Nonlinear Systems and Controls

Jürgen Adamy

Nonlinear Systems and Controls

Second Edition

Springer

Jürgen Adamy
Fachgebiet Regelungsmethoden und
Intelligente Systeme
Technische Universität Darmstadt
Darmstadt, Germany

ISBN 978-3-662-68689-8 ISBN 978-3-662-68690-4 (eBook)
https://doi.org/10.1007/978-3-662-68690-4

This Springer imprint is published by the registered company Springer-Verlag GmbH, DE, part of Springer Nature.
The registered company address is: Heidelberger Platz 3, 14197 Berlin, Germany

Paper in this product is recyclable.

For Claudia, Ursula, and Timm

Preface

Nonlinear dynamic systems and controls play an important role in science, mathematics, economics, and engineering. The regulation of human body temperature, the regulation of the money supply in an economy, and the autopilot of an airplane are examples. A book about these systems can be written from the point of view of the scientist, the mathematician, or the engineer. A book of this kind must orient itself towards the objective of the discipline in question: a book focused on the natural sciences would explain the dynamic behavior of natural phenomena, such as planetary motion or the blood pressure regulation in mammals. A mathematical approach would focus on theorems and their proofs, and a book addressed to engineers would describe the theory and its practical application, such as how to model and control nonlinear processes and systems like power plants and rockets.

Karl Popper's three-world theory illustrates these three perspectives: it is a model of the scientific process that divides the world into the world of matter, the world of the subjective ego, and the world of objective knowledge, as shown in Figure P.1. The material world includes all kinds of matter including flora, fauna, and the human body. The world of the subjective ego includes the perceptions, thoughts, and ideas of an individual human being. In contrast, the objective world consists of, among other things, scientific theories, works of art, laws and agreements that are available to all or many people to examine and change.

Scientists observe the material world and use the subjective ego to develop a theory of how it works: they make discoveries. A discovery is published in a scientific journal and is thus made part of the objective world. In this way, a theory is broadly accessible and can be tested by other scientists and either confirmed or falsified. In contrast to the classical natural sciences (physics, chemistry, and biology), mathematics is a method, or we could also say a language, with which natural and technological processes can be quantitatively described. This is not possible using our everyday language. Mathematicians are constantly expanding this language and exploring new ways of describing facts using formulas and solving problems mathematically: they draft math-

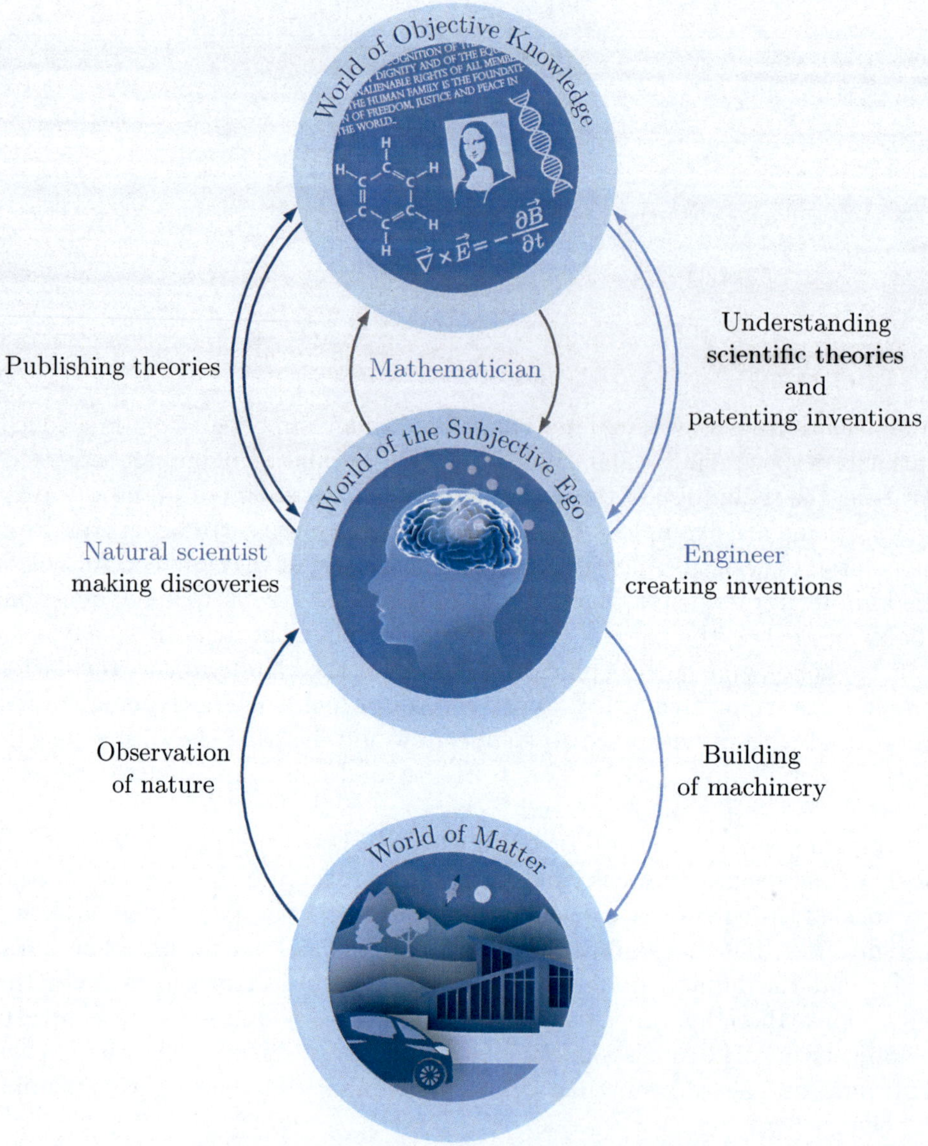

Fig. P.1: Karl Popper's three-world ontology

ematical hypotheses and prove them. Engineers make use both of scientific theories, especially physical theories, and mathematics, the language in which these theories are formulated. They use this knowledge of the objective world to create, within the world of their subjective ego, the idea for a new machine or a new technical process: they invent something new. By realizing these inventions, they change the material world. They build skyscrapers, airplanes, power plants, and robots. Construction plans and system descriptions published in patents and journals become part of the objective world.

Although these three disciplines are sharply delineated above, there are a large number of fields in which they overlap. These include experimental physics, in which scientists build special equipment for their research such as particle accelerators, and control engineering, in which new mathematical theories, such as that of flat systems, are developed to solve engineering problems.

This book is primarily oriented towards the engineer's point of view. The aim of the book is therefore to convey the knowledge of nonlinear systems and controls that is important for an engineer, and to illustrate its applicability. In doing so, the book describes the current state of the theory of these systems. For this purpose, the theorems of this mathematical theory which have proved useful in engineering practice have been selected. Since these theorems and their proofs are available in the literature, the proofs have not been shown here in the classical theorem-proof sequence: the emphasis in this book is on teaching the theory and its application rather than proving of the theorems.

Literature Recommendations

To learn more about a particular field of research in detail, it is useful to consult various books which address that field. In the case of the theory of nonlinear systems and controls, for example, the works of H. K. Khalil, E. D. Sontag, S. Sastry, H. Nijmeijer and A. van der Schaft, J.-J. E. Slotine and W. Li, H. J. Marquez, and A. Isidori are recommended. C. M. Kellett and P. Braun cover both linear and nonlinear control systems in their book. An introduction to the subject is given in the compact book by G. Chen.

To understand the theory of nonlinear systems, the theory of linear systems is useful or necessary in many cases. Describing this theory would require a similarly extensive volume, so it cannot be addressed in detail in this book and it is assumed that the reader is already familiar with it. Suitable works on this topic have been written by G. F. Franklin, J. D. Powell, and A. Emami-Naeini, as well as R. C. Dorf and R. H. Bishop, among others.

Acknowledgements

A large number of scholars have proofread the book for comprehensibility, errors, and mistakes, and have contributed to its layout. For the first as well as the second English edition, I would like to especially thank K. Olhofer-Karova for patiently preparing the LaTeX source code and Matlab simulations, and V. Ansel for preparing the color images. My thanks go to the American translator H. Krehbiel for the linguistic revision of the text.

Jürgen Adamy, Technische Universität Darmstadt, 2023

Contents

1

Fundamentals of Nonlinear Systems

1.1 System Description and System Behavior

1.1.1 Linear and Nonlinear Systems

Only some of the processes and systems that are regularly encountered in nature or industrial practice can be adequately described by linear models and linear systems theory. A significant number of processes and systems are nonlinear and must be represented by nonlinear models. Linear systems theory is not generally applicable to nonlinear systems. Exceptions are nonlinear systems, which can be sufficiently well approximated by a linear model.

Therefore, nonlinear systems and control loops require their own analysis and design methods. Because nonlinear systems are mathematically much more complex than linear systems, so far such methods have only been developed for certain classes of nonlinear systems and for specific applications. This is quite different in the case of linear systems, for which an almost completely elaborated systems theory exists with only a few unexplored areas [133, 152]. Figure 1.1 illustrates this different level of knowledge.

It is not entirely correct to speak of a nonlinear systems theory, since the term actually refers to a collection of incoherent methods and theories. Rather than a single theory, as shown schematically in Figure 1.1, there are a number of theories for different classes of nonlinear systems and controllers [77, 203, 232, 338, 402, 465]. Despite this diversity and depending on the focus, it is common to subsume these theories under the terms *nonlinear control* or *nonlinear systems*. One of the most important and most developed theories is that of control-affine systems, which depend nonlinearly on the system states but linearly on the input variables. The theory of such systems is almost completely elaborated and is also of practical importance.

The most important great methods of nonlinear control engineering, which are also most relevant to industrial practice, are the subject of this book. To begin with, in this chapter, the system dynamics and the solution of nonlinear differential equations are addressed. This is followed by a description

J. Adamy, *Nonlinear Systems and Controls*,
https://doi.org/10.1007/978-3-662-68690-4_1

Fig. 1.1: Level of knowledge regarding linear and nonlinear systems. Blue represents the known areas; white represents the unexplored areas.

of the stability theory of nonlinear systems and feedback controls in Chapter 2. Chapter 3 deals with the system property of controllability, which is the prerequisite for any kind of feedback and feedforward control, which are also known as open-loop and closed-loop control, respectively. It further describes the flatness property of a system, which is related to its controllability, and contains an explanation of state-space transformations and diffeomorphisms.

This fundamental knowledge is the starting point of a series of design methods for nonlinear controllers of linear plants, i. e. of linear controlled systems, as described in Chapter 4, and nonlinear plants, which are discussed in Chapter 5. An important part of this chapter is the theory of control-affine systems and their controls. Design methods for controllers that can be used both in linear and in nonlinear plants follow in Chapter 6. Since most nonlinear control methods require the state vector of the controlled system, which is often not fully measurable, observers for nonlinear control loops are dealt with in Chapter 7.

1.1.2 System Description and Nonlinear Control Loops

Nonlinear systems can generally be described by a vector differential equation

$$\dot{\boldsymbol{x}} = \boldsymbol{f}(\boldsymbol{x}, \boldsymbol{u}) \tag{1.1}$$

and an output equation

$$\boldsymbol{y} = \boldsymbol{g}(\boldsymbol{x}, \boldsymbol{u}).$$

Here, $\boldsymbol{x}$ is the n - dimensional state vector, $\boldsymbol{u}$ is the m - dimensional *actuator* or *input variable vector*, also called the *control variable vector*, $\boldsymbol{f}$ is the n - dimensional *system function vector*, $\boldsymbol{y}$ is the r - dimensional *output variable vector*, and $\boldsymbol{g}$ is the r - dimensional *output function vector*, or *output vector* for short. In most cases, the output variable vector $\boldsymbol{y}$ does not depend on the input variable vector $\boldsymbol{u}$, i. e.

$$\boldsymbol{y} = \boldsymbol{g}(\boldsymbol{x}).$$

If the output equation depends on $\boldsymbol{u}$, this dependency is referred to as *feedthrough*.

In the case of linear systems

$$\begin{aligned}\dot{\boldsymbol{x}} &= \boldsymbol{A}\boldsymbol{x} + \boldsymbol{B}\boldsymbol{u},\\ \boldsymbol{y} &= \boldsymbol{C}\boldsymbol{x} + \boldsymbol{D}\boldsymbol{u},\end{aligned}$$

we call $\boldsymbol{A} \in \mathbb{R}^{n\times n}$ the *system matrix*, $\boldsymbol{B} \in \mathbb{R}^{n\times m}$ the *input matrix* or the *input vector* if $m = 1$, $\boldsymbol{C} \in \mathbb{R}^{r\times n}$ the *output matrix* or the *output vector* if $r = 1$, and $\boldsymbol{D} \in \mathbb{R}^{r\times m}$ the *feedthrough* or *direct transmission matrix*.

In contrast to linear systems, which are defined on the entire real coordinate space $\mathbb{R}^n$, nonlinear systems, more precisely the functions $\boldsymbol{f}$ and $\boldsymbol{g}$, are sometimes defined only on a subset. This subset, the *system's domain of definition*, is the intersection of the functions $\boldsymbol{f}$ and $\boldsymbol{g}$'s domains of definition. It can be represented as the Cartesian product $D_{\boldsymbol{x},\mathrm{def}} \times D_{\boldsymbol{u},\mathrm{def}}$ of the sets $D_{\boldsymbol{x},\mathrm{def}} \subseteq \mathbb{R}^n$ of vectors $\boldsymbol{x}$ and $D_{\boldsymbol{u},\mathrm{def}} \subseteq \mathbb{R}^m$ of vectors $\boldsymbol{u}$, where $D_{\boldsymbol{x},\mathrm{def}}$ and $D_{\boldsymbol{u},\mathrm{def}}$ are the sets on which the two functions $\boldsymbol{f}(\boldsymbol{x},\boldsymbol{u})$ and $\boldsymbol{g}(\boldsymbol{x},\boldsymbol{u})$ are both defined. For simplicity, we will often only refer to the subspace $D_{\boldsymbol{x},\mathrm{def}}$ as the system's domain of definition. In the following, if no domain of definition is explicitly mentioned, we will assume $D_{\boldsymbol{x},\mathrm{def}} \times D_{\boldsymbol{u},\mathrm{def}}$.

Sometimes, especially when a system possesses only a single-input and single-output variable, this system can be described by an explicit higher order differential equation

$$y^{(n)} = h(y, \dot{y}, \ldots, y^{(n-1)}, u_1, \ldots, u_m). \tag{1.2}$$

By introducing the variables

$$\begin{aligned}x_1 &= y,\\ x_2 &= \dot{y},\\ &\;\vdots\\ x_n &= y^{(n-1)},\end{aligned}$$

equation (1.2) can also be represented as the vector differential equation

$$\dot{\boldsymbol{x}} = \boldsymbol{f}(\boldsymbol{x},\boldsymbol{u}) = \begin{bmatrix} x_2\\ x_3\\ \vdots\\ x_n\\ h(x_1,\ldots,x_n,u_1,\ldots,u_m)\end{bmatrix}.$$

Systems with a single-input variable u and a single-output variable y are referred to as *SISO systems*. Systems with multiple-input variables and multiple-output variables are called *MIMO systems*. The system represented by equation (1.1) is said to be *autonomous* if the function $\boldsymbol{f}$ does not depend on a time-varying input variable $\boldsymbol{u}(t)$, or – which is essentially the same when solving the differential equation – if $\boldsymbol{f}$ is not directly time-dependent, i. e.

$$\dot{\boldsymbol{x}} = \boldsymbol{f}(\boldsymbol{x}).$$

If a system is explicitly time-dependent, i. e.

$$\dot{\boldsymbol{x}} = \boldsymbol{f}(\boldsymbol{x}, \boldsymbol{u}, t)$$

holds and $\boldsymbol{f}$ is defined for $t \in D_{t,\mathrm{def}} \subseteq I\!R$, we call it *time-variant*. If it is not, i. e. when

$$\dot{\boldsymbol{x}} = \boldsymbol{f}(\boldsymbol{x}, \boldsymbol{u}) \tag{1.3}$$

holds, it is called *time-invariant*. If the input signal $\boldsymbol{u}(t)$ of a system is equal to zero or the input signal is missing, i. e. if

$$\dot{\boldsymbol{x}} = \boldsymbol{f}(\boldsymbol{x}, t, \boldsymbol{u} = \boldsymbol{0}) \quad \text{or} \quad \dot{\boldsymbol{x}} = \boldsymbol{f}(\boldsymbol{x}, t)$$

holds, we call it a *free system*. A function $\boldsymbol{x}(t)$ that fulfills the differential equation (1.3) for a given initial vector $\boldsymbol{x}(0)$ and a given input signal $u(t)$ is called *solution* or *trajectory*.

In control theory, three classes of nonlinear control systems are considered to be most relevant. The first class consists of the linear controllers for nonlinear plants, as depicted in Figure 1.2. They are often associated with linear controller design methods, since in this case the plant is linearizable and the

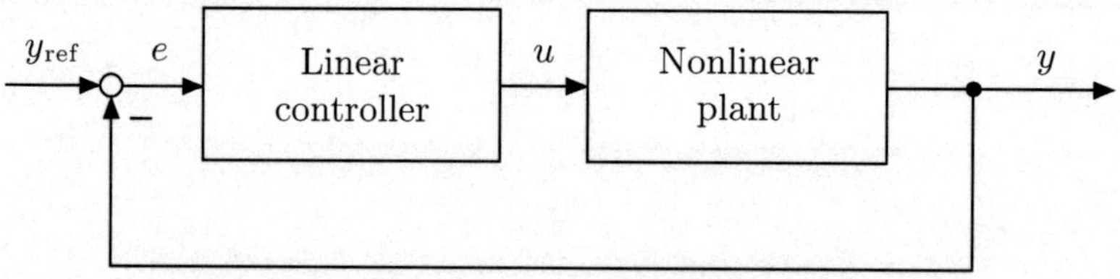

Fig. 1.2: Nonlinear plant with a linear controller

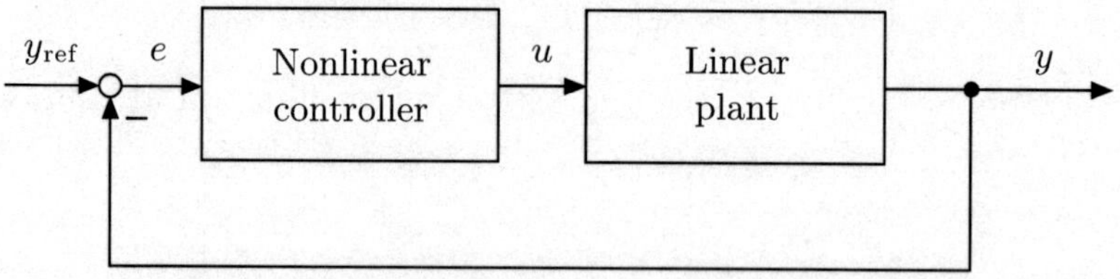

Fig. 1.3: Linear plant with a nonlinear controller

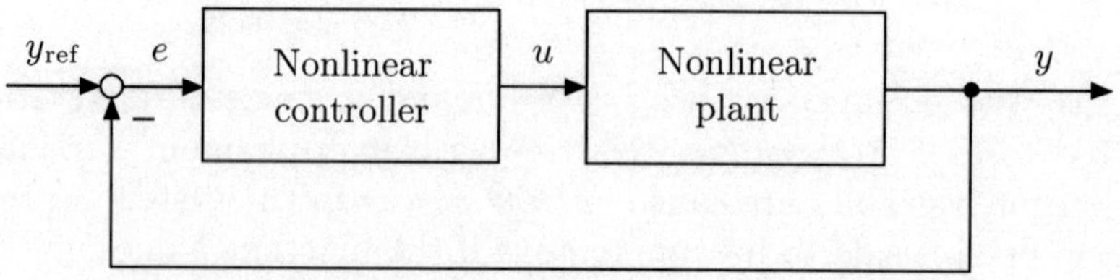

Fig. 1.4: Nonlinear plant with a nonlinear controller

linearized model is sufficiently accurate. Thus linear systems theory can be applied. The reference variable and the control error are denoted by y_{ref} and e, respectively. In the case of systems with multiple-input and multiple-output variables, we use the bold letters $\boldsymbol{e}$ and $\boldsymbol{y}_{\text{ref}}$.

The second class consists of nonlinear controllers for linear plants, as shown in Figure 1.3. Often, simple nonlinear controllers are applied to linear plants for technical reasons or because of their low cost. A common example is the temperature control in electric irons using a bimetal. The bimetal has the characteristic of a switch with hysteresis, i. e. it is a nonlinear system. Nonlinear controllers with a more complex design are also used to control linear plants in order to achieve better control results compared to those that can be achieved using linear controllers.

The third class consists of nonlinear controllers for nonlinear plants, as illustrated in Figure 1.4. Nonlinear plants are often very complex in terms of their behavior. For such cases, linear controllers are often not able to provide the desired quality, and nonlinear controllers must be designed. For example, it is sometimes possible to combine a nonlinear plant with a nonlinear controller in such a way that the resulting control loop is linear. Due to the linearity, its behavior is easy to comprehend.

1.1.3 Equilibrium Points of Nonlinear Systems

An equilibrium point is a state of a dynamic system which does not change with time, i. e. it will remain constant for all future time. One of the central objectives of control engineering is to drive the state variables of a plant to an equilibrium point and to keep it there. For example, the aim is to steer an aircraft to a certain altitude using an autopilot, to heat up the water in a boiler to a desired temperature, or to roll a sheet of metal to a predetermined thickness, and then to hold it there. Therefore to design a suitable controller, an appropriate equilibrium point must first be found. This raises the question of how to determine the equilibrium point of a nonlinear system. Before this question is examined in more detail, the term *equilibrium point* should be clearly defined.

Definition 1 (Equilibrium Point). *Consider the system*

$$\dot{\boldsymbol{x}} = \boldsymbol{f}(\boldsymbol{x}, \boldsymbol{u}).$$

A point $\boldsymbol{x}_{\text{eq}}$ within the state space is called an equilibrium point if

$$\dot{\boldsymbol{x}} = \boldsymbol{f}(\boldsymbol{x}_{\text{eq}}, \boldsymbol{0}) = \boldsymbol{0}$$

holds.

In this definition, it was assumed that

$$\boldsymbol{u} = \boldsymbol{0}.$$

Of course, m - dimensional input variable vectors

$$\boldsymbol{u} = \boldsymbol{c} \neq \boldsymbol{0}$$

may exist, where $\boldsymbol{c} \in I\!R^m$ is a constant vector, such that

$$\dot{\boldsymbol{x}} = \boldsymbol{f}(\boldsymbol{x}, \boldsymbol{u} = \boldsymbol{c}) = \boldsymbol{0} \tag{1.4}$$

holds. Often, depending on $\boldsymbol{c}$, systems have an infinite number of equilibrium points. This is why, according to the above definition, we always refer to the equilibrium points of the free system, i. e. $\dot{\boldsymbol{x}} = \boldsymbol{f}(\boldsymbol{x}, \boldsymbol{u} = \boldsymbol{0})$, when speaking of an *equilibrium point* without further specification. The case of equation (1.4) is covered by the above definition when the transformation

$$\boldsymbol{u} = \tilde{\boldsymbol{u}} + \boldsymbol{c}$$

is performed. Then we must analyze the system

$$\dot{\boldsymbol{x}} = \boldsymbol{f}(\boldsymbol{x}, \tilde{\boldsymbol{u}} + \boldsymbol{c}) = \tilde{\boldsymbol{f}}(\boldsymbol{x}, \tilde{\boldsymbol{u}})$$

which has an equilibrium point at $\boldsymbol{x}_{\text{eq}}$ for $\tilde{\boldsymbol{u}} = \boldsymbol{0}$. This implies that

$$\dot{\boldsymbol{x}} = \tilde{\boldsymbol{f}}(\boldsymbol{x}_{\text{eq}}, \boldsymbol{0}) = \boldsymbol{0}$$

holds.

Determining an equilibrium point of a nonlinear system is often not an easy task to accomplish. This is because we must solve the implicit equation

$$\dot{\boldsymbol{x}} = \boldsymbol{f}(\boldsymbol{x}_{\text{eq}}, \boldsymbol{0}) = \boldsymbol{0}$$

for $\boldsymbol{x}_{\text{eq}}$. One, zero, or multiple solutions may exist. A continuum of solutions is also possible, or there may be none at all. This is illustrated in Figure 1.5 for a one-dimensional function f.

When calculating the equilibrium points $\boldsymbol{x}_{\text{eq}}$, the following three cases may occur. In the first case, the implicit equation $\boldsymbol{f}(\boldsymbol{x}_{\text{eq}}, \boldsymbol{0}) = \boldsymbol{0}$ is explicitly solvable for $\boldsymbol{x}_{\text{eq}}$, i. e.

$$\boldsymbol{x}_{\text{eq}} = \boldsymbol{f}^{-1}(\boldsymbol{0})$$

holds.

In the second case, the implicit equation is transcendental. Here we must resort to numerical methods, such as the multidimensional Newton's method. This can be difficult because we often face the problem of not knowing how many equilibrium points exist, or even if there are any equilibrium points at all.

The third case applies to many technical systems: we can often surmise an equilibrium point $\boldsymbol{x}_{\text{eq}}$, either from intuition or from knowledge that is available about the system. Substituting this estimate into the equation system allows us to verify our assumption.

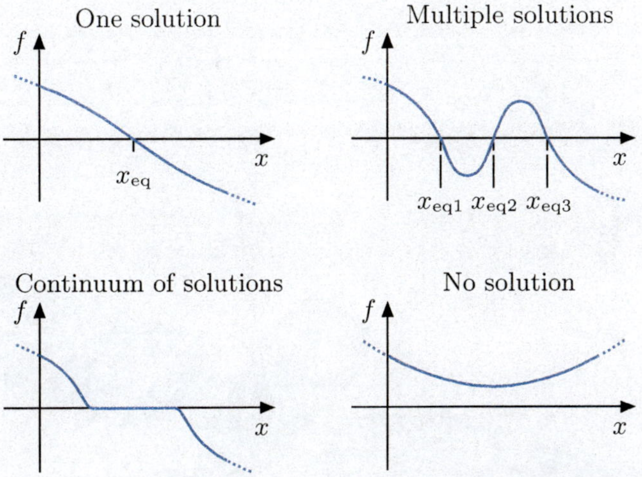

Fig. 1.5: Possible solutions of the equilibrium point equation $f(x_{\mathrm{eq}}, 0) = 0$

1.1.4 Example: Satellite

Let us consider the rotation of a satellite or a space probe that can be rotated around its axes via control jets. Figure 1.6 shows such a satellite with its corresponding body-fixed coordinate system (x, y, z) and a space-fixed coordinate system $(\tilde{x}, \tilde{y}, \tilde{z})$.

For the angular momentum vector $\boldsymbol{L}$ of the satellite which rotates with its corresponding body-fixed coordinate system and angular velocity vector $\boldsymbol{\omega}$ around the space-fixed coordinate system, using the inertia tensor $\boldsymbol{J}$, the relation

$$\boldsymbol{L} = \boldsymbol{J}\boldsymbol{\omega}$$

holds. The equation of motion is derived from

$$\frac{d\boldsymbol{L}}{dt} = \boldsymbol{M}, \tag{1.5}$$

wherein the torque vector $\boldsymbol{M}$ contains the torques caused by the control jets. For the derivative of the angular momentum $\boldsymbol{L}$ with respect to time, we obtain

$$\frac{d\boldsymbol{L}}{dt} = \boldsymbol{J}\dot{\boldsymbol{\omega}} + \boldsymbol{\omega} \times (\boldsymbol{J}\boldsymbol{\omega}), \tag{1.6}$$

where the cross product $\boldsymbol{\omega} \times (\boldsymbol{J}\boldsymbol{\omega})$ results from the rotation of the body-fixed coordinate system around the space-fixed coordinate system with velocity $\boldsymbol{\omega}$. From equation (1.5) and equation (1.6), it follows that

$$\boldsymbol{J}\dot{\boldsymbol{\omega}} = -\boldsymbol{\omega} \times (\boldsymbol{J}\boldsymbol{\omega}) + \boldsymbol{M}. \tag{1.7}$$

If the axes of the satellite's body-fixed coordinate system are identical to its principal axes of inertia, it holds that

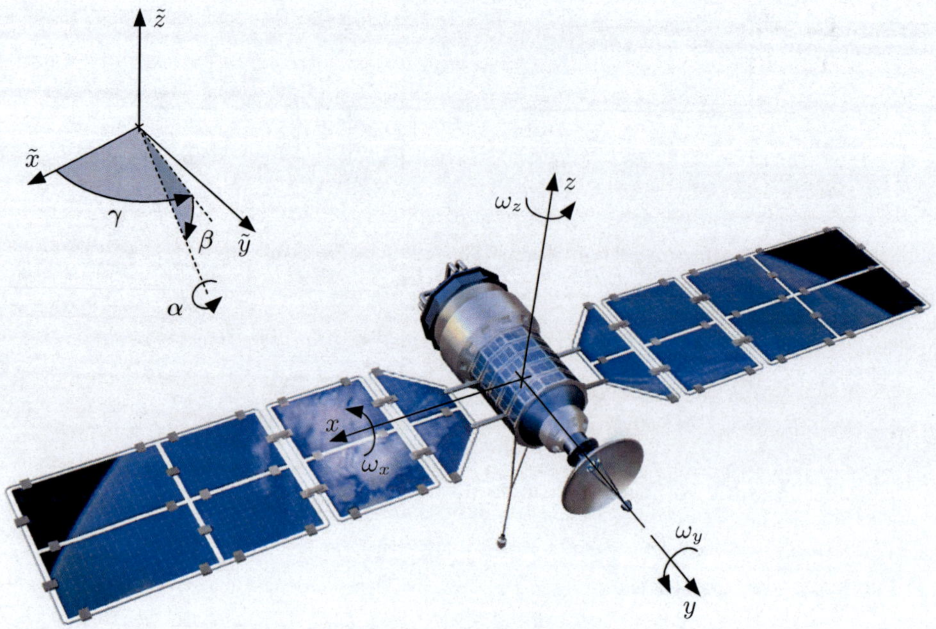

Fig. 1.6: Satellite with a body-fixed coordinate system (x, y, z) and a space-fixed coordinate system $(\tilde{x}, \tilde{y}, \tilde{z})$. The orientation of the satellite relative to the space-fixed coordinate system is specified by the Euler angles α, β, γ.

$$\boldsymbol{J} = \begin{bmatrix} J_x & 0 & 0 \\ 0 & J_y & 0 \\ 0 & 0 & J_z \end{bmatrix}.$$

In this case, Euler's rotation equations

$$\begin{aligned} J_x\dot{\omega}_x &= -(J_z - J_y)\omega_y\omega_z + M_x, \\ J_y\dot{\omega}_y &= -(J_x - J_z)\omega_x\omega_z + M_y, \\ J_z\dot{\omega}_z &= -(J_y - J_x)\omega_x\omega_y + M_z \end{aligned} \tag{1.8}$$

follow from equation (1.7) as the satellite's equations of motion.

Let us now examine when $\dot{\omega} = 0$ holds. The satellite has equilibrium points only if at least two of the angular velocities ω_x, ω_y, and ω_z are equal to zero. We assume that the actuating variables M_x, M_y, and M_z are also equal to zero. This results in three equilibrium sets, namely

$$\boldsymbol{\omega}_{\text{eq1}} = \begin{bmatrix} \omega_1 \\ 0 \\ 0 \end{bmatrix}, \quad \boldsymbol{\omega}_{\text{eq2}} = \begin{bmatrix} 0 \\ \omega_2 \\ 0 \end{bmatrix}, \quad \boldsymbol{\omega}_{\text{eq3}} = \begin{bmatrix} 0 \\ 0 \\ \omega_3 \end{bmatrix}.$$

Here the angular velocities ω_1, ω_2, $\omega_3 \in I\!R$ may take arbitrary values. Obviously, an infinite number of equilibrium points exists. Note that

$$\omega_1 = \omega_2 = \omega_3 = 0$$

is also possible. Thus $\boldsymbol{\omega} = \mathbf{0}$ is also an equilibrium point.

This example also illustrates that the term *equilibrium point* does not necessarily imply the absence of movement, in this case of a rigid body. Instead, the term *equilibrium point* signifies that the states of the system under consideration are not going to change with time.

Now we will describe the orientation of the satellite in the space-fixed coordinate system. This is done by introducing the Euler angles α, β, and γ [195, 221], which are depicted in Figure 1.6. By defining the vector

$$\boldsymbol{\varphi} = [\alpha \ \beta \ \gamma]^T,$$

the motion equations of the satellite are given by

$$\begin{bmatrix} \dot{\boldsymbol{\varphi}} \\ \dot{\boldsymbol{\omega}} \end{bmatrix} = \begin{bmatrix} \boldsymbol{\Lambda}(\boldsymbol{\varphi})\boldsymbol{\omega} \\ -\boldsymbol{J}^{-1}[\boldsymbol{\omega} \times (\boldsymbol{J}\boldsymbol{\omega})] \end{bmatrix} + \begin{bmatrix} \mathbf{0} \\ \boldsymbol{J}^{-1} \end{bmatrix} \boldsymbol{M}, \tag{1.9}$$

where the matrix

$$\boldsymbol{\Lambda}(\boldsymbol{\varphi}) = \begin{bmatrix} 1 & \sin(\alpha)\tan(\beta) & \cos(\alpha)\tan(\beta) \\ 0 & \cos(\alpha) & -\sin(\alpha) \\ 0 & \sin(\alpha)/\cos(\beta) & \cos(\alpha)/\cos(\beta) \end{bmatrix}$$

transforms the angular velocity vector $\boldsymbol{\omega}$ to the angular velocity vector $\dot{\boldsymbol{\varphi}}$ of the space-fixed coordinate system. The transformation

$$\dot{\boldsymbol{\varphi}} = \boldsymbol{\Lambda}(\boldsymbol{\varphi})\boldsymbol{\omega}$$

is dependent on the Euler angles, i. e. on the vector $\boldsymbol{\varphi}$. Obviously, the system of equation (1.9) again possesses infinitely many equilibrium points

$$\begin{bmatrix} \boldsymbol{\varphi}_{\text{eq}} \\ \boldsymbol{\omega}_{\text{eq}} = \mathbf{0} \end{bmatrix}, \quad \boldsymbol{\varphi}_{\text{eq}} \in \mathbb{R}^3,$$

if $\boldsymbol{M} = \mathbf{0}$ holds for the actuator vector. In this case, however, the satellite does not spin around any of its axes.

1.1.5 Equilibrium Points of Linear Systems

The linear systems

$$\begin{aligned} \dot{\boldsymbol{x}} &= \boldsymbol{A}\boldsymbol{x} + \boldsymbol{B}\boldsymbol{u}, \\ \boldsymbol{y} &= \boldsymbol{C}\boldsymbol{x} + \boldsymbol{D}\boldsymbol{u} \end{aligned}$$

are a particular case of the general system description

$$\begin{aligned} \dot{\boldsymbol{x}} &= \boldsymbol{f}(\boldsymbol{x}, \boldsymbol{u}), \\ \boldsymbol{y} &= \boldsymbol{g}(\boldsymbol{x}, \boldsymbol{u}) \end{aligned}$$

and will be considered briefly below to demonstrate an important difference between them and nonlinear systems.

The equilibrium points of linear systems are easily determined by

$$\dot{\boldsymbol{x}} = \boldsymbol{A}\boldsymbol{x}_{\text{eq}} = \boldsymbol{0}$$

if $\boldsymbol{u} = \boldsymbol{0}$ holds. The following cases may occur: if $\det(\boldsymbol{A}) \neq 0$ holds, only the equilibrium point $\boldsymbol{x}_{\text{eq}} = \boldsymbol{0}$ exists. If $\det(\boldsymbol{A}) = 0$, the system $\boldsymbol{A}$ has zero-valued eigenvalues and there is a linear subspace of vectors $\boldsymbol{x}_{\text{eq}}$, for which

$$\boldsymbol{A}\boldsymbol{x}_{\text{eq}} = \boldsymbol{0}$$

holds. This means that a continuum of equilibrium points exists. A simple example is the system

$$\dot{\boldsymbol{x}} = \begin{bmatrix} 0 & 1 \\ 0 & 0 \end{bmatrix} \boldsymbol{x} + \begin{bmatrix} 0 \\ 1 \end{bmatrix} u,$$

which consists of two integrators and is therefore described by the transfer function $1/s^2$ that is shown in Figure 1.7. Obviously, all states

$$\boldsymbol{x}_{\text{eq}} = \begin{bmatrix} a \\ 0 \end{bmatrix}, \quad a \in \mathbb{R},$$

are equilibrium points. This means that the x_1-axis forms a continuum of equilibrium points, as illustrated in Figure 1.7.

If $\boldsymbol{u} = \boldsymbol{c}$ and $\det(\boldsymbol{A}) = 0$ holds, it is possible that no equilibrium point exists. This could occur when

$$\boldsymbol{A}\boldsymbol{x} = -\boldsymbol{B}\boldsymbol{c}$$

is an overdetermined system of equations. Therefore, a linear system has either one, zero, or a continuum of equilibrium points. The case of multiple isolated equilibrium points, which is possible for nonlinear systems, does not occur in linear systems.

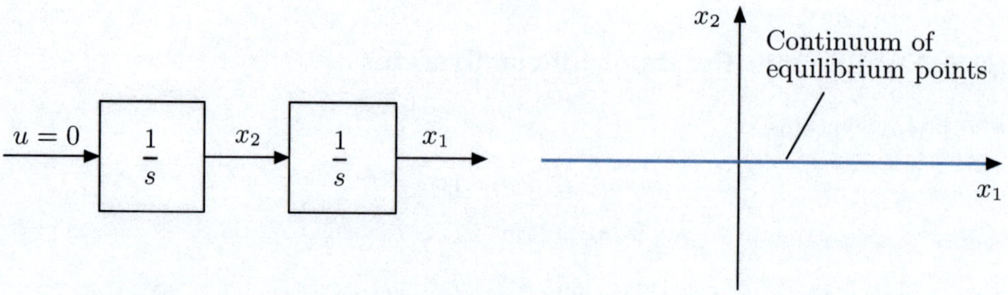

Fig. 1.7: The system $1/s^2$ and its continuum of equilibrium points $\boldsymbol{x}_{\text{eq}} = [a\ 0]^T$ with $a \in \mathbb{R}$

1.1.6 Stability and Asymptotic Stability

An equilibrium point is said to be *stable* if all trajectories $\boldsymbol{x}(t)$ that begin in a neighborhood of the equilibrium point converge to this point in the course of time and remain there. In a weaker sense, the term *stable* is still used, even if the trajectories do not terminate at the equilibrium point, but remain in a neighborhood of the equilibrium point.

Let us consider a few examples of stable and unstable equilibrium points below, in order to gain a preliminary impression and to develop a basic understanding of the different kinds of stability. The easiest way to achieve this goal is by considering linear systems with the associated equilibrium point $\boldsymbol{x}_{\text{eq}} = \boldsymbol{0}$. Since all solutions $\boldsymbol{x}(t)$ of the linear differential equation

$$\dot{\boldsymbol{x}} = \boldsymbol{A}\boldsymbol{x}$$

only involve terms given by

$$e^{\lambda_i t} \quad \text{and} \quad t^k e^{\lambda_j t} \quad \text{with} \quad k \in \{1, 2, 3, \dots\},$$

all trajectories $\boldsymbol{x}(t)$ of the system obviously lead to the equilibrium point $\boldsymbol{x}_{\text{eq}} = \boldsymbol{0}$ for $t \to \infty$ if the inequality

$$\operatorname{Re}\{\lambda_i\} < 0$$

holds for every eigenvalue λ_i of the system. In this case, both the equilibrium point and the linear system are referred to as *stable*. Figure 1.8 shows the trajectories of such a linear system.

However, if

$$\operatorname{Re}\{\lambda_i\} > 0$$

holds for at least one of the eigenvalues of the system, the equilibrium point is unstable. Then there are trajectories aiming away from the equilibrium point and tending to infinity. Such a linear system is referred to as *unstable*. Figure 1.9 provides an example of the trajectories of an unstable system.

If a linear second-order system has a pair of complex conjugate eigenvalues $\lambda_{1/2} = \pm j$, the system is called a *simple harmonic oscillator*. Its trajectories are shown in Figure 1.10. Obviously, the trajectories do not tend to the equilibrium point $\boldsymbol{x}_{\text{eq}} = \boldsymbol{0}$ or to infinity. In this case, the equilibrium point $\boldsymbol{x}_{\text{eq}} = \boldsymbol{0}$ still possesses stability to a certain extent. A similar case is that of a second-order system with one eigenvalue $\lambda_1 = 0$ and one eigenvalue $\lambda_2 < 0$, as shown in Figure 1.11. Although all trajectories tend to an equilibrium point on the x_1-axis, none of the equilibrium points attracts all trajectories. On the other hand, none of the trajectories tends to infinity. Therefore such systems are also referred to as stable systems.

To develop a preliminary impression of stability conditions for nonlinear systems, the system defined by

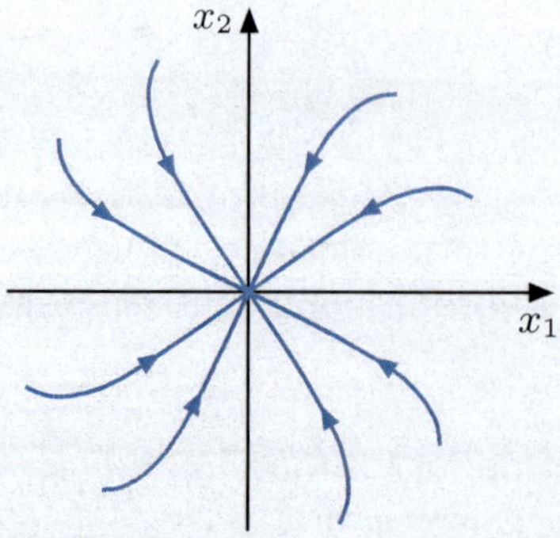

Fig. 1.8: Trajectories of a stable linear system with eigenvalues λ_1, λ_2 with $\text{Re}\{\lambda_1\} < 0$, $\text{Re}\{\lambda_2\} < 0$

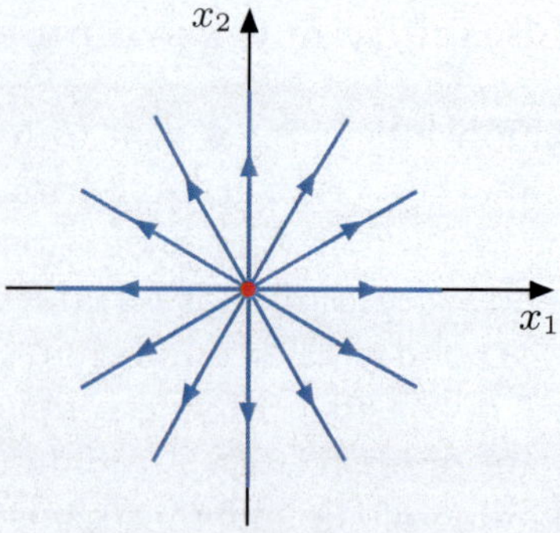

Fig. 1.9: Trajectories of an unstable linear system with eigenvalues λ_1, λ_2 with $\text{Re}\{\lambda_1\} > 0$, $\text{Re}\{\lambda_2\} > 0$

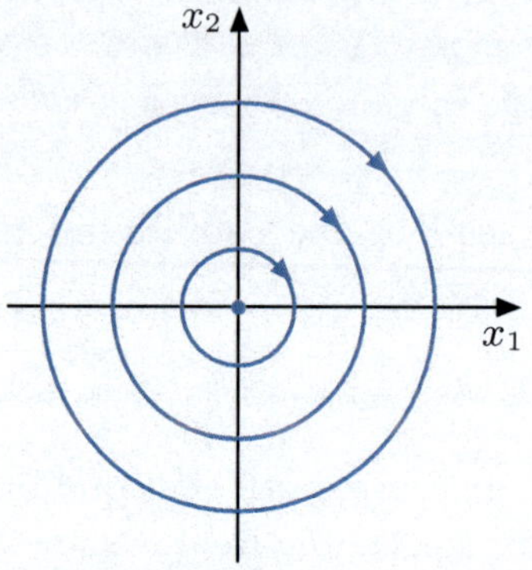

Fig. 1.10: Trajectories of an oscillator with eigenvalues $\lambda_{1/2} = \pm j$

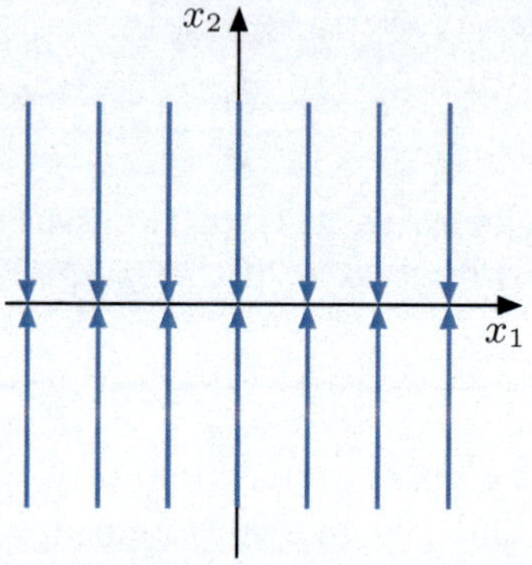

Fig. 1.11: Trajectories of a system with $\lambda_1 = 0$ and $\lambda_2 < 0$

$$\begin{aligned} \dot{x}_1 &= x_1(x_2 - 1), \\ \dot{x}_2 &= x_2(x_1 - 1) \end{aligned} \tag{1.10}$$

is regarded as an example. It possesses two equilibrium points at

$$\boldsymbol{x}_{\text{eq1}} = \begin{bmatrix} 0 \\ 0 \end{bmatrix} \qquad \text{and} \qquad \boldsymbol{x}_{\text{eq2}} = \begin{bmatrix} 1 \\ 1 \end{bmatrix}.$$

This demonstrates the previously mentioned difference regarding linear systems, because there are two isolated equilibrium points. This is not possible for linear systems. They either have one equilibrium point at $\boldsymbol{x}_{\text{eq}} = \boldsymbol{0}$ or a continuum of equilibrium points.

Figure 1.12 depicts the course of the trajectories of the system that is described by equation (1.10) in the vicinity of the equilibrium points. Near the origin, all trajectories tend toward the equilibrium point $\boldsymbol{x}_{\text{eq1}} = \boldsymbol{0}$. This equilibrium point can be referred to as being stable. The trajectories belonging

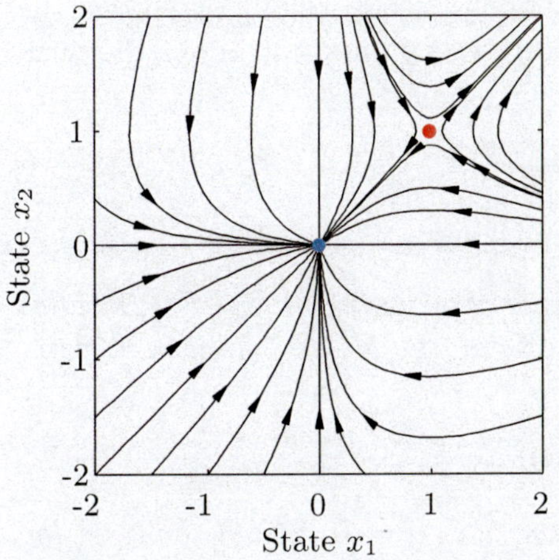

Fig. 1.12: Course of the trajectories of the system that is described by equation (1.10).

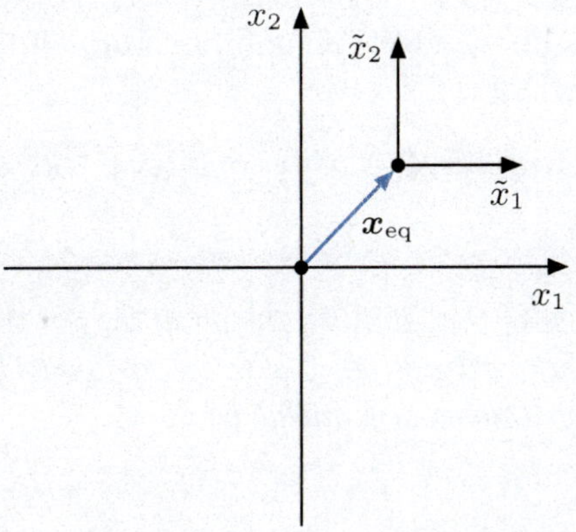

Fig. 1.13: Transformation, more precisely translation, of an equilibrium point $\boldsymbol{x}_{\text{eq}}$ into the origin

to the equilibrium point $\boldsymbol{x}_{\text{eq2}} = [1\ 1]^T$ tend away from this point to infinity, which means that this equilibrium point must be considered unstable.

This example demonstrates that nonlinear systems cannot generally be classified as being stable or unstable, which is possible for linear systems. Rather, we must consider the stability behavior of the system in the neighborhood of an equilibrium point, i. e. the stability behavior of the equilibrium point. If multiple equilibrium points exist, the term stability only refers to a specifically considered equilibrium point and not to the entire system. In this context there is a need for clarification of

(1) the behavior of trajectories in the neighborhood of an equilibrium point,
(2) the size of the region surrounding an equilibrium point, in which all trajectories starting within the region tend to the equilibrium point, and
(3) the mathematical definition of the stability of an equilibrium point.

Before the above three issues are clarified, a simplification of the analysis must be conducted. If an equilibrium point $\boldsymbol{x}_{\text{eq}}$ has been determined for a system, it can be shifted by an appropriate transformation

$$\boldsymbol{x} = \boldsymbol{x}_{\text{eq}} + \tilde{\boldsymbol{x}}$$

into the origin, i. e. to $\tilde{\boldsymbol{x}} = \boldsymbol{0}$. The system equations are then given by

$$\begin{aligned}\dot{\tilde{\boldsymbol{x}}} &= \boldsymbol{f}(\boldsymbol{x}_{\text{eq}}+\tilde{\boldsymbol{x}}, \boldsymbol{u}) = \tilde{\boldsymbol{f}}(\tilde{\boldsymbol{x}}, \boldsymbol{u}),\\ \boldsymbol{y} &= \boldsymbol{g}(\boldsymbol{x}_{\text{eq}}+\tilde{\boldsymbol{x}}, \boldsymbol{u}) = \tilde{\boldsymbol{g}}(\tilde{\boldsymbol{x}}, \boldsymbol{u}).\end{aligned}$$

Since this transformation is always possible, we assume hereafter that the equilibrium point of interest has been shifted to zero if it is not already at this point. Figure 1.13 illustrates this transformation.

In order to characterize the behavior of the trajectories of a system in the neighborhood of an equilibrium point $\boldsymbol{x}_{\text{eq}} = \boldsymbol{0}$, we will first introduce the term *attractivity*.

Definition 2 (Attractivity). *Let a system*

$$\dot{\boldsymbol{x}} = \boldsymbol{f}(\boldsymbol{x}, \boldsymbol{u})$$

possess the equilibrium point $\boldsymbol{x}_{\text{eq}} = \boldsymbol{0}$. The equilibrium point $\boldsymbol{x}_{\text{eq}} = \boldsymbol{0}$ is called locally attractive if a neighborhood $U(\boldsymbol{0})$ exists around the equilibrium point such that every initial vector

$$\boldsymbol{x}(0) \in U(\boldsymbol{0})$$

leads to a trajectory $\boldsymbol{x}(t)$ of the free system, i. e. $\boldsymbol{u} = \boldsymbol{0}$, which tends to the equilibrium point $\boldsymbol{x}_{\text{eq}} = \boldsymbol{0}$ for $t \to \infty$. If every trajectory of the free system tends to zero for $t \to \infty$, the equilibrium point is called globally attractive.

The set of initial points $\boldsymbol{x}(0)$ which lead to trajectories all tending to the equilibrium point $\boldsymbol{x}_{\text{eq}} = \boldsymbol{0}$ is called the *region of attraction.*

Figure 1.14 illustrates the concept of attractivity. The attractivity of an equilibrium point therefore ensures that every trajectory that begins at $U(\boldsymbol{0})$ tends to the equilibrium point. However, the term attractivity does not tell us how far the trajectory departs from the equilibrium point $\boldsymbol{x}_{\text{eq}} = \boldsymbol{0}$. From a practical point of view, this can be problematic. Basically, for real systems, we would like to know which possibly dangerously large values the state $\boldsymbol{x}$ of the system can take before converging to the equilibrium point. The subsequent stability term is more detailed in this respect.

Definition 3 (Lyapunov Stability). *Let a system*

$$\dot{\boldsymbol{x}} = \boldsymbol{f}(\boldsymbol{x}, \boldsymbol{u})$$

possess the equilibrium point $\boldsymbol{x}_{\text{eq}} = \boldsymbol{0}$. This equilibrium point is called locally Lyapunov stable, or Lyapunov stable for short, if for every ε-neighborhood

$$U_\varepsilon(\boldsymbol{0}) = \{\boldsymbol{x} \in I\!R^n \mid |\boldsymbol{x}| < \varepsilon\}$$

a δ-neighborhood

$$U_\delta(\boldsymbol{0}) = \{\boldsymbol{x} \in I\!R^n \mid |\boldsymbol{x}| < \delta\}$$

exists such that all trajectories $\boldsymbol{x}(t)$ of the free system which start within the δ-neighborhood, i. e.

$$\boldsymbol{x}(0) \in U_\delta(\boldsymbol{0}),$$

remain in the ε-neighborhood along their further course, i. e.

$$\boldsymbol{x}(t) \in U_\varepsilon(\boldsymbol{0}) \quad \text{for all} \quad t > 0.$$

If, in addition, there is an ϵ-neighborhood $U_\varepsilon(\boldsymbol{0})$ for every δ-neighborhood $U_\delta(\boldsymbol{0})$, the equilibrium point $\boldsymbol{x}_{\text{eq}} = \boldsymbol{0}$ is called globally Lyapunov stable.

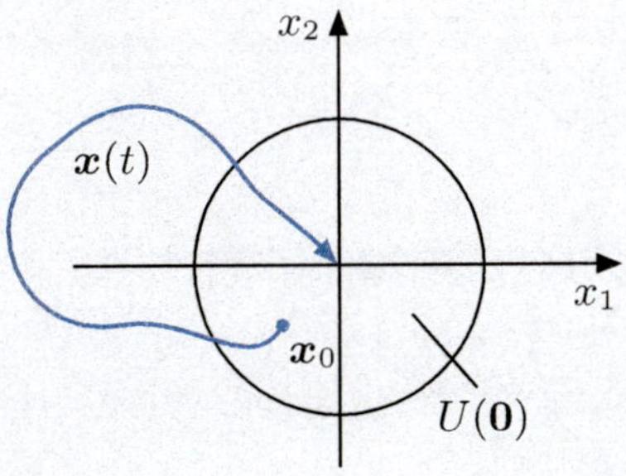

Fig. 1.14: Attractive equilibrium point

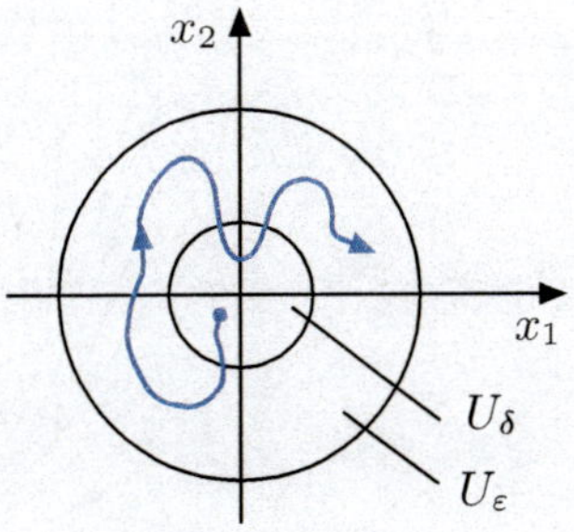

Fig. 1.15: Illustration of the definition of Lyapunov stability

Figure 1.15 illustrates the above definition of stability by Lyapunov. Note that the trajectories $\boldsymbol{x}(t)$ do not necessarily have to tend to the equilibrium point $\boldsymbol{x}_{\text{eq}} = \mathbf{0}$ for an equilibrium point to be Lyapunov stable[1]. A specific example of this case is the harmonic oscillator

$$\dot{\boldsymbol{x}} = \begin{bmatrix} 0 & 1 \\ -1 & 0 \end{bmatrix} \boldsymbol{x},$$

whose trajectories $\boldsymbol{x}(t)$ we already saw in Figure 1.10 and whose equilibrium point is globally Lyapunov stable.

As we know, an attractive equilibrium point is not necessarily Lyapunov stable. This is illustrated by the system

$$\begin{aligned} \dot{x}_1 &= x_1\left(1 - \sqrt{x_1^2 + x_2^2}\right) - \frac{x_2}{2}\left(1 - \frac{x_1}{\sqrt{x_1^2 + x_2^2}}\right), \\ \dot{x}_2 &= x_2\left(1 - \sqrt{x_1^2 + x_2^2}\right) + \frac{x_1}{2}\left(1 - \frac{x_1}{\sqrt{x_1^2 + x_2^2}}\right). \end{aligned} \tag{1.11}$$

It is not defined in $\boldsymbol{x} = \mathbf{0}$, but can be continuously continued there by $\dot{\boldsymbol{x}} = \mathbf{0}$. This continued system possesses two equilibrium points

$$\boldsymbol{x}_{\text{eq1}} = \begin{bmatrix} 0 \\ 0 \end{bmatrix} \quad \text{and} \quad \boldsymbol{x}_{\text{eq2}} = \begin{bmatrix} 1 \\ 0 \end{bmatrix},$$

neither of which is stable. However, in contrast to the equilibrium point $\boldsymbol{x}_{\text{eq1}}$, the equilibrium point $\boldsymbol{x}_{\text{eq2}}$ is locally attractive.

Figure 1.16 illustrates this. All trajectories which begin in a neighborhood of $\boldsymbol{x}_{\text{eq1}}$ tend away from this equilibrium point. On the other hand, all trajectories with initial values

$$\boldsymbol{x}(0) \in \mathbb{R}^n \backslash \{\mathbf{0}\}$$

[1] The terms *stable* and *stable in the sense of Lyapunov* are used as synonyms for the term *Lyapunov stable*.

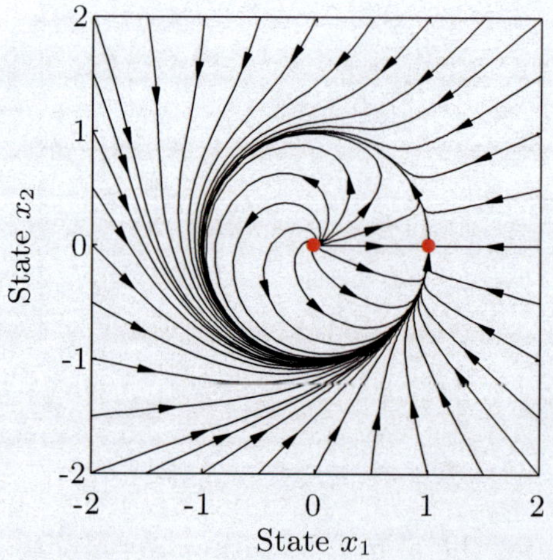

Fig. 1.16: A system for which the equilibrium point $\boldsymbol{x}_{\text{eq2}}^T = [1\ 0]$ is attractive but not Lyapunov stable

converge to the attractive equilibrium point $\boldsymbol{x}_{\text{eq2}}$ for $t \to \infty$. Notably, a special trajectory $\boldsymbol{x}_{\text{uc}}(t)$, which runs counterclockwise on the unit circle, also tends to $\boldsymbol{x}_{\text{eq2}}$. In particular, we consider the case in which this trajectory, which is given by

$$\boldsymbol{x}_{\text{uc}}(t) = \begin{bmatrix} 1 - \dfrac{2}{1+z^2(t)} \\ -\dfrac{2z(t)}{1+z^2(t)} \end{bmatrix}, \qquad z(t) = \frac{t}{2} - \operatorname{sgn}(x_2(0)) \cdot \sqrt{\frac{1+x_1(0)}{1-x_1(0)}},$$

starts on the unit circle arbitrarily close to $\boldsymbol{x}_{\text{eq2}}$ with positive initial values $x_1(0)$ and $x_2(0)$. Under such conditions, $\boldsymbol{x}_{\text{uc}}(t)$ initially depart away from the equilibrium point $\boldsymbol{x}_{\text{eq2}}$, and then, running on the unit circle, it tends to the equilibrium point $\boldsymbol{x}_{\text{eq2}}$ for $t \to \infty$. Since the trajectory first departs far from $\boldsymbol{x}_{\text{eq2}}$ even if it starts arbitrarily close to $\boldsymbol{x}_{\text{eq2}}$ in the positive quadrant, it follows that it is not possible for every ε-neighborhood $U_\epsilon(\boldsymbol{x}_{\text{eq2}})$ of $\boldsymbol{x}_{\text{eq2}}$ to define a δ-neighborhood $U_\delta(\boldsymbol{x}_{\text{eq2}})$, for which all trajectories that start within $U_\delta(\boldsymbol{x}_{\text{eq2}})$ remain in $U_\epsilon(\boldsymbol{x}_{\text{eq2}})$ along their further course. Therefore, the conditions of Definition 3 are not fulfilled and therefore the equilibrium point $\boldsymbol{x}_{\text{eq2}}$ is not Lyapunov stable.

For practical applications in control, equilibrium points which are both Lyapunov stable and attractive are of central importance. This is because the combination of these two properties ensures that a system can permanently maintain a specific nominal state, i. e. an equilibrium point. Accordingly, the term *asymptotic stability* is defined in

Definition 4 (Asymptotic Stability). *If an equilibrium point* $\boldsymbol{x}_{\text{eq}} = \mathbf{0}$ *is locally (globally) attractive and locally (globally) Lyapunov stable, it is called locally (globally) asymptotically stable.*

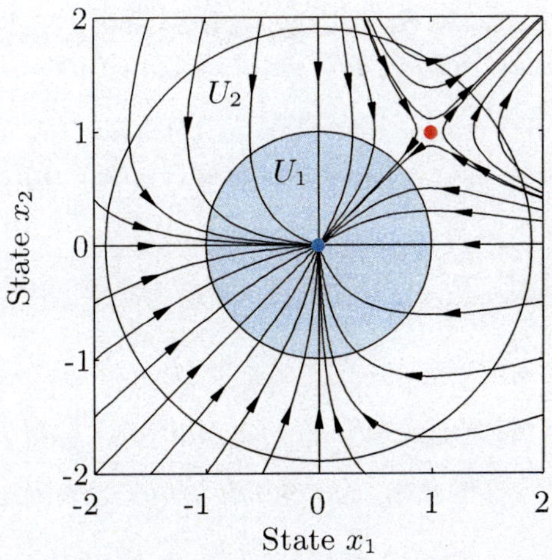

Fig. 1.17: Asymptotically stable equilibrium point $\boldsymbol{x}_{\mathrm{eq}} = \mathbf{0}$ and a corresponding catchment region U_1 (blue). The neighborhood U_2 is not a catchment region.

For an asymptotically stable equilibrium point $\boldsymbol{x}_{\mathrm{eq}}$, the neighborhood $U(\boldsymbol{x}_{\mathrm{eq}})$ within which all starting trajectories tend to the equilibrium point is also of interest. Not every neighborhood has this property, as illustrated by the example of equation (1.10) and the accompanying Figure 1.17. In the neighborhood U_1 of the equilibrium point $\boldsymbol{x}_{\mathrm{eq}} = \mathbf{0}$, all trajectories $\boldsymbol{x}(t)$ tend to the origin. For the neighborhood U_2, it is clearly visible that this is not the case.

For such situations, we define the term *catchment region* as follows:

Definition 5 (Catchment Region). *A neighborhood of an asymptotically stable equilibrium point is called a catchment region if all trajectories that begin within this neighborhood both remain within it and tend to the equilibrium point in their further course.*

Definition 6 (Region of Asymptotic Stability). *The largest catchment region is called the region of asymptotic stability.*

The region of asymptotic stability is also often called the *basin of the equilibrium point.*

If there is only one globally asymptotically stable equilibrium point, then the entire state space is the catchment region. From the engineering point of view, this is the desirable case, and engineers attempt to achieve it in control-loop design and other applications. In any case, the size of an equilibrium point's region of asymptotic stability is of great importance in practical applications.

So far, we have discussed the stability of equilibria. However, we are also interested in the system's stability behavior as a whole. In the case of a system with a globally asymptotically stable equilibrium point, this is straightforward, and we can formulate

Definition 7 (Asymptotic Stability of a System). *A system with a globally asymptotically stable equilibrium point is called asymptotically stable.*

However, the situation is more complex in the case of multiple equilibrium points. This is why we will now introduce a more generalized definition of the stability of a system:

Definition 8 (Stability of a System). *Let us consider a system*

$$\dot{\boldsymbol{x}} = \boldsymbol{f}(\boldsymbol{x}, \boldsymbol{u}).$$

We call the system stable if the following conditions hold for the free system:

(1) There is a bounded set S in which all trajectories remain during their course.
(2) There is a bounded set $H \supset S$ for every bounded set $G \supset S$ such that all trajectories beginning in H remain in G during their course.
(3) There is a set $G \supset S$ for every set H.

This definition enables us to estimate the behavior of the system as a whole and not just its behavior around its equilibrium points. Notice that Definition 8 provides us with a global definition of stability which is similar to Lyapunov's global definition of stability. This kind of stability of a system ensures that there is no solution tending to infinity. However, it does not ensure that all trajectories converge to a single equilibrium point. It is also possible that trajectories do not run into an equilibrium point and therefore $\dot{\boldsymbol{x}}(t)$ does not become zero. The simple harmonic oscillator, for example, is a stable system.

If we want to ensure that the trajectories of a stable system all converge to a bounded set, we have to formulate a stricter version of Definition 8. We then obtain

Definition 9 (Strict Stability of a System). *We call a stable system $\dot{\boldsymbol{x}} = \boldsymbol{f}(\boldsymbol{x}, \boldsymbol{u})$ strictly stable if there is a bounded set S such that $\boldsymbol{x}(t \to \infty) \in S$ holds for all trajectories $\boldsymbol{x}(t)$ of the free system.*

It is worth noting that the size of the set S is not specified in Definition 9. It can be of any size. In practical applications, however, the size of the set S does matter. In most cases, it should be as small as possible. As in Definition 8, the trajectories $\boldsymbol{x}(t)$ within the set S do not have to converge or tend to stable equilibrium points. System (1.11), for example, is strictly stable even though its two equilibrium points are unstable.

1.1.7 Exponential Stability of Equilibrium Points

The term *asymptotically stable* signifies that the trajectories $\boldsymbol{x}(t)$ of a system for $t \to \infty$ tend to the corresponding equilibrium point. However, it does not provide us with a measure of how quickly this happens. For example, let us consider the linear asymptotically stable system

$$\dot{x} = -\lambda x, \quad x(0) = x_0, \quad \lambda > 0.$$

The system's solution is given by

$$x(t) = x_0 e^{-\lambda t}.$$

We can see that the solution of the system not only tends asymptotically to zero but also decreases exponentially fast.

Accordingly, for all systems whose solutions decrease toward zero exponentially or faster than exponentially, we can now define the term *exponential stability*.

Definition 10 (Exponential Stability). *If a system $\dot{\boldsymbol{x}} = \boldsymbol{f}(\boldsymbol{x}, \boldsymbol{u})$ possesses an asymptotically stable equilibrium point at $\boldsymbol{x} = \boldsymbol{0}$, this equilibrium point is called locally exponentially stable if positive constants m, α, and δ exist such that*

$$|\boldsymbol{x}(t)| \leq m e^{-\alpha t} |\boldsymbol{x}(0)| \tag{1.12}$$

holds for the free system for all $|\boldsymbol{x}(0)| < \delta$ and all $t \geq 0$. If, additionally, the above inequality is fulfilled for every initial vector $\boldsymbol{x}(0)$ of the system, the equilibrium point is called globally exponentially stable.

In this context, we will point out that an exponentially stable equilibrium point is always asymptotically stable, i. e.

exponentially stable $\Rightarrow$ asymptotically stable.

Furthermore, the asymptotically stable equilibrium point $\boldsymbol{x}_{\text{eq}} = \boldsymbol{0}$ of a linear system is exponentially stable, since all eigen-motions, i. e. the solutions of the linear system for $\boldsymbol{u} = \boldsymbol{0}$, decay exponentially. The largest possible constant α in equation (1.12) is called the *convergence rate* of the system.

As an example, let us consider the system

$$\dot{x} = -x^3, \quad x(0) = x_0,$$

and its equilibrium point $x_{\text{eq}} = 0$. The equilibrium point is asymptotically stable but not exponentially stable. We can deduce this from the system solution

$$x(t) = \frac{x_0}{\sqrt{2x_0^2 t + 1}}$$

because there is a time t_0 for every $m, \alpha > 0$ so that the inequality

$$|x(t)| = \frac{|x_0|}{\sqrt{2x_0^2 t + 1}} > m e^{-\alpha t} |x_0|, \quad t > t_0, \tag{1.13}$$

is fulfilled. This is because equation (1.13) is equivalent to

$$\frac{1}{m}e^{\alpha t} > \sqrt{2x_0^2 t + 1}, \quad t > t_0.$$

Hence, the system's equilibrium point $x_{\mathrm{eq}} = 0$ is not exponentially stable because equation (1.12) from Definition 10 is not fulfilled.

We can determine that for linear systems, asymptotic stability entails exponential stability. On the other hand, there are nonlinear systems with asymptotically stable equilibrium points that are not exponentially stable.

1.1.8 Instability of Equilibrium Points

Intuitively, we expect that an equilibrium point which is not stable, or more precisely formulated, which is not Lyapunov stable, should be called unstable. Such an equilibrium point is characterized by at least one trajectory which begins in an arbitrarily small neighborhood of the equilibrium point and then departs from it. Accordingly, we can formulate

Definition 11 (Instability). *An equilibrium point is called unstable if it is not Lyapunov stable.*

However, we must point out that unstable equilibrium points may well be attractive, as illustrated by our example of system (1.11). Its two equilibrium points are unstable; nevertheless, its equilibrium point

$$\boldsymbol{x}_{\mathrm{eq2}} = \begin{bmatrix} 1 \\ 0 \end{bmatrix}$$

is attractive. This means that all trajectories which start within an arbitrarily small neighborhood around it finally converge to it. In their course, the trajectories are bounded and never depart from the region that lies inside a circle with a diameter greater than one after they have reached it. Thus no trajectory tends to infinity.

However, systems exist with an equilibrium point for which trajectories which start within an arbitrarily small neighborhood around it not only depart from the equilibrium point, but also tend to infinity. From an engineering perspective, such behavior is more dangerous for the functionality of a system than the case of an unstable equilibrium point, for which no trajectory tends to infinity. It is thus reasonable to introduce an instability definition that goes beyond the concept given in Definition 11.

Definition 12 (Strict Instability). *Let a system $\dot{\boldsymbol{x}} = \boldsymbol{f}(\boldsymbol{x}, \boldsymbol{u})$ possess an equilibrium point at $\boldsymbol{x} = \boldsymbol{0}$ for $\boldsymbol{u} = \boldsymbol{0}$. This equilibrium point is called strictly unstable if, for every neighborhood U of the equilibrium point, there is at least one trajectory $\boldsymbol{x}(t)$ of the free system which begins at $\boldsymbol{x}(0) \in U$ and has the limit*

$$\lim_{t \to t_{\inf}} |\boldsymbol{x}(t)| = \infty$$

for a time $t_{\inf} \in (0, \infty]$.

Of course, any strictly unstable equilibrium point is also unstable, but not vice versa. For example, the equilibrium points of system (1.11) are unstable, but not strictly unstable. Examples of strictly unstable equilibrium points are the equilibrium points $\boldsymbol{x}_{\text{eq}} = \boldsymbol{0}$ of linear systems with eigenvalues that possess a positive real part.

As a further example, let us consider the system

$$\begin{aligned} \dot{x}_1 &= x_1^2 - x_2^2, \\ \dot{x}_2 &= 2x_1x_2 \end{aligned} \tag{1.14}$$

with the initial values $x_1(0) = x_{10}$ and $x_2(0) = x_{20}$. It possesses the solution

$$\begin{aligned} x_1(t) &= \frac{x_{10} - \left(x_{10}^2 + x_{20}^2\right) t}{\left(x_{10}^2 + x_{20}^2\right) t^2 - 2x_{10}t + 1}, \\ x_2(t) &= \frac{x_{20}}{\left(x_{10}^2 + x_{20}^2\right) t^2 - 2x_{10}t + 1} \end{aligned} \tag{1.15}$$

and a single equilibrium point at $\boldsymbol{x} = \boldsymbol{0}$. Its trajectories $\boldsymbol{x}(t)$ are in circular shape and all tend to the equilibrium point $\boldsymbol{x}_{\text{eq}} = \boldsymbol{0}$. They are shown in Figure 1.18.

Although all trajectories tend to the equilibrium point $\boldsymbol{x}_{\text{eq}} = \boldsymbol{0}$, the system is unstable, since it does not fulfill the definition of Lyapunov stability that was given in Definition 3. In addition, the equilibrium point is also strictly unstable, since all trajectories beginning on the positive x_1-axis, i. e. trajectories with initial values $x_{10} > 0$ and $x_{20} = 0$, are given by

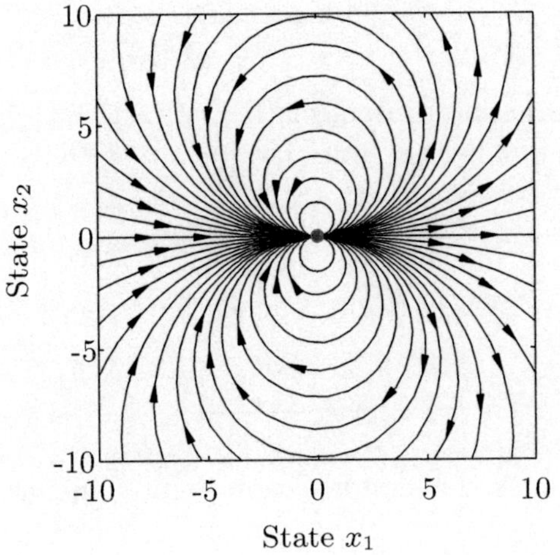

Fig. 1.18: Trajectories of the system given in equation (1.14)

$$\boldsymbol{x}(t) = \begin{bmatrix} \dfrac{x_{10}}{1 - x_{10}t} \\ 0 \end{bmatrix}. \tag{1.16}$$

For

$$t \to t_{\text{inf}} = \frac{1}{x_{10}},$$

they tend to infinity. If we now consider the time progression of the state x_1 for $x_{10} > 0$, as shown in Figure 1.19, we observe that the state $x_1(t)$ first reaches infinity for $t = 1/x_{10}$, and subsequently returns from infinity and tends asymptotically to zero via the negative x_1-axis for

$$t > t_{\text{inf}} = \frac{1}{x_{10}}.$$

If we include the point infinity in the progression of the trajectories, it follows that every trajectory (1.16) ultimately tends to zero. Below we will examine this behavior in more detail.

First we will consider the trajectory points

$$\boldsymbol{x}(t) = \begin{bmatrix} 0 \\ x_2(t) \end{bmatrix}$$

for all trajectories. We compute them for the initial vector

$$\boldsymbol{x}_0 = \begin{bmatrix} x_{10} \\ x_{20} \end{bmatrix} = \begin{bmatrix} 1 \\ x_{20} \end{bmatrix} \quad \text{with} \quad x_{20} > 0$$

and thereby obtain the trajectory points

$$\boldsymbol{x}_{\text{s}} = \begin{bmatrix} 0 \\ \dfrac{x_{20}^2 + 1}{x_{20}} \end{bmatrix},$$

at which the trajectories $\boldsymbol{x}(t)$ intersect the x_2-axis.

We note that the smaller the initial value x_{20} with

$$0 < x_{20} \leq 1$$

becomes, i. e. the more closely it approaches the x_1-axis, the larger the intersection point

$$x_{\text{s}2} = \frac{x_{20}^2 + 1}{x_{20}}$$

with the x_2-axis becomes. This intersection point coincides with the diameter

$$d = x_{\text{s}2}$$

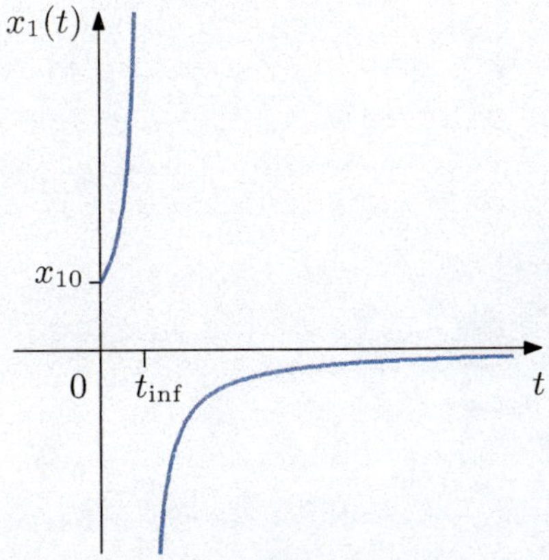

Fig. 1.19: Time course of the solution $x_1(t)$ for $x_{10} > 0$ and $x_{20} = 0$

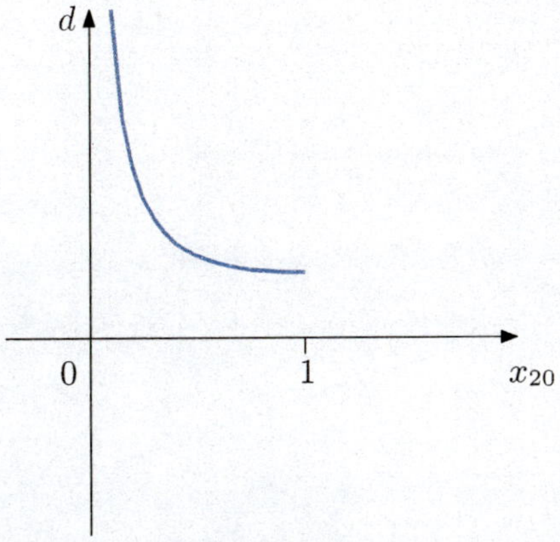

Fig. 1.20: Diameter d of the trajectory with dependence on x_{20}

of the associated circular trajectory. Figure 1.20 illustrates this progression. For the limit $x_{20} \to 0$, the diameter and the intersection point of the associated trajectory with the x_2-axis become infinite. We can therefore envision this trajectory as starting on the x_1-axis with the initial values $x_{10} = 1$ and $x_{20} = 0$ and progressing along this axis to infinity. It follows the other trajectories in their circular path, but passes through infinity from the positive real axis to the negative real axis and then to the equilibrium point $\boldsymbol{x}_{\text{eq}} = \mathbf{0}$. Thus this trajectory progresses along a circle of infinite radius. However, we have not taken into account the fact that infinity does not belong to the set of real numbers. This means that the trajectory $\boldsymbol{x}(t)$, $t \in I\!R$, is not defined for

$$t = t_{\text{inf}} = \frac{1}{x_0}$$

in our example.

In the following, we will therefore be mathematically exact and utilize a result from topology. We extend the real plane through a so-called *one-point compactification* [56] by including infinity. This can be illustrated as a stereographic projection which uniquely maps the real plane, which has been extended to include infinity, onto a sphere. Here the sphere is tangent to the plane that lies beneath it. Figure 1.21 shows this. From the north pole of the sphere, we can draw a line to every point $\boldsymbol{x}_{\text{p}}$ on the plane and obtain an intersection point $\boldsymbol{x}_{\text{c}}$ with the surface of the sphere. The point $\boldsymbol{x}_{\text{c}}$ is a mapping of the point $\boldsymbol{x}_{\text{p}}$ lying on the plane. In this way, every point on the plane can be bijectively mapped to a point on the surface of the sphere. The north pole corresponds to infinity in the extended plane, while the south pole corresponds to the origin of the extended plane. The trajectory that starts in the plane on the x_1-axis at $x_{10} > 0$ is projected into the north pole of the sphere when it progresses to infinity. From there, it progresses further down to the opposite side of the sphere to its south pole, which corresponds to the

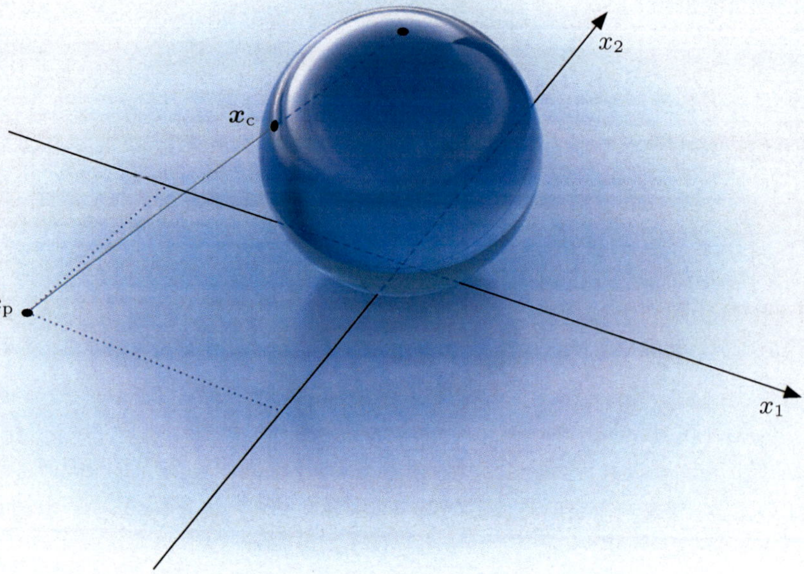

Fig. 1.21: Illustration of the one-point compactification of $I\!R^2$. The north pole corresponds to infinity in the real plane.

progression of the trajectory on the negative real axis to the equilibrium point $\boldsymbol{x}_{\text{eq}} = \mathbf{0}$. In conclusion, all trajectories (1.15) tend to the equilibrium point $\boldsymbol{x}_{\text{eq}} = \mathbf{0}$ for $t \to \infty$. The equilibrium point is therefore globally attractive. As noted previously, however, it is also strictly unstable.

Thus systems with attractive, but unstable, or even strictly unstable equilibrium points also exist. We will conclude our descriptions of stability and instability with a visual summary of their relations. For this purpose, Figure 1.22 illustrates the sets

M_{us} = set of unstable equilibrium points,

M_{sus}= set of strictly unstable equilibrium points, $M_{\text{sus}} \subset M_{\text{us}}$,

M_{Ls} = set of Lyapunov stable equilibrium points, $M_{\text{Ls}} \cap M_{\text{us}} = \emptyset$,

M_{as} = set of asymptotically stable equilibrium points, $M_{\text{as}} \subset M_{\text{Ls}}$,

M_{es} = set of exponentially stable equilibrium points, $M_{\text{es}} \subset M_{\text{as}}$,

M_{att}= set of attractive equilibrium points, $M_{\text{as}} \subset M_{\text{att}}$, $M_{\text{att}} \cap M_{\text{sus}} \neq \emptyset$.

At this point, it becomes clear once again that the term *attractivity* is unsuited to describe the stability behavior of an equilibrium point.

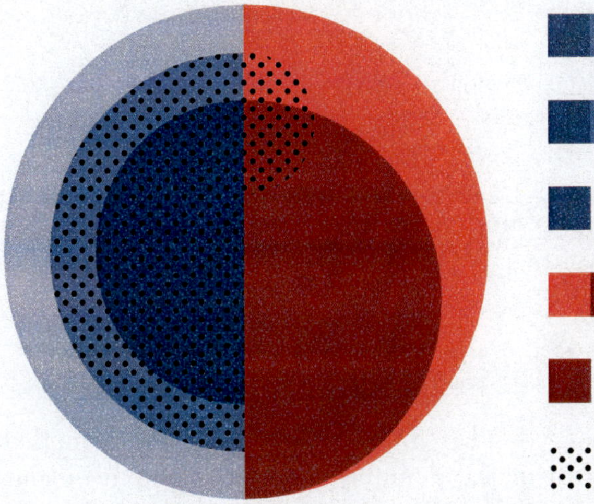

Fig. 1.22: The sets describing different types of stable, unstable, and attractive equilibrium points and the relations among them

1.1.9 Stability in the Case of Variable Input Signals

So far we have dealt with the stability of the equilibrium point $\boldsymbol{x}_{\mathrm{eq}} = \mathbf{0}$ of a system

$$\begin{aligned}\dot{\boldsymbol{x}} &= \boldsymbol{f}(\boldsymbol{x}, \boldsymbol{u}),\\ \boldsymbol{y} &= \boldsymbol{g}(\boldsymbol{x}, \boldsymbol{u})\end{aligned}$$

for $\boldsymbol{u} = \mathbf{0}$. In addition, we are also interested in how the state vector $\boldsymbol{x}(t)$ and the output variable vector $\boldsymbol{y}(t)$, or more precisely their absolute values $|\boldsymbol{x}(t)|$ and $|\boldsymbol{y}(t)|$, change with dependence on the input signal $\boldsymbol{u}(t)$. From the engineer's practical point of view, it is desirable for bounded input signals to produce bounded output signals and bounded state variables.

In simple words, we term a system input-to-state stable [430] if the course of the state vector $\boldsymbol{x}(t)$ remains bounded for a bounded course of the input variable vector $\boldsymbol{u}(t)$. For mathematical purposes, we will formulate this more precisely in

Definition 13 (Input-to-State Stability). *A system*

$$\dot{\boldsymbol{x}} = \boldsymbol{f}(\boldsymbol{x}, \boldsymbol{u})$$

is called input-to-state stable if, for all its initial state vectors $\boldsymbol{x}_0 = \boldsymbol{x}(0)$ and for every bounded input function $\boldsymbol{u}(t)$,

(1) there exists a function $\beta(|\boldsymbol{x}_0|, t)$ with $\beta(0, t) = 0$ for all $t \geq 0$ and $\beta(|\boldsymbol{x}_0|, t \to \infty) = 0$ for all $\boldsymbol{x}_0 \in \mathbb{R}^n$, which is a strictly increasing continuous function of its first argument $|\boldsymbol{x}_0|$ and a strictly decreasing continuous function of its second argument t, and

(2) there exists a strictly increasing continuous function γ with $\gamma(0) = 0$ such that

$$|\boldsymbol{x}(t)| \leq \beta(|\boldsymbol{x}_0|, t) + \gamma(\sup_{t\geq 0}(|\boldsymbol{u}(t)|))$$

holds for all $t \geq 0$.

In the literature, input-to-state stability is usually abbreviated to ISS. The above definition of stability ensures that the state vector $\boldsymbol{x}(t)$ takes values that lie within a hypersphere with radius

$$\beta(|\boldsymbol{x}_0|, t) + \gamma(\sup_{t\geq 0}(|\boldsymbol{u}(t)|)),$$

therefore remaining bounded for bounded input signals $\boldsymbol{u}(t)$ and bounded initial values $\boldsymbol{x}_0$. If we set $\boldsymbol{u} = \boldsymbol{0}$ in the defining equation of input-output stability, we see that all trajectories $\boldsymbol{x}(t)$ of an input-to-state stable system tend to zero for $t \to \infty$. Consequently, an input-to-state stable system has a globally asymptotically stable equilibrium point. However, in general, the reverse conclusion does not hold.

Below we will determine the functions β and γ for a linear system

$$\dot{\boldsymbol{x}} = \boldsymbol{A}\boldsymbol{x} + \boldsymbol{B}\boldsymbol{u} \tag{1.17}$$

using the system's solution [133]

$$\boldsymbol{x}(t) = e^{\boldsymbol{A}t}\boldsymbol{x}_0 + \int_0^t e^{\boldsymbol{A}(t-\tau)}\boldsymbol{B}\boldsymbol{u}(\tau)d\tau. \tag{1.18}$$

Based on this, we obtain an upper bound of the absolute value $\boldsymbol{x}(t)$ using

$$|\boldsymbol{x}(t)| = |e^{\boldsymbol{A}t}\boldsymbol{x}_0 + \int_0^t e^{\boldsymbol{A}(t-\tau)}\boldsymbol{B}\boldsymbol{u}(\tau)d\tau| \leq |e^{\boldsymbol{A}t}\boldsymbol{x}_0| + |\int_0^t e^{\boldsymbol{A}(t-\tau)}\boldsymbol{B}\boldsymbol{u}(\tau)d\tau|. \tag{1.19}$$

Firstly, the inequality

$$|e^{\boldsymbol{A}t}\boldsymbol{x}_0| \leq me^{-\alpha t} \cdot |\boldsymbol{x}_0| = \beta(|\boldsymbol{x}_0|, t), \quad \text{for some } m, \alpha > 0\ , \tag{1.20}$$

holds if system (1.17) is asymptotically stable and is thus also exponentially stable, since all asymptotically stable linear systems are exponentially stable. Here m and α must be chosen appropriately. The function

$$\beta(|\boldsymbol{x}_0|, t) = me^{-\alpha t} \cdot |\boldsymbol{x}_0|$$

is a strictly increasing continuous function of its first argument $|\boldsymbol{x}_0|$ and a strictly decreasing continuous function of its second argument t. We have therefore found an appropriate function β according to Definition 13.

Secondly, the following holds for the integral in equation (1.18)

$$\left|\int_0^t e^{\boldsymbol{A}(t-\tau)}\boldsymbol{B}\boldsymbol{u}(\tau)d\tau\right| \leq \int_0^t |e^{\boldsymbol{A}(t-\tau)}\boldsymbol{B}\boldsymbol{u}(\tau)|d\tau \leq \int_0^t me^{-\alpha(t-\tau)}|\boldsymbol{B}\boldsymbol{u}(\tau)|d\tau. \tag{1.21}$$

With μ denoting the largest eigenvalue of the matrix $\boldsymbol{B}^T\boldsymbol{B}$, inequality

$$|\boldsymbol{B}\boldsymbol{u}(t)| \leq \sqrt{\mu}\cdot \sup_{t\geq 0}(|\boldsymbol{u}(t)|)$$

holds [40], hence

$$\begin{aligned}
\int_0^t me^{-\alpha(t-\tau)}|\boldsymbol{B}\boldsymbol{u}(\tau)|d\tau &\leq m\sqrt{\mu}\sup_{t\geq 0}(|\boldsymbol{u}(t)|)\cdot \int_0^t e^{\alpha(\tau-t)}d\tau \\
&= m\sqrt{\mu}\sup_{t\geq 0}(|\boldsymbol{u}(t)|)\frac{1}{\alpha}(1-e^{-\alpha t}) \\
&\leq \frac{m\sqrt{\mu}}{\alpha}\sup_{t\geq 0}(|\boldsymbol{u}(t)|) = \gamma(\sup_{t\geq 0}(|\boldsymbol{u}(t)|)).
\end{aligned}$$

Using the bounds given in equations (1.19), (1.20), and (1.21), we obtain inequality

$$|\boldsymbol{x}(t)| \leq \underbrace{me^{-\alpha t}\cdot|\boldsymbol{x}_0|}_{\beta(|\boldsymbol{x}_0|,t)} + \underbrace{\frac{m\sqrt{\mu}}{\alpha}\sup_{t\geq 0}(|\boldsymbol{u}(t)|)}_{\gamma(\sup_{t\geq 0}(|\boldsymbol{u}(t)|))}.$$

Using Definition 13, from the asymptotic stability of a linear system (1.17), i. e. $\alpha > 0$, we can now conclude its input-to-state stability. Thus we arrive at

Theorem 1 (Input-to-State Stability of Linear Systems). *A linear system is input-to-state stable if and only if it is asymptotically stable.*

As mentioned above, this equivalence does not generally apply to nonlinear systems; additionally, the functions β and γ usually cannot be determined for nonlinear systems. To illustrate the first, let us consider the system

$$\dot{x} = -x(1-u).$$

For the input value $u = 0$, this system possesses a globally asymptotically stable equilibrium point. For values $u > 1$, however, this system becomes strictly unstable, which means that it is not input-to-state stable.

Analogously to the input-to-state stability, we can define the *input-output stability* as follows [432, 433]:

Definition 14 (Input-Output Stability). *A system*

$$\dot{\boldsymbol{x}} = \boldsymbol{f}(\boldsymbol{x}, \boldsymbol{u}),$$
$$\boldsymbol{y} = \boldsymbol{g}(\boldsymbol{x}, \boldsymbol{u})$$

is called input-output stable if, for all its initial state vectors $\boldsymbol{x}_0 = \boldsymbol{x}(0)$ and for every bounded input function $\mathbf{u}(t)$, there exist two functions β and γ according to Definition 13, so that

$$|\boldsymbol{y}(t)| \leq \beta(|\boldsymbol{x}_0|, t) + \gamma(\sup_{t\geq 0}(|\boldsymbol{u}(t)|))$$

holds for all $t \geq 0$.

Again, the case of linear systems is simple. We obtain the only sufficient

Theorem 2 (Input-Output Stability of Linear Systems). *An asymptotically stable linear system is input-output stable.*

However, if a linear system has a coprime transfer function, the input-to-state stable, input-output stable, and asymptotically stable properties are equivalent to each other [427]. For the case of linear systems, the term *Bounded Input-Bounded Output stable (BIBO stable)* is commonly used as a synonym for input-output stability.

The simple example of the nonlinear system

$$\dot{x} = -x + u,$$
$$y = \frac{1}{x}$$

illustrates that not every system with a globally asymptotically stable equilibrium point $x_{\text{eq}} = 0$ is also input-output stable. If the trajectory $x(t)$ of this example tends to the globally asymptotically stable equilibrium point $x_{\text{eq}} = 0$, the output variable $y(t)$ becomes infinitely large.

1.1.10 Limit Cycles

In nonlinear systems, just as in linear systems, sustained oscillations may occur. For these oscillations, the system states are repeated periodically and the trajectory of a sustained oscillation is a closed curve. These oscillations are called *limit cycles.*

As an example, Figure 1.23 illustrates the sustained oscillation of the Van der Pol differential equation

$$\ddot{x}_1 - (1 - x_1^2)\dot{x}_1 + x_1 = 0.$$

The state-space model

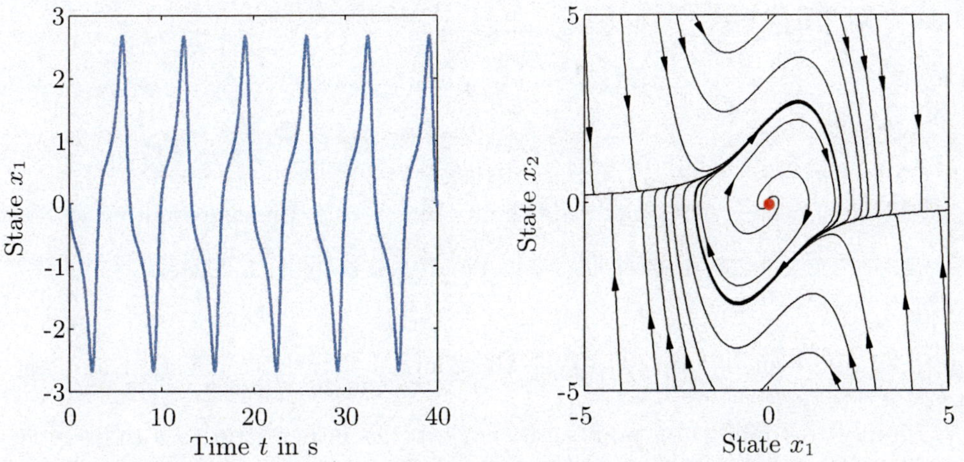

Fig. 1.23: The graph on the left shows the temporal variation of $x_1(t)$; on the right, the trajectories $\boldsymbol{x}(t)$ and the limit cycle of the Van der Pol differential equation.

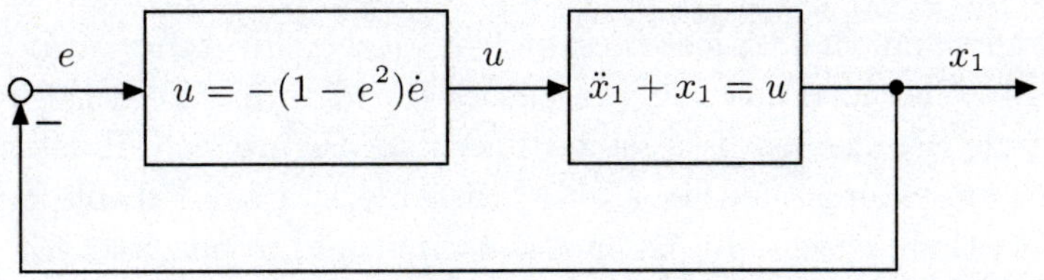

Fig. 1.24: Van der Pol oscillator, represented as a control loop

$$\dot{x}_1 = x_2,$$
$$\dot{x}_2 = -x_1 + (1 - x_1^2)x_2$$

is equivalent to the differential equation above.

The Van der Pol differential equation describes situations such as the behavior of an electric oscillating circuit for radio stations consisting of a triode, a capacitor, a coil, and a resistor. Here the term $(1 - x_1^2)$ acts as a nonlinear attenuator. We can also represent a Van der Pol oscillator as a control loop with a control error given by $e = -x_1$ and a nonlinear characteristic curve

$$u = f(e, \dot{e}) = -(1 - e^2)\dot{e}$$

as the control law. The linear differential equation

$$\ddot{x}_1 + x_1 = u$$

is the plant, as shown in Figure 1.24. The purpose of the control loop, in this case, is not to control the trajectory into an equilibrium point, but to maintain an oscillation. In this example, the limit cycle is produced deliberately. Normally, however, limit cycles are undesired in control loops. Generally, the

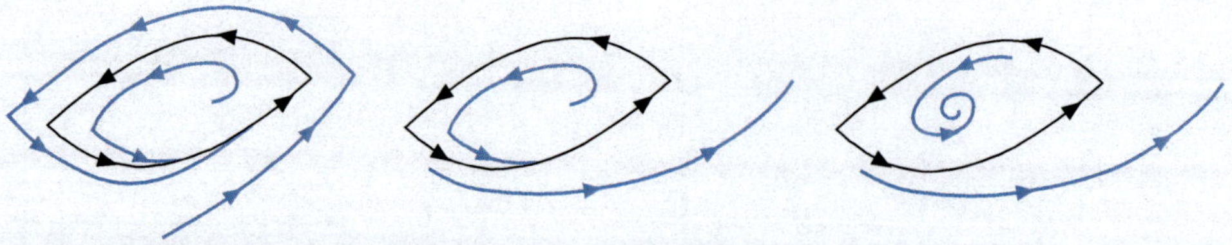

Fig. 1.25: Limit cycles (black) and their stability behavior

purpose of a control loop is to keep the controlled variable constant and not to generate oscillations.

Similar to equilibrium points, the trajectories either tend to a limit cycle or away from it. The term stability can thus be applied to limit cycles. Three cases can be distinguished. Firstly, there are asymptotically stable limit cycles, toward which all trajectories in the closer vicinity converge. Secondly, there are semi-stable limit cycles for which all trajectories on one side of the limit cycle converge and the trajectories on the other side tend away from the limit cycle. In the third case, all trajectories originating from the vicinity of the limit cycle diverge, so it is called unstable. Figure 1.25 illustrates these cases. In linear systems, neither stable, unstable, nor semi-stable limit cycles can occur. In these cases only harmonic oscillators are possible, for which an infinite number of closed trajectories exist. No other trajectories approach or tend away from these trajectories.

Unstable and semi-stable limit cycles are of no practical significance, since the trajectory always departs from the limit cycle when very small disturbances, which are always present in a real system, occur. However, what is most relevant to control theory is the stable limit cycle. In general, as mentioned before, it is undesired. To detect limit cycles in control loops, we typically use the method of harmonic balance, which will be discussed in detail in Section 2.1.

1.1.11 Sliding Modes

Apart from limit cycles, nonlinear systems exhibit further behavior which does not exist in linear systems. *Sliding modes* belong to this category of phenomena. They occur in systems with unsteady behavior, e. g. in systems $\dot{\boldsymbol{x}} = \boldsymbol{f}(\boldsymbol{x}, \boldsymbol{u})$ with discontinuous functions $\boldsymbol{f}$.

To explain the sliding mode phenomenon, we will consider the plant

$$\dot{\boldsymbol{x}} = \begin{bmatrix} 0 & 1 \\ -2 & -3 \end{bmatrix} \boldsymbol{x} + \begin{bmatrix} 0 \\ 1 \end{bmatrix} u,$$
$$y = \begin{bmatrix} 0.5 & 1 \end{bmatrix} \boldsymbol{x},$$

which possesses the transfer function

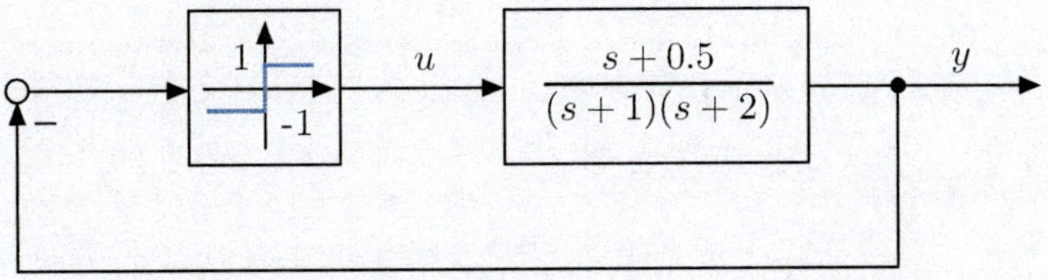

Fig. 1.26: Control loop with two-position controller

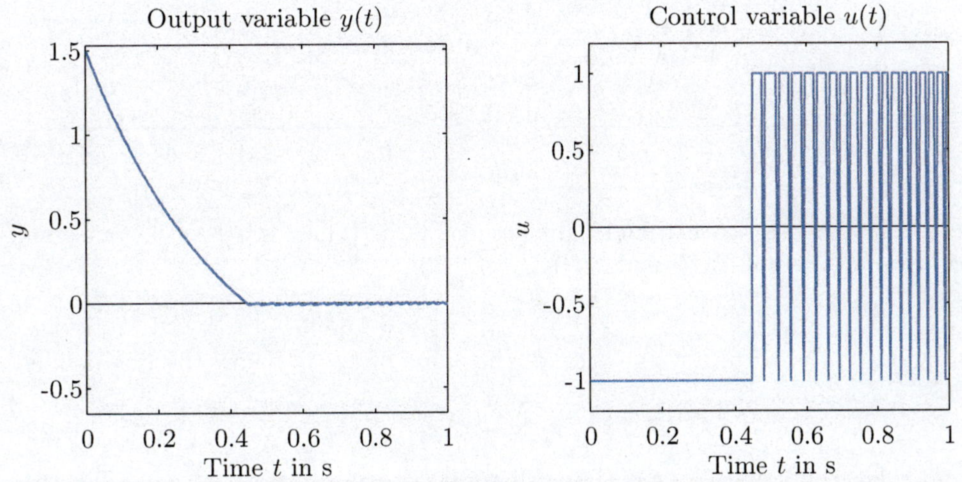

Fig. 1.27: Temporal course of the output variable $y(t)$ and control variable $u(t)$

$$G(s) = \frac{s + 0.5}{(s + 1)(s + 2)}.$$

As a controller, we use a two-position controller

$$u = \begin{cases} 1, & y \leq 0, \\ -1, & y > 0, \end{cases}$$

which is sometimes termed a *bang-bang controller*. Figure 1.26 shows the corresponding control loop. Simulating the system for the initial values $x_1(0) = 1$ and $x_2(0) = 1$, we obtain the courses of the output variable y and the control variable u which are plotted in Figure 1.27. Obviously, the output variable y tends to zero. The control variable u, on the other hand, from a certain point onward does not remain constant but switches back and forth between the values $u = 1$ and $u = -1$ with a high frequency.

This behavior can be explained by considering the trajectories $\boldsymbol{x}(t)$ of the system in the phase plane, as depicted in Figure 1.28, and analyzing the control law

$$u = \begin{cases} 1, & x_2 \leq -0.5x_1, \\ -1, & x_2 > -0.5x_1. \end{cases}$$

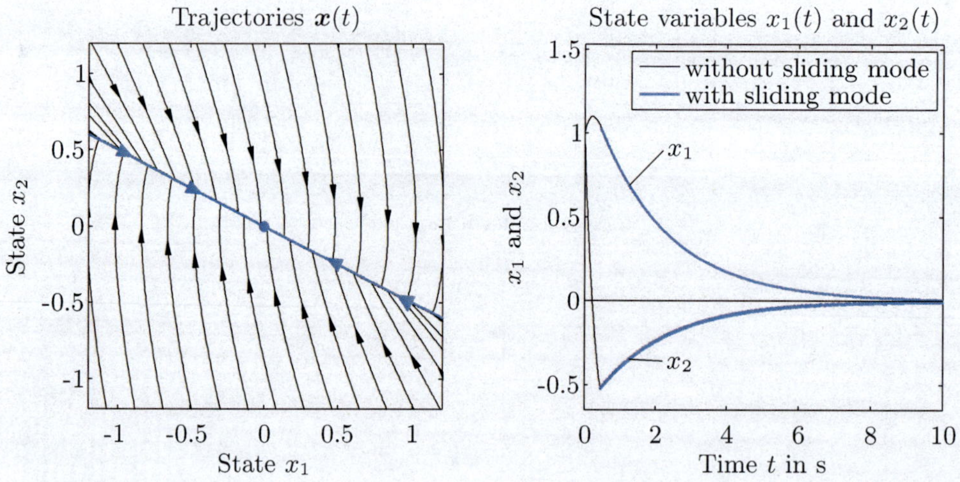

Fig. 1.28: Trajectories $\boldsymbol{x}(t)$ and sliding mode on the (blue) switching line and time courses of the states $x_1(t)$ and $x_2(t)$

Obviously, the actuator signal takes the value of $u = 1$ below the line defined by

$$x_2 = -0.5x_1 \tag{1.22}$$

and takes the value $u = -1$ above this line.

The trajectories $\boldsymbol{x}(t)$ converge from both sides onto the straight line described by equation (1.22). If a trajectory impinges onto the line, it briefly switches to the other side and the actuator signal jumps from $u = 1$ to $u = -1$, or vice versa. The trajectory then again impinges onto the line, briefly switches sides, and the process starts all over again. This explains the previously observed high-frequency switching of the actuator signal. The trajectory itself slides along the switching line, accompanied by high-frequency switching of the actuator signal, into the equilibrium point

$$\boldsymbol{x}_{\mathrm{eq}} = \boldsymbol{0}.$$

The term *sliding mode* is derived from this sliding behavior.

The behavior described above has the disadvantage that the actuator, which can be a valve or some other mechanical actuator, is heavily stressed and quickly wears out. Therefore sliding modes are usually undesired and should be avoided.

However, the sliding state also has an advantage. It can be shown that the sliding of the trajectory $\boldsymbol{x}(t)$ into the equilibrium point is robust against changes of the parameters of the plant. This means that the control system that is in a sliding mode always has the same dynamic, even if the plant changes. This behavior can therefore be exploited for the design of a certain class of robust controllers, the sliding mode controllers. We will examine this topic in Section 6.2.

1.1.12 Chaos

Chaos occurs in biological, meteorological, economic, and technical systems [50, 233, 451]. Specific examples of this are economic cycles, single-mode lasers, micromechanical oscillators, and the development of populations in ecological systems. The most important characteristic of a chaotic system is that we cannot say exactly how its state variables will develop in future. This is somewhat surprising, since chaotic systems can be described by ordinary differential equations with a deterministic behavior. In this context, the term *deterministic* excludes any kind of stochastic influences on the system.

Intuitively explained, chaotic behavior can be characterized by the following three properties:

(1) The trajectories run aperiodically, i. e. they do not tend to limit cycles.
(2) The trajectories do not tend to an equilibrium point or to infinity.
(3) Arbitrarily close initial values or vectors lead to very different progressions of trajectories.

As an example of a chaotic system, we will consider the double pendulum, which is shown in Figure 1.29. The masses m_1 and m_2 are appended from two pendulum rods of lengths l_1 and l_2, respectively. The joints allow for a free rotation of the pendulum; that is, the rotation angles Θ_1 and Θ_2 are not limited by stops. We will assume ideal conditions so that there is no friction

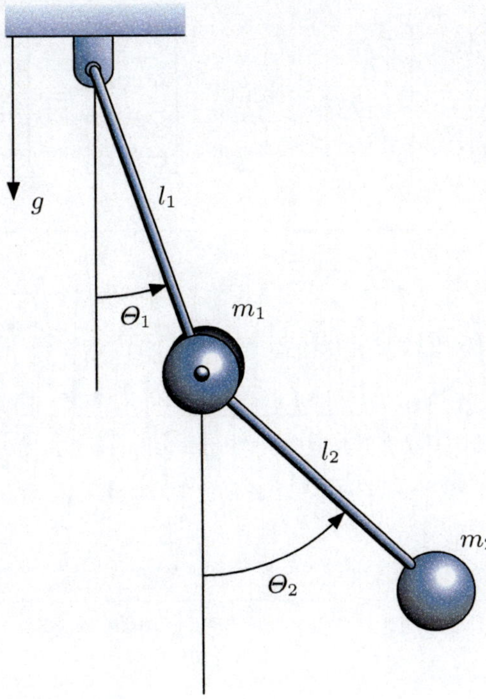

Fig. 1.29: Double pendulum with chaotic behavior

or other energy losses. Due to the gravitational acceleration $g = 9.81\,\mathrm{m\,s^{-2}}$, which acts upon the pendulum masses, the lossless double pendulum has a Lyapunov stable, but not asymptotically stable, equilibrium point for the angles

$$\Theta_1 = \Theta_2 = 0.$$

The system is described by the second-order differential equations [333]

$$\ddot{\Theta}_1 = \frac{g\,[\sin(\Theta_2)\cos(\Delta\Theta) - \mu\sin(\Theta_1)] - \sin(\Delta\Theta)\left[l_2\dot{\Theta}_2^2 + l_1\dot{\Theta}_1^2\cos(\Delta\Theta)\right]}{l_1\,[\mu - \cos^2(\Delta\Theta)]},$$

$$\ddot{\Theta}_2 = \frac{\mu g\,[\sin(\Theta_1)\cos(\Delta\Theta) - \sin(\Theta_2)] + \sin(\Delta\Theta)\left[\mu l_1\dot{\Theta}_1^2 + l_2\dot{\Theta}_2^2\cos(\Delta\Theta)\right]}{l_2\,[\mu - \cos^2(\Delta\Theta)]}$$

with

$$\Delta\Theta = \Theta_1 - \Theta_2,$$
$$\mu = 1 + \frac{m_1}{m_2}.$$

For a simulation, shown in Figure 1.30, we choose $m_1 = m_2 = 1\,\mathrm{kg}$ and $l_1 = l_2 = 1\,\mathrm{m}$ as the system's parameter values. The chaotic behavior of the double pendulum becomes evident if we consider two closely spaced initial vectors

$$\begin{bmatrix}\Theta_1(0)\\ \dot{\Theta}_1(0)\\ \Theta_2(0)\\ \dot{\Theta}_2(0)\end{bmatrix} = \begin{bmatrix}\pi/2\\ 0\\ -\pi/2\\ 0.01\end{bmatrix} \quad \text{and} \quad \begin{bmatrix}\tilde{\Theta}_1(0)\\ \dot{\tilde{\Theta}}_1(0)\\ \tilde{\Theta}_2(0)\\ \dot{\tilde{\Theta}}_2(0)\end{bmatrix} = \begin{bmatrix}\pi/2\\ 0\\ -\pi/2\\ 0.0101\end{bmatrix}$$

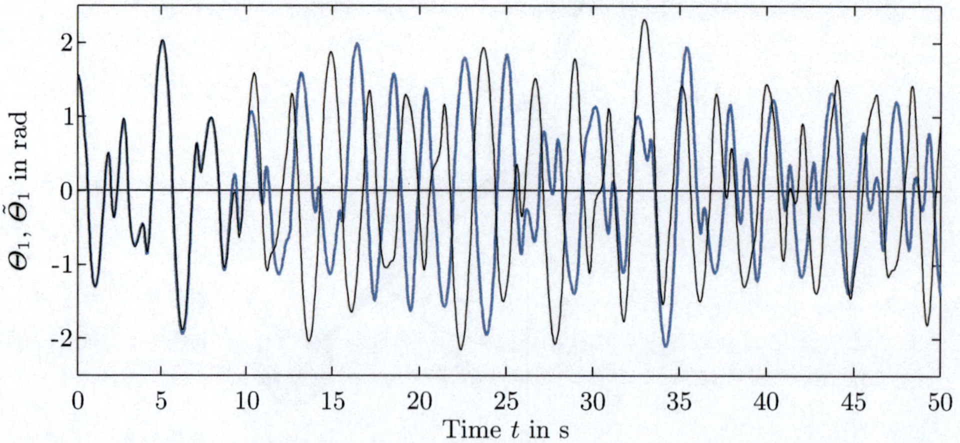

Fig. 1.30: Time courses of the angles Θ_1 (black curve) and $\tilde{\Theta}_1$ (blue curve) of the double pendulum with chaotic behavior

and compare the two resulting progressions of the angles $\Theta_1(t)$ and $\tilde{\Theta}_1(t)$. It can be seen from Figure 1.30 that, after a while, the angles progress entirely differently and without any recognizable pattern, even though their initial conditions were nearly identical.

Technical systems very rarely exhibit chaotic behavior. This is because the developers and users of technical systems are generally interested in designing their devices, processes, or vehicles in such a way that they exhibit predictable behavior. In Section 5.6, we will consider another chaotic technical system, a fluid system, and how to control it in such a way that it has a globally asymptotically stable equilibrium point and, consequently, no longer exhibits chaotic behavior. It turns out that a chaotic system is no more difficult to control than a conventional one. A detailed description of chaotic systems can be found in [19, 20, 77, 92, 354, 436, 452].

1.1.13 Discrete-Time Systems

Apart from continuous-time systems, in control engineering and systems theory, we also encounter systems for which the state variables and the signals of the inputs and outputs are not observed or measured continuously over time, but only at certain, meaning discrete instants k [143, 200, 201, 231, 259, 341]. Such discrete-time systems are often the result of sampling continuous-time systems, which is typically the case with feedforward and feedback control systems in industrial plants where the control is implemented via a programmable logic controller. The discrete-time sequence, however, can also be an inherent property of the system. An example are radars that only detect an object after a full revolution.

These discrete-time systems are described by difference equations

$$\begin{aligned} \boldsymbol{x}(k+1) &= \boldsymbol{f}(\boldsymbol{x}(k), \boldsymbol{u}(k)), \\ \boldsymbol{y}(k) &= \boldsymbol{g}(\boldsymbol{x}(k), \boldsymbol{u}(k)) \end{aligned} \tag{1.23}$$

with $k = 0, 1, 2, \ldots$ Analogously to the continuous-time systems, $\boldsymbol{x}$ is the n - dimensional state vector, $\boldsymbol{u}$ denotes the m - dimensional input variable vector, and $\boldsymbol{y}$ is the r - dimensional output variable vector.

The equilibrium points $\boldsymbol{x}_{\text{eq}}$ of a discrete-time system are defined by the equation

$$\boldsymbol{x}_{\text{eq}} = \boldsymbol{f}(\boldsymbol{x}_{\text{eq}}, \boldsymbol{0}).$$

In this, as in the continuous-time case, without loss of generality, $\boldsymbol{u} = \boldsymbol{0}$ is assumed.

The stability of an equilibrium point can be defined in a manner similar to the continuous-time case. Assuming again that the equilibrium point lies at $\boldsymbol{x} = \boldsymbol{0}$, we can formulate the following definition:

Definition 15 (Stability of Equilibrium Points of Discrete-Time Systems). *Let a system*

$$\boldsymbol{x}(k+1) = \boldsymbol{f}(\boldsymbol{x}(k), \boldsymbol{u}(k))$$

possess the equilibrium point $\boldsymbol{x}_{\mathrm{eq}} = \mathbf{0}$. *This equilibrium point is called Lyapunov stable if, for every ε-neighborhood*

$$U_\varepsilon(\mathbf{0}) = \{\boldsymbol{x} \in I\!R^n \mid |\boldsymbol{x}| < \varepsilon\},$$

there exists a δ-neighborhood

$$U_\delta(\mathbf{0}) = \{\boldsymbol{x} \in I\!R^n \mid |\boldsymbol{x}| < \delta\}$$

such that all sequences of states $\boldsymbol{x}(k)$ *of the free system which begin in the δ-neighborhood, i. e.*

$$\boldsymbol{x}(0) \in U_\delta(\mathbf{0}),$$

in their further progression, remain within the ε-neighborhood, i. e.

$$\boldsymbol{x}(k) \in U_\varepsilon(\mathbf{0}) \quad \textit{for} \quad k > 0.$$

If the condition

$$\lim_{k\to\infty} \boldsymbol{x}(k) = \mathbf{0}$$

is additionally fulfilled for all $\boldsymbol{x}(0) \in U_\delta(\mathbf{0})$, *the equilibrium point is called asymptotically stable.*

Limit cycles can occur in discrete-time systems, just as in continuous-time systems. This is the case if for a sequence of state vectors $\boldsymbol{x}(0),\ \boldsymbol{x}(1), \ldots,\ \boldsymbol{x}(l)$, the relation

$$\boldsymbol{x}(0) = \boldsymbol{x}(l)$$

is fulfilled for a constant input variable vector $\boldsymbol{u}(k)$. Since the limit cycle

$$\boldsymbol{x}(0),\ \boldsymbol{x}(1),\ \ldots,\ \boldsymbol{x}(l) = \boldsymbol{x}(0)$$

in total comprises $l+1$ states, this is referred to as a *cycle* or *periodic orbit* of length $l+1$.

Chaotic behavior is also possible. For example, chaos occurs in a discrete-time system of order one if a cycle

$$x(0),\ x(1),\ x(2) = x(0),$$

i. e. a cycle of length three, exists [276]. Of course, the cycle itself is not chaotic. However, as soon as the state departs marginally from the values $x(0)$, $x(1)$, $x(2)$ of the cycle, the course $x(k)$ becomes chaotic. It is worth mentioning that chaos can occur in discrete-time systems of order $n = 1$ or larger. In autonomous continuous-time systems, however, chaos can only arise in systems of order $n = 3$ or larger.

In most cases, it is impossible to model a sampled continuous-time nonlinear system exactly using a difference equation (1.23). This is completely different from the situation with linear systems, where the discrete-time model that results from the sampling of the continuous-time model can be calculated from the continuous-time model in a simple and efficient way. This simple calculation results from the knowledge of the solution of the linear differential equation.

In the nonlinear case, we do not usually have the system solution available, and therefore we cannot determine the discrete-time model, at least not exactly. However, an approximation is possible. Using the numerical solutions of nonlinear differential equations, it is possible to determine such approximate models for sampled nonlinear systems. The focus of the next sections is on solving nonlinear differential equations.

1.2 Solution of Nonlinear Differential Equations

1.2.1 Existence of Solutions

For linear systems and for their associated initial-value problems

$$\begin{aligned}\dot{\boldsymbol{x}} &= \boldsymbol{A}\boldsymbol{x} + \boldsymbol{B}\boldsymbol{u}\\ \boldsymbol{x}(0) &= \boldsymbol{x}_0,\end{aligned}$$

we know the solution. It can be explicitly stated and, as we know from linear control theory [133], it is given by

$$\boldsymbol{x}(t) = e^{\boldsymbol{A}t}\boldsymbol{x}_0 + \int_0^t e^{\boldsymbol{A}(t-\tau)}\boldsymbol{B}\boldsymbol{u}(\tau)d\tau.$$

In nonlinear systems, the task of determining a solution becomes significantly more difficult. In most cases, the solution cannot even be specified explicitly. If it is nonetheless possible, we can attempt this by trial and error, transforming or substituting variables, and integrating the differential equation. For some differential equations, such as the Bernoulli and the d'Alembert differential equations, solutions or solution methods are known. We will deal with the most well-known methods in the exercises. An extensive collection of analytically solvable differential equations and their solutions can be found in [364]. However, there is no systematic general method for determining an analytical solution. For this reason, we have to use numerical solution methods more often than not. Before we deal with these, we will clarify the conditions sufficient for an ordinary differential equation to be solvable.

We will consider the non-autonomous differential equations

$$\dot{\boldsymbol{x}} = \tilde{\boldsymbol{f}}(\boldsymbol{x}(t), \boldsymbol{u}(t)). \tag{1.24}$$

The time-varying input function $\boldsymbol{u}(t)$ can also be interpreted, in its effect, as a time-dependence of the function $\boldsymbol{f}$, i. e.

$$\boldsymbol{f}(\boldsymbol{x}(t), t) = \tilde{\boldsymbol{f}}(\boldsymbol{x}(t), \boldsymbol{u}(t)).$$

Equations of the (1.24) type are thus included in

$$\dot{\boldsymbol{x}} = \boldsymbol{f}(\boldsymbol{x}(t), t), \tag{1.25}$$

provided the input signal $\boldsymbol{u}(t)$ is known.

Determining a solution of the differential equation (1.25) requires that a solution actually exists. Whether a solution exists is shown by [327]

Theorem 3 (Peano Existence Theorem). *Let*

$$\begin{aligned} &\dot{\boldsymbol{x}} = \boldsymbol{f}(\boldsymbol{x}, t), \quad \boldsymbol{x} \in D_{\boldsymbol{x},\text{def}} \subseteq I\!R^n, \quad t \in D_{t,\text{def}} \subseteq I\!R, \\ &\boldsymbol{x}(t_0) = \boldsymbol{x}_0, \end{aligned}$$

be a given initial-value problem. If the function $\boldsymbol{f}$ is continuous in a neighborhood of the point $(\boldsymbol{x}_0, t_0)$, the initial-value problem possesses at least one solution $\boldsymbol{x}(t)$ in an open interval around t_0.

Peano's existence theorem guarantees the solvability of a differential equation. However, it does not guarantee the uniqueness of the solution. This means that multiple solutions may well exist. To illustrate this, we can consider the differential equation

$$\dot{x} = \sqrt[3]{x^2} \tag{1.26}$$

with the initial value

$$x(0) = 0 \tag{1.27}$$

as an example. All functions

$$x = \begin{cases} \left(\dfrac{1}{3}t - a\right)^3, & t < 3a, \\ 0, & 3a \le t \le 3b, \\ \left(\dfrac{1}{3}t - b\right)^3, & t > 3b, \end{cases}$$

with arbitrary values $a \le 0$, $b \ge 0$ fulfill the above initial-value problem (1.26), (1.27). Hence infinitely many solutions exist, i. e. the solution is not unique.

If the Peano existence theorem is extended by the condition of Lipschitz continuity of the function $\boldsymbol{f}$, we arrive at the Picard-Lindelöf theorem[2]. This theorem guarantees both the existence of a solution and its uniqueness. We thus require Lipschitz continuity, which is defined by

[2] The Picard-Lindelöf theorem is also known as Picard's existence theorem or the Cauchy-Lipschitz theorem.

Definition 16 (Lipschitz Continuity). *A function $\boldsymbol{f}(\boldsymbol{x})$ is called globally Lipschitz continuous if the so-called Lipschitz condition*

$$|\boldsymbol{f}(\boldsymbol{x}_1) - \boldsymbol{f}(\boldsymbol{x}_2)| \leq L|\boldsymbol{x}_1 - \boldsymbol{x}_2|$$

is fulfilled for a real number $L \geq 0$ for all $\boldsymbol{x}_1, \boldsymbol{x}_2 \in D_{\boldsymbol{x},\mathrm{def}} \subseteq I\!R^n$. If a Lipschitz condition is only fulfilled within a neighborhood of each point $\boldsymbol{x} \in D_{\boldsymbol{x}} \subseteq D_{\boldsymbol{x},\mathrm{def}}$, the function $\boldsymbol{f}(\boldsymbol{x})$ is called locally Lipschitz continuous.

Definition 17 (Lipschitz Continuity of Multivariate Functions). *A function $\boldsymbol{f}(\boldsymbol{x},t)$ is called globally Lipschitz continuous in $\boldsymbol{x}$ if the Lipschitz condition*

$$|\boldsymbol{f}(\boldsymbol{x}_1,t) - \boldsymbol{f}(\boldsymbol{x}_2,t)| \leq L|\boldsymbol{x}_1 - \boldsymbol{x}_2|$$

is fulfilled for a real number $L \geq 0$ for all $t \in I\!R$ and all $\boldsymbol{x}_1, \boldsymbol{x}_2 \in D_{\boldsymbol{x},\mathrm{def}} \subseteq I\!R^n$. If a Lipschitz condition is only fulfilled within a neighborhood of each point $\boldsymbol{x} \in D_{\boldsymbol{x}} \subseteq D_{\boldsymbol{x},\mathrm{def}}$ and for all t in an interval $(t_l, t_u) \subseteq I\!R$, the function $\boldsymbol{f}(\boldsymbol{x},t)$ is called locally Lipschitz continuous in $\boldsymbol{x}$.

For example, the function

$$f(x) = x^2$$

is locally, but not globally, Lipschitz continuous. The following chain of deductions holds:

globally Lipschitz continuous $\Rightarrow$ locally Lipschitz continuous
$\Rightarrow$ locally continuous.

On the one hand, Lipschitz continuity is therefore a strict form of continuity. On the other hand, it provides us with a measure, a maximum rate L, for the change in the function values, which resembles the maximum value of a function's derivative. In particular, continuously differentiable functions are locally Lipschitz continuous and continuously differentiable functions with a bounded first derivative are globally Lipschitz continuous. However, not all Lipschitz continuous functions are differentiable, but all differentiable globally Lipschitz continuous functions have bounded derivatives.

Let us consider two examples. The first is the absolute value function $f(x) = |x|$, which is globally Lipschitz continuous but not differentiable. The second is the differential equation (1.26), whose right side is not globally Lipschitz continuous and also not locally Lipschitz continuous around $x = 0$, since its derivative

$$\frac{\partial f(x)}{\partial x} = \frac{\partial \sqrt[3]{x^2}}{\partial x} = \frac{2}{3}\frac{1}{\sqrt[3]{x}}$$

becomes infinite at the point $x = 0$. Therefore, it is not possible to specify a number L for any neighborhood of $x = 0$.

So equipped, we can formulate [327]

Theorem 4 (Picard-Lindelöf Local Existence Theorem). *Let*

$$\begin{aligned} \dot{\boldsymbol{x}} &= \boldsymbol{f}(\boldsymbol{x}, t), \quad \boldsymbol{x} \in D_{\boldsymbol{x},\mathrm{def}} \subseteq I\!R^n, \quad t \in D_{t,\mathrm{def}} \subseteq I\!R, \\ \boldsymbol{x}(t_0) &= \boldsymbol{x}_0 \end{aligned}$$

be a given initial-value problem. If the function $\boldsymbol{f}$ is locally Lipschitz continuous in $\boldsymbol{x}$ and continuous in t in a neighborhood of the point $(\boldsymbol{x}_0, t_0)$, there exists exactly one solution $\boldsymbol{x}(t)$ that is defined in an open interval around t_0.

It is worth noting that this version of the Picard-Lindelöf existence theorem guarantees the solution's uniqueness but does not guarantee its existence for all $t \in I\!R$. To see this, we will consider the example

$$\begin{aligned} \dot{x} &= -\frac{1}{2x}, \\ x(0) &= 1. \end{aligned} \tag{1.28}$$

Obviously, the function

$$f(x) = -1/(2x)$$

is locally Lipschitz continuous in the neighborhood of the point

$$(x(0), t_0) = (1, 0).$$

It is not globally Lipschitz continuous, since the derivative of $f(x)$ becomes infinite as x approaches zero. In this case, it is not possible to specify a constant L that satisfies the Lipschitz condition. However, the conditions of the local existence theorem are fulfilled. The only solution to the initial-value problem (1.28) is

$$x(t) = \sqrt{1 - t}.$$

We notice that no solution exists for values $t > 1$. The time $t \in I\!R$, for which $x(t) \to \infty$ (in this case $t = 1$) is called *finite escape time*. In contrast to some nonlinear systems, unstable linear systems cannot run to infinity in finite time.

If we want to guarantee the existence of a solution for all values $t \in I\!R$, we must resort to [264]

Theorem 5 (Picard-Lindelöf Global Existence Theorem). *Let*

$$\begin{aligned} \dot{\boldsymbol{x}} &= \boldsymbol{f}(\boldsymbol{x}, t), \quad \boldsymbol{x} \in D_{\boldsymbol{x},\mathrm{def}} = I\!R^n, \quad t \in D_{t,\mathrm{def}} = I\!R, \\ \boldsymbol{x}(t_0) &= \boldsymbol{x}_0 \end{aligned}$$

be a given initial-value problem. If the function $\boldsymbol{f}$ is globally Lipschitz continuous in $\boldsymbol{x}$ and continuous in t, there exists exactly one solution $\boldsymbol{x}(t)$ for every $\boldsymbol{x}_0 \in I\!R^n$, which is defined for all $t \in I\!R$.

A very well-known example of this global existence theorem is the linear initial-value problem

$$\begin{aligned}\dot{\boldsymbol{x}} &= \boldsymbol{A}\boldsymbol{x},\\ \boldsymbol{x}(0) &= \boldsymbol{x}_0.\end{aligned}$$

It is known that the unique solution is given by

$$\boldsymbol{x}(t) = e^{\boldsymbol{A}t}\boldsymbol{x}_0,$$

which is defined for all $t \in \mathbb{R}$. This result is consistent with our global existence theorem, since the linear function

$$\boldsymbol{f}(\boldsymbol{x}) = \boldsymbol{A}\boldsymbol{x}$$

is globally Lipschitz continuous, and thus a solution exists for all $t \in \mathbb{R}$.

1.2.2 Numerical Solution and Euler Method

We have already mentioned in the previous section that unlike linear differential equations, solving nonlinear differential equations with initial values $\boldsymbol{x}(t_0)$, i. e. initial-value problems

$$\begin{aligned}\dot{\boldsymbol{x}}(t) &= \boldsymbol{f}(\boldsymbol{x}(t), \boldsymbol{u}(t)),\\ \boldsymbol{x}(t_0) &= \boldsymbol{x}_0,\end{aligned}$$

is only numerically possible in most cases. For this purpose, a number of numerical integration methods have been developed [172, 173, 423].

They are based on the following principle, which will be illustrated without loss of generality for the case of a single-state variable x, i. e.

$$\dot{x}(t) = f(x(t), u(t)). \tag{1.29}$$

By integrating equation (1.29), we obtain the solution

$$x(t) = x(t_0) + \int_{t_0}^{t} f\left(x(\tau), u(\tau)\right) d\tau \tag{1.30}$$

of the differential equation. Note that this equation is implicit with respect to $x(t)$, and therefore it is usually not analytically solvable. The integration methods thus solve the integral in equation (1.30) numerically via approximations, which are more or less suitable. Accuracy and computational complexity depend on the choice of the numerical integration method.

To obtain a numerical solution, we discretize equation (1.30). As shown in Figure 1.31, the time-axis is divided by a set of equidistant points

$$t_i = t_0 + h \cdot i, \qquad i = 0, \ldots, k,$$

where the sampling interval h is called the *step size*. Now it is possible to write the solution of the differential equation (1.29) at the time instances t_i as a recursive formula

$$x(t_{i+1}) = \underbrace{x(t_0) + \int\limits_{t_0}^{t_i} f(x,u)\,dt}_{x(t_i)} + \int\limits_{t_i}^{t_i+1} f(x,u)\,dt,$$

which is equivalent to

$$x(t_{i+1}) = x(t_i) + \int\limits_{t_i}^{t_{i+1}} f(x,u)\,dt.$$

The points $x(t_i)$ are termed *integration points.*

The aim of the integration methods is to find a good approximation for the integral. In the simplest case, as shown in Figure 1.32, the area

$$F = \int\limits_{t_i}^{t_{i+1}} f(x,u)dt$$

between t_i and t_{i+1} is approximated by a rectangle whose area is

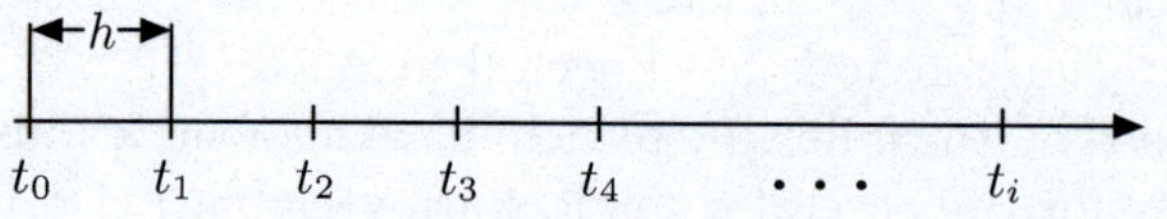

Fig. 1.31: Time instances t_i and step size h of the numerical solution

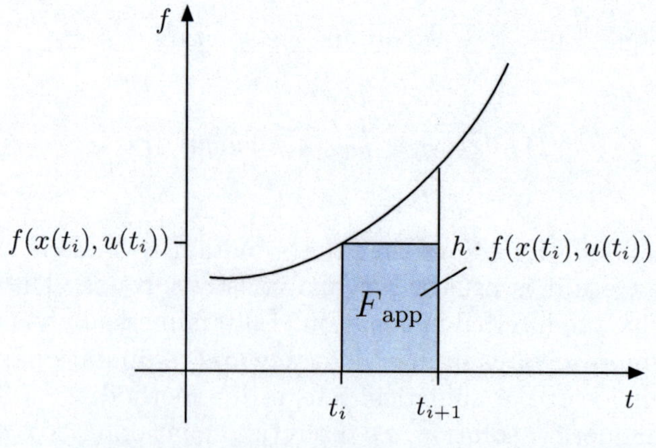

Fig. 1.32: Approximate solution of the integral

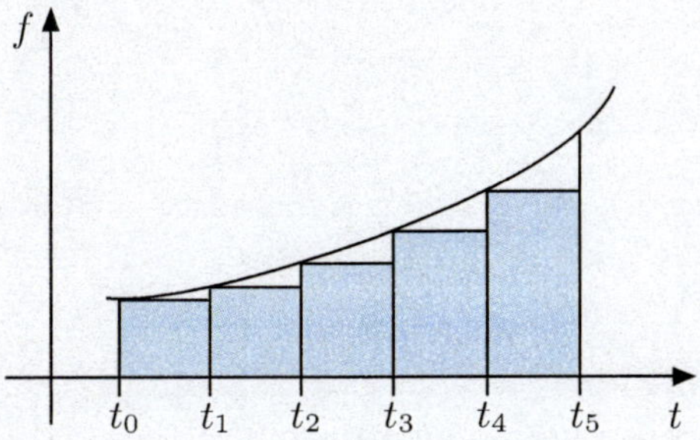

Fig. 1.33: Area calculation using the Euler method

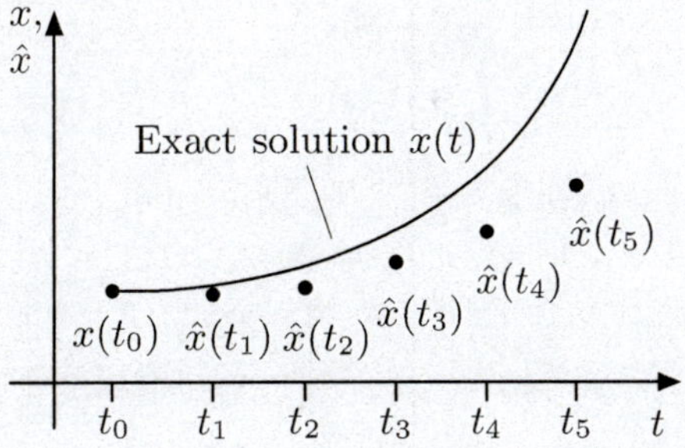

Fig. 1.34: Comparison between the exact solution $x(t)$ and its approximation $\hat{x}(t_i)$

$$F_{\text{app}} = h \cdot f\left(x(t_i), u(t_i)\right).$$

This yields the value

$$\hat{x}(t_{i+1}) = \hat{x}(t_i) + h \cdot f\left(\hat{x}(t_i), u(t_i)\right)$$

as an approximation for $x(t_{i+1})$. In the multidimensional case, using the abbreviations $\hat{\boldsymbol{x}}(t_i) = \hat{\boldsymbol{x}}_i$ and $\boldsymbol{u}(t_i) = \boldsymbol{u}_i$,

$$\hat{\boldsymbol{x}}_{i+1} = \hat{\boldsymbol{x}}_i + h \cdot \boldsymbol{f}(\hat{\boldsymbol{x}}_i, \boldsymbol{u}_i)$$

holds.

Here $\hat{\boldsymbol{x}}_0 = \boldsymbol{x}_0$ serves as the initial value for this recursion. The calculation above is referred to as the *Euler method.* It is simple but inaccurate. Figure 1.33 illustrates the approximation that is yielded by this procedure.

1.2.3 Accuracy of the Numerical Solution

The approximation of the area under the curve of the differential equation's solution by a series of rectangles has a direct influence on the accuracy of the solution, as illustrated in Figure 1.34. The error at step i, the cumulative error

$$\varepsilon_i = |\boldsymbol{x}(t_i) - \hat{\boldsymbol{x}}(t_i)| = \sqrt{(x_1(t_i) - \hat{x}_1(t_i))^2 + \ldots + (x_n(t_i) - \hat{x}_n(t_i))^2},$$

is dependent on the overall number of previous steps and on the step size h. Using the Euler method, the estimation

$$\varepsilon_k \le \alpha \cdot h$$

holds for the error ε_k after a fixed integration time $T = k \cdot h$ with k steps. The method error ε_k therefore decreases for decreasing step sizes. In general, α is an unknown constant.

For integration methods that are more accurate than the Euler method, the error ε_k of the approximate solution can be decreased much more quickly by reducing the step size h. In general, it holds that

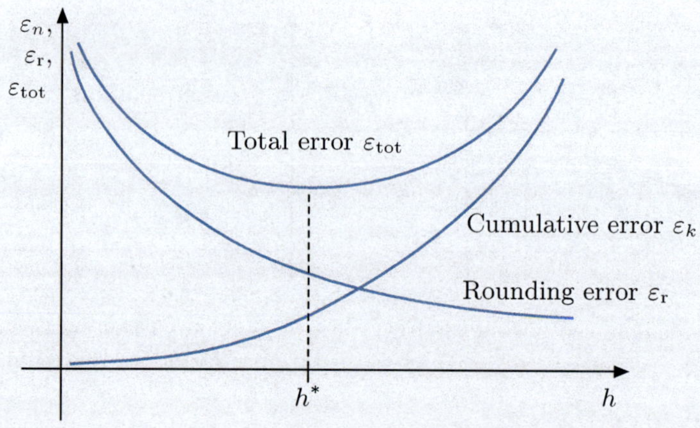

Fig. 1.35: The total error ε_{tot} of a numerical integration method is made up of the method error ε_k and the rounding error ε_r.

$$\varepsilon_k \leq \alpha \cdot h^q.$$

The parameter q is called the *order of accuracy* or *error order* of an integration method. The value of q defines how quickly the error ε_k converges to zero as h tends to zero. Here a larger value of q indicates the greater accuracy of a given integration method.

Regarding the precision of a numerical solution of differential equations, it therefore seems appropriate to choose very small step sizes h. However, as h is decreased, the number of required integration points increases, and with it the computation time of the simulation.

There is another disadvantage to very small step sizes h. Although the cumulative error ε_k decreases when the step size h is reduced, the computer's rounding error ε_r increases, as shown in Figure 1.35. Therefore an optimal step size h^* exists for which the total error ε_{tot} is minimal. Inconveniently, the optimal step size h^* generally cannot be determined.

1.2.4 The Modified Euler Method

The Euler method is very imprecise because of the chosen rectangular approximation $h \cdot f(\hat{x}_i, u_i)$. It can be improved by not choosing the value

$$f(\hat{x}(t_i), u(t_i))$$

at position $\hat{x}(t_i) = \hat{x}_i$ as the height, but rather

$$f\left(\hat{x}\left(t_i + \frac{h}{2}\right), u\left(t_i + \frac{h}{2}\right)\right)$$

at position

$$\hat{x}_{i+1/2} = \hat{x}(t_i + \frac{h}{2})$$

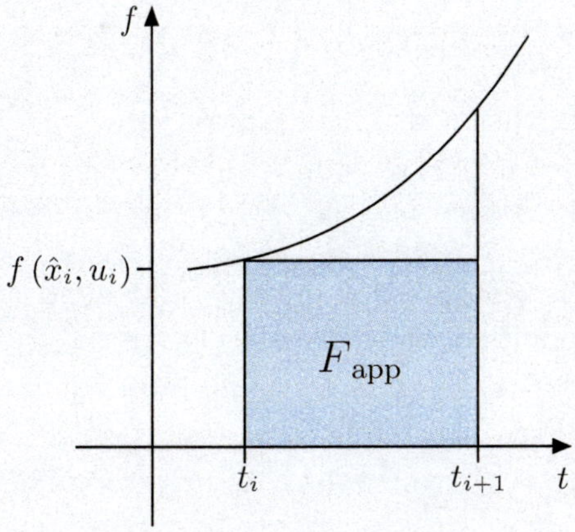

Fig. 1.36: Euler method

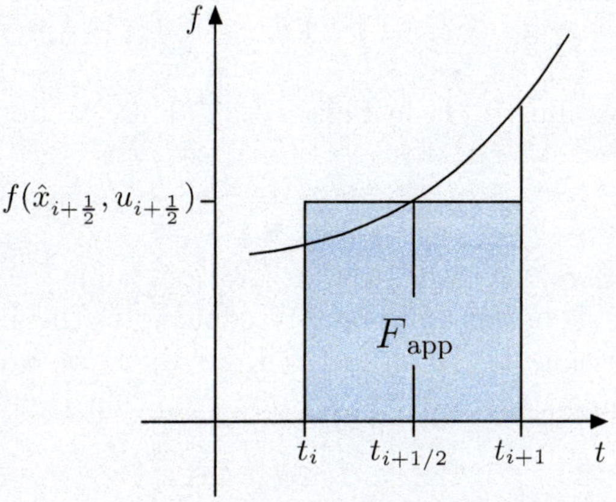

Fig. 1.37: Modified Euler method

and

$$u_{i+1/2} = u(t_i + \frac{h}{2}).$$

Figures 1.36 and 1.37 illustrate these conditions.

The integral of the modified Euler method, also known as the *midpoint method*, is calculated as

$$F_{\text{app}} = h \cdot f\left(\hat{x}_{i+1/2}, u_{i+1/2}\right).$$

In this way we obtain the following algorithm for the improved method in the general case of multidimensional system equations:

(1) $\hat{\boldsymbol{x}}_{i+1/2} = \hat{\boldsymbol{x}}_i + \frac{h}{2} \cdot \boldsymbol{f}\left(\hat{\boldsymbol{x}}_i, \boldsymbol{u}_i\right),$

(2) $\hat{\boldsymbol{x}}_{i+1} = \hat{\boldsymbol{x}}_i + h \cdot \boldsymbol{f}\left(\hat{\boldsymbol{x}}_{i+1/2}, \boldsymbol{u}_{i+1/2}\right).$

In contrast to the Euler method with the order of accuracy $q = 1$, the modified Euler method has the order of accuracy $q = 2$.

1.2.5 The Heun and Simpson Methods

Further improvements in accuracy compared to the numerical methods described above can be achieved if the integral

$$F = \int_{t_i}^{t_{i+1}} f(x(t), u(t))dt$$

is more accurately approximated by means of a trapezoid area. For this purpose, the Heun method replaces the rectangle with a trapezoid to calculate the approximation of the integral

$$F_{\text{app}} = \frac{h}{2}\left[f(x_i, u_i) + f(x_{i+1}, u_{i+1})\right],$$

as illustrated in Figure 1.38. Thus we obtain the recurrence relation

$$\hat{x}_{i+1} = \hat{x}_i + \frac{h}{2}\left[f\left(\hat{x}_i, u_i\right) + f\left(\hat{x}_{i+1}, u_{i+1}\right)\right].$$

Here, however, the integration point $\hat{x}_{i+1}$ is also present on the right-hand side of the equation. We could solve this implicit equation iteratively for $\hat{x}_{i+1}$. Using the Heun method, however, we proceed differently and compute $\hat{x}_{i+1}$ approximately via a Euler step that is given by

$$\hat{x}_{i+1} \approx \tilde{x}_{i+1} = \hat{x}_i + h \cdot f\left(\hat{x}_i, u_i\right).$$

In summary, we arrive at the Heun method for the general multidimensional case:

(1) $\tilde{\boldsymbol{x}}_{i+1} = \hat{\boldsymbol{x}}_i + h \cdot \boldsymbol{f}\left(\hat{\boldsymbol{x}}_i, \boldsymbol{u}_i\right)$,

(2) $\hat{\boldsymbol{x}}_{i+1} = \hat{\boldsymbol{x}}_i + \dfrac{h}{2}\left[\boldsymbol{f}\left(\hat{\boldsymbol{x}}_i, \boldsymbol{u}_i\right) + \boldsymbol{f}(\tilde{\boldsymbol{x}}_{i+1}, \boldsymbol{u}_{i+1})\right]$.

Heun's method is a so-called *predictor-corrector method*. The result of its first equation yields the predictor with an approximation of the solution. This is improved upon by the second equation, which is called the corrector. It is worth noting that the Heun method, like the modified Euler method, has the order of accuracy $q = 2$, and therefore does not improve upon the modified Euler method in terms of accuracy.

If we use a parabola for the approximation of the area

$$F = \int\limits_{t_i}^{t_{i+1}} f(x, u)dt$$

as the interpolation function for the function f on the interval $[t_i, t_{i+1}]$, the accuracy can be further increased. By using the parabola, as depicted in Figure 1.39, we arrive at Simpson's rule. To compute the parameters of the parabola, an additional point at

$$t_{i+1/2} = t_i + \frac{h}{2},$$

which lies between the points $(t_i, f(\hat{x}_i, u_i))$ and $(t_{i+1}, f(\hat{x}_{i+1}, u_{i+1}))$ of the Heun method, and the intermediate value

$$u_{i+1/2} = u(t_i + \frac{h}{2})$$

of the control signal are required. The integration of the parabola on the interval $[t_i, t_{i+1}]$ then provides the estimate F_{app} of the area to be integrated.

The resulting calculation rule is called the Simpson method and is summarized as follows:

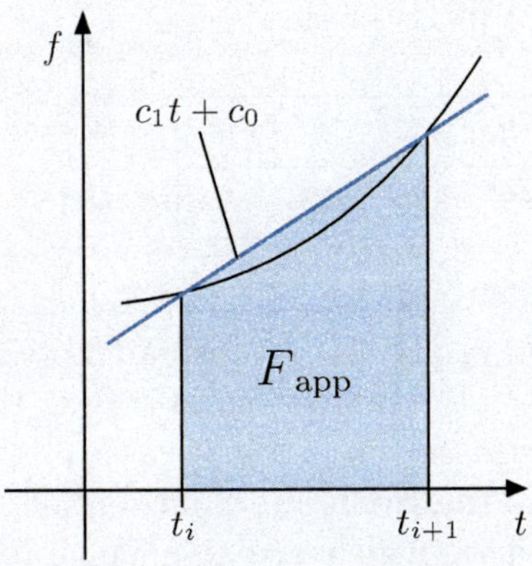

Fig. 1.38: Heun method

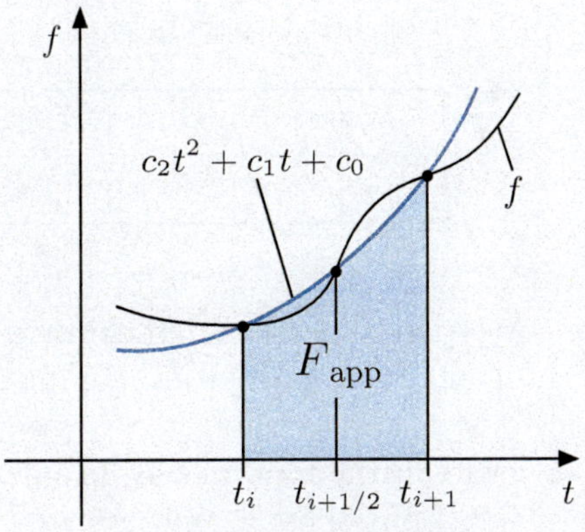

Fig. 1.39: Simpson method

(1) $\boldsymbol{k}_1 = \boldsymbol{f}\,(\hat{\boldsymbol{x}}_i, \boldsymbol{u}_i)$,

(2) $\boldsymbol{k}_2 = \boldsymbol{f}(\hat{\boldsymbol{x}}_i + \frac{h}{2}\boldsymbol{k}_1, \boldsymbol{u}_{i+1/2})$,

(3) $\boldsymbol{k}_3 = \boldsymbol{f}\,(\hat{\boldsymbol{x}}_i - h\boldsymbol{k}_1 + 2h\boldsymbol{k}_2, \boldsymbol{u}_{i+1})$,

(4) $\hat{\boldsymbol{x}}_{i+1} = \boldsymbol{x}_i + \frac{h}{6}(\boldsymbol{k}_1 + 4\boldsymbol{k}_2 + \boldsymbol{k}_3)$.

It has the order of accuracy $q = 3$. The method is also known as the third-order Runge-Kutta method.

1.2.6 The Runge-Kutta Methods

The Euler methods, the Heun method, and the Simpson method are special cases of the *Runge-Kutta methods*. All of these calculation rules are called *one-step methods*, since they determine a new integration point $\hat{\boldsymbol{x}}_{i+1}$ based only on a previous point $\hat{\boldsymbol{x}}_i$. The Runge-Kutta methods have the following general form:

(1) $\boldsymbol{k}_1 = \boldsymbol{f}\,(\hat{\boldsymbol{x}}_i, \boldsymbol{u}_i)$,

(2) $\boldsymbol{k}_2 = \boldsymbol{f}\,(\hat{\boldsymbol{x}}_i + h\alpha_{21}\boldsymbol{k}_1, \boldsymbol{u}(t_i + \beta_2 h))$,

(3) $\boldsymbol{k}_3 = \boldsymbol{f}\,(\hat{\boldsymbol{x}}_i + h\,(\alpha_{31}\boldsymbol{k}_1 + \alpha_{32}\boldsymbol{k}_2)\,, \boldsymbol{u}(t_i + \beta_3 h))$,

$$\vdots$$

(m) $\boldsymbol{k}_m = \boldsymbol{f}\,(\hat{\boldsymbol{x}}_i + h\,(\alpha_{m1}\boldsymbol{k}_1 + \ldots + \alpha_{m,m-1}\boldsymbol{k}_{m-1})\,, \boldsymbol{u}(t_i + \beta_m h))$,

(m+1) $\hat{\boldsymbol{x}}_{i+1} = \hat{\boldsymbol{x}}_i + h \cdot \sum_{j=1}^{m} \gamma_j \boldsymbol{k}_j$.

Table 1.1: Orders of accuracy q of the Runge-Kutta methods

m	1	2	3	4	5	6	7	8	9
q	1	2	3	4	4	5	6	6	7

The corresponding orders of accuracy are listed in Table 1.1. With an increasing number of approximation computations, i. e. the number m of stages, the accuracy q of the method being used also increases. The special cases of the Euler method, the modified Euler method, and the Heun and Simpson methods result from the parameters m, γ, β, and α, which are given in Table 1.2.

Table 1.2: Special cases of the Runge-Kutta methods

	m	γ	β	α_{ij}
Euler	$m = 1$	$\gamma_1 = 1$	–	–
Modified Euler	$m = 2$	$\gamma_1 = 0$ $\gamma_2 = 1$	$\beta_2 = \frac{1}{2}$	$\alpha_{21} = \frac{1}{2}$
Heun	$m = 2$	$\gamma_1 = \frac{1}{2}$ $\gamma_2 = \frac{1}{2}$	$\beta_2 = 1$	$\alpha_{21} = 1$
Simpson	$m = 3$	$\gamma_1 = \frac{1}{6}$ $\gamma_2 = \frac{4}{6}$ $\gamma_3 = \frac{1}{6}$	$\beta_2 = \frac{1}{2}$ $\beta_3 = 1$	$\alpha_{21} = \frac{1}{2}$ $\alpha_{31} = -1$ $\alpha_{32} = 2$

Several variants exist for the same value of $m \geq 2$, of which the classical Runge-Kutta method with number of stages $m = 4$ is the most common. On the one hand, this is because the computational effort is limited in this case. On the other hand, it is because the order of accuracy q, as shown in Table 1.1, does not increase when the number of stages increases from $m = 4$ to $m = 5$, whereas the computational effort becomes greater. Also, the achievable accuracy is high in proportion to the computational effort. Another advantage is the simplicity of the Runge-Kutta equations; they are easy to remember. A good mnemonic for the methods' parameters provides the Butcher tableau [23]. The most commonly used and best-known Runge-Kutta method is

(1) $\boldsymbol{k}_1 = \boldsymbol{f}\left(\hat{\boldsymbol{x}}_i, \boldsymbol{u}_i\right),$

(2) $\boldsymbol{k}_2 = \boldsymbol{f}(\hat{\boldsymbol{x}}_i + \frac{h}{2}\boldsymbol{k}_1, \boldsymbol{u}_{i+1/2}),$

(3) $\boldsymbol{k}_3 = \boldsymbol{f}(\hat{\boldsymbol{x}}_i + \frac{h}{2}\boldsymbol{k}_2, \boldsymbol{u}_{i+1/2}),$

(4) $\boldsymbol{k}_4 = \boldsymbol{f}\left(\hat{\boldsymbol{x}}_i + h\boldsymbol{k}_3, \boldsymbol{u}_{i+1}\right),$

(5) $\hat{\boldsymbol{x}}_{i+1} = \hat{\boldsymbol{x}}_i + \frac{h}{6}(\boldsymbol{k}_1 + 2\boldsymbol{k}_2 + 2\boldsymbol{k}_3 + \boldsymbol{k}_4)\,.$

The method has an order of accuracy $q = 4$. Runge-Kutta methods of higher orders can be found in [172].

Due to the advantageous characteristics mentioned above, the fourth-order Runge-Kutta method and the Dormand-Prince 4/5 method [99, 172], another very advantageous Runge-Kutta method, are the best-known integration methods and are very frequently used.

1.2.7 Adaptation of the Step Size

So far, the step size h has been kept constant during the recursion. This is not always appropriate. If the differential equation inhibits parts of strongly differing dynamics, as illustrated in Figure 1.40, the integration methods that we have discussed so far lead to inaccurate or computationally intensive solutions. This is because we would have to choose a constant step size h to be small enough to produce a good approximation of the course in the region of the oscillations ($t < 5s$). In the regions that do not contain oscillations, i. e. ($t > 5s$), the simulation would require an unnecessarily large amount of steps, since the step size is too small in that region. Therefore, during the simulation

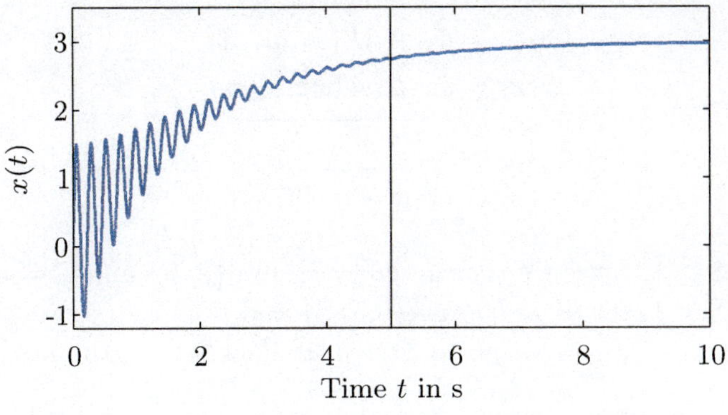

Fig. 1.40: Differential equation with strongly differing dynamic components

it is advisable to adapt h to the obtained progression of the solutions $\hat{\boldsymbol{x}}(t)$, i. e. we choose small step sizes for fast dynamic progressions and large step sizes for slow ones. With this step-size adaptation, the effort and duration of the simulation can be reduced.

A simple way of adapting the step size is provided by the following algorithm, where Φ denotes the function of the applied integration method, which is used to compute an integration step with the approximation result $\hat{\boldsymbol{x}}_i$, and ε is a prespecified error:

Step 1: Compute two recursion steps with h:

$$\begin{aligned}\hat{\boldsymbol{x}}_{i+1} &= \hat{\boldsymbol{x}}_i + h\Phi(\hat{\boldsymbol{x}}_i, \boldsymbol{u}_i, h),\\ \hat{\boldsymbol{x}}_{i+2} &= \hat{\boldsymbol{x}}_{i+1} + h\Phi(\hat{\boldsymbol{x}}_{i+1}, \boldsymbol{u}_{i+1}, h).\end{aligned}$$

Step 2: Compute a recursion step with $2h$:

$$\tilde{\boldsymbol{x}}_{i+2} = \hat{\boldsymbol{x}}_i + 2h\Phi(\hat{\boldsymbol{x}}_i, \boldsymbol{u}_i, 2h).$$

Step 3: If

$$|\hat{\boldsymbol{x}}_{i+2} - \tilde{\boldsymbol{x}}_{i+2}| > \varepsilon,$$

set $h^* = h/2$ and $i^* = i$. If

$$|\hat{\boldsymbol{x}}_{i+2} - \tilde{\boldsymbol{x}}_{i+2}| \leq 0.1\varepsilon,$$

set $h^* = 2h$ and $i^* = i + 2$, otherwise set $h^* = h$ and $i^* = i + 2$. Set $i = i^*$, $h = h^*$, go to Step 1 and start from the beginning.

A much more effective step-size adaptation than the intuitive one mentioned above is as follows. We choose two one-step integration methods, i. e. a method Γ with the order of accuracy q and a method Φ with $q + 1$. Let the ε be a prespecified error. Then the following algorithm [108] applies:

Step 1: Compute

$$\begin{aligned}\breve{\boldsymbol{x}}_{i+1} &= \hat{\boldsymbol{x}}_i + h\Gamma(\hat{\boldsymbol{x}}_i, \boldsymbol{u}_i, h),\\ \tilde{\boldsymbol{x}}_{i+1} &= \hat{\boldsymbol{x}}_i + h\Phi(\hat{\boldsymbol{x}}_i, \boldsymbol{u}_i, h),\\ S &= \sqrt[q]{\frac{h \cdot \varepsilon}{|\breve{\boldsymbol{x}}_{i+1} - \tilde{\boldsymbol{x}}_{i+1}|}}.\end{aligned}$$

Step 2: If $S \geq 1$, set $\hat{\boldsymbol{x}}_{i+1} = \tilde{\boldsymbol{x}}_{i+1}$, $h^* = h \cdot \min\{2; S\}$ and $i^* = i + 1$. If $S < 1$, set $h^* = h \cdot \max\{0.5; S\}$ and $i^* = i$. Set $i = i^*$, $h = h^*$, go to Step 1 and start from the beginning.

Particularly for complex systems with dynamics that are difficult to assess, it is advisable to always use an adaptive step size.

1.2.8 The Adams-Bashforth Methods

So far we have only dealt with one-step methods, i. e. methods for which only one preceding value $\hat{x}_i$ is used to estimate the area F and thereby the value x_{i+1}. To approximate the integral

$$F = \int_{t_i}^{t_{i+1}} f(x(t), u(t))dt$$

even more precisely, it is useful to fit f using an interpolation polynomial through a set of interpolation points

$$\begin{aligned}(t_{i-k}, f(\hat{x}_{i-k}, u_{i-k})) &= (t_{i-k}, f_{i-k}),\\ &\vdots\\ (t_{i-1}, f(\hat{x}_{i-1}, u_{i-1})) &= (t_{i-1}, f_{i-1}),\\ (t_i, f(\hat{x}_i, u_i)) &= (t_i, f_i),\end{aligned}$$

and possibly

$$(t_{i+1}, f(\hat{x}_{i+1}, u_{i+1})) = (t_{i+1}, f_{i+1}).$$

This latter interpolation point is precisely the one that we wish to calculate, meaning it is not yet known. Figure 1.41 illustrates this. Since more than one interpolation point is used for the area approximation, these methods are called *multi-step methods*.

The case in which the point (t_{i+1}, f_{i+1}) is not used as an interpolation point results in the *Adams-Bashforth* methods. Where four interpolation points are used, i. e. a polynomial of order three, the Adams-Bashforth method is given by

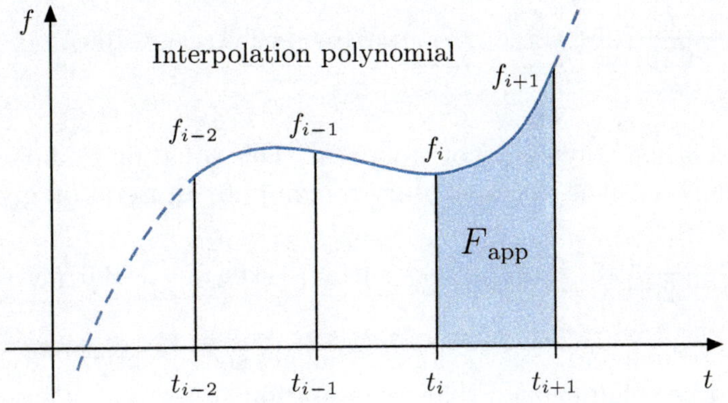

Fig. 1.41: Adams-Bashforth method

$$\hat{\boldsymbol{x}}_{i+1} = \hat{\boldsymbol{x}}_i + \frac{h}{24}(55\boldsymbol{f}_i - 59\boldsymbol{f}_{i-1} + 37\boldsymbol{f}_{i-2} - 9\boldsymbol{f}_{i-3})$$

with the order of accuracy $q = 4$. Note that the first three values $\hat{\boldsymbol{x}}_1$, $\hat{\boldsymbol{x}}_2$, $\hat{\boldsymbol{x}}_3$ and $\boldsymbol{f}_1$, $\boldsymbol{f}_2$, $\boldsymbol{f}_3$, beginning from $\hat{\boldsymbol{x}}_0$ and $\boldsymbol{f}_0$, must be computed using a one-step method.

An advantage of the Adams-Bashforth methods is that we only need to calculate one new function value in each step. A disadvantage is that the interpolation polynomial uses only the interpolation points

$$(t_{i-k}, f_{i-k}), \ldots, (t_i, f_i)$$

but not

$$(t_{i+1}, f_{i+1}).$$

However, the approximation of F by F_{app} is performed on the interval $[t_i, t_{i+1}]$. Now, the problem is that an interpolation polynomial outside its interpolation interval, which is here $[t_{i-k}, t_i]$, tends to infinity for $x \to \infty$. Consequently, the interpolation polynomial deviates more and more from $f(x)$ with increasing stepsize and the approximation error is larger than desired. To compensate for this disadvantage in the Adams-Bashforth methods, we will improve upon them in the following section.

1.2.9 The Adams-Moulton Predictor-Corrector Methods

The *Adams-Moulton methods* apply an Adams-Bashforth method as a predictor and improve the result by computing a correction term. This term is based on an interpolation polynomial that includes the unknown interpolation point

$$(t_{i+1}, f_{i+1})$$

in addition to the four interpolation points of the Adams-Bashforth method. This yields the relation

$$\hat{x}_{i+1} = \hat{x}_i + \frac{h}{720}\Big(251\underbrace{f(\hat{x}_{i+1}, u_{i+1})}_{f_{i+1}} + 646f_i - 264f_{i-1} + 106f_{i-2} - 19f_{i-3}\Big)$$

for an interpolation polynomial of order four. This equation is implicit in $\hat{x}_{i+1}$. To determine $\hat{x}_{i+1}$, it is thus necessary to compute an iteration given by

$$\hat{x}_{i+1}^{(l+1)} = \hat{x}_i + \frac{h}{720}\Big(251f(\hat{x}_{i+1}^{(l)}, u_{i+1}) + 646f_i - 264f_{i-1} + 106f_{i-2} - 19f_{i-3}\Big).$$

This has to be done until $\hat{x}_{i+1}^{(l)}$ no longer changes significantly. The chosen step size h should be small enough that two iteration steps $l = 1, 2$ are sufficient.

In summary, in the multidimensional case this yields the Adams-Moulton predictor-corrector method with the two recursions

(1) $\hat{\boldsymbol{x}}_{i+1}^{(0)} = \hat{\boldsymbol{x}}_i + \dfrac{h}{24}\big(55\boldsymbol{f}_i - 59\boldsymbol{f}_{i-1} + 37\boldsymbol{f}_{i-2} - 9\boldsymbol{f}_{i-3}\big),$

(2) $\hat{\boldsymbol{x}}_{i+1}^{(l+1)} = \hat{\boldsymbol{x}}_i + \dfrac{h}{720}\Big(251\boldsymbol{f}(\hat{\boldsymbol{x}}_{i+1}^{(l)}, \boldsymbol{u}_{i+1}) + 646\boldsymbol{f}_i - 264\boldsymbol{f}_{i-1} + 106\boldsymbol{f}_{i-2} - 19\boldsymbol{f}_{i-3}\Big)$

and the order of accuracy $q = 5$. In this case, the first three points $\hat{\boldsymbol{x}}_i$ after $\boldsymbol{x}_0$ must be computed again using a one-step method. It is advisable to use a one-step method of the same order of accuracy $q = 5$ as the Adams-Moulton multi-step method has.

The step size can also be adapted in the case of multi-step methods. However, this is more complex than for one-step methods [63].

1.2.10 Stability of Numerical Integration Methods

The accuracy of an integration method is significant, but so is its stability. Below we will analyze the stability behavior of the integration methods we have discussed. This requires answering the question of whether the error ε_n of the numerical solution remains bounded or whether it exceeds all bounds for an increasing number of steps and an increasing simulation duration. The answer depends on the step size h, the differential equation $\dot{\boldsymbol{x}} = \boldsymbol{f}(\boldsymbol{x}, \boldsymbol{u})$ itself, and the chosen integration method. If the error ε_n remains bounded, the method is called stable.

Unfortunately the above question generally cannot be answered precisely due to the nonlinearity of $\boldsymbol{f}$. For the simple linear test case

$$\dot{x} = -\lambda x \quad \text{with} \quad \lambda > 0 \tag{1.31}$$

and the initial value

$$x(0) = x_0,$$

however, the stability range of h can be determined. Since this is possible for all the methods we have discussed, the methods can be compared based on their stability behavior, which gives us a basic insight into the circumstances.

For the Euler method, we obtain

$$\hat{x}_{i+1} = \hat{x}_i + h(-\lambda \hat{x}_i) = (1 - h\lambda)\hat{x}_i. \tag{1.32}$$

Obviously, this difference equation is asymptotically stable if the inequality

$$|1 - h\lambda| < 1, \qquad \text{i.e.} \qquad 0 < h\lambda < 2,$$

is fulfilled. This result can also be obtained by determining the zero of the system's characteristic polynomial from equation (1.32),

$$P(z) = z - (1 - h\lambda),$$

which must lie within the interval $[-1, 1]$ for the difference equation (1.32) to be Lyapunov stable and within the interval $(-1, 1)$ to be asymptotically

Table 1.3: Stability ranges

Euler	$h\lambda < 2$
Modified Euler	$h\lambda < 2$
Heun	$h\lambda < 2$
Simpson	$h\lambda < 2.5359$
Runge-Kutta (4th order)	$h\lambda < 2.7853$
Adams-Bashforth (4th order)	$h\lambda < 0.3$
Adams-Moulton (5th order)	$h\lambda < 1.8367$

stable. For $h\lambda < 2$ the solution of the difference equation (1.32) thus tends to the same value as the solution of the differential equation (1.31). The progression of $\hat{x}_i$, however, can deviate substantially from $x(t)$.

In a way similar to the Euler method, stability for the test case of $\dot{x} = -\lambda x$ can also be determined for other methods. In these methods, the respective characteristic polynomial has an order greater than one. If its roots lie within the unit circle, the method is stable. Taking this into account, we obtain the stability ranges $h\lambda$ that are provided in Table 1.3. Note that $h\lambda > 0$ must hold.

The Adams-Bashforth method has a very small range of stability. This can be explained by the fact that the interpolation on the interval $[t_i, t_{i+1}]$ is computed via a polynomial that is based on the previous interpolation points $(t_{i-k}, x_{i-k}), \ldots, (t_i, x_i)$, as mentioned previously. Interpolation polynomials are only sufficiently exact between the interpolation points. Outside the interpolation range, they often differ significantly from the course that they should approximate. This is why the computation of the integral becomes inaccurate.

To illustrate the stability behavior of the methods, Figure 1.42 plots different simulations which have been carried out for the test case

$$\dot{x} = -\lambda x$$

with $\lambda = 1$ and the initial value $x(0) = 1$ for different step sizes h. The analytical solution of this differential equation is

$$x(t) = e^{-t}.$$

As we can see from Figure 1.42, stable step sizes h do not always lead to numerically determined solutions with sufficiently small simulation errors. Figure 1.43 provides the relative error as a percentage for the example of $\dot{x} = -x$, i. e.

$$\varepsilon_n^{\mathrm{rel}} = \frac{|\varepsilon_n|}{|x_n|} \cdot 100\% = \frac{|x_n - \hat{x}_n|}{|x_n|} \cdot 100\%,$$

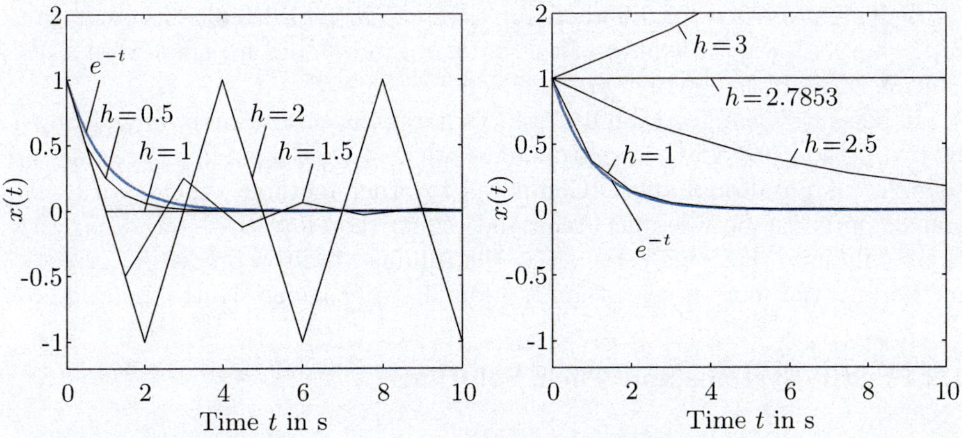

Fig. 1.42: Solution of $\dot{x} = -x$ with the Euler method (left) and the fourth-order Runge-Kutta method (right) for different step sizes h

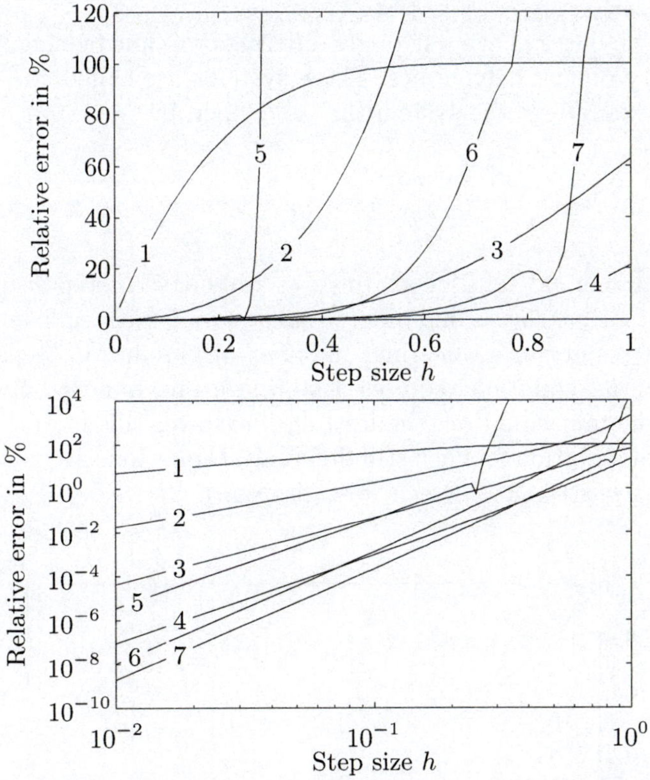

Fig. 1.43: Relative error (in percentage points) in linear representation and in double logarithmic representation for the following methods: 1. Euler, 2. modified Euler, 3. Simpson, 4. fourth-order Runge-Kutta, 5. Adams-Bashforth with the order of accuracy $q = 4$, 6. Adams-Moulton with $q = 5$ (one iteration), 7. Adams-Moulton with $q = 5$ (ten iterations)

at the instant $t = 10s$ for different step sizes h. The logarithmic representation of the error shows the course of the error in more detail for small step sizes, contrasted with the linear representation.

It becomes clear from Figure 1.43 that the classical fourth-order Runge-Kutta method provides an approximate solution with a small error and reasonable computational effort. Compared to other methods, this is an excellent compromise between accuracy and effort; therefore this method is very commonly used. For small step sizes, the Adams-Moulton predictor-corrector method is even more precise than the fourth-order Runge-Kutta method.

1.2.11 Stiff Systems and Their Solutions

The one-step and multi-step methods that we have outlined so far are often not suitable for systems

$$\dot{\boldsymbol{x}} = \boldsymbol{f}(\boldsymbol{x}),$$

whose state variables x_i have a strongly differing dynamic behavior, or contain parts with very different dynamics. These systems are called *stiff systems*. A simple linear example is the system that is defined by

$$\begin{aligned} \dot{x}_1 &= -x_1, \\ \dot{x}_2 &= -100x_2. \end{aligned}$$

Obviously, a numerical solution requires us to choose a step size h which is small enough to solve the second differential equation with sufficient accuracy. However, this results in a very high number of simulation steps, since the second differential equation requires a step size one hundred times smaller than that of the first equation. Figure 1.44 illustrates this fact.

One possible solution for such stiff differential equations are one-step methods with adaptive step sizes, which were discussed in Section 1.2.7. In the case

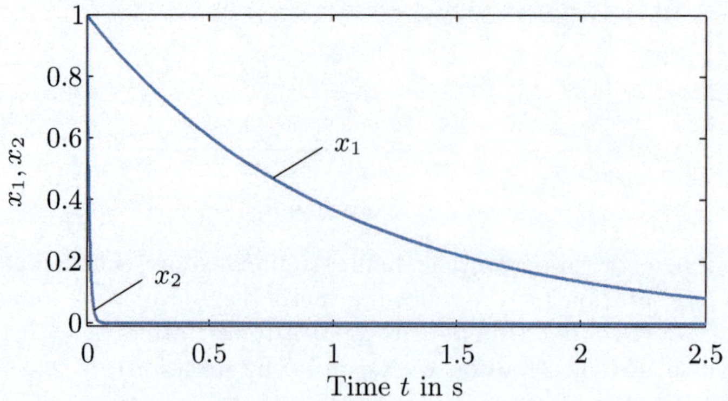

Fig. 1.44: Example of a stiff system

of multi-step methods, specially developed integration methods are used which are based on implicit recursive formulas and which exhibit particularly good stability properties.

If we take the Euler method, for example, it is possible to use

$$F = \int_{t_i}^{t_{i+1}} f(x,u)dt \approx hf(x_{i+1}, u_{i+1})$$

instead of

$$F \approx hf(x_i, u_i).$$

This yields the approximation equation

$$\hat{x}_{i+1} = \hat{x}_i + hf(\hat{x}_{i+1}, u_{i+1}).$$

The implicit character of this equation requires the computation of multiple iterations for each simulation step i. As in the correction formula of the Adams-Moulton method, this yields an iteratively solvable equation

$$\hat{x}_{i+1}^{(l+1)} = \hat{x}_i + hf(\hat{x}_{i+1}^{(l)}, u_{i+1}).$$

This method is called the *implicit Euler method*, which is also the simplest of an entire class of integration methods for stiff differential equations: the *Gear methods*. All these methods are implicit calculation rules. Table 1.4 lists the Gear formulas up to the order of accuracy $q = 4$ for the multidimensional case. Here M denotes the number of interpolation points. As initialization for the corrector equations, the simple predictor

$$\hat{\boldsymbol{x}}_{i+1}^{0} = \hat{\boldsymbol{x}}_i$$

Table 1.4: Gear formulas

M	Gear formulas	q
1	$\hat{\boldsymbol{x}}_{i+1}^{(l+1)} = \hat{\boldsymbol{x}}_i + h\boldsymbol{f}_{i+1}^{(l)}$	1
2	$\hat{\boldsymbol{x}}_{i+1}^{(l+1)} = \frac{1}{3}\left(4\hat{\boldsymbol{x}}_i - \hat{\boldsymbol{x}}_{i-1} + 2h\boldsymbol{f}_{i+1}^{(l)}\right)$	2
3	$\hat{\boldsymbol{x}}_{i+1}^{(l+1)} = \frac{1}{11}\left(18\hat{\boldsymbol{x}}_i - 9\hat{\boldsymbol{x}}_{i-1} + 2\hat{\boldsymbol{x}}_{i-2} + 6h\boldsymbol{f}_{i+1}^{(l)}\right)$	3
4	$\hat{\boldsymbol{x}}_{i+1}^{(l+1)} = \frac{1}{25}\left(48\hat{\boldsymbol{x}}_i - 36\hat{\boldsymbol{x}}_{i-1} + 16\hat{\boldsymbol{x}}_{i-2} - 3\hat{\boldsymbol{x}}_{i-3} + 12h\boldsymbol{f}_{i+1}^{(l)}\right)$	4

is used, and typically three iterations are computed.

An important characteristic of the Gear methods is their large range of stability. Once again, let us select the test example

$$\dot{x} = -\lambda x, \quad \lambda > 0.$$

Its stability range is given by

$$h\lambda < \infty.$$

This stability range applies to all Gear methods. There are also adaptive step-size variants of the Gear methods and other methods which can be used to solve stiff differential equations [173, 224]. An overview of integration methods for stiff differential equations can be found in [173].

1.3 Exercises

Exercise 1.1 For the system

$$\begin{aligned}\dot{x}_1 &= x_1^3 x_2^3 + x_1 u,\\ \dot{x}_2 &= -x_1^2 x_2,\\ y &= x_2,\end{aligned}$$

calculate the differential equation which is only dependent on the output variable y and its time derivatives, as well as on the input variable u.

Exercise 1.2 Let us consider the system

$$\begin{aligned}\dot{x}_1 &= x_1(1 - x_1 - x_2),\\ \dot{x}_2 &= x_2(1 - x_1 - 2x_2).\end{aligned}$$

Calculate the zero set of $\dot{x}_1$, i.e. the set of values x_1 and x_2 for which

$$x_1(1 - x_1 - x_2) = 0$$

applies. Next, calculate the zero set of $\dot{x}_2$. Using the zero sets, determine the system's equilibrium points.

Exercise 1.3 Calculate the equilibrium points of the systems

(a) $\dot{x} = x^2 + 1,$

(b) $\dot{x}_1 = (x_2 - 1)^2 \sin(x_1),$
$\dot{x}_2 = x_1^2(1 - \cos^2(x_2)),$

(c) $\dot{x}_1 = x_2,$
$\dot{x}_2 = -x_1 + \dfrac{x_1^3}{6} - x_2,$

(d) $\dot{x}_1 = -x_1 + x_2,$
$\dot{x}_2 = 0.1x_1 - 2x_2 - x_1^2 - 0.1x_1^3,$

(e) $\dot{x}_1 = x_2,$
$\dot{x}_2 = -x_1 + x_2(1 - 3x_1^2 - 2x_2^2),$

(f) $\dot{x}_1 = -x_1 + x_2(1 + x_1),$
$\dot{x}_2 = -x_1(1 + x_1),$

(g) $\dot{x}_1 = (x_1 - x_2)(x_1^2 + x_2^2 - 1),$
$\dot{x}_2 = (x_1 + x_2)(x_1^2 + x_2^2 - 1),$

(h) $\dot{x}_1 = x_1^2 + x_2^2 + x_3^2 - 1,$
$\dot{x}_2 = x_1 x_2 x_3,$
$\dot{x}_3 = x_1 - x_2^2.$

Exercise 1.4 Let us examine the contagion model of an infectious disease such as measles with x_1 representing the number of healthy, but potentially vulnerable persons, x_2 the number of infected patients, and x_3 the number of recovered patients who now have lifelong immunity. The epidemiological model can be described by the equations

$$\begin{aligned}\dot{x}_1 &= \alpha N - r\beta \frac{x_2}{N} x_1 - \mu x_1,\\ \dot{x}_2 &= r\beta \frac{x_2}{N} x_1 - \gamma x_2 - \mu x_2,\\ \dot{x}_3 &= \gamma x_2 - \mu x_3\end{aligned}$$

with

α representing the birth rate,
N the population size, which we assume to be constant,
r the number of contacts between people per unit of time,
β the probability of infection per contact,
μ the death rate, and
γ the recovery rate.

(a) Determine the model's equilibrium points.
(b) Which situation is represented by each of the equilibrium points?

Let us now examine the case of an incurable infectious disease such as AIDS. In this case, no recovered patients exist and thus x_3 is not included in our calculation. Let us assume that the population size $N = x_1 + x_2$ remains constant. We will use the simplified model

$$\begin{aligned}\dot{x}_1 &= -r\beta \frac{x_2}{N} x_1,\\ \dot{x}_2 &= r\beta \frac{x_2}{N} x_1.\end{aligned}$$

(c) Reduce the system of differential equations to one differential equation with x_2 as the variable.
(d) Determine the equilibrium points.
(e) Calculate the solution to the differential equation. For this purpose, transform the differential equation using $x_2 = z^{-1}$.
(f) Which equilibrium point does the system tend to if one or more persons are contagious?
(g) What is the shortcoming of the model from (c), and how can it be remedied?

Exercise 1.5 Let us examine the system

$$\dot{x}_1 = -x_1 + \frac{1}{3}x_1^4 x_2^2, \quad x_{10} = x_1(0),$$
$$\dot{x}_2 = -x_2, \quad x_{20} = x(0),$$

whose solution is

$$x_1(t) = x_{10}\sqrt[3]{\frac{5e^{2t}}{x_{10}^3 x_{20}^2 + (5 - x_{10}^3 x_{20}^2)e^{5t}}}, \quad x_2(t) = x_{20}e^{-t},$$

and whose only equilibrium point is $\boldsymbol{x}_{\text{eq}} = \mathbf{0}$.
(a) Show that the equilibrium point is not globally but locally attractive.
(b) Show that the equilibrium point is asymptotically stable.
(c) Show that the equilibrium point is not globally asymptotically stable.
(d) Determine the region of attraction G_{ROA}.
(e) Determine the region of asymptotic stability G_{RAS}.

Exercise 1.6 Let us examine Chua's circuit, which is shown in Figure 1.45.
(a) Create a mathematical model of the circuit. Use the capacitor voltages u_{C1} and u_{C2} and the inductor current i_{L} as the state variables. Begin by calculating the nonlinear curve $i(u_{\text{C1}})$, assuming that the diodes and the operational amplifier are ideal components.
(b) What type of nonlinearity characterizes the circuit?
(c) Now, take

$$x_1 = \frac{u_{\text{C1}}}{u_0}, \quad x_2 = \frac{u_{\text{C2}}}{u_0}, \quad x_3 = \frac{R}{u_0}i_{\text{L}} \quad \text{with } u_0 = \frac{R_2}{R_2 + R_3}u_{\text{d}}$$

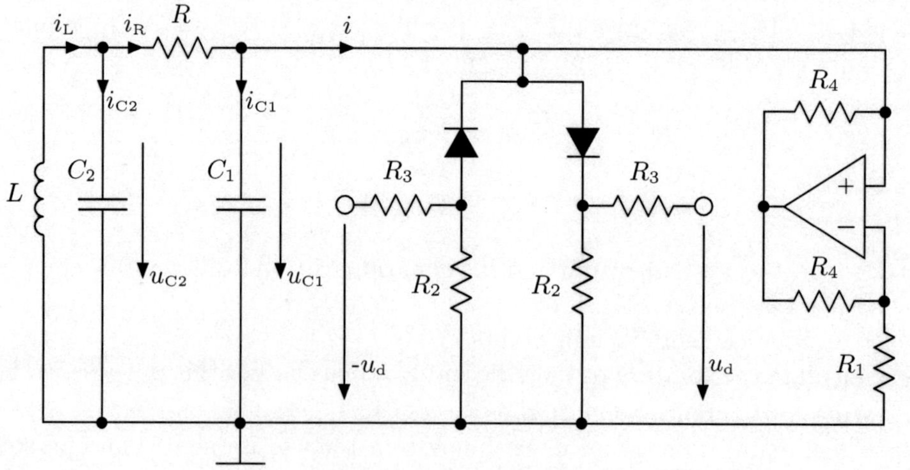

Fig. 1.45: Chua's circuit

as the new state variables and rescale the time using

$$t = RC_2\tau.$$

In addition, use the abbreviations

$$\alpha = \frac{C_2}{C_1}, \quad \beta = \frac{R^2 C_2}{L}, \quad a = -\frac{R}{R_1}, \quad b = R\frac{R_2 + R_3}{R_2 R_3} - \frac{R}{R_1}.$$

Based on this, calculate the normalized equations of the model.

(d) What type does this normalization take, and why do we perform it?

Exercise 1.7 Let us consider the system

$$\begin{aligned} \dot{x}_1 &= \alpha\left(-x_1 + x_2 - g(x_1)\right), \\ \dot{x}_2 &= x_1 - x_2 + x_3, \\ \dot{x}_3 &= -\beta x_2, \end{aligned} \qquad \text{with} \quad g(x_1) = \begin{cases} bx_1 + a - b, & x_1 \geq 1, \\ ax_1, & |x_1| < 1, \\ bx_1 + b - a, & x_1 \leq -1. \end{cases}$$

Here, $a < b < 0$ and $a < -1$, as well as $\alpha > 0$ and $\beta > 0$ apply.

(a) Determine the equilibrium points of the system.

(b) Now

$$a = -8/7, \quad b = -5/7, \quad \alpha = 44/5, \quad \text{and} \quad \beta = 15$$

apply. For each equilibrium point, determine whether it is asymptotically stable, Lyapunov stable, or unstable.

(c) The system has no limit cycle, and no trajectory tends to infinity. What do you conclude from this, taking into account the results from (b)?

Exercise 1.8 If linear systems are connected in series, it is possible to change their order without changing anything in the overall transfer behavior. This does not apply if nonlinear components are connected in series. Let us examine two examples shown in Fig. 1.46. An oscillation of the type $u(t) = A\cos(\omega t)$, $A > a$, is applied at the inputs of both circuits at $t = 0$. What form do the output signals y_1 and y_2 take? Make a sketch.

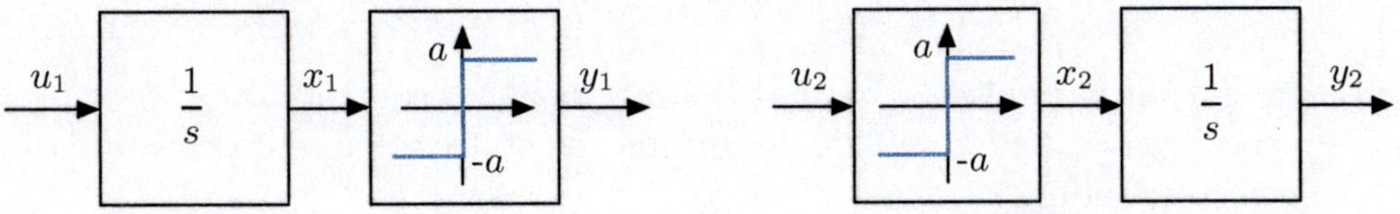

Fig. 1.46: Serial connections of a linear and a nonlinear system

Exercise 1.9 Consider the control loop shown in Figure 1.47.

(a) Draw the output signal $y(t)$ corresponding to the input signal $y_{\text{ref}}(t)$ shown in Figure 1.48, and for the time interval $0 \leq t \leq 40$ and $x(0) = 0$.

(b) How great is the steady-state error e_∞ when $y_{\text{ref}} = 3$ and $y_{\text{ref}} = -3$?

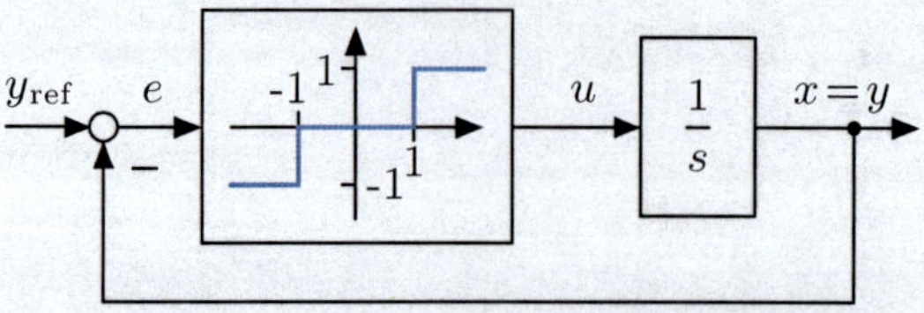

Fig. 1.47: Loop with three-position control

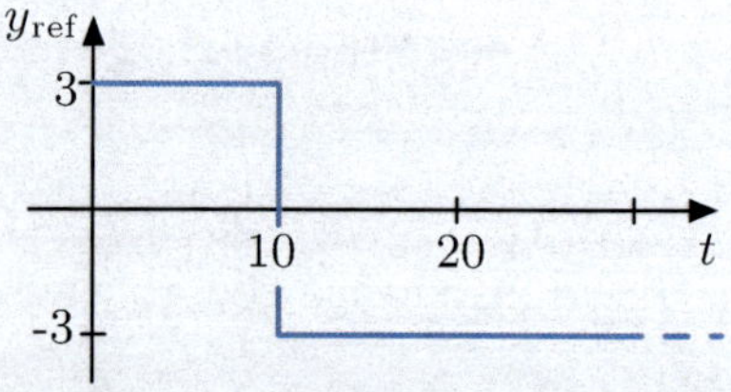

Fig. 1.48: Input signal $y_{ref}(t)$

Exercise 1.10 We will now analyze the stability behavior of the control loop shown in Figure 1.49.

(a) What are the eigenvalues of the plant?

(b) Calculate all the possible trajectories and sketch them on the x_1x_2-plane.

(c) Where do equilibrium points occur, and what are the stability properties of each?

(d) How does the choice of the initial vector $\boldsymbol{x}(0)$ influence the stability behavior?

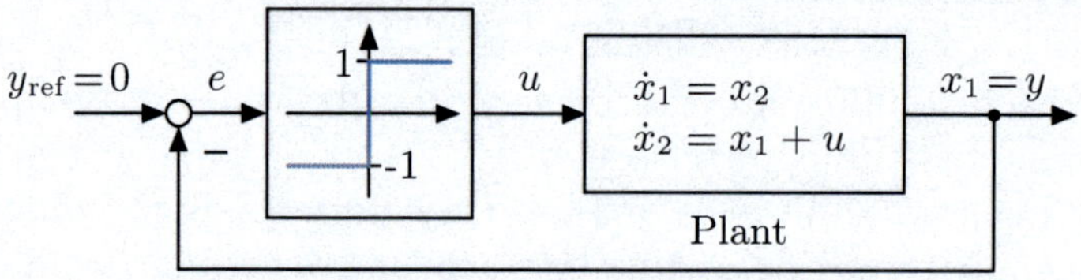

Fig. 1.49: Control loop with two-position control and unstable plant

Exercise 1.11 Let us examine the plant

$$\dot{x}_1 = x_2,$$
$$\dot{x}_2 = u$$

with the output variable $y = x_1$ and the proportional control $u = -ky$.

(a) For both $k = 4$ and $k = 0.25$, determine the state-space representation of the corresponding two control loops.

(b) Specify the stability behavior of both control loops.

(c) Calculate the time solution $x_1(t)$ and $x_2(t)$ for $k = 4$ and $k = 0.25$ for the initial values $x_1(0) = 0$ and $x_2(0) \in \mathbb{R}$. What shape do the trajectories $\boldsymbol{x}(t)$ take for the two control loops? Draw them both.

(d) We will now apply the switching control law

$$u = \begin{cases} -4y, & x_1x_2 \geq 0, \\ -0.25y, & x_1x_2 < 0. \end{cases}$$

Draw the trajectories of the control loop. What equilibrium points does it have, and what is their stability behavior?

(e) We will now change the switching control law to the following:

$$u = \begin{cases} -4y, & x_1 x_2 < 0, \\ -0.25y, & x_1 x_2 \geq 0. \end{cases}$$

Now draw the trajectories of the control loop again. What equilibrium points does it have, and what is their stability behavior?

Exercise 1.12 Prove that a *gradient system*

$$\dot{\boldsymbol{x}} = -\left(\frac{\partial h(\boldsymbol{x})}{\partial \boldsymbol{x}}\right)^T$$

has no limit cycles.

Exercise 1.13 Commercial fishing has a significant impact on the population dynamics of fish species. This example deals with tuna and mackerel. Both are prey of human beings; however, the mackerel is also the prey of tuna. The population variable x_1 of the mackerel can be described by

$$\dot{x}_1 = a_1 x_1 - a_2 x_1 x_2 - b_1 x_1 u_1$$

[36, 400]. The change $\dot{x}_1$ in the population size x_1 of the mackerel depends on $a_1 x_1$, where a_1 is the growth coefficient, the number of predators x_2, and the catch rate u_1 by human fishers. The increase $\dot{x}_2$ in the number of tuna depends on the number x_1 of their prey, but also on the competition $-a_3 x_2$ between them and their catch rate u_2 by human fishers. It is given by

$$\dot{x}_2 = -a_3 x_2 + a_4 x_1 x_2 - b_2 x_2 u_2,$$

where $a_1, a_2, a_3, a_4, b_1, b_2 > 0$ and $0 \leq u_1 \leq 1$ and $0 \leq u_2 \leq 1$ apply.

(a) Determine the equilibrium points of the system for constant catch rates u_1 and u_2.

(b) Linearize the system around the equilibrium point for which $x_1 \neq 0$, $x_2 \neq 0$, and $u_1 = u_2 = 0.5$ apply, where Δx_1 and Δx_2 denote the variations in the mackerel and tuna populations.

Assume $a_1 > 0.5b_1$ in the following.

(c) What eigenvalues does the linearized model have?

(d) Assume that the catch rate of the mackerel is $u_1 = 0.5$. Can the populations x_1 and x_2 around the equilibrium point from (b) be stabilized if we use the control $\Delta u_2 = -k\Delta x_2$ based only on the variation Δx_2 in the tuna population?

Exercise 1.14 Let us now examine a distribution process for granular material which is transported into two silos via a conveyer belt. The silos are filled in turn using a swiveling tube. The identical silos each have volume V

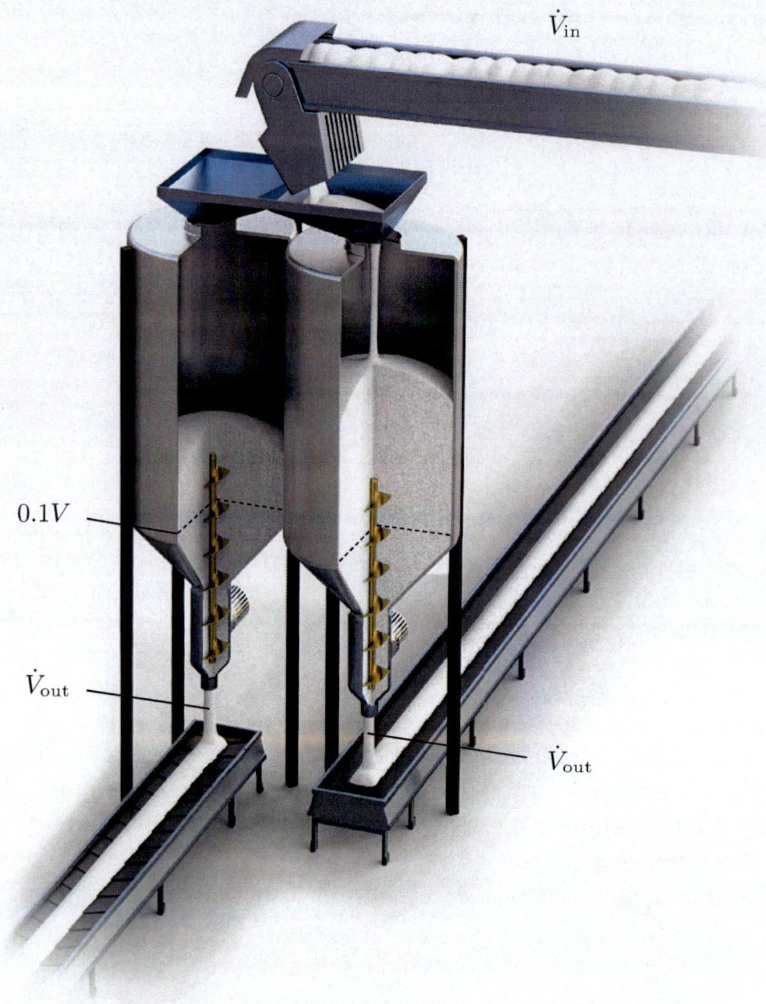

Fig. 1.50: Distribution system for bulk material

and serve as buffer storage units for the subsequent removal and continued transport of the granular material via two conveyer belts which carry it on to processing plants. The volume flows $\dot{V}_{1,\text{out}}$ and $\dot{V}_{2,\text{out}}$ exiting the silos can be regulated using adjustable unloaders. Figure 1.50 shows how this process works.

A control mechanism is required to swivel the filling tube from the left-hand to the right-hand silo and vice versa, ensuring that neither of the silos is emptied entirely. The control mechanism in this example is such that the tube swings out over the silo whose contents have fallen to 10% and pours the material into that silo.

In the special case examined in the following, the constant volume flow $\dot{V}_{\text{in}}$ of the material being transported amounts to 9/50 of a silo's volume V per hour; furthermore, the relation

$$\frac{10}{9}\dot{V}_{\text{in}} = \dot{V}_{1,\text{out}} + \dot{V}_{2,\text{out}}$$

holds. The two volume flows exiting the silos are constant and equal, i.e. $\dot{V}_{1,\text{out}} = \dot{V}_{2,\text{out}} = \dot{V}_{\text{out}}$ applies. For safety reasons, the process is stopped when one of the silos has contents lower than $0.1V$.

(a) Determine the state-space model of the system with the control mechanism described above. As state variables, use the filled volume V_1 and V_2 of the silos, which depend on the volume flow $\dot{V}_{\text{in}}$ being transported into each silo, and as a switching variable z. If the swiveling tube fills the left silo, $z = 1$ holds; if it fills the right one, then $z = 0$ applies.
(b) Calculate the time intervals Δt_i between switchovers of the swiveling tube from one silo to the other. At time $t = 0$, $V_1 = 0.1V$ and $V_2 = 0.5V$ apply.
(c) Calculate the switching times t_i at which the supplying mechanism swivels from one silo to the other.
(d) Calculate the sum t_{tot} of the times Δt_i of all switching intervals.
(e) Assess the control mechanism described here in terms of its applicability.
(f) What problem occurs if we simulate the system using an integration method, e. g. a Runge-Kutta method?

Exercise 1.15 Let us examine the ordinary differential equation

$$\dot{x} = p(t)x + q(t)x^{\alpha}, \quad \alpha \in I\!R,$$

which is known as the *Bernoulli differential equation*.

(a) Execute the transformation $z = x^{1-\alpha}$.
(b) What type is the transformed Bernoulli differential equation?

Exercise 1.16 Determine the solution to the differential equation

$$\dot{x} = ax\left(1 - \frac{x}{b}\right),$$

known as the *logistic differential equation*.

Exercise 1.17 Differential equations of the form

$$\dot{x} = f(x)g(t)$$

are called *separable*, since we can separate the variables x and t such that the left side of the equation depends only on x and the right one only on t. Thus we obtain

$$\frac{1}{f(x)}dx = g(t)dt.$$

Determine the general solution of a *separable differential equation*.

Exercise 1.18 Determine the analytical solutions $x(t)$ of the following differential equations:

(a) $\dot{x} = nx^{(n-1)/n}, \quad x(0) = 0,$

(b) $\dot{x} = -\dfrac{2(1+x^3)}{3x^2}, \quad x(0) \in \mathbb{R}, \quad$ hint: transform $z = 1 + x^3$,

(c) $\dot{x} = -x + x^2, \quad x(0) \in \mathbb{R},$

(d) $\dot{x} = t^2\sqrt{1-x^2}, \quad x(0) \in [-1, 1], \quad$ and

(e) $\ddot{x} = \sqrt{1+\dot{x}^2}, \quad x(0) \in \mathbb{R}, \quad \dot{x}(0) \in \mathbb{R}.$

Exercise 1.19 Consider the differential equation

$$\dot{x} = -\operatorname{sgn}(x)\sqrt{|x|}.$$

(a) Show that $x(t) = \begin{cases} \operatorname{sgn}(x_0)\left(\sqrt{|x_0|} - 0.5t\right)^2, & 0 \le t < 2\sqrt{|x_0|}, \\ 0, & t \ge 2\sqrt{|x_0|} \end{cases}$

for $x_0 \in \mathbb{R}\backslash\{0\}$ is the only solution to the differential equation.

(b) Show that the equilibrium point $x_{\text{eq}} = 0$ is globally asymptotically stable.

(c) What time t_{eq} does it take for the solutions $x(t)$ to reach the equilibrium point $x_{\text{eq}} = 0$?

Exercise 1.20 The aim of this exercise is to determine the computing effort of integration methods with different orders of accuracy q_i for the ordinary differential equation $\dot{x} = -x$. For this purpose, take into account the functional relationship between the relative error $\varepsilon^{\text{rel}} = \tilde{\alpha}h^q$ and the step size h for the various integration methods for the differential equation $\dot{x} = -x$ in Figure 1.51, which shows a magnified area of Figure 1.43 in Section 1.2.10 on p. 55. Display the results in table form.

(a) What is the relationship between the slopes and the order of accuracy q_i of each integration method? What are the slopes m_i and the parameters $\tilde{\alpha}_i$ of the curves shown in the double logarithm plot in Figure 1.43 on p. 55? To determine this, take only the approximately linear parts of the curves into account.

(b) In each of the integration methods, what approximate step size h should we choose for the example $\dot{x} = -x$ such that the relative error is limited to $\varepsilon^{\text{rel}} = 10^{-6}$? If necessary, extrapolate the curves shown in the graph.

(c) The time required to calculate the function $f(x, u)$ normally takes up most of the time required for the simulation of a common differential equation. For the methods shown in Figure 1.43, estimate how often the function $f(x, u)$ must be calculated to determine the solution of the differential equation $\dot{x} = -x$ within the time interval $[0, 10]$ if we wish to keep the relative error below 10^{-6}.

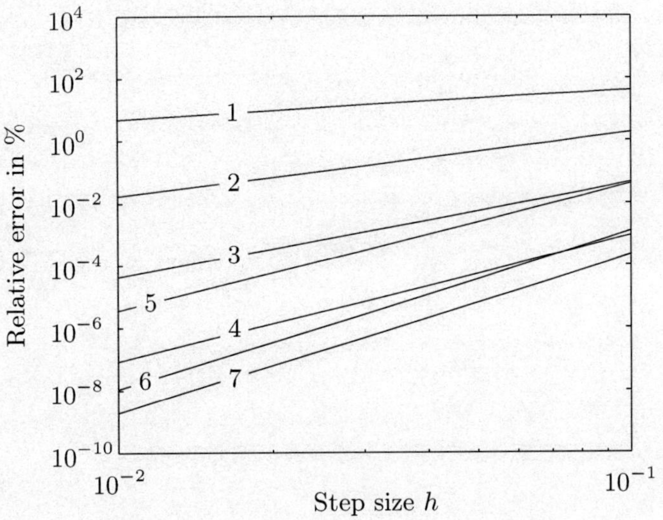

Fig. 1.51: Relative error in double logarithmic representation for the following methods: 1. Euler, 2. modified Euler, 3. Simpson, 4. fourth-order Runge-Kutta, 5. Adams-Bashforth with $q = 4$, 6. Adams-Moulton with $q = 5$ (one iteration), 7. Adams-Moulton with $q = 5$ (ten iterations)

(d) Why is it impossible, from a comparison of the integration methods' orders of accuracy q alone, to determine which method is the most suitable? What is the most suitable one-step method? Based on the analysis above, which integration method would you use for similar problems?

Exercise 1.21 Using the z-transformation, calculate the stability range of the step size for the differential equation $\dot{x} = -\lambda x$ with $\lambda > 0$ for the following integration methods:

(1) Simpson,
(2) Runge-Kutta, 4th order,
(3) Adams-Moulton, 5th order,
(4) Gear, 4th order.

2

Limit Cycles and Stability Criteria

2.1 The Describing Function Method

2.1.1 Idea behind the Method

The describing function method, also called the *method of harmonic balance*, is used to detect limit cycles in nonlinear control loops, which are structured as depicted in Figure 2.1. It can also be used for nonlinear control loops that have been transformed into such a structure. The control loop shown is referred to as the *nonlinear standard control loop*. It consists of a linear system, represented here by its Laplace transfer function $G(s)$, and a nonlinear characteristic curve $u = f(e)$, representing a controller or a plant's nonlinearity, for example.

The absence of a reference variable y_{ref} does not pose a serious problem, since a constant reference variable can be shifted to zero via a transformation. In any case, a limit cycle must also be ruled out for $y_{\text{ref}} = 0$. Thus it is adequate to consider only the case $y_{\text{ref}} = 0$.

Nonlinear standard control loops are frequently encountered in practice, either because nonlinear controllers are deliberately deployed, or because nonlinear characteristic curves are included as undesirable elements within the control-loop structure, e. g. in the form of the limiting characteristic curve of the actuator. Typical characteristic curves are illustrated in Figure 2.2.

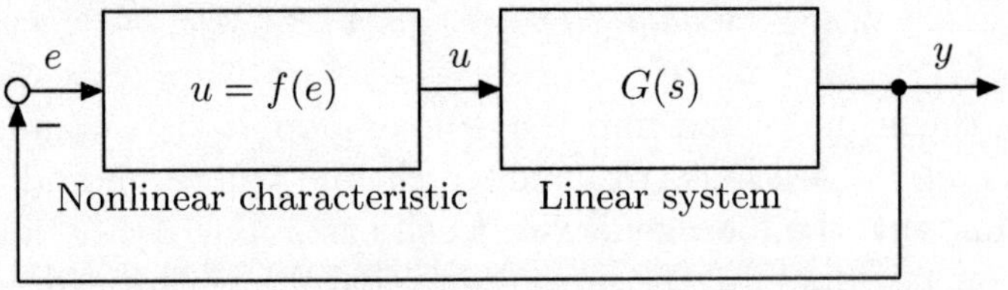

Fig. 2.1: Nonlinear standard control loop

J. Adamy, *Nonlinear Systems and Controls*,
https://doi.org/10.1007/978-3-662-68690-4_2

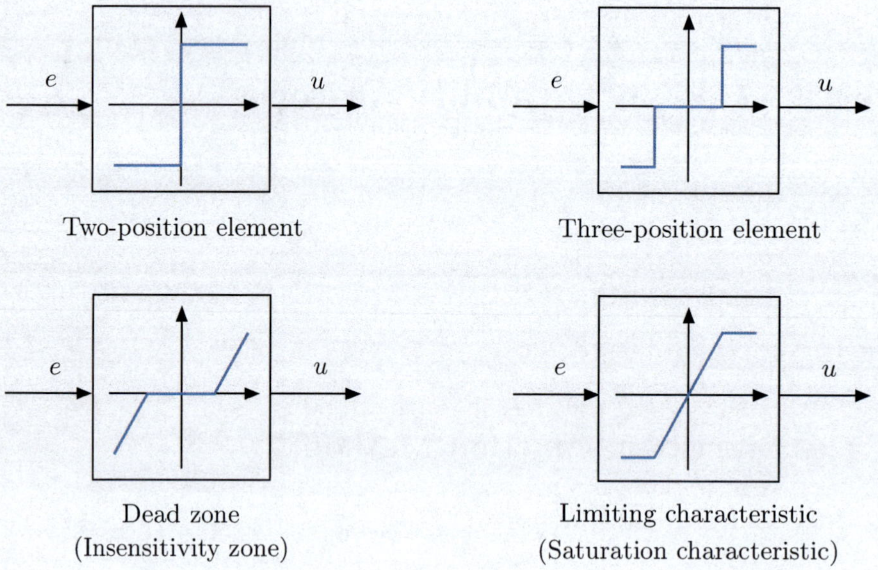

Fig. 2.2: Typical characteristic curves in control loops

A question that arises is whether limit cycles can occur in the above control loop. To gradually approach the solution of this problem, we will first consider the special case of a linear characteristic curve

$$u = f(e) = K \cdot e.$$

In this case, the control loop takes the form shown in Figure 2.3. A permanent oscillation, i. e. one that is self-sustaining within the control loop, arises if a harmonic signal

$$e(t) = A \cdot \sin(\omega_0 t) \tag{2.1}$$

is fed into the control loop at the input of the linear characteristic and is phase-shifted by π, i. e. 180°, and of amplitude A at the output of the linear plant $G(s)$, i. e.

$$y(t) = A \cdot \sin(\omega_0 t - \pi) = -A \cdot \sin(\omega_0 t);$$

because then this signal is fed into the control loop at the summation point, the new signal $e(t) = -y(t)$ corresponds to the old one (2.1), and the process repeats. In this way, the loop generates a self-sustaining oscillation.

In the frequency domain, the above condition for a sustained oscillation is formulated as

$$A \cdot e^{j(\omega_0 t - \pi)} = K \cdot G(j\omega_0) \cdot A \cdot e^{j\omega_0 t}$$

or

$$K \cdot G(j\omega_0) = -1. \tag{2.2}$$

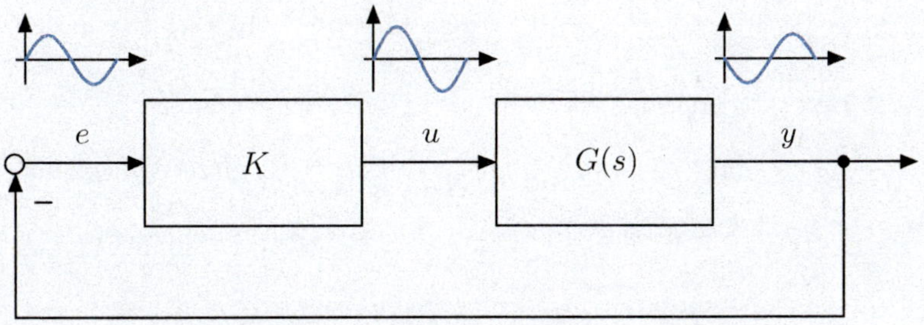

Fig. 2.3: Control loop with a linear controller

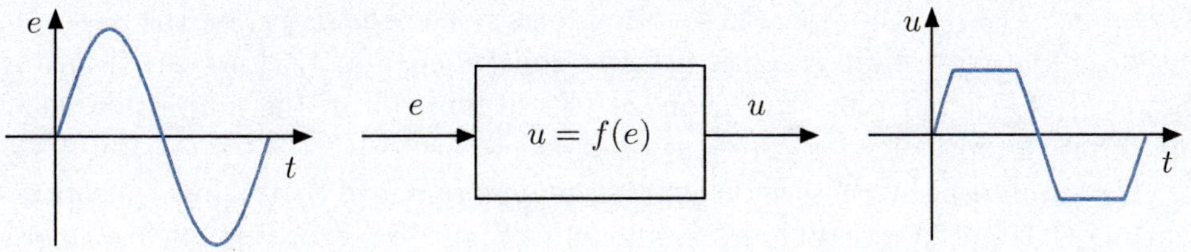

Fig. 2.4: Distortion of the input signal by the nonlinearity

This perpetuation of the sinusoidal oscillations at the input and the output of the open loop is referred to as the state of *harmonic balance.*

From the above, we can now deduce the nonlinear situation as follows. If we feed a sinusoidal signal

$$e(t) = A \cdot \sin(\omega_0 t)$$

into the nonlinearity, we obtain a distorted sinusoidal signal at the output, as illustrated in Figure 2.4.

The output signal u can be represented by a Fourier series as

$$u(t) = c_0(A) + \sum_{i=1}^{\infty} c_i(A) \cdot \sin(i \cdot \omega_0 t + \varphi_i(A)).$$

If the nonlinearity is of the kind such that

$$c_0(A) = 0 \qquad \text{and} \qquad c_i \ll c_1, \quad i = 2, 3, \ldots$$

holds, i. e. the constant component is zero and the amplitudes of the harmonics c_i are small compared to the amplitude c_1 of the fundamental wave, the output $u(t)$ of the controller can be approximated by

$$u(t) \approx c_1(A) \cdot \sin(\omega_0 t + \varphi_1(A)).$$

The condition $c_0(A) = 0$ is fulfilled if the characteristic curve is point-symmetric to the origin.

With the above approximation, we have linearized the nonlinearity f, and the linear approximation has an amplitude-dependent gain. It is obtained from the input and the output signal

$$e(t) = A \cdot \sin(\omega_0 t) \qquad \text{and} \qquad u(t) = c_1(A) \cdot \sin(\omega_0 t + \varphi_1(A))$$

or in phasor representations

$$\underline{e}(t) = A \cdot e^{j(\omega_0 t - \frac{\pi}{2})} \qquad \text{and} \qquad \underline{u}(t) = c_1(A) \cdot e^{j(\omega_0 t - \frac{\pi}{2} + \varphi_1(A))}$$

as

$$N(A) = \frac{\underline{u}(t)}{\underline{e}(t)} = \frac{c_1(A)}{A} \cdot \mathrm{e}^{j\,\varphi_1(A)}\,. \tag{2.3}$$

This gain $N(A)$ of the linearized characteristic element is called the *describing function*. Note that $N(A)$ is not frequency-dependent. However, the gain factor $c_1(A)/A$ and phase rotation $\varphi_1(A)$ do depend on the amplitude A of the input signal.

The nonlinear characteristic element is now replaced by its linear approximation (2.3) in the control loop, as shown in Figure 2.5. For this linear control loop the previously derived condition (2.2) for the state of harmonic balance, i. e. a self-sustaining oscillation, is given by

$$N(A) \cdot G(j\omega) = -1$$

or

$$G(j\omega) = -\frac{1}{N(A)}. \tag{2.4}$$

Of course, the above equation only holds if the neglected harmonics occurring at the frequencies $2\omega_0$, $3\omega_0$, $4\omega_0, \ldots$ are sufficiently damped by the plant $G(s)$. This means that the transfer function $G(s)$ must exhibit a sufficiently strong low-pass behavior.

Summarizing the results above and generalizing them to the characteristic curves

$$u = f(e, \dot{e}),$$

we obtain the following heuristic.

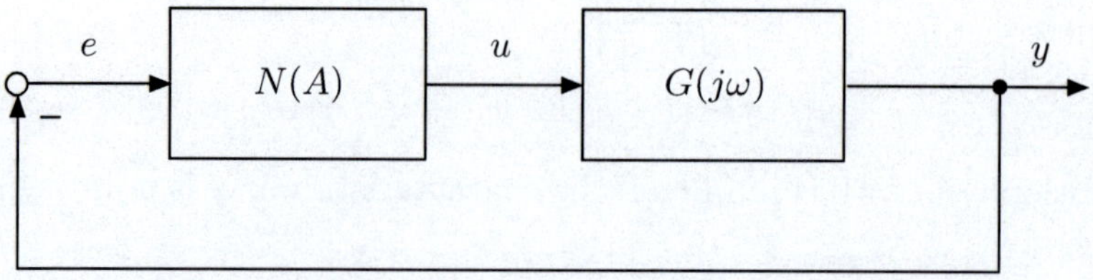

Fig. 2.5: Linearized characteristic element and linear plant

Heuristic 1 (Harmonic Balance). *Consider a nonlinear standard control loop*

$$\begin{aligned} Y(s) &= G(s)U(s), \\ e &= -y, \\ u &= f(e, \dot{e}). \end{aligned}$$

Let the characteristic curve $u = f(e, \dot{e})$ *be point-symmetric with respect to the origin, i. e.* $f(-e, -\dot{e}) = -f(e, \dot{e})$ *holds. Further, assume that the plant* $G(s)$ *possesses a sufficiently strong low-pass characteristic. If there exist values* ω *and* A*, such that the equality*

$$G(j\omega) = -\frac{1}{N(A)}$$

is fulfilled, presumably a sustained oscillation occurs which has approximately the frequency ω *and the amplitude* A*.*

The describing function $N(A)$ is real-valued if the point-symmetric nonlinearity only depends on e. If it is also a function of $\dot{e}$, $N(A)$ usually has an imaginary part as well.

Condition (2.4) can be evaluated graphically in the complex plane. To this end, we will draw the Nyquist plot of $G(j\omega)$ and the locus $-1/N(A)$ of the describing function, which we will call *nonlinear locus* for short. If an intersection point exists, equation (2.4) is fulfilled, and presumably a sustained oscillation occurs. The frequency and amplitude of the presumed sustained oscillation can also be approximately computed with the help of the intersection point.

2.1.2 Illustrative Example

To illustrate the method described above, we will consider the plant

$$G(s) = \frac{9}{s\,(s+1)\,(s+9)},$$

which we control with a two-position element or relay characteristic

$$u = -b \cdot \operatorname{sgn}(y).$$

Figure 2.6 shows the corresponding control loop.

First we determine the describing function $N(A)$ of the two-position controller. For every sinusoidal signal input, its output function is a sequence of square-wave signals. The associated Fourier series of the sequence of square-wave signals is

$$u(t) = \frac{4b}{\pi}\left(\sin(\omega_0 t) + \frac{\sin(3\omega_0 t)}{3} + \frac{\sin(5\omega_0 t)}{5} + \ldots\right).$$

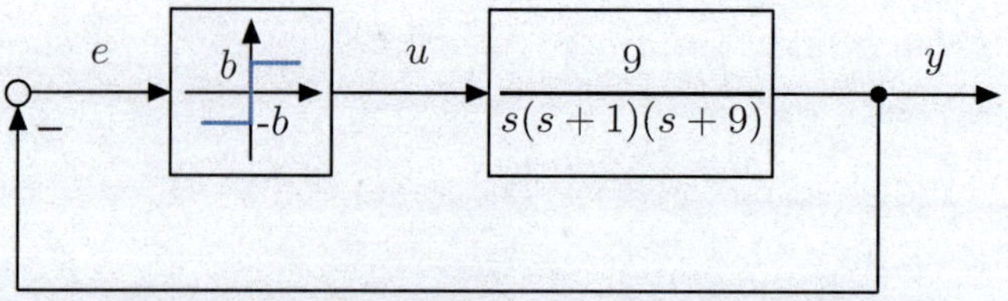

Fig. 2.6: Control loop with two-position controller

Approximating

$$u(t) \approx \frac{4b}{\pi}\sin(\omega_0 t) = c_1(A)\cdot\sin(\omega_0 t),$$

we obtain

$$N(A) = \frac{c_1(A)}{A} = \frac{4b}{\pi A}.$$

To detect a sustained oscillation, both sides of the equation of the harmonic balance

$$G(j\omega) = -\frac{1}{N(A)},$$

i. e.

$$\frac{9}{j\omega(j\omega+1)(j\omega+9)} = -\frac{\pi A}{4b}, \tag{2.5}$$

are graphically illustrated. This is shown in Figure 2.7. Since an intersection point of the Nyquist plot of $G(j\omega)$ and the nonlinear locus $-1/N(A)$ exists, equation (2.5) is fulfilled for a value pair (ω, A), and we can conclude that a sustained oscillation exists. Its amplitude and frequency are determined for $b = 1$ from equation (2.5), i. e. from

$$-\frac{4}{\pi A} = -\frac{10}{9}\omega^2 + j\frac{9\omega - \omega^3}{9},$$

to be

$$\omega = 3 \quad \text{and} \quad A = \frac{2}{5\pi} \approx 0.127.$$

The approximative nature of the method becomes apparent when we compare these values to a simulation, from which the values

$$\omega = 2.5 \quad \text{and} \quad A = 0.195$$

are obtained.

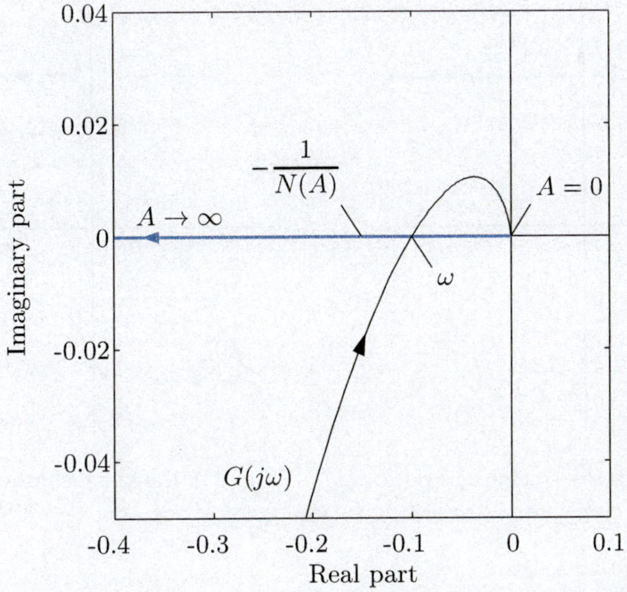

Fig. 2.7: Nyquist plot of $G(j\omega)$ and the nonlinear locus $-1/N(A)$

2.1.3 Characteristic Curves and Their Describing Functions

The method of harmonic balance is also applicable if the nonlinear characteristic curve not only depends on e, but also on $\dot{e}$. One of the most important characteristic curves of this kind is the hysteresis curve

$$\begin{aligned} u &= \begin{cases} b \cdot \operatorname{sgn}(e + a), & \dot{e} < 0, \\ b \cdot \operatorname{sgn}(e - a), & \dot{e} > 0 \end{cases} \\ &= b \cdot \operatorname{sgn}(e - a \cdot \operatorname{sgn}(\dot{e})) \end{aligned}$$

shown in Figure 2.8. Hysteresis curves can be found in temperature-control loops such as those in irons, among other applications. In these cases, the heater is turned on using a bimetal. When this bimetal has heated up to a sufficient temperature, it deforms in such a way that the contact opens, turning off the heating current. After cooling down, the bimetal relaxes into its original form, turns the heating current back on, and the cycle begins anew.

Backlash behavior is another frequently occurring nonlinearity which depends both on e and on $\dot{e}$. This behavior is described by the characteristic curve shown in Figure 2.9. Backlash, as depicted in Figure 2.10, occurs in mechanical systems as play between gear wheels, carriers, linkages, etc. Note that, depending on the sign of $\dot{e}$, the horizontal branches in Figure 2.9 are passed through from both directions. The horizontal branches can occur for every value of u. Backlash behavior can be modeled by

$$\dot{u} = m\dot{e}\Big[H(me - u - ma)H(\dot{e}) + H(-me + u - ma)H(-\dot{e})\Big], \tag{2.6}$$

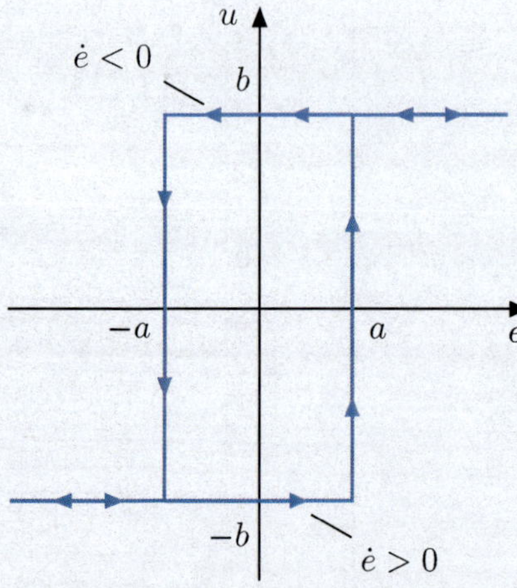

Fig. 2.8: Hysteresis characteristic curve

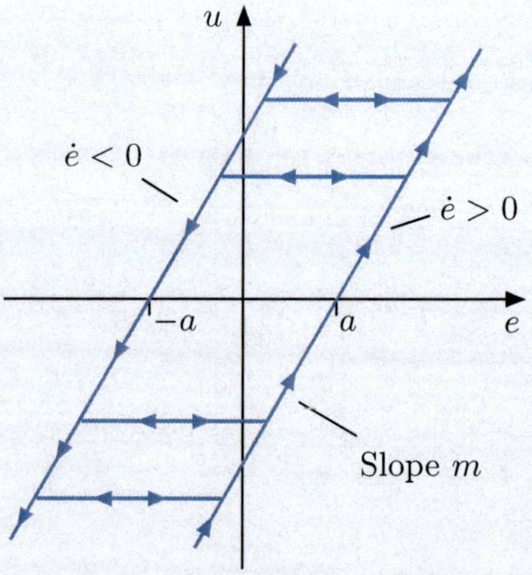

Fig. 2.9: Backlash characteristic curve

where the initial values $u(0)$ and $e(0)$ are restricted by

$$me(0) - ma \leq u(0) \leq me(0) + ma.$$

The parameter m is the backlash's slope, the clearance is $2a$, and

$$H(x) = \begin{cases} 0, & x < 0, \\ 1, & x \geq 0, \end{cases}$$

is the Heaviside function. For most systems, especially mechanical ones, the slope is $m = 1$.

The describing functions $N(A)$ of the hysteresis and backlash characteristic curves are also determined by choosing a sine function as the input signal and representing the output as a Fourier series. The describing function $N(A)$ then results again from the fundamental wave. The describing functions $N(A)$ of the hysteresis, backlash, and other important characteristic curves are given in Table 2.1.

Many characteristic curves can be represented as an additive combination of standard characteristic curves. This summation of characteristic curves corresponds to a parallel connection of their function blocks. Examples are step curves which are point-symmetric. They can be formed by adding three-position characteristic curves. In this way, the step curve of an A/D converter can be emulated, for example.

Since the relation

$$u = N_1(A)e + N_2(A)e = (N_1(A) + N_2(A))\,e$$

holds, the total describing function $N_{\text{tot}}(A)$ of k characteristic curves which are connected in parallel, i. e.

Fig. 2.10: Examples of systems with backlash behavior

$$u = \sum_{i=1}^{k} f_i(e, \dot{e}),$$

is given by the sum of the describing functions $N_i(A)$ of the individual nonlinearities f_i, according to

$$N_{\text{tot}}(A) = \sum_{i=1}^{k} N_i(A).$$

An important example of a composite characteristic curve $N_{\text{tot}}(A)$ is a piecewise linear continuous characteristic curve, i. e. a polygonal line. This polygonal line can be constructed as a summation of dead zones or limiting elements. In each of the $2k$ intervals

$$[\pm a_i, \pm a_{i+1}), \qquad i = 0, \dots, k,$$

of the polygon, there is one straight line with corresponding slope m_i. Further, the polygon line is point-symmetric to the origin and consequently passes through zero, so we obtain

$$u = \begin{cases} m_1 e, & e \in [0, \pm a_1), \\ m_2 e \pm (m_1 - m_2) a_1, & e \in [\pm a_1, \pm a_2), \\ m_3 e \pm (m_1 - m_2) a_1 \pm (m_2 - m_3) a_2, & e \in [\pm a_2, \pm a_3), \\ \vdots & \\ m_k e \pm \displaystyle\sum_{i=1}^{k-1} (m_i - m_{i+1}) a_i, & e \in [\pm a_{k-1}, \pm \infty). \end{cases}$$

We can construct the corresponding describing function of the polygon line by using the describing function of a dead zone, which can be found in Table 2.1.

Table 2.1: Characteristic curves and their describing functions

<table>
<tr><th>Nonlinearity</th><th>Describing Function $N(A)$ and Nonlinear Locus $-1/N(A)$</th></tr>
<tr><td>Two-position element
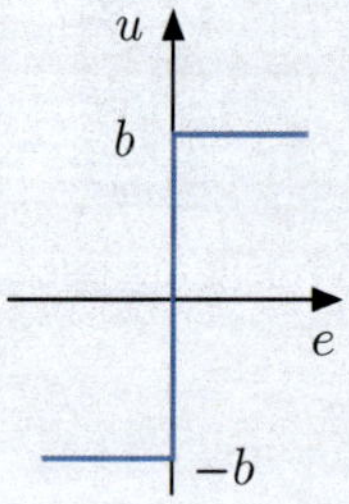</td><td>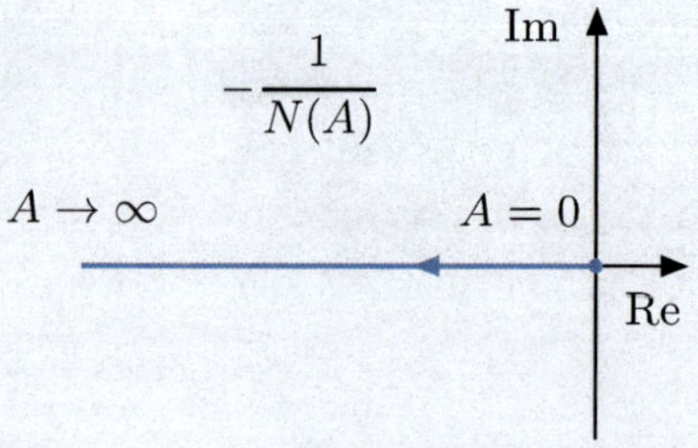
$N(A) = \dfrac{4b}{\pi A},\ A \geq 0$</td></tr>
<tr><td>Preload
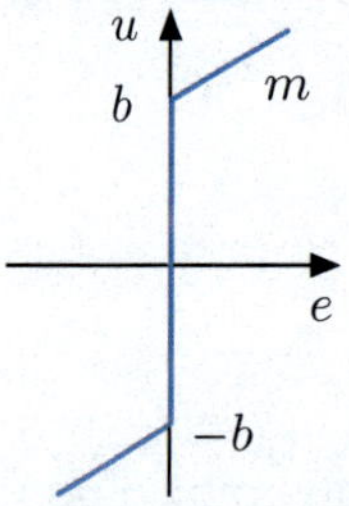
Slope: m</td><td>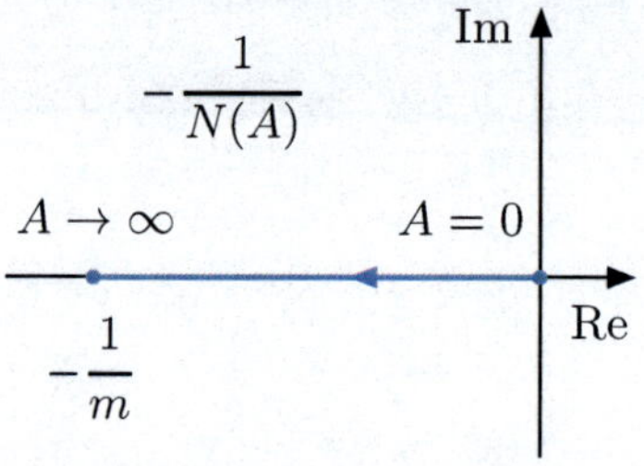
$N(A) = \dfrac{4b}{\pi A} + m,\ A \geq 0$</td></tr>
<tr><td>Three-position element
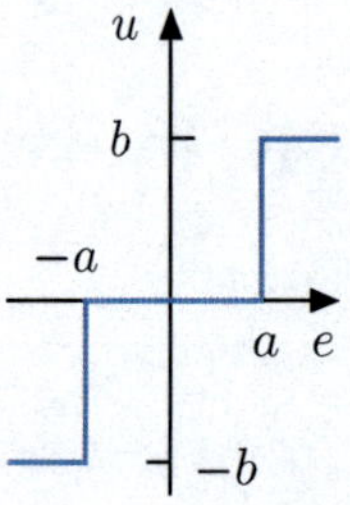</td><td>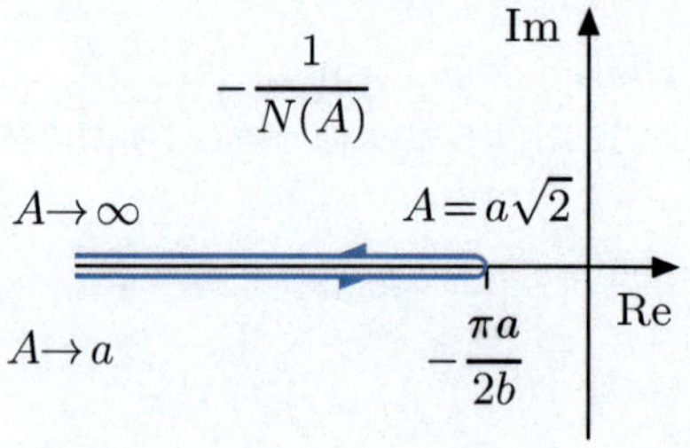
$N(A) = \dfrac{4b}{\pi A}\sqrt{1-\left(\dfrac{a}{A}\right)^2},\ A \geq a$</td></tr>
</table>

Table 2.1: Characteristic curves and their describing functions - continued

Nonlinearity	Describing Function $N(A)$ and Nonlinear Locus $-1/N(A)$
Dead zone 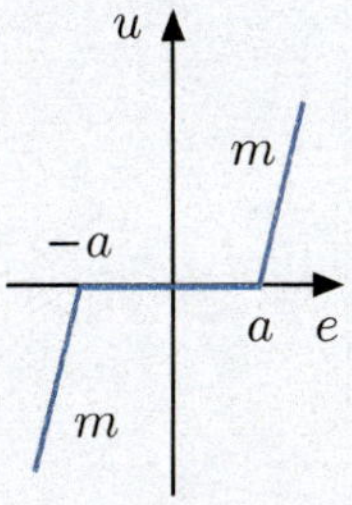 Slope: m	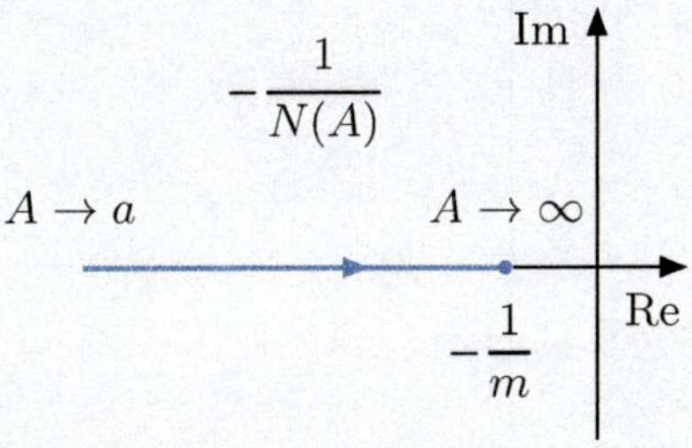 $N(A) = m\left(1-\frac{2}{\pi}\arcsin\left(\frac{a}{A}\right)-\frac{2}{\pi}\frac{a}{A}\sqrt{1-\left(\frac{a}{A}\right)^2}\right),\ A \geq a$
Saturation 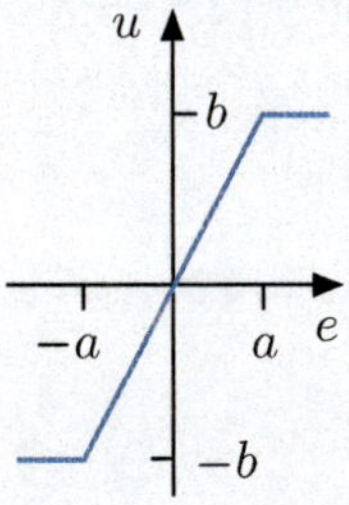 Slope: $m = \frac{b}{a}$	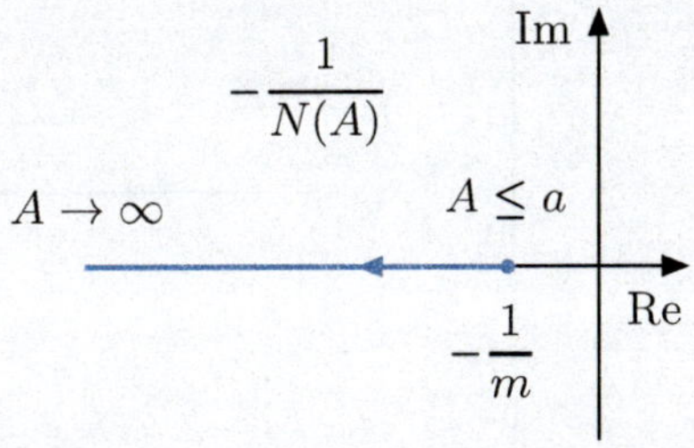 $N(A) = \begin{cases} m, & 0 \leq A \leq a \\ \frac{2m}{\pi}\left(\arcsin\left(\frac{a}{A}\right)+\frac{a}{A}\sqrt{1-\left(\frac{a}{A}\right)^2}\right), & A > a \end{cases}$
Power 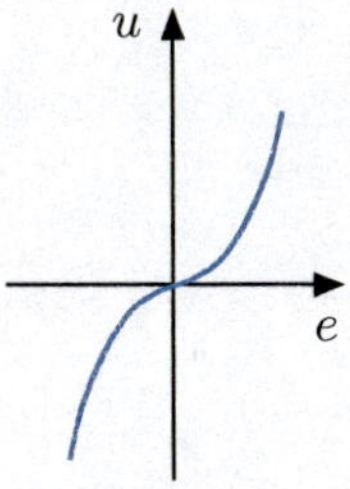 $u = e\,\lvert e\rvert,$ $u = e^3$	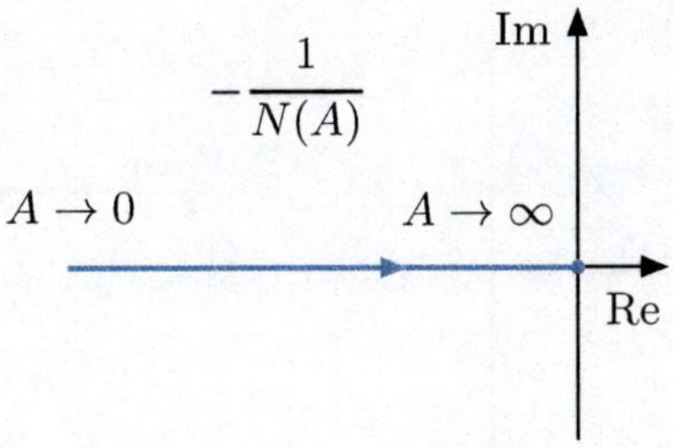 $N(A) = \frac{8A}{3\pi},\quad$ for $\quad u = e\,\lvert e\rvert \quad$ and $\quad A \geq 0$ $N(A) = \frac{3}{4}A^2,\quad$ for $\quad u = e^3 \quad$ and $\quad A \geq 0$

Table 2.1: Characteristic curves and their describing functions - continued

Nonlinearity	Describing Function $N(A)$ and Nonlinear Locus $-1/N(A)$
Roots 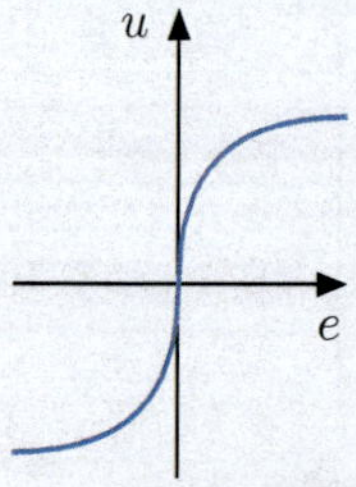 $u = \mathrm{sgn}(e)\sqrt{\lvert e \rvert}$, $u = \sqrt[3]{e}$	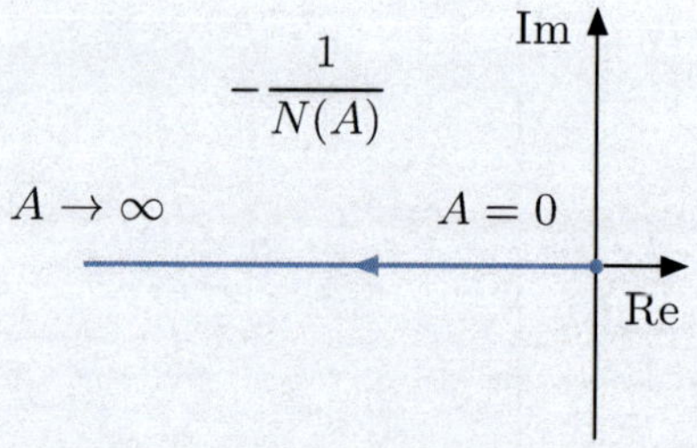 $N(A) = 1.11A^{-1/2}$, for $u = \mathrm{sgn}(e)\sqrt{\lvert e \rvert}$ and $A \geq 0$ $N(A) = 1.16A^{-2/3}$, for $u = \sqrt[3]{e}$ and $A \geq 0$
Dry friction 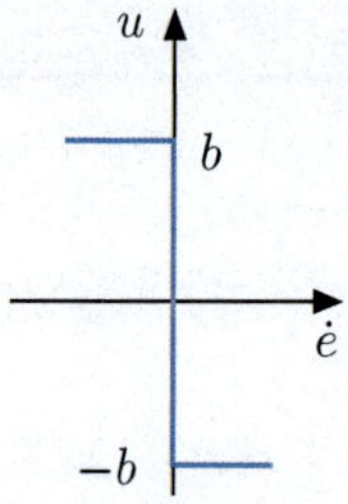 Force of friction: b $u = -b\,\mathrm{sgn}\,(\dot{e})$	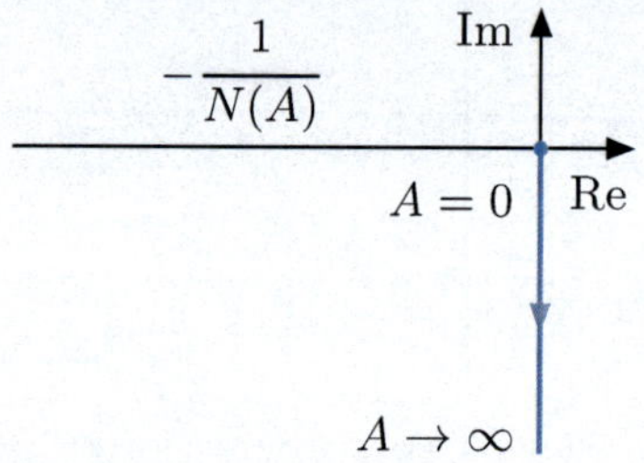 $N(A) = -j\,\frac{4\,b}{\pi A}$, $A \geq 0$
Backlash 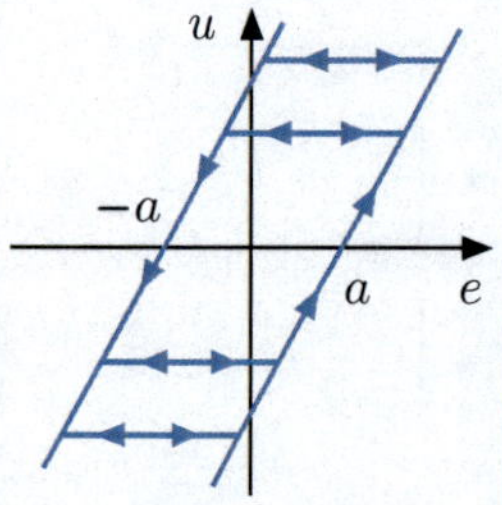 Slope: m	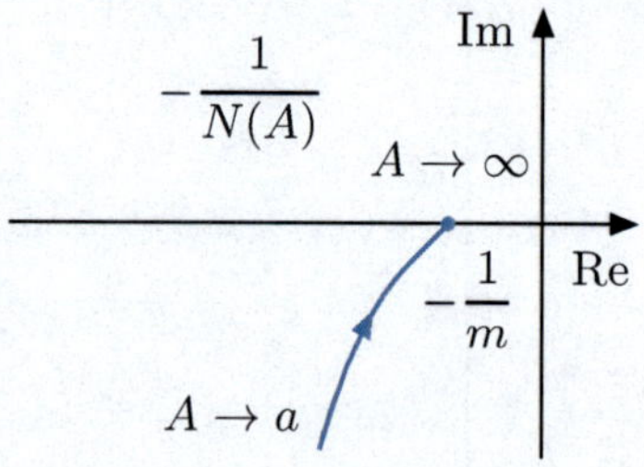 $N(A) = \frac{m}{2} + \frac{m}{\pi}\left(\arcsin(\alpha) + \alpha\sqrt{1-\alpha^2}\right) - j\frac{m}{\pi}\,(1-\alpha^2)$ with $\alpha = 1 - \frac{2a}{A}$, $A > a$

Table 2.1: Characteristic curves and their describing functions - continued

Nonlinearity	Describing Function $N(A)$ and Nonlinear Locus $-1/N(A)$
Sharp hysteresis 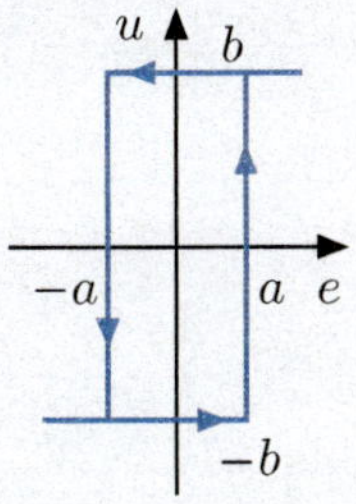	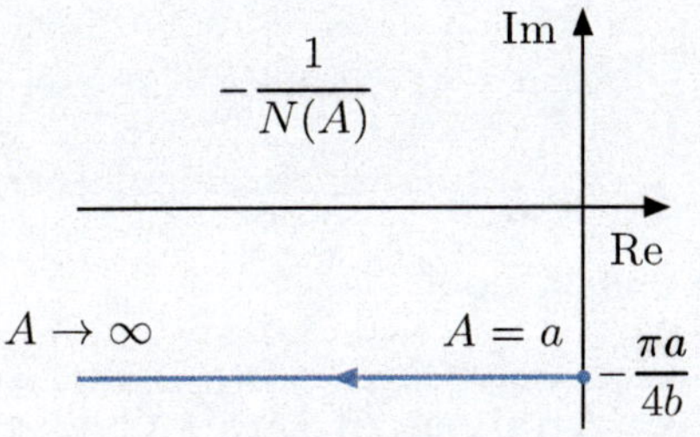 $N(A) = \frac{4b}{\pi A}\sqrt{1-\left(\frac{a}{A}\right)^2} - j\frac{4ab}{\pi A^2},\ A \geq a$
Elastic hysteresis 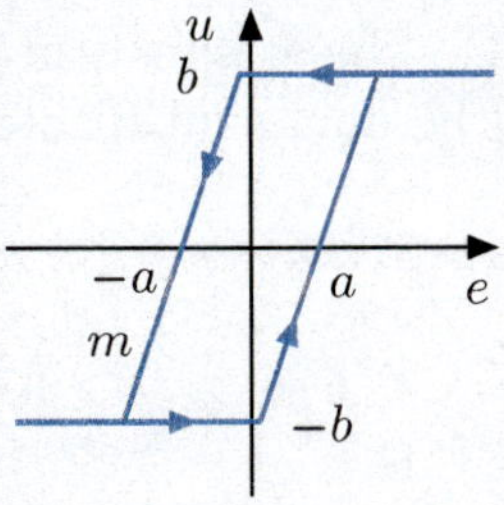 Slope: m	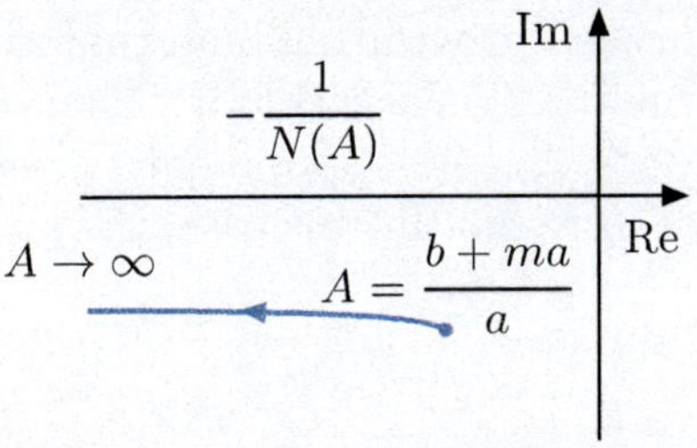 $N(A) = \frac{m}{\pi}\left(\mu\left(\frac{b+ma}{mA}\right) + \mu\left(\frac{b-ma}{mA}\right)\right) - j\frac{4ba}{\pi A^2},$ $\mu(x) = \arcsin(x) + x\sqrt{1-x^2},\ A \geq \frac{b+ma}{m}$
Three-position element with hysteresis 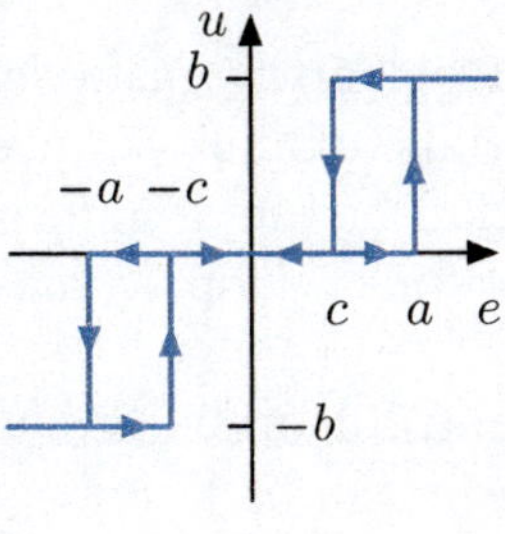	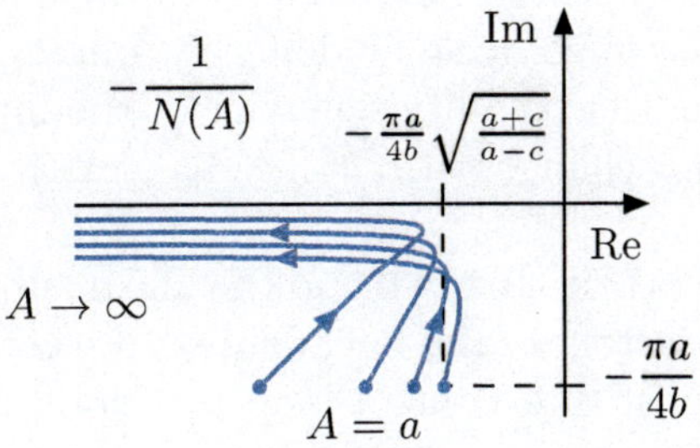 $N(A) = \frac{2b}{\pi A}\left(\sqrt{1-\left(\frac{c}{A}\right)^2} + \sqrt{1-\left(\frac{a}{A}\right)^2}\right) - j\frac{2b(a-c)}{\pi A^2},$ $A \geq a$

For the polygonal line, we can then calculate

$$N_{\mathrm{tot}}(A) = m_1 + \sum_{i=1}^{k-1} (m_{i+1} - m_i)\left(1 - \frac{2}{\pi}\mu\left(\frac{a_i}{A}\right)\right)$$

with

$$\mu(x) = \arcsin(x) + x\sqrt{1 - x^2}.$$

Here $A \geq a_{k-1}$ must hold.

2.1.4 Stability Analysis of Limit Cycles

Based on the Nyquist plot of $G(j\omega)$ and the nonlinear locus $-1/N(A)$, it is not only possible to determine whether a limit cycle could exist; it is also possible to deduce its stability behavior. Thus we can ascertain whether the limit cycle is stable, semi-stable, or unstable.

For this purpose, we will assume that the limit cycle of amplitude A_{lc} has been determined. Then within a vicinity of the limit cycle, the behavior of the control loop can be approximately described by the linear substitute loop shown in Figure 2.3. In this case, the gain factor of the linear substitute controller is

$$K = N(A).$$

If the amplitude is changed slightly by ΔA to

$$A = A_{\mathrm{lc}} + \Delta A,$$

K also changes only slightly. With this change in the amplitude A, we have left the limit cycle and have to answer the question of whether the trajectory returns to the limit cycle or departs from it. This question can be answered by examining the stability behavior of the linear substitute control loop (see Figure 2.5) for changes in $K = N(A_{\mathrm{lc}} + \Delta A)$. Four cases are possible; see also Figure 2.11.

For Case (a), let $\Delta A > 0$ and the linear substitute control loop be unstable. Thus the trajectory of the linear substitute control loop tends to infinity. Since the linear substitute control loop is a good approximation of the nonlinear control loop, we can conclude that the trajectory of the nonlinear loop departs from the limit cycle.

For Case (b), let $\Delta A > 0$ and the linear substitute control loop be stable, so that the amplitude A decreases over time. Consequently, the trajectory of the nonlinear control loop tends to the limit cycle.

For Case (c), let $\Delta A < 0$ and the linear substitute control loop be unstable, which is why the amplitude A increases over time and the trajectory of the nonlinear control loop tends to the limit cycle.

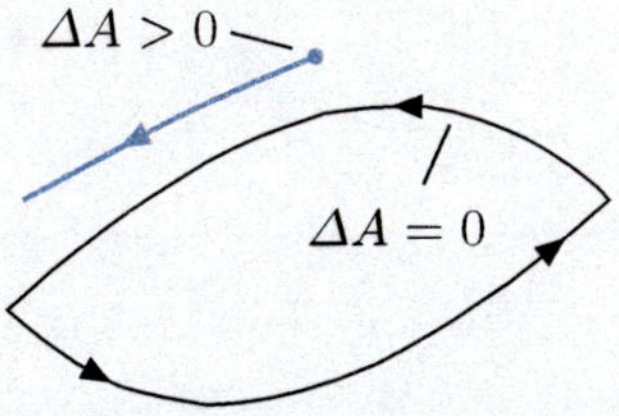

Case (a): $\Delta A > 0$ and unstable linear substitute control loop

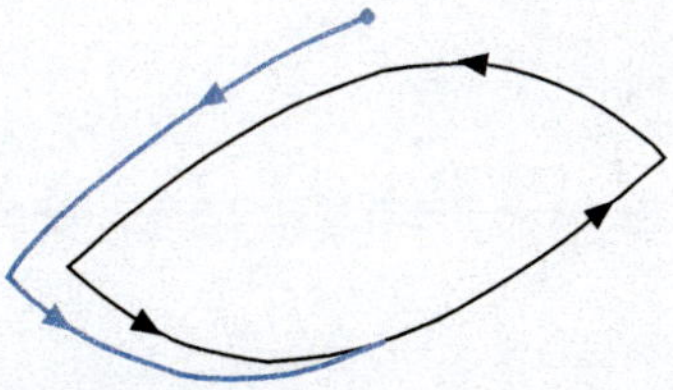

Case (b): $\Delta A > 0$ and stable linear substitute control loop

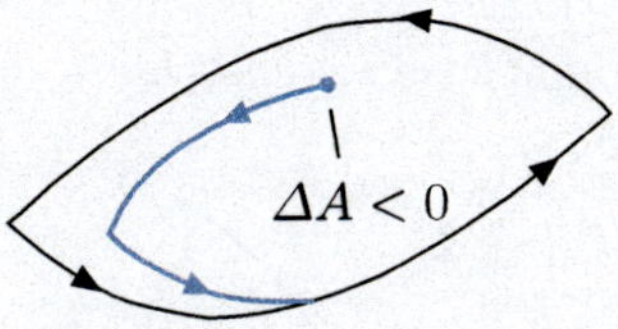

Case (c): $\Delta A < 0$ and unstable linear substitute control loop

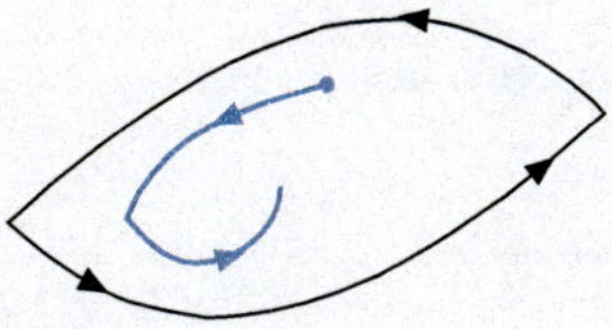

Case (d): $\Delta A < 0$ and stable linear substitute control loop

Fig. 2.11: Stability behavior for amplitude changes ΔA in the limit cycle

For Case (d), let $\Delta A < 0$ and the linear substitute control loop be stable. As a consequence, the trajectory of the nonlinear control loop departs from the limit cycle.

Because both of the cases $\Delta A > 0$ and $\Delta A < 0$ occur in the control loop, the stability of the limit cycle can be inferred from the following situations:

Situation 1: Cases (a) and (c) : semi-stable limit cycle,
Situation 2: Cases (a) and (d) : unstable limit cycle,
Situation 3: Cases (b) and (c) : stable limit cycle,
Situation 4: Cases (b) and (d) : semi-stable limit cycle.

Based on the simplified Nyquist criterion [130], we will decide whether the linear substitute control loop becomes stable or unstable with a change in ΔA and therefore a change in

$$K = N(A_{\mathrm{lc}} + \Delta A).$$

In this way we can find out which of the above situations arises. From linear systems theory, we know that the simplified Nyquist criterion is applicable to an open loop with a transfer function $K \cdot G(s)$ that has exclusively poles λ_i with $\mathrm{Re}\{\lambda_i\} < 0$ and at most two poles at $s = 0$. If, in this case, the Nyquist plot of $G(j\omega)$ does not enclose or touch the critical point

$$-\frac{1}{K} = -\frac{1}{N(A_{\mathrm{lc}} + \Delta A)}$$

for $0 \le \omega \le \infty$, which is especially so when the Nyquist plot passes the critical point on its right-hand side, the control loop is asymptotically stable.

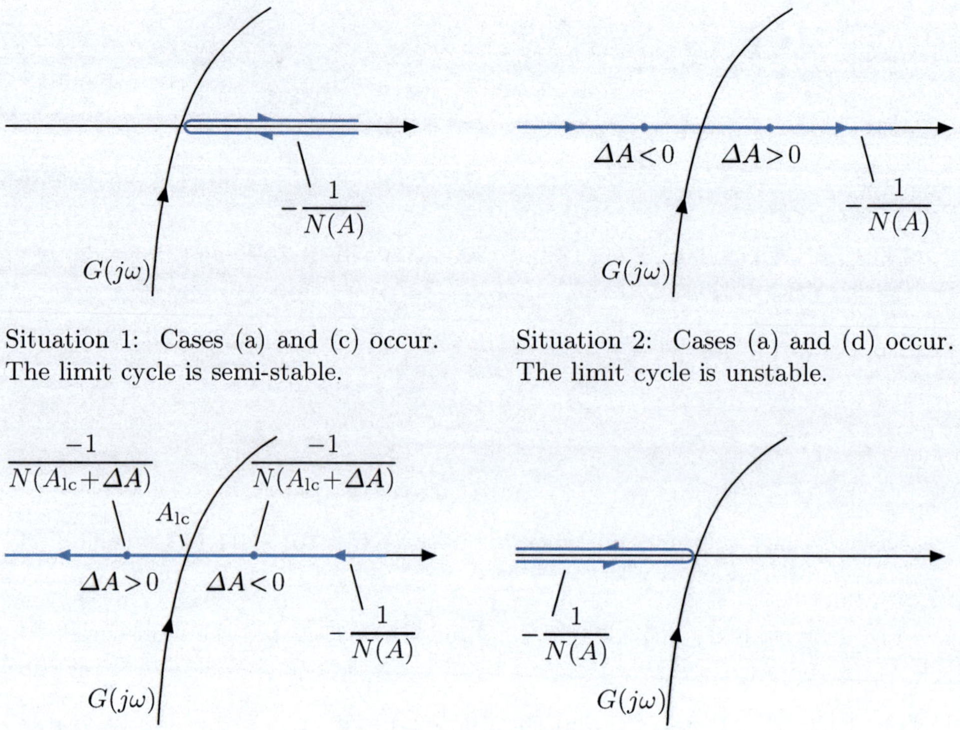

Situation 1: Cases (a) and (c) occur. The limit cycle is semi-stable.

Situation 2: Cases (a) and (d) occur. The limit cycle is unstable.

Situation 3: Cases (b) and (c) occur. The limit cycle is stable.

Situation 4: Cases (b) and (d) occur. The limit cycle is semi-stable.

Fig. 2.12: Possible situations for the stability of limit cycles

Otherwise it is unstable. Thus, each of the four situations can be identified by analyzing the Nyquist plot and the nonlinear locus, as shown in Figure 2.12.

For example, consider Situation 3 in Figure 2.12. For $\Delta A > 0$, the Nyquist plot of $G(j\omega)$ passes by the point $-1/N(A_{\mathrm{lc}}+\Delta A)$ on its right. Thus the linear substitute control loop is stable. This situation corresponds to Case (b) from Figure 2.11, i. e. the trajectory tends to the limit cycle from the outside. For $\Delta A < 0$, on the other hand, the linear substitute control loop is unstable, since the Nyquist plot of $G(j\omega)$ passes by the point $-1/N(A_{\mathrm{lc}} + \Delta A)$ on its left. This leads to Case (c) in Figure 2.11, and the trajectory tends to the limit cycle from the inside. The limit cycle is therefore stable.

The calculations above are the basis for the following criterion for the stability of limit cycles. Like the method of harmonic balance itself, it does not provide a determination, only an indication of possible limit cycles' stability or instability.

Heuristic 2 (Stability of Limit Cycles). *For a nonlinear standard control loop with a plant* $G(s)$ *having at most two poles at* $s = 0$ *and otherwise only poles* λ_i *with* $\mathrm{Re}\{\lambda_i\} < 0$*, a limit cycle is presumably*

(1) *stable if the nonlinear locus crosses the Nyquist plot from right to left at the corresponding intersection point,*
(2) *semi-stable if the nonlinear locus is tangential to the Nyquist plot at the corresponding intersection point,*
(3) *unstable if the nonlinear locus crosses the Nyquist plot from left to right at the corresponding intersection point.*

In this context, the directions *left* and *right* refer to the directions resulting from moving along the Nyquist plot beginning at $\omega = 0$.

We cannot normally observe unstable or semi-stable limit cycles in practical applications. Furthermore, an unstable or semi-stable limit cycle is often not seriously critical for a control loop if the system has an asymptotically stable equilibrium point. This is the case because the trajectory departs from the limit cycle, even for the smallest disturbances, and then, for example, tends to the equilibrium point. Functions $-1/N(A)$ with a nonlinear locus which, as shown in Situation 2 of Figure 2.12, aim toward the origin usually do not lead to stable limit cycles. This is due to the fact that the nonlinear locus $-1/N(A)$ crosses the Nyquist plot of $G(j\omega)$ from left to right if the Nyquist plot of $G(j\omega)$ progresses in a clockwise rotating manner. The latter is usually the case.

For the harmonic balance method, there are a number of extensions for control loops with multiple characteristic curves, asymmetric characteristic curves, and discrete controllers [88, 147, 319, 446].

2.1.5 Example: Power-Assisted Steering System

Now we will consider a power-assisted steering system for motor vehicles which utilizes the principle of angle superposition [225, 243, 244]. In a vehicle's steering system, a high steering transmission ratio is used to reduce the steering wheel torque for the driver. A resulting disadvantage is the large steering wheel angle. A motor-powered superposition gear therefore reduces the steering wheel angle to increase driving comfort. To do this, the superposition gear generates a supplementary angle δ_2 which is superposed onto the steering wheel angle δ_1. Both together produce the output angle δ_{y}. Figure 2.13 shows the principle structure.

Steering systems with angle superposition are also used for active steering. At low speeds, active steering reduces the steering wheel angle that the driver must set for cornering. In this way, a driver can drive along narrow curves with small movements of the steering wheel. At high speeds, the superposition motor counter-steers. Hence large steering wheel rotations lead only to small steering angles. This increases driving safety in this velocity range.

Normally the driver holds onto the steering wheel and specifies a steering angle δ_1. When the driver releases the steering wheel, this specification of the steering angle disappears, and limit cycles can occur in the steering system, which of course are undesired.

Fig. 2.13: Power-assisted steering system which uses the principle of angle superposition

In order to detect possible limit cycles, we must examine the steering control loop for this particular case where the driver has released the steering wheel. It is shown in Figure 2.14. The servomotor with a current control loop and differential drive constitutes the plant to be controlled, whose input variable is the required torque value M of the servomotor. One output variable of this plant is the output angle

$$\delta_y = \delta_1 + \delta_2.$$

On the other hand, the steering wheel angle δ_1 is also an output variable, since the driver does not hold onto the steering wheel. The PD controller for the supplementary angle δ_2 has the parameters

$$K_{PD} = 3000\,\mathrm{N\,m\,rad^{-1}} \quad \text{and} \quad T_D = 0.02\,\mathrm{s}.$$

The servomotor including the current control and superposition gear is described by the linear state-space model

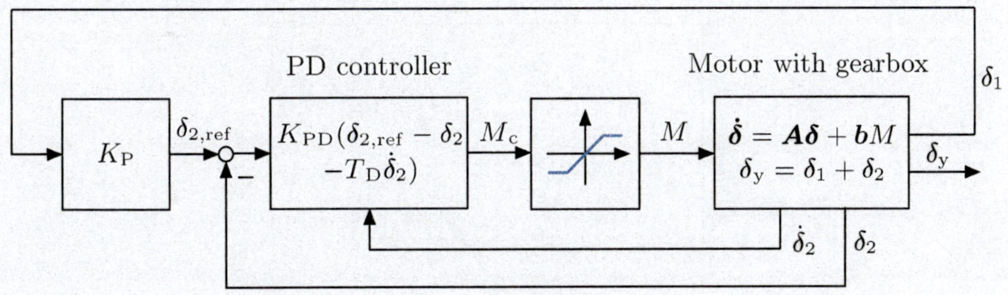

Fig. 2.14: Control loop of the steering system for a steering wheel that is not held by the driver

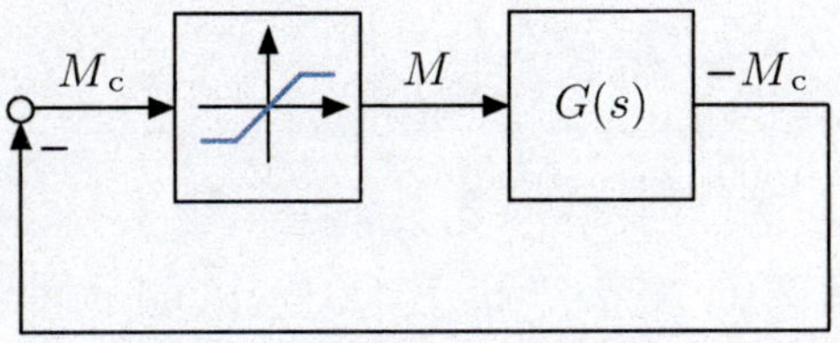

Fig. 2.15: Steering system's nonlinear standard control loop, which is derived by transformation of the control loop depicted in Figure 2.14

$$\dot{\boldsymbol{\delta}} = \begin{bmatrix} 0 & 0 & 1 & 0 \\ 0 & 0 & 0 & 1 \\ -67.3568 & -67.3568 & -11.3988 & -11.3988 \\ -24.1480 & -24.1480 & -4.0866 & -4.0866 \end{bmatrix} \boldsymbol{\delta} + \begin{bmatrix} 0 \\ 0 \\ -4.0123 \cdot 10^{-2} \\ 1.8977 \end{bmatrix} M.$$

Here the state vector is given by

$$\boldsymbol{\delta} = \left[\delta_1 \; \delta_2 \; \dot{\delta}_1 \; \dot{\delta}_2\right]^T.$$

The torque M is limited to $|M| \leq 21\,\mathrm{N\,m}$. Thus, if the PD controller

$$M_c = K_{PD}(\delta_{2,\mathrm{ref}} - \delta_2 - T_D\dot{\delta}_2)$$

specifies torques M_c, which exceed these limits, they are limited by the saturation characteristic

$$M = \begin{cases} M_{\max}, & M_c > M_{\max}, \\ M_c, & |M_c| \leq M_{\max}, \\ -M_{\max}, & M_c < -M_{\max}. \end{cases}$$

The steering wheel angle δ_1 specifies the reference value $\delta_{2,\mathrm{ref}}$ of the supplementary angle δ_2 for a given vehicular velocity by using the factor $K_P = 1.5$. Since the driver is not holding onto the steering wheel, the superposition gear impacts directly upon the steering wheel angle δ_1, as shown in Figure 2.14.

The control loop in Figure 2.14 can be transformed into a nonlinear standard control loop, as shown in Figure 2.15. Taking into account the dynamics

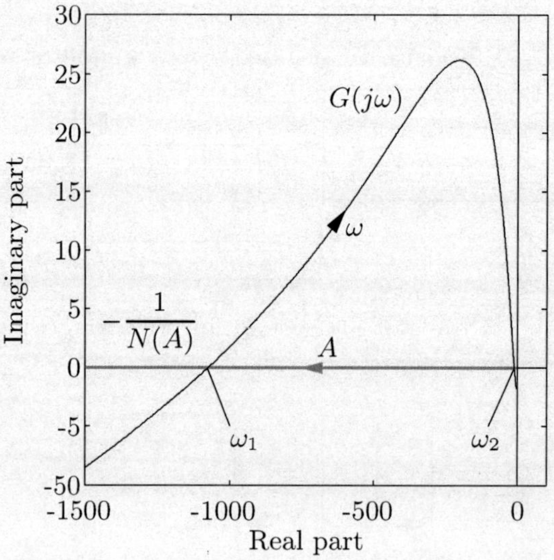

Fig. 2.16: Nyquist plot of $G(j\omega)$ and nonlinear locus $-1/N(A)$

of the current-controlled servomotors with a superposition gear, the transfer function $G(s)$ is obtained as

$$G(s) = -\frac{M_{\mathrm{c}}(s)}{M(s)} = \frac{113.9s^3 + 7181s^2 + 171200s + 965900}{s^2(s^2 + 15.49s + 91.50)}.$$

Applying the describing function method results in two intersecting points of the Nyquist plot of $G(j\omega)$ with the nonlinear locus $-1/N(A)$ of the saturation element's describing function, as shown in Figure 2.16. We therefore expect two limit cycles with frequencies close to ω_1 and ω_2. For the left-hand intersecting point in Figure 2.16, the Nyquist plot of $G(j\omega)$ is crossed from right to left by the nonlinear locus curve. This limit cycle with frequency ω_1 is stable. The second limit cycle, on the other hand, is unstable and thus not of practical significance, in contrast to the stable limit cycle, which is of interest. The latter's period length is determined by solving the equation of harmonic balance $G(j\omega) = -1/N(A)$, to yield

$$T_{\mathrm{harm}} = \frac{2\pi}{\omega_1} = 1.97\,\mathrm{s}.$$

Because of the approximating nature of the harmonic balance, this value is imprecise. By simulating the system as shown in Figure 2.17, the period length can be determined to be $T_{\mathrm{sim}} = 2.42\,\mathrm{s}$.

As can be seen from Figure 2.17, the oscillation has a high amplitude, which is, of course, out of the question for a motor vehicle. The limit cycle can be eliminated with an extension of the control system [244].

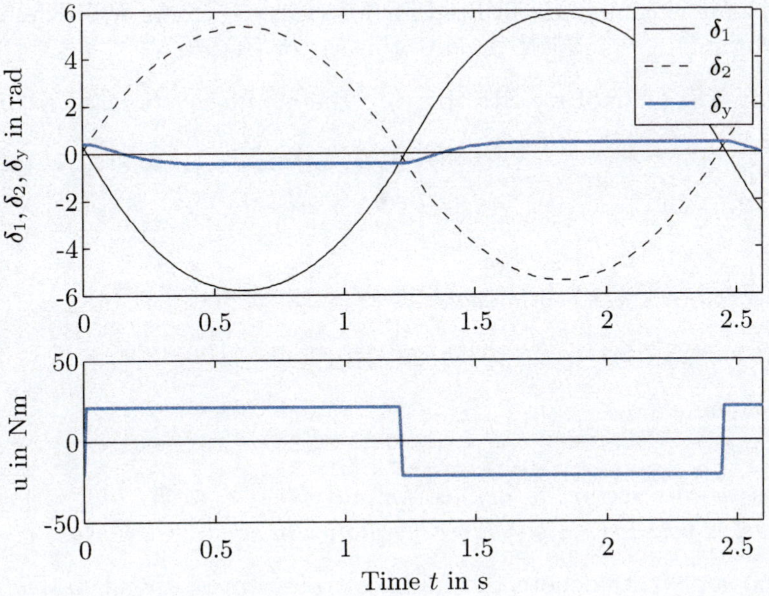

Fig. 2.17: Time courses of the angles $\delta_1, \delta_2, \delta_\mathrm{y}$, and the actuating variable u of the limit cycle

2.2 Absolute Stability

2.2.1 The Concept of Absolute Stability

In the previous section, we examined the nonlinear standard control loop, which is illustrated again in Figure 2.18, for limit cycles. Of course, it is also relevant to analyze the stability behavior in the absence of limit cycles. For example, it is useful for practical purposes if we can determine for which characteristic curves f the control loop has a globally asymptotically stable equilibrium point.

We will address this question below, focusing our attention on characteristic curves which lie in a sector bounded by two lines

$$u = K_1 e \quad \text{and} \quad u = K_2 e.$$

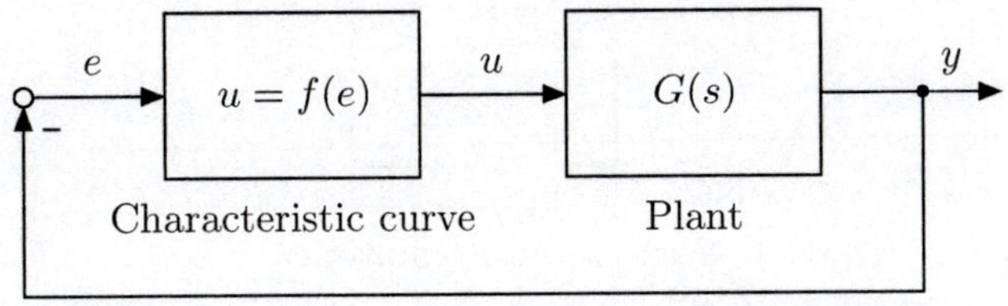

Fig. 2.18: Nonlinear standard control loop

Figure 2.19 plots this sector, for which we will use the short-hand notation $[K_1, K_2]$, as it is commonly utilized for intervals. Next, we will define a new stability term [10] which is related to this sector $[K_1, K_2]$.

Definition 18 (Absolute Stability). *The nonlinear standard control loop*

$$Y(s) = G(s)U(s),$$
$$e = -y,$$
$$u = f(e)$$

is called absolutely stable in a sector

$$[K_1, K_2]$$

if for every function

$$u = f(e)$$

which lies in this sector, is defined for all values $e \in \mathbb{R}$*, and is piecewise continuous, it possesses a globally asymptotically stable equilibrium point.*

Here, $Y(s)$ and $U(s)$ denote the Laplace transform of the signals $y(t)$ and $u(t)$, respectively.

It is already possible to initially estimate the size of the sector of absolute stability[1]. Taking only linear characteristics into account, i. e. functions

$$u = f(e) = K \cdot e,$$

we can determine the maximum range of parameters

$$(K_{\mathrm{H1}}, K_{\mathrm{H2}}) \tag{2.7}$$

which result in an asymptotically stable linear control loop by using the Routh criterion or the root locus method. The sector (2.7) is called the *Hurwitz sector*[2]. Obviously, the maximum sector of absolute stability is always smaller than or equal to the Hurwitz sector. Figure 2.20 illustrates the situation.

---

[1] The problem of finding conditions which guarantee that the nonlinear standard control loop has a globally stable equilibrium point $\boldsymbol{x}_{\mathrm{eq}}$ is known as the *Lur'e problem*. In this context, and when the nonlinearity has been shifted into the feedback connection, the nonlinear standard control loop is usually called a *Lur'e system*, which is shown in the figure below. A. I. Lur'e was the first to examine the stability of this class of systems [290, 291].

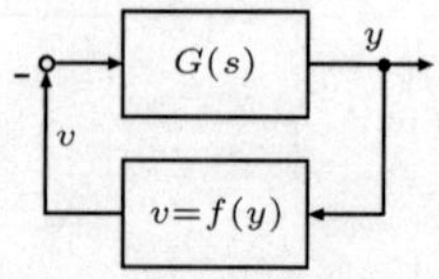

[2] Only gains $K \in (K_{\mathrm{H1}}, K_{\mathrm{H2}})$ lead to an asymptotically stable control loop. The non-included interval endpoints K_{H1} and K_{H2} lead to unstable or Lyapunov-stable control loops. This is similar in the case of the largest sector of absolute stability.

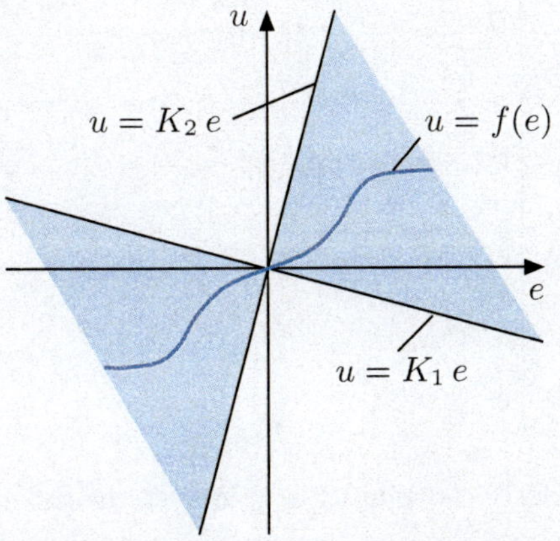

Fig. 2.19: Sector of characteristic curves bounded by $u = K_1 e$ and $u = K_2 e$

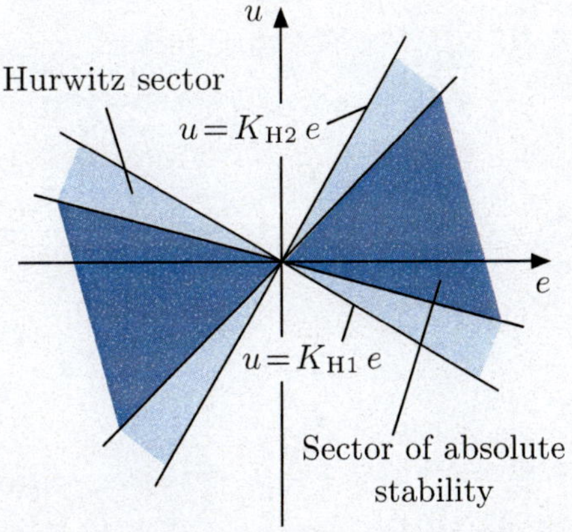

Fig. 2.20: Hurwitz sector and sector of absolute stability

2.2.2 The Popov Criterion and Its Application

A criterion for showing absolute stability was developed by V. M. Popov in 1959 [278, 365, 366, 367, 368, 369, 370, 371, 372]. It is easy to use and allows graphical interpretation. For its formulation, it is necessary to first define the concept of *strict marginal stability*.

Definition 19 (Marginal Stability and Strict Marginal Stability). *A linear system with the transfer function $G(s)$ is called marginally stable if it has only poles p_i with*

$$\mathrm{Re}\,\{p_i\} \leq 0,$$

where $\mathrm{Re}\,\{p_i\} = 0$ holds for at least one pole. We call a marginally stable linear system strictly marginally stable if the linear control loop

$$G_\varepsilon(s) = \frac{G(s)}{1 + \varepsilon\, G(s)}$$

is stable for every arbitrarily small $\varepsilon > 0$.

Viewed in graphical terms, the strict marginal stability of a system means that the branches of the root locus, which begin on the imaginary axis, tend to the left for increasing ε. This is illustrated in Figure 2.21 using the plant

$$G(s) = \frac{s + b}{s^2 + a}, \quad a > 0, \; b > 0,$$

as an example.

Popov's criterion is formulated as follows.

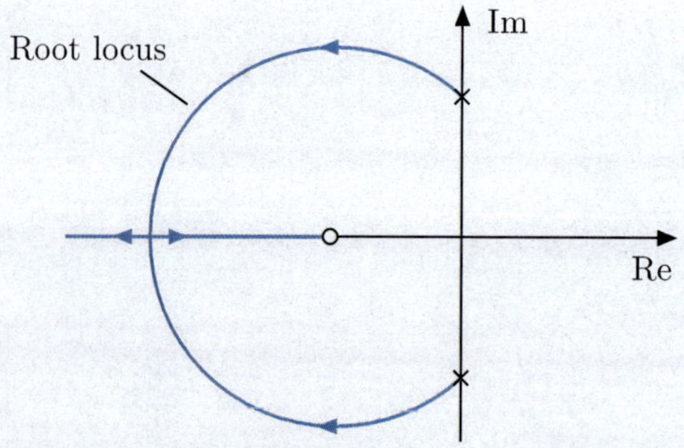

Fig. 2.21: Example of a root locus of a strictly marginally stable system

Theorem 6 (Popov Criterion). *Let us consider the nonlinear standard control loop*

$$\begin{aligned} Y(s) &= G(s)U(s), \\ e &= -y, \\ u &= f(e) \end{aligned}$$

with the asymptotically stable or strictly marginally stable plant $G(s)$. *Further, let the degree* m *of the numerator of* $G(s)$ *be less than the degree* n *of the denominator, and let the function* $u = f(e)$ *be defined for all* $e \in \mathbb{R}$, *piecewise continuous, and pass through the origin. Then the control loop defined above is absolutely stable*

(1) in the sector $[0, K]$ *in the case of an asymptotically stable* $G(s)$,
(2) in the sector $[\varepsilon, K]$ *with an arbitrarily small* $\varepsilon > 0$ *in the case of a strictly marginally stable* $G(s)$

if it is possible to find a real number q *such that the Popov inequality*

$$\text{Re}\{(1 + q \cdot j\omega) \cdot G(j\omega)\} > -\frac{1}{K}$$

is fulfilled for all $\omega \geq 0$.

Two aspects of this criterion require some explanation. On the one hand, we have to consider the applicability being limited to the sectors $[0, K]$ and $[\varepsilon, K]$ and, on the other hand, the Popov inequality requires further discussion.

First we will elaborate on the sector of applicability. The distinction between the sectors $[0, K]$ for stable plants $G(s)$ and $[\varepsilon, K]$ for strictly marginally stable plants $G(s)$ can be explained by the fact that a strictly marginally stable plant $G(s)$ and a characteristic curve

$$u = 0 \cdot e$$

result in a control loop which is not globally asymptotically stable. At the least, an amplification of ε, i. e.

$$u = \varepsilon \cdot e$$

is required to stabilize the control system.

The distinction between the sectors $[0, K]$ and $[\varepsilon, K]$ also has an effect on the characteristic curves $u = f(e)$ being discussed. For instance, there are characteristic curves which lie in $[0, K]$ but not in $[\varepsilon, K]$. This is also the case although $\varepsilon > 0$ is allowed to be arbitrarily small. For purposes of illustration, we will look at two examples. The first example is depicted in Figure 2.22. Here the characteristic curve $u = f(e)$ tends to zero for $e \to \infty$, but never reaches this value. Obviously, there is no sector $[\varepsilon, K]$ for which the line $u = \varepsilon \cdot e$ does not intersect the characteristic curve. In the second example, which is illustrated in Figure 2.23, the function

$$u = \operatorname{sgn}(e)\sqrt{|e|}$$

tends to infinity for $e \to \infty$, but it does so more weakly than any line $u = \varepsilon \cdot e$, so that there is always an intersection point. Thus in this example as well, the characteristic curve $u = \operatorname{sgn}(e)\sqrt{|e|}$ will not lie in a sector $[\varepsilon, K]$.

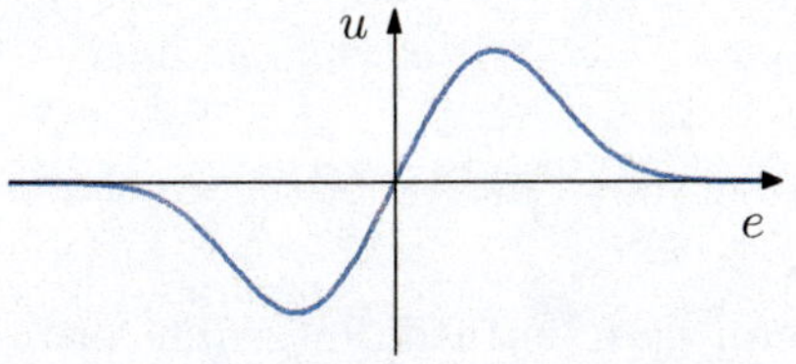

Fig. 2.22: Characteristic curve which tends to zero for $e \to \infty$

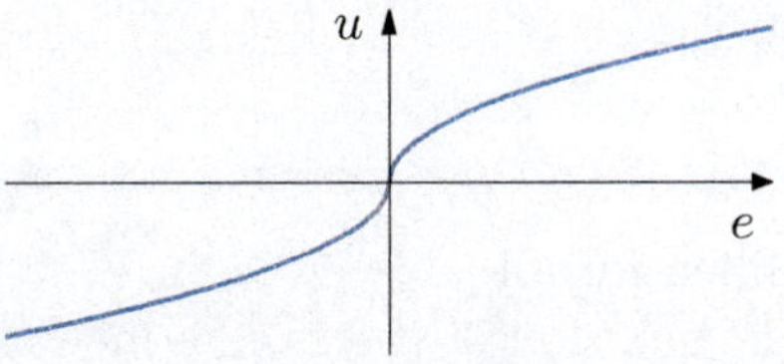

Fig. 2.23: Characteristic curve which tends to ∞ for $e \to \infty$

The limitation to sectors $[0, K]$ instead of $[K_1, K_2]$ is only seemingly restrictive. By rearranging the control loop from Figure 2.18, for which we will consider the sector $[K_1, K_2]$, $K_1 = 0$ can be obtained. This is done by adding two proportional elements with a gain of K_1, as shown in Figure 2.24, to the control loop. Obviously, the two proportional elements in effect cancel each other out, so that the control loop remains unchanged. By combining the two subsystems, we obtain the control loop depicted in Figure 2.25. For the rearranged control loop, we have to look at the transformed sector

$$[K_1 - K_1, K_2 - K_1] = [0, K = K_2 - K_1].$$

Furthermore, unstable control loops can be stabilized by these transformations, so that the Popov criterion is also applicable to such plants.

Now that the Popov criterion's area of applicability has been clarified, let us consider its application. In particular, we would like to determine whether it is possible to find a real number q such that the Popov inequality

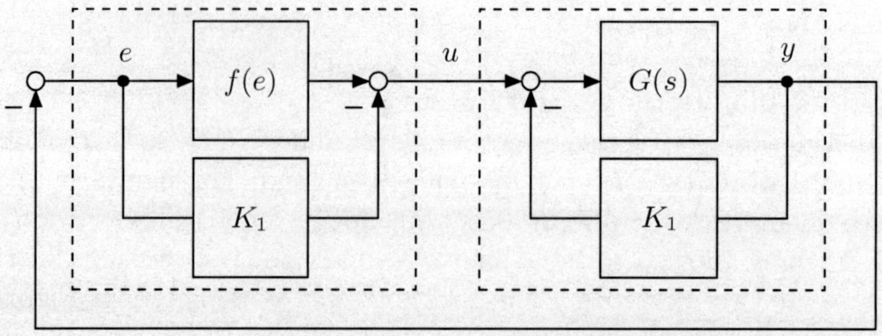

Fig. 2.24: Inserting a factor K_1 into the nonlinear standard control loop

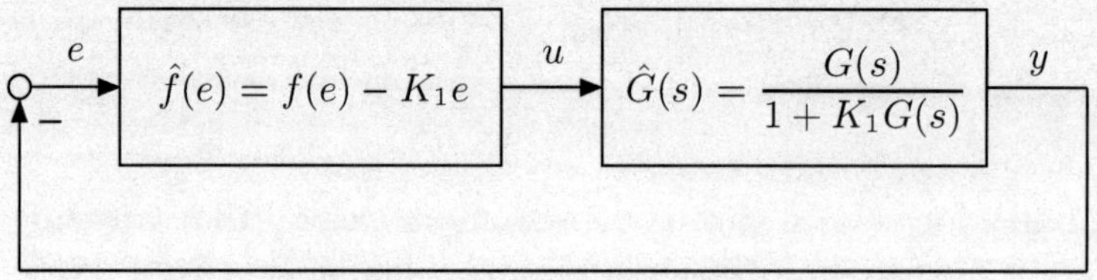

Fig. 2.25: A standard nonlinear control loop which is an equivalent version of the control loop shown in Figure 2.24

$$\mathrm{Re}\,\{(1 + q \cdot j\omega) \cdot G(j\omega)\} > -\frac{1}{K}$$

is fulfilled for all $\omega \geq 0$.

The solution of the Popov inequality can be illustrated in graph form. To do this, we first need to rearrange the inequality to obtain

$$\underbrace{\mathrm{Re}\,\{G(j\omega)\}}_{X(\omega)} - q \cdot \underbrace{\omega\ \mathrm{Im}\,\{G(j\omega)\}}_{Y(\omega)} > -\frac{1}{K}.$$

This leads to the inequality

$$X(\omega) - q \cdot Y(\omega) + \frac{1}{K} > 0, \tag{2.8}$$

which is parameterized in terms of ω. The above inequality is fulfilled for all points (X, Y) which lie to the right of the line defined by

$$X - q \cdot Y + \frac{1}{K} = 0.$$

The slope of this line is equal to $1/q$, and its intersection with the real axis is at $-1/K$. Figure 2.26 illustrates this. The line is referred to as the *Popov line*.

We do not need to check for all values of X and Y to see if they satisfy equation (2.8), since they depend on ω via

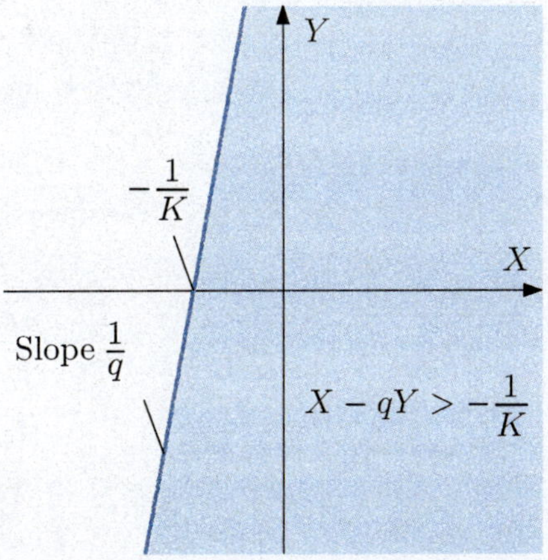

Fig. 2.26: Sector (blue), in which $X - qY > -1/K$ is fulfilled

$$\begin{aligned} X(\omega) &= \mathrm{Re}\,\{G(j\omega)\}, \\ Y(\omega) &= \omega\ \mathrm{Im}\,\{G(j\omega)\}. \end{aligned} \tag{2.9}$$

The curve defined by equation (2.9) can be pictured in the coordinate plane with X and Y as its coordinates. It is similar to a Nyquist plot if we set

$$\tilde{G}(j\omega) = X(\omega) + jY(\omega) = \mathrm{Re}\,\{G(j\omega)\} + j\omega\ \mathrm{Im}\,\{G(j\omega)\}.$$

The related locus is termed the *Popov plot*. It results from the Nyquist plot of $G(j\omega)$ when we multiply the imaginary part of $G(j\omega)$ by ω. The graphic interpretation of this is that the Nyquist plot of $G(j\omega)$ is modified in the direction of the imaginary axis. However, the real part is left unchanged, as we see in Figure 2.27.

The possible values for X and Y are thus given by equation (2.9): they are the values of the Popov plot. This result can be depicted in a graph and utilized as follows. Popov's inequality is fulfilled if the Popov plot lies on the right-hand side of a line with the arbitrary slope $1/q$ which intersects the X-axis, i. e. the real axis at $-1/K$.

With the above method, it is possible to solve the Popov inequality graphically. Obviously, all lines to the left of the Popov plot fulfill the inequality. From these, the value of $-1/K$ and thus the sectors of absolute stability

$$[0, K] \qquad \text{or} \qquad [\varepsilon, K]$$

are directly visible. Clearly, we are interested in determining the largest possible sector. To do this, the Popov line is shifted to the Popov plot such that it is its tangent, which results in the largest possible $K = K_{\mathrm{Po}}$, as shown in Figure 2.28. The corresponding sector

$$[0, K_{\mathrm{Po}}) \qquad \text{or} \qquad [\varepsilon, K_{\mathrm{Po}})$$

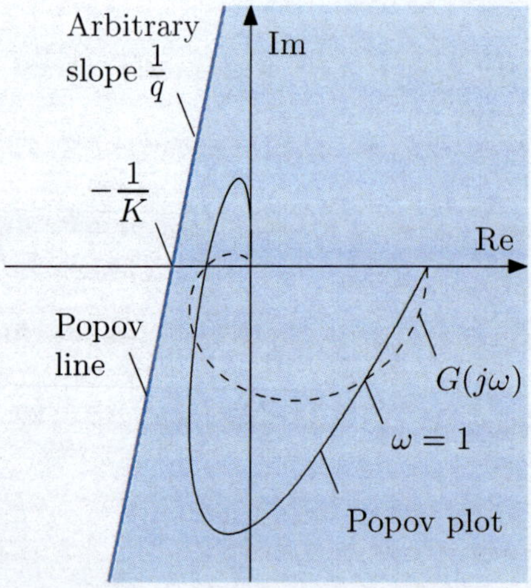

Fig. 2.27: The Nyquist plot of $G(j\omega)$, the Popov plot, a possible Popov line, and a region (blue) where Popov's inequality is fulfilled

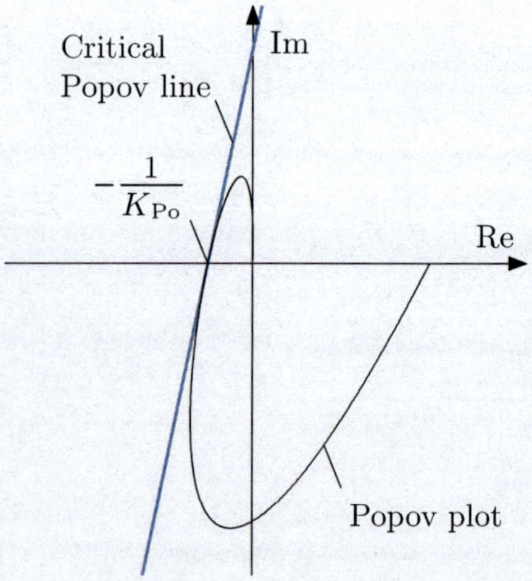

Fig. 2.28: The critical Popov line is the tangent to the Popov plot which produces the largest Popov sector.

is called the *Popov sector*. For the gain K_{Po} of the Popov sector $[0, K_{\mathrm{Po}})$, stability is not proven by the Popov criterion, since the Popov inequality requires values $Re\{G(j\omega)\} - q\omega \mathrm{Im}\{G(j\omega)\}$ to be greater than, not greater than or equal to $-1/K$. However, this distinction is not relevant in practice, since, in any practical application, a safety distance from the critical value K_{Po} is always maintained.

It should be noted that the Popov criterion is of course also applicable if the control loop takes the form

$$\dot{\boldsymbol{x}} = \boldsymbol{A}\boldsymbol{x} - \boldsymbol{b} f(\boldsymbol{c}^T \boldsymbol{x})$$

and its linear plant is controllable and observable.

Below we will consider two extensions of the Popov criterion. The Popov criterion is also applicable if $G(s)$ takes the form

$$G(s) = \hat{G}(s) \cdot \mathrm{e}^{-Ts}$$

and $\hat{G}(s)$ is asymptotically stable. In this case, however, only positive values of q are allowed in the Popov inequality and the nonlinearity is required to be continuous [277, 373].

In the case that the nonlinearity is time-variant, i. e. if $u = f(e, t)$ holds, the Popov criterion is also applicable. However, in this case, the following additional restrictions apply [391]:

(1) $0 < \dfrac{f(e,t)}{e} < K, \;\; e \neq 0$, and $f(e,t) = 0$ for all $t \in I\!R$,
(2) $G(s)$ has no more than one pole at $s = 0$, and otherwise only poles s_i with $\mathrm{Re}\{s_i\} < 0$, and

(3) $q = 0$.

The requirement $q = 0$ implies that the Popov line is vertical.

2.2.3 The Aizerman and Kalman Conjectures

For many nonlinear standard control loops, the application of the Popov criterion is unnecessary. This is because for these systems, Aizerman's conjecture [9] holds. It states that the largest sector of absolute stability is identical to the Hurwitz sector (2.7), which is much easier to determine. Even though the conjecture does not hold for cases in general [93, 122], it is true for all first- and second-order[3] plants with $m < n$. It is also true for all third-order plants without zeros and only real poles [10, 37, 139]. If we restrict ourselves to sectors $[0, K_{\mathrm{H2}})$ and $[\varepsilon, K_{\mathrm{H2}})$, i. e. if we only consider the nonnegative part of the Hurwitz sector, Aizerman's conjecture also holds for the following plants [404, 478]:

(1) asymptotically stable third-order systems with

$$G(s) = \frac{b_1 s + b_0}{s^3 + a_2 s^2 + a_1 s + a_0}$$

or strictly marginally stable systems with

$$G(s) = \frac{1}{s} \cdot \frac{b_1 s + b_0}{s^2 + a_1 s + a_0},$$

(2) fourth-order systems without zeros, with at most one pole at $s = 0$, and otherwise only real negative poles.

Similar to Aizerman's conjecture, Kalman [217] guesses equality of the Hurwitz sector and the sector of absolute stability if the nonlinearity's derivative is bounded by

$$K_1 < \frac{\partial f(e)}{\partial e} < K_2. \tag{2.10}$$

Note that this restriction is stricter than Aizerman's requirement of a sector

$$K_1 < \frac{f(e)}{e} < K_2, \quad e \neq 0. \tag{2.11}$$

In both cases above, we assume that $f(0) = 0$ holds. Thus, every function f which fulfills inequality (2.10) satisfies inequality (2.11).

The Kalman conjecture is also false in general [255, 271]. However, Kalman's conjecture is valid in some cases not covered by the cases in which Aizerman's conjecture was proven to be true. For example, it holds for all

[3] There is a special case [10, 57, 380] in which Aizerman's conjecture does not hold for second-order systems, but it has no practical significance in engineering tasks.

linear systems of the third order or below with $m < n$; thus, for these, the sector of absolute stability and the Hurwitz sector are identical if inequality (2.10) holds [28, 270].

Furthermore, similar to the Kalman conjecture, we can easily calculate a sector of absolute stability

$$[0, K_{\mathrm{H}2}) \quad \text{or} \quad [\varepsilon, K_{\mathrm{H}2})$$

for nonlinear standard control loops which may have an integrator and a delay. Here, the gain $K_{\mathrm{H}2} > 0$ is again the upper bound of the Hurwitz sector $(K_{\mathrm{H}1}, K_{\mathrm{H}2})$, which results when we use only linear characteristics $u = Ke$. The gain $K_{\mathrm{H}2}$ is also known as the *Nyquist value* of a linear plant, since it results from the Nyquist criterion and indicates the control loop's gain where the system's behavior changes from stability to instability. We use the results from [498] and obtain

Theorem 7 (Absolute Stability of Control Loops with Delay). *Consider a nonlinear standard control loop with a plant*

$$G_1(s) = \frac{be^{-Ts}}{D(s)} \quad \text{or} \quad G_2(s) = \frac{be^{-Ts}}{sD(s)}$$

where $b > 0$ and $T \geq 0$, and a nonlinear controller

$$u = f(e) \quad \text{with} \quad f(0) = 0.$$

If

(1) the polynomial $D(s)$ has only roots λ_i with

$$\operatorname{Re}\{\lambda_i\} < 0 \quad \text{and} \quad |\operatorname{Im}\{\lambda_i\}| < |\operatorname{Re}\{\lambda_i\}|$$

and

(2) the nonlinearity's derivative is bounded by

$$0 \leq \frac{\partial f(e)}{\partial e} < K_{\mathrm{H}2},$$

where $K_{\mathrm{H}2}$ is the Nyquist value of $G_1(s)$ or $G_2(s)$,

then the nonlinear standard control loop is absolutely stable in the sector $[0, K_{\mathrm{H}2})$ for $G_1(s)$ and in the sector $[\varepsilon, K_{\mathrm{H}2})$ for $G_2(s)$.

At least for the plants above, it is possible to simply determine the largest possible sector of absolute stability or its nonnegative part by calculating the Hurwitz sector $(K_{\mathrm{H}1}, K_{\mathrm{H}2})$ or its subsector $[0, K_{\mathrm{H}2})$.

Next we will look at an example to complement the above considerations. The plant is given by

$$G(s) = \frac{1}{(s^2 + 0.1s + 10)(s^2 + 0.2s + 20)}.$$

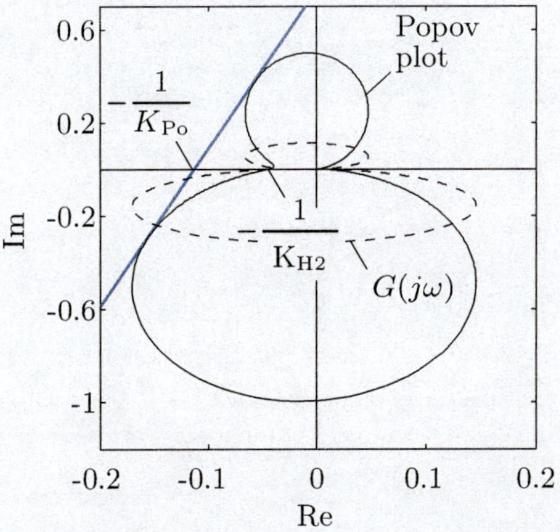

Fig. 2.29: Example of a system for which the Hurwitz sector is larger than the Popov sector

Figure 2.29 shows the corresponding Nyquist and Popov plots, so that the Popov sector can be graphically calculated as $[0, K_{\mathrm{Po}} \approx 8.7)$. From the Nyquist plot, it also becomes clear that the nonnegative Hurwitz sector with the maximum slope K_{H2} is larger than the Popov sector, since it holds that

$$K_{\mathrm{Po}} \approx 8.7 < \frac{1012}{45} = K_{\mathrm{H2}}.$$

However, the fact that the Popov sector is smaller than the Hurwitz sector is not sufficient for us to conclude that the sector of absolute stability is smaller than the Hurwitz sector. This is because the proof [431] of the Popov criterion only shows its sufficiency, and we do not know if it delivers the largest sector of absolute stability. Thus, Aizerman's conjecture could still hold for the above system. As mentioned previously, it does not hold for every system [93, 122, 246, 362, 363]. However, this is difficult to prove in specific cases. The following general rule deserves emphasis:

Popov sector ≤ largest sector of absolute stability ≤ Hurwitz sector.

So far, we have not addressed the question whether the Popov sector $[0, K_{\mathrm{Po}})$ is always equal to the largest nonnegative sector of absolute stability $[0, K_{\mathrm{as}})$. It has been shown that this is not the case [281, 377].

2.2.4 Example: Controlling a Ship

As an example application, we will look at the control of the course of the US Coast Guard ship *Tampa*, which is shown in Figure 2.30. The controlling mechanism should be designed in such a way that the change in the angle of

Fig. 2.30: Ship *Tampa*

the course φ of 10° requires a maximum settling time of 50 s. In this case, the control variable ϑ should not exceed ±10°. The yaw rate $\dot{\varphi}$ is measured with a gyroscope.

The *Tampa*'s transfer function $G(s)$ which relates the rudder angle ϑ to the yaw rate $\dot{\varphi}$ has been experimentally determined [133] to be

$$G(s) = \frac{\Omega(s)}{\Theta(s)} = \frac{-0.0184(s + 0.0068)}{(s + 0.2647)(s + 0.0063)},$$

where $\Omega(s)$ and $\Theta(s)$ are the Laplace transforms of $\omega = \dot{\varphi}$ and ϑ, respectively. The variable to be controlled is φ. For this reason, the yaw rate $\dot{\varphi}$, which is measured by the gyroscope, is integrated to obtain φ. A cascade control with two proportional controllers for the yaw rate and the course angle are used. The integral part of the plant ensures that no permanent control error exists. To limit the rudder angle to ±10°, a saturation element is added in front of the plant. The control loop is shown in Figure 2.31. The controls are designed in such a way that a 10° change in the course corresponds to a maximum control variable of 10°. In this case, the control loop is linear. For larger changes in the course, the actuating variable saturates, i. e. the control loop becomes nonlinear.

Now the question arises whether the control loop is also stable in nonlinear cases. To investigate this using the Popov criterion, the control loop is first

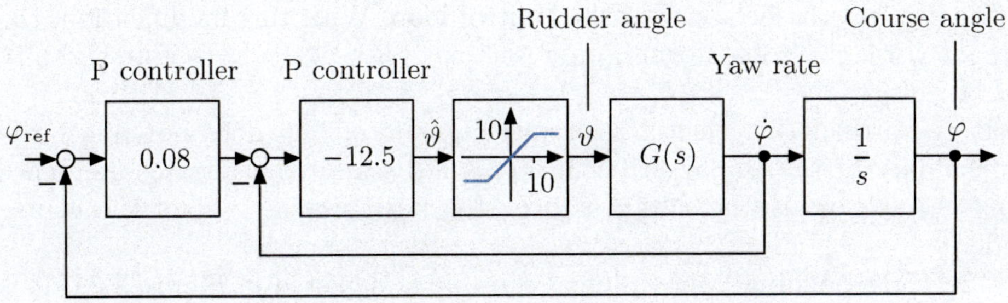

Fig. 2.31: Course control loop of the *Tampa*

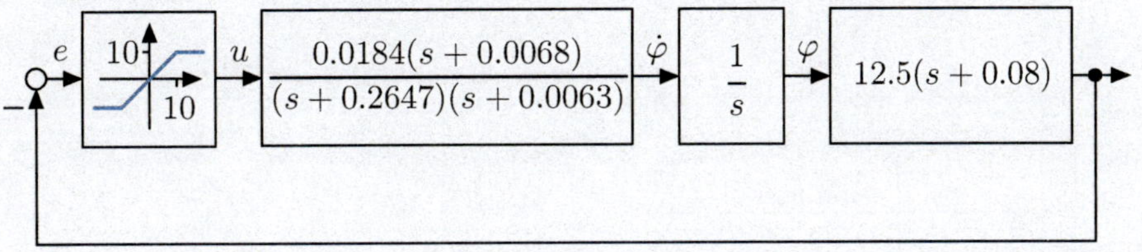

Fig. 2.32: Control loop of the ship in the form of a nonlinear standard control loop

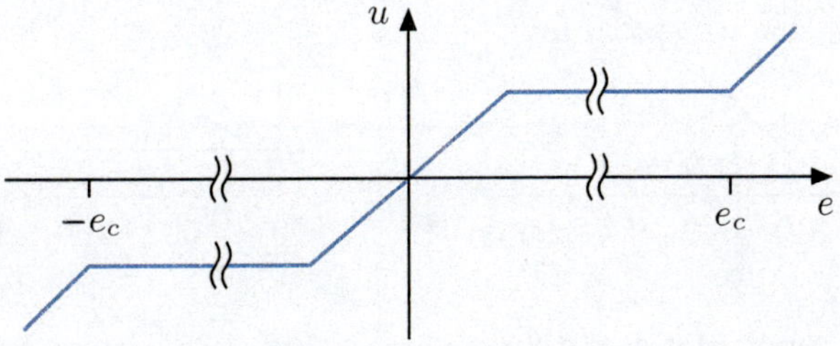

Fig. 2.33: Modified saturation characteristic curve

transformed into the nonlinear standard control loop. Figure 2.32 shows the result of this transformation. Since the plant contains an integrator and is strictly marginally stable, only sectors $[\varepsilon, K]$ can be obtained using the Popov criterion. The sector $[\varepsilon, K > 1]$, however, does not include the saturation characteristic curve, since its parallel to the abscissa always intersects any arbitrary line with slope $\varepsilon > 0$. However, for small values of ε, this only happens for large values of e. The saturation characteristic curve thus does not lie in the sector $[\varepsilon, K > 1]$, and the Popov criterion is hence inapplicable in this case.

Because we wish to apply it anyway, we will resort to the following remedy. We can note that it is not possible for arbitrarily large values of the control error e to occur in a practically operated control loop. For this reason, the saturation characteristic curve may be allowed to increase further beyond some determinable value e_c, which is never exceeded in a practically operated system. This change in the characteristic curve, as depicted in Figure 2.33, does not affect the behavior of the control loop. With this modification, the characteristic curve lies in the Popov sector $[\varepsilon, K > 1]$ and absolute stability can be ensured.

It is worth noting that a reference input $\varphi_{\text{ref}} \neq 0$ does not affect the applicability of the Popov criterion. This is because the negated reference input $-\varphi_{\text{ref}}$ can also be interpreted as the initial value $\varphi(0)$ of the course angle.

The corresponding Popov plot of the ship is depicted in Figure 2.34. It is evident that a line can be shifted from the left such that it touches the Popov plot and passes through the origin. The system is thus absolutely stable in the sector $[\varepsilon, \infty)$.

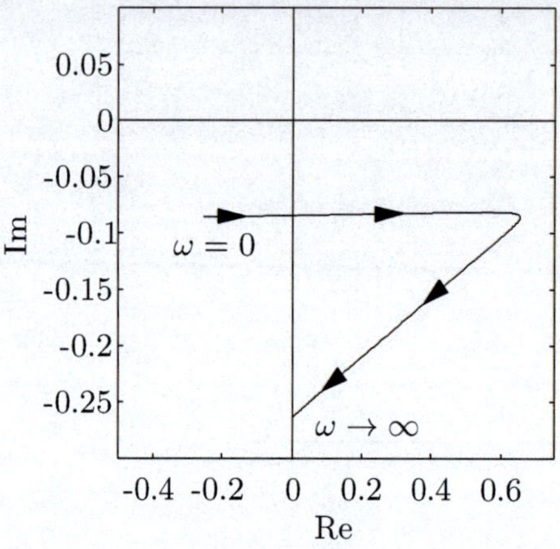

Fig. 2.34: Popov plot of the ship

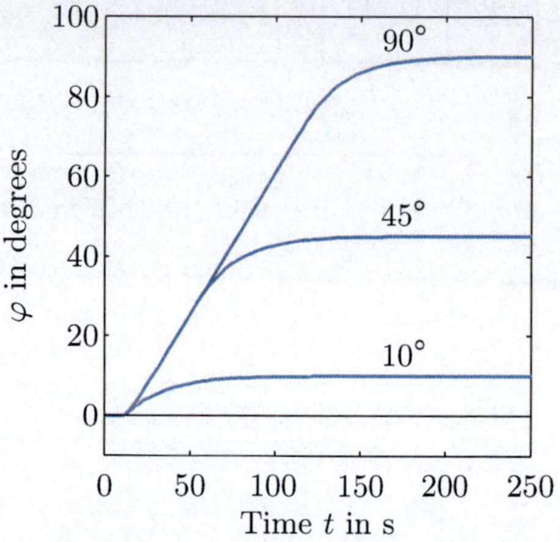

Fig. 2.35: Course angle φ of the ship

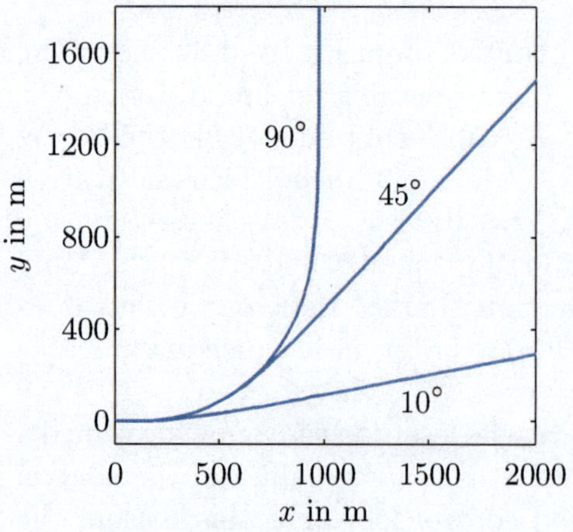

Fig. 2.36: View of the ship's course from the top

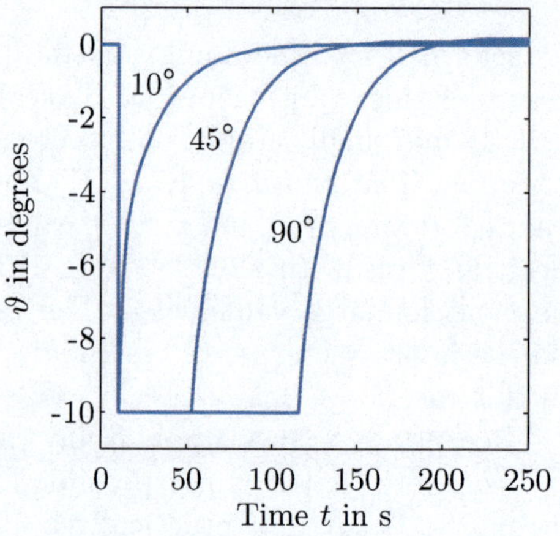

Fig. 2.37: Control signal, i. e. rudder angle ϑ

Figures 2.35, 2.36, and 2.37 show the simulated trajectories of the course control for course angle changes of $10°$, $45°$, and $90°$ beginning at $t = 10\,\text{s}$. Here the velocity of the ship is assumed to be $10\,\text{m}\,\text{s}^{-1}$. Figure 2.37 illustrates that, for a course change exceeding $10°$, the control signal reaches saturation. The ship's rudder then has its maximum deflection. It also becomes clear from these figures that, due to the saturation effect, large changes in the course are carried out more slowly than small changes. This is reasonable, since large curves should not be navigated as rapidly as small ones. Figure 2.36 shows the curves that were navigated and the ship's course changes from the top, i. e. in the xy-plane.

2.2.5 The Circle Criterion

Like the Popov criterion, the circle criterion allows us to analyze nonlinear standard control loops, such as the one shown in Figure 2.38, regarding absolute stability. The nonlinear function

$$u = f(e,t)$$

can be time-varying and lies in a slope sector $[K_1, K_2]$ with

$$0 \leq K_1 < K_2,$$

i. e. it holds that

$$K_1 \leq \frac{f(e,t)}{e} \leq K_2, \quad e \neq 0.$$

Consequently,

$$f(0,t) = 0$$

must be fulfilled. Figure 2.39 illustrates this case.

The circle criterion [330, 401, 402, 491, 492] confirms whether absolute stability of the standard nonlinear control loop is guaranteed in a sector $[K_1, K_2]$. Like the Popov criterion, however, the circle criterion is only a sufficient condition. Thus, in general, we do not know whether we have determined the maximum possible sector of absolute stability.

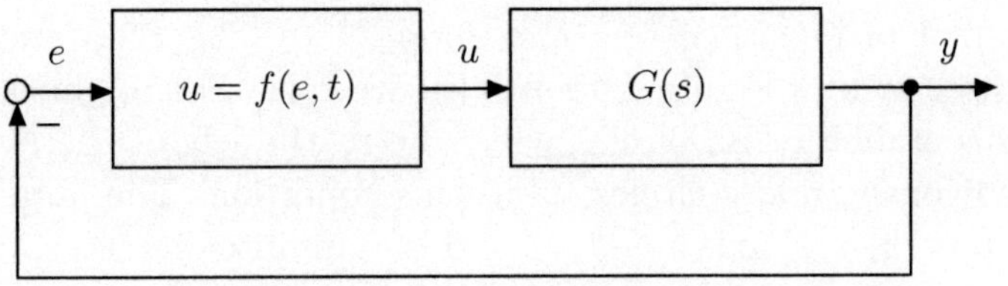

Fig. 2.38: Nonlinear standard control loop with a time-varying characteristic curve

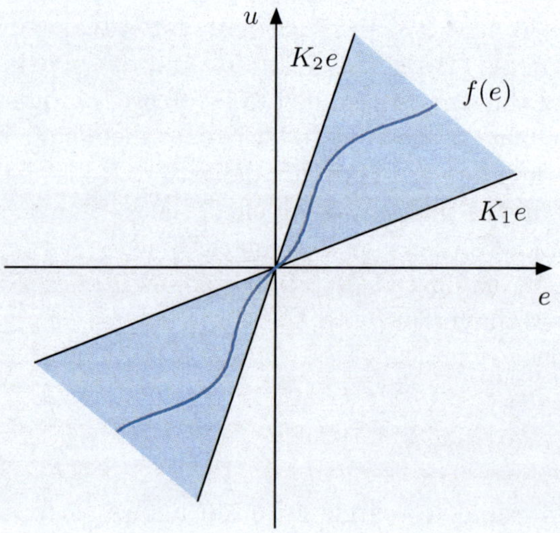

Fig. 2.39: Slope sector $[K_1, K_2]$ of the circle criterion

Theorem 8 (Circle Criterion for Asymptotically Stable Systems). *Let the nonlinear standard control loop*

$$\begin{aligned} Y(s) &= G(s)U(s), \\ e &= -y, \\ u &= f(e,t), \quad e \in I\!R,\ t \in I\!R, \end{aligned}$$

be given whose open-loop transfer function $G(s)$ *is coprime, has only poles* λ_i *with* $\mathrm{Re}\{\lambda_i\} < 0$*, and has a degree* m *of the numerator which is less than the degree* n *of the denominator. Let the function* $u = f(e,t)$ *lie in the sector* $[K_1, K_2]$ *with* $0 \le K_1 < K_2$*, and let* $f(0,t) = 0$ *hold for all* $t \in I\!R$*. If the Nyquist plot of the frequency response* $G(j\omega)$ *does not encompass, intersect, or touch the circle* $D(K_1, K_2)$ *which has its center on the real axis of the complex plane and passes through the points*

$$-\frac{1}{K_1} \quad \text{and} \quad -\frac{1}{K_2},$$

then the control loop is absolutely stable in the sector $[K_1, K_2]$.

The application of the circle criterion is simple. We need merely sketch the Nyquist plot of $G(j\omega)$ of the open loop for $0 \le \omega < \infty$ and the circle $D(K_1, K_2)$, as shown in Figure 2.40 and Figure 2.41. The nonlinear standard control loop is stable for all circles which lie to the left of the Nyquist plot of $G(j\omega)$. Obviously, many circles fulfill this condition. The largest of these circles intersects the real axis at $-1/K$ and is of infinite diameter. In this case, we obtain the sector $[0, K]$. Note that the circle is not allowed to intersect or touch the Nyquist plot. Sectors of different circles must not be combined to obtain a larger sector of absolute stability.

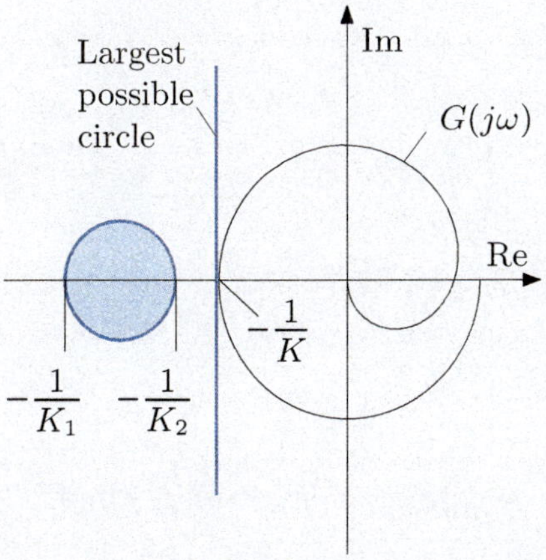

Fig. 2.40: Application of the circle criterion

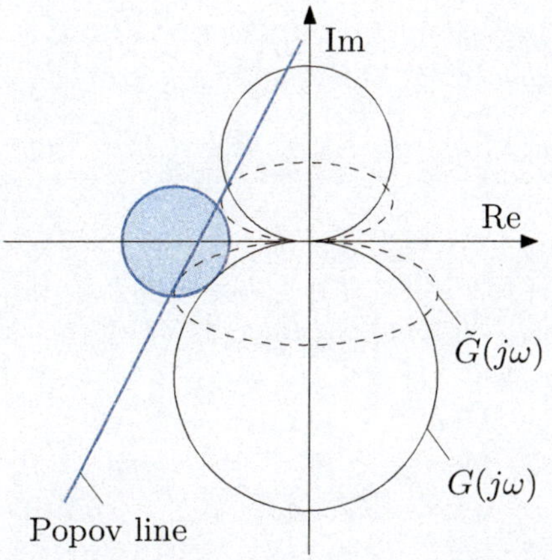

Fig. 2.41: Comparison of the Popov and circle criteria

The circle criterion is somewhat easier to apply than the Popov criterion, since we only need the Nyquist plot of $G(j\omega)$ and not the Popov plot. In specific cases, however, the circle criterion yields a different sector of absolute stability. Figure 2.41 illustrates such a case, where the Popov plot is denoted with $\tilde{G}(j\omega)$.

In the case of an unstable plant or sectors with negative K_1, the sector transformation we introduced for the Popov criterion in Section 2.2.2 is applied. In this way, we obtain an asymptotically stable plant or a transformed sector with $K_1 \geq 0$ and the circle criterion is applicable again. As an alternative to this approach, the circle criterion can be formulated in such a way [308, 402, 465] that the sector transformation is not required and the criterion encompasses the cases mentioned above. Then the following criterion for systems without poles on the imaginary axis holds.

Theorem 9 (Circle Criterion for Systems without Poles on the Imaginary Axis). *Let the nonlinear standard control loop*

$$
\begin{aligned}
Y(s) &= G(s)U(s),\\
e &= -y,\\
u &= f(e,t), \quad e \in \mathbb{R},\ t \in \mathbb{R},
\end{aligned}
$$

be given whose open-loop transfer function $G(s)$ is coprime and has a degree m of the numerator which is less than the degree n of the denominator; let $G(s)$ have $\nu \in \{0,1,\ldots,n\}$ poles λ_i with $\operatorname{Re}\{\lambda_i\} > 0$ and no poles λ_i with $\operatorname{Re}\{\lambda_i\} = 0$. Let the function $f(e,t)$ lie in the sector $[K_1,K_2]$, let $f(0,t) = 0$ for all $t \in \mathbb{R}$, and let $D(K_1,K_2)$ represent the circle in the complex plane which passes through the points

$$
-\frac{1}{K_1} \quad \textit{and} \quad -\frac{1}{K_2}
$$

and which has its center on the real axis. In this case, the control loop is absolutely stable in the sector $[K_1, K_2]$

(1) for $0 < K_1 < K_2$ *if the Nyquist plot of* $G(j\omega)$ *for frequencies* ω *ranging from* $-\infty$ *to* ∞ *does not intersect or touch the circle* $D(K_1, K_2)$*, and orbits it* ν *times in a counterclockwise direction,*

(2) for $0 = K_1 < K_2$ *and* $\nu = 0$ *if the Nyquist plot is located to the right of the vertical line crossing the real axis at* $-1/K_2$*,*

(3) for $K_1 < 0 < K_2$ *and* $\nu = 0$ *if the Nyquist plot lies inside the circle* $D(K_1, K_2)$*, and*

(4) for $K_1 < K_2 \leq 0$ *if, after replacing* $G(j\omega)$ *with* $-G(j\omega)$*, and* K_1 *and* K_2 *with* $-K_1$ *and* $-K_2$*, respectively, either Condition (1) or (2) is fulfilled.*

Figure 2.42 illustrates the four cases of the above theorem in graphic form.The subplots are shown as the areas colored in blue through which the Nyquist plot must not pass if absolute stability is to be proven. If, on the other hand,

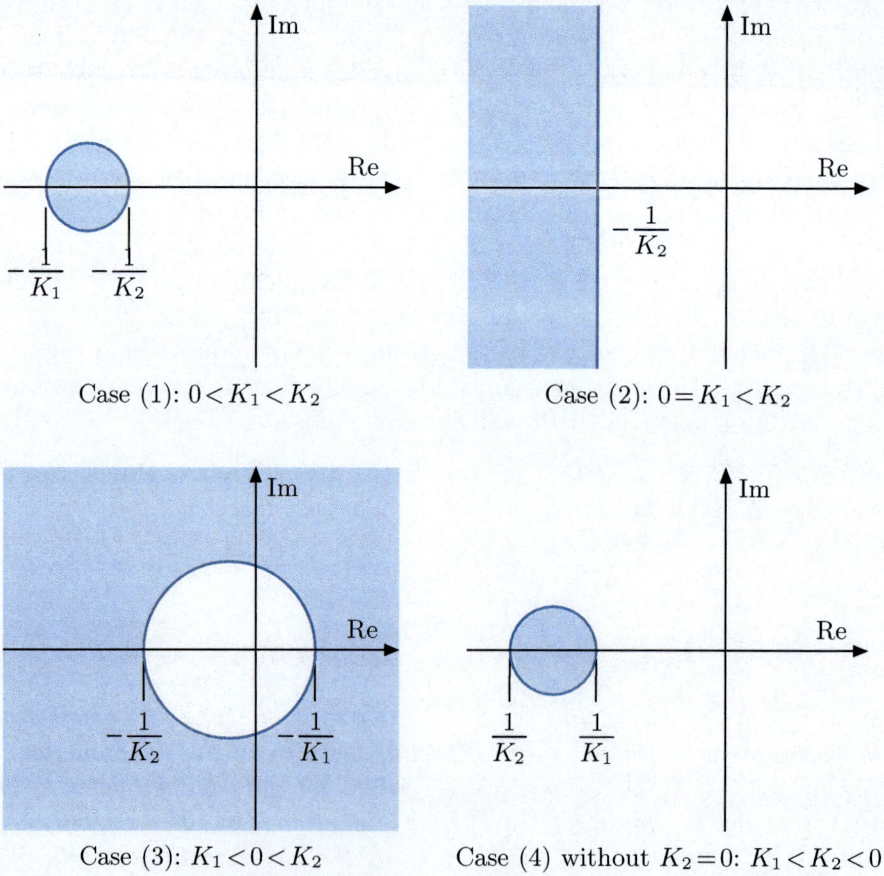

Fig. 2.42: To achieve absolute stability in the sector $[K_1, K_2]$ using Theorem 9, the Nyquist plot of $G(j\omega)$ must lie entirely in the white areas.

the Nyquist plot runs through a blue area, absolute stability of the control loop cannot be proven.

Note that in the case of Condition (1), it is absolutely necessary to use the Nyquist plot of $G(j\omega)$ with $-\infty < \omega < \infty$. In all other cases and in Theorem 8, it is irrelevant whether the Nyquist plot is drawn for all frequencies $-\infty < \omega < \infty$, or only for nonnegative frequencies $0 \leq \omega < \infty$. This is because of the mirror symmetry of the Nyquist plot with $-\infty < \omega < \infty$ with respect to the real axis.

In the following, we will examine a specific example [213] for Case (1) of Theorem 9. The plant is chosen as

$$G(s) = \frac{2}{(s-1)(s+2)}. \tag{2.12}$$

It has a stable pole at $s_1 = -2$ and an unstable pole at $s_2 = 1$. Therefore, $\nu = 1$ holds. Figure 2.43 shows the corresponding Nyquist plot. The Nyquist plot orbits the circle $D(K_1, K_2)$ once in a counterclockwise direction, so we can deduce that the nonlinear standard control loop is absolutely stable in the sector $[K_1, K_2]$.

The above circle criterion enables us to address unstable systems whose poles lie in the open right half-plane of the complex numbers. However, it is limited to systems which do not have poles on the imaginary axis[4]. It can admittedly be extended to systems with poles on the imaginary axis [274]. However, this can be laborious. This is unsatisfactory because many systems in practical applications have such poles. Therefore, it would be useful if we were also able to apply the circle criterion to such systems in a simple way.

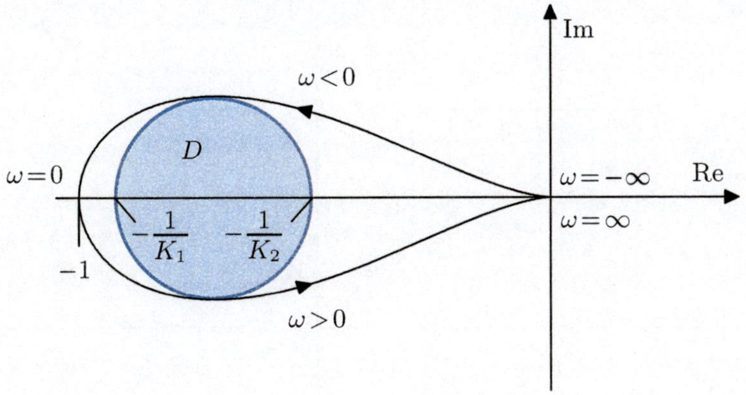

Fig. 2.43: Nyquist plot of $G(j\omega)$ of system (2.12) and application of Case (1) of the circle criterion. Here are $K_1 = 1.083$ and $K_2 = 2.003$.

[4] An illustrative example of this restriction is the system $G(s) = 1/s^4$. If poles on the imaginary axis were allowed in Theorem 9, it would seem as if there were a sector $[\varepsilon, \infty)$ of absolute stability. However, there is no Hurwitz sector for this system and, therefore, no sector of absolute stability.

To do this, we need the *continuous angle variation*, which we define as follows [130, 483]:

Definition 20 (Continuous Angle Variation). *Let the Nyquist plot of a frequency response $G(j\omega)$ for $\omega \in [0, \infty]$ and a point p on the negative real axis of the complex plane be given. For each frequency ω, there is a phasor pointing from the point p to the point $G(j\omega)$. The continuous variation of this phasor's angle $\varphi(\omega)$ which results when $G(j\omega)$ starts at $\omega = 0$ and passes through all frequencies up to $\omega = \infty$ is called the continuous angle variation Ω with respect to the point p.*

In some cases the Nyquist plot is piecewise continuous, i. e. the Nyquist plot has jump discontinuities between its continuous segments. If this is the case, the continuous angle variation Ω consists of the sum of all continuous angle variations of the segments.

Two examples of a continuous angle variation Ω are shown in Figure 2.44 and 2.45. For the first example

$$G(s) = \frac{1}{(s+1)(s+2)(s+3)},$$

we can easily read the continuous angle variation $\Omega = 0$ from the Nyquist plot. The second example

$$G(s) = \frac{4s^2 + 3s + 1}{s^2(s-1)}$$

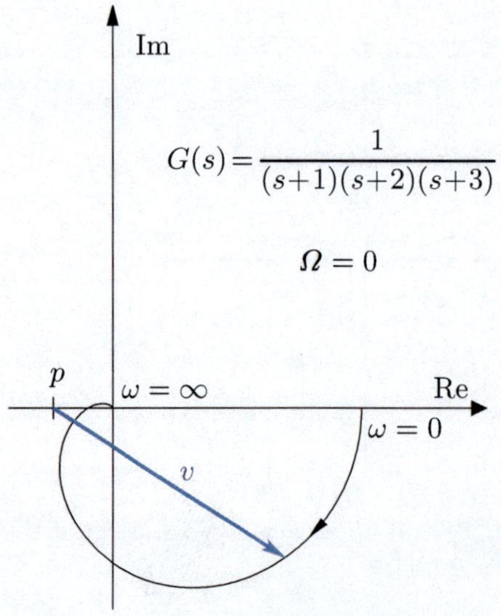

Fig. 2.44: Nyquist plot of $G(j\omega)$ for $\omega \in [0, \infty]$ and a continuous angle variation $\Omega = 0$ of the phasor v

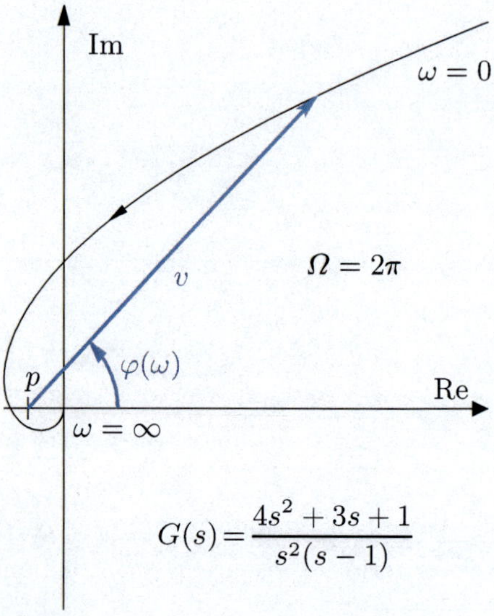

Fig. 2.45: Nyquist plot of $G(j\omega)$ for $\omega \in [0, \infty]$ and a continuous angle variation $\Omega = 2\pi$ of the phasor v

requires a bit of calculation since we do not know the phasor's angle $\varphi(\omega)$ of $G(j\omega)$ for $\omega = 0$ in advance. Using

$$\varphi(\omega) = \arctan\left(\frac{\operatorname{Im}\{G(j\omega)\}}{\operatorname{Re}\{G(j\omega)\} - p}\right) = \arctan\left(4\frac{\omega^3 - \omega}{7\omega^2 - 1 + p\omega^2(1+\omega^2)}\right),$$

we obtain $\varphi(0) = 0$. With this result, we can read a continuous angle variation of $\Omega = 2\pi$ from the Nyquist plot. In general, Ω is easy to determine.

Using the continuous angle variation, we obtain the following general circle criterion, the proof of which is given in Appendix A.1.

Theorem 10 (General Circle Criterion). *Let the nonlinear standard control loop*

$$\begin{aligned} Y(s) &= G(s)U(s), \\ e &= -y, \\ u &= f(e,t), \quad e \in \mathbb{R},\ t \in \mathbb{R}, \end{aligned}$$

be given whose open-loop transfer function $G(s)$ is coprime and has a degree m of the numerator which is less then the degree n of the denominator; let $G(s)$ have ν poles λ_i with $\operatorname{Re}\{\lambda_i\} > 0$ and μ poles λ_i with $\operatorname{Re}\{\lambda_i\} = 0$. Let the function $f(e,t)$ lie in the sector $[K_1, K_2]$, let $f(0,t) = 0$ for all $t \in \mathbb{R}$, and let $D(K_1, K_2)$ represent the circle in the complex plane which passes through the points

$$-\frac{1}{K_1} \quad \textit{and} \quad -\frac{1}{K_2}$$

and which has its center on the real axis. In this case, the control loop is absolutely stable in the sector $[K_1, K_2]$

(1) for $0 < K_1 < K_2$ if the Nyquist plot of $G(j\omega)$ does not intersect or touch the circle $D(K_1, K_2)$ and if the continuous angle variation of $G(j\omega)$ with respect to the point $-1/K_1$ is

$$\Omega = \nu\pi + \mu\frac{\pi}{2}, \tag{2.13}$$

(2) for $0 = K_1 < K_2$ and $\nu = \mu = 0$ if the Nyquist plot is located to the right of the vertical line crossing the real axis at $-1/K_2$,

(3) for $K_1 < 0 < K_2$ and $\nu = \mu = 0$ if the Nyquist plot lies inside the circle $D(K_1, K_2)$, and

(4) for $K_1 < K_2 \le 0$ if, after replacing $G(j\omega)$ with $-G(j\omega)$, and K_1 and K_2 with $-K_1$ and $-K_2$, respectively, either Condition (1) or (2) is fulfilled.

As an example, we look again at system (2.12) and its Nyquist plot, which is shown in Figure 2.43. The circle $D(K_1, K_2)$ is not intersected or touched by the Nyquist plot of $G(j\omega)$, and the continuous angle variation Ω with respect to the point $-1/K_1$ is $\Omega = \pi$. Since $\nu = 1$ and $\mu = 0$ hold for system (2.12), condition (2.13) is fulfilled. Thus, the control loop is absolutely stable in the sector $[K_1 = 1.083,\ K_2 = 2.003]$.

If $K_1 = K_2$, the sector collapses to a line, the characteristic curve is reduced to a linear function $u = f(e) = K_1 e$, and the circle becomes a point located at $-1/K_1$. In this case, the circle criterion is equivalent to the Nyquist criterion.

2.2.6 The Tsypkin Criterion for Discrete-Time Systems

When considering sampled control loops with a sampling period T and a z-domain transfer function $G(z)$, the concept of absolute stability can be defined in a way analogous to the continuous-time case. Again, let us examine nonlinear static characteristic curves f. To determine the absolute stability of such discrete-time nonlinear control loops, as shown in Figure 2.46, Ya. Z. Tsypkin derived theorems comparable to the Popov criterion [260, 455, 456, 503, 502]. It will become clear that their application is similarly simple. However, like the Popov criterion, the Tsypkin criteria are only sufficient. Therefore, we cannot determine with certainty the largest sector of absolute stability by applying these criteria. First we will look at the following simple criterion.

Theorem 11 (Basic Tsypkin Criterion). *Let the nonlinear standard control loop*

$$
\begin{aligned}
Y(z) &= G(z)U(z),\\
e &= -y,\\
u &= f(e)
\end{aligned}
$$

with the plant $G(z)$ be given. Let the function $u = f(e)$ be defined for all values $e \in \mathbb{R}$, piecewise continuous, pass through zero, and fulfill $f(e \to \infty) \neq 0$. Then the control loop is absolutely stable

(1) in the sector $[0, K]$ for asymptotically stable transfer functions $G(z)$ and
(2) in the sector $[\varepsilon, K]$ with an arbitrarily small $\varepsilon > 0$ for transfer functions $G(z)$ with only one pole λ_i with $\lambda_i = 1$ and otherwise only poles λ_i with $|\lambda_i| < 1$

if the inequality

$$
\operatorname{Re}\{G(z = e^{j\omega T})\} > -\frac{1}{K}
$$

is fulfilled for all $0 \le \omega T \le \pi$.

The criterion is still valid if the nonlinearity is time-varying [260]. It can be simply represented geometrically. We need the Nyquist plot for $G(z = e^{j\omega T})$, as shown in Figure 2.47, and then we just need to move a vertical line from the left to the right through the complex plane toward the Nyquist plot until

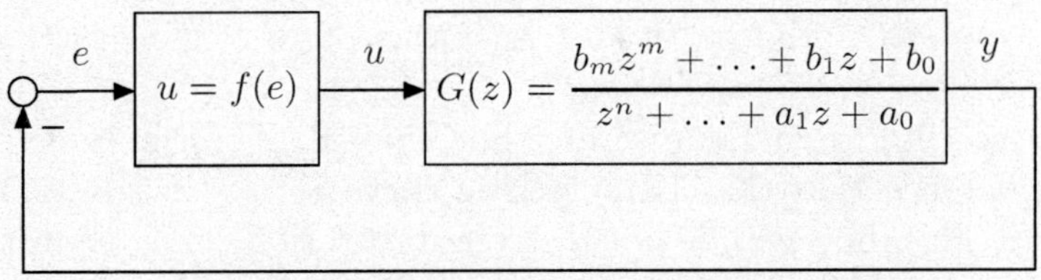

Fig. 2.46: Discrete-time nonlinear standard control loop

the line reaches the plot but does not yet touch it. The intersection $-1/K$ with the real axis yields the sector $[0, K]$ or $[\varepsilon, K]$, of absolute stability. The advantage of this criterion is its simple applicability. Its disadvantage is that, in some cases, it does not provide a good estimate of the sector of absolute stability. A criterion providing a better estimation is

Theorem 12 (Extended Tsypkin Criterion). *Let the nonlinear standard control loop*

$$\begin{aligned} Y(z) &= G(z)U(z), \\ e &= -y, \\ u &= f(e) \end{aligned}$$

with an asymptotically stable plant $G(z)$ be given. Let the function $u = f(e)$ be defined for all values $e \in \mathbb{R}$, piecewise continuous, pass through zero, and be monotonous. Then the control loop is absolutely stable in the sector $[0, K]$ if there exists a real number $q \geq 0$ for which the Tsypkin inequality

$$\operatorname{Re}\left\{\left(1 + q(1 - e^{-j\omega T})\right) \cdot G(z = e^{j\omega T})\right\} > -\frac{1}{K}$$

is fulfilled for all $0 \leq \omega T \leq \pi$.

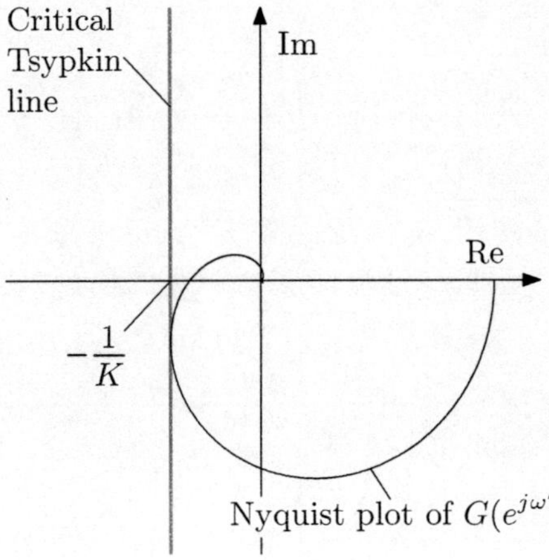

Fig. 2.47: Application of the basic Tsypkin criterion

Here, $Y(z)$ and $U(z)$ denote the z-transforms of the signals $y(t)$ and $u(t)$, respectively.

The above criterion is the discrete-time equivalent to the Popov criterion with the difference that the characteristic curve $u = f(e)$ is required to be monotonous. This, however, is not a strong restriction, since almost all characteristic curves which exist in practice fulfill this additional requirement. Further, $q \geq 0$ must hold.

The extended Tsypkin criterion can be calculated in graphic terms in a fashion similar to the Popov criterion. The Tsypkin inequality

$$\operatorname{Re}\{G(e^{j\omega T}) + q(1 - e^{-j\omega T}) \cdot G(e^{j\omega T})\} > -\frac{1}{K}$$

is rearranged to

$$\underbrace{\operatorname{Re}\{G(e^{j\omega T})\}}_{U(\omega)} - q\underbrace{\left(\operatorname{Re}\{e^{-j\omega T} G(e^{j\omega T})\} - \operatorname{Re}\{G(e^{j\omega T})\}\right)}_{V(\omega)} > -\frac{1}{K}.$$

As in the case of the Popov criterion, this results in the parameterized inequality

$$U(\omega) - qV(\omega) + \frac{1}{K} > 0.$$

Again, an artificial frequency response, or more precisely the corresponding locus curve, the *Tsypkin plot*, is defined by

$$\tilde{G}(e^{j\omega T}) = U(\omega) + jV(\omega).$$

If the Tsypkin plot lies to the right-hand side, or below a line with slope $1/q$ which intersects the real axis at $-1/K$, the control loop is absolutely stable. Figure 2.48 illustrates this. Again, a line is moved until it touches the Tsypkin

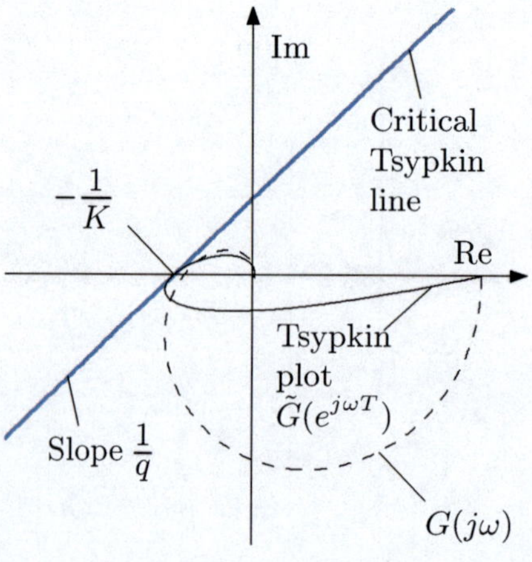

Fig. 2.48: Application of the extended Tsypkin criterion

plot, resulting in the largest possible sector of absolute stability that can be determined using the Tsypkin criterion.

For discrete-time systems, it is also possible to apply the circle criteria, more precisely Theorem 8 and Theorem 9 [465, 504]. In this case, the Nyquist plot of $G(j\omega)$ of the continuous-time system must be replaced with the Nyquist plot of $G(z = e^{j\omega T})$, where $-\pi \leq \omega T \leq \pi$ holds. The variable ν again represents the number of unstable poles of the plant, i. e. the number of poles λ_i with $|\lambda_i| > 1$.

2.3 Lyapunov's Stability Theory

2.3.1 The Concept and the Direct Method

In the previous sections, we dealt with methods for analyzing the stability of nonlinear standard control loops, i. e. a limited class of systems. These include the describing function method, the circle criterion, and the Popov criterion. The methods mentioned above are important to control theory, since they deal with types of control loops that frequently occur in practice.

Unfortunately, these methods of stability analysis are not applicable to arbitrary nonlinear systems. A generic method for analyzing the stability of nonlinear systems was introduced in 1892 by A. M. Lyapunov[5] [279, 292, 293, 355]. This method has been extended in many ways [26, 162, 170, 389]. In principle, we can analyze whether an equilibrium point of a dynamical system is stable or not using the Lyapunov method. It will become apparent, however, that this is often not possible in practice. Hence the Lyapunov method does not entirely solve the problem of how to analyze the stability of nonlinear systems.

To illustrate the key principle behind the Lyapunov method, we will look at different cases of possible stability behavior using the example of a ball with friction influenced by the acceleration of gravity g. Figure 2.49 shows the mechanical arrangements used for this purpose; only the left one exhibits a stable equilibrium point.

The potential energy

$$E_{\text{pot}} = mgy$$

of the ball with mass m is proportional to its height y. Each of the arrangements in question forces the ball to move along a specific track, so that the y-coordinate is a function f of x and z. For the potential energy, we therefore obtain

$$E_{\text{pot}} = mgf(x, z).$$

Obviously an equilibrium point is only stable if the potential energy has a minimum at the equilibrium point, i. e. in this example, the function f must have a minimum.

[5] The name is also sometimes spelled Liapunov.

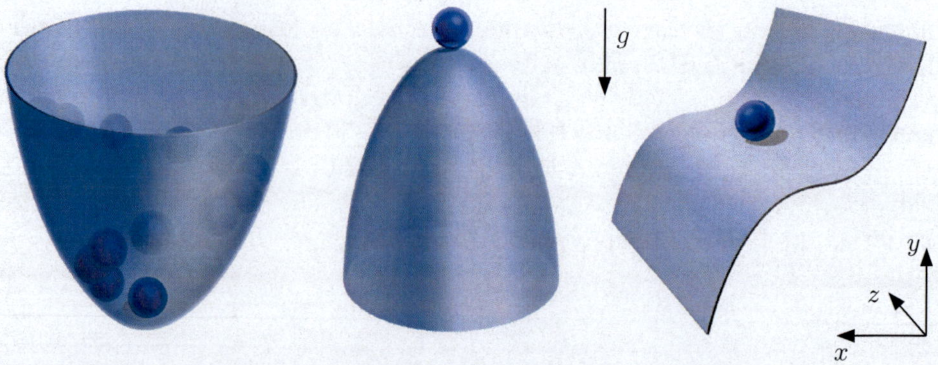

Fig. 2.49: Stability and instability of a ball influenced by gravity

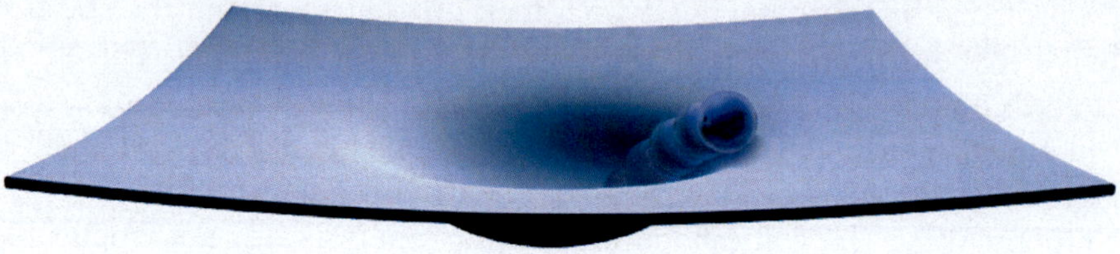

Fig. 2.50: Ball with energy source and force driving an upswing

The following example demonstrates that this alone, however, does not suffice to ensure stability. Now imagine that the ball has an energy source which causes it to swing up and move upwards, as illustrated in Figure 2.50. Thus the equilibrium point is not stable, even though the potential energy has a minimum. This is due to the fact that the system has an internal source of energy.

In addition to the requirement that the potential energy has a minimum, a further condition must obviously be satisfied to ensure stability of the equilibrium point. We therefore require the potential energy along all trajectories in the neighborhood of the equilibrium point to decrease or at least stay constant.

Looking at the problem more closely, it seems possible to move away from our starting point of a potential energy function and to generalize the above idea. To assess the stability of an equilibrium point, it seems to be sufficient to use an arbitrary function for which the following two conditions are fulfilled:

(1) The function must have a minimum at the equilibrium point.
(2) The function must decrease within a neighborhood of the equilibrium point along all of its trajectories.

This is the basic concept behind the *direct Lyapunov method*, which is also referred to as his *second method*. A. M. Lyapunov proved the following theorem [170, 279, 389], which is essential to the stability analysis of dynamic systems.

Theorem 13 (Lyapunov's Direct Method). *Let the differential equation* $\dot{\boldsymbol{x}} = \boldsymbol{f}(\boldsymbol{x})$ *with the equilibrium point* $\boldsymbol{x}_{\mathrm{eq}} = \boldsymbol{0}$ *possess a continuous and unique solution for every initial state vector within a neighborhood* $U_1(\boldsymbol{0})$ *of the origin. If a function* $V(\boldsymbol{x})$ *exists which possesses continuous partial derivatives and fulfills the conditions*

(1) $V(\boldsymbol{0}) = 0$,
(2) $V(\boldsymbol{x}) > 0$ *for* $\boldsymbol{x} \neq \boldsymbol{0}$,
(3) $\dot{V}(\boldsymbol{x}) < 0$ *for* $\boldsymbol{x} \neq \boldsymbol{0}$ *(or* $\dot{V}(\boldsymbol{x}) \leq 0$*)*

within a neighborhood $U_2(\boldsymbol{0}) \subseteq U_1(\boldsymbol{0})$*, the equilibrium point* $\boldsymbol{x}_{\mathrm{eq}} = \boldsymbol{0}$ *is asymptotically stable (or Lyapunov stable).*

A function V which fulfills Conditions (1) and (2) is called *positive definite* and has a minimum at $\boldsymbol{x} = \boldsymbol{0}$. Condition (3) means that V decreases or stays constant over time along all trajectories $\boldsymbol{x}(t)$ starting from $U_2(\boldsymbol{0})$. It is worth noting, once again, that the assumption of an equilibrium point at $\boldsymbol{x} = \boldsymbol{0}$ is not a loss of generality, since every equilibrium point can be transformed to $\boldsymbol{x} = \boldsymbol{0}$.

Theorem 13 allows us to verify whether an equilibrium point is Lyapunov stable or asymptotically stable. This depends on whether $\dot{V}(\boldsymbol{x}) \leq 0$ or $\dot{V}(\boldsymbol{x}) < 0$ holds. Figure 2.51 illustrates a case in which $\dot{V}(\boldsymbol{x}) \leq 0$. Where $\dot{V}(\boldsymbol{x}) = 0$ holds, it is possible that trajectories $\boldsymbol{x}(t)$ which do not tend to the equilibrium point $\boldsymbol{x}_{\mathrm{eq}} = \boldsymbol{0}$ fulfill the inequality $\dot{V}(\boldsymbol{x}) \leq 0$, i. e. the equilibrium point is only Lyapunov stable. Figure 2.52, on the other hand, illustrates a case of asymptotic stability. In this case the function $V(\boldsymbol{x})$ decreases along all trajectories $\boldsymbol{x}(t)$ except the trivial trajectory $\boldsymbol{x}(t) = \boldsymbol{0}$, i. e. it holds for the time derivative that $\dot{V}(\boldsymbol{x}) < 0$ and $\boldsymbol{x} \neq \boldsymbol{0}$.

If Conditions (2) and (3) are fulfilled for the entire state space, and additionally

$$V(\boldsymbol{x}) \to \infty \quad \text{whenever} \quad |\boldsymbol{x}| \to \infty \tag{2.14}$$

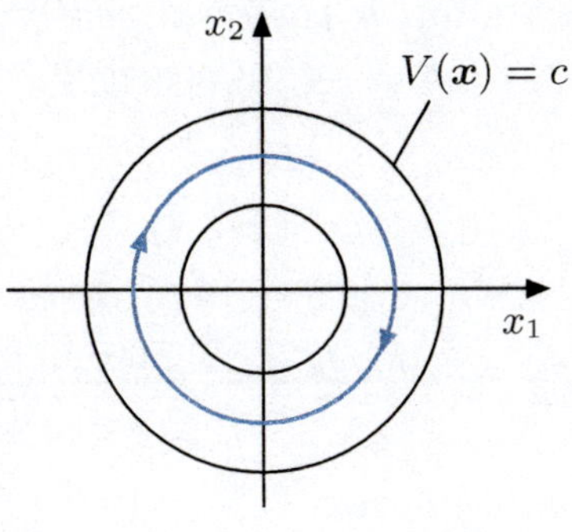

Fig. 2.51: Lyapunov stability

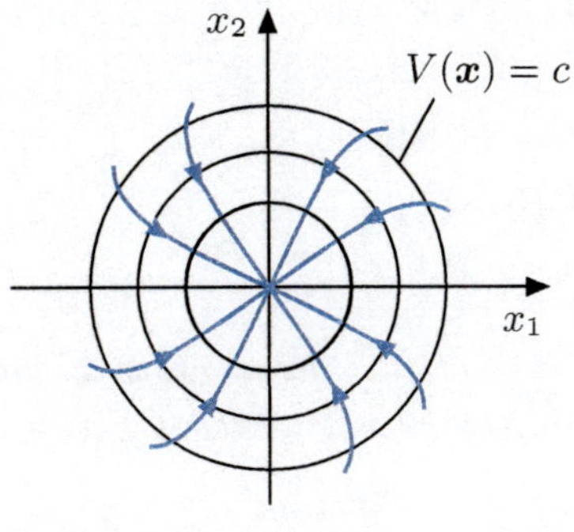

Fig. 2.52: An example of asymptotic stability

holds, the equilibrium point is globally asymptotically stable (or globally Lyapunov stable). A function with the property (2.14) is called *radially unbounded*. Functions $V(\boldsymbol{x})$ which satisfy the conditions of the Lyapunov stability theorem are called *Lyapunov functions* in the case in which $\dot{V}(\boldsymbol{x}) \leq 0$ and *strict Lyapunov functions* in the case in which $\dot{V}(\boldsymbol{x}) < 0$. In practical use, the time derivative of the Lyapunov functions $V(\boldsymbol{x})$ is determined by the gradient and the time derivative of the state to be

$$\dot{V}(\boldsymbol{x}) = \dot{\boldsymbol{x}}^T \operatorname{grad}(V(\boldsymbol{x})) = \sum_{i=1}^{n} \dot{x}_i \frac{\partial V}{\partial x_i}. \tag{2.15}$$

The derivative of the state vector

$$\dot{\boldsymbol{x}} = \boldsymbol{f}(\boldsymbol{x})$$

can now be inserted into equation (2.15). Then we must check if $\dot{V}(\boldsymbol{x}) \leq 0$ or if $\dot{V}(\boldsymbol{x}) < 0$ holds for $\boldsymbol{x} \neq \boldsymbol{0}$. This is illustrated by Figure 2.53. The solution of the differential equation, which for nonlinear systems in many cases cannot be computed analytically, is not required for us to apply Theorem 13. The name *direct method* stems from the direct application of the differential equation for the calculation of $\dot{V}(\boldsymbol{x})$.

Next we will address the case in which we know a Lyapunov function V for a system $\dot{\boldsymbol{x}} = \boldsymbol{f}(\boldsymbol{x})$ for which we can only show that $\dot{V}(\boldsymbol{x}) \leq 0$. Therefore, using Theorem 13, we can only prove Lyapunov stability and not asymptotic stability. However, if no trajectory $\boldsymbol{x}(t)$ exists which starts at some initial value $\boldsymbol{x}(0)$ and along which the derivative $\dot{V}(\boldsymbol{x}(t))$ is continuously identical to zero, the function V decreases along all trajectories that start within the neighborhood of the equilibrium point $\boldsymbol{x}_{\text{eq}} = \boldsymbol{0}$. Thus the asymptotic stability of the equilibrium point can also be proven for this case [29]. This is more precisely formulated in

Theorem 14 (Barbashin and Krasovskii Theorem). *Let the differential equation $\dot{\boldsymbol{x}} = \boldsymbol{f}(\boldsymbol{x})$ with the equilibrium point $\boldsymbol{x}_{\text{eq}} = \boldsymbol{0}$ possess a continuous and unique solution for every initial state vector within a neighborhood $U_1(\boldsymbol{0})$ of the origin. Let a function $V(\boldsymbol{x})$ exist which possesses continuous partial derivatives and fulfills the following conditions within a neighborhood $U_2(\boldsymbol{0}) \subseteq U_1(\boldsymbol{0})$:*

(1) $V(\boldsymbol{0}) = 0$,
(2) $V(\boldsymbol{x}) > 0$ *for* $\boldsymbol{x} \neq \boldsymbol{0}$,
(3) $\dot{V}(\boldsymbol{x}) \leq 0$.
(4) The set of state vectors $\boldsymbol{x}$ for which $\dot{V}(\boldsymbol{x}) = 0$ holds does not include a trajectory $\boldsymbol{x}(t)$, except $\boldsymbol{x}(t) = \boldsymbol{0}$.

Then the equilibrium point $\boldsymbol{x}_{\text{eq}} = \boldsymbol{0}$ is asymptotically stable.

If the Lyapunov function of the above theorem is also radially unbounded and $U_2(\boldsymbol{0}) = I\!R^n$, the equilibrium point is globally asymptotically stable.

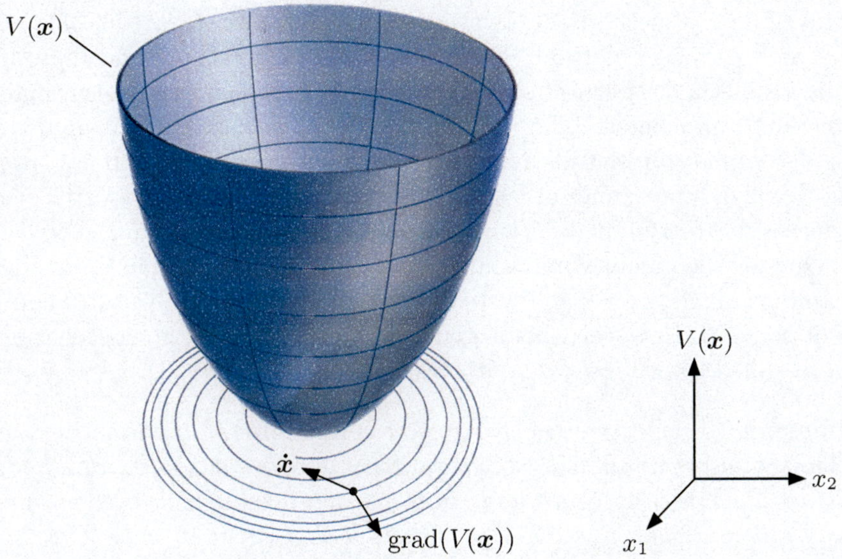

Fig. 2.53: Illustration of the equation $\dot{V}(\boldsymbol{x}) = \dot{\boldsymbol{x}}^T \operatorname{grad}(V(\boldsymbol{x})) < 0$

Seen in terms of graphics, Condition (4) means that no trajectory $\boldsymbol{x}(t)$ may exist which progresses in such a way that $V(\boldsymbol{x}(t))$ is always constant along its course. Then the trajectory would run along a contour line of V and therefore would never reach the equilibrium point $\boldsymbol{x}_{\text{eq}} = \boldsymbol{0}$. The trajectory runs precisely along a contour line if and only if its direction $\dot{\boldsymbol{x}}(t)$ is orthogonal to the gradient $(\partial V/\partial \boldsymbol{x})^T$. In this case,

$$\dot{V}(\boldsymbol{x}(t)) = \left(\frac{\partial V}{\partial \boldsymbol{x}}\right)^T \dot{\boldsymbol{x}}(t) = 0 \tag{2.16}$$

holds. Condition (4), however, does not rule out the existence of individual points $\boldsymbol{x}$ of the trajectory for which equation (2.16) is fulfilled. It requires that equation (2.16) does not hold for the entire trajectory $\boldsymbol{x}(t)$.

Theorem 14 is often useful in practice when it is only possible to find a Lyapunov function with $\dot{V}(\boldsymbol{x}) \leq 0$. Condition (4) can be verified by determining the set of vectors $\boldsymbol{x}$, for which $\dot{V}(\boldsymbol{x}) = 0$ holds. These vectors are inserted into $\dot{\boldsymbol{x}} = \boldsymbol{f}(\boldsymbol{x})$. If the set contains a solution $\boldsymbol{x}(t)$ of the differential equation, which is different from the equilibrium point $\boldsymbol{x}_{\text{eq}} = \boldsymbol{0}$, Condition (4) is not fulfilled. In most cases, Condition (4) is fulfilled, since it only rarely happens that a trajectory $\boldsymbol{x}(t)$ continuously runs along a contour line of the Lyapunov function V.

The problem with applying the above stability theorems mainly consists of finding a Lyapunov function $V(\boldsymbol{x})$. For linear systems and for some special cases of nonlinear systems, as discussed below, it is easy to find a Lyapunov function based on intuition aided by graphs. Generally, however, it turns out to be very difficult to determine a Lyapunov function. Although there are a series of design methods for Lyapunov functions [30, 151, 164, 339, 356, 405] such as Aizerman's method, Schultz and Gibson's method, and Ingwerson's or Zubow's method, they are only applicable to special cases and in many cases difficult to implement. The method that is most successful in many cases is the determination of the energy function of the system and its usage as a possible Lyapunov function. Ultimately, however, in the majority of cases it is necessary to try out different approaches for $V(\boldsymbol{x})$.

Due to this difficulty in finding a solution, we might suspect that there is no Lyapunov function for some systems with an asymptotically stable equilibrium point $\boldsymbol{x}_{\mathrm{eq}} = \mathbf{0}$. However, this is generally not the case. An existence theorem, the so-called *converse Lyapunov theorem* [300, 301, 311, 312, 486], is given by

Theorem 15 (Converse Lyapunov Theorem). *If the system $\dot{\boldsymbol{x}} = \boldsymbol{f}(\boldsymbol{x})$ possesses an asymptotically stable equilibrium point at $\boldsymbol{x} = \mathbf{0}$, and the function $\boldsymbol{f}$ is locally Lipschitz continuous in a neighborhood of $\boldsymbol{x} = \mathbf{0}$, then a continuously differentiable function $V(\boldsymbol{x})$ exists with $V(\mathbf{0}) = 0$, with $V(\boldsymbol{x}) > 0$, and $\dot{V}(\boldsymbol{x}) < 0$ for $\boldsymbol{x} \neq \mathbf{0}$.*

If the condition of the theorem above is fulfilled, the existence of a Lyapunov function is assured. The problem is finding it, as mentioned previously. In the case of an equilibrium point $\boldsymbol{x}_{\mathrm{eq}} = \mathbf{0}$ which is not asymptotically stable, but only Lyapunov stable, only the existence of a time-dependent Lyapunov function $V(\boldsymbol{x}, t)$ with $V(\mathbf{0}, t) = 0$ as well as $V(\boldsymbol{x}, t) > 0$ and

$$\dot{V}(\boldsymbol{x}, t) = \frac{\partial V}{\partial \boldsymbol{x}} \dot{\boldsymbol{x}} + \frac{\partial V}{\partial t} \leq 0$$

for $\boldsymbol{x} \neq \mathbf{0}$ is guaranteed [448].

2.3.2 Illustrative Example

Let us return once again to the system (1.10) on p. 12, i. e.

$$\begin{aligned} \dot{x}_1 &= x_1(x_2 - 1), \\ \dot{x}_2 &= x_2(x_1 - 1), \end{aligned}$$

with the equilibrium points $\boldsymbol{x}_{\mathrm{eq1}} = \mathbf{0}$ and $\boldsymbol{x}_{\mathrm{eq2}} = [1 \;\; 1]^T$. Its trajectories are shown in Figure 2.54. A candidate for a Lyapunov function to prove the stability of the equilibrium point $\boldsymbol{x}_{\mathrm{eq1}} = \mathbf{0}$ is

$$V(\boldsymbol{x}) = x_1^2 + x_2^2, \tag{2.17}$$

since it holds that $V(\mathbf{0}) = 0$ and $V(\boldsymbol{x}) > 0$ otherwise. The contour lines of the function V are circles. As we remember, the equilibrium point $\boldsymbol{x}_{\mathrm{eq2}}$ is unstable.

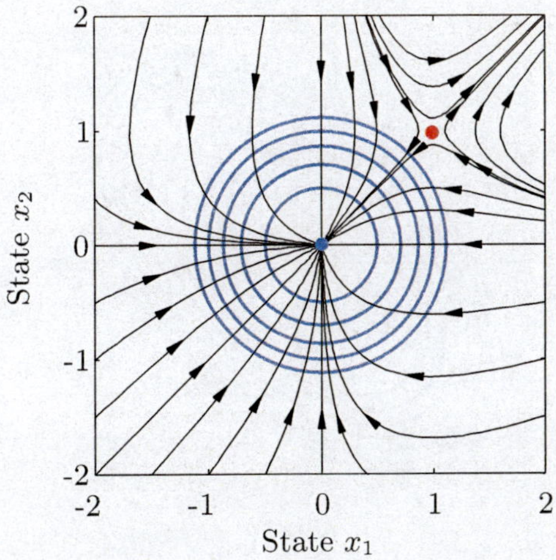

Fig. 2.54: Trajectories $\boldsymbol{x}(t)$ and circular contour lines (blue) of the Lyapunov function $V(\boldsymbol{x}) = x_1^2 + x_2^2$

Now it is necessary to determine whether the function (2.17) decreases along all system trajectories $\boldsymbol{x}(t)$ in the neighborhood of the equilibrium point. For this purpose, we calculate

$$\dot{V}(\boldsymbol{x}) = \dot{\boldsymbol{x}}^T \operatorname{grad}(V(\boldsymbol{x})) = \begin{bmatrix}\dot{x}_1 & \dot{x}_2\end{bmatrix} \begin{bmatrix}2x_1 \\ 2x_2\end{bmatrix} = 2x_1^2(x_2 - 1) + 2x_2^2(x_1 - 1).$$

It holds that

$$\dot{V}(\boldsymbol{x}) < 0 \quad \text{for} \quad x_1 < 1 \quad \text{and} \quad x_2 < 1,$$

so that the equilibrium point $\boldsymbol{x}_{\text{eq}} = \boldsymbol{0}$ is asymptotically stable and the function (2.17) is a strict Lyapunov function. Figure 2.54 illustrates these results, with the contour lines $V(\boldsymbol{x})$ forming circles as mentioned earlier.

2.3.3 Quadratic Lyapunov Functions

Based on graphs supporting intuition, functions with circular or ellipsoidal contour lines seem to be suitable candidates for Lyapunov functions for different systems. Their general form is determined by the positive definite quadratic form

$$V(\boldsymbol{x}) = \boldsymbol{x}^T \boldsymbol{R} \boldsymbol{x}.$$

Figure 2.55 shows the contour lines of such a function.

Condition (1) of Theorem 13,

$$V(\boldsymbol{0}) = 0,$$

is obviously fulfilled; if $\boldsymbol{R}$ is a positive definite matrix, Condition (2), i. e.

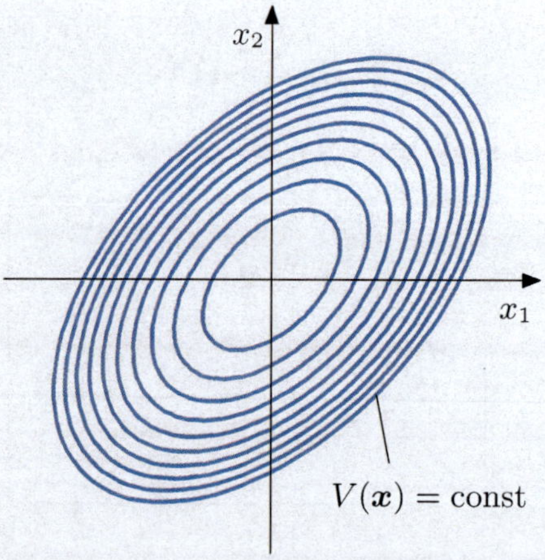

Fig. 2.55: Contour lines of a positive definite quadratic function

$$V(\boldsymbol{x}) > 0 \quad \text{for} \quad \boldsymbol{x} \neq \boldsymbol{0},$$

is also fulfilled. It remains necessary to verify Condition (3), i. e.

$$\dot{V}(\boldsymbol{x}) \leq 0 \quad \text{or} \quad \dot{V}(\boldsymbol{x}) < 0,$$

for the system in question.

We will investigate the extent to which the approach of choosing quadratic forms $\boldsymbol{x}^T\boldsymbol{R}\boldsymbol{x}$ for a Lyapunov function is effective using linear systems

$$\dot{\boldsymbol{x}} = \boldsymbol{A}\boldsymbol{x}. \tag{2.18}$$

With

$$V(\boldsymbol{x}) = \boldsymbol{x}^T\boldsymbol{R}\boldsymbol{x},$$

it holds for $\dot{V}(\boldsymbol{x})$ that

$$\dot{V}(\boldsymbol{x}) = \boldsymbol{x}^T\boldsymbol{R}\dot{\boldsymbol{x}} + \dot{\boldsymbol{x}}^T\boldsymbol{R}\boldsymbol{x}.$$

Inserting the differential equation (2.18) yields

$$\begin{aligned}\dot{V}(\boldsymbol{x}) &= \boldsymbol{x}^T\boldsymbol{R}\boldsymbol{A}\boldsymbol{x} + \boldsymbol{x}^T\boldsymbol{A}^T\boldsymbol{R}\boldsymbol{x}\\ &= \boldsymbol{x}^T\underbrace{\left(\boldsymbol{R}\boldsymbol{A} + \boldsymbol{A}^T\boldsymbol{R}\right)}_{-\boldsymbol{Q}}\boldsymbol{x}.\end{aligned}$$

For asymptotic stability, we have to fulfill the inequality

$$\dot{V}(\boldsymbol{x}) = -\boldsymbol{x}^T\boldsymbol{Q}\boldsymbol{x} < 0.$$

Thus the matrix $\boldsymbol{Q}$ is required to be positive definite.

If the matrix equation

$$\boldsymbol{A}^T\boldsymbol{R} + \boldsymbol{R}\boldsymbol{A} = -\boldsymbol{Q} \tag{2.19}$$

provides a positive definite matrix $\boldsymbol{Q}$, $V(\boldsymbol{x}) = \boldsymbol{x}^T\boldsymbol{R}\boldsymbol{x}$ is a strict Lyapunov function and the system $\dot{\boldsymbol{x}} = \boldsymbol{A}\boldsymbol{x}$ is asymptotically stable. It is also possible to proceed inversely by specifying an arbitrary positive definite matrix $\boldsymbol{Q}$ and – if system (2.18) is asymptotically stable – determining a positive definite matrix $\boldsymbol{R}$ and thus a Lyapunov function. Equation (2.19) is called the *Lyapunov equation*. The following theorem [258] holds.

Theorem 16 (Lyapunov Equation). *The equilibrium point* $\boldsymbol{x}_{\mathrm{eq}} = \boldsymbol{0}$ *of the linear system* $\dot{\boldsymbol{x}} = \boldsymbol{A}\boldsymbol{x}$ *is Lyapunov stable (asymptotically stable) if and only if, for an arbitrary real-valued symmetric, positive semidefinite (positive definite) matrix* $\boldsymbol{Q}$*, a positive definite matrix* $\boldsymbol{R}$ *exists such that*

$$\boldsymbol{A}^T\boldsymbol{R} + \boldsymbol{R}\boldsymbol{A} = -\boldsymbol{Q}$$

holds. Then the function $V = \boldsymbol{x}^T\boldsymbol{R}\boldsymbol{x}$ *is a (strict) Lyapunov function for the system.*

For stable linear systems, it is therefore always possible to find quadratic Lyapunov functions. At least in this case, the approach of using quadratic Lyapunov functions proves to be very suitable.

Theorem 16 is actually not significant in the stability analysis of linear systems, which can be more easily determined using the eigenvalues of the system. Its importance instead lies in the design theory of many nonlinear control systems and also in the stability analysis of linearized nonlinear systems. We will address these topics later on.

2.3.4 Example: Mutualism

Let us consider a dynamic ecological system. In ecological systems there are many and often very different dependencies between species. The best-known are the predator-prey relationships, which are often modeled by Lotka-Volterra equations. Interrelations between two species where both species profit from the relationship can also be described by differential equations that are closely related to the Lotka-Volterra equations. The type of cohabitation that is beneficial for both species is called mutualism.

An example of this is the mutualism between the ocellaris clownfish (*Amphiprion ocellaris*) and the magnificent sea anemone (*Heteractis magnifica*), which is depicted in Figure 2.56. On the one hand, the anemone protects the clownfish against predators with its poisonous tentacles; on the other hand, the fish also protects the anemone against predators such as filefish (*Monacanthidae*). Another example is the mutualism between humans and wheat. Mutualism of this kind is modeled by the equations

$$\begin{aligned} \dot{x}_1 &= ax_1 - cx_1^2 + ex_1x_2, \\ \dot{x}_2 &= bx_2 - dx_2^2 + fx_1x_2, \end{aligned} \tag{2.20}$$

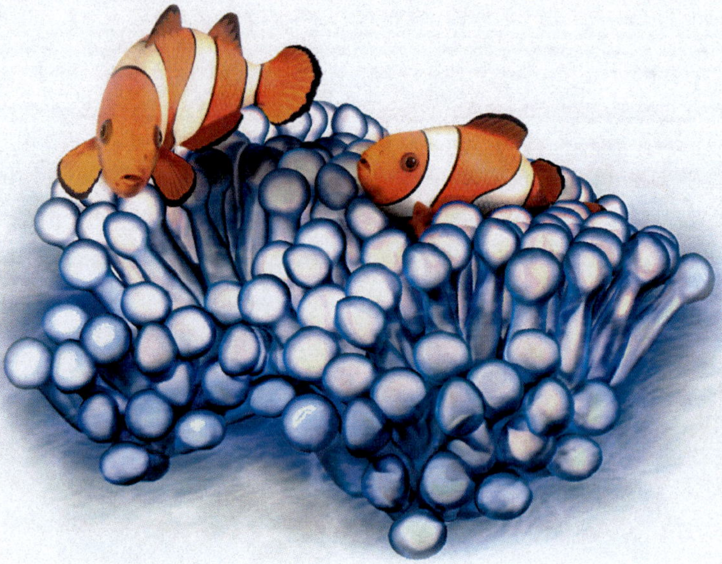

Fig. 2.56: Mutualism between ocellaris clownfish and anemone

in which x_1 denotes the number of individuals of the first species and x_2 that of the other species. The values a, b, c, d, e, and f are constant positive parameters.

In equation (2.20), the terms $\dot{x}_1 = ax_1$ and $\dot{x}_2 = bx_2$ describe linear growth laws, for which the population growth increases linearly depending on their sizes. The term $-cx_1^2$, on the other hand, inhibits the growth of population x_1 for an increasing population size, because of rivalry for food within one species, for example. This is also called intraspecific competition. The term $-dx_2^2$ has the same effect. The components ex_1x_2 and fx_1x_2 lead to a mutual promotion of growth in both populations. These two components thus describe the mutualism within the system.

The model (2.20) possesses the equilibrium points

$$\boldsymbol{x}_{\text{eq1}} = \begin{bmatrix} 0 \\ 0 \end{bmatrix}, \quad \boldsymbol{x}_{\text{eq2}} = \begin{bmatrix} a/c \\ 0 \end{bmatrix}, \quad \boldsymbol{x}_{\text{eq3}} = \begin{bmatrix} 0 \\ b/d \end{bmatrix}.$$

If the inequality $ef < cd$ is additionally fulfilled, a fourth equilibrium point occurs at

$$\boldsymbol{x}_{\text{eq4}} = \begin{bmatrix} \dfrac{be + ad}{cd - ef} \\ \dfrac{bc + af}{cd - ef} \end{bmatrix},$$

which is created by mutualism. In the case where $ef > cd$, no equilibrium point with positive coordinates exists, since the mutualism is much stronger than the intraspecific competition and the populations x_1 and x_2 grow infinitely large.

We will look at the following specific case

$$\begin{aligned}\dot{x}_1 &= x_1 - 10^{-3}x_1^2 + 0.5\cdot 10^{-3}x_1x_2,\\ \dot{x}_2 &= x_2 - 10^{-3}x_2^2 + 0.5\cdot 10^{-3}x_1x_2\end{aligned} \tag{2.21}$$

with the equilibrium points

$$\boldsymbol{x}_{\text{eq1}} = \begin{bmatrix}0\\0\end{bmatrix}, \quad \boldsymbol{x}_{\text{eq2}} = \begin{bmatrix}1000\\0\end{bmatrix}, \quad \boldsymbol{x}_{\text{eq3}} = \begin{bmatrix}0\\1000\end{bmatrix}, \quad \boldsymbol{x}_{\text{eq4}} = \begin{bmatrix}2000\\2000\end{bmatrix}.$$

Figure 2.57 depicts the courses of the trajectories in this ecological system. We are especially interested in the equilibrium point $\boldsymbol{x}_{\text{eq4}}$ caused by mutualism. To prove its stability, we will first transform $\boldsymbol{x}_{\text{eq4}}$ via

$$\boldsymbol{x} = \boldsymbol{z} + \boldsymbol{x}_{\text{eq4}}$$

into the origin, which yields

$$\begin{aligned}\dot{z}_1 &= (z_1 + 2000) - 10^{-3}(z_1 + 2000)^2 + 0.5\cdot 10^{-3}(z_1 + 2000)(z_2 + 2000),\\ \dot{z}_2 &= (z_2 + 2000) - 10^{-3}(z_2 + 2000)^2 + 0.5\cdot 10^{-3}(z_1 + 2000)(z_2 + 2000)\end{aligned}$$

for the system (2.21) after transformation. The courses of the trajectories in the transformed system are shown in Figure 2.58.

Now we choose

$$V(\boldsymbol{z}) = z_1^2 + z_2^2$$

as potential Lyapunov function, and we arrive at

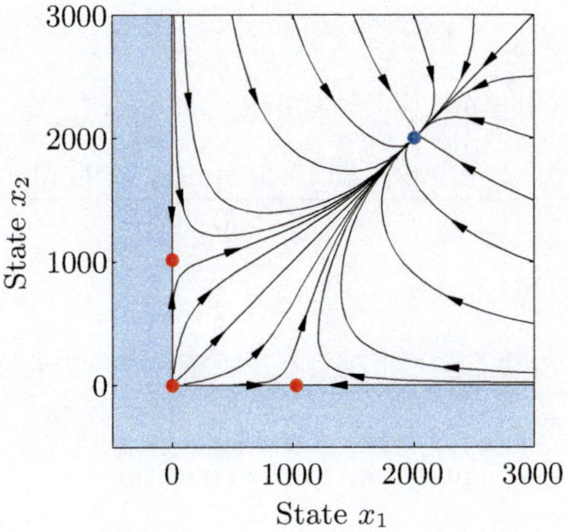

Fig. 2.57: Trajectories of a mutualistic system. For a negative number of individuals x_1 and x_2 (blue area), no trajectories occur in reality.

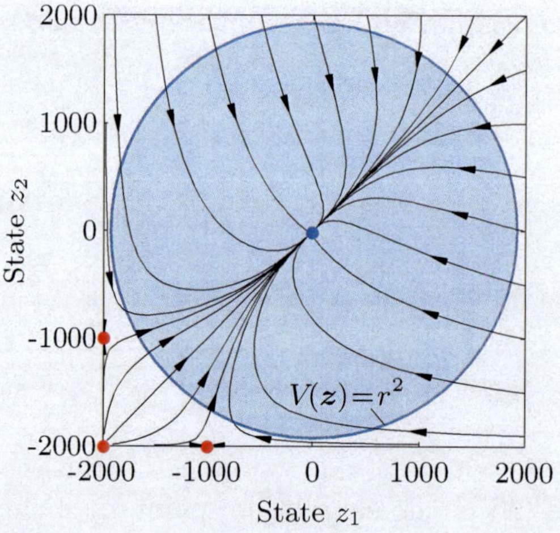

Fig. 2.58: A circular catchment region with radius $r = 1924$ (blue) of the mutualistic equilibrium point, which has been transformed to $\boldsymbol{z} = \boldsymbol{0}$

$$\begin{aligned}\dot{V}(\boldsymbol{z}) &= 2z_1\dot{z}_1 + 2z_2\dot{z}_2\\ &= -4(z_1^2+z_2^2) - 2\cdot 10^{-3}(z_1^3+z_2^3) + 10^{-3}z_1z_2(z_1+z_2) + 4z_1z_2.\end{aligned}\tag{2.22}$$

In order to determine when $\dot{V}(\boldsymbol{z}) < 0$ holds, we will define the polar coordinates

$$\begin{aligned}z_1 &= r\cos(\varphi),\\ z_2 &= r\sin(\varphi).\end{aligned}$$

Here r denotes the radius, φ represents the angle of the polar coordinates, and r^2 further represents the level of a contour line of the Lyapunov function

$$V(\boldsymbol{z}) = z_1^2 + z_2^2 = r^2.$$

Thus it follows from equation (2.22) that

$$\dot{V} = -r^2\frac{16000 + 5r\cos(\varphi) + 3r\cos(3\varphi) + 5r\sin(\varphi) - 8000\sin(2\varphi) - 3r\sin(3\varphi)}{4000},$$

which is obviously negative if

$$16000 - 8000\sin(2\varphi) + r\left[5\cos(\varphi) + 3\cos(3\varphi) + 5\sin(\varphi) - 3\sin(3\varphi)\right] > 0 \tag{2.23}$$

holds. Equation (2.23) is equivalent to the condition

$$\frac{1}{r} > \frac{3\sin(3\varphi) - 3\cos(3\varphi) - 5\cos(\varphi) - 5\sin(\varphi)}{8000(2-\sin(2\varphi))},$$

which is certainly fulfilled if

$$\frac{1}{r} > \max\left\{\frac{3\sin(3\varphi) - 3\cos(3\varphi) - 5\cos(\varphi) - 5\sin(\varphi)}{8000(2-\sin(2\varphi))}\right\} \approx \frac{1}{1924}$$

holds.

For all values $r < 1924$, i. e. the ones within a circle of radius $r = 1924$, it holds that $\dot{V}(\boldsymbol{z}) < 0$. The equilibrium point $\boldsymbol{x}_{\mathrm{eq4}}$ is therefore asymptotically stable and the circle with $r = 1924$ around its center at $\boldsymbol{x}_{\mathrm{eq4}}$ is a catchment region, sometimes called a *Lyapunov region*, since the region is outlined by a contour line from a Lyapunov function. However, as shown in Figure 2.58, this Lyapunov region does not provide the maximum catchment region. The latter would consist of the part of the state plane in Figure 2.57 where x_1 and x_2 are positive. Here the coordinate axes do not belong to the maximum catchment region, since the trajectories which start here do not tend to the non-mutualistic equilibrium points.

2.3.5 The Direct Method for Discrete-Time Systems

Similar to continuous-time systems, the direct Lyapunov method can also be applied to discrete-time systems

$$\boldsymbol{x}_{k+1} = \boldsymbol{f}(\boldsymbol{x}_k).$$

The first two conditions of the Stability Theorem 13, $V(\boldsymbol{0}) = 0$ and $V(\boldsymbol{x}) > 0$ for all $\boldsymbol{x} \neq \boldsymbol{0}$, hold unchanged. Only the third condition,

$$\dot{V}(\boldsymbol{x}) < 0 \quad \text{for all} \quad \boldsymbol{x} \neq \boldsymbol{0},$$

which we are only calculating for the asymptotically stable case at this point, has to be replaced by the condition

$$\Delta V_k = V(\boldsymbol{x}_{k+1}) - V(\boldsymbol{x}_k) < 0 \quad \text{for all} \quad \boldsymbol{x}_k \neq \boldsymbol{0}.$$

In the case of linear systems

$$\boldsymbol{x}_{k+1} = \boldsymbol{\Phi}\boldsymbol{x}_k$$

and with quadratic Lyapunov functions

$$V(\boldsymbol{x}_k) = \boldsymbol{x}_k^T \boldsymbol{R} \boldsymbol{x}_k,$$

we arrive at

$$\begin{aligned}\Delta V_k &= \boldsymbol{x}_{k+1}^T \boldsymbol{R} \boldsymbol{x}_{k+1} - \boldsymbol{x}_k^T \boldsymbol{R} \boldsymbol{x}_k \\ &= \boldsymbol{x}_k^T \boldsymbol{\Phi}^T \boldsymbol{R} \boldsymbol{\Phi} \boldsymbol{x}_k - \boldsymbol{x}_k^T \boldsymbol{R} \boldsymbol{x}_k \\ &= \boldsymbol{x}_k^T (\boldsymbol{\Phi}^T \boldsymbol{R} \boldsymbol{\Phi} - \boldsymbol{R}) \boldsymbol{x}_k < 0.\end{aligned}$$

The above inequality is obviously fulfilled if the matrix $\boldsymbol{Q}$ in the equation

$$\boldsymbol{\Phi}^T \boldsymbol{R}\boldsymbol{\Phi} - \boldsymbol{R} = -\boldsymbol{Q} \tag{2.24}$$

is positive definite. Equation (2.24) is called the *discrete Lyapunov equation*. In contrast to the Lyapunov equation for the continuous-time case, the Lyapunov equation (2.24) depends on a quadratic function of the system matrix $\boldsymbol{\Phi}$.

2.3.6 The Indirect Method

The above results for continuous-time linear systems are the basis for the stability analysis of nonlinear systems of the form

$$\dot{\boldsymbol{x}} = \boldsymbol{A}\boldsymbol{x} + \boldsymbol{g}(\boldsymbol{x}) \tag{2.25}$$

with the associated equilibrium point $\boldsymbol{x}_{\text{eq}} = \boldsymbol{0}$. Here $\boldsymbol{g}(\boldsymbol{x})$ should tend faster to $\boldsymbol{x} = \boldsymbol{0}$ than $|\boldsymbol{x}|$ for $|\boldsymbol{x}| \to 0$. This is the case when

$$\lim_{|\boldsymbol{x}| \to 0} \frac{\boldsymbol{g}(\boldsymbol{x})}{|\boldsymbol{x}|} = \boldsymbol{0} \tag{2.26}$$

holds. For these systems, Lyapunov's stability theory allows us to prove the following theorem [320], which is also referred to as *Lyapunov's indirect method*, *Lyapunov's first method*, or *Lyapunov's linearization method*.

Theorem 17 (Lyapunov's Indirect Method of Stability). *Let the system*

$$\dot{\boldsymbol{x}} = \boldsymbol{A}\boldsymbol{x} + \boldsymbol{g}(\boldsymbol{x}),$$

where $\boldsymbol{g}(x)$ is a continuous function, possess an equilibrium point at $\boldsymbol{x} = \boldsymbol{0}$ and a continuous and unique solution for every initial state vector within a neighborhood of $\boldsymbol{x} = \boldsymbol{0}$. Let it further hold that

$$\lim_{|\boldsymbol{x}| \to 0} \frac{\boldsymbol{g}(\boldsymbol{x})}{|\boldsymbol{x}|} = \boldsymbol{0}.$$

If, in this case, the matrix $\boldsymbol{A}$ has only eigenvalues with a negative real part, the equilibrium point $\boldsymbol{x}_{\text{eq}} = \boldsymbol{0}$ is asymptotically stable and a Lyapunov function

$$V(\boldsymbol{x}) = \boldsymbol{x}^T \boldsymbol{R}\boldsymbol{x}$$

will always exist for $\dot{\boldsymbol{x}} = \boldsymbol{A}\boldsymbol{x} + \boldsymbol{g}(\boldsymbol{x})$, whose matrix $\boldsymbol{R}$ results from

$$\boldsymbol{A}^T \boldsymbol{R} + \boldsymbol{R}\boldsymbol{A} = -\boldsymbol{Q}$$

with an arbitrary positive definite matrix $\boldsymbol{Q}$. If $\boldsymbol{A}$ has one or more eigenvalues with a positive real part, the equilibrium point is unstable. If $\boldsymbol{A}$ does not have any eigenvalues λ_i with a positive real part, but at least one with $\operatorname{Re}\{\lambda_i\} = 0$, the equilibrium point is stable or unstable depending on the form of $\boldsymbol{g}$.

The class of nonlinear systems (2.25) is particularly interesting, because most of the nonlinear systems

$$\dot{\boldsymbol{x}} = \boldsymbol{f}(\boldsymbol{x})$$

with an equilibrium point $\boldsymbol{x}_{\mathrm{eq}} = \mathbf{0}$ can be represented by a Taylor polynomial

$$\dot{\boldsymbol{x}} = \boldsymbol{f}(\boldsymbol{x}) = \underbrace{\boldsymbol{f}(\mathbf{0})}_{\mathbf{0}} + \underbrace{\left.\frac{\partial \boldsymbol{f}}{\partial \boldsymbol{x}}\right|_{\boldsymbol{x}=\mathbf{0}}}_{\boldsymbol{A}} \boldsymbol{x} + \boldsymbol{g}(\boldsymbol{x})$$

with residual $\boldsymbol{g}(\boldsymbol{x})$ and can be defined in terms of equation (2.25) in this way. In these cases, equation (2.26) always holds for the residual $\boldsymbol{g}(\boldsymbol{x})$ of a function $\boldsymbol{f}$ which is representable by a Taylor series. Thus using the above theorem, we can deduce the stability behavior of the equilibrium point $\boldsymbol{x}_{\mathrm{eq}} = \mathbf{0}$ of a differential equation $\dot{\boldsymbol{x}} = \boldsymbol{f}(\boldsymbol{x})$ by its corresponding linearization

$$\dot{\boldsymbol{x}} = \boldsymbol{A}\boldsymbol{x} \quad \text{with} \quad \boldsymbol{A} = \left.\frac{\partial \boldsymbol{f}}{\partial \boldsymbol{x}}\right|_{\boldsymbol{x}=\mathbf{0}}$$

around the equilibrium point $\boldsymbol{x}_{\mathrm{eq}} = \mathbf{0}$. For this reason, the above theorem is of exceptional importance, since it states that the stability behavior of a large class of nonlinear systems can be easily determined based on the corresponding linearized systems. However, it does not guarantee a large catchment region.

2.3.7 Determining Exponential Stability

To prove that a system

$$\dot{\boldsymbol{x}} = \boldsymbol{f}(\boldsymbol{x})$$

has an exponentially stable equilibrium point $\boldsymbol{x}_{\mathrm{eq}} = \mathbf{0}$, positive constants m and α must exist, so that the inequality

$$|\boldsymbol{x}(t)| \leq m e^{-\alpha t} |\boldsymbol{x}(0)| \tag{2.27}$$

is fulfilled. We know this from Definition 10, p. 19. The solution $\boldsymbol{x}(t)$ of the differential equation $\dot{\boldsymbol{x}} = \boldsymbol{f}(\boldsymbol{x})$ tends to the equilibrium point $\boldsymbol{x}_{\mathrm{eq}} = \mathbf{0}$ faster than or equally as fast as the exponential function $m e^{-\alpha t}|\boldsymbol{x}(0)|$. For simplicity's sake, we will write $\boldsymbol{x}$ for $\boldsymbol{x}(t)$ and $\boldsymbol{x}_0$ for the initial value $\boldsymbol{x}(0)$ in the following.

For nonlinear differential equations $\dot{\boldsymbol{x}} = \boldsymbol{f}(\boldsymbol{x})$ it is generally very difficult, if not impossible, to prove exponential stability using the defining equation (2.27), since in most cases we cannot determine the solution $\boldsymbol{x}(t)$ of the differential equation $\dot{\boldsymbol{x}} = \boldsymbol{f}(\boldsymbol{x})$ analytically. Also, in this case, Lyapunov functions often allow us to prove stability. However, the conditions on the Lyapunov function are stricter in the case of exponential stability than in the case of asymptotic stability. The requirements that

$$V(\mathbf{0}) = 0$$

and

$$V(\boldsymbol{x}) > 0, \quad \boldsymbol{x} \neq \mathbf{0},$$

from the Stability Theorem 13 on p. 115, are replaced by the stricter requirements

$$k_1|\boldsymbol{x}|^p \leq V(\boldsymbol{x}) \leq k_2|\boldsymbol{x}|^p, \tag{2.28}$$

where k_1, k_2 and p are positive constants which remain to be determined. Below we will assume that V is continuously differentiable. Now we will replace the condition

$$\dot{V}(\boldsymbol{x}) < 0, \quad \boldsymbol{x} \neq \mathbf{0},$$

of Theorem 13 with the stricter condition

$$\dot{V}(\boldsymbol{x}) \leq -k_3|\boldsymbol{x}|^p, \tag{2.29}$$

where k_3 is once again a positive constant which remains to be determined. From inequality (2.29) and using inequality (2.28), it follows that

$$\dot{V}(\boldsymbol{x}) \leq -\frac{k_3}{k_2}V(\boldsymbol{x}) = -aV(\boldsymbol{x}), \qquad a = \frac{k_3}{k_2}. \tag{2.30}$$

At this point, we will utilize the differentiable form of Grönwall's Lemma[6]. It allows us to deduce the inequality

$$V(\boldsymbol{x}) \leq V(\boldsymbol{x}_0)e^{-at} \tag{2.31}$$

from the differential inequality (2.30).

We will again apply inequality (2.28), which along with inequality (2.31) yields the relationships

$$|\boldsymbol{x}|^p \leq \frac{V(\boldsymbol{x})}{k_1} \leq \frac{V(\boldsymbol{x}_0)}{k_1}e^{-at} \leq \frac{k_2}{k_1}|\boldsymbol{x}_0|^p e^{-at}$$

[6] Grönwall's Lemma states that if a nonnegative, continuously differentiable function $\eta(t)$ fulfills the differential inequality

$$\dot{\eta}(t) \leq c\eta(t) + \Psi(t), \qquad t \in [0, T \in \mathbb{R}],$$

with a constant $c \in \mathbb{R}$ and a nonnegative integrable function Ψ, the Grönwall inequality

$$\eta(t) \leq \left(\eta(0) + \int_0^t \Psi(s)e^{-cs}ds\right)e^{ct}$$

holds for all $t \in [0, T]$ [357]. Note that in the case discussed here, $\boldsymbol{x}$ and therefore V are functions of time t, and it holds that $\Psi(t) = 0$.

and thus

$$|\boldsymbol{x}| \leq \sqrt[p]{\frac{k_2}{k_1}}|\boldsymbol{x}_0|e^{-\frac{a}{p}t} = m|\boldsymbol{x}_0|e^{-\alpha t}, \qquad m = \sqrt[p]{\frac{k_2}{k_1}},\ \alpha = \frac{a}{p}.$$

The defining expression (2.27) for exponential stability is therefore fulfilled by the above inequality if condition (2.28) and condition (2.29) are fulfilled. This leads to the following theorem.

Theorem 18 (Exponential Stability). *Let the system* $\dot{\boldsymbol{x}} = \boldsymbol{f}(\boldsymbol{x})$ *with the equilibrium point* $\boldsymbol{x}_{\text{eq}} = \boldsymbol{0}$ *possess a continuous and unique solution for every initial state vector within a neighborhood* $U_1(\boldsymbol{0})$ *of the origin. If a function* $V(\boldsymbol{x})$ *exists which is continuous in a neighborhood* $U_2(\boldsymbol{0}) \subseteq U_1(\boldsymbol{0})$, *possesses continuous partial derivatives, and positive constants* k_1, k_2, k_3, *and* p *exist such that the inequalities*

(1) $k_1|\boldsymbol{x}|^p \leq V(\boldsymbol{x}) \leq k_2|\boldsymbol{x}|^p$,
(2) $\dot{V}(\boldsymbol{x}) \leq -k_3|\boldsymbol{x}|^p$

are fulfilled at $U_2(\boldsymbol{0})$, *then the equilibrium point* $\boldsymbol{x}_{\text{eq}} = \boldsymbol{0}$ *is exponentially stable.*

If the Conditions (1) and (2) of the theorem are satisfied in the entire state space, the equilibrium point is globally exponentially stable.

2.3.8 Example: Underwater Glider

Underwater gliders are a type of autonomous underwater robot used for long-term missions, such as to measure environmental data or salinity in oceans and seas [178, 183, 263, 294]. Figure 2.59 illustrates a possible structure for such a glider. They consume very little energy and can operate autonomously for years. These gliders are about 2 m long or longer and have wings on the sides of their hull, similar to sailplanes. Their principle of movement is also similar to that of sailplanes. However, the cause of the glider's movement is different. It is based on a density change within the glider. To this end, for example, a hydraulic liquid is pumped back and forth between an internal swim bladder, i. e. one located inside the hull of the glider, and an external one located outside the impermeable part of the hull at the stern.

For example, if the underwater glider is at the surface and the hydraulic liquid is pumped from the external swim bladder to the internal one, the volume of the glider is reduced due to the contraction of the external bladder, and the specific density of the underwater vehicle becomes greater than that of the surrounding water. Simultaneously, the center of gravity of the underwater vehicle moves toward the bow and the glider begins to sink, bow first. The center of gravity can be fine-tuned by a motor which changes the position of the accumulator providing the energy supply for the glider. Due to the wings, which have a similar effect as the wings of an aircraft, the glider does

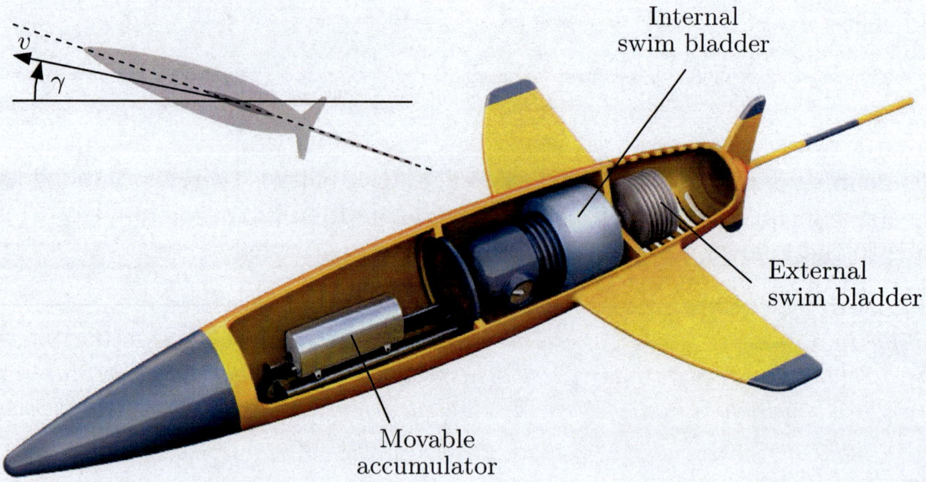

Fig. 2.59: Structure of an underwater glider

not sink vertically; it glides on a slanting course into the depths, as shown in Figure 2.60. After the glider reaches a prespecified depth, the hydraulic liquid is pumped back into the external bladder at the stern. Consequently, the volume of the glider increases while its weight remains constant and the glider begins to rise. This takes place along a slanting upward path with a gliding angle of γ. When the glider has reached the surface, radio frequency signals can be transmitted and then the glider descends again. The resulting motion as a function of time is a sawtooth-like course of movement. A typical velocity for such a glider is approximately $1\,\mathrm{km\,h^{-1}}$.

Below we will present a simplified model of the underwater glider [45, 46]. It describes the relative deviation

$$x_1 = \frac{v - v_\mathrm{s}}{v_\mathrm{s}}$$

of the glider's velocity v from the stationary velocity v_s and the deviation

$$x_2 = \gamma - \gamma_\mathrm{s}$$

of the gliding angle γ from the stationary gliding angle γ_s via

$$\begin{aligned}
\dot{x}_1 &= -\frac{1}{a}\left[a\,(1+x_1)^2 + mg\sin(x_2+\gamma_\mathrm{s})\right], \quad & a &= -mg\sin(\gamma_\mathrm{s}),\\
\dot{x}_2 &= \frac{1}{a(1+x_1)}\left[b\,(1+x_1)^2 - mg\cos(x_2+\gamma_\mathrm{s})\right], \quad & b &= mg\cos(\gamma_\mathrm{s}),
\end{aligned} \tag{2.32}$$

and has an equilibrium point at $\boldsymbol{x}_\mathrm{eq} = \mathbf{0}$. Here m is the mass of the glider and g is the gravitational acceleration. For the deviation, it holds that

$$x_1 \in (x_{1,\min}, x_{1,\max}) \quad \text{with} \quad x_{1,\min} > -1 \quad \text{and} \quad x_2 \in [-\pi, \pi).$$

For the above system, the Lyapunov function

$$V(\boldsymbol{x}) = \frac{2}{3} - (1 + x_1)\cos(x_2) + \frac{1}{3}(1 + x_1)^3 \qquad (2.33)$$

was found in [45]. See also [46]. Using this Lyapunov function, the exponential stability of the equilibrium point $\boldsymbol{x}_{\text{eq}} = \mathbf{0}$ can be proven. For this purpose, we will first consider Condition (1) of Theorem 18, namely

$$k_1|\boldsymbol{x}|^p \leq V(\boldsymbol{x}) \leq k_2|\boldsymbol{x}|^p,$$

where we choose $p = 2$. When we apply the Taylor series of the cosine function,

$$\cos(\alpha) = 1 - \frac{\alpha^2}{2!} + \frac{\alpha^4}{4!} - \frac{\alpha^6}{6!} + \frac{\alpha^8}{8!} - \ldots,$$

equation (2.33) becomes

Fig. 2.60: Sawtooth-like course of movement of an underwater glider

$$
\begin{aligned}
V(\boldsymbol{x}) &= \left(1+\frac{x_1}{3}\right)x_1^2 - (1+x_1)\left(-\frac{x_2^2}{2!}+\frac{x_2^4}{4!}-\frac{x_2^6}{6!}+\frac{x_2^8}{8!}-\ldots\right) \\
&= \left(1+\frac{x_1}{3}\right)x_1^2 + \frac{1+x_1}{2}x_2^2\left[\left(1-\frac{x_2^2}{3\cdot 4}\right)+\frac{2x_2^4}{6!}\left(1-\frac{x_2^2}{7\cdot 8}\right)+\ldots\right].
\end{aligned} \tag{2.34}
$$

Within $-\pi \leq x_2 < \pi$, it holds for the elements of the above series that

$$1-\frac{x_2^2}{3\cdot 4} > 0,\quad 1-\frac{x_2^2}{7\cdot 8} > 0,\quad \cdots,$$

such that

$$V(\boldsymbol{x}) \geq \left(1+\frac{x_1}{3}\right)x_1^2 + \frac{1+x_1}{2}\left(1-\frac{x_2^2}{3\cdot 4}\right)x_2^2 \geq k_1(x_1^2+x_2^2)$$

holds, where the positive constant k_1 is given by

$$k_1 = \min\left\{1+\frac{x_{1,\min}}{3}, \frac{1+x_{1,\min}}{2}\left(1-\frac{\pi^2}{12}\right)\right\},\quad x_{1,\min} > -1.$$

Furthermore, we obtain

$$V(\boldsymbol{x}) = \left(1+\frac{x_1}{3}\right)x_1^2 + \frac{1+x_1}{2}x_2^2 - \frac{1+x_1}{4!}x_2^4\left[\left(1-\frac{x_2^2}{5\cdot 6}\right)+\frac{4!x_2^4}{8!}\left(1-\frac{x_2^2}{9\cdot 10}\right)+\ldots\right]$$

by utilizing equation (2.34). Similar to the case above, it therefore follows for $-\pi \leq x_2 \leq \pi$ that

$$V(\boldsymbol{x}) \leq \left(1+\frac{x_1}{3}\right)x_1^2 + \frac{1+x_1}{2}x_2^2 \leq k_2\left(x_1^2+x_2^2\right)$$

with

$$k_2 = \max\left\{1+\frac{x_{1,\max}}{3}, \frac{1+x_{1,\max}}{2}\right\}.$$

The first condition of Theorem 18 is therefore fulfilled.

Next, the second condition of this theorem, namely

$$\dot{V}(\boldsymbol{x}) \leq -k_3|\boldsymbol{x}|^2, \tag{2.35}$$

is verified with $p = 2$. It holds that

$$
\begin{aligned}
\dot{V}(\boldsymbol{x}) &= \frac{\partial V(\boldsymbol{x})}{\partial \boldsymbol{x}}\dot{\boldsymbol{x}} \\
&= \begin{bmatrix} (1+x_1)^2-\cos(x_2) \\ (1+x_1)\sin(x_2) \end{bmatrix}^T \begin{bmatrix} -\dfrac{1}{a}\left[a(1+x_1)^2+mg\sin(x_2+\gamma_{\mathrm{s}})\right] \\ \dfrac{1}{a(1+x_1)}\left[b(1+x_1)^2-mg\cos(x_2+\gamma_{\mathrm{s}})\right] \end{bmatrix} \\
&= \frac{1}{a}\Big[-a(1+x_1)^4+a(1+x_1)^2\cos(x_2)-mg(1+x_1)^2\sin(x_2+\gamma_{\mathrm{s}}) \\
&\quad + b(1+x_1)^2\sin(x_2) + mg\underbrace{(\cos(x_2)\sin(x_2+\gamma_{\mathrm{s}})-\cos(x_2+\gamma_{\mathrm{s}})\sin(x_2))}_{\sin(\gamma_{\mathrm{s}})}\Big],
\end{aligned}
$$

and with $a = -mg\sin(\gamma_s)$ and $b = mg\cos(\gamma_s)$ from equation (2.32), it follows that

$$\begin{aligned}\dot{V}(\boldsymbol{x}) &= \frac{1}{a}\left[-a(1+x_1)^4 + a(1+x_1)^2\cos(x_2)\right.\\ &\qquad\left. + mg(1+x_1)^2\underbrace{(\cos(\gamma_s)\sin(x_2) - \sin(x_2+\gamma_s))}_{-\sin(\gamma_s)\cos(x_2)} - a\right]\\ &= \frac{1}{a}\left[-a(1+x_1)^4 + 2a(1+x_1)^2\cos(x_2) - a\right]\\ &= -1 + 2(1+x_1)^2\cos(x_2) - (1+x_1)^4. \end{aligned} \tag{2.36}$$

In the following, we make use of

$$\sin^2\left(\frac{\alpha}{2}\right) = \frac{1}{2}(1-\cos(\alpha)) \quad \text{and} \quad \cos(\alpha) = 1 - 2\sin^2\left(\frac{\alpha}{2}\right)$$

in equation (2.36) to obtain

$$\begin{aligned}\dot{V}(\boldsymbol{x}) &= -1 + 2(1+x_1)^2 - (1+x_1)^4 - 4(1+x_1)^2\sin^2\left(\frac{x_2}{2}\right)\\ &= -\left[x_1(x_1+2)\right]^2 - 4(1+x_1)^2\sin^2\left(\frac{x_2}{2}\right). \end{aligned} \tag{2.37}$$

Obviously $\dot{V}(\boldsymbol{x}) < 0$ holds for $-\pi \le x_2 < \pi$; thus the equilibrium point is asymptotically stable according to Lyapunov's stability theorem on p. 115. To prove exponential stability, however, the stricter condition (2.35) must be met.

Taking into account that

$$\sin^2\left(\frac{x_2}{2}\right) = \sin^2\left(\frac{|x_2|}{2}\right)$$

holds, we will first evaluate the Taylor series for $-\pi \le x_2 < \pi$

$$\begin{aligned}\sin\left(\frac{|x_2|}{2}\right) &= \frac{|x_2|}{2} - \frac{1}{3!}\frac{|x_2|^3}{2^3} + \frac{1}{5!}\frac{|x_2|^5}{2^5}\left(1 - \frac{|x_2|^2}{2^2\cdot 6\cdot 7}\right) + \ldots\\ &\ge \frac{|x_2|}{2} - \frac{1}{3!}\frac{|x_2|^3}{2^3} = \frac{|x_2|}{2}\left(1 - \frac{|x_2|^2}{24}\right)\\ &\ge \frac{|x_2|}{2}\left(1 - \frac{\pi^2}{24}\right) \end{aligned}$$

and deduce that

$$\sin^2\left(\frac{x_2}{2}\right) \ge \frac{1}{4}\left(1 - \frac{\pi^2}{24}\right)^2 x_2^2.$$

Thus $\dot{V}(\boldsymbol{x})$ from equation (2.37) can be estimated by

$$\dot{V}(\boldsymbol{x}) \leq -(2+x_1)^2 x_1^2 - \left(1 - \frac{\pi^2}{24}\right)^2 (1+x_1)^2 x_2^2$$
$$\leq \underbrace{-\left(1 - \frac{\pi^2}{24}\right)^2 (1+x_{1,\min})^2}_{k_3} |\boldsymbol{x}|^2,$$

and the second condition for exponential stability, the inequality (2.35), is also fulfilled. Therefore the glider always tends exponentially from an initial deflection or a disturbance to the equilibrium point $\boldsymbol{x}_{\mathrm{eq}} = \boldsymbol{0}$, i. e. it dives on a path at a stationary velocity v_{s} and stationary gliding angle γ_{s}.

2.3.9 Catchment Regions

So far the stability of an equilibrium point has been analyzed. In practice, whether an equilibrium point is stable is not the only important question. We are also interested in the largest region containing the equilibrium point for which all trajectories beginning in this region tend to this point. This region, the *maximal catchment region*, is also called the *region of asymptotic stability* or *basin* of an equilibrium point. Figure 2.61 depicts this region for the system (1.10) that was discussed in Section 1.1.6, p. 11 et seq., and Section 2.3.2, p. 118. The curves, or for higher dimensions the hypersurfaces, which separate the stable courses of trajectories from unstable ones are termed *separatrix*. In Figure 2.61 the separatrix is defined by the two trajectories which tend to the unstable equilibrium point

$$\boldsymbol{x}_{\mathrm{eq2}} = \begin{bmatrix} 1 \\ 1 \end{bmatrix}.$$

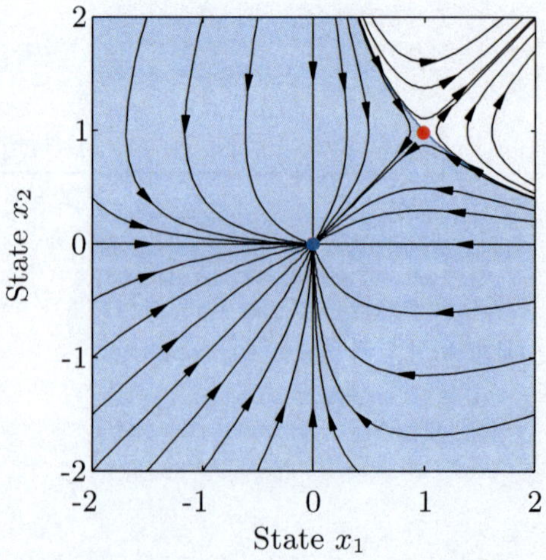

Fig. 2.61: Maximal catchment region (blue) of the exemplary system (1.10) on p. 12

If the region of asymptotic stability of a locally asymptotically stable equilibrium point is very small, the equilibrium point cannot be called asymptotically stable in practice. The proof of stability of an equilibrium point is therefore not sufficient in practice. We must also consider the vicinity of the equilibrium point, and the maximal catchment region must be large enough to include the starting points of all trajectories that are of interest.

In general, the region of asymptotic stability of an equilibrium point cannot be determined analytically. However, subsets of the region of asymptotic stability can be determined if we have found a Lyapunov function. Such a subset is a catchment region if all trajectories which start in this region never leave it and tend to the equilibrium point $\boldsymbol{x}_{\mathrm{eq}} = \boldsymbol{0}$. These catchment regions are bounded by the contour lines of a Lyapunov function, for example. As mentioned earlier, they are often referred to as *Lyapunov sets*, *Lyapunov regions*, or *contractive positively invariant sets*. More specifically, we obtain [462]

Theorem 19 (Catchment Region). *If $V(\boldsymbol{x})$ is a Lyapunov function for the system $\dot{\boldsymbol{x}} = \boldsymbol{f}(\boldsymbol{x})$ with the asymptotically stable equilibrium point $\boldsymbol{x}_{\mathrm{eq}} = \boldsymbol{0}$, the region*

$$G = \{\boldsymbol{x} \in I\!R^n \mid V(\boldsymbol{x}) < c\}, \quad c \in I\!R^+,$$

is bounded, and

$$\dot{V}(\boldsymbol{x}) < 0$$

holds everywhere in $G \setminus \{\boldsymbol{0}\}$, then G is a catchment region of the equilibrium point $\boldsymbol{x}_{\mathrm{eq}} = \boldsymbol{0}$.

With no loss of generality, the open set G in Theorem 19 can be replaced by a closed set

$$G = \{\boldsymbol{x} \in I\!R^n \mid V(\boldsymbol{x}) \le c\}.$$

Theorem 19 maintains its validity in this case.

As can be seen in Figure 2.53, the condition $\dot{V}(\boldsymbol{x}) < 0$ ensures that V decreases along all trajectories. Thus none of the trajectories can leave the region G. Since $\dot{V}(\boldsymbol{x}) < 0$ holds everywhere in this region and G is bounded, they also tend to zero.

The condition that the region G must be bounded is essential. If it is not fulfilled, G is not necessarily a catchment region, as shown in the following example. We will examine the function

$$V(\boldsymbol{x}) = \frac{x_1^2}{1 + x_1^2} + 2x_2^2. \tag{2.38}$$

This function is not radially unbounded, i. e. it does not hold that $V(\boldsymbol{x}) \to \infty$ for $|\boldsymbol{x}| \to \infty$, and thus there are also regions

$$G = \{\boldsymbol{x} \in I\!R^2 \mid V(\boldsymbol{x}) < c\}$$

which are unbounded, i. e. the contour lines are not closed in every case; rather, they tend to infinity for values $c \geq 1$. The graph of the contour lines of this function is shown in Figure 2.62.

There are systems with an equilibrium point $\boldsymbol{x}_{\text{eq}} = \boldsymbol{0}$ for which the function (2.38) is a Lyapunov function V and $\dot{V}(\boldsymbol{x}) < 0$ holds everywhere in $\mathbb{R}^n \backslash \{\boldsymbol{0}\}$. Nevertheless, trajectories may tend to infinity along the contour lines of $V(\boldsymbol{x})$ with $c > 1$, even though along the contour lines it holds that $\dot{V}(\boldsymbol{x}) < 0$. Figure 2.62 shows such a trajectory. Obviously, in this case, G is not a catchment region of the equilibrium point $\boldsymbol{x}_{\text{eq}} = \boldsymbol{0}$.

Such a system is [44]

$$\begin{aligned} \dot{x}_1 &= -x_1 + 2x_1^3 x_2^2, \\ \dot{x}_2 &= -x_2. \end{aligned} \tag{2.39}$$

Its trajectories are shown in Figure 2.63. For this system, which has a single equilibrium point at $\boldsymbol{x}_{\text{eq}} = \boldsymbol{0}$, and the function (2.38), the inequality

$$\dot{V}(\boldsymbol{x}) = -\frac{2x_1^2 + 4x_2^2\left(1 + 2x_1^2\right)}{\left(1 + x_1^2\right)^2} < 0$$

holds for all $\boldsymbol{x} \in \mathbb{R}^2 \backslash \{\boldsymbol{0}\}$. Thus, the function (2.38) is a Lyapunov function for system (2.39), and the equilibrium point $\boldsymbol{x}_{\text{eq}} = \boldsymbol{0}$ is asymptotically stable. However, as we will soon see, it is not globally asymptotically stable. Note that the latter cannot be verified by means of the Lyapunov function (2.38), because it is not radially unbounded. Radial unboundedness, however, is a necessary condition of a Lyapunov function V to prove global stability, as required in equation (2.14) on p. 115.

The solutions of the system (2.39) are given by

$$\begin{aligned} x_1(t) &= \frac{x_{10}e^t}{\sqrt{x_{10}^2 x_{20}^2 + \left(1 - x_{10}^2 x_{20}^2\right)e^{4t}}}, \\ x_2(t) &= x_{20}e^{-t}, \end{aligned} \tag{2.40}$$

where $x_{10} = x_1(0)$ and $x_{20} = x_2(0)$. From (2.40), we can directly deduce that the trajectory $\boldsymbol{x}(t)$ tend to the equilibrium point $\boldsymbol{x}_{\text{eq}} = \boldsymbol{0}$ for $x_{10}^2 x_{20}^2 < 1$ and to infinity for $x_{10}^2 x_{20}^2 \geq 1$. The initial values which fulfill the equation $x_{10}^2 x_{20}^2 = 1$, i. e.

$$x_{20} = \pm\frac{1}{x_{10}},$$

make up the separatrix, which represents the boundary between stable and unstable trajectories. The trajectories that lie on the separatrix itself are unstable and are given by

$$\begin{aligned} x_1(t) &= x_{10}e^t, \\ x_2(t) &= x_{20}e^{-t}. \end{aligned}$$

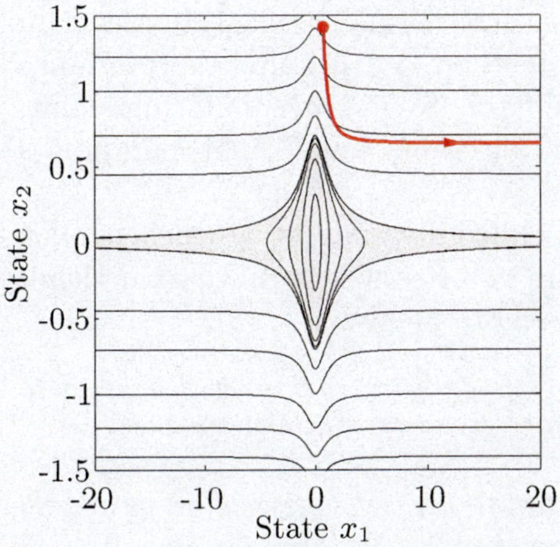

Fig. 2.62: The contour lines of the Lyapunov function (2.38). The region with closed contour lines is gray, an unstable trajectory of the system (2.39) is shown in red.

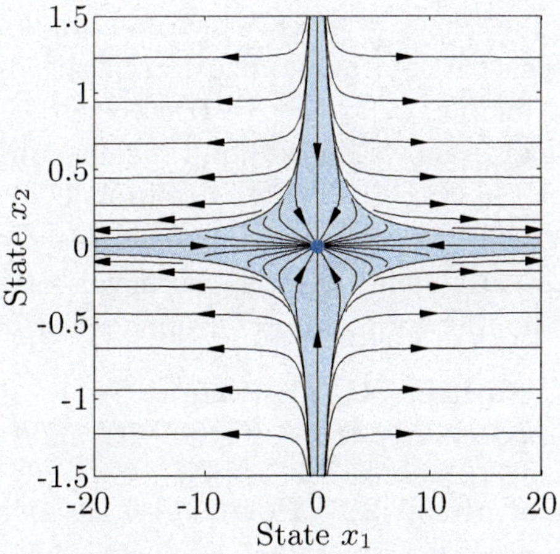

Fig. 2.63: Trajectories $\boldsymbol{x}(t)$ of the system (2.39). The region of asymptotic stability $G_{\max}$ is shown in blue. All trajectories outside of $G_{\max}$ are unstable.

In the case where $x_{10}^2 x_{20}^2 > 1$, all trajectories tend to infinity in finite time. The denominator of equation (2.40), in this case, for the finite escape time

$$t_{\mathrm{e}} = \frac{1}{4} \ln \left(\frac{x_{10}^2 x_{20}^2}{x_{10}^2 x_{20}^2 - 1} \right)$$

becomes identical to zero[7] and therefore x_1 becomes infinitely large.

The maximal catchment region, the region of asymptotic stability, is thus given by

$$G_{\mathrm{RAS}} = \left\{ \boldsymbol{x} \in I\!R^2 \mid x_1^2 x_2^2 < 1 \right\}.$$

Figure 2.63 not only illustrates the course of the trajectories $\boldsymbol{x}(t)$ but also the region G_{RAS}, whose boundary is the separatrix of the system. Figure 2.62 also shows an unstable trajectory, which has its starting point at $x_{10} = x_{20} = 1.5$. This illustrates the previously mentioned case in which $V(\boldsymbol{x})$ decreases along a trajectory, which nevertheless tends to infinity.

2.3.10 LaSalle's Invariance Principle

In addition to the direct Lyapunov method, there is a more general and more powerful method for analyzing the convergence of the solutions of a differential

[7] The solution (2.40) of the system (2.39) is only defined for the time interval $[-\infty, t_{\mathrm{e}})$. For times $t \geq t_{\mathrm{e}}$ no solution exists. This is because the right side of the differential equation (2.39) is locally but not globally Lipschitz continuous. See Theorems 4 and 5 on p. 40.

equation $\dot{\boldsymbol{x}} = \boldsymbol{f}(\boldsymbol{x})$, which is LaSalle's invariance principle [261]. In a certain sense, it is a generalization of the direct method. With LaSalle's invariance principle, we are not restricted to the stability analysis of an equilibrium point; it also allows for a simultaneous consideration of multiple equilibrium points or the stability analysis of limit cycles.

The invariance principle is based on so-called invariant sets, which, similar to catchment regions, are never left by trajectories which run within them. The following two definitions describe this more precisely.

Definition 21 (Invariant Set). *A set G is called invariant with respect to a system $\dot{\boldsymbol{x}} = \boldsymbol{f}(\boldsymbol{x})$ if $\boldsymbol{x}(t) \in G$ holds for all $\boldsymbol{x}(0) \in G$ and all $-\infty < t < \infty$.*

Definition 22 (Positively Invariant Set). *A set G is called positively invariant with respect to a system $\dot{\boldsymbol{x}} = \boldsymbol{f}(\boldsymbol{x})$ if $\boldsymbol{x}(t) \in G$ holds for all $\boldsymbol{x}(0) \in G$ and $t \geq 0$.*

As specified in Definition 21, the trajectories whose courses lie within an invariant set are inside this set for the past, $t < 0$, the present, $t = 0$, and the future, $t > 0$. Trajectories thus never enter an invariant set; they start inside it. In contrast to invariant sets, trajectories may enter positively invariant sets for $t \leq 0$, which they will no longer leave. Every invariant set is also a positively invariant set. An invariant or positively invariant set for which all trajectories inside it tend to an equilibrium point is a catchment region of the equilibrium point, synonymously called a *contractive positively invariant set*, as mentioned before. We are using the term catchment region here, since it is shorter and more intuitive.

For example, if V is a Lyapunov function for an equilibrium point $\boldsymbol{x}_{\mathrm{eq}} = \boldsymbol{0}$ of a system $\dot{\boldsymbol{x}} = \boldsymbol{f}(\boldsymbol{x})$, it follows that

$$G = \{\boldsymbol{x} \in I\!R^n \mid V(\boldsymbol{x}) < c\}$$

is a positively invariant set.

Building upon the concept of invariant and positively invariant sets, we can state [261]

Theorem 20 (LaSalle's Invariance Principle). *Let $\dot{\boldsymbol{x}} = \boldsymbol{f}(\boldsymbol{x})$ be a system with a compact positively invariant set Ω, and let $V_{\mathrm{LaSalle}}(\boldsymbol{x})$ be a continuously differentiable function with*

$$\dot{V}_{\mathrm{LaSalle}}(\boldsymbol{x}) \leq 0$$

for all $\boldsymbol{x} \in \Omega$. Further, let N denote the set of all points $\boldsymbol{x} \in \Omega$ with

$$\dot{V}_{\mathrm{LaSalle}}(\boldsymbol{x}) = 0$$

and let M denote the largest invariant set in N. In this case all solutions $\boldsymbol{x}(t)$ that start within Ω tend to the set M for $t \to \infty$.

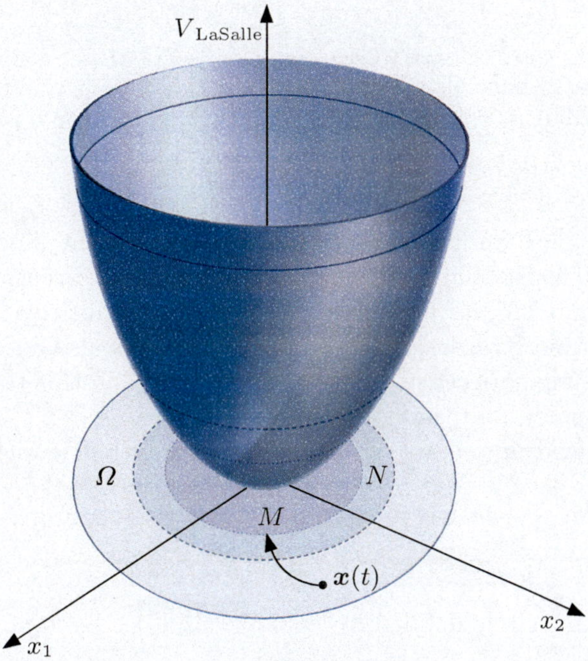

Fig. 2.64: The function V_{LaSalle}, the positively invariant set Ω, the set N, and the invariant set M, to which the trajectory $\boldsymbol{x}(t)$ tends

Figure 2.64 illustrates the sets Ω, N, and M, as well as the LaSalle function V_{LaSalle}. The invariance principle does not require the function V_{LaSalle} to be positive definite and $V_{\text{LaSalle}}(\mathbf{0}) = 0$ to hold, as was the case for Lyapunov's direct method. However, the approach using a positive definite function V_{LaSalle} is reasonable, since it then holds that

$$\Omega = \{\boldsymbol{x} \in I\!R^n \mid V_{\text{LaSalle}}(\boldsymbol{x}) \leq c\} \tag{2.41}$$

is a positively invariant set of the system, assuming that the inequality

$$\dot{V}_{\text{LaSalle}}(\boldsymbol{x}) \leq 0$$

holds. Note that the set (2.41) is closed, but not necessarily bounded. It is compact only if it is closed and bounded. Since this property is required in Theorem 20, we have to verify the boundedness of set (2.41) in the use of this theorem.

A special case of the invariance principle occurs when only the solution

$$\boldsymbol{x}(t) = \mathbf{0} \in \Omega$$

fulfills the equation

$$\dot{V}_{\text{LaSalle}}(\boldsymbol{x}) = 0,$$

i. e. when

$$N = \{\mathbf{0}\}$$

holds, and everywhere else $\dot{V}_{\text{LaSalle}}(\boldsymbol{x}) < 0$ is fulfilled. Obviously, the trivial solution $\boldsymbol{x}(t) = \mathbf{0}$ of the differential equation in this case is also the largest invariant set, i. e.

$$M = N = \{\mathbf{0}\}.$$

Thus $\boldsymbol{x} = \mathbf{0}$ is an asymptotically stable equilibrium point, since all trajectories tend to zero. This special case equates to Lyapunov's direct method, given in Theorem 13 on p. 115, if we assume that V_{LaSalle} is a positive definite function.

Barbashin and Krasovskii's Theorem on p. 116 is also a special case of the invariance principle, because $M = \{\mathbf{0}\}$ is stipulated in this theorem, although it does not usually hold that $N = \{\mathbf{0}\}$.

As a further example, we will examine the limit cycle of the system

$$\begin{aligned} \dot{x}_1 &= -x_2 + x_1(1 - x_1^2 - x_2^2), \\ \dot{x}_2 &= x_1 + x_2(1 - x_1^2 - x_2^2), \end{aligned} \tag{2.42}$$

which is the unit circle defined by

$$x_1^2 + x_2^2 = 1. \tag{2.43}$$

It is depicted in Figure 2.65. We can show that the unit circle is a limit cycle of the system by inserting equation (2.43) into equation (2.42). The derivatives of x_1 and x_2 on the unit circle are then

$$\begin{aligned} \dot{x}_1 &= -x_2, \\ \dot{x}_2 &= x_1. \end{aligned} \tag{2.44}$$

We can prove that the unit circle is a closed trajectory by means of the normal vectors

$$\boldsymbol{n} = \frac{\partial(x_1^2 + x_2^2)}{\partial \boldsymbol{x}} = 2\begin{bmatrix} x_1 \\ x_2 \end{bmatrix}, \quad \boldsymbol{x} \in \left\{\boldsymbol{x} \in I\!R^2 \mid x_1^2 + x_2^2 = 1\right\},$$

of the circle, i. e. the vectors which are perpendicular to the boundary of the circle. Then, applying equation (2.44), we obtain

$$\boldsymbol{n}^T \dot{\boldsymbol{x}} = 0,$$

so that $\dot{\boldsymbol{x}}$ is perpendicular to $\boldsymbol{n}$. Thus the circle forms a trajectory. This means the unit circle is an invariant set for the example in question.

To determine the sets Ω, N, and M, we will choose the function

$$V_{\text{LaSalle}}(\boldsymbol{x}) = (x_1^2 + x_2^2 - 1)^2,$$

for which

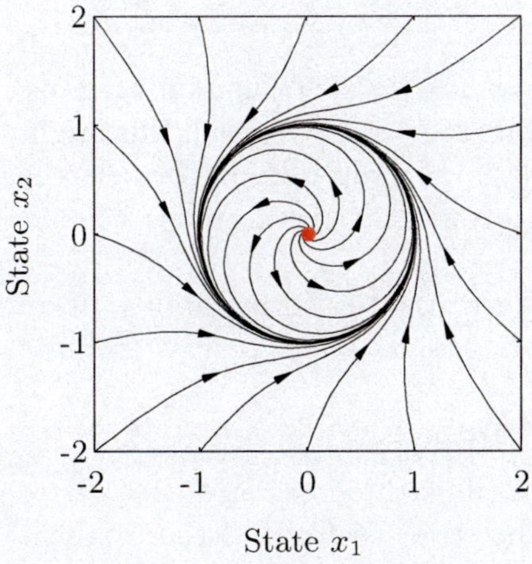

Fig. 2.65: Trajectories and limit cycle of the system (2.42)

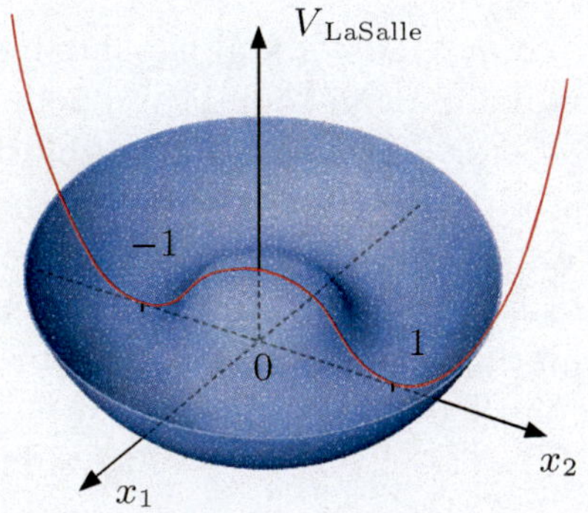

Fig. 2.66: Shape of the LaSalle function $V_{\text{LaSalle}}(\boldsymbol{x}) = (x_1^2 + x_2^2 - 1)^2$

$$V_{\text{LaSalle}}(\boldsymbol{x}) \geq 0$$

holds, but not

$$V_{\text{LaSalle}}(\mathbf{0}) = 0.$$

Instead of the latter, all points of the unit circle satisfy the equation $V_{\text{LaSalle}}(\boldsymbol{x}) = 0$, i. e.

$$V_{\text{LaSalle}}(\boldsymbol{x}) = 0 \text{ for all } \boldsymbol{x} \in \{\boldsymbol{x} \in \mathbb{R}^2 \mid x_1^2 + x_2^2 - 1 = 0\}.$$

Furthermore, the function V_{LaSalle} shown in Figure 2.66 has its minima on the unit circle. For the time derivative, we obtain

$$\dot{V}_{\text{LaSalle}}(\boldsymbol{x}) = -4(x_1^2 + x_2^2)(x_1^2 + x_2^2 - 1)^2.$$

On the unit circle and in $\boldsymbol{x} = \mathbf{0}$, the derivative of the LaSalle function becomes

$$\dot{V}_{\text{LaSalle}}(\boldsymbol{x}) = 0.$$

We are leaving the point $\boldsymbol{x} = \mathbf{0}$ out of our calculations. Taking this exception into account, we obtain the sets

$$\begin{aligned}
&\Omega = \{\boldsymbol{x} \in \mathbb{R}^2 \mid V_{\text{LaSalle}}(\boldsymbol{x}) = (x_1^2 + x_2^2 - 1)^2 \leq 1 - \varepsilon\}, \quad 1 \gg \varepsilon > 0,\\
&N = \{\boldsymbol{x} \in \mathbb{R}^2 \mid x_1^2 + x_2^2 = 1\},\\
&M = N,
\end{aligned}$$

where ε is an arbitrarily small positive number.

By means of Theorem 20 and the result above, we can calculate that all trajectories $\boldsymbol{x}(t)$ which begin outside of N tend to the set $M = N$, i. e. the unit circle, in its further progression. We have thus verified that the unit circle is an asymptotically stable limit cycle.

2.3.11 Instability Criterion

All the methods of stability analysis we have described so far aim to prove the stability of an equilibrium point. If stability cannot be verified, this might be because the method of stability analysis is not suitable or because the equilibrium point is unstable. Although the latter is a trivial reason, it is not always an immediately obvious one.

In such a case, a useful approach is to investigate the instability of an equilibrium point. This is possible by showing that, for a positive definite function $V(\boldsymbol{x})$,

$$\dot{V}(\boldsymbol{x}) = \dot{\boldsymbol{x}}^T \operatorname{grad}(V(\boldsymbol{x})) > 0$$

holds around the equilibrium point $\boldsymbol{x}_{\text{eq}} = \boldsymbol{0}$, as illustrated in Figure 2.67. Here we realize that the temporal change $\dot{\boldsymbol{x}}$ of the state vector and the gradient of the function V both point away from the equilibrium point $\boldsymbol{x}_{\text{eq}} = \boldsymbol{0}$ in all cases. The trajectories $\boldsymbol{x}(t)$ therefore tend away from the equilibrium point. Accordingly, we can formulate the following theorem [479], which is a reversal of Lyapunov's stability theorem.

Theorem 21 (Instability). *Let the differential equation* $\dot{\boldsymbol{x}} = \boldsymbol{f}(\boldsymbol{x})$ *with the equilibrium point* $\boldsymbol{x}_{\text{eq}} = \boldsymbol{0}$ *possess a continuous and unique solution for every initial state vector within a neighborhood* $U_1(\boldsymbol{0})$ *of the origin. If a function* $V(\boldsymbol{x})$ *exists which possesses continuous partial derivatives and fulfills the conditions*

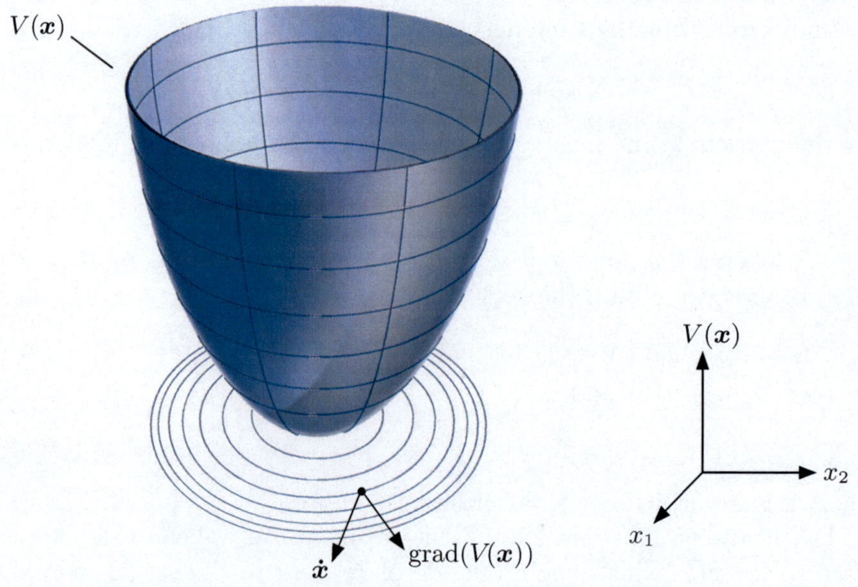

Fig. 2.67: Illustration of the equation $\dot{V}(\boldsymbol{x}) = \dot{\boldsymbol{x}}^T \operatorname{grad}(V(\boldsymbol{x})) > 0$

(1) $V(\mathbf{0}) = 0$,
(2) $V(\boldsymbol{x}) > 0$ *for* $\boldsymbol{x} \neq \mathbf{0}$,
(3) $\dot{V}(\boldsymbol{x}) > 0$ *for* $\boldsymbol{x} \neq \mathbf{0}$

in a neighborhood $U_2(\mathbf{0}) \subseteq U_1(\mathbf{0})$, *then the equilibrium point* $\boldsymbol{x}_{\mathrm{eq}} = \mathbf{0}$ *is unstable.*

This theorem, however, cannot prove the instability of an equilibrium point if in addition to the trajectories progressing away from it, there are trajectories that tend to it. Theorems that allow for such an analysis can be found in [170, 389]. These are not often used in practice, however.

2.4 Passivity and Stability

2.4.1 Passive Systems

Passive systems, as indicated by their name, do not have an internal energy source such as a battery; they are only excitable via their inputs. Such systems frequently occur in practice. For example, mechanical systems without propulsion or electrical circuits consisting only of resistors, capacitors, and coils are always passive systems. In order to define the term *passivity* of a system more precisely and at the same time provide an illustrative explanation, let us consider a simple linear system consisting of a coil L and a resistor R, as shown in Figure 2.68.

The energy balance of the system is of the form

$$\underbrace{\int_0^t R i_{\mathrm{R}}^2(\tau) d\tau}_{\substack{\text{consumed}\\ \text{energy}}} + \underbrace{\frac{1}{2} L i_{\mathrm{L}}^2(t) - \frac{1}{2} L i_{\mathrm{L}}^2(0)}_{\substack{\text{stored}\\ \text{energy}}} = \underbrace{\int_0^t u(\tau) i(\tau) d\tau}_{\substack{\text{supplied}\\ \text{energy}}}.$$

Since the consumed energy always has to be positive (or in the ideal but not real case equal to zero), obviously the law

$$\text{stored energy} \leq \text{supplied energy}$$

holds. If we introduce the storage function

$$S(t) = \frac{1}{2} L i_{\mathrm{L}}^2(t)$$

for the stored energy, it holds with $y = i$ that

$$S(t) - S(0) \leq \int_0^t u(\tau) y(\tau) d\tau.$$

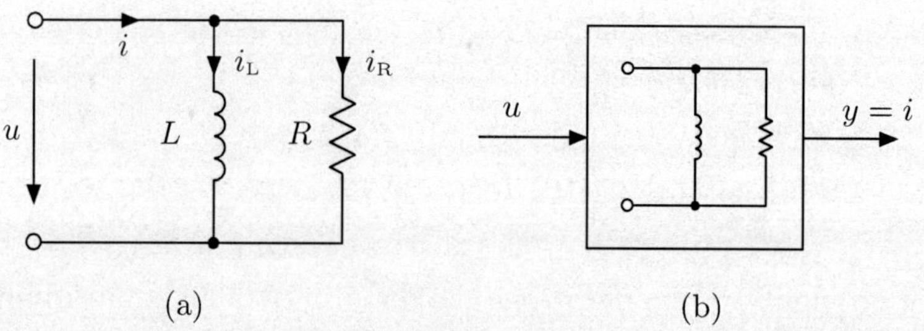

Fig. 2.68: RL circuit as a) a circuit diagram and b) a block diagram with input variable u and output variable $y = i$

We can now generalize the above law, which states that the stored energy always has to be smaller than or equal to the energy supplied, to arbitrary systems including MIMO systems

$$\begin{aligned}\dot{\boldsymbol{x}} &= \boldsymbol{f}(\boldsymbol{x}, \boldsymbol{u}),\\ \boldsymbol{y} &= \boldsymbol{g}(\boldsymbol{x}, \boldsymbol{u})\end{aligned}$$

with the restriction that $\dim(\boldsymbol{y}) = \dim(\boldsymbol{u})$, i.e. it holds that

$$S(\boldsymbol{x}(t)) - S(\boldsymbol{x}(0)) \leq \int_0^t \boldsymbol{u}(\tau)^T \boldsymbol{y}(\tau)\, d\tau, \tag{2.45}$$

where $S(\boldsymbol{x}(t))$ describes the total energy contained in the system. Since this is positive, S is a function with $S(\boldsymbol{x}(t)) \geq 0$, where with no loss of generality we can assume $S(\boldsymbol{0}) = 0$. In this context, we use

Definition 23 (Positive and Negative Definite Functions). *A function* $v(\boldsymbol{x})$ *with* $v(\boldsymbol{0}) = 0$ *is called*

(1) negative semidefinite if $v(\boldsymbol{x}) \leq 0$,
(2) negative definite if $v(\boldsymbol{x}) < 0$ *for all* $\boldsymbol{x} \neq \boldsymbol{0}$,
(3) positive semidefinite if $v(\boldsymbol{x}) \geq 0$,
(4) positive definite if $v(\boldsymbol{x}) > 0$ *for all* $\boldsymbol{x} \neq \boldsymbol{0}$.

The storage function S is therefore at least positive semidefinite. Lyapunov functions, which we described in the previous section and, as we will soon see, are closely related to storage functions, are positive definite. Below we keep in mind that the variables depend on the time, and we will write them in simplified form without explicitly referring to the time t, e.g. $\boldsymbol{x}$ instead of $\boldsymbol{x}(t)$.

We can now drop the restriction which states that the function S must describe the total energy of the system and regard S as a general positive semidefinite function.

The inequality (2.45) can be transformed to

$$\dot{S}(\boldsymbol{x}) = \frac{\partial S(\boldsymbol{x})}{\partial \boldsymbol{x}} \dot{\boldsymbol{x}} \leq \boldsymbol{u}^T \boldsymbol{y},$$

since equation (2.45) must hold for all $t \geq 0$. Combining this with the above results, we arrive at the following definition of passivity, strict passivity, and losslessness.

Definition 24 (Passivity, Strict Passivity, and Losslessness). *Let a system be defined by*

$$\begin{aligned} \dot{\boldsymbol{x}} &= \boldsymbol{f}(\boldsymbol{x}, \boldsymbol{u}), \\ \boldsymbol{y} &= \boldsymbol{g}(\boldsymbol{x}, \boldsymbol{u}) \end{aligned}$$

with $m = \dim(\boldsymbol{u}) = \dim(\boldsymbol{y})$. *If a continuously differentiable, positive semidefinite function* $S(\boldsymbol{x})$ *exists, and where required a positive definite function* $R(\boldsymbol{x})$, *such that for all* $\boldsymbol{x} \in I\!R^n$ *and* $\boldsymbol{u} \in I\!R^m$

(1) $\dot{S}(\boldsymbol{x}) \leq \boldsymbol{u}^T \boldsymbol{y}$ *holds, the system is called passive;*
(2) $\dot{S}(\boldsymbol{x}) + R(\boldsymbol{x}) \leq \boldsymbol{u}^T \boldsymbol{y}$ *holds, the system is called strictly passive;*
(3) $\dot{S}(\boldsymbol{x}) = \boldsymbol{u}^T \boldsymbol{y}$ *holds, the system is called lossless.*

It follows from the definition that strictly passive systems and lossless systems are also always passive.

If we examine the special case of a static system

$$\boldsymbol{y} = \boldsymbol{g}(\boldsymbol{u}),$$

the state vector $\boldsymbol{x}$ is not present. In this case, we can see that $S(\boldsymbol{x})$ is independent of $\boldsymbol{x}$ and therefore it is constant. Thus it holds that

$$\dot{S}(\boldsymbol{x}) = 0$$

and a static system, i. e. a characteristic curve or a characteristic diagram, is passive if

$$0 \leq \boldsymbol{u}^T \boldsymbol{y}$$

holds.

For the scalar case, this equation simplifies to

$$0 \leq uy, \tag{2.46}$$

from which it immediately follows that

$$\operatorname{sgn}(u) = \operatorname{sgn}(y).$$

All characteristic curves which satisfy equation (2.46) therefore lie within the first and third quadrants and pass through the origin, as illustrated in Figure 2.69. The limiting element, the dead zone, and the three-position element are passive, for example.

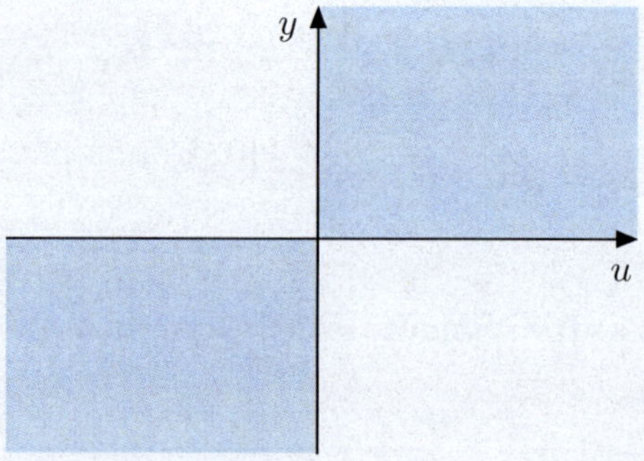

Fig. 2.69: Characteristic elements $y = g(u)$, whose characteristic curves lie entirely in the blue shaded sector, are passive.

2.4.2 Stability of Passive Systems

The closeness of the storage function $S(\boldsymbol{x})$ in the definition of passivity to a Lyapunov function of the passive system is obvious: inserting $\boldsymbol{u} = \boldsymbol{0}$, which we have required for an equilibrium point at $\boldsymbol{x} = \boldsymbol{0}$, it follows from Definition 24 of passivity that

$$\dot{S}(\boldsymbol{x}) \leq 0.$$

This is precisely the central requirement of Theorem 13, i. e. Lyapunov's stability theorem on p. 115. However, there is a small but important difference: the storage function $S(\boldsymbol{x})$ from Definition 24 need only be positive semidefinite, while a Lyapunov function is required to be positive definite. If we also demand that the latter be fulfilled for the storage function, the conditions of Theorem 13 are fulfilled, and we obtain

Theorem 22 (Stability of Passive Systems). *A passive system with a positive definite storage function S possesses an equilibrium point $\boldsymbol{x}_{\mathrm{eq}} = \boldsymbol{0}$ which is Lyapunov stable.*

If the storage function is radially unbounded, the equilibrium point is even globally Lyapunov stable. If the storage function is not positive definite, its stability is not guaranteed. A passive system can thus also be unstable. An example of such a case is the system

$$\dot{\boldsymbol{x}} = \begin{bmatrix} -1 & 0 \\ 0 & 1 \end{bmatrix} \boldsymbol{x} + \begin{bmatrix} 1 \\ 0 \end{bmatrix} u,$$
$$y = \begin{bmatrix} 1 \; 0 \end{bmatrix} \boldsymbol{x},$$

for which the state variable $x_2(t)$ is unobservable and unstable. A positive semidefinite storage function for this system is

$$S(\boldsymbol{x}) = \frac{1}{2} \boldsymbol{x}^T \begin{bmatrix} 1 & 0 \\ 0 & 0 \end{bmatrix} \boldsymbol{x}.$$

In this case, the passivity inequality

$$\dot{S}(\boldsymbol{x}) = \boldsymbol{x}^T \begin{bmatrix} 1 & 0 \\ 0 & 0 \end{bmatrix} \dot{\boldsymbol{x}} = x_1 \dot{x}_1 = -x_1^2 + u x_1 \le u y = u x_1$$

from Definition 24, which is equivalent to

$$-x_1^2 \le 0,$$

is fulfilled. The system is therefore passive.

We will now return to the general case. If, when using a positive definite storage function, we require that the set

$$\{\boldsymbol{x} \in I\!R^n \mid \dot{S}(\boldsymbol{x}) = 0\}$$

does not contain a trajectory $\boldsymbol{x}(t)$ except for $\boldsymbol{x}(t) = \boldsymbol{0}$, this means that the conditions from the Barbashin and Krasovskii stability theorem, i. e. Theorem 14 on p. 116, are fulfilled. In this case $\boldsymbol{x} = \boldsymbol{0}$ is an asymptotically stable equilibrium point, and we obtain

Theorem 23 (Asymptotic Stability of Passive Systems). *A passive system with a positive definite storage function S possesses an asymptotically stable equilibrium point at $\boldsymbol{x} = \boldsymbol{0}$ if no trajectory $\boldsymbol{x}(t)$ other than $\boldsymbol{x}(t) = \boldsymbol{0}$ is contained within the set $\{\boldsymbol{x} \in I\!R^n \mid \dot{S}(\boldsymbol{x}) = 0\}$.*

For strictly passive systems, similar to the passive case above, the inequality

$$\dot{S}(\boldsymbol{x}) + R(\boldsymbol{x}) \le 0$$

holds with $\boldsymbol{u} = \boldsymbol{0}$. Since the function $R(\boldsymbol{x})$ is positive definite, by rearranging the inequality to

$$\dot{S}(\boldsymbol{x}) \le -R(\boldsymbol{x}), \tag{2.47}$$

we can first deduce that $S(\boldsymbol{x})$ is positive definite and not only positive semidefinite. This is because if $S(\boldsymbol{x})$ were only positive semidefinite, points $\boldsymbol{x} \neq \boldsymbol{0}$ with $S(\boldsymbol{x}) = 0$ would exist. Such points are inevitably minima of a positive semidefinite function. Consequently, the gradient of $S(\boldsymbol{x})$ would have to fulfill the equation

$$\left(\frac{\partial S(\boldsymbol{x})}{\partial \boldsymbol{x}}\right)^T = \boldsymbol{0}$$

for these points. The latter, however, is impossible, since because of equation (2.47), i. e.

$$\dot{S}(\boldsymbol{x}) = \frac{\partial S(\boldsymbol{x})}{\partial \boldsymbol{x}} \dot{\boldsymbol{x}} \le -R(\boldsymbol{x}) < 0 \quad \text{for} \quad \boldsymbol{x} \neq \boldsymbol{0}, \tag{2.48}$$

the derivative

$$\frac{\partial S(\boldsymbol{x})}{\partial \boldsymbol{x}}$$

must be different from zero. Thus we can conclude that there is no minimum, except for $\boldsymbol{x} = \boldsymbol{0}$ with $S(\boldsymbol{0}) = 0$. Therefore, it follows that S is positive definite.

Furthermore, because of the positive definiteness of $R(\boldsymbol{x})$ and equation (2.48), the inequality

$$\dot{S}(\boldsymbol{x}) < 0$$

is also fulfilled for $\boldsymbol{x} \neq \boldsymbol{0}$. Thus the conditions of Lyapunov's stability theorem for asymptotic stability are fulfilled, the storage function $S(\boldsymbol{x})$ is a Lyapunov function, and we obtain

Theorem 24 (Asymptotic Stability of Strictly Passive Systems). *A strictly passive system possesses an asymptotically stable equilibrium point at* $\boldsymbol{x} = \boldsymbol{0}$.

If the storage function is radially unbounded, the equilibrium point is globally asymptotically stable. The term *strict passivity*, compared to *passivity*, contains the property of asymptotic stability of the equilibrium point $\boldsymbol{x}_{\text{eq}} = \boldsymbol{0}$. In contrast to this, passivity – assuming the existence of a positive definite storage function – only ensures its Lyapunov stability.

2.4.3 Passivity of Connected Systems

When passive systems are connected, in many cases, a passive system again results. We will first illustrate this by examining two strictly passive systems which are connected in parallel. Figure 2.70 illustrates the parallel connection of both strictly passive systems and their input and output variables.

For the strict passivity of the individual systems, it holds that

$$\dot{S}_i(\boldsymbol{x}_i) + R_i(\boldsymbol{x}_i) \leq \boldsymbol{u}_i^T \boldsymbol{y}_i, \qquad i = 1, 2. \tag{2.49}$$

A strictly passive overall system requires the existence of a positive semidefinite storage function $S(\boldsymbol{x})$ and a positive definite function $R(\boldsymbol{x})$ so that

$$\dot{S}(\boldsymbol{x}) + R(\boldsymbol{x}) \leq \boldsymbol{u}^T \boldsymbol{y} \quad \text{with } \boldsymbol{x}^T = \left[\boldsymbol{x}_1^T \ \boldsymbol{x}_2^T\right] \tag{2.50}$$

is fulfilled. With the connection scheme of Figure 2.70 and with the equations

$$\begin{aligned} S(\boldsymbol{x}) &= S_1(\boldsymbol{x}_1) + S_2(\boldsymbol{x}_2), \\ R(\boldsymbol{x}) &= R_1(\boldsymbol{x}_1) + R_2(\boldsymbol{x}_2), \\ \boldsymbol{y} &= \boldsymbol{y}_1 + \boldsymbol{y}_2, \\ \boldsymbol{u} &= \boldsymbol{u}_1 = \boldsymbol{u}_2, \end{aligned}$$

it follows from equation (2.50) that

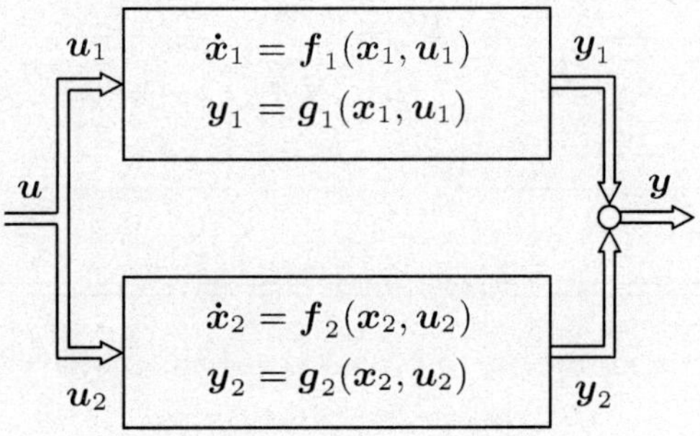

Fig. 2.70: Parallel connection of two passive systems

$$\dot{S}_1(\boldsymbol{x}_1) + R_1(\boldsymbol{x}_1) + \dot{S}_2(\boldsymbol{x}_2) + R_2(\boldsymbol{x}_2) \leq \boldsymbol{u}^T \boldsymbol{y} = \boldsymbol{u}_1^T \boldsymbol{y}_1 + \boldsymbol{u}_2^T \boldsymbol{y}_2.$$

Provided that equation (2.49) is valid, this inequality is obviously fulfilled. Since the statements above also hold for

$$R_1(\boldsymbol{x}_1) = 0 \quad \text{and} \quad R_2(\boldsymbol{x}_2) = 0,$$

i. e. the case of passivity, we can formulate

Theorem 25 (Parallel Connection of Passive Systems). *Let there be two (strictly) passive systems*

$$\begin{aligned} \dot{\boldsymbol{x}}_1 &= \boldsymbol{f}_1(\boldsymbol{x}_1, \boldsymbol{u}_1), & \dot{\boldsymbol{x}}_2 &= \boldsymbol{f}_2(\boldsymbol{x}_2, \boldsymbol{u}_2), \\ \boldsymbol{y}_1 &= \boldsymbol{g}_1(\boldsymbol{x}_1, \boldsymbol{u}_1), & \boldsymbol{y}_2 &= \boldsymbol{g}_2(\boldsymbol{x}_2, \boldsymbol{u}_2). \end{aligned}$$

Their parallel connection

$$\begin{aligned} \dot{\boldsymbol{x}}_1 &= \boldsymbol{f}_1(\boldsymbol{x}_1, \boldsymbol{u}), \quad \boldsymbol{u} = \boldsymbol{u}_1 = \boldsymbol{u}_2, \\ \dot{\boldsymbol{x}}_2 &= \boldsymbol{f}_2(\boldsymbol{x}_2, \boldsymbol{u}), \\ \boldsymbol{y} &= \boldsymbol{g}_1(\boldsymbol{x}_1, \boldsymbol{u}) + \boldsymbol{g}_2(\boldsymbol{x}_2, \boldsymbol{u}) \end{aligned}$$

is a (strictly) passive system.

However, the series connection of two passive (or strictly passive) systems does not always result in a passive (or strictly passive) system.

Of special interest and importance in control theory are control loops with passive or strictly passive subsystems, as shown in Figure 2.71. In particular, we also want to prove the strict passivity of the connected system in the case in which the subsystems are strictly passive. We will choose the sum $S(\boldsymbol{x}) = S_1(\boldsymbol{x}_1) + S_2(\boldsymbol{x}_2)$ of the storage functions S_1 and S_2 of the subsystems as the storage function S of the overall system; for the function R, correspondingly, $R(\boldsymbol{x}) = R_1(\boldsymbol{x}_1) + R_2(\boldsymbol{x}_2)$. Then equation (2.50) with

$$\begin{aligned} \boldsymbol{e}_1 &= \boldsymbol{u}_1 - \boldsymbol{y}_2, \\ \boldsymbol{e}_2 &= \boldsymbol{u}_2 + \boldsymbol{y}_1, \end{aligned} \qquad \boldsymbol{u} = \begin{bmatrix} \boldsymbol{u}_1 \\ \boldsymbol{u}_2 \end{bmatrix}, \qquad \boldsymbol{y} = \begin{bmatrix} \boldsymbol{y}_1 \\ \boldsymbol{y}_2 \end{bmatrix} \tag{2.51}$$

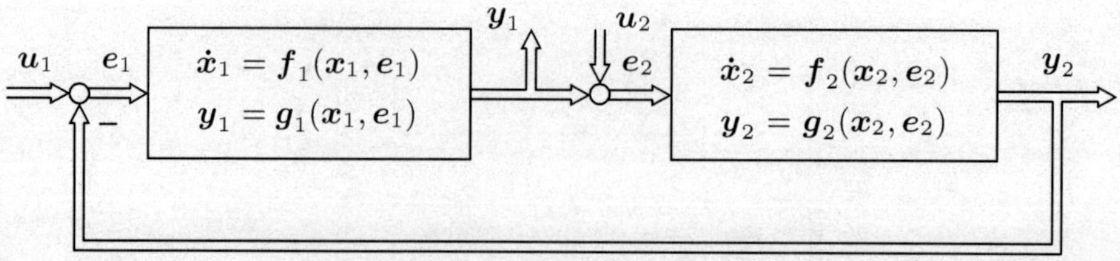

Fig. 2.71: A control loop consisting of two passive systems

takes the form

$$\dot{S}_1(\boldsymbol{x}_1) + R_1(\boldsymbol{x}_1) + \dot{S}_2(\boldsymbol{x}_2) + R_2(\boldsymbol{x}_2) \le \boldsymbol{u}^T\boldsymbol{y} = \boldsymbol{u}_1^T\boldsymbol{y}_1 + \boldsymbol{u}_2^T\boldsymbol{y}_2.$$

Using equation (2.51) once again, we can rearrange this to

$$\dot{S}_1(\boldsymbol{x}_1)+R_1(\boldsymbol{x}_1)+\dot{S}_2(\boldsymbol{x}_2)+R_2(\boldsymbol{x}_2)\le(\boldsymbol{e}_1+\boldsymbol{y}_2)^T\boldsymbol{y}_1+(\boldsymbol{e}_2-\boldsymbol{y}_1)^T\boldsymbol{y}_2=\boldsymbol{e}_1^T\boldsymbol{y}_1+\boldsymbol{e}_2^T\boldsymbol{y}_2. \tag{2.52}$$

Because of the strict passivity of the subsystems, i. e. because of

$$\dot{S}_i(\boldsymbol{x}_i) + R_i(\boldsymbol{x}_i) \le \boldsymbol{e}_i^T\boldsymbol{y}_i, \qquad i = 1, 2,$$

the inequality (2.52) is fulfilled and the control loop is also strictly passive. With the results above, we have also derived the passivity of the control loop for the case of two passive subsystems, since for this derivation, we only have to set $R_1(\boldsymbol{x}) = 0$ and $R_2(\boldsymbol{x}) = 0$.

Thus we obtain

Theorem 26 (Control Loop with Passive Subsystems). *Let there be two (strictly) passive systems*

$$\begin{aligned} \dot{\boldsymbol{x}}_1 &= \boldsymbol{f}_1(\boldsymbol{x}_1, \boldsymbol{e}_1), & \dot{\boldsymbol{x}}_2 &= \boldsymbol{f}_2(\boldsymbol{x}_2, \boldsymbol{e}_2), \\ \boldsymbol{y}_1 &= \boldsymbol{g}_1(\boldsymbol{x}_1, \boldsymbol{e}_1), & \boldsymbol{y}_2 &= \boldsymbol{g}_2(\boldsymbol{x}_2, \boldsymbol{e}_2). \end{aligned}$$

The control loop that is composed of these systems with $\boldsymbol{e}_1 = \boldsymbol{u}_1 - \boldsymbol{y}_2$ *and* $\boldsymbol{e}_2 = \boldsymbol{u}_2 + \boldsymbol{y}_1$, *i. e.*

$$\begin{aligned} \dot{\boldsymbol{x}}_1 &= \boldsymbol{f}_1(\boldsymbol{x}_1, \boldsymbol{u}_1 - \boldsymbol{y}_2), \\ \dot{\boldsymbol{x}}_2 &= \boldsymbol{f}_2(\boldsymbol{x}_2, \boldsymbol{u}_2 + \boldsymbol{y}_1), \\ \boldsymbol{y}_1 &= \boldsymbol{g}_1(\boldsymbol{x}_1, \boldsymbol{u}_1 - \boldsymbol{y}_2), \\ \boldsymbol{y}_2 &= \boldsymbol{g}_2(\boldsymbol{x}_2, \boldsymbol{u}_2 + \boldsymbol{y}_1) \end{aligned}$$

is a (strictly) passive system.

Combined with the stability theorems from the previous section, the theorem above provides a very useful tool for designing controllers. This is because, according to the above theorem and Figure 2.71, combining a strictly passive plant with a strictly passive controller leads to a control loop with a globally asymptotically stable equilibrium point $\boldsymbol{x}_{\text{eq}} = \boldsymbol{0}$. This always requires that the storage functions of the subsystems are radially unbounded and positive definite.

In summary, passive systems with positive definite storage functions are Lyapunov stable, while strictly passive systems are even asymptotically stable. (More precisely: their equilibrium point $\boldsymbol{x}_{\text{eq}} = \boldsymbol{0}$ is Lyapunov stable.) Since suitably connected passive systems are themselves passive, the connected systems also have a stable equilibrium point at $\boldsymbol{x} = \boldsymbol{0}$. Thus, such connected passive systems always yield stable systems. Showing that a system is passive, or even better strictly passive, is therefore very advantageous.

2.4.4 Passivity of Linear Systems

It is useful to investigate which conditions are necessary and sufficient for a linear system to be passive or strictly passive. This is because, according to the results obtained in the previous section, we can connect passive or strictly passive linear plants to passive or strictly passive nonlinear controllers and thus obtain a stable control loop.

For our analysis, we will consider the MIMO systems

$$\begin{aligned} \dot{\boldsymbol{x}} &= \boldsymbol{A}\boldsymbol{x} + \boldsymbol{B}\boldsymbol{u}, \quad && \boldsymbol{x} \in I\!R^n,\ \boldsymbol{u} \in I\!R^m, \\ \boldsymbol{y} &= \boldsymbol{C}\boldsymbol{x} + \boldsymbol{D}\boldsymbol{u}, \quad && \boldsymbol{y} \in I\!R^m, \end{aligned} \tag{2.53}$$

with feedthrough $\boldsymbol{D}\boldsymbol{u}$. In the following, we will restrict ourselves to positive definite storage functions, even though the definition of passivity only requires positive semidefinite storage functions. But we have already established that passive systems with positive semidefinite storage functions do not have to be stable. It is therefore appropriate to rule out this case right from the start. For linear systems that we want to examine for passivity and stability, we can then assume a positive definite quadratic storage function

$$S(\boldsymbol{x}) = \frac{1}{2}\boldsymbol{x}^T\boldsymbol{R}\boldsymbol{x}$$

with no loss of generality, as shown in [448, 477]. This yields the passivity inequality

$$\dot{S}(\boldsymbol{x}) = \frac{1}{2}\left(\dot{\boldsymbol{x}}^T\boldsymbol{R}\boldsymbol{x} + \boldsymbol{x}^T\boldsymbol{R}\dot{\boldsymbol{x}}\right) \le \boldsymbol{u}^T\boldsymbol{y} \quad \text{for all}\ \ \boldsymbol{x} \in I\!R^n,\ \boldsymbol{u} \in I\!R^m.$$

From this it follows with equation (2.53) that

$$\frac{1}{2}\boldsymbol{x}^T\left(\boldsymbol{A}^T\boldsymbol{R} + \boldsymbol{R}\boldsymbol{A}\right)\boldsymbol{x} + \boldsymbol{u}^T\boldsymbol{B}^T\boldsymbol{R}\boldsymbol{x} \le \boldsymbol{u}^T\boldsymbol{C}\boldsymbol{x} + \boldsymbol{u}^T\boldsymbol{D}\boldsymbol{u}. \tag{2.54}$$

It is possible to decompose every matrix $\boldsymbol{M}$ into a skew-symmetric component and a symmetric component according to

$$\boldsymbol{M} = \underbrace{\frac{1}{2}\left(\boldsymbol{M} - \boldsymbol{M}^T\right)}_{\text{skew-symmetric}} + \underbrace{\frac{1}{2}\left(\boldsymbol{M} + \boldsymbol{M}^T\right)}_{\text{symmetric}}.$$

For the skew-symmetric matrix, it holds that

$$\boldsymbol{z}^T\left(\boldsymbol{M} - \boldsymbol{M}^T\right)\boldsymbol{z} = 0 \quad \text{for all } \boldsymbol{z} \in I\!R^n.$$

Thus we can formulate

$$\boldsymbol{u}^T\boldsymbol{D}\boldsymbol{u} = \frac{1}{2}\boldsymbol{u}^T\left(\boldsymbol{D} + \boldsymbol{D}^T\right)\boldsymbol{u}.$$

Equivalent to inequality (2.54), we arrive at the inequality

$$\frac{1}{2}\begin{bmatrix}\boldsymbol{x}\\ \boldsymbol{u}\end{bmatrix}^T\begin{bmatrix}\boldsymbol{A}^T\boldsymbol{R}+\boldsymbol{R}\boldsymbol{A} & \boldsymbol{R}\boldsymbol{B}-\boldsymbol{C}^T\\ \boldsymbol{B}^T\boldsymbol{R}-\boldsymbol{C} & -\boldsymbol{D}-\boldsymbol{D}^T\end{bmatrix}\begin{bmatrix}\boldsymbol{x}\\ \boldsymbol{u}\end{bmatrix} \le 0,$$

which leads to

Theorem 27 (Passivity of Linear Systems). *A linear system*

$$\begin{aligned}\dot{\boldsymbol{x}} &= \boldsymbol{A}\boldsymbol{x} + \boldsymbol{B}\boldsymbol{u}, \quad \boldsymbol{x} \in I\!R^n,\ \boldsymbol{u} \in I\!R^m,\\ \boldsymbol{y} &= \boldsymbol{C}\boldsymbol{x} + \boldsymbol{D}\boldsymbol{u}, \quad \boldsymbol{y} \in I\!R^m,\end{aligned}$$

with a positive definite storage function is passive if and only if a positive definite $n \times n$ matrix $\boldsymbol{R}$ exists such that the $(n+m) \times (n+m)$ matrix

$$\begin{bmatrix}\boldsymbol{A}^T\boldsymbol{R}+\boldsymbol{R}\boldsymbol{A} & \boldsymbol{R}\boldsymbol{B}-\boldsymbol{C}^T\\ \boldsymbol{B}^T\boldsymbol{R}-\boldsymbol{C} & -\boldsymbol{D}-\boldsymbol{D}^T\end{bmatrix} \tag{2.55}$$

is negative semidefinite.

The matrix above can now be represented by a matrix $\boldsymbol{L}$ and a matrix $\boldsymbol{W}$ as

$$\begin{bmatrix}\boldsymbol{A}^T\boldsymbol{R}+\boldsymbol{R}\boldsymbol{A} & \boldsymbol{R}\boldsymbol{B}-\boldsymbol{C}^T\\ \boldsymbol{B}^T\boldsymbol{R}-\boldsymbol{C} & -\boldsymbol{D}-\boldsymbol{D}^T\end{bmatrix} = -\begin{bmatrix}\boldsymbol{L}^T\\ \boldsymbol{W}^T\end{bmatrix}\begin{bmatrix}\boldsymbol{L} & \boldsymbol{W}\end{bmatrix}, \tag{2.56}$$

where the matrix on the right-hand side of the equation above is always negative semidefinite[8].

Taking into account the symmetry of the matrix, the matrix equation (2.56) leads to the three so-called Kalman-Yakubovich-Popov equations (KYP equations)

[8] An initial and detailed treatment of this topic can be found in J. G. Willems [477] and further information is provided in [163].

$$\begin{aligned} \boldsymbol{A}^T\boldsymbol{R} + \boldsymbol{R}\boldsymbol{A} &= -\boldsymbol{L}^T\boldsymbol{L}, \\ \boldsymbol{R}\boldsymbol{B} - \boldsymbol{C}^T &= -\boldsymbol{L}^T\boldsymbol{W}, \\ \boldsymbol{D} + \boldsymbol{D}^T &= \boldsymbol{W}^T\boldsymbol{W}. \end{aligned} \tag{2.57}$$

It is important to note that the requirement of a negative semidefinite matrix (2.55), the fulfillment of equation (2.56), and thus (2.57) are equivalent. This is because, by choosing $\boldsymbol{L}$ as a matrix of dimension $(n+m) \times n$ and $\boldsymbol{W}$ as a matrix of dimension $(n+m) \times m$, creating the matrix

$$\begin{bmatrix} \boldsymbol{L}^T \\ \boldsymbol{W}^T \end{bmatrix} [\boldsymbol{L}\ \boldsymbol{W}], \tag{2.58}$$

it is possible to create any arbitrary positive semidefinite and also positive definite symmetric $(n+m) \times (n+m)$ matrix.

This circumstance draws our attention to the dimensions of the matrices $\boldsymbol{L}$ and $\boldsymbol{W}$. Obviously we can vary their dimensions if we want to create a positive semidefinite matrix (2.58). For example, either

$$\boldsymbol{L} \in I\!R^{n\times n} \quad \text{and} \quad \boldsymbol{W} \in I\!R^{n\times m}$$

or

$$\boldsymbol{L} \in I\!R^{m\times n} \quad \text{and} \quad \boldsymbol{W} \in I\!R^{m\times m} \tag{2.59}$$

would fit together. Thus, the dimensions of the matrices $\boldsymbol{L}$ and $\boldsymbol{W}$ are not predetermined. When choosing them, however, their compatibility with each other and with the other matrices which occur in the Kalman-Yakubovich-Popov equations (2.57) must be maintained. The choice of matrices (2.59), which allows for low dimensions of the matrices, is sufficient for the solvability of the KYP equations [477].

Using the above results, we formulate

Theorem 28 (Passivity and the KYP Equations). *A linear system*

$$\begin{aligned} \dot{\boldsymbol{x}} &= \boldsymbol{A}\boldsymbol{x} + \boldsymbol{B}\boldsymbol{u}, \quad \boldsymbol{x} \in I\!R^n,\ \boldsymbol{u} \in I\!R^m \\ \boldsymbol{y} &= \boldsymbol{C}\boldsymbol{x} + \boldsymbol{D}\boldsymbol{u}, \quad \boldsymbol{y} \in I\!R^m, \end{aligned}$$

is passive with a positive definite storage function if and only if the Kalman-Yakubovich-Popov equations

$$\begin{aligned} \boldsymbol{A}^T\boldsymbol{R} + \boldsymbol{R}\boldsymbol{A} &= -\boldsymbol{L}^T\boldsymbol{L}, \\ \boldsymbol{L}^T\boldsymbol{W} &= \boldsymbol{C}^T - \boldsymbol{R}\boldsymbol{B}, \\ \boldsymbol{W}^T\boldsymbol{W} &= \boldsymbol{D} + \boldsymbol{D}^T \end{aligned}$$

can be solved by a positive definite symmetric $n \times n$ matrix $\boldsymbol{R}$, a matrix $\boldsymbol{L}$, and a matrix $\boldsymbol{W}$.

In the case of strictly passive linear systems (2.53), we can proceed similarly as in the case of passivity above. We will examine the defining inequality

$$\dot{S}(\boldsymbol{x}) + R(\boldsymbol{x}) \leq \boldsymbol{u}^T \boldsymbol{y} \tag{2.60}$$

for strict passivity. Once again, we will choose a positive definite storage function

$$S(\boldsymbol{x}) = \frac{1}{2}\boldsymbol{x}^T \boldsymbol{R} \boldsymbol{x} \tag{2.61}$$

with the $n \times n$ matrix $\boldsymbol{R}$ and also a quadratic function

$$R(\boldsymbol{x}) = \mu \frac{1}{2}\boldsymbol{x}^T \boldsymbol{P} \boldsymbol{x} \tag{2.62}$$

with a positive, real-valued constant μ and a positive definite matrix $\boldsymbol{P}$. At this point, we will note that choosing a general quadratic form (2.62) for the function $R(\boldsymbol{x})$ does not restrict the generality of our further deductions.

If we now insert the functions (2.61) and (2.62) into equation (2.60), we obtain – similarly to the case of passivity – the inequality (2.60) for strict passivity in the form of

$$\frac{1}{2}\boldsymbol{x}^T \left(\boldsymbol{A}^T\boldsymbol{R} + \boldsymbol{R}\boldsymbol{A} + \mu\boldsymbol{P}\right)\boldsymbol{x} + \boldsymbol{u}^T\left(\boldsymbol{B}^T\boldsymbol{R} - \boldsymbol{C}\right)\boldsymbol{x} - \frac{1}{2}\boldsymbol{u}^T\left(\boldsymbol{D} + \boldsymbol{D}^T\right)\boldsymbol{u} \leq 0$$

or, equivalently,

$$\frac{1}{2}\begin{bmatrix}\boldsymbol{x}\\ \boldsymbol{u}\end{bmatrix}^T \begin{bmatrix}\boldsymbol{A}^T\boldsymbol{R} + \boldsymbol{R}\boldsymbol{A} + \mu\boldsymbol{P} & \boldsymbol{R}\boldsymbol{B} - \boldsymbol{C}^T \\ \boldsymbol{B}^T\boldsymbol{R} - \boldsymbol{C} & -\boldsymbol{D}^T - \boldsymbol{D}\end{bmatrix}\begin{bmatrix}\boldsymbol{x}\\ \boldsymbol{u}\end{bmatrix} \leq 0. \tag{2.63}$$

The linear system (2.53) is thus strictly passive if and only if positive definite matrices $\boldsymbol{R}$ and $\mu\boldsymbol{P}$ exist such that the matrix of the quadratic form (2.63) is negative semidefinite.

This is equivalently formulated as

Theorem 29 (Strict Passivity and the KYP Equations). *A linear system*

$$\begin{aligned}\dot{\boldsymbol{x}} &= \boldsymbol{A}\boldsymbol{x} + \boldsymbol{B}\boldsymbol{u}, \quad \boldsymbol{x} \in I\!R^n, \quad \boldsymbol{u} \in I\!R^m,\\ \boldsymbol{y} &= \boldsymbol{C}\boldsymbol{x} + \boldsymbol{D}\boldsymbol{u}, \quad \boldsymbol{y} \in I\!R^m,\end{aligned}$$

is strictly passive if and only if the Kalman-Yakubovich-Popov equations

$$\begin{aligned}\boldsymbol{A}^T\boldsymbol{R} + \boldsymbol{R}\boldsymbol{A} &= -\boldsymbol{L}^T\boldsymbol{L} - \mu\boldsymbol{P},\\ \boldsymbol{L}^T\boldsymbol{W} &= \boldsymbol{C}^T - \boldsymbol{R}\boldsymbol{B},\\ \boldsymbol{W}^T\boldsymbol{W} &= \boldsymbol{D} + \boldsymbol{D}^T\end{aligned}$$

can be fulfilled by positive definite symmetric $n \times n$ matrices $\boldsymbol{R}$ and $\mu\boldsymbol{P}$, a matrix $\boldsymbol{L}$, and a matrix $\boldsymbol{W}$.

In cases without a feedthrough $\boldsymbol{D}$, the Kalman-Yakubovich-Popov equations from Theorem 29 can be simplified, because, from $\boldsymbol{D} = \boldsymbol{0}$, it follows that $\boldsymbol{W} = \boldsymbol{0}$. Furthermore, we can add the matrix $\boldsymbol{L}^T\boldsymbol{L}$, which is freely selectable in this case, and the positive definite matrix $\mu\boldsymbol{P}$, resulting in a positive definite matrix $\boldsymbol{Q}$, which yields

$$\begin{aligned} \boldsymbol{A}^T\boldsymbol{R} + \boldsymbol{R}\boldsymbol{A} &= -\boldsymbol{Q}, \\ \boldsymbol{C} &= \boldsymbol{B}^T\boldsymbol{R}. \end{aligned}$$

In the case of Theorem 28, the matrix

$$\boldsymbol{Q} = \boldsymbol{L}^T\boldsymbol{L}$$

only needs to be positive semidefinite. An explicit solution method for the Kalman-Yakubovich-Popov equations can be found in [398].

2.4.5 Example: Transporting System for Material Webs

In the industrial production of paper, plastic foil, and textile fabrics, the material webs are transported over a roller system, either during production or subsequent treatment, such as coating. For this purpose, pairs of rollers located between the dispenser and the rewinder press down onto the material web, so that it is moved along by them. Figure 2.72 illustrates the basic process.

The speed of the material web must be equal everywhere, i. e. the angular velocity ω_i of rollers with the same radius must be equal. From this requirement, it follows as a further condition that the differences

$$F_i - F_{i-1}$$

between tractions F_i affecting the web at its contact points with the rollers should be zero or at least not too large, since otherwise the material web will tear or flutter [324, 488]. We now want to determine the torque equilibria at the individual rollers. Here M_i is the driving torque of the ith roller, F_i is the tension force which impacts the material web between the rollers i and $i+1$, r denotes the roller radius, J the moment of inertia of the roller elements, and $d \cdot \omega_i$, with d being the friction coefficient, describes the sliding friction of the gears. Simplifying, we assume that the radius of the rolls, including the wound web, is the same.

For the dispenser, it holds that

$$J\dot{\omega}_1 = rF_1 - d \cdot \omega_1 + M_1. \tag{2.64}$$

For the second and third roller elements, we obtain

$$J\dot{\omega}_i = r(F_i - F_{i-1}) - d \cdot \omega_i + M_i, \quad i = 2, 3, \tag{2.65}$$

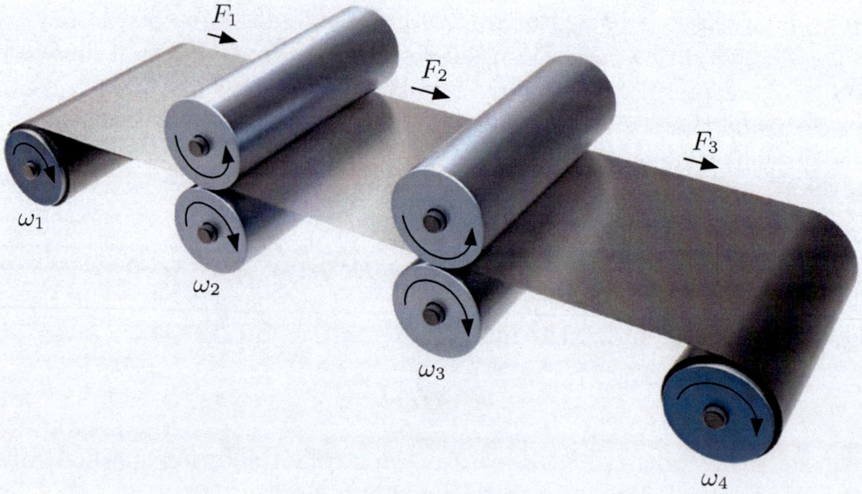

Fig. 2.72: Transporting system for material webs; the roller elements are constructed in such a way that they all have the same inertia moment J.

and for the rewinder

$$J\dot{\omega}_4 = -rF_3 - d \cdot \omega_4 + M_4 \tag{2.66}$$

in a similar way. Further, we must take into account that, between the rollers, the material web acts like a spring, with c denoting the spring constant. According to Hooke's law, it follows for the change in the tension between the rollers i and $i+1$ that

$$\dot{F}_i = c(v_{i+1} - v_i) = c\,r(\omega_{i+1} - \omega_i), \quad i = 1, \ldots, 3 \tag{2.67}$$

holds.

Using equations (2.64), (2.65), (2.66), and (2.67), we thus obtain seven differential equations in total, which we subsume with the state vector

$$\boldsymbol{x} = \begin{bmatrix} \omega_1 & F_1 & \omega_2 & F_2 & \omega_3 & F_3 & \omega_4 \end{bmatrix}^T$$

and the input variable vector

$$\boldsymbol{u} = \begin{bmatrix} M_1 & M_2 & M_3 & M_4 \end{bmatrix}^T$$

in the linear system description

$$\begin{aligned} \dot{\boldsymbol{x}} &= \boldsymbol{A}\boldsymbol{x} + \boldsymbol{B}\boldsymbol{u}, \\ \boldsymbol{y} &= \begin{bmatrix} x_1 & x_3 & x_5 & x_7 \end{bmatrix}^T = \begin{bmatrix} \omega_1 & \omega_2 & \omega_3 & \omega_4 \end{bmatrix}^T \end{aligned} \tag{2.68}$$

with

$$\boldsymbol{A} = \begin{bmatrix} -\frac{d}{J} & \frac{r}{J} & 0 & 0 & 0 & 0 & 0 \\ -cr & 0 & cr & 0 & 0 & 0 & 0 \\ 0 & -\frac{r}{J} & -\frac{d}{J} & \frac{r}{J} & 0 & 0 & 0 \\ 0 & 0 & -cr & 0 & cr & 0 & 0 \\ 0 & 0 & 0 & -\frac{r}{J} & -\frac{d}{J} & \frac{r}{J} & 0 \\ 0 & 0 & 0 & 0 & -cr & 0 & cr \\ 0 & 0 & 0 & 0 & 0 & -\frac{r}{J} & -\frac{d}{J} \end{bmatrix}, \qquad \boldsymbol{B} = \frac{1}{J}\begin{bmatrix} 1 & 0 & 0 & 0 \\ 0 & 0 & 0 & 0 \\ 0 & 1 & 0 & 0 \\ 0 & 0 & 0 & 0 \\ 0 & 0 & 1 & 0 \\ 0 & 0 & 0 & 0 \\ 0 & 0 & 0 & 1 \end{bmatrix}.$$

To verify the system's passivity, we will choose

$$S(\boldsymbol{x}) = \frac{1}{2}\boldsymbol{x}^T \boldsymbol{R}\boldsymbol{x}, \quad \boldsymbol{R} = \begin{bmatrix} J & 0 & 0 & 0 & 0 & 0 & 0 \\ 0 & \frac{1}{c} & 0 & 0 & 0 & 0 & 0 \\ 0 & 0 & J & 0 & 0 & 0 & 0 \\ 0 & 0 & 0 & \frac{1}{c} & 0 & 0 & 0 \\ 0 & 0 & 0 & 0 & J & 0 & 0 \\ 0 & 0 & 0 & 0 & 0 & \frac{1}{c} & 0 \\ 0 & 0 & 0 & 0 & 0 & 0 & J \end{bmatrix} \tag{2.69}$$

as the storage function. For the temporal derivative of the storage function, it holds that

$$\begin{aligned} \dot{S}(\boldsymbol{x}) &= \frac{1}{2}\boldsymbol{x}^T(\boldsymbol{A}^T\boldsymbol{R} + \boldsymbol{R}\boldsymbol{A})\boldsymbol{x} + \boldsymbol{u}^T\boldsymbol{B}^T\boldsymbol{R}\boldsymbol{x} \\ &= -d(x_1^2 + x_3^2 + x_5^2 + x_7^2) + u_1x_1 + u_2x_3 + u_3x_5 + u_4x_7 \\ &= -d(x_1^2 + x_3^2 + x_5^2 + x_7^2) + u_1y_1 + u_2y_2 + u_3y_3 + u_4y_4. \end{aligned} \tag{2.70}$$

From the passivity inequality

$$\dot{S}(\boldsymbol{x}) \le \boldsymbol{u}^T\boldsymbol{y} = u_1y_1 + u_2y_2 + u_3y_3 + u_4y_4,$$

inserting equation (2.70), we obtain the inequality

$$-d(x_1^2 + x_3^2 + x_5^2 + x_7^2) \le 0,$$

which is obviously fulfilled. The system is thus passive.

This conclusion is also reached when verifying the Kalman-Yakubovich-Popov equations from Theorem 29,

$$\boldsymbol{A}^T\boldsymbol{R} + \boldsymbol{R}\boldsymbol{A} = -\boldsymbol{Q}, \tag{2.71}$$

$$\boldsymbol{C}^T = \boldsymbol{R}\boldsymbol{B}, \tag{2.72}$$

in the case that there is no feedthrough. Equation (2.71) is fulfilled with the matrix $\boldsymbol{R}$ from equation (2.69), where

$$\boldsymbol{Q} = 2\begin{bmatrix} d & 0 & 0 & 0 & 0 & 0 & 0 \\ 0 & 0 & 0 & 0 & 0 & 0 & 0 \\ 0 & 0 & d & 0 & 0 & 0 & 0 \\ 0 & 0 & 0 & 0 & 0 & 0 & 0 \\ 0 & 0 & 0 & 0 & d & 0 & 0 \\ 0 & 0 & 0 & 0 & 0 & 0 & 0 \\ 0 & 0 & 0 & 0 & 0 & 0 & d \end{bmatrix}$$

is positive semidefinite. Equation (2.72) is also fulfilled, since

$$\boldsymbol{C} = \boldsymbol{B}^T\boldsymbol{R} = \begin{bmatrix} 1 & 0 & 0 & 0 & 0 & 0 & 0 \\ 0 & 0 & 1 & 0 & 0 & 0 & 0 \\ 0 & 0 & 0 & 0 & 1 & 0 & 0 \\ 0 & 0 & 0 & 0 & 0 & 0 & 1 \end{bmatrix}$$

holds. This corresponds to the output matrix $\boldsymbol{C}$ of the system (2.68) in the output equation $\boldsymbol{y} = \boldsymbol{C}\boldsymbol{x}$.

2.4.6 Positive Real Transfer Functions

If we wish to establish passivity conditions using the transfer function or the transfer matrix

$$\boldsymbol{G}(s) = \boldsymbol{C}(s\boldsymbol{I} - \boldsymbol{A})^{-1}\boldsymbol{B} + \boldsymbol{D} \tag{2.73}$$

rather than the state-space representation, we use the concept of *positive realness* [58, 307, 329, 445]. We will later see that the passivity of a linear state-space model and the positive realness of its transfer function are equivalent system properties. The positive realness is therefore the frequency-domain analogon of passivity in the state space.

First we will define positive realness for SISO systems, and then for MIMO systems.

Definition 25 (Positive Realness of Transfer Functions). *A transfer function*

$$G(s) = k\frac{(s - n_m)\cdots(s - n_1)}{(s - p_n)\cdots(s - p_1)}$$

is called positive real if, for all $s \in \mathbb{C}$ *with* $\operatorname{Re}\{s\} > 0$, *the inequality*

$$\operatorname{Re}\{G(s)\} \geq 0$$

holds. It is called strictly positive real if $G(s - \epsilon)$ *for some* $\epsilon > 0$ *is positive real.*

Here $\mathbb{C}$ denotes the set of complex numbers. The term *positive real*[9],[10] is derived from the fact that the real part $\operatorname{Re}\{G(s)\}$ of the transfer function must always be positive or equal to zero.

Some properties of positive real transfer functions can be deduced from the definition itself. First of all, we can easily conclude for

$$\operatorname{Re}\{s\} > 0$$

and with $s = \sigma + j\omega$ that

$$\lim_{\sigma \to 0} \operatorname{Re}\{G(\sigma + j\omega)\} = \operatorname{Re}\{G(j\omega)\} \geq 0.$$

The Nyquist plot of $G(j\omega)$ of positive real or strictly positive real systems must therefore lie within the closed or open right half-plane of the complex plane. Figure 2.73 illustrates this.

From the fact that the Nyquist plot of $G(j\omega)$ must lie within the right complex half-plane, for a positive real transfer function $G(s)$ it follows that

$$n - m = 0 \quad \text{or} \quad n - m = 1$$

holds for its degree m of the nominator and its degree n of the denominator. Otherwise the phase would be less than $-90°$.

Since the requirements of Definition 25 are generally difficult to verify, the following necessary and sufficient criteria are useful. We start with [58]

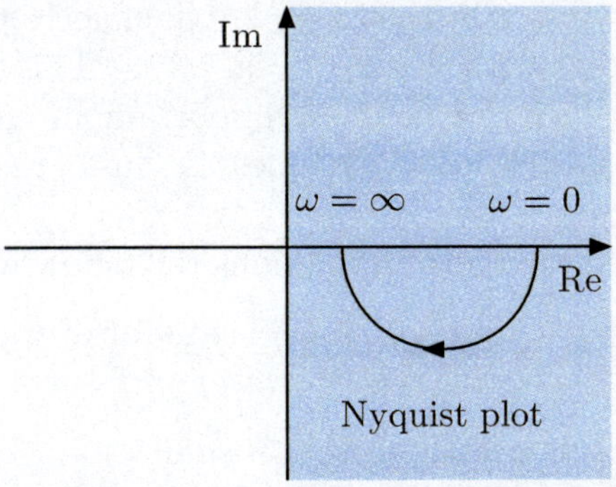

Fig. 2.73: Positive real transfer functions have a Nyquist plot in the closed right half-plane.

[9] Both the term *positive real* and in particular the term *strictly positive real* are defined differently and inconsistently in the literature [214, 228, 240, 288, 445]. Of the known definitions, we use here those from which it follows that passivity and strict passivity are equivalent to positive realness and strict positive realness, respectively.

[10] Closely related to the properties of *passivity* and *positive realness* is the property of *hyperstability*, which was introduced by V. M. Popov [374, 375]. However, in the literature the term *passivity* has prevailed over *hyperstability*. Detailed information on hyperstability can be found in [16, 375].

Theorem 30 (Positive Real Transfer Functions I). *A transfer function $G(s)$ is positive real if and only if*

(1) $G(s)$ only has poles p_i with $\operatorname{Re}\{p_i\} \le 0$, and
(2) $\operatorname{Re}\{G(j\omega)\} \ge 0$ holds for every $\omega \ge 0$, for which $j\omega$ is not a pole of $G(s)$, and
(3) possible poles $j\omega_0$ are single poles and have nonnegative real residuals

$$\operatorname*{Res}_{s=j\omega_0}(G(s)) = \lim_{s\to j\omega_0}(s - j\omega_0)G(s).$$

The conditions of the above sufficient and necessary theorem are also often used as an alternative to Definition 25, to define the term *positive realness* [58, 232]. Furthermore we can use [78, 473]

Theorem 31 (Positive Real Transfer Functions II). *A coprime transfer function*

$$G(s) = \frac{N(s)}{D(s)}$$

is positive real if and only if

(1) the polynomial $N(s) + D(s)$ only has zeros with a negative real part, and
(2) $\operatorname{Re}\{G(j\omega)\} \ge 0$ holds for every $\omega \ge 0$, for which $j\omega$ is not a pole of $G(s)$.

The property of positive realness at most ensures the Lyapunov stability of a system and not asymptotic stability, since the real parts of the poles are allowed to be equal to zero. On the other hand, strictly positive real transfer functions only have poles with negative real parts. This property guarantees asymptotic stability, which is extremely important in control loops. From a practical point of view, strict positive realness is thus of much greater relevance than positive realness.

For strictly positive real transfer functions the following holds [84, 85, 232]:

Theorem 32 (Strictly Positive Real Transfer Functions). *A transfer function $G(s)$ is strictly positive real if and only if*

(1) $G(s)$ only has poles p_i with $\operatorname{Re}\{p_i\} < 0$, and
(2) for every ω with $0 \le \omega < \infty$ the inequality $\operatorname{Re}\{G(j\omega)\} > 0$ holds, and
(3) either

$$G(\infty) > 0$$

or

$$G(\infty) = 0$$

is fulfilled and, in the latter case,

$$\lim_{\omega\to\infty} \omega^2 \operatorname{Re}\{G(j\omega)\} > 0$$

holds.

In addition to the above sufficient and necessary conditions for positive or strict positive realness, the following necessary conditions [58], which we partially described at the beginning of this section, also apply.

Theorem 33 (Necessary Conditions for Positive Realness). *For a (strictly) positive real transfer function $G(s)$ it holds that*

(1) only poles p_i with $\operatorname{Re}\{p_i\} \leq 0$ ($\operatorname{Re}\{p_i\} < 0$) exist,
(2) only zeros n_i with $\operatorname{Re}\{n_i\} \leq 0$ ($\operatorname{Re}\{n_i\} < 0$) exist,
(3) the difference $n-m$ between the numerator degree m and the denominator degree n is either zero or one,
(4) the Nyquist plot of $G(j\omega)$ lies completely within the closed (open) right half-plane of the complex plane.

The positive realness of a transfer matrix

$$\boldsymbol{G}(s) = \begin{bmatrix} G_{11}(s) & \cdots & G_{1m}(s) \\ \vdots & \ddots & \vdots \\ G_{m1}(s) & \cdots & G_{mm}(s) \end{bmatrix}$$

of the MIMO systems

$$\boldsymbol{Y}(s) = \boldsymbol{G}(s) \cdot \boldsymbol{U}(s)$$

is defined similarly to the positive realness of a transfer function in

Definition 26 (Positive Realness of Transfer Matrices). *A transfer matrix $\boldsymbol{G}(s)$ is called positive real if, for all $s \in \mathbb{C}$ with $\operatorname{Re}\{s\} > 0$, the inequality*

$$\operatorname{Re}\{\boldsymbol{z}^*\boldsymbol{G}(s)\boldsymbol{z}\} \geq 0 \quad \textit{for all} \;\; \boldsymbol{z} \in \mathbb{C}^m \tag{2.74}$$

holds. It is called strictly positive real if $\boldsymbol{G}(s-\epsilon)$ is positive real for a constant $\epsilon > 0$.

Here $\boldsymbol{z}^*$ denotes the complex conjugate transpose of the vector $\boldsymbol{z}$ and $\mathbb{C}^m$ is the m - dimensional space of complex numbers.

The defining equation (2.74) is equivalent to the requirement that the matrix

$$\boldsymbol{G}(s) + \boldsymbol{G}^*(s) \tag{2.75}$$

be positive semidefinite for all $s \in \mathbb{C}$ with $\operatorname{Re}\{s\} > 0$. Here

$$\boldsymbol{G}^*(s) = \boldsymbol{G}^T(\overline{s})$$

is the conjugate transpose, i. e. the hermitian transpose matrix of $\boldsymbol{G}(s)$, and $\overline{s}$ is the complex conjugate of s.

For transfer matrices $\boldsymbol{G}(s)$, a necessary and sufficient criterion also exists to determine positive realness [116, 232, 442]. This is

Theorem 34 (Positive Real Transfer Matrices). *The transfer matrix $\boldsymbol{G}(s)$ is positive real if and only if*

(1) all transfer functions $G_{ij}(s)$ only have poles p_i with $\operatorname{Re}\{p_i\} \leq 0$, and
(2) for every $\omega \geq 0$, for which $j\omega$ is not a pole of $\boldsymbol{G}(s)$,

$$\boldsymbol{G}(j\omega) + \boldsymbol{G}^*(j\omega)$$

is a positive semidefinite matrix, and
(3) possible poles $j\omega_0$ are single poles and the residual matrix

$$\operatorname*{Res}_{s=j\omega_0}(\boldsymbol{G}(s)) = \lim_{s\to j\omega_0}(s - j\omega_0)\boldsymbol{G}(s)$$

is a hermitian and positive semidefinite matrix.

A quadratic $n \times n$ matrix $\boldsymbol{M}$ is hermitian if all its diagonal elements are real and the relationship $m_{ik} = \overline{m}_{ki}$ holds for all symmetric off-diagonal elements m_{ik} and m_{ki}. Here $\overline{m}_{ki}$ is the complex conjugate of m_{ki}. Thus a hermitian matrix $\boldsymbol{M}$ takes the form

$$\boldsymbol{M} = \begin{bmatrix} m_{11} & m_{12} & m_{13} & \cdots & m_{1n} \\ \overline{m}_{12} & m_{22} & m_{23} & \cdots & m_{2n} \\ \overline{m}_{13} & \overline{m}_{23} & m_{33} & \cdots & m_{3n} \\ \vdots & \vdots & \vdots & \ddots & \vdots \\ \overline{m}_{1n} & \overline{m}_{2n} & \overline{m}_{3n} & \cdots & m_{nn} \end{bmatrix}.$$

Note in this context that

$$\boldsymbol{G}^*(j\omega) = \boldsymbol{G}^T(-j\omega)$$

holds.

Analogous to the above Theorem 34, there is a counterpart to strictly positive real transfer matrices [83, 84, 118], namely

Theorem 35 (Strictly Positive Real Transfer Matrices). *The $m \times m$ transfer matrix $\boldsymbol{G}(s)$ is strictly positive real if and only if*

(1) all transfer functions $G_{ij}(s)$ only have poles p_i with $\operatorname{Re}\{p_i\} < 0$, and
(2) for every ω with $0 \leq \omega < \infty$, the matrix

$$\boldsymbol{G}(j\omega) + \boldsymbol{G}^*(j\omega)$$

is positive definite, and
(3) the inequality

$$\lim_{|\omega|\to\infty} \omega^{2(m-\mu)} \det(\boldsymbol{G}(j\omega) + \boldsymbol{G}^*(j\omega)) > 0, \quad \mu = \operatorname{rank}(G(\infty) + G^*(\infty)),$$

is fulfilled.

If we have a look at the state-space representation belonging to the transfer function (2.73), i. e.

$$\begin{aligned}\dot{\boldsymbol{x}} &= \boldsymbol{A}\boldsymbol{x} + \boldsymbol{B}\boldsymbol{u},\\ \boldsymbol{y} &= \boldsymbol{C}\boldsymbol{x} + \boldsymbol{D}\boldsymbol{u},\end{aligned} \tag{2.76}$$

it holds for the feedthrough matrix $\boldsymbol{D}$ that

$$\boldsymbol{D} = \boldsymbol{G}(\infty) \quad \text{and} \quad \boldsymbol{D}^T = \boldsymbol{D}^* = \boldsymbol{G}^*(\infty).$$

The rank μ of the matrix $\boldsymbol{G}(\infty) + \boldsymbol{G}^*(\infty)$ from the theorem above can thus also be determined via

$$\mu = \operatorname{rank}(\boldsymbol{D} + \boldsymbol{D}^T).$$

In the large majority of cases $\boldsymbol{D} = \boldsymbol{0}$ is given, and thus it holds that $\mu = 0$ in Condition (3) of Theorem 35, since only very few practical systems (2.76) have a direct feedthrough of the control signal vector $\boldsymbol{u}$ to the output signal vector $\boldsymbol{y}$. Some further results if $\boldsymbol{D} \neq \boldsymbol{0}$ can be found in [475].

2.4.7 Equivalence of Positive Realness and Passivity

The requirements of Theorems 34 and 35 for MIMO systems are generally very difficult to verify. Based on the Kalman-Yakubovich-Popov equations

$$\boldsymbol{A}^T\boldsymbol{R} + \boldsymbol{R}\boldsymbol{A} = -\boldsymbol{L}^T\boldsymbol{L}, \tag{2.77}$$

$$\boldsymbol{L}^T\boldsymbol{W} = \boldsymbol{C}^T - \boldsymbol{R}\boldsymbol{B}, \tag{2.78}$$

$$\boldsymbol{W}^T\boldsymbol{W} = \boldsymbol{D} + \boldsymbol{D}^T, \tag{2.79}$$

we can derive criteria for positive or strict positive realness which are based on the state-space representation. These are often easier to utilize than the corresponding frequency-domain criteria. Our starting point is the defining inequality (2.74) for positive real transfer functions, or the equivalent requirement that the matrix (2.75), i. e.

$$\boldsymbol{G}(s) + \boldsymbol{G}^*(s) = \boldsymbol{G}(s) + \boldsymbol{G}^T(\overline{s}), \tag{2.80}$$

must be positive semidefinite for all $s \in \mathbb{C}$ with $\operatorname{Re}\{s\} > 0$. We assume that the system in question

$$\begin{aligned}\dot{\boldsymbol{x}} &= \boldsymbol{A}\boldsymbol{x} + \boldsymbol{B}\boldsymbol{u},\\ \boldsymbol{y} &= \boldsymbol{C}\boldsymbol{x} + \boldsymbol{D}\boldsymbol{u}\end{aligned} \tag{2.81}$$

is controllable and observable. We then calculate its transfer matrix

$$\boldsymbol{G}(s) = \boldsymbol{C}(s\boldsymbol{I} - \boldsymbol{A})^{-1}\boldsymbol{B} + \boldsymbol{D}. \tag{2.82}$$

In this context, we recall linear system theory and the relationship between transfer matrix and state-space representation. A consequence of the controllability and observability of a MIMO state-space model (2.81) with degree n is that its transfer matrix (2.82) has the same degree and coprime transfer functions $G_{ij}(s)$ as elements. Conversely, a state-space model (2.81) is a controllable and observable realization of the transfer matrix (2.82) if and only if all $G_{ij}(s)$ are coprime and the degree of this state-space model is equal to the degree of the transfer matrix[11] [73].

We now insert equation (2.82) into equation (2.80). This yields the expression

$$\boldsymbol{G}(s)+\boldsymbol{G}^T(\overline{s}) = \boldsymbol{C}(s\boldsymbol{I}-\boldsymbol{A})^{-1}\boldsymbol{B}+\boldsymbol{D}+\boldsymbol{B}^T[(\overline{s}\boldsymbol{I}-\boldsymbol{A})^{-1}]^T\boldsymbol{C}^T+\boldsymbol{D}^T$$

for equation (2.80). By applying the KYP equations (2.78) and (2.79), i. e.

$$\boldsymbol{C} = \boldsymbol{W}^T\boldsymbol{L}+\boldsymbol{B}^T\boldsymbol{R} \quad \text{and} \quad \boldsymbol{D}+\boldsymbol{D}^T = \boldsymbol{W}^T\boldsymbol{W},$$

the equation

$$\begin{aligned}\boldsymbol{G}(s)+\boldsymbol{G}^T(\overline{s}) &= \\ &= (\boldsymbol{W}^T\boldsymbol{L}+\boldsymbol{B}^T\boldsymbol{R})(s\boldsymbol{I}-\boldsymbol{A})^{-1}\boldsymbol{B}+\boldsymbol{B}^T(\overline{s}\boldsymbol{I}-\boldsymbol{A}^T)^{-1}(\boldsymbol{L}^T\boldsymbol{W}+\boldsymbol{R}\boldsymbol{B})+\boldsymbol{W}^T\boldsymbol{W} \\ &= \boldsymbol{W}^T\boldsymbol{L}(s\boldsymbol{I}-\boldsymbol{A})^{-1}\boldsymbol{B}+\boldsymbol{B}^T(\overline{s}\boldsymbol{I}-\boldsymbol{A}^T)^{-1}\boldsymbol{L}^T\boldsymbol{W}+\boldsymbol{W}^T\boldsymbol{W} \\ &\quad \underbrace{+\,\boldsymbol{B}^T\left[\boldsymbol{R}(s\boldsymbol{I}-\boldsymbol{A})^{-1}+(\overline{s}\boldsymbol{I}-\boldsymbol{A}^T)^{-1}\boldsymbol{R}\right]\boldsymbol{B}}_{\boldsymbol{F}} \end{aligned} \tag{2.83}$$

follows. In the last summand $\boldsymbol{F}$ of the equation above, we now factor out the matrix $(s\boldsymbol{I}-\boldsymbol{A})^{-1}$ and its complex conjugate variant $(\overline{s}\boldsymbol{I}-\boldsymbol{A}^T)^{-1}$. This last summand $\boldsymbol{F}$ then takes the form

$$\begin{aligned}\boldsymbol{F} &= \boldsymbol{B}^T(\overline{s}\boldsymbol{I}-\boldsymbol{A}^T)^{-1}\left[(\overline{s}\boldsymbol{I}-\boldsymbol{A}^T)\boldsymbol{R}+\boldsymbol{R}(s\boldsymbol{I}-\boldsymbol{A})\right](s\boldsymbol{I}-\boldsymbol{A})^{-1}\boldsymbol{B} \\ &= \boldsymbol{B}^T(\overline{s}\boldsymbol{I}-\boldsymbol{A}^T)^{-1}\left[(s+\overline{s})\boldsymbol{R}-\boldsymbol{A}^T\boldsymbol{R}-\boldsymbol{R}\boldsymbol{A}\right](s\boldsymbol{I}-\boldsymbol{A})^{-1}\boldsymbol{B} \\ &= \boldsymbol{B}^T(\overline{s}\boldsymbol{I}-\boldsymbol{A}^T)^{-1}\left[2\mathrm{Re}\{s\}\cdot\boldsymbol{R}+\boldsymbol{L}^T\boldsymbol{L}\right](s\boldsymbol{I}-\boldsymbol{A})^{-1}\boldsymbol{B}, \end{aligned} \tag{2.84}$$

where the last conversion is based on the KYP equation (2.77). Equation (2.84) is now inserted into equation (2.83), which yields

$$\begin{aligned}\boldsymbol{G}(s)+\boldsymbol{G}^T(\overline{s}) &= \left[\boldsymbol{W}^T+\boldsymbol{B}^T(\overline{s}\boldsymbol{I}-\boldsymbol{A}^T)^{-1}\boldsymbol{L}^T\right]\left[\boldsymbol{W}+\boldsymbol{L}(s\boldsymbol{I}-\boldsymbol{A})^{-1}\boldsymbol{B}\right] \\ &\quad +2\mathrm{Re}\{s\}\cdot\boldsymbol{B}^T(\overline{s}\boldsymbol{I}-\boldsymbol{A}^T)^{-1}\boldsymbol{R}(s\boldsymbol{I}-\boldsymbol{A})^{-1}\boldsymbol{B}.\end{aligned}$$

[11] The degree n of a transfer matrix $\boldsymbol{G}(s)$ is defined as the degree of the lowest common denominator of all minors of $\boldsymbol{G}(s)$. This denominator is called the characteristic polynomial of the transfer matrix.

When the equation is more compactly formulated, i. e.

$$\begin{aligned} \boldsymbol{G}(s)+\boldsymbol{G}^T(\overline{s}) &= \boldsymbol{M}^*\boldsymbol{M}+2\mathrm{Re}\,\{s\}\cdot\boldsymbol{N}^*\boldsymbol{R}\boldsymbol{N}, \\ \boldsymbol{M} &= \boldsymbol{W}+\boldsymbol{L}(s\boldsymbol{I}-\boldsymbol{A})^{-1}\boldsymbol{B}, \\ \boldsymbol{N} &= (s\boldsymbol{I}-\boldsymbol{A})^{-1}\boldsymbol{B}, \end{aligned}$$

we immediately recognize that $\boldsymbol{G}(s)+\boldsymbol{G}^T(\overline{s})$ is positive semidefinite for all $s \in \mathbb{C}$ with $\mathrm{Re}\,\{s\} > 0$. This is because the matrix $\boldsymbol{M}^*\boldsymbol{M}$ is positive semidefinite and so is the matrix $\boldsymbol{N}^*\boldsymbol{R}\boldsymbol{N}$, since $\boldsymbol{R}$ is positive definite. The summation

$$\boldsymbol{M}^*\boldsymbol{M}+2\mathrm{Re}\,\{s\}\cdot\boldsymbol{N}^*\boldsymbol{R}\boldsymbol{N}$$

of both matrices is also positive semidefinite only, and not positive definite, since the determinant

$$\det\left(\boldsymbol{M}^*\boldsymbol{M}+2\mathrm{Re}\,\{s\}\cdot\boldsymbol{N}^*\boldsymbol{R}\boldsymbol{N}\right)$$

is a polynomial in s and can take the value zero. The KYP inequalities (2.77), (2.78), and (2.79) are thus sufficient for the positive realness of the transfer matrix (2.82). As shown in [17, 58] they are also necessary, so the following theorem holds:

Theorem 36 (Positive Realness and the KYP Equations). *Let there be a controllable and observable system*

$$\begin{aligned} \dot{\boldsymbol{x}} &= \boldsymbol{A}\boldsymbol{x}+\boldsymbol{B}\boldsymbol{u}, \quad \boldsymbol{x}\in I\!R^n, \quad \boldsymbol{u}\in I\!R^m, \\ \boldsymbol{y} &= \boldsymbol{C}\boldsymbol{x}+\boldsymbol{D}\boldsymbol{u}, \quad \boldsymbol{y}\in I\!R^m, \end{aligned}$$

with the associated transfer matrix

$$\boldsymbol{G}(s) = \boldsymbol{C}(s\boldsymbol{I}-\boldsymbol{A})^{-1}\boldsymbol{B}+\boldsymbol{D}.$$

The transfer matrix $\boldsymbol{G}(s)$ is positive real if and only if the Kalman-Yakubovich-Popov equations

$$\begin{aligned} \boldsymbol{A}^T\boldsymbol{R}+\boldsymbol{R}\boldsymbol{A} &= -\boldsymbol{L}^T\boldsymbol{L}, \\ \boldsymbol{L}^T\boldsymbol{W} &= \boldsymbol{C}^T-\boldsymbol{R}\boldsymbol{B}, \\ \boldsymbol{W}^T\boldsymbol{W} &= \boldsymbol{D}+\boldsymbol{D}^T \end{aligned}$$

are solvable by a positive definite symmetric $n\times n$ matrix $\boldsymbol{R}$, a matrix $\boldsymbol{L}$, and a matrix $\boldsymbol{W}$.

Further, we obtain the KYP equations for the strictly positive real case if we combine the KYP equation

$$\boldsymbol{A}^T\boldsymbol{R}+\boldsymbol{R}\boldsymbol{A} = -\boldsymbol{L}^T\boldsymbol{L} \tag{2.85}$$

of the above theorem with the requirement from Definition 26 of strict positive realness, i. e. that

$$\begin{aligned} \boldsymbol{G}(s-\epsilon) &= \boldsymbol{C}[(s-\epsilon)\boldsymbol{I}-\boldsymbol{A}]^{-1}\boldsymbol{B}+\boldsymbol{D} \\ &= \boldsymbol{C}[s\boldsymbol{I}-(\boldsymbol{A}+\epsilon\boldsymbol{I})]^{-1}\boldsymbol{B}+\boldsymbol{D} \end{aligned}$$

must be positive real. Accordingly, in equation (2.85) instead of the matrix $\boldsymbol{A}$ we use the matrix $\boldsymbol{A}+\epsilon\boldsymbol{I}$ and arrive at

$$(\boldsymbol{A}+\epsilon\boldsymbol{I})^T\boldsymbol{R}+\boldsymbol{R}(\boldsymbol{A}+\epsilon\boldsymbol{I}) = -\boldsymbol{L}^T\boldsymbol{L}.$$

From this we obtain the equation

$$\boldsymbol{A}^T\boldsymbol{R}+\boldsymbol{R}\boldsymbol{A} = -\boldsymbol{L}^T\boldsymbol{L}-2\epsilon\boldsymbol{R},$$

which we can represent with no loss of generality using

$$\mu = 2\epsilon > 0$$

as

$$\boldsymbol{A}^T\boldsymbol{R}+\boldsymbol{R}\boldsymbol{A} = -\boldsymbol{L}^T\boldsymbol{L}-\mu\boldsymbol{R}.$$

From this follows [465]

Theorem 37 (Strict Positive Realness and the KYP Equations). *Let there be a controllable and observable system*

$$\begin{aligned} \dot{\boldsymbol{x}} &= \boldsymbol{A}\boldsymbol{x}+\boldsymbol{B}\boldsymbol{u}, \quad \boldsymbol{x}\in I\!R^n,\ \boldsymbol{u}\in I\!R^m, \\ \boldsymbol{y} &= \boldsymbol{C}\boldsymbol{x}+\boldsymbol{D}\boldsymbol{u}, \quad \boldsymbol{y}\in I\!R^m, \end{aligned}$$

with the associated transfer matrix

$$\boldsymbol{G}(s) = \boldsymbol{C}(s\boldsymbol{I}-\boldsymbol{A})^{-1}\boldsymbol{B}+\boldsymbol{D}.$$

The transfer matrix $\boldsymbol{G}(s)$ is strictly positive real if and only if the Kalman-Yakubovich-Popov equations

$$\begin{aligned} \boldsymbol{A}^T\boldsymbol{R}+\boldsymbol{R}\boldsymbol{A} &= -\boldsymbol{L}^T\boldsymbol{L}-\mu\boldsymbol{R}, \\ \boldsymbol{L}^T\boldsymbol{W} &= \boldsymbol{C}^T-\boldsymbol{R}\boldsymbol{B}, \\ \boldsymbol{W}^T\boldsymbol{W} &= \boldsymbol{D}+\boldsymbol{D}^T \end{aligned}$$

are solvable by a positive definite symmetric $n\times n$ matrix $\boldsymbol{R}$, a matrix $\boldsymbol{L}$, a matrix $\boldsymbol{W}$, and a positive constant $\mu > 0$.

By comparing Theorem 36 and Theorem 28, we can deduce the equivalence of the terms *passive* and *positive real*. However, if we compare Theorem 37 and Theorem 29, equivalence between strictly passive systems and strictly positive real transfer functions seems not to be present. This is because the KYP equation

$$\boldsymbol{A}^T\boldsymbol{R} + \boldsymbol{R}\boldsymbol{A} = -\boldsymbol{L}^T\boldsymbol{L} - \mu\boldsymbol{P} \tag{2.86}$$

from Theorem 29 and the KYP equation

$$\boldsymbol{A}^T\boldsymbol{R} + \boldsymbol{R}\boldsymbol{A} = -\boldsymbol{L}^T\boldsymbol{L} - \mu\boldsymbol{R} \tag{2.87}$$

from Theorem 37 differ. Obviously, equation (2.87) from Theorem 37 is more restrictive, since in equation (2.86) from Theorem 29 we are not restricted in our choice of the matrix $\boldsymbol{P}$ to the particular matrix $\boldsymbol{R}$. However, it is possible to state that in the case of a positive definite solution matrix $\boldsymbol{R}$ in equation (2.87), if we choose

$$\boldsymbol{P} = \boldsymbol{R},$$

equation (2.86) is also fulfilled. Thus we have so far only proven that systems with strictly positive real transfer functions are strictly passive.

However, it is possible to show that the reverse is also true [297], because in the case of positive definite solution matrices $\boldsymbol{R}$ and $\boldsymbol{P}$ for equation (2.86), equation (2.87) is also fulfilled in every case. The system properties of strict passivity and strict positive realness are thus equivalent to each other. They describe the same circumstance: one is based on the system representation in the state space, the other on the system representation in the frequency domain.

Summarizing all of this, we can formulate [297]

Theorem 38 (Passivity and Positive Realness). *A controllable and observable system*

$$\begin{aligned}\dot{\boldsymbol{x}} &= \boldsymbol{A}\boldsymbol{x} + \boldsymbol{B}\boldsymbol{u},\\ \boldsymbol{y} &= \boldsymbol{C}\boldsymbol{x} + \boldsymbol{D}\boldsymbol{u}\end{aligned}$$

is (strictly) passive with a positive definite storage function if and only if its transfer matrix is (strictly) positive real.

Via the theorems on positive realness, which are based on transfer functions, we obtain an alternate way of evaluating whether a system is passive or strictly passive. Here the criteria for SISO systems are easy to utilize, while the criteria for MIMO systems are often very laborious.

All theorems that are based on the KYP equations involve searching for a matrix $\boldsymbol{R}$ which solves them. Especially for the determination of strict passivity, a criterion would be useful which provides a statement on the strict passivity or strict positive realness of a system without requiring the search for a matrix $\boldsymbol{R}$. For the most common case, i. e. for systems without a feedthrough, this is provided by [85]

Theorem 39 (Corless and Shorten's Theorem). *A transfer function*

$$\boldsymbol{G}(s) = \boldsymbol{C}(s\boldsymbol{I} - \boldsymbol{A})^{-1}\boldsymbol{B}, \quad \boldsymbol{C} \in I\!R^{m\times n},\ \boldsymbol{B} \in I\!R^{n\times m},$$

is strictly positive real if and only if

(1) all poles of $\boldsymbol{G}(s)$ have negative real parts,
(2) the $m \times m$ matrix $\boldsymbol{R} = -\boldsymbol{CAB} - (\boldsymbol{CAB})^T$ is positive definite, and
(3) the matrix

$$\begin{bmatrix} \boldsymbol{A} + \boldsymbol{ABR}^{-1}\boldsymbol{CA} & \boldsymbol{ABR}^{-1}\boldsymbol{B}^T\boldsymbol{A}^T \\ -\boldsymbol{A}^T\boldsymbol{C}^T\boldsymbol{R}^{-1}\boldsymbol{CA} & -\boldsymbol{A}^T - \boldsymbol{A}^T\boldsymbol{C}^T\boldsymbol{R}^{-1}\boldsymbol{B}^T\boldsymbol{A}^T \end{bmatrix}$$

has $2m$ eigenvalues at zero, and no further eigenvalues on the imaginary axis.

Since strict positive realness and strict passivity, according to Theorem 38, are equivalent to each other for controllable and observable systems, Theorem 39 is also a necessary and sufficient criterion for the strict passivity of a controllable and observable system.

2.4.8 Lossless Hamiltonian Systems

A class of passive systems common in control engineering is the class of mechanical systems, in particular multi-body systems such as cranes, satellites, and industrial robots. We can determine these systems' equations of motion by their kinetic and potential energies E_{kin} and E_{pot} via the Lagrange equations [184]

$$\frac{d}{dt}\frac{\partial L}{\partial \dot{q}_i} - \frac{\partial L}{\partial q_i} = \tau_i, \quad i = 1, \ldots, k,$$

or the corresponding vectorial form

$$\frac{d}{dt}\frac{\partial L}{\partial \dot{\boldsymbol{q}}} - \frac{\partial L}{\partial \boldsymbol{q}} = \boldsymbol{\tau}^T. \tag{2.88}$$

Here

$$L = E_{\text{kin}} - E_{\text{pot}} \tag{2.89}$$

is the Lagrange function; the variables q_i written as vector

$$\boldsymbol{q} = \begin{bmatrix} q_1 \\ \vdots \\ q_k \end{bmatrix}$$

denote generalized coordinates, i. e. lengths or angles, and the values τ_i written as vector

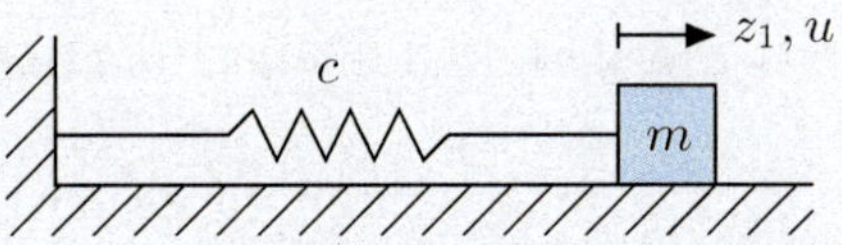

Fig. 2.74: Spring-mass system

$$\boldsymbol{\tau} = \begin{bmatrix} \tau_1 \\ \vdots \\ \tau_k \end{bmatrix}$$

denote forces or torques.

For illustration purposes, we will look at a frictionless spring-mass system, as depicted in Figure 2.74. By applying an external force u that extends the spring, the mass m can be displaced by a stretch z_1. The spring with its associated stiffness c counteracts this action. For the kinetic and potential energy, it holds that

$$E_{\text{kin}} = \frac{1}{2} m \dot{z}_1^2,$$
$$E_{\text{pot}} = \frac{1}{2} c z_1^2,$$

and therefore

$$L = \frac{1}{2}\left(m\dot{z}_1^2 - cz_1^2\right).$$

Inserting this into the Lagrange equation yields

$$\frac{1}{2}\left[\frac{d}{dt}\left(\frac{\partial(m\dot{z}_1^2 - cz_1^2)}{\partial \dot{z}_1}\right) - \frac{\partial(m\dot{z}_1^2 - cz_1^2)}{\partial z_1}\right] = u.$$

From this, we obtain

$$m\ddot{z}_1 + cz_1 = u.$$

Using the state vector

$$\boldsymbol{z} = \begin{bmatrix} z_1 \\ z_2 \end{bmatrix} = \begin{bmatrix} z_1 \\ \dot{z}_1 \end{bmatrix},$$

the state-space model of the system becomes

$$\dot{\boldsymbol{z}} = \begin{bmatrix} 0 & 1 \\ -\dfrac{c}{m} & 0 \end{bmatrix} \boldsymbol{z} + \begin{bmatrix} 0 \\ \dfrac{1}{m} \end{bmatrix} u.$$

Aside from the Lagrange method, there is a further method from which a description of the system can be derived and which is convenient in control theory. This is one in which the passivity of the system is immediately apparent. It is based on a generalized energy function, which is called the *Hamiltonian function.*

We can derive this convenient system description by using the Lagrange function as follows. First we will define the new coordinates

$$\boldsymbol{p} = \left(\frac{\partial L}{\partial \dot{\boldsymbol{q}}}\right)^T. \tag{2.90}$$

Our next step is to define the Hamiltonian function

$$H(\boldsymbol{q},\boldsymbol{p}) = \dot{\boldsymbol{q}}^T\boldsymbol{p} - L(\boldsymbol{q},\dot{\boldsymbol{q}}). \tag{2.91}$$

Now we differentiate equation (2.91) on both sides with respect to time and arrive at

$$\frac{dH(\boldsymbol{q},\boldsymbol{p})}{dt} = \frac{d\dot{\boldsymbol{q}}^T\boldsymbol{p}}{dt} - \frac{dL(\boldsymbol{q},\dot{\boldsymbol{q}})}{dt},$$

which yields

$$\frac{\partial H}{\partial \boldsymbol{q}}\dot{\boldsymbol{q}} + \frac{\partial H}{\partial \boldsymbol{p}}\dot{\boldsymbol{p}} = \ddot{\boldsymbol{q}}^T\boldsymbol{p} + \dot{\boldsymbol{q}}^T\dot{\boldsymbol{p}} - \frac{\partial L}{\partial \boldsymbol{q}}\dot{\boldsymbol{q}} - \frac{\partial L}{\partial \dot{\boldsymbol{q}}}\ddot{\boldsymbol{q}}. \tag{2.92}$$

Now we insert equation (2.90) into equation (2.92), which yields

$$\frac{\partial H}{\partial \boldsymbol{q}}\dot{\boldsymbol{q}} + \frac{\partial H}{\partial \boldsymbol{p}}\dot{\boldsymbol{p}} = -\frac{\partial L}{\partial \boldsymbol{q}}\dot{\boldsymbol{q}} + \dot{\boldsymbol{q}}^T\dot{\boldsymbol{p}}. \tag{2.93}$$

From the Lagrange equation (2.88) and using equation (2.90), it follows that

$$\frac{\partial L}{\partial \boldsymbol{q}} = \frac{d}{dt}\frac{\partial L}{\partial \dot{\boldsymbol{q}}} - \boldsymbol{\tau}^T = \dot{\boldsymbol{p}}^T - \boldsymbol{\tau}^T.$$

We insert this into equation (2.93) and arrive at

$$\frac{\partial H}{\partial \boldsymbol{q}}\dot{\boldsymbol{q}} + \frac{\partial H}{\partial \boldsymbol{p}}\dot{\boldsymbol{p}} = (-\dot{\boldsymbol{p}} + \boldsymbol{\tau})^T\dot{\boldsymbol{q}} + \dot{\boldsymbol{q}}^T\dot{\boldsymbol{p}}. \tag{2.94}$$

By comparing both sides of the equation, we obtain the system equations, which are known as the *Hamiltonian canonical form*

$$\begin{aligned} \dot{\boldsymbol{q}} &= \left(\frac{\partial H}{\partial \boldsymbol{p}}\right)^T, \\ \dot{\boldsymbol{p}} &= -\left(\frac{\partial H}{\partial \boldsymbol{q}}\right)^T + \boldsymbol{\tau}. \end{aligned} \tag{2.95}$$

They are also referred to as the *Hamiltonian equations*, which are $2k$ first-order differential equations, while the corresponding equivalent Lagrange equations are k second-order differential equations.

Below we will explain the meaning of the Hamiltonian function H. Here we will take into account that, in conventional mechanical systems, the kinetic energy takes the form

$$E_{\text{kin}} = \frac{1}{2}\dot{\boldsymbol{q}}^T \boldsymbol{M}(\boldsymbol{q})\dot{\boldsymbol{q}}. \tag{2.96}$$

For example, in the simple case of linear motion it holds that

$$E_{\text{kin}} = \frac{1}{2}mv^2.$$

In equation (2.96), the matrix $\boldsymbol{M}(\boldsymbol{q})$ is a generalized mass matrix whose elements depend on the position $\boldsymbol{q}$; an example is the joint angles q_i of an industrial robot. This generalized mass matrix is positive definite and thus also regular. In contrast to the kinetic energy E_{kin}, the potential energy E_{pot} is only a function of the position $\boldsymbol{q}$ and not of the velocity $\dot{\boldsymbol{q}}$.

From equations (2.89) and (2.90) it follows with equation (2.96) that

$$\boldsymbol{p} = \left(\frac{\partial L}{\partial \dot{\boldsymbol{q}}}\right)^T = \left(\frac{\partial E_{\text{kin}}(\boldsymbol{q},\dot{\boldsymbol{q}})}{\partial \dot{\boldsymbol{q}}} - \frac{\partial E_{\text{pot}}(\boldsymbol{q})}{\partial \dot{\boldsymbol{q}}}\right)^T = \boldsymbol{M}(\boldsymbol{q})\dot{\boldsymbol{q}}. \tag{2.97}$$

Inserting this into equation (2.91), along with equations (2.89), (2.96), and (2.97), yields

$$\begin{aligned} H(\boldsymbol{q},\boldsymbol{p}) &= \dot{\boldsymbol{q}}^T \boldsymbol{M}(\boldsymbol{q})\dot{\boldsymbol{q}} - (E_{\text{kin}}(\boldsymbol{q},\dot{\boldsymbol{q}}) - E_{\text{pot}}(\boldsymbol{q})) \\ &= \frac{1}{2}\dot{\boldsymbol{q}}^T \boldsymbol{M}(\boldsymbol{q})\dot{\boldsymbol{q}} + E_{\text{pot}}(\boldsymbol{q}) \end{aligned}$$

and thus

$$H(\boldsymbol{q},\boldsymbol{p}) = E_{\text{kin}}(\boldsymbol{q},\boldsymbol{p}) + E_{\text{pot}}(\boldsymbol{q}).$$

The Hamiltonian function therefore corresponds to the total amount of energy in the system.

We can now derive a further important property of the Hamiltonian function. For this purpose, we will take a further look at equation (2.94),

$$\underbrace{\frac{\partial H}{\partial \boldsymbol{q}}\dot{\boldsymbol{q}} + \frac{\partial H}{\partial \boldsymbol{p}}\dot{\boldsymbol{p}}}_{\dfrac{dH}{dt}} = \boldsymbol{\tau}^T\dot{\boldsymbol{q}}\underbrace{-\dot{\boldsymbol{p}}^T\dot{\boldsymbol{q}} + \dot{\boldsymbol{q}}^T\dot{\boldsymbol{p}}}_{=0}.$$

It immediately follows that

$$\frac{dH}{dt} = \boldsymbol{\tau}^T\dot{\boldsymbol{q}}. \tag{2.98}$$

The interpretation of this equation is useful for our understanding. Two cases can be distinguished:

(1) No external forces $\boldsymbol{\tau}$ influence the system. It therefore holds that

$$\frac{dH}{dt} = 0,$$

which means that H, and thus the total energy of the system, is constant. This case describes lossless systems without an external energy input, such as ideal pendulum clocks, meaning clocks without friction.

(2) An external force $\boldsymbol{\tau}$ influences the system. In this case,

$$\frac{dH}{dt} = \boldsymbol{\tau}^T \dot{\boldsymbol{q}}$$

defines the supplied power. An example of this is an industrial robot, for which motors induce forces onto the joints.

What specific steps do we take to determine the Hamiltonian equations? The following five steps describe the procedure:

Step 1: Write down the Lagrange function $L = E_{\text{kin}} - E_{\text{pot}}$.

Step 2: Determine the generalized impulses

$$\boldsymbol{p} = \left(\frac{\partial L}{\partial \dot{\boldsymbol{q}}}\right)^T = \boldsymbol{M}(\boldsymbol{q})\dot{\boldsymbol{q}}.$$

Step 3: By inverting the equation from the previous step, determine the generalized velocities

$$\dot{\boldsymbol{q}} = \boldsymbol{M}^{-1}(\boldsymbol{q})\boldsymbol{p}.$$

Step 4: Write down the Hamiltonian function

$$H(\boldsymbol{q},\boldsymbol{p}) = E_{\text{kin}}(\boldsymbol{q},\boldsymbol{p}) + E_{\text{pot}}(\boldsymbol{q}).$$

Step 5: Form the Hamiltonian equations (2.95).

Determining the inverse function $\dot{\boldsymbol{q}} = \boldsymbol{M}^{-1}(\boldsymbol{q})\boldsymbol{p}$ of Step 3 can be especially complicated and time-consuming. This is one of the reasons why the Hamiltonian equations are often more difficult to derive than the motion equations via the Lagrange equations. Furthermore, the Hamiltonian equations are often of a more complex form.

As an example, we will again examine the spring-mass system from Figure 2.74. With

$$q = z_1 \quad \text{and} \quad p = \frac{\partial L}{\partial \dot{q}} = \frac{1}{2}\frac{\partial\left(m\dot{z}_1^2 - cz_1^2\right)}{\partial \dot{z}_1} = m\dot{z}_1,$$

the Hamiltonian function becomes

$$H = E_{\text{kin}} + E_{\text{pot}} = \frac{1}{2}m\dot{z}_1^2 + \frac{1}{2}cz_1^2 = \frac{1}{2m}p^2 + \frac{c}{2}q^2, \tag{2.99}$$

which yields

$$\begin{aligned}\dot{q} &= \frac{\partial H}{\partial p} = \frac{p}{m},\\ \dot{p} &= -\frac{\partial H}{\partial q} + u = -cq + u\end{aligned} \tag{2.100}$$

for the Hamiltonian equations (2.95). In this example, the Hamiltonian equations have a simple form.

Now we will investigate the Hamiltonian equations (2.95) for passivity. We subsume the coordinates $\boldsymbol{q}$ and $\boldsymbol{p}$ to a state vector

$$\boldsymbol{x} = \begin{bmatrix} \boldsymbol{q} \\ \boldsymbol{p} \end{bmatrix}$$

so that we can formulate the Hamiltonian equations as

$$\dot{\boldsymbol{x}} = \boldsymbol{J}_{\text{H}} \left(\frac{\partial H}{\partial \boldsymbol{x}}\right)^T + \boldsymbol{B} \cdot \boldsymbol{\tau}, \quad \boldsymbol{J}_{\text{H}} = \begin{bmatrix} \boldsymbol{0} & \boldsymbol{I} \\ -\boldsymbol{I} & \boldsymbol{0} \end{bmatrix}, \quad \boldsymbol{B} = \begin{bmatrix} \boldsymbol{0} \\ \boldsymbol{I} \end{bmatrix}, \tag{2.101}$$

where $\boldsymbol{I}$ is the k-dimensional identity matrix.

If we choose

$$\boldsymbol{y} = \left(\frac{\partial H}{\partial \boldsymbol{p}}\right)^T = \dot{\boldsymbol{q}} \tag{2.102}$$

as the output vector, set $\boldsymbol{u} = \boldsymbol{\tau}$, and assume a storage function $S(\boldsymbol{x})$, it holds for the passivity inequality that

$$\dot{S}(\boldsymbol{x}) \le \boldsymbol{u}^T \boldsymbol{y} = \boldsymbol{\tau}^T \dot{\boldsymbol{q}}. \tag{2.103}$$

This equation can obviously be solved if we choose

$$S(\boldsymbol{x}) = H(\boldsymbol{x})$$

as the storage function, since, using equation (2.98), the identity

$$\dot{S} = \dot{H} = \boldsymbol{\tau}^T \dot{\boldsymbol{q}} \tag{2.104}$$

follows. We know if H is a positive semidefinite function, i. e. it holds that $H(\boldsymbol{0}) = 0$ and $H(\boldsymbol{x}) \ge 0$ for $\boldsymbol{x} \neq 0$, the system is passive. According to the equal sign in equation (2.103) and equation (2.104), it is also lossless.

In our spring-mass example, H is a positive definite function, since

$$H(p,q) = 0 \quad \text{for} \quad [q \;\; p]^T = [z_1 \;\; m\dot{z}_1]^T = \boldsymbol{0}$$

holds and $H(p,q)$ is positive everywhere else, both of which we can see from equation (2.99). Further, with equation (2.99) and equation (2.100) it holds that

$$\dot{H} = \frac{p\dot{p}}{m} + cq\dot{q} = \frac{up}{m} = u\dot{q} = uy,$$

so that equation (2.103) is fulfilled, and thus the spring-mass system is passive.

In the following, we generalize the Hamiltonian equations (2.101) with the output function (2.102) to

$$\begin{aligned} \dot{\boldsymbol{x}} &= \boldsymbol{J}(\boldsymbol{x}) \left(\frac{\partial S}{\partial \boldsymbol{x}}\right)^T + \boldsymbol{B}(\boldsymbol{x}) \cdot \boldsymbol{u}, \\ \boldsymbol{y} &= \boldsymbol{B}^T(\boldsymbol{x}) \left(\frac{\partial S}{\partial \boldsymbol{x}}\right)^T \end{aligned} \tag{2.105}$$

by considering the matrices $\boldsymbol{J}$ and $\boldsymbol{B}$ to be dependent on the state vector $\boldsymbol{x}$. Here arbitrary skew-symmetric matrices

$$\boldsymbol{J}(\boldsymbol{x}) = -\boldsymbol{J}^T(\boldsymbol{x})$$

are allowed. They describe the lossless energy exchange within the system. The positive definite function $S(\boldsymbol{x})$ we are now using instead of the Hamiltonian function $H(\boldsymbol{x})$ also constitutes a generalization. It may be a Hamiltonian function, but could also be a differently obtained positive definite function. However, we will restrict ourselves to positive definite functions $S(\boldsymbol{x})$, since only these functions, when used as storage functions, guarantee stability in connection with passivity. The generalized system (2.105), which is described by the generalized Hamiltonian equations, is called an *input* or *port-controlled Hamiltonian system (PCH system)*. The term *port-controlled* results from the fact that we analyze systems with input variables $\boldsymbol{u}$ or $\boldsymbol{\tau}$.

The system (2.105) is also passive and lossless, and consequently stable if $S(\boldsymbol{x})$ is positive definite, since it holds that

$$\dot{S}(\boldsymbol{x}) = \frac{\partial S}{\partial \boldsymbol{x}}\dot{\boldsymbol{x}} = \frac{\partial S}{\partial \boldsymbol{x}}\boldsymbol{J}(\boldsymbol{x})\left(\frac{\partial S}{\partial \boldsymbol{x}}\right)^T + \frac{\partial S}{\partial \boldsymbol{x}}\boldsymbol{B}(\boldsymbol{x}) \cdot \boldsymbol{u}.$$

Based on the skew-symmetry of $\boldsymbol{J}$, we obtain

$$\frac{\partial S}{\partial \boldsymbol{x}}\boldsymbol{J}\left(\frac{\partial S}{\partial \boldsymbol{x}}\right)^T = 0,$$

and thus

$$\dot{S}(\boldsymbol{x}) = \frac{\partial S}{\partial \boldsymbol{x}}\boldsymbol{B}(\boldsymbol{x}) \cdot \boldsymbol{u} = \boldsymbol{u}^T \boldsymbol{y}.$$

The passivity inequality (2.103) is therefore fulfilled, and the function S is the storage function that we were searching for.

2.4.9 Example: Self-Balancing Vehicle

As an example of a PCH system, we will look at a self-balancing vehicle with one axis and one or two wheels, as shown in Figure 2.75. Industrial applications of such systems are the one-wheeled Honda U3-X [174], the Segway [220, 322], and the Toyota Winglet [275], which both have two wheels. One or two motors in the chassis drive the wheel (or wheels) with the generated torque M.

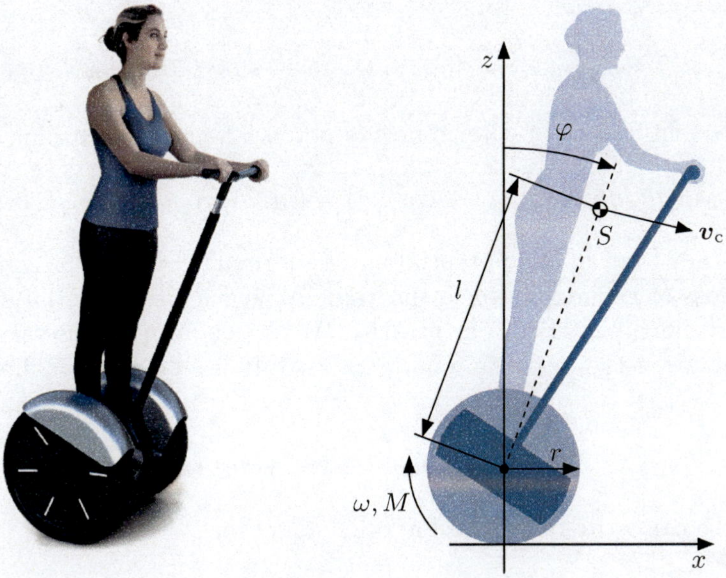

Fig. 2.75: Self-balancing vehicle

To set up the dynamic equations in PCH form, we will first determine the kinetic energy of the wheels

$$E_{\text{kin,w}} = \frac{1}{2}J_{\text{w}}\omega^2 + \frac{1}{2}m_{\text{w}}r^2\omega^2 = \frac{1}{2}\frac{J_{\text{w}}}{r^2}\dot{x}^2 + \frac{1}{2}m_{\text{w}}\dot{x}^2,$$

where ω denotes the angular velocity of the wheels, which we consider to be equal for both wheels. The velocity of the vehicle is given by $\dot{x} = r\omega$. The sum of the inertia moment of the wheels is denoted as J_{w}. Their mass and radius are denoted by m_{w} and r, respectively. The driving force that acts upon the wheels is the torque M, which is generated by the motor.

Similarly, for the combined kinetic energy of the driver and chassis mounted between the wheels, it holds that

$$E_{\text{kin,c}} = \frac{1}{2}J_{\text{c}}\dot{\varphi}^2 + \frac{1}{2}m_{\text{c}}|\boldsymbol{v}_{\text{c}}|^2, \tag{2.106}$$

where φ is the angle between the vertical z-coordinate axis and the line passing through the center of gravity S and the wheel axis, $\boldsymbol{v}_{\text{c}}$ is the velocity of the center of gravity S, J_c is the inertia moment of the driver and chassis with respect to the wheel axis as the axis of rotation, and m_c is the mass of the driver and chassis.

For the velocity $\boldsymbol{v}_\mathrm{c}$ of the center of gravity S, it holds that

$$\boldsymbol{v}_\mathrm{c} = \begin{bmatrix} r\omega + l\dot{\varphi}\cos(\varphi) \\ -l\dot{\varphi}\sin(\varphi) \end{bmatrix} = \begin{bmatrix} \dot{x} + l\dot{\varphi}\cos(\varphi) \\ -l\dot{\varphi}\sin(\varphi) \end{bmatrix}$$

and thus for the kinetic energy (2.106) of the driver and chassis it follows that

$$E_\mathrm{kin,c} = \frac{1}{2}J_\mathrm{c}\dot{\varphi}^2 + \frac{1}{2}m_\mathrm{c}\left(\dot{x}^2 + 2l\dot{x}\dot{\varphi}\cos(\varphi) + l^2\dot{\varphi}^2\right).$$

Subsequently, we will determine the potential energy according to

$$E_\mathrm{pot} = m_\mathrm{w}gr + m_\mathrm{c}g(r + l\cos(\varphi)) = (m_\mathrm{w} + m_\mathrm{c})gr + m_\mathrm{c}gl\cos(\varphi).$$

The constant term in the potential energy formula can be arbitrarily chosen with no loss of generality, since the value at which the potential energy becomes zero can always be set arbitrarily. We choose $m_\mathrm{c}gl$ as this value, so that the minimum of the potential energy is zero. It lies at $\varphi = \pi$. Thus it holds that

$$E_\mathrm{pot} = m_\mathrm{c}gl\cos(\varphi) + m_\mathrm{c}gl.$$

Now we can write down the Lagrange function

$$L = E_\mathrm{kin,w} + E_\mathrm{kin,c} - E_\mathrm{pot}$$

from which, in the subsequent second step, we will be able to determine the generalized impulses

$$\boldsymbol{p} = \left(\frac{\partial L}{\partial \dot{\boldsymbol{q}}}\right)^T.$$

In the present case, we have

$$\boldsymbol{q} = \begin{bmatrix} x \\ \varphi \end{bmatrix},$$

and thus

$$\boldsymbol{p} = \left(\frac{\partial L}{\partial \dot{\boldsymbol{q}}}\right)^T = \begin{bmatrix} \dfrac{\partial L}{\partial \dot{x}} \\ \dfrac{\partial L}{\partial \dot{\varphi}} \end{bmatrix} = \underbrace{\begin{bmatrix} m_\mathrm{w} + m_\mathrm{c} + \dfrac{J_\mathrm{w}}{r^2} & m_\mathrm{c}l\cos(\varphi) \\ m_\mathrm{c}l\cos(\varphi) & m_\mathrm{c}l^2 + J_\mathrm{c} \end{bmatrix}}_{\boldsymbol{M}(\boldsymbol{q})} \underbrace{\begin{bmatrix} \dot{x} \\ \dot{\varphi} \end{bmatrix}}_{\dot{\boldsymbol{q}}} \tag{2.107}$$

holds.

In the third step, we can determine from equation (2.107), i. e. from

$$\boldsymbol{p} = \boldsymbol{M}(\boldsymbol{q})\dot{\boldsymbol{q}},$$

the vector

$$\dot{\boldsymbol{q}} = \boldsymbol{M}^{-1}(\boldsymbol{q})\boldsymbol{p},$$

which yields

$$\dot{\boldsymbol{q}} = \begin{bmatrix} \dot{x} \\ \dot{\varphi} \end{bmatrix} = \frac{1}{ab - c^2\cos^2(\varphi)} \begin{bmatrix} b & -c\cdot\cos(\varphi) \\ -c\cdot\cos(\varphi) & a \end{bmatrix} \boldsymbol{p} \tag{2.108}$$

with the abbreviations

$$a = m_\mathrm{w} + m_\mathrm{c} + \frac{J_\mathrm{w}}{r^2}, \qquad b = m_\mathrm{c} l^2 + J_\mathrm{c}, \qquad c = m_\mathrm{c} l.$$

The fourth step provides the Hamiltonian function

$$\begin{aligned} H(\boldsymbol{q},\boldsymbol{p}) =& E_\mathrm{kin,w} + E_\mathrm{kin,c} + E_\mathrm{pot} \\ =& \frac{1}{2}\left(m_\mathrm{w} + m_\mathrm{c} + \frac{J_\mathrm{w}}{r^2}\right)\dot{x}^2 + \frac{1}{2}\left(m_\mathrm{c} l^2 + J_\mathrm{c}\right)\dot{\varphi}^2 + c\dot{x}\dot{\varphi}\cos(\varphi) + cg\cos(\varphi) + cg \\ =& \frac{1}{2}a\dot{x}^2 + \frac{1}{2}b\dot{\varphi}^2 + c\dot{x}\dot{\varphi}\cos(\varphi) + cg\cos(\varphi) + cg. \end{aligned}$$

In this equation, regarding

$$[\dot{x}\ \ \dot{\varphi}]^T = \dot{\boldsymbol{q}},$$

we insert equation (2.108), where

$$\boldsymbol{p} = [p_1\ \ p_2]^T.$$

This results in

$$H(\boldsymbol{q},\boldsymbol{p}) = \frac{ap_2^2 + bp_1^2 - 2cp_1p_2\cos(\varphi)}{2\left(ab - c^2\cos^2(\varphi)\right)} + cg\cos(\varphi) + cg. \tag{2.109}$$

The Hamiltonian function for $\dot{x} = 0$ and $\dot{\varphi} = 0$, i. e. $p_1 = 0$ and $p_2 = 0$ due to equation (2.107), and for $\varphi = \pi$ as well as $x \in \mathbb{R}$, possesses minima with

$$H([x\ \pi]^T, \boldsymbol{0}) = 0.$$

The angle $\varphi = \pi$ obviously does not appear in the real system; however, in the mathematical system it provides stable equilibrium points for all states $x \in \mathbb{R}$ and $\boldsymbol{p} = \boldsymbol{0}$. It is plausible that the equilibrium points at $\varphi = 0$, $x \in \mathbb{R}$, and $\boldsymbol{p} = \boldsymbol{0}$ are unstable.

In the fifth step, we obtain the system dynamics in Hamiltonian form as

$$\dot{\boldsymbol{q}} = \begin{bmatrix} \dot{x} \\ \dot{\varphi} \end{bmatrix} = \left(\frac{\partial H}{\partial \boldsymbol{p}}\right)^T = \frac{1}{ab - c^2\cos^2(\varphi)} \begin{bmatrix} bp_1 - cp_2\cos(\varphi) \\ ap_2 - cp_1\cos(\varphi) \end{bmatrix},$$

$$\dot{\boldsymbol{p}} = \begin{bmatrix} \dot{p}_1 \\ \dot{p}_2 \end{bmatrix} = -\left(\frac{\partial H}{\partial \boldsymbol{q}}\right)^T + \boldsymbol{\tau} = -\begin{bmatrix} \dfrac{\partial H}{\partial x} \\ \dfrac{\partial H}{\partial \varphi} \end{bmatrix} + \boldsymbol{\tau}$$

$$= \begin{bmatrix} 0 \\ c\cdot\sin(\varphi)\left[\dfrac{c\cos(\varphi)\left(ap_2^2 + bp_1^2 - cp_1p_2\cos(\varphi)\right) - abp_1p_2}{\left(ab - c^2\cos^2(\varphi)\right)^2} + g\right] \end{bmatrix} + \boldsymbol{\tau}.$$

The external forces or torques $\boldsymbol{\tau}$ result from the torque M, i. e.

$$\boldsymbol{\tau} = \begin{bmatrix} \dfrac{M}{r} \\ -M \end{bmatrix},$$

so that we finally arrive at the Hamiltonian equations

$$\begin{bmatrix} \dot{x} \\ \dot{\varphi} \\ \dot{p}_1 \\ \dot{p}_2 \end{bmatrix} = \begin{bmatrix} \dfrac{bp_1 - cp_2\cos(\varphi)}{ab - c^2\cos^2(\varphi)} \\ \dfrac{ap_2 - cp_1\cos(\varphi)}{ab - c^2\cos^2(\varphi)} \\ 0 \\ c\cdot\sin(\varphi)\left[\dfrac{c\cos(\varphi)\left(ap_2^2 + bp_1^2 - cp_1p_2\cos(\varphi)\right) - abp_1p_2}{\left(ab - c^2\cos^2(\varphi)\right)^2} + g\right] \end{bmatrix} + \begin{bmatrix} 0 \\ 0 \\ \dfrac{1}{r} \\ -1 \end{bmatrix} M. \tag{2.110}$$

For the output variables, we subsequently choose

$$\boldsymbol{y} = \left(\frac{\partial H}{\partial \boldsymbol{p}}\right)^T = \dot{\boldsymbol{q}} = \begin{bmatrix} \dot{x} \\ \dot{\varphi} \end{bmatrix}. \tag{2.111}$$

It holds that $H(\mathbf{0},\mathbf{0}) = 2cg$, and otherwise $H(\boldsymbol{q},\boldsymbol{p}) \geq 0$. Because of the inequality $H(\mathbf{0},\mathbf{0}) \neq 0$, the Hamiltonian function in the form from equation (2.109) is not positive semidefinite. Therefore we cannot use it to establish passivity for the vehicle. However, if we do not choose the angle φ as a state, but rather the angle $\Theta = \pi - \varphi$, we arrive at the modified Hamiltonian function

$$\begin{aligned}\hat{H}([x\ \Theta]^T,\boldsymbol{p}) &= H([x\ \varphi=\pi-\Theta]^T,\boldsymbol{p})\\ &= \frac{ap_2^2+bp_1^2+2cp_1p_2\cos(\Theta)}{2(ab-c^2\cos^2(\Theta))}+cg\,(1-\cos(\Theta)),\end{aligned}$$

which is positive semidefinite and with which we can finally prove the passivity of the system (2.110), (2.111).

For the sake of comparison, from the Lagrange equation

$$\frac{d}{dt}\frac{\partial L}{\partial \dot{\boldsymbol{q}}}-\frac{\partial L}{\partial \boldsymbol{q}}=\boldsymbol{\tau}^T,$$

which is the common way of deducing a system's dynamics, we can derive the equations

$$\begin{aligned}\ddot{x} &= \frac{c}{a}\left(\dot{\varphi}^2\sin(\varphi)-\ddot{\varphi}\cos(\varphi)\right)+\frac{M}{ar},\\ \ddot{\varphi} &= \frac{c}{b}\left(g\sin(\varphi)-\ddot{x}\cos(\varphi)\right)-\frac{M}{b},\end{aligned}$$

where the state vector is given by

$$[q_1\ \dot{q}_1\ q_2\ \dot{q}_2]^T=[x\ \dot{x}\ \varphi\ \dot{\varphi}]^T.$$

They are not only easier to determine, but are also less complex than the Hamiltonian equations (2.110). However, the passivity of the system cannot be immediately determined from these equations of dynamics and equation (2.111).

2.4.10 Dissipative Hamiltonian Systems

So far we have studied lossless Hamiltonian systems such as the spring-mass system shown in Figure 2.74 on p. 169. If we wish to consider losses as well, such as those due to friction, we must integrate an energy-consuming, i.e. dissipative element into our considerations.

To this end, we will introduce a dissipation term $\boldsymbol{D}(\boldsymbol{x})$ into the PCH system (2.105) which describes the system's internal energy losses, thus obtaining

$$\begin{aligned}\dot{\boldsymbol{x}} &= (\boldsymbol{J}(\boldsymbol{x})-\boldsymbol{D}(\boldsymbol{x}))\left(\frac{\partial V}{\partial \boldsymbol{x}}\right)^T+\boldsymbol{B}(\boldsymbol{x})\cdot\boldsymbol{u}, \qquad \boldsymbol{J}(\boldsymbol{x})=-\boldsymbol{J}^T(\boldsymbol{x}),\\ \boldsymbol{y} &= \boldsymbol{B}^T(\boldsymbol{x})\left(\frac{\partial V}{\partial \boldsymbol{x}}\right)^T.\end{aligned} \tag{2.112}$$

Here $\boldsymbol{D}(\boldsymbol{x})$ is a symmetric and positive semidefinite matrix. Such systems are termed *port-controlled Hamiltonian systems with dissipation* (*PCHD systems*). In this case V is no longer a Hamiltonian function in the conventional sense and does not necessarily have to represent the energy of the system. However, $V(\boldsymbol{x})$ should be a positive definite function. With

$$\dot{V}(\boldsymbol{x}) = \frac{\partial V}{\partial \boldsymbol{x}}\dot{\boldsymbol{x}} = \underbrace{\frac{\partial V}{\partial \boldsymbol{x}}\boldsymbol{J}(\boldsymbol{x})\left(\frac{\partial V}{\partial \boldsymbol{x}}\right)^T}_{=0} - \frac{\partial V}{\partial \boldsymbol{x}}\boldsymbol{D}(\boldsymbol{x})\left(\frac{\partial V}{\partial \boldsymbol{x}}\right)^T + \frac{\partial V}{\partial \boldsymbol{x}}\boldsymbol{B}(\boldsymbol{x})\cdot\boldsymbol{u}$$

we obtain the passivity inequality

$$\dot{V}(\boldsymbol{x}) = \boldsymbol{u}^T\boldsymbol{y} - \frac{\partial V}{\partial \boldsymbol{x}}\boldsymbol{D}(\boldsymbol{x})\left(\frac{\partial V}{\partial \boldsymbol{x}}\right)^T \le \boldsymbol{u}^T\boldsymbol{y}, \tag{2.113}$$

which is obviously fulfilled because of the positive semidefiniteness of $\boldsymbol{D}(\boldsymbol{x})$.

PCHD systems are thus always passive. For this reason, the system representation (2.112) is also referred to as the *passivity canonical form* [253, 419]. Since the storage function V is assumed to be positive definite, and because

$$\dot{V}(\boldsymbol{x}) \le 0$$

holds for $\boldsymbol{u} = \boldsymbol{0}$, V is a Lyapunov function of the passive system. This is also the reason why we no longer use the letter S but the letter V for the symbol of the storage function. The latter typically represents Lyapunov functions.

Furthermore, if the dissipation matrix $\boldsymbol{D}(\boldsymbol{x})$ is positive definite, we can fulfill the defining inequality for strict passivity

$$\dot{V}(\boldsymbol{x}) + R(\boldsymbol{x}) \le \boldsymbol{u}^T\boldsymbol{y},$$

for example by choosing

$$R(\boldsymbol{x}) = \frac{1}{2}\frac{\partial V}{\partial \boldsymbol{x}}\boldsymbol{D}(\boldsymbol{x})\left(\frac{\partial V}{\partial \boldsymbol{x}}\right)^T.$$

It subsequently holds that

$$\dot{V}(\boldsymbol{x}) + R(\boldsymbol{x}) = \boldsymbol{u}^T\boldsymbol{y} - \frac{1}{2}\frac{\partial V}{\partial \boldsymbol{x}}\boldsymbol{D}(\boldsymbol{x})\left(\frac{\partial V}{\partial \boldsymbol{x}}\right)^T \le \boldsymbol{u}^T\boldsymbol{y}.$$

A PCHD system with a positive definite matrix $\boldsymbol{D}(\boldsymbol{x})$ is thus strictly passive. According to Theorem 24 on p. 148, a strictly passive system possesses an asymptotically stable equilibrium point at $\boldsymbol{x} = \boldsymbol{0}$. This brings us to

Theorem 40 (Stability of PCHD Systems). *A PCHD system*

$$\dot{\boldsymbol{x}} = (\boldsymbol{J}(\boldsymbol{x}) - \boldsymbol{D}(\boldsymbol{x}))\left(\frac{\partial V}{\partial \boldsymbol{x}}\right)^T + \boldsymbol{B}(\boldsymbol{x})\cdot\boldsymbol{u},$$

$$\boldsymbol{y} = \boldsymbol{B}^T(\boldsymbol{x})\left(\frac{\partial V}{\partial \boldsymbol{x}}\right)^T$$

is strictly passive and has an asymptotically stable equilibrium point at $\boldsymbol{x} = \boldsymbol{0}$ *if* $\boldsymbol{D}(\boldsymbol{x})$ *is positive definite.*

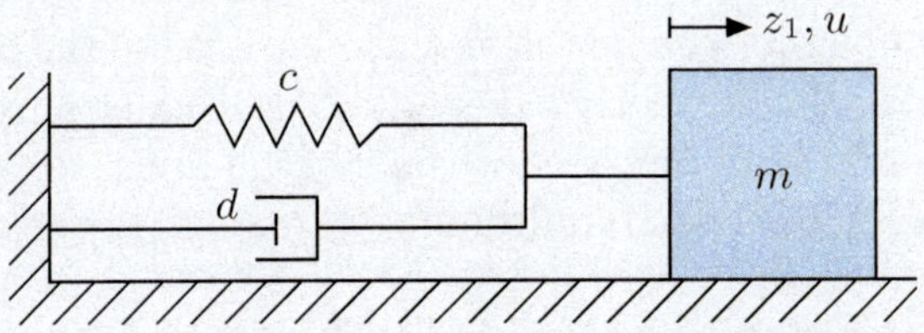

Fig. 2.76: Spring-mass system with damping element

As an illustrative example, we will again consider the spring-mass system and add a damping term to it as an energy consuming element, as shown in Figure 2.76. For the equilibrium of forces, it holds that

$$m\ddot{z}_1 = -cz_1 - d\dot{z}_1 + u.$$

The equations of motion are therefore given by

$$\begin{aligned} \dot{z}_1 &= z_2, \\ \dot{z}_2 &= -\frac{c}{m}z_1 - \frac{d}{m}z_2 + \frac{u}{m}. \end{aligned} \tag{2.114}$$

For the PCHD representation, we again set

$$q = z_1 \qquad \text{and} \qquad p = m\dot{z}_1 = mz_2.$$

Using the Hamiltonian function as the storage function V, i. e.

$$V(q,p) = E_{\text{kin}} + E_{\text{pot}} = \frac{1}{2}m\dot{z}_1^2 + \frac{1}{2}cz_1^2 = \frac{1}{2}\frac{p^2}{m} + \frac{1}{2}cq^2,$$

and $\boldsymbol{x} = [q \;\; p]^T$, we can represent system (2.114) as

$$\dot{\boldsymbol{x}} = \begin{bmatrix} \dot{q} \\ \dot{p} \end{bmatrix} = \begin{bmatrix} \dfrac{\partial V}{\partial p} \\ -\dfrac{\partial V}{\partial q} \end{bmatrix} - \begin{bmatrix} 0 \\ \dfrac{d}{m}p \end{bmatrix} + \begin{bmatrix} 0 \\ u \end{bmatrix} = \left[\begin{bmatrix} 0 & 1 \\ -1 & 0 \end{bmatrix} - \begin{bmatrix} 0 & 0 \\ 0 & d \end{bmatrix}\right]\left(\frac{\partial V}{\partial \boldsymbol{x}}\right)^T + \begin{bmatrix} 0 \\ 1 \end{bmatrix} u$$

and thus the PCHD representation as

$$\begin{bmatrix} \dot{q} \\ \dot{p} \end{bmatrix} = \left[\begin{bmatrix} 0 & 1 \\ -1 & 0 \end{bmatrix} - \begin{bmatrix} 0 & 0 \\ 0 & d \end{bmatrix}\right]\begin{bmatrix} cq \\ \dfrac{p}{m} \end{bmatrix} + \begin{bmatrix} 0 \\ 1 \end{bmatrix} u \tag{2.115}$$

with the output value

$$y = \frac{1}{m}p.$$

From this example, it becomes clear that, because of the additional damping term, we can no longer derive the differential equations for PCHD systems from the Hamiltonian function or the storage function by using equation (2.95), as we can for lossless Hamiltonian systems. This especially holds for the dissipation matrix $\boldsymbol{D}$.

Furthermore, the PCHD representation of a system is not unique. Rather, there are many different PCHD representations whose forms depend on the choice of the storage function V. It is not possible to prove asymptotic stability of a strictly passive system's equilibrium point $\boldsymbol{x}_{\text{eq}} = \boldsymbol{0}$ using every PCHD representation, since the matrix $\boldsymbol{D}$ need not be positive definite for a strictly passive system, as the above example illustrates. It depends on the choice of the function V whether a PCHD representation has a positive or positive semidefinite matrix $\boldsymbol{D}$. In Section 2.4.12 we will continue with example (2.114) and determine a PCHD representation with a positive definite matrix $\boldsymbol{D}$.

It is worth noting that many physical systems can be modeled as PCHD systems (2.112). In particular, passive mechanical and electrotechnical systems can be represented in this form. However, it is not always easy to find a PCHD representation for a strictly passive system.

2.4.11 Example: Separately Excited Direct-Current Machine

As a further example of a PCHD system, we will consider the separately excited direct-current machine. Its stator consists of an electromagnet, at which the stator voltage u_{s} is applied and through which the stator current i_{s} flows. It generates the magnetic flux linkage

$$\Psi_{\text{s}} = L_{\text{s}} i_{\text{s}},$$

where L_{s} is the inductance of the stator winding. Figure 2.77 shows the corresponding circuit diagram. Since the equation

$$u_{\text{sw}} = L_{\text{s}} \dot{i}_{\text{s}} = \dot{\Psi}_{\text{s}}$$

holds for the voltage u_{sw} which is induced in the stator winding, we obtain

$$u_{\text{s}} = u_{\text{sw}} + R_{\text{s}} i_{\text{s}} = \dot{\Psi}_{\text{s}} + \frac{R_{\text{s}}}{L_{\text{s}}} \Psi_{\text{s}} \tag{2.116}$$

for the stator voltage u_{s} as the sum of the induced voltage u_{sw} and the voltage at the ohmic stator resistor R_{s}.

The voltage induced in the rotor winding can be calculated using

$$u_{\text{rw}} = k\omega\Psi_{\text{s}}.$$

It depends bilinearly on the angular velocity ω and the flux linkage Ψ_{s} via the machine constant k. This yields the differential equation

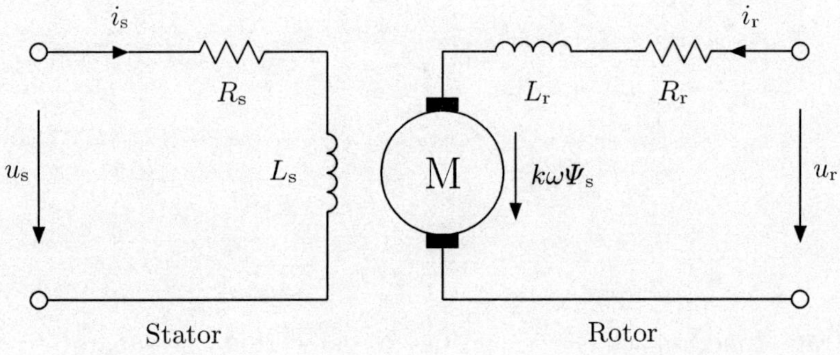

Fig. 2.77: Separately excited direct-current motor

$$u_r = R_r i_r + L_r \dot{i}_r + k\omega\Psi_s \tag{2.117}$$

for the rotor current i_r, where u_r is the terminal voltage, L_r is the rotor inductance, and R_r is the rotor resistance.

The machine generates the torque

$$M_{DCM} = k i_r \Psi_s,$$

so that we obtain

$$J\dot{\omega} = k i_r \Psi_s - d \cdot \omega - M_L \tag{2.118}$$

for the balance of torques, where $d \cdot \omega$ is the sliding-friction torque with d being the friction coefficient, M_L denotes the load torque, and J is the inertia moment of the rotor.

With the state vector $\boldsymbol{x}^T = [i_r \;\; \Psi_s \;\; \omega]$ and equations (2.116), (2.117), and (2.118), the state-space model

$$\dot{\boldsymbol{x}} = \begin{bmatrix} -\dfrac{R_r}{L_r}x_1 - \dfrac{k}{L_r}x_2x_3 \\ -\dfrac{R_s}{L_s}x_2 \\ \dfrac{k}{J}x_1x_2 - \dfrac{d}{J}x_3 \end{bmatrix} + \begin{bmatrix} \dfrac{1}{L_r} & 0 & 0 \\ 0 & 1 & 0 \\ 0 & 0 & -\dfrac{1}{J} \end{bmatrix} \begin{bmatrix} u_r \\ u_s \\ M_L \end{bmatrix}, \tag{2.119}$$

$$\boldsymbol{y} = \begin{bmatrix} x_1 & \dfrac{1}{L_s}x_2 & -x_3 \end{bmatrix}^T$$

can be derived. By adding up the energies that are stored in the inductances

$$E_{ind} = \frac{1}{2}L_r i_r^2 + \frac{1}{2}\frac{1}{L_s}\Psi_s^2$$

and the rotation energy

$$E_{\mathrm{rot}} = \frac{1}{2} J \omega^2,$$

we obtain

$$\begin{aligned} V(\boldsymbol{x}) = E_{\mathrm{ind}} + E_{\mathrm{rot}} &= \frac{1}{2} L_{\mathrm{r}} i_{\mathrm{r}}^2 + \frac{1}{2}\frac{1}{L_{\mathrm{s}}} \Psi_{\mathrm{s}}^2 + \frac{1}{2} J \omega^2 \\ &= \frac{1}{2} L_{\mathrm{r}} x_1^2 + \frac{1}{2}\frac{1}{L_{\mathrm{s}}} x_2^2 + \frac{1}{2} J x_3^2 \end{aligned}$$

as a storage function which we can use to show that the separately excited direct-current machine is passive. For the passivity inequality, we obtain

$$\dot{V}(\boldsymbol{x}) = L_{\mathrm{r}} x_1 \dot{x}_1 + \frac{1}{L_{\mathrm{s}}} x_2 \dot{x}_2 + J x_3 \dot{x}_3 \leq \boldsymbol{u}^T \boldsymbol{y},$$

where the input variable vector $\boldsymbol{u}$ is given by

$$\boldsymbol{u} = [u_{\mathrm{r}} \;\; u_{\mathrm{s}} \;\; M_{\mathrm{L}}]^T.$$

From this, along with equation (2.119), it follows that

$$\begin{aligned} \dot{V}(\boldsymbol{x}) &= -R_{\mathrm{r}} x_1^2 - \frac{R_{\mathrm{s}}}{L_{\mathrm{s}}^2} x_2^2 - d x_3^2 + x_1 u_{\mathrm{r}} + \frac{1}{L_{\mathrm{s}}} x_2 u_{\mathrm{s}} - x_3 M_{\mathrm{L}} \\ &\leq u_{\mathrm{r}} x_1 + \frac{1}{L_{\mathrm{s}}} u_{\mathrm{s}} x_2 - M_{\mathrm{L}} x_3, \end{aligned}$$

and from the above equation, we immediately arrive at

$$-R_{\mathrm{r}} x_1^2 - \frac{R_{\mathrm{s}}}{L_{\mathrm{s}}^2} x_2^2 - d x_3^2 \leq 0.$$

Since this inequality holds for all $\boldsymbol{x} \neq 0$, the separately excited direct-current machine is passive. It is even strictly passive, since we can define a positive definite function $R(\boldsymbol{x})$ in such a way that

$$\dot{V}(\boldsymbol{x}) + R(\boldsymbol{x}) \leq \boldsymbol{u}^T \boldsymbol{y}$$

holds. For example, we can choose

$$R(\boldsymbol{x}) = \frac{1}{2}\left[R_{\mathrm{r}} x_1^2 + \frac{R_{\mathrm{s}}}{L_{\mathrm{s}}^2} x_2^2 + d x_3^2 \right].$$

The passivity of the machine can also be directly deduced from its PCHD representation. For this purpose, we rearrange equation (2.119). This yields

$$\dot{\boldsymbol{x}} = (\boldsymbol{J}(\boldsymbol{x}) - \boldsymbol{D}(\boldsymbol{x})) \left(\frac{\partial V(\boldsymbol{x})}{\partial \boldsymbol{x}} \right)^T + \boldsymbol{B}\boldsymbol{u}, \tag{2.120}$$

where the matrices are

$$\boldsymbol{J}(\boldsymbol{x}) = \begin{bmatrix} 0 & 0 & -\dfrac{k}{JL_{\mathrm{r}}}x_2 \\ 0 & 0 & 0 \\ \dfrac{k}{JL_{\mathrm{r}}}x_2 & 0 & 0 \end{bmatrix}, \quad \boldsymbol{D}(\boldsymbol{x}) = \begin{bmatrix} \dfrac{R_{\mathrm{r}}}{L_{\mathrm{r}}^2} & 0 & 0 \\ 0 & R_{\mathrm{s}} & 0 \\ 0 & 0 & \dfrac{d}{J^2} \end{bmatrix},$$

$$\boldsymbol{B}(\boldsymbol{x}) = \begin{bmatrix} \dfrac{1}{L_{\mathrm{r}}} & 0 & 0 \\ 0 & 1 & 0 \\ 0 & 0 & -\dfrac{1}{J} \end{bmatrix}.$$

The dissipation matrix $\boldsymbol{D}$ describes the energy loss resulting from the machine's ohmic resistances and friction. Since this matrix is positive definite, according to Theorem 40 the system is strictly passive and its equilibrium point $\boldsymbol{x}_{\mathrm{eq}} = \boldsymbol{0}$ is asymptotically stable.

2.4.12 Linear Hamiltonian Systems

Considering the linear systems

$$\begin{aligned} \dot{\boldsymbol{x}} &= \boldsymbol{A}\boldsymbol{x} + \boldsymbol{B}\boldsymbol{u}, \\ \boldsymbol{y} &= \boldsymbol{C}\boldsymbol{x}, \end{aligned} \tag{2.121}$$

we can ask in which case such a system can be represented as a PCH system

$$\begin{aligned} \dot{\boldsymbol{x}} &= \boldsymbol{J}\left(\frac{\partial V}{\partial \boldsymbol{x}}\right)^T + \boldsymbol{B}\boldsymbol{u}, \\ \boldsymbol{y} &= \boldsymbol{B}^T\left(\frac{\partial V}{\partial \boldsymbol{x}}\right)^T \end{aligned} \tag{2.122}$$

with the associated storage function $V(\boldsymbol{x})$. As a condition for the storage function, by comparing equation (2.121) with equation (2.122), we obtain the identity

$$\boldsymbol{C}\boldsymbol{x} = \boldsymbol{B}^T\left(\frac{\partial V}{\partial \boldsymbol{x}}\right)^T. \tag{2.123}$$

This means that the positive definite storage function V must be a quadratic form

$$V(\boldsymbol{x}) = \frac{1}{2}\boldsymbol{x}^T\boldsymbol{R}\boldsymbol{x}, \qquad \boldsymbol{R} = \boldsymbol{R}^T, \tag{2.124}$$

with a positive definite matrix $\boldsymbol{R}$. Thus equation (2.123) is equivalent to

$$\boldsymbol{C} = \boldsymbol{B}^T \boldsymbol{R}. \tag{2.125}$$

Furthermore, the system given in equation (2.121) must be lossless and consequently passive to be representable in PCH form (2.122). This means that $\dot{V} = \boldsymbol{u}^T \boldsymbol{y}$ must hold, i. e.

$$\dot{V}(\boldsymbol{x}) = \frac{\partial V}{\partial \boldsymbol{x}} \dot{\boldsymbol{x}} = \frac{1}{2} \boldsymbol{x}^T \left(\boldsymbol{A}^T \boldsymbol{R} + \boldsymbol{R}\boldsymbol{A} \right) \boldsymbol{x} + \boldsymbol{u}^T \boldsymbol{B}^T \boldsymbol{R} \boldsymbol{x} = \boldsymbol{u}^T \boldsymbol{y} = \boldsymbol{u}^T \boldsymbol{B}^T \boldsymbol{R} \boldsymbol{x}.$$

It follows that

$$\boldsymbol{A}^T \boldsymbol{R} + \boldsymbol{R}\boldsymbol{A} = \boldsymbol{0}, \tag{2.126}$$

which is equivalent to

$$\dot{V}(\boldsymbol{x}) = 0 \quad \text{for} \quad \boldsymbol{u} = \boldsymbol{0}.$$

This in turn is equivalent to the fact that the storage function V is a Lyapunov function and that all the system's trajectories run along the ellipsoidal contour lines of the function V. The system (2.121) is thus Lyapunov stable, but not asymptotically stable. Consequently, the system may not have more than one eigenvalue at $\lambda_i = 0$, and otherwise has only complex conjugate eigenvalues $\lambda_i = j\omega_i$ and $\lambda_{i+1} = -j\omega_i$ with $\omega_i \in I\!R \backslash \{0\}$.

For us to be able to represent the system (2.121) in PCH form (2.122), it must further hold that

$$\boldsymbol{A}\boldsymbol{x} = \boldsymbol{J} \left(\frac{\partial V}{\partial \boldsymbol{x}} \right)^T = \boldsymbol{J}\boldsymbol{R}\boldsymbol{x},$$

or

$$\boldsymbol{J} = \boldsymbol{A}\boldsymbol{R}^{-1}. \tag{2.127}$$

Since $\boldsymbol{J}$ is a skew-symmetric matrix and $\boldsymbol{R}$ is symmetric, it must hold that

$$\boldsymbol{A}\boldsymbol{R}^{-1} = -\left(\boldsymbol{A}\boldsymbol{R}^{-1} \right)^T = -\boldsymbol{R}^{-1}\boldsymbol{A}^T$$

and thus, after multiplying both sides by $\boldsymbol{R}$, it follows that

$$\boldsymbol{A}^T \boldsymbol{R} + \boldsymbol{R}\boldsymbol{A} = \boldsymbol{0}.$$

This equation is identical to equation (2.126) and is therefore fulfilled, so that we can choose equation (2.127) for $\boldsymbol{J}$.

Equations (2.125) and (2.126) are the Kalman-Yakubovich-Popov equations from Theorem 28 on p. 153 for the case $\boldsymbol{L} = \boldsymbol{0}$. They are the necessary and sufficient conditions for the system (2.121) being lossless.

Summarizing the results above, we obtain

Theorem 41 (Linear PCH Systems). *A linear system*

$$\begin{aligned}\dot{\boldsymbol{x}} &= \boldsymbol{A}\boldsymbol{x} + \boldsymbol{B}\boldsymbol{u},\\ \boldsymbol{y} &= \boldsymbol{C}\boldsymbol{x}\end{aligned}$$

is lossless and representable as a PCH system

$$\begin{aligned}\dot{\boldsymbol{x}} &= \boldsymbol{J}\left(\frac{\partial V}{\partial \boldsymbol{x}}\right)^T + \boldsymbol{B}\boldsymbol{u},\\ \boldsymbol{y} &= \boldsymbol{B}^T\left(\frac{\partial V}{\partial \boldsymbol{x}}\right)^T\end{aligned}$$

with a positive definite storage function

$$V(\boldsymbol{x}) = \frac{1}{2}\boldsymbol{x}^T\boldsymbol{R}\boldsymbol{x}$$

and $\boldsymbol{J} = \boldsymbol{A}\boldsymbol{R}^{-1}$ *if and only if a positive definite matrix* $\boldsymbol{R}$ *exists such that*

$$\begin{aligned}\boldsymbol{A}^T\boldsymbol{R} + \boldsymbol{R}\boldsymbol{A} &= \boldsymbol{0},\\ \boldsymbol{B}^T\boldsymbol{R} &= \boldsymbol{C}\end{aligned}$$

holds.

Next we will examine PCHD systems

$$\begin{aligned}\dot{\boldsymbol{x}} &= (\boldsymbol{J} - \boldsymbol{D})\left(\frac{\partial V}{\partial \boldsymbol{x}}\right)^T + \boldsymbol{B}\boldsymbol{u},\\ \boldsymbol{y} &= \boldsymbol{B}^T\left(\frac{\partial V}{\partial \boldsymbol{x}}\right)^T,\end{aligned} \tag{2.128}$$

intending to analyze which conditions allow the linear system (2.121) to be represented as a PCHD system with the associated storage function V. We remind ourselves that the dissipation matrix $\boldsymbol{D}$ is positive semidefinite in this case. This property of $\boldsymbol{D}$ will be needed later.

We will assume that equations (2.123), (2.124), and (2.125) hold again. If the system (2.121) is to be a passive PCHD system with a positive definite storage function (2.124), the passivity inequality

$$\dot{V}(\boldsymbol{x}) = \frac{1}{2}\boldsymbol{x}^T\left(\boldsymbol{A}^T\boldsymbol{R} + \boldsymbol{R}\boldsymbol{A}\right)\boldsymbol{x} + \boldsymbol{u}^T\boldsymbol{B}^T\boldsymbol{R}\boldsymbol{x} \le \boldsymbol{u}^T\boldsymbol{y} = \boldsymbol{u}^T\boldsymbol{B}^T\boldsymbol{R}\boldsymbol{x}$$

must hold, i. e. the equation

$$\boldsymbol{A}^T\boldsymbol{R} + \boldsymbol{R}\boldsymbol{A} = -\boldsymbol{Q} \tag{2.129}$$

with the positive semidefinite matrix $\boldsymbol{Q}$ must be fulfilled.

Furthermore, if the inequality

$$\dot{V}(\boldsymbol{x}) + R(\boldsymbol{x}) \leq \boldsymbol{u}^T \boldsymbol{y},$$

i. e.

$$\frac{1}{2}\boldsymbol{x}^T \left(\boldsymbol{A}^T\boldsymbol{R} + \boldsymbol{R}\boldsymbol{A}\right)\boldsymbol{x} + R(\boldsymbol{x}) \leq 0, \tag{2.130}$$

is fulfilled with $R(\boldsymbol{x})$ being a positive definite function, it follows that the system (2.121) is strictly passive. Since $R(\boldsymbol{x})$ is positive definite, the matrix

$$\boldsymbol{A}^T\boldsymbol{R} + \boldsymbol{R}\boldsymbol{A}$$

must be negative definite in this case, so that equation (2.130) can be fulfilled. Since we are free to choose $R(\boldsymbol{x})$ as we wish, we will set

$$R(\boldsymbol{x}) = -\frac{1}{4}\boldsymbol{x}^T \left(\boldsymbol{A}^T\boldsymbol{R} + \boldsymbol{R}\boldsymbol{A}\right)\boldsymbol{x} = \frac{1}{4}\boldsymbol{x}^T\boldsymbol{Q}\boldsymbol{x} > 0 \quad \text{for} \quad \boldsymbol{x} \neq \boldsymbol{0},$$

which for equation (2.130) yields

$$\frac{1}{4}\boldsymbol{x}^T \left(\boldsymbol{A}^T\boldsymbol{R} + \boldsymbol{R}\boldsymbol{A}\right)\boldsymbol{x} \leq 0.$$

This inequality is obviously fulfilled. Thus if the matrix $\boldsymbol{Q}$ from equation (2.129) is positive definite, it follows that the system (2.121) is strictly passive. According to Theorem 16 on p. 121, the storage function $V(\boldsymbol{x})$ is thus a strict Lyapunov function of the system (2.121) and the system is, in contrast to PCH systems, asymptotically stable.

With the storage function (2.124) the differential equation of system (2.128) takes the form

$$\dot{\boldsymbol{x}} = (\boldsymbol{J} - \boldsymbol{D})\,\boldsymbol{R}\boldsymbol{x} + \boldsymbol{B}\boldsymbol{u}.$$

Comparing this equation to (2.121) yields

$$\boldsymbol{A} = (\boldsymbol{J} - \boldsymbol{D})\,\boldsymbol{R}. \tag{2.131}$$

Similar to the previous case involving PCH systems, we will utilize the fact that

$$\boldsymbol{J} = \boldsymbol{A}\boldsymbol{R}^{-1} + \boldsymbol{D}$$

is skew-symmetric. It must therefore hold that

$$\boldsymbol{A}\boldsymbol{R}^{-1} + \boldsymbol{D} = -\left(\boldsymbol{A}\boldsymbol{R}^{-1} + \boldsymbol{D}\right)^T.$$

It follows that

$$AR^{-1} + R^{-1}A^T = -2D.$$

After multiplying both sides of this equation by R and using the matrix

$$Q = 2RDR,$$

we finally arrive at the Lyapunov equation

$$A^T R + RA = -Q. \tag{2.132}$$

Since R is positive definite and D is positive semidefinite due to the system's passivity, it follows that Q is a positive semidefinite matrix. Equation (2.132) is identical to equation (2.129), which is fulfilled because of the system's passivity.

According to Theorem 16 on p. 121, the Lyapunov equation (2.129) with the positive semidefinite matrix Q is only solvable if the matrix A does not possess any eigenvalues with a positive real part, i. e. the system (2.121) must be Lyapunov stable or asymptotically stable.

How do we obtain the matrices J and D? To do this, we will use equation (2.131) and rearrange it to

$$J - D = AR^{-1}.$$

The matrix J is skew-symmetric and D is symmetric. We will take advantage of this, since every matrix M can be decomposed into a skew-symmetric first part and a symmetric second part using

$$M = \frac{1}{2}\left(M - M^T\right) + \frac{1}{2}\left(M + M^T\right).$$

For the skew-symmetric part, we thus obtain

$$J = \frac{1}{2}\left(AR^{-1} - R^{-1}A^T\right)$$

and for the symmetric positive semidefinite part, we obtain

$$D = -\frac{1}{2}\left(AR^{-1} + R^{-1}A^T\right).$$

With these matrices, the system

$$\begin{aligned} \dot{x} &= Ax + Bu, \\ y &= Cx \end{aligned} \tag{2.133}$$

can be represented in PCHD form (2.128) under the condition that equation (2.125) and equation (2.132) hold. Again, these two equations are the conditions of the Kalman-Yakubovich-Popov lemma, i. e. Theorem 28 on p. 153, or Theorem 29 on p. 154. Since this is sufficient and necessary for passivity, we can formulate

Theorem 42 (Linear PCHD Systems). *A linear system*

$$\dot{\boldsymbol{x}} = \boldsymbol{A}\boldsymbol{x} + \boldsymbol{B}\boldsymbol{u},$$
$$\boldsymbol{y} = \boldsymbol{C}\boldsymbol{x}$$

is (strictly) passive and can be represented as a PCHD system

$$\dot{\boldsymbol{x}} = (\boldsymbol{J} - \boldsymbol{D})\left(\frac{\partial V}{\partial \boldsymbol{x}}\right)^T + \boldsymbol{B}\boldsymbol{u},$$
$$\boldsymbol{y} = \boldsymbol{B}^T\left(\frac{\partial V}{\partial \boldsymbol{x}}\right)^T$$

with a positive definite storage function

$$V(\boldsymbol{x}) = \frac{1}{2}\boldsymbol{x}^T\boldsymbol{R}\boldsymbol{x} \tag{2.134}$$

and

$$\boldsymbol{J} = \frac{1}{2}\left(\boldsymbol{A}\boldsymbol{R}^{-1} - \boldsymbol{R}^{-1}\boldsymbol{A}^T\right),$$
$$\boldsymbol{D} = -\frac{1}{2}\left(\boldsymbol{A}\boldsymbol{R}^{-1} + \boldsymbol{R}^{-1}\boldsymbol{A}^T\right)$$

if and only if a positive definite matrix $\boldsymbol{R}$ *and a positive semidefinite (positive definite) matrix* $\boldsymbol{Q}$ *exist such that*

$$\boldsymbol{A}^T\boldsymbol{R} + \boldsymbol{R}\boldsymbol{A} = -\boldsymbol{Q}, \tag{2.135}$$
$$\boldsymbol{B}^T\boldsymbol{R} = \boldsymbol{C}$$

holds.

Similar to the case of PCH systems, it follows that a linear system (2.133) can be represented as a PCHD system if and only if it is passive.

The theorem above once again highlights why the system representation of PCHD systems is referred to as the *passivity canonical form*. This results from the fact that passive linear systems can always be written in the passivity canonical form, i. e. as PCHD systems. As already mentioned, it may well be that there is a strictly passive system for which the storage function (2.134) only leads to a positive semidefinite matrix $\boldsymbol{Q}$ in equation (2.135). However, choosing a different appropriate storage function, or more precisely, a matrix $\boldsymbol{R}$, we can prove strict passivity.

For illustrative purposes, we will again consider the damped spring-mass system (2.114), or (2.115) with

$$c = 1\,\mathrm{N\,m^{-1}},$$
$$d = 1\,\mathrm{kg\,s^{-1}},$$
$$m = 0.5\,\mathrm{kg},$$

i. e.

$$\dot{\boldsymbol{z}} = \underbrace{\begin{bmatrix} 0 & 1 \\ -2 & -2 \end{bmatrix}}_{\boldsymbol{A}} \boldsymbol{z} + \underbrace{\begin{bmatrix} 0 \\ 2 \end{bmatrix}}_{\boldsymbol{b}} u. \tag{2.136}$$

Accordingly, a PCHD system representation is given by

$$\begin{aligned} \dot{\boldsymbol{z}} &= \Bigg(\underbrace{\begin{bmatrix} 0 & 1 \\ -1 & 0 \end{bmatrix}}_{\boldsymbol{J}} - \underbrace{\begin{bmatrix} 0 & 0 \\ 0 & 2 \end{bmatrix}}_{\boldsymbol{D}} \Bigg) \left(\frac{\partial V}{\partial \boldsymbol{z}} \right)^T + \underbrace{\begin{bmatrix} 0 \\ 2 \end{bmatrix}}_{\boldsymbol{b}} u, \\ y &= \boldsymbol{b}^T \boldsymbol{R}_1 \boldsymbol{z} \end{aligned} \tag{2.137}$$

with

$$V(\boldsymbol{z}) = \frac{1}{2} \boldsymbol{z}^T \underbrace{\begin{bmatrix} 2 & 0 \\ 0 & 1 \end{bmatrix}}_{\boldsymbol{R}_1} \boldsymbol{z}. \tag{2.138}$$

This PCHD representation differs from the one given in equation (2.115), since we are not using the Hamiltonian coordinates

$$q = z_1$$

and

$$p = m z_2,$$

but the original coordinates z_1 and z_2. In this way, we do not have to perform a change of coordinates, and we can use the same state vector $\boldsymbol{z}$ in both system representations (2.136) and (2.137). The storage function (2.138), which is also a Lyapunov function, leads – as already shown in equation (2.113) – to the passivity inequality

$$\dot{V}(\boldsymbol{z}) = uy - \frac{\partial V}{\partial \boldsymbol{z}} \boldsymbol{D} \left(\frac{\partial V}{\partial \boldsymbol{z}} \right)^T \leq uy, \tag{2.139}$$

i. e. to

$$-\frac{\partial V}{\partial \boldsymbol{z}} \boldsymbol{D} \left(\frac{\partial V}{\partial \boldsymbol{z}} \right)^T \leq 0$$

with the positive semidefinite matrix $\boldsymbol{D}$. The matrix

$$\boldsymbol{A}^T \boldsymbol{R}_1 + \boldsymbol{R}_1 \boldsymbol{A} = -\begin{bmatrix} 0 & 0 \\ 0 & 4 \end{bmatrix}$$

is also only negative semidefinite. Since $\boldsymbol{D}$ is only positive semidefinite and not positive definite, in equation (2.139) the equality is fulfilled for some specific cases. So with equation (2.139), we have only provided evidence for the passivity of the system.

If, on the other hand, we choose

$$V(\boldsymbol{z}) = \frac{1}{2}\boldsymbol{z}^T \underbrace{\begin{bmatrix} 2 & 1 \\ 1 & 1 \end{bmatrix}}_{\boldsymbol{R}_2} \boldsymbol{z}, \tag{2.140}$$

as the storage function, we arrive at the PCHD representation

$$\begin{aligned} \dot{\boldsymbol{z}} &= \left(\underbrace{\begin{bmatrix} 0 & 1 \\ -1 & 0 \end{bmatrix}}_{\boldsymbol{J}} - \underbrace{\begin{bmatrix} 1 & -1 \\ -1 & 2 \end{bmatrix}}_{\boldsymbol{D}} \right) \left(\frac{\partial V}{\partial \boldsymbol{z}} \right)^T + \underbrace{\begin{bmatrix} 0 \\ 2 \end{bmatrix}}_{\boldsymbol{b}} u, \\ y &= \boldsymbol{b}^T \boldsymbol{R}_2 \boldsymbol{z}. \end{aligned} \tag{2.141}$$

For strict passivity, the inequality

$$\dot{V}(\boldsymbol{z}) + R(\boldsymbol{z}) = -\frac{\partial V}{\partial \boldsymbol{z}} \boldsymbol{D} \left(\frac{\partial V}{\partial \boldsymbol{z}} \right)^T + R(\boldsymbol{z}) + uy \le uy, \tag{2.142}$$

must be satisfied, where $R(\boldsymbol{z})$ must be a positive definite function. With the derivative

$$\frac{\partial V}{\partial \boldsymbol{z}} = \boldsymbol{R}_2 \boldsymbol{z},$$

equation (2.142) is equivalent to

$$-\boldsymbol{z}^T \boldsymbol{R}_2 \boldsymbol{D} \boldsymbol{R}_2 \boldsymbol{z} + R(\boldsymbol{z}) \le 0. \tag{2.143}$$

Since the matrix $\boldsymbol{D}$ and thus the matrix $\boldsymbol{R}_2 \boldsymbol{D} \boldsymbol{R}_2$ are positive definite, we can choose

$$R(\boldsymbol{z}) = \frac{1}{2} \boldsymbol{z}^T \boldsymbol{R}_2 \boldsymbol{D} \boldsymbol{R}_2 \boldsymbol{z}$$

as a positive definite function which fulfills equation (2.142) or, equivalently, equation (2.143). Furthermore, it is also equivalent to the result given in equation (2.143), that the matrix

$$\boldsymbol{A}^T \boldsymbol{R}_2 + \boldsymbol{R}_2 \boldsymbol{A} = -\begin{bmatrix} 4 & 2 \\ 2 & 2 \end{bmatrix}$$

is negative definite. We have thus proven the strict passivity of the system (2.141) with the associated storage function (2.140). At this point, we note that different PCHD models of a system have different output variables and that the output is irrelevant to stability.

In conclusion, we will note that there is an infinite number of PCHD representations of the same system. Whether only the passivity of the system can be proven based on the storage function V, or whether the strict passivity can be proven as well, depends on whether the dissipation matrix $\boldsymbol{D}$ is positive semidefinite or positive definite.

2.5 Exercises

Exercise 2.1 Calculate the describing functions for the four functions in Figure 2.78.

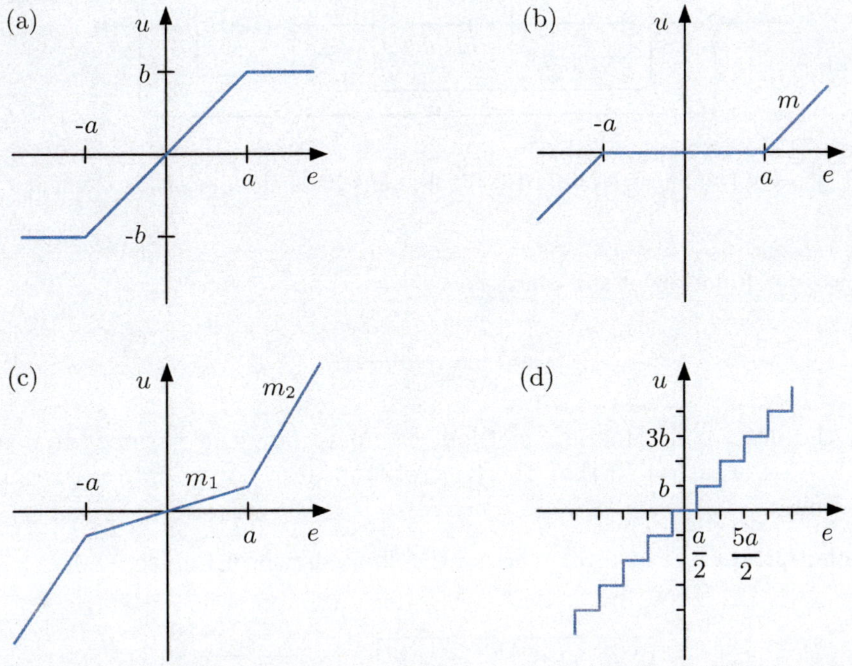

Fig. 2.78: Nonlinear functions; slopes are denoted by m, m_1, and m_2.

Exercise 2.2 Calculate the describing functions for the following two functions:

(a) $f(e) = k_1 e + k_2 e^3 + k_3 e^5$, (b) $f(e) = \sum_{i=1}^{n} k_i e^{2i-1}$.

Exercise 2.3 As in Exercises 1.6 and 1.7, we will now view Chua's circuit, here in the normalized form already calculated in Exercise 1.6 on p. 60 and analyzed in Exercise 1.7 on p. 61:

$$\dot{\boldsymbol{x}} = \begin{bmatrix} -\frac{44}{5} & \frac{44}{5} & 0 \\ 1 & -1 & 1 \\ 0 & -15 & 0 \end{bmatrix} \boldsymbol{x} + \begin{bmatrix} 1 \\ 0 \\ 0 \end{bmatrix} (-g(x_1)), \quad g(x_1) = \frac{1}{7} \begin{cases} -5x_1 - 3, & x_1 \geq 1, \\ -8x_1, & |x_1| < 1, \\ -5x_1 + 3, & x_1 \leq -1, \end{cases}$$

with the output variable $y = x_1$.

(a) Determine the transfer function $G(s) = Y(s)/U(s)$ with $u = -g(x_1)$.

(b) Represent Chua's circuit as a nonlinear standard control loop.

(c) Determine the describing function of $-g(x_1)$. Note that $g(-x_1) = -g(x_1)$.

(d) Demonstrate that the harmonic balance does not give any indication of a limit cycle.

Exercise 2.4 Let us examine the nonlinear standard control loop shown in Figure 2.79.

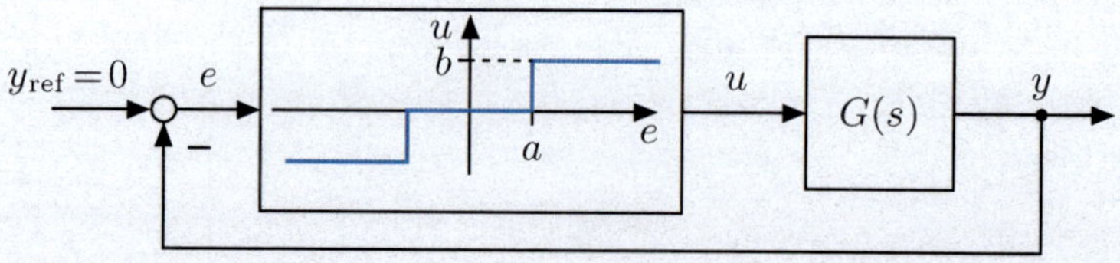

Fig. 2.79: Nonlinear standard control loop with a three-position element

The transfer function of the plant is

$$G = \frac{4}{s(s^2 + 4s + 4)}.$$

Using the describing function method, calculate how the parameters a and b need to be adjusted so that the control loop does not go into a sustained oscillation.

Exercise 2.5 Let us examine the control loop shown in Figure 2.80.

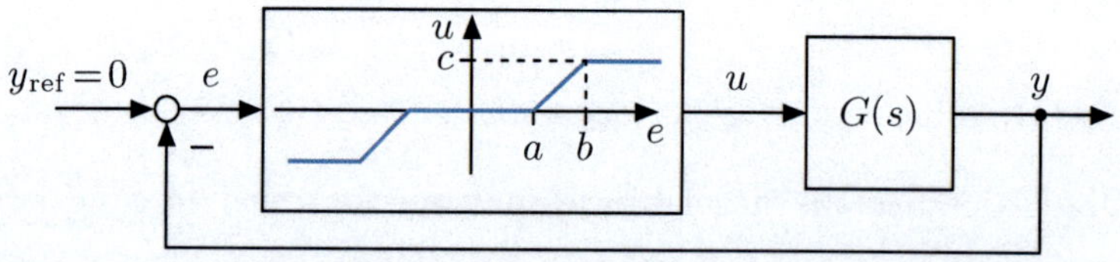

Fig. 2.80: Nonlinear standard control loop with a limited dead zone

(a) Calculate and sketch the describing function of the nonlinear element.

Let us now examine the special case $a = 1$, $b = 3$, $c = 2$.

(b) Assume the plant to be

$$G(s) = \frac{K}{0.25s^3 + 0.375s^2 + s + 1}.$$

Approximately calculate the amplitudes of the two limit circles that can occur and determine their frequency for $K = 1$.

(c) For which value $K = K_s$ do we obtain a semi-stable limit circle? How would you describe its stability behavior for varying input amplitudes A?

(d) What types of stability behavior do the limit cycles have for values K which are smaller or greater than K_s?

Exercise 2.6 Now we will consider the blood-pressure regulation system in humans and other mammals [91, 287, 384]. The blood pressure y is essentially determined by the heart's pumping force, which varies according to the patient's physical and emotional state. To control it, the brain stem, or more accurately, the *medulla oblongata*, generates a reference value of the blood pressure y_{ref}. This is compared to the actual blood pressure y, which – as shown in Figure 2.81 – is measured by so-called baroreceptors within the artery walls, or more accurately in the aortic arch and in the carotid artery.

Baroreceptors are pressure-sensitive neurons whose readings are electrically transmitted to the *medulla oblongata* via their axons. Here, the control error $y_{\mathrm{ref}} - y$ is calculated and the control signal generated. The latter is transmitted via the nervous system to the sinoatrial node of the heart. The sinoatrial node in turn triggers the contraction, i.e. the pumping process, by means of electrical signals. Figure 2.82 shows the simplified control loop in block diagram form. It must be noted that the transmission of signals along the nerve fibers takes place with a delay such that a time lag τ_{a} exists. In addition, the neuronal activity is subjected to a sigmoidal saturation, as shown in Figure 2.83, and which we will model here as

$$\tilde{u} = f(\tilde{e}) = b \cdot \arctan(m(\tilde{e} - \tilde{e}_c)) + \tilde{u}_c.$$

We will use the parameters $b = 57.89$, $\tilde{e}_c = 121$, $\tilde{u}_c = 139$, and $m = 0.06$ of a patient with high blood pressure. The vascular system can be modeled using a first-order transfer function with a time delay component

$$G(s) = \frac{e^{-\tau_{\mathrm{e}} s}}{1 + \tau_\nu s}, \qquad \tau_{\mathrm{e}} = 2.98\,\mathrm{s}, \;\; \tau_\nu = 5\,\mathrm{s}.$$

The measurement and the signal transmission – as mentioned above – cause an additional time delay of

$$H(s) = e^{-\tau_{\mathrm{a}} s}, \qquad \tau_{\mathrm{a}} = 0.01\,\mathrm{s}.$$

The regulation can be approximated by a P controller with the amplification factor $k = 1.65$.

(a) Convert the control loop shown in Figure 2.82 into a nonlinear standard control loop with a point-symmetric sigmoidal curve $u = f(e)$ which passes through zero.

(b) Prove that the describing function $N(A)$ of the sigmoidal saturation curve $u = b\arctan(me)$ can be stated as

$$N(A) = \frac{2b}{mA^2}\left(\sqrt{m^2A^2 + 1} - 1\right).$$

In doing so, use the definite integral

$$\int_0^\pi \frac{1}{1 + m^2A^2\sin^2(\alpha)}\, d\alpha = \frac{\pi}{\sqrt{m^2A^2 + 1}}.$$

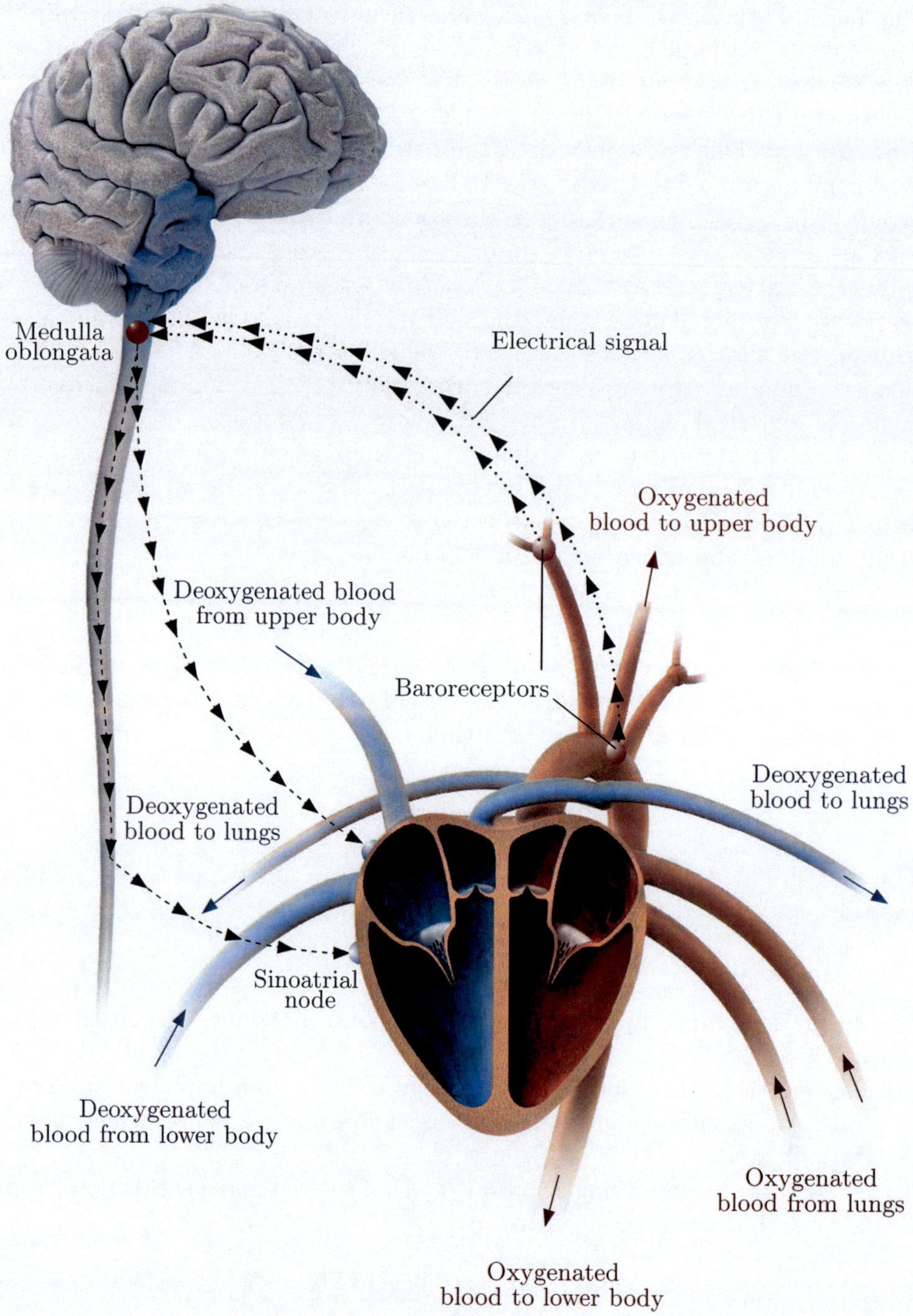

Fig. 2.81: The circulatory system of humans and mammals and its control

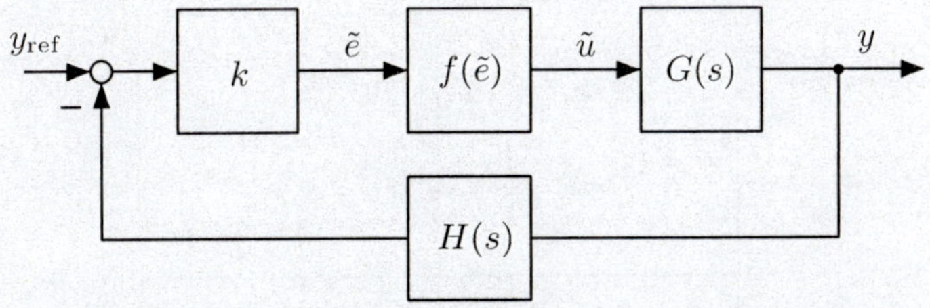

Fig. 2.82: Simplified blood-pressure control loop

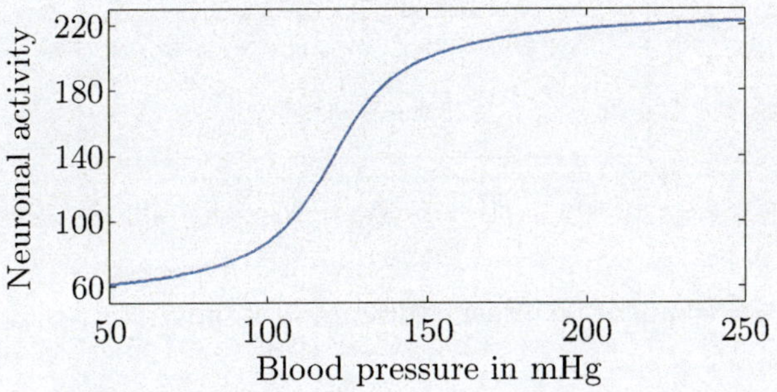

Fig. 2.83: Neuronal activity, given as a dimensionless measure

(c) Determine whether a limit cycle can exist in this case, and if so, what its frequency is.
(d) What statement can you make about the stability behavior of the limit cycle?

Exercise 2.7 Now we will examine a nonlinear standard control loop with the plant

$$G(s) = \frac{0.1111}{s^4 + 0.5s^3 + 10.06s^2 + 2.1s + 9}$$

and an unspecified nonlinear element. Figure 2.84 shows the Nyquist plot (dashed line) and the Popov plot (solid line).
(a) Using the Popov criterion, identify the maximum possible sector $[0, K_{\text{Po}})$ of absolute stability which can be determined using this criterion.
(b) In which sector can a linear curve lie so that asymptotic stability is ensured? What controller do we obtain in this case?
(c) Let us examine the nonlinearity

$$u = e^3.$$

What maximum control errors e are permissible if absolute stability is to be ensured?

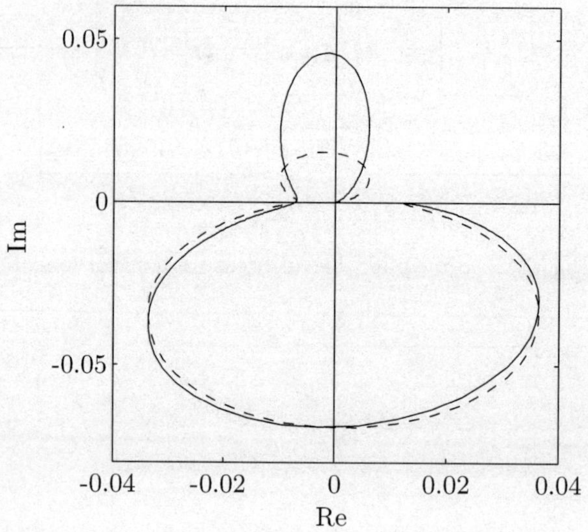

Fig. 2.84: Nyquist plot (dashed line) and Popov plot (solid line)

Exercise 2.8 For optimal functioning of a catalytic converter meant to reduce the emissions from a combustion engine, the fuel must be entirely burnt; otherwise carbon monoxide will result. Thus, oxygen sufficient for combustion must be present. The amount of air necessary to ensure this without leaving any oxygen behind is called the stoichiometric amount of air m_{st}. A measure of the completeness of the combustion is the ratio

$$\lambda = \frac{m_{\mathrm{air}}}{m_{\mathrm{st}}}$$

between the amount of air m_{air} available and the stoichiometric amount. Where stoichiometric combustion occurs, the ratio is $\lambda = 1$. To ensure this, a control system is used, the principal elements of which are shown in Figure 2.85.

The plant consists of a fuel injector with the duration v of the fuel injection as the control signal, as well as the combustion process, which is characterized by a time delay and a linear differential equation of the first order. Normally, a PI controller is used for regulation. Figure 2.86 shows this control loop. Because the oxygen sensor already includes the minus sign of the feedback, the measured value is added to the reference value, instead of being subtracted from it as is usual in a control loop. In our example, we have $T_{\mathrm{e}} = 0.2\,\mathrm{s}$, $T_{\mathrm{d}} = 0.5\,\mathrm{s}$, and $K_{\mathrm{e}} = 25$. In the control loop, all variables are shown as deviations from the operating point $\lambda = 1$, so that the curve of the sensor for this value passes through the origin and the reference value of the control loop is zero. Within the linear region, the slope of the sensor's curve $u(\lambda)$ has a minimum value of $m = du/d\lambda = -10$. The sensor curve is shown in Figure 2.87.

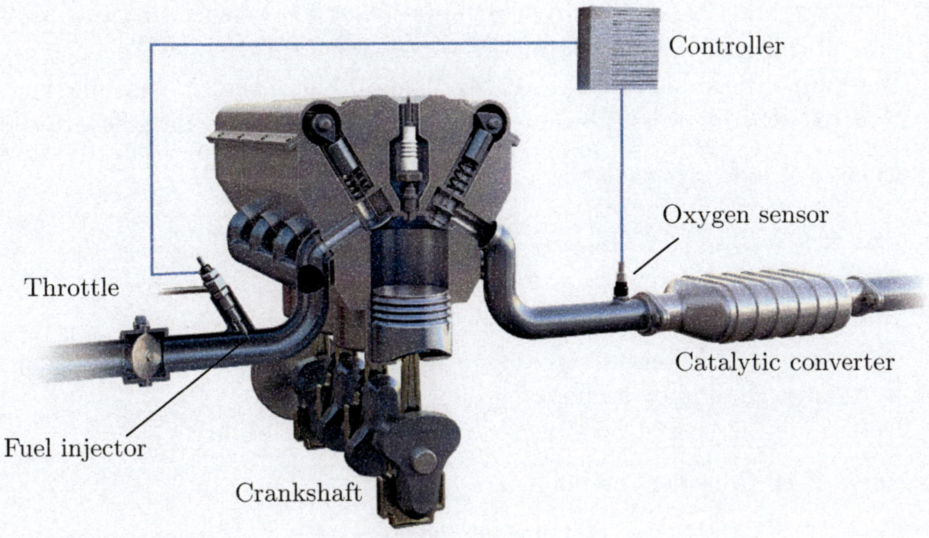

Fig. 2.85: Combustion engine

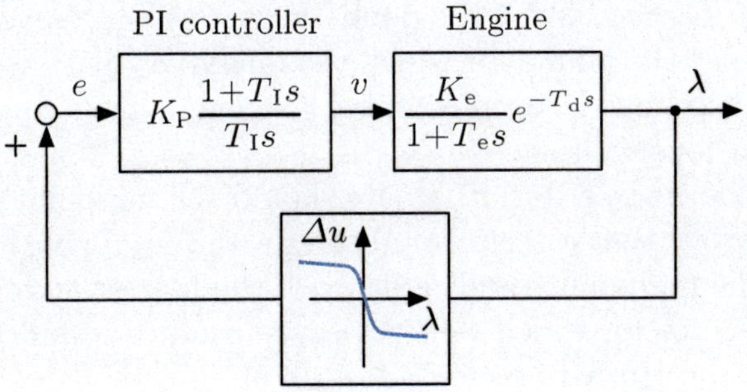

Fig. 2.86: Control loop

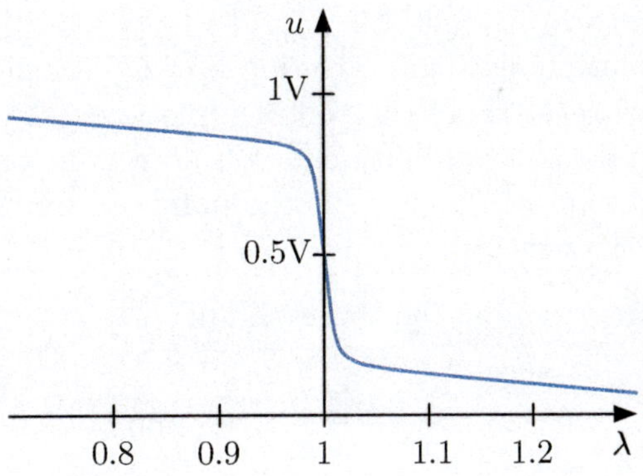

Fig. 2.87: Curve of an oxygen sensor with output voltage u depending on λ

(a) Configure the PI controller $K_{\mathrm{PI}} = K_{\mathrm{P}}(1+T_{\mathrm{I}}s)/(T_{\mathrm{I}}s))$ such that the linear part of the open control loop has the transfer function $ce^{-T_{\mathrm{d}}s}/s$.
(b) Determine the value c and thus also the gain K_{P} of the PI controller such that the control loop is absolutely stable in the sector $[\varepsilon, 12)$ where $\varepsilon > 0$.

Exercise 2.9 Let us consider

$$G(s) = \frac{1}{(s+1)(s+2)(s+3)}.$$

(a) Determine the largest sector of absolute stability that can be obtained using the circle criterion.
(b) Determine the Popov sector.
(c) Determine the largest possible sector of absolute stability.

Exercise 2.10 Consider the plant

$$G(s) = \frac{30}{(s-1)(s+2)(s+3)}.$$

(a) Sketch the Nyquist plot of $G(jw)$ for $\omega \in [-\infty, \infty]$. To do this, use the points $G(j0)$, $G(\pm j\infty)$, and the points for which $\operatorname{Im}\{G(j\omega)\} = 0$ applies. In addition, use the points for which the imaginary part assumes its maxima and/or minima. Determine these extreme values approximately by trying out various values ω.
(b) If we apply the circle criterion, where should the midpoint of the circle lie in order to prove that a sector of absolute stability exists?
(c) Determine the midpoint c and radius r of the largest possible circle. How large is the sector for which we can prove absolute stability?
(d) How large is the Hurwitz sector?

Exercise 2.11 Consider the plant

$$G(s) = \frac{s+1}{s^2+1}.$$

(a) Determine the continuous angle variation Ω of $G(j\omega)$ in relation to a point p on the negative real axis of the complex plane.
(b) Which sector of absolute stability can you determine using the circle criterion if the midpoint of the circle is the point -1 on the real axis?
(c) Identify the Hurwitz sector.

Exercise 2.12 Let us examine the three plants

$$G_1(s) = \frac{s+1}{s^2}, \quad G_2(s) = \frac{2s}{s^2-2s+4}, \quad \text{and} \quad G_3(s) = \frac{s^2-1}{s^3}.$$

(a) Sketch the Nyquist plots. Use the circle criterion. Which sector of absolute stability can you determine in each case?

(b) Determine the Hurwitz sector for each case.

Exercise 2.13 Let us examine the system

$$\begin{aligned}\dot{x}_1 &= x_2,\\ \dot{x}_2 &= -2x_1 - 2x_2 - 4x_1^3\end{aligned}$$

and the Lyapunov function

$$V(\boldsymbol{x}) = 4x_1^2 + 4x_1^4 + 2x_2^2.$$

Show that the equilibrium point $\boldsymbol{x}_{\mathrm{eq}} = \boldsymbol{0}$ is Lyapunov stable.

Exercise 2.14 We will examine the system

$$\begin{aligned}\dot{x}_1 &= x_2 + x_1\left(x_1^2 + x_2^2\right),\\ \dot{x}_2 &= -x_1 + x_2\left(x_1^2 + x_2^2\right).\end{aligned}$$

(a) Determine the equilibrium points of the system.
(b) Using the function $V(\boldsymbol{x}) = x_1^2 + x_2^2$, analyze the stability of the equilibrium point or points.

Exercise 2.15 Given the system

$$\ddot{x} + d\dot{x}^3 + kx = 0$$

with $d, k > 0$, demonstrate that

$$V(\boldsymbol{x}) = \frac{1}{2}(kx^2 + \dot{x}^2)$$

is a Lyapunov function of the system's equilibrium point. What stability property can you prove using the function $V(\boldsymbol{x})$?

Exercise 2.16 We will now analyze the system

$$\begin{aligned}\dot{x}_1 &= -x_1 - x_2,\\ \dot{x}_2 &= x_1 + (x_1^2 - 1)x_2.\end{aligned}$$

(a) Using Lyapunov's indirect method, show that the origin is asymptotically stable.
(b) Show that $V(\boldsymbol{x}) = \boldsymbol{x}^T\boldsymbol{x}$ is a Lyapunov function of the linearized system.
(c) What is the maximum catchment region we can calculate using the Lyapunov function $V(\boldsymbol{x}) = \boldsymbol{x}^T\boldsymbol{x}$?

Exercise 2.17 Let us examine the system

$$\begin{aligned}\dot{x}_1 &= x_2,\\ x_2 &= -x_1.\end{aligned}$$

(a) Show that the function

$$V(\boldsymbol{x}) = 0.1(x_1^2 + x_2^2)\left[2 + \cos\left(\sqrt{x_1^2 + x_2^2}\right)\right]$$

is a Lyapunov function for the equilibrium point $\boldsymbol{x}_{\text{eq}} = \boldsymbol{0}$ and that this equilibrium point is Lyapunov stable.
(b) What form do the level curves of the function $V(\boldsymbol{x})$ take?
(c) Sketch the function $V(\boldsymbol{x})$ for $x_1 \in [-50, 50]$ and $x_2 = 0$.
(d) What shape do the trajectories $\boldsymbol{x}(t)$ of the system take, and what is a system of this kind called?

Exercise 2.18 Consider the control loop shown in Figure 2.88 with the parameters

$$\boldsymbol{A} = \begin{bmatrix} -1 & 6 \\ -1 & -2 \end{bmatrix}, \quad \boldsymbol{b} = \begin{bmatrix} 1 \\ 1 \end{bmatrix}, \quad \boldsymbol{k}^T = \begin{bmatrix} 3 & 21 \end{bmatrix},$$

and the nonlinear controller $u = e + e^3$.

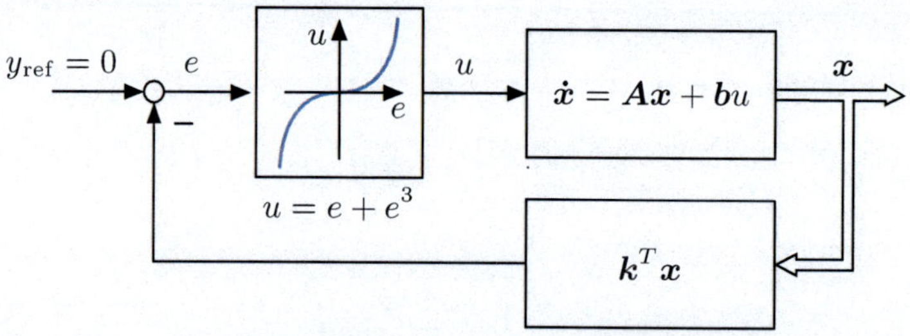

Fig. 2.88: Nonlinear control loop

(a) Formulate the differential equations of the closed control loop.
(b) Determine the equilibrium point $\boldsymbol{x}_{\text{eq}}$ of the control loop.
(c) Using Lyapunov's direct method, show that the equilibrium point of the control loop is globally asymptotically stable. To do this, use the Lyapunov function

$$V(\boldsymbol{x}) = (\boldsymbol{x} - \boldsymbol{x}_{\text{eq}})^T(\boldsymbol{x} - \boldsymbol{x}_{\text{eq}}).$$

Exercise 2.19 Consider the system

$$\begin{aligned}
\dot{x}_1 &= (1 - x_1^2 - x_2^2 - x_3^2)x_1, \\
\dot{x}_2 &= (1 - x_1^2 - x_2^2 - x_3^2)x_2, \\
\dot{x}_3 &= (1 - x_1^2 - x_2^2 - x_3^2)x_3.
\end{aligned}$$

(a) Determine the system's equilibrium points.
(b) Analyze the stability behavior of the equilibrium points.

Exercise 2.20 Let us examine the stability of the linear system

$$\dot{\boldsymbol{x}} = \left(\boldsymbol{A}_0 + \sum_{i=1}^{k} p_i \boldsymbol{A}_i \right) \boldsymbol{x}$$

with parameters p_i which are either unknown or change only very slowly over time. We will assume that the parameters p_i are limited by the hyperrectangle

$$P = \{p_i \in I\!R \mid p_i \in [p_{i,\min}, p_{i,\max}], \quad i = 1, ..., k\}.$$

Prove that the system for all $\boldsymbol{p} = [p_1 \; p_2 \; ... \; p_k]^T \in P$ has a globally asymptotically stable equilibrium point at $\boldsymbol{x} = \boldsymbol{0}$ with a Lyapunov function $V(\boldsymbol{x}) = \boldsymbol{x}^T \boldsymbol{R} \boldsymbol{x}$ if

$$\dot{V}(\boldsymbol{x}) < 0$$

applies at all 2^k vertices $\boldsymbol{p}_{\mathrm{v},i}$ of the hyperrectangle P.

Exercise 2.21 Let us examine the passivity of the satellite model (1.8) from Section 1.1.4 on p. 8.

(a) Choose a suitable storage function $S(\boldsymbol{\omega})$, and show that the model is passive and lossless where the input variable vector $\boldsymbol{u} = \boldsymbol{M}$ and the output variable vector $\boldsymbol{y} = \boldsymbol{\omega}$.

(b) What property must the matrix $\boldsymbol{K}$ have in the control law $\boldsymbol{u} = -\boldsymbol{K}\boldsymbol{\omega}$ such that the closed control loop has a globally asymptotically stable equilibrium point at $\boldsymbol{\omega} = \boldsymbol{0}$?

(c) Represent the satellite model as a PCH system. For this purpose, select suitable new coordinates.

(d) Now choose a control law with the moment vector $\boldsymbol{M}$ as the control variable vector such that a PCHD system with a globally asymptotically stable equilibrium point at the origin results.

Exercise 2.22 Is a PID controller passive, strictly passive, or neither?

Exercise 2.23 If a system

$$\dot{\boldsymbol{x}} = \boldsymbol{A}\boldsymbol{x} + \boldsymbol{b}u, \qquad y = \boldsymbol{c}^T \boldsymbol{x}$$

fulfills the Kalman-Yakubovich-Popov equations

$$\boldsymbol{A}^T \boldsymbol{R} + \boldsymbol{R}\boldsymbol{A} = -\boldsymbol{Q}, \qquad \boldsymbol{c}^T = \boldsymbol{b}^T \boldsymbol{R}$$

(here $\boldsymbol{Q}$ is positive definite), then it is strictly passive. Using the Lyapunov function $V(\boldsymbol{x}) = \boldsymbol{x}^T \boldsymbol{R} \boldsymbol{x}$, show that in this case it is absolutely stable in the sector $[0, \infty]$.

Exercise 2.24 Consider the system

$$G(s) = \frac{s + 0.5}{s^2 + s + 1}. \tag{2.144}$$

(a) Is the system strictly positive real, only positive real, or neither?
(b) Is the system strictly passive, only passive, or lossless?
(c) For the general case, show that the nonlinear standard control loop is strictly passive if the plant is strictly passive and the nonlinearity is passive. Which nonlinearities are possible in this case if we wish the control loop to have a globally asymptotic equilibrium point?
(d) How large is the largest sector of absolute stability in the case of system (2.144)?

Exercise 2.25 State the condition for the series connection of a system with a strictly positive real transfer function $G(s)$ and a PI controller

$$H(s) = P + \frac{I}{s}$$

to be a passive system.

Exercise 2.26 Consider a mathematical pendulum, i.e. an ideal pendulum with a massless rod of length l, point mass m, and no friction losses. It is shown in Figure 2.89.

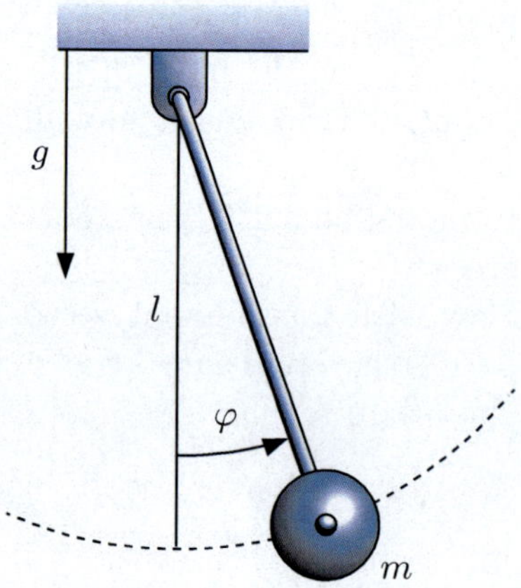

Fig. 2.89: Pendulum

(a) Determine the Lagrangian function L.
(b) Formulate the Lagrangian equations and use them to calculate the equations of motion for the pendulum.
(c) Determine the Hamiltonian function.
(d) Determine the Hamiltonian equations.

3

Controllability and Flatness

3.1 Controllability

3.1.1 Definition of Controllability

The objective of feedforward or feedback control is to influence a plant such that a specified state $\boldsymbol{x}_{\mathrm{e}}$ is reached or – stated in a more far-reaching way – a specified trajectory $\boldsymbol{x}_{\mathrm{ref}}(t)$ is followed. To ensure that at least the first objective is fulfilled, the plant

$$\dot{\boldsymbol{x}} = \boldsymbol{f}(\boldsymbol{x}, \boldsymbol{u})$$

must be controllable. That means it must be possible to transform an arbitrary state $\boldsymbol{x}_0$ to any other state $\boldsymbol{x}_{\mathrm{e}}$ via a suitable course of the actuating variable $\boldsymbol{u}(t)$. More precisely, we define the term *controllability* by

Definition 27 (Controllability). *Let a system*

$$\dot{\boldsymbol{x}} = \boldsymbol{f}(\boldsymbol{x}, \boldsymbol{u})$$

be defined for $\boldsymbol{x} \in D_{\boldsymbol{x},\mathrm{def}} \subseteq \mathbb{R}^n$ *and* $\boldsymbol{u} \in D_{\boldsymbol{u},\mathrm{def}} \subseteq \mathbb{R}^m$. *Let* $D_{\boldsymbol{x}} \subseteq D_{\boldsymbol{x},\mathrm{def}}$ *be a path-connected set and* $D_{\boldsymbol{u}} \subseteq D_{\boldsymbol{u},\mathrm{def}}$. *Then the system is called controllable if any state* $\boldsymbol{x}_0 \in D_{\boldsymbol{x}}$ *can be transferred to any state* $\boldsymbol{x}_{\mathrm{e}} \in D_{\boldsymbol{x}}$ *by an input function* $\boldsymbol{u}$ *with function values* $\boldsymbol{u}(t) \in D_{\boldsymbol{u}}$ *along a trajectory* $\boldsymbol{x}(t) \in D_{\boldsymbol{x}}$ *within a finite time interval* $[0, T]$.

Obviously, the controllability of a system is a fundamental requirement for the realization and proper functioning of a feedforward or feedback control system. This is intuitively clear when thinking of an airplane that the pilot or autopilot cannot maneuver into the desired flight altitude. For technical systems, the controllability is usually given, since the designer certainly regards it as a major feature of the system and consequently considers it during the design process. Therefore, if we find out that a technical system is not

J. Adamy, *Nonlinear Systems and Controls*,
https://doi.org/10.1007/978-3-662-68690-4_3

controllable, the question arises whether the modeling has been performed correctly.

An example of an uncontrollable system is

$$\dot{x} = u^2,$$

since no initial value x_0 can be transferred to a state x_e if $x_e < x_0$, independent of the course of the input function $u(t)$.

The controllability of a system implies that we can reach any state from any other state. However, it does not necessarily imply that we can do this via any trajectory $\boldsymbol{x}(t)$. As a matter of fact, trajectories may exist which a controllable system cannot execute regardless of the choice of the input function $\boldsymbol{u}(t)$. To illustrate this circumstance, we will consider the system

$$\begin{aligned} \dot{x}_1 &= u, \\ \dot{x}_2 &= x_1. \end{aligned} \tag{3.1}$$

The direction $[\dot{x}_1 \; \dot{x}_2]^T$ in which a trajectory can move is limited by $\dot{x}_2 = x_1$. For positive values of x_1, the state variable x_2 always increases, whereas for negative values, it always decreases. Figure 3.1 illustrates the possible directions of the trajectories. The figure also shows a realizable and an unrealizable trajectory.

However, the above system is controllable. We can show this using the controllability criterion [152, 98] for linear systems. A linear system

$$\dot{\boldsymbol{x}} = \boldsymbol{A}\boldsymbol{x} + \boldsymbol{B}\boldsymbol{u}$$

is controllable if and only if the $n \times mn$ controllability matrix

$$\boldsymbol{M}_{\text{contr}} = \begin{bmatrix} \boldsymbol{B} & \boldsymbol{A}\boldsymbol{B} & \boldsymbol{A}^2\boldsymbol{B} & \ldots & \boldsymbol{A}^{n-1}\boldsymbol{B} \end{bmatrix} \tag{3.2}$$

is of rank n. This is the case here, since with

$$\boldsymbol{A} = \begin{bmatrix} 0 & 0 \\ 1 & 0 \end{bmatrix} \quad \text{and} \quad \boldsymbol{B} = \begin{bmatrix} 1 \\ 0 \end{bmatrix}$$

it holds that

$$\text{rank}(\boldsymbol{M}_{\text{contr}}) = \text{rank}\left(\begin{bmatrix} 1 & 0 \\ 0 & 1 \end{bmatrix}\right) = 2.$$

The example of system (3.1) also illustrates that, even though controllability implies that any point $\boldsymbol{x}$ can be reached, it does not necessarily mean that we can remain at this point. Here, such points are $[x_1 \neq 0 \;\; x_2]$ since it holds that $\dot{x}_2 = x_1 \neq 0$.

Next we will discuss a stricter definition of controllability which gives us new possibilities of influencing the system. First, let us consider the set of all points $\boldsymbol{x}_e = \boldsymbol{x}(t)$ which the trajectories $\boldsymbol{x}(t)$ of a system reach for all times t with $0 \leq t \leq T$ and all possible $\boldsymbol{u}(t)$ beginning at $\boldsymbol{x}_0 = \boldsymbol{x}(0)$. This set, which

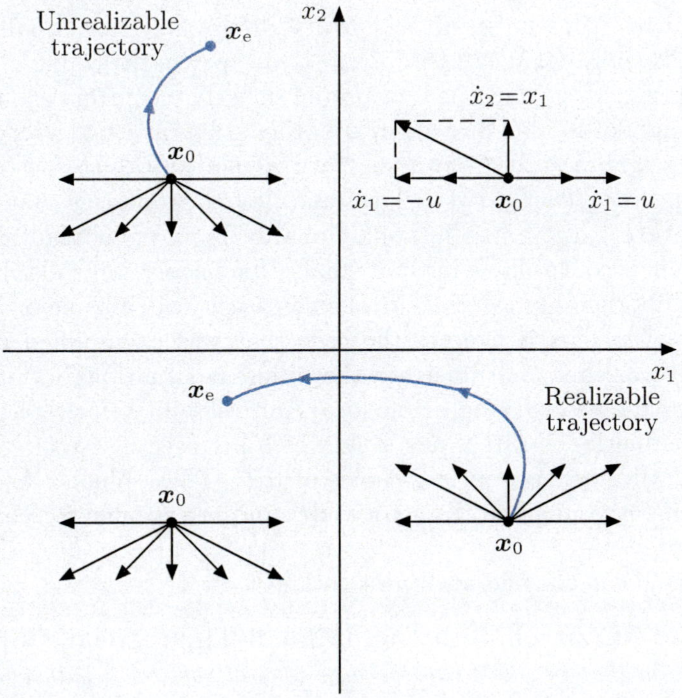

Fig. 3.1: Directions of realizable and unrealizable trajectories of the system (3.1)

we call the *reachable set*, is denoted by $R(\boldsymbol{x}_0, T)$. In short, $R(\boldsymbol{x}_0, T)$ is the set of all states that can be reached from $\boldsymbol{x}_0$ within a finite time T or a shorter period of time. The above definition of the set $R(\boldsymbol{x}_0, T)$ also implies that all trajectories $\boldsymbol{x}(t)$ with $0 \leq t \leq T$ take their course within $R(\boldsymbol{x}_0, T)$. Using $R(\boldsymbol{x}_0, T)$ we can define the term *small-time local controllability (STLC)*:

Definition 28 (Small-Time Local Controllability). *Let a system*

$$\dot{\boldsymbol{x}} = \boldsymbol{f}(\boldsymbol{x}, \boldsymbol{u})$$

be defined for $\boldsymbol{x} \in D_{\boldsymbol{x},\text{def}} \subseteq I\!R^n$ *and* $\boldsymbol{u} \in D_{\boldsymbol{u},\text{def}} \subset I\!R^m$. *Let* $D_{\boldsymbol{x}} \subseteq D_{\boldsymbol{x},\text{def}}$ *be an open set and* $D_{\boldsymbol{u}} \subseteq D_{\boldsymbol{u},\text{def}}$ *a closed and bounded set. Then the system is called small-time locally controllable if for every state* $\boldsymbol{x}_0 \in D_{\boldsymbol{x}}$, *for* $\boldsymbol{u}(t) \in D_{\boldsymbol{u}}$ *where* $0 < t \leq T$, *and for all* $0 < T < \infty$ *the sets* $R(\boldsymbol{x}_0, T)$ *are neighborhoods of* $\boldsymbol{x}_0$.

Obviously, as stated above, $D_{\boldsymbol{x}}$ must be an open set if the system is to be small-time locally controllable on this set, since the boundary points $\boldsymbol{x}_\text{b}$ of a closed set $D_{\boldsymbol{x}}$ do not possess a neighborhood $R(\boldsymbol{x}_\text{b}, T) \subset D_{\boldsymbol{x}}$.

According to the above definition, every set $R(\boldsymbol{x}_0, T)$ is a neighborhood of $\boldsymbol{x}_0$. Small-time local controllability thus means that we can reach every point $\boldsymbol{x}_\text{e}$ located within an arbitrarily small neighborhood $R(\boldsymbol{x}_0, T)$ on a trajectory whose course, beginning from $\boldsymbol{x}_0$, lies entirely within this neighborhood $R(\boldsymbol{x}_0, T)$ in small time. This, however, does not imply that trajectories beginning at $\boldsymbol{x}_0$ can aim in any arbitrary direction. Thus, small-time local controllability does not mean that every trajectory is realizable.

The term *local* here must be interpreted such that a system can reach every state in a neighborhood of any arbitrary other state on a trajectory which lies within this neighborhood, i. e. a trajectory which is local. It does not refer to the common distinction between local and global properties that is made in control theory, e. g. local stability and global stability of an equilibrium point.

If a system is controllable but not small-time locally controllable, $R(\boldsymbol{x}_0, T)$ is not a neighborhood of $\boldsymbol{x}_0$, since in such a case $\boldsymbol{x}_0$ can also lie on the boundary of $R(\boldsymbol{x}_0, T)$. This is exactly the case that was exemplified in equation (3.1). Figures 3.2 and 3.3 illustrate the different situations for the cases in which controllability and small-time local controllability exist. The blue highlighted areas mark the directions $\dot{\boldsymbol{x}}$ in which a trajectory $\boldsymbol{x}(t)$ can progress starting at $\boldsymbol{x}_0$ for an appropriate choice of $\boldsymbol{u}(t)$. These blue areas of starting directions change along the trajectory $\boldsymbol{x}(t)$ and are specific for each point of the trajectory.

From the above consideration, we obtain

Theorem 43 (Controllability and Small-Time Local Controllability). *A system that is small-time locally controllable on a path-connected set $D_{\boldsymbol{x}}$ is controllable on this set.*

The requirement that $D_{\boldsymbol{x}}$ is a path-connected set is essential and usually fulfilled. Small-time local controllability – in contrast to controllability – can hold for a set that is not path-connected, since in this case it is not required that any point $\boldsymbol{x}_{\mathrm{e}} \in D_{\boldsymbol{x}}$ is reachable from any other point $\boldsymbol{x}_0 \in D_{\boldsymbol{x}}$. It is precisely the latter that is an essential characteristic of controllability. The most important difference between controllable and small-time locally controllable systems is that there might be a long detour to arrive at $\boldsymbol{x}_{\mathrm{e}}$ from $\boldsymbol{x}_0$ for a system that is only controllable. For a small-time locally controllable system,

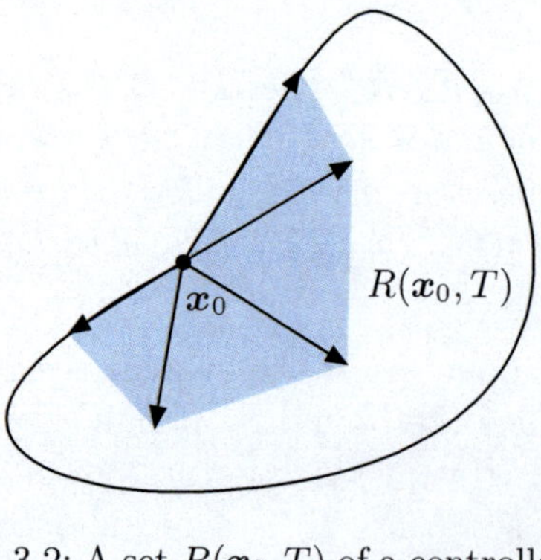

Fig. 3.2: A set $R(\boldsymbol{x}_0, T)$ of a controllable system which can be reached within the time T starting from the point $\boldsymbol{x}_0$. Here, $R(\boldsymbol{x}_0, T)$ is not a neighborhood of $\boldsymbol{x}_0$.

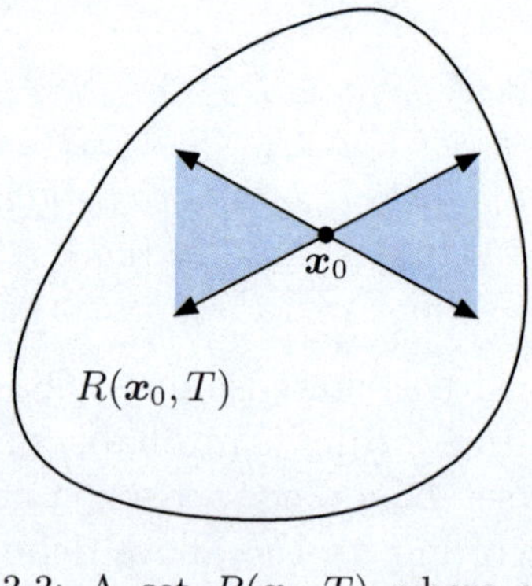

Fig. 3.3: A set $R(\boldsymbol{x}_0, T)$ where small-time local controllability exists. The trajectories, in general, cannot progress from the point $\boldsymbol{x}_0$ in all directions.

on the other hand, this route is very short and lies within a neighborhood close to $\boldsymbol{x}_0$ and $\boldsymbol{x}_e$.

For some systems, the controlling possibilities allow the realization of any arbitrary trajectory $\boldsymbol{x}(t)$. For this, the trajectories must be able to progress from $\boldsymbol{x}$ in all directions, i. e. the vector $\dot{\boldsymbol{x}}$ representing the direction of motion must be omnidirectionally controllable by the actuating variable $\boldsymbol{u}$. Accordingly, we can now introduce the term *omnidirectional controllability*, which is a stricter form of controllability than the two controllability definitions given previously.

Definition 29 (Omnidirectional Controllability). *Let a system*

$$\dot{\boldsymbol{x}} = \boldsymbol{f}(\boldsymbol{x}, \boldsymbol{u})$$

be defined for $\boldsymbol{x} \in D_{\boldsymbol{x},\text{def}} \subseteq \mathbb{R}^n$ and $\boldsymbol{u} \in D_{\boldsymbol{u},\text{def}} \subseteq \mathbb{R}^m$. Let $D_{\boldsymbol{x}} \subseteq D_{\boldsymbol{x},\text{def}}$ be a path-connected set and $D_{\boldsymbol{u}} \subseteq D_{\boldsymbol{u},\text{def}}$. Then the system is called omnidirectionally controllable if any state $\boldsymbol{x}_0 \in D_{\boldsymbol{x}}$ can be transferred to any state $\boldsymbol{x}_e \in D_{\boldsymbol{x}}$ by an input function $\boldsymbol{u}$ with function values $\boldsymbol{u}(t) \in D_{\boldsymbol{u}}$ along any arbitrary defined trajectory $\boldsymbol{x}(t) \in D_{\boldsymbol{x}}$ within a finite time interval $[0, T < \infty]$.

A weak form of omnidirectional controllability holds if an arbitrarily defined progression of a trajectory $\boldsymbol{x}_{\text{ref}}(\tau)$ can be followed exactly, but only at a slower velocity. This means that for the realizable progressions of trajectories $\boldsymbol{x}(t)$, the relation

$$\boldsymbol{x}(t) = \boldsymbol{x}_{\text{ref}}(c\tau) \qquad \text{with} \quad 0 < c < 1$$

holds. This may occur if the temporal change $\dot{\boldsymbol{x}}$ of the trajectories $\boldsymbol{x}(t)$ is limited, i. e. $|\dot{\boldsymbol{x}}| < a$, $a > 0$. For example, bounded input variables $\boldsymbol{u}$ cause this.

For the case of omnidirectional controllability, the reachable set $R(\boldsymbol{x}_0, T)$ is also a neighborhood of the state $\boldsymbol{x}_0$, as was the case for the small-time local controllability. In contrast to the latter, as illustrated in Figure 3.4, a trajectory that starts at $\boldsymbol{x}_0$ can progress in every direction. Accordingly, any arbitrary trajectory can be realized. From an engineering perspective, omnidirectional controllability is thus a more valuable property than controllability or small-time local controllability.

For linear systems

$$\dot{\boldsymbol{x}} = \boldsymbol{A}\boldsymbol{x} + \boldsymbol{B}\boldsymbol{u},$$

omnidirectional controllability is given if and only if the control matrix $\boldsymbol{B}$ has rank n. This means that an equal number of input and state variables exist. In this case only, the state vector's temporal derivative $\dot{\boldsymbol{x}}$ and thus a trajectory $\boldsymbol{x}(t)$ can be defined arbitrarily via $\boldsymbol{B}\boldsymbol{u}$. In practice, however, linear systems only rarely possess n input variables and are therefore only rarely omnidirectionally controllable. This is different for nonlinear systems. Their

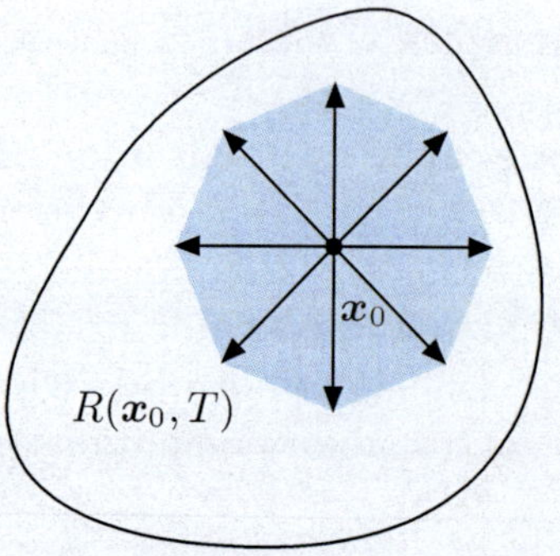

Fig. 3.4: The set $R(\boldsymbol{x}_0, T)$ in the case of omnidirectional controllability. The trajectories can progress starting from $\boldsymbol{x}_0$ in all directions.

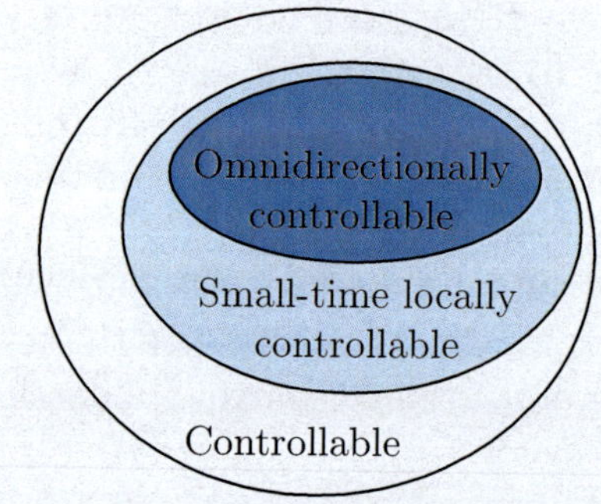

Fig. 3.5: The sets of controllable, small-time locally, and omnidirectionally controllable systems in case of path-connected open sets $D_{\boldsymbol{x}}$

complex dynamics are often accompanied by a larger number of input variables when compared to the linear case. These are introduced by the designer to allow the system to be more easily influenced, meaning that nonlinear systems which are omnidirectionally controllable are more common.

If we compare Definitions 28 and 29, we immediately obtain

Theorem 44 (Small-Time Local and Omnidirectional Controllability). *A system that is omnidirectionally controllable on an open set* $D_{\boldsymbol{x}}$ *is also small-time locally controllable on this set.*

Figure 3.5 illustrates the relationship between the different kinds of controllability, assuming that $D_{\boldsymbol{x}}$ is a path-connected open set. If $D_{\boldsymbol{x}}$ is not path-connected, the set of small-time locally controllable systems is not a subset of the controllable systems. This is because it is possible for a system to be small-time locally controllable on a set that is not path-connected. However, it is never controllable on a set of this kind, since not every point can be reached from every other point. It is worth noting that this problem disappears if we consider only path-connected sets. The latter should be the case in almost all real-world applications.

Figure 3.6 shows an example of an omnidirectionally controllable system, a ship with transverse thrusters in the bow and stern. Transverse thrusters are tunnels that are mounted perpendicular to the direction of travel and contain an internal propeller drive. Using the transverse thrusters, the ship can be steered transversely to the normal direction of travel. The simultaneous usage of the ship's regular propulsion and transverse thrusters enables it to move in any arbitrary direction, as illustrated in Figure 3.7. Furthermore, the ship can be rotated on the spot, i. e. it can change its orientation. Examples are drilling ships which must be exactly positioned over a drill hole, cruise ships, large container ships, and ferries which must maneuver in harbors with an accuracy of about one meter.

Fig. 3.6: Ship with transverse thrusters in the bow and stern, shown here in a service ship which provides maintenance for submarine boreholes

Fig. 3.7: The ship's transverse and rotational movement functions

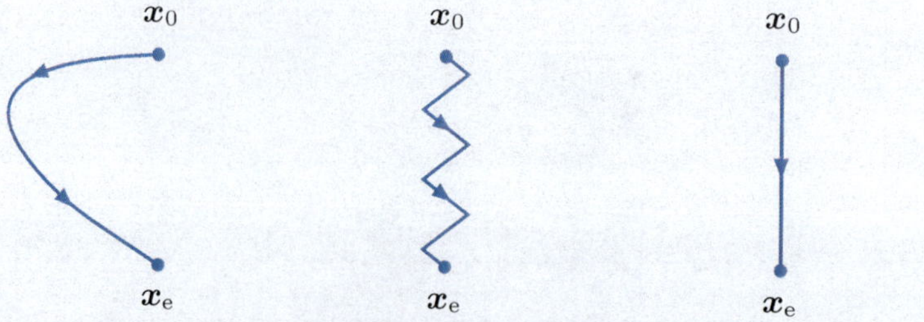

Fig. 3.8: The shortest trajectories from x_0 to x_e for all three kinds of controllability

The shortest trajectories from one point x_0 to a different point x_e for the cases of controllability, small-time local controllability, and omnidirectional controllability are illustrated in Figure 3.8. This figure also illustrates the meaning of the different kinds of controllability; for example, where a system is to be controlled such that it follows a predefined trajectory. For controllable systems, this is only possible with restrictions since their trajectories can only move in specific directions. An improvement is provided by small-time locally controllable systems which, inside a small neighborhood of a point $x(t)$, can change direction $\dot{x}$ within a wider range than controllable systems normally can. However, only omnidirectionally controllable systems allow any predefined trajectory to be followed exactly.

An example of the abovementioned restrictions would be a motor vehicle without a reverse gear, with position and orientation as state variables, such as a motorbike. It is merely controllable, but not small-time locally controllable. This is why motorbikes cannot be turned around in a small space by shunting. In contrast to this, motor vehicles with a reverse gear and tracked vehicles are small-time locally controllable. However, neither is omnidirectionally controllable, since they cannot be moved orthogonally to their direction of travel, in contrast to the ship discussed above.

In the next section, we will distinguish between locally and globally controllable systems.

3.1.2 Global and Local Controllability

If they are controllable, linear systems are always controllable in the entire state space $\mathbb{R}^n$. Therefore, the controllability is – similar to the stability of the equilibrium point $x_{eq} = \mathbf{0}$ – a global system characteristic of linear systems. For nonlinear systems, this is not always the case. Nonlinear systems exist which are only controllable on subsets D_x of their domain of definition $D_{x,\text{def}}$, i. e. they are only locally controllable. Furthermore, there are also nonlinear systems whose controllability characteristic varies, e. g. a system which is small-time locally controllable on one subset and omnidirectionally controllable on another. Consequently, we can define the terms *global* and *local controllability*.

Definition 30 (Global Controllability): *Let a system*

$$\dot{\boldsymbol{x}} = \boldsymbol{f}(\boldsymbol{x}, \boldsymbol{u})$$

be defined for $\boldsymbol{x} \in D_{\boldsymbol{x},\mathrm{def}} \subseteq \mathbb{R}^n$ *and for* $\boldsymbol{u} \in D_{\boldsymbol{u},\mathrm{def}} \subseteq \mathbb{R}^m$. *If the system is (small-time locally or omnidirectionally) controllable on its domain of definition* $D_{\boldsymbol{x},\mathrm{def}}$ *by means of* $\boldsymbol{u} \in D_{\boldsymbol{u}} \subseteq D_{\boldsymbol{u},\mathrm{def}}$, *the system is called globally (small-time locally or omnidirectionally) controllable by* $D_{\boldsymbol{u}}$.

In this context, we recall that the set on which a system is controllable or omnidirectional controllable must be path-connected.

Definition 31 (Local Controllability). *Let a system*

$$\dot{\boldsymbol{x}} = \boldsymbol{f}(\boldsymbol{x}, \boldsymbol{u})$$

be defined for $\boldsymbol{x} \in D_{\boldsymbol{x},\mathrm{def}} \subseteq \mathbb{R}^n$ *and for* $\boldsymbol{u} \in D_{\boldsymbol{u},\mathrm{def}} \subseteq \mathbb{R}^m$. *If the system is only (small-time locally or omnidirectionally) controllable on a proper subset* $D_{\boldsymbol{x}}$ *of the system's domain of definition* $D_{\boldsymbol{x},\mathrm{def}}$ *by means of* $\boldsymbol{u} \in D_{\boldsymbol{u}} \subseteq D_{\boldsymbol{u},\mathrm{def}}$, *it is called locally (small-time locally or omnidirectionally) controllable on* $D_{\boldsymbol{x}}$ *by* $D_{\boldsymbol{u}}$.

It should be noted that in practical applications, the set $D_{\boldsymbol{u}}$ of admissible or available control variable vectors $\boldsymbol{u}$ is often not identical to the set $D_{\boldsymbol{u},\mathrm{def}}$, on which the system is defined. Since in some cases controllability depends not only on the size of the set $D_{\boldsymbol{x}}$ but also on that of the set $D_{\boldsymbol{u}} \subseteq D_{\boldsymbol{u},\mathrm{def}}$, as in the definitions above, it is useful to describe this dependence as *controllable on* $D_{\boldsymbol{x}}$ *by* $D_{\boldsymbol{u}}$.

For illustration purposes, as an initial example, we will consider the system

$$\dot{x} = \frac{u}{x}$$

with the domain of definition $D_{x,\mathrm{def}} = \mathbb{R}\backslash\{0\}$. It is directly apparent from the differential equation that it is not possible to reach any negative endpoint x_{e} from any positive starting point x_0, or vice versa. For this, the trajectory would have to run through the point $x = 0$, for which the system is not defined. This is also evident from the solution

$$x(t) = \operatorname{sgn}(x_0) \cdot \sqrt{x_0^2 + 2\int\limits_0^t u(\tau)d\tau}, \quad x_0 \neq 0, \tag{3.3}$$

of the differential equation, because the sign of the solution $x(t)$ can never change.

Even though the system is not controllable on $D_x = D_{x,\text{def}}$, i. e. it is not globally controllable, it is locally controllable on the two subsets $D_x^+ = I\!R^+$ and $D_x^- = I\!R^-$. The local controllability is directly apparent from its solution (3.3). Furthermore, the system is globally small-time locally controllable.

Let us now consider a second example to further our understanding of global and local controllability. In particular, we will consider the system

$$\dot{x} = xu \tag{3.4}$$

which is defined on $D_{x,\text{def}} = I\!R$. Its system solution is

$$x(t) = x_0 e^{\int_0^t u(\tau)d\tau} . \tag{3.5}$$

Although the system is defined at $x = 0$, it is not controllable there. This is because we cannot reach any point $x_\text{e} \neq 0$ from $x_0 = 0$. Furthermore, the system solution (3.5) also shows that the sign of the solution $x(t)$ can never change. The system is thus not globally controllable. However, it is locally controllable on $D_x^+ = I\!R^+$ and $D_x^- = I\!R^-$.

The system (3.4) is also not globally small-time locally controllable, since for the point $x_0 = 0$ no reachable set

$$R(x_0, T) \neq \{0\},$$

i. e. no neighborhood $R(x_0, T)$ of x_0, exists. On the set $D_x^+ = I\!R^+$ of positive real numbers and on the set $D_x^- = I\!R^-$ of negative real numbers, however, the system (3.4) is locally small-time locally controllable[1].

As a final example, we will consider the system

$$\dot{\boldsymbol{x}} = \begin{bmatrix} x_2 \\ 0 \end{bmatrix} + \begin{bmatrix} 0 \\ 1 \end{bmatrix} u_1 + \begin{bmatrix} g(x_2) \\ 0 \end{bmatrix} u_2 \quad \text{with} \quad g(x_2) = \begin{cases} x_2^2, & x_2 > 0, \\ 0, & x_2 \leq 0. \end{cases}$$

It is globally controllable, since the linear system component is controllable. For $x_2 > 0$, the system is also omnidirectionally controllable, since for this case

$$g(x_2) \neq 0$$

holds, and using u_1 and u_2, we can specify any arbitrary direction via $\dot{\boldsymbol{x}}$ and thus also any arbitrary trajectory $\boldsymbol{x}(t)$. Since, for $x_2 > 0$, the system is locally omnidirectionally controllable, according to Theorem 44 it is also locally small-time locally controllable.

[1] If a system is small-time locally controllable only on a subset of its domain of definition, then we have to call it *locally small-time locally controllable.* Although this phrase is meaningful, it may seem confusing due to the double use of the term *locally.* A non-confusing term would be *locally small-time proximally controllable.* However, the term *small-time proximally controllable* is not established in the literature.

3.1.3 Proving Controllability

Now that we have introduced and defined the different concepts of controllability, which are of fundamental importance for feedforward and feedback control systems, the question arises how to verify the controllability of a specific system.

For linear systems, we know the answer: we check if the rank of the controllability matrix (3.2) equals n. For general nonlinear systems, no such easily applicable criterion exists. This unsatisfactory situation arises because it is necessary – with some exceptions – to know the solution of the system's differential equation to determine whether it is possible to transform any point $\boldsymbol{x}_0$ into any other point $\boldsymbol{x}_\mathrm{e}$ by means of an input signal within finite time. However, these solutions are rarely analytically determinable for nonlinear systems.

For a special class of nonlinear systems, however, the controllability can be examined. These are systems that are nonlinear in their state variables x_i, but linear in their input variables u_i, i. e. they can be represented by the differential equation

$$\dot{\boldsymbol{x}} = \boldsymbol{a}(\boldsymbol{x}) + \boldsymbol{B}(\boldsymbol{x}) \cdot \boldsymbol{u} = \boldsymbol{a}(\boldsymbol{x}) + \sum_{i=1}^{m} \boldsymbol{b}_i(\boldsymbol{x}) \cdot u_i \tag{3.6}$$

with $\boldsymbol{x} \in D_{\boldsymbol{x},\mathrm{def}} \subseteq I\!R^n$ and $\boldsymbol{u} \in I\!R^m$. Such systems are called *control-affine systems*[2]. In practice, this class of systems is of great importance because many technological systems are control-affine.

We will now consider a particular control-affine system of the form

$$\begin{bmatrix} \dot{x}_1 \\ \dot{x}_2 \\ \vdots \\ \dot{x}_{n-1} \\ \dot{x}_n \end{bmatrix} = \begin{bmatrix} x_2 \\ x_3 \\ \vdots \\ x_n \\ \alpha(\boldsymbol{x}) \end{bmatrix} + \begin{bmatrix} 0 \\ 0 \\ \vdots \\ 0 \\ \beta(\boldsymbol{x}) \end{bmatrix} u \tag{3.7}$$

with

$$\boldsymbol{x} = \begin{bmatrix} x_1 \ x_2 \ \cdots \ x_n \end{bmatrix}^T,$$

which is referred to as the *nonlinear controller canonical form*. Figure 3.9 shows the associated block diagram.

Assuming that

$$\beta(\boldsymbol{x}) \neq 0$$

holds for all $\boldsymbol{x} \in D_{\boldsymbol{x}} \subseteq D_{\boldsymbol{x},\mathrm{def}}$, it follows that a new control or input variable v can be used via

[2] Such systems are also termed *input-linear systems*, *input-affine systems*, or *affine-in-control systems* in the literature.

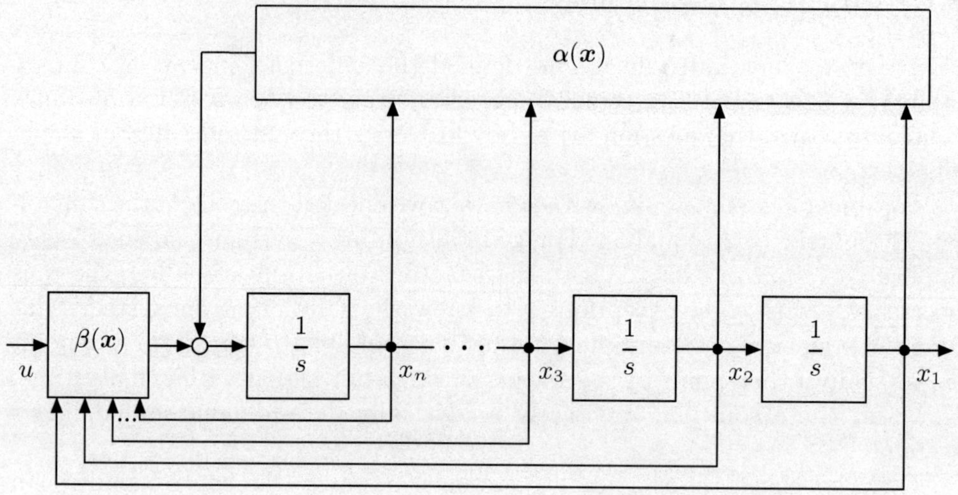

Fig. 3.9: System in nonlinear controller canonical form

$$u = \frac{-\alpha(\boldsymbol{x}) + v}{\beta(\boldsymbol{x})}. \tag{3.8}$$

With equation (3.8) the system (3.7) takes the form of a linear system

$$\begin{bmatrix} \dot{x}_1 \\ \dot{x}_2 \\ \vdots \\ \dot{x}_{n-1} \\ \dot{x}_n \end{bmatrix} = \begin{bmatrix} 0 & 1 & 0 \cdots 0 \\ 0 & 0 & 1 \cdots 0 \\ \vdots & \vdots & \vdots \ddots \vdots \\ 0 & 0 & 0 \cdots 1 \\ 0 & 0 & 0 \cdots 0 \end{bmatrix} \begin{bmatrix} x_1 \\ x_2 \\ \vdots \\ x_{n-1} \\ x_n \end{bmatrix} + \begin{bmatrix} 0 \\ 0 \\ \vdots \\ 0 \\ 1 \end{bmatrix} v \tag{3.9}$$

in linear controller canonical form. This is controllable, as quickly becomes evident from its controllability matrix

$$\boldsymbol{M}_{\text{contr}} = \begin{bmatrix} \boldsymbol{b} & \boldsymbol{A}\boldsymbol{b} & \boldsymbol{A}^2\boldsymbol{b} & \ldots & \boldsymbol{A}^{n-1}\boldsymbol{b} \end{bmatrix} = \begin{bmatrix} 0 & 0 & \cdots & 0 & 1 \\ 0 & 0 & \cdots & 1 & 0 \\ \vdots & \vdots & \ddots & \vdots & \vdots \\ 0 & 1 & \cdots & 0 & 0 \\ 1 & 0 & \cdots & 0 & 0 \end{bmatrix}$$

which is of rank n. Since the linear system (3.9) is controllable, so is the nonlinear system (3.7). It follows that a system that is given in nonlinear controller canonical form with $\beta(\boldsymbol{x}) \neq 0$ is controllable. This holds for all systems that can be transformed into the nonlinear controller canonical form using a bijective coordinate transformation. Such a transformation $\boldsymbol{z} = \boldsymbol{q}(\boldsymbol{x})$ assigns exactly one state $\boldsymbol{z}$ to each state $\boldsymbol{x}$, and vice versa. Here it should be at least once continuously differentiable. Such a transformation is referred to as a *diffeomorphism* [425]. The proposition above holds because the controllability is a system characteristic that is invariant to transformations by diffeomorphisms. Therefore, the controllability of a system does not change if the system is represented in different state coordinates.

For MIMO systems with n state variables and m input variables, the nonlinear controller canonical form of m subsystems is of the form

$$\begin{bmatrix} \dot{x}_1 \\ \vdots \\ \dot{x}_{n_1-1} \\ \dot{x}_{n_1} \\ \dot{x}_{n_1+1} \\ \vdots \\ \dot{x}_{n_2-1} \\ \dot{x}_{n_2} \\ \vdots \\ \dot{x}_{n_{m-1}+1} \\ \vdots \\ \dot{x}_{n-1} \\ \dot{x}_n \end{bmatrix} = \begin{bmatrix} x_2 \\ \vdots \\ x_{n_1} \\ \alpha_1(\boldsymbol{x}) \\ x_{n_1+2} \\ \vdots \\ x_{n_2} \\ \alpha_2(\boldsymbol{x}) \\ \vdots \\ x_{n_{m-1}+2} \\ \vdots \\ x_n \\ \alpha_m(\boldsymbol{x}) \end{bmatrix} + \begin{bmatrix} 0 & 0 & 0 & \cdots & 0 & 0 \\ \vdots & \vdots & \vdots & \ddots & \vdots & \vdots \\ 0 & 0 & 0 & \cdots & 0 & 0 \\ \beta_{1,1}(\boldsymbol{x}) & \beta_{1,2}(\boldsymbol{x}) & \beta_{1,3}(\boldsymbol{x}) & \cdots & \beta_{1,m-1}(\boldsymbol{x}) & \beta_{1,m}(\boldsymbol{x}) \\ 0 & 0 & 0 & \cdots & 0 & 0 \\ \vdots & \vdots & \vdots & \ddots & \vdots & \vdots \\ 0 & 0 & 0 & \cdots & 0 & 0 \\ 0 & \beta_{2,2}(\boldsymbol{x}) & \beta_{2,3}(\boldsymbol{x}) & \cdots & \beta_{2,m-1}(\boldsymbol{x}) & \beta_{2,m}(\boldsymbol{x}) \\ \vdots & \vdots & \vdots & \ddots & \vdots & \vdots \\ 0 & 0 & 0 & \cdots & 0 & 0 \\ \vdots & \vdots & \vdots & \ddots & \vdots & \vdots \\ 0 & 0 & 0 & \cdots & 0 & 0 \\ 0 & 0 & 0 & \cdots & 0 & \beta_{m,m}(\boldsymbol{x}) \end{bmatrix} \begin{bmatrix} u_1 \\ u_2 \\ u_3 \\ \vdots \\ u_{m-1} \\ u_m \end{bmatrix}. \tag{3.10}$$

The individual subsystems[3] possess the nonlinear controller canonical form. With the upper $m \times m$ triangular matrix

$$\boldsymbol{D}(\boldsymbol{x}) = \begin{bmatrix} \beta_{1,1}(\boldsymbol{x}) & \beta_{1,2}(\boldsymbol{x}) & \cdots & \beta_{1,m}(\boldsymbol{x}) \\ 0 & \beta_{2,2}(\boldsymbol{x}) & \cdots & \beta_{2,m}(\boldsymbol{x}) \\ \vdots & \vdots & \ddots & \vdots \\ 0 & 0 & \cdots & \beta_{m,m}(\boldsymbol{x}), \end{bmatrix},$$

similar to the SISO case, we can use a new input variable vector $\boldsymbol{v}$ via

$$\boldsymbol{u} = \boldsymbol{D}^{-1}(\boldsymbol{x}) \cdot \left(- \begin{bmatrix} \alpha_1(\boldsymbol{x}) \\ \alpha_2(\boldsymbol{x}) \\ \vdots \\ \alpha_m(\boldsymbol{x}) \end{bmatrix} + \begin{bmatrix} v_1 \\ v_2 \\ \vdots \\ v_m \end{bmatrix} \right). \tag{3.11}$$

We call the matrix $\boldsymbol{D}$ the *decoupling matrix*. Here it is assumed that $\beta_{i,i}(\boldsymbol{x}) \neq 0$ holds for all $\boldsymbol{x} \in D_{\boldsymbol{x}}$ and $i = 1, \ldots, m$, so that

$$\det(\boldsymbol{D}(\boldsymbol{x})) = \beta_{1,1}(\boldsymbol{x}) \cdot \beta_{2,2}(\boldsymbol{x}) \cdot \ \ldots \ \cdot \beta_{m,m}(\boldsymbol{x}) \neq 0$$

and consequently the inverse $\boldsymbol{D}^{-1}(\boldsymbol{x})$ exists.

[3] Notice that the degrees of the m subsystems are n_1, $n_2 - n_1$, $n_3 - n_2, \ldots, n - n_{m-1}$.

Inserting equation (3.11) into the system's representation (3.10), we arrive at the specific linear controller canonical form

$$\begin{bmatrix} \dot{x}_1 \\ \vdots \\ \dot{x}_{n_1-1} \\ \dot{x}_{n_1} \\ \dot{x}_{n_1+1} \\ \vdots \\ \dot{x}_{n_2-1} \\ \dot{x}_{n_2} \\ \vdots \\ \dot{x}_{n_{m-1}+1} \\ \vdots \\ \dot{x}_{n-1} \\ \dot{x}_n \end{bmatrix} = \begin{bmatrix} x_2 \\ \vdots \\ x_{n_1} \\ 0 \\ x_{n_1+2} \\ \vdots \\ x_{n_2} \\ 0 \\ \vdots \\ x_{n_{m-1}+2} \\ \vdots \\ x_n \\ 0 \end{bmatrix} + \begin{bmatrix} 0 & 0 & 0 & \cdots & 0 & 0 \\ \vdots & \vdots & \vdots & \ddots & \vdots & \vdots \\ 0 & 0 & 0 & \cdots & 0 & 0 \\ 1 & 0 & 0 & \cdots & 0 & 0 \\ 0 & 0 & 0 & \cdots & 0 & 0 \\ \vdots & \vdots & \vdots & \ddots & \vdots & \vdots \\ 0 & 0 & 0 & \cdots & 0 & 0 \\ 0 & 1 & 0 & \cdots & 0 & 0 \\ \vdots & \vdots & \vdots & \ddots & \vdots & \vdots \\ 0 & 0 & 0 & \cdots & 0 & 0 \\ \vdots & \vdots & \vdots & \ddots & \vdots & \vdots \\ 0 & 0 & 0 & \cdots & 0 & 0 \\ 0 & 0 & 0 & \cdots & 0 & 1 \end{bmatrix} \begin{bmatrix} v_1 \\ v_2 \\ v_3 \\ \vdots \\ v_{m-1} \\ v_m \end{bmatrix},$$

which is also referred to as the *Brunovsky canonical form* [60]. Analogous to the SISO case, we can deduce that since the linear MIMO system in controller canonical form is controllable, so is the corresponding system in nonlinear controller canonical form. We can thus formulate the following important theorem.

Theorem 45 (Controllability of Nonlinear Systems). *A system*

$$\dot{\boldsymbol{x}} = \boldsymbol{f}(\boldsymbol{x}, \boldsymbol{u})$$

with $\boldsymbol{x} \in D_{\boldsymbol{x},\mathrm{def}} \subseteq \mathbb{R}^n$ *and* $\boldsymbol{u} \in \mathbb{R}^m$, *which is in nonlinear controller canonical form, or which can be transformed into this form by means of a continuously differentiable, bijective coordinate transformation, is controllable if the decoupling matrix* $\boldsymbol{D}(\boldsymbol{x})$ *fulfills*

$$\det(\boldsymbol{D}(\boldsymbol{x})) \neq 0$$

for all $\boldsymbol{x}$ *of a path-connected set* $D_{\boldsymbol{x}} \subseteq D_{\boldsymbol{x},\mathrm{def}}$.

Of course, this theorem also includes the SISO case, in which

$$\det(\boldsymbol{D}(\boldsymbol{x})) = \beta(\boldsymbol{x})$$

holds.

The above theorem is of great importance because quite a few nonlinear systems exist which can be transformed into the nonlinear controller canonical form. The most important of these belong to the class of control-affine systems. How these systems can be transformed into the nonlinear controller canonical form, and how to control them, is discussed in Section 5.2.

3.1.4 Example: Industrial Robot

We will consider an industrial robot with three degrees of freedom, two rotational degrees and one translational. The rotation angle Θ_1 of a vertical arm A_1 with associated mass m_1 constitutes the first degree of freedom of movement with respect to the foundation, as shown in Figure 3.10. Attached to the endpoint of the arm is a joint which realizes the second degree of freedom by means of a rotation with angle Θ_2. The third degree of freedom is obtained through a shiftably mounted arm A_2 within the rotary joint with associated mass m_2. The arm allows for a translational movement along the stretch r, which defines the distance of the center of gravity of A_2 to the rotary joint.

In order to derive the dynamic model of the robot, we first compute the robot's kinetic and potential energy, E_{kin} and E_{pot}. The difference between the two is the Lagrange function

$$L = E_{\text{kin}} - E_{\text{pot}}, \tag{3.12}$$

which we insert into the Lagrange equations [184]

$$\begin{aligned} M_i &= \frac{d}{dt}\frac{\partial L}{\partial \dot{\Theta}_i} - \frac{\partial L}{\partial \Theta_i}, \quad i = 1,2, \\ F &= \frac{d}{dt}\frac{\partial L}{\partial \dot{r}} - \frac{\partial L}{\partial r} \end{aligned} \tag{3.13}$$

and thereby determine the torques M_i and the force F. The torques M_i and the force F are the robot's control variables, and are proportional to the currents of the electric motors which adjust the joints' angles Θ_i and the translation r.

The kinetic energy of the arm A_1 is

$$E_{\text{kin},A_1} = \frac{1}{2} J_z \dot{\Theta}_1^2,$$

where J_z is the moment of inertia of arm A_1 with respect to its longitudinal axis z.

For arm A_2, the kinetic energy is composed of its rotational energy E_{rot,A_2} and the energy E_{trans,A_2} from the translational movement. With the angular velocity vector $\boldsymbol{\omega}_{A_2}$ and the inertia matrix

$$\boldsymbol{J} = \begin{bmatrix} J_{\tilde{x}} & 0 & 0 \\ 0 & J_{\tilde{y}} & 0 \\ 0 & 0 & J_{\tilde{z}} \end{bmatrix}$$

of arm A_2, it holds that

$$E_{\text{rot},A_2} = \frac{1}{2} \boldsymbol{\omega}_{A_2}^T \boldsymbol{J} \, \boldsymbol{\omega}_{A_2}. \tag{3.14}$$

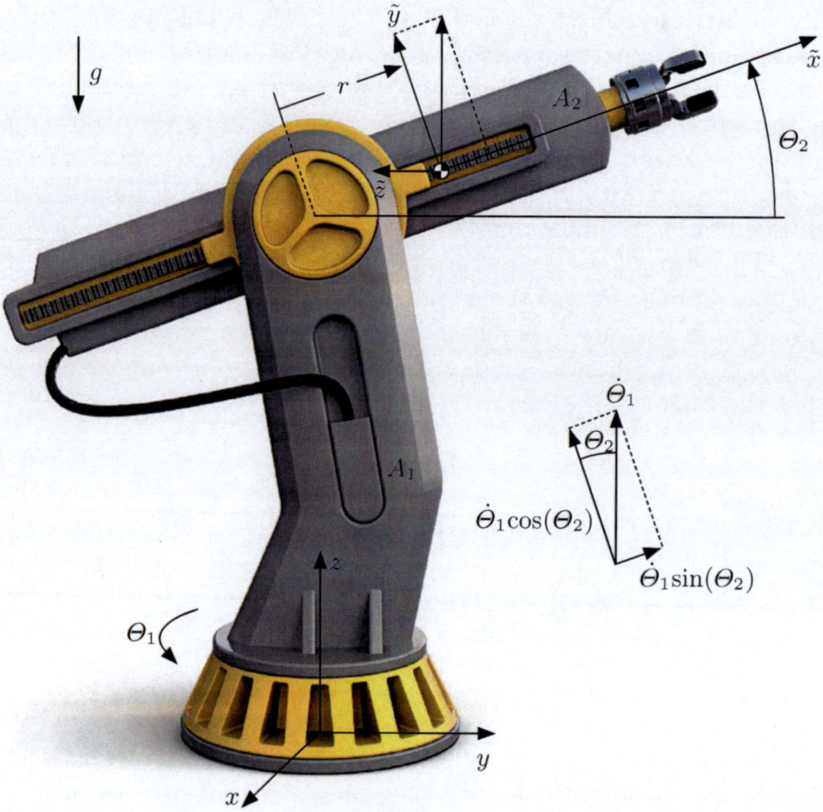

Fig. 3.10: Industrial robot with three degrees of freedom. The $\tilde{x}, \tilde{y}, \tilde{z}$-coordinate system has its origin in the center of gravity of the robot arm A_2.

With the unit vectors of the space-fixed coordinate system (x, y, z) and the arm-coordinate system $(\tilde{x}, \tilde{y}, \tilde{z})$

$$\boldsymbol{e}_x = \begin{bmatrix}1\\0\\0\end{bmatrix},\ \boldsymbol{e}_y = \begin{bmatrix}0\\1\\0\end{bmatrix},\ \boldsymbol{e}_z = \begin{bmatrix}0\\0\\1\end{bmatrix} \quad \text{and} \quad \boldsymbol{e}_{\tilde{\boldsymbol{x}}} = \begin{bmatrix}1\\0\\0\end{bmatrix},\ \boldsymbol{e}_{\tilde{\boldsymbol{y}}} = \begin{bmatrix}0\\1\\0\end{bmatrix},\ \boldsymbol{e}_{\tilde{\boldsymbol{z}}} = \begin{bmatrix}0\\0\\1\end{bmatrix},$$

and the decomposition of the angular velocity shown in Figure 3.10

$$\dot{\Theta}_1 \boldsymbol{e}_z = \dot{\Theta}_1 \sin(\Theta_2)\boldsymbol{e}_{\tilde{x}} + \dot{\Theta}_1 \cos(\Theta_2)\boldsymbol{e}_{\tilde{y}},$$

as well as the angular velocity $\dot{\Theta}_2 \boldsymbol{e}_{\tilde{z}}$ of the arm A_2 rotating around the $\tilde{z}$-joint axis, it holds that

$$\boldsymbol{\omega}_{A_2} = \dot{\Theta}_1 \sin(\Theta_2)\boldsymbol{e}_{\tilde{x}} + \dot{\Theta}_1 \cos(\Theta_2)\boldsymbol{e}_{\tilde{y}} + \dot{\Theta}_2 \boldsymbol{e}_{\tilde{z}}. \tag{3.15}$$

Inserting equation (3.15) into equation (3.14) yields

$$E_{\text{rot},A_2} = \frac{1}{2}\left[(J_{\tilde{x}}\sin^2(\Theta_2) + J_{\tilde{y}}\cos^2(\Theta_2))\dot{\Theta}_1^2 + J_{\tilde{z}}\dot{\Theta}_2^2\right].$$

For the computation of the translational energy E_{trans,A_2}, we require the velocity vector $\boldsymbol{v}_{A_2}$ of A_2's center of mass. It is composed of the radial velocity $\dot{r}\boldsymbol{e}_{\tilde{x}}$ as well as the tangential velocities $\dot{\Theta}_2 r\boldsymbol{e}_{\tilde{y}}$ and $\dot{\Theta}_1 r\cos(\Theta_2)\boldsymbol{e}_{\tilde{z}}$ and is given by the equation

$$\boldsymbol{v}_{A_2} = \dot{r}\boldsymbol{e}_{\tilde{x}} + \dot{\Theta}_2 r\boldsymbol{e}_{\tilde{y}} - \dot{\Theta}_1 r\cos(\Theta_2)\boldsymbol{e}_{\tilde{z}}.$$

We then obtain

$$E_{\text{trans},A_2} = \frac{1}{2}m_2\,\boldsymbol{v}_{A_2}^T\boldsymbol{v}_{A_2} = \frac{1}{2}m_2\left[\dot{r}^2 + r^2\left(\dot{\Theta}_2^2 + \dot{\Theta}_1^2\cos^2(\Theta_2)\right)\right].$$

Furthermore, we have to determine the robot's potential energy, for which we obtain

$$E_{\text{pot}} = m_2\,g\,r\sin(\Theta_2).$$

It then follows according to equation (3.12) that

$$\begin{aligned} L &= \underbrace{E_{\text{kin},A_1} + E_{\text{rot},A_2} + E_{\text{trans},A_2}}_{E_{\text{kin}}} - E_{\text{pot}} \\ &= \frac{1}{2}\Big[\left(J_z + J_{\tilde{x}}\sin^2(\Theta_2) + (J_{\tilde{y}} + m_2r^2)\cos^2(\Theta_2)\right)\dot{\Theta}_1^2 + (J_{\tilde{z}} + m_2r^2)\dot{\Theta}_2^2 \\ &\quad + m_2\dot{r}^2 - 2m_2\,g\,r\sin(\Theta_2)\Big] \end{aligned} \tag{3.16}$$

holds.

We now insert equation (3.16) into the Lagrange equations (3.13) and obtain the expressions for the torques M_1, M_2, and the force F as

$$\begin{aligned} M_1 &= \left[J_z + J_{\tilde{x}}\sin^2(\Theta_2) + (J_{\tilde{y}} + m_2r^2)\cos^2(\Theta_2)\right]\ddot{\Theta}_1 \\ &\quad + 2m_2\,r\,\dot{r}\,\dot{\Theta}_1\cos^2(\Theta_2) - (J_{\tilde{y}} - J_{\tilde{x}} + m_2r^2)\dot{\Theta}_1\dot{\Theta}_2\sin(2\Theta_2), \\ M_2 &= (J_{\tilde{z}} + m_2r^2)\ddot{\Theta}_2 + 2m_2\,r\,\dot{r}\,\dot{\Theta}_2 \\ &\quad + \frac{1}{2}(J_{\tilde{y}} - J_{\tilde{x}} + m_2r^2)\dot{\Theta}_1^2\sin(2\Theta_2) + m_2\,g\,r\cos(\Theta_2), \\ F &= m_2\ddot{r} - m_2\,r(\dot{\Theta}_1^2\cos^2(\Theta_2) + \dot{\Theta}_2^2) + m_2\,g\sin(\Theta_2). \end{aligned} \tag{3.17}$$

We can reformulate equation (3.17) by using the state vector $\boldsymbol{x}$, the input variable vector $\boldsymbol{u}$, and the output variable vector $\boldsymbol{y}$,

$$\boldsymbol{x} = \left[\Theta_1\;\dot{\Theta}_1\;\Theta_2\;\dot{\Theta}_2\;r\;\dot{r}\right]^T,\quad \boldsymbol{u} = \left[M_1\;M_2\;F\right]^T,\quad \boldsymbol{y} = \left[\Theta_1\;\Theta_2\;r\right]^T,$$

yielding the dynamic model

$$\begin{aligned}\dot{\boldsymbol{x}} &= \boldsymbol{a}(\boldsymbol{x}) + \boldsymbol{B}(\boldsymbol{x}) \cdot \boldsymbol{u}, \\ \boldsymbol{y} &= \boldsymbol{c}(\boldsymbol{x})\end{aligned} \tag{3.18}$$

with

$$\boldsymbol{a}(\boldsymbol{x}) = \begin{bmatrix} x_2 \\ -\dfrac{2m_2\, x_2\, x_5\, x_6 \cos^2(x_3) - (J_{\tilde{y}} - J_{\tilde{x}} + m_2\, x_5^2)x_2\, x_4 \sin(2x_3)}{J_z + J_{\tilde{x}} \sin^2(x_3) + (J_{\tilde{y}} + m_2\, x_5^2)\cos^2(x_3)} \\ x_4 \\ -\dfrac{2m_2\, x_4\, x_5\, x_6 + \frac{1}{2}(J_{\tilde{y}} - J_{\tilde{x}} + m_2\, x_5^2)x_2^2 \sin(2x_3) + m_2\, g\, x_5 \cos(x_3)}{J_{\tilde{z}} + m_2\, x_5^2} \\ x_6 \\ x_5(x_2^2 \cos^2(x_3) + x_4^2) - g \sin(x_3) \end{bmatrix},$$

$$\boldsymbol{B}(\boldsymbol{x}) = \begin{bmatrix} 0 & 0 & 0 \\ \dfrac{1}{J_z + J_{\tilde{x}} \sin^2(x_3) + (J_{\tilde{y}} + m_2 x_5^2)\cos^2(x_3)} & 0 & 0 \\ 0 & 0 & 0 \\ 0 & \dfrac{1}{J_{\tilde{z}} + m_2 x_5^2} & 0 \\ 0 & 0 & 0 \\ 0 & 0 & \dfrac{1}{m_2} \end{bmatrix},$$

$$\boldsymbol{c}(\boldsymbol{x}) = \begin{bmatrix} x_1 \\ x_3 \\ x_5 \end{bmatrix}.$$

The system (3.18) is in the nonlinear controller canonical form

$$\begin{aligned}\dot{\boldsymbol{x}} &= \begin{bmatrix} x_2 \\ \alpha_1(\boldsymbol{x}) \\ x_4 \\ \alpha_2(\boldsymbol{x}) \\ x_6 \\ \alpha_3(\boldsymbol{x}) \end{bmatrix} + \begin{bmatrix} 0 & 0 & 0 \\ \beta_1(\boldsymbol{x}) & 0 & 0 \\ 0 & 0 & 0 \\ 0 & \beta_2(\boldsymbol{x}) & 0 \\ 0 & 0 & 0 \\ 0 & 0 & \beta_3(\boldsymbol{x}) \end{bmatrix} \boldsymbol{u}, \\ \boldsymbol{y} &= \begin{bmatrix} x_1 & x_3 & x_5 \end{bmatrix}^T\end{aligned} \tag{3.19}$$

and it holds that $\beta_i(\boldsymbol{x}) \neq 0$ for $i = 1, 2, 3$. The matrix

$$\boldsymbol{D}(\boldsymbol{x}) = \begin{bmatrix} \beta_1(\boldsymbol{x}) & 0 & 0 \\ 0 & \beta_2(\boldsymbol{x}) & 0 \\ 0 & 0 & \beta_3(\boldsymbol{x}) \end{bmatrix}$$

with the associated determinant

$$\det(\boldsymbol{D}(\boldsymbol{x})) = \beta_1(\boldsymbol{x})\beta_2(\boldsymbol{x})\beta_3(\boldsymbol{x})$$

is different from zero, since $\beta_i(\boldsymbol{x}) \neq 0$ holds for $i = 1, 2, 3$. Thus the robot is controllable.

The nonlinear controller canonical form allows for controller design which is especially easy to handle, hence its name. For this design, we use the control law

$$\boldsymbol{u} = -\begin{bmatrix} \beta_1^{-1}(\boldsymbol{x}) & 0 & 0 \\ 0 & \beta_2^{-1}(\boldsymbol{x}) & 0 \\ 0 & 0 & \beta_3^{-1}(\boldsymbol{x}) \end{bmatrix} \begin{bmatrix} \alpha_1(\boldsymbol{x}) + a_{1,1}x_2 + a_{1,0}x_1 - V_1 \cdot y_{\mathrm{ref},1} \\ \alpha_2(\boldsymbol{x}) + a_{2,1}x_4 + a_{2,0}x_3 - V_2 \cdot y_{\mathrm{ref},2} \\ \alpha_3(\boldsymbol{x}) + a_{3,1}x_6 + a_{3,0}x_5 - V_3 \cdot y_{\mathrm{ref},3} \end{bmatrix} =$$

$$\begin{bmatrix} 2m_2x_2x_5x_6\cos^2(x_3) - (J_\Delta + m_2x_5^2)x_2x_4\sin(2x_3) - \dfrac{a_{1,1}x_2 + a_{1,0}x_1 - V_1y_{\mathrm{ref},1}}{\beta_1(\boldsymbol{x})} \\ 2m_2x_4x_5x_6 + \dfrac{(J_\Delta + m_2x_5^2)x_2^2\sin(2x_3)}{2} + m_2gx_5\cos(x_3) - \dfrac{a_{2,1}x_4 + a_{2,0}x_3 - V_2y_{\mathrm{ref},2}}{\beta_2(\boldsymbol{x})} \\ -m_2x_5(x_2^2\cos^2(x_3) + x_4^2) + m_2g\sin(x_3) - \dfrac{a_{3,1}x_6 + a_{3,0}x_5 - V_3y_{\mathrm{ref},3}}{\beta_3(\boldsymbol{x})} \end{bmatrix}, \tag{3.20}$$

where $y_{\mathrm{ref},1}$, $y_{\mathrm{ref},2}$, and $y_{\mathrm{ref},3}$ are the reference variables and

$$J_\Delta = J_{\tilde{y}} - J_{\tilde{x}}.$$

The coefficients $a_{i,j}$ and V_i are freely selectable. Inserted into equation (3.18) or (3.19), the control law (3.20) yields a linear control loop with the system equations

$$\begin{aligned} \ddot{y}_1 + a_{1,1}\dot{y}_1 + a_{1,0}y_1 &= V_1 \cdot y_{\mathrm{ref},1}, \\ \ddot{y}_2 + a_{2,1}\dot{y}_2 + a_{2,0}y_2 &= V_2 \cdot y_{\mathrm{ref},2}, \\ \ddot{y}_3 + a_{3,1}\dot{y}_3 + a_{3,0}y_3 &= V_3 \cdot y_{\mathrm{ref},3}. \end{aligned}$$

It is possible to design such linear control loops not only for the industrial robot described here, but for all other nonlinear plants that can be represented in nonlinear controller canonical form as well. We will utilize this property in Section 5.2.

3.1.5 Small-Time Local Controllability of Driftless Systems

In the following, we will address control-affine systems (3.6) which have no free dynamics $\boldsymbol{a}(\boldsymbol{x})$, sometimes called *drift*, i. e. systems characterized by

$$\dot{\boldsymbol{x}} = \boldsymbol{B}(\boldsymbol{x}) \cdot \boldsymbol{u}.$$

Accordingly, these systems are called *driftless systems*. For the special case of a driftless linear system

$$\dot{\boldsymbol{x}} = \boldsymbol{B}\boldsymbol{u}, \tag{3.21}$$

controllability exists if and only if the control matrix $\boldsymbol{B}$ is of rank n, i. e. if

$$\det(\boldsymbol{B}) \neq 0$$

holds. On the one hand, this is plausible, since all n derivatives must be manipulated to change a state $\boldsymbol{x}_0$ into a different state $\boldsymbol{x}_\mathrm{e}$. On the other hand, controllability also follows from the full rank of the controllability matrix of the system, i. e. from

$$\begin{aligned}\operatorname{rank}(\boldsymbol{M}_\mathrm{contr}) &= \operatorname{rank}\left(\begin{bmatrix}\boldsymbol{B} & \boldsymbol{A}\boldsymbol{B} & \boldsymbol{A}^2\boldsymbol{B} & \cdots & \boldsymbol{A}^{n-1}\boldsymbol{B}\end{bmatrix}\right)\\ &= \operatorname{rank}(\boldsymbol{B}) = n.\end{aligned}$$

Additionally, the system is also omnidirectionally controllable.

Driftless control-affine systems, in contrast to the linear case described above, can also be controllable if $\boldsymbol{B}(\boldsymbol{x})$ is not of rank n, i. e. if the number m of input variables is smaller than the number n of states. This is not directly plausible, since an $n \times m$ matrix $\boldsymbol{B}(\boldsymbol{x})$ with $m < n$ does not span the entire $I\!R^n$, and therefore the controllability matrix

$$\boldsymbol{M}_\mathrm{contr} = \begin{bmatrix}\boldsymbol{B}(\boldsymbol{x}) & \boldsymbol{0} & \cdots & \boldsymbol{0}\end{bmatrix}$$

is not of rank n for a fixed $\boldsymbol{x}$. Accordingly, the corresponding driftless linear system (3.21) with

$$\operatorname{rank}(\boldsymbol{B}) < n$$

would not be controllable.

At second glance, however, it becomes clear that the columns $\boldsymbol{b}_i(\boldsymbol{x})$ of a nonlinear system's matrix $\boldsymbol{B}(\boldsymbol{x})$ change along a trajectory $\boldsymbol{x}(t)$, i. e. the direction $\boldsymbol{b}_i(\boldsymbol{x})$, in which the control input u_i is effective, varies with $\boldsymbol{x}$. After a movement from $\boldsymbol{x}_\mathrm{a}$ to $\boldsymbol{x}_\mathrm{b}$, it is thus possible to steer the trajectory in a new, previously impossible direction: $\boldsymbol{B}(\boldsymbol{x}_\mathrm{b}) \cdot \boldsymbol{u}$.

As an example, we will view the system

$$\begin{bmatrix}\dot{x}_1\\ \dot{x}_2\\ \dot{x}_3\end{bmatrix} = \underbrace{\begin{bmatrix}0\\0\\1\end{bmatrix}}_{\boldsymbol{b}_1} u_1 + \underbrace{\begin{bmatrix}\cos(x_3)\\ \sin(x_3)\\ 0\end{bmatrix}}_{\boldsymbol{b}_2(x_3)} u_2, \tag{3.22}$$

which describes a mobile robot with two individually driven wheels and a freely movable stabilizing wheel, as shown in Figure 3.11. The control variable u_1 is the robot's angular velocity and the control variable u_2 is its translational velocity in the direction of motion. The state variables x_1 and x_2 describe the position of the robot and x_3 denotes the angular orientation of the robot in space.

Obviously, the direction on the xy-plane where the robot can move, i. e. $\boldsymbol{b}_2(x_3)$, depends on its orientation x_3. Figure 3.12 illustrates this dependency. Choosing a suitable sequence of rotations u_1 and longitudinal movements u_2, it is possible to reach any arbitrary point $\boldsymbol{x}_{\mathrm{e}}$ from any other arbitrary point $\boldsymbol{x}_0$, although $\boldsymbol{b}_1$ and $\boldsymbol{b}_2(x_3)$ only span $I\!R^2$ for a fixed x_3. The red line in Figure 3.12 shows such a trajectory. To generate this trajectory, we use the control signal sequence

$$\begin{bmatrix} u_1 \\ u_2 \end{bmatrix} = \begin{bmatrix} 0 \\ -1 \end{bmatrix}, \quad \begin{bmatrix} u_1 \\ u_2 \end{bmatrix} = \begin{bmatrix} 1 \\ 0 \end{bmatrix}, \quad \begin{bmatrix} u_1 \\ u_2 \end{bmatrix} = \begin{bmatrix} 0 \\ 1 \end{bmatrix}$$

to arrive at the sequence

$$\boldsymbol{x}_0 = \boldsymbol{x}(0) \to \boldsymbol{x}(\Delta t) \to \boldsymbol{x}(2\Delta t) \to \boldsymbol{x}(3\Delta t) = \boldsymbol{x}_{\mathrm{e}}$$

of the state vectors. According to the geometric illustration, it is plausible that the system is controllable.

We will now consider the more general case of a system

$$\dot{\boldsymbol{x}} = \boldsymbol{b}_1(\boldsymbol{x}) \cdot u_1 + \boldsymbol{b}_2(\boldsymbol{x}) \cdot u_2 \tag{3.23}$$

with $\boldsymbol{x} \in I\!R^n$. Similar to the case of the mobile robot above, we utilize the sequence

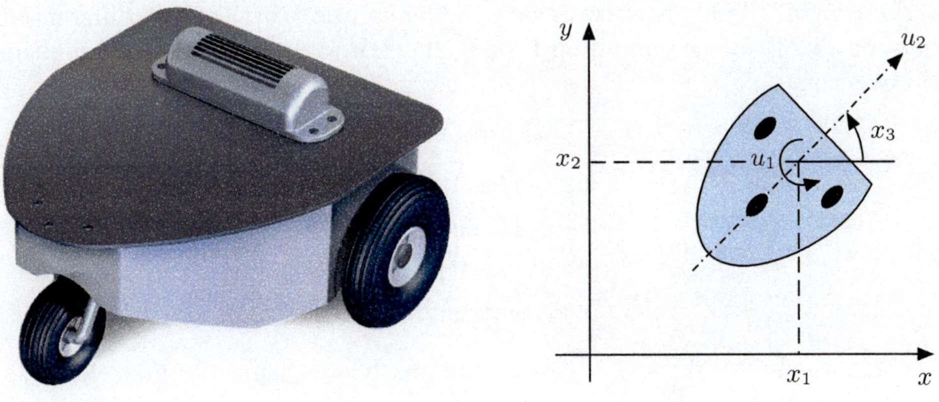

Fig. 3.11: Mobile robot with two individually driven wheels and a stabilizing wheel

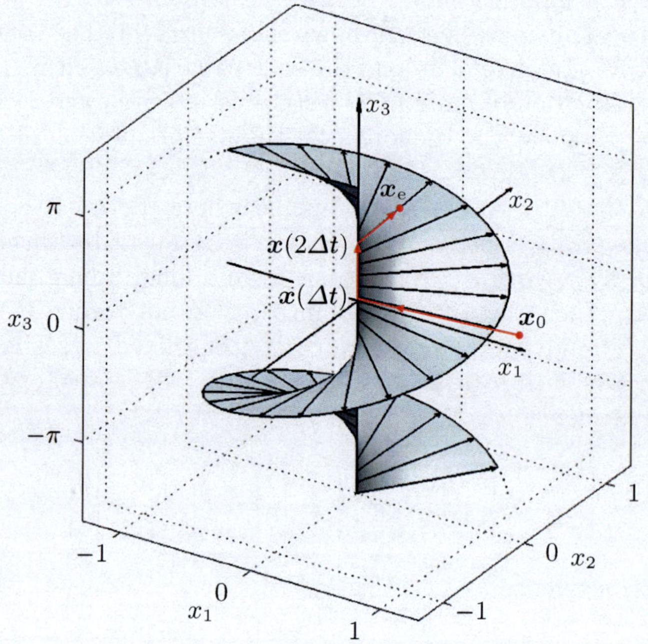

Fig. 3.12: Directions of the vector $\boldsymbol{b}_2(x_3)$ for $u_2 = 1$ and for the case in which the robot is in the state $[x_1\ x_2\ x_3] = [0\ 0\ \int_0^t u_1(\tau)d\tau]$. A possible trajectory is depicted by the red line.

$$\boldsymbol{u}(t) = \begin{cases} \boldsymbol{u}^T = [\ \ 1\ \ \ 0], & 0 \le t < \Delta t, \\ \boldsymbol{u}^T = [\ \ 0\ \ \ 1], & \Delta t \le t < 2\Delta t, \\ \boldsymbol{u}^T = [-1\ \ \ 0], & 2\Delta t \le t < 3\Delta t, \\ \boldsymbol{u}^T = [\ \ 0\ -1], & 3\Delta t \le t < 4\Delta t, \end{cases} \tag{3.24}$$

as the control signal. Next we examine the solution $\boldsymbol{x}(t)$ of the differential equation (3.23) for the points in time $0, \Delta t$, $2\Delta t$, $3\Delta t$, and $4\Delta t$, making the abbreviations

$$\begin{aligned} \boldsymbol{x}_0 &= \boldsymbol{x}(0), \\ \boldsymbol{x}_1 &= \boldsymbol{x}(\Delta t), \\ \boldsymbol{x}_2 &= \boldsymbol{x}(2\Delta t), \\ \boldsymbol{x}_3 &= \boldsymbol{x}(3\Delta t), \\ \boldsymbol{x}_4 &= \boldsymbol{x}(4\Delta t) = \boldsymbol{x}_e. \end{aligned}$$

To compute these values, we then approximate $\boldsymbol{x}(t)$ using the first elements of a Taylor series

$$\boldsymbol{x}((i+1)\Delta t) \approx \boldsymbol{x}(i\Delta t) + \Delta t \dot{\boldsymbol{x}}(i\Delta t) + \frac{1}{2}\Delta t^2 \ddot{\boldsymbol{x}}(i\Delta t),$$

which can be more briefly formulated as

$$\boldsymbol{x}_{i+1} \approx \boldsymbol{x}_i + \Delta t \dot{\boldsymbol{x}}_i + \frac{1}{2} \Delta t^2 \ddot{\boldsymbol{x}}_i.$$

Using the differential equation (3.23) and the course of the control signal (3.24), it follows that

$$\boldsymbol{x}_1 \approx \boldsymbol{x}_0 + \Delta t\, \boldsymbol{b}_1(\boldsymbol{x}_0) + \frac{1}{2}\Delta t^2 \left.\frac{\partial \boldsymbol{b}_1}{\partial \boldsymbol{x}}\right|_{\boldsymbol{x}_0} \cdot \boldsymbol{b}_1(\boldsymbol{x}_0), \tag{3.25}$$

$$\boldsymbol{x}_2 \approx \boldsymbol{x}_1 + \Delta t\, \boldsymbol{b}_2(\boldsymbol{x}_1) + \frac{1}{2}\Delta t^2 \left.\frac{\partial \boldsymbol{b}_2}{\partial \boldsymbol{x}}\right|_{\boldsymbol{x}_1} \cdot \boldsymbol{b}_2(\boldsymbol{x}_1), \tag{3.26}$$

$$\boldsymbol{x}_3 \approx \boldsymbol{x}_2 - \Delta t\, \boldsymbol{b}_1(\boldsymbol{x}_2) + \frac{1}{2}\Delta t^2 \left.\frac{\partial \boldsymbol{b}_1}{\partial \boldsymbol{x}}\right|_{\boldsymbol{x}_2} \cdot \boldsymbol{b}_1(\boldsymbol{x}_2), \tag{3.27}$$

$$\boldsymbol{x}_4 \approx \boldsymbol{x}_3 - \Delta t\, \boldsymbol{b}_2(\boldsymbol{x}_3) + \frac{1}{2}\Delta t^2 \left.\frac{\partial \boldsymbol{b}_2}{\partial \boldsymbol{x}}\right|_{\boldsymbol{x}_3} \cdot \boldsymbol{b}_2(\boldsymbol{x}_3). \tag{3.28}$$

In the following we will first insert equation (3.25) into equation (3.26) and subsequently insert the result into equation (3.27). The result obtained from this is in turn inserted into equation (3.28). We thus obtain $\boldsymbol{x}_4$ in dependence on the starting point $\boldsymbol{x}_0$, so that we can deduce the direction which points from $\boldsymbol{x}_0$ to $\boldsymbol{x}_4$ due to the control signal sequence (3.24).

As stated above, we will begin by inserting equation (3.25) into equation (3.26) and obtain

$$\begin{aligned}\boldsymbol{x}_2 \approx{}& \boldsymbol{x}_0 + \Delta t\, \boldsymbol{b}_1(\boldsymbol{x}_0) + \frac{1}{2}\Delta t^2 \left.\frac{\partial \boldsymbol{b}_1}{\partial \boldsymbol{x}}\right|_{\boldsymbol{x}_0} \cdot \boldsymbol{b}_1(\boldsymbol{x}_0) + \Delta t\, \boldsymbol{b}_2(\boldsymbol{x}_0 + \Delta t \boldsymbol{b}_1(\boldsymbol{x}_0) + \ldots)\\ &+ \frac{1}{2}\Delta t^2 \left.\frac{\partial \boldsymbol{b}_2}{\partial \boldsymbol{x}}\right|_{\boldsymbol{x}_0 + \Delta t \boldsymbol{b}_1(\boldsymbol{x}_0) + \ldots} \cdot \boldsymbol{b}_2(\boldsymbol{x}_0 + \Delta t \boldsymbol{b}_1(\boldsymbol{x}_0) + \ldots).\end{aligned} \tag{3.29}$$

In the following, just as above, we will only include the terms up to the power of two, i. e. Δt^2, in all approximations, since we are only interested in small values of Δt. Again, we approximate

$$\boldsymbol{b}_2(\boldsymbol{x}_0 + \Delta t\, \boldsymbol{b}_1(\boldsymbol{x}_0) + \frac{1}{2}\Delta t^2 \left.\frac{\partial \boldsymbol{b}_1}{\partial \boldsymbol{x}}\right|_{\boldsymbol{x}_0} \cdot \boldsymbol{b}_1(\boldsymbol{x}_0)) \approx \boldsymbol{b}_2(\boldsymbol{x}_0) + \Delta t \left.\frac{\partial \boldsymbol{b}_2}{\partial \boldsymbol{x}}\right|_{\boldsymbol{x}_0} \cdot \boldsymbol{b}_1(\boldsymbol{x}_0)$$

using the Taylor series and insert the result into equation (3.29), at which point we can make the approximation

$$\boldsymbol{x}_1 = \boldsymbol{x}_0 + \Delta t \boldsymbol{b}_1(\boldsymbol{x}_0) + \ldots \approx \boldsymbol{x}_0$$

for the differentiation point $\boldsymbol{x}_0 + \Delta t \boldsymbol{b}_1(\boldsymbol{x}_0) + \ldots$ in the last term of equation (3.29). It follows that

$$\boldsymbol{x}_2 \approx \boldsymbol{x}_0 + \Delta t\,[\boldsymbol{b}_1(\boldsymbol{x}_0) + \boldsymbol{b}_2(\boldsymbol{x}_0)]$$
$$+ \Delta t^2 \left[\frac{1}{2}\frac{\partial \boldsymbol{b}_1}{\partial \boldsymbol{x}}\bigg|_{\boldsymbol{x}_0} \cdot \boldsymbol{b}_1(\boldsymbol{x}_0) + \frac{\partial \boldsymbol{b}_2}{\partial \boldsymbol{x}}\bigg|_{\boldsymbol{x}_0} \cdot \boldsymbol{b}_1(\boldsymbol{x}_0) + \frac{1}{2}\frac{\partial \boldsymbol{b}_2}{\partial \boldsymbol{x}}\bigg|_{\boldsymbol{x}_0} \cdot \boldsymbol{b}_2(\boldsymbol{x}_0) \right].$$

We then apply this result in equation (3.27) and obtain

$$\boldsymbol{x}_3 \approx \boldsymbol{x}_0 + \Delta t\,\boldsymbol{b}_2(\boldsymbol{x}_0) + \Delta t^2 \left[\frac{\partial \boldsymbol{b}_2}{\partial \boldsymbol{x}}\bigg|_{\boldsymbol{x}_0} \cdot \boldsymbol{b}_1(\boldsymbol{x}_0) - \frac{\partial \boldsymbol{b}_1}{\partial \boldsymbol{x}}\bigg|_{\boldsymbol{x}_0} \cdot \boldsymbol{b}_2(\boldsymbol{x}_0) + \frac{1}{2}\frac{\partial \boldsymbol{b}_2}{\partial \boldsymbol{x}}\bigg|_{\boldsymbol{x}_0} \cdot \boldsymbol{b}_2(\boldsymbol{x}_0) \right].$$

Finally, inserting the result into equation (3.28) yields

$$\boldsymbol{x}_4 \approx \boldsymbol{x}_0 + \Delta t^2 \left[\frac{\partial \boldsymbol{b}_2}{\partial \boldsymbol{x}}\bigg|_{\boldsymbol{x}_0} \cdot \boldsymbol{b}_1(\boldsymbol{x}_0) - \frac{\partial \boldsymbol{b}_1}{\partial \boldsymbol{x}}\bigg|_{\boldsymbol{x}_0} \cdot \boldsymbol{b}_2(\boldsymbol{x}_0) \right]. \tag{3.30}$$

Based on equation (3.30) we can deduce that, at least for small time intervals Δt, we can move from $\boldsymbol{x}_0$ in the direction

$$\boldsymbol{b}_3(\boldsymbol{x}) = \frac{\partial \boldsymbol{b}_2}{\partial \boldsymbol{x}}\bigg|_{\boldsymbol{x}_0} \cdot \boldsymbol{b}_1(\boldsymbol{x}_0) - \frac{\partial \boldsymbol{b}_1}{\partial \boldsymbol{x}}\bigg|_{\boldsymbol{x}_0} \cdot \boldsymbol{b}_2(\boldsymbol{x}_0) = [\boldsymbol{b}_1, \boldsymbol{b}_2]$$

to reach $\boldsymbol{x}_4 = \boldsymbol{x}_\mathrm{e}$ by means of the control signal sequence (3.24). The short form of the expression above is $[\boldsymbol{b}_1, \boldsymbol{b}_2]$. Such a differential operator

$$[\boldsymbol{f}_1, \boldsymbol{f}_2] = \frac{\partial \boldsymbol{f}_2(\boldsymbol{x})}{\partial \boldsymbol{x}}\boldsymbol{f}_1(\boldsymbol{x}) - \frac{\partial \boldsymbol{f}_1(\boldsymbol{x})}{\partial \boldsymbol{x}}\boldsymbol{f}_2(\boldsymbol{x})$$

is termed the *Lie bracket* after M. S. Lie.

Therefore, in addition to the directions $\boldsymbol{b}_1$ and $\boldsymbol{b}_2$, it is also possible to move in the direction $\boldsymbol{b}_3$ if the control signal sequence is chosen accordingly. If one chooses control signal sequences which are different and longer than the ones in equation (3.24), where $u_i \in \{-1, 0, 1\}$ holds, this results in additional directions

$$\begin{gathered}
[\boldsymbol{b}_1, [\boldsymbol{b}_1, \boldsymbol{b}_2]], \\
[\boldsymbol{b}_2, [\boldsymbol{b}_1, \boldsymbol{b}_2]], \\
[\boldsymbol{b}_1, [\boldsymbol{b}_1, [\boldsymbol{b}_1, \boldsymbol{b}_2]]], \\
\vdots \\
[[\boldsymbol{b}_1, \boldsymbol{b}_2], [\boldsymbol{b}_1, [\boldsymbol{b}_1, [\boldsymbol{b}_1, \boldsymbol{b}_2]]]], \\
\vdots
\end{gathered}$$

into which the system state $\boldsymbol{x}$ can proceed.

The considerations above (see also [80, 81, 179, 285, 338, 435]) bring us to

Theorem 46 (Chow's Theorem). *Let the system*

$$\dot{\boldsymbol{x}} = \sum_{i=1}^{m} \boldsymbol{b}_i(\boldsymbol{x}) \cdot u_i$$

be defined for $\boldsymbol{x} \in D_{\boldsymbol{x},\text{def}} \subseteq I\!R^n$. Let $D_{\boldsymbol{x}} \subseteq D_{\boldsymbol{x},\text{def}}$ be an open set, and assume that $u_i \in [u_{\text{min},i}, u_{\text{max},i}]$ holds with $0 > u_{\text{min},i} \in I\!R$ and $0 < u_{\text{max},i} \in I\!R$ for $i = 1, \ldots, m$. Then, if n of the vectors

$$\boldsymbol{b}_1, \ldots, \boldsymbol{b}_m, [\boldsymbol{b}_1, \boldsymbol{b}_2], \ldots, [\boldsymbol{b}_{m-1}, \boldsymbol{b}_m], [\boldsymbol{b}_1, [\boldsymbol{b}_1, \boldsymbol{b}_2]], \ldots, [[\boldsymbol{b}_1, \boldsymbol{b}_2], [\boldsymbol{b}_2, \boldsymbol{b}_3]], \ldots$$

are linearly independent for all $\boldsymbol{x} \in D_{\boldsymbol{x}}$, the system is small-time locally controllable.

Below we will discuss the details of the theorem above. The requirement that positive as well as negative control values u_i exist is essential. If u_i is only either positive or negative, generally not all directions can be accessed. Note once again that the controllability of the system is derived from the small-time local controllability according to Theorem 43 on p. 208 for a path-connected set $D_{\boldsymbol{x}}$ only. The small-time local controllability is more useful in control applications than the controllability, since it allows for shorter trajectories from $\boldsymbol{x}_0$ to $\boldsymbol{x}_\text{e}$.

When calculating a Lie bracket in the above theorem, we begin with the simple brackets $[\boldsymbol{b}_i, \boldsymbol{b}_j]$ and determine whether it is possible to derive a matrix from them and the vectors $\boldsymbol{b}_i$ with a determinant that is different from zero. If this is not possible, higher, i. e. nested Lie brackets must be included. In this case, all combinations of Lie brackets and vectors $\boldsymbol{b}_i$, and combinations of combinations, must be included.

Let us return to the mobile robot and its system description (3.22). Its controllability is plausible due to its design. Since

$$\boldsymbol{M} = \begin{bmatrix} \boldsymbol{b}_1 & \boldsymbol{b}_2 & [\boldsymbol{b}_1, \boldsymbol{b}_2] \end{bmatrix} = \begin{bmatrix} 0 & \cos(x_3) & -\sin(x_3) \\ 0 & \sin(x_3) & \cos(x_3) \\ 1 & 0 & 0 \end{bmatrix}$$

holds additionally, and therefore for the determinant of this matrix it follows that

$$\det(\boldsymbol{M}) = \cos^2(x_3) + \sin^2(x_3) = 1,$$

the conditions of Theorem 46 are fulfilled and the robot is also small-time locally controllable.

It is easy to see that the robot is controllable. However, its small-time local controllability is not apparent. If we consider a small neighborhood around a point, such as $\boldsymbol{x}_0 = \boldsymbol{0}$, all directions in which the robot can move are severely limited. Figure 3.13 shows an example.

Nevertheless, we can reach the point $\boldsymbol{x}_\text{e}$ which lies outside the blue highlighted area in Figure 3.13 by executing the following steps. First, we will proceed toward the state

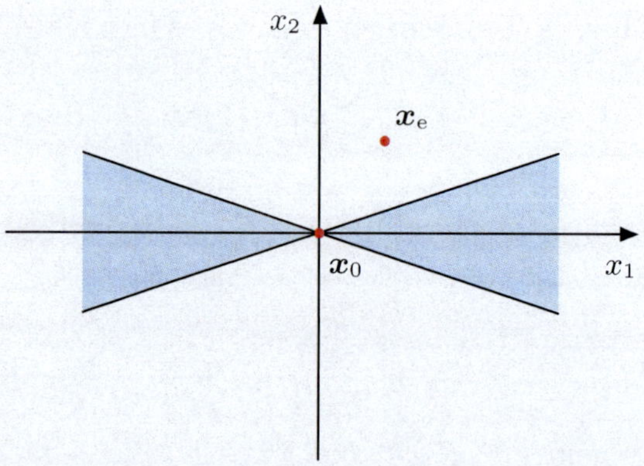

Fig. 3.13: Directions on the x_1x_2-plane in which the robot can move if only small changes in x_3 around zero are included

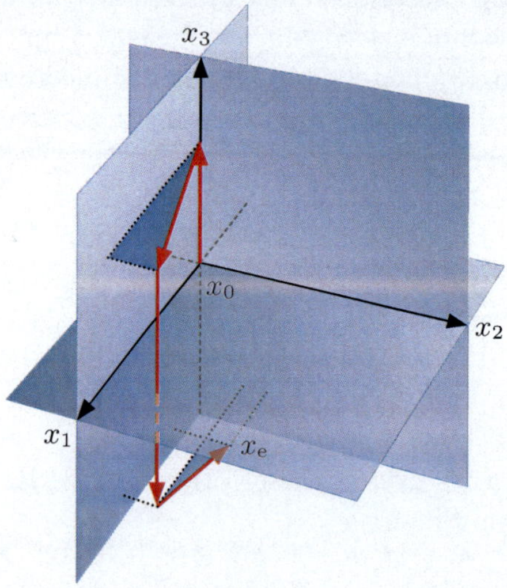

Fig. 3.14: Trajectory of the robot from $\boldsymbol{x}_0$ to $\boldsymbol{x}_{\mathrm{e}} = \boldsymbol{x}(4\Delta t)$

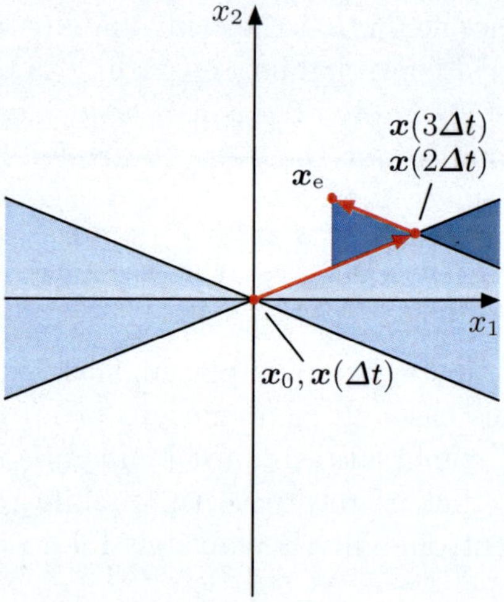

Fig. 3.15: Trajectory of the robot viewed from above

$$\boldsymbol{x}(\Delta t) = \begin{bmatrix} 0 \\ 0 \\ x_3(\Delta t) \end{bmatrix}$$

by means of $\boldsymbol{u} = [1\ 0]^T$ in time Δt. This only requires a small change in x_3. In the second step, we use $\boldsymbol{u} = [0\ 1]^T$ and steer toward the point $\boldsymbol{x}(2\Delta t)$, as shown in Figures 3.14 and 3.15. From point $\boldsymbol{x}(2\Delta t)$, we can maneuver in approximately the same directions as we can from point $\boldsymbol{x}_0 = \boldsymbol{0}$ if, again, we are in a small environment of $x_3 = 0$. In the next step, we define the control variable as $\boldsymbol{u} = [-2\ 0]^T$, so that we only need to change x_3, thus reaching

$\boldsymbol{x}(3\Delta t)$. From here, we move to $\boldsymbol{x}_\mathrm{e} = \boldsymbol{x}(4\Delta t)$ using the control signal vector $\boldsymbol{u} = [0 \ \ -1]^T$.

By taking a zigzag trajectory, it is therefore possible to maneuver to any arbitrary position. This explains why, for small-time locally controllable systems, it is possible to create trajectories which remain within an arbitrarily small neighborhood of a point $\boldsymbol{x}_0$ and reach any point there.

3.1.6 Example: Motor Vehicle with Trailer

We will describe a higher-dimensional example of a driftless control-affine system: a front-wheel drive motor vehicle with a trailer, as illustrated in Figure 3.16. The motor vehicle with a trailer is described by the model [31, 257]

$$\dot{\boldsymbol{x}} = \underbrace{\begin{bmatrix} \cos(x_3)\cos(x_4) \\ \cos(x_3)\sin(x_4) \\ 0 \\ \dfrac{1}{l}\sin(x_3) \\ \dfrac{1}{d}\cos(x_3)\sin(x_4 - x_5) \end{bmatrix}}_{\boldsymbol{b}_1(\boldsymbol{x})} u_1 + \underbrace{\begin{bmatrix} 0 \\ 0 \\ 1 \\ 0 \\ 0 \end{bmatrix}}_{\boldsymbol{b}_2(\boldsymbol{x})} u_2 .$$

Here x_1 and x_2 denote the position of the vehicle, or, more precisely, the center point of its rear axle. The state variable x_3 is the steering angle of the vehicle's front wheels, x_4 is the angle of the vehicle's longitudinal axis to the x-axis of the space-fixed coordinate system, and x_5 is the angle of the trailer axle to the x-axis. The velocity of the vehicle u_1 and the velocity of the steering angle of the vehicle's wheel u_2 are the control variables. Figure 3.16 illustrates this. The distance between the vehicle's front and rear axles, and the difference between the center points of the trailer axle and the vehicle's rear axle are denoted by l and d, respectively. As a simplification, it is assumed that the trailer hitch lies at the center point of the vehicle's rear axle.

The admissible range of definition $D_{\boldsymbol{x}}$ of possible system states x_i is limited. On the one hand, there are limitations to the steering angle of the vehicle's wheel x_3 which are given by

$$-\frac{\pi}{4} < x_3 < \frac{\pi}{4}.$$

On the other hand, there are also limitations which arise due to the fact that the trailer's longitudinal axis may not form too great an angle Θ to the vehicle's longitudinal axis. In this case, we require

$$-\frac{\pi}{2} < \Theta = x_5 - x_4 < \frac{\pi}{2}.$$

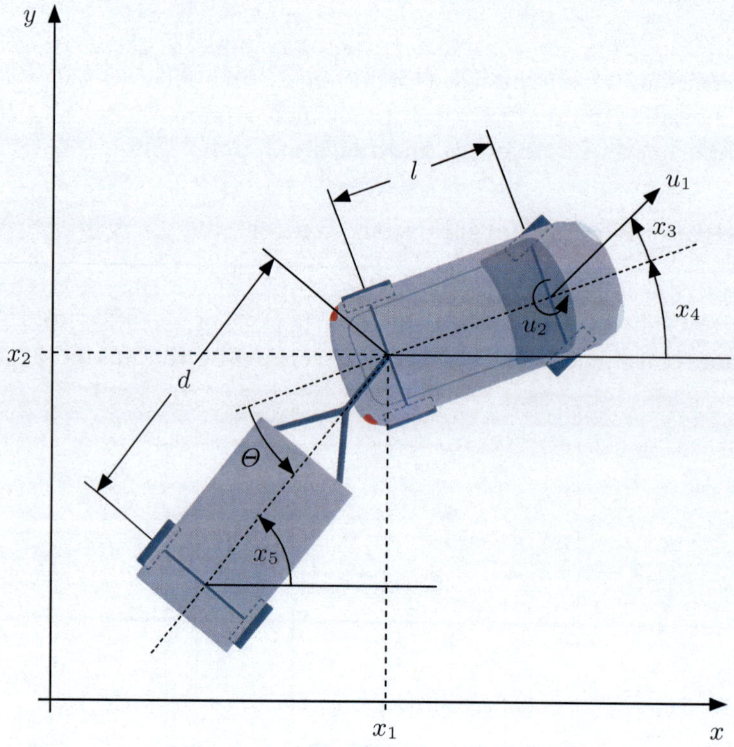

Fig. 3.16: Schematic view from above of the motor vehicle with its trailer in the space-fixed xy-coordinate system

This yields

$$D_{\boldsymbol{x}} = \{\boldsymbol{x} \in I\!R^5 \mid x_1, x_2 \in I\!R, |x_3| < \frac{\pi}{4}, |x_5 - x_4| < \frac{\pi}{2}\}$$

as the admissible domain of the state variables.

We will now examine the system's controllability, and for this purpose we need to compute the Lie brackets

$$[\boldsymbol{b}_1, \boldsymbol{b}_2] = \begin{bmatrix} \sin(x_3)\cos(x_4) \\ \sin(x_3)\sin(x_4) \\ 0 \\ -\dfrac{1}{l}\cos(x_3) \\ \dfrac{1}{d}\sin(x_3)\sin(x_4 - x_5) \end{bmatrix}, \quad [\boldsymbol{b}_1, [\boldsymbol{b}_1, \boldsymbol{b}_2]] = \begin{bmatrix} -\dfrac{1}{l}\sin(x_4) \\ \dfrac{1}{l}\cos(x_4) \\ 0 \\ 0 \\ \dfrac{1}{ld}\cos(x_4 - x_5) \end{bmatrix},$$

and

$$[\boldsymbol{b}_1,[\boldsymbol{b}_1,[\boldsymbol{b}_1,\boldsymbol{b}_2]]] = \begin{bmatrix} -\dfrac{1}{l^2}\sin(x_3)\cos(x_4) \\ -\dfrac{1}{l^2}\sin(x_3)\sin(x_4) \\ 0 \\ 0 \\ \dfrac{l\cos(x_3) - d\sin(x_3)\sin(x_4 - x_5)}{l^2 d^2} \end{bmatrix}.$$

Along with the vectors $\boldsymbol{b}_1(\boldsymbol{x})$ and $\boldsymbol{b}_2(\boldsymbol{x})$, we generate the matrix

$$\boldsymbol{M} = \begin{bmatrix} \boldsymbol{b}_1 & \boldsymbol{b}_2 & [\boldsymbol{b}_1,\boldsymbol{b}_2] & [\boldsymbol{b}_1,[\boldsymbol{b}_1,\boldsymbol{b}_2]] & [\boldsymbol{b}_1,[\boldsymbol{b}_1,[\boldsymbol{b}_1,\boldsymbol{b}_2]]] \end{bmatrix}$$

and compute its determinant as

$$\det(\boldsymbol{M}) = -\frac{\cos(x_3)}{l^3 d^2}.$$

Since the steering angle of the vehicle's wheel x_3 is limited to the interval $(-\pi/4, \pi/4)$, the determinant is different from zero. The system is therefore small-time locally controllable.

3.1.7 Omnidirectional Controllability

The controllability of a control-affine system with an equal number of input variables as state variables, i. e. $m = n$, can be examined in a simple fashion. We will thus consider a system

$$\dot{\boldsymbol{x}} = \boldsymbol{a}(\boldsymbol{x}) + \boldsymbol{B}(\boldsymbol{x}) \cdot \boldsymbol{u} \tag{3.31}$$

with $\boldsymbol{x} \in D_{\boldsymbol{x},\mathrm{def}} \subseteq I\!R^n$ and $\boldsymbol{u} \in I\!R^m$. If $m = n$ and

$$\det(\boldsymbol{B}(\boldsymbol{x})) \neq 0$$

holds for all $\boldsymbol{x}$ of a set $D_{\boldsymbol{x}} \subseteq D_{\boldsymbol{x},\mathrm{def}}$, using the new input variable vector $\boldsymbol{v}$, we can express the control variable $\boldsymbol{u}$ as

$$\boldsymbol{u} = -\boldsymbol{B}^{-1}(\boldsymbol{x})\,(\boldsymbol{a}(\boldsymbol{x}) - \boldsymbol{v})$$

and represent the system (3.31) as

$$\dot{\boldsymbol{x}} = \boldsymbol{v}.$$

Obviously, it is possible to specify any arbitrary trajectory $\boldsymbol{x}(t)$ such that this trajectory can begin or continue in any arbitrary direction $\dot{\boldsymbol{x}}$ if and only if $\boldsymbol{B}(\boldsymbol{x})$ is regular or

$$\operatorname{rank}(\boldsymbol{B}(\boldsymbol{x})) = n$$

holds in the case of an $n \times m$ matrix $\boldsymbol{B}(\boldsymbol{x})$ with $m > n$. Correspondingly, every trajectory $\boldsymbol{x}(t) \in D_{\boldsymbol{x}}$ between an initial point $\boldsymbol{x}_0 \in D_{\boldsymbol{x}}$ and every destination point $\boldsymbol{x}_\mathrm{e} \in D_{\boldsymbol{x}}$ is possible given a suitable choice of $\boldsymbol{v}(t) \in I\!R^n$. Thus we obtain

Theorem 47 (Omnidirectional Controllability of Control-Affine Systems). *A control-affine system*

$$\dot{\boldsymbol{x}} = \boldsymbol{a}(\boldsymbol{x}) + \boldsymbol{B}(\boldsymbol{x}) \cdot \boldsymbol{u}$$

with $\boldsymbol{x} \in D_{\boldsymbol{x},\mathrm{def}} \subseteq I\!R^n$ *and* $\boldsymbol{u} \in I\!R^m$ *is omnidirectionally controllable if and only if*

$$\mathrm{rank}(\boldsymbol{B}(\boldsymbol{x})) = n$$

holds for all $\boldsymbol{x}$ *of a path-connected set* $D_{\boldsymbol{x}} \subseteq D_{\boldsymbol{x},\mathrm{def}}$.

Let us illustrate the difference between small-time local controllability and omnidirectional controllability once again using two mobile robots. The first is the robot powered by two wheels which was discussed in Section 3.1.5. It is small-time locally, but not omnidirectionally controllable. The second is an omnidirectional robot, which, as the name suggests, is omnidirectionally controllable. It possesses three so-called *Swedish wheels* which not only allow for a movement perpendicular to the wheel's axle, but also along the wheel's axle itself. Figure 3.17 displays these wheels and the corresponding robot. The system representation

$$\begin{bmatrix} \dot{x}_1 \\ \dot{x}_2 \\ \dot{x}_3 \end{bmatrix} = \underbrace{\begin{bmatrix} 0 & \cos(x_3) & -\sin(x_3) \\ 0 & \sin(x_3) & \cos(x_3) \\ 1 & 0 & 0 \end{bmatrix}}_{\boldsymbol{B}(\boldsymbol{x})} \begin{bmatrix} u_1 \\ u_2 \\ u_3 \end{bmatrix} \tag{3.32}$$

follows from the robot's kinematics. Here x_1 denotes its x-position and x_2 denotes its y-position in space. The orientation of the robot, i. e. the angle between its main axis and the x-axis of the space-fixed coordinate system, is denoted by x_3. The control variable u_1 represents the angular velocity with respect to the robot's vertical axis, the control variable u_2 is the velocity in the longitudinal direction, and u_3 is the velocity in the transverse direction. Since

$$\det(\boldsymbol{B}(\boldsymbol{x})) = 1$$

holds, the robot is omnidirectionally controllable. This is also intuitively plausible based on its design.

For the general nonlinear system

$$\dot{\boldsymbol{x}} = \boldsymbol{f}(\boldsymbol{x}, \boldsymbol{u})$$

with $\boldsymbol{x} \in I\!R^n$ and n input variables u_i, i. e. $\boldsymbol{u} \in I\!R^n$, we can derive a theorem that is similar to Theorem 47. The existence of the inverse function

$$\boldsymbol{u} = \boldsymbol{f}^{-1}(\boldsymbol{x}, \dot{\boldsymbol{x}}) \tag{3.33}$$

is sufficient to confirm the omnidirectional controllability of the system. Thus it is possible to specify an arbitrary direction $\dot{\boldsymbol{x}}$ for every point $\boldsymbol{x}$, and to

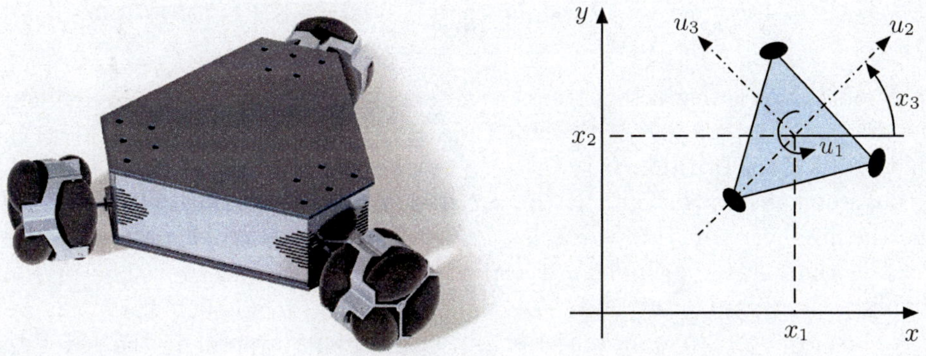

Fig. 3.17: Omnidirectional robot with three individually powered wheel axles and Swedish wheels

deduce the required input variable vector $\boldsymbol{u}$ from equation (3.33), which is interpreted as the control law. The requirement that the inverse function $\boldsymbol{f}^{-1}$ exists, however, need not be fulfilled for the omnidirectional controllability in every case, since, in some cases, $\dot{\boldsymbol{x}}$ can be generated not only by a single $\boldsymbol{u}$, but also by several different vectors $\boldsymbol{u}$. In the latter case, however, an inverse function would not exist.

Computing the inverse function $\boldsymbol{f}^{-1}$ is often difficult or impossible, so it would be useful if the verification of the omnidirectional controllability could be simplified. This is achieved as follows. According to the fundamental theorem on implicit functions [167], the inverse function $\boldsymbol{f}^{-1}$ exists if the Jacobian matrix

$$\frac{\partial \boldsymbol{f}(\boldsymbol{x},\boldsymbol{u})}{\partial \boldsymbol{u}} \tag{3.34}$$

is of rank n. This is also sufficient if $\dim(\boldsymbol{u}) > n$ holds. This leads to the general

Theorem 48 (Omnidirectional Controllability). *A system*

$$\dot{\boldsymbol{x}} = \boldsymbol{f}(\boldsymbol{x},\boldsymbol{u})$$

with $\boldsymbol{x} \in D_{\boldsymbol{x},\mathrm{def}} \subseteq I\!R^n$ *and* $\boldsymbol{u} \in I\!R^m$ *is omnidirectionally controllable if*

$$\operatorname{rank}\left(\frac{\partial \boldsymbol{f}(\boldsymbol{x},\boldsymbol{u})}{\partial \boldsymbol{u}}\right) = n$$

holds for all $\boldsymbol{x}$ *of a path-connected set* $D_{\boldsymbol{x}} \subseteq D_{\boldsymbol{x},\mathrm{def}}$ *and* $\boldsymbol{u} \in I\!R^m$.

Alternatively, this theorem can also be proven by calculating the Taylor series expansion for $\boldsymbol{f}$ around the point $(\boldsymbol{x}_\mathrm{p}, \boldsymbol{u}_\mathrm{p})$, and thereby representing the system as

$$\dot{\boldsymbol{x}} = \boldsymbol{f}(\boldsymbol{x}_\mathrm{p}, \boldsymbol{u}_\mathrm{p}) + \left.\frac{\partial \boldsymbol{f}}{\partial \boldsymbol{x}}\right|_{\boldsymbol{x}_\mathrm{p}, \boldsymbol{u}_\mathrm{p}} \cdot (\boldsymbol{x} - \boldsymbol{x}_\mathrm{p}) + \left.\frac{\partial \boldsymbol{f}}{\partial \boldsymbol{u}}\right|_{\boldsymbol{x}_\mathrm{p}, \boldsymbol{u}_\mathrm{p}} \cdot (\boldsymbol{u} - \boldsymbol{u}_\mathrm{p}) + \boldsymbol{remainder}. \quad (3.35)$$

If the remainder is neglected, i. e. the system is linearized, it becomes evident that any arbitrary $\dot{\boldsymbol{x}}$ can be specified via $\boldsymbol{u}$ if the Jacobian matrix (3.34) in equation (3.35) is of rank n.

Theorem 48 is also valid if the number m of control variables is larger than the number n of state variables. However, the case in which more control variables than state variables are available does not occur often. The reason for this is the redundancy of $m - n$ control variables u_i in such cases.

An example of an omnidirectionally controllable system is the satellite discussed in Section 1.1.4, if the state vector is only composed of its angular velocity vector. The evaporation plant for syrup production which we will encounter in Section 6.1.6 is also omnidirectionally controllable. In the following section, we will describe a further example of such a system.

3.1.8 Example: Steam Generator

Among other applications, steam generators are used in power plants to generate hot steam for a turbine. For this purpose, gas, oil, or pulverized coal are burnt in a combustion chamber. Water flowing in a pipe system contained in this combustion chamber is heated up. As illustrated in Figure 3.18, the hot water ascends in the pipes and reaches the steam boiler. Here in the boiler, steam rises up and is channeled into the turbine. Because colder water has a higher specific gravity than hot water, colder water streams out of the steam boiler into the lower pipe system in the combustion chamber. There it is heated up anew and subsequently rises into the steam boiler again. The fresh water is also fed into the steam boiler, either directly or via a preheater which heats the water using the exhaust gas from the burner.

The steam thus generated is fed into a superheater which heats the steam past the evaporation temperature so that the remaining liquid droplets are dissolved. On the one hand, this serves to prevent damage to the turbine blades by liquid droplets striking them. On the other hand, it increases the degree of efficiency of the process. Downstream of the superheater, there is a desuperheater. It compensates for the rise in temperature that occurs when the load of the turbine decreases. For this purpose, demineralized water is finely atomized and fed into the steam current. Using a valve, the injected water can be controlled such that the desired steam temperature is reached when the steam enters the turbine. Before this is done, the blend of steam and water vapor is subsequently dried again in a second superheater, i. e. liquid droplets are reheated so that they evaporate. After leaving the superheater the amount of steam that is finally injected into the turbine can be controlled by another valve.

The three state variables, i. e. the boiler pressure p in $\mathrm{kg\,cm^{-2}}$, the mass flow q of the steam at the entrance of the turbine in $\mathrm{kg\,s^{-1}}$, and the water level

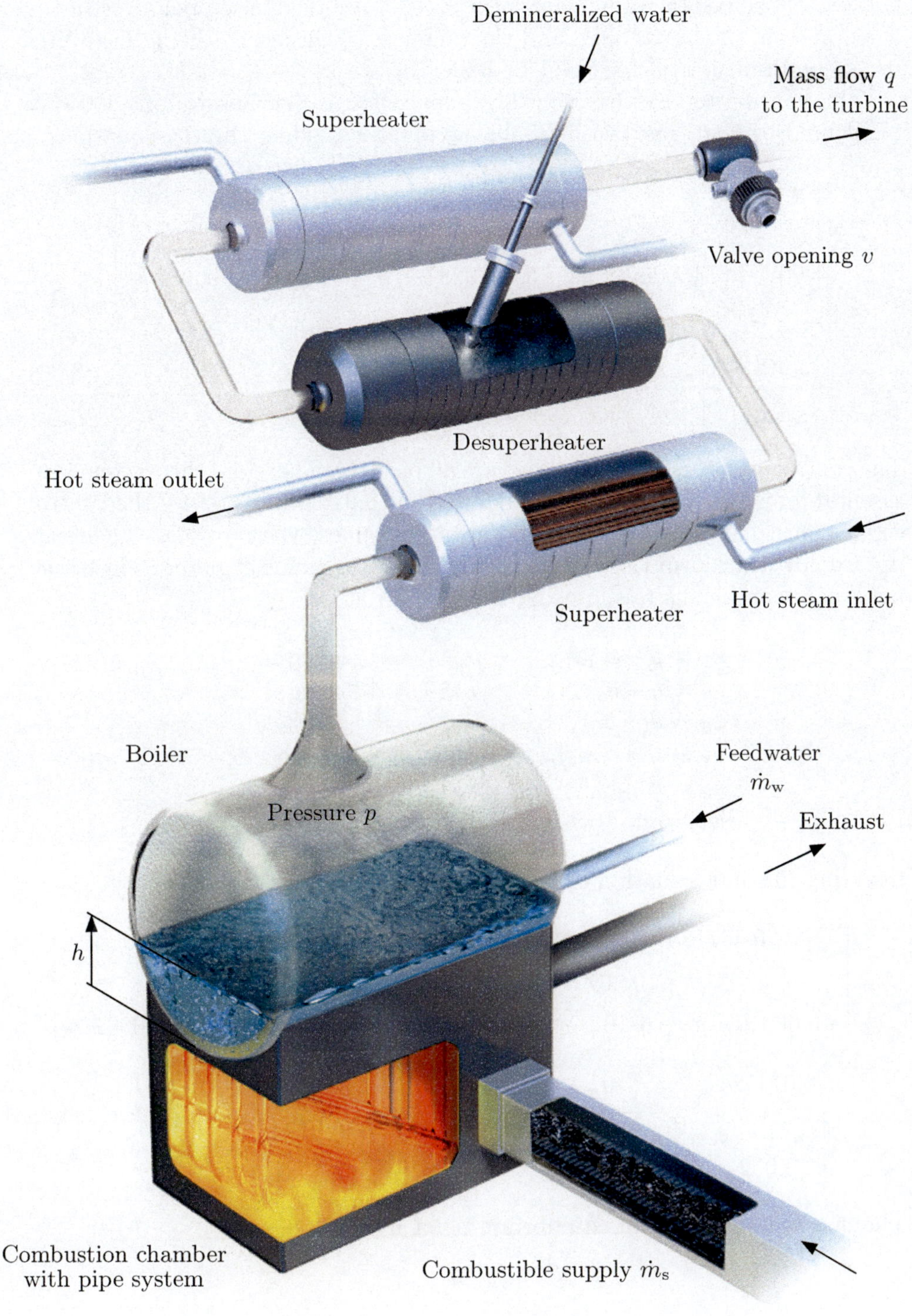

Fig. 3.18: Steam generator

h in cm can be influenced by three control variables. The latter are the mass flow $\dot{m}_\mathrm{s}$ of the combustible material in $\mathrm{kg\,s^{-1}}$, the degree of opening v of the valve in front of the turbine, and the amount $\dot{m}_\mathrm{w}$ in $\mathrm{kg\,s^{-1}}$ of the feedwater per second that is injected into the boiler.

As a specific application, we will examine the steam generator of a 200 MW plant fired by coal, gas, or oil [383]. The plant is modeled by the equations

$$\begin{bmatrix} \dot{p} \\ \dot{q} \\ \dot{h} \end{bmatrix} = \begin{bmatrix} -0.00193\, q\sqrt[8]{p} + 0.00121\, h \\ -0.785716\, q \\ -0.000006\, p^2 - 0.007328\, q - 0.00914\, h - 0.000082\, h^2 \end{bmatrix} + \begin{bmatrix} 0.014524 & 0 & -0.000736 \\ 0 & 10\sqrt{p} & 0 \\ 0.002 & 0.463 & 0.00863 \end{bmatrix} \begin{bmatrix} \dot{m}_\mathrm{s} \\ v \\ \dot{m}_\mathrm{w} \end{bmatrix}. \tag{3.36}$$

The control variables $\dot{m}_\mathrm{s}$, v, and $\dot{m}_\mathrm{w}$ are all positive. Therefore this system representation is not appropriate to analyze the controllability, since the control signals must have both positive and negative values. Without loss of generality, we now transform the system to the operating point at 60 percent of the maximum continuous rating using the new variables

$$\begin{aligned} x_1 &= p - p_\mathrm{op}, & p_\mathrm{op} &= 175.8\,\mathrm{kg\,cm^{-2}}, \\ x_2 &= q - q_\mathrm{op}, & q_\mathrm{op} &= 135.0\,\mathrm{kg\,s^{-1}}, \\ x_3 &= h - h_\mathrm{op}, & h_\mathrm{op} &= 64\,\mathrm{cm}, \\ u_1 &= \dot{m}_\mathrm{s} - \dot{m}_\mathrm{s,op}, & \dot{m}_\mathrm{s,op} &= 38.577\,\mathrm{kg\,s^{-1}}, \\ u_2 &= v - v_\mathrm{op}, & v_\mathrm{op} &= 0.8, \\ u_3 &= \dot{m}_\mathrm{w} - \dot{m}_\mathrm{w,op}, & \dot{m}_\mathrm{w,op} &= 190.961\,\mathrm{kg\,s^{-1}}. \end{aligned}$$

Inserting this into equation (3.36) yields the transformed model

$$\dot{\boldsymbol{x}} = \begin{bmatrix} 0.497185 - 0.00193\sqrt[8]{175.8 + x_1}(135 + x_2) + 0.00121\, x_3 \\ -106.072 + 8\sqrt{175.8 + x_1} - 0.785716\, x_2 \\ -0.0021096\, x_1 - 6\cdot 10^{-6} x_1^2 - 0.007328\, x_2 - 0.019636\, x_3 - 0.000082\, x_3^2 \end{bmatrix} + \begin{bmatrix} 0.014524 & 0 & -0.000736 \\ 0 & 10\sqrt{175.8 + x_1} & 0 \\ 0.002 & 0.463 & 0.00863 \end{bmatrix} \boldsymbol{u}. \tag{3.37}$$

The operating point is an equilibrium point if

$$\boldsymbol{u} = \boldsymbol{0}.$$

The control variables u_1, u_2, and u_3 can take on both positive and negative values. For the determinant of $\boldsymbol{B}(\boldsymbol{x})$, we obtain

$$\det(\boldsymbol{B}(\boldsymbol{x})) = 0.001268\sqrt{175.8 + x_1},$$

i. e. it is different from zero for all $x_1 > -175.8$. Since $p > 0$ holds, only values $x_1 > -175.8$ are possible, the determinant is always positive, and the mathematical model (3.37) of the steam generator is omnidirectionally controllable. Since all control variables u_i and states x_i are limited in a real-world system, the real system is only locally weakly omnidirectionally controllable in practice.

3.2 Flatness

3.2.1 Basic Concept and Definition of Flatness

The term *flatness* signifies that we can rearrange the dynamic equations of a system in such a way that all input variables and state variables of a system can be represented by functions that depend only on the output variable vector and its derivatives [124, 125, 127, 273]. The practical use of this system property is obvious: if a system is flat, by specifying the trajectory of the output variable vector it is possible to directly calculate the required course of the input variables, i. e. the appropriate feedforward control.

As an example, we will consider the system of linear differential equations

$$\begin{aligned} \dot{x}_1 &= x_2, \\ \dot{x}_2 &= -x_1 - x_2 + u, \\ y &= x_1 \end{aligned} \tag{3.38}$$

with the state variables x_1, x_2, the input variable u, and the output variable y. We can describe the state variables as a function of the output variable and its derivative as

$$\boldsymbol{x} = \begin{bmatrix} x_1 \\ x_2 \end{bmatrix} = \begin{bmatrix} y \\ \dot{y} \end{bmatrix}. \tag{3.39}$$

The system of differential equations (3.38) can be reformulated as the differential equation

$$\ddot{y} + \dot{y} + y = u. \tag{3.40}$$

With this representation, we also know the relationship which explicitly describes u in dependence on y, $\dot{y}$, and $\ddot{y}$.

If we know $y(t)$, $\dot{y}(t)$, and $\ddot{y}(t)$ or if we specify their progressions, it is not only possible to compute the progression of $u(t)$ associated with $y(t)$ using equation (3.40), but also to determine the associated state progressions $x_1(t)$ and $x_2(t)$ using equation (3.39). Here the input variable u and the state variables $x_1(t)$ and $x_2(t)$ can be directly determined from the output variable y and its derivatives by evaluating a function. The output y is then called a *flat output* and the system is called *flat*.

The above example of a flat system illustrates why the property of *flatness* is useful. The result obtained for this specific example can be generalized to all flat systems. As mentioned previously, this allows for a simple determination of the control signal $u(t)$, which generates a desired trajectory

$$(y(t), \dot{y}(t), \ldots, y^{(\beta)}(t)).$$

For the system characteristic of flatness, it is irrelevant whether the output in question really exists or not. To verify flatness, we can come up with any suitable system output. An output which does not exist in reality is called *fictitious.* If only fictitious flat outputs exist for the implementation of a control, they must be converted to the real output or vice versa.

Based on the description above, we will define the term *flatness* [127]. In doing so, we will only consider the systems

$$\dot{\boldsymbol{x}} = \boldsymbol{f}(\boldsymbol{x}, \boldsymbol{u}) \tag{3.41}$$

with $m \leq n$ for which no control variable u_i can be represented as a function of other control variables $u_{j \neq i}$. This is fulfilled if the Taylor series expansion of $\boldsymbol{f}$ in equation (3.41),

$$\dot{\boldsymbol{x}} = \boldsymbol{f}(\boldsymbol{x}_0, \boldsymbol{u}_0) + \left.\frac{\partial \boldsymbol{f}}{\partial \boldsymbol{x}}\right|_{\boldsymbol{x}_0, \boldsymbol{u}_0} \cdot (\boldsymbol{x} - \boldsymbol{x}_0) + \left.\frac{\partial \boldsymbol{f}}{\partial \boldsymbol{u}}\right|_{\boldsymbol{x}_0, \boldsymbol{u}_0} \cdot (\boldsymbol{u} - \boldsymbol{u}_0) + \boldsymbol{remainder},$$

has a matrix

$$\left.\frac{\partial \boldsymbol{f}}{\partial \boldsymbol{u}}\right|_{\boldsymbol{x}_0, \boldsymbol{u}_0}$$

of rank m for all $\boldsymbol{x}_0$ and $\boldsymbol{u}_0$. We will now specify the concept of flatness more precisely in

Definition 32 (Flatness). *Let a system*

$$\dot{\boldsymbol{x}} = \boldsymbol{f}(\boldsymbol{x}, \boldsymbol{u})$$

be defined for $\boldsymbol{x} \in D_{\boldsymbol{x},\mathrm{def}} \subseteq I\!R^n$ *and* $\boldsymbol{u} \in D_{\boldsymbol{u},\mathrm{def}} \subseteq I\!R^m$ *with* $m \leq n$ *and let*

$$\operatorname{rank}\left(\frac{\partial \boldsymbol{f}(\boldsymbol{x}, \boldsymbol{u})}{\partial \boldsymbol{u}}\right) = m$$

hold. The system is called flat if a real or fictitious output variable vector

$$\boldsymbol{y} = \boldsymbol{h}(\boldsymbol{x}, \boldsymbol{u}, \dot{\boldsymbol{u}}, \ldots, \boldsymbol{u}^{(\alpha)})$$

with a finite value for $\alpha \in I\!N$ *exists, so that*

(1) the state vector $\boldsymbol{x}$ *can be represented as a function of* $\boldsymbol{y}$ *and a finite number* β *of derivatives* $\boldsymbol{y}^{(i)}$ *as*

$$\boldsymbol{x} = \boldsymbol{\Psi}_1(\boldsymbol{y}, \dot{\boldsymbol{y}}, \ldots, \boldsymbol{y}^{(\beta)}), \tag{3.42}$$

(2) the input variable vector $\boldsymbol{u}$ *can be represented as a function*

$$\boldsymbol{u} = \boldsymbol{\Psi}_2(\boldsymbol{y}, \dot{\boldsymbol{y}}, \ldots, \boldsymbol{y}^{(\beta+1)}), \tag{3.43}$$

and

(3) it holds for the input and output variable vector that

$$\dim(\boldsymbol{y}) = \dim(\boldsymbol{u})\,.$$

In this case, the output variable vector $\boldsymbol{y}$ *is called flat output.*

For SISO systems, it holds that $y = h(\boldsymbol{x})$ and $\beta = n - 1$. On the other hand, the value of β is not known *a priori* for MIMO systems [467]. Usually, not only in SISO systems but in MIMO systems as well, the flat output $\boldsymbol{y}$ does not directly depend on the control variable $\boldsymbol{u}$ or one of its derivatives $\boldsymbol{u}^{(i)}$. It therefore often holds that $\boldsymbol{y} = \boldsymbol{h}(\boldsymbol{x})$.

In general, we can distinguish between *local* and *global flatness*, depending on whether Conditions (1), (2), and (3) of the above definition are only fulfilled for a proper subset of the domain of definition of $\boldsymbol{f}(\boldsymbol{x}, \boldsymbol{u})$, or whether they are fulfilled for its entire domain of definition.

The term *differentially flat* is often used synonymously with the term *flat*. This is because only the derivatives of $\boldsymbol{y}$, but no integrals of $\boldsymbol{y}$, are used to determine $\boldsymbol{u}$ and $\boldsymbol{x}$ in the functions $\boldsymbol{\Psi}_1$ and $\boldsymbol{\Psi}_2$.

The Conditions (2) and (3) of Definition 32 are equivalent to stating that the function $\boldsymbol{y}(t)$ does not fulfill any differential equation

$$\varphi(\boldsymbol{y}, \dot{\boldsymbol{y}}, \ldots, \boldsymbol{y}^{(\gamma)}) = 0, \quad \gamma \in \{0, 1, 2, \ldots\}. \tag{3.44}$$

If, for example, $\dim(\boldsymbol{y}) > \dim(\boldsymbol{u})$, cases would exist for which equation (3.44) would be fulfilled. The example

$$u = y_1 + 2\dot{y}_1 + y_2 + \dot{y}_2$$

with $\dim(\boldsymbol{u}) = 1$ and $\dim(\boldsymbol{y}) = 2$ illustrates this. Obviously, for this case, we can choose y_2 and $\dot{y}_2$ such that

$$y_2 + \dot{y}_2 = 0,$$

which is a differential equation of the form (3.44), and u is determined only from y_1 and $\dot{y}_1$. If, on the other hand, the inequality $\dim(\boldsymbol{y}) < \dim(\boldsymbol{u})$ holds, then not all u_i are independent of each other. We can illustrate this by the example

$$u_1 = y_1 + \dot{y}_1,$$
$$u_2 = y_1,$$

from which

$$u_1 = u_2 + \dot{u}_2$$

follows. As described earlier, the independence of the control variables u_i is a reasonable assumption, since some of the variables u_i would otherwise be unnecessary to generate the output trajectory $\boldsymbol{y}(t)$. Assuming this independence, equation (3.43) consists of $m = \dim(\boldsymbol{u}) = \dim(\boldsymbol{y})$ independent differential equations for $y_1, \ldots, y_m$, which depend on the input variables u_i; thus, these equations are not of the form (3.44).

If equation (3.44) is not fulfilled, the outputs y_i are called *differentially independent*. This property ensures that no output y_i or any of its derivatives is determined by one or more of the others. Consequently, there is no function which describes a relationship between the variables

$$y_i, \dot{y}_i, \ldots, y_i^{(\beta+1)}.$$

We are therefore completely free in choosing the output trajectory

$$(\boldsymbol{y}, \dot{\boldsymbol{y}}, \ddot{\boldsymbol{y}}, \ldots, \boldsymbol{y}^{(\beta+1)}),$$

and can also realize this trajectory using the control input from equation (3.43).

It is most useful if a real output $\boldsymbol{y}$ is flat. On the one hand, in this case the flatness can be easily verified using equation (3.42) and equation (3.43). On the other hand, we can directly specify the required control (3.43) from the trajectory $(\boldsymbol{y}, \dot{\boldsymbol{y}}, \ddot{\boldsymbol{y}}, \ldots, \boldsymbol{y}^{(\beta+1)})$.

Unfortunately, the real output is frequently not flat. In this case, it is necessary to search for a fictitious flat output. Similar to the case of the Lyapunov function, no practically applicable general method exists to find or construct a fictitious flat output. Therefore we have to try out different outputs to determine a flat output. It has proven useful to start by considering outputs $\boldsymbol{y}$ which exhibit a derivative $\boldsymbol{y}^{(k)}$ which depends on the input variable vector $\boldsymbol{u}$ as the first of the derivatives $\boldsymbol{y}^{(i)}$, $i = 1, \ldots, k$, and has an order k which is as high as possible. However, for a flat system there is not only one single flat output, but an infinite number of flat outputs. All these outputs can be converted into each other.

After determining a candidate $\boldsymbol{y}$ for a flat output, we can verify whether it is truly flat. To do this, the function

$$\boldsymbol{y} = \boldsymbol{h}(\boldsymbol{x}, \boldsymbol{u}, \dot{\boldsymbol{u}}, \ldots, \boldsymbol{u}^{(\alpha)})$$

is derived multiple times, which yields

$$\begin{aligned}
\dot{\boldsymbol{y}} &= \frac{d\boldsymbol{h}(\boldsymbol{x}, \boldsymbol{u}, \dot{\boldsymbol{u}}, \ldots, \boldsymbol{u}^{(\alpha)})}{dt}, \\
\ddot{\boldsymbol{y}} &= \frac{d^2\boldsymbol{h}(\boldsymbol{x}, \boldsymbol{u}, \dot{\boldsymbol{u}}, \ldots, \boldsymbol{u}^{(\alpha)})}{dt^2}, \\
&\ \ \vdots
\end{aligned}$$

until an algebraic system of equations for the variables $\boldsymbol{x}, \boldsymbol{u}, \dot{\boldsymbol{u}}, \ldots, \boldsymbol{u}^{(\alpha)}$ is obtained which can be solved. From this system, we can determine the functions

$$\begin{aligned} \boldsymbol{x} &= \boldsymbol{\Psi}_1(\boldsymbol{y}, \dot{\boldsymbol{y}}, \ldots, \boldsymbol{y}^{(\beta)}), \\ \boldsymbol{u} &= \boldsymbol{\Psi}_2(\boldsymbol{y}, \dot{\boldsymbol{y}}, \ldots, \boldsymbol{y}^{(\beta+1)}) \end{aligned}$$

of Definition 32, and thereby prove that the output is flat.

In certain cases, a flat system representation results directly from the derivation of the physical equations. For example, this applies to mechanical systems such as the industrial robot from Section 3.1.4 if they are modeled using their kinetic and potential energies, E_{kin} and E_{pot}, and the Lagrange equations

$$\frac{d}{dt}\frac{\partial L}{\partial \dot{q}_i} - \frac{\partial L}{\partial q_i} = \tau_i, \quad i = 1, \ldots, k,$$

or their vectorial form

$$\frac{d}{dt}\frac{\partial L}{\partial \dot{\boldsymbol{q}}} - \frac{\partial L}{\partial \boldsymbol{q}} = \boldsymbol{\tau}^T.$$

Here

$$L = E_{\text{kin}} - E_{\text{pot}}$$

is the Lagrange function, the vector $\boldsymbol{q}^T = [q_1 \ldots q_k]$ contains the generalized coordinates, i. e. the distances or angles, and the vector $\boldsymbol{\tau}^T = [\tau_1 \ldots \tau_k]$ contains the forces and torques. From the Lagrange equations, we directly obtain a flat system description with

$$\begin{aligned} \boldsymbol{y} &= \begin{bmatrix} q_1 & \cdots & q_k \end{bmatrix}^T, \\ \boldsymbol{u} &= \begin{bmatrix} \tau_1 & \cdots & \tau_k \end{bmatrix}^T = \begin{bmatrix} \dfrac{d}{dt}\dfrac{\partial L}{\partial \dot{\boldsymbol{q}}} - \dfrac{\partial L}{\partial \boldsymbol{q}} \end{bmatrix}^T = \boldsymbol{\Psi}_1(\boldsymbol{y}, \dot{\boldsymbol{y}}, \ddot{\boldsymbol{y}}), \\ \boldsymbol{x} &= \begin{bmatrix} q_1 & \cdots & q_k & \dot{q}_1 & \cdots & \dot{q}_k \end{bmatrix}^T = \begin{bmatrix} y_1 & \cdots & y_k & \dot{y}_1 & \cdots & \dot{y}_k \end{bmatrix}^T = \boldsymbol{\Psi}_2(\boldsymbol{y}, \dot{\boldsymbol{y}}), \end{aligned}$$

and the flat output $\boldsymbol{y}$.

3.2.2 The Lie-Bäcklund Transformation

In the state-space representation

$$\dot{\boldsymbol{x}} = \boldsymbol{f}(\boldsymbol{x}, \boldsymbol{u}) \tag{3.45}$$

of the system with the flat output

$$\boldsymbol{y} = \boldsymbol{h}(\boldsymbol{x}, \boldsymbol{u}, \dot{\boldsymbol{u}}, \ldots, \boldsymbol{u}^{(\alpha)}), \tag{3.46}$$

we use the state vector $\boldsymbol{x}$ and its time course $\boldsymbol{x}(t)$ to describe the system's behavior. In order to calculate the course of the state vector $\boldsymbol{x}(t)$ and the output variable vector $\boldsymbol{y}(t)$, we need to solve the differential equation. As mentioned, the output function $\boldsymbol{h}$ in equation (3.46) only depends on $\boldsymbol{x}$ in the majority of cases, i. e. there is no feedthrough.

If, on the other hand, we use the flat system representation

$$\boldsymbol{x} = \boldsymbol{\Psi}_1(\boldsymbol{y}, \dot{\boldsymbol{y}}, \ldots, \boldsymbol{y}^{(\beta)}), \tag{3.47}$$

$$\boldsymbol{u} = \boldsymbol{\Psi}_2(\boldsymbol{y}, \dot{\boldsymbol{y}}, \ldots, \boldsymbol{y}^{(\beta+1)}) \tag{3.48}$$

and the flat output variable vector $\boldsymbol{y}$, we do not need to solve a differential equation to determine $\boldsymbol{x}$. The flat coordinates

$$\boldsymbol{z}_\mathrm{f} = \begin{bmatrix} \boldsymbol{y} \\ \dot{\boldsymbol{y}} \\ \vdots \\ \boldsymbol{y}^{(\beta)} \end{bmatrix}$$

are not subject to any dynamics since they are differentially independent of each other, i. e. they do not have a mutual dependence in the form of a differential equation (3.44). Rather, the time course of the m output variables y_i as well as all the associated derivatives

$$y_i^{(k)}(t), \quad k = 1, 2, \ldots,$$

can be chosen arbitrarily. They therefore represent a system without dynamics, which is referred to as a *trivial system*. The two system descriptions (3.45), (3.46) and (3.47), (3.48) are equivalent to each other. They can be bijectively converted into each other via a transformation that is called the *Lie-Bäcklund transformation* or *Lie-Bäcklund isomorphism* [18, 126]. The equations (3.47), (3.48), and

$$\begin{aligned}
\dot{\boldsymbol{u}} &= \dot{\boldsymbol{\Psi}}_2(\boldsymbol{y}, \dot{\boldsymbol{y}}, \ldots, \boldsymbol{y}^{(\beta+1)}), \\
\ddot{\boldsymbol{u}} &= \ddot{\boldsymbol{\Psi}}_2(\boldsymbol{y}, \dot{\boldsymbol{y}}, \ldots, \boldsymbol{y}^{(\beta+1)}), \\
&\vdots
\end{aligned}$$

make up the Lie-Bäcklund transformations and the equations

$$\begin{aligned}
\boldsymbol{y} &= \boldsymbol{h}(\boldsymbol{x}, \boldsymbol{u}, \dot{\boldsymbol{u}}, \ldots, \boldsymbol{u}^{(\alpha)}), \\
\dot{\boldsymbol{y}} &= \dot{\boldsymbol{h}}(\boldsymbol{x}, \boldsymbol{u}, \dot{\boldsymbol{u}}, \ldots, \boldsymbol{u}^{(\alpha)}), \\
\ddot{\boldsymbol{y}} &= \ddot{\boldsymbol{h}}(\boldsymbol{x}, \boldsymbol{u}, \dot{\boldsymbol{u}}, \ldots, \boldsymbol{u}^{(\alpha)}), \\
&\vdots
\end{aligned}$$

the associated inverse transformation. Figure 3.19 illustrates both coordinate spaces with their coordinate vectors $\boldsymbol{x}$ and $\boldsymbol{z}_\mathrm{f}$ which consists of $\boldsymbol{y}^{(i)}$, $i = 0, 1, \ldots, \beta + 1$, and the Lie-Bäcklund transformation.

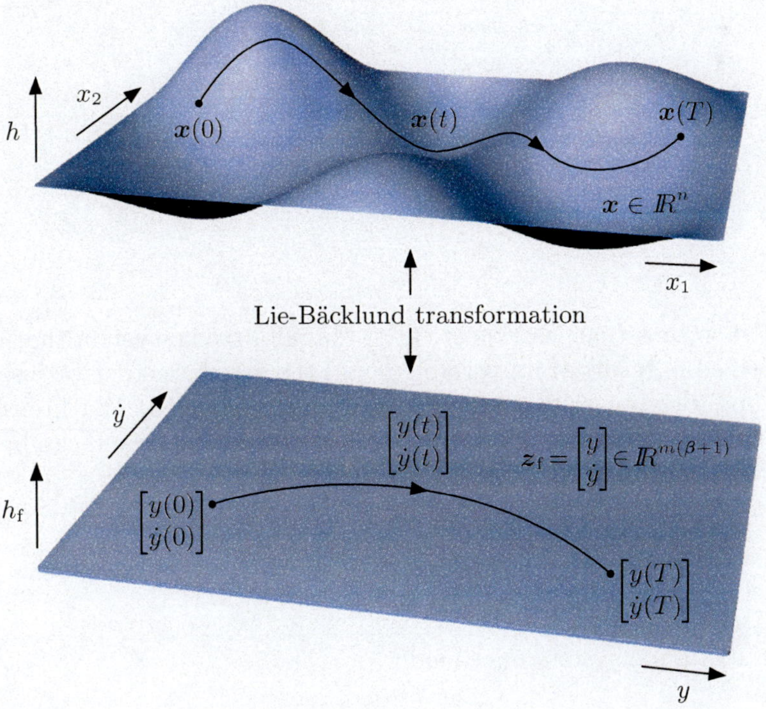

Fig. 3.19: The Lie-Bäcklund transformation transforms the system representation from the state-space coordinate system to the space of flat coordinates. For illustration purposes, the potential h of a so-called gradient system $\dot{\boldsymbol{x}} = \boldsymbol{f}(\boldsymbol{x}, \boldsymbol{u})$ with its gradient $\operatorname{grad}(h) = \boldsymbol{f}(\boldsymbol{x}, \boldsymbol{0})$ and the potential function h_f of the trivial system in flat system representation with $\operatorname{grad}(h_f) = \boldsymbol{0}$ is shown here.

In general, the Lie-Bäcklund transformation, even though it is bijective, does not maintain the dimension of the original coordinate system, since it holds that $\boldsymbol{x} \in I\!R^n$, but $\boldsymbol{z}_f \in I\!R^{m(\beta+1)}$. The dimension of the input space $I\!R^m$, on the other hand, is not affected by the transformation.

As a simple example, we will look at system (3.38) once again. The Lie-Bäcklund transformation is given here by

$$\begin{aligned} \boldsymbol{x} &= \boldsymbol{\Psi}_1(y, \dot{y}) = \begin{bmatrix} y \\ \dot{y} \end{bmatrix}, \\ u &= \Psi_2(y, \dot{y}, \ddot{y}) = \ddot{y} + \dot{y} + y. \end{aligned} \tag{3.49}$$

The inverse transformation is

$$\begin{aligned} y &= x_1, \\ \dot{y} &= x_2, \\ \ddot{y} &= -x_1 - x_2 + u. \end{aligned} \tag{3.50}$$

If the transformation rule (3.49) is inserted into the system equations (3.38), we obtain

$$\begin{aligned}\dot{x}_1 &= x_2 = \dot{y},\\ \dot{x}_2 &= -x_1 - x_2 + u = \ddot{y},\\ y &= x_1 = y.\end{aligned}$$

Incorporating $\dot{x}_1 = \dot{y}$ and $\dot{x}_2 = \ddot{y}$ from equation (3.50), the trivial relation

$$\begin{aligned}\dot{y} &= \dot{y},\\ \ddot{y} &= \ddot{y},\\ y &= y\end{aligned}$$

follows. This means that, as expected, we obtain a trivial system that can be freely specified in terms of its variable y and the associated derivatives.

If we apply the inverse transformation equations (3.50) and insert them into the trivial system, for which y, $\dot{y}$, and $\ddot{y}$ are independent and can be freely chosen, we get

$$\begin{aligned}\dot{y} = \dot{x}_1 &= x_2,\\ \ddot{y} = \dot{x}_2 &= -x_1 - x_2 + u,\\ y = x_1&,\end{aligned}$$

which is the original system once again.

3.2.3 Example: VTOL Aircraft

As an example, we will consider a vertical take-off and landing (VTOL) aircraft. Such an aircraft can hover as well as take off and land vertically [129, 309]. For this purpose engines are mounted to the ends of the wings and can be rotated into a vertical position for take-off and landing. For regular flight, on the other hand, the thrusts are oriented horizontally. This allows much higher flight speeds than is possible for helicopters. Such VTOL aircraft include the Dornier Do 31, the Hawker Siddeley Harrier, and the Bell-Boeing V-22 Osprey. The illustration of the AgustaWestland AW609, which is based on the Bell-Boeing V-22 Osprey, is shown in Figure 3.20.

For vertical take-off and landing, the corresponding motion equations for the horizontal position z_1, as well as the vertical position z_2 of the center of gravity S, and the rotation φ in relation to the aircraft's roll axis can be derived from the balance of forces

$$\begin{aligned}m\ddot{z}_1 &= -(F_\mathrm{l} + F_\mathrm{r})\cos(\alpha)\sin(\varphi) + (F_\mathrm{l} - F_\mathrm{r})\sin(\alpha)\cos(\varphi),\\ m\ddot{z}_2 &= -mg + (F_\mathrm{l} + F_\mathrm{r})\cos(\alpha)\cos(\varphi) + (F_\mathrm{l} - F_\mathrm{r})\sin(\alpha)\sin(\varphi)\end{aligned} \tag{3.51}$$

with the jet-engine forces denoted by F_l and F_r, and from the torque equation

$$J\ddot{\varphi} = (F_\mathrm{l} - F_\mathrm{r})(b\cos(\alpha) - a\sin(\alpha)). \tag{3.52}$$

Here, m denotes the mass of the airplane, J is the inertia moment with respect to the roll axis, α is the angle of inclination of an engine with respect to the

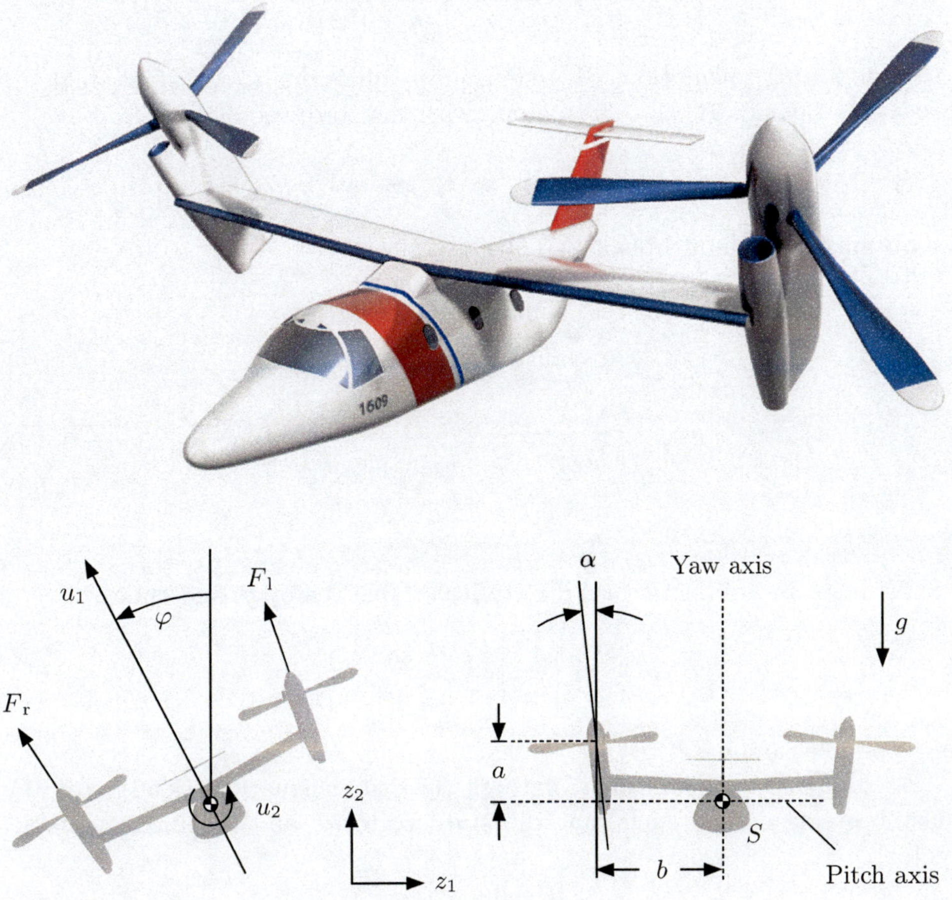

Fig. 3.20: The AgustaWestland AW609, an example of a VTOL aircraft. The two smaller diagrams illustrate the front view of the aircraft and the relevant parameters.

yaw axis, and a and b represent the distances of an engine to the center of gravity S in the direction of the yaw and pitch axis, respectively. For the gravitational acceleration, the common notation g is used.

For simplification, we will define the control variable u_1 as the acceleration in relation to the direction of the aircraft's yaw axis, i. e.

$$u_1 = \frac{\cos(\alpha)}{m}(F_\mathrm{l} + F_\mathrm{r}), \tag{3.53}$$

and the control variable u_2 as the angular acceleration in relation to the roll axis, i. e.

$$u_2 = \frac{b\cos(\alpha) - a\sin(\alpha)}{J}(F_\mathrm{l} - F_\mathrm{r}). \tag{3.54}$$

For the sake of convenience, we will use the artificial length

$$\varepsilon = \frac{J\sin(\alpha)}{m(b\cos(\alpha) - a\sin(\alpha))} \tag{3.55}$$

in the following calculations. It assumes low values due to the small angle α. From equations (3.51) to (3.55), along with the state vector

$$\boldsymbol{x} = \begin{bmatrix} \dot{z}_1 & \dot{z}_2 & \dot{\varphi} & z_1 & z_2 & \varphi \end{bmatrix}^T,$$

we obtain for the model of the VTOL aircraft

$$\dot{x}_1 = -u_1\sin(x_6) + \varepsilon u_2\cos(x_6), \tag{3.56}$$
$$\dot{x}_2 = u_1\cos(x_6) + \varepsilon u_2\sin(x_6) - g, \tag{3.57}$$
$$\dot{x}_3 = u_2, \tag{3.58}$$
$$\dot{x}_4 = x_1, \tag{3.59}$$
$$\dot{x}_5 = x_2, \tag{3.60}$$
$$\dot{x}_6 = x_3. \tag{3.61}$$

We now aim to identify a flat output variable vector $\boldsymbol{y}$, and choose

$$\boldsymbol{y} = \begin{bmatrix} y_1 \\ y_2 \end{bmatrix} = \begin{bmatrix} x_4 - \varepsilon\sin(x_6) \\ x_5 + \varepsilon\cos(x_6) \end{bmatrix} \tag{3.62}$$

as a possible candidate.

According to the definition of flatness, the system equations (3.56) to (3.61) must be reformulated such that the state vector $\boldsymbol{x}$ and the input variable vector

$$\boldsymbol{u} = [u_1 \; u_2]^T$$

are represented as functions

$$\boldsymbol{x} = \boldsymbol{\Psi}_1(\boldsymbol{y}, \dot{\boldsymbol{y}}, \ldots, \boldsymbol{y}^{(\beta)}),$$
$$\boldsymbol{u} = \boldsymbol{\Psi}_2(\boldsymbol{y}, \dot{\boldsymbol{y}}, \ldots, \boldsymbol{y}^{(\beta+1)}).$$

To achieve this, we convert equation (3.62) into

$$x_4 = y_1 + \varepsilon\sin(x_6), \tag{3.63}$$
$$x_5 = y_2 - \varepsilon\cos(x_6). \tag{3.64}$$

The next task is the elimination of the state variable x_6 in equations (3.63) and (3.64). To do this, we compute the second derivative of equation (3.63) which yields

$$\ddot{y}_1 = \ddot{x}_4 - \varepsilon\ddot{x}_6\cos(x_6) + \varepsilon\dot{x}_6^2\sin(x_6).$$

In this equation, we incorporate equations (3.56), (3.58), (3.59), and (3.61) so that

$$\ddot{y}_1 = (\varepsilon\dot{x}_6^2 - u_1)\sin(x_6) \tag{3.65}$$

follows. Similarly, we obtain

$$\ddot{y}_2 = -(\varepsilon\dot{x}_6^2 - u_1)\cos(x_6) - g. \tag{3.66}$$

Dividing equation (3.66) by equation (3.65) results in

$$x_6 = -\arctan\left(\frac{\ddot{y}_1}{\ddot{y}_2 + g}\right). \tag{3.67}$$

By applying

$$\arctan\left(\frac{p}{q}\right) = \arcsin\left(\frac{p}{\sqrt{p^2+q^2}}\right),$$

it follows from equation (3.63) and equation (3.67) that

$$x_4 = y_1 - \varepsilon\frac{\ddot{y}_1}{\sqrt{\ddot{y}_1^2 + (\ddot{y}_2 + g)^2}}. \tag{3.68}$$

In a comparable way, we obtain

$$x_5 = y_2 - \varepsilon\frac{\ddot{y}_2 + g}{\sqrt{\ddot{y}_1^2 + (\ddot{y}_2 + g)^2}} \tag{3.69}$$

from equation (3.64) and equation (3.67).

We have now represented x_4, x_5, and x_6 in dependence on the output variables y_1 and y_2, and their derivatives, to determine the function $\boldsymbol{\Psi}_1$. In the next step, we will focus our attention on the state variables x_1, x_2, and x_3.

It holds that

$$x_1 = \dot{x}_4 = \dot{y}_1 + \varepsilon\frac{\ddot{y}_1\dddot{y}_2(\ddot{y}_2 + g) - \dddot{y}_1(\ddot{y}_2 + g)^2}{[\ddot{y}_1^2 + (\ddot{y}_2 + g)^2]^{3/2}} \tag{3.70}$$

and

$$x_2 = \dot{x}_5 = \dot{y}_2 + \varepsilon\frac{\ddot{y}_1\dddot{y}_1(\ddot{y}_2 + g) - \ddot{y}_1^2\dddot{y}_2}{[\ddot{y}_1^2 + (\ddot{y}_2 + g)^2]^{3/2}} \tag{3.71}$$

as well as

$$x_3 = \dot{x}_6 = \frac{\ddot{y}_1\dddot{y}_2 - \dddot{y}_1(\ddot{y}_2 + g)}{\ddot{y}_1^2 + (\ddot{y}_2 + g)^2}. \tag{3.72}$$

Now that we have derived the function $\boldsymbol{\Psi}_1$ with

$$\beta = 3$$

from equation (3.67) to equation (3.72), we still have to prove the existence of

$$\boldsymbol{u} = \boldsymbol{\Psi}_2(\boldsymbol{y}, \dot{\boldsymbol{y}}, \ldots, \boldsymbol{y}^{(4)})$$

to verify flatness. We begin with the control variable u_1 and derive

$$u_1 = \varepsilon \dot{x}_6^2 - \frac{\ddot{y}_1}{\sin(x_6)}$$

from equation (3.65). Inserting equations (3.67) and (3.72) into this equation yields

$$u_1 = \varepsilon \left[\frac{\dddot{y}_1 \ddot{y}_2 - \dddot{y}_1 (\ddot{y}_2 + g)}{\ddot{y}_1^2 + (\ddot{y}_2 + g)^2} \right]^2 + \sqrt{\ddot{y}_1^2 + (\ddot{y}_2 + g)^2}.$$

For the control variable u_2, with equation (3.58) and equation (3.72) we obtain

$$u_2 = \ddot{x}_3 = \frac{\ddot{y}_1 y_2^{(4)} - y_1^{(4)}(\ddot{y}_2 + g)}{\ddot{y}_1^2 + (\ddot{y}_2 + g)^2} - 2\frac{[\ddot{y}_1 \dddot{y}_2 - \dddot{y}_1(\ddot{y}_2 + g)][\ddot{y}_1 \dddot{y}_1 + (\ddot{y}_2 + g)\dddot{y}_2]}{[\ddot{y}_1^2 + (\ddot{y}_2 + g)^2]^2},$$

so that we have also determined the function $\boldsymbol{\Psi}_2$. Thus, the VTOL aircraft model (3.56) to (3.61) is flat.

3.2.4 Flatness and Controllability

Below we will examine a flat linear system that is given in controller canonical form

$$\dot{\boldsymbol{x}} = \begin{bmatrix} 0 & 1 & 0 & \cdots & 0 \\ 0 & 0 & 1 & \cdots & 0 \\ \vdots & \vdots & \vdots & \ddots & \vdots \\ 0 & 0 & 0 & \cdots & 1 \\ -a_0 & -a_1 & -a_2 & \cdots & -a_{n-1} \end{bmatrix} \boldsymbol{x} + \begin{bmatrix} 0 \\ 0 \\ \vdots \\ 0 \\ 1 \end{bmatrix} u, \tag{3.73}$$
$$y = \begin{bmatrix} b_0 & b_1 & \cdots & b_m & 0 & \cdots & 0 \end{bmatrix} \boldsymbol{x},$$

or has been transformed into this form. We can assume without loss of generality that the system is in this canonical form, since, like controllability, flatness is invariant with respect to bijective state transformations, i. e. a system characteristic which does not change.

The flatness of the system (3.73) can be quickly verified by introducing the fictitious output

$$y_\mathrm{f} = x_1.$$

This is because it holds that

$$\begin{aligned} x_1 &= y_\mathrm{f}, \\ x_2 &= \dot{x}_1 = \dot{y}_\mathrm{f}, \\ x_3 &= \dot{x}_2 = \ddot{y}_\mathrm{f}, \\ &\vdots \\ x_n &= \dot{x}_{n-1} = y_\mathrm{f}^{(n-1)}, \end{aligned}$$

i. e.

$$\boldsymbol{x} = \begin{bmatrix} y_\mathrm{f} & \dot{y}_\mathrm{f} & \ddot{y}_\mathrm{f} & \cdots & y_\mathrm{f}^{(n-1)} \end{bmatrix}^T = \boldsymbol{\Psi}_1(y_\mathrm{f}, \dot{y}_\mathrm{f}, \ldots, y_\mathrm{f}^{(n-1)}),$$

and

$$\begin{aligned} u &= a_0 x_1 + a_1 x_2 + \ldots + a_{n-2} x_{n-1} + a_{n-1} x_n + \dot{x}_n \\ &= a_0 y_\mathrm{f} + a_1 \dot{y}_\mathrm{f} + \ldots + a_{n-2} y_\mathrm{f}^{(n-2)} + a_{n-1} y_\mathrm{f}^{(n-1)} + y_\mathrm{f}^{(n)} \\ &= \Psi_2(y_\mathrm{f}, \dot{y}_\mathrm{f}, \ldots, y_\mathrm{f}^{(n)}). \end{aligned}$$

Therefore, the fictitious output $y_\mathrm{f} = x_1$ is flat.

Since every controllable linear system can be transformed into the controller canonical form (3.73), a controllable linear system is always flat. On the other hand, a flat system is also always controllable [127]. Both hold for MIMO systems as well. Accordingly, we are able to formulate [273]

Theorem 49 (Controllability and Flatness of Linear Systems). *A linear system $\dot{\boldsymbol{x}} = \boldsymbol{A}\boldsymbol{x} + \boldsymbol{B}\boldsymbol{u}$ with $\boldsymbol{x} \in I\!R^n$ and $\boldsymbol{u} \in I\!R^m$ is flat if and only if it is controllable.*

The above criterion is necessary and sufficient. A similar, merely sufficient theorem [127] also exists for the case of nonlinear systems:

Theorem 50 (Controllability and Flatness of Nonlinear Systems). *A nonlinear system $\dot{\boldsymbol{x}} = \boldsymbol{f}(\boldsymbol{x}, \boldsymbol{u})$ with $\boldsymbol{x} \in D_{\boldsymbol{x},\mathrm{def}} \subseteq I\!R^n$ and $\boldsymbol{u} \in D_{\boldsymbol{u},\mathrm{def}} \subseteq I\!R^m$ is controllable if it is flat.*

There is a close relationship between controllability and flatness. The set of flat systems is a subset of the controllable systems. In particular, in Theorem 50 we now have a criterion to determine whether a nonlinear system is controllable.

In summary, we can deduce that the controllability of a system allows us to transform its state from any starting point to any end point if we choose suitable control variables. Similarly, flatness allows us to generate every desired output variable signal which possesses a derivative of sufficiently high order using a suitable control variable. Both system properties allow for a systematic manipulation of the system via a control variable.

3.2.5 Flat Outputs of Linear Systems

Although it is not possible to directly specify the set of flat outputs of a system in general, it can easily be done for linear systems

$$\dot{\boldsymbol{x}} = \boldsymbol{A}\boldsymbol{x} + \boldsymbol{b}u. \tag{3.74}$$

This makes sense, since flatness is an important and useful property of linear systems as well. As is the case for nonlinear systems, it allows us to design a control system for a specified output variable signal $y_\mathrm{ref}(t)$.

In the previous section, we have already determined a flat output for the special case of linear systems in controller canonical form. After multiplication by an arbitrary constant different from zero, this output generates the set of all flat outputs in controller canonical form. Consequently, a possible method of determining the flat outputs of a linear system (3.74) is to transform it to the controller canonical form. The flat outputs of the controller canonical form are then transformed back to the original coordinates of the system (3.74). Below, we will discuss an alternative possibility.

To determine the flat outputs

$$y_{\mathrm{f}} = \boldsymbol{c}^T \boldsymbol{x}$$

of a linear system in the general form (3.74), we will first derive the equation

$$\boldsymbol{x} = \boldsymbol{\Psi}_1(y_{\mathrm{f}}, \dot{y}_{\mathrm{f}}, \ldots, y_{\mathrm{f}}^{(n-1)}). \tag{3.75}$$

For this purpose, we will determine

$$\begin{aligned}
y_{\mathrm{f}} &= \boldsymbol{c}^T \boldsymbol{x},\\
\dot{y}_{\mathrm{f}} &= \boldsymbol{c}^T \dot{\boldsymbol{x}} = \boldsymbol{c}^T \boldsymbol{A}\boldsymbol{x} + \boldsymbol{c}^T \boldsymbol{b} u,\\
\ddot{y}_{\mathrm{f}} &= \boldsymbol{c}^T \boldsymbol{A}\dot{\boldsymbol{x}} + \boldsymbol{c}^T \boldsymbol{b}\dot{u} = \boldsymbol{c}^T \boldsymbol{A}^2\boldsymbol{x} + \boldsymbol{c}^T \boldsymbol{A}\boldsymbol{b} u + \boldsymbol{c}^T \boldsymbol{b}\dot{u},\\
&\ \vdots\\
y_{\mathrm{f}}^{(n-1)} &= \boldsymbol{c}^T \boldsymbol{A}^{n-2}\dot{\boldsymbol{x}} \ldots = \boldsymbol{c}^T \boldsymbol{A}^{n-1}\boldsymbol{x} + \boldsymbol{c}^T \boldsymbol{A}^{n-2}\boldsymbol{b} u + \boldsymbol{c}^T \boldsymbol{A}^{n-3}\boldsymbol{b}\dot{u} + \ldots + \boldsymbol{c}^T \boldsymbol{b} u^{(n-2)}.
\end{aligned}$$

To obtain a relationship consistent with equation (3.75), the above derivatives must be independent of $u, \dot{u}, \ldots, u^{(n-2)}$, i. e. it must hold that

$$\begin{aligned}
\boldsymbol{c}^T \boldsymbol{b} &= 0,\\
\boldsymbol{c}^T \boldsymbol{A}\boldsymbol{b} &= 0,\\
\boldsymbol{c}^T \boldsymbol{A}^2\boldsymbol{b} &= 0,\\
&\ \vdots\\
\boldsymbol{c}^T \boldsymbol{A}^{n-2}\boldsymbol{b} &= 0.
\end{aligned} \tag{3.76}$$

It follows that

$$\begin{bmatrix} y_{\mathrm{f}} \\ \dot{y}_{\mathrm{f}} \\ \ddot{y}_{\mathrm{f}} \\ \vdots \\ y_{\mathrm{f}}^{(n-1)} \end{bmatrix} = \underbrace{\begin{bmatrix} \boldsymbol{c}^T \\ \boldsymbol{c}^T \boldsymbol{A} \\ \boldsymbol{c}^T \boldsymbol{A}^2 \\ \vdots \\ \boldsymbol{c}^T \boldsymbol{A}^{n-1} \end{bmatrix}}_{\boldsymbol{M}_{\mathrm{obs}}} \boldsymbol{x},$$

where $\boldsymbol{M}_{\text{obs}}$ is the observability matrix of the linear system (3.74) with the output $y_{\text{f}} = \boldsymbol{c}^T \boldsymbol{x}$. Assuming the regularity of $\boldsymbol{M}_{\text{obs}}$, i. e. the observability of the linear system with output y_{f}, it holds that

$$\boldsymbol{x} = \boldsymbol{M}_{\text{obs}}^{-1} \begin{bmatrix} y_{\text{f}} \\ \dot{y}_{\text{f}} \\ \ddot{y}_{\text{f}} \\ \vdots \\ y_{\text{f}}^{(n-1)} \end{bmatrix} = \boldsymbol{\Psi}_1(y_{\text{f}}, \dot{y}_{\text{f}}, \ldots, y^{(n-1)}).$$

We can therefore deduce that a system (3.74) with a flat output $y_{\text{f}} = \boldsymbol{c}^T \boldsymbol{x}$ is observable.

Furthermore, a flat output y_{f} must satisfy an equation of the form

$$u = \Psi_2(y_{\text{f}}, \dot{y}_{\text{f}}, \ldots, y_{\text{f}}^{(n)}).$$

The derivative

$$y_{\text{f}}^{(n)} = \boldsymbol{c}^T \boldsymbol{A}^n \boldsymbol{x} + \boldsymbol{c}^T \boldsymbol{A}^{n-1} \boldsymbol{b} u + \boldsymbol{c}^T \boldsymbol{A}^{n-2} \boldsymbol{b} \dot{u} + \ldots + \boldsymbol{c}^T \boldsymbol{b} u^{(n-1)},$$

which reduces to

$$y_{\text{f}}^{(n)} = \boldsymbol{c}^T \boldsymbol{A}^n \boldsymbol{x} + \boldsymbol{c}^T \boldsymbol{A}^{n-1} \boldsymbol{b} u \tag{3.77}$$

due to equation (3.76), must therefore be dependent on u. This means that it is required that

$$\boldsymbol{c}^T \boldsymbol{A}^{n-1} \boldsymbol{b} = \alpha \tag{3.78}$$

holds for an arbitrary $\alpha \in I\!R \backslash \{0\}$.

We combine equation (3.76) and equation (3.78) and arrive at

$$\boldsymbol{c}^T \underbrace{\left[\boldsymbol{b} \;\; \boldsymbol{A}\boldsymbol{b} \cdots \boldsymbol{A}^{n-2}\boldsymbol{b} \;\; \boldsymbol{A}^{n-1}\boldsymbol{b}\right]}_{\boldsymbol{M}_{\text{contr}}} = \left[0 \; 0 \cdots 0 \; \alpha\right].$$

Here $\boldsymbol{M}_{\text{contr}}$ is the controllability matrix of system (3.74). If we assume controllability, i. e. regularity of $\boldsymbol{M}_{\text{contr}}$, all flat outputs $y_{\text{f}} = \boldsymbol{c}^T \boldsymbol{x}$ are parameterized by

$$\boldsymbol{c}^T = \left[0 \cdots 0 \; \alpha\right] \boldsymbol{M}_{\text{contr}}^{-1}.$$

Additionally, we can introduce new coordinates

$$\boldsymbol{z} = \begin{bmatrix} y_\mathrm{f} \\ \dot{y}_\mathrm{f} \\ \vdots \\ y_\mathrm{f}^{(n-1)} \end{bmatrix} = \begin{bmatrix} \boldsymbol{c}^T \\ \boldsymbol{c}^T\boldsymbol{A} \\ \vdots \\ \boldsymbol{c}^T\boldsymbol{A}^{n-1} \end{bmatrix} \boldsymbol{x} = \underbrace{\alpha \begin{bmatrix} [0 \cdots 0\;1]\,\boldsymbol{M}_\mathrm{contr}^{-1} \\ [0 \cdots 0\;1]\,\boldsymbol{M}_\mathrm{contr}^{-1}\boldsymbol{A} \\ \vdots \\ [0 \cdots 0\;1]\,\boldsymbol{M}_\mathrm{contr}^{-1}\boldsymbol{A}^{n-1} \end{bmatrix}}_{\boldsymbol{T}} \boldsymbol{x}.$$

The coordinate transformation $\boldsymbol{z} = \boldsymbol{T}\boldsymbol{x}$ transforms system (3.74) into the form

$$\dot{\boldsymbol{z}} = \begin{bmatrix} \dot{y}_\mathrm{f} \\ \ddot{y}_\mathrm{f} \\ \vdots \\ y_\mathrm{f}^{(n-1)} \\ y_\mathrm{f}^{(n)} \end{bmatrix} = \begin{bmatrix} z_2 \\ z_3 \\ \vdots \\ z_n \\ \boldsymbol{c}^T\boldsymbol{A}^n\boldsymbol{x} + \boldsymbol{c}^T\boldsymbol{A}^{n-1}\boldsymbol{b}u \end{bmatrix} = \begin{bmatrix} z_2 \\ z_3 \\ \vdots \\ z_n \\ \boldsymbol{c}^T\boldsymbol{A}^n\boldsymbol{T}^{-1}\boldsymbol{z} \end{bmatrix} + \begin{bmatrix} 0 \\ 0 \\ \vdots \\ 0 \\ \alpha \end{bmatrix} u \qquad (3.79)$$

with the flat output

$$y_\mathrm{f} = \boldsymbol{c}^T\boldsymbol{x} = z_1.$$

For this transformation, it is not necessary to explicitly compute the transformation $\boldsymbol{z} = \boldsymbol{T}\boldsymbol{x}$. The system equations (3.79) directly follow from

$$\dot{z}_i = y_\mathrm{f}^{(i)} = z_{i+1}, \quad i = 1, \ldots, n-1,$$

and from equation (3.77).

Without loss of generality, we multiply equation (3.79) by the factor α^{-1} on both sides, and rescale the coordinates z_i to

$$\tilde{z}_i = \alpha^{-1} z_i, \quad i = 1, \ldots, n.$$

In this way we obtain

$$\begin{aligned} \dot{\tilde{\boldsymbol{z}}} &= \begin{bmatrix} 0 & 1 & 0 & \cdots & 0 \\ 0 & 0 & 1 & \cdots & 0 \\ \vdots & \vdots & \vdots & \ddots & \vdots \\ 0 & 0 & 0 & \cdots & 1 \\ -a_0 & -a_1 & -a_2 & \cdots & -a_{n-1} \end{bmatrix} \tilde{\boldsymbol{z}} + \begin{bmatrix} 0 \\ 0 \\ \vdots \\ 0 \\ 1 \end{bmatrix} u, \\ y_\mathrm{f} &= \alpha \tilde{z}_1 \end{aligned} \qquad (3.80)$$

for the transformed system. Here the values for $a_0, \ldots, a_{n-1}$ are the coefficients of the characteristic polynomial $P(s) = \det(s\boldsymbol{I} - \boldsymbol{A})$ of $\boldsymbol{A}$. The form in equation (3.80) is a special case of the controller canonical form, for which the corresponding transfer function does not include zeros.

Every system in the controller canonical form equation (3.80) is thus flat with $y_\mathrm{f} = \alpha\tilde{z}_1$ as the flat output. Conversely, any controllable linear system (3.74) with a flat output

$$y_{\mathrm{f}} = \boldsymbol{c}^T \boldsymbol{x} = \alpha[0 \cdots 1]\boldsymbol{M}_{\mathrm{contr}}^{-1}\boldsymbol{x}$$

can be transformed into the controller canonical form (3.80). The output $y_{\mathrm{f}} = \alpha \tilde{z}_1$ with an arbitrary $\alpha \neq 0$ already describes all flat outputs. Equation (3.80) can therefore be called the *flat canonical form* of a linear system.

3.2.6 Verification of Flatness

In the case of linear systems, flatness is easily verifiable, since, according to Theorem 49, a controllable linear system is flat and vice versa. Therefore, we only need to examine the controllability of a linear system in order to verify its flatness. Verifying the flatness of nonlinear systems is not a simple task as in the linear case. Nevertheless, a flatness criterion also exists for nonlinear systems [273], although it is not easily applicable. Its conditions require the solution of partial differential equations. Thus, the original problem is transformed into a different one which is similarly difficult to solve. Flatness can be systematically analyzed and evaluated with a reasonable effort for particular classes of systems only.

Members of this class of systems are control-affine systems

$$\dot{\boldsymbol{x}} = \boldsymbol{a}(\boldsymbol{x}) + \boldsymbol{B}(\boldsymbol{x}) \cdot \boldsymbol{u}$$

or, more generally, nonlinear systems

$$\dot{\boldsymbol{x}} = \boldsymbol{f}(\boldsymbol{x}, \boldsymbol{u}),$$

which are either in nonlinear controller canonical form or can be transformed into this form using a bijective transformation which is continuously differentiable. From the nonlinear controller canonical form

$$\begin{bmatrix} \dot{x}_1 \\ \vdots \\ \dot{x}_{n_1-1} \\ \dot{x}_{n_1} \\ \dot{x}_{n_1+1} \\ \vdots \\ \dot{x}_{n_2-1} \\ \dot{x}_{n_2} \\ \vdots \\ \dot{x}_{n_{m-1}+1} \\ \vdots \\ \dot{x}_{n-1} \\ \dot{x}_n \end{bmatrix} = \begin{bmatrix} x_2 \\ \vdots \\ x_{n_1} \\ \alpha_1(\boldsymbol{x}) \\ x_{n_1+2} \\ \vdots \\ x_{n_2} \\ \alpha_2(\boldsymbol{x}) \\ \vdots \\ x_{n_{m-1}+2} \\ \vdots \\ x_n \\ \alpha_m(\boldsymbol{x}) \end{bmatrix} + \begin{bmatrix} 0 & 0 & 0 & \cdots & 0 & 0 \\ \vdots & \vdots & \vdots & \ddots & \vdots & \vdots \\ 0 & 0 & 0 & \cdots & 0 & 0 \\ \beta_{1,1}(\boldsymbol{x}) & \beta_{1,2}(\boldsymbol{x}) & \beta_{1,3}(\boldsymbol{x}) & \cdots & \beta_{1,m-1}(\boldsymbol{x}) & \beta_{1,m}(\boldsymbol{x}) \\ 0 & 0 & 0 & \cdots & 0 & 0 \\ \vdots & \vdots & \vdots & \ddots & \vdots & \vdots \\ 0 & 0 & 0 & \cdots & 0 & 0 \\ 0 & \beta_{2,2}(\boldsymbol{x}) & \beta_{2,3}(\boldsymbol{x}) & \cdots & \beta_{2,m-1}(\boldsymbol{x}) & \beta_{2,m}(\boldsymbol{x}) \\ \vdots & \vdots & \vdots & \ddots & \vdots & \vdots \\ 0 & 0 & 0 & \cdots & 0 & 0 \\ \vdots & \vdots & \vdots & \ddots & \vdots & \vdots \\ 0 & 0 & 0 & \cdots & 0 & 0 \\ 0 & 0 & 0 & \cdots & 0 & \beta_{m,m}(\boldsymbol{x}) \end{bmatrix} \begin{bmatrix} u_1 \\ u_2 \\ u_3 \\ \vdots \\ u_{m-1} \\ u_m \end{bmatrix}, \tag{3.81}$$

the flat output can be directly determined as

$$\boldsymbol{y} = \begin{bmatrix} x_1 \\ x_{n_1+1} \\ \vdots \\ x_{n_{m-1}+1} \end{bmatrix}.$$

Its flatness can be verified as follows. With δ_i denoting the orders of the m subsystems and

$$\begin{aligned} n_1 &= \delta_1, \\ n_2 &= \delta_1 + \delta_2, \\ n_3 &= \delta_1 + \delta_2 + \delta_3, \\ &\vdots \\ n_{m-1} &= \delta_1 + \ldots + \delta_{m-1}, \\ n &= \delta_1 + \ldots + \delta_m, \end{aligned}$$

the derivatives of the output signals can be stated as

$$\begin{array}{llll} y_1 = x_1, & y_2 = x_{n_1+1}, & \ldots, & y_m = x_{n_{m-1}+1}, \\ \dot{y}_1 = \dot{x}_1 = x_2, & \dot{y}_2 = x_{n_1+2}, & \ldots, & \dot{y}_m = x_{n_{m-1}+2}, \\ \ddot{y}_1 = \dot{x}_2 = x_3, & \ddot{y}_2 = x_{n_1+3}, & \ldots, & \ddot{y}_m = x_{n_{m-1}+3}, \\ \vdots & \vdots & & \vdots \\ y_1^{(\delta_1-1)} = x_{n_1}, & y_2^{(\delta_2-1)} = x_{n_2}, & \ldots, & y_m^{(\delta_m-1)} = x_n, \end{array}$$

and, with

$$\delta_{\max} = \max\{\delta_i, i = 1, \ldots, m\},$$

this yields

$$\begin{aligned} \boldsymbol{x} &= [y_1 \;\; \dot{y}_1 \;\; \cdots \;\; y_1^{(\delta_1-1)} \;\; \cdots \;\; y_m \;\; \dot{y}_m \;\; \cdots \;\; y_m^{(\delta_m-1)}]^T \\ &= \boldsymbol{\Psi}_1(\boldsymbol{y}, \dot{\boldsymbol{y}}, \ldots, \boldsymbol{y}^{(\delta_{\max}-1)}). \end{aligned} \tag{3.82}$$

If the decoupling matrix

$$\boldsymbol{D}(\boldsymbol{x}) = \begin{bmatrix} \beta_{1,1}(\boldsymbol{x}) & \beta_{1,2}(\boldsymbol{x}) & \beta_{1,3}(\boldsymbol{x}) & \cdots & \beta_{1,m}(\boldsymbol{x}) \\ 0 & \beta_{2,2}(\boldsymbol{x}) & \beta_{2,3}(\boldsymbol{x}) & \cdots & \beta_{2,m}(\boldsymbol{x}) \\ 0 & 0 & \beta_{3,3}(\boldsymbol{x}) & \cdots & \beta_{3,m}(\boldsymbol{x}) \\ \vdots & \vdots & \vdots & \ddots & \vdots \\ 0 & 0 & 0 & \cdots & \beta_{m,m}(\boldsymbol{x}) \end{bmatrix}$$

composed of the elements $\beta_{i,j}(\boldsymbol{x})$ used in equation (3.81) is regular for all $\boldsymbol{x}$, we obtain

$$\boldsymbol{u} = \boldsymbol{D}^{-1}(\boldsymbol{x})\left(-\begin{bmatrix}\alpha_1(\boldsymbol{x})\\ \alpha_2(\boldsymbol{x})\\ \vdots\\ \alpha_m(\boldsymbol{x})\end{bmatrix}+\begin{bmatrix}y_1^{(\delta_1)}\\ y_2^{(\delta_2)}\\ \vdots\\ y_m^{(\delta_m)}\end{bmatrix}\right)$$

for the system's input signal. With equation (3.82), this can also be written as

$$\boldsymbol{u}{=}\boldsymbol{D}^{-1}\left(\boldsymbol{\Psi}_1(\boldsymbol{y},\dot{\boldsymbol{y}},\dots,\boldsymbol{y}^{(\delta_{\max}-1)})\right)\left(\begin{bmatrix}y_1^{(\delta_1)}\\ y_2^{(\delta_2)}\\ \vdots\\ y_m^{(\delta_m)}\end{bmatrix}-\begin{bmatrix}\alpha_1(\boldsymbol{\Psi}_1(\boldsymbol{y},\dot{\boldsymbol{y}},\dots,\boldsymbol{y}^{(\delta_{\max}-1)}))\\ \alpha_2(\boldsymbol{\Psi}_1(\boldsymbol{y},\dot{\boldsymbol{y}},\dots,\boldsymbol{y}^{(\delta_{\max}-1)}))\\ \vdots\\ \alpha_m(\boldsymbol{\Psi}_1(\boldsymbol{y},\dot{\boldsymbol{y}},\dots,\boldsymbol{y}^{(\delta_{\max}-1)}))\end{bmatrix}\right)$$
$$= \boldsymbol{\Psi}_2(\boldsymbol{y},\dot{\boldsymbol{y}},\dots,\boldsymbol{y}^{(\delta_{\max})}). \tag{3.83}$$

Based on Definition 32 and using equations (3.82) and (3.83), we have shown that $\boldsymbol{y}$ is a flat output. Now we can formulate the following sufficient criterion for MIMO systems. For SISO systems, it is not only sufficient but also necessary [420].

Theorem 51 (Flatness of the Controller Canonical Form). *A system*

$$\dot{\boldsymbol{x}} = \boldsymbol{f}(\boldsymbol{x},\boldsymbol{u})$$

with $\boldsymbol{x} \in D_{\boldsymbol{x},\text{def}} \subseteq I\!R^n$ and $\boldsymbol{u} \in D_{\boldsymbol{u},\text{def}} \subseteq I\!R^m$ ($u \in D_{u,\text{def}} \subseteq I\!R$) is flat if (if and only if) it is available in nonlinear controller canonical form, or if it can be transformed into this form by means of a continuously differentiable, bijective coordinate transformation, and if the decoupling matrix $\boldsymbol{D}(\boldsymbol{x})$ fulfills

$$\det\left(\boldsymbol{D}(\boldsymbol{x})\right) \neq 0$$

for all $\boldsymbol{x} \in I\!R^n$.

Since there are quite a few systems which can be transformed into the nonlinear controller canonical form, the above theorem is of great importance. This is all the more true since there is no general, practically applicable criterion for proving flatness. However, a necessary condition [127, 310, 390, 420] exists. It is based on lines lying in a hypersurface. With this condition, it is at least possible to show that a system is not flat. The following theorem which allows such a verification is also known as the *ruled manifold criterion*.

Theorem 52 (Necessary Condition for Flatness). *Let a system be given by*

$$\dot{\boldsymbol{x}} = \boldsymbol{f}(\boldsymbol{x}, \boldsymbol{u}), \quad \boldsymbol{x} \in D_{\boldsymbol{x},\mathrm{def}} \subseteq I\!R^n, \quad \boldsymbol{u} \in D_{\boldsymbol{u},\mathrm{def}} \subseteq I\!R^m,$$

and let the underdetermined system resulting from the elimination of $\boldsymbol{u}$ *from this system be given by*

$$\boldsymbol{h}(\boldsymbol{x}, \dot{\boldsymbol{x}}) = \boldsymbol{0}$$

with $\dim(\boldsymbol{h}) = n - m$. *If the system* $\dot{\boldsymbol{x}} = \boldsymbol{f}(\boldsymbol{x}, \boldsymbol{u})$ *is flat, a real vector* $\boldsymbol{a} \neq \boldsymbol{0}$ *with* $\boldsymbol{a} \in I\!R^n$ *exists such that the algebraic equation*

$$\boldsymbol{h}(\boldsymbol{p}, \boldsymbol{q} + \lambda \boldsymbol{a}) = \boldsymbol{0}$$

is fulfilled for all real numbers λ *and all* $\boldsymbol{p}, \boldsymbol{q} \in I\!R^n$, *for which* $\boldsymbol{h}(\boldsymbol{p}, \boldsymbol{q}) = \boldsymbol{0}$ *holds.*

We will consider the example

$$\dot{x}_1 = u, \tag{3.84}$$

$$\dot{x}_2 = -x_1 + u^3. \tag{3.85}$$

Inserting equation (3.84) into equation (3.85) yields the underdetermined system of equations

$$h(\boldsymbol{x}, \dot{\boldsymbol{x}}) = -x_1 + \dot{x}_1^3 - \dot{x}_2 = 0$$

and the algebraic equation

$$h(\boldsymbol{p}, \boldsymbol{q} + \lambda \boldsymbol{a}) = -p_1 + (q_1 + \lambda a_1)^3 - (q_2 + \lambda a_2) = 0$$

with $\boldsymbol{a} = [a_1 \;\; a_2]^T$, $\boldsymbol{p} = [p_1 \;\; p_2]^T$, and $\boldsymbol{q} = [q_1 \;\; q_2]^T$. Multiplying out the brackets, we arrive at

$$a_1^3 \lambda^3 + 3a_1^2 q_1 \lambda^2 + (3a_1 q_1^2 - a_2)\lambda + q_1^3 - p_1 - q_2 = 0. \tag{3.86}$$

We then insert

$$h(\boldsymbol{p}, \boldsymbol{q}) = q_1^3 - p_1 - q_2 = 0$$

into equation (3.86) and obtain the equation

$$a_1^3 \lambda^3 + 3a_1^2 q_1 \lambda^2 + (3a_1 q_1^2 - a_2)\lambda = 0$$

as a necessary condition for the flatness of the system, which must be fulfilled for all $\lambda \in I\!R$. This is only possible if

$$a_1 = a_2 = 0$$

holds. Obviously, no $\boldsymbol{a} \neq \boldsymbol{0}$ exists for which the condition of Theorem 52 is fulfilled. Consequently, the system (3.84), (3.85) is not flat.

We will conclude with the autonomous systems

$$\dot{\boldsymbol{x}} = \boldsymbol{f}(\boldsymbol{x}),$$

i. e. systems which do not explicitly depend on an input variable vector $\boldsymbol{u}(t)$. Assuming that we have found a flat output $\boldsymbol{y}$ for such a system, we obtain

$$\boldsymbol{x} = \boldsymbol{\Psi}_1(\boldsymbol{y}, \dot{\boldsymbol{y}}, \ldots, \boldsymbol{y}^{(\beta)})$$

according to the definition of flatness: Definition 32 on p. 240. If this equation is inserted into the differential equation $\dot{\boldsymbol{x}} = \boldsymbol{f}(\boldsymbol{x})$, it follows that

$$\frac{\partial \boldsymbol{\Psi}_1}{\partial \boldsymbol{y}}\dot{\boldsymbol{y}} + \frac{\partial \boldsymbol{\Psi}_1}{\partial \dot{\boldsymbol{y}}}\ddot{\boldsymbol{y}} + \ldots + \frac{\partial \boldsymbol{\Psi}_1}{\partial \boldsymbol{y}^{(\beta)}}\boldsymbol{y}^{(\beta+1)} = \boldsymbol{f}(\boldsymbol{\Psi}_1(\boldsymbol{y}, \dot{\boldsymbol{y}}, \ldots, \boldsymbol{y}^{(\beta)})).$$

This equation, however, contradicts the requirement (3.44) of differential independence of a flat output on p. 241, i. e. the requirement that no differential equation

$$\boldsymbol{\varphi}(\boldsymbol{y}, \dot{\boldsymbol{y}}, \ldots, \boldsymbol{y}^{(\gamma)}) = \boldsymbol{0}$$

exists. Thus we obtain

Theorem 53 (Nonexistent Flatness of Autonomous Systems). *An autonomous system $\dot{\boldsymbol{x}} = \boldsymbol{f}(\boldsymbol{x})$ is not flat.*

This theorem can also be directly deduced from the fact that no flat output $\boldsymbol{y}$ can exist for autonomous systems, because Requirement (3) of Definition 32, $\dim(\boldsymbol{y}) = \dim(\boldsymbol{u})$, is not fulfilled.

We will apply flatness theory to the design of feedforward and feedback control systems in Section 5.4. For supplementary literature on flatness theory, we refer the interested reader to [1, 72, 127, 128, 129, 273, 420, 485], among others.

3.3 Nonlinear State Transformations

3.3.1 Transformations and Transformed System Equations

In linear systems theory, it is often useful to transform the representation of a system

$$\dot{\boldsymbol{x}} = \boldsymbol{A}\boldsymbol{x} + \boldsymbol{B}\boldsymbol{u} \tag{3.87}$$

into a different one. This is done via a coordinate transformation

$$\boldsymbol{x} = \boldsymbol{T}\boldsymbol{z} \tag{3.88}$$

where the vector $\boldsymbol{z}$ represents the new coordinates. The matrix $\boldsymbol{T}$ is a regular $n \times n$ matrix. The new system is then represented by

$$\dot{\boldsymbol{z}} = \boldsymbol{T}^{-1}\boldsymbol{A}\boldsymbol{T}\boldsymbol{z} + \boldsymbol{T}^{-1}\boldsymbol{B}\boldsymbol{u}. \tag{3.89}$$

Often the transformation matrix $\boldsymbol{T}$ is chosen such that the new system matrix $\boldsymbol{T}^{-1}\boldsymbol{A}\boldsymbol{T}$ is diagonal or takes the form of a companion matrix

$$\boldsymbol{T}^{-1}\boldsymbol{A}\boldsymbol{T} = \begin{bmatrix} 0 & 1 & 0 & \cdots & 0 \\ 0 & 0 & 1 & \cdots & 0 \\ \vdots & \vdots & \vdots & & \vdots \\ -a_0 & -a_1 & -a_2 & \cdots & -a_{n-1} \end{bmatrix}.$$

These representations are useful for the controller design or for directly identifying system properties.

Similarly, for nonlinear systems

$$\dot{\boldsymbol{x}} = \boldsymbol{f}(\boldsymbol{x}, \boldsymbol{u}), \tag{3.90}$$

the transformations

$$\boldsymbol{z} = \boldsymbol{q}(\boldsymbol{x}) \quad \text{or} \quad \boldsymbol{x} = \boldsymbol{q}^{-1}(\boldsymbol{z}) \tag{3.91}$$

can be useful as well if it is possible to transform the system description into a suitable form this way for purposes such as controller design. Furthermore, as discussed in the previous section, to verify the controllability of a control-affine system, it is very helpful to be able to transform a system representation into the controller canonical form.

Usually, transformations (3.91) are required to be continuously differentiable and bijective. The latter means that only a single $\boldsymbol{z}$ can be assigned to every $\boldsymbol{x}$ and vice versa. As already mentioned, such a continuously differentiable transformation is called *diffeomorphism*. The transformed representation results from inserting the transformation equation (3.91) into the system description (3.90). This yields

$$\frac{d\boldsymbol{q}^{-1}(\boldsymbol{z})}{dt} = \boldsymbol{f}(\boldsymbol{q}^{-1}(\boldsymbol{z}), \boldsymbol{u}),$$

from which

$$\frac{\partial \boldsymbol{q}^{-1}(\boldsymbol{z})}{\partial \boldsymbol{z}} \cdot \dot{\boldsymbol{z}} = \boldsymbol{f}(\boldsymbol{q}^{-1}(\boldsymbol{z}), \boldsymbol{u})$$

and, finally, the transformed system representation

$$\dot{\boldsymbol{z}} = \left(\frac{\partial \boldsymbol{q}^{-1}(\boldsymbol{z})}{\partial \boldsymbol{z}}\right)^{-1} \boldsymbol{f}(\boldsymbol{q}^{-1}(\boldsymbol{z}), \boldsymbol{u}) = \hat{\boldsymbol{f}}(\boldsymbol{z}, \boldsymbol{u}) \tag{3.92}$$

follows. Similarly, for the retransformation of the $\boldsymbol{z}$-coordinates to the $\boldsymbol{x}$-coordinates, it holds that

$$\dot{\boldsymbol{x}} = \left(\frac{\partial \boldsymbol{q}(\boldsymbol{x})}{\partial \boldsymbol{x}}\right)^{-1} \hat{\boldsymbol{f}}(\boldsymbol{q}(\boldsymbol{x}), \boldsymbol{u}) = \boldsymbol{f}(\boldsymbol{x}, \boldsymbol{u}). \tag{3.93}$$

To calculate the inverse of the Jacobian matrix of $\boldsymbol{q}^{-1}$ used in equation (3.92), we apply the rule of derivation for inverse functions, obtaining

$$\left(\frac{\partial \boldsymbol{q}^{-1}(\boldsymbol{z})}{\partial \boldsymbol{z}}\right)^{-1} = \left.\frac{\partial \boldsymbol{q}(\boldsymbol{x})}{\partial \boldsymbol{x}}\right|_{\boldsymbol{x}=\boldsymbol{q}^{-1}(\boldsymbol{z})}.$$

In many cases, the application of this identity simplifies the computation of the transformed system in equation (3.92).

The calculations above make it clear why the transformation functions $\boldsymbol{q}(\boldsymbol{x})$ and $\boldsymbol{q}^{-1}(\boldsymbol{z})$ have to be continuously differentiable. Clearly, if $\boldsymbol{q}(\boldsymbol{x})$ and $\boldsymbol{q}^{-1}(\boldsymbol{z})$ are continuously differentiable, the elements of the matrices

$$\left(\frac{\partial \boldsymbol{q}^{-1}(\boldsymbol{z})}{\partial \boldsymbol{z}}\right)^{-1} \qquad \text{and} \qquad \frac{\partial \boldsymbol{q}(\boldsymbol{x})}{\partial \boldsymbol{x}}$$

are continuous. If this is not the case, the right side of the differential equations (3.92) and (3.93) is discontinuous as a result of the transformation. This, however, would be an unreasonable situation, since the left side of the differential equations $\dot{\boldsymbol{z}}$ or $\dot{\boldsymbol{x}}$ is the time derivative of a – logically – differentiable function, which cannot jump, i. e. is continuous. In such a case, the transformed differential equation, in contrast to the original differential equation, would not be solvable.

For a linear system (3.87) and a linear transformation (3.88), it is easy to show that the transformed system representation (3.89) follows from equation (3.92). This is because the equations

$$\boldsymbol{x} = \boldsymbol{q}^{-1}(\boldsymbol{z}) = \boldsymbol{T}\boldsymbol{z}$$

and

$$\frac{\partial \boldsymbol{q}^{-1}(\boldsymbol{z})}{\partial \boldsymbol{z}} = \boldsymbol{T}$$

hold.

Next we will derive an important theorem on diffeomorphisms, i. e. bijective and continuously differentiable coordinate transformations $\boldsymbol{q}(\boldsymbol{x})$. It holds that

$$\boldsymbol{q}^{-1}(\boldsymbol{q}(\boldsymbol{x})) = \boldsymbol{x}$$

and for the derivative of this identity

$$\frac{\partial \boldsymbol{q}^{-1}(\boldsymbol{z})}{\partial \boldsymbol{z}} \cdot \frac{\partial \boldsymbol{q}(\boldsymbol{x})}{\partial \boldsymbol{x}} = \boldsymbol{I}.$$

Here $\boldsymbol{I}$ is the identity matrix. Accordingly, the Jacobian matrix $\partial \boldsymbol{q}(\boldsymbol{x})/\partial \boldsymbol{x}$ of a diffeomorphism $\boldsymbol{q}(\boldsymbol{x})$ must be regular.

The regularity of the Jacobian matrix, however, is not only necessary for a diffeomorphism; it is also sufficient. If the Jacobian matrix of $\boldsymbol{q}(\boldsymbol{x})$ is regular at a point $\boldsymbol{x}_0$, according to the implicit function theorem [167], a continuously differentiable inverse function $\boldsymbol{q}^{-1}$ exists in a neighborhood of $\boldsymbol{x}_0$ for the continuously differentiable transformation function $\boldsymbol{q}$. If the Jacobian matrix is not only regular at a point $\boldsymbol{x}_0$ but for an open set, there is at least a subset where the inverse function $\boldsymbol{q}^{-1}$ exists for all points in this set. If not only a diffeomorphism $\boldsymbol{q}$ but also its inverse $\boldsymbol{q}^{-1}$ bijectively maps $\mathbb{R}^n$ into $\mathbb{R}^n$, we call it global. If this is not the case, we call the diffeomorphism local. Summarizing the results above, we can use the following theorem to verify whether a mapping $\boldsymbol{z} = \boldsymbol{q}(\boldsymbol{x})$ is a local diffeomorphism or not.

Theorem 54 (Local Diffeomorphism). *Let a function $\boldsymbol{z} = \boldsymbol{q}(\boldsymbol{x})$ be given which is defined on a set $D_{\boldsymbol{x},\mathrm{def}} \subseteq \mathbb{R}^n$ and maps $D_{\boldsymbol{x},\mathrm{def}}$ into $\mathbb{R}^n$. If and only if*

(1) $\boldsymbol{q}$ is continuously differentiable on a set $D_{\boldsymbol{x}} \subseteq D_{\boldsymbol{x},\mathrm{def}}$ and

(2) the inequality $\det(\frac{\partial \boldsymbol{q}(\boldsymbol{x})}{\partial \boldsymbol{x}}) \neq 0$ holds for all $\boldsymbol{x} \in D_{\boldsymbol{x}}$,

then a set $S \subseteq D_{\boldsymbol{x}}$ exists where $\boldsymbol{z} = \boldsymbol{q}(\boldsymbol{x})$ is a local diffeomorphism.

The above theorem guarantees the existence of the diffeomorphism only for a set S – which generally has to be identified – and not for the domain of definition $D_{\boldsymbol{x},\mathrm{def}}$ itself. An illustrative example [481] is

$$\boldsymbol{z} = \boldsymbol{q}(\boldsymbol{x}) = \begin{bmatrix} e^{x_1}\cos(x_2) \\ e^{x_1}\sin(x_2) \end{bmatrix}, \qquad D_{\boldsymbol{x},\mathrm{def}} = \mathbb{R}^2.$$

The determinant

$$\det(\frac{\partial \boldsymbol{q}(\boldsymbol{x})}{\partial \boldsymbol{x}}) = \det(\begin{bmatrix} e^{x_1}\cos(x_2) & -e^{x_1}\sin(x_2) \\ e^{x_1}\sin(x_2) & e^{x_1}\cos(x_2) \end{bmatrix}) = e^{2x_1}$$

is unequal to zero for all $\boldsymbol{x} \in \mathbb{R}^2$. Nevertheless, $\boldsymbol{q}$ is not a global diffeomorphism because $\boldsymbol{z} = \boldsymbol{q}(\boldsymbol{x})$ is not a one-to-one and onto, i. e. not a bijective, mapping for all $\boldsymbol{x} \in \mathbb{R}^2$. For example, the two vectors

$$\boldsymbol{x}_{\mathrm{a}} = \boldsymbol{0} \quad \text{and} \quad \boldsymbol{x}_{\mathrm{b}} = \begin{bmatrix} 0 \\ 2\pi \end{bmatrix}$$

both result in

$$\boldsymbol{q}(\boldsymbol{x}_{\mathrm{a}}) = \boldsymbol{q}(\boldsymbol{x}_{\mathrm{b}}) = \begin{bmatrix} 1 \\ 0 \end{bmatrix}.$$

In some cases, we are interested in having a global diffeomorphism to map $\mathbb{R}^n$ into $\mathbb{R}^n$ and vice versa. The best, but mostly very laborious way to verify whether a mapping $\boldsymbol{q}$ is a global diffeomorphism is calculating its inverse $\boldsymbol{q}^{-1}$. The following theorem [154, 245, 394, 481], Hadamard's global inverse function theorem, provides us with a less laborious way, but does not yield the inverse.

Theorem 55 (Global Diffeomorphism). *A function* $\boldsymbol{z} = \boldsymbol{q}(\boldsymbol{x})$ *which maps* $\mathbb{R}^n$ *into* $\mathbb{R}^n$ *is a global diffeomorphism if and only if*

(1) $\boldsymbol{q}$ *is continuously differentiable everywhere on* $\mathbb{R}^n$,

(2) the inequality $\det(\dfrac{\partial \boldsymbol{q}(\boldsymbol{x})}{\partial \boldsymbol{x}}) \neq 0$ *holds for all* $\boldsymbol{x} \in \mathbb{R}^n$, *and*

(3) the limit $|\boldsymbol{q}(\boldsymbol{x})| \to \infty$ *results whenever* $|\boldsymbol{x}| \to \infty$.

In general, the transformation equations may also be dependent on the input variable vector $\boldsymbol{u}$ and possibly on some temporal derivatives $\boldsymbol{u}^{(j)}$, $j = 1, \ldots, i$, i.e. they can be stated as

$$\boldsymbol{z} = \boldsymbol{q}(\boldsymbol{x}, \boldsymbol{u}, \dot{\boldsymbol{u}}, \ldots, \boldsymbol{u}^{(i)})$$

or

$$\boldsymbol{x} = \boldsymbol{q}^{-1}(\boldsymbol{z}, \boldsymbol{u}, \dot{\boldsymbol{u}}, \ldots, \boldsymbol{u}^{(i)}). \tag{3.94}$$

If we now transform the system

$$\dot{\boldsymbol{x}} = \boldsymbol{f}(\boldsymbol{x}, \boldsymbol{u})$$

to $\boldsymbol{z}$-coordinates, we obtain

$$\frac{d\boldsymbol{q}^{-1}(\boldsymbol{z},\boldsymbol{u},\ldots,\boldsymbol{u}^{(i)})}{dt} = \frac{\partial \boldsymbol{q}^{-1}(\boldsymbol{z},\boldsymbol{u},\ldots,\boldsymbol{u}^{(i)})}{\partial \boldsymbol{z}} \cdot \dot{\boldsymbol{z}} + \sum_{j=0}^{i} \frac{\partial \boldsymbol{q}^{-1}(\boldsymbol{z},\boldsymbol{u},\ldots,\boldsymbol{u}^{(i)})}{\partial \boldsymbol{u}^{(j)}} \cdot \boldsymbol{u}^{(j+1)}$$
$$= \boldsymbol{f}(\boldsymbol{q}^{-1}(\boldsymbol{z},\boldsymbol{u},\ldots,\boldsymbol{u}^{(i)}), \boldsymbol{u})$$

after inserting equation (3.94) into the system equation $\dot{\boldsymbol{x}} = \boldsymbol{f}(\boldsymbol{x}, \boldsymbol{u})$. The transformed system equation follows as

$$\dot{\boldsymbol{z}} = \left(\frac{\partial \boldsymbol{q}^{-1}(\boldsymbol{z},\boldsymbol{u},\ldots,\boldsymbol{u}^{(i)})}{\partial \boldsymbol{z}}\right)^{-1} \cdot \left(\boldsymbol{f}(\boldsymbol{q}^{-1}(\boldsymbol{z},\boldsymbol{u},\ldots,\boldsymbol{u}^{(i)}),\boldsymbol{u}) - \sum_{j=0}^{i} \frac{\partial \boldsymbol{q}^{-1}(\boldsymbol{z},\boldsymbol{u},\ldots,\boldsymbol{u}^{(i)})}{\partial \boldsymbol{u}^{(j)}} \boldsymbol{u}^{(j+1)}\right). \tag{3.95}$$

The equation for the retransformation can be derived in a comparable fashion.

It generally requires some effort to determine a transformation

$$\boldsymbol{z} = \boldsymbol{q}(\boldsymbol{x})$$

or

$$\boldsymbol{z} = \boldsymbol{q}(\boldsymbol{x},\boldsymbol{u},\dot{\boldsymbol{u}},\ldots,\boldsymbol{u}^{(i)})$$

which induces a change of coordinates resulting in the desired system representation. This is already the case for linear systems when diagonalizing a system, for example. To do this, the eigenvectors which make up the columns of the transformation matrix $\boldsymbol{T}$ must be determined. For the nonlinear case, the effort required is often much greater. We will discuss this problem in Section 3.3.4.

3.3.2 Illustrative Example

Next we will consider a nonlinear example. Let a system be described by

$$\dot{x} = \frac{1}{x}. \tag{3.96}$$

This differential equation is not defined for $x = 0$. We will therefore exclude the point $x = 0$ from our considerations in the following. We select

$$z = q(x) = e^{\operatorname{sgn}(x)x^2/2} \tag{3.97}$$

as the transformation in order to simplify the representation of the differential equation and to calculate its solution. The retransformation is given by

$$x = q^{-1}(z) = \operatorname{sgn}\left(\ln\left(z^2\right)\right)\left|\ln\left(z^2\right)\right|^{1/2}. \tag{3.98}$$

The transformation rules (3.97) and (3.98) are bijective, continuously differentiable, and q maps the space of real numbers $I\!R$ to the interval $(0, \infty)$. Figure 3.21 shows the graph of $q(x)$. We are now able to calculate the derivative

$$\frac{\partial q^{-1}(z)}{\partial z} = \frac{\left|\ln\left(z^2\right)\right|^{-1/2}}{z}$$

and, consistent with equation (3.92), we obtain

$$\dot{z} = \left(\frac{\left|\ln\left(z^2\right)\right|^{-1/2}}{z}\right)^{-1} \cdot \left(\frac{1}{\operatorname{sgn}\left(\ln\left(z^2\right)\right)\left|\ln\left(z^2\right)\right|^{1/2}}\right)$$

for the transformed system. Simplifying the latter, we arrive at

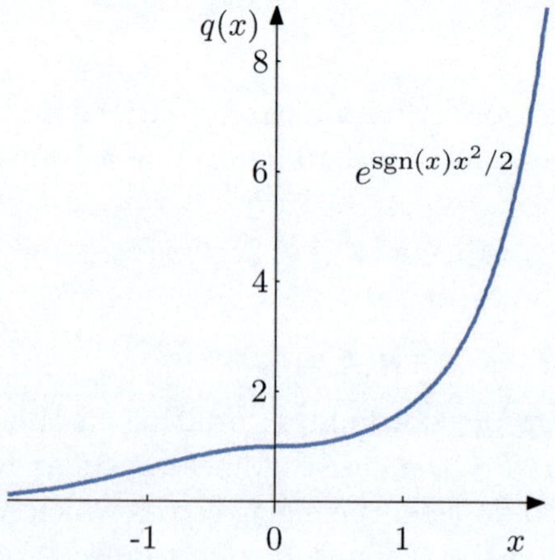

Fig. 3.21: Graph of the function $z = q(x)$

$$\dot{z} = z\,\mathrm{sgn}\left(\ln\left(z^2\right)\right) = \begin{cases} z, & z > 1, \\ -z, & 0 < z < 1. \end{cases} \tag{3.99}$$

By means of the transformation (3.97), we have managed to transform the nonlinear system description (3.96) into one, or, more precisely, into two linear system descriptions. Note that the two system descriptions (3.96) and (3.99) are equivalent, since the transformation is bijective.

Finally, we solve the differential equation (3.96) by first determining the solution of equation (3.99). Its solution is

$$z(t) = \begin{cases} z_0 e^t, & z > 1, \\ z_0 e^{-t}, & 0 < z < 1, \end{cases}$$

with an initial value $z_0 = z(0)$. After retransformation according to equation (3.98), we obtain the solution of the nonlinear system (3.96) as

$$x(t) = q^{-1}(z(t)) = \begin{cases} \mathrm{sgn}\left(\ln\left(z_0^2 e^{2t}\right)\right) \cdot \left|\ln\left(z_0^2 e^{2t}\right)\right|^{1/2}, & z > 1, \\ \mathrm{sgn}\left(\ln\left(z_0^2 e^{-2t}\right)\right) \cdot \left|\ln\left(z_0^2 e^{-2t}\right)\right|^{1/2}, & 0 < z < 1. \end{cases} \tag{3.100}$$

Now we replace the initial value z_0 with the initial value $x_0 = x(0)$, obtaining

$$x(0) = \mathrm{sgn}\left(\ln\left(z_0^2\right)\right) \cdot \left|\ln\left(z_0^2\right)\right|^{1/2},$$

from which

$$\mathrm{sgn}(x_0) = \mathrm{sgn}\left(\ln\left(z_0^2\right)\right)$$

results. Using equation (3.97), the relation

$$\begin{aligned} x_0^2 &= \ln\left(z_0^2\right), \quad z_0 > 1, \\ -x_0^2 &= \ln\left(z_0^2\right), \quad 0 < z_0 < 1, \end{aligned}$$

between the initial values x_0 and z_0 follows. Inserting this result into equation (3.100) yields

$$\begin{aligned} x(t) &= \begin{cases} \mathrm{sgn}\left(x_0\right) \cdot \left|x_0^2 + 2t\right|^{1/2}, & x_0 > 0, \\ \mathrm{sgn}\left(x_0\right) \cdot \left|-x_0^2 - 2t\right|^{1/2}, & x_0 < 0, \end{cases} \\ &= \mathrm{sgn}\left(x_0\right) \cdot \left(x_0^2 + 2t\right)^{1/2}, \quad x \neq 0, \end{aligned}$$

as the solution of the differential equation $\dot{x} = 1/x$. We will calculate this solution again in Exercise 3.16, using a simpler transformation equation than equation (3.98).

3.3.3 Example: Park Transformation

One of the most important nonlinear transformations used in electrical engineering is the Park transformation. It allows us to simplify the model equations of synchronous machines and induction machines (asynchronous machines) significantly. In the following, we will consider induction machines, for which the rotors are provided with a three-phase current of adjustable frequency $\Delta\omega$ via slip rings and a current converter. These are called doubly fed induction machines. They are often used as generators for wind turbines. In this application, they have a set of advantages compared to other generator designs: they only require a small current converter, which merely converts the rotor current which is fed into the rotor coils via the slip rings. The rotor voltage not only allows us to control the effective power but also the reactive power. It is also an advantage that the stator voltage $\boldsymbol{u}_{\mathrm{sabc}}$ can be set to be the line voltage. Furthermore, the frequency of the stator currents can remain constant and consistent with the line frequency, even for varying angular velocities of the rotor blade due to varying wind velocities. A disadvantage, however, is the slip rings, since they wear out and require maintenance. Figure 3.22 illustrates the structure of a wind turbine with a doubly fed induction generator.

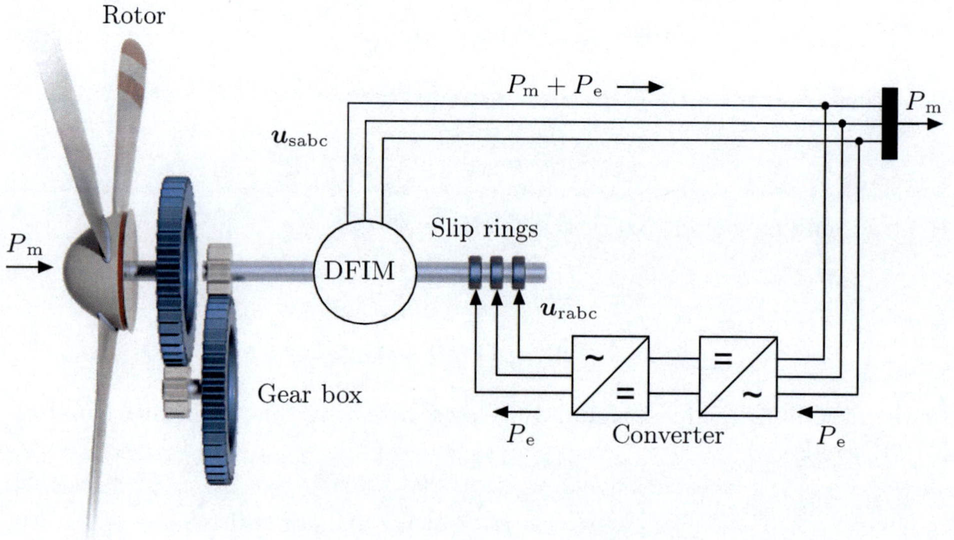

Fig. 3.22: Wind turbine with a doubly fed induction machine (DFIM). The graph also shows the energy flow associated with the absorbed wind power P_m and the power of the rotor circuit P_e, where losses are assumed to be zero. The vectors $\boldsymbol{u}_\mathrm{sabc}$ and $\boldsymbol{u}_\mathrm{rabc}$ correspond to the stator and rotor voltages of the phases a, b, c.

First we will derive the model of the doubly fed induction machine [247, 254]. Figure 3.23 shows the stator and rotor coils, as well as their positions relative to each other. The rotor of the induction generator, powered by the turbine, turns with the mechanical frequency

$$\omega_{\mathrm{m}} = \dot{\Theta}.$$

Here the geometric angle Θ denotes the angle between the rotor coil in phase a_{r} and the stator coil in phase a_{s}, as shown in Figure 3.23. The indices r and s stand for the rotor and stator, respectively. The rotor currents i_{ra}, i_{rb}, and i_{rc} of frequency $\Delta\omega$ are fed into the rotor coils via the slip rings and the current converter. For a machine with p pole pairs, the magnetic field of the rotor revolves with the frequency

$$\frac{\Delta\omega}{p} = \dot{\varphi}$$

around it. Here φ is the rotation angle of the rotor's magnetic field. With respect to the stator, the magnetic field therefore rotates at a frequency of

$$\omega_{\mathrm{s}} = \omega_{\mathrm{m}} + \frac{\Delta\omega}{p}.$$

The rotor field induces a magnetic field in the stator coils that also rotates at

$$\omega_{\mathrm{s}} = \dot{\beta} = \dot{\Theta} + \dot{\varphi},$$

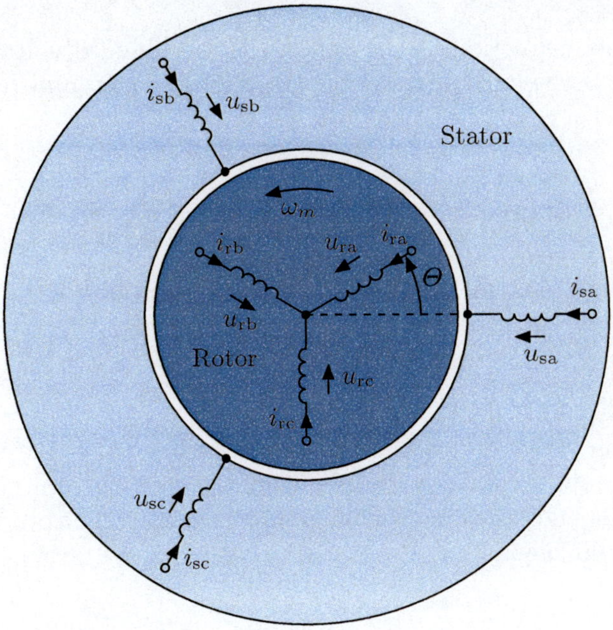

Fig. 3.23: Rotor and stator coils and their relative positions

where $\beta = \Theta + \varphi$ is the rotation angle of the stator field. The frequency of the stator field $\dot{\beta}$ therefore corresponds to the sum of the mechanical rotation frequency $\dot{\Theta}$ and the rotational frequency $\dot{\varphi}$ of the rotor field. Since the induction machine has p pole pairs, a voltage of frequency

$$\omega_{\text{electric}} = p \cdot \omega_{\text{s}}$$

is produced at the stator terminals. The relative difference

$$s = \frac{\omega_{\text{s}} - \omega_{\text{m}}}{\omega_{\text{s}}} = \frac{\dot{\varphi}}{\omega_{\text{s}}} \tag{3.101}$$

between the frequency ω_{s} of the stator field and the mechanical frequency ω_{m} of the rotor is called *slip*. Using the current converter, we can adjust the rotor frequency $\Delta\omega$. Thus we are able to regulate the generator frequency

$$\omega_{\text{electric}} = p \cdot \omega_{\text{s}} = p \cdot \omega_{\text{m}} + \Delta\omega$$

via the rotor frequency $\Delta\omega$ such that it is identical to the line frequency even if the wind turbine has a varying rotational frequency ω_{m}.

For the stator voltages of the phases a_{s}, b_{s}, c_{s}, we obtain

$$\underbrace{\begin{bmatrix} u_{\text{sa}} \\ u_{\text{sb}} \\ u_{\text{sc}} \end{bmatrix}}_{\boldsymbol{u}_{\text{sabc}}} = \underbrace{\begin{bmatrix} \dot{\Psi}_{\text{sa}} \\ \dot{\Psi}_{\text{sb}} \\ \dot{\Psi}_{\text{sc}} \end{bmatrix}}_{\dot{\boldsymbol{\Psi}}_{\text{sabc}}} + \underbrace{\begin{bmatrix} R_{\text{s}} & 0 & 0 \\ 0 & R_{\text{s}} & 0 \\ 0 & 0 & R_{\text{s}} \end{bmatrix}}_{\boldsymbol{R}_{\text{s}}} \underbrace{\begin{bmatrix} i_{\text{sa}} \\ i_{\text{sb}} \\ i_{\text{sc}} \end{bmatrix}}_{\boldsymbol{i}_{\text{sabc}}}, \tag{3.102}$$

where Ψ_{sa} is the magnetic flux linkage, $\dot{\Psi}_{\text{sa}}$ is the induced voltage of the coil of phase a_{s}, i_{sa} is the current that flows through the coil, and R_{s} is its ohmic resistance. Similarly, for the rotor voltages, we obtain

$$\underbrace{\begin{bmatrix} u_{\text{ra}} \\ u_{\text{rb}} \\ u_{\text{rc}} \end{bmatrix}}_{\boldsymbol{u}_{\text{rabc}}} = \underbrace{\begin{bmatrix} \dot{\Psi}_{\text{ra}} \\ \dot{\Psi}_{\text{rb}} \\ \dot{\Psi}_{\text{rc}} \end{bmatrix}}_{\dot{\boldsymbol{\Psi}}_{\text{rabc}}} + \underbrace{\begin{bmatrix} R_{\text{r}} & 0 & 0 \\ 0 & R_{\text{r}} & 0 \\ 0 & 0 & R_{\text{r}} \end{bmatrix}}_{\boldsymbol{R}_{\text{r}}} \underbrace{\begin{bmatrix} i_{\text{ra}} \\ i_{\text{rb}} \\ i_{\text{rc}} \end{bmatrix}}_{\boldsymbol{i}_{\text{rabc}}} \tag{3.103}$$

with the resistance R_{r} of each rotor coil, and the corresponding voltages, currents, and flux linkages.

The stator and the rotor coils are coupled via the magnetic fluxes $\boldsymbol{\Psi}_{\text{rabc}}$ and $\boldsymbol{\Psi}_{\text{sabc}}$. Since the rotor rotates with frequency $\omega_{\text{m}} = \dot{\Theta}$, the mutual inductance of the rotor coils and stator coils changes according to the rotor angle Θ.

For the flux linkage of the stator coil of phase a_{s}, we obtain

$$\Psi_{\text{sa}} = L_{\text{ss}} i_{\text{sa}} + L_{\text{ms}}(i_{\text{sb}} + i_{\text{sc}}) + L_{\text{msr}}\left(i_{\text{ra}} \cos(\Theta) + i_{\text{rb}} \cos(\Theta + \frac{2\pi}{3}) + i_{\text{rc}} \cos(\Theta - \frac{2\pi}{3})\right), \tag{3.104}$$

while for the rotor coil of phase a_r

$$\Psi_{ra}=L_{sr}i_{ra}+L_{mr}(i_{rb}+i_{rc})+L_{msr}\left(i_{sa}\cos(\Theta)+i_{sb}\cos(\Theta-\frac{2\pi}{3})+i_{sc}\cos(\Theta+\frac{2\pi}{3})\right) \tag{3.105}$$

holds.

Here L_{ss} is the self-inductance of the stator coils, L_{sr} is that of the rotor coils, L_{ms} is the mutual inductance of two stator coils, while L_{mr} is that of two rotor coils and L_{msr} is the mutual inductance of the stator coil of phase a_s with respect to the rotor coil of phase a_r at $\Theta = 0$.

Using

$$i_{sa} + i_{sb} + i_{sc} = 0, \quad i_{ra} + i_{rb} + i_{rc} = 0,$$

the abbreviations

$$L_s = L_{ss} - L_{ms}, \quad L_r = L_{sr} - L_{mr},$$

and equations (3.104) and (3.105), the equations

$$\Psi_{sa} = L_s i_{sa} + L_{msr}\left(i_{ra}\cos(\Theta) + i_{rb}\cos(\Theta + \frac{2\pi}{3}) + i_{rc}\cos(\Theta - \frac{2\pi}{3})\right),$$

$$\Psi_{ra} = L_r i_{ra} + L_{msr}\left(i_{sa}\cos(\Theta) + i_{sb}\cos(\Theta - \frac{2\pi}{3}) + i_{sc}\cos(\Theta + \frac{2\pi}{3})\right)$$

follow. Including phases b and c as well, we obtain

$$\underbrace{\begin{bmatrix}\Psi_{sa}\\ \Psi_{sb}\\ \Psi_{sc}\end{bmatrix}}_{\boldsymbol{\Psi}_{sabc}} = \underbrace{\begin{bmatrix}L_s & 0 & 0\\ 0 & L_s & 0\\ 0 & 0 & L_s\end{bmatrix}}_{\boldsymbol{L}_s}\underbrace{\begin{bmatrix}i_{sa}\\ i_{sb}\\ i_{sc}\end{bmatrix}}_{\boldsymbol{i}_{sabc}} + L_{msr}\underbrace{\begin{bmatrix}\cos(\Theta) & \cos(\Theta+\frac{2\pi}{3}) & \cos(\Theta-\frac{2\pi}{3})\\ \cos(\Theta-\frac{2\pi}{3}) & \cos(\Theta) & \cos(\Theta+\frac{2\pi}{3})\\ \cos(\Theta+\frac{2\pi}{3}) & \cos(\Theta-\frac{2\pi}{3}) & \cos(\Theta)\end{bmatrix}}_{\boldsymbol{\Gamma}_s(\Theta)}\underbrace{\begin{bmatrix}i_{ra}\\ i_{rb}\\ i_{rc}\end{bmatrix}}_{\boldsymbol{i}_{rabc}} \tag{3.106}$$

for the flux linkages of the stator, and

$$\underbrace{\begin{bmatrix}\Psi_{ra}\\ \Psi_{rb}\\ \Psi_{rc}\end{bmatrix}}_{\boldsymbol{\Psi}_{rabc}} = \underbrace{\begin{bmatrix}L_r & 0 & 0\\ 0 & L_r & 0\\ 0 & 0 & L_r\end{bmatrix}}_{\boldsymbol{L}_r}\underbrace{\begin{bmatrix}i_{ra}\\ i_{rb}\\ i_{rc}\end{bmatrix}}_{\boldsymbol{i}_{rabc}} + L_{msr}\underbrace{\begin{bmatrix}\cos(\Theta) & \cos(\Theta-\frac{2\pi}{3}) & \cos(\Theta+\frac{2\pi}{3})\\ \cos(\Theta+\frac{2\pi}{3}) & \cos(\Theta) & \cos(\Theta-\frac{2\pi}{3})\\ \cos(\Theta-\frac{2\pi}{3}) & \cos(\Theta+\frac{2\pi}{3}) & \cos(\Theta)\end{bmatrix}}_{\boldsymbol{\Gamma}_r(\Theta)}\underbrace{\begin{bmatrix}i_{sa}\\ i_{sb}\\ i_{sc}\end{bmatrix}}_{\boldsymbol{i}_{sabc}} \tag{3.107}$$

for the flux linkage of the rotor. In vectorial form, we arrive at

$$\begin{bmatrix} \dot{\boldsymbol{\Psi}}_{\text{sabc}} \\ \dot{\boldsymbol{\Psi}}_{\text{rabc}} \end{bmatrix} = \begin{bmatrix} \boldsymbol{u}_{\text{sabc}} \\ \boldsymbol{u}_{\text{rabc}} \end{bmatrix} - \begin{bmatrix} \boldsymbol{R}_{\text{s}} & \boldsymbol{0} \\ \boldsymbol{0} & \boldsymbol{R}_{\text{r}} \end{bmatrix} \begin{bmatrix} \boldsymbol{i}_{\text{sabc}} \\ \boldsymbol{i}_{\text{rabc}} \end{bmatrix} \tag{3.108}$$

and

$$\begin{bmatrix} \boldsymbol{\Psi}_{\text{sabc}} \\ \boldsymbol{\Psi}_{\text{rabc}} \end{bmatrix} = \begin{bmatrix} \boldsymbol{L}_{\text{s}} & \boldsymbol{0} \\ \boldsymbol{0} & \boldsymbol{L}_{\text{r}} \end{bmatrix} \begin{bmatrix} \boldsymbol{i}_{\text{sabc}} \\ \boldsymbol{i}_{\text{rabc}} \end{bmatrix} + L_{\text{msr}} \begin{bmatrix} \boldsymbol{0} & \boldsymbol{\Gamma}_{\text{s}}(\Theta) \\ \boldsymbol{\Gamma}_{\text{r}}(\Theta) & \boldsymbol{0} \end{bmatrix} \begin{bmatrix} \boldsymbol{i}_{\text{sabc}} \\ \boldsymbol{i}_{\text{rabc}} \end{bmatrix} \tag{3.109}$$

from equations (3.102), (3.103), (3.106), and (3.107).

In the above formula, we need to know the angle Θ of the rotor position. It can be calculated from the mechanical equation of motion

$$\ddot{\Theta} = \frac{M}{J}, \tag{3.110}$$

where M is the rotor shaft's effective torque, i. e. the difference between the drive torque and the machine torque, and J is the rotation components' moment of inertia. Combined, equations (3.108), (3.109), and (3.110) make up the model of the induction generator.

In equation (3.109), the mutual inductances $L_{\text{msr}}\boldsymbol{\Gamma}_{\text{s}}(\Theta)$ and $L_{\text{msr}}\boldsymbol{\Gamma}_{\text{r}}(\Theta)$ depend on the rotor angle $\Theta(t)$ and are therefore time-dependent. In order to eliminate the flux linkage in equation (3.108) via equation (3.109), it would be necessary to calculate the time derivative of equation (3.109), and thus the time derivatives of the mutual inductances $L_{\text{msr}}\boldsymbol{\Gamma}_{\text{s}}(\Theta)$ and $L_{\text{msr}}\boldsymbol{\Gamma}_{\text{r}}(\Theta)$ as well. This would involve laborious calculations and a very complex system description. Therefore, the system variables are transformed to a coordinate space in which the mutual inductances are constant by applying the Park transformation [254]. The latter is a nonlinear state transformation. The resulting coordinate system, which is referred to as the dq0-coordinate system, rotates around the stator-fixed abc-coordinate system with synchronous frequency ω_{s}. The q-axis precedes the d-axis by 90°. Below only the electric and magnetic state variables $\boldsymbol{u}$, $\boldsymbol{i}$, and $\boldsymbol{\Psi}$ will be transformed, not the mechanical quantities.

The transformation equations can be stated as

$$\begin{aligned} \begin{bmatrix} \boldsymbol{u}_{\text{sdq0}} \\ \boldsymbol{u}_{\text{rdq0}} \end{bmatrix} &= \begin{bmatrix} \boldsymbol{T}(\beta) & \boldsymbol{0} \\ \boldsymbol{0} & \boldsymbol{T}(\varphi) \end{bmatrix} \begin{bmatrix} \boldsymbol{u}_{\text{sabc}} \\ \boldsymbol{u}_{\text{rabc}} \end{bmatrix}, \\ \begin{bmatrix} \boldsymbol{i}_{\text{sdq0}} \\ \boldsymbol{i}_{\text{rdq0}} \end{bmatrix} &= \begin{bmatrix} \boldsymbol{T}(\beta) & \boldsymbol{0} \\ \boldsymbol{0} & \boldsymbol{T}(\varphi) \end{bmatrix} \begin{bmatrix} \boldsymbol{i}_{\text{sabc}} \\ \boldsymbol{i}_{\text{rabc}} \end{bmatrix}, \\ \begin{bmatrix} \boldsymbol{\Psi}_{\text{sdq0}} \\ \boldsymbol{\Psi}_{\text{rdq0}} \end{bmatrix} &= \begin{bmatrix} \boldsymbol{T}(\beta) & \boldsymbol{0} \\ \boldsymbol{0} & \boldsymbol{T}(\varphi) \end{bmatrix} \begin{bmatrix} \boldsymbol{\Psi}_{\text{sabc}} \\ \boldsymbol{\Psi}_{\text{rabc}} \end{bmatrix} \end{aligned} \tag{3.111}$$

with

$$\boldsymbol{T}(\alpha) = \frac{2}{3}\begin{bmatrix} \cos(\alpha) & \cos(\alpha - \frac{2\pi}{3}) & \cos(\alpha + \frac{2\pi}{3}) \\ -\sin(\alpha) & -\sin(\alpha - \frac{2\pi}{3}) & -\sin(\alpha + \frac{2\pi}{3}) \\ \frac{1}{2} & \frac{1}{2} & \frac{1}{2} \end{bmatrix},$$

$$\boldsymbol{T}^{-1}(\alpha) = \begin{bmatrix} \cos(\alpha) & -\sin(\alpha) & 1 \\ \cos(\alpha - \frac{2\pi}{3}) & -\sin(\alpha - \frac{2\pi}{3}) & 1 \\ \cos(\alpha + \frac{2\pi}{3}) & -\sin(\alpha + \frac{2\pi}{3}) & 1 \end{bmatrix}.$$

The transformation (3.111) depends on the angles $\beta(t)$ and $\varphi(t)$, and therefore also on the state variable

$$\Theta = \beta - \varphi.$$

It is thus a transformation consistent with equation (3.91), i. e. a diffeomorphism which nonlinearly depends on the state of the system.

By inserting the transformation equations (3.111) into the system equation (3.108), we arrive at

$$\begin{aligned} \frac{d}{dt}\left(\begin{bmatrix} \boldsymbol{T}^{-1}(\beta) & \boldsymbol{0} \\ \boldsymbol{0} & \boldsymbol{T}^{-1}(\varphi) \end{bmatrix}\begin{bmatrix} \boldsymbol{\Psi}_{\mathrm{sdq0}} \\ \boldsymbol{\Psi}_{\mathrm{rdq0}} \end{bmatrix}\right) &= \begin{bmatrix} \boldsymbol{T}^{-1}(\beta) & \boldsymbol{0} \\ \boldsymbol{0} & \boldsymbol{T}^{-1}(\varphi) \end{bmatrix}\begin{bmatrix} \boldsymbol{u}_{\mathrm{sdq0}} \\ \boldsymbol{u}_{\mathrm{rdq0}} \end{bmatrix} \\ &\quad - \begin{bmatrix} \boldsymbol{R}_{\mathrm{s}} & \boldsymbol{0} \\ \boldsymbol{0} & \boldsymbol{R}_{\mathrm{r}} \end{bmatrix}\begin{bmatrix} \boldsymbol{T}^{-1}(\beta) & \boldsymbol{0} \\ \boldsymbol{0} & \boldsymbol{T}^{-1}(\varphi) \end{bmatrix}\begin{bmatrix} \boldsymbol{i}_{\mathrm{sdq0}} \\ \boldsymbol{i}_{\mathrm{rdq0}} \end{bmatrix}, \end{aligned}$$

from which, with $\boldsymbol{T}\boldsymbol{R}_{\mathrm{s}}\boldsymbol{T}^{-1} = \boldsymbol{R}_{\mathrm{s}}$ and $\boldsymbol{T}\boldsymbol{R}_{\mathrm{r}}\boldsymbol{T}^{-1} = \boldsymbol{R}_{\mathrm{r}}$,

$$\begin{bmatrix} \dot{\boldsymbol{\Psi}}_{\mathrm{sdq0}} \\ \dot{\boldsymbol{\Psi}}_{\mathrm{rdq0}} \end{bmatrix} + \begin{bmatrix} \boldsymbol{T}(\beta) & \boldsymbol{0} \\ \boldsymbol{0} & \boldsymbol{T}(\varphi) \end{bmatrix}\begin{bmatrix} \frac{d\boldsymbol{T}^{-1}(\beta)}{dt} & \boldsymbol{0} \\ \boldsymbol{0} & \frac{d\boldsymbol{T}^{-1}(\varphi)}{dt} \end{bmatrix}\begin{bmatrix} \boldsymbol{\Psi}_{\mathrm{sdq0}} \\ \boldsymbol{\Psi}_{\mathrm{rdq0}} \end{bmatrix} = \begin{bmatrix} \boldsymbol{u}_{\mathrm{sdq0}} \\ \boldsymbol{u}_{\mathrm{rdq0}} \end{bmatrix} - \begin{bmatrix} \boldsymbol{R}_{\mathrm{s}} & \boldsymbol{0} \\ \boldsymbol{0} & \boldsymbol{R}_{\mathrm{r}} \end{bmatrix}\begin{bmatrix} \boldsymbol{i}_{\mathrm{sdq0}} \\ \boldsymbol{i}_{\mathrm{rdq0}} \end{bmatrix} \tag{3.112}$$

follows. Using

$$\frac{d\boldsymbol{T}^{-1}(\alpha)}{dt} = \dot{\alpha}\begin{bmatrix} -\sin(\alpha) & -\cos(\alpha) & 0 \\ -\sin(\alpha - \frac{2\pi}{3}) & -\cos(\alpha - \frac{2\pi}{3}) & 0 \\ -\sin(\alpha + \frac{2\pi}{3}) & -\cos(\alpha + \frac{2\pi}{3}) & 0 \end{bmatrix},$$

we obtain

$$\boldsymbol{T}(\alpha) \cdot \frac{d\boldsymbol{T}^{-1}(\alpha)}{dt} = \dot{\alpha} \underbrace{\begin{bmatrix} 0 & -1 & 0 \\ 1 & 0 & 0 \\ 0 & 0 & 0 \end{bmatrix}}_{\boldsymbol{K}}.$$

With $\dot{\beta} = \omega_{\mathrm{s}}$ and $\dot{\varphi} = s\omega_{\mathrm{s}}$, equation (3.112) leads to

$$\begin{bmatrix} \dot{\boldsymbol{\Psi}}_{\mathrm{sdq0}} \\ \dot{\boldsymbol{\Psi}}_{\mathrm{rdq0}} \end{bmatrix} + \begin{bmatrix} \omega_{\mathrm{s}}\boldsymbol{K} & \boldsymbol{0} \\ \boldsymbol{0} & s\omega_{\mathrm{s}}\boldsymbol{K} \end{bmatrix} \begin{bmatrix} \boldsymbol{\Psi}_{\mathrm{sdq0}} \\ \boldsymbol{\Psi}_{\mathrm{rdq0}} \end{bmatrix} = \begin{bmatrix} \boldsymbol{u}_{\mathrm{sdq0}} \\ \boldsymbol{u}_{\mathrm{rdq0}} \end{bmatrix} - \begin{bmatrix} \boldsymbol{R}_{\mathrm{s}} & \boldsymbol{0} \\ \boldsymbol{0} & \boldsymbol{R}_{\mathrm{r}} \end{bmatrix} \begin{bmatrix} \boldsymbol{i}_{\mathrm{sdq0}} \\ \boldsymbol{i}_{\mathrm{rdq0}} \end{bmatrix}. \tag{3.113}$$

This is independent of the angle Θ.

The algebraic equation (3.109) still has to be transformed via equation (3.111), a calculation which yields

$$\begin{bmatrix} \boldsymbol{\Psi}_{\mathrm{sdq0}} \\ \boldsymbol{\Psi}_{\mathrm{rdq0}} \end{bmatrix} = \begin{bmatrix} \boldsymbol{L}_{\mathrm{s}} & \boldsymbol{0} \\ \boldsymbol{0} & \boldsymbol{L}_{\mathrm{r}} \end{bmatrix} \begin{bmatrix} \boldsymbol{i}_{\mathrm{sdq0}} \\ \boldsymbol{i}_{\mathrm{rdq0}} \end{bmatrix} + L_{\mathrm{msr}} \begin{bmatrix} \boldsymbol{T}(\beta) & \boldsymbol{0} \\ \boldsymbol{0} & \boldsymbol{T}(\varphi) \end{bmatrix} \begin{bmatrix} \boldsymbol{0} & \boldsymbol{\Gamma}_{\mathrm{s}}(\Theta) \\ \boldsymbol{\Gamma}_{\mathrm{r}}(\Theta) & \boldsymbol{0} \end{bmatrix} \begin{bmatrix} \boldsymbol{T}^{-1}(\beta) & \boldsymbol{0} \\ \boldsymbol{0} & \boldsymbol{T}^{-1}(\varphi) \end{bmatrix} \begin{bmatrix} \boldsymbol{i}_{\mathrm{sdq0}} \\ \boldsymbol{i}_{\mathrm{rdq0}} \end{bmatrix}. \tag{3.114}$$

We now determine

$$\begin{bmatrix} \boldsymbol{T}(\beta) & \boldsymbol{0} \\ \boldsymbol{0} & \boldsymbol{T}(\varphi) \end{bmatrix} \begin{bmatrix} \boldsymbol{0} & \boldsymbol{\Gamma}_{\mathrm{s}}(\Theta) \\ \boldsymbol{\Gamma}_{\mathrm{r}}(\Theta) & \boldsymbol{0} \end{bmatrix} \begin{bmatrix} \boldsymbol{T}^{-1}(\beta) & \boldsymbol{0} \\ \boldsymbol{0} & \boldsymbol{T}^{-1}(\varphi) \end{bmatrix} = \frac{3}{2} \begin{bmatrix} \boldsymbol{0} & \boldsymbol{U} \\ \boldsymbol{U} & \boldsymbol{0} \end{bmatrix}$$

with

$$\boldsymbol{U} = \begin{bmatrix} 1 & 0 & 0 \\ 0 & 1 & 0 \\ 0 & 0 & 0 \end{bmatrix}.$$

Using this result, and the abbreviation $L_{\mathrm{m}} = 3L_{\mathrm{msr}}/2$, we then simplify equation (3.114) in order to get the transformed equation

$$\begin{bmatrix} \boldsymbol{\Psi}_{\mathrm{sdq0}} \\ \boldsymbol{\Psi}_{\mathrm{rdq0}} \end{bmatrix} = \begin{bmatrix} \boldsymbol{L}_{\mathrm{s}} & \boldsymbol{0} \\ \boldsymbol{0} & \boldsymbol{L}_{\mathrm{r}} \end{bmatrix} \begin{bmatrix} \boldsymbol{i}_{\mathrm{sdq0}} \\ \boldsymbol{i}_{\mathrm{rdq0}} \end{bmatrix} + L_{\mathrm{m}} \begin{bmatrix} \boldsymbol{0} & \boldsymbol{U} \\ \boldsymbol{U} & \boldsymbol{0} \end{bmatrix} \begin{bmatrix} \boldsymbol{i}_{\mathrm{sdq0}} \\ \boldsymbol{i}_{\mathrm{rdq0}} \end{bmatrix}, \tag{3.115}$$

i. e. an equation independent of the mechanical rotation angle Θ. The equations (3.113) and (3.115) represent the electrical system's equations for the doubly fed induction machine.

Having arrived at the transformed system equations (3.113) and (3.115), we have actually reached our goal of making the mutual inductivities of the original system equations (3.108) and (3.109) independent of the rotation angle Θ by applying the Park transformation (3.111). In addition, we will now insert equation (3.115) into equation (3.113) to eliminate the magnetic flux linkage. With

$$\boldsymbol{K}\boldsymbol{U} = \boldsymbol{K},$$

we achieve

$$\begin{bmatrix} \boldsymbol{L}_{\mathrm{s}} & L_{\mathrm{m}}\boldsymbol{U} \\ L_{\mathrm{m}}\boldsymbol{U} & \boldsymbol{L}_{\mathrm{r}} \end{bmatrix} \begin{bmatrix} \dot{\boldsymbol{i}}_{\mathrm{sdq0}} \\ \dot{\boldsymbol{i}}_{\mathrm{rdq0}} \end{bmatrix} + \begin{bmatrix} \omega_{\mathrm{s}}\boldsymbol{K}\boldsymbol{L}_{\mathrm{s}} + \boldsymbol{R}_{\mathrm{s}} & \omega_{\mathrm{s}}L_{\mathrm{m}}\boldsymbol{K} \\ s\omega_{\mathrm{s}}L_{\mathrm{m}}\boldsymbol{K} & s\omega_{\mathrm{s}}\boldsymbol{K}\boldsymbol{L}_{\mathrm{r}} + \boldsymbol{R}_{\mathrm{r}} \end{bmatrix} \begin{bmatrix} \boldsymbol{i}_{\mathrm{sdq0}} \\ \boldsymbol{i}_{\mathrm{rdq0}} \end{bmatrix} = \begin{bmatrix} \boldsymbol{u}_{\mathrm{sdq0}} \\ \boldsymbol{u}_{\mathrm{rdq0}} \end{bmatrix}$$

for the electrical equations of the doubly fed induction machine, which are independent of the flux linkage. For the mechanical equation of the machine, equation (3.110), we obtain

$$\ddot{\Theta} = -\dot{s}\omega_{\mathrm{s}} = \frac{M}{J} \qquad \text{or} \qquad \dot{s} = -\frac{\ddot{\Theta}}{\omega_{\mathrm{s}}} = -\frac{M}{\omega_{\mathrm{s}}J},$$

assuming a constant stator frequency ω_{s} in equation (3.101). Similarly, for synchronous machines, the Park transformation allows for a simplification of the system equations, i. e. an independence of the mutual inductances relative to the rotor position.

3.3.4 Determining the Transformation Rule

So far, we have assumed that the transformation rules

$$\boldsymbol{z} = \boldsymbol{q}(\boldsymbol{x}) = \boldsymbol{p}^{-1}(\boldsymbol{x}) \quad \text{and} \quad \boldsymbol{x} = \boldsymbol{q}^{-1}(\boldsymbol{z}) = \boldsymbol{p}(\boldsymbol{z}) \tag{3.116}$$

which transform a system

$$\dot{\boldsymbol{x}} = \boldsymbol{f}(\boldsymbol{x}, \boldsymbol{u}) \tag{3.117}$$

into the desired form

$$\dot{\boldsymbol{z}} = \hat{\boldsymbol{f}}(\boldsymbol{z}, \boldsymbol{u}) \tag{3.118}$$

and back again are known. However, generally this is not the case. In fact the desired system representation (3.118) is usually given, and we need to determine the corresponding transformation equation, the diffeomorphism (3.116).

For this purpose, we first insert equation (3.116) into equation (3.117) which allows us to obtain

$$\dot{\boldsymbol{z}} = \left(\frac{\partial \boldsymbol{p}(\boldsymbol{z})}{\partial \boldsymbol{z}}\right)^{-1} \cdot \boldsymbol{f}(\boldsymbol{p}(\boldsymbol{z}), \boldsymbol{u}). \tag{3.119}$$

Next, we equate (3.119) with the desired form (3.118) of the system, yielding the conditional equation

$$\frac{\partial \boldsymbol{p}(\boldsymbol{z})}{\partial \boldsymbol{z}} \cdot \hat{\boldsymbol{f}}(\boldsymbol{z}, \boldsymbol{u}) - \boldsymbol{f}(\boldsymbol{p}(\boldsymbol{z}), \boldsymbol{u}) = \boldsymbol{0} \tag{3.120}$$

for the diffeomorphism $\boldsymbol{x} = \boldsymbol{p}(\boldsymbol{z})$ which we have been seeking.

This first-order partial differential equation, however, is only analytically solvable in a few cases, which largely restricts its practical use. Nevertheless, this result is helpful, since it demonstrates the difficulty of finding suitable nonlinear state transformations. Later, in Section 5.2, we will see that, at least for the class of control-affine systems

$$\dot{\boldsymbol{x}} = \boldsymbol{a}(\boldsymbol{x}) + \boldsymbol{B}(\boldsymbol{x}) \cdot \boldsymbol{u},$$

these transformations can often be determined.

3.3.5 Illustration Using Linear Systems

As a simple example of an application of equation (3.120), we will take a linear system

$$\dot{\boldsymbol{x}} = \boldsymbol{A}\boldsymbol{x} + \boldsymbol{b} \cdot u, \tag{3.121}$$

which is to be transformed into the desired form

$$\dot{\boldsymbol{z}} = \hat{\boldsymbol{A}}\boldsymbol{z} + \hat{\boldsymbol{b}} \cdot u. \tag{3.122}$$

Using equation (3.120), the required transformation rule is determined from

$$\frac{\partial \boldsymbol{p}(\boldsymbol{z})}{\partial \boldsymbol{z}}(\hat{\boldsymbol{A}}\boldsymbol{z} + \hat{\boldsymbol{b}} \cdot u) - (\boldsymbol{A}\boldsymbol{p}(\boldsymbol{z}) + \boldsymbol{b} \cdot u) = \boldsymbol{0}.$$

It follows that

$$\frac{\partial \boldsymbol{p}(\boldsymbol{z})}{\partial \boldsymbol{z}}\hat{\boldsymbol{A}}\boldsymbol{z} - \boldsymbol{A}\boldsymbol{p}(\boldsymbol{z}) + \left(\frac{\partial \boldsymbol{p}(\boldsymbol{z})}{\partial \boldsymbol{z}}\hat{\boldsymbol{b}} - \boldsymbol{b}\right)u = \boldsymbol{0}.$$

Since the above equation must hold for all $\boldsymbol{z} \in I\!\!R^n$ and $u \in I\!\!R$, we obtain

$$\begin{aligned} \frac{\partial \boldsymbol{p}(\boldsymbol{z})}{\partial \boldsymbol{z}}\hat{\boldsymbol{A}}\boldsymbol{z} - \boldsymbol{A}\boldsymbol{p}(\boldsymbol{z}) &= \boldsymbol{0}, \\ \frac{\partial \boldsymbol{p}(\boldsymbol{z})}{\partial \boldsymbol{z}}\hat{\boldsymbol{b}} - \boldsymbol{b} &= \boldsymbol{0} \end{aligned} \tag{3.123}$$

from the previous equation. Our approach to solving this partial differential equation utilizes the linear transformation (3.88), i. e.

$$\boldsymbol{x} = \boldsymbol{T}\boldsymbol{z} = \boldsymbol{p}(\boldsymbol{z}) \tag{3.124}$$

with the $n \times n$ matrix $\boldsymbol{T}$ which was previously used in Section 3.3.1. Inserting this transformation into equation (3.123) yields

$$\begin{aligned} \frac{\partial \boldsymbol{T}\boldsymbol{z}}{\partial \boldsymbol{z}}\hat{\boldsymbol{A}}\boldsymbol{z} - \boldsymbol{A}\boldsymbol{T}\boldsymbol{z} = \boldsymbol{0} \quad &\Leftrightarrow \quad \boldsymbol{T}\hat{\boldsymbol{A}} - \boldsymbol{A}\boldsymbol{T} = \boldsymbol{0}, \\ \frac{\partial \boldsymbol{T}\boldsymbol{z}}{\partial \boldsymbol{z}}\hat{\boldsymbol{b}} - \boldsymbol{b} = \boldsymbol{0} \quad &\Leftrightarrow \quad \boldsymbol{T}\hat{\boldsymbol{b}} - \boldsymbol{b} = \boldsymbol{0}. \end{aligned} \tag{3.125}$$

We note that, according to Theorem 55, the Jacobian matrix

$$\frac{\partial \boldsymbol{p}(\boldsymbol{z})}{\partial \boldsymbol{z}} = \boldsymbol{T}$$

must be regular. The known result from linear systems theory follows as

$$\begin{aligned} \hat{\boldsymbol{A}} &= \boldsymbol{T}^{-1}\boldsymbol{A}\boldsymbol{T}, \\ \hat{\boldsymbol{b}} &= \boldsymbol{T}^{-1}\boldsymbol{b}. \end{aligned} \tag{3.126}$$

From equation (3.126) we can conclude that the matrix $\hat{\boldsymbol{A}}$ of the desired system (3.122) must possess the same eigenvalues as the system matrix $\boldsymbol{A}$. If this is not the case, no diffeomorphism (3.124) exists to transform the system (3.121) into the form of equation (3.122).

It remains to be determined how the matrix $\boldsymbol{T}$ can be computed from $\boldsymbol{A}$ and $\hat{\boldsymbol{A}}$, or, more precisely, from equation (3.125),

$$\boldsymbol{T}\hat{\boldsymbol{A}} - \boldsymbol{A}\boldsymbol{T} = \boldsymbol{0}. \tag{3.127}$$

First, note that the matrix $\boldsymbol{T}$ is regular if and only if $\boldsymbol{A}$ and $\hat{\boldsymbol{A}}$ can be transformed to the same Jordan canonical form $\boldsymbol{J}$ [258], meaning that the matrices $\boldsymbol{A}$ and $\hat{\boldsymbol{A}}$ must have the same eigenvalues with the same associated algebraic and geometric multiplicities[1]. The matrices $\boldsymbol{V}$ and $\hat{\boldsymbol{V}}$ transform the matrices $\boldsymbol{A}$ and $\hat{\boldsymbol{A}}$ into the Jordan canonical form $\boldsymbol{J}$ according to

$$\boldsymbol{V}^{-1}\boldsymbol{A}\boldsymbol{V} = \boldsymbol{J},$$
$$\hat{\boldsymbol{V}}^{-1}\hat{\boldsymbol{A}}\hat{\boldsymbol{V}} = \boldsymbol{J}.$$

The solution of equation (3.127) is [144, 258]

$$\boldsymbol{T} = \boldsymbol{V}\boldsymbol{K}\hat{\boldsymbol{V}}^{-1}, \tag{3.128}$$

where $\boldsymbol{K}$ is a block diagonal matrix. Let the matrices $\boldsymbol{A}$ and $\hat{\boldsymbol{A}}$ possess the eigenvalues $\lambda_1, \ldots, \lambda_k$. Each eigenvalue λ_i is of algebraic multiplicity r_i, and of geometric multiplicity s_i. Then the associated Jordan canonical form [155] is consistent with the block diagonal matrix

$$\boldsymbol{J} = \begin{bmatrix} \boldsymbol{J}_{1,1} \cdots & \boldsymbol{0} & \boldsymbol{0} & \cdots & \boldsymbol{0} & \cdots & \boldsymbol{0} & \cdots & \boldsymbol{0} \\ \vdots & \ddots & \vdots & \vdots & \ddots & \vdots & \ddots & \vdots & \ddots & \vdots \\ \boldsymbol{0} & \cdots \boldsymbol{J}_{1,s_1} & \boldsymbol{0} & \cdots & \boldsymbol{0} & \cdots & \boldsymbol{0} & \cdots & \boldsymbol{0} \\ \boldsymbol{0} & \cdots & \boldsymbol{0} & \boldsymbol{J}_{2,1} \cdots & \boldsymbol{0} & \cdots & \boldsymbol{0} & \cdots & \boldsymbol{0} \\ \vdots & \ddots & \vdots & \vdots & \ddots & \vdots & \ddots & \vdots & \ddots & \vdots \\ \boldsymbol{0} & \cdots & \boldsymbol{0} & \boldsymbol{0} & \cdots \boldsymbol{J}_{2,s_2} \cdots & \boldsymbol{0} & \cdots & \boldsymbol{0} \\ \vdots & \ddots & \vdots & \vdots & \ddots & \vdots & \ddots & \vdots & \ddots & \vdots \\ \boldsymbol{0} & \cdots & \boldsymbol{0} & \boldsymbol{0} & \cdots & \boldsymbol{0} & \cdots \boldsymbol{J}_{k,1} \cdots & \boldsymbol{0} \\ \vdots & \ddots & \vdots & \vdots & \ddots & \vdots & \ddots & \vdots & \ddots & \vdots \\ \boldsymbol{0} & \cdots & \boldsymbol{0} & \boldsymbol{0} & \cdots & \boldsymbol{0} & \cdots & \boldsymbol{0} & \cdots \boldsymbol{J}_{k,s_k} \end{bmatrix}, \text{ where } \boldsymbol{J}_{i,j} = \begin{bmatrix} \lambda_i & 1 & 0 & \cdots & 0 & 0 \\ 0 & \lambda_i & 1 & \cdots & 0 & 0 \\ \vdots & \vdots & \vdots & \ddots & \vdots & \vdots \\ 0 & 0 & 0 & \cdots & \lambda_i & 1 \\ 0 & 0 & 0 & \cdots & 0 & \lambda_i \end{bmatrix},$$

and where the matrices $\boldsymbol{J}_{i,j}$ are $p_{ij} \times p_{ij}$ matrices. It holds that

[1] The algebraic multiplicity r_i of an eigenvalue λ_i of matrix $\boldsymbol{X}$ determines the number of linear factors $(s - \lambda_i)$ of the characteristic polynomial of $\boldsymbol{X}$. The number of linearly independent eigenvectors of an eigenvalue λ_i corresponds to its geometric multiplicity s_i.

$$1 \leq p_{ij} \leq r_i \quad \text{and} \quad p_{i1} + p_{i2} + \ldots + p_{is_i} = r_i.$$

The matrix $\boldsymbol{K}$ in equation (3.128) is of the particular block diagonal form

$$\boldsymbol{K} = \begin{bmatrix} \boldsymbol{K}_{1,1} \cdots & \mathbf{0} & \mathbf{0} & \cdots & \mathbf{0} & \cdots & \mathbf{0} & \cdots & \mathbf{0} \\ \vdots & \ddots & \vdots & \vdots & \ddots & \vdots & \ddots & \vdots & \ddots & \vdots \\ \mathbf{0} & \cdots \boldsymbol{K}_{1,s_1} & \mathbf{0} & \cdots & \mathbf{0} & \cdots & \mathbf{0} & \cdots & \mathbf{0} \\ \mathbf{0} & \cdots & \mathbf{0} & \boldsymbol{K}_{2,1} \cdots & \mathbf{0} & \cdots & \mathbf{0} & \cdots & \mathbf{0} \\ \vdots & \ddots & \vdots & \vdots & \ddots & \vdots & \ddots & \vdots & \ddots & \vdots \\ \mathbf{0} & \cdots & \mathbf{0} & \mathbf{0} & \cdots \boldsymbol{K}_{2,s_2} \cdots & \mathbf{0} & \cdots & \mathbf{0} \\ \vdots & \ddots & \vdots & \vdots & \ddots & \vdots & \ddots & \vdots & \ddots & \vdots \\ \mathbf{0} & \cdots & \mathbf{0} & \mathbf{0} & \cdots & \mathbf{0} & \cdots \boldsymbol{K}_{k,1} \cdots & \mathbf{0} \\ \vdots & \ddots & \vdots & \vdots & \ddots & \vdots & \ddots & \vdots & \ddots & \vdots \\ \mathbf{0} & \cdots & \mathbf{0} & \mathbf{0} & \cdots & \mathbf{0} & \cdots & \mathbf{0} & \cdots \boldsymbol{K}_{k,s_k} \end{bmatrix}$$

with the submatrices

$$\boldsymbol{K}_{i,j} = \begin{bmatrix} k_1(i,j) & k_2(i,j) & k_3(i,j) & \cdots & k_{p_{ij}}(i,j) \\ 0 & k_1(i,j) & k_2(i,j) & \cdots & k_{p_{ij}-1}(i,j) \\ \vdots & \vdots & \vdots & \ddots & \vdots \\ 0 & 0 & 0 & \cdots & k_2(i,j) \\ 0 & 0 & 0 & \cdots & k_1(i,j) \end{bmatrix}.$$

The submatrices $\boldsymbol{K}_{i,j}$ are upper $p_{ij} \times p_{ij}$ triangular Toeplitz matrices with p_{ij} arbitrary parameters $k_l(i,j)$, where $l = 1, \ldots, p_{ij}$. Because of the freely selectable parameters, the solution (3.128) is not unique. Hence, an infinite number of transformations exist which transform the system $\dot{\boldsymbol{x}} = \boldsymbol{A}\boldsymbol{x} + \boldsymbol{b}u$ into a form consistent with $\dot{\boldsymbol{z}} = \hat{\boldsymbol{A}}\boldsymbol{z} + \hat{\boldsymbol{b}}u$.

Where $\boldsymbol{A}$ and $\hat{\boldsymbol{A}}$ possess exactly n different eigenvalues, the Jordan canonical form is diagonal. This means the matrix $\boldsymbol{K}$ is a diagonal matrix and the solution (3.128) of the conditional equation (3.127) can be easily determined from $\boldsymbol{T}$. In this case, the matrices $\boldsymbol{V}$ and $\hat{\boldsymbol{V}}$ are made up of the eigenvectors of the matrices $\boldsymbol{A}$ and $\hat{\boldsymbol{A}}$.

Based on this example of a linear system, it seems plausible that for nonlinear systems as well, the choice of the transformed system, i. e. the system $\dot{\boldsymbol{z}} = \hat{\boldsymbol{f}}(\boldsymbol{z}, \boldsymbol{u})$ into which we wish to transform the original system $\dot{\boldsymbol{x}} = \boldsymbol{f}(\boldsymbol{x}, \boldsymbol{u})$, is severely limited. (In the linear case described above, this limitation is due to the requirement of identical eigenvalues and $\hat{\boldsymbol{A}}$ and $\boldsymbol{A}$ having an identical Jordan canonical form.) Consequently, a diffeomorphism does not exist for any choice of $\dot{\boldsymbol{z}} = \hat{\boldsymbol{f}}(\boldsymbol{z}, \boldsymbol{u})$; and, even if it did exist, determining it could become very involved.

3.4 Exercises

Exercise 3.1 Analyze to what extent the system $\dot{x} = xu$ is controllable.

Exercise 3.2 Explain why the system

$$\dot{\boldsymbol{x}} = \begin{bmatrix} x_2 \\ \alpha(\boldsymbol{x}) \\ x_3 \end{bmatrix} + \begin{bmatrix} 0 \\ \beta(\boldsymbol{x}) \\ 0 \end{bmatrix} u$$

is not controllable.

Exercise 3.3 Let us examine the service ship from Section 3.1.1, p. 211, in the Cartesian coordinate system, shown in Figure 3.24.

(a) Create the state-space model of the ship with the position coordinates x_1 and x_2 and the course angle x_3 as state variables. The input variables are the ship speeds v_1 and v_2 in the longitudinal and sideways directions, respectively, and the change in time v_3 of the course angle x_3.

(b) Show that the model from (a) is omnidirectionally controllable and flat.

(c) The speed vector $\boldsymbol{v} = \begin{bmatrix} v_1 \; v_2 \; v_3 \end{bmatrix}^T$ results from the rotational speeds of the transverse thrusters u_2 and u_3 in the bow and stern and the rotational speed u_1 of the ship's propeller. This relationship can be modeled by

$$\boldsymbol{M}\dot{\boldsymbol{v}} + \boldsymbol{D}\boldsymbol{v} = \boldsymbol{u}, \qquad \boldsymbol{u} = \begin{bmatrix} u_1 \; u_2 \; u_3 \end{bmatrix}^T. \tag{3.129}$$

Determine the overall model with $[\boldsymbol{x}^T \; \boldsymbol{v}^T]^T$ as the state vector resulting from the model from (a) above and equation (3.129). Examine whether

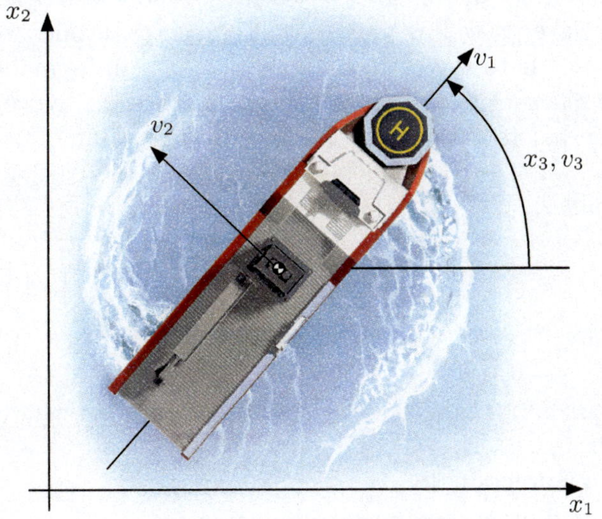

Fig. 3.24: State and input variables of the service ship model

the overall model is omnidirectionally controllable, only controllable, or neither of the above. The matrix $\boldsymbol{M}$ is regular.

(d) Give a flat representation of the overall system.

Exercise 3.4 Given the differentiable vector functions $\boldsymbol{g}_1(\boldsymbol{x}), \boldsymbol{g}_2(\boldsymbol{x}), \boldsymbol{g}_3(\boldsymbol{x})$, and $\boldsymbol{g}_4(\boldsymbol{x})$, prove

(a) $[\boldsymbol{g}_1, \boldsymbol{g}_1] = 0$,

(b) $[\boldsymbol{g}_1, \boldsymbol{g}_2] = -[\boldsymbol{g}_2, \boldsymbol{g}_1]$ (antisymmetry),

(c) $[\boldsymbol{g}_1, \boldsymbol{g}_2] = -[-\boldsymbol{g}_1, \boldsymbol{g}_2] = -[\boldsymbol{g}_1, -\boldsymbol{g}_2]$,

(d) $[\boldsymbol{g}_1, c_1\boldsymbol{g}_2 + c_2\boldsymbol{g}_3] = c_1[\boldsymbol{g}_1, \boldsymbol{g}_2] + c_2[\boldsymbol{g}_1, \boldsymbol{g}_3], \qquad c_1, c_2 \in I\!R$,

(e) $[[\boldsymbol{g}_1, \boldsymbol{g}_2], [\boldsymbol{g}_3, \boldsymbol{g}_4]] = [\boldsymbol{g}_1, [\boldsymbol{g}_2, [\boldsymbol{g}_3, \boldsymbol{g}_4]]] - [\boldsymbol{g}_2, [\boldsymbol{g}_1, [\boldsymbol{g}_3, \boldsymbol{g}_4]]]$; use the Jacobi identity $[\boldsymbol{g}_1, [\boldsymbol{g}_2, \boldsymbol{g}_3]] + [\boldsymbol{g}_3, [\boldsymbol{g}_1, \boldsymbol{g}_2]] + [\boldsymbol{g}_2, [\boldsymbol{g}_3, \boldsymbol{g}_1]] = 0$,

(f) $[\boldsymbol{g}_1, [\boldsymbol{g}_2, [\boldsymbol{g}_2, \boldsymbol{g}_1]]] = [\boldsymbol{g}_2, [\boldsymbol{g}_1, [\boldsymbol{g}_2, \boldsymbol{g}_1]]]$,

(g)
$$\begin{aligned}[\lambda(\boldsymbol{x})\boldsymbol{g}_1, \mu(\boldsymbol{x})\boldsymbol{g}_2] =& \lambda(\boldsymbol{x})\mu(\boldsymbol{x})[\boldsymbol{g}_1, \boldsymbol{g}_2] + \lambda(\boldsymbol{x})\left(\frac{\partial \mu(\boldsymbol{x})}{\partial \boldsymbol{x}}\boldsymbol{g}_1(\boldsymbol{x})\right)\boldsymbol{g}_2(\boldsymbol{x}) \\ &- \mu(\boldsymbol{x})\left(\frac{\partial \lambda(\boldsymbol{x})}{\partial \boldsymbol{x}}\boldsymbol{g}_2(\boldsymbol{x})\right)\boldsymbol{g}_1\end{aligned}$$

with $\lambda(\boldsymbol{x})$ and $\mu(\boldsymbol{x})$ being differentiable scalar functions.

Exercise 3.5 A polymer electrolyte membrane (PEM) fuel cell has two electrodes which are separated by a polymer membrane; see Figure 3.25. At the anode, the hydrogen molecules H_2 being supplied split into two protons H^+, setting two electrons free. The protons diffuse through the membrane and at the cathode they encounter the oxygen molecules O_2 being supplied to that point, where they bind together into water molecules H_2O. In the course of this, each water molecule binds two electrons. Before this reaction takes place, the electrons have passed as an external current i through the load and the line from the anode to the cathode. This process can be modeled [328] as below using the relative gas pressures x_1, x_2, and x_3 of the hydrogen, the oxygen, and the steam as state variables, along with the volume flows u_1 and u_2 of the hydrogen and the oxygen being supplied and the electric current $i = u_3$ as input variables:

$$\begin{aligned}\dot{x}_1 &= a(p - x_1)u_1 - 2ac(p - x_1)u_3, \\ \dot{x}_2 &= b(p - x_2)u_2 - bc(p - x_2)u_3, \\ \dot{x}_3 &= -bx_3u_2 + 2bc(p - x_3)u_3.\end{aligned}$$

The parameters a, b, c, and p are constants. Determine whether the model of the fuel cell is

(a) locally omnidirectionally controllable,

(b) globally omnidirectionally controllable on $I\!R^3$,

(c) locally small-time locally controllable,

(d) globally small-time locally controllable on $I\!R^3$.

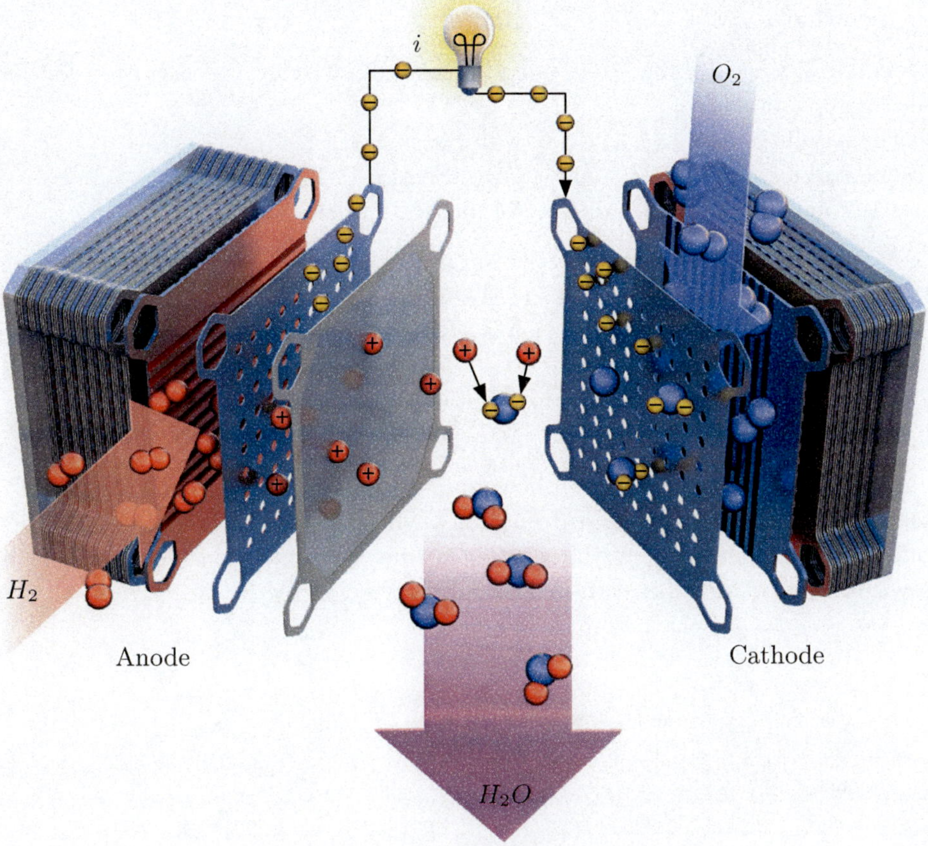

Fig. 3.25: Polymer electrolyte membrane fuel cell

Exercise 3.6 Let us consider the system

$$\begin{aligned}\dot{x}_1 &= f_1(x_1) + h_1(x_1)x_2\\ \dot{x}_2 &= f_2(x_1,x_2) + h_2(x_1,x_2)x_3\\ \dot{x}_3 &= f_3(x_1,x_2,x_3) + h_3(x_1,x_2,x_3)u.\end{aligned}$$

(a) Identify the diffeomorphism which transforms the system into the nonlinear controller canonical form.
(b) Identify the vector $\boldsymbol{b}(\boldsymbol{z})$ of the transformed system $\dot{\boldsymbol{z}} = \boldsymbol{a}(\boldsymbol{z}) + \boldsymbol{b}(\boldsymbol{z})u$, i.e. of the nonlinear controller canonical form for the case in question.
(c) Is the system always controllable?

Exercise 3.7 Let us consider the linear system

$$\dot{\boldsymbol{x}} = \begin{bmatrix} 0 & 1\\ -1 & 0\end{bmatrix}\boldsymbol{x} + \begin{bmatrix}0\\1\end{bmatrix}u.$$

(a) Identify a flat output and the corresponding flat system representation.
(b) Determine the input signal $u(t)$ of a feedforward control for the system such that $y = 1 - e^{-t} - te^{-t}$ applies.

Exercise 3.8 Determine a flat output and the flat system description for the following systems:
(a) the satellite model (1.8) on p. 8 in Section 1.1.4,
(b) the robot model (3.22) on p. 224 in Section 3.1.5, and
(c) the robot model (3.32) on p. 234 in Section 3.1.7.

Exercise 3.9 Let us examine the separately excited direct-current motor (2.119) on p. 183 in Section 2.4.11. We will assume that the load torque M_{L} is constant. In this case the motor has only the two input variables u_{r} and u_{s}.
(a) Identify a flat output.
(b) Determine the flat system representation.

Exercise 3.10 Let us examine an active suspension system of the type used in automobiles, among other applications. Because the suspension and shock-absorbing components of an automobile's wheels have an identical construction, it is sufficient to view the suspension mechanism of only one wheel, as shown in Figure 3.26. Its state-space model is given by

$$
\begin{aligned}
\dot{x}_1 &= x_2,\\
\dot{x}_2 &= -\frac{1}{m_{\mathrm{c}}}\left[F_{\mathrm{s}}(x_1,x_3) + F_{\mathrm{d}}(x_2,x_4)\right] + \frac{1}{m_{\mathrm{c}}}u,\\
\dot{x}_3 &= x_4,\\
\dot{x}_4 &= -\frac{c_{\mathrm{t}}}{m_{\mathrm{s}}}x_3 + \frac{1}{m_{\mathrm{s}}}\left[F_{\mathrm{s}}(x_1,x_3) + F_{\mathrm{d}}(x_2,x_4)\right] - \frac{1}{m_{\mathrm{s}}}u + \frac{c_{\mathrm{t}}}{m_{\mathrm{s}}}x_{\mathrm{d}}
\end{aligned}
$$

with the spring and damping forces as follows:

$$
\begin{aligned}
F_{\mathrm{s}}(x_1,x_3) &= c_{\mathrm{s}}(x_1 - x_3) + c_{\mathrm{nls}}(x_1 - x_3)^3,\\
F_{\mathrm{d}}(x_2,x_4) &= d(x_2 - x_4) + d_{\mathrm{nl}}(x_2 - x_4)^2\,\mathrm{sgn}(x_2 - x_4).
\end{aligned}
$$

Here x_1 and x_3 are the spring travels and x_2 and x_4 are the corresponding speeds. In addition, m_{c} is one-fourth of the mass of the automobile without the mass of the mounting rod and wheel; m_{s} is the mass of the mounting rod including the wheel; c_{s}, c_{nls}, and c_{t} are the spring constants; and d and d_{nl} are the damping coefficients. An additional force u introduced into the system serves as the control variable. Assume that the unevenness of the road, which is represented by x_{d}, is measured by a radar sensor, and x_{d} is therefore known.
(a) Demonstrate that $y = m_{\mathrm{c}}x_1 + m_{\mathrm{s}}x_3$ is a flat output and give the flat system description.
(b) Is the system controllable and if so, why?
(c) Now introduce the flat output and its derivatives as the new coordinates

$$
\boldsymbol{z}^T = \left[y\ \dot{y}\ \ddot{y}\ \dddot{y}\right].
$$

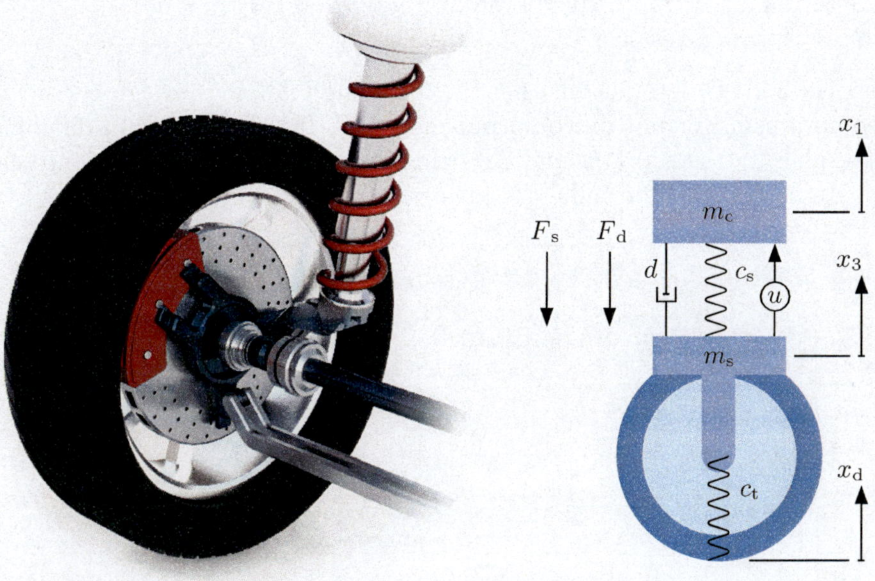

Fig. 3.26: Active suspension of a wheel

Formulate $\boldsymbol{z} = \boldsymbol{t}(\boldsymbol{x})$ and calculate $\dot{\boldsymbol{z}} = \dot{\boldsymbol{t}}(\boldsymbol{x})$. Here, replace $\boldsymbol{x}$ with the new coordinates $\boldsymbol{z}$. Which system representation $\dot{\boldsymbol{z}} = \boldsymbol{f}(\boldsymbol{z})$ results, and what is it called?

Exercise 3.11 Determine the inverse function of

$$\boldsymbol{z} = \boldsymbol{q}(\boldsymbol{x}) = \begin{bmatrix} e^{x_1}\cos(x_2) \\ e^{x_1}\sin(x_2) \end{bmatrix}.$$

Exercise 3.12 Let the differential equation

$$\dot{x} = x - x^2$$

with $x_0 = x(0)$ be given.

(a) Identify the transformation $x = p(z)$ which transforms the above differential equation into the linear differential equation $\dot{z} = -z + 1$, $z_0 = z(0)$.

(b) State the solution of the differential equation $\dot{x} = x - x^2$ for $x_0 = x(0)$.

Exercise 3.13 Let us examine the system

$$\dot{x}_1 = \sin(x_2),$$
$$\dot{x}_2 = -x_1 + u.$$

(a) Calculate the inverse $\boldsymbol{x} = \boldsymbol{q}^{-1}(\boldsymbol{z})$ of the diffeomorphism $\boldsymbol{z} = \boldsymbol{q}(\boldsymbol{x})$ where $z_1 = x_1$ and $z_2 = \sin(x_2)$.
(b) Determine the transformed system description $\dot{\boldsymbol{z}} = \boldsymbol{f}(\boldsymbol{z}, u)$.
(c) What form does $\dot{\boldsymbol{z}} = \boldsymbol{f}(\boldsymbol{z}, u)$ take?

Exercise 3.14 Determine whether the following mappings are global or local diffeomorphisms, or not diffeomorphisms at all. If there is a local diffeomorphism, identify the sets $M_{\boldsymbol{x}}$ and $M_{\boldsymbol{z}}$ which are bijectively mapped onto each other. In each case, determine the inverse mapping if possible.

(a) $x = z^3$,

(b) $\boldsymbol{x} = \boldsymbol{T}\boldsymbol{z} + \boldsymbol{a}, \quad \boldsymbol{T} \in I\!R^{n\times n}, \boldsymbol{a} \in I\!R^n$,

(c) $\boldsymbol{x} = \begin{bmatrix} z_1\cos(z_2) \\ z_1\sin(z_2) \end{bmatrix}$ (polar coordinates),

(d) $\boldsymbol{x} = \begin{bmatrix} z_1^3 + z_2 \\ z_2^3 - z_1 \end{bmatrix}$,

(e) $\boldsymbol{x} = \begin{bmatrix} e^{-z_1 z_2} \\ e^{z_1 z_2} \end{bmatrix}$,

(f) $\boldsymbol{x} = \begin{bmatrix} (R + z_1\cos(z_2))\cos(z_3) \\ (R + z_1\cos(z_2))\sin(z_3) \\ z_1\sin(z_2) \end{bmatrix}$ with $0 < z_1 < R$ (torus coordinates).

Figure 3.27 shows the polar and torus coordinates.

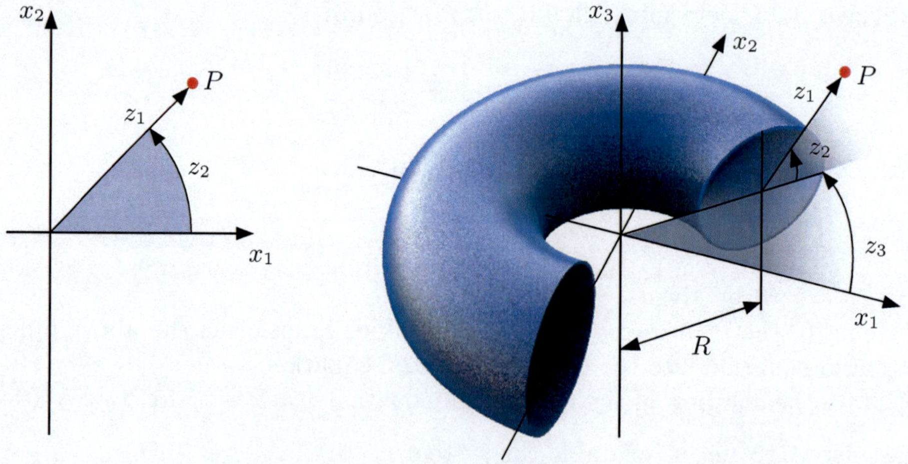

Fig. 3.27: Polar coordinates (left) and torus coordinates (right)

Exercise 3.15 Let us examine the system (1.11) on p. 15 from Section 1.1.6.
(a) Transform the system (1.11) into polar coordinates using

$$x_1 = z_1 \cos(z_2),$$
$$x_2 = z_1 \sin(z_2).$$

(b) Using the system description in polar coordinates, describe the system's equilibrium points and trajectories.

Exercise 3.16 Let us consider the differential equation

$$\dot{x} = \frac{1}{x}. \tag{3.130}$$

from Section 3.3.2, p. 264 et seq., once again.
(a) Transform the differential equation (3.130) using $z = \operatorname{sgn}(x)x^2$. What is the result?
(b) Determine the differential equation's solution.
(c) Compare the approach here with that of Section 3.3.2 on p. 264. What is your finding?
(d) What type is the differential equation (3.130)?

4

Nonlinear Control of Linear Systems

4.1 Control with Anti-Windup

4.1.1 The Windup Effect

Every real actuator of a control loop has a limitation in terms of the control variable u because its maximum control power is finite. Consequently, every real technical system possesses input constraints. This is illustrated in Figure 4.1, which shows a control loop with a plant $G(s)$, a controller $K(s)$, and a limiting element. In this case, the complete control element consists of the controller and the limiting element. The limitation of the control variable is described by the saturation characteristic

$$u = \operatorname{sat}(u_c) = \begin{cases} u_{\max}, & u_c > u_{\max}, \\ u_c, & u_{\min} \le u_c \le u_{\max}, \\ u_{\min}, & u_c < u_{\min}. \end{cases}$$

Examples of such limitations of the control variable u_c are a servomotor's maximum torque, a ship's maximum rudder angle, and the maximum aperture of a valve. Often the saturation characteristic curve has symmetric limiting points, i. e.

$$u_{\min} = -u_{\max}$$

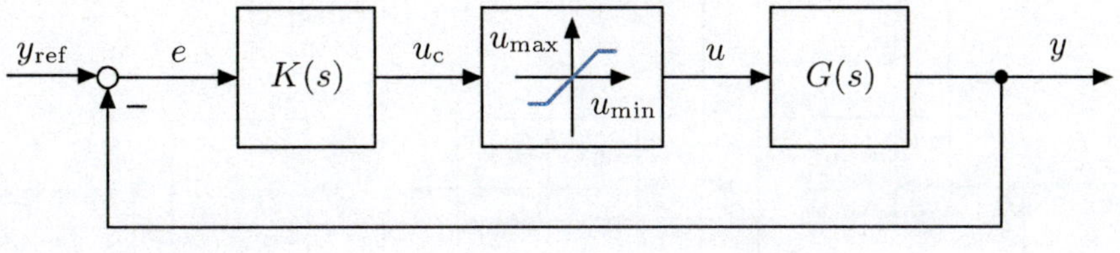

Fig. 4.1: Control loop with a linear plant, a linear controller $K(s)$, and a limitation of the control variable u_c

J. Adamy, *Nonlinear Systems and Controls*,
https://doi.org/10.1007/978-3-662-68690-4_4

holds.

If the control variable u_c exceeds the limitations, the control loop is no longer linear. In many cases, this is disadvantageous to its stability and control behavior. We will examine this below.

As an important case in practice, we will consider the general structure of a PID controller,

$$K(s) = K_{\mathrm{P}}\left(1 + \frac{1}{T_{\mathrm{I}}s} + T_{\mathrm{D}}s\right) = K_{\mathrm{PD}}(s) + \frac{K_{\mathrm{I}}}{s}, \quad K_{\mathrm{I}} = \frac{K_{\mathrm{P}}}{T_{\mathrm{I}}},$$

in the control loop of Figure 4.1. The first term on the right side of the above equation, $K_{\mathrm{PD}}(s)$, can represent a P, PD, or D controller. Where only an I controller exists, $K_{\mathrm{PD}}(s)$ is not present.

Where $u = u_c$, i.e. in the non-saturated case, the control loop is obviously linear. In the saturated case, on the other hand,

$$u = u_{\max} \quad \text{or} \quad u = u_{\min}$$

holds, i.e. a constant control variable u acts upon the plant $G(s)$.

Because of the constant control variable u in the case of saturation, no control is possible. In practical terms, the control loop is interrupted. Figure 4.2 illustrates this. Obviously, the interruption causes the integrator to continually integrate the control error e such that the integrator's content increases until the control error changes its sign. This process is called *windup*. Summing up the control error e is not only useless in this case; it is also harmful, since a value of

$$u_c < u_{\min} \quad \text{or} \quad u_c > u_{\max}$$

cannot act upon the plant. The harm would be caused by the high value of the integrator, which would only go down bit by bit after the control error e changed its sign. Thus, the high value of the integrator prevents the needed change in the control variable's sign for a certain time. The control loop is therefore interrupted until the integrator value is reduced such that the control

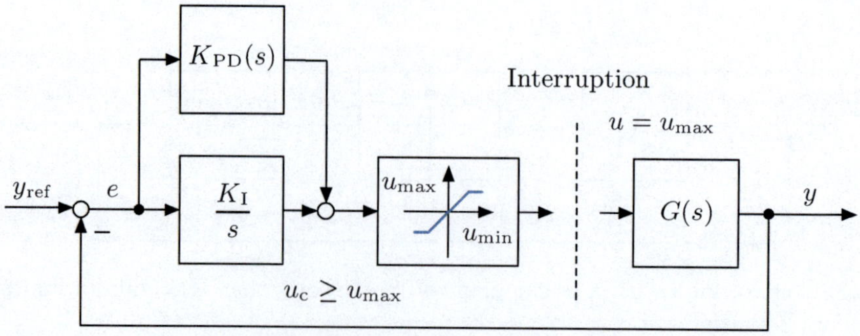

Fig. 4.2: Control loop in the saturated case

variable u_c lies within the saturation limits. At this point, a feedback control becomes possible again.

The windup behavior negatively affects the control behavior and it can lead to increased overshooting or even to instability. In simplified form, the situation can be interpreted as follows: if the control variable is in saturation such that the control loop is interrupted, the integration part of the controller is an unstable element of the open loop. Obviously, this is not a desirable situation.

4.1.2 PID Controller with Anti-Windup Element

It is possible to avoid the windup described above by means of an anti-windup structure, which ensures that the controller's output variable u_c does not exceed the limits u_{min} and u_{max}. Figure 4.3 shows a PID controller with anti-windup. The characteristic curve of the anti-windup element is a dead zone described by

$$v = \begin{cases} m(u_c - u_{max}), & u_c > u_{max}, \\ 0, & u_{min} \le u_c \le u_{max}, \\ m(u_c - u_{min}), & u_c < u_{min}, \end{cases}$$

with a very large positive value $m \gg 1$ for its slope.

The dead zone creates a negative feedback from the control output to the input of the integral component. If the controller's output value u_c is within the limits of the control variable u_{min} and u_{max}, the feedback is not active, i. e. the controller behaves like a normal PID controller. If u_c exceeds one of the limits u_{min} or u_{max}, the dead zone acts on the integral component as a strong negative feedback. The negative feedback instantaneously reduces the integral component to such a small value that u_c does not exceed the control variable limitation. Since the slope m is large but finite, the limits u_{min} and

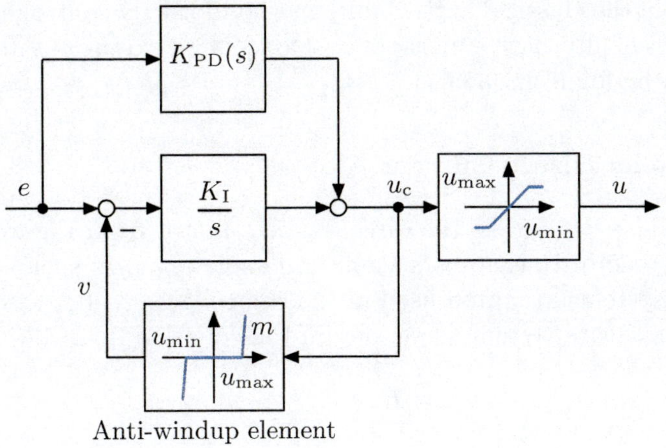

Fig. 4.3: PID controller with anti-windup element

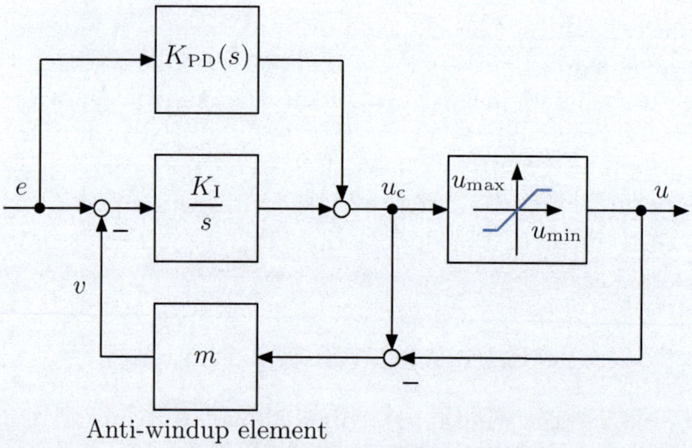

Fig. 4.4: Alternative implementation of the anti-windup structure shown in Figure 4.3

$u_{\max}$ can be marginally exceeded by the control variable u_{c}. In practice, this does not matter.

An implementation equivalent to the anti-windup structure in Figure 4.3 which does not require a dead zone is shown in Figure 4.4. The advantage of this anti-windup structure is that only one nonlinear characteristic curve is required. When there is a difference between u_{c} and u, the integrator state is reduced by the feedback $m(u_{\mathrm{c}} - u)$. Since $m \gg 1$ holds, this happens rapidly. The transfer behavior between u and u_{c} is given by

$$G_{\mathrm{aw}} = \frac{U_{\mathrm{c}}(s)}{U(s)} = \frac{m \cdot K_{\mathrm{I}}}{s + m \cdot K_{\mathrm{I}}}$$

if $K_{\mathrm{PD}}(s)$ is neglected. As is easily apparent, the reaction time of the anti-windup element can be varied by m.

Apart from the classical anti-windup method for PID controllers discussed above, a series of alternative methods and associated extensions also exist. An overview can be found in [359].

4.1.3 Example: Direct-Current Motor

As an example, we will take the direct-current motor from Figure 4.5, which acts as a servomotor that adjusts a rotation angle φ_{load}. In such servomotors, anti-windup systems are often used in the controller.

For the armature voltage u, we obtain the equation

$$u = Ri + L\dot{i} + k_1\omega \tag{4.1}$$

with the armature current i and the voltage $k_1\omega$ induced in the armature winding. Here, R is the resistance, L is the inductance, φ is the rotation

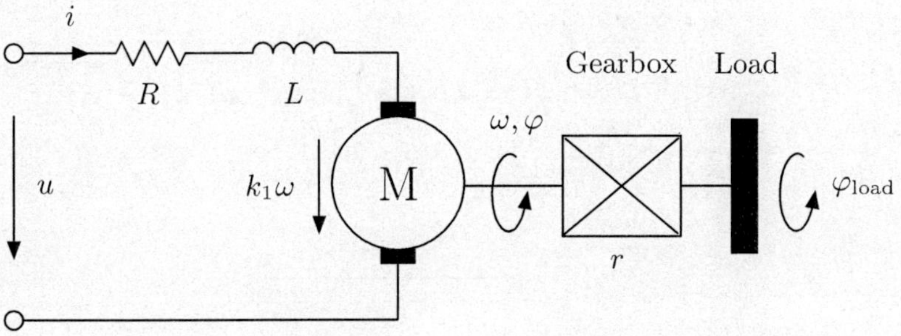

Fig. 4.5: Direct-current motor with gearbox and load

angle, and $\omega = \dot{\varphi}$ the armature's angular velocity. The generated torque can be computed from

$$M = k_2 i$$

which is equal to

$$M = J\dot{\omega} + k_3\omega,$$

where $k_3\omega$ is the friction term proportional to the velocity. It follows that

$$k_2 i = J\dot{\omega} + k_3\omega. \tag{4.2}$$

With the armature's inertia moment J_{a}, and the load's inertia moment J_{load}, which is converted via the gearbox with the transmission ratio r into the inertia moment belonging to the armature shaft, we obtain the total inertia moment

$$J = J_{\mathrm{a}} + \frac{1}{r^2} J_{\mathrm{load}}.$$

The values k_1, k_2, and k_3 are motor parameters.

From equations (4.1) and (4.2), along with the Laplace transformed variables $U(s)$, $I(s)$, $\Phi(s)$ of u, i, φ, we calculate

$$\begin{aligned} U(s) &= RI(s) + LI(s) \cdot s + k_1 \Phi(s) \cdot s, \\ I(s) &= \frac{J}{k_2}\Phi(s) \cdot s^2 + \frac{k_3}{k_2}\Phi(s) \cdot s \end{aligned}$$

if all initial values are identical to zero. Combining the last two equations yields the transfer function

$$\frac{\Phi(s)}{U(s)} = \frac{1}{s\left[\dfrac{LJ}{k_2}s^2 + \left(\dfrac{RJ}{k_2} + \dfrac{Lk_3}{k_2}\right)s + \left(\dfrac{Rk_3}{k_2} + k_1\right)\right]}.$$

If we also take into account that $\varphi = r\varphi_{\mathrm{load}}$, it follows that

$$G(s) = \frac{\Phi_{\mathrm{load}}(s)}{U(s)} = \frac{\dfrac{k_2}{rLJ}}{s\left[s^2 + \left(\dfrac{R}{L} + \dfrac{k_3}{J}\right)s + \left(\dfrac{Rk_3}{LJ} + \dfrac{k_1 k_2}{LJ}\right)\right]}.$$

Inserting the motor and load parameters

$$R = 8.9\,\Omega, \quad L = 0.1\,\mathrm{H}, \quad J = 0.1\,\mathrm{Nm\,s^2\,rad^{-1}}, \quad r = 10,$$

$$k_1 = 1.7775\,\mathrm{V\,s\,rad^{-1}}, \quad k_2 = 4\,\mathrm{Nm\,A^{-1}}, \quad k_3 = 0.1\,\mathrm{Nm\,s\,rad^{-1}}$$

yields

$$G(s) = \frac{40}{s(s^2 + 90s + 800)}.$$

The armature voltage u, the control variable, is subject to the symmetric limitations

$$-100\,\mathrm{V} \le u \le 100\,\mathrm{V}.$$

We now use a PI controller with

$$H(s) = K_{\mathrm{P}} + K_{\mathrm{I}} \cdot \frac{1}{s}$$

and the parameters $K_{\mathrm{P}} = 90$ and $K_{\mathrm{I}} = 150$.

The simulations in Figure 4.6 show a small step response of $\varphi_{\mathrm{load}} = 0\,\mathrm{rad}$ to $\varphi_{\mathrm{load}} = 1\,\mathrm{rad} = 57.3\,°$ with a linear regulation, for which the control variable does not saturate, and a large response of $\varphi_{\mathrm{load}} = 0\,\mathrm{rad}$ to $\varphi_{\mathrm{load}} = 10\,\mathrm{rad} = 573.0\,°$ with a nonlinear regulation. In the latter case of a PI controller with no anti-windup element for which the control variable saturates, the negative

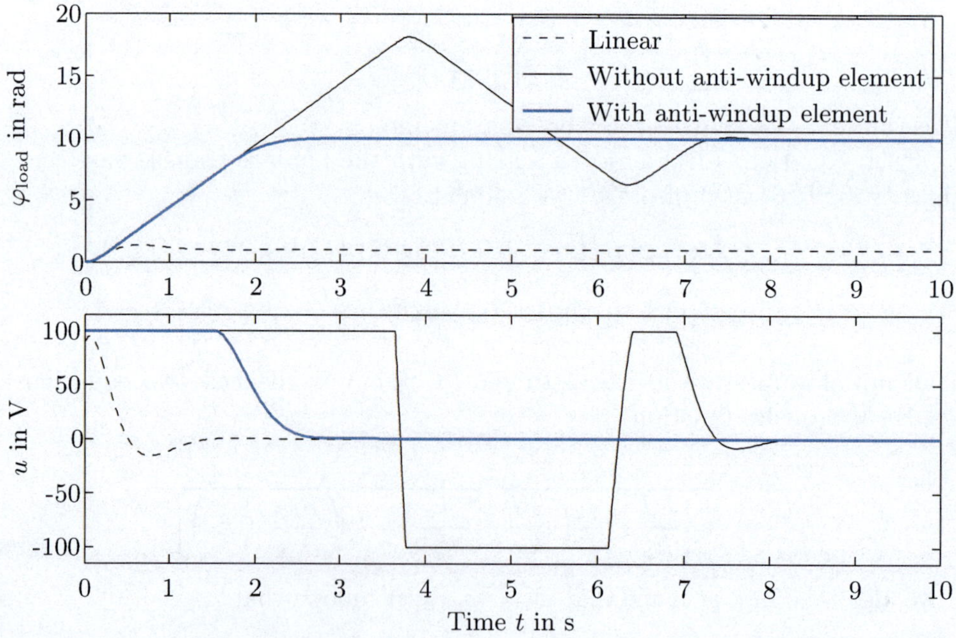

Fig. 4.6: Rotation angle φ_{load} and armature voltage u in the linear case, the case with an anti-windup element, and the case without an anti-windup element, in which the controller is in saturation

effect of the windup of the I controller, i. e. the oscillation of the angle φ_{load}, is obvious. The positive effect of the anti-windup element, which removes the oscillation occurring due to the unnecessary windup of the integrator, is equally evident.

4.1.4 A General Anti-Windup Method

Not all controllers are PID controllers. So the question arises which anti-windup procedures could be used for a more general controller

$$U_{\text{c}}(s) = H(s) \cdot E(s).$$

If the controller transfer function $H(s)$ has unstable poles or poles with a real part equal to zero, windup of the controller states and thus the controller output u_{c} occurs for limited control variables. Figure 4.7 shows the structure of this more general control loop. For such controllers $H(s)$, various anti-windup methods exist [103, 142, 153, 222, 239, 424, 443, 444, 457, 489].

A very plausible anti-windup structure for general cases is obtained if we take a small detour, as we will discuss in the next few pages. While doing so, we will not only obtain a general anti-windup method for arbitrary linear controllers but a simple method for handling saturated control variables of state-space control loops with an observer as well [185].

Now we will examine a state-control loop with an observer, also known as a *control observer*, as shown in Figure 4.8. The limitation of the control variable is part of the plant and the pre-filter G_{pf} acts as compensation for the steady-state error. In the linear operation mode, meaning

$$u_{\min} \le u_{\text{c}} \le u_{\max}, \tag{4.3}$$

the estimated state vector $\tilde{\boldsymbol{x}}$ corresponds to the plant's state vector $\boldsymbol{x}$ after a certain settling time. Note that if the condition (4.3) for u_{c} is not fulfilled, the control loop is nonlinear.

Below we will first address the linear case. The linear state-control loop with an observer has $2n$ eigenvalues. As we know from the theory of linear observers and from the separation principle [98], these are the n eigenvalues of the plant which is controlled by

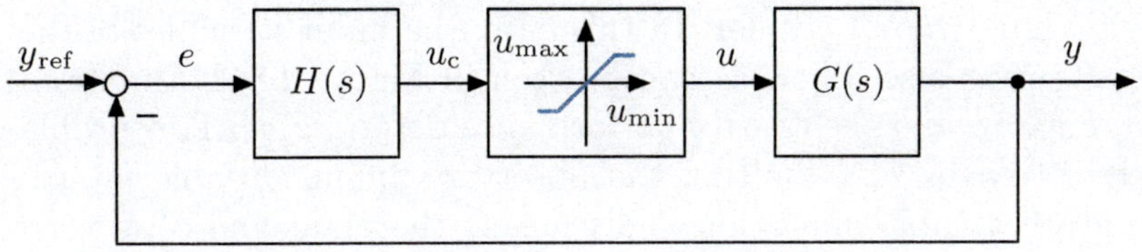

Fig. 4.7: Control loop with a saturation characteristic curve and an arbitrary linear controller $H(s)$

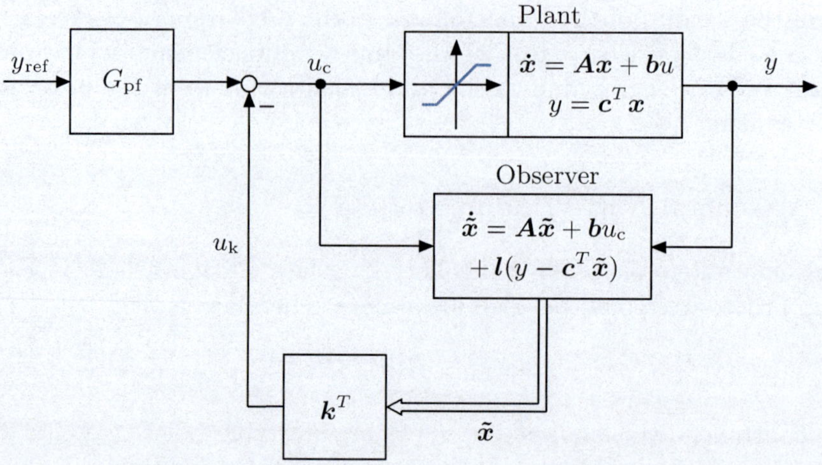

Fig. 4.8: State-control loop with an observer

$$\boldsymbol{u} = -\boldsymbol{k}^T\boldsymbol{x},$$

i. e. the eigenvalues of the matrix

$$\hat{\boldsymbol{A}} = \boldsymbol{A} - \boldsymbol{b}\boldsymbol{k}^T,$$

as well as the n eigenvalues of the observer matrix

$$\boldsymbol{F} = \boldsymbol{A} - \boldsymbol{l}\boldsymbol{c}^T.$$

The characteristic polynomial of the control loop with a control observer is thus represented by

$$P(s) = \underbrace{\det\left(s\boldsymbol{I} - \boldsymbol{A} + \boldsymbol{b}\boldsymbol{k}^T\right)}_{\delta(s)}\underbrace{\det\left(s\boldsymbol{I} - \boldsymbol{A} + \boldsymbol{l}\boldsymbol{c}^T\right)}_{\Delta(s)},$$

which means it consists of the control loop's characteristic polynomial $\delta(s)$ and the observer's characteristic polynomial $\Delta(s)$. Here, due to the separation principle, the control loop's and the observer's eigenvalues can be specified independently by the state-feedback gain vector $\boldsymbol{k}$ and the observer gain vector $\boldsymbol{l}$, respectively.

If the control variable u_c violates the limitation, the control variable u_c and the plant's input signal u differ. In this case, the input variables of the plant and the observer also differ, as can be seen in Figure 4.8. Estimation errors $\boldsymbol{e} = \boldsymbol{x} - \tilde{\boldsymbol{x}}$ follow and the control performance deteriorates. This problem can be easily solved by also limiting the observer's input variable u_{c} using the same control variable limitation. This means the plant and observer always have the same control variable. Figure 4.9 displays the modified structure.

Estimation errors resulting from the violation of the control variable limitation no longer occur using the additional saturation element. The simple

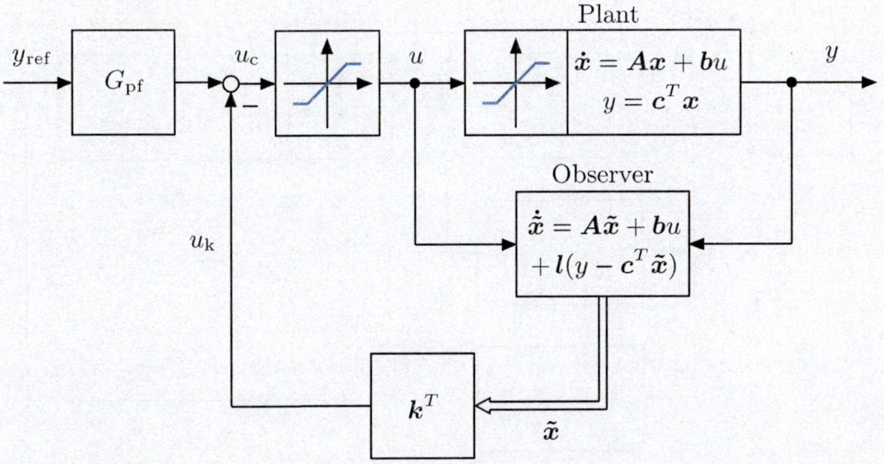

Fig. 4.9: Control loop with an additional saturation characteristic curve to avoid the negative effects of the control variable limitation

structure shown in Figure 4.9 also provides good control performance in many cases if the control loop is temporarily operated in saturation. This particularly holds for stable plants.

To find the anti-windup structure for the control loop in Figure 4.7 which we are attempting to identify, in the first step we will describe the observer's behavior

$$\dot{\tilde{\boldsymbol{x}}} = \left(\boldsymbol{A} - \boldsymbol{l}\boldsymbol{c}^T\right)\tilde{\boldsymbol{x}} + \boldsymbol{b}u + \boldsymbol{l}y = \boldsymbol{F}\tilde{\boldsymbol{x}} + \boldsymbol{b}u + \boldsymbol{l}y$$

using transfer functions. Utilizing the Laplace transformed variables $\tilde{\boldsymbol{X}}(s)$, $U(s)$, and $Y(s)$, we obtain

$$s\tilde{\boldsymbol{X}}(s) - \tilde{\boldsymbol{x}}(0) = \boldsymbol{F} \cdot \tilde{\boldsymbol{X}}(s) + \boldsymbol{b} \cdot U(s) + \boldsymbol{l} \cdot Y(s).$$

Assuming

$$\tilde{\boldsymbol{x}}(0) = \boldsymbol{0},$$

it follows that

$$\tilde{\boldsymbol{X}}(s) = (s\boldsymbol{I} - \boldsymbol{F})^{-1}\boldsymbol{b} \cdot U(s) + (s\boldsymbol{I} - \boldsymbol{F})^{-1}\boldsymbol{l} \cdot Y(s).$$

Along with the state controller

$$U_{\text{k}}(s) = \boldsymbol{k}^T\tilde{\boldsymbol{X}}(s),$$

we obtain

$$U_{\text{k}}(s) = \boldsymbol{k}^T(s\boldsymbol{I} - \boldsymbol{F})^{-1}\boldsymbol{b} \cdot U(s) + \boldsymbol{k}^T(s\boldsymbol{I} - \boldsymbol{F})^{-1}\boldsymbol{l} \cdot Y(s).$$

Next we will calculate the transfer functions

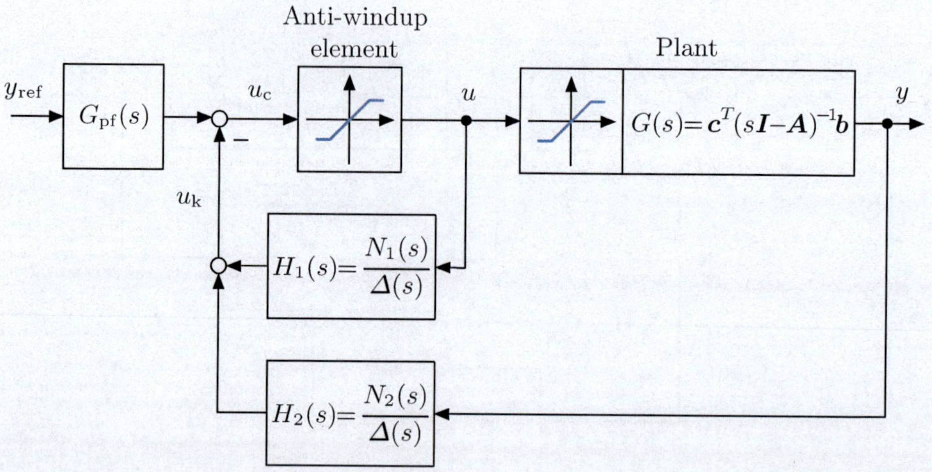

Fig. 4.10: State-control loop with a controller and an observer, represented by transfer functions and anti-windup element

$$H_1(s) = \frac{U_{\mathrm{k}}(s)}{U(s)} = \boldsymbol{k}^T \left(s\boldsymbol{I} - \boldsymbol{F}\right)^{-1} \boldsymbol{b} = \frac{N_1(s)}{\Delta(s)}$$

and

$$H_2(s) = \frac{U_{\mathrm{k}}(s)}{Y(s)} = \boldsymbol{k}^T \left(s\boldsymbol{I} - \boldsymbol{F}\right)^{-1} \boldsymbol{l} = \frac{N_2(s)}{\Delta(s)},$$

which describe the relationship between u_{k} and u or y, respectively. The transfer functions $H_1(s)$ and $H_2(s)$ have the same denominator:

$$\Delta(s) = \det(s\boldsymbol{I} - \boldsymbol{F}).$$

The control loop with an observer can thus be represented by transfer functions, as shown in Figure 4.10. Now we will interpret the additional saturation characteristic curve $u = \mathrm{sat}(u_{\mathrm{c}})$ as a kind of anti-windup element for the transfer functions $H_1(s)$ and $H_2(s)$, which represent the controller. This is because the additional saturation element prevents the undesired effects of the control variable limitation.

It should be noted that the degree n of the denominators of $H_1(s)$ and $H_2(s)$ is identical to the order of $\boldsymbol{A}$, since we have fully reconstructed the state vector $\boldsymbol{x}$. For a system with a reduced-order observer, the denominators of $H_1(s)$ and $H_2(s)$ have a lower system order consistent with that of the observer.

We have also generalized the structure of the control loop in Figure 4.10 by replacing the pre-filter G_{pf} with a general transfer function $G_{\mathrm{pf}}(s)$. The functionality of the anti-windup element and the state controller with an observer remains unaffected by this generalization.

In our second step, we will again examine the classical control loop with the plant and the controller

$$G(s) = \frac{N(s)}{D(s)} \quad \text{and} \quad H(s) = \frac{N_c(s)}{D_c(s)},$$

respectively, seeking its anti-windup element. This control loop, which is shown in Figures 4.7 and 4.11, has the transfer function

$$G_{\text{loop}}(s) = \frac{N_c(s)N(s)}{D_c(s)D(s) + N_c(s)N(s)}$$

for the linear, i. e. the unsaturated case. The control loop from Figure 4.11 can be rearranged so that it is consistent with the structure shown in Figure 4.12. At this point, the polynomial $\Delta(s)$ is inserted in addition. This means that the structure of Figure 4.12, except for the anti-windup element, complies with that of the state-control loop with an observer, as shown in Figure 4.10. The anti-windup element can now be integrated into the control loop, as shown in Figure 4.12. Figure 4.13 depicts the resulting control loop.

In this way, after the process we went through to avoid saturation effects for the state controller with an observer, we have obtained an anti-windup element for the nonlinear standard control loop with an arbitrary linear controller $H(s)$. This control-loop structure, which is shown in Figure 4.13, thus provides a solution to the anti-windup problem of the nonlinear standard control loop from Figures 4.7 and 4.11.

4.1.5 Dimensioning the General Anti-Windup Controller

The question of choosing the polynomial $\Delta(s)$ which we posed in the previous section still remains to be answered. In resolving this issue, we realize once again that the following three control-loop structures are identical for the linear case:

(1) the state controller with an observer and anti-windup element from Figure 4.9,
(2) the standard control loop from Figure 4.11,
(3) the standard control loop with an anti-windup element from Figure 4.13.

For the characteristic polynomial of the control loop with an observer, we have already established that

$$\delta(s) \cdot \Delta(s) = \det(s\boldsymbol{I} - \hat{\boldsymbol{A}}) \det(s\boldsymbol{I} - \boldsymbol{F}) \,. \tag{4.4}$$

The characteristic polynomial of the standard control loop from Figure 4.11 and that of the control loop shown in Figure 4.13 are the same and can be stated as

$$P(s) = D_c(s)D(s) + N_c(s)N(s). \tag{4.5}$$

For the control-loop structures from Figure 4.10 and Figure 4.13 to be identical, their characteristic polynomials (4.4) and (4.5) must be identical as well. Therefore, the equation

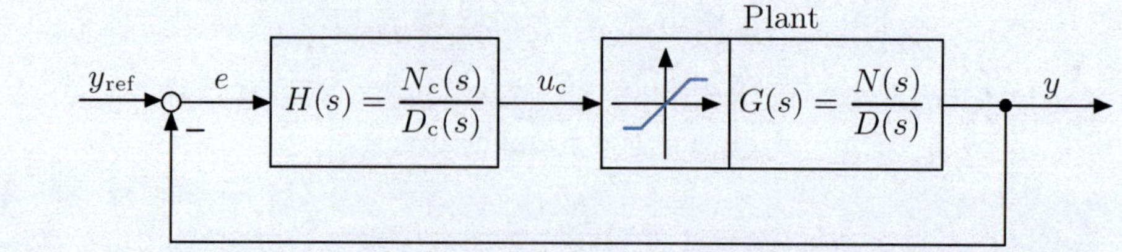

Fig. 4.11: Control loop with a control variable limitation and a general linear controller $H(s)$

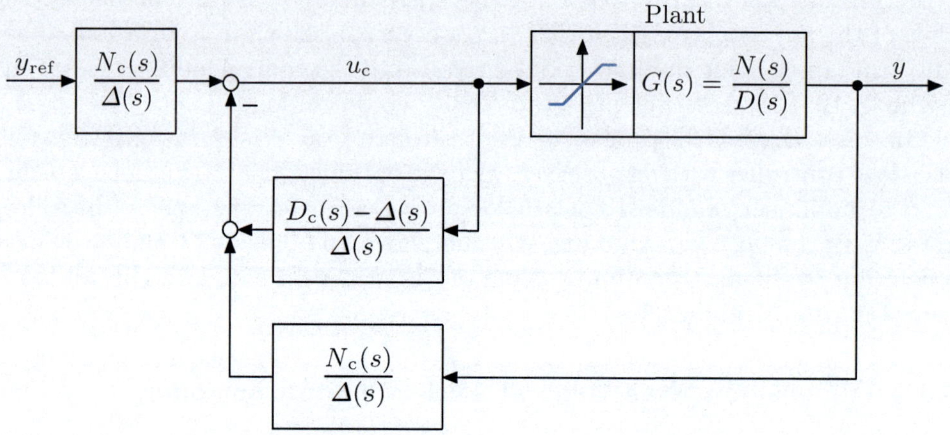

Fig. 4.12: Control-loop structure equivalent to the structure shown in Figure 4.11

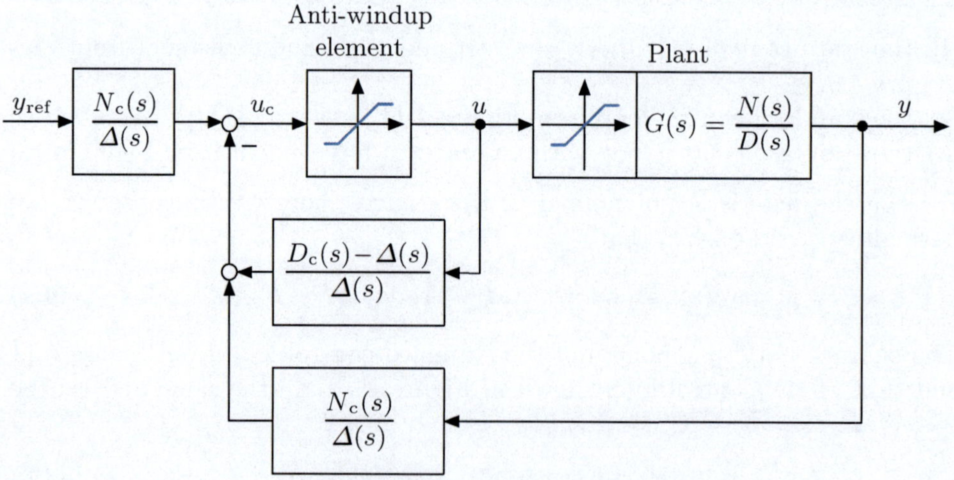

Fig. 4.13: Control loop with saturation, a general controller, and an anti-windup element

$$P(s) = D_c(s)D(s) + N_c(s)N(s) = \delta(s) \cdot \Delta(s)$$

has to hold. All roots of the polynomial $\Delta(s)$ being sought must therefore be roots of the characteristic polynomial $P(s)$ of the standard control loop shown in Figure 4.11. Note that the order of $P(s)$ is $n + k$, where n is the order of the plant and k is the order of the controller. The polynomial $\delta(s)$ is of degree n and the polynomial $\Delta(s)$ is of degree k.

Where $n = k$, $\Delta(s)$ represents the characteristic polynomial of a full-state observer for all states x_i of the plant. Where $k < n$, the polynomial $\Delta(s)$ has fewer than n roots and represents the characteristic polynomial of a reduced-order observer. The case in which $k < n$ is the usual case, since the controller $H(s)$ has the same order n as the plant in special cases only.

Note that the state-control loop with an observer does not have to be designed in order to determine the desired characteristic polynomial $\Delta(s)$. The equivalence between the state-control loop with an observer and the standard control loop only serves to explain how the anti-windup is designed.

The results are summarized in the following theorem.

Theorem 56 (General Anti-Windup Structure). *Let the standard control loop shown below*

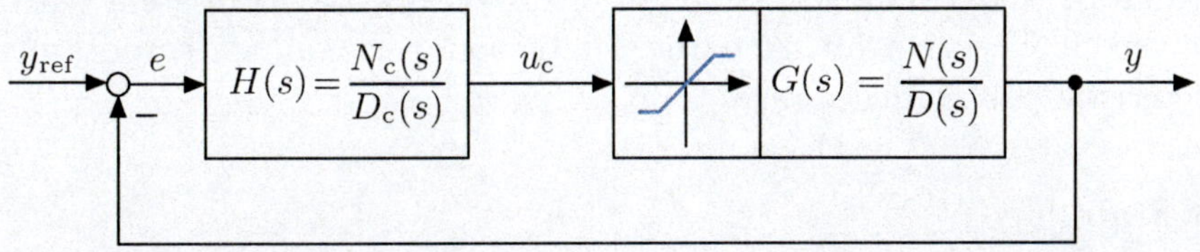

consist of the controller $N_c(s)/D_c(s)$ of order k and the plant $N(s)/D(s)$ of order n. The control loop shown below

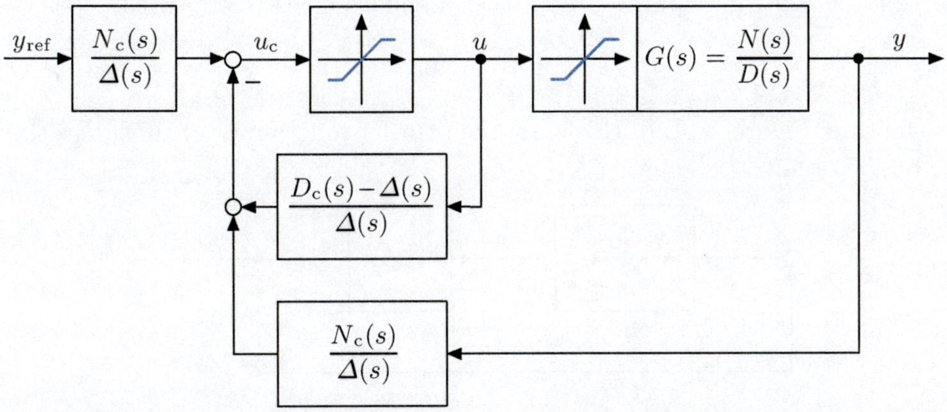

displays the same linear control behavior as the standard control loop, and has an anti-windup element in addition. The k roots of the polynomial $\Delta(s)$ are chosen such that they correspond to k roots of the control loop's characteristic polynomial

$$P(s) = D_c(s)D(s) + N_c(s)N(s).$$

When applying the above theorem, the k roots of the polynomial $\Delta(s)$ can be freely specified, as long as each one corresponds to one of the $n+k$ zeros of the polynomial $P(s)$. A suitable strategy is to choose them in a way that yields good control behavior. Unfortunately, no rule exists for choosing them. Instead, we are dependent on trial and error based on simulations to verify the behavior.

In infrequent specific cases, the choice of the k zeros of $\Delta(s)$ may be difficult. This is the case when the polynomial $P(s)$ has only complex conjugate roots and the degree k of the polynomial $\Delta(s)$ is odd. Obviously, $\Delta(s)$ then possesses a real root. A real root, however, does not exist in the set of complex roots of $P(s)$. This problem is solved by choosing the complex conjugate pair of roots of $P(s)$ with the largest damping coefficient d. Associated with this pair is the polynomial

$$s^2 + 2d\omega_0 s + \omega_0^2.$$

We approximate this polynomial via

$$(s+\omega_0)^2$$

and choose ω_0 as a real zero of $\Delta(s)$. The remaining $k-1$ roots of the polynomial $\Delta(s)$, whose number is even, are chosen in a way consistent with the rule given in Theorem 56 from the set of complex conjugate roots of the characteristic polynomial $P(s)$.

4.1.6 Stability

Finally, we will discuss the stability of a control loop with an anti-windup element. For a constant reference variable y_{ref}, this can be addressed by converting the control loop into a nonlinear standard control loop with

$$y_{\text{ref}}(t) = 0,$$

as shown in Figure 4.14. In this case, the transfer function $\tilde{G}(s)$ subsumes the

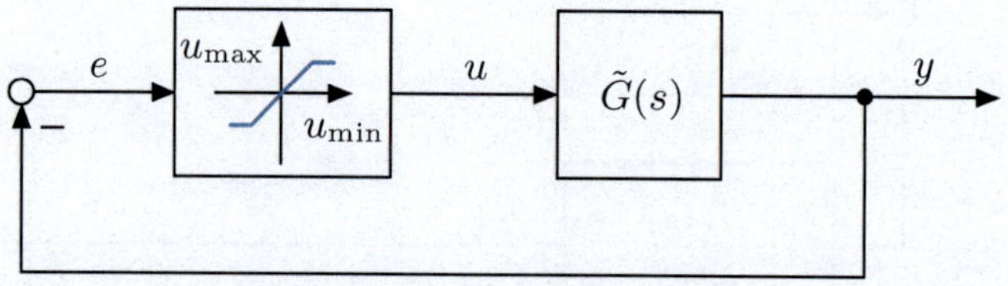

Fig. 4.14: Nonlinear standard control loop

controller and plant transfer functions. Once represented in the form of the nonlinear standard control loop, the circle criterion, among other methods, can be used for stability analysis.

4.2 Time-Optimal Control

4.2.1 Fundamentals and Fel'dbaum's Theorem

In practice, the design of nonlinear controllers is often based on heuristics. This is particularly true for the design of controllers which consist of elements with nonlinear characteristic curves. Examples are many two-position controllers and frequently controllers with an anti-windup element. Controllers of this kind are designed based on intuition, prior knowledge, and assumptions about the plant and its control-loop behavior with a selected controller. After designing such nonlinear controls, simulations are typically conducted to verify the control performance. In addition, one of the methods discussed in Chapter 2 should be applied to ensure stability. Consequently, this approach can be divided into three steps: heuristic controller design, stability analysis of the control loop, and simulation.

The reasons for a heuristic approach of this kind may be the fact that for many problems analytic design methods do not exist or that the design problems are very complex. In contrast, a further possible reason may be that the design problem is very simple and the implementation must be inexpensive. For example, a temperature controller for electric irons or coffee machines can be designed using an element with a hysteresis characteristic curve, e. g. a bimetal.

Such an approach does not succeed if the control performance requirements are high or the plant is extremely complex. These cases require appropriate controller design methods which aim to achieve better control performance than would be possible with a linear controller, or even to achieve the optimal control performance with respect to a given quality measure. One class of optimal feedforward and feedback controllers are time-optimal controllers. As indicated by the name, this class of controllers provides a feedforward or feedback compensation from an initial state $\boldsymbol{x}_0$ to the final state $\boldsymbol{x}_\mathrm{e} = \boldsymbol{0}$ within the shortest possible time t_e.

As plants, we will select the linear SISO systems

$$\dot{\boldsymbol{x}} = \boldsymbol{A}\boldsymbol{x} + \boldsymbol{b}u. \tag{4.6}$$

The starting point for the controller design is the requirement that the system's trajectory $\boldsymbol{x}(t)$ is steered to the equilibrium state $\boldsymbol{x}_\mathrm{eq} = \boldsymbol{0}$ in minimum time t_e. That means the performance index

$$J = t_\mathrm{e} \tag{4.7}$$

must be minimized by an appropriately chosen control signal $u(t)$. In this context, we need to remember that the control signal is limited by

$$-u_\mathrm{max} \leq u \leq u_\mathrm{max}. \tag{4.8}$$

Consequently, the following optimization problem must be solved: find the feedforward control function $u(t)$ for the system (4.6) with a control variable limitation (4.8), such that the performance index (4.7) is minimized for a given initial displacement $\boldsymbol{x}_0$. Closely related to this problem is that of determining the time-optimal feedback control law $u(\boldsymbol{x})$.

The above problems can be solved using Pontryagin's maximum principle [186, 237, 286, 453]. In general, the time-optimal feedback controls $u(\boldsymbol{x})$ are very difficult to determine and extremely complex to implement. That is the reason why time-optimal feedback controllers in industrial practice, with few exceptions [269, 353, 471], are found infrequently. Time-optimal feedforward controllers $u(t)$, on the other hand, are often easier to derive and implement. They are used in various applications, such as [25, 35, 59, 79, 86, 136, 476]. We will not discuss the maximum principle, since most of the practically relevant cases, particularly time-optimal feedforward controllers, can be computed without it.

The course of the control variable $u(t)$ of a time-optimal control is a very simple one, since $u(t)$ merely jumps from $-u_{\max}$ to $u_{\max}$ and vice versa, as depicted in Figure 4.15. In this way, a series of switching operations between $-u_{\max}$ and $u_{\max}$ is performed. The difficulty in designing a time-optimal control is the determination of the switching times

$$t_1, t_2, t_3, \ldots, t_{\mathrm{e}}$$

for a feedforward control, and the determination of the feedback control law $u(\boldsymbol{x})$ for a feedback control. The feedback control law $u(\boldsymbol{x})$ depends on the state vector $\boldsymbol{x}$, unlike the feedforward control sequence $u(t)$ which is only a function of time.

However, an important special case exists for which the switching times are relatively easy to determine. These are plants which possess only real eigenvalues. Without applying the maximum principle, time-optimal control functions $u(t)$ can then be deduced [112, 113, 114] using

Theorem 57 (Fel'dbaum's Theorem). *If the controllable system*

$$\dot{\boldsymbol{x}} = \boldsymbol{A}\boldsymbol{x} + \boldsymbol{b}u$$

of order n has only real eigenvalues, the course of the time-optimal control function $u(t)$ consists of a maximum of n switching intervals, for which $u(t)$ alternates between $-u_{\max}$ and $u_{\max}$.

As illustrated in Figure 4.15 for a fourth-order system with only real eigenvalues, switching between $-u_{\max}$ and $u_{\max}$ occurs a maximum of three times, i. e. we have four switching intervals. For a system with complex conjugate eigenvalues, the number of switching intervals can be greater than n.

Note that the time-optimal control sequence does not exist for all $\boldsymbol{x} \in I\!R^n$ where plants are unstable. This is because, for limited control variables u, not all initial states $\boldsymbol{x}(0)$ can be brought to the equilibrium point $\boldsymbol{x}_{\mathrm{eq}} = \boldsymbol{0}$. The limited control power is not sufficient in this case. On the other hand,

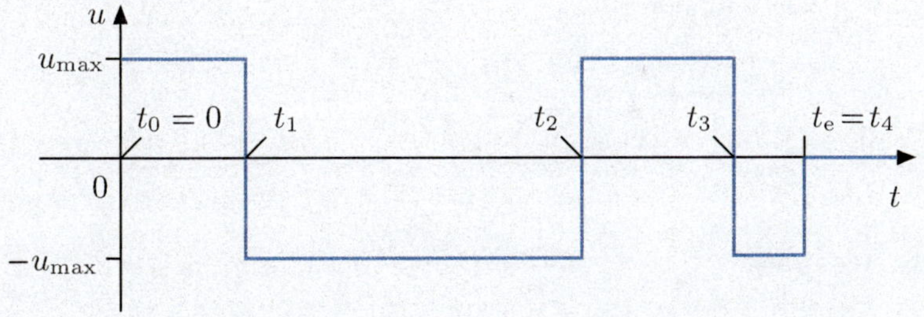

Fig. 4.15: Time course of the control variable of a time-optimal feedforward or feedback controller

for controllable plants which do not have any eigenvalues with positive real parts, time-optimal feedforward control sequences $u(t)$ and feedback control laws $u(\boldsymbol{x})$ exist for all $\boldsymbol{x} \in I\!R^n$.

4.2.2 Computation of Time-Optimal Controls

For systems with only real eigenvalues, the switching times t_i can be determined as follows. For a linear system's differential equation

$$\dot{\boldsymbol{x}} = \boldsymbol{A}\boldsymbol{x} + \boldsymbol{b}u,$$

the solution

$$\boldsymbol{x}(t) = \mathrm{e}^{\boldsymbol{A}t}\boldsymbol{x}_0 + \int\limits_0^t \mathrm{e}^{\boldsymbol{A}(t-\tau)}\boldsymbol{b}u(\tau)\,d\tau \tag{4.9}$$

is well-known. In the first switching interval, the sign of the trajectory of $u(t)$ equals

$$\alpha = 1 \quad \text{or} \quad \alpha = -1.$$

Thus,

$$u(t) = (-1)^{i-1}\alpha \cdot u_{\max} \qquad \text{for} \quad t \in [t_{i-1}, t_i)$$

holds for $i = 1, \ldots, n$. At present, we will leave the question open whether $\alpha = 1$ or $\alpha = -1$ and go ahead calculating the solution of the system by inserting the switching sequence $u(t)$ into equation (4.9). We obtain

$$\boldsymbol{x}(t_e = t_n) = \mathrm{e}^{\boldsymbol{A}t_n}\boldsymbol{x}_0 + \sum_{i=1}^{n} \int\limits_{t_{i-1}}^{t_i} \mathrm{e}^{\boldsymbol{A}(t_n-\tau)}\boldsymbol{b}(-1)^{i-1}\alpha \cdot u_{\max}\,d\tau$$

and with the final state

$$\boldsymbol{x}(t_e = t_n) = \boldsymbol{0}$$

the equation

$$\mathbf{0} = e^{\boldsymbol{A}t_n}\boldsymbol{x}_0 + \alpha \cdot u_{\max} \sum_{i=1}^{n} \int_{t_{i-1}}^{t_i} e^{\boldsymbol{A}(t_n-\tau)}\boldsymbol{b}(-1)^{i-1}\,d\tau. \tag{4.10}$$

Taking into account

$$e^{\boldsymbol{A}(t_n-\tau)} = \mathrm{e}^{\boldsymbol{A}t_n}\,\mathrm{e}^{-\boldsymbol{A}\tau},$$

equation (4.10) multiplied by $e^{-\boldsymbol{A}t_n}$ results in

$$\boldsymbol{x}_0 = -\alpha \cdot u_{\max} \sum_{i=1}^{n} (-1)^{i-1} \underbrace{\int_{t_{i-1}}^{t_i} e^{-\boldsymbol{A}\tau}\boldsymbol{b}\,d\tau}_{\boldsymbol{w}(t_i)-\boldsymbol{w}(t_{i-1})}. \tag{4.11}$$

Here, $\boldsymbol{w}(\tau)$ is the antiderivative vector of $e^{-\boldsymbol{A}\tau}\boldsymbol{b}$. Using equation (4.11), we obtain

$$-\frac{\boldsymbol{x}_0}{\alpha u_{\max}} = [\boldsymbol{w}(t_1)-\boldsymbol{w}(t_0)]-[\boldsymbol{w}(t_2)-\boldsymbol{w}(t_1)]+\ldots+(-1)^{n-1}[\boldsymbol{w}(t_n)-\boldsymbol{w}(t_{n-1})].$$

This equation leads to the nonlinear system of equations

$$\boldsymbol{w}(t_1) - \boldsymbol{w}(t_2) + \boldsymbol{w}(t_3) - \ldots + \frac{1}{2}(-1)^{n-1}\boldsymbol{w}(t_n) = \frac{1}{2}\boldsymbol{w}(0) - \frac{\boldsymbol{x}_0}{2\alpha u_{\max}} \tag{4.12}$$

with n equations and n unknowns, i. e. $t_1, t_2, \ldots, t_n$ for a starting time $t_0 = 0$.

In principle, α is also unknown. Since it is not known *a priori* whether $u_{\max}$ or $-u_{\max}$ holds for the first switching interval, both values $\alpha = 1$ and $\alpha = -1$ are tried. The correspondent signals $u(t)$ are shown in Figure 4.16. For one of the two cases the system of equations has a solution, whereas for the other, it does not. The system of equations is only analytically solvable

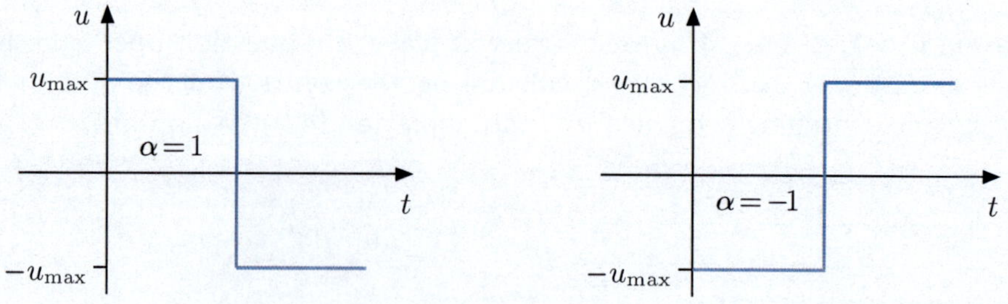

Fig. 4.16: The first switching interval of a time-optimal control sequence begins with $u_{\max}$ or $-u_{\max}$.

for lower-order systems. Otherwise, it is transcendental and must be solved numerically.

The time-optimal switching times of systems $\dot{\boldsymbol{x}} = \boldsymbol{A}\boldsymbol{x} + \boldsymbol{b}u$ with complex conjugate eigenvalues which satisfy the system of equations (4.12) can also be computed. However, we do not know whether n switching intervals are sufficient. If no solution is found, we need to try $n+1, n+2, \ldots$ intervals. This can lead to additional solutions which do not provide time-optimal switching times.

4.2.3 Example $1/s^2$

As a classic example [213], we will consider the plant $1/s^2$. It occurs in situations such as when a mass is accelerated. The corresponding state-space representation is

$$\dot{\boldsymbol{x}} = \begin{bmatrix} 0 & 1 \\ 0 & 0 \end{bmatrix} \boldsymbol{x} + \begin{bmatrix} 0 \\ 1 \end{bmatrix} u,$$

and we will assume that the control variable is symmetrically limited by

$$-u_{\max} \leq u \leq u_{\max}.$$

Thus, u is the acceleration, x_2 is the velocity, and x_1 is the distance traveled. Figure 4.17 shows the corresponding block diagram.

According to Fel'dbaum's theorem, the time-optimal control $u(t)$ has a maximum of two switching intervals, meaning one switching operation between $-u_{\max}$ and $u_{\max}$. The switching times t_1 and t_2 are determined from

$$\boldsymbol{w}(t_1) - \frac{1}{2}\boldsymbol{w}(t_2) = \frac{1}{2}\boldsymbol{w}(0) - \frac{\boldsymbol{x}_0}{2\alpha u_{\max}} \tag{4.13}$$

with

$$\boldsymbol{w}(\tau) = \int e^{-\boldsymbol{A}\tau}\, d\tau \cdot \boldsymbol{b} \quad \text{and} \quad \boldsymbol{x}_0 = \begin{bmatrix} x_{10} \\ x_{20} \end{bmatrix} = \begin{bmatrix} x_1(0) \\ x_2(0) \end{bmatrix}.$$

First, using the inverse Laplace transformation $\mathcal{L}^{-1}$ the state-transition matrix

$$e^{-\boldsymbol{A}t} = \mathcal{L}^{-1}\left\{(s\boldsymbol{I} + \boldsymbol{A})^{-1}\right\} = \mathcal{L}^{-1}\left\{\begin{bmatrix} s & 1 \\ 0 & s \end{bmatrix}^{-1}\right\} = \begin{bmatrix} 1 & -t \\ 0 & 1 \end{bmatrix}$$

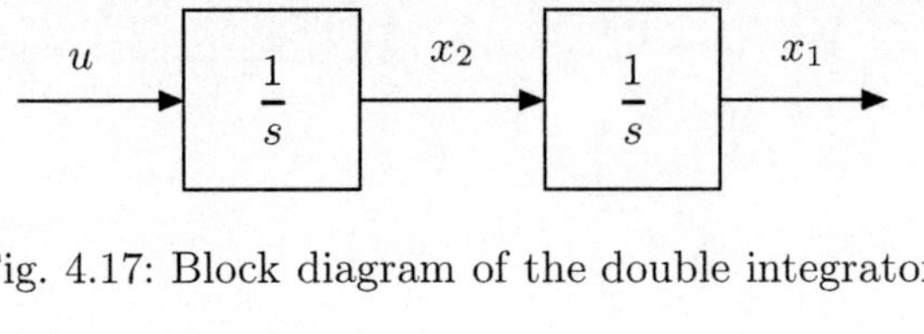

Fig. 4.17: Block diagram of the double integrator

is derived. Using the latter, we calculate

$$\boldsymbol{w}(\tau) = \int e^{-\boldsymbol{A}\tau}\, d\tau \cdot \boldsymbol{b} = \int \begin{bmatrix} -\tau \\ 1 \end{bmatrix} d\tau = \begin{bmatrix} -\dfrac{1}{2}\tau^2 + C_1 \\ \tau + C_2 \end{bmatrix}. \tag{4.14}$$

Furthermore, equation (4.14) substituted into equation (4.13) yields

$$-2t_1^2 + t_2^2 = -2\frac{x_{10}}{\alpha u_{\max}},$$
$$2t_1 - t_2 = -\frac{x_{20}}{\alpha u_{\max}}.$$

This nonlinear system of equations is easily solved. The resulting solutions are

$$t_1 = -\frac{x_{20}}{\alpha u_{\max}} \pm \sqrt{\frac{1}{2}\left(\frac{x_{20}}{\alpha u_{\max}}\right)^2 - \frac{x_{10}}{\alpha u_{\max}}}, \tag{4.15}$$

$$t_2 = -\frac{x_{20}}{\alpha u_{\max}} \pm 2\sqrt{\frac{1}{2}\left(\frac{x_{20}}{\alpha u_{\max}}\right)^2 - \frac{x_{10}}{\alpha u_{\max}}}. \tag{4.16}$$

The question remains whether $\alpha = 1$ or $\alpha = -1$ holds, i. e. whether the control sequence begins with $u_{\max}$ or with $-u_{\max}$. To find this out, we will examine equation (4.15) more closely. Obviously

$$t_1 = -\frac{x_{20}}{\alpha u_{\max}} \pm \sqrt{\frac{1}{2}\left(\frac{x_{20}}{u_{\max}}\right)^2 - \frac{x_{10}}{\alpha u_{\max}}} \geq 0 \tag{4.17}$$

must hold. Our first task is to determine which values x_{10} and x_{20} lead to a value $t_1 \geq 0$ if $\alpha = 1$ holds. In this case, equation (4.17) multiplied by $u_{\max}$ is of the form

$$-x_{20} \pm \sqrt{\frac{1}{2}x_{20}^2 - u_{\max}x_{10}} \geq 0. \tag{4.18}$$

A requirement is that the term in the root is positive or equal to zero, i. e.

$$x_{10} \leq \frac{1}{2u_{\max}}x_{20}^2 \tag{4.19}$$

must hold. Furthermore, we need only consider cases in which the sign in front of the root in equation (4.18) is positive, since the associated solution set also contains all solutions which result from a negative sign. As a further condition, we therefore obtain

$$x_{20} \leq \sqrt{\frac{1}{2}x_{20}^2 - u_{\max}x_{10}}\,. \tag{4.20}$$

Two cases must be distinguished for this inequality. We will start with the case $x_{20} \leq 0$. In this case, equation (4.20) is obviously always fulfilled as long as equation (4.19) holds. In the other case, $x_{20} > 0$ holds. We now square both sides and obtain

$$x_{20}^2 \leq \frac{1}{2}x_{20}^2 - u_{\max}x_{10} \quad \text{for} \quad x_{20} > 0$$

or rather

$$x_{10} \leq -\frac{1}{2u_{\max}}x_{20}^2 \quad \text{for} \quad x_{20} > 0. \tag{4.21}$$

The set of initial values

$$\boldsymbol{x}_0 = \begin{bmatrix} x_{10} \; x_{20} \end{bmatrix}^T$$

for which equation (4.15) is solvable with $\alpha = 1$ is thus given by the inequalities (4.19) for $x_{20} \leq 0$ and (4.21). It is bounded by branches of the two parabolas defined by the equal sign in these inequalities. Figure 4.18 illustrates this region in blue. For $\alpha = -1$, a similar analysis leads to the result that $u = -u_{\max}$ holds for the region above the branches of the parabolas, i. e. the white area in Figure 4.18.

In summary, the regions of the state space for which $\alpha = 1$ or $\alpha = -1$ holds, meaning $u = u_{\max}$ or $u = -u_{\max}$ also holds, are separated by

$$x_{10} = \begin{cases} \dfrac{x_{20}^2}{2u_{\max}}, & x_{20} \leq 0, \\ -\dfrac{x_{20}^2}{2u_{\max}}, & x_{20} > 0. \end{cases}$$

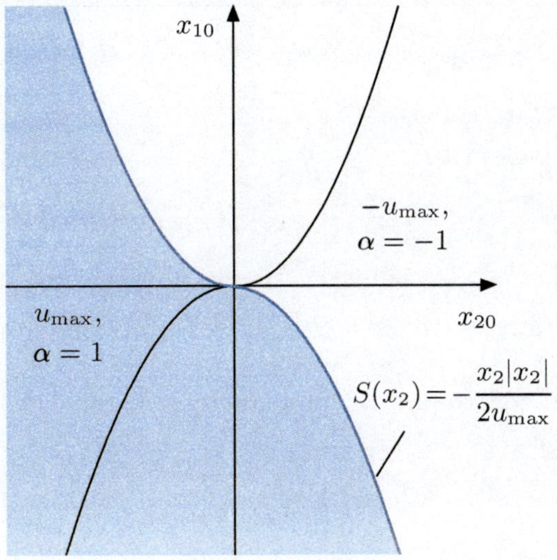

Fig. 4.18: The phase plane is separated into two halves by the switching curve $x_1 = S(x_2)$, which is shown in blue.

Together, these parabola branches form the switching curve

$$S(x_2) = x_1 = -\frac{x_2|x_2|}{2u_{\max}}.$$

Below the switching curve $S(x_2)$, $u = u_{\max}$ holds, whereas above it $u = -u_{\max}$ holds.

It should be noticed that the minus sign in front of the roots in equations (4.15) and (4.16) can be omitted, as this leads to negative or irrelevant switching times. With the results above, the switching times t_i are computed as

$$t_1 = -\frac{x_{20}}{\alpha u_{\max}} + \sqrt{\frac{1}{2}\left(\frac{x_{20}}{u_{\max}}\right)^2 - \frac{x_{10}}{\alpha u_{\max}}},$$

$$t_2 = -\frac{x_{20}}{\alpha u_{\max}} + 2\sqrt{\frac{1}{2}\left(\frac{x_{20}}{u_{\max}}\right)^2 - \frac{x_{10}}{\alpha u_{\max}}}$$

with

$$\alpha = \begin{cases} 1, & x_{10} < S(x_{20}) \text{ or } x_{10} = S(x_{20}) > 0, \\ -1, & x_{10} > S(x_{20}) \text{ or } x_{10} = S(x_{20}) < 0. \end{cases}$$

Using these equations, we are able to calculate the time-optimal feedforward control.

The time-optimal feedback control can now also be determined. Above the switching curve $S(x_2)$ the actuator signal $-u_{\max}$ is used and below it $u_{\max}$, i. e. the control law is

$$u(\boldsymbol{x}) = \begin{cases} u_{\max}, & x_1 - S(x_2) < 0, \\ -u_{\max}, & x_1 - S(x_2) > 0. \end{cases}$$

The latter is equivalent to

$$\begin{aligned} u &= \begin{cases} u_{\max}, & \operatorname{sgn}(x_1 - S(x_2)) < 0, \\ -u_{\max}, & \operatorname{sgn}(x_1 - S(x_2)) > 0 \end{cases} \\ &= -u_{\max} \cdot \operatorname{sgn}(x_1 - S(x_2)) = u_{\max} \cdot \operatorname{sgn}(S(x_2) - x_1). \end{aligned}$$

The time-optimal control law of the plant $1/s^2$ can eventually be represented as

$$u = u_{\max} \cdot \operatorname{sgn}\left(-\frac{x_2|x_2|}{2u_{\max}} - x_1\right). \tag{4.22}$$

The control law (4.22) provides the value $u = 0$ on the switching curve, i. e. for $x_1 = S(x_2)$. To be correct, in this case, $u = -u_{\max}$ for $x_1 < 0$ and $u = u_{\max}$ for $x_1 > 0$ should hold. In practice, however, this is irrelevant since

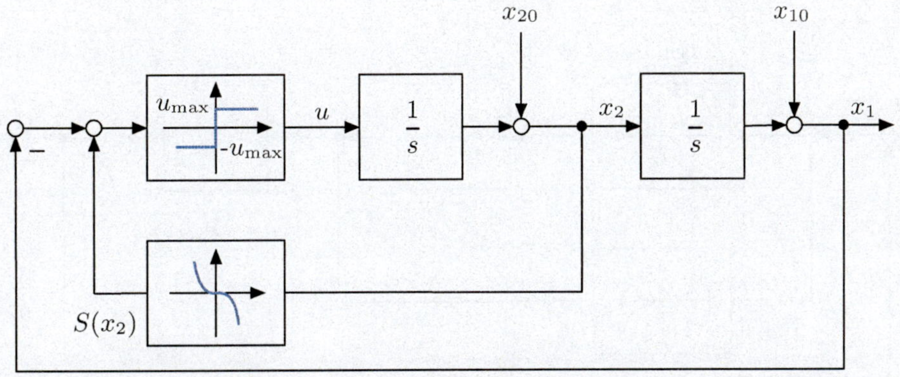

Fig. 4.19: Time-optimal feedback control for the plant $1/s^2$

the trajectory can never run precisely along the switching curve due to noise. The corresponding controller is shown in Figure 4.19.

The control law above and further time-optimal control laws for second-order plants with real eigenvalues can be relatively easily derived by calculating all trajectories generated by $u = u_{\max}$ and $u = -u_{\max}$ in the phase plane. Parts of these trajectories obviously form the set of all trajectories of the time-optimally controlled system. In particular, the switching curve S is identical to the parts of the two trajectories which pass through the origin $\boldsymbol{x} = \boldsymbol{0}$ for $u = u_{\max}$ or $u = -u_{\max}$, respectively. Therefore, to derive the time-optimal control law, it is sufficient to compute these two kinds of trajectories and to formulate the switching curve S from parts of these trajectories based on geometrical considerations in the phase plane.

4.2.4 Time-Optimal Control of Low-Order Systems

In the following, we will address second- and third-order systems with the control variable limitation

$$|u| \leq u_{\max}$$

and their associated time-optimal control laws. First, let us turn to plants

$$\dot{\boldsymbol{x}} = \begin{bmatrix} 0 & 1 \\ 0 & -a \end{bmatrix} \boldsymbol{x} + \begin{bmatrix} 0 \\ 1 \end{bmatrix} u, \tag{4.23}$$

which have one eigenvalue at zero and one eigenvalue at $\lambda = -a < 0$. Note that all controllable plants with the above eigenvalue configuration can be transformed into a form consistent with (4.23), i. e. the controller canonical form.

The time-optimal feedback control law is obtained as

$$u = u_{\max} \operatorname{sgn}(S(x_2) - x_1)$$

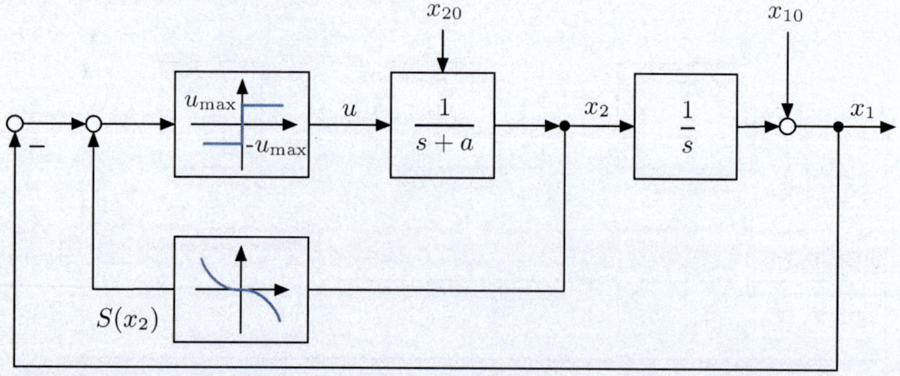

Fig. 4.20: Time-optimal feedback control system for the plant $1/(s(s+a))$

with the switching curve

$$S(x_2) = -\frac{1}{a}x_2 + \frac{u_{max}}{a^2}\operatorname{sgn}(x_2)\ln\left(1 + \frac{a|x_2|}{u_{max}}\right).$$

Figure 4.20 shows the corresponding block diagram.

The second case we will address is stable second-order plants with real non-zero eigenvalues $\lambda_1 < \lambda_2 < 0$. Here we assume that the system description is in, or can be transformed into, a form consistent with

$$\dot{\boldsymbol{x}} = \begin{bmatrix}\lambda_1 & 0\\ 0 & \lambda_2\end{bmatrix}\boldsymbol{x} + \begin{bmatrix}\lambda_1\\ \lambda_2\end{bmatrix}u. \tag{4.24}$$

Then the time-optimal feedback control law is stated as

$$u = u_{max}\operatorname{sgn}(S(x_2) - x_1)$$

with

$$S(x_2) = u_{max}\operatorname{sgn}(x_2)\left[\left(1 + \frac{|x_2|}{u_{max}}\right)^{\lambda_1/\lambda_2} - 1\right].$$

The time-optimal control law can also be computed for third-order systems with two eigenvalues at zero and a negative eigenvalue $\lambda = -a$ if they are stated in controller canonical form, or can be transformed into this form [24]. However, here it is useful to transform the controller canonical form

$$\dot{\tilde{\boldsymbol{x}}} = \begin{bmatrix}0 & 1 & 0\\ 0 & 0 & 1\\ 0 & 0 & -a\end{bmatrix}\tilde{\boldsymbol{x}} + \begin{bmatrix}0\\0\\1\end{bmatrix}u$$

by means of

$$\tilde{\boldsymbol{x}} = \frac{1}{a^3}\begin{bmatrix} 1 & 0 & 1 \\ 0 & a & -a \\ 0 & 0 & a^2 \end{bmatrix}\boldsymbol{x}$$

into the form

$$\dot{\boldsymbol{x}} = \begin{bmatrix} 0 & a & 0 \\ 0 & 0 & 0 \\ 0 & 0 & -a \end{bmatrix}\boldsymbol{x} + \begin{bmatrix} -a \\ a \\ a \end{bmatrix} u. \tag{4.25}$$

For the time-optimal control of the system (4.25), we obtain [24]

$$u = u_{\max}\operatorname{sgn}(S(x_1, x_2) - x_3) \tag{4.26}$$

with

$$\begin{aligned} S(x_1, x_2) &= u_{\max} d\left[e^c \cdot (2 - e^{\sqrt{b}}) - 1\right], \\ d &= \operatorname{sgn}\left(x_1 + x_2 + \frac{x_2|x_2|}{2u_{\max}}\right), \\ b &= \frac{x_2^2}{2u_{\max}^2} + \frac{d}{u_{\max}}(x_1 + x_2), \\ c &= \frac{d \cdot x_2}{u_{\max}} + \sqrt{b}. \end{aligned}$$

If the plant was originally not in one of the corresponding state-space representations (4.23), (4.24), or (4.25) but was transformed into such a form, the corresponding control law must be transformed back to the plant's original coordinates. The following section shows an example.

The time-optimal feedback control laws for some third-order systems can be found in [141, 395] and for certain fourth-order systems in [396]. In general, for higher-order plants no closed-form expressions for the control law can be derived. However, for such stable plants with exclusively real eigenvalues, a control law is obtained for which a transcendental nonlinear system of equations must be solved to determine u [24]. In practice, apart from a few exceptions with a low system order [24, 336, 399, 414, 474], the calculation and implementation of a time-optimal feedback control law are no longer possible for systems with complex conjugate eigenvalues.

4.2.5 Example: Submarine

Submarines can dive both dynamically and statically. For dynamic diving, the depth rudder is adjusted during travel such that a downward force is generated. This allows the submarine to dive into deeper waters, even though it is not heavier than the water it displaces. On the other hand, for static diving, ocean water is inserted into ballast tanks so that the submarine becomes heavier and sinks. When the submarine rises, the water is expelled with

Fig. 4.21: Submarine

compressed air from the tanks shown in Figure 4.21. In the following, we will design a time-optimal depth control for static diving.

We assume that the submarine is balanced such that it floats in a certain depth h and has an associated mass m. Starting from the ocean surface, the depth h is measured by means of the water pressure. If additional water of mass Δm is inserted into or ejected from the ballast tank, a vertical force

$$F = g \cdot \Delta m$$

is generated which acts upon the submarine. Subsequently, we assume that $\Delta m \ll m$. The ballast water mass accelerates the submarine with a total mass of $m + \Delta m$ as follows:

$$\ddot{h} = \frac{g \cdot \Delta m}{m + \Delta m} \approx \frac{g}{m} \cdot \Delta m, \tag{4.27}$$

so that the submarine rises or sinks statically. For $\Delta m < 0$ it rises, whereas for $\Delta m > 0$ it sinks.

The ballast water mass Δm_{ref} to be inserted or expelled from the submarine is stated as a reference value to a secondary controller which is subordinated to the time-optimal controller and is assumed to be known. This secondary controller can be described by the differential equation

$$(\Delta m)^{\cdot} + a\Delta m = a\Delta m_{\text{ref}}. \tag{4.28}$$

Inserting equation (4.27) into equation (4.28) yields

$$\dddot{h} + a\ddot{h} = \frac{a \cdot g}{m} \cdot \Delta m_{\text{ref}}.$$

In the next calculation, we use

$$\tilde{u} = \frac{g}{m} \cdot \Delta m_{\text{ref}}$$

as the control variable for the depth controller, which is the primary controller and is yet to be designed. For the state vector, we choose

$$\tilde{\boldsymbol{x}} = \begin{bmatrix} h \\ \dot{h} \\ \ddot{h} \end{bmatrix}.$$

The parameters $a = 0.005$ and $m = 1962\,\text{t}$, as well as the units of the variables, are chosen so as to model a Swedish submarine by the manufacturer Kockumation AB [166]; we then obtain

$$\dot{\tilde{\boldsymbol{x}}} = \begin{bmatrix} 0 & 1 & 0 \\ 0 & 0 & 1 \\ 0 & 0 & -0.005 \end{bmatrix} \tilde{\boldsymbol{x}} + \begin{bmatrix} 0 \\ 0 \\ 0.005 \end{bmatrix} \tilde{u}$$

with the diving depth as the output variable $y = \tilde{x}_1$. Furthermore, we set $\tilde{u}_{\max} = 0.005$ to be consistent with [166].

To obtain the time-optimal control law (4.26) as the depth control, we transform the control variable $\tilde{u}$ via $\tilde{u} = 200u$, and thus obtain the system description

$$\begin{aligned} \dot{\tilde{\boldsymbol{x}}} &= \begin{bmatrix} 0 & 1 & 0 \\ 0 & 0 & 1 \\ 0 & 0 & -0.005 \end{bmatrix} \tilde{\boldsymbol{x}} + \begin{bmatrix} 0 \\ 0 \\ 1 \end{bmatrix} u, \\ y &= \tilde{x}_1 \end{aligned} \tag{4.29}$$

with the input constraints

$$|u| \le u_{\max} = 2.5 \cdot 10^{-5}.$$

We can now apply the control law (4.26) for the plant (4.29). However, we need to use the transformed state variable vector $\boldsymbol{x}$ in equation (4.26), i. e.

$$\boldsymbol{x} = a^3 \begin{bmatrix} 1 & 0 & 1 \\ 0 & a & -a \\ 0 & 0 & a^2 \end{bmatrix}^{-1} \cdot \tilde{\boldsymbol{x}} = \begin{bmatrix} a^3 & 0 & -a \\ 0 & a^2 & a \\ 0 & 0 & a \end{bmatrix} \cdot \tilde{\boldsymbol{x}},$$

to compute u. Inserting the transformation equation above into equation (4.26) yields

$$u = u_{\max} \operatorname{sgn}(\tilde{S}(\tilde{x}_1, \tilde{x}_2, \tilde{x}_3) - a\tilde{x}_3)$$

as the time-optimal control law for the submarine at the original coordinates $\tilde{\boldsymbol{x}}$, where

$$\tilde{S}(\tilde{x}_1, \tilde{x}_2, \tilde{x}_3) = u_{\max} d \left[e^c \cdot (2 - e^{\sqrt{b}}) - 1 \right]$$

with

$$d = \operatorname{sgn}\left(a\tilde{x}_1 + \tilde{x}_2 + \frac{(a\tilde{x}_2 + \tilde{x}_3)|a\tilde{x}_2 + \tilde{x}_3|}{2u_{\max}} \right),$$
$$b = \frac{a^2(a\tilde{x}_2 + \tilde{x}_3)^2}{2u_{\max}^2} + \frac{da^2(a\tilde{x}_1 + \tilde{x}_2)}{u_{\max}},$$
$$c = \frac{ad(a\tilde{x}_2 + \tilde{x}_3)}{u_{\max}} + \sqrt{b}.$$

As an example, we will describe how the submarine surfaces from a depth of 100 meters, so the initial vector is:

$$\tilde{\boldsymbol{x}}(0) = \begin{bmatrix} 100 \\ 0 \\ 0 \end{bmatrix}.$$

Figure 4.22 shows the progressions of the depth $\tilde{x}_1(t)$ and the control variable $\tilde{u}(t)$ which the time-optimal control generates.

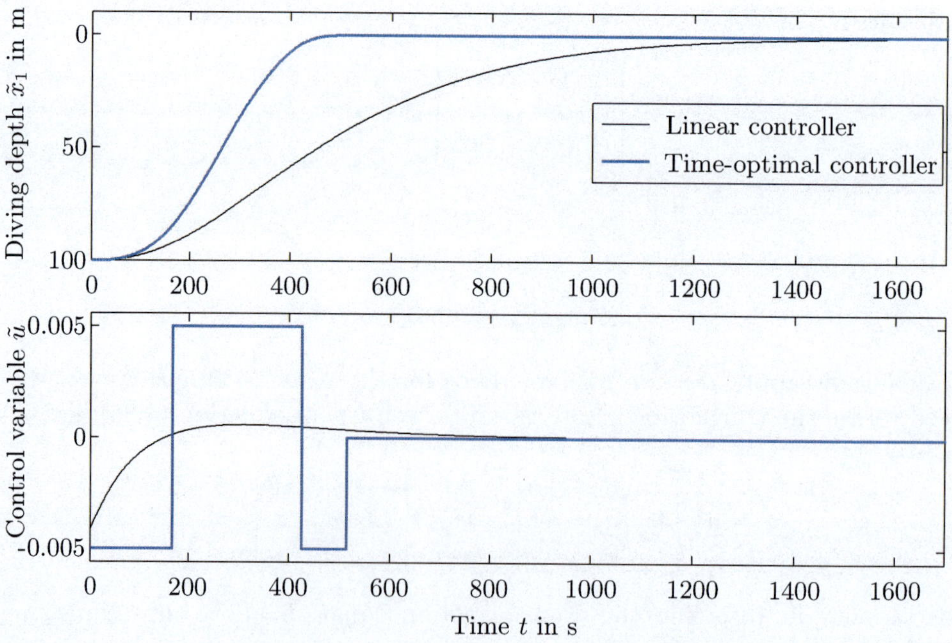

Fig. 4.22: Time courses of the diving depth $\tilde{x}_1(t) = h(t)$ and the control variables $\tilde{u}(t)$ of the submarine's linear (black) and time-optimal (blue) controllers

As a comparison, the trajectories for a very good linear controller,

$$\tilde{u} = 200u = -\begin{bmatrix} 4.164 \cdot 10^{-5} & 2.128 \cdot 10^{-2} & 2.592 \end{bmatrix} \tilde{\boldsymbol{x}},$$

are also shown. This controller was designed to adhere to the control variable limitation while exhibiting a short settling time and no overshooting. The substantially shorter settling time of the time-optimal controller is evident. However, its disadvantages are the discontinuous course of the control variable and its high amplitude, which leads to increased energy consumption.

4.2.6 Time-Optimal Pilot Control

For the second-order and third-order plants with real eigenvalues we are viewing, determining the time-optimal feedback control law $u(\boldsymbol{x})$ is relatively simple. For higher-order plants, analytically identifying time-optimal feedback control laws is no longer feasible. As mentioned previously, the design and implementation become impossible or extremely complex. Time-optimal feedforward controllers, on the other hand, are usually determinable and can be implemented in practice. In particular, their application is appropriate in systems which progress from one point $\boldsymbol{x}_0$ to a point $\boldsymbol{x}_{\mathrm{e}}$ again and again, i. e. systems which repeatedly take the same trajectory, as shown in Figure 4.23. To eliminate disturbances, a linear feedback controller can be used additionally, as shown in Figure 4.24. This feedback controller corrects the feedforward control, which is called a *pilot control* or *pre-control* in this context.

Experience has shown that existing algorithms for the calculation of the switching times for time-optimal feedforward controllers [109, 158, 159] exhibit numerical problems for higher-order plants with complex eigenvalues. In such cases, time-optimal solutions often cannot be determined. A solution to this problem is the calculation of the feedforward control sequence of discrete-time systems. In this approach, the continuous-time system is transformed into a discrete-time system. The time-optimal feedforward control sequences for discrete-time linear systems can be relatively easily calculated using linear

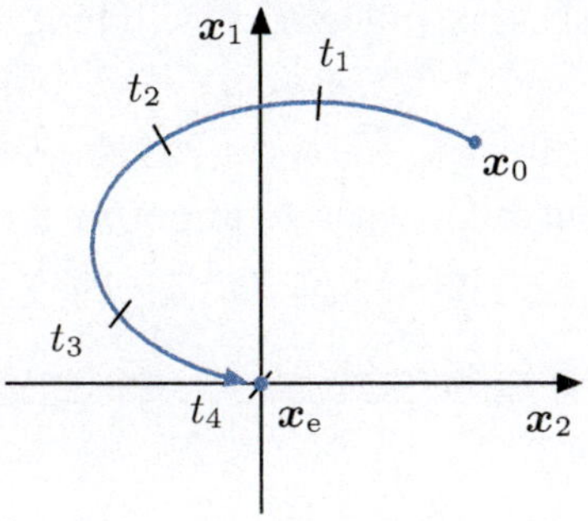

Fig. 4.23: Time-optimal trajectory with switching times t_1,. . . ,t_4

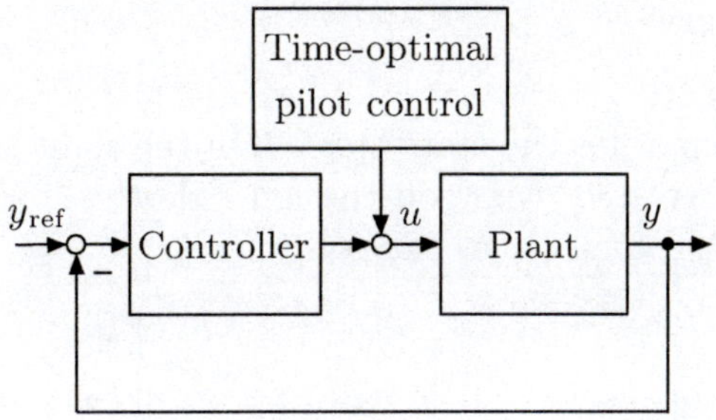

Fig. 4.24: Control loop with time-optimal pilot control

programming methods [33, 49, 74, 82, 227, 236, 295, 347, 409]. For this purpose, the shorter the chosen sampling time, the better the step-optimal feedforward control sequence approximates the time-optimal feedforward control function of the continuous-time system.

Comparing time-optimal controllers with linear controllers leads to the following dilemma: time-optimal controllers have a short settling time, but they are in general difficult or even impossible to design and implement. In comparison, linear controllers are slow but simple to design and implement. Evidently, it is impossible to achieve both very good regulation by the controller and simplicity of design and implementation. However, this dilemma can be resolved, because a trade-off exists between these two extremes. This is the subject of the following section, which addresses variable structure control systems.

4.3 Variable Structure Control Without Sliding Mode

4.3.1 Fundamentals of Variable Structure Control

Variable structure controllers can be divided into different classes. An important class consists of the *parameter-* and *structure-switching controllers*. For these, switching between different controllers is mostly performed in dependence on the state vector $\boldsymbol{x}$. As plants, we will consider the linear systems

$$\dot{\boldsymbol{x}} = \boldsymbol{A}\boldsymbol{x} + \boldsymbol{b}u$$

where the control variable u is limited according to

$$-u_{\max} \leq u \leq u_{\max}.$$

Figure 4.25 shows the general structure of a control loop such as this. A detailed description of switching control systems can be found in [280, 403].

In switching controllers, we can distinguish between two types of dynamic behavior, which are illustrated using an example in the following. We will examine switching between two P controllers and a second-order plant, as shown in Figure 4.26. The switching strategy works as follows: a switching curve

$$s(\boldsymbol{x}) = \boldsymbol{r}^T\boldsymbol{x} = 0$$

separates the state space. On the right side the controller with a gain of k_1 is activated, while on the left side the controller with a gain of k_2 is activated. The control law takes the form

$$u = \begin{cases} k_1 e, & s(\boldsymbol{x}) \geq 0, \\ k_2 e, & s(\boldsymbol{x}) < 0. \end{cases}$$

Depending on the choice of k_1 and k_2, different trajectories can be generated. We can now distinguish between two general cases. In the first, the trajectories

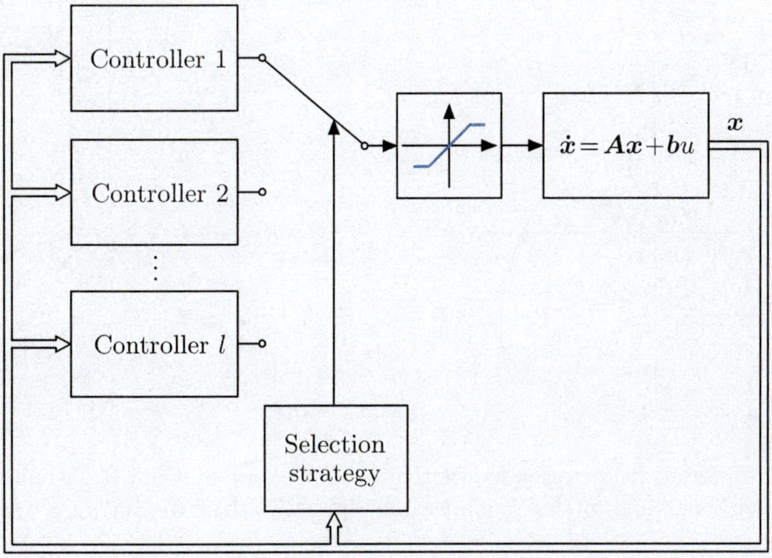

Fig. 4.25: Variable structure control system

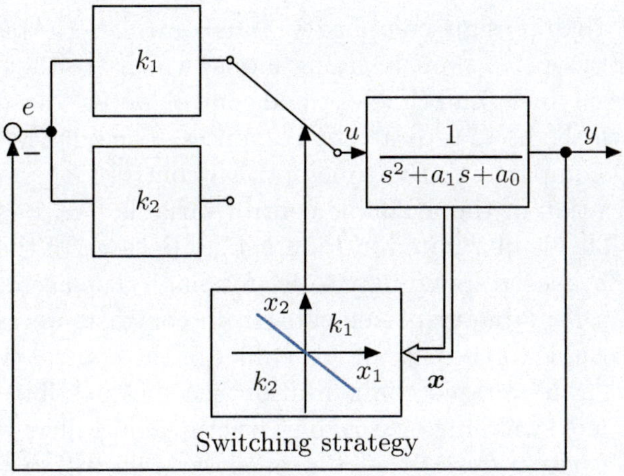

Fig. 4.26: Control loop that switches between two P controllers with gains k_1 and k_2

switch from one region into the other and continue their course there. This is illustrated in the graph on the left of Figure 4.27.

In the second case, the trajectories $\boldsymbol{x}(t)$ tend to the switching curve $s(\boldsymbol{x})$ from both sides, as shown in the upper left quadrant of the right-hand graph in Figure 4.27. Once it has reached the switching curve, the trajectory repeatedly switches from one side to the other. While maintaining this continuous switching behavior, it eventually reaches the equilibrium point. This repeated switching, which we already discussed in Section 1.1.11, is termed *sliding mode*. Sliding mode controllers make up a special class of variable structure

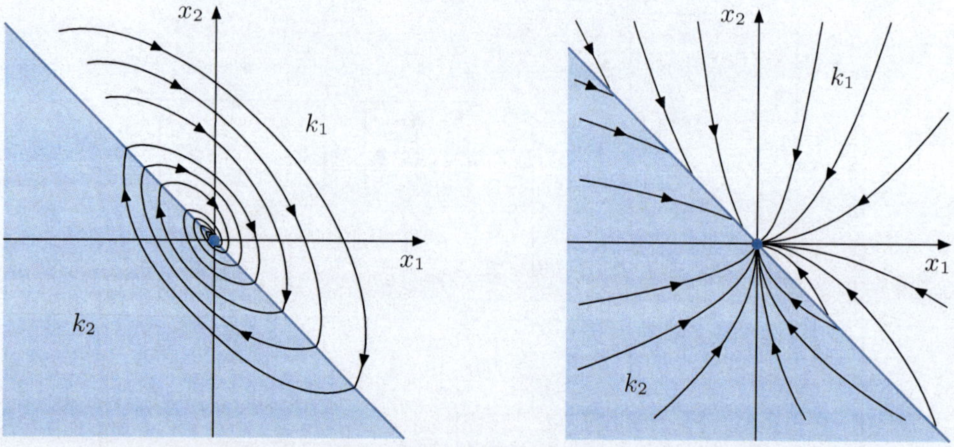

Fig. 4.27: Different trajectories for switching controllers: on the left without and on the right with a sliding mode. The blue line represents the switching curve $s(\boldsymbol{x}) = 0$.

controllers and are not addressed here, but will be discussed in detail in Section 6.2.

In terms of their design complexity and control performance, variable structure control loops without a sliding mode, which are the subject of this section, lie between linear and time-optimal control loops. An example is given in Figure 4.28 which shows typical step responses. The quality loss in the linear control loop compared to the time-optimal control loop is caused by the insufficient utilization of the available control variable $-u_{\max} \leq u \leq u_{\max}$ in the linear case. This is illustrated in Figure 4.29. Because of the linearity, the control variable u is correspondingly lower for small displacements $\boldsymbol{x}$ or control errors e than for large displacements. If, in contrast, we use high values of the control variable in the case of low values of the control error e or small displacements $\boldsymbol{x}$, a more rapid compensation becomes possible.

The basic idea of switching or variable structure controllers without sliding mode is to make better use of the controlled variable within the admissible range $-u_{\max} \leq u \leq u_{\max}$ than can be done using linear controllers. For this purpose, several linear controllers are used. We switch between them by using a selection strategy which normally depends on the system's state vector $\boldsymbol{x}$. This is done in such a way that successively more effective controllers, i. e. controllers making the control loop faster and faster, are used during the control process. Figure 4.25 shows the corresponding control loop.

For variable structure controls, linear systems are mostly considered to be controlled systems. The individual controllers can be either linear or nonlinear. However, they are usually linear. In any case, because of the switching operation, the resulting control loop is nonlinear.

The performance of a parameter or structure switching control is defined by three factors: the number l of controllers, the controllers' type and design, and the switching strategy. Provided that the controllers and the selection

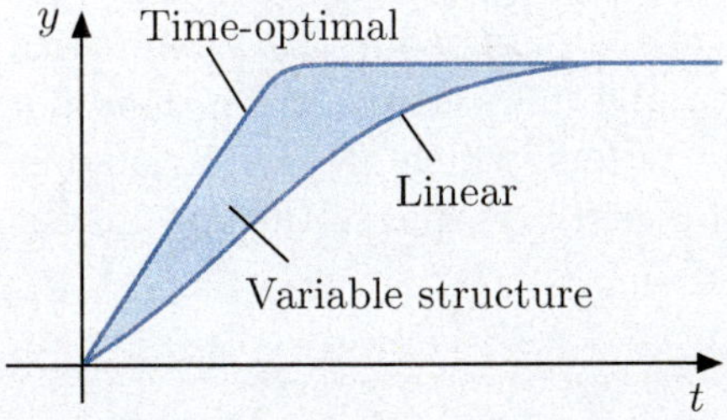

Fig. 4.28: Comparison of the step responses of linear, variable structure, and time-optimal control systems

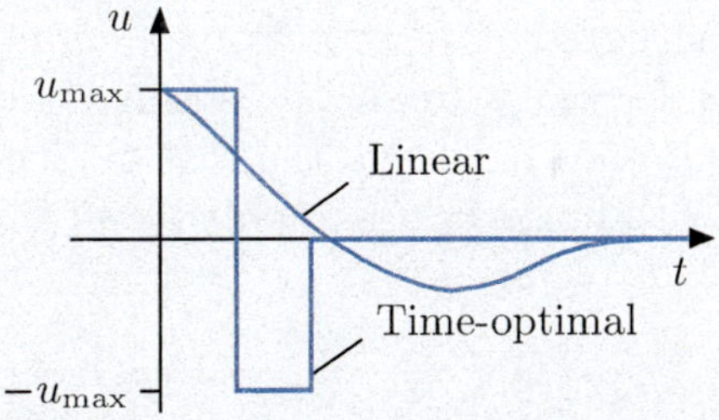

Fig. 4.29: Typical control variable signals of linear and time-optimal controllers

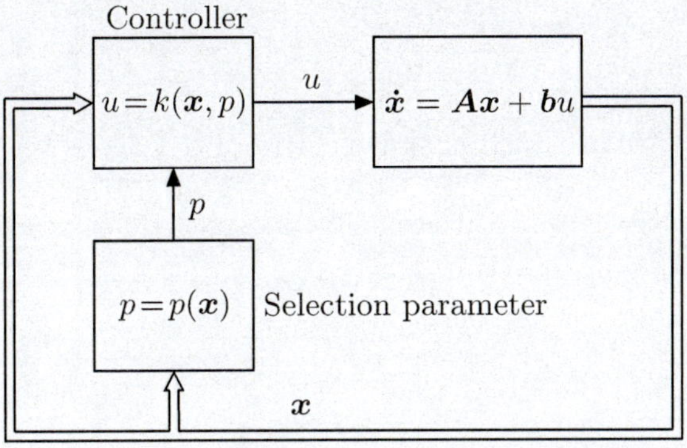

Fig. 4.30: Soft variable structure controller

strategy are adequately chosen, the achievable control performance increases for an increasing number of controllers used in maximally close sequence.

The switching need not be based on a switching line, as discussed for the simple example given above. It can also follow other types of curves or, for higher-dimensional spaces, other types of surfaces. For example, in the next section, we will discuss a controller with a selection strategy that switches based on ellipsoids.

As mentioned previously, the achievable control performance of a switching controller without a sliding mode increases with the number of controllers used. So an obvious approach is to use as many controllers as possible and ultimately an infinite number. Then the controller change does not occur due to switching between the controllers; instead it is due to a continuous change of the control parameters in the control law

$$u = k(\boldsymbol{x}, p),$$

where the variation of the control law depends on a selection parameter

$$p = p(\boldsymbol{x}),$$

which is continuous in $\boldsymbol{x}$. These controllers, which are shown in Figure 4.30, are called *soft variable structure controllers* [6, 61, 132, 192]. They can be seen as a systematic evolution and improvement of switching controllers. The term *soft* results from the continuous control variable trajectory, which does not jump as it does in the case of switching controllers, which are discontinuous. A sliding mode is impossible in this kind of controllers due to the continuous control variable.

4.3.2 Piecewise Linear Control

In the following, we will describe a switching controller without a sliding mode for linear plants with a control variable limitation [235, 480]

$$\dot{\boldsymbol{x}} = \boldsymbol{A}\boldsymbol{x} + \boldsymbol{b}u,$$
$$|u| \leq u_{\max},$$

the general structure of which is shown in Figure 4.25. A set of l controllers

$$u = -\boldsymbol{k}_1^T\boldsymbol{x},$$
$$u = -\boldsymbol{k}_2^T\boldsymbol{x},$$
$$\vdots$$
$$u = -\boldsymbol{k}_l^T\boldsymbol{x},$$

is used, designed such that the feedback control system

$$\dot{\boldsymbol{x}} = \left(\boldsymbol{A} - \boldsymbol{b}\boldsymbol{k}_i^T\right)\boldsymbol{x} = \hat{\boldsymbol{A}}_i\boldsymbol{x}, \quad i = 1, \ldots, l, \tag{4.30}$$

becomes faster with the increasing index i, i. e. the settling time of the corresponding linear subcontrol loop decreases for increasing index values i. For example, this can be achieved by calculating each $\boldsymbol{k}_i$ as the parameter vector of a linear-quadratic controller based on the performance integral

$$J = \int_0^\infty \left(p_i\boldsymbol{x}^T\boldsymbol{P}\boldsymbol{x} + \frac{1}{p_i}ru^2\right)\, dt.$$

Here the matrix $\boldsymbol{P}$ is positive definite or positive semidefinite. The factor p_i must be chosen so that the control loop (4.30) utilizes increasingly faster controllers with higher control variable values for increasing p_i.

A second simple possibility exists for selecting the controller vectors $\boldsymbol{k}_i$. We can specify them using an eigenvalue placement such that the eigenvalues λ_{ji}, $j = 1, \ldots, n$, of the closed control loop (4.30) take on lower and lower real parts for increasing index values i. Among other methods, we can do this by shifting the eigenvalues λ_{ji} on the rays

$$\lambda_{j(i+1)} = \alpha_i \cdot \lambda_{ji}, \quad \alpha_i > 1,$$

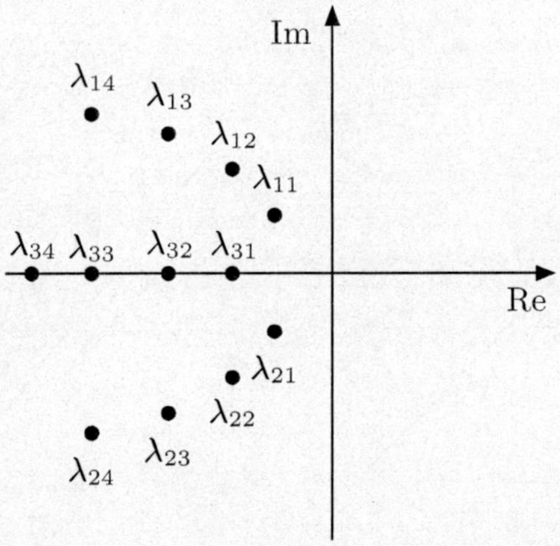

Fig. 4.31: Eigenvalues λ_{ji} of the parameter switching control loop

further and further to the left in the negative half-plane for increasing values i. Other eigenvalue progressions are also possible, as shown in Figure 4.31.

For each of the l control loops (4.30), we determine a catchment region

$$G_i = \left\{ \boldsymbol{x} \in I\!R^n \mid \boldsymbol{x}^T \boldsymbol{R}_i \boldsymbol{x} < c_i \right\},$$

where $\boldsymbol{R}_i$ results from the Lyapunov equation

$$\hat{\boldsymbol{A}}_i^T \boldsymbol{R}_i + \boldsymbol{R}_i \hat{\boldsymbol{A}}_i = -\boldsymbol{Q}_i.$$

The matrices $\boldsymbol{Q}_i$ must be chosen as suitable positive definite matrices.

The scaling factors c_i determine the expansion of the catchment regions G_i. Using the Lagrange multiplier method to maximize a region G_i so that its boundary is tangent to the hyperplanes

$$|u| = |\boldsymbol{k}_i^T \boldsymbol{x}| = u_{\max},$$

as shown in Figure 4.32, we can calculate the scaling factor c_i, obtaining

$$c_i = \frac{u_{\max}^2}{\boldsymbol{k}_i^T \boldsymbol{R}_i^{-1} \boldsymbol{k}_i}.$$

This ensures that no trajectory $\boldsymbol{x}(t)$ beginning in G_i violates any control variable limitation

$$|u| = |\boldsymbol{k}_i^T \boldsymbol{x}| \le u_{\max},$$

since G_i is a catchment region, and thus the trajectory $\boldsymbol{x}(t)$ does not leave it.

The next step is verifying whether all the catchment regions G_i are nested within each other, i. e. if

$$G_l \subset G_{l-1} \subset \ldots \subset G_2 \subset G_1$$

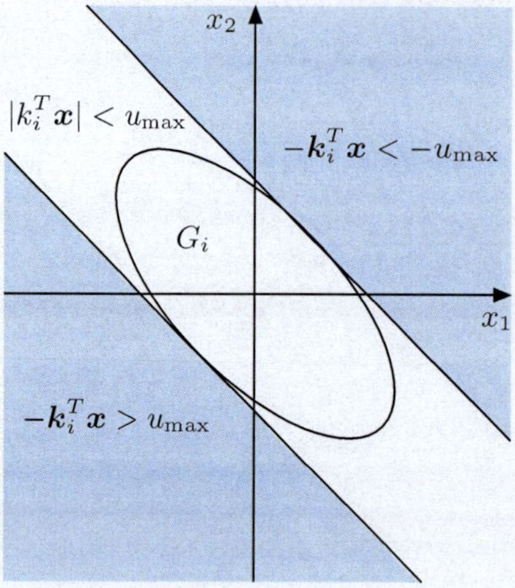

Fig. 4.32: Catchment regions G_i and the control variable limitation

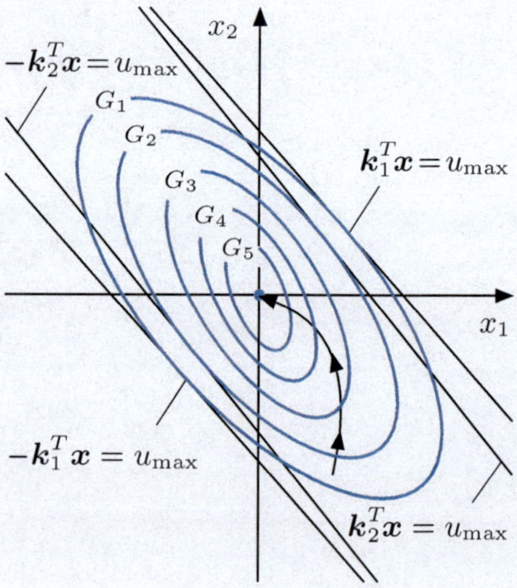

Fig. 4.33: A series of nested catchment regions G_i

holds. Figure 4.33 illustrates this nesting.

If the nesting condition is fulfilled, each trajectory moves from a larger region G_i into the smaller region G_{i+1}. Since the switching is always to a better, meaning a faster controller $\boldsymbol{k}_{i+1}$, the compensation becomes more rapid for each region.

The nesting of the regions G_i is simple to verify, since the requirement $G_{i+1} \subset G_i$ is fulfilled if and only if

$$\frac{\boldsymbol{x}^T \boldsymbol{R}_i \boldsymbol{x}}{c_i} < \frac{\boldsymbol{x}^T \boldsymbol{R}_{i+1} \boldsymbol{x}}{c_{i+1}}$$

or

$$0 < \boldsymbol{x}^T \left(\frac{\boldsymbol{R}_{i+1}}{c_{i+1}} - \frac{\boldsymbol{R}_i}{c_i} \right) \boldsymbol{x}$$

holds. The latter is fulfilled if the matrix

$$\boldsymbol{S}_i = \frac{\boldsymbol{R}_{i+1}}{c_{i+1}} - \frac{\boldsymbol{R}_i}{c_i} \tag{4.31}$$

is positive definite. The nesting of all regions G_i therefore exists if all matrices $\boldsymbol{S}_i$, for $i = 1, \ldots, l-1$, are positive definite. If the nesting condition is not fulfilled for a specified set of controller vectors $\boldsymbol{k}_i$ and catchment regions G_i, new vectors $\boldsymbol{k}_i$ and regions G_i with greater distances between them are chosen such that condition (4.31) holds for all $i = 1, \ldots, l-1$.

The control law is given by

$$u = -\boldsymbol{k}_i^T \boldsymbol{x}$$

with

$$i = \begin{cases} 1 & \text{if } \boldsymbol{x} \in G_1 \backslash G_2, \\ 2 & \text{if } \boldsymbol{x} \in G_2 \backslash G_3, \\ \vdots & \qquad \vdots \\ l-1 & \text{if } \boldsymbol{x} \in G_{l-1} \backslash G_l, \\ l & \text{if } \boldsymbol{x} \in G_l. \end{cases}$$

Here the set differences $G_i \backslash G_{i+1}$ make up the zones between two regions G_i. A controller $\boldsymbol{k}_i$ is always active if the state vector $\boldsymbol{x}$ lies within the associated ringlike zone $G_i \backslash G_{i+1}$. Since, for a progressing compensation, the trajectory $\boldsymbol{x}(t)$ subsequently enters smaller regions G_i and faster controllers $\boldsymbol{k}_i$ are activated, the compensation compared to a classical linear state controller such as $u = -\boldsymbol{k}_1^T \boldsymbol{x}$ becomes increasingly rapid. Figure 4.33 illustrates this procedure.

In summary, the following design scheme results:

Step 1: For the plant

$$\dot{\boldsymbol{x}} = \boldsymbol{A}\boldsymbol{x} + \boldsymbol{b}u \quad \text{with } |u| \leq u_{\max},$$

we select a family of l controllers

$$u = -\boldsymbol{k}_i^T \boldsymbol{x}, \quad i = 1, \ldots, l.$$

Step 2: For every subcontrol loop

$$\dot{\boldsymbol{x}} = \left(\boldsymbol{A} - \boldsymbol{b}\boldsymbol{k}_i^T\right) \boldsymbol{x} = \hat{\boldsymbol{A}}_i \boldsymbol{x}, \quad i = 1, \ldots, l,$$

we determine a catchment region

$$G_i = \left\{ \boldsymbol{x} \in I\!R^n \mid \; \boldsymbol{x}^T \boldsymbol{R}_i \boldsymbol{x} < c_i = \frac{u_{\max}^2}{\boldsymbol{k}_i^T \boldsymbol{R}_i^{-1} \boldsymbol{k}_i} \right\}$$

from

$$\hat{\boldsymbol{A}}_i^T \boldsymbol{R}_i + \boldsymbol{R}_i \hat{\boldsymbol{A}} = -\boldsymbol{Q}_i, \quad i = 1, \ldots, l$$

with an arbitrary positive definite matrix $\boldsymbol{Q}_i$.

Step 3: We verify whether the matrices

$$\boldsymbol{S}_i = \frac{\boldsymbol{R}_{i+1}}{c_{i+1}} - \frac{\boldsymbol{R}_i}{c_i}$$

are positive definite for all $i = 1, \ldots, l-1$. If this is not the case for an index i, a new design iteration must be started by returning to Step 1.

If instead of a finite number l of controllers $\boldsymbol{k}_i$ we select an infinite number with infinitely many catchment regions G_i, a soft variable structure control results [6, 192, 234, 267]. The regions G_i are then infinitesimally densely nested, that is, without any space in between. This means the settling time can be further improved and becomes nearly time-optimal. Furthermore, the control signal is continuous, since the controller parameters are continuously changed in dependence on the system's state, and, thus, there is no switching between any controllers.

4.3.3 Example: Ship-to-Shore Gantry Crane

As an example, we will consider a ship-to-shore gantry crane [303], as is used for loading and unloading container ships. Here, the container is suspended from a system of ropes under the crane's trolley, which transports the container from the dock to the ship, or vice-versa, as shown in Figure 4.34.

To achieve the fastest possible dispatching of the ship in the harbor, the cranes should load, and unload, the maximum possible number of containers per hour. This requires a fast back and forth movement of the trolley, which can lead to oscillations of the containers. Evidently, these are undesired, since they render the accurate placement of the containers impossible. A suitable controller enables the trolley to move at high speed. Moreover, it prevents the container from oscillating over its target position of placement.

The dynamics of the crane equipment are described by the model

$$\dot{\boldsymbol{x}} = \begin{bmatrix} 0 & 1 & 0 & 0 & 0 \\ 0 & 0 & \dfrac{-\beta\delta}{\alpha\gamma-\beta^2} & 0 & \dfrac{\gamma}{\alpha\gamma-\beta^2} \\ 0 & 0 & 0 & 1 & 0 \\ 0 & 0 & \dfrac{-\alpha\delta}{\alpha\gamma-\beta^2} & 0 & \dfrac{\beta}{\alpha\gamma-\beta^2} \\ 0 & 0 & 0 & 0 & -\omega_{\mathrm{s}} \end{bmatrix} \boldsymbol{x} + \begin{bmatrix} 0 \\ 0 \\ 0 \\ 0 \\ K \end{bmatrix} u.$$

Here, the position s of the trolley, its velocity $\dot{s}$, the container's swing angle φ, the angular velocity $\dot{\varphi}$, and the force F with which the trolley is accelerated make up the elements of the state vector

$$\boldsymbol{x} = [s \quad \dot{s} \quad \varphi \quad \dot{\varphi} \quad F]^T.$$

The current of the trolley's electric motor is the input variable u. For the parameters of the crane, we use the abbreviations

$$\begin{aligned} \alpha &= m_{\mathrm{t}} + m_{\mathrm{l}}, \\ \beta &= m_{\mathrm{l}}(f-e), \end{aligned}$$

Fig. 4.34: Ship-to-shore gantry crane

$$\gamma = m_{\mathrm{l}}(f-e)^2 + J,$$

$$\delta = m_{\mathrm{l}} \cdot g \cdot \left(e + \frac{b}{d-b} f + \frac{(d-b)d}{4l^2-(d-b)^2} f \right),$$

$$f = \frac{bl}{d-b}\sqrt{1-\frac{1}{4}\left(\frac{d-b}{l}\right)^2}.$$

Here m_{t} is the mass of the trolley, m_{l} is the mass of the container and its mount, l is the rope length, b is the distances between the container's rope attachments, d is the distances between the trolley's rope attachments, e is the vertical distance of the container's rope attachment to its center of gravity, J is the container's inertia moment, and $g = 9.81 \mathrm{m\,s^{-2}}$ is the gravitational

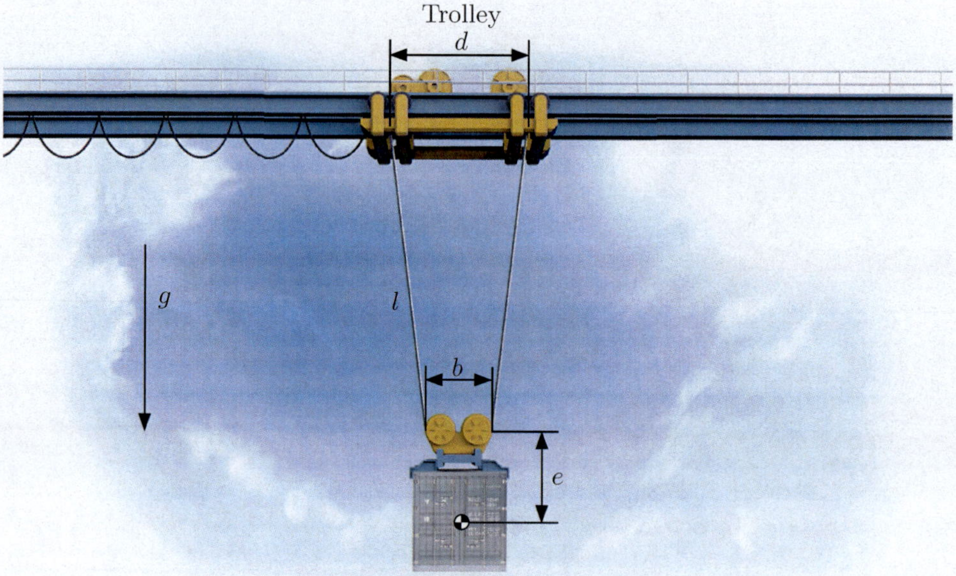

Fig. 4.35: Trolley and container

acceleration, as shown in Figure 4.35. The constants K and ω_s are parameters of the electric motor.

In the following, we will consider a design with

$$\begin{aligned} m_t &= 32\,\text{t}, & m_l &= 18.8\,\text{t}, & J &= 20.5\,\text{t}\,\text{m}^2,\\ d &= 3\,\text{m}, & b &= 2\,\text{m}, & e &= 1.6\,\text{m},\\ l &= 15\,\text{m}, & \omega_s &= 0.91\,\text{s}^{-1}, & K &= 0.94\,\text{kN}\,\text{s}^{-1}\text{A}^{-1}, \end{aligned}$$

and a control variable limitation of

$$u_{\max} = 100\,\text{A}.$$

With the above parameter specifications, we arrive at

$$\dot{\boldsymbol{x}} = \begin{bmatrix} 0 & 1 & 0 & 0 & 0\\ 0 & 0 & -12.4949 & 0 & 0.0312\\ 0 & 0 & 0 & 1 & 0\\ 0 & 0 & -1.1895 & 0 & 0.0011\\ 0 & 0 & 0 & 0 & -0.9100 \end{bmatrix} \boldsymbol{x} + \begin{bmatrix} 0\\ 0\\ 0\\ 0\\ 0.94 \end{bmatrix} u \qquad (4.32)$$

for the model. The set of possible initial displacements $\boldsymbol{x}_0 = \boldsymbol{x}(0)$ is

$$X_0 = \{\boldsymbol{x}_0 \in I\!R^5 \mid |s| \leq 20\,\text{m},\ |\dot{s}| \leq 3\,\text{m}\,\text{s}^{-1},\ |\varphi| \leq 5^\circ,\ |\dot{\varphi}| \leq 2^\circ\text{s}^{-1},\ F = 0\,\text{kN}\}.$$

Note that, for clarity, the angle φ and the angular velocity $\dot{\varphi}$ are given in degrees and degrees per second, respectively. In the differential equation (4.32), on the other hand, radians and radians per second must be used.

Next we will select ten state controllers with vectors $\boldsymbol{k}_i$ for the design of the switching controller consisting of these subcontrollers $\boldsymbol{k}_i$ and the selection strategy. Then we determine ten adequate catchment regions

$$G_i = \left\{ \boldsymbol{x} \in I\!R^5 \mid \boldsymbol{x}^T \boldsymbol{R}_i \boldsymbol{x} < c_i \right\}.$$

The largest of these regions, G_1, is specified such that it contains the initial set X_0. The parameters of the subcontrollers $\boldsymbol{k}_i$ and the matrices $\boldsymbol{R}_i$ can be found in Appendix A.2. The eigenvalues of the subcontrol loops

$$\dot{\boldsymbol{x}} = (\boldsymbol{A} - \boldsymbol{b}\boldsymbol{k}_i^T)\boldsymbol{x}$$

are shown in Figure 4.36. As an initial displacement, we will take

$$\boldsymbol{x}_0 = \begin{bmatrix} s \\ \dot{s} \\ \varphi \\ \dot{\varphi} \\ F \end{bmatrix} = \begin{bmatrix} 20 \\ 0 \\ 0 \\ 0 \\ 0 \end{bmatrix}.$$

The simulation of the state variables $x_1 = s$, $x_3 = \varphi$, $x_5 = F$, and the control variable u is shown in Figure 4.37. As a comparison, the graphs also show the time courses for the linear state controller with gain $\boldsymbol{k}_1$ and the time-optimal feedforward controller. As expected, the control signal $u(t)$ of the switching controller, i. e. the motor current, contains discontinuities. These can be tolerated in this case, since the force F, which results from the motor current u and which drives the crane, has a continuous progression.

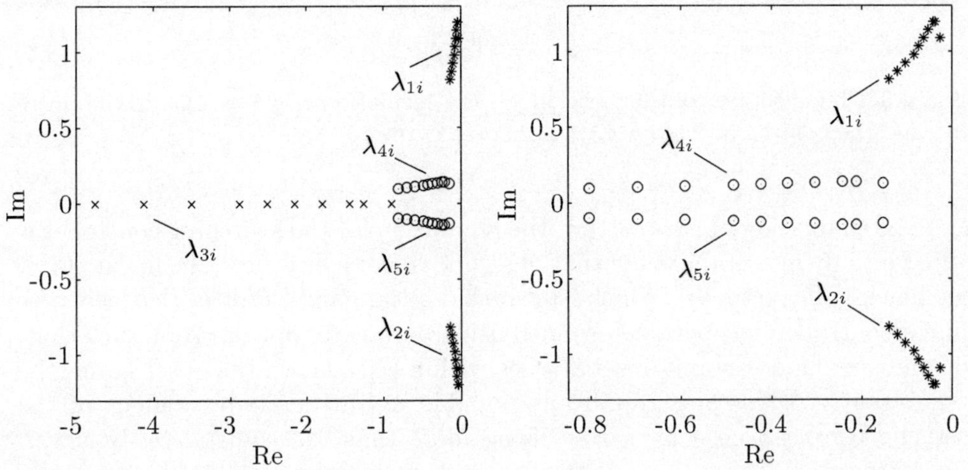

Fig. 4.36: The left plot shows all eigenvalues of the ten subcontrol loops. The right plot shows a detailed view of the area of interest on the left plot so as to better display the values taken on by λ_{1i} and λ_{2i} as well as λ_{4i} and λ_{5i}.

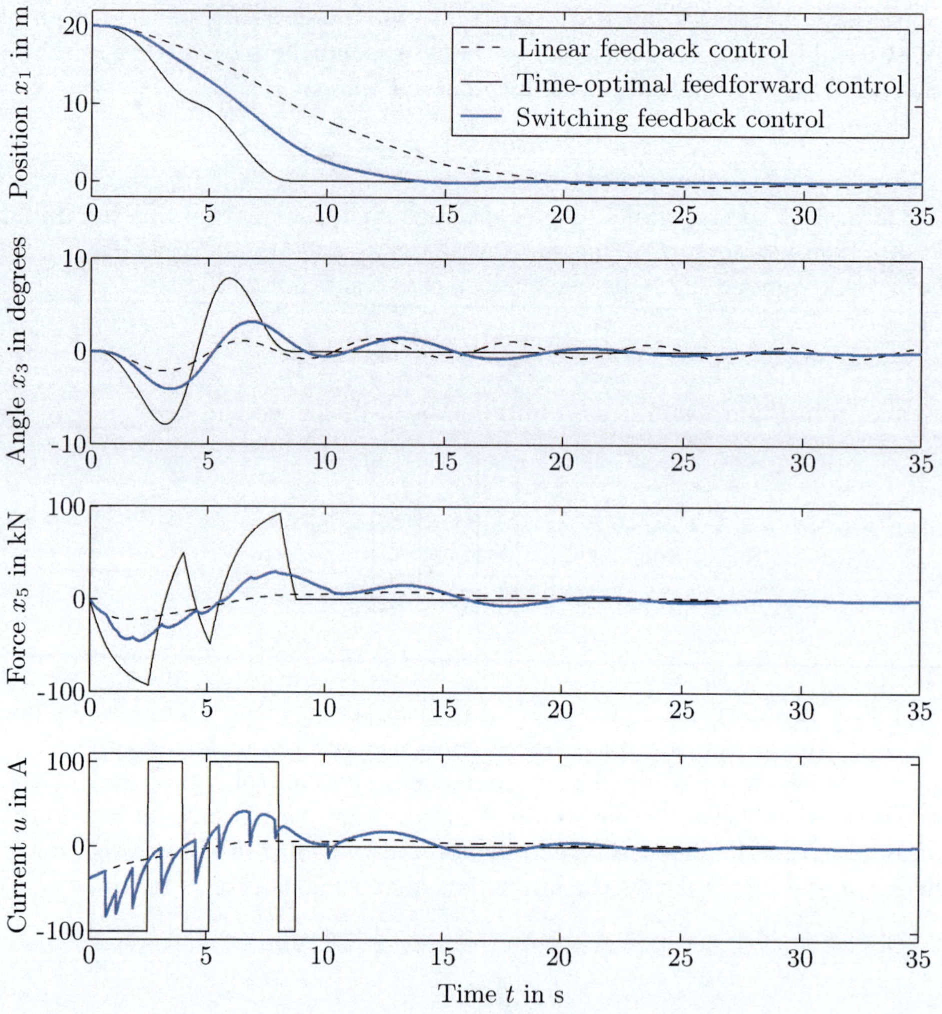

Fig. 4.37: Time courses of the position x_1, the oscillation angle x_3, the driving force x_5, and the control variable u, i. e. the motor current

The simulation illustrates that the regulation for the switching controller is substantially more rapid than that of a good linear controller. The linear state-feedback gain vector $\boldsymbol{k}_1$, which is also the initial gain vector of the piecewise linear controller, has been determined using parameter optimization such that, on the one hand, the initial set X_0 lies within G_1 and, on the other hand, the regulation progresses as rapidly as possible. A further improvement in the settling time can be achieved by using more than ten controllers. However, for an increasing number of controllers, the design complexity also increases. Note that the improvement in the control variable utilization compared to the linear controller comes at the cost of a higher control energy consumption.

4.4 Saturation Controllers

4.4.1 Basics and Stability

Again we will consider a linear plant

$$\dot{\boldsymbol{x}} = \boldsymbol{A}\boldsymbol{x} + \boldsymbol{b}u$$

with the input constraints

$$|u| \le u_{\max}. \tag{4.33}$$

As a controller, we will use a state controller $u = -\boldsymbol{k}^T\boldsymbol{x}$ which is allowed to violate the limitation (4.33). We thus obtain the control law

$$u = \operatorname{sat}(-\boldsymbol{k}^T\boldsymbol{x}) = \begin{cases} u_{\max}, & -\boldsymbol{k}^T\boldsymbol{x} > u_{\max}, \\ -\boldsymbol{k}^T\boldsymbol{x}, & -u_{\max} \le -\boldsymbol{k}^T\boldsymbol{x} \le u_{\max}, \\ -u_{\max}, & -\boldsymbol{k}^T\boldsymbol{x} < -u_{\max}. \end{cases} \tag{4.34}$$

Figure 4.38 shows the structure of the resulting control loop. In the following, it will become apparent that this simple nonlinear controller structure not only ensures the stability of the control loop but can also achieve fast regulation for certain plants.

Here a set X_0 of possible initial displacements $\boldsymbol{x}(0)$ is given. All trajectories beginning in X_0 need to be stabilized so that they finally tend to the equilibrium point $\boldsymbol{x}_{\mathrm{eq}} = \boldsymbol{0}$. For unstable plants, this stabilization is not possible for every set X_0. The reason for this is the control variable limitation. If $u_{\max}$ is too low or X_0 is too large, the available maximal control variable $u_{\max}$ does not suffice for large initial displacements $\boldsymbol{x}(0)$ to stabilize the trajectory $\boldsymbol{x}(t)$ which starts at $\boldsymbol{x}(0)$.

In order to ensure stability, the set X_0 is assumed to lie within an ellipsoid-shaped catchment region

$$G = \left\{\boldsymbol{x} \in I\!R^n \mid \boldsymbol{x}^T\boldsymbol{R}\boldsymbol{x} < c\right\},$$

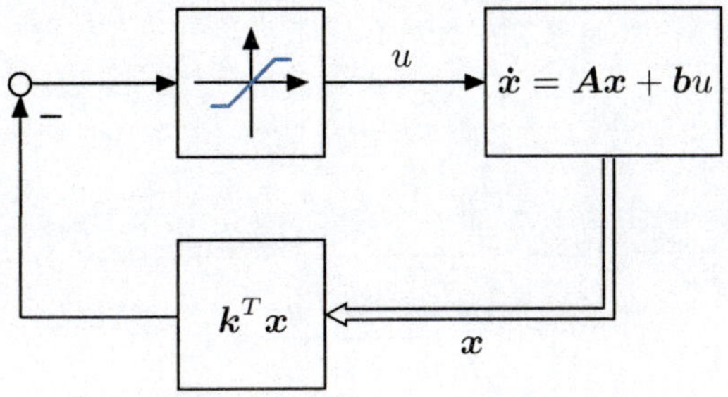

Fig. 4.38: Control loop with saturation controller $u = -\operatorname{sat}(\boldsymbol{k}^T\boldsymbol{x})$

i. e. within a region which no trajectory of the control loop leaves. The matrix $\boldsymbol{R}$ that defines this region will be determined in the course of the stability analysis.

For the stability analysis, we will now transform the control law (4.34) so that it takes on the form

$$u = -\operatorname{sat}(\boldsymbol{k}^T\boldsymbol{x}) = -\boldsymbol{h}^T\boldsymbol{x} - p(\boldsymbol{k}^T\boldsymbol{x} - \boldsymbol{h}^T\boldsymbol{x}) \tag{4.35}$$

with

$$p = \frac{\operatorname{sat}(\boldsymbol{k}^T\boldsymbol{x}) - \boldsymbol{h}^T\boldsymbol{x}}{\boldsymbol{k}^T\boldsymbol{x} - \boldsymbol{h}^T\boldsymbol{x}}. \tag{4.36}$$

Here $\boldsymbol{h}$ is configured in such a way that the catchment region G is a subset of the set

$$Z_{\boldsymbol{h}} = \{\boldsymbol{x} \in \mathbb{R}^n \mid |\boldsymbol{h}^T\boldsymbol{x}| \le u_{\max}\}.$$

The two hyperplanes

$$-\boldsymbol{h}^T\boldsymbol{x} = \pm u_{\max}$$

can be tangent to the ellipse G, but must not intersect it. Figure 4.39 illustrates this.

The state-feedback gain $\boldsymbol{k}$ is chosen such that it leads to regulation which is significantly more rapid than the one we would obtain if we were to use $\boldsymbol{h}$ instead of $\boldsymbol{k}$. Its control signal $u(t)$ is correspondingly larger and the set

$$Z_{\boldsymbol{k}} = \left\{\boldsymbol{x} \in \mathbb{R}^n \mid |\boldsymbol{k}^T\boldsymbol{x}| \le u_{\max}\right\}$$

intersects the catchment region G, as shown in Figure 4.40.

We will now address all state vectors $\boldsymbol{x} \in G$. These will be regulated to the equilibrium point $\boldsymbol{x}_{\text{eq}} = \boldsymbol{0}$ using the control law (4.34) or, equivalently, (4.35). This yields different cases; we will first view the case for which

$$\boldsymbol{x} \in Z_{\boldsymbol{k}} \quad \text{and} \quad \boldsymbol{x} \in G$$

hold, i. e. the state $\boldsymbol{x}$ is in the dark blue shaded set shown in Figure 4.40. Obviously, for such vectors $\boldsymbol{x}$ the inequalities

$$|\boldsymbol{h}^T\boldsymbol{x}| \le u_{\max} \quad \text{and} \quad |\boldsymbol{k}^T\boldsymbol{x}| \le u_{\max}$$

are fulfilled, so that with $\operatorname{sat}(\boldsymbol{k}^T\boldsymbol{x}) = \boldsymbol{k}^T\boldsymbol{x}$ we obtain the factor

$$p = \frac{\boldsymbol{k}^T\boldsymbol{x} - \boldsymbol{h}^T\boldsymbol{x}}{\boldsymbol{k}^T\boldsymbol{x} - \boldsymbol{h}^T\boldsymbol{x}} = 1 \tag{4.37}$$

from equation (4.36). The control law (4.35) is linear for these values of $\boldsymbol{x}$, so that

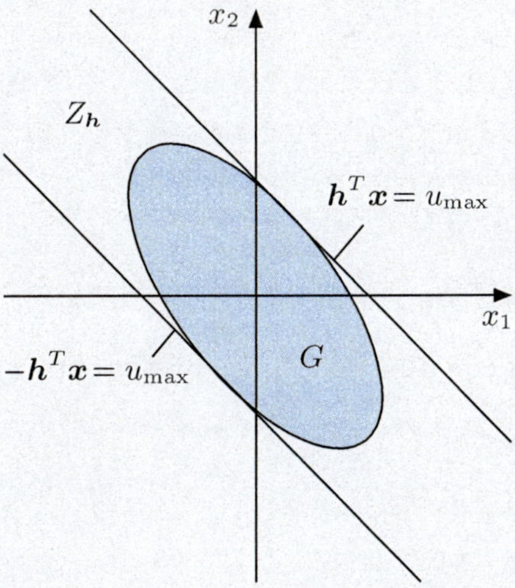

Fig. 4.39: Region G, set Z_h, and the hyperplanes $|h^T x| = u_{max}$

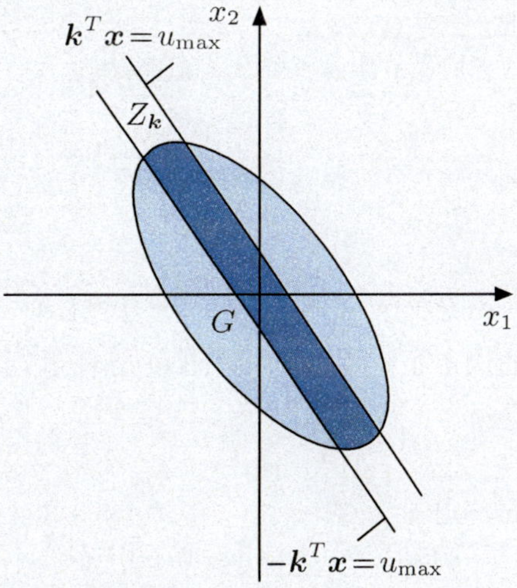

Fig. 4.40: Region G, set Z_k, and the hyperplanes $|k^T x| = u_{max}$

$$u = -\boldsymbol{k}^T \boldsymbol{x}$$

applies.

As the next vector $\boldsymbol{x}$ to be adjusted, we will select an arbitrary vector $\boldsymbol{x}$ which lies within G but not within $Z_{\boldsymbol{k}}$. This set $G \backslash Z_{\boldsymbol{k}}$ is highlighted in light blue in Figure 4.40. For the states $\boldsymbol{x} \in G \backslash Z_{\boldsymbol{k}}$, the inequality

$$|\boldsymbol{k}^T \boldsymbol{x}| > u_{max}$$

holds, i. e. the control is in saturation. Therefore, the control law

$$u = -\operatorname{sat}(\boldsymbol{k}^T \boldsymbol{x}) = -u_{max} \operatorname{sgn}(\boldsymbol{k}^T \boldsymbol{x}) \quad \text{for} \quad \boldsymbol{x} \in G \backslash Z_{\boldsymbol{k}}$$

follows. Using

$$\begin{aligned}\boldsymbol{k}^T \boldsymbol{x} &= |\boldsymbol{k}^T \boldsymbol{x}| \operatorname{sgn}(\boldsymbol{k}^T \boldsymbol{x}),\\ \boldsymbol{h}^T \boldsymbol{x} &= |\boldsymbol{h}^T \boldsymbol{x}| \operatorname{sgn}(\boldsymbol{h}^T \boldsymbol{x}),\end{aligned}$$

this yields

$$p = \frac{u_{max} \operatorname{sgn}(\boldsymbol{k}^T \boldsymbol{x}) - |\boldsymbol{h}^T \boldsymbol{x}| \operatorname{sgn}(\boldsymbol{h}^T \boldsymbol{x})}{|\boldsymbol{k}^T \boldsymbol{x}| \operatorname{sgn}(\boldsymbol{k}^T \boldsymbol{x}) - |\boldsymbol{h}^T \boldsymbol{x}| \operatorname{sgn}(\boldsymbol{h}^T \boldsymbol{x})}$$

for equation (4.36). We now multiply both the numerator and denominator by $\operatorname{sgn}(\boldsymbol{k}^T \boldsymbol{x})$ and arrive at

$$p = \frac{u_{max} - |\boldsymbol{h}^T \boldsymbol{x}| \operatorname{sgn}(\boldsymbol{h}^T \boldsymbol{x}) \operatorname{sgn}(\boldsymbol{k}^T \boldsymbol{x})}{|\boldsymbol{k}^T \boldsymbol{x}| - |\boldsymbol{h}^T \boldsymbol{x}| \operatorname{sgn}(\boldsymbol{h}^T \boldsymbol{x}) \operatorname{sgn}(\boldsymbol{k}^T \boldsymbol{x})} = \frac{u_{max} \pm |\boldsymbol{h}^T \boldsymbol{x}|}{|\boldsymbol{k}^T \boldsymbol{x}| \pm |\boldsymbol{h}^T \boldsymbol{x}|}$$

for $\boldsymbol{x} \in G \backslash Z_{\boldsymbol{k}}$.

From this result, and with

$$|\boldsymbol{h}^T \boldsymbol{x}| \leq u_{\max} < |\boldsymbol{k}^T \boldsymbol{x}| \quad \text{for } \boldsymbol{x} \in G \backslash Z_{\boldsymbol{k}},$$

we obtain

$$0 \leq p < 1 \tag{4.38}$$

for all $\boldsymbol{x} \in G \backslash Z_{\boldsymbol{k}}$. With equation (4.38) and equation (4.37), we finally obtain

$$0 \leq p \leq 1$$

for all $\boldsymbol{x} \in G$.

To guarantee the stability of the control loop

$$\dot{\boldsymbol{x}} = \boldsymbol{A}\boldsymbol{x} - \boldsymbol{b}\,\mathrm{sat}(\boldsymbol{k}^T \boldsymbol{x}) = \left[\boldsymbol{A} - \boldsymbol{b}\boldsymbol{h}^T - p\boldsymbol{b}\left(\boldsymbol{k}^T - \boldsymbol{h}^T\right)\right]\boldsymbol{x}$$

for all $\boldsymbol{x} \in G$ and to specify G as a catchment region, we choose

$$V(\boldsymbol{x}) = \boldsymbol{x}^T \boldsymbol{R}\boldsymbol{x}$$

as a Lyapunov function. For the time derivative of V, we obtain

$$\dot{V}(\boldsymbol{x}) = \boldsymbol{x}^T\left[\left(\boldsymbol{A} - \boldsymbol{b}\boldsymbol{h}^T - p\boldsymbol{b}\left(\boldsymbol{k}^T - \boldsymbol{h}^T\right)\right)^T \boldsymbol{R} + \boldsymbol{R}\left(\boldsymbol{A} - \boldsymbol{b}\boldsymbol{h}^T - p\boldsymbol{b}\left(\boldsymbol{k}^T - \boldsymbol{h}^T\right)\right)\right]\boldsymbol{x} < 0. \tag{4.39}$$

Obviously, $\dot{V}(\boldsymbol{x})$ is a linear function of p, so that $\dot{V}(\boldsymbol{x})$ takes on its maximum value, either in $p = 0$, or in $p = 1$ for $p \in [0, 1]$. The requirement (4.39) is thus fulfilled for all $p \in [0, 1]$ if

$$\boldsymbol{x}^T\left[\left(\boldsymbol{A} - \boldsymbol{b}\boldsymbol{h}^T\right)^T \boldsymbol{R} + \boldsymbol{R}\left(\boldsymbol{A} - \boldsymbol{b}\boldsymbol{h}^T\right)\right]\boldsymbol{x} < 0$$

holds for $p = 0$, and

$$\boldsymbol{x}^T\left[\left(\boldsymbol{A} - \boldsymbol{b}\boldsymbol{k}^T\right)^T \boldsymbol{R} + \boldsymbol{R}\left(\boldsymbol{A} - \boldsymbol{b}\boldsymbol{k}^T\right)\right]\boldsymbol{x} < 0$$

holds for $p = 1$. Using the results above, we can now derive the following stability theorem for the state control with saturation [191].

Theorem 58 (Hu and Lin's Stability Theorem). *The control loop*

$$\dot{\boldsymbol{x}} = \boldsymbol{A}\boldsymbol{x} - \boldsymbol{b}\,\mathrm{sat}(\boldsymbol{k}^T \boldsymbol{x})$$

possesses the asymptotically stable equilibrium point $\boldsymbol{x}_{\mathrm{eq}} = \boldsymbol{0}$ *with the catchment region*

$$G = \{\boldsymbol{x} \in I\!R^n \mid \boldsymbol{x}^T \boldsymbol{R} \boldsymbol{x} < c\}$$

if a vector $\boldsymbol{h}$ *exists, so that*

$$G \subseteq \{\boldsymbol{x} \in I\!R^n \mid |\boldsymbol{h}^T \boldsymbol{x}| \leq u_{\max}\} \tag{4.40}$$

holds, and the matrices

$$\left(\boldsymbol{A} - \boldsymbol{b}\boldsymbol{h}^T\right)^T \boldsymbol{R} + \boldsymbol{R}\left(\boldsymbol{A} - \boldsymbol{b}\boldsymbol{h}^T\right) = -\boldsymbol{Q}_{\boldsymbol{h}},$$
$$\left(\boldsymbol{A} - \boldsymbol{b}\boldsymbol{k}^T\right)^T \boldsymbol{R} + \boldsymbol{R}\left(\boldsymbol{A} - \boldsymbol{b}\boldsymbol{k}^T\right) = -\boldsymbol{Q}_{\boldsymbol{k}}$$

are negative definite.

Note that the vector $\boldsymbol{h}$ does not influence the control law $u = -\operatorname{sat}(\boldsymbol{k}^T\boldsymbol{x})$. However, it is of fundamental importance for the stability statement of Theorem 58. Requirement (4.40) is equivalent to the inequality

$$c \cdot \boldsymbol{h}^T \boldsymbol{R}^{-1} \boldsymbol{h} \leq u_{\max}^2.$$

This inequality can be easily verified by inserting $\boldsymbol{h}$ and $\boldsymbol{R}^{-1}$.

4.4.2 Design in Multiple Steps

Before describing the design of the saturation controller, we will first review the tasks involved in its design. The control loop is stated as

$$\dot{\boldsymbol{x}} = \boldsymbol{A}\boldsymbol{x} - \boldsymbol{b}\operatorname{sat}(\boldsymbol{k}^T\boldsymbol{x})$$

where a state-feedback gain $\boldsymbol{k}$ is sought such that all initial displacements $\boldsymbol{x}(0) \in X_0$ are regulated to the equilibrium point $\boldsymbol{x}_{\text{eq}} = \boldsymbol{0}$. Thus stability is ensured. The set of possible initial displacements X_0 is often a hypercube

$$X_0 = \{\boldsymbol{x} \in I\!R^n \mid -\alpha_i \leq x_i \leq \beta_i,\ i = 1, \ldots, n\}.$$

Let us now consider the design procedure, which can be divided into six steps:

Step 1: Determine a vector $\boldsymbol{h}$ such that

$$\dot{\boldsymbol{x}} = (\boldsymbol{A} - \boldsymbol{b}\boldsymbol{h}^T)\boldsymbol{x}$$

is asymptotically stable.

Step 2: Choose a positive definite matrix $\boldsymbol{Q}_{\boldsymbol{h}}$ and determine the matrix $\boldsymbol{R}$ of the catchment region $G = \{\boldsymbol{x} \in I\!R^n \mid \boldsymbol{x}^T\boldsymbol{R}\boldsymbol{x} < c\}$ from the Lyapunov equation

$$\left(\boldsymbol{A} - \boldsymbol{b}\boldsymbol{h}^T\right)^T \boldsymbol{R} + \boldsymbol{R}\left(\boldsymbol{A} - \boldsymbol{b}\boldsymbol{h}^T\right) = -\boldsymbol{Q}_{\boldsymbol{h}}.$$

Step 3: Set

$$c = \frac{u_{\max}^2}{\boldsymbol{h}^T \boldsymbol{R}^{-1} \boldsymbol{h}}$$

so that the two hyperplanes $|\boldsymbol{h}^T \boldsymbol{x}| = u_{\max}$ are tangent to the catchment region G.

Step 4: Using the vertices $\boldsymbol{x}_{\mathrm{v},i}$ of X_0, verify whether $X_0 \subseteq G$ holds, i. e. examine whether

$$\boldsymbol{x}_{\mathrm{v},i}^T \boldsymbol{R} \boldsymbol{x}_{\mathrm{v},i} < c$$

holds for all $i = 1, \ldots, 2^n$. If this is the case, proceed with Step 5. Otherwise, go back to Step 1 and choose a controller $\boldsymbol{h}$ with a smaller control-variable amplitude, or go to Step 2 and vary $\boldsymbol{Q}_{\boldsymbol{h}}$.

Step 5: Choose a state-feedback gain $\boldsymbol{k}$ and examine whether the matrix

$$\left(\boldsymbol{A} - \boldsymbol{b}\boldsymbol{k}^T\right)^T \boldsymbol{R} + \boldsymbol{R}\left(\boldsymbol{A} - \boldsymbol{b}\boldsymbol{k}^T\right) = -\boldsymbol{Q}_{\boldsymbol{k}}$$

is negative definite. If this is not the case, repeat Step 5.

Step 6: Simulate the control loop

$$\dot{\boldsymbol{x}} = \boldsymbol{A}\boldsymbol{x} - \boldsymbol{b}\,\mathrm{sat}(\boldsymbol{k}^T \boldsymbol{x}).$$

If the control behavior is satisfactory, finish the design. If it is not, go to Step 1 or Step 5, and repeat the design from there.

The disadvantage of the above design procedure lies in its heuristic components. The vectors $\boldsymbol{h}$ and $\boldsymbol{k}$, as well as the matrix $\boldsymbol{Q}_{\boldsymbol{h}}$ must be chosen based on guesswork. There is also no guarantee at the outset that $X_0 \subseteq G$ will hold. Instead, different design steps must potentially be repeated multiple times, and the procedure is based on trial and error.

The design procedure consisting of the six steps described above can also be done on a computer and then run automatically. Furthermore, a one-step design is possible [191] if all design requirements are summarized into a system of linear matrix inequalities [55]. The disadvantages of the six-step design procedure as described above do not occur in this case. Saturation controllers can also be used for MIMO plants [191].

4.4.3 Example: Helicopter

As an example, we will control a rescue helicopter using an autopilot. Helicopters of this kind are used by the coast guards of various countries. In a maritime rescue, as shown in Figure 4.41, the aim is for the helicopter to hover in a stationary position over the surface of the water. In situations such as gusty winds resulting in position displacements, the controller of the autopilot brings the helicopter back to the stationary target position. Aiding the pilot in this way is of great importance for emergency missions in stormy weather.

Below we will focus on the horizontal position displacement x_4 of the helicopter, measured in meters, relative to its rescue position. As part of the autopilot function, we aim to compensate for this position deviation to zero by using a state controller with saturation. Furthermore, we aim to correct displacements caused by gusty winds or other forces to zero as quickly as possible. The position change can be described by the model [98]

Fig. 4.41: Emergency rescue helicopter

$$\dot{\boldsymbol{x}} = \begin{bmatrix} -0.415 & 0 & -0.0111 & 0 \\ 1 & 0 & 0 & 0 \\ -1.43 & 9.8 & -0.0198 & 0 \\ 0 & 0 & 1 & 0 \end{bmatrix} \boldsymbol{x} + \begin{bmatrix} 6.27 \\ 0 \\ 9.8 \\ 0 \end{bmatrix} u,$$

$$y = \begin{bmatrix} 0\,0\,0\,1 \end{bmatrix} \boldsymbol{x},$$

where x_1 is the pitch angle rate, x_2 is the pitch angle, x_3 is the horizontal velocity, and x_4 is the position, or as mentioned above, the position deviation. The control variable u is the adjustable rotor angle, measured in radians. The rotor angle is limited by

$$-0.11 \leq u \leq 0.11.$$

For the set of possible disturbances, or initial position displacements, we will select

$$X_0 = \left\{ \boldsymbol{x} \in I\!R^4 \mid |x_1| \leq 0.1,\, |x_2| \leq 0.2,\, |x_3| \leq 1,\, |x_4| \leq 10 \right\}.$$

For a saturation controller

$$u = -\operatorname{sat}(\boldsymbol{k}^T \boldsymbol{x})$$

which corrects all disturbances of the helicopter's position within X_0, taking the control variable limitation

$$|u| \leq 0.11$$

into account, we determine

$$\boldsymbol{k}^T = \begin{bmatrix} 0.16226 & 0.20860 & 0.0089414 & 0.0020975 \end{bmatrix}.$$

Figure 4.42 shows the time courses of the correction for the disturbance displacement

$$\boldsymbol{x}_0 = \begin{bmatrix} 0.1 \\ 0.2 \\ 1 \\ 10 \end{bmatrix}.$$

The time courses of a linear controller

$$u = -\boldsymbol{k}_{\text{lin}}{}^T \boldsymbol{x}$$

with

$$\boldsymbol{k}_{\text{lin}}{}^T = \begin{bmatrix} 1.7719 & 2.8852 & 0.15954 & 0.038056 \end{bmatrix}$$

and the time-optimal feedforward controller are also shown for comparison. The substantially more rapid regulation by the saturation controller compared to the linear controller is plainly evident. However, this comes at the cost of a higher control signal velocity $\dot{u}$.

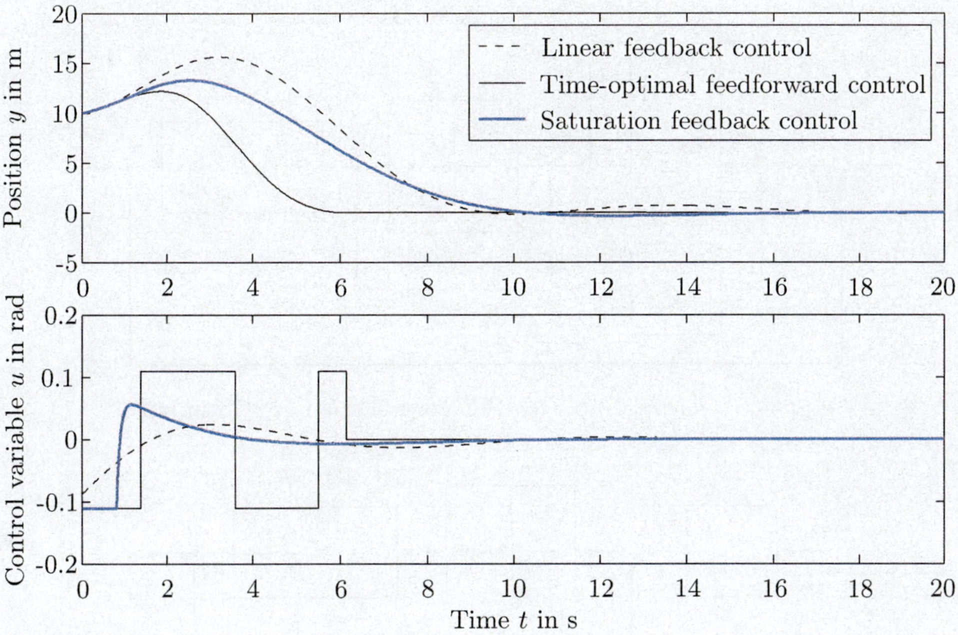

Fig. 4.42: Time courses of the position y and the control variable u of the helicopter

4.5 Exercises

Exercise 4.1 Let us examine the control loop with a PID controller and anti-windup from Figure 4.43. The PID controller's transfer function is

$$G_{\mathrm{PID}}(s) = \frac{N_{\mathrm{PID}}(s)}{s} = G_{\mathrm{PD}}(s) + \frac{K_{\mathrm{I}}}{s}.$$

(a) Convert the control loop from Figure 4.43 such that it takes the same form as the control loop from Figure 4.44 and thus makes possible the application of the Popov criterion and the circle criterion. In doing this, determine the transfer functions $Q(s)$, $R(s)$, and $G(s)R(s)$ with dependence on m, K_{I}, $N(s)$, $D(s)$, and $N_{\mathrm{PID}}(s)$.

Let us now examine the case of the plant

$$G(s) = \frac{1}{s}$$

and PI controller

$$G_{\mathrm{PI}}(s) = \frac{K_{\mathrm{P}}s + K_{\mathrm{I}}}{s}, \qquad K_{\mathrm{P}} = 2,\; K_{\mathrm{I}} = 1.$$

(b) Calculate the transfer function $G(s)G_{\mathrm{PI}}(s)$ of the open loop in the linear case, i. e. without anti-windup, its poles, and the transfer function $G(s)R(s)$ of the open loop with anti-windup and $m = 1000$.

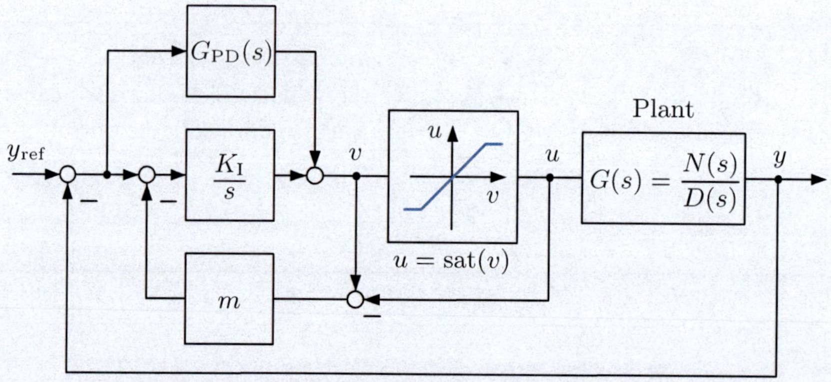

Fig. 4.43: Control loop with PID controller and anti-windup

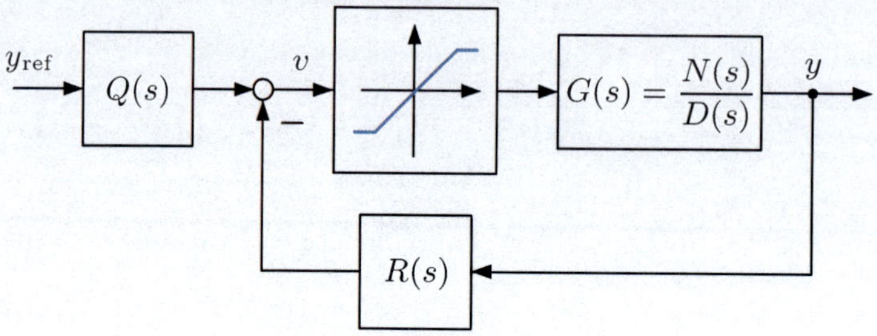

Fig. 4.44: Control loop

(c) Are the Popov criterion and the circle criterion applicable?
(d) Draw the Nyquist plots and the Popov plots of $G(s)G_{\mathrm{PI}}(s)$ and $G(s)R(s)$.
(e) If possible, identify the sectors of absolute stability for the control loop without anti-windup and the one with anti-windup, such that the global asymptotic stability of the origin is ensured.
(f) Calculate the Hurwitz sectors for the control loop with and without anti-windup. Is Aizerman's conjecture correct in these cases?

Exercise 4.2 Show that in the formulas (4.15) and (4.16) of the optimal switching times of the double integrator on p. 304, only a positive sign in front of the square root yields the correct switching times.

Exercise 4.3 Calculate the time-optimal control feedback law for the plant

$$\dot{\boldsymbol{x}} = \begin{bmatrix} 0 & 1 \\ 0 & 0 \end{bmatrix} \boldsymbol{x} + \begin{bmatrix} 0 \\ 1 \end{bmatrix} u, \qquad |u| \leq u_{\max},$$

using the following steps:
(a) Determine the trajectories of the plant for the two control values $u = u_{\max}$ and $u = -u_{\max}$.

(b) Determine the trajectories from (a) which pass through $\boldsymbol{x} = \mathbf{0}$. The parts of these two trajectories which move in the direction of and towards $\boldsymbol{x} = \mathbf{0}$ make up the switching curve $S(\boldsymbol{x}) = x_1$.

(c) Using the trajectories from (b), determine the time-optimal feedback control law $u(\boldsymbol{x})$ for all $\boldsymbol{x} \in I\!R^2$.

Exercise 4.4 For an unstable linear system

$$\dot{\boldsymbol{x}} = \boldsymbol{A}\boldsymbol{x} + \boldsymbol{b}u, \quad \boldsymbol{x} \in I\!R^n,$$

with a control variable limitation $|u| \leq u_{\max}$, no controller $u(\boldsymbol{x})$ exists with which all initial values $\boldsymbol{x}(0) \in I\!R^n$ can be transitioned to the equilibrium point $\boldsymbol{x}_{\text{eq}} = \mathbf{0}$, i.e. the system cannot be stabilized for all $\boldsymbol{x}(0) \in I\!R^n$. However, this is possible for a subset of $I\!R^n$. The largest of these sets is called *the null-controllable region* X_{null}. Determine the null-controllable region X_{null} for the unstable first-order system

$$\dot{x} = \lambda x + bu, \quad |u| \leq u_{\max}, \ \lambda > 0.$$

Exercise 4.5 Determine the switching time t_1 of the time-optimal feedforward controls and time-optimal feedback control laws $u(x)$ for the plants

(a) $\dot{x} = u, \quad |u| \leq u_{\max},$

(b) $\dot{x} = \lambda x + bu, \quad |u| \leq u_{\max}, \quad \lambda \in I\!R \setminus \{0\},\ b > 0.$

Exercise 4.6 Determine the time-optimal feedback control law for the system

$$\dot{\boldsymbol{x}} = \begin{bmatrix} \lambda & 1 \\ 0 & \lambda \end{bmatrix} \boldsymbol{x} + \begin{bmatrix} 0 \\ \lambda \end{bmatrix} u, \quad \lambda < 0, \quad |u| \leq u_{\max}.$$

In doing so, take into account that the switching curve is identical to sections of the system's trajectories for which $u = u_{\max}$ and $u = -u_{\max}$ apply and which pass through the origin $\boldsymbol{x} = \mathbf{0}$.

Exercise 4.7 The time-optimal feedback control of the triple integrator [96, 111, 141]

$$\dot{\boldsymbol{x}} = \begin{bmatrix} 0 & 1 & 0 \\ 0 & 0 & 1 \\ 0 & 0 & 0 \end{bmatrix} \boldsymbol{x} + \begin{bmatrix} 0 \\ 0 \\ 1 \end{bmatrix} u$$

with the control variable limitation $|u| \leq 1$ is given by

$$\begin{aligned} u &= -\operatorname{sgn}(a|a| + b^3), \\ a &= x_1 + \frac{1}{3}x_3^3 + x_2 x_3 \operatorname{sgn}(x_2 + \frac{1}{2}x_3|x_3|), \\ b &= x_2 + \frac{1}{2}x_3^2 \operatorname{sgn}(x_2 + \frac{1}{2}x_3|x_3|). \end{aligned}$$

State the control law for the case $|u| \leq u_{\max}$.

Exercise 4.8 Nuclear fusion reactors make possible the controlled fusion of deuterium and tritium nuclei. The heat that results can be used to produce electricity. The Tokamak is one design for a plant in which an electricity-conducting plasma with a temperature of up to 150 million degrees Celsius is generated in a torus-shaped chamber. This plasma is shaped and held within the chamber by means of magnetic fields of up to 10 T such that it does not touch the walls of the chamber. If this were not the case, the plasma would cool and the nuclear fusion would stop. The magnetic fields are generated by means of active inductors, i.e. inductors to which an external variable voltage u_{a} is applied and through which an electrical current i_{a} flows, and passive inductors in which an electrical current i_{v} is induced by the magnetic fields. These passive inductors have no external electricity supply. Using these inductors and their magnetic fields, the plasma, or more accurately, its center line, can be positioned within the torus; see Figure 4.45. The vertical shift z of the midpoint of the plasma can be represented by the simple model [407]

$$\begin{bmatrix} \dot{i}_a \\ \dot{i}_v \end{bmatrix} = \frac{1}{1-k_{\mathrm{av}}} \begin{bmatrix} -\dfrac{R_{\mathrm{aa}}}{L_{\mathrm{aa}}} & k_{\mathrm{av}}\dfrac{R_{\mathrm{vv}}}{L_{\mathrm{av}}} \\ k_{\mathrm{av}}\dfrac{R_{\mathrm{aa}}}{L_{\mathrm{av}}} & -\dfrac{R_{\mathrm{vv}}(k_{\mathrm{av}}-M_{\mathrm{vp}})}{L_{\mathrm{vv}}(1-M_{\mathrm{vp}})} \end{bmatrix} \underbrace{\begin{bmatrix} i_{\mathrm{a}} \\ i_{\mathrm{v}} \end{bmatrix}}_{\boldsymbol{i}} + \begin{bmatrix} \dfrac{1}{L_{\mathrm{aa}}(1-k_{\mathrm{av}})} \\ -\dfrac{k_{\mathrm{av}}}{L_{\mathrm{av}}(1-k_{\mathrm{av}})} \end{bmatrix} u_{\mathrm{a}}$$

$$z = \frac{1}{i_{\mathrm{p}}} \begin{bmatrix} -\dfrac{L_{\mathrm{ap}}}{A_{\mathrm{pp}}} & -\dfrac{L_{\mathrm{vp}}}{A_{\mathrm{pp}}} \end{bmatrix} \boldsymbol{i}, \quad \boldsymbol{i} = \begin{bmatrix} i_{\mathrm{a}} \; i_{\mathrm{v}} \end{bmatrix}^T. \tag{4.41}$$

Here i_{p} is the plasma flow. In the case of the joint European torus (JET), we have the following parameters:

$R_{\mathrm{aa}} = 35.0 \cdot 10^{-3}\,\Omega$	(Resistance of the active inductors),
$R_{\mathrm{vv}} = 2.56 \cdot 10^{-3}\,\Omega$	(Resistance of the passive inductors),
$L_{\mathrm{aa}} = 42.5 \cdot 10^{-3}\,\mathrm{H}$	(Self-inductance of the active inductors),
$L_{\mathrm{av}} = 0.432 \cdot 10^{-3}\,\mathrm{H}$	(Mutual inductance of the active and passive inductors),
$L_{\mathrm{vv}} = 0.012 \cdot 10^{-3}\,\mathrm{H}$	(Self-inductance of the passive inductors),
$L_{\mathrm{ap}} = 115 \cdot 10^{-6}\,\mathrm{Hm}^{-1}$	(Mutual inductance between active inductors and plasma per unit of length),
$L_{\mathrm{vp}} = 3.2 \cdot 10^{-6}\,\mathrm{Hm}^{-1}$	(Mutual inductance between passive inductors and plasma per unit of length),
$A_{\mathrm{pp}} = 0.5 \cdot 10^{-6}\,\mathrm{Hm}^{-2}$.	

In addition, the following apply:

Fig. 4.45: Tokamak-type nuclear fusion reactor

$$k_{\mathrm{av}} = \frac{L_{\mathrm{av}}^2}{L_{\mathrm{aa}}L_{\mathrm{vv}}} \quad \text{and} \quad M_{\mathrm{vp}} = \frac{A_{\mathrm{pp}}L_{\mathrm{vv}}}{L_{\mathrm{vp}}^2}.$$

For the maximum inductor voltage u_{a}, the limitation

$$|u_{\mathrm{a}}| \leq 10\,\mathrm{kV}$$

applies, while the plasma current is $i_{\mathrm{p}} = 400 \cdot 10^3\,\mathrm{A}$.

(a) Determine the model (4.41) of the JET and its eigenvalues λ_1 and λ_2 numerically.

(b) Transform the model of the JET calculated in (a) into the form

$$\dot{\boldsymbol{x}} = \begin{bmatrix} \lambda_1 & 0 \\ 0 & \lambda_2 \end{bmatrix} \boldsymbol{x} + \begin{bmatrix} \lambda_1 \\ \lambda_2 \end{bmatrix} u, \quad \lambda_1 > 0, \quad \lambda_2 < 0, \tag{4.42}$$

with $u = u_{\mathrm{a}}$. State the transformation $\boldsymbol{i} = \boldsymbol{T}\boldsymbol{x}$.

(c) For a general system (4.42) with $|u| \leq u_{\max}$, determine the time-optimal feedback control. For this purpose, determine the trajectories of the system for the control signal values $-u_{\max}$ and $u_{\max}$, and put together the switching curve of the time-optimal control using sections of the trajectories that pass through the equilibrium point $\boldsymbol{x}_{\mathrm{eq}} = \mathbf{0}$.

(d) Calculate the region $G \subseteq \mathbb{R}^2$ in $\boldsymbol{x}$-coordinates for which the JET can be stably controlled to zero, taking into account the control variable limitation $|u| \le u_{\max} = 10\,\mathrm{kV}$.

Exercise 4.9 Let the plant

$$\dot{\boldsymbol{x}} = \begin{bmatrix} 0 & 1 \\ 0 & 0 \end{bmatrix} \boldsymbol{x} + \begin{bmatrix} 0 \\ 1 \end{bmatrix} u \tag{4.43}$$

with the control variable limitation $|u| \le u_{\max}$ be given.

(a) Determine the time-optimal feedback control law $u_{\mathrm{to}}(\boldsymbol{x})$ which calculates the correct actuator value $u_{\mathrm{to}}(\boldsymbol{x})$ for all $\boldsymbol{x} \in \mathbb{R}^2$.

(b) In addition, determine a linear state controller $u = -\boldsymbol{k}^T\boldsymbol{x}$ for the plant (4.43) such that both eigenvalues of the control loop are $\lambda_{1,2} = -1$.

(c) Explain why the function

$$V(\boldsymbol{x}) = \boldsymbol{x}^T\boldsymbol{x}$$

is a Lyapunov function of the linear control loop with the controller from (b), and why it ensures asymptotic stability of the equilibrium point $\boldsymbol{x}_{\mathrm{eq}}=\boldsymbol{0}$.

(d) Examine the controller

$$u(\boldsymbol{x}) = \begin{cases} -\boldsymbol{k}^T\boldsymbol{x}, & \text{if } \boldsymbol{x}^T\boldsymbol{x} \le 0.01, \\ u_{\mathrm{to}}(\boldsymbol{x}), & \text{if } \boldsymbol{x}^T\boldsymbol{x} > 0.01, \end{cases} \tag{4.44}$$

which starts the control process with a time-optimal controller and then switches to a linear state controller in the final phase. Such a controller with two different modes is called a *dual-mode controller*. In the state plane, draw the areas in which $u = u_{\max}$, $u = -u_{\max}$, and $u = -\boldsymbol{k}^T\boldsymbol{x}$ apply.

(e) Why is the equilibrium point $\boldsymbol{x}_{\mathrm{eq}} = \boldsymbol{0}$ of the control loop (4.43), (4.44) globally asymptotically stable?

(f) What advantage does the control loop with the control law (4.44) have compared to the time-optimal control loop?

Exercise 4.10 Let us examine the discrete-time first-order system

$$x(k+1) = ax(k) + bu(k), \qquad 0 < a < 1,\ b > 0,$$

with the control variable limitation $|u| \le u_{\max}$.

(a) Calculate $x(k)$ in dependence on the initial value $x(0)$ and the input values $u(0), u(1), \ldots, u(k-1)$.

(b) Demonstrate that

$$x(k)a^{-k} = x(0) + \sum_{i=1}^{k} a^{-i} bu(i-1)$$

applies.

(c) Determine the minimum number N of steps k with which the system, starting at $x(0) \in \mathbb{R}$, can be brought to the origin, i.e. $x(N) = 0$.
(d) Now determine the control sequence $u(0), u(1), \ldots, u(N-1)$ with the optimal number of steps which brings the system from $x(0)$ to the origin in the minimum number of steps N.

Exercise 4.11 Let us examine the switching system

$$\dot{\boldsymbol{x}} = \boldsymbol{A}_i \boldsymbol{x} \text{ with } \boldsymbol{A}_i = \begin{cases} \boldsymbol{A}_1, & 2i\Delta t \le t < (2i+1)\Delta t, \quad i = 0, 1, 2, \ldots, \\ \boldsymbol{A}_2, & (2i+1)\Delta t \le t < (2i+2)\Delta t, \quad i = 0, 1, 2, \ldots \end{cases}$$

(a) Show that if $\boldsymbol{A}_1 + \boldsymbol{A}_2$ only has stable eigenvalues and a sufficiently small value of Δt is selected, this is sufficient for the solution $\boldsymbol{x}(t)$ for all initial values $\boldsymbol{x}_0 = \boldsymbol{x}(0) \in \mathbb{R}^n$ to tend to zero for $t \to \infty$.
(b) State a sufficient and necessary criterion for global asymptotic stability of the switching system above.
(c) Determine whether the switching system for

$$\boldsymbol{A}_1 = \begin{bmatrix} -1 & -4 \\ 0 & -1 \end{bmatrix} \quad \text{and} \quad \boldsymbol{A}_2 = \begin{bmatrix} -1 & 0 \\ 9 & -1 \end{bmatrix}$$

is stable if in a first case study $\Delta t = \ln(1.4)$ and in a second $\Delta t = \ln(3)$ apply.

Exercise 4.12 Let us consider a linear system

$$\dot{\boldsymbol{x}} = \boldsymbol{A}\boldsymbol{x} + \boldsymbol{b}u, \quad \boldsymbol{x} \in \mathbb{R}^n, \tag{4.45}$$

and a piecewise constant control

$$u = u_i \text{ for } t_{i-1} \le t < t_i \text{ with } i = 1, \ldots, N, \; N \in \mathbb{N} \text{ and } t_0 = 0, \tag{4.46}$$

with the values u_i each being constant.
(a) Let $N = 2$. Calculate the values u_1 and u_2 for

$$\boldsymbol{A}=\begin{bmatrix} 0 & -1 \\ 1 & 0 \end{bmatrix}, \; \boldsymbol{b}=\begin{bmatrix} 0 \\ 1 \end{bmatrix}, \; \boldsymbol{x}(0)=\begin{bmatrix} -2 \\ -4 \end{bmatrix}, \; \boldsymbol{x}(t_2)=\begin{bmatrix} 0 \\ 0 \end{bmatrix}, \; N=2, \; t_1=\frac{\pi}{2}, \text{ and } t_2=\frac{3}{2}\pi.$$

(b) Repeat the calculation of u_1 and u_2 for the switching times $t_1 = \pi$ and $t_2 = 2\pi$. What result do you obtain?
(c) Derive a necessary and sufficient criterion for the existence of a control (4.46) of the general system (4.45) which is capable of driving any initial state $\boldsymbol{x}(0)$ into any other state $\boldsymbol{x}(t_N)$ in time t_N.

Exercise 4.13 Below we will attempt to design a piecewise linear control. Let us take a controllable and observable linear plant which is given without restriction of generality in the controller canonical form

$$\dot{\boldsymbol{x}} = \begin{bmatrix} 0 & 1 & \cdots & 0 \\ 0 & 0 & \cdots & 0 \\ \vdots & \vdots & \ddots & \vdots \\ 0 & 0 & \cdots & 1 \\ -a_0 & -a_1 & \cdots & -a_{n-1} \end{bmatrix} \boldsymbol{x} + \begin{bmatrix} 0 \\ 0 \\ \vdots \\ 0 \\ 1 \end{bmatrix} u$$

and has a parameter-dependent state controller

$$u = -\boldsymbol{k}^T(q)\boldsymbol{x} \quad \text{with} \quad \boldsymbol{k}(q) = \boldsymbol{D}^{-1}(q)\hat{\boldsymbol{a}} - \boldsymbol{a}$$

with the diagonal matrix

$$\boldsymbol{D}(q) = \operatorname{diag}(q^n, q^{n-1}, \ldots, q), \quad q \in (0, 1],$$

and the vectors

$$\boldsymbol{a} = \begin{bmatrix} a_0 \; a_1 \; \cdots \; a_{n-1} \end{bmatrix} \quad \text{and} \quad \hat{\boldsymbol{a}} = \begin{bmatrix} \hat{a}_0 \; \hat{a}_1 \; \cdots \; \hat{a}_{n-1} \end{bmatrix}.$$

Here the values $\hat{a}_i$ represent the coefficients of the characteristic polynomial of the closed control loop when $q = 1$. The control variable u is limited by

$$|u| \leq u_{\max}.$$

(a) Demonstrate that the closed control loop

$$\dot{\boldsymbol{x}} = (\boldsymbol{A} - \boldsymbol{b}\boldsymbol{k}^T(q))\boldsymbol{x} \tag{4.47}$$

has the system matrix

$$\hat{\boldsymbol{A}}(q) = \frac{1}{q}\boldsymbol{D}(q)\hat{\boldsymbol{A}}_1\boldsymbol{D}^{-1}(q), \qquad \hat{\boldsymbol{A}}_1 = \hat{\boldsymbol{A}}(1) = \boldsymbol{A} - \boldsymbol{b}\boldsymbol{k}^T(1).$$

(b) Show that for the eigenvalues $\lambda_i(q)$ of $\hat{\boldsymbol{A}}(q)$

$$\lambda_i(q) = \frac{\lambda_{1i}}{q}, \qquad \lambda_{1i} = \lambda_i(1),$$

holds.

(c) Demonstrate that each set

$$G(q) = \left\{\boldsymbol{x} \in I\!R^n \mid e(q)\boldsymbol{x}^T\boldsymbol{D}^{-1}(q)\boldsymbol{R}\boldsymbol{D}^{-1}(q)\boldsymbol{x} \leq 1\right\} \tag{4.48}$$

is a catchment region of the equilibrium point $\boldsymbol{x}_{\text{eq}} = \boldsymbol{0}$ of the control loop (4.47) if the matrix

$$\hat{\boldsymbol{A}}_1^T\boldsymbol{R} + \boldsymbol{R}\hat{\boldsymbol{A}}_1 = -\boldsymbol{Q}$$

is negative definite. Here $\boldsymbol{R} \in I\!R^{n\times n}$ is a positive definite matrix and $e(q)$ is a positive function with which the size of the regions $G(q)$ can be varied.

(d) Show that each of the areas $G(q)$ is tangent to the corresponding hyperplanes

$$|\boldsymbol{k}^T(q)\boldsymbol{x}| = u_{\max}$$

if

$$e(q) = \frac{\boldsymbol{k}^T(q)\boldsymbol{D}(q)\boldsymbol{R}^{-1}\boldsymbol{D}(q)\boldsymbol{k}(q)}{u_{\max}^2}$$

applies.

(e) Parameter-dependent convex areas

$$G(q) = \{\boldsymbol{x} \in I\!R^n \mid g(\boldsymbol{x}, q) \le 1\}$$

are nested within one another without touching if

$$\frac{\partial g(\boldsymbol{x}, q)}{\partial q} < 0$$

applies for all $\boldsymbol{x} \in I\!R^n$. Demonstrate that the areas (4.48) are nested within one another if

$$\frac{\partial e(q)}{\partial q} \le 0$$

applies and the matrix $\boldsymbol{S}$ in the equation

$$\boldsymbol{N}\boldsymbol{R} + \boldsymbol{R}\boldsymbol{N} = -\boldsymbol{S}, \qquad \boldsymbol{N} = \operatorname{diag}(-n, -(n-1), \ldots, 1),$$

is positive definite.

(f) Create a procedure by which the above results can be used to design a piecewise linear control in accordance with Section 4.3.2, p. 318 et seq.

(g) Based on the above results, determine a soft variable structure controller. The latter consists of the control law $u(\boldsymbol{x}, q)$ and the equation to calculate q. Note that, in this case, the catchment regions $G(q)$ are infinitesimally densely nested and the parameters of the controller vector $\boldsymbol{k}(q)$ are continuously varied. State the three requisite design equations.

(h) Explain why $V(\boldsymbol{x}) = q$ is a Lyapunov function of this soft variable structure control.

Exercise 4.14 What characteristic does the control law

$$u = -\operatorname{sat}(\boldsymbol{h}^T\boldsymbol{x}) \quad \text{for } \boldsymbol{x} \in G$$

in Hu and Lin's theorem on p. 330 have for $\boldsymbol{k} = \boldsymbol{h}$?

Exercise 4.15 Let us consider the linear plant

$$\dot{\boldsymbol{x}} = \begin{bmatrix} 0 & 1 \\ -1 & 0 \end{bmatrix} \boldsymbol{x} + \begin{bmatrix} 0 \\ 1 \end{bmatrix} u$$

with the control variable limitation

$$|u| \le u_{\max} = 6.$$

Design a saturation controller based on Hu and Lin's theorem on p. 330. Proceed as follows:

(a) What eigenvalues does the plant have?

(b) Determine a state controller $u = -\boldsymbol{h}^T\boldsymbol{x}$ such that the linear control loop $\boldsymbol{x} = (\boldsymbol{A} - \boldsymbol{b}\boldsymbol{h}^T)\boldsymbol{x}$ has the eigenvalues $s_{1,2} = -1$.

(c) For the diagonal matrix $\boldsymbol{Q}_h = \operatorname{diag}(2, 2)$, determine the matrix $\boldsymbol{R}$ from the Lyapunov equation

$$(\boldsymbol{A} - \boldsymbol{b}\boldsymbol{h}^T)^T\boldsymbol{R} + \boldsymbol{R}(\boldsymbol{A} - \boldsymbol{b}\boldsymbol{h}^T) = -\boldsymbol{Q}_h.$$

(d) Determine the catchment region

$$G = \{\boldsymbol{x} \in I\!R^2 \mid \boldsymbol{x}^T\boldsymbol{R}\boldsymbol{x} < c\}$$

such that it is tangent to the hyperplanes of the control variable limitation $|u| = |\boldsymbol{h}^T\boldsymbol{x}| = u_{\text{max}}$.

(e) Select the controller $u = -\boldsymbol{k}^T\boldsymbol{x}$ such that the eigenvalues of the linear control loop $\dot{\boldsymbol{x}} = (\boldsymbol{A} - \boldsymbol{b}\boldsymbol{k}^T)\boldsymbol{x}$ are $\lambda_{1,2} = -2$.

(f) Is the equilibrium point $\boldsymbol{x}_{\text{eq}} = \boldsymbol{0}$ of the nonlinear control loop

$$\dot{\boldsymbol{x}} = (\boldsymbol{A} - \boldsymbol{b}\,\text{sat}(\boldsymbol{k}^T\boldsymbol{x}))$$

asymptotically stable for all $\boldsymbol{x} \in G$? Give reasons for your answer.

5

Nonlinear Control of Nonlinear Systems

5.1 Gain-Scheduling Control

5.1.1 Mode of Operation and Design

Gain-scheduling methods allow for the design of relatively simple field-proven controllers for nonlinear plants based on linear systems theory. For this reason, gain-scheduling controls are widespread. Typical applications are flight control systems, controllers in the chemical processing industry, and automotive engine controls.

For this class of controllers, the plant is linearized for different operating points. This results in a series of linear submodels, for each of which a linear controller is designed. In rare cases, nonlinear controllers are chosen. During operation, the controller whose associated plant model corresponds best to the current situation is activated or, alternatively, one suitably interpolates between the controllers. Based on the linearized plant, it is possible to apply the design methods of linear systems theory to nonlinear plants. This can greatly simplify the design of control laws for nonlinear systems, which is certainly one reason for the popularity of this design method.

The starting point of gain scheduling is the nonlinear plant

$$\begin{aligned}
\dot{\boldsymbol{x}} &= \boldsymbol{f}(\boldsymbol{x},\boldsymbol{u},\boldsymbol{d}), \quad \boldsymbol{x} \in \mathbb{R}^n,\ \boldsymbol{u} \in \mathbb{R}^m,\ \boldsymbol{d} \in \mathbb{R}^s,\\
\boldsymbol{y} &= \boldsymbol{g}(\boldsymbol{x}), \quad \boldsymbol{y} \in \mathbb{R}^r.
\end{aligned}$$

The vector $\boldsymbol{d}$ describes possible external factors or disturbances. The system's equilibrium points $\boldsymbol{x}_{\mathrm{eq}}$ are determined from the equation

$$\dot{\boldsymbol{x}} = \boldsymbol{f}(\boldsymbol{x}_{\mathrm{eq}},\boldsymbol{u}_{\mathrm{eq}},\boldsymbol{d}_{\mathrm{eq}}) = \boldsymbol{0}.$$

They depend on the input variable vector $\boldsymbol{u}_{\mathrm{eq}}$, and possibly on external factors $\boldsymbol{d}_{\mathrm{eq}}$, such as disturbances. These equilibrium points or operating points of the system can commonly be represented as functions of a parameter vector $\boldsymbol{\rho}$ in the form

J. Adamy, *Nonlinear Systems and Controls*,
https://doi.org/10.1007/978-3-662-68690-4_5

$$\begin{aligned}\boldsymbol{u}_{\text{eq}} &= \boldsymbol{u}_{\text{eq}}(\boldsymbol{\rho}),\\ \boldsymbol{x}_{\text{eq}} &= \boldsymbol{x}_{\text{eq}}(\boldsymbol{\rho}),\\ \boldsymbol{d}_{\text{eq}} &= \boldsymbol{d}_{\text{eq}}(\boldsymbol{\rho}),\end{aligned}$$

and

$$\boldsymbol{y}_{\text{eq}} = \boldsymbol{g}(\boldsymbol{x}_{\text{eq}}(\boldsymbol{\rho})) = \boldsymbol{y}_{\text{eq}}(\boldsymbol{\rho}).$$

In the following, the dimension of the parameter vector $\boldsymbol{\rho}$ is indicated by l.

From the set

$$E = \{\boldsymbol{u}_{\text{eq}}(\boldsymbol{\rho}), \boldsymbol{x}_{\text{eq}}(\boldsymbol{\rho}), \boldsymbol{d}_{\text{eq}}(\boldsymbol{\rho}), \boldsymbol{y}_{\text{eq}}(\boldsymbol{\rho}), \quad \boldsymbol{\rho} \in M \subset \mathbb{R}^l\} \subset \mathbb{R}^{m+n+s+r}$$

of equilibrium points, a subset of p equilibrium points (or operating points)

$$(\boldsymbol{u}_{\text{eq}}(\boldsymbol{\rho}_i), \boldsymbol{x}_{\text{eq}}(\boldsymbol{\rho}_i), \boldsymbol{d}_{\text{eq}}(\boldsymbol{\rho}_i), \boldsymbol{y}_{\text{eq}}(\boldsymbol{\rho}_i)), \qquad i = 1, ..., p,$$

is selected. Around each of these, a linearized model

$$\begin{aligned}\Delta\dot{\boldsymbol{x}}_i &= \boldsymbol{A}(\boldsymbol{\rho}_i)\Delta\boldsymbol{x}_i + \boldsymbol{B}(\boldsymbol{\rho}_i)\Delta\boldsymbol{u}_i + \boldsymbol{S}(\boldsymbol{\rho}_i)\Delta\boldsymbol{d}_i,\\ \Delta\boldsymbol{y}_i &= \boldsymbol{C}(\boldsymbol{\rho}_i)\Delta\boldsymbol{x}_i\end{aligned}$$

is derived, where

$$\begin{aligned}\Delta\boldsymbol{x}_i &= \boldsymbol{x} - \boldsymbol{x}_{\text{eq}}(\boldsymbol{\rho}_i),\\ \Delta\boldsymbol{u}_i &= \boldsymbol{u} - \boldsymbol{u}_{\text{eq}}(\boldsymbol{\rho}_i),\\ \Delta\boldsymbol{d}_i &= \boldsymbol{d} - \boldsymbol{d}_{\text{eq}}(\boldsymbol{\rho}_i),\\ \Delta\boldsymbol{y}_i &= \boldsymbol{y} - \boldsymbol{y}_{\text{eq}}(\boldsymbol{\rho}_i),\end{aligned}$$

and

$$\begin{aligned}\boldsymbol{A}(\boldsymbol{\rho}_i) &= \left.\frac{\partial \boldsymbol{f}(\boldsymbol{x}, \boldsymbol{u}_{\text{eq}}(\boldsymbol{\rho}_i), \boldsymbol{d}_{\text{eq}}(\boldsymbol{\rho}_i))}{\partial \boldsymbol{x}}\right|_{\boldsymbol{x}=\boldsymbol{x}_{\text{eq}}(\boldsymbol{\rho}_i)},\\ \boldsymbol{B}(\boldsymbol{\rho}_i) &= \left.\frac{\partial \boldsymbol{f}(\boldsymbol{x}_{\text{eq}}(\boldsymbol{\rho}_i), \boldsymbol{u}, \boldsymbol{d}_{\text{eq}}(\boldsymbol{\rho}_i))}{\partial \boldsymbol{u}}\right|_{\boldsymbol{u}=\boldsymbol{u}_{\text{eq}}(\boldsymbol{\rho}_i)},\\ \boldsymbol{S}(\boldsymbol{\rho}_i) &= \left.\frac{\partial \boldsymbol{f}(\boldsymbol{x}_{\text{eq}}(\boldsymbol{\rho}_i), \boldsymbol{u}_{\text{eq}}(\boldsymbol{\rho}_i), \boldsymbol{d})}{\partial \boldsymbol{d}}\right|_{\boldsymbol{d}=\boldsymbol{d}_{\text{eq}}(\boldsymbol{\rho}_i)},\\ \boldsymbol{C}(\boldsymbol{\rho}_i) &= \left.\frac{\partial \boldsymbol{g}(\boldsymbol{x})}{\partial \boldsymbol{x}}\right|_{\boldsymbol{x}=\boldsymbol{x}_{\text{eq}}(\boldsymbol{\rho}_i)}\end{aligned}$$

hold. Below, operating points are indicated by $\boldsymbol{\rho}_i$ for simplicity instead of using the operating point

$$(\boldsymbol{u}_{\text{eq}}(\boldsymbol{\rho}_i), \boldsymbol{x}_{\text{eq}}(\boldsymbol{\rho}_i), \boldsymbol{d}_{\text{eq}}(\boldsymbol{\rho}_i), \boldsymbol{y}_{\text{eq}}(\boldsymbol{\rho}_i))$$

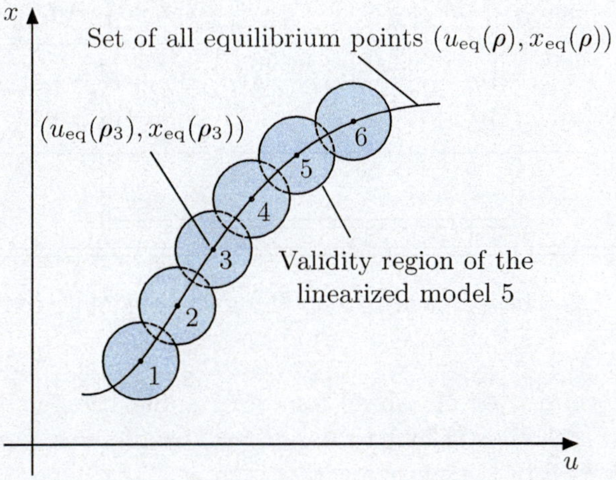

Fig. 5.1: Validity regions of the parameterized linearization family

itself. The set of linearized models is called the *parameterized linearization family.* These p linearized models have only a limited validity range around the respective operating points. The regions of validity should be adjacent to each other without gaps and should overlap, as illustrated in Figure 5.1. The parameterized linearization family often does not cover the entire space of the input and state variables with its validity ranges, so that this model family no longer reflects the behavior of the nonlinear system sufficiently well outside these regions.

For each of the p linearized models with the associated parameter vector $\boldsymbol{\rho}_i$, a controller

$$\begin{aligned}\dot{\boldsymbol{z}}_i &= \boldsymbol{h}_i(\boldsymbol{z}_i, \boldsymbol{\Delta d}_i, \boldsymbol{\Delta x}_i, \boldsymbol{\Delta y}_i, \boldsymbol{\Delta y}_{\mathrm{ref},i}),\\ \boldsymbol{\Delta u}_i &= \boldsymbol{k}_i(\boldsymbol{z}_i, \boldsymbol{\Delta d}_i, \boldsymbol{\Delta x}_i, \boldsymbol{\Delta y}_i, \boldsymbol{\Delta y}_{\mathrm{ref},i})\end{aligned}$$

is designed whose optional dynamic is determined by a differential equation with controller states $\boldsymbol{z}_i$. Here,

$$\boldsymbol{\Delta y}_{\mathrm{ref},i} = \boldsymbol{y}_{\mathrm{ref}} - \boldsymbol{y}_{\mathrm{ref,eq}}(\boldsymbol{\rho}_i)$$

is the deviation of the reference value vector $\boldsymbol{y}_{\mathrm{ref}}$ from the operating point $\boldsymbol{y}_{\mathrm{ref,eq}}(\boldsymbol{\rho}_i)$. Figure 5.2 shows the i-th of these p control loops, which consists of the i-th subcontroller and the plant which is linearized at the operating point ρ_i. The subcontrollers can be of various types, such as linear state controllers, PID controllers, etc.

After designing the p controllers, we combine them into an overall controller. This combining can be performed in different ways. It can be done, for example, so that only one controller is activated. A more suitable method is the interpolation between two or more controllers. Ideally, the selection and the interpolation depend on the parameter vector $\boldsymbol{\rho}(t)$. However, the vector $\boldsymbol{\rho}$

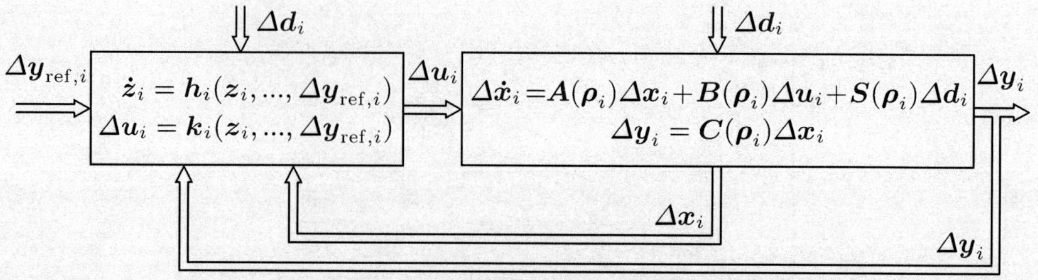

Fig. 5.2: The i-th control loop of a gain-scheduling controller

cannot be determined or measured for every application. It is then replaced by a vector $\boldsymbol{\beta}(t)$ which is referred to as the *scheduling vector*, whose elements $\beta_k(t)$ are composed of known quantities such as

(1) the state variables x_k,
(2) the output variables y_k,
(3) the control variables u_k,
(4) the controller states z_k, or
(5) the reference values $y_{\mathrm{ref},k}$.

A requirement is for

$$l = \dim(\boldsymbol{\rho}) = \dim(\boldsymbol{\beta})$$

to hold and for the steady-state case

$$\boldsymbol{\beta}_{\mathrm{s}} = \boldsymbol{\beta}(t \to \infty) = \boldsymbol{\rho} \tag{5.1}$$

to exist.

The simplest way to generate the control signal $\boldsymbol{u}$ is to activate a single controller i with the control signal

$$\boldsymbol{u} = \boldsymbol{u}_{\mathrm{eq}}(\boldsymbol{\rho}_i) + \Delta \boldsymbol{u}_i. \tag{5.2}$$

The controller i is determined by

$$i = \arg\min_k |\boldsymbol{\beta} - \boldsymbol{\rho}_k|. \tag{5.3}$$

Thus, the controller i whose parameter vector $\boldsymbol{\rho}_i$ is closest to the current value of the scheduling vector $\boldsymbol{\beta}$ is active.

The control law (5.2) is simple, but has the disadvantage of switching between controllers by means of the selection law (5.3). This results in undesired jumps in the control signal which most of the actuators cannot perform or withstand. It is therefore sensible to use a control law which interpolates between the controllers, or more precisely, between their control variables. The actuating variable $\boldsymbol{u}$ is then composed of the values $\boldsymbol{u}_{\mathrm{eq}}(\boldsymbol{\rho}_i)$ and $\Delta \boldsymbol{u}_i$. This is achieved by interpolating between the controllers i in whose validity region the scheduling vector $\boldsymbol{\beta}$ lies.

The interpolation between the control variables $\boldsymbol{u}_{\mathrm{eq}}(\boldsymbol{\rho}_i) + \boldsymbol{\Delta u}_i$ of the individual controllers is carried out based on factors such as the weighted arithmetic mean, which then results in the equation of the gain-scheduling controller

$$\boldsymbol{u} = \frac{\sum\limits_{i=1}^{p} \mu_i(\boldsymbol{\beta}) \cdot (\boldsymbol{u}_{\mathrm{eq}}(\boldsymbol{\rho}_i) + \boldsymbol{\Delta u}_i)}{\sum\limits_{i=1}^{p} \mu_i(\boldsymbol{\beta})}. \tag{5.4}$$

The weight functions μ_i satisfy

$$0 \leq \mu_i(\boldsymbol{\beta}) \leq 1,$$

are equal to one at the operating point

$$\boldsymbol{\beta} = \boldsymbol{\rho}_i,$$

and continuously decrease to zero or approximately to zero at the boundary of the validity region which belongs to this operating point. One function with this property is the multidimensional Gaussian function

$$\mu_i(\boldsymbol{\beta}) = e^{-|\boldsymbol{\Sigma}(\boldsymbol{\beta}-\boldsymbol{\rho}_i)|^2} = e^{-(\boldsymbol{\beta}-\boldsymbol{\rho}_i)^T \boldsymbol{\Sigma}^T \boldsymbol{\Sigma}(\boldsymbol{\beta}-\boldsymbol{\rho}_i)}. \tag{5.5}$$

The $l \times l$ matrix $\boldsymbol{\Sigma}$ can be a diagonal matrix such as

$$\boldsymbol{\Sigma} = \begin{bmatrix} \dfrac{1}{\sqrt{2}\sigma_1} & \cdots & 0 \\ \vdots & \ddots & \vdots \\ 0 & \cdots & \dfrac{1}{\sqrt{2}\sigma_l} \end{bmatrix}.$$

The diagonal matrix gives each coordinate direction of the scheduling vector $\boldsymbol{\beta}$ its own scaling with $1/(2\sigma_j^2)$. Thus, in addition to circles as shown in Figure 5.1, ellipses can also be formed as contour lines of the function (5.5). In this case, the axes of the ellipses are Cartesian oriented. If we use non-diagonal positive definite matrices $\boldsymbol{\Sigma}^T\boldsymbol{\Sigma}$, we can orient the ellipses arbitrarily in space.

Normally, the functions $\mu_i(\boldsymbol{\beta})$ are designed using an appropriate choice of the parameters σ_i, $i = 1, \ldots, l$, such that each function $\mu_i(\boldsymbol{\beta})$ is nearly zero for $\boldsymbol{\beta} = \boldsymbol{\rho}_j$ if $j \neq i$ and equal to one if $j = i$. Then, for $\boldsymbol{\beta} = \boldsymbol{\rho}_i$, equation (5.4) reduces to

$$\boldsymbol{u} \approx \frac{\mu_i(\boldsymbol{\rho}_i) \cdot (\boldsymbol{u}_{\mathrm{eq}}(\boldsymbol{\rho}_i) + \boldsymbol{\Delta u}_i)}{\mu_i(\boldsymbol{\rho}_i)} = \boldsymbol{u}_{\mathrm{eq}}(\boldsymbol{\rho}_i) + \boldsymbol{\Delta u}_i \tag{5.6}$$

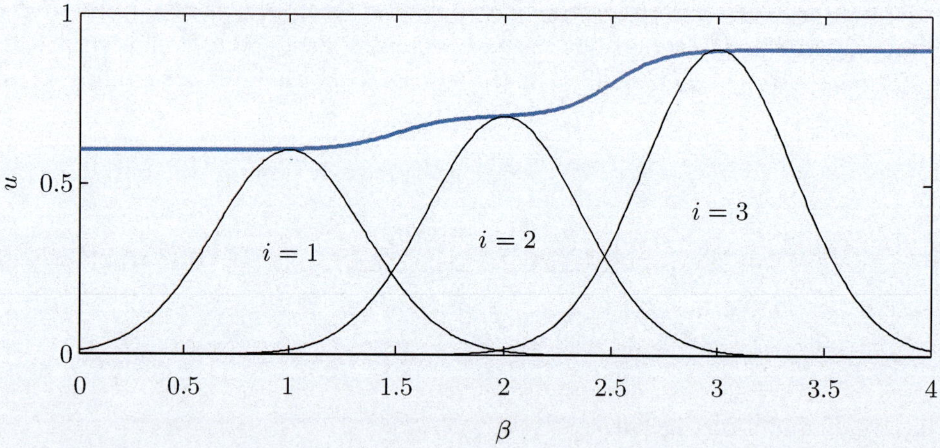

Fig. 5.3: The interpolation u of the controller output values $u_{\text{eq}}(\rho_i) + \Delta u_i$ based on the weighted arithmetic mean is shown as the blue curve. The black curves depict the functions $\mu_i(\beta) \cdot (u_{\text{eq}}(\rho_i) + \Delta u_i)$ for $i = 1, \ldots, 3$.

such that only the associated controller $\boldsymbol{\Delta u}_i$ becomes active at the operating point $\boldsymbol{\rho}_i$. Between the operating points, a weighted average of the individual controller output variables $\boldsymbol{u}_{\text{eq}}(\boldsymbol{\rho}_i) + \boldsymbol{\Delta u}_i$ is used. Figure 5.3 illustrates this.

The interpolation law (5.4) has the often favorable property that, if a vector $\boldsymbol{\beta}$ lies outside the valid operating regions in question, the controller of the operating points closest to $\boldsymbol{\beta}$ remains active. This can be seen from equation (5.4), because for vectors $\boldsymbol{\beta}$ of this type,

$$\mu_i(\boldsymbol{\beta}) \ll 1$$

holds for all i. Thus the highest of these values $\mu_i(\boldsymbol{\beta})$, meaning the one with the smallest Euclidean distance $|\boldsymbol{\beta} - \boldsymbol{\rho}_i|$, has the greatest effect on equation (5.4). In this case, equation (5.6) applies as well. Figure 5.3 illustrates this property.

Overall, the design of a gain-scheduling control can be implemented in four steps:

Step 1: Calculate the parameterized linearization family.

Step 2: Design the linear subcontrollers using linear design methods.

Step 3: Define the selection or interpolation law for the subcontrollers, which depends on the scheduling parameters.

Step 4: Simulate the entire control loop and check stability and control quality.

Step 4 in particular requires attention, as stability generally cannot be ensured using analytical means. The stability of each of the linear subcontrol loops is not sufficient in this case, since the entire system is nonlinear.

Additional and supplementary material on gain-scheduling control can be found in [21, 97, 107, 265, 468, 499]. An overview can be found in [266, 393, 472], among other sources.

5.1.2 Illustrative Example

Let us consider the simple example

$$\dot{x} = -x^3 + u.$$

The system has the equilibrium point

$$x_{\mathrm{eq}} = \sqrt[3]{u_{\mathrm{eq}}}.$$

Choosing

$$\rho = u_{\mathrm{eq}}$$

results in the parameterized linearization family

$$\begin{aligned}\Delta\dot{x} = A\Delta x + b\Delta u &= \left.\frac{\partial f(x, u_{\mathrm{eq}})}{\partial x}\right|_{x_{\mathrm{eq}}} \cdot \Delta x + \left.\frac{\partial f(x_{\mathrm{eq}}, u)}{\partial u}\right|_{u_{\mathrm{eq}}} \cdot \Delta u \\ &= -3x_{\mathrm{eq}}^2\Delta x + \Delta u = -3\sqrt[3]{u_{\mathrm{eq}}^2}\Delta x + \Delta u.\end{aligned}$$

Furthermore, five operating points with

$$\begin{aligned}
\rho_1 &= u_{\mathrm{eq1}} = -2, & x_{\mathrm{eq1}} &= -\sqrt[3]{2} \approx -1.26,\\
\rho_2 &= u_{\mathrm{eq2}} = -1, & x_{\mathrm{eq2}} &= -\sqrt[3]{1} = -1,\\
\rho_3 &= u_{\mathrm{eq3}} = 0, & x_{\mathrm{eq3}} &= \sqrt[3]{0} = 0,\\
\rho_4 &= u_{\mathrm{eq4}} = 1, & x_{\mathrm{eq4}} &= \sqrt[3]{1} = 1,\\
\rho_5 &= u_{\mathrm{eq5}} = 2, & x_{\mathrm{eq5}} &= \sqrt[3]{2} \approx 1.26,
\end{aligned}$$

and

$$\Delta u_i = u - u_{\mathrm{eq},i} \quad \text{and} \quad \Delta x_i = x - x_{\mathrm{eq},i}$$

are selected. Thus we obtain five linearized models with

$$\begin{aligned}
\rho_1 &= u_{\mathrm{eq1}} = -2 & &\text{and} & \Delta\dot{x}_1 &= -4.76\Delta x_1 + \Delta u_1,\\
\rho_2 &= u_{\mathrm{eq2}} = -1 & &\text{and} & \Delta\dot{x}_2 &= -3.00\Delta x_2 + \Delta u_2,\\
\rho_3 &= u_{\mathrm{eq3}} = 0 & &\text{and} & \Delta\dot{x}_3 &= \Delta u_3,\\
\rho_4 &= u_{\mathrm{eq4}} = 1 & &\text{and} & \Delta\dot{x}_4 &= -3.00\Delta x_4 + \Delta u_4,\\
\rho_5 &= u_{\mathrm{eq5}} = 2 & &\text{and} & \Delta\dot{x}_5 &= -4.76\Delta x_5 + \Delta u_5.
\end{aligned}$$

For each model, a proportional controller with the control error

$$\Delta e_i = \Delta y_{\mathrm{ref},i} - \Delta x_i$$

and

$$\Delta u_i = k_i \Delta e_i$$

is designed, so that the corresponding control loops have the eigenvalues

$$\lambda_1 = -12, \quad \lambda_2 = -4.5, \quad \lambda_3 = -1.5, \quad \lambda_4 = -4.5, \quad \lambda_5 = -12.$$

The subcontrollers $\Delta u_i = k_i \Delta e_i$ have the parameters

$$k_1 = 7.24, \quad k_2 = 1.50, \quad k_3 = 1.50, \quad k_4 = 1.50, \quad k_5 = 7.24.$$

The switching rule from equations (5.2) and (5.3) is applied here and results in the overall control law

$$u = u_{\mathrm{eq}}(\rho_i) + \Delta u_i = u_{\mathrm{eq},i} + \Delta u_i = u_{\mathrm{eq},i} + k_i \cdot \Delta e_i \tag{5.7}$$

with

$$i = \arg \min_{k=1,\ldots,5} \left|\beta - u_{\mathrm{eq},k}\right|. \tag{5.8}$$

If we select

$$\beta = \rho = u$$

as the scheduling parameter in the control law (5.7), (5.8), the control law is implicit in u. This is unfavorable, because in this case computing u is a complex and time-consuming task. A more suitable option using the equilibrium equation

$$\dot{x} = -x^3 + u = 0 \quad \Leftrightarrow \quad u = x^3,$$

results in choosing

$$\beta = x^3.$$

Obviously in this case, condition (5.1), i. e. $\beta_{\mathrm{s}} = \rho$ for the steady-state case, is fulfilled.

Figure 5.4 shows the entire control loop. We have

$$y_{\mathrm{ref,eq},i} = x_{\mathrm{eq},i}$$

and

$$\Delta e_i = \Delta y_{\mathrm{ref},i} - \Delta x_i = y_{\mathrm{ref}} - x = \Delta e.$$

Figure 5.5 shows the simulation of the control-loop behavior for a step in the reference variable from

$$y_{\mathrm{ref}} = 0$$

to

$$y_{\mathrm{ref}} = 1.5$$

at time $t = 0$. Since we are only using P controllers, there is a permanent control error. The discontinuities in the course of u are caused by a change in the index i, meaning a change in $u_{\mathrm{eq},i}$ and Δu_i according to

$$u = u_{\mathrm{eq},i} + \Delta u_i.$$

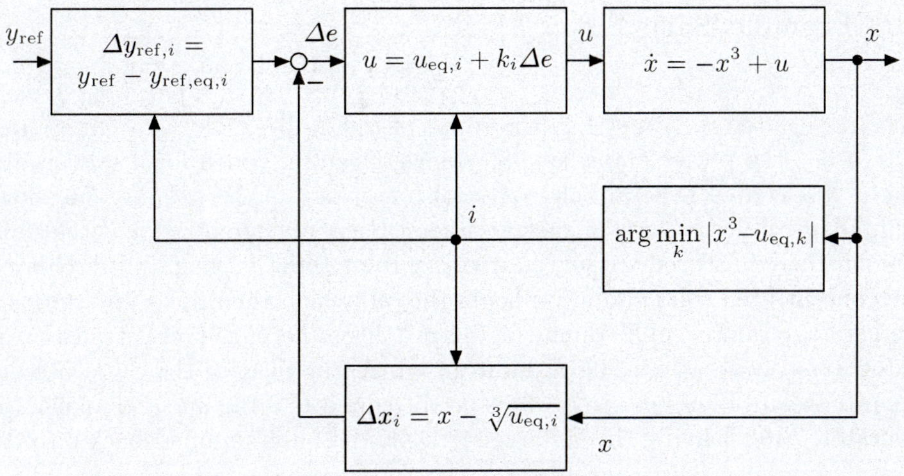

Fig. 5.4: Block diagram of the entire control loop

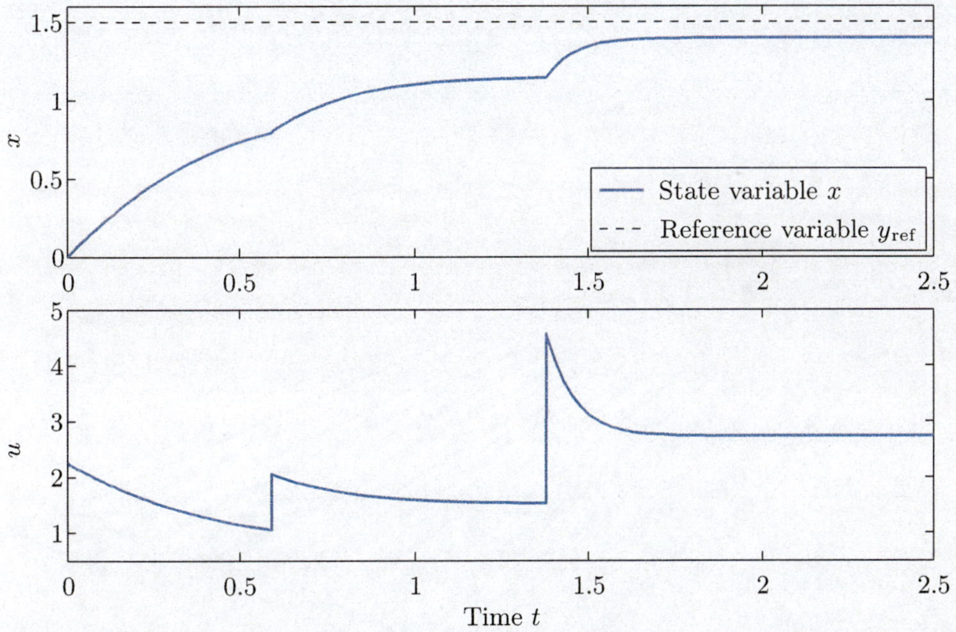

Fig. 5.5: Simulation of state variable x and actuating variable u from the gain-scheduling control of the plant $\dot{x} = -x^3 + u$

As mentioned above, this is often undesirable, because not every actuator can perform such jumps. But the abrupt change in the course of the state variable x is also undesirable if we consider a flight controller, for example. In many cases, a smooth progression of actuating variables and state variables is preferable. In the next section, we will look at an example in which interpolation takes place between the controllers so that no jumps occur in the control signal.

5.1.3 Example: Solar Power Plant

Below we will describe the solar power plant Acurex with a parabolic trough collector, located in Almería in the south of Spain [67, 209, 392]. On an area of 2674 m^2, the power plant has 480 solar collectors, which for a solar radiation of 900 $W\,m^{-2}$ generate an electrical power of 1.2 MW. Figure 5.6 shows the fundamental structure of such a system. Parabolic troughs focus the sun's rays via their mirrored surface in a translucent pipe in which oil circulates. The concentrated solar radiation heats the oil, which then flows into an insulated storage tank with a volume of 140 m^3. From here, the oil is transported into a steam generator using a pump in which the heat of the oil is used to generate steam. Finally, this steam is used to operate a turbine and an electric generator. After leaving the steam generator, the oil is pumped back into the

Fig. 5.6: Solar power station

collector array's pipes at a rate of 2 to 10 $\mathrm{l\,s^{-1}}$. The oil is therefore constantly circulating throughout the plant.

The oil's outlet temperature T_o is a measure of the collected energy. The dynamic behavior of the outlet temperature when leaving the collector field can be approximately described by the power balance

$$C\dot{T}_\mathrm{o} = \eta_0 SJ - qp_\mathrm{cp}(T_\mathrm{o} - T_\mathrm{e}). \tag{5.9}$$

Here, T_e is the temperature of the oil when entering the collector field. It is either constant or varies only very slowly and slightly. The solar radiation intensity is taken into account by the variable J, where $0 \leq J \leq 1\ \mathrm{kW\,m^{-2}}$ holds.

Unlike the intensity J of the solar radiation, the volume flow q of oil can be set between 2 and 10 $\mathrm{l\,s^{-1}}$ by adjusting the pump power. Furthermore, η_0 is the optical efficiency factor of the collectors, S is the effective collector area, C the heat capacity of the oil, and p_cp a system constant. For the Acurex plant in Almería, the parameter set

$$\eta_0 S = 1322\ \mathrm{m^2}, \quad C = 2267\ \mathrm{kJ\ {}^\circ C^{-1}}, \quad p_\mathrm{cp} = 1.924\ \mathrm{kJ\ {}^\circ C^{-1}\ l^{-1}}$$

applies.

The oil temperature T_o must not exceed the temperature of 305°C, and in particular the difference between the inlet temperature T_e and exit temperature T_o must not be higher than 100°C. Normal values for this difference are around 70°C. If the temperature difference $T_\mathrm{o} - T_\mathrm{e}$ is too great, the oil pressure in the pipes of the collector field rises too high. Leaks can be the result.

Therefore, our aim is to maintain the temperature difference

$$\Delta T = T_\mathrm{o} - T_\mathrm{e}$$

at the constant value of 70°C. This is done using a control system which affects the temperature T_o and thus ΔT by controlling the volume flow q of the oil in equation (5.9) or in the alternative equation

$$C(\Delta T)^{\cdot} + C\dot{T}_\mathrm{e} = \eta_0 SJ - qp_\mathrm{cp}\Delta T. \tag{5.10}$$

Since the inlet temperature T_e is approximately constant or varies only very slowly, i. e. $\dot{T}_\mathrm{e} \approx 0$, we obtain the approximation

$$C(\Delta T)^{\cdot} = \eta_0 SJ - qp_\mathrm{cp}\Delta T \tag{5.11}$$

for equation (5.10).

To be in conformity with the rules of common denomination, we then specify the state ΔT of system (5.11) as x and its control variable q as u. We thus obtain

$$\dot{x} = aJ - bxu \tag{5.12}$$

with

$$a = \eta_0 S C^{-1} = 0.5831^\circ\mathrm{C}\,\mathrm{m}^2\mathrm{kJ}^{-1}$$

and

$$b = p_{\mathrm{cp}} C^{-1} = 0.8487 \cdot 10^{-3} \mathrm{l}^{-1}.$$

The intensity J of the solar radiation, from a control engineering point of view, is a disturbance variable.

The equilibrium points x_{eq} of system (5.10) are given by

$$x_{\mathrm{eq}} = \frac{aJ_{\mathrm{eq}}}{bu_{\mathrm{eq}}}. \tag{5.13}$$

The temperature difference x_{eq} in the steady state is therefore proportional to the measurable intensity J of the solar radiation. In the case of a constant temperature difference $x_{\mathrm{eq}} = 70^\circ\mathrm{C}$, we obtain

$$u_{\mathrm{eq}} = 9.82 J_{\mathrm{eq}}$$

by using equation (5.13).

In a manner similar to [209], we wish to control the system (5.12) by means of gain-scheduling. For this purpose, we will first determine the parameterized linearization family

$$\Delta\dot{x} = \left.\frac{\partial aJ}{\partial J}\right|_{J=J_{\mathrm{eq}}} \cdot \Delta J - \left.\frac{\partial bxu_{\mathrm{eq}}}{\partial x}\right|_{x=x_{\mathrm{eq}}} \cdot \Delta x - \left.\frac{\partial bx_{\mathrm{eq}}u}{\partial u}\right|_{u=u_{\mathrm{eq}}} \cdot \Delta u \tag{5.14}$$

of the system, where the variables

$$\Delta J = J - J_{\mathrm{eq}}, \quad \Delta x = x - x_{\mathrm{eq}}, \quad \Delta u = u - u_{\mathrm{eq}}$$

denote the deviations from the values at the equilibrium points.

The linearized model

$$\Delta\dot{x} = a\Delta J - bu_{\mathrm{eq}}\Delta x - \frac{aJ_{\mathrm{eq}}}{u_{\mathrm{eq}}}\Delta u$$

follows from equation (5.13) and equation (5.14), where the parameter vector $\boldsymbol{\rho}$ is

$$\boldsymbol{\rho} = [u_{\mathrm{eq}}\ J_{\mathrm{eq}}]^T.$$

From the possible ranges of parameters,

$$2\,\mathrm{l\,s}^{-1} \le u_{\mathrm{eq}} \le 10\,\mathrm{l\,s}^{-1} \quad \text{and} \quad 0\,\mathrm{kW\,m}^{-2} \le J_{\mathrm{eq}} \le 1\,\mathrm{kW\,m}^{-2},$$

we select six operating points

$$\boldsymbol{\rho}_i = \begin{bmatrix} u_{\mathrm{eq},i} \\ J_{\mathrm{eq},i} \end{bmatrix} \in \left\{ \begin{bmatrix} 4 \\ 0.5 \end{bmatrix}, \begin{bmatrix} 6 \\ 0.5 \end{bmatrix}, \begin{bmatrix} 8 \\ 0.5 \end{bmatrix}, \begin{bmatrix} 4 \\ 0.8 \end{bmatrix}, \begin{bmatrix} 6 \\ 0.8 \end{bmatrix}, \begin{bmatrix} 8 \\ 0.8 \end{bmatrix} \right\}$$

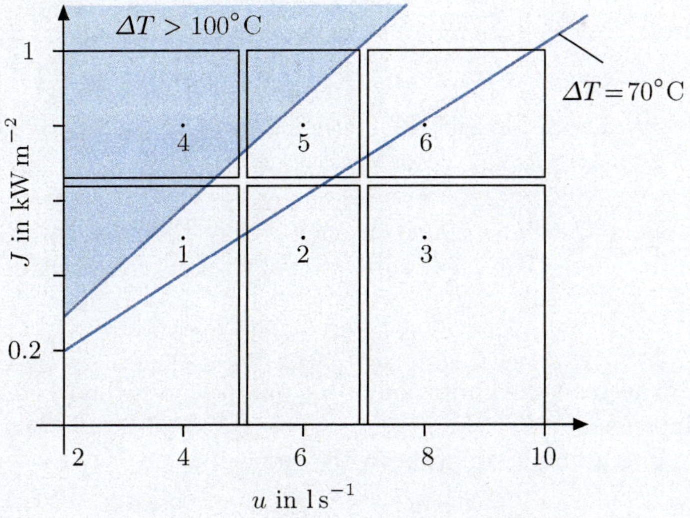

Fig. 5.7: Operating points (1, 2, 3, 4, 5, and 6) and validity regions of the linear submodels in which the respective Gaussian function of an operating point dominates

as illustrated in Figure 5.7. The region above the straight line

$$J_{\mathrm{eq}} = 0.1455 u_{\mathrm{eq}}$$

yielded by equation (5.13) includes temperature differences of $\Delta T > 100°\mathrm{C}$. This region, in which the control loop must not be operated, is shown in blue in Figure 5.7. The desired steady-state conditions with $x_{\mathrm{ref}} = \Delta T = 70°\mathrm{C}$, which depend on the intensity J of the solar radiation, are represented by a straight line that is consistent with

$$J_{\mathrm{eq}} = 0.1019 u_{\mathrm{eq}}.$$

We thus obtain

$$\begin{aligned}
i = 1: &\quad \Delta\dot{x}_1 = 0.5831 \cdot \Delta J_1 - 3.395 \cdot 10^{-3} \cdot \Delta x_1 - 72.89 \cdot 10^{-3} \cdot \Delta u_1,\\
i = 2: &\quad \Delta\dot{x}_2 = 0.5831 \cdot \Delta J_2 - 5.092 \cdot 10^{-3} \cdot \Delta x_2 - 48.60 \cdot 10^{-3} \cdot \Delta u_2,\\
i = 3: &\quad \Delta\dot{x}_3 = 0.5831 \cdot \Delta J_3 - 6.790 \cdot 10^{-3} \cdot \Delta x_3 - 36.44 \cdot 10^{-3} \cdot \Delta u_3,\\
i = 4: &\quad \Delta\dot{x}_4 = 0.5831 \cdot \Delta J_4 - 3.395 \cdot 10^{-3} \cdot \Delta x_4 - 116.6 \cdot 10^{-3} \cdot \Delta u_4,\\
i = 5: &\quad \Delta\dot{x}_5 = 0.5831 \cdot \Delta J_5 - 5.092 \cdot 10^{-3} \cdot \Delta x_5 - 77.75 \cdot 10^{-3} \cdot \Delta u_5,\\
i = 6: &\quad \Delta\dot{x}_6 = 0.5831 \cdot \Delta J_6 - 6.790 \cdot 10^{-3} \cdot \Delta x_6 - 58.31 \cdot 10^{-3} \cdot \Delta u_6
\end{aligned}$$

for the linear submodels at the i-th operating point. We then design a PI controller

$$\Delta u_i = K_i \left(x_{\mathrm{ref}} - x\right) + K_i \omega_i \int_0^t \left(x_{\mathrm{ref}} - x\right) d\tau$$

for each of the submodels with the respective controller coefficients

$$\begin{aligned}
K_1 &= -1.3254, & K_4 &= -0.8284, & \omega_1 &= 0.0248, & \omega_4 &= 0.0248,\\
K_2 &= -1.9532, & K_5 &= -1.2207, & \omega_2 &= 0.0253, & \omega_5 &= 0.0253,\\
K_3 &= -2.5577, & K_6 &= -1.5985, & \omega_3 &= 0.0257, & \omega_6 &= 0.0257.
\end{aligned}$$

The controllers are dimensioned in such a way that the control loops they form with the respective linear submodel always have eigenvalues at

$$\lambda_1 = -0.04 \quad \text{and} \quad \lambda_2 = -0.06.$$

To take into account the limitation of the pump power, the PI controllers have anti-windup elements.

Taking into account equation (5.13), we will use

$$\boldsymbol{\beta} = \begin{bmatrix} u = \dfrac{aJ}{bx} \\ J \end{bmatrix} \neq \boldsymbol{\rho}$$

as the scheduling vector, where $\boldsymbol{\beta}(t \to \infty) = \boldsymbol{\rho}$ holds. The control variables

$$u_i = u_{\mathrm{eq},i} + \Delta u_i$$

are averaged using the weighted arithmetic mean (5.4) to obtain the gain-scheduling controller

$$u = \frac{\sum\limits_{i=1}^{6} e^{-|\boldsymbol{\Sigma}(\boldsymbol{\beta}-\boldsymbol{\rho}_i)|^2} u_i}{\sum\limits_{i=1}^{6} e^{-|\boldsymbol{\Sigma}(\boldsymbol{\beta}-\boldsymbol{\rho}_i)|^2}},$$

where we choose

$$\boldsymbol{\Sigma} = \begin{bmatrix} 2 & 0 \\ 0 & 0.3 \end{bmatrix}.$$

We then simulate the gain-scheduling control loop of the solar power plant for the assumed daily solar radiation curve shown in Figure 5.8. The progressions of the pump flow u and the temperature difference $x = \Delta T$ between the oil's outlet temperature and the inlet temperature are also shown in Figure 5.8. As can be seen, this does not violate the constraints of the pump flow rate, $2\,\mathrm{l\,s^{-1}} \leq u \leq 10\,\mathrm{l\,s^{-1}}$. The temperature difference x is also kept at the desired value of $x = 70°\mathrm{C}$. However, the temperature difference falls below $70°\mathrm{C}$ if the solar radiation J is less than $0.2\,\mathrm{kW\,m^{-2}}$, since in this case the heating of the oil by the sun is too low to produce a temperature difference of $70°\mathrm{C}$ with a minimum oil flow of $2\,\mathrm{l\,s^{-1}}$.

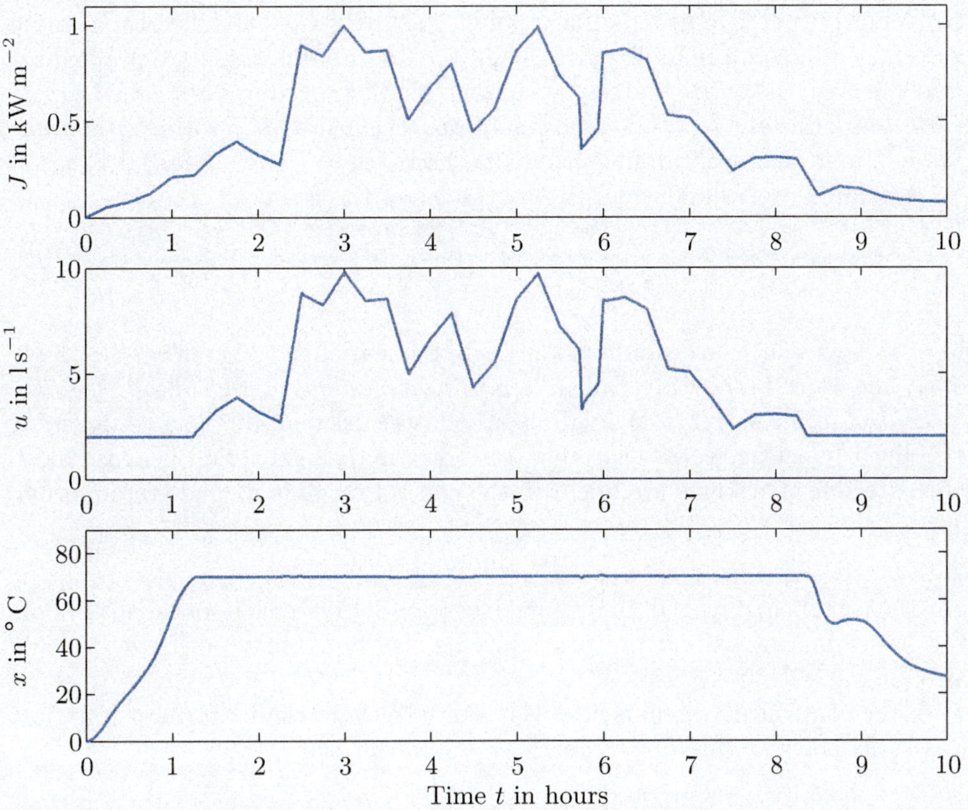

Fig. 5.8: Intensity J of the solar radiation, volume flow $u = q$ of the pump, and temperature difference $x = \Delta T$ of the oil between the inlet and outlet of the collector field

5.2 Input-Output Linearization

5.2.1 Basic Concept and Nonlinear Controller Canonical Form

In this section, we will describe a method which enables us to design controllers for nonlinear plants directly, not utilizing a linear approximation of the nonlinear plant. The basic concept is to design a nonlinear controller so that it completely compensates for the nonlinearity of the plant, thereby creating a linear control loop. This has given rise to the terms *feedback linearization* or *exact linearization* for this approach.

The methods of feedback linearization can be divided into two categories. On the one hand, these are procedures that bring about a linearization of the system behavior between input and output. We call this *input-output linearization* or, more precisely, *input-output exact linearization*[1]. In this case,

[1] The modifier *exact* emphasizes that we have a linearization of the system everywhere within a set of the input and output variables and not only an approximate linearization around an operating point.

the input-output behavior is linear, but the state-space model of the control loop may still contain nonlinearities. On the other hand, these are procedures that linearize the entire state-space model. These procedures are called *input-state linearization*, *state-space exact linearization* or *full-state linearization*. We will first deal with input-output linearization.

As plants, we will consider SISO systems of the form

$$\begin{aligned} \dot{\boldsymbol{x}} &= \boldsymbol{a}(\boldsymbol{x}) + \boldsymbol{b}(\boldsymbol{x}) \cdot u, \\ y &= c(\boldsymbol{x}), \end{aligned} \tag{5.15}$$

i. e. systems which are nonlinear in $\boldsymbol{x}$ but linear in u. Therefore, as previously mentioned, they are called *input-linear systems*, *input-affine systems*, or *control-affine systems*. We will use the latter designation. The assumption of a linearly acting control variable u is not a major restriction because most of the technical systems are linear in the control variable u. The system order is

$$n = \dim(\boldsymbol{x})\,.$$

For the time being, we will limit ourselves to SISO systems in order to explain the basic concept behind the procedure as clearly as possible. Later we will also look at MIMO systems.

If the controlled system is now represented as a particular form of equation (5.15), namely the *nonlinear controller canonical form*

$$\begin{aligned} \begin{bmatrix} \dot{x}_1 \\ \vdots \\ \dot{x}_{n-1} \\ \dot{x}_n \end{bmatrix} &= \begin{bmatrix} x_2 \\ \vdots \\ x_n \\ \alpha(\boldsymbol{x}) \end{bmatrix} + \begin{bmatrix} 0 \\ \vdots \\ 0 \\ \beta(\boldsymbol{x}) \end{bmatrix} u, \\ y &= x_1, \end{aligned} \tag{5.16}$$

the design objective of an overall linear control loop as described above can be easily achieved. As the control law, we will select

$$u = -\frac{\alpha(\boldsymbol{x}) + \sum\limits_{i=1}^{n} a_{i-1}x_i}{\beta(\boldsymbol{x})} + \frac{V}{\beta(\boldsymbol{x})} y_{\mathrm{ref}} \tag{5.17}$$

with the reference variable y_{ref} as well as the freely selectable coefficients a_{i-1}, $i = 1, \ldots, n$, and V, yielding

$$\begin{aligned} \dot{\boldsymbol{x}} &= \begin{bmatrix} 0 & 1 & 0 & \cdots & 0 \\ 0 & 0 & 1 & \cdots & 0 \\ \vdots & \vdots & \vdots & \ddots & \vdots \\ 0 & 0 & 0 & \cdots & 1 \\ -a_0 & -a_1 & -a_2 & \cdots & -a_{n-1} \end{bmatrix} \boldsymbol{x} + \begin{bmatrix} 0 \\ 0 \\ \vdots \\ 0 \\ V \end{bmatrix} y_{\mathrm{ref}}, \\ y &= x_1 \end{aligned}$$

for the closed loop, which is a linear system in controller canonical form[2]. Its eigenvalues can be freely specified via the n coefficients a_i. Throughout these calculations, we assume that $\beta(\boldsymbol{x}) \neq 0$ holds. This is usually the case in technical applications, since otherwise the control variable (5.17) would be infinitely large and the plant would not be controllable.

Naturally, if we model real systems using their original state variables, only few will be in the nonlinear controller canonical form (5.16). But if we succeed in bijectively transforming a system (5.15) into this form, we can apply the control law above and obtain a linear control loop. This is the basic concept of feedback linearization.

We will now attempt to identify the bijective state coordinate transformation which transforms the system (5.15) into the nonlinear controller canonical form (5.16). For this purpose, we will use the Lie derivative, which is defined as the gradient of a scalar function $h(\boldsymbol{x})$ multiplied by a vector field $\boldsymbol{f}(\boldsymbol{x})$, i. e.

$$L_{\boldsymbol{f}} h(\boldsymbol{x}) = \frac{\partial h(\boldsymbol{x})}{\partial \boldsymbol{x}} \boldsymbol{f}(\boldsymbol{x}) = \operatorname{grad}^T(h(\boldsymbol{x})) \cdot \boldsymbol{f}(\boldsymbol{x}).$$

The Lie derivative can be illustrated geometrically: if it is positive, the vector field $\boldsymbol{f}$ points in the direction of increasing values of the function h. If it is negative, it is the other way round and the vector field $\boldsymbol{f}$ points towards decreasing function values $h(\boldsymbol{x})$. The latter, for example, is the case for a Lyapunov function V for an asymptotically stable equilibrium point $\boldsymbol{x}_{\text{eq}} = \boldsymbol{0}$ of a system $\dot{\boldsymbol{x}} = \boldsymbol{f}(\boldsymbol{x})$. We remember that in this case

$$\dot{V}(\boldsymbol{x}) = \frac{\partial V(\boldsymbol{x})}{\partial \boldsymbol{x}} \dot{\boldsymbol{x}} = \frac{\partial V(\boldsymbol{x})}{\partial \boldsymbol{x}} \boldsymbol{f}(\boldsymbol{x}) = L_{\boldsymbol{f}} V(\boldsymbol{x}) < 0$$

holds. However, the situations above do not play a role in the input-output linearization, and we use the Lie derivative here to simplify the calculations procedures.

For the transformation we are looking for, we need, in particular, the Lie derivatives

$$L_{\boldsymbol{a}} c(\boldsymbol{x}) = \frac{\partial c(\boldsymbol{x})}{\partial \boldsymbol{x}} \boldsymbol{a}(\boldsymbol{x}) \quad \text{and} \quad L_{\boldsymbol{b}} c(\boldsymbol{x}) = \frac{\partial c(\boldsymbol{x})}{\partial \boldsymbol{x}} \boldsymbol{b}(\boldsymbol{x}).$$

We start by calculating the time derivative of the output variable y and obtain

$$\dot{y} = \frac{dc(\boldsymbol{x})}{dt} = \frac{\partial c(\boldsymbol{x})}{\partial x_1} \dot{x}_1 + \ldots + \frac{\partial c(\boldsymbol{x})}{\partial x_n} \dot{x}_n = \frac{\partial c(\boldsymbol{x})}{\partial \boldsymbol{x}} \dot{\boldsymbol{x}}.$$

Substituting

[2] The term *nonlinear controller canonical form* indicates that the controller design is simple if the plant is in this canonical form. This is similar to the case of linear systems, where the term *controller canonical form* also indicates that the design of a linear state controller by means of pole placement is especially simple.

$$\dot{\boldsymbol{x}} = \boldsymbol{a}(\boldsymbol{x}) + \boldsymbol{b}(\boldsymbol{x}) \cdot u$$

into the equation above yields

$$\dot{y} = \frac{\partial c(\boldsymbol{x})}{\partial \boldsymbol{x}} \boldsymbol{a}(\boldsymbol{x}) + \frac{\partial c(\boldsymbol{x})}{\partial \boldsymbol{x}} \boldsymbol{b}(\boldsymbol{x}) \cdot u$$

or, using Lie derivatives,

$$\dot{y} = L_{\boldsymbol{a}} c(\boldsymbol{x}) + L_{\boldsymbol{b}} c(\boldsymbol{x}) \cdot u.$$

For most technical systems, in the above equation

$$L_{\boldsymbol{b}} c(\boldsymbol{x}) = \frac{\partial c(\boldsymbol{x})}{\partial \boldsymbol{x}} \boldsymbol{b}(\boldsymbol{x}) = 0 \tag{5.18}$$

holds so that we obtain

$$\dot{y} = L_{\boldsymbol{a}} c(\boldsymbol{x}).$$

Systems for which this does not apply will be addressed later.

Linear systems in the controller canonical form

$$\dot{\boldsymbol{x}} = \boldsymbol{A}\boldsymbol{x} + \boldsymbol{b}u = \begin{bmatrix} 0 & 1 & 0 & \cdots & 0 \\ 0 & 0 & 1 & \cdots & 0 \\ \vdots & \vdots & \vdots & \ddots & \vdots \\ 0 & 0 & 0 & \cdots & 1 \\ -a_0 & -a_1 & -a_2 & \cdots & -a_{n-1} \end{bmatrix} \boldsymbol{x} + \begin{bmatrix} 0 \\ 0 \\ \vdots \\ 0 \\ 1 \end{bmatrix} u,$$

$$y = \boldsymbol{c}^T \boldsymbol{x} = \begin{bmatrix} b_0 \ b_1 \ \cdots \ b_m \ 0 \cdots 0 \end{bmatrix} \boldsymbol{x}$$

with $m < n-1$ provide an example of systems consistent with equation (5.18). For these systems, the equation

$$L_{\boldsymbol{b}} c(\boldsymbol{x}) = \frac{\partial \boldsymbol{c}^T \boldsymbol{x}}{\partial \boldsymbol{x}} \boldsymbol{b}(\boldsymbol{x}) = \boldsymbol{c}^T \boldsymbol{b} = 0$$

obviously holds.

Next, the second time derivative $\ddot{y}$ is to be determined. Starting with $\dot{y} = L_{\boldsymbol{a}} c(\boldsymbol{x})$, we obtain the expression

$$\ddot{y} = \frac{dL_{\boldsymbol{a}} c(\boldsymbol{x})}{dt} = \frac{\partial L_{\boldsymbol{a}} c(\boldsymbol{x})}{\partial \boldsymbol{x}} \dot{\boldsymbol{x}} = \underbrace{\frac{\partial L_{\boldsymbol{a}} c(\boldsymbol{x})}{\partial \boldsymbol{x}} \boldsymbol{a}(\boldsymbol{x})}_{L_{\boldsymbol{a}} L_{\boldsymbol{a}} c(\boldsymbol{x})} + \underbrace{\frac{\partial L_{\boldsymbol{a}} c(\boldsymbol{x})}{\partial \boldsymbol{x}} \boldsymbol{b}(\boldsymbol{x})}_{L_{\boldsymbol{b}} L_{\boldsymbol{a}} c(\boldsymbol{x})} \cdot u.$$

For the first term in the latter equation, the abbreviation

$$L_{\boldsymbol{a}} L_{\boldsymbol{a}} c(\boldsymbol{x}) = L_{\boldsymbol{a}}^2 c(\boldsymbol{x})$$

is used, since the Lie derivative $L_{\boldsymbol{a}}$ has been applied twice in a row. For the second term, the identity

$$L_{\boldsymbol{b}}L_{\boldsymbol{a}}(\boldsymbol{x}) = \frac{\partial L_{\boldsymbol{a}}c(\boldsymbol{x})}{\partial \boldsymbol{x}}\boldsymbol{b}(\boldsymbol{x}) = 0$$

often applies once again, yielding

$$\ddot{y} = L_{\boldsymbol{a}}^2 c(\boldsymbol{x}).$$

When the higher derivatives are also calculated, we finally obtain

$$\begin{aligned} y &= c(\boldsymbol{x}), \\ \dot{y} &= L_{\boldsymbol{a}}c(\boldsymbol{x}), \\ \ddot{y} &= L_{\boldsymbol{a}}^2 c(\boldsymbol{x}), \\ &\vdots \\ y^{(\delta-1)} &= L_{\boldsymbol{a}}^{\delta-1}c(\boldsymbol{x}), \\ y^{(\delta)} &= L_{\boldsymbol{a}}^{\delta}c(\boldsymbol{x}) + L_{\boldsymbol{b}}L_{\boldsymbol{a}}^{\delta-1}c(\boldsymbol{x}) \cdot u. \end{aligned} \tag{5.19}$$

In this sequence of time derivatives, the equation

$$L_{\boldsymbol{b}}L_{\boldsymbol{a}}^i c(\boldsymbol{x}) = \frac{\partial L_{\boldsymbol{a}}^i c(\boldsymbol{x})}{\partial \boldsymbol{x}}\boldsymbol{b}(\boldsymbol{x}) = 0$$

holds for all indices

$$i = 0, \ldots, \delta - 2.$$

The Lie derivative $L_{\boldsymbol{b}}L_{\boldsymbol{a}}^{\delta-1}c(\boldsymbol{x})$ is not equal to zero only for indices $i \geq \delta - 1$. We refer to δ as the *difference degree* or mostly as the *relative degree* of the system. The relative degree indicates the degree of the output variable y's derivative at which it directly depends on the control variable u for the first time. For linear systems, the relative degree is equal to the difference

$$\delta = n - m$$

between the denominator degree n and the numerator degree m of the transfer function. For $\delta = n$ the numerator polynomial has the order zero, i. e. the numerator is a constant.

We will first examine the case

$$\delta = n,$$

i. e. the case in which the system order $n = \dim(\boldsymbol{x})$ is equal to the relative degree δ. Then the new state coordinates

$$\boldsymbol{z} = \begin{bmatrix} z_1 \\ z_2 \\ z_3 \\ \vdots \\ z_n \end{bmatrix} = \begin{bmatrix} y \\ \dot{y} \\ \ddot{y} \\ \vdots \\ y^{(n-1)} \end{bmatrix} = \begin{bmatrix} c(\boldsymbol{x}) \\ L_{\boldsymbol{a}}c(\boldsymbol{x}) \\ L_{\boldsymbol{a}}^2 c(\boldsymbol{x}) \\ \vdots \\ L_{\boldsymbol{a}}^{n-1}c(\boldsymbol{x}) \end{bmatrix} = \boldsymbol{t}(\boldsymbol{x}) \tag{5.20}$$

can be introduced. If the function $\boldsymbol{t}(\boldsymbol{x})$ is continuously differentiable and the inverse function $\boldsymbol{t}^{-1}$ exists, i. e. if

$$\boldsymbol{t}^{-1}(\boldsymbol{t}(\boldsymbol{x})) = \boldsymbol{x}$$

holds, and furthermore if $\boldsymbol{t}^{-1}$ is continuously differentiable, the mapping

$$\boldsymbol{t} : \mathbb{R}^n \to \mathbb{R}^n$$

is referred to as a *diffeomorphism*, as previously described in Section 3.3.1, p. 260 et seq. It forms the required bijective coordinate transformation and transforms the system

$$\begin{aligned}\dot{\boldsymbol{x}} &= \boldsymbol{a}(\boldsymbol{x}) + \boldsymbol{b}(\boldsymbol{x}) \cdot u,\\ y &= c(\boldsymbol{x})\end{aligned}$$

by differentiating the transformation equation (5.20), i. e.

$$\dot{\boldsymbol{z}} = \dot{\boldsymbol{t}}(\boldsymbol{x}) = \frac{\partial \boldsymbol{t}(\boldsymbol{x})}{\partial \boldsymbol{x}} \dot{\boldsymbol{x}} = \begin{bmatrix} L_{\boldsymbol{a}} c(\boldsymbol{x}) \\ L_{\boldsymbol{a}}^2 c(\boldsymbol{x}) \\ \vdots \\ L_{\boldsymbol{a}}^{n-1} c(\boldsymbol{x}) \\ L_{\boldsymbol{a}}^n c(\boldsymbol{x}) + L_{\boldsymbol{b}} L_{\boldsymbol{a}}^{n-1} c(\boldsymbol{x}) \cdot u \end{bmatrix},$$

and replacing $\boldsymbol{x}$ by $\boldsymbol{x} = \boldsymbol{t}^{-1}(\boldsymbol{z})$ into the nonlinear controller canonical form

$$\begin{aligned}\begin{bmatrix} \dot{z}_1 \\ \vdots \\ \dot{z}_{n-1} \\ \dot{z}_n \end{bmatrix} &= \begin{bmatrix} z_2 \\ \vdots \\ z_n \\ L_{\boldsymbol{a}}^n c(\boldsymbol{x}) \end{bmatrix} + \begin{bmatrix} 0 \\ \vdots \\ 0 \\ L_{\boldsymbol{b}} L_{\boldsymbol{a}}^{n-1} c(\boldsymbol{x}) \end{bmatrix} u = \begin{bmatrix} z_2 \\ \vdots \\ z_n \\ L_{\boldsymbol{a}}^n c(\boldsymbol{t}^{-1}(\boldsymbol{z})) \end{bmatrix} + \begin{bmatrix} 0 \\ \vdots \\ 0 \\ L_{\boldsymbol{b}} L_{\boldsymbol{a}}^{n-1} c(\boldsymbol{t}^{-1}(\boldsymbol{z})) \end{bmatrix} u,\\ y &= z_1.\end{aligned} \tag{5.21}$$

We have thus achieved our goal and in the following section can use the system in nonlinear controller canonical form for purposes of controller design.

5.2.2 Nonlinear Controller and Linear Control Loop

We already know the control law (5.17) for the nonlinear controller canonical form which yields a linear control loop. Using

$$\alpha(\boldsymbol{x}) = L_{\boldsymbol{a}}^n c(\boldsymbol{x}) \quad \text{and} \quad \beta(\boldsymbol{x}) = L_{\boldsymbol{b}} L_{\boldsymbol{a}}^{n-1} c(\boldsymbol{x}),$$

we can write it as

$$u(\boldsymbol{x}, y_{\text{ref}}) = -r(\boldsymbol{x}) + v(\boldsymbol{x}) \cdot y_{\text{ref}} \tag{5.22}$$

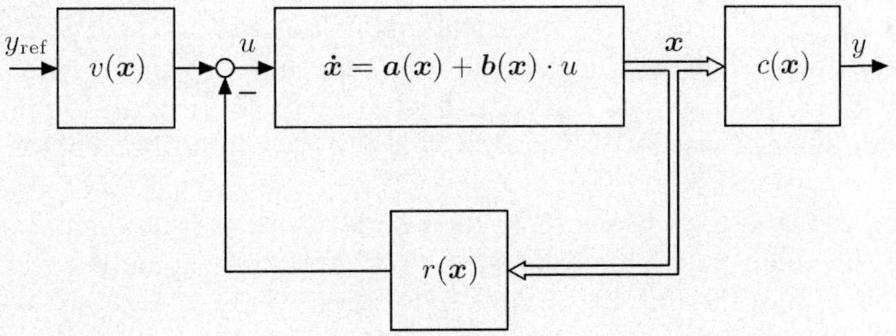

Fig. 5.9: Control loop with linear dynamics between input y_{ref} and output y

with

$$r(\boldsymbol{x}) = \frac{L_{\boldsymbol{a}}^n c(\boldsymbol{x}) + \boldsymbol{k}^T \boldsymbol{z}}{L_{\boldsymbol{b}} L_{\boldsymbol{a}}^{n-1} c(\boldsymbol{x})}, \qquad \boldsymbol{k}^T = [a_0 \;\; a_1 \;\; \cdots \;\; a_{n-1}]$$

as a controller and

$$v(\boldsymbol{x}) = \frac{V}{L_{\boldsymbol{b}} L_{\boldsymbol{a}}^{n-1} c(\boldsymbol{x})}$$

as a pre-filter. Figure 5.9 illustrates the corresponding block diagram. The variable y_{ref} acts as the reference variable of the control loop. Obviously the inequality $L_{\boldsymbol{b}} L_{\boldsymbol{a}}^{n-1} c(\boldsymbol{x}) \neq 0$ must be fulfilled for all states $\boldsymbol{x}$ in which the control is to be executed. If $L_{\boldsymbol{b}} L_{\boldsymbol{a}}^{n-1} \boldsymbol{c}(\boldsymbol{x}) = 0$, according to equation (5.21) the control variable would have no effect on the controlled system. The general controller given by equation (5.22) is equal to the special control law (5.17) from our initial calculation in which the controlled system is *a priori* in nonlinear controller canonical form.

We can also interpret the control law

$$u = -\frac{L_{\boldsymbol{a}}^n c(\boldsymbol{x}) + \boldsymbol{k}^T \boldsymbol{z}}{L_{\boldsymbol{b}} L_{\boldsymbol{a}}^{n-1} c(\boldsymbol{x})} + \frac{V}{L_{\boldsymbol{b}} L_{\boldsymbol{a}}^{n-1} c(\boldsymbol{x})} y_{\text{ref}} \tag{5.23}$$

as a state-dependent input variable transformation if we replace the input variable u with the new input variable y_{ref}.

Inserted in the nonlinear controller canonical form (5.21), the control law (5.22) results in a linear control loop in controller canonical form

$$\begin{bmatrix} \dot{z}_1 \\ \dot{z}_2 \\ \vdots \\ \dot{z}_{n-1} \\ \dot{z}_n \end{bmatrix} = \begin{bmatrix} 0 & 1 & 0 & \cdots & 0 \\ 0 & 0 & 1 & \cdots & 0 \\ \vdots & \vdots & \vdots & \ddots & \vdots \\ 0 & 0 & 0 & \cdots & 1 \\ -a_0 & -a_1 & -a_2 & \cdots & -a_{n-1} \end{bmatrix} \begin{bmatrix} z_1 \\ z_2 \\ \vdots \\ z_{n-1} \\ z_n \end{bmatrix} + \begin{bmatrix} 0 \\ 0 \\ \vdots \\ 0 \\ V \end{bmatrix} y_{\text{ref}}, \tag{5.24}$$

$$y = z_1.$$

We are free to choose the n coefficients a_i in the vector $\boldsymbol{k}$ so that we can impose any desired eigenvalue configuration and thus any linear dynamics on the control loop. The value V is also freely selectable. In practice, however, the limitation of the control variable, $|u| \leq u_{\max}$, limits our choice of the parameters a_i and V.

Using the transformations (5.20) and (5.23), we were successful in transforming the nonlinear system representation into a linear one. For the differential equation of the output variable y, we thus obtain

$$y^{(n)} + a_{n-1}y^{(n-1)} + \ldots + a_1\dot{y} + a_0 y = V \cdot y_{\text{ref}}$$

from equation (5.24). As already indicated, the term *input-output exact linearization* and its shortened version *input-output linearization* have their origin in the linear behavior between the input variable y_{ref} and the output variable y.

To make the control law (5.23) independent of $\boldsymbol{z}$, we insert

$$\boldsymbol{z} = \boldsymbol{t}(\boldsymbol{x}) = \begin{bmatrix} c(\boldsymbol{x}) \\ L_{\boldsymbol{a}}c(\boldsymbol{x}) \\ \vdots \\ L_{\boldsymbol{a}}^{n-1}c(\boldsymbol{x}) \end{bmatrix}$$

into equation (5.23) and arrive at the *nonlinear Ackermann formula*

$$u = -\frac{L_{\boldsymbol{a}}^{n}c(\boldsymbol{x}) + a_{n-1}L_{\boldsymbol{a}}^{n-1}c(\boldsymbol{x}) + \ldots + a_1 L_{\boldsymbol{a}}c(\boldsymbol{x}) + a_0 c(\boldsymbol{x})}{L_{\boldsymbol{b}}L_{\boldsymbol{a}}^{n-1}c(\boldsymbol{x})} + \frac{V \cdot y_{\text{ref}}}{L_{\boldsymbol{b}}L_{\boldsymbol{a}}^{n-1}c(\boldsymbol{x})}$$

as the control law, which only depends on the original state $\boldsymbol{x}$.

We can summarize all these results in the following theorem.

Theorem 59 (Input-Output Linearization at the Maximum Relative Degree). *Let the plant*

$$\begin{aligned} \dot{\boldsymbol{x}} &= \boldsymbol{a}(\boldsymbol{x}) + \boldsymbol{b}(\boldsymbol{x}) \cdot u, \quad \boldsymbol{x} \in D_{\boldsymbol{x},\text{def}} \subseteq I\!R^n, \quad u \in I\!R, \\ y &= c(\boldsymbol{x}), \end{aligned}$$

with the relative degree $\delta = n$ and the control law

$$u = -r(\boldsymbol{x}) + v(\boldsymbol{x}) \cdot y_{\text{ref}}$$

be given. If the controller has the form

$$r(\boldsymbol{x}) = \frac{L_{\boldsymbol{a}}^{n}c(\boldsymbol{x}) + a_{n-1}L_{\boldsymbol{a}}^{n-1}c(\boldsymbol{x}) + \ldots + a_1 L_{\boldsymbol{a}}c(\boldsymbol{x}) + a_0 c(\boldsymbol{x})}{L_{\boldsymbol{b}}L_{\boldsymbol{a}}^{n-1}c(\boldsymbol{x})},$$

and the pre-filter is given by

$$v(\boldsymbol{x}) = \frac{V}{L_{\boldsymbol{b}} L_{\boldsymbol{a}}^{n-1} c(\boldsymbol{x})}, \qquad V \in I\!R,$$

with

$$L_{\boldsymbol{b}} L_{\boldsymbol{a}}^{n-1} c(\boldsymbol{x}) \neq 0,$$

then the control loop has linear dynamic behavior described by

$$y^{(n)} + a_{n-1} y^{(n-1)} + \ldots + a_1 \dot{y} + a_0 y = V \cdot y_{\text{ref}}.$$

We refer to the procedure in Theorem 59 as *input-output linearization at the maximum relative degree*, since it is tailored to the highest possible value $\delta = n$. The case of $\delta < n$ is discussed in Section 5.2.4.

The following fact should be emphasized again: for input-output linearization to be possible on the system's domain of definition $D_{\boldsymbol{x},\text{def}}$, the condition

$$L_{\boldsymbol{b}} L_{\boldsymbol{a}}^{n-1} c(\boldsymbol{x}) \neq 0$$

must hold for all $\boldsymbol{x} \in D_{\boldsymbol{x},\text{def}}$. If this is the case, we designate the relative degree as *well-defined*. In the opposite case it is termed *not well-defined* and input-output linearization is only possible on a proper subset $D_{\boldsymbol{x}} \subset D_{\boldsymbol{x},\text{def}}$. Depending on whether the relative degree is well-defined or not, the system is globally or locally linearizable.

From the above it follows that a control-affine system with a relative degree $\delta = n$ is always controllable (as long as $\beta(\boldsymbol{x}) = L_{\boldsymbol{b}} L_{\boldsymbol{a}}^{n-1} c(\boldsymbol{x}) \neq 0$ holds), since it can be bijectively transformed into the nonlinear controller canonical form (5.21). The controllability of such systems has already been addressed in Theorem 45 on p. 218. For controllable and observable plants, it further holds that the inequality $\delta \leq n$ is fulfilled. However, in the case of uncontrollable and unobservable systems, $L_{\boldsymbol{b}} L_{\boldsymbol{a}}^{n-1} c(\boldsymbol{x}) = 0$ may be true and thus the input signal u has no influence on the output signal y. If we calculate δ in this case, we obtain $\delta = \infty$. We can only refer to δ as a relative degree if $\delta \leq n$ holds.

Literature on input-output linearization can also be found in [203, 232, 308, 338, 402].

5.2.3 Example: Magnetic Bearing

We will now describe an active magnetic bearing [316, 454] as used in electric generators and motors, pumps, and turbines. It enables frictionless bearing of the machine shaft and thus saves energy. The schematic diagram of a magnetic bearing and one of its pairs of electromagnets are shown in Figure 5.10. The electromagnets hold the shaft in position $h = 0$.

The magnetic fluxes Φ_1 and Φ_2, generated by the currents i_1 and i_2 in the electromagnets, generate the forces F_1 and F_2 which act on the shaft. These forces are adjusted in such a way that the bearing is mounted without contact.

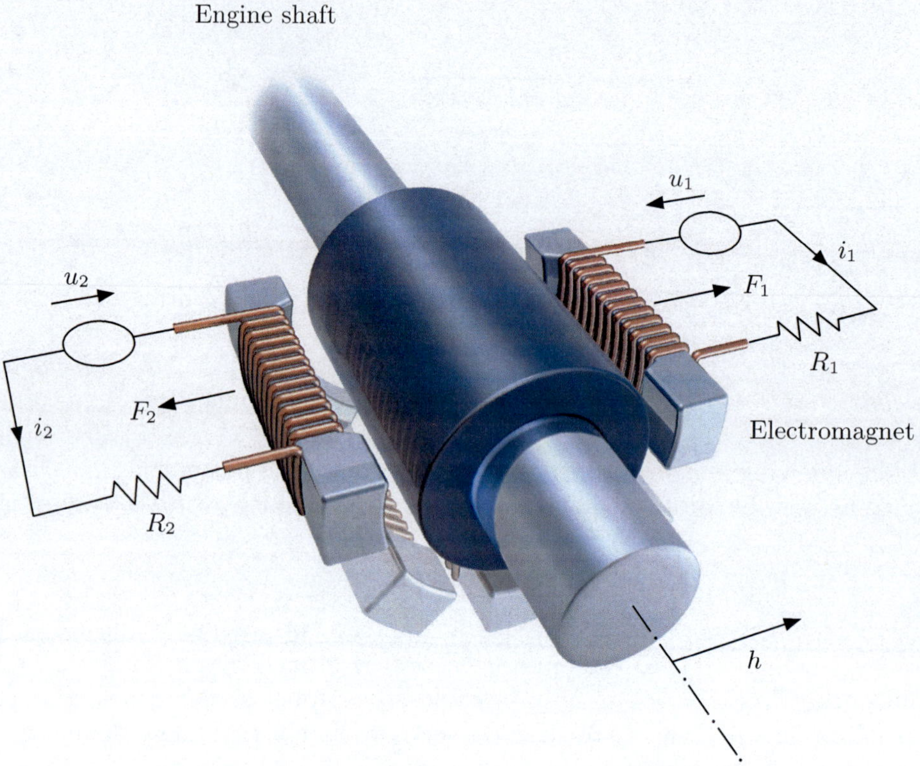

Fig. 5.10: Schematic diagram of an active magnetic bearing with one degree of freedom in the direction h. Further degrees of freedom not addressed here are produced by additional electromagnets.

To save electrical energy, the electromagnets can be alternately switched on and off. This operating mode is called a zero-bias or a low-bias operation.

The forces acting on the shaft are given by

$$F_i = \frac{1}{\mu_0 A} \Phi_i^2, \qquad i = 1, 2. \tag{5.25}$$

They variably depend on the magnetic flux Φ_i of the respective electromagnets i. Here, μ_0 is the magnetic field constant and A the surface of one pole of the electromagnets. The magnetic flux

$$\Phi_i = \Phi_0 + \phi_i, \qquad i = 1, 2, \tag{5.26}$$

consists of two components, where $\Phi_0 \ll \phi_i$ holds. The flux component Φ_0 is constant, i. e. it makes up the bias. The flux components ϕ_i include the components which change over time. Neglecting the coil resistances R_1 and R_2, we obtain for the relationship between the voltage u_i of the electromagnet i and the flux Φ_i

$$\dot{\Phi}_i = \dot{\phi}_i = \frac{u_i}{N}, \qquad i = 1, 2. \tag{5.27}$$

In this case, u_i is the voltage applied to the i-th electromagnet and N is the number of the coil's turns. We now define

$$\phi = \phi_1 - \phi_2 \tag{5.28}$$

and switch the voltages of the two electromagnets on and off according to

$$u_1 = \begin{cases} v & \text{for } \phi < 0, \\ 0 & \text{for } \phi \geq 0 \end{cases} \tag{5.29}$$

and

$$u_2 = \begin{cases} 0 & \text{for } \phi < 0, \\ -v & \text{for } \phi \geq 0. \end{cases} \tag{5.30}$$

The voltage v is variable and makes up the control signal

$$u = \dot{\phi} = \frac{v}{N}$$

of the controller which is yet to be designed.

With equation (5.25), the total force acting on the shaft can be stated as

$$F_{\text{tot}} = F_1 - F_2 = \frac{1}{\mu_0 A}(\Phi_1^2 - \Phi_2^2).$$

For the difference $\Phi_1^2 - \Phi_2^2$ arising from equations (5.26), (5.27), and (5.28), and taking into account switching strategies (5.29) and (5.30), the relation

$$\Phi_1^2 - \Phi_2^2 = 2\bar{\Phi}_0\phi + \phi|\phi|$$

with

$$\bar{\Phi}_0 = \Phi_0 + \min\{\phi_1(0), \phi_2(0)\}$$

follows after an intermediate calculation [454].

Thus, for the acceleration $\ddot{h}$ of the driving shaft with mass m, we obtain

$$\ddot{h} = \frac{1}{\mu_0 A m}(2\bar{\Phi}_0\phi + \phi|\phi|).$$

We define $x_1 = h$, $x_2 = \dot{h}$, and $x_3 = \phi$ as state variables of the system, and the voltage $u = v/N$ as the control variable. We then obtain the control-affine state-space model

$$\dot{\boldsymbol{x}} = \underbrace{\begin{bmatrix} x_2 \\ \alpha_1 x_3 + \alpha_2 x_3 |x_3| \\ 0 \end{bmatrix}}_{\boldsymbol{a}(\boldsymbol{x})} + \underbrace{\begin{bmatrix} 0 \\ 0 \\ 1 \end{bmatrix}}_{\boldsymbol{b}(\boldsymbol{x})} u$$

with $\boldsymbol{x} = [x_1 \;\; x_2 \;\; x_3]^T$ as the state vector and $u = \dot{\phi}$ as the control variable. The parameters are given by

$$\alpha_1 = \frac{2\bar{\Phi}_0}{\mu_0 Am} \quad \text{and} \quad \alpha_2 = \frac{1}{\mu_0 Am}.$$

The output variable y of the magnetic bearing is the displacement h, i. e.

$$y = c(\boldsymbol{x}) = x_1.$$

We can now design a controller using input-output linearization according to Theorem 59. We will take into account the fact that the function

$$f(x) = x|x|$$

is differentiable and that its derivative is

$$f'(x) = 2|x|.$$

We can now proceed in the following steps:

Step 1: The Lie derivatives $L_{\boldsymbol{a}}^{i} c(\boldsymbol{x})$ are

$$\begin{aligned}
i = 0: &\quad L_{\boldsymbol{a}}^{0} c(\boldsymbol{x}) = c(\boldsymbol{x}) = x_1,\\
i = 1: &\quad L_{\boldsymbol{a}} c(\boldsymbol{x}) = \frac{\partial c(\boldsymbol{x})}{\partial \boldsymbol{x}} \boldsymbol{a}(\boldsymbol{x}) = x_2,\\
i = 2: &\quad L_{\boldsymbol{a}}^{2} c(\boldsymbol{x}) = \frac{\partial L_{\boldsymbol{a}} c(\boldsymbol{x})}{\partial \boldsymbol{x}} \boldsymbol{a}(\boldsymbol{x}) = \alpha_1 x_3 + \alpha_2 x_3 |x_3|,\\
i = 3: &\quad L_{\boldsymbol{a}}^{3} c(\boldsymbol{x}) = \frac{\partial L_{\boldsymbol{a}}^{2} c(\boldsymbol{x})}{\partial \boldsymbol{x}} \boldsymbol{a}(\boldsymbol{x}) = 0.
\end{aligned}$$

Step 2: With the results from Step 1, we are able to calculate the terms

$$\begin{aligned}
i = 0: &\quad L_{\boldsymbol{b}} c(\boldsymbol{x}) = \frac{\partial x_1}{\partial \boldsymbol{x}} \boldsymbol{b}(\boldsymbol{x}) = 0,\\
i = 1: &\quad L_{\boldsymbol{b}} L_{\boldsymbol{a}} c(\boldsymbol{x}) = \frac{\partial x_2}{\partial \boldsymbol{x}} \boldsymbol{b}(\boldsymbol{x}) = 0,\\
i = 2: &\quad L_{\boldsymbol{b}} L_{\boldsymbol{a}}^{2} c(\boldsymbol{x}) = \frac{\partial (\alpha_1 x_3 + \alpha_2 x_3 |x_3|)}{\partial \boldsymbol{x}} \boldsymbol{b}(\boldsymbol{x}) = \alpha_1 + 2\alpha_2 |x_3|.
\end{aligned}$$

Thus for $i = 2$ the term $L_{\boldsymbol{b}} L_{\boldsymbol{a}}^{i} c(\boldsymbol{x})$ is unequal to zero and $\delta = 3 = n$ follows.

Step 3: It holds that

$$L_{\boldsymbol{b}} L_{\boldsymbol{a}}^{2} c(\boldsymbol{x}) = \alpha_1 + 2\alpha_2 |x_3| \neq 0$$

for all $\boldsymbol{x} \in I\!R^3$, because $\alpha_1 > 0$, $\alpha_2 > 0$, and $|x_3| \geq 0$.

Step 4: Since the relative degree is $\delta = 3$, the transfer function $G(s)$ of the control loop also takes an order of three, i.e.

$$G(s) = \frac{V}{s^3 + a_2 s^2 + a_1 s + a_0}.$$

We now select three real poles $s_1 = s_2 = s_3 = -\lambda$ with $\lambda > 0$ and, choosing $V = \lambda^3$, we obtain the transfer function

$$G(s) = \frac{\lambda^3}{s^3 + 3\lambda s^2 + 3\lambda^2 s + \lambda^3}.$$

Step 5: Including the results of Steps 1, 2, and 4 yields the controller

$$\begin{aligned} r(\boldsymbol{x}) &= \frac{L_{\boldsymbol{a}}^3 c(\boldsymbol{x}) + a_2 L_{\boldsymbol{a}}^2 c(\boldsymbol{x}) + a_1 L_{\boldsymbol{a}} c(\boldsymbol{x}) + a_0 c(\boldsymbol{x})}{L_{\boldsymbol{b}} L_{\boldsymbol{a}}^2 c(\boldsymbol{x})} \\ &= \frac{3\lambda(\alpha_1 x_3 + \alpha_2 x_3 |x_3|) + 3\lambda^2 x_2 + \lambda^3 x_1}{\alpha_1 + 2\alpha_2 |x_3|}. \end{aligned}$$

Step 6: For the pre-filter, we obtain

$$v(x) = \frac{V}{L_{\boldsymbol{b}} L_{\boldsymbol{a}}^2 c(\boldsymbol{x})} = \frac{\lambda^3}{\alpha_1 + 2\alpha_2 |x_3|}.$$

The control loop thus calculated is shown in Figure 5.11. The dynamics of the control loop, especially the control variable amplitude, and the required actuator energy can be influenced by means of the triple pole at $-\lambda$, which has not yet been specified. This makes it possible to keep the control value u within prescribed limits, such as $-1 \leq u \leq 1$. However, ensuring that the limits of the control variable are not exceeded is a difficult task in case of exact linearization [165, 223, 302, 352].

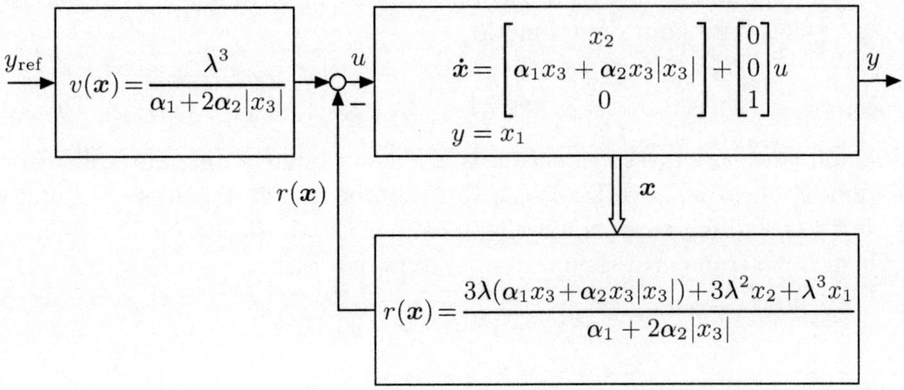

Fig. 5.11: Control of the active magnetic bearing using input-output linearization

5.2.4 Plants with Internal Dynamics

Below we will address the case in which

$$\delta < n,$$

i. e. the case in which the relative degree δ is lower than the system order n. We recall that the sequence of derivatives (5.19) without dependence on the control signal u stops at $y^{(\delta-1)}$. In this case as well, a nonlinear bijective transformation rule which is continuously differentiable, a diffeomorphism

$$\boldsymbol{z} = \boldsymbol{t}(\boldsymbol{x}), \tag{5.31}$$

can be found to arrive at a system representation which is favorable to controller design. Based on this, we can again design a control loop with linear input-output behavior.

Since $\delta < n$ holds, only the first δ components $t_1, \ldots, t_\delta$ of the diffeomorphism (5.31) can be used for the new state variables $z_1, \ldots, z_\delta$ in the same form as in the case in which $\delta = n$ holds. It holds that

$$\boldsymbol{z} = \begin{bmatrix} y \\ \dot{y} \\ \vdots \\ y^{(\delta-1)} \\ z_{\delta+1} \\ \vdots \\ z_n \end{bmatrix} = \begin{bmatrix} z_1 \\ z_2 \\ \vdots \\ z_\delta \\ z_{\delta+1} \\ \vdots \\ z_n \end{bmatrix} = \boldsymbol{t}(\boldsymbol{x}) = \begin{bmatrix} c(\boldsymbol{x}) \\ L_{\boldsymbol{a}} c(\boldsymbol{x}) \\ \vdots \\ L_{\boldsymbol{a}}^{\delta-1} c(\boldsymbol{x}) \\ t_{\delta+1}(\boldsymbol{x}) \\ \vdots \\ t_n(\boldsymbol{x}) \end{bmatrix}. \tag{5.32}$$

A comparison with the case $\delta = n$ in equation (5.20) demonstrates the identical procedure for the first δ elements of $\boldsymbol{z}$. The functions $t_{\delta+1}, \ldots, t_n$ can be chosen arbitrarily, as long as it is certain that $\boldsymbol{t}$ is a diffeomorphism. For this reason, the following requirements must be met: that $\boldsymbol{t}$ is continuously differentiable, that the inverse function

$$\boldsymbol{x} = \boldsymbol{t}^{-1}(\boldsymbol{z})$$

exists, and that this inverse function is also continuously differentiable. From Theorem 54 on p. 262, we know that all the above requirements are fulfilled and that $\boldsymbol{t}$ is a diffeomorphism if the Jacobian matrix $\partial \boldsymbol{t}(\boldsymbol{x})/\partial \boldsymbol{x}$ is regular.

Then, if we transform the original system description

$$\begin{aligned} \dot{\boldsymbol{x}} &= \boldsymbol{a}(\boldsymbol{x}) + \boldsymbol{b}(\boldsymbol{x}) \cdot u, \\ y &= c(\boldsymbol{x}) \end{aligned}$$

using the diffeomorphism (5.32), the result is

$$\dot{\boldsymbol{z}} = \begin{bmatrix} \dot{z}_1 \\ \dot{z}_2 \\ \vdots \\ \dot{z}_{\delta-1} \\ \dot{z}_\delta \\ \dot{z}_{\delta+1} \\ \vdots \\ \dot{z}_n \end{bmatrix} = \frac{d}{dt} \begin{bmatrix} c(\boldsymbol{x}) \\ L_{\boldsymbol{a}}c(\boldsymbol{x}) \\ \vdots \\ L_{\boldsymbol{a}}^{\delta-2}c(\boldsymbol{x}) \\ L_{\boldsymbol{a}}^{\delta-1}c(\boldsymbol{x}) \\ t_{\delta+1}(\boldsymbol{x}) \\ \vdots \\ t_n(\boldsymbol{x}) \end{bmatrix} = \begin{bmatrix} L_{\boldsymbol{a}}c(\boldsymbol{x}) \\ L_{\boldsymbol{a}}^2c(\boldsymbol{x}) \\ \vdots \\ L_{\boldsymbol{a}}^{\delta-1}c(\boldsymbol{x}) \\ L_{\boldsymbol{a}}^{\delta}c(\boldsymbol{x}) + L_{\boldsymbol{b}}L_{\boldsymbol{a}}^{\delta-1}c(\boldsymbol{x}) \cdot u \\ \dot{t}_{\delta+1}(\boldsymbol{x}) \\ \vdots \\ \dot{t}_n(\boldsymbol{x}) \end{bmatrix}.$$

Utilizing equation (5.32) once again leads to

$$\dot{\boldsymbol{z}} = \begin{bmatrix} \dot{z}_1 \\ \dot{z}_2 \\ \vdots \\ \dot{z}_{\delta-1} \\ \dot{z}_\delta \\ \dot{z}_{\delta+1} \\ \vdots \\ \dot{z}_n \end{bmatrix} = \begin{bmatrix} z_2 \\ z_3 \\ \vdots \\ z_\delta \\ L_{\boldsymbol{a}}^{\delta}c(\boldsymbol{x}) + L_{\boldsymbol{b}}L_{\boldsymbol{a}}^{\delta-1}c(\boldsymbol{x}) \cdot u \\ \dot{t}_{\delta+1}(\boldsymbol{x}) \\ \vdots \\ \dot{t}_n(\boldsymbol{x}) \end{bmatrix}, \tag{5.33}$$

while

$$y = z_1$$

still holds. In contrast to the case where

$$\delta = n,$$

only the first δ rows of equation (5.33) are in controller canonical form.

Let us take a closer look at the functions $t_{\delta+1}, \dots, t_n$. Using the transformation

$$\boldsymbol{x} = \boldsymbol{t}^{-1}(\boldsymbol{z})$$

for $i = \delta + 1, \dots, n$, the derivatives

$$\begin{aligned} \dot{t}_i(\boldsymbol{x}) &= \frac{\partial t_i(\boldsymbol{x})}{\partial \boldsymbol{x}}\dot{\boldsymbol{x}} = \frac{\partial t_i(\boldsymbol{x})}{\partial \boldsymbol{x}}(\boldsymbol{a}(\boldsymbol{x}) + \boldsymbol{b}(\boldsymbol{x}) \cdot u) \\ &= L_{\boldsymbol{a}}t_i(\boldsymbol{x}) + L_{\boldsymbol{b}}t_i(\boldsymbol{x}) \cdot u \\ &= \hat{q}_i(\boldsymbol{x}, u) = \hat{q}_i(\boldsymbol{t}^{-1}(\boldsymbol{z}), u) = q_i(\boldsymbol{z}, u) \end{aligned} \tag{5.34}$$

can be determined. If the functions t_i have been appropriately chosen or have been calculated using the method specified in [388], so that

$$L_{\boldsymbol{b}}t_i(\boldsymbol{x}) = \frac{\partial t_i(\boldsymbol{x})}{\partial \boldsymbol{x}}\boldsymbol{b}(\boldsymbol{x}) = 0 \tag{5.35}$$

holds, equation (5.34) can be simplified. The dependency on u is now dropped, and

$$\dot{t}_i(\boldsymbol{x}) = L_{\boldsymbol{a}} t_i(\boldsymbol{x}) = \hat{q}_i(\boldsymbol{x}) = q_i(\boldsymbol{z}) \tag{5.36}$$

applies. The elimination of the dependence on u simplifies the representation of the transformed system greatly, since the internal dynamics now no longer depend on u.

Equation (5.35) is a partial differential equation whose solution $t_i(\boldsymbol{x})$ usually exists, but is difficult to determine in many cases. In the following, and for reasons of simplification as described above, we will assume that t_i fulfills equation (5.35) for all $i = \delta+1, \ldots, n$, and insert the functions $\dot{t}_i$ from equation (5.36) into equation (5.33). We then obtain the so-called *Byrnes-Isidori canonical form*

$$\left.\begin{aligned} \begin{bmatrix} \dot{z}_1 \\ \vdots \\ \dot{z}_{\delta-1} \\ \dot{z}_\delta \\ \dot{z}_{\delta+1} \\ \vdots \\ \dot{z}_n \end{bmatrix} &= \begin{bmatrix} z_2 \\ \vdots \\ z_\delta \\ \alpha(\boldsymbol{z}) \\ q_{\delta+1}(\boldsymbol{z}) \\ \vdots \\ q_n(\boldsymbol{z}) \end{bmatrix} + \begin{bmatrix} 0 \\ \vdots \\ 0 \\ \beta(\boldsymbol{z}) \\ 0 \\ \vdots \\ 0 \end{bmatrix} u, \\ y &= z_1 \end{aligned}\right\} \begin{array}{l} \text{external dynamics} \\ \text{internal dynamics} \\ \text{output variable} \end{array} \tag{5.37}$$

as the system description, where

$$\alpha(\boldsymbol{z}) = L_{\boldsymbol{a}}^{\delta} c(\boldsymbol{t}^{-1}(\boldsymbol{z})) \quad \text{and} \quad \beta(\boldsymbol{z}) = L_{\boldsymbol{b}} L_{\boldsymbol{a}}^{\delta-1} c(\boldsymbol{t}^{-1}(\boldsymbol{z}))$$

hold.

As shown in equation (5.37), the system dynamics can be divided into an external and an internal part. The external dynamics of the system, describing the changes in the states $z_1, \ldots, z_\delta$, can be linearized by transforming the input variable u into the new input variable y_{ref}, as in the case in which $\delta = n$. Based on equation (5.33) or equation (5.37), we can determine this transformation as

$$u = -\frac{L_{\boldsymbol{a}}^{\delta} c(\boldsymbol{x}) + \boldsymbol{k}^T \boldsymbol{z}}{L_{\boldsymbol{b}} L_{\boldsymbol{a}}^{\delta-1} c(\boldsymbol{x})} + \frac{V}{L_{\boldsymbol{b}} L_{\boldsymbol{a}}^{\delta-1} c(\boldsymbol{x})} \cdot y_{\text{ref}} \tag{5.38}$$

with

$$\boldsymbol{k}^T = [a_0 \ \cdots \ a_{\delta-1} \ 0 \ \cdots \ 0].$$

If this is inserted into equation (5.37), the dynamic equations of the transformed system are obtained as follows:

$$\begin{bmatrix} \dot{z}_1 \\ \dot{z}_2 \\ \vdots \\ \dot{z}_{\delta-1} \\ \dot{z}_\delta \end{bmatrix} = \begin{bmatrix} 0 & 1 & 0 & \cdots & 0 \\ 0 & 0 & 1 & \cdots & 0 \\ \vdots & \vdots & \vdots & \ddots & \vdots \\ 0 & 0 & 0 & \cdots & 1 \\ -a_0 & -a_1 & -a_2 & \cdots & -a_{\delta-1} \end{bmatrix} \begin{bmatrix} z_1 \\ z_2 \\ \vdots \\ z_{\delta-1} \\ z_\delta \end{bmatrix} + \begin{bmatrix} 0 \\ 0 \\ \vdots \\ 0 \\ V \end{bmatrix} y_{\text{ref}}, \left.\vphantom{\begin{bmatrix} 0 \\ 0 \\ \vdots \\ 0 \\ V \end{bmatrix}}\right\} \begin{matrix}\text{external} \\ \text{dynamics}\end{matrix} \tag{5.39}$$

$$\begin{bmatrix} \dot{z}_{\delta+1} \\ \vdots \\ \dot{z}_n \end{bmatrix} = \begin{bmatrix} q_{\delta+1}(\boldsymbol{z}) \\ \vdots \\ q_n(\boldsymbol{z}) \end{bmatrix}, \left.\vphantom{\begin{bmatrix} 0 \\ \vdots \\ 0 \end{bmatrix}}\right\} \begin{matrix}\text{internal} \\ \text{dynamics}\end{matrix} \tag{5.40}$$

$$y = z_1.$$

After applying the transformation (5.38) of the input variable, the external dynamics are independent of the state variables $z_{\delta+1}, \ldots, z_n$ of the internal dynamics. In particular, the state variables of the internal dynamics have no influence on the output variable y. As a result, the internal dynamics (5.40) are not observable. The term *internal* symbolizes this property. This is why the control loop as a whole loses its observability due to the linearization. In practice, however, the latter is of little importance.

Obviously, the procedure above does not linearize the internal dynamics. Fortunately, as already stated, it does not influence the output variable $y = z_1$ or the controlled external dynamics. The internal dynamics possess the input variables $z_1, \ldots, z_\delta$ and no output variable. We can interpret the internal dynamics as an independent system with the associated state variables $z_{\delta+1}, \ldots, z_n$. This is illustrated by Figure 5.12.

Provided that

$$L_{\boldsymbol{b}} L_{\boldsymbol{a}}^{\delta-1} c(\boldsymbol{t}^{-1}(\boldsymbol{z})) \neq 0$$

holds, the external dynamics (5.37) are always controllable because they are present in nonlinear controller canonical form. We know this fact from Theorem 45 on p. 218. In this case, again, we can interpret the transformation equation (5.38) as a control law which has a state controller in the first summand and a pre-filter for the reference variable y_{ref} in the second. To make equation (5.38) independent of the artificial state vector $\boldsymbol{z}$, we can formulate

$$\boldsymbol{k}^T \boldsymbol{z} = a_{\delta-1} L_{\boldsymbol{a}}^{\delta-1} c(\boldsymbol{x}) + \ldots + a_1 L_{\boldsymbol{a}} c(\boldsymbol{x}) + a_0 c(\boldsymbol{x})$$

using

$$z_1 = c(\boldsymbol{x}), \quad z_2 = L_{\boldsymbol{a}} c(\boldsymbol{x}), \quad \ldots, \quad z_\delta = L_{\boldsymbol{a}}^{\delta-1} c(\boldsymbol{x}),$$

which follows from equation (5.32). Thus we obtain

$$u = -\frac{L_{\boldsymbol{a}}^{\delta} c(\boldsymbol{x}) + a_{n-1} L_{\boldsymbol{a}}^{\delta-1} c(\boldsymbol{x}) + \ldots + a_1 L_{\boldsymbol{a}} c(\boldsymbol{x}) + a_0 c(\boldsymbol{x})}{L_{\boldsymbol{b}} L_{\boldsymbol{a}}^{\delta-1} c(\boldsymbol{x})} + \frac{V \cdot y_{\text{ref}}}{L_{\boldsymbol{b}} L_{\boldsymbol{a}}^{\delta-1} c(\boldsymbol{x})}$$

as the control law.

The controlled external dynamics, meaning the control loop (5.39) itself, have the same structure as in the case in which $\delta = n$, i. e. as shown in Figure 5.9. Similar to the case in which $\delta = n$, the output variable y is given by a linear differential equation

$$y^{(\delta)} + a_{\delta-1}y^{(\delta-1)} + \ldots + a_1\dot{y} + a_0 y = V \cdot y_{\mathrm{ref}}. \tag{5.41}$$

However, it only has the relative degree $\delta < n$. We can also see from equation (5.41) that the lower the relative degree δ, the more directly the input variable u affects the output variable y.

At this point, the additional significance of the relative degree δ to the behavior of the control loop becomes clear: it determines the order of the linear differential equation of the control loop.

Internal dynamics do not affect the dynamics of the linear control loop, but they do affect the dynamics of the overall system. It is obvious that unstable internal dynamics[3] lead to an unstable overall system. Therefore, we need to analyze the internal dynamics' stability, which is a prerequisite if we are to obtain a stable control loop.

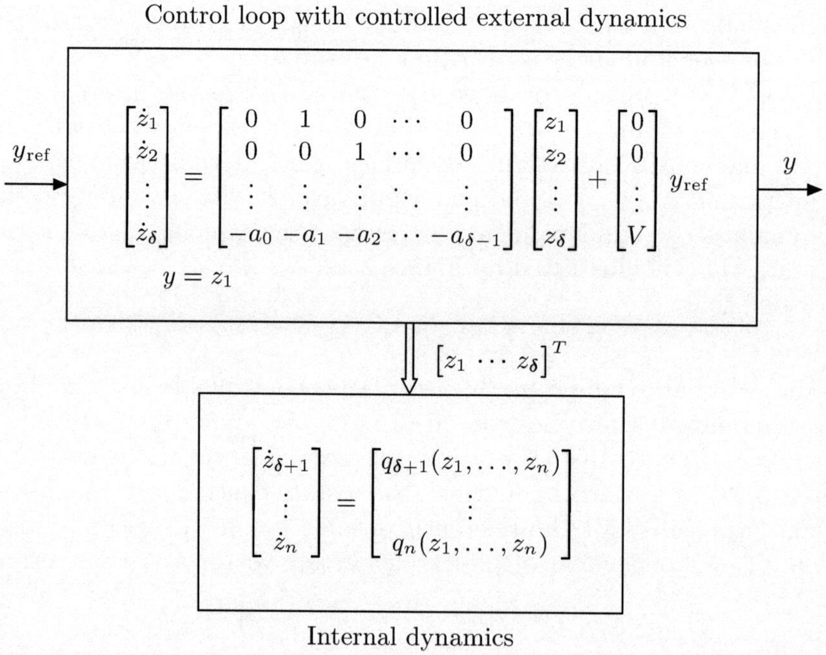

Fig. 5.12: Block diagram of the control loop with controlled external dynamics and unobservable internal dynamics

[3] The internal dynamics may have several equilibria, not merely a single one. In this case the equilibria which are relevant to practical operation must be stable. In this context, see also Definition 8 on p. 18, and the explanations of stability and internal dynamics in Section 5.2.10, p. 398, for a deeper understanding.

Again, we can summarize these results in a theorem.

Theorem 60 (Input-Output Linearization with a Reduced Relative Degree). *Let the plant*

$$\begin{aligned}\dot{\boldsymbol{x}} &= \boldsymbol{a}(\boldsymbol{x}) + \boldsymbol{b}(\boldsymbol{x}) \cdot u, \quad \boldsymbol{x} \in D_{\boldsymbol{x},\mathrm{def}} \subseteq I\!R^n, \quad u \in I\!R,\\ y &= c(\boldsymbol{x})\end{aligned}$$

with the relative degree $\delta < n$ and the control law

$$u = -r(\boldsymbol{x}) + v(\boldsymbol{x}) \cdot y_{\mathrm{ref}}$$

be given. If the controller has the form

$$r(\boldsymbol{x}) = \frac{L_{\boldsymbol{a}}^{\delta} c(\boldsymbol{x}) + a_{\delta-1} L_{\boldsymbol{a}}^{\delta-1} c(\boldsymbol{x}) + \ldots + a_1 L_{\boldsymbol{a}} c(\boldsymbol{x}) + a_0 c(\boldsymbol{x})}{L_{\boldsymbol{b}} L_{\boldsymbol{a}}^{\delta-1} c(\boldsymbol{x})}$$

and the pre-filter is given by

$$v(\boldsymbol{x}) = \frac{V}{L_{\boldsymbol{b}} L_{\boldsymbol{a}}^{\delta-1} c(\boldsymbol{x})}, \qquad V \in I\!R,$$

with

$$L_{\boldsymbol{b}} L_{\boldsymbol{a}}^{\delta-1} c(\boldsymbol{x}) \neq 0,$$

then the control loop has linear dynamic behavior described by

$$y^{(\delta)} + a_{\delta-1} y^{(\delta-1)} + \ldots + a_1 \dot{y} + a_0 y = V \cdot y_{\mathrm{ref}}.$$

In addition, the control loop has the internal dynamics

$$\begin{bmatrix} \dot{z}_{\delta+1} \\ \vdots \\ \dot{z}_n \end{bmatrix} = \begin{bmatrix} q_{\delta+1}(\boldsymbol{z}) \\ \vdots \\ q_n(\boldsymbol{z}) \end{bmatrix}$$

with

$$\boldsymbol{z} = \boldsymbol{t}(\boldsymbol{x}).$$

The internal dynamics do not influence the output y. If both the internal and the linear external dynamics of the control loop have a globally asymptotically stable equilibrium point at the origin and the internal dynamics with its input variable vector $[z_1 \ \cdots \ z_\delta]$ are input-to-state stable, then the equilibrium point $\boldsymbol{z}_{\mathrm{eq}} = \boldsymbol{0}$ of the whole system is globally asymptotically stable.

Theorem 59 and Theorem 60 are very similar and differ only in that internal dynamics are included in Theorem 60. We will discuss the stability of an input-output linearized system in more detail in Section 5.2.10. It is important to note that although the control loop has unobservable internal dynamics, the controlled system is normally observable. Therefore, the state vector $\boldsymbol{x}$ required for the control can be estimated using an observer.

In the linear transfer behavior of the control loop

$$G(s) = \frac{V}{s^{\delta} + a_{\delta-1}s^{\delta-1} + \ldots + a_1 s + a_0}$$

the coefficients

$$a_0, a_1, \ldots, a_{\delta-1}$$

and the factor V of the pre-filter $v(\boldsymbol{x})$ are arbitrarily selectable for both cases, $\delta = n$ as well as $\delta < n$. The dynamics of the transfer function $G(s)$ can also be freely defined using the parameters $a_0, \ldots, a_{\delta-1}$ and V. In practice, however, the limitation of the control variable given by the inequality

$$|u| \leq u_{\max}$$

again limits the available flexibility.

5.2.5 Design Procedure

The following general procedure, which applies to $\delta = n$ as well as $\delta < n$, describes how to calculate the controller and the pre-filter. We have already followed a similar procedure for the special case of the magnetic bearing in Section 5.2.3.

Step 1: Determine the Lie derivatives

$$L_{\boldsymbol{a}}^{i} c(\boldsymbol{x}), \quad i = 0, \ldots, n = \dim(\boldsymbol{x}) .$$

Step 2: Determine the Lie derivatives

$$L_{\boldsymbol{b}} L_{\boldsymbol{a}}^{i} c(\boldsymbol{x}) = \frac{\partial L_{\boldsymbol{a}}^{i} c(\boldsymbol{x})}{\partial \boldsymbol{x}} \boldsymbol{b}(\boldsymbol{x}) \tag{5.42}$$

with $i = 0, 1, 2, \ldots$ in ascending order. The lowest index i for which the above term (5.42) is not equal to zero yields the relative degree

$$\delta = i + 1.$$

The corresponding term in equation (5.42) reads as

$$L_{\boldsymbol{b}} L_{\boldsymbol{a}}^{\delta-1} c(\boldsymbol{x}) = \frac{\partial L_{\boldsymbol{a}}^{\delta-1} c(\boldsymbol{x})}{\partial \boldsymbol{x}} \boldsymbol{b}(\boldsymbol{x}).$$

Step 3: For all occurring $\boldsymbol{x}$, the inequality

$$L_{\boldsymbol{b}} L_{\boldsymbol{a}}^{\delta-1} c(\boldsymbol{x}) \neq 0$$

must hold, i. e. the relative degree δ must be well-defined.

Step 4: Select the linear dynamics of the control loop

$$G(s) = \frac{V}{s^{\delta} + a_{\delta-1}s^{\delta-1} + \ldots + a_1 s + a_0}$$

by setting the freely selectable parameters V and $a_0, a_1, \ldots, a_{\delta-1}$.

Step 5: Utilize the results from Steps 1, 2, and 4 to determine the controller

$$r(\boldsymbol{x}) = \frac{L_{\boldsymbol{a}}^{\delta} c(\boldsymbol{x}) + a_{\delta-1} L_{\boldsymbol{a}}^{\delta-1} c(\boldsymbol{x}) + \ldots + a_0 c(\boldsymbol{x})}{L_{\boldsymbol{b}} L_{\boldsymbol{a}}^{\delta-1} c(\boldsymbol{x})}.$$

Step 6: The pre-filter is determined according to

$$v(\boldsymbol{x}) = \frac{V}{L_{\boldsymbol{b}} L_{\boldsymbol{a}}^{\delta-1} c(\boldsymbol{x})}.$$

Step 7: For $\delta < n$, determine the internal dynamics and check them for stability.

Checking the internal dynamics for stability can be complicated, since this dynamic component is principally nonlinear.

5.2.6 Example: Lunar Module

In the following, we will look at an example with internal dynamics: a lander [318] such as the lunar module Eagle of the Apollo 11 mission [22, 343], the Phoenix Mars probe, and the lunar lander Altair from NASA's Constellation Program. The landing module of a potentially manned Mars mission might have a similar design. Figure 5.13 shows the lunar module Eagle, which we will consider in the following.

The main purpose of propulsion control for a landing module of the type above is to ensure a soft landing. For this purpose, the engine's vertically acting thrust force is used to decelerate the gravity-induced approach toward the lunar surface.

The following equation

$$m\ddot{h} = -v\dot{m} - mg \tag{5.43}$$

describes the balance of forces, where m is the mass of the lunar module including the fuel supply, h denotes the altitude, v is the speed of the gas ejected from the engine, and $g = 1.62\,\mathrm{m\,s^{-2}}$ is the gravitational acceleration of the moon. The term $-v\dot{m}$ represents the thrust of the engine, where $\dot{m} < 0$ applies because the fuel mass decreases during landing. The Eagle had a thrust of $4.56 \ldots 45.04\,\mathrm{kN}$ during the landing phase. The engine's exit velocity was $v = 3050\,\mathrm{m\,s^{-1}}$.

It should also be noted that the fuel mass makes up a substantial part of the total mass m. The total weight of the Eagle was $m = 15264\,\mathrm{kg}$, of which

Fig. 5.13: Lunar module Eagle of the Apollo 11 mission

8165 kg of fuel was available for the landing phase alone. The decrease in the mass m due to the consumption of the fuel supply is thus an essential aspect of the equation of the shuttle's motion (5.43).

Utilizing the state variables

$$x_1 = h, \;\; x_2 = \dot{h}, \;\; x_3 = m,$$

the control variable

$$u = -\dot{m},$$

and equation (5.43), we arrive at the model

$$\begin{bmatrix} \dot{x}_1 \\ \dot{x}_2 \\ \dot{x}_3 \end{bmatrix} = \underbrace{\begin{bmatrix} x_2 \\ -g \\ 0 \end{bmatrix}}_{\boldsymbol{a}(\boldsymbol{x})} + \underbrace{\begin{bmatrix} 0 \\ \dfrac{v}{x_3} \\ -1 \end{bmatrix}}_{\boldsymbol{b}(\boldsymbol{x})} u, \tag{5.44}$$

$$y = c(\boldsymbol{x}) = x_1$$

for the Eagle's landing phase. Note that $u = -\dot{m}$ is always positive.

Consistent with the design procedure from Section 5.2.5, we determine

$$\begin{aligned} L_{\boldsymbol{a}}^0 c(\boldsymbol{x}) &= c(\boldsymbol{x}) = x_1, \\ L_{\boldsymbol{a}}^1 c(\boldsymbol{x}) &= \frac{\partial c(\boldsymbol{x})}{\partial \boldsymbol{x}} \boldsymbol{a}(\boldsymbol{x}) = x_2, \\ L_{\boldsymbol{a}}^2 c(\boldsymbol{x}) &= \frac{\partial L_{\boldsymbol{a}} c(\boldsymbol{x})}{\partial \boldsymbol{x}} \boldsymbol{a}(\boldsymbol{x}) = -g. \end{aligned}$$

Further, we can compute

$$\begin{aligned} L_{\boldsymbol{b}} c(\boldsymbol{x}) &= \frac{\partial c(\boldsymbol{x})}{\partial \boldsymbol{x}} \boldsymbol{b}(\boldsymbol{x}) = 0, \\ L_{\boldsymbol{b}} L_{\boldsymbol{a}} c(\boldsymbol{x}) &= \frac{\partial L_{\boldsymbol{a}} c(\boldsymbol{x})}{\partial \boldsymbol{x}} \boldsymbol{b}(\boldsymbol{x}) = \frac{v}{x_3}. \end{aligned}$$

Since $L_{\boldsymbol{b}} L_{\boldsymbol{a}} c(\boldsymbol{x}) \neq 0$, the system has the relative degree $\delta = 2$. Because the system order is $n = 3 > \delta$, the system has internal dynamics. The mass $x_3 = m$ of the shuttle is always positive. Thus,

$$L_{\boldsymbol{b}} L_{\boldsymbol{a}} c(\boldsymbol{x}) = \frac{v}{x_3} > 0$$

holds and the relative degree is well-defined. We can therefore design a controller using input-output linearization. We must also ensure the stability of the internal dynamics to guarantee the stability of the control system as a whole.

The diffeomorphism can be written as

$$\boldsymbol{z} = \boldsymbol{t}(\boldsymbol{x}) = \begin{bmatrix} c(\boldsymbol{x}) \\ L_{\boldsymbol{a}} c(\boldsymbol{x}) \\ t_3(\boldsymbol{x}) \end{bmatrix} = \begin{bmatrix} x_1 \\ x_2 \\ t_3(\boldsymbol{x}) \end{bmatrix}, \tag{5.45}$$

where the function $t_3(\boldsymbol{x})$ is yet to be selected. This is done in such a way that the partial differential equation (5.35) is fulfilled, i. e. in this case

$$L_{\boldsymbol{b}} t_3(\boldsymbol{x}) = \frac{\partial t_3(\boldsymbol{x})}{\partial \boldsymbol{x}} \boldsymbol{b}(\boldsymbol{x}) = 0.$$

We remember: if we do it this way, the internal dynamics do not depend on the input variable u. Specifically, we obtain

$$L_{\boldsymbol{b}} t_3(\boldsymbol{x}) = \frac{\partial t_3(\boldsymbol{x})}{\partial x_2} \cdot \frac{v}{x_3} - \frac{\partial t_3(\boldsymbol{x})}{\partial x_3} = 0. \tag{5.46}$$

This partial differential equation has the solution

$$t_3(\boldsymbol{x}) = x_3 \cdot e^{x_2/v},$$

which can easily be checked by inserting it into equation (5.46). So the diffeomorphism (5.45) is of the form

$$\boldsymbol{z} = \boldsymbol{t}(\boldsymbol{x}) = \begin{bmatrix} x_1 \\ x_2 \\ x_3 e^{x_2/v} \end{bmatrix} \quad \text{or} \quad \boldsymbol{x} = \boldsymbol{t}^{-1}(\boldsymbol{z}) = \begin{bmatrix} z_1 \\ z_2 \\ z_3 e^{-z_2/v} \end{bmatrix}. \tag{5.47}$$

Using equation (5.44) and equation (5.47), we obtain the system description with external and internal dynamics as

$$\begin{bmatrix} \dot{z}_1 \\ \dot{z}_{\delta=2} \\ \dot{z}_3 \end{bmatrix} = \begin{bmatrix} L_{\boldsymbol{a}} c(\boldsymbol{x}) \\ L_{\boldsymbol{a}}^2 c(\boldsymbol{x}) + L_{\boldsymbol{b}} L_{\boldsymbol{a}} c(\boldsymbol{x}) u \\ \dot{t}_3(x_2, x_3) \end{bmatrix} = \begin{bmatrix} x_2 \\ -g + \dfrac{v}{x_3} u \\ -v^{-1} g x_3 e^{x_2/v} \end{bmatrix}$$

$$= \begin{bmatrix} z_2 \\ -g + \dfrac{v}{x_3} u \\ -v^{-1} g z_3 \end{bmatrix}. \quad \begin{matrix} \left.\vphantom{\begin{matrix} a \\ b \end{matrix}}\right\} \text{external dynamics} \\ \} \text{ internal dynamics} \end{matrix} \tag{5.48}$$

We can now design the control law according to Theorem 60 using

$$\begin{aligned} u &= -\frac{L_{\boldsymbol{a}}^2 c(\boldsymbol{x}) + a_1 L_{\boldsymbol{a}} c(\boldsymbol{x}) + a_0 c(\boldsymbol{x})}{L_{\boldsymbol{b}} L_{\boldsymbol{a}} c(\boldsymbol{x})} \\ &= -\frac{-g + a_1 x_2 + a_0 x_1}{v} x_3 = -\frac{-g + a_1 z_2 + a_0 z_1}{v} x_3. \end{aligned} \tag{5.49}$$

A reference variable y_{ref} is not required here, since the control adjusts the state variables $x_1 = z_1$ and $x_2 = z_2$ to the equilibrium point

$$\begin{bmatrix} x_{1,\text{eq}} \\ x_{2,\text{eq}} \end{bmatrix} = \boldsymbol{0}.$$

Therefore, the pre-filter $v(\boldsymbol{x})$ is not necessary.

With equations (5.47), (5.48), and (5.49) we can now represent the entire control system by

$$\begin{bmatrix} \dot{z}_1 \\ \dot{z}_2 \\ \dot{z}_3 \end{bmatrix} = \begin{bmatrix} z_2 \\ -a_1 z_2 - a_0 z_1 \\ -v^{-1} g z_3 \end{bmatrix}, \tag{5.50}$$

$$y = z_1.$$

As can be seen from equation (5.50), the internal and external dynamics are decoupled. The internal dynamics especially cause no problems because they are stable. It is also apparent that the internal dynamics are not observable, since z_3 does not affect $y = z_1$.

We will now transform the above controlled system description back to the original coordinates $\boldsymbol{x}$, and, with equation (5.50) and diffeomorphism (5.47), we thus arrive at the representation

$$\begin{bmatrix} \dot{x}_1 \\ \dot{x}_2 \\ \dot{x}_3 \end{bmatrix} = \begin{bmatrix} x_2 \\ -a_1 x_2 - a_0 x_1 \\ v^{-1} x_3 (a_1 x_2 + a_0 x_1 - g) \end{bmatrix}$$

of the control loop. We can also obtain this result by inserting the control law (5.49) into the system description (5.44). The above equation shows that the internal dynamics do not influence the controlled external dynamics. However, we need the internal state variable x_3 in the control law (5.49). This does not pose a problem, since $x_3 = m$ holds and m is measurable.

As a concrete numerical example, we will continue with the lunar module Eagle from the Apollo 11 mission and simulate the approach and landing phase on the moon using the designed controller (5.49). We select the coefficients $a_0 = 0.02$ and $a_1 = 1.1$, so that the eigenvalues of the controlled external dynamics are

$$\lambda_1 = -0.0185 \quad \text{and} \quad \lambda_2 = -1.0815.$$

So we can describe the controlled lunar module by

$$\begin{bmatrix} \dot{x}_1 \\ \dot{x}_2 \\ \dot{x}_3 \end{bmatrix} = \begin{bmatrix} x_2 \\ -1.1 x_2 - 0.02 x_1 \\ x_3 (3.61 \cdot 10^{-4} x_2 + 6.56 \cdot 10^{-6} x_1 - 5.31 \cdot 10^{-4}) \end{bmatrix}$$

with $g = 1.62\,\mathrm{m\,s^{-2}}$ and $v = 3050\,\mathrm{m\,s^{-1}}$.

We begin the approach and landing phase [22] at an altitude of $x_1(0) = 2450\,\mathrm{m}$ and a sinking velocity of $x_2(0) = -45\,\mathrm{m\,s^{-1}}$. The fuel supply for the landing is still 1633 kg at this point. The total mass is $x_3(0) = 8732$ kg. We assume that the movement of the lunar module in this phase is approximately

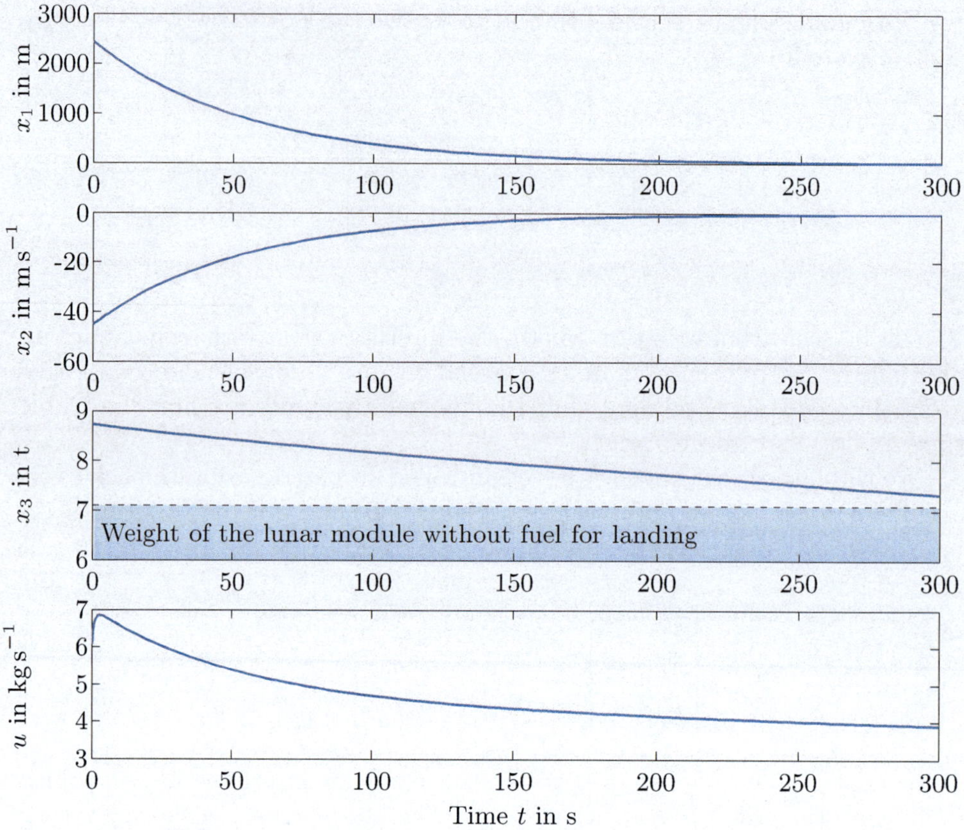

Fig. 5.14: Height x_1, velocity x_2, total mass x_3, and control variable u

vertical to the lunar surface[4]. The time courses for the height x_1, the speed x_2, and the remaining total mass x_3 are shown in Figure 5.14.

5.2.7 Input-Output Linearization of General SISO Systems

We can extend input-output linearization, discussed in previous sections for control-affine systems, to general nonlinear SISO systems

$$\begin{aligned} \dot{\boldsymbol{x}} &= \boldsymbol{f}(\boldsymbol{x}, u), \\ y &= g(\boldsymbol{x}). \end{aligned} \tag{5.51}$$

Similar to the control-affine case, we will utilize the Lie derivative of g with respect to $\boldsymbol{f}$, i. e.

$$L_{\boldsymbol{f}} g(\boldsymbol{x}) = \frac{\partial g(\boldsymbol{x})}{\partial \boldsymbol{x}} \boldsymbol{f}(\boldsymbol{x}, u)$$

[4] In reality, the Eagle lunar module's trajectory was not completely vertical during the final phase of the landing due to the orbit of curved descent around a part of the moon.

and the multiple Lie derivative

$$L^i_{\boldsymbol{f}} g(\boldsymbol{x}) = \frac{\partial L^{i-1}_{\boldsymbol{f}} g(\boldsymbol{x})}{\partial \boldsymbol{x}} \boldsymbol{f}(\boldsymbol{x}, u).$$

We can now calculate

$$\dot{y} = \frac{\partial g(\boldsymbol{x})}{\partial \boldsymbol{x}} \dot{\boldsymbol{x}} = \frac{\partial g(\boldsymbol{x})}{\partial \boldsymbol{x}} \boldsymbol{f}(\boldsymbol{x}, u) = L_{\boldsymbol{f}} g(\boldsymbol{x}) \tag{5.52}$$

and

$$\ddot{y} = \frac{\partial L_{\boldsymbol{f}} g(\boldsymbol{x})}{\partial \boldsymbol{x}} \boldsymbol{f}(\boldsymbol{x}, u) + \frac{\partial L_{\boldsymbol{f}} g(\boldsymbol{x})}{\partial u} \dot{u}.$$

If

$$\frac{\partial L_{\boldsymbol{f}} g(\boldsymbol{x})}{\partial u} \dot{u} = 0 \tag{5.53}$$

holds, obviously $L_{\boldsymbol{f}} g(\boldsymbol{x})$ and thus $\dot{y}$ in equation (5.52) do not depend on u. We assume that equation (5.53) is valid so that we obtain

$$\ddot{y} = L^2_{\boldsymbol{f}} g(\boldsymbol{x})$$

next. We can now compute

$$\dddot{y} = \frac{\partial L^2_{\boldsymbol{f}} g(\boldsymbol{x})}{\partial \boldsymbol{x}} \boldsymbol{f}(\boldsymbol{x}, u) + \frac{\partial L^2_{\boldsymbol{f}} g(\boldsymbol{x})}{\partial u} \dot{u}.$$

If

$$\frac{\partial L^2_{\boldsymbol{f}} g(\boldsymbol{x})}{\partial u} \dot{u} = 0$$

holds, which is similar to the case described above, the double Lie derivative $L^2_{\boldsymbol{f}} g(\boldsymbol{x})$ is not dependent on u either. We then continue calculating further derivatives $y^{(i)}$ as long as

$$\frac{\partial L^{i-1}_{\boldsymbol{f}} g(\boldsymbol{x})}{\partial u} = 0$$

holds, until we arrive at two equations of the form

$$\begin{aligned} y^{(\delta)} &= \underbrace{\frac{\partial L^{\delta-1}_{\boldsymbol{f}} g(\boldsymbol{x})}{\partial \boldsymbol{x}} \boldsymbol{f}(\boldsymbol{x}, u)}_{L^{\delta}_{\boldsymbol{f}} g(\boldsymbol{x})} + \underbrace{\frac{\partial L^{\delta-1}_{\boldsymbol{f}} g(\boldsymbol{x})}{\partial u}}_{0} \dot{u} = L^{\delta}_{\boldsymbol{f}} g(\boldsymbol{x}) = \varphi(\boldsymbol{x}, u), \\ y^{(\delta+1)} &= \frac{\partial L^{\delta}_{\boldsymbol{f}} g(\boldsymbol{x})}{\partial \boldsymbol{x}} \boldsymbol{f}(\boldsymbol{x}, u) + \underbrace{\frac{\partial L^{\delta}_{\boldsymbol{f}} g(\boldsymbol{x})}{\partial u}}_{\neq 0} \dot{u}. \end{aligned} \tag{5.54}$$

Since now

$$\frac{\partial L_{\boldsymbol{f}}^{\delta} g(\boldsymbol{x})}{\partial u} \neq 0 \tag{5.55}$$

holds in equation (5.54), the function $L_{\boldsymbol{f}}^{\delta} g(\boldsymbol{x})$ depends on u. Since all derivatives $y^{(i)} = L_{\boldsymbol{f}}^{i} g(\boldsymbol{x})$ with $i = 0, \ldots, \delta - 1$ are independent of u, we obtain

$$\begin{aligned} y &= g(\boldsymbol{x}), \\ \dot{y} &= L_{\boldsymbol{f}} g(\boldsymbol{x}), \\ &\vdots \\ y^{(\delta-1)} &= L_{\boldsymbol{f}}^{\delta-1} g(\boldsymbol{x}), \\ y^{(\delta)} &= L_{\boldsymbol{f}}^{\delta} g(\boldsymbol{x}) = \varphi(\boldsymbol{x}, u). \end{aligned} \tag{5.56}$$

As in the case of control-affine systems, the index δ is referred to as the *relative degree.* Note that the δth Lie derivative $L_{\boldsymbol{f}}^{\delta} g(\boldsymbol{x})$ is used in equation (5.55). Compared to this, for a control-affine system

$$\dot{\boldsymbol{x}} = \boldsymbol{a}(\boldsymbol{x}) + \boldsymbol{b}(\boldsymbol{x}) \cdot u, \quad y = c(\boldsymbol{x}),$$

the Lie derivative $L_{\boldsymbol{b}} L_{\boldsymbol{a}}^{\delta-1} c(\boldsymbol{x})$ is used to determine the relative degree δ.

In equation (5.56), we can replace the control variable u with a new control variable

$$v = \varphi(\boldsymbol{x}, u) \tag{5.57}$$

so that we obtain the relation

$$y^{(\delta)} = v.$$

This procedure assumes that the implicit equation (5.57) is uniquely solvable for u; more precisely, that the function φ is bijective, meaning the inverse function $u = \varphi^{-1}(\boldsymbol{x}, v)$ exists. Here it will often be the case that the inverse function φ^{-1} can only be determined numerically.

If we select the control law with the new control variable v such that

$$v = -a_{\delta-1} y^{(\delta-1)} - \ldots - a_1 \dot{y} - a_0 y + V \cdot y_{\text{ref}} \tag{5.58}$$

holds, where y_{ref} is the reference variable of the controlled system, we obtain the linear control-loop dynamics according to

$$y^{(\delta)} + a_{\delta-1} y^{(\delta-1)} + \ldots + a_1 \dot{y} + a_0 y = V \cdot y_{\text{ref}}.$$

Again, the coefficients a_i are freely selectable. The input-output exact linearization of general nonlinear systems has thus been achieved.

Let us now transform system (5.51) into the so-called *generalized Byrnes-Isidori canonical form* using a suitable diffeomorphism $\boldsymbol{t}(\boldsymbol{x})$. As in the control-affine case, we choose

$$\boldsymbol{z} = \boldsymbol{t}(\boldsymbol{x}) = \begin{bmatrix} g(\boldsymbol{x}) \\ L_{\boldsymbol{f}} g(\boldsymbol{x}) \\ L_{\boldsymbol{f}}^2 g(\boldsymbol{x}) \\ \vdots \\ L_{\boldsymbol{f}}^{\delta-1} g(\boldsymbol{x}) \\ t_{\delta+1}(\boldsymbol{x}) \\ \vdots \\ t_n(\boldsymbol{x}) \end{bmatrix}.$$

By deriving $\boldsymbol{z} = \boldsymbol{t}(\boldsymbol{x})$, we then calculate

$$\dot{\boldsymbol{z}} = \begin{bmatrix} L_{\boldsymbol{f}} g(\boldsymbol{x}) \\ L_{\boldsymbol{f}}^2 g(\boldsymbol{x}) \\ \vdots \\ L_{\boldsymbol{f}}^{\delta-1} g(\boldsymbol{x}) \\ L_{\boldsymbol{f}}^{\delta} g(\boldsymbol{x}) \\ \dot{t}_{\delta+1}(\boldsymbol{x}) \\ \vdots \\ \dot{t}_n(\boldsymbol{x}) \end{bmatrix} = \begin{bmatrix} z_2 \\ z_3 \\ \vdots \\ z_\delta \\ \hat{\varphi}(\boldsymbol{z}, u) \\ L_{\boldsymbol{f}} t_{\delta+1}(\boldsymbol{x}) \\ \vdots \\ L_{\boldsymbol{f}} t_n(\boldsymbol{x}) \end{bmatrix}, \quad \begin{matrix} \left.\right\} \text{external dynamics} \\ \left.\right\} \text{internal dynamics} \end{matrix}$$

$$y = z_1,$$

where $\hat{\varphi}(\boldsymbol{z}, u) = \varphi(\boldsymbol{t}^{-1}(\boldsymbol{z}), u)$ holds, and then note that

$$\dot{t}_i(\boldsymbol{x}) = L_{\boldsymbol{f}} t_i(\boldsymbol{x}) = \frac{\partial t_i(\boldsymbol{x})}{\partial \boldsymbol{x}} \boldsymbol{f}(\boldsymbol{x}, u), \quad i = \delta+1, \ldots, n, \tag{5.59}$$

depends on the control variable u. The dependence of the functions $\dot{t}_i$ on the control variable u is similar to the control-affine case. However, this dependence is not as simple to eliminate as in control-affine systems, where only equation (5.35) on p. 373 needs to be solved. For equation (5.59) to be independent of a non-constant control variable u, the partial differential equation

$$\frac{\partial \dot{t}_i(\boldsymbol{x})}{\partial u} = \frac{\partial}{\partial u}\left(\frac{\partial t_i(\boldsymbol{x})}{\partial \boldsymbol{x}} \boldsymbol{f}(\boldsymbol{x}, u)\right) = \frac{\partial t_i(\boldsymbol{x})}{\partial \boldsymbol{x}} \cdot \frac{\partial \boldsymbol{f}(\boldsymbol{x}, u)}{\partial u} = 0$$

must hold for all $i = \delta + 1, \ldots, n$. So we need to find $n - \delta$ functions t_i such that

$$\sum_{k=1}^{n} \frac{\partial t_i(\boldsymbol{x})}{\partial x_k} \cdot \frac{\partial f_k(\boldsymbol{x}, u)}{\partial u} = 0, \qquad i = \delta + 1, \ldots, n, \tag{5.60}$$

holds.

When choosing or determining functions t_i as solutions of equation (5.60), we have to keep in mind that a diffeomorphism $\boldsymbol{z} = \boldsymbol{t}(\boldsymbol{x})$ must result, i. e. according to Theorem 54, p. 262,

$$\det\left(\frac{\partial \boldsymbol{t}(\boldsymbol{x})}{\partial \boldsymbol{x}}\right) \neq 0$$

must hold. If we succeed in doing all the above, we then derive $\boldsymbol{z} = \boldsymbol{t}(\boldsymbol{x})$ with respect to time and obtain a similar transformed system representation as in the control-affine case in equation (5.37). This is the *generalized Byrnes-Isidori canonical form*

$$\dot{\boldsymbol{z}} = \begin{bmatrix} L_{\boldsymbol{f}} g(\boldsymbol{x}) \\ L_{\boldsymbol{f}}^2 g(\boldsymbol{x}) \\ \vdots \\ L_{\boldsymbol{f}}^{\delta-1} g(\boldsymbol{x}) \\ L_{\boldsymbol{f}}^{\delta} g(\boldsymbol{x}) \\ \dot{t}_{\delta+1}(\boldsymbol{x}) \\ \vdots \\ \dot{t}_n(\boldsymbol{x}) \end{bmatrix} = \begin{bmatrix} z_2 \\ z_3 \\ \vdots \\ z_\delta \\ \hat{\varphi}(\boldsymbol{z}, u) \\ q_{\delta+1}(\boldsymbol{z}) \\ \vdots \\ q_n(\boldsymbol{z}) \end{bmatrix}, \quad \begin{matrix} \left.\begin{matrix} \\ \\ \\ \\ \\ \end{matrix}\right\} \text{external dynamics} \\ \left.\begin{matrix} \\ \\ \\ \end{matrix}\right\} \text{internal dynamics} \end{matrix} \tag{5.61}$$

$$y = z_1.$$

Here, the nonlinear part of the external dynamics is

$$\dot{z}_\delta = \hat{\varphi}(\boldsymbol{z}, u) = v$$

and the internal dynamics are

$$\dot{z}_i = q_i(\boldsymbol{z}) = \dot{t}_i(\boldsymbol{x}) = \dot{t}_i(\boldsymbol{t}^{-1}(\boldsymbol{z})), \quad i = \delta + 1, \ldots, n.$$

In summarizing, using

$$u = \hat{\varphi}^{-1}(\boldsymbol{z}, v) \qquad \text{or} \qquad u = \varphi^{-1}(\boldsymbol{x}, v)$$

and the control law (5.58), we can control the external dynamics of the transformed system (5.61) and impose any linear dynamics on the control loop.

5.2.8 Relative Degree and Internal Dynamics of Linear Systems

In the following, simply to deepen our understanding of the input-output linearization method, we will apply input-output linearization to the controllable and observable linear systems

$$\begin{aligned}\dot{\boldsymbol{x}} &= \boldsymbol{A}\boldsymbol{x} + \boldsymbol{b}\cdot u,\\ y &= \boldsymbol{c}^T\boldsymbol{x}.\end{aligned} \tag{5.62}$$

Without any restriction of generality, we can assume that the system (5.62) is present in controller canonical form or has been transformed into that form, i. e. in the following we can assume that the system's description is given by

$$\boldsymbol{A} = \begin{bmatrix} 0 & 1 & 0 & \cdots & 0 \\ 0 & 0 & 1 & \cdots & 0 \\ \vdots & \vdots & \vdots & \ddots & \vdots \\ 0 & 0 & 0 & \cdots & 1 \\ -a_0 & -a_1 & -a_2 & \cdots & -a_{n-1} \end{bmatrix}, \quad \boldsymbol{b} = \begin{bmatrix} 0 \\ 0 \\ \vdots \\ 0 \\ 1 \end{bmatrix}, \tag{5.63}$$

$$\boldsymbol{c}^T = \begin{bmatrix} b_0 & b_1 \cdots b_m & 0 \cdots 0 \end{bmatrix}, \quad b_m \neq 0.$$

The associated transfer function is

$$G(s) = \frac{b_m s^m + \ldots + b_1 s + b_0}{s^n + a_{n-1}s^{n-1} + \ldots + a_1 s + a_0}. \tag{5.64}$$

First, we will calculate the relative degree δ. In the derivatives

$$y^{(i)} = L_{\boldsymbol{a}}^i c(\boldsymbol{x}) + L_{\boldsymbol{b}} L_{\boldsymbol{a}}^{i-1} c(\boldsymbol{x}) \cdot u = \boldsymbol{c}^T\boldsymbol{A}^i\boldsymbol{x} + \boldsymbol{c}^T\boldsymbol{A}^{i-1}\boldsymbol{b}\cdot u,$$

we use the identity

$$L_{\boldsymbol{b}} L_{\boldsymbol{a}}^{i-1} c(\boldsymbol{x}) = \boldsymbol{c}^T\boldsymbol{A}^{i-1}\boldsymbol{b} = 0 \tag{5.65}$$

for $i = 1, \ldots, \delta - 1$. The inequality

$$\boldsymbol{c}^T\boldsymbol{A}^i\boldsymbol{b} \neq 0 \tag{5.66}$$

is fulfilled only for $i \geq \delta - 1$. Using the Leverrier-Faddeev-Souriau-Frame equation [135, 216]

$$(s\boldsymbol{I} - \boldsymbol{A})^{-1} = \frac{1}{D(s)} \sum_{k=0}^{n-1} \left(\sum_{l=0}^{k} a_{n-k+l}\boldsymbol{A}^l \right) s^{n-1-k},$$

where $a_n = 1$ and

$$D(s) = \det(s\boldsymbol{I} - \boldsymbol{A}) = s^n + a_{n-1}s + \ldots + a_1 s + a_0$$

is the characteristic polynomial of $\boldsymbol{A}$, we obtain the expression

$$G(s) = \boldsymbol{c}^T(s\boldsymbol{I} - \boldsymbol{A})^{-1}\boldsymbol{b} = \frac{1}{D(s)}\sum_{k=0}^{n-1}\left(\sum_{l=0}^{k} a_{n-k+l}\,\boldsymbol{c}^T\boldsymbol{A}^l\boldsymbol{b}\right)s^{n-1-k}$$

for the transfer function of the system. Since $\boldsymbol{c}^T\boldsymbol{A}^l\boldsymbol{b} = 0$ holds for $l < \delta - 1$, the transfer function can be written as

$$G(s) = \frac{1}{D(s)}\sum_{k=\delta-1}^{n-1}\left(\sum_{l=\delta-1}^{k} a_{n-k+l}\,\boldsymbol{c}^T\boldsymbol{A}^l\boldsymbol{b}\right)s^{n-1-k}.$$

Using

$$b_{n-1-k} = \sum_{l=\delta-1}^{k} a_{n-k+l}\,\boldsymbol{c}^T\boldsymbol{A}^l\boldsymbol{b}, \qquad k = \delta-1,\ldots,n-1 \text{ and } a_n = 1,$$

the transfer function finally becomes

$$G(s) = \frac{b_{n-\delta}s^{n-\delta} + b_{n-\delta-1}s^{n-\delta-1} + \ldots + b_1 s + b_0}{s^n + a_{n-1}s^{n-1} + \ldots + a_1 s + a_0}. \tag{5.67}$$

If we compare equation (5.67) to equation (5.64), we see that the relative degree of a linear system is

$$\delta = n - m.$$

This means that the relative degree, also referred to as the difference degree, is equal to the difference between the denominator degree and the numerator degree.

The diffeomorphism (5.32) and the new coordinates $\boldsymbol{z}$, which allow us to formulate the system (5.62) in Byrnes-Isidori canonical form (5.37), are given by

$$\boldsymbol{z} = \begin{bmatrix} z_1 \\ z_2 \\ \vdots \\ z_\delta \\ z_{\delta+1} \\ \vdots \\ z_n \end{bmatrix} = \begin{bmatrix} c(\boldsymbol{x}) \\ L_{\boldsymbol{a}}c(\boldsymbol{x}) \\ \vdots \\ L_{\boldsymbol{a}}^{\delta-1}c(\boldsymbol{x}) \\ t_{\delta+1}(\boldsymbol{x}) \\ \vdots \\ t_n(\boldsymbol{x}) \end{bmatrix} = \begin{bmatrix} \boldsymbol{c}^T\boldsymbol{x} \\ \boldsymbol{c}^T\boldsymbol{A}\boldsymbol{x} \\ \vdots \\ \boldsymbol{c}^T\boldsymbol{A}^{\delta-1}\boldsymbol{x} \\ x_1 \\ \vdots \\ x_{n-\delta} \end{bmatrix} = \boldsymbol{t}(\boldsymbol{x}) \tag{5.68}$$

in this case. The functions

$$t_{\delta+i}(\boldsymbol{x}) = x_i, \quad i = 1,\ldots,n-\delta,$$

satisfy the partial differential equations (5.35) from Section 5.2.4 on p. 373, which here take the form

$$\frac{\partial t_{\delta+i}(\boldsymbol{x})}{\partial \boldsymbol{x}}\,\boldsymbol{b}(\boldsymbol{x}) = \frac{\partial x_i}{\partial \boldsymbol{x}}\,\boldsymbol{b} = 0, \quad i = 1,\ldots,n-\delta.$$

At this point, recall that these differential equations ensure the independence of the internal dynamics from the control variable u.

Now our aim is to transform the system (5.62) using the diffeomorphism (5.68), which is linear here, i. e.

$$\boldsymbol{t}(\boldsymbol{x}) = \boldsymbol{T}\boldsymbol{x} = \boldsymbol{z}, \tag{5.69}$$

so that it is represented in $\boldsymbol{z}$-coordinates and the internal dynamics become visible. To this end, we will consider the upper δ elements of the vector in equation (5.68), for which

$$\begin{aligned}
\boldsymbol{c}^T &= \begin{bmatrix} b_0 & b_1 \cdots b_m & 0 \cdots 0 \end{bmatrix}, \\
\boldsymbol{c}^T\boldsymbol{A} &= \begin{bmatrix} 0 & b_0 & b_1 \cdots b_m & 0 \cdots 0 \end{bmatrix}, \\
\boldsymbol{c}^T\boldsymbol{A}^2 &= \begin{bmatrix} 0 & 0 & b_0 & b_1 \cdots b_m & 0 \cdots 0 \end{bmatrix}, \\
&\;\;\vdots \\
\boldsymbol{c}^T\boldsymbol{A}^{\delta-1} &= \begin{bmatrix} 0 \cdots 0 & b_0 & b_1 \cdots b_m \end{bmatrix}
\end{aligned} \tag{5.70}$$

holds. For the $n \times n$ matrix $\boldsymbol{T}$ of the transformation equation (5.69), we thus obtain

$$\boldsymbol{T} = \begin{bmatrix}
b_0 & b_1 & b_2 & \cdots & b_{m-1} & b_m & 0 & 0 & \cdots & 0 & 0 \\
0 & b_0 & b_1 & \cdots & b_{m-2} & b_{m-1} & b_m & 0 & \cdots & 0 & 0 \\
\vdots & \vdots & \vdots & \ddots & \vdots & \vdots & \vdots & \vdots & \ddots & \vdots & \vdots \\
0 & 0 & 0 & \cdots & \cdots & \cdots & \cdots & \cdots & \cdots & 0 & 0 \\
0 & 0 & 0 & \cdots & \cdots & \cdots & \cdots & \cdots & \cdots & b_m & 0 \\
0 & 0 & 0 & \cdots & \cdots & \cdots & \cdots & \cdots & \cdots & b_{m-1} & b_m \\
1 & 0 & 0 & \cdots & 0 & 0 & 0 & 0 & \cdots & 0 & 0 \\
0 & 1 & 0 & \cdots & 0 & 0 & 0 & 0 & \cdots & 0 & 0 \\
0 & 0 & 1 & \cdots & 0 & 0 & 0 & 0 & \cdots & 0 & 0 \\
\vdots & \vdots & \vdots & \ddots & \vdots & \vdots & \vdots & \vdots & \ddots & \vdots & \vdots \\
0 & 0 & 0 & \cdots & 1 & 0 & 0 & 0 & \cdots & 0 & 0
\end{bmatrix}
\begin{matrix}
\left.\vphantom{\begin{matrix}a\\a\\a\\a\\a\\a\end{matrix}}\right\}\delta \text{ rows} \\
\left.\vphantom{\begin{matrix}a\\a\\a\\a\\a\end{matrix}}\right\}\begin{matrix} m=n-\delta \\ \text{rows}\end{matrix}
\end{matrix},$$

$$\underbrace{\hspace{8em}}_{m\,=\,n\,-\,\delta \text{ columns}}\quad\underbrace{\hspace{10em}}_{\delta \text{ columns}}$$

which is identical to the observability matrix

$$\boldsymbol{M}_{\text{obs}} = \begin{bmatrix} \boldsymbol{c}^T \\ \boldsymbol{c}^T\boldsymbol{A} \\ \boldsymbol{c}^T\boldsymbol{A}^2 \\ \vdots \\ \boldsymbol{c}^T\boldsymbol{A}^{n-1} \end{bmatrix}$$

consisting of the n row vectors (5.70) if $\delta = n$. The Matrix $\boldsymbol{T}$ can be divided into two parts according to

$$\boldsymbol{T} = \begin{bmatrix} \boldsymbol{M} \\ \boldsymbol{Z} \end{bmatrix}, \quad \boldsymbol{M} \in \mathbb{R}^{\delta \times n}, \quad \boldsymbol{Z} \in \mathbb{R}^{(n-\delta)\times n},$$

where $\boldsymbol{M}$ is equal to the first δ rows of $\boldsymbol{T}$ and $\boldsymbol{Z}$ matches the last $n - \delta$ rows.

The matrix $\boldsymbol{T}$ is regular and so is the transformation, because

$$\det(\boldsymbol{T}) = \det \begin{bmatrix} \boldsymbol{M} \\ \boldsymbol{Z} \end{bmatrix} = (-1)^{n-\delta} \det \begin{bmatrix} \boldsymbol{Z} \\ \boldsymbol{M} \end{bmatrix} = (-1)^{n-\delta} b_m^\delta$$

holds, since the swapping of $\boldsymbol{M}$ and $\boldsymbol{Z}$ results in a lower triangular matrix whose determinant is equal to the product of its diagonal elements [40].

To obtain the transformed system description, we will first calculate the derivative of the transformation equation (5.69)

$$\dot{\boldsymbol{z}} = \boldsymbol{T}\dot{\boldsymbol{x}} = \begin{bmatrix} \boldsymbol{c}^T\boldsymbol{A}\boldsymbol{x} \\ \boldsymbol{c}^T\boldsymbol{A}^2\boldsymbol{x} \\ \vdots \\ \boldsymbol{c}^T\boldsymbol{A}^\delta\boldsymbol{x} + \boldsymbol{c}^T\boldsymbol{A}^{\delta-1}\boldsymbol{b}\cdot u \\ \dot{x}_1 \\ \vdots \\ \dot{x}_{n-\delta} \end{bmatrix} = \begin{bmatrix} z_2 \\ z_3 \\ \vdots \\ \boldsymbol{c}^T\boldsymbol{A}^\delta\boldsymbol{T}^{-1}\boldsymbol{z} + \boldsymbol{c}^T\boldsymbol{A}^{\delta-1}\boldsymbol{b}\cdot u \\ x_2 \\ \vdots \\ x_{n-\delta+1} \end{bmatrix}, \tag{5.71}$$
$$y = \boldsymbol{c}^T\boldsymbol{x} = z_1$$

based on equation (5.68) as well as equations (5.62) and (5.63). Furthermore, according to equation (5.68), the identities

$$x_1 = z_{\delta+1}, \quad x_2 = z_{\delta+2}, \quad x_3 = z_{\delta+3}, \quad \ldots, \quad x_{n-\delta} = z_n \tag{5.72}$$

apply, and, using

$$c^T x = b_0\, x_1 + b_1\, x_2 + \ldots + b_{n-\delta}\, x_{n-\delta+1} = z_1$$

and $m = n - \delta$, we obtain the relation

$$\begin{aligned} x_{n-\delta+1} &= \frac{z_1 - b_0 x_1 - b_1 x_2 - \ldots - b_{n-\delta-1} x_{n-\delta}}{b_{n-\delta}} \\ &= \frac{z_1 - b_0 z_{\delta+1} - b_1 z_{\delta+2} - \ldots - b_{m-1} z_n}{b_m}. \end{aligned} \tag{5.73}$$

If we insert equations (5.72) and (5.73) into equation (5.71), the system description eventually yields

$$\begin{bmatrix} \dot z_1 \\ \vdots \\ \dot z_{\delta-1} \\ \dot z_\delta \\ \dot z_{\delta+1} \\ \vdots \\ \dot z_{n-1} \\ \dot z_n \end{bmatrix} = \begin{bmatrix} z_2 \\ \vdots \\ z_\delta \\ \boldsymbol{c}^T \boldsymbol{A}^\delta \boldsymbol{T}^{-1} \boldsymbol{z} + \boldsymbol{c}^T \boldsymbol{A}^{\delta-1} \boldsymbol{b} \cdot u \\ z_{\delta+2} \\ \vdots \\ z_n \\ -\dfrac{b_0}{b_m} z_{\delta+1} - \dfrac{b_1}{b_m} z_{\delta+2} - \ldots - \dfrac{b_{m-1}}{b_m} z_n + \dfrac{1}{b_m} z_1 \end{bmatrix}, \quad \begin{matrix} \left.\begin{matrix} \\ \\ \\ \\ \end{matrix}\right\} \text{external dynamics} \\ \left.\begin{matrix} \\ \\ \\ \\ \end{matrix}\right\} \text{internal dynamics} \end{matrix}$$

$$y = z_1. \tag{5.74}$$

This is the Byrnes-Isidori canonical form of a linear system (5.62).

Considering only the internal dynamics

$$\begin{bmatrix} \dot z_{\delta+1} \\ \dot z_{\delta+2} \\ \vdots \\ \dot z_{n-1} \\ \dot z_n \end{bmatrix} = \begin{bmatrix} 0 & 1 & 0 & \cdots & 0 & 0 \\ 0 & 0 & 1 & \cdots & 0 & 0 \\ \vdots & \vdots & \vdots & & \vdots & \vdots \\ 0 & 0 & 0 & \cdots & 0 & 1 \\ -\dfrac{b_0}{b_m} & -\dfrac{b_1}{b_m} & -\dfrac{b_2}{b_m} & \cdots & -\dfrac{b_{m-2}}{b_m} & -\dfrac{b_{m-1}}{b_m} \end{bmatrix} \begin{bmatrix} z_{\delta+1} \\ z_{\delta+2} \\ \vdots \\ z_{n-1} \\ z_n \end{bmatrix} + \begin{bmatrix} 0 \\ 0 \\ \vdots \\ 0 \\ \dfrac{1}{b_m} \end{bmatrix} z_1,$$

the state variable z_1 acts as the input variable. As we can immediately see in the companion matrix above, the internal system dynamics possess the characteristic polynomial $P(s)/b_m$ with

$$P(s) = b_m s^m + b_{m-1} s^{m-1} + \ldots + b_1 s + b_0.$$

The eigenvalues of the internal dynamics are therefore identical to the zeros of the linear system. Of course, the transformed system (5.74) as a whole has the same eigenvalues as the original system (5.62).

In summary, we can formulate

Theorem 61 (Relative Degree and Internal Dynamics of Linear Systems). *A controllable and observable linear system with the transfer function*

$$G(s) = \frac{b_m s^m + b_{m-1} s^{m-1} + \ldots + b_1 s + b_0}{s^n + a_{n-1} s^{n-1} + \ldots + a_1 s + a_0}$$

has the relative degree $\delta = n - m$. The eigenvalues of the internal dynamics are identical to the zeros of the transfer function.

We will now address the exceptional case of an uncontrollable and unobservable system. If equation (5.66) is not fulfilled for any $i = 1, \ldots, n-1$, i. e. if $\boldsymbol{c}^T \boldsymbol{A}^{n-1} \boldsymbol{b} = 0$, then $\delta > n$ applies; more precisely, $\delta = \infty$. To determine that a system such as this is uncontrollable and unobservable, we calculate

$$[\boldsymbol{c}^T \boldsymbol{b} \quad \boldsymbol{c}^T \boldsymbol{A} \boldsymbol{b} \quad \boldsymbol{c}^T \boldsymbol{A}^2 \boldsymbol{b} \quad \cdots \quad \boldsymbol{c}^T \boldsymbol{A}^{n-1} \boldsymbol{b}] = \boldsymbol{c}^T \boldsymbol{M}_{\text{contr}} = (\boldsymbol{M}_{\text{obs}} \boldsymbol{b})^T = \boldsymbol{0}^T$$

using equation (5.65). The expression above can only be identical to zero if both the controllability matrix

$$\boldsymbol{M}_{\text{contr}} = [\boldsymbol{b} \;\; \boldsymbol{A}\boldsymbol{b} \;\; \cdots \;\; \boldsymbol{A}^{n-1}\boldsymbol{b}]$$

and the observability matrix

$$\boldsymbol{M}_{\text{obs}} = \begin{bmatrix} \boldsymbol{c}^T \\ \boldsymbol{c}^T \boldsymbol{A} \\ \boldsymbol{c}^T \boldsymbol{A}^2 \\ \vdots \\ \boldsymbol{c}^T \boldsymbol{A}^{n-1} \end{bmatrix}$$

are singular. Therefore, the system is neither controllable nor observable if $\delta = \infty$.

In the following we will turn to a further issue and draw an analogy to nonlinear systems. In linear and nonlinear systems, the internal dynamics can be represented by

$$\dot{\boldsymbol{z}}_{\text{int}} = \boldsymbol{q}(\boldsymbol{z}_{\text{ext}}, \boldsymbol{z}_{\text{int}})$$

with

$$\boldsymbol{z}_{\text{ext}} = [z_1 \; \cdots \; z_\delta]^T \quad \text{and} \quad \boldsymbol{z}_{\text{int}} = [z_{\delta+1} \; \cdots \; z_n]^T .$$

Inserting the expression

$$u = -\frac{\boldsymbol{c}^T \boldsymbol{A}^\delta \boldsymbol{T}^{-1} \boldsymbol{z}}{\boldsymbol{c}^T \boldsymbol{A}^{\delta-1} \boldsymbol{b}}$$

into equation (5.74) and setting the initial values of $\boldsymbol{z}_{\text{ext}}(t)$ to $\boldsymbol{z}_{\text{ext}}(0) = \boldsymbol{0}$ yields

$$\boldsymbol{z}_{\text{ext}}(t) = \boldsymbol{0} \qquad \text{for } t \geq 0,$$

and, correspondingly, $y(t) = z_1 = 0$. The internal dynamics are now unaffected by the state vector $\boldsymbol{z}_{\text{ext}}$, or more precisely the state z_1, which result from the external dynamics. The variable z_1 can be interpreted as the input variable of the internal dynamics. Thus,

$$\dot{\boldsymbol{z}}_{\text{int}} = \boldsymbol{q}(\boldsymbol{0}, \boldsymbol{z}_{\text{int}}), \tag{5.75}$$

and in the case of linear systems the internal dynamics' behavior are only determined by the zeros of the system. This special case of internal dynamics is referred to as *zero dynamics*, which is also defined for nonlinear systems by (5.75). The zero dynamics of nonlinear systems thus have a similar meaning as the zeros in linear systems.

If the zero dynamics (5.75), or more precisely their equilibrium point of interest, $\boldsymbol{z}_{\text{int,eq}} = \boldsymbol{0}$, are asymptotically stable, the nonlinear system is often described as *minimum phase*[5]. This is analogous to minimum-phase linear systems, which are designated as minimum phase if they have only poles and zeros s_i with $\operatorname{Re}\{s_i\} < 0$ [199, 342]. Where the equilibrium point $\boldsymbol{z}_{\text{int,eq}}$ of the zero dynamics is not asymptotically stable, only Lyapunov stable, the system is designated as *weakly minimum phase*. Since the zero dynamics of nonlinear systems are normally nonlinear, more than one equilibrium point may exist. Consequently, discussing its stability can be exceptionally complicated in this case.

5.2.9 Control Law for the Linear Case

In addition to the previous section, we will revisit the input-output linearization method for the linear plant (5.62), (5.63) and determine its control law. From a practical point of view, of course, this is not necessary, since a linear system does not require linearization. But this gives us two further interesting insights into the method.

We obtain

$$u = -\frac{\boldsymbol{c}^T\boldsymbol{A}^{\delta}\boldsymbol{T}^{-1}\boldsymbol{z} + \boldsymbol{k}^T\boldsymbol{z}}{\boldsymbol{c}^T\boldsymbol{A}^{\delta-1}\boldsymbol{b}} + \frac{V}{\boldsymbol{c}^T\boldsymbol{A}^{\delta-1}\boldsymbol{b}}y_{\text{ref}}, \quad \boldsymbol{k}^T = [\hat{a}_0 \ \cdots \ \hat{a}_{\delta-1} \ 0 \ \cdots \ 0] \tag{5.76}$$

[5] The term *minimum phase* was introduced by H.W. Bode [52] and is mainly used in communications engineering [342]. Its definition is not consistent in the literature. Among other fields, it is used in filter design, for which the lowest possible phase shift is important. The common usage of the term as a stability description for internal or zero dynamics seems to be misleading, because the stability of a nonlinear system does not correspond to a phase response, which a nonlinear system does not have at all. In addition, nonlinear zero dynamics can have several equilibrium points with different stability behavior. Therefore, in this book the terms *asymptotic stability* and *Lyapunov stable* are used in connection with the internal dynamics and zero dynamics, because they are more appropriate. See also [496].

for the control law of the input-output linearization. Inserting this into the Byrnes-Isidori canonical form (5.74) of a linear plant yields

$$\begin{bmatrix} \dot{z}_1 \\ \vdots \\ \dot{z}_{\delta-1} \\ \dot{z}_\delta \\ \dot{z}_{\delta+1} \\ \vdots \\ \dot{z}_{n-1} \\ \dot{z}_n \end{bmatrix} = \underbrace{\begin{bmatrix} 0 & 1 & 0 & \cdots & 0 & 0 & 0 & 0 & \cdots & 0 \\ \vdots & \vdots & \vdots & & \vdots & \vdots & \vdots & \vdots & \ddots & \vdots \\ 0 & 0 & 0 & \cdots & 1 & 0 & 0 & 0 & \cdots & 0 \\ -\hat{a}_0 & -\hat{a}_1 & -\hat{a}_2 & \cdots & -\hat{a}_{\delta-1} & 0 & 0 & 0 & \cdots & 0 \\ 0 & 0 & 0 & \cdots & 0 & 0 & 1 & 0 & \cdots & 0 \\ \vdots & \vdots & \vdots & \ddots & \vdots & \vdots & \vdots & \vdots & & \vdots \\ 0 & 0 & 0 & \cdots & 0 & 0 & 0 & 0 & \cdots & 1 \\ \frac{1}{b_m} & 0 & 0 & \cdots & 0 & -\frac{b_0}{b_m} & -\frac{b_1}{b_m} & -\frac{b_2}{b_m} & \cdots & -\frac{b_{m-1}}{b_m} \end{bmatrix}}_{\hat{\boldsymbol{A}}} \begin{bmatrix} z_1 \\ \vdots \\ z_{\delta-1} \\ z_\delta \\ z_{\delta+1} \\ \vdots \\ z_{n-1} \\ z_n \end{bmatrix} + \underbrace{\begin{bmatrix} 0 \\ \vdots \\ 0 \\ V \\ 0 \\ \vdots \\ 0 \\ 0 \end{bmatrix}}_{\hat{\boldsymbol{b}}} y_{\text{ref}},$$

$$y = \underbrace{\begin{bmatrix} 1 & 0 & \cdots & 0 \end{bmatrix}}_{\hat{\boldsymbol{c}}^T} \boldsymbol{z} \tag{5.77}$$

as equations for the control loop. This illustrates again that only the first δ equations are relevant for the input-output behavior.

Next we will determine the transfer function

$$G(s) = \frac{Y(s)}{W(s)} = \hat{\boldsymbol{c}}^T(s\boldsymbol{I} - \hat{\boldsymbol{A}})^{-1}\hat{\boldsymbol{b}}.$$

The system matrix $\hat{\boldsymbol{A}}$ and thus $s\boldsymbol{I} - \hat{\boldsymbol{A}}$ are block lower triangular matrices. Therefore, we can easily calculate $(s\boldsymbol{I} - \hat{\boldsymbol{A}})^{-1}$, because for the inverse of a block lower triangular matrix

$$\boldsymbol{X} = \begin{bmatrix} \boldsymbol{X}_{11} & \boldsymbol{0} \\ \boldsymbol{X}_{21} & \boldsymbol{X}_{22} \end{bmatrix},$$

we obtain [40]

$$\boldsymbol{X}^{-1} = \begin{bmatrix} \boldsymbol{X}_{11}^{-1} & \boldsymbol{0} \\ -\boldsymbol{X}_{22}^{-1}\boldsymbol{X}_{21}\boldsymbol{X}_{11}^{-1} & \boldsymbol{X}_{22}^{-1} \end{bmatrix}.$$

Next we will partition $\hat{\boldsymbol{A}}$, $\hat{\boldsymbol{b}}$, and $\hat{\boldsymbol{c}}^T$ correspondingly, resulting in

$$\hat{\boldsymbol{A}} = \begin{bmatrix} \hat{\boldsymbol{A}}_{\text{ext}} & \boldsymbol{0} \\ \hat{\boldsymbol{A}}_{\text{v}} & \hat{\boldsymbol{A}}_{\text{int}} \end{bmatrix}, \quad \hat{\boldsymbol{b}} = \begin{bmatrix} \hat{\boldsymbol{b}}_{\text{ext}} \\ \boldsymbol{0} \end{bmatrix}, \quad \text{and} \quad \hat{\boldsymbol{c}}^T = \begin{bmatrix} \hat{\boldsymbol{c}}_{\text{ext}} & \boldsymbol{0} \end{bmatrix}, \tag{5.78}$$

where $\hat{\boldsymbol{A}}_{\text{ext}} \in \mathbb{R}^{\delta\times\delta}$, $\hat{\boldsymbol{A}}_{\text{v}} \in \mathbb{R}^{m\times\delta}$, $\hat{\boldsymbol{A}}_{\text{int}} \in \mathbb{R}^{m\times m}$, and $\hat{\boldsymbol{c}}_{\text{ext}}$, $\hat{\boldsymbol{b}}_{\text{ext}} \in \mathbb{R}^{\delta}$ hold. Note that $\hat{\boldsymbol{c}}^T$ has only one non-zero element $\hat{c}_1 = 1$, so it is not necessary to include the bottom row of $\hat{\boldsymbol{A}}$ in equation (5.78) in the calculation of the transfer function. We thus arrive at

$$
\begin{aligned}
G(s) &= \hat{\boldsymbol{c}}^T(s\boldsymbol{I}-\hat{\boldsymbol{A}})^{-1}\hat{\boldsymbol{b}} = \hat{\boldsymbol{c}}_{\text{ext}}^T(s\boldsymbol{I}-\hat{\boldsymbol{A}}_{\text{ext}})^{-1}\hat{\boldsymbol{b}}_{\text{ext}} = \\
&= \begin{bmatrix} 1\,0\cdots 0\end{bmatrix}\left(s\boldsymbol{I}-\begin{bmatrix} 0 & 1 & 0 & \cdots & 0\\ \vdots & \vdots & \vdots & & \vdots\\ 0 & 0 & 0 & \cdots & 1\\ -\hat{a}_0 & -\hat{a}_1 & -\hat{a}_2 & \cdots & -\hat{a}_{\delta-1}\end{bmatrix}\right)^{-1}\begin{bmatrix}0\\ \vdots\\ 0\\ V\end{bmatrix}\\
&= \frac{V}{s^\delta+\hat{a}_{\delta-1}s^{\delta-1}+\ldots+\hat{a}_1 s+\hat{a}_0}.
\end{aligned}\qquad (5.79)
$$

The above result is not surprising, since the transfer behavior of a control loop with a relative degree δ resulting from an input-output linearization is always of the form (5.79).

Once again it is clear that the controller (5.76) leads to a reduction in the system order from the order n of the plant (5.74) to the order δ of the transfer function (5.79). However, this is only the case if the system matrix $\boldsymbol{A}$, the input vector $\boldsymbol{b}$, and the output vector $\boldsymbol{c}$ are identical to the model parameters $\boldsymbol{A}_{\text{m}}$, $\boldsymbol{b}_{\text{m}}$, and $\boldsymbol{c}_{\text{m}}$ of the plant in the control law (5.76). If the model parameters $\boldsymbol{A}_{\text{m}}$, $\boldsymbol{b}_{\text{m}}$, and $\boldsymbol{c}_{\text{m}}$ deviate from the true system parameters $\boldsymbol{A}$, $\boldsymbol{b}$, and $\boldsymbol{c}$, the control law (5.76) no longer compensates for the eigenmovements

$$
\boldsymbol{c}^T\boldsymbol{A}^\delta\boldsymbol{T}^{-1}\boldsymbol{z}
$$

of the external dynamics in equation (5.74). That is, after inserting the control law (5.76), which now takes the form

$$
u = -\frac{\boldsymbol{c}_{\text{m}}{}^T\boldsymbol{A}_{\text{m}}{}^\delta\boldsymbol{T}^{-1}\boldsymbol{z}+\boldsymbol{k}^T\boldsymbol{z}}{\boldsymbol{c}_{\text{m}}{}^T\boldsymbol{A}_{\text{m}}{}^{\delta-1}\boldsymbol{b}_{\text{m}}} + \frac{V}{\boldsymbol{c}_{\text{m}}{}^T\boldsymbol{A}_{\text{m}}{}^{\delta-1}\boldsymbol{b}_{\text{m}}}y_{\text{ref}},
$$

into equation (5.74), we no longer obtain the differential equation

$$
\dot{z}_\delta = \boldsymbol{k}^T\boldsymbol{z}+V\cdot y_{\text{ref}} = -\hat{a}_{\delta-1}z_\delta - \hat{a}_{\delta-2}z_{\delta-1} - \ldots - \hat{a}_1 z_2 - \hat{a}_0 z_1 + V\cdot y_{\text{ref}};
$$

instead we obtain

$$
\dot{z}_\delta = \left[\boldsymbol{c}^T\boldsymbol{A}^\delta\boldsymbol{T}^{-1} - \frac{\boldsymbol{c}^T\boldsymbol{A}^{\delta-1}\boldsymbol{b}}{\boldsymbol{c}_{\text{m}}{}^T\boldsymbol{A}_{\text{m}}{}^{\delta-1}\boldsymbol{b}_{\text{m}}}(\boldsymbol{c}_{\text{m}}{}^T\boldsymbol{A}_{\text{m}}{}^\delta\boldsymbol{T}^{-1}+\boldsymbol{k}^T)\right]\boldsymbol{z} + \frac{\boldsymbol{c}^T\boldsymbol{A}^{\delta-1}\boldsymbol{b}V\cdot y_{\text{ref}}}{\boldsymbol{c}_{\text{m}}{}^T\boldsymbol{A}_{\text{m}}{}^{\delta-1}\boldsymbol{b}_{\text{m}}}.
$$

Now the derivative $\dot{z}_\delta$ depends on the entire state vector $\boldsymbol{z}$ and thus on the states $z_{\delta+1}, \ldots, z_n$ of the internal dynamics. As a result, the system matrix $\hat{\boldsymbol{A}}$ of the control loop no longer exhibits the shape of a block lower triangular matrix as in equation (5.77); rather it is fully populated and the transfer function is of order n and not of order δ as in the case of exact modeling. So if the modeling is imprecise, an increase in the system order of the transfer function (5.79) occurs and the states of the internal dynamics now also influence the input-output behavior. The resulting problem is obvious. For example, weakly damped or slow eigenmovements caused by the imprecise linearization may now disturb the originally planned transfer behavior (5.77). In the nonlinear case, these robustness problems are similar. However, they are even more challenging, because if the model deviates from the real controlled system, the control loop's input-output behavior is no longer linear.

Note that input-output linearization requires precise modeling if the planned linear transfer behavior between the reference variable y_{ref} and the output variable y is actually to be achieved.

A further insight emerges: it is now clear that the controller (5.76) of the input-output linearization functions as a compensation controller, because poles are canceled against zeros. This means that the control loop is no longer observable. The unobservability is related to the fact that the transfer function (5.79) has an order reduced by the value $n-\delta$ compared to the state-space representation of the plant (5.77). From linear systems theory, we know that a state-space model of order n has a transfer function with an order of less than n if and only if it is not controllable or not observable, or both. Based on the observability matrix

$$\boldsymbol{M}_{\text{obs}} = \begin{bmatrix} \hat{\boldsymbol{c}}^T \\ \hat{\boldsymbol{c}}^T\hat{\boldsymbol{A}} \\ \hat{\boldsymbol{c}}^T\hat{\boldsymbol{A}}^2 \\ \vdots \\ \hat{\boldsymbol{c}}^T\hat{\boldsymbol{A}}^{n-1} \end{bmatrix} = \begin{bmatrix} \hat{\boldsymbol{c}}_{\text{ext}}^T & \boldsymbol{0} \\ \hat{\boldsymbol{c}}_{\text{ext}}^T\hat{\boldsymbol{A}}_{\text{ext}} & \boldsymbol{0} \\ \hat{\boldsymbol{c}}_{\text{ext}}^T\hat{\boldsymbol{A}}_{\text{ext}}^2 & \boldsymbol{0} \\ \vdots & \vdots \\ \hat{\boldsymbol{c}}_{\text{ext}}^T\hat{\boldsymbol{A}}_{\text{ext}}^{n-1} & \boldsymbol{0} \end{bmatrix}$$

of the control loop (5.77), we can show its unobservability, because the last $n-\delta$ columns of the observability matrix are zero vectors. Therefore, $\boldsymbol{M}_{\text{obs}}$ is not regular. However, the regularity of $\boldsymbol{M}_{\text{obs}}$ is necessary and sufficient for the observability of the control loop. Note that the plant (5.62) is generally observable.

5.2.10 Stability of Internal and Zero Dynamics

In the previous sections, we have only briefly touched on the importance of the initial dynamics for the control loop's stability. If we take a closer look, various questions arise: what kind of stability is required for the internal dynamics? Is input-to-state stability a necessary prerequisite? What is the significance of zero dynamics in this context?

First of all, we must emphasize that the internal dynamics

$$\begin{bmatrix} \dot{z}_{\delta+1} \\ \vdots \\ \dot{z}_n \end{bmatrix} = \begin{bmatrix} q_{\delta+1}(z_1, \ldots, z_\delta, z_{\delta+1}, \ldots, z_n) \\ \vdots \\ q_n(z_1, \ldots, z_\delta, z_{\delta+1}, \ldots, z_n) \end{bmatrix}$$

are generally nonlinear. In the following, we will formulate it more concisely as

$$\dot{\boldsymbol{z}}_{\mathrm{int}} = \boldsymbol{q}(\boldsymbol{z}_{\mathrm{ext}}, \boldsymbol{z}_{\mathrm{int}}), \tag{5.80}$$

using

$$\boldsymbol{z}_{\mathrm{ext}} = \begin{bmatrix} z_1 \\ \vdots \\ z_\delta \end{bmatrix} \quad \text{and} \quad \boldsymbol{z}_{\mathrm{int}} = \begin{bmatrix} z_{\delta+1} \\ \vdots \\ z_n \end{bmatrix}.$$

Due to its nonlinearity, it might have more than one equilibrium point. Therefore, in general it is sometimes misleading to refer to the stability of the internal dynamics; it is more accurate to refer to the stability of their equilibrium points. In consequence, we must distinguish between local and global stability or we must examine the system's stability as defined in Definition 8 or 9 on p. 18 as a whole. The controlled external dynamics of the input-output linearized system, on the other hand, are linear. Thus, generally speaking, there is only one asymptotically stable equilibrium point at zero.

Since the stability analysis of an equilibrium point is always based on vanishing – or at least constant – input values, the input values of the internal dynamics

$$z_1, \ldots, z_\delta$$

must be set at zero for this investigation. We thus obtain the zero dynamics

$$\dot{\boldsymbol{z}}_{\mathrm{int}} = \boldsymbol{q}(\boldsymbol{0}, \boldsymbol{z}_{\mathrm{int}})$$

alone; that is, we can analyze the stability of the equilibrium point or points of the internal dynamics using the zero dynamics. The simplest situation is that involving a single globally asymptotically stable equilibrium point. If, on the other hand, several equilibrium points exist, at most local stability is possible. In the latter case, it must be ensured that this local stability is sufficient for control-loop operation.

A single globally asymptotically stable equilibrium point of the internal dynamics is a necessary condition for the global asymptotical stability of the equilibrium point of the control loop. But is this, along with a globally asymptotically stable equilibrium point of the external dynamics, also sufficient to ensure the global asymptotic stability of the equilibrium of the control loop consisting of external and internal dynamics? This question is entirely justified because the external dynamics influence the internal dynamics via the variables $z_1, \dots, z_\delta$. Let us denote the external dynamics of the input-output linearized control loop as

$$\dot{\boldsymbol{z}}_{\text{ext}} = \hat{\boldsymbol{A}}_{\text{ext}} \boldsymbol{z}_{\text{ext}} + \hat{\boldsymbol{b}}_{\text{ext}} \cdot y_{\text{ref}},$$

which, along with the internal dynamics (5.80), yields a composite system of the form

$$\begin{aligned} \dot{\boldsymbol{z}}_{\text{ext}} &= \hat{\boldsymbol{A}}_{\text{ext}} \boldsymbol{z}_{\text{ext}} + \hat{\boldsymbol{b}}_{\text{ext}} \cdot y_{\text{ref}}, &&\} \text{ external dynamics} \\ \dot{\boldsymbol{z}}_{\text{int}} &= \boldsymbol{q}(\boldsymbol{z}_{\text{ext}}, \boldsymbol{z}_{\text{int}}) &&\} \text{ internal dynamics} \end{aligned} \tag{5.81}$$

with the associated equilibrium point

$$\boldsymbol{z}_{\text{eq}} = \begin{bmatrix} \boldsymbol{z}_{\text{ext,eq}} \\ \boldsymbol{z}_{\text{int,eq}} \end{bmatrix} = \boldsymbol{0}.$$

We can assume, without loss of generality, that the equilibrium point or, where several equilibrium points exist, one of the equilibrium points of the internal dynamics lies in $\boldsymbol{z}_{\text{int}} = \boldsymbol{0}$ or was transformed there.

System (5.81) is a cascaded system, whose stability was analyzed in several works [66, 115, 351, 410, 427, 429, 464]. Using the results of [428, 429], and [464], we obtain

Theorem 62 (Stability of Input-Output Linearized Systems). *The equilibrium point* $\boldsymbol{z}_{\text{eq}} = \boldsymbol{0}$ *of an input-output linearized system*

$$\dot{\boldsymbol{z}} = \begin{bmatrix} \dot{\boldsymbol{z}}_{\text{ext}} \\ \dot{\boldsymbol{z}}_{\text{int}} \end{bmatrix} = \begin{bmatrix} \hat{\boldsymbol{A}}_{\text{ext}} \boldsymbol{z}_{\text{ext}} \\ \boldsymbol{q}(\boldsymbol{z}_{\text{ext}}, \boldsymbol{z}_{\text{int}}) \end{bmatrix} + \begin{bmatrix} \hat{\boldsymbol{b}}_{\text{ext}} \\ \boldsymbol{0} \end{bmatrix} y_{\text{ref}}$$

whose matrix $\hat{\boldsymbol{A}}_{\text{ext}}$ *we assume to possess only eigenvalues with negative real parts is*

(1) asymptotically stable if and only if the equilibrium point $\boldsymbol{z}_{\text{int,eq}} = \boldsymbol{0}$ *of the zero dynamics* $\dot{\boldsymbol{z}}_{\text{int}} = \boldsymbol{q}(\boldsymbol{0}, \boldsymbol{z}_{\text{int}})$ *is asymptotically stable and all functions* q_i *have bounded derivatives in a neighborhood of the origin,*

(2) globally asymptotically stable if the equilibrium point $\boldsymbol{z}_{\text{int,eq}} = \boldsymbol{0}$ *of the zero dynamics* $\dot{\boldsymbol{z}}_{\text{int}} = \boldsymbol{q}(\boldsymbol{0}, \boldsymbol{z}_{\text{int}})$ *is globally asymptotically stable and the internal dynamics* $\dot{\boldsymbol{z}}_{\text{int}} = \boldsymbol{q}(\boldsymbol{z}_{\text{ext}}, \boldsymbol{z}_{\text{int}})$ *with its input variable vector* $\boldsymbol{z}_{\text{ext}}$ *are input-to-state stable,*

(3) globally exponentially stable if and only if the equilibrium point $\boldsymbol{z}_{\text{int,eq}} = \boldsymbol{0}$ *of the zero dynamics* $\dot{\boldsymbol{z}}_{\text{int}} = \boldsymbol{q}(\boldsymbol{0}, \boldsymbol{z}_{\text{int}})$ *is globally exponentially stable and all functions* q_i *have bounded derivatives.*

With the result above, we have established that the stability of the equilibrium point $\boldsymbol{z}_{\text{eq}} = \boldsymbol{0}$ of the input-output linearized system is essentially determined by the zero dynamics. Notice that the requirement for bounded derivatives of the functions q_i in Statement (1) is satisfied if the functions q_i are continuously differentiable in a neighborhood of the origin.

5.2.11 Input-Output Linearization of MIMO Systems

In the following our aim is to apply input-output linearization to MIMO systems

$$\begin{aligned}\dot{\boldsymbol{x}} &= \boldsymbol{a}(\boldsymbol{x}) + \sum_{i=1}^{m} \boldsymbol{b}_i(\boldsymbol{x})u_i,\\ \boldsymbol{y} &= \boldsymbol{c}(\boldsymbol{x})\end{aligned} \tag{5.82}$$

with n state variables x_j, m input variables u_i, and the same number of m output variables y_i.

Requiring that the input and output variables both amount to m in number appears to be a restriction. If we wish to include systems with more output than input variables, this is definitely the case. To add additional input variables involves a constructive extension of the system and leads to additional costs. Therefore, this option is usually not considered. If, on the other hand, we wish to address systems with more inputs than outputs, we can adapt the dimension of the output variable vector $\boldsymbol{y}$ to that of the input variable vector $\boldsymbol{u}$ by adding further output variables. Such additional output variables y_i can be state variables x_j, which do not occur in the output vector $\boldsymbol{c}(\boldsymbol{x})$. One could counter that additional output variables imply that they require measurement and thus mean additional effort. However, all state variables must be measured in any case in order to implement a controller based on input-output linearization.

As in the SISO case, we aim to design a controller and a pre-filter which linearize the dynamic behavior occurring between the input and output variables. For this, we will begin with a single output variable $y_i = c_i(\boldsymbol{x})$ of the system and differentiate it according to

$$\begin{aligned}
y_i &= c_i(\boldsymbol{x}),\\
\dot{y}_i &= L_{\boldsymbol{a}}c_i(\boldsymbol{x}) + \sum_{k=1}^{m} \underbrace{L_{\boldsymbol{b}_k}c_i(\boldsymbol{x})}_{0}\,u_k,\\
\ddot{y}_i &= L_{\boldsymbol{a}}^2c_i(\boldsymbol{x}) + \sum_{k=1}^{m} \underbrace{L_{\boldsymbol{b}_k}L_{\boldsymbol{a}}c_i(\boldsymbol{x})}_{0}\,u_k,\\
&\;\;\vdots\\
y_i^{(\delta_i-1)} &= L_{\boldsymbol{a}}^{\delta_i-1}c_i(\boldsymbol{x}) + \sum_{k=1}^{m} \underbrace{L_{\boldsymbol{b}_k}L_{\boldsymbol{a}}^{\delta_i-2}c_i(\boldsymbol{x})}_{0}\,u_k,\\
y_i^{(\delta_i)} &= L_{\boldsymbol{a}}^{\delta_i}c_i(\boldsymbol{x}) + \sum_{k=1}^{m} \underbrace{L_{\boldsymbol{b}_k}L_{\boldsymbol{a}}^{\delta_i-1}c_i(\boldsymbol{x})}_{\substack{\neq 0 \text{ for at}\\ \text{least one } k}}\,u_k.
\end{aligned} \tag{5.83}$$

In this sequence of derivatives, all Lie derivatives $L_{\boldsymbol{b}_k} L_{\boldsymbol{a}}^i c_i(\boldsymbol{x})$ with respect to the vectors $\boldsymbol{b}_k$, $k = 1, \ldots, m$, are equal to zero. Only in the derivative $y_i^{(\delta_i)}$, at least one of these Lie derivatives is unequal to zero. Therefore, in a manner similar to the SISO case, we will define the relative degree δ_i belonging to the output value y_i, such that

$$L_{\boldsymbol{b}_k} L_{\boldsymbol{a}}^{l-1} c_i(\boldsymbol{x}) = 0, \quad l = 1, \ldots, \delta_i - 1$$

and

$$L_{\boldsymbol{b}_k} L_{\boldsymbol{a}}^{\delta_i - 1} c_i(\boldsymbol{x}) \neq 0 \tag{5.84}$$

hold for at least one $k \in \{1, \ldots, m\}$.

Now we will perform this computation for all output variables $y_i, \ldots, y_m$, and thus determine the corresponding relative degrees $\delta_i, \ldots, \delta_m$ as well as the m equations

$$\begin{aligned}
y_1^{(\delta_1)} &= L_{\boldsymbol{a}}^{\delta_1} c_1(\boldsymbol{x}) + \sum_{k=1}^{m} L_{\boldsymbol{b}_k} L_{\boldsymbol{a}}^{\delta_1 - 1} c_1(\boldsymbol{x}) u_k, \\
y_2^{(\delta_2)} &= L_{\boldsymbol{a}}^{\delta_2} c_2(\boldsymbol{x}) + \sum_{k=1}^{m} L_{\boldsymbol{b}_k} L_{\boldsymbol{a}}^{\delta_2 - 1} c_2(\boldsymbol{x}) u_k, \\
&\ \ \vdots \\
y_m^{(\delta_m)} &= L_{\boldsymbol{a}}^{\delta_m} c_m(\boldsymbol{x}) + \sum_{k=1}^{m} L_{\boldsymbol{b}_k} L_{\boldsymbol{a}}^{\delta_m - 1} c_m(\boldsymbol{x}) u_k.
\end{aligned} \tag{5.85}$$

The equations (5.85) can be written more concisely in vectorial form as

$$\mathring{\boldsymbol{y}} = \mathring{\boldsymbol{c}}(\boldsymbol{x}) + \boldsymbol{D}(\boldsymbol{x}) \cdot \boldsymbol{u} \tag{5.86}$$

with the m - dimensional vectors

$$\mathring{\boldsymbol{y}} = \begin{bmatrix} y^{(\delta_1)} \\ y^{(\delta_2)} \\ \vdots \\ y^{(\delta_m)} \end{bmatrix}, \qquad \mathring{\boldsymbol{c}}(\boldsymbol{x}) = \begin{bmatrix} L_{\boldsymbol{a}}^{\delta_1} c_1(\boldsymbol{x}) \\ L_{\boldsymbol{a}}^{\delta_2} c_2(\boldsymbol{x}) \\ \vdots \\ L_{\boldsymbol{a}}^{\delta_m} c_m(\boldsymbol{x}) \end{bmatrix}$$

and the $m \times m$ decoupling matrix

$$\boldsymbol{D}(\boldsymbol{x}) = \begin{bmatrix} L_{\boldsymbol{b}_1} L_{\boldsymbol{a}}^{\delta_1 - 1} c_1(\boldsymbol{x}) & L_{\boldsymbol{b}_2} L_{\boldsymbol{a}}^{\delta_1 - 1} c_1(\boldsymbol{x}) & \cdots & L_{\boldsymbol{b}_m} L_{\boldsymbol{a}}^{\delta_1 - 1} c_1(\boldsymbol{x}) \\ L_{\boldsymbol{b}_1} L_{\boldsymbol{a}}^{\delta_2 - 1} c_2(\boldsymbol{x}) & L_{\boldsymbol{b}_2} L_{\boldsymbol{a}}^{\delta_2 - 1} c_2(\boldsymbol{x}) & \cdots & L_{\boldsymbol{b}_m} L_{\boldsymbol{a}}^{\delta_2 - 1} c_2(\boldsymbol{x}) \\ \vdots & \vdots & \ddots & \vdots \\ L_{\boldsymbol{b}_1} L_{\boldsymbol{a}}^{\delta_m - 1} c_m(\boldsymbol{x}) & L_{\boldsymbol{b}_2} L_{\boldsymbol{a}}^{\delta_m - 1} c_m(\boldsymbol{x}) & \cdots & L_{\boldsymbol{b}_m} L_{\boldsymbol{a}}^{\delta_m - 1} c_m(\boldsymbol{x}) \end{bmatrix}.$$

The circles on the vectors $\mathring{\boldsymbol{y}}$ and $\mathring{\boldsymbol{c}}$ symbolize the multiple derivatives $y^{(\delta_i)}$ and $L_{\boldsymbol{a}}^{\delta_i} c_i(\boldsymbol{x})$ of the vector elements.

The relationship (5.86) is the key equation for the exact input-output linearization of MIMO systems. Obviously, if the decoupling matrix $\boldsymbol{D}(\boldsymbol{x})$ is regular, we can introduce a new input variable vector $\boldsymbol{v} \in I\!\!R^m$ with

$$\boldsymbol{u} = -\boldsymbol{D}^{-1}(\boldsymbol{x}) \cdot (\mathring{\boldsymbol{c}}(\boldsymbol{x}) - \boldsymbol{v}), \tag{5.87}$$

which, if inserted into equation (5.86), not only linearizes the input-output behavior of the resulting system to

$$\mathring{\boldsymbol{y}} = \boldsymbol{v},$$

i. e.

$$\begin{aligned} y_1^{(\delta_1)} &= v_1, \\ y_2^{(\delta_2)} &= v_2, \\ &\;\vdots \\ y_m^{(\delta_m)} &= v_m; \end{aligned} \tag{5.88}$$

it also completely decouples the effect of the input variables v_i on the derivatives $y_i^{(\delta_i)}$ and thus the output variables y_i. This means that each input variable v_i influences only one output variable y_i.

From equation (5.88), the output variables y_i are obtained by the δ_i-fold integration of the new control variables v_i. Each of the m equations thus represents an integrator chain of length δ_i. If we select the actuating variables v_i as control laws

$$v_i = -a_{i,\delta_i-1} y_i^{(\delta_i-1)} - a_{i,\delta_i-2} y_i^{(\delta_i-2)} - \ldots - a_{i,0} y_i + V_i \cdot y_{\mathrm{ref},i}, \quad i = 1, \ldots, m \tag{5.89}$$

with the reference variables $y_{\mathrm{ref},i}$, the result is control loops, to which any dynamics

$$y_i^{(\delta_i)} + a_{i,\delta_i-1} y_i^{(\delta_i-1)} + \ldots + a_{i,1} \dot{y}_i + a_{i,0} y_i = V_i \cdot y_{\mathrm{ref},i}, \quad i = 1, \ldots, m$$

can be assigned by freely choosing the coefficients $a_{i,j}$. This corresponds to the SISO case.

If we also insert the relations obtained from equation (5.83)

$$y_i = c_i(\boldsymbol{x}), \quad \dot{y}_i = L_{\boldsymbol{a}} c_i(\boldsymbol{x}), \quad \ldots, \quad y_i^{(\delta_i-1)} = L_{\boldsymbol{a}}^{\delta_i-1} c_i(\boldsymbol{x})$$

into equation (5.89) and subsequently equation (5.87), we obtain the control law

$$
\boldsymbol{u}=-\boldsymbol{D}^{-1}(\boldsymbol{x})\begin{bmatrix} L_{\boldsymbol{a}}^{\delta_1}c_1(\boldsymbol{x})+\ldots+a_{1,1}L_{\boldsymbol{a}}c_1(\boldsymbol{x})+a_{1,0}c_1(\boldsymbol{x})-V_1\cdot y_{\mathrm{ref},1}\\ L_{\boldsymbol{a}}^{\delta_2}c_2(\boldsymbol{x})+\ldots+a_{2,1}L_{\boldsymbol{a}}c_2(\boldsymbol{x})+a_{2,0}c_2(\boldsymbol{x})-V_2\cdot y_{\mathrm{ref},2}\\ \vdots\\ L_{\boldsymbol{a}}^{\delta_m}c_m(\boldsymbol{x})+\ldots+a_{m,1}L_{\boldsymbol{a}}c_m(\boldsymbol{x})+a_{m,0}c_m(\boldsymbol{x})-V_m\cdot y_{\mathrm{ref},m} \end{bmatrix}, \tag{5.90}
$$

which is dependent on the state vector $\boldsymbol{x}$ and is structured analogously to the control laws from Theorems 59 on p. 366 and 60 on p. 377 for the SISO case.

As we can see from the derivation above, we can perform input-output linearization without transforming the system into the nonlinear controller canonical form via a diffeomorphism. However, in this case, information about the remaining system behavior will be lacking, i. e. information about the unobservable internal dynamics which are described by the transformed state variables.

A prerequisite for the procedure above is the regularity of the decoupling matrix $\boldsymbol{D}(\boldsymbol{x})$. Only in this case is complete decoupling between input variables v_i or $y_{\mathrm{ref},i}$ and output variables y_i possible. As we will see, internal dynamics can also occur. Whether such dynamics are present or not depends once again on the relative degree of the MIMO system. We will define it in

Definition 33 (Vector Relative Degree of MIMO Systems). *Let there be a system*

$$
\begin{aligned}
\dot{\boldsymbol{x}} &= \boldsymbol{a}(\boldsymbol{x}) + \sum_{i=1}^{m} \boldsymbol{b}_i(\boldsymbol{x})u_i,\\
\boldsymbol{y} &= \boldsymbol{c}(\boldsymbol{x})
\end{aligned}
$$

with $\boldsymbol{x} \in D_{\boldsymbol{x},\mathrm{def}} \subseteq I\!R^n$ and $\boldsymbol{y} \in I\!R^m$. Then

$$
\boldsymbol{\delta} = \begin{bmatrix}\delta_1 \; \delta_2 \; \cdots \; \delta_m\end{bmatrix}
$$

is called the vector relative degree of the system if the following applies:

(1) $L_{\boldsymbol{b}_k} L_{\boldsymbol{a}}^{l-1} c_i(\boldsymbol{x}) = 0, \quad i,k = 1,\ldots,m, \; l = 1,\ldots,\delta_i - 1.$

(2) The $m\times m$ decoupling matrix $\boldsymbol{D}(\boldsymbol{x})$ is regular.

The fulfillment of requirement (5.84) that, for each $i = 1,\ldots,m$, at least one $k \in \{1,\ldots,m\}$ with

$$
L_{\boldsymbol{b}_k} L_{\boldsymbol{a}}^{\delta_i - 1} c_i(\boldsymbol{x}) \neq 0
$$

exists is ensured by the regularity of the matrix $\boldsymbol{D}(\boldsymbol{x})$.

The sum of the values δ_i,

$$\delta = \delta_1 + \delta_2 + \ldots + \delta_m,$$

is referred to as the *total relative degree* of a MIMO system. If the decoupling matrix $\boldsymbol{D}(\boldsymbol{x})$ is regular for all $\boldsymbol{x} \in I\!R^n$, the relative degree, as in the SISO case, is *well-defined*. The following holds [459]:

Theorem 63 (Maximum Total Relative Degree). *For the total relative degree δ of a controllable and observable control-affine MIMO system with the system order n, the relation*

$$\delta = \delta_1 + \delta_2 + \ldots + \delta_m \le n$$

holds.

In the SISO case, the total relative degree is equal to the relative degree. For the exceptional situation where a system is uncontrollable and unobservable, we obtain

$$\delta = \infty,$$

as in the SISO case. This means we cannot perform input-output linearization.

5.2.12 MIMO Control Loops in State-Space Representation

We will transform the control-affine MIMO system (5.82) into the Byrnes-Isidori canonical form and thus obtain the state description of the transformed system. For this purpose, we will use the diffeomorphism

$$\boldsymbol{z} = \begin{bmatrix} y_1 \\ \dot{y}_1 \\ \vdots \\ y_1^{(\delta_1-1)} \\ \vdots \\ y_m \\ \dot{y}_m \\ \vdots \\ y_m^{(\delta_m-1)} \\ z_{\delta+1} \\ z_{\delta+2} \\ \vdots \\ z_n \end{bmatrix} = \boldsymbol{t}(\boldsymbol{x}) = \begin{bmatrix} c_1(\boldsymbol{x}) \\ L_{\boldsymbol{a}} c_1(\boldsymbol{x}) \\ \vdots \\ L_{\boldsymbol{a}}^{\delta_1-1} c_1(\boldsymbol{x}) \\ \vdots \\ c_m(\boldsymbol{x}) \\ L_{\boldsymbol{a}} c_m(\boldsymbol{x}) \\ \vdots \\ L_{\boldsymbol{a}}^{\delta_m-1} c_m(\boldsymbol{x}) \\ t_{\delta+1}(\boldsymbol{x}) \\ t_{\delta+2}(\boldsymbol{x}) \\ \vdots \\ t_n(\boldsymbol{x}) \end{bmatrix}. \tag{5.91}$$

The functions

$$t_{\delta+1}, \ldots, t_n$$

must be selected in such a way that the mapping

$$\boldsymbol{z} = \boldsymbol{t}(\boldsymbol{x})$$

is bijective and both $\boldsymbol{t}$ and $\boldsymbol{t}^{-1}$ are continuously differentiable, i. e. $\boldsymbol{t}(\boldsymbol{x})$ is a diffeomorphism. Where

$$\delta = n,$$

the system has a maximum total relative degree and the functions $t_{\delta+1}, \ldots, t_n$ are omitted. In this case, the system has no internal dynamics. Because the functions $t_{\delta+1}, \ldots, t_n$ need not be determined in this case, the calculation of the diffeomorphism is simplified considerably.

We will analyze the general case with internal dynamics, meaning $\delta < n$, and determine the functions $t_i(\boldsymbol{x})$. But first we will derive the transformed system description, which will lead to a determination method for the functions $t_i(\boldsymbol{x})$. The procedure is similar to that used in the SISO case.

Accordingly, we will transform the original system description (5.82) by differentiating the diffeomorphism (5.91) with respect to time, which results in

$$\dot{\boldsymbol{z}} = \frac{d\boldsymbol{t}(\boldsymbol{x})}{dt} = \begin{bmatrix} L_{\boldsymbol{a}} c_1(\boldsymbol{x}) \\ L_{\boldsymbol{a}}^2 c_1(\boldsymbol{x}) \\ \vdots \\ L_{\boldsymbol{a}}^{\delta_1} c_1(\boldsymbol{x}) + \sum\limits_{k=1}^{m} L_{\boldsymbol{b}_k} L_{\boldsymbol{a}}^{\delta_1 - 1} c_1(\boldsymbol{x}) u_k \\ \vdots \\ L_{\boldsymbol{a}} c_m(\boldsymbol{x}) \\ L_{\boldsymbol{a}}^2 c_m(\boldsymbol{x}) \\ \vdots \\ L_{\boldsymbol{a}}^{\delta_m} c_m(\boldsymbol{x}) + \sum\limits_{k=1}^{m} L_{\boldsymbol{b}_k} L_{\boldsymbol{a}}^{\delta_m - 1} c_m(\boldsymbol{x}) u_k \\ \dot{t}_{\delta+1}(\boldsymbol{x}) \\ \vdots \\ \dot{t}_n(\boldsymbol{x}) \end{bmatrix}.$$

From this, the Byrnes-Isidori canonical form

$$\begin{bmatrix} \dot{z}_1 \\ \dot{z}_2 \\ \vdots \\ \dot{z}_{\delta_1 - 1} \\ \dot{z}_{\delta_1} \\ \vdots \\ \dot{z}_{\delta_1+\ldots+\delta_{m-1}+1} \\ \dot{z}_{\delta_1+\ldots+\delta_{m-1}+2} \\ \vdots \\ \dot{z}_{\delta-1} \\ \dot{z}_{\delta} \\ \dot{z}_{\delta+1} \\ \vdots \\ \dot{z}_n \end{bmatrix} = \begin{bmatrix} z_2 \\ z_3 \\ \vdots \\ z_{\delta_1} \\ L_{\boldsymbol{a}}^{\delta_1} c_1(\boldsymbol{x}) + \sum\limits_{k=1}^{m} L_{\boldsymbol{b}_k} L_{\boldsymbol{a}}^{\delta_1 - 1} c_1(\boldsymbol{x}) u_k \\ \vdots \\ z_{\delta_1+\ldots+\delta_{m-1}+2} \\ z_{\delta_1+\ldots+\delta_{m-1}+3} \\ \vdots \\ z_{\delta} \\ L_{\boldsymbol{a}}^{\delta_m} c_m(\boldsymbol{x}) + \sum\limits_{k=1}^{m} L_{\boldsymbol{b}_k} L_{\boldsymbol{a}}^{\delta_m - 1} c_m(\boldsymbol{x}) u_k \\ \dot{t}_{\delta+1}(\boldsymbol{x}) \\ \vdots \\ \dot{t}_n(\boldsymbol{x}) \end{bmatrix} \tag{5.92}$$

follows. At this point, recall that the total relative degree δ in the equation above satisfies the relation

$$\delta = \delta_1 + \delta_2 + \ldots + \delta_m.$$

The output variable vector in the transformed system from equation (5.92) above is given by

$$\boldsymbol{y} = \begin{bmatrix} z_1 \\ z_{\delta_1+1} \\ z_{\delta_1+\delta_2+1} \\ \vdots \\ z_{\delta_1+\delta_2+\ldots+\delta_{m-1}+1} \end{bmatrix}.$$

For the derivatives $\dot{t}_i(\boldsymbol{x})$, we obtain

$$\dot{t}_i(\boldsymbol{x}) = \frac{\partial t_i(\boldsymbol{x})}{\partial \boldsymbol{x}} \cdot \dot{\boldsymbol{x}} = L_{\boldsymbol{a}} t_i(\boldsymbol{x}) + \sum_{k=1}^{m} L_{\boldsymbol{b}_k} t_i(\boldsymbol{x}) u_k, \quad i = \delta + 1, \ldots, n. \tag{5.93}$$

In contrast to the SISO case, which we discussed in Section 5.2.4, these $n - \delta$ equations generally cannot be made independent of all actuating variables u_k,

$k = 1, \ldots, m$, by a suitable choice of functions $t_i(\boldsymbol{x})$. This means that the $m(n-\delta)$ partial differential equations

$$L_{\boldsymbol{b}_k} t_i(\boldsymbol{x}) = \frac{\partial t_i(\boldsymbol{x})}{\partial \boldsymbol{x}} \boldsymbol{b}_k(\boldsymbol{x}) = 0, \qquad i = \delta+1, \ldots, n, \qquad k = 1, \ldots, m, \tag{5.94}$$

in many cases cannot all be fulfilled. The reason for this problem is that there are $m(n-\delta)$ coupled partial differential equations for $n-\delta$ functions t_i instead of $n-\delta$ single partial differential equations in the SISO case.

Inserting

$$\boldsymbol{x} = \boldsymbol{t}^{-1}(\boldsymbol{z}) \tag{5.95}$$

into equation (5.92), using

$$\varphi_i(\boldsymbol{z}, \boldsymbol{u}) = L_{\boldsymbol{a}}^{\delta_i} c_i(\boldsymbol{t}^{-1}(\boldsymbol{z})) + \sum_{k=1}^{m} L_{\boldsymbol{b}_k} L_{\boldsymbol{a}}^{\delta_i - 1} c_i(\boldsymbol{t}^{-1}(\boldsymbol{z})) u_k, \quad i = 1, \ldots, m, \tag{5.96}$$

as an abbreviation, and inserting diffeomorphism (5.95) into equation (5.93), which results in

$$q_i(\boldsymbol{z}, \boldsymbol{u}) = \dot{t}_i(\boldsymbol{x}) = L_{\boldsymbol{a}} t_i(\boldsymbol{t}^{-1}(\boldsymbol{z})) + \sum_{k=1}^{m} L_{\boldsymbol{b}_k} t_i(\boldsymbol{t}^{-1}(\boldsymbol{z})) u_k, \quad i = \delta+1, \ldots, n,$$

yields a transformation of the system to the Byrnes-Isidori canonical form

$$\begin{bmatrix} \dot{z}_1 \\ \dot{z}_2 \\ \vdots \\ \dot{z}_{\delta_1 - 1} \\ \dot{z}_{\delta_1} \\ \vdots \\ \dot{z}_{\delta_1 + \ldots + \delta_{m-1} + 1} \\ \dot{z}_{\delta_1 + \ldots + \delta_{m-1} + 2} \\ \vdots \\ \dot{z}_{\delta - 1} \\ \dot{z}_{\delta} \\ \dot{z}_{\delta+1} \\ \vdots \\ \dot{z}_n \end{bmatrix} = \begin{bmatrix} z_2 \\ z_3 \\ \vdots \\ z_{\delta_1} \\ \varphi_1(\boldsymbol{z}, \boldsymbol{u}) \\ \vdots \\ z_{\delta_1 + \ldots + \delta_{m-1} + 2} \\ z_{\delta_1 + \ldots + \delta_{m-1} + 3} \\ \vdots \\ z_{\delta} \\ \varphi_m(\boldsymbol{z}, \boldsymbol{u}) \\ q_{\delta+1}(\boldsymbol{z}, \boldsymbol{u}) \\ \vdots \\ q_n(\boldsymbol{z}, \boldsymbol{u}) \end{bmatrix} \begin{matrix} \left.\vphantom{\begin{matrix} 1\\1\\1\\1\\1\\1\\1\\1\\1\\1\\1 \end{matrix}}\right\} \text{external dynamics} \\ \left.\vphantom{\begin{matrix} 1\\1\\1 \end{matrix}}\right\} \text{internal dynamics} \end{matrix} . \tag{5.97}$$

The functions $\varphi_i(\boldsymbol{z},\boldsymbol{u})$ in equation (5.96) are identical to the derivatives $y_i^{(\delta_i)}$ in equation (5.85). From equation (5.85), we have already derived the control law (5.90) for the input-output linearization which we will also apply in this case. It linearizes the m nonlinear differential equations in equation (5.97), i. e.

$$
\begin{aligned}
\dot{z}_{\delta_1} &= \varphi_1(\boldsymbol{z},\boldsymbol{u}) = L_{\boldsymbol{a}}^{\delta_1}\boldsymbol{c}_1(\boldsymbol{x}) + \sum_{k=1}^{m} L_{\boldsymbol{b}_k} L_{\boldsymbol{a}}^{\delta_1-1}\boldsymbol{c}_1(\boldsymbol{x})u_k,\\
\dot{z}_{\delta_1+\delta_2} &= \varphi_2(\boldsymbol{z},\boldsymbol{u}) = L_{\boldsymbol{a}}^{\delta_2}\boldsymbol{c}_2(\boldsymbol{x}) + \sum_{k=1}^{m} L_{\boldsymbol{b}_k} L_{\boldsymbol{a}}^{\delta_2-1}\boldsymbol{c}_2(\boldsymbol{x})u_k,\\
&\ \ \vdots\\
\dot{z}_{\delta_1+\ldots+\delta_m} &= \varphi_m(\boldsymbol{z},\boldsymbol{u}) = L_{\boldsymbol{a}}^{\delta_m}\boldsymbol{c}_m(\boldsymbol{x}) + \sum_{k=1}^{m} L_{\boldsymbol{b}_k} L_{\boldsymbol{a}}^{\delta_m-1}\boldsymbol{c}_m(\boldsymbol{x})u_k.
\end{aligned}
$$

Using the diffeomorphism (5.91) in addition to the control law (5.90), we obtain the state-space model of the input-output linearized and completely decoupled control loop

$$
\begin{bmatrix}
\dot{z}_1\\ \dot{z}_2\\ \vdots\\ \dot{z}_{\delta_1-1}\\ \dot{z}_{\delta_1}\\ \vdots\\ \dot{z}_{\delta_1+\ldots+\delta_{m-1}+1}\\ \dot{z}_{\delta_1+\ldots+\delta_{m-1}+2}\\ \vdots\\ \dot{z}_{\delta-1}\\ \dot{z}_{\delta}\\ \dot{z}_{\delta+1}\\ \vdots\\ \dot{z}_n
\end{bmatrix}
=
\begin{bmatrix}
z_2\\ z_3\\ \vdots\\ z_{\delta_1}\\ -a_{1,0}z_1 - a_{1,1}z_2 - \ldots - a_{1,\delta_1-1}z_{\delta_1} + V_1 y_{\mathrm{ref},1}\\ \vdots\\ z_{\delta_1+\ldots+\delta_{m-1}+2}\\ z_{\delta_1+\ldots+\delta_{m-1}+3}\\ \vdots\\ z_{\delta}\\ -a_{m,0}z_{\delta_1+\ldots+\delta_{m-1}+1} - \ldots - a_{m,\delta_m-1}z_{\delta} + V_m y_{\mathrm{ref},m}\\ q_{\delta+1}(\boldsymbol{z},\boldsymbol{u}) = \hat{q}_{\delta+1}(\boldsymbol{z},\boldsymbol{y}_{\mathrm{ref}})\\ \vdots\\ q_n(\boldsymbol{z},\boldsymbol{u}) = \hat{q}_{\delta+1}(\boldsymbol{z},\boldsymbol{y}_{\mathrm{ref}})
\end{bmatrix}
$$

with the reference variable vector

$$\boldsymbol{y}_{\text{ref}} = \begin{bmatrix} y_{\text{ref},1} \\ \vdots \\ y_{\text{ref},m} \end{bmatrix}.$$

As expected, the external dynamics of this control loop are stated in the linear controller canonical form of a MIMO system. The last $n - \delta$ differential equations

$$\dot{\boldsymbol{z}}_{\text{int}} = \hat{\boldsymbol{q}}(\boldsymbol{z}, \boldsymbol{y}_{\text{ref}})$$

which make up the internal dynamics are usually nonlinear, as in the SISO case. If we satisfy condition (5.94), all $\hat{q}_i$ are independent of $\boldsymbol{y}_{\text{ref}}$.

In the case of a total relative degree

$$\delta = n$$

there are no internal dynamics. For

$$\delta < n,$$

we must also include the internal dynamics in our stability analysis.

5.2.13 Example: Combustion Engine

We will use a combustion engine as an example [361, 376]. Our objective is to design a control for the idle speed. The available control variables u_1 and u_2 are measures of the ignition timing and a variable representing the position of the throttle valve, respectively. The latter controls the air-mass flow necessary for combustion, as shown schematically in Figure 5.15. With the rotational speed x_1, a measure x_2 for the air-mass flow, and the air pressure x_3 in the intake manifold as state variables, as well as the disturbance variable d, the model can be stated as

$$\begin{aligned} \dot{\boldsymbol{x}} &= \boldsymbol{a}(\boldsymbol{x}) + \boldsymbol{B}(\boldsymbol{x}) \cdot \boldsymbol{u} + \boldsymbol{s} \cdot d, \\ \boldsymbol{y} &= \boldsymbol{c}(\boldsymbol{x}) \end{aligned}$$

with

$$\boldsymbol{a}(\boldsymbol{x}) = \begin{bmatrix} a_0 + a_1 x_1 + a_2 x_1^2 + a_3 x_2 \\ a_4 x_1 x_2 + a_5 x_1^2 x_3 + a_6 x_1^2 x_3^2 \\ a_7 x_1 x_3 + a_8 x_1 x_3^2 \end{bmatrix}, \quad \boldsymbol{B}(\boldsymbol{x}) = \begin{bmatrix} 1 & 0 \\ 0 & 0 \\ 0 & b(x_3) \end{bmatrix},$$

$$\boldsymbol{s} = \begin{bmatrix} -a_d \\ 0 \\ 0 \end{bmatrix}, \quad \boldsymbol{c}(\boldsymbol{x}) = \begin{bmatrix} x_1 \\ x_2 \end{bmatrix},$$

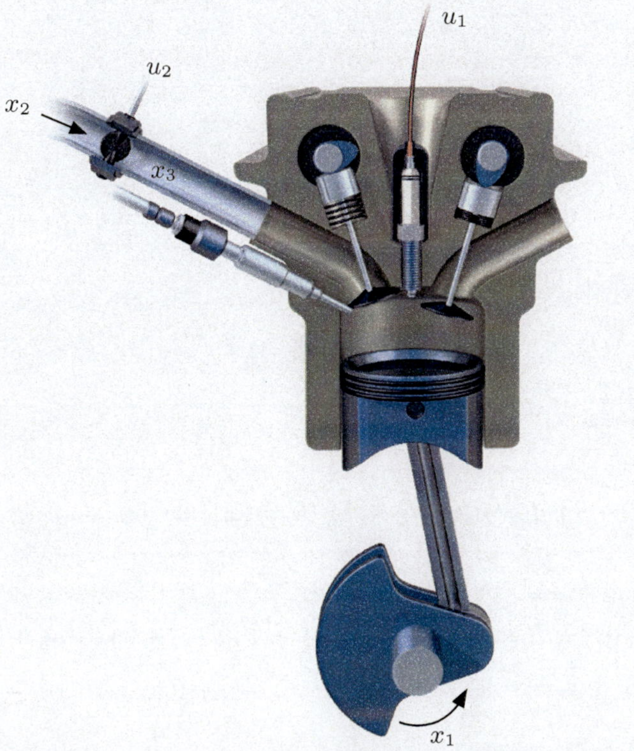

Fig. 5.15: Combustion engine with variables u_1 and u_2 representing the ignition timing and the throttle position as control variables, as well as the rotational speed x_1, a variable x_2 representing the air-mass flow, and the air pressure x_3 in the intake manifold as state variables

and the continuous function

$$b(x_3) = \begin{cases} a_9, & x_3 \leq 0.5P_\mathrm{a}, \\ \dfrac{2a_9}{P_\mathrm{a}}\sqrt{x_3 P_\mathrm{a} - x_3^2}, & x_3 > 0.5P_\mathrm{a}. \end{cases}$$

Here P_a is the atmospheric air pressure. The disturbance d includes all disturbances as they may be caused by the power consumption of the servo hydraulics, the lighting, the radio system, the air conditioning system, etc.

We can determine

$$\begin{aligned} \dot{y}_1 &= L_{\boldsymbol{a}} c_1(\boldsymbol{x}) + L_{\boldsymbol{b}_1} c_1(\boldsymbol{x}) \cdot u_1 + L_{\boldsymbol{b}_2} c_1(\boldsymbol{x}) \cdot u_2 + L_{\boldsymbol{s}} c_1(\boldsymbol{x}) \cdot d \\ &= \dot{x}_1 = a_0 + a_1 x_1 + a_2 x_1^2 + a_3 x_2 + u_1 - a_d d \end{aligned}$$

with

$$L_{\boldsymbol{a}}c_1(\boldsymbol{x}) = a_0 + a_1x_1 + a_2x_1^2 + a_3x_2,$$

$$L_{\boldsymbol{b}_1}c_1(\boldsymbol{x}) = 1,$$

$$L_{\boldsymbol{b}_2}c_1(\boldsymbol{x}) = 0,$$

$$L_{\boldsymbol{s}}c_1(\boldsymbol{x}) = -a_d,$$

in which we treat the disturbance d as an input variable. Since $\dot{y}_1$ depends on u_1, $\delta_1 = 1$ follows.

Further, it holds that

$$\begin{aligned}\dot{y}_2 &= L_{\boldsymbol{a}}c_2(\boldsymbol{x}) + L_{\boldsymbol{b}_1}c_2(\boldsymbol{x}) \cdot u_1 + L_{\boldsymbol{b}_2}c_2(\boldsymbol{x}) \cdot u_2 + L_{\boldsymbol{s}}c_2(\boldsymbol{x}) \cdot d = L_{\boldsymbol{a}}c_2(\boldsymbol{x})\\ &= \dot{x}_2 = a_4x_1x_2 + a_5x_1^2x_3 + a_6x_1^2x_3^2\end{aligned}$$

and

$$\ddot{y}_2 = L_{\boldsymbol{a}}^2c_2(\boldsymbol{x}) + L_{\boldsymbol{b}_1}L_{\boldsymbol{a}}c_2(\boldsymbol{x}) \cdot u_1 + L_{\boldsymbol{b}_2}L_{\boldsymbol{a}}c_2(\boldsymbol{x}) \cdot u_2 + L_{\boldsymbol{s}}L_{\boldsymbol{a}}c_2(\boldsymbol{x}) \cdot d$$

with

$$\begin{aligned}L_{\boldsymbol{a}}^2c_2(\boldsymbol{x}) = x_1^3x_3(a_5 + 2a_6x_3)(a_7 + a_8x_3) + a_4x_1^2[a_4x_2 + x_1x_3(a_5 + a_6x_3)]\\ + (a_0 + a_1x_1 + a_2x_1^2 + a_3x_2)[a_4x_2 + 2x_1x_3(a_5 + a_6x_3)],\end{aligned}$$

$$L_{\boldsymbol{b}_1}L_{\boldsymbol{a}}c_2(\boldsymbol{x}) = a_4x_2 + 2a_5x_1x_3 + 2a_6x_1x_3^2,$$

$$L_{\boldsymbol{b}_2}L_{\boldsymbol{a}}c_2(\boldsymbol{x}) = b(x_3)(a_5x_1^2 + 2a_6x_1^2x_3),$$

and

$$L_{\boldsymbol{s}}L_{\boldsymbol{a}}c_2(\boldsymbol{x}) = -a_d(a_4x_2 + 2a_5x_1x_3 + 2a_6x_1x_3^2).$$

This means the relative degree of the second state variable is $\delta_2 = 2$ and the vector relative degree becomes

$$\boldsymbol{\delta} = [1 \quad 2].$$

With these results, we obtain

$$\begin{bmatrix}\dot{y}_1\\ \ddot{y}_2\end{bmatrix} = \begin{bmatrix}L_{\boldsymbol{a}}c_1(\boldsymbol{x})\\ L_{\boldsymbol{a}}^2c_2(\boldsymbol{x})\end{bmatrix} + \begin{bmatrix}L_{\boldsymbol{b}_1}c_1(\boldsymbol{x}) & 0\\ L_{\boldsymbol{b}_1}L_{\boldsymbol{a}}c_2(\boldsymbol{x}) & L_{\boldsymbol{b}_2}L_{\boldsymbol{a}}c_2(\boldsymbol{x})\end{bmatrix}\boldsymbol{u} + \begin{bmatrix}-a_d\\ L_{\boldsymbol{s}}L_{\boldsymbol{a}}c_2(\boldsymbol{x})\end{bmatrix}d.$$

Using the control law (5.87), which takes the form

$$\boldsymbol{u} = -\frac{1}{L_{\boldsymbol{b}_1}c_1(\boldsymbol{x})L_{\boldsymbol{b}_2}L_{\boldsymbol{a}}c_2(\boldsymbol{x})} \cdot \begin{bmatrix}L_{\boldsymbol{b}_2}L_{\boldsymbol{a}}c_2(\boldsymbol{x}) & 0\\ -L_{\boldsymbol{b}_1}L_{\boldsymbol{a}}c_2(\boldsymbol{x}) & L_{\boldsymbol{b}_1}c_1(\boldsymbol{x})\end{bmatrix}\begin{bmatrix}L_{\boldsymbol{a}}c_1(\boldsymbol{x}) - v_1\\ L_{\boldsymbol{a}}^2c_2(\boldsymbol{x}) - v_2\end{bmatrix} \tag{5.98}$$

in this case, and the two new input variables v_1 and v_2 yields

$$\begin{aligned}\dot{y}_1 &= v_1 - a_d d,\\ \ddot{y}_2 &= v_2 - a_d(a_4 x_2 + 2a_5 x_1 x_3 + 2a_6 x_1 x_3^2)d.\end{aligned} \tag{5.99}$$

Here, formula (5.99) deviates from that of equation (5.88) due to the additional disturbance d. Using

$$\begin{aligned}v_1 &= -a_{1,0}y_1 + V_1 y_{\text{ref},1} = -a_{1,0}x_1 + V_1 y_{\text{ref},1},\\ v_2 &= -a_{2,1}\dot{y}_2 - a_{2,0}y_2 + V_2 y_{\text{ref},2} = -a_{2,1}\dot{x}_2 - a_{2,0}x_2 + V_2 y_{\text{ref},2}\\ &= -a_{2,1}(a_4 x_1 x_2 + a_5 x_1^2 x_3 + a_6 x_1^2 x_3^2) - a_{2,0}x_2 + V_2 y_{\text{ref},2}\end{aligned}$$

in the control law (5.98), we can impose any desired input-output dynamics

$$\begin{aligned}\dot{y}_1 + a_{1,0}y_1 &= V_1 y_{\text{ref},1} - a_d d,\\ \ddot{y}_2 + a_{2,1}\dot{y}_2 + a_{2,0}y_2 &= V_2 y_{\text{ref},2} - a_d(a_4 x_2 + 2a_5 x_1 x_3 + 2a_6 x_1 x_3^2)d\end{aligned}$$

on the system using the freely selectable parameters $a_{1,0}$, $a_{2,1}$, and $a_{2,0}$. The disturbance d affects both differential equations.

5.3 Full-State Linearization

5.3.1 Full-State Linearization of SISO Systems

We will begin by briefly recapitulating the basic concept of input-output linearization from Section 5.2: by means of a suitable state-space transformation and a state-dependent control variable transformation, which we can also interpret as a controller, we had linearized the system behavior between input and output. In general, however, we were not able to do this for the system's behavior as a whole; the internal dynamics, if present, remain nonlinear. For the input-output linearization of SISO systems, we started with the system

$$\begin{aligned}\dot{\boldsymbol{x}} &= \boldsymbol{a}(\boldsymbol{x}) + \boldsymbol{b}(\boldsymbol{x}) \cdot u,\\ y &= c(\boldsymbol{x})\end{aligned} \tag{5.100}$$

with the relative degree δ and calculated the derivatives

$$\begin{aligned}\dot{y} &= L_{\boldsymbol{a}}c(\boldsymbol{x}) + \underbrace{L_{\boldsymbol{b}}c(\boldsymbol{x})}_{0}\cdot u,\\ \ddot{y} &= L_{\boldsymbol{a}}^2c(\boldsymbol{x}) + \underbrace{L_{\boldsymbol{b}}L_{\boldsymbol{a}}c(\boldsymbol{x})}_{0}\cdot u,\\ &\;\;\vdots\\ y^{(\delta-1)} &= L_{\boldsymbol{a}}^{\delta-1}c(\boldsymbol{x}) + \underbrace{L_{\boldsymbol{b}}L_{\boldsymbol{a}}^{\delta-2}c(\boldsymbol{x})}_{0}\cdot u,\\ y^{(\delta)} &= L_{\boldsymbol{a}}^{\delta}c(\boldsymbol{x}) + \underbrace{L_{\boldsymbol{b}}L_{\boldsymbol{a}}^{\delta-1}c(\boldsymbol{x})}_{\neq 0}\cdot u,\end{aligned} \tag{5.101}$$

which enables us to linearize the input-output behavior. If the relative degree is $\delta = n$, we can use the diffeomorphism

$$\boldsymbol{z} = \boldsymbol{t}(\boldsymbol{x}) = \begin{bmatrix} c(\boldsymbol{x}) \\ L_{\boldsymbol{a}}c(\boldsymbol{x}) \\ L_{\boldsymbol{a}}^2 c(\boldsymbol{x}) \\ \vdots \\ L_{\boldsymbol{a}}^{n-1}c(\boldsymbol{x}) \end{bmatrix} \tag{5.102}$$

to transform the system description into the nonlinear controller canonical form

$$\dot{\boldsymbol{z}} = \begin{bmatrix} z_2 \\ \vdots \\ z_n \\ \alpha(\boldsymbol{x}) + \beta(\boldsymbol{x}) \cdot u \end{bmatrix}.$$

If

$$\beta(\boldsymbol{x}) = L_{\boldsymbol{b}} L_{\boldsymbol{a}}^{\delta-1} c(\boldsymbol{x}) \neq 0$$

holds, the system can be transformed via the new input variable v and the control variable transformation

$$u = \frac{-\,\alpha(\boldsymbol{x}) + v}{\beta(\boldsymbol{x})}$$

into the Brunovsky canonical form

$$\dot{\boldsymbol{z}} = \boldsymbol{A}\boldsymbol{z} + \boldsymbol{b} \cdot v,$$

where

$$\boldsymbol{A} = \begin{bmatrix} 0 & 1 & 0 & \cdots & 0 \\ 0 & 0 & 1 & \cdots & 0 \\ \vdots & \vdots & \vdots & \ddots & \vdots \\ 0 & 0 & 0 & \cdots & 1 \\ 0 & 0 & 0 & \cdots & 0 \end{bmatrix}, \quad \boldsymbol{b} = \begin{bmatrix} 0 \\ 0 \\ \vdots \\ 0 \\ 1 \end{bmatrix}.$$

Its block diagram is shown in Figure 5.16. The output variable y can be stated as

$$y = \boldsymbol{c}^T \boldsymbol{z}, \quad \boldsymbol{c}^T = \begin{bmatrix} 1 \; 0 \cdots 0 \end{bmatrix}.$$

If the above transformations are feasible, which is the case if $\delta = n$ and $\beta(\boldsymbol{x}) \neq 0$, we can thus describe the nonlinear system using an equivalent linear state-space model. The following definition summarizes this.

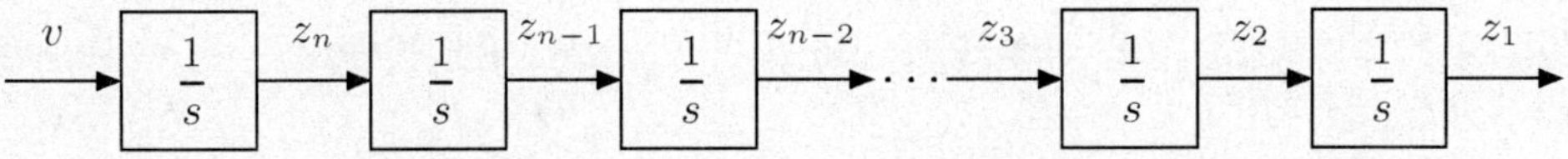

Fig. 5.16: Block diagram of the Brunovsky canonical form

Definition 34 (Full-State Linearizability). *A system*

$$\dot{\boldsymbol{x}} = \boldsymbol{a}(\boldsymbol{x}) + \boldsymbol{b}(\boldsymbol{x}) \cdot u, \quad \boldsymbol{x} \in D_{\boldsymbol{x},\mathrm{def}} \subseteq I\!R^n, \quad u \in I\!R,$$

is called full-state linearizable if a diffeomorphism $\boldsymbol{z} = \boldsymbol{t}(\boldsymbol{x})$ and an input transformation $u = p(\boldsymbol{x}, v)$ exist, so that the system can be transformed into the Brunovsky canonical form.

If a system has the relative degree $\delta = n$, it is full-state linearizable. However, if the relative order δ is less than the system order n, the transformation to the Brunovsky canonical form and the linearization in general cannot be performed completely for all states. This results in internal dynamics, and these are usually nonlinear. Here the question arises whether we can achieve full-state linearization with a different choice of the output variable $y = c(\boldsymbol{x})$. In this case, we must select $c(\boldsymbol{x})$ such that $\delta = n$ holds, and we thus obtain a diffeomorphism (5.102), i. e. an input-output linearization, which consequently has no internal dynamics and therefore no nonlinear dynamic components. The question of whether such a choice of y is possible is especially relevant to systems without an output variable $y = c(\boldsymbol{x})$: can we achieve linearization as described above by choosing a suitable output variable y?

Undoubtedly, we can achieve this goal if we are able to find an artificial output variable

$$y = \lambda(\boldsymbol{x})$$

so that the system has the relative degree $\delta = n$. This is equivalent to the requirements

$$\begin{aligned} L_{\boldsymbol{b}}\lambda(\boldsymbol{x}) &= 0, \\ L_{\boldsymbol{b}}L_{\boldsymbol{a}}\lambda(\boldsymbol{x}) &= 0, \\ &\vdots \\ L_{\boldsymbol{b}}L_{\boldsymbol{a}}^{n-2}\lambda(\boldsymbol{x}) &= 0, \end{aligned} \tag{5.103}$$

and

$$L_{\boldsymbol{b}}L_{\boldsymbol{a}}^{n-1}\lambda(\boldsymbol{x}) \neq 0, \tag{5.104}$$

as is clear if we consider the sequence of derivatives in equation (5.101).

Assuming that equations (5.103) and (5.104) are fulfilled, we can refer to $y = \lambda(\boldsymbol{x})$ as the *linearizing output*. This output does not necessarily have to be of practical importance. To determine it, we must solve equations (5.103).

These are homogeneous linear partial differential equations of higher order; an example is

$$L_{\boldsymbol{b}} L_{\boldsymbol{a}}^2 \lambda(\boldsymbol{x}) = \frac{\partial}{\partial \boldsymbol{x}} \left[\frac{\partial}{\partial \boldsymbol{x}} \left(\frac{\partial \lambda(\boldsymbol{x})}{\partial \boldsymbol{x}} \boldsymbol{a}(\boldsymbol{x}) \right) \boldsymbol{a}(\boldsymbol{x}) \right] \boldsymbol{b}(\boldsymbol{x}) = 0.$$

We can simplify the differential equations (5.103) by converting them into homogeneous linear partial differential equations of the first-order.

For this purpose, we need the Lie bracket

$$[\boldsymbol{f}, \boldsymbol{g}] = \frac{\partial \boldsymbol{g}(\boldsymbol{x})}{\partial \boldsymbol{x}} \boldsymbol{f}(\boldsymbol{x}) - \frac{\partial \boldsymbol{f}(\boldsymbol{x})}{\partial \boldsymbol{x}} \boldsymbol{g}(\boldsymbol{x})$$

of two vector functions $\boldsymbol{f}(\boldsymbol{x})$ and $\boldsymbol{g}(\boldsymbol{x})$. The Lie bracket is a special derivative operator for vector functions, which we have already encountered in Section 3.1.5 on p. 228. If we nest Lie brackets, we can abbreviate this by writing

$$\mathrm{ad}_{\boldsymbol{f}}^i \boldsymbol{g} = [\boldsymbol{f}, \mathrm{ad}_{\boldsymbol{f}}^{i-1} \boldsymbol{g}] \qquad \text{with } \mathrm{ad}_{\boldsymbol{f}}^0 \boldsymbol{g} = \boldsymbol{g} \ \text{ for } \ i = 1, 2, \ldots .$$

This leads to

$$\begin{aligned} \mathrm{ad}_{\boldsymbol{f}}^0 \boldsymbol{g} &= \boldsymbol{g}, \\ \mathrm{ad}_{\boldsymbol{f}} \boldsymbol{g} &= [\boldsymbol{f}, \boldsymbol{g}], \\ \mathrm{ad}_{\boldsymbol{f}}^2 \boldsymbol{g} &= [\boldsymbol{f}, [\boldsymbol{f}, \boldsymbol{g}]], \\ \mathrm{ad}_{\boldsymbol{f}}^3 \boldsymbol{g} &= [\boldsymbol{f}, [\boldsymbol{f}, [\boldsymbol{f}, \boldsymbol{g}]]], \\ &\ \vdots \end{aligned}$$

The ad operator will make our calculations below much simpler. In addition to this operator and the Lie bracket, we also need the identity[6]

$$L_{\mathrm{ad}_{\boldsymbol{f}} \boldsymbol{g}} h(\boldsymbol{x}) = L_{[\boldsymbol{f}, \boldsymbol{g}]} h(\boldsymbol{x}) = L_{\boldsymbol{f}} L_{\boldsymbol{g}} h(\boldsymbol{x}) - L_{\boldsymbol{g}} L_{\boldsymbol{f}} h(\boldsymbol{x}) \tag{5.105}$$

of a function $h(\boldsymbol{x})$.

So equipped, we can achieve our goal to transform equation (5.103) into a system of homogeneous linear partial differential equations of the first order. Using the identity (5.105) in the form

$$L_{\mathrm{ad}_{\boldsymbol{a}} \boldsymbol{b}} \lambda(\boldsymbol{x}) = L_{\boldsymbol{a}} L_{\boldsymbol{b}} \lambda(\boldsymbol{x}) - L_{\boldsymbol{b}} L_{\boldsymbol{a}} \lambda(\boldsymbol{x}),$$

and

[6] This identity is sometimes called the Jacobi identity [422]. However, the common form of the Jacobi identity is

$$[\boldsymbol{f}, [\boldsymbol{g}, \boldsymbol{h}] + [\boldsymbol{h}, [\boldsymbol{f}, \boldsymbol{g}]] + [\boldsymbol{g}, [\boldsymbol{h}, \boldsymbol{f}]]] = \boldsymbol{0}.$$

It can be derived from the identity (5.105).

$$L_{\boldsymbol{b}}\lambda(\boldsymbol{x}) = 0,$$
$$L_{\boldsymbol{b}}L_{\boldsymbol{a}}\lambda(\boldsymbol{x}) = 0,$$

which results from equation (5.103), we obtain

$$L_{\mathrm{ad}_{\boldsymbol{a}}\boldsymbol{b}}\lambda(\boldsymbol{x}) = 0.$$

Again using the identity (5.105), we then compute

$$\begin{aligned} L_{\mathrm{ad}^2_{\boldsymbol{a}}\boldsymbol{b}}\lambda(\boldsymbol{x}) &= L_{[\boldsymbol{a},[\boldsymbol{a},\boldsymbol{b}]]}\lambda(\boldsymbol{x}) = L_{\boldsymbol{a}}L_{[\boldsymbol{a},\boldsymbol{b}]}\lambda(\boldsymbol{x}) - L_{[\boldsymbol{a},\boldsymbol{b}]}L_{\boldsymbol{a}}\lambda(\boldsymbol{x}) \\ &= L_{\boldsymbol{a}}(L_{\boldsymbol{a}}L_{\boldsymbol{b}}\lambda(\boldsymbol{x}) - L_{\boldsymbol{b}}L_{\boldsymbol{a}}\lambda(\boldsymbol{x})) - (L_{\boldsymbol{a}}L_{\boldsymbol{b}}L_{\boldsymbol{a}}\lambda(\boldsymbol{x}) - L_{\boldsymbol{b}}L_{\boldsymbol{a}}L_{\boldsymbol{a}}\lambda(\boldsymbol{x})) \\ &= L^2_{\boldsymbol{a}}L_{\boldsymbol{b}}\lambda(\boldsymbol{x}) - 2L_{\boldsymbol{a}}L_{\boldsymbol{b}}L_{\boldsymbol{a}}\lambda(\boldsymbol{x}) + L_{\boldsymbol{b}}L^2_{\boldsymbol{a}}\lambda(\boldsymbol{x}). \end{aligned}$$

Inserting

$$L_{\boldsymbol{b}}\lambda(\boldsymbol{x}) = 0,$$
$$L_{\boldsymbol{b}}L_{\boldsymbol{a}}\lambda(\boldsymbol{x}) = 0,$$
$$L_{\boldsymbol{b}}L^2_{\boldsymbol{a}}\lambda(\boldsymbol{x}) = 0$$

into the latter, we then obtain

$$L_{\mathrm{ad}^2_{\boldsymbol{a}}\boldsymbol{b}}\lambda(\boldsymbol{x}) = 0.$$

This calculation can be continued with

$$L_{\mathrm{ad}^3_{\boldsymbol{a}}\boldsymbol{b}}\lambda(\boldsymbol{x}),\ L_{\mathrm{ad}^4_{\boldsymbol{a}}\boldsymbol{b}}\lambda(\boldsymbol{x}),\ldots$$

As an equivalent condition to equation (5.103), we then obtain the desired system of homogeneous first-order linear partial differential equations

$$\begin{aligned} L_{\boldsymbol{b}}\lambda(\boldsymbol{x}) &= \frac{\partial\lambda(\boldsymbol{x})}{\partial\boldsymbol{x}}\boldsymbol{b} = 0, \\ L_{\mathrm{ad}_{\boldsymbol{a}}\boldsymbol{b}}\lambda(\boldsymbol{x}) &= \frac{\partial\lambda(\boldsymbol{x})}{\partial\boldsymbol{x}}[\boldsymbol{a},\boldsymbol{b}] = 0, \\ L_{\mathrm{ad}^2_{\boldsymbol{a}}\boldsymbol{b}}\lambda(\boldsymbol{x}) &= \frac{\partial\lambda(\boldsymbol{x})}{\partial\boldsymbol{x}}[\boldsymbol{a},[\boldsymbol{a},\boldsymbol{b}]] = 0, \\ &\vdots \\ L_{\mathrm{ad}^{n-2}_{\boldsymbol{a}}\boldsymbol{b}}\lambda(\boldsymbol{x}) &= \frac{\partial\lambda(\boldsymbol{x})}{\partial\boldsymbol{x}}[\boldsymbol{a},[\boldsymbol{a},\ldots[\boldsymbol{a},\boldsymbol{b}]\ldots]] = 0 \end{aligned} \tag{5.106}$$

and

$$L_{\mathrm{ad}^{n-1}_{\boldsymbol{a}}\boldsymbol{b}}\lambda(\boldsymbol{x}) = (-1)^{n-1}L_{\boldsymbol{b}}L^{n-1}_{\boldsymbol{a}}\lambda(\boldsymbol{x}) \neq 0 \tag{5.107}$$

for equation (5.104). We have shown so far that a solution of equations (5.103), (5.104) always fulfills equations (5.106), (5.107) as well. The reverse conclusion is also correct [203, 422, 465]. Thus, equations (5.103), (5.104) and equations (5.106), (5.107) are equivalent.

For us to solve the partial differential equations (5.106), it is very useful if they are linear and of the first order. In practice, however, these differential equations are often not simple to solve. In addition, their solutions are often very complex and extremely lengthy when the system (5.100) is of higher order.

We can examine the solvability of the above system of partial differential equations (5.106) as well as the fulfillment of condition (5.107) and the associated full-state linearizability [197, 198, 203, 206, 232], i. e. the existence of a linearizing output with the relative degree $\delta = n$, using

Theorem 64 (Full-State Linearizability of SISO Systems). *A control-affine system $\dot{\boldsymbol{x}} = \boldsymbol{a}(\boldsymbol{x}) + \boldsymbol{b}(\boldsymbol{x}) \cdot u$ with $\boldsymbol{x} \in D_{\boldsymbol{x},\mathrm{def}} \subseteq I\!R^n$ and $u \in I\!R$ is full-state linearizable if and only if*

(1) $\operatorname{rank}\left(\begin{bmatrix} \boldsymbol{b} & \operatorname{ad}_{\boldsymbol{a}}\boldsymbol{b} & \operatorname{ad}_{\boldsymbol{a}}^2\boldsymbol{b} & \cdots & \operatorname{ad}_{\boldsymbol{a}}^{n-1}\boldsymbol{b}\end{bmatrix}\right) = n$ *holds and*

(2) all Lie brackets

$$[\operatorname{ad}_{\boldsymbol{a}}^i \boldsymbol{b}, \operatorname{ad}_{\boldsymbol{a}}^k \boldsymbol{b}] \quad \text{with} \quad i,k \in \{0,1,\ldots,n-2\}$$

are linear combinations of the vectors

$$\boldsymbol{b},\ \operatorname{ad}_{\boldsymbol{a}}\boldsymbol{b}, \ldots, \operatorname{ad}_{\boldsymbol{a}}^{n-2}\boldsymbol{b}.$$

The set of vector functions $\{\boldsymbol{b}, \operatorname{ad}_{\boldsymbol{a}}\boldsymbol{b}, \operatorname{ad}_{\boldsymbol{a}}^2\boldsymbol{b}, \ldots, \operatorname{ad}_{\boldsymbol{a}}^{n-2}\boldsymbol{b}\}$ is termed *involutive* if the functions satisfy Condition (2) of the theorem. Note that the conditions of Theorem 64 do not depend on an artificial output $y = \lambda(\boldsymbol{x})$.

Theorem 64 provides a statement about the full-state linearizability of the system

$$\dot{\boldsymbol{x}} = \boldsymbol{a}(\boldsymbol{x}) + \boldsymbol{b}(\boldsymbol{x}) \cdot u,$$

but it is not constructive. That is, it does not provide us with a procedure for determining a linearizing output $y = \lambda(\boldsymbol{x})$. Summarizing our results above, we obtain the following theorem which enables us to determine a linearizing output and the diffeomorphism needed to transform a system into the Brunovsky canonical form.

Theorem 65 (Full-State Linearization of SISO Systems). *A full-state linearizable system $\dot{\boldsymbol{x}} = \boldsymbol{a}(\boldsymbol{x}) + \boldsymbol{b}(\boldsymbol{x}) \cdot u$ with $\boldsymbol{x} \in D_{\boldsymbol{x},\mathrm{def}} \subseteq I\!R^n$ and $u \in I\!R$ is transformed into the Brunovsky canonical form by the diffeomorphism*

$$\boldsymbol{z} = \boldsymbol{t}(\boldsymbol{x}) = \begin{bmatrix} \lambda(\boldsymbol{x}) \\ L_{\boldsymbol{a}}\lambda(\boldsymbol{x}) \\ L_{\boldsymbol{a}}^2\lambda(\boldsymbol{x}) \\ \vdots \\ L_{\boldsymbol{a}}^{n-1}\lambda(\boldsymbol{x}) \end{bmatrix}$$

and the input variable transformation

$$u = \frac{v - L_{\boldsymbol{a}}^{n}\lambda(\boldsymbol{x})}{L_{\boldsymbol{b}}L_{\boldsymbol{a}}^{n-1}\lambda(\boldsymbol{x})}$$

with the new input variable v if the artificial output $y = \lambda(\boldsymbol{x})$ fulfills the partial differential equations

$$\begin{aligned} L_{\boldsymbol{b}}\lambda(\boldsymbol{x}) &= 0, \\ L_{\mathrm{ad}_{\boldsymbol{a}}\boldsymbol{b}}\lambda(\boldsymbol{x}) &= 0, \\ L_{\mathrm{ad}_{\boldsymbol{a}}^{2}\boldsymbol{b}}\lambda(\boldsymbol{x}) &= 0, \\ &\vdots \\ L_{\mathrm{ad}_{\boldsymbol{a}}^{n-2}\boldsymbol{b}}\lambda(\boldsymbol{x}) &= 0 \end{aligned}$$

and the inequality

$$L_{\boldsymbol{b}}L_{\boldsymbol{a}}^{n-1}\lambda(\boldsymbol{x}) \neq 0.$$

Note that the requirement stated in equation (5.107) and the last equation in Theorem 65 are equivalent. Furthermore, the conditions of Theorem 65 can always be fulfilled, since it presupposes full-state linearizability.

As mentioned above, verifying the full-state linearizability of higher order systems using Theorem 64 can be very complex and often futile. In particular, Condition (2) from Theorem 64, the involution condition, restricts the number of full-state linearizable systems and is often time-consuming to verify.

If we have succeeded in transforming a full-state linearizable system into the Brunovsky canonical form, we can immediately design a linear control for the transformed system. This is especially simple because the Brunovsky canonical form consists of a chain of n integrators.

5.3.2 Example: Drilling Rig

Deep boreholes in the earth's crust are mainly used to access oil, gas, and geothermal sources. The drilling technique is similar in all cases. The rotary drilling technique uses a drilling rig, at the lower end of which a revolving drill bit is installed as shown in Figure 5.17. A drilling fluid is pumped through the interior of the drill pipe to the drill bit. This liquid cools and lubricates the drill bit. At the same time, the liquid streams up into the borehole transporting the milled rock to the surface. It is flushed up in the annular gap between the drill pipes and the borehole wall. The borehole wall is covered with steel pipes so it does not collapse. The drill pipe is turned by a carrier rod by means of a rotary table driven by an electric motor. The rotary table has a large flywheel mass in order to ensure that the speed of the drill pipe is as constant as possible. From time to time, extensions approximately 10 m in length must be added to the line of drill rods.

Such boreholes typically reach depths between 3000 m and 5000 m. At these drilling depths, the drill string will twist. Therefore, the angle Θ_1 at the top of the drill pipe and the angle Θ_2 at its end can differ considerably. For example, at a depth of 3500 m, the drill pipe will twist 360 degrees about twenty times. Since the drill pipe acts as a torsion spring, torsional vibrations of the drill head can occur. These vibrations are triggered by stick-slip friction

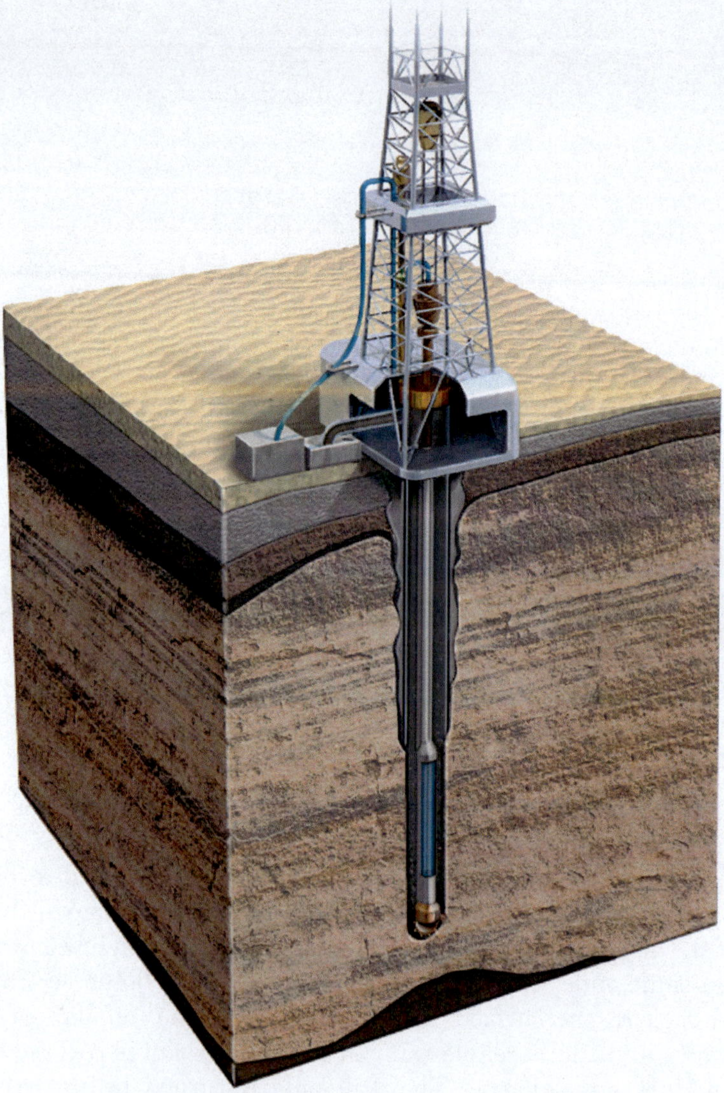

Fig. 5.17: Drilling rig

on the drill bit and are undesirable because they harm the bit and the drill pipe and, in the worst case, can cause parts of the drill string to shear off. Therefore, vibrations of this kind are suppressed by applying a control system [2, 69, 412].

We can approximately model the drill pipe with two flywheels connected by a torsion spring, as shown in Figure 5.18. The flywheels' movements are viscously damped by the drilling fluid, where d_1 and d_2 are the damping constants. Here M_{drive} is the driving torque at the rig. The load torque M_{load} at the bit is caused by the stick-slip friction on the rock during the milling process. The torque caused by the friction can be modeled by

$$M_{\text{fr}}(\dot{\Theta}_2) = \frac{2}{\pi} M_{\text{c}}(k_1 \dot{\Theta}_2 e^{-k_2|\dot{\Theta}_2|} + \arctan(k_3 \dot{\Theta}_2))$$

with $M_{\text{c}} = 0.5\,\text{kN m}$, $k_1 = 9.5\,\text{s rad}^{-1}$, $k_2 = 2.2\,\text{s rad}^{-1}$, and $k_3 = 35.0\,\text{s rad}^{-1}$. Furthermore, an unknown, time-dependent torque $M_{\text{d}}(t)$ is added to the friction torque to formulate the load torque

$$M_{\text{load}} = M_{\text{fr}}(\dot{\Theta}_2) + M_{\text{d}}(t).$$

We view the torque $M_{\text{d}}(t)$ as a disturbance which depends on the type of rock, the temperature of the drill bit, etc. Figure 5.19 shows the curve of the stick-slip friction $M_{\text{fr}}(\dot{\Theta}_2)$.

The equations of motion can be stated as

$$\begin{aligned} J_1 \ddot{\Theta}_1 &= -c(\Theta_1 - \Theta_2) - d_1 \dot{\Theta}_1 + M_{\text{drive}}, \\ J_2 \ddot{\Theta}_2 &= c(\Theta_1 - \Theta_2) - d_2 \dot{\Theta}_2 - M_{\text{fr}}(\dot{\Theta}_2) - M_{\text{d}}(t), \end{aligned} \tag{5.108}$$

where J_1 summarizes the moments of inertia of the rotary table, the electric motor, and its gear, and J_2 summarizes the moment of inertia of the lower drill pipe and that of the bit. The term

$$c(\Theta_1 - \Theta_2)$$

describes the spring force, where $c = 473\,\text{N m rad}^{-1}$ is the spring constant. The viscous friction forces

$$-d_1 \dot{\Theta}_1 \quad \text{and} \quad -d_2 \dot{\Theta}_2$$

have the damping constants $d_1 = 425\,\text{N m s rad}^{-1}$ and $d_2 = 50\,\text{N m s rad}^{-1}$, respectively. The moments of inertia take on the values $J_1 = 2122\,\text{kg m}^2$ and $J_2 = 374\,\text{kg m}^2$.

With

$$\boldsymbol{x} = \begin{bmatrix} \dot{\Theta}_1 & \Theta_1 - \Theta_2 & \dot{\Theta}_2 \end{bmatrix}^T$$

as the state vector and the driving torque $u = M_{\text{drive}}$ as the input variable, we obtain the model

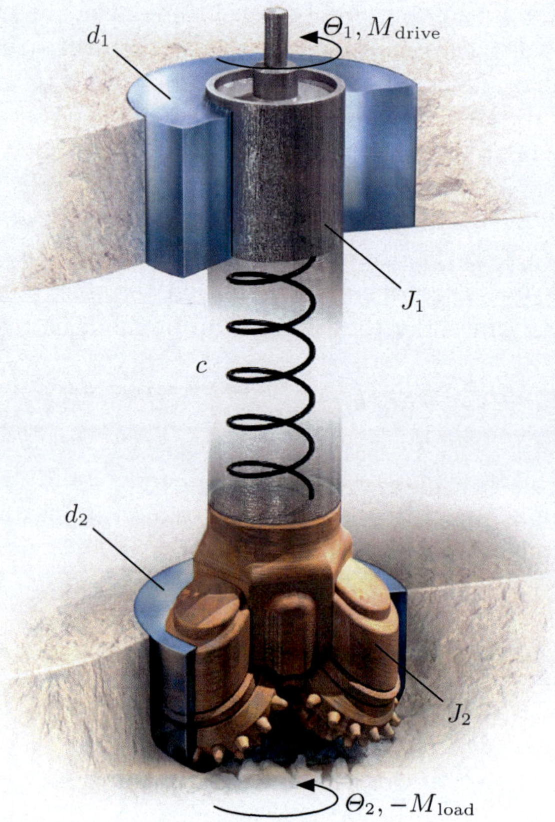

Fig. 5.18: Simplified and abstracted model of the drill pipe with two concentrated masses, their moments of inertia J_1 and J_2, and a torsion spring with spring constant c

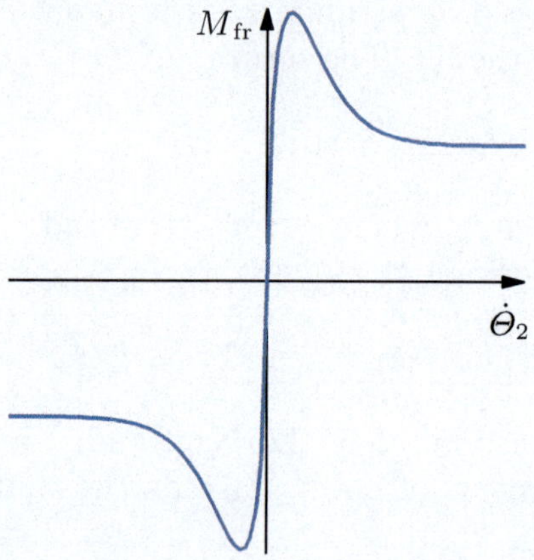

Fig. 5.19: Curve of the stick-slip friction $M_{\text{fr}}(\dot{\Theta}_2)$

$$\dot{\boldsymbol{x}} = \boldsymbol{a}(\boldsymbol{x}) + \boldsymbol{b} \cdot u + \boldsymbol{s} \cdot M_{\mathrm{d}} \tag{5.109}$$

from equation (5.108), where the disturbance is denoted by $M_{\mathrm{d}}(t)$ and

$$\boldsymbol{a}(\boldsymbol{x}) = \begin{bmatrix} -a_2 x_1 - a_1 x_2 \\ x_1 - x_3 \\ a_3 x_2 - a_4 x_3 \underbrace{-a_5 x_3 e^{-k_2|x_3|} - a_6 \arctan(k_3 x_3)}_{-\,s M_{\mathrm{fr}}(x_3)} \end{bmatrix},$$

$$\boldsymbol{b} = \begin{bmatrix} b \\ 0 \\ 0 \end{bmatrix}, \quad \boldsymbol{s} = \begin{bmatrix} 0 \\ 0 \\ -s \end{bmatrix}.$$

The system's parameters are

$$a_1 = \frac{c}{J_1}, \qquad a_2 = \frac{d_1}{J_1}, \qquad a_3 = \frac{c}{J_2}, \qquad a_4 = \frac{d_2}{J_2},$$

$$a_5 = \frac{2M_{\mathrm{c}} k_1}{\pi J_2}, \qquad a_6 = \frac{2M_{\mathrm{c}}}{\pi J_2}, \qquad b = \frac{1}{J_1}, \qquad s = \frac{1}{J_2}.$$

To determine whether the system (5.109) is full-state linearizable, we will apply Theorem 64 and begin by calculating

$$\begin{aligned} \mathrm{ad}_{\boldsymbol{a}} \boldsymbol{b} &= [\boldsymbol{a}, \boldsymbol{b}] = \frac{\partial \boldsymbol{b}}{\partial \boldsymbol{x}} \boldsymbol{a} - \frac{\partial \boldsymbol{a}}{\partial \boldsymbol{x}} \boldsymbol{b} \\ &= - \begin{bmatrix} -a_2 & -a_1 & 0 \\ 1 & 0 & -1 \\ 0 & a_3 & -a_4 - s M_{\mathrm{fr}}'(x_3) \end{bmatrix} \begin{bmatrix} b \\ 0 \\ 0 \end{bmatrix} = \begin{bmatrix} a_2 b \\ -b \\ 0 \end{bmatrix}, \end{aligned}$$

$$\begin{aligned} \mathrm{ad}_{\boldsymbol{a}}^2 \boldsymbol{b} &= [\boldsymbol{a}, [\boldsymbol{a}, \boldsymbol{b}]] = \frac{\partial [\boldsymbol{a}, \boldsymbol{b}]}{\partial \boldsymbol{x}} \boldsymbol{a} - \frac{\partial \boldsymbol{a}}{\partial \boldsymbol{x}} [\boldsymbol{a}, \boldsymbol{b}] \\ &= - \begin{bmatrix} -a_2 & -a_1 & 0 \\ 1 & 0 & -1 \\ 0 & a_3 & -a_4 - s M_{\mathrm{fr}}'(x_3) \end{bmatrix} \begin{bmatrix} a_2 b \\ -b \\ 0 \end{bmatrix} = \begin{bmatrix} a_2^2 b - a_1 b \\ -a_2 b \\ a_3 b \end{bmatrix}, \end{aligned}$$

where

$$M_{\mathrm{fr}}'(x_3) = \frac{\partial M_{\mathrm{fr}}(x_3)}{\partial x_3}$$

holds. This yields

$$\mathrm{rank}([\boldsymbol{b} \ \ \mathrm{ad}_{\boldsymbol{a}} \boldsymbol{b} \ \ \mathrm{ad}_{\boldsymbol{a}}^2 \boldsymbol{b}]) = \mathrm{rank} \left(b \begin{bmatrix} 1 & a_2 & a_2^2 - a_1 \\ 0 & -1 & -a_2 \\ 0 & 0 & a_3 l \end{bmatrix} \right) = 3,$$

and Condition (1) of Theorem 64 is fulfilled. Since the vectors $\boldsymbol{b}$, $\mathrm{ad}_{\boldsymbol{a}}\,\boldsymbol{b}$, and $\mathrm{ad}_{\boldsymbol{a}}^2\,\boldsymbol{b}$ are constant, all Lie brackets result in

$$[\mathrm{ad}_{\boldsymbol{a}}^i\,\boldsymbol{b}, \mathrm{ad}_{\boldsymbol{a}}^k\,\boldsymbol{b}] = \begin{bmatrix} 0 \\ 0 \\ 0 \end{bmatrix} \quad \text{for } i,k \in \{0,1\},$$

i. e. they are trivial linear combinations of the vectors $\boldsymbol{b}$ and $\mathrm{ad}_{\boldsymbol{a}}\,\boldsymbol{b}$. Thus Condition (2) of Theorem 64 is also fulfilled and the system can be full-state linearized.

It is interesting to note that in the case of a disappearing spring constant c, i. e. where $a_3 = 0$, the drilling system is no longer full-state linearizable. In this unrealistic case, the spring is omitted and the connection between the flywheels is broken. Then the control variable u has no influence on the state x_3, which is why the plant is not controllable and therefore is not full-state linearizable.

We can use the partial differential equations

$$\begin{aligned} L_{\boldsymbol{b}}\lambda(\boldsymbol{x}) &= \frac{\partial \lambda}{\partial \boldsymbol{x}}\boldsymbol{b} = \frac{\partial \lambda}{\partial x_1} b = 0, \\ L_{\mathrm{ad}_{\boldsymbol{a}}\,\boldsymbol{b}}\lambda(\boldsymbol{x}) &= \frac{\partial \lambda}{\partial \boldsymbol{x}}\,\mathrm{ad}_{\boldsymbol{a}}\,\boldsymbol{b} = \frac{\partial \lambda}{\partial x_1} a_2 b - \frac{\partial \lambda}{\partial x_2} b = 0 \end{aligned} \tag{5.110}$$

from Theorem 65 to determine the linearizing output $y = \lambda(\boldsymbol{x})$. In this case the task is straightforward, because

$$y = \lambda(\boldsymbol{x}) = x_3$$

obviously satisfies the differential equations (5.110).

Thus, the diffeomorphism

$$\boldsymbol{z} = \boldsymbol{t}(\boldsymbol{x}) = \begin{bmatrix} \lambda(\boldsymbol{x}) \\ L_{\boldsymbol{a}}\lambda(\boldsymbol{x}) \\ L_{\boldsymbol{a}}^2\lambda(\boldsymbol{x}) \end{bmatrix}$$

which transforms the system (5.109) into the Brunovsky canonical form can be formulated as

$$\boldsymbol{z} = \boldsymbol{t}(\boldsymbol{x}) = \begin{bmatrix} x_3 \\ a_3 x_2 - a_4 x_3 - s M_{\mathrm{fr}}(x_3) \\ a_3(x_1 - x_3) - (a_4 + s M_{\mathrm{fr}}'(x_3))(a_3 x_2 - a_4 x_3 - s M_{\mathrm{fr}}(x_3)) \end{bmatrix}.$$

The related retransformation is

$$\boldsymbol{x} = \boldsymbol{t}^{-1}(\boldsymbol{z}) = \begin{bmatrix} \dfrac{1}{a_3}[z_3 + (a_4 + sM'_{\text{fr}}(z_1))\, z_2] + z_1 \\ \dfrac{1}{a_3}[z_2 + a_4 z_1 + sM_{\text{fr}}(z_1)] \\ z_1 \end{bmatrix}.$$

By differentiating the equation $\boldsymbol{z} = \boldsymbol{t}(\boldsymbol{x})$ with respect to time, we obtain the transformed system representation, whose general form is already known to us from Section 3.3.1 on p. 260, supplemented by the perturbation $M_{\text{d}}(t)$, as

$$\dot{\boldsymbol{z}} = \frac{\partial \boldsymbol{t}(\boldsymbol{x})}{\partial \boldsymbol{x}} \dot{\boldsymbol{x}} = \left.\frac{\partial \boldsymbol{t}(\boldsymbol{x})}{\partial \boldsymbol{x}}\right|_{\boldsymbol{x}=\boldsymbol{t}^{-1}(\boldsymbol{z})} \cdot \left(\boldsymbol{a}(\boldsymbol{t}^{-1}(\boldsymbol{z})) + \boldsymbol{b}(\boldsymbol{t}^{-1}(\boldsymbol{z})) \cdot u + \boldsymbol{s}(\boldsymbol{t}^{-1}(\boldsymbol{z})) \cdot M_{\text{d}}\right),$$

where

$$\left.\frac{\partial \boldsymbol{t}(\boldsymbol{x})}{\partial \boldsymbol{x}}\right|_{\boldsymbol{x}=\boldsymbol{t}^{-1}(\boldsymbol{z})} = \begin{bmatrix} 0 & 0 & 1 \\ 0 & a_3 & -(a_4 + sM'_{\text{fr}}(z_1)) \\ a_3 & -a_3(a_4+sM'_{\text{fr}}(z_1)) & -a_3+(a_4+sM'_{\text{fr}}(z_1))^2 - sM''_{\text{fr}}(z_1)z_2 \end{bmatrix}$$

is the Jacobi matrix of the diffeomorphism $\boldsymbol{t}(\boldsymbol{x})$. Here the abbreviation

$$M''_{\text{fr}}(z_1) = \frac{\partial^2 M'_{\text{fr}}(z_1)}{\partial z_1^2}$$

is used. Thus, we obtain

$$\dot{\boldsymbol{z}} = \begin{bmatrix} z_2 \\ z_3 \\ -(a_1a_4 + a_2a_3)z_1 - (a_1 + a_2a_4 + a_3)z_2 - (a_2 + a_4)z_3 - h(\boldsymbol{z}) \end{bmatrix} + \begin{bmatrix} 0 \\ 0 \\ ba_3 \end{bmatrix} u$$
$$+ \left.\frac{\partial \boldsymbol{t}(\boldsymbol{x})}{\partial \boldsymbol{x}}\right|_{\boldsymbol{x}=\boldsymbol{t}^{-1}(\boldsymbol{z})} \cdot \boldsymbol{s}(\boldsymbol{t}^{-1}(\boldsymbol{z})) \cdot M_{\text{d}} \tag{5.111}$$

for the transformed system, where

$$h(\boldsymbol{z}) = a_1 sM_{\text{fr}}(z_1) + (z_3 + a_2z_2)sM'_{\text{fr}}(z_1) + sz_2^2 M''_{\text{fr}}(z_1).$$

Using the new input variable v and applying

$$u = \frac{(a_1a_4 + a_2a_3)z_1 + (a_1 + a_2a_4 + a_3)z_2 + (a_2 + a_4)z_3 + h(\boldsymbol{z})}{ba_3} + \frac{v}{ba_3},$$

we can transform system (5.111) into the Brunovsky canonical form. So we obtain

$$\dot{\boldsymbol{z}} = \begin{bmatrix} 0 & 1 & 0 \\ 0 & 0 & 1 \\ 0 & 0 & 0 \end{bmatrix} \boldsymbol{z} + \begin{bmatrix} 0 \\ 0 \\ 1 \end{bmatrix} v + s \begin{bmatrix} -1 \\ a_4 + sM'_{\text{fr}}(z_1) \\ a_3 - (a_4 + sM'_{\text{fr}}(z_1))^2 + sM''_{\text{fr}}(z_1)z_2 \end{bmatrix} M_{\text{d}}.$$

The nonlinear term represents the disturbance.

5.3.3 Full-State Linearization of MIMO Systems

We can extend Definition 34 of full-state linearizability to control-affine MIMO systems

$$\dot{\boldsymbol{x}} = \boldsymbol{a}(\boldsymbol{x}) + \sum_{i=1}^{m} \boldsymbol{b}_i(\boldsymbol{x}) \cdot u_i. \tag{5.112}$$

As in the SISO case, system (5.112) is referred to as full-state linearizable if a diffeomorphism $\boldsymbol{z} = \boldsymbol{t}(\boldsymbol{x})$ and its input variable transformation $\boldsymbol{u} = \boldsymbol{p}(\boldsymbol{x}, \boldsymbol{v})$ exist, so that the system can be transformed into the MIMO Brunovsky canonical form

$$\dot{\boldsymbol{x}} = \begin{bmatrix} \boldsymbol{A}_1 & \boldsymbol{0} & \cdots & \boldsymbol{0} \\ \boldsymbol{0} & \boldsymbol{A}_2 & \cdots & \boldsymbol{0} \\ \vdots & \vdots & \ddots & \vdots \\ \boldsymbol{0} & \boldsymbol{0} & \cdots & \boldsymbol{A}_m \end{bmatrix} \boldsymbol{x} + \begin{bmatrix} \boldsymbol{e}_1 & \boldsymbol{0} & \cdots & \boldsymbol{0} \\ \boldsymbol{0} & \boldsymbol{e}_2 & \cdots & \boldsymbol{0} \\ \vdots & \vdots & \ddots & \vdots \\ \boldsymbol{0} & \boldsymbol{0} & \cdots & \boldsymbol{e}_m \end{bmatrix} \boldsymbol{v}$$

with the $\delta_i \times \delta_i$ matrices $\boldsymbol{A}_i$ and the δ_i - dimensional vectors $\boldsymbol{e}_i$ of the form

$$\boldsymbol{A}_i = \begin{bmatrix} 0 & 1 & 0 & \cdots & 0 \\ 0 & 0 & 1 & \cdots & 0 \\ \vdots & \vdots & \vdots & \ddots & \vdots \\ 0 & 0 & 0 & \cdots & 1 \\ 0 & 0 & 0 & \cdots & 0 \end{bmatrix} \quad \text{and} \quad \boldsymbol{e}_i = \begin{bmatrix} 0 \\ 0 \\ \vdots \\ 0 \\ 1 \end{bmatrix}.$$

Then $[\delta_1 \cdots \delta_m]$ forms the vector relative degree of the system. The total relative degree is $\delta = \delta_1 + \ldots + \delta_m = n$.

Similarly to the SISO case, the following applies with regard to linearizability [197, 203, 206, 402]:

Theorem 66 (Full-State Linearizability of MIMO Systems). *A control-affine system*

$$\dot{\boldsymbol{x}} = \boldsymbol{a}(\boldsymbol{x}) + \sum_{i=1}^{m} \boldsymbol{b}_i(\boldsymbol{x}) \cdot u_i$$

with $\boldsymbol{x} \in D_{\boldsymbol{x},\text{def}} \subseteq I\!R^n$, $u_i \in I\!R$, $i = 1, \ldots, m$, *and a matrix* $\boldsymbol{M}_0 = \begin{bmatrix} \boldsymbol{b}_1 \cdots \boldsymbol{b}_m \end{bmatrix}$ *of rank* m *can be full-state linearized if and only if*

(1) all matrices

$$\boldsymbol{M}_1 = \left[\boldsymbol{b}_1 \cdots \boldsymbol{b}_m \;\; \mathrm{ad}_{\boldsymbol{a}} \boldsymbol{b}_1 \cdots \mathrm{ad}_{\boldsymbol{a}} \boldsymbol{b}_m\right],$$

$$\vdots$$

$$\boldsymbol{M}_{n-2} = \left[\boldsymbol{b}_1 \cdots \boldsymbol{b}_m \;\; \mathrm{ad}_{\boldsymbol{a}} \boldsymbol{b}_1 \cdots \mathrm{ad}_{\boldsymbol{a}} \boldsymbol{b}_m \cdots \mathrm{ad}_{\boldsymbol{a}}^{n-2} \boldsymbol{b}_1 \cdots \mathrm{ad}_{\boldsymbol{a}}^{n-2} \boldsymbol{b}_m\right]$$

are of constant rank,
(2) the matrix

$$\boldsymbol{M}_{n-1} = \left[\boldsymbol{b}_1 \cdots \boldsymbol{b}_m \;\; \mathrm{ad}_{\boldsymbol{a}} \boldsymbol{b}_1 \cdots \mathrm{ad}_{\boldsymbol{a}} \boldsymbol{b}_m \cdots \mathrm{ad}_{\boldsymbol{a}}^{n-1} \boldsymbol{b}_1 \cdots \mathrm{ad}_{\boldsymbol{a}}^{n-1} \boldsymbol{b}_m\right]$$

is of rank n, and
(3) all sets

$$\left\{\boldsymbol{b}_1, \ldots, \boldsymbol{b}_m\right\},$$
$$\left\{\boldsymbol{b}_1, \ldots, \boldsymbol{b}_m, ad_{\boldsymbol{a}}\boldsymbol{b}_1, \ldots, ad_{\boldsymbol{a}}\boldsymbol{b}_m\right\},$$
$$\vdots$$
$$\left\{\boldsymbol{b}_1, \ldots, \boldsymbol{b}_m, \mathrm{ad}_{\boldsymbol{a}} \boldsymbol{b}_1, \ldots, ad_{\boldsymbol{a}}\boldsymbol{b}_m, \ldots, ad_{\boldsymbol{a}}^{n-2}\boldsymbol{b}_1, \ldots, \mathrm{ad}_{\boldsymbol{a}}^{n-2} \boldsymbol{b}_m\right\}$$

are involutive.

Recall that a set

$$\{\boldsymbol{h}_1(\boldsymbol{x}), \boldsymbol{h}_2(\boldsymbol{x}), \ldots, \boldsymbol{h}_m(\boldsymbol{x})\}$$

of vector functions $\boldsymbol{h}_k(\boldsymbol{x})$ is referred to as *involutive* if the Lie brackets

$$[\boldsymbol{h}_i, \boldsymbol{h}_j], \; i, j \in \{1, \ldots, m\}$$

of all possible combinations (i, j) can be represented as linear combinations

$$\left[\boldsymbol{h}_i, \boldsymbol{h}_j\right] = \sum_{k=1}^{m} \alpha_k(\boldsymbol{x}) \boldsymbol{h}_k(\boldsymbol{x})$$

of the functions $\boldsymbol{h}_k(\boldsymbol{x})$. The coefficients $\alpha_k(\boldsymbol{x})$ are scalar functions which must be appropriately determined.

As in Theorem 64, which deals with the SISO case, Theorem 66 is not constructive. However, we can once again derive a theorem in a way similar to the SISO case, one which allows us to calculate the linearizing outputs $\lambda_1(\boldsymbol{x}), \ldots, \lambda_m(\boldsymbol{x})$ and the associated diffeomorphism. This is

Theorem 67 (Full-State Linearization of MIMO Systems). *A full-state linearizable system*

$$\dot{\boldsymbol{x}} = \boldsymbol{a}(\boldsymbol{x}) + \sum_{i=1}^{m} \boldsymbol{b}_i(\boldsymbol{x}) \cdot u_i$$

with $\boldsymbol{x} \in D_{\boldsymbol{x},\mathrm{def}} \subseteq I\!R^n$ *and* $u_i \in I\!R$, $i = 1, \ldots, m$, *is transformed into the Brunovsky canonical form by means of the diffeomorphism*

$$\boldsymbol{z} = \boldsymbol{t}(\boldsymbol{x}) = \begin{bmatrix} \lambda_1(\boldsymbol{x}) \\ L_{\boldsymbol{a}}\lambda_1(\boldsymbol{x}) \\ \vdots \\ L_{\boldsymbol{a}}^{\delta_1-1}\lambda_1(\boldsymbol{x}) \\ \vdots \\ \lambda_m(\boldsymbol{x}) \\ L_{\boldsymbol{a}}\lambda_m(\boldsymbol{x}) \\ \vdots \\ L_{\boldsymbol{a}}^{\delta_m-1}\lambda_m(\boldsymbol{x}) \end{bmatrix}$$

with $\delta = \delta_1 + \delta_2 + \ldots + \delta_m = n$, *and the input variable transformation*

$$\boldsymbol{u} = -\boldsymbol{D}^{-1}(\boldsymbol{x}) \cdot (\mathring{\boldsymbol{c}}(\boldsymbol{x}) - \boldsymbol{v})$$

with the new input variable vector $\boldsymbol{v}$ *and the associated decoupling matrix* $\boldsymbol{D}$, *stated as*

$$\boldsymbol{D}(\boldsymbol{x}) = \begin{bmatrix} L_{\boldsymbol{b}_1}L_{\boldsymbol{a}}^{\delta_1-1}\lambda_1(\boldsymbol{x}) & \cdots & L_{\boldsymbol{b}_m}L_{\boldsymbol{a}}^{\delta_1-1}\lambda_1(\boldsymbol{x}) \\ L_{\boldsymbol{b}_1}L_{\boldsymbol{a}}^{\delta_2-1}\lambda_2(\boldsymbol{x}) & \cdots & L_{\boldsymbol{b}_m}L_{\boldsymbol{a}}^{\delta_2-1}\lambda_2(\boldsymbol{x}) \\ \vdots & \ddots & \vdots \\ L_{\boldsymbol{b}_1}L_{\boldsymbol{a}}^{\delta_m-1}\lambda_m(\boldsymbol{x}) & \cdots & L_{\boldsymbol{b}_m}L_{\boldsymbol{a}}^{\delta_m-1}\lambda_m(\boldsymbol{x}) \end{bmatrix}, \quad \mathring{\boldsymbol{c}}(\boldsymbol{x}) = \begin{bmatrix} L_{\boldsymbol{a}}^{\delta_1}\lambda_1(\boldsymbol{x}) \\ L_{\boldsymbol{a}}^{\delta_2}\lambda_2(\boldsymbol{x}) \\ \vdots \\ L_{\boldsymbol{a}}^{\delta_m}\lambda_m(\boldsymbol{x}) \end{bmatrix},$$

if the artificial output variable vector $\boldsymbol{\lambda} = [\lambda_1 \cdots \lambda_m]^{\mathrm{T}}$ *fulfills the partial differential equations*

$$\begin{aligned} L_{\boldsymbol{b}_i}\lambda_j &= 0, \\ L_{\mathrm{ad}_{\boldsymbol{a}}\boldsymbol{b}_i}\lambda_j &= 0, \\ L_{\mathrm{ad}_{\boldsymbol{a}}^2\boldsymbol{b}_i}\lambda_j &= 0, \\ &\vdots \\ L_{\mathrm{ad}_{\boldsymbol{a}}^{\delta_j-2}\boldsymbol{b}_i}\lambda_j &= 0 \end{aligned}$$

for all $i = 1, \ldots, m$ *and all* $j = 1, \ldots, m$, *and the matrix* $\boldsymbol{D}(\boldsymbol{x})$ *is regular.*

Since the control-affine system in the above theorem is assumed to be full-state linearizable, $\delta = n$ holds for the vector relative degree $[\delta_1 \ \cdots \ \delta_m]$ of the system. To carry out a full-state linearization, the numbers $\delta_1, \ldots, \delta_m$ must be calculated in advance to determine the linearizing output variable vector $\boldsymbol{\lambda} = [\lambda_1 \cdots \lambda_m]^{\mathrm{T}}$. This is possible due to the following [272, 305]:

Theorem 68 (Vector Relative Degree). *A full-state linearizable system*

$$\dot{\boldsymbol{x}} = \boldsymbol{a}(\boldsymbol{x}) + \sum_{i=1}^{m} \boldsymbol{b}_i(\boldsymbol{x}) \cdot u_i$$

with $\boldsymbol{x} \in D_{\boldsymbol{x},\mathrm{def}} \subseteq I\!R^n$ *and a matrix* $\boldsymbol{M}_0 = \begin{bmatrix} \boldsymbol{b}_1 \cdots \boldsymbol{b}_m \end{bmatrix}$ *of rank* m *possesses an artificial output variable vector* $\boldsymbol{\lambda}(\boldsymbol{x})$ *with the vector relative degree*

$$\boldsymbol{\delta} = \begin{bmatrix} \delta_1 \ \cdots \ \delta_m \end{bmatrix} \quad \textit{with} \ \ \delta_1 + \ldots + \delta_m = n$$

where a value δ_i, $i = 1, \ldots, m$, *is equal to the number of values* $r_k \geq i$, $k = 0, \ldots, n-1$, *and*

$$\begin{aligned} r_0 &= \operatorname{rank}(\boldsymbol{M}_0) = m, \\ r_1 &= \operatorname{rank}(\boldsymbol{M}_1) - \operatorname{rank}(\boldsymbol{M}_0), \\ &\ \vdots \\ r_{n-1} &= \operatorname{rank}(\boldsymbol{M}_{n-1}) - \operatorname{rank}(\boldsymbol{M}_{n-2}) \end{aligned}$$

holds.

Note that it is not necessary to know the artificial output variable vector $\boldsymbol{\lambda}(\boldsymbol{x})$ when applying Theorem 68. A direct result of the theorem above is the sequence

$$\delta_1 \geq \delta_2 \geq \cdots \geq \delta_m.$$

To perform a full-state linearization of a specific system, we follow the procedure below:

Step 1: Using Theorem 66, check whether the system is full-state linearizable.

Step 2: Based on Theorem 68, determine the vector relative degree $\boldsymbol{\delta}$.

Step 3: Solve the partial differential equations from Theorem 67 and obtain the linearizing output variable vector $\boldsymbol{\lambda}$.

Step 4: By means of the output variable vector $\boldsymbol{\lambda}$, determine the diffeomorphism $\boldsymbol{z} = \boldsymbol{t}(\boldsymbol{x})$ from Theorem 67 and the input variable transformation $\boldsymbol{u} = -\boldsymbol{D}^{-1}(\boldsymbol{x}) \cdot (\mathring{\boldsymbol{c}}(\boldsymbol{x}) - \boldsymbol{v})$.

Step 3 is the most difficult and complex step, because the partial differential equations can often be solved only with great effort, as is the case for SISO systems.

Theorems 64 and 66 provide us with criteria we can use to determine whether a control-affine system can be full-state linearized and thus transformed into nonlinear controller canonical form. From Theorem 45, p. 218, we know that systems in nonlinear controller canonical form are always controllable. Therefore, Theorems 64 and 66 also enable us to determine the controllability of a control-affine system.

5.3.4 Flatness of Full-State Linearizable Systems

From Theorem 51 on p. 257, we know that SISO systems $\dot{\boldsymbol{x}} = \boldsymbol{f}(\boldsymbol{x}, u)$ are flat if and only if they are in nonlinear controller canonical form, or if they are transformable into such a form, and if $\beta(\boldsymbol{x}) \neq 0$ holds. In this case, a flat or fictitious flat output of the nonlinear controller canonical form is $y = x_1$. Furthermore, we know from Section 5.3.1 that control-affine systems

$$\dot{\boldsymbol{x}} = \boldsymbol{a}(\boldsymbol{x}) + \boldsymbol{b}(\boldsymbol{x}) \cdot u \tag{5.113}$$

can be full-state linearized if and only if they can be transformed into the nonlinear controller canonical form and $\beta(\boldsymbol{x}) \neq 0$ applies. Thus we obtain

Theorem 69 (Flatness of Full-State Linearizable SISO Systems). *A system*

$$\dot{\boldsymbol{x}} = \boldsymbol{a}(\boldsymbol{x}) + \boldsymbol{b}(\boldsymbol{x}) \cdot u, \quad \boldsymbol{x} \in D_{\boldsymbol{x},\mathrm{def}} \subseteq \mathbb{R}^n, \quad u \in \mathbb{R},$$

is flat if and only if it is full-state linearizable. Furthermore, if the system has the relative degree n for an output y, it follows that y is a flat output.

Therefore the conditions from Theorem 64 on p. 418, which are sufficient and necessary for the full-state linearizability of control-affine SISO systems, are also sufficient and necessary for the flatness of these systems. At this point, we remember that the properties of flatness and state linearizability may not hold for all values $\boldsymbol{x} \in D_{\boldsymbol{x},\mathrm{def}}$ and, in such cases, are only locally valid.

The flat system representation of a control-affine SISO system with the output variable $y = c(\boldsymbol{x})$ and relative degree $\delta = n$ can be derived using the diffeomorphism

$$\boldsymbol{z} = \boldsymbol{t}(\boldsymbol{x}) = \begin{bmatrix} c(\boldsymbol{x}) \\ L_{\boldsymbol{a}} c(\boldsymbol{x}) \\ \vdots \\ L_{\boldsymbol{a}}^{n-1} c(\boldsymbol{x}) \end{bmatrix} = \begin{bmatrix} y \\ \dot{y} \\ \vdots \\ y^{(n-1)} \end{bmatrix}. \tag{5.114}$$

Thus, the inverse function provides the first equation

$$\begin{aligned}\boldsymbol{x} &= \boldsymbol{t}^{-1}(\boldsymbol{z}) = \boldsymbol{t}^{-1}\left(y, \dot{y}, \ldots, y^{(n-1)}\right) \\ &= \boldsymbol{\Psi}_1\left(y, \dot{y}, \ldots, y^{(n-1)}\right)\end{aligned} \tag{5.115}$$

of the flat system representation.

For the calculation of u as a function of y and its derivatives, we will now transform the system (5.113) by differentiating the diffeomorphism (5.114) into the form

$$\dot{\boldsymbol{z}} = \begin{bmatrix} \dot{y} \\ \ddot{y} \\ \vdots \\ y^{(n-1)} \\ y^{(n)} \end{bmatrix} = \begin{bmatrix} z_2 \\ z_3 \\ \vdots \\ z_n \\ L_{\boldsymbol{a}}^n c(\boldsymbol{x}) + L_{\boldsymbol{b}} L_{\boldsymbol{a}}^{n-1} c(\boldsymbol{x}) \cdot u \end{bmatrix}.$$

This results in

$$\begin{aligned} u &= \frac{y^{(n)} - L_{\boldsymbol{a}}^n c(\boldsymbol{x})}{L_{\boldsymbol{b}} L_{\boldsymbol{a}}^{n-1} c(\boldsymbol{x})} = \frac{y^{(n)} - L_{\boldsymbol{a}}^n c\left(\boldsymbol{t}^{-1}\left(y, \dot{y}, \ldots, y^{(n-1)}\right)\right)}{L_{\boldsymbol{b}} L_{\boldsymbol{a}}^{n-1} c\left(\boldsymbol{t}^{-1}\left(y, \dot{y}, \ldots, y^{(n-1)}\right)\right)} \\ &= \Psi_2\left(y, \dot{y}, \ldots, y^{(n)}\right). \end{aligned} \tag{5.116}$$

Therefore, with equations (5.115) and (5.116), we have obtained the flat system representation.

In the case of control-affine SISO systems, the property of flatness does not reveal any significant new findings. This is because the control-affine system is flat if it can be full-state linearized and vice versa. In particular, as shown above, the flat system representation can be determined directly from the solution of the full-state linearization problem.

However, in the case of control-affine MIMO systems, flatness and state linearizability are no longer equivalent to each other, as is the case for SISO systems. This is because the following theorem, which we can readily derive from the nonlinear controller canonical form and Theorem 51 on p. 257, applies merely sufficiently.

Theorem 70 (Flatness of Full-State Linearizable MIMO Systems). *A system*

$$\dot{\boldsymbol{x}} = \boldsymbol{a}(\boldsymbol{x}) + \sum_{i=1}^{m} \boldsymbol{b}_i(\boldsymbol{x}) \cdot u_i, \quad \boldsymbol{x} \in D_{\boldsymbol{x},\text{def}} \subseteq I\!R^n, \quad u_i \in I\!R,\ i = 1, \ldots, m,$$

is flat if it is full-state linearizable. If the system has the total relative degree n for an output $\boldsymbol{y}$, then $\boldsymbol{y}$ is a flat output.

Consequently, it may be that a control-affine MIMO system cannot be full-state linearized, but is flat nevertheless.

5.3.5 Example: Rocket

A rocket [129, 420] is an example of a flat system which is not full-state linearizable. As shown in Figure 5.20, the rocket is placed in a fixed reference coordinate system with specified x- and z-coordinates for the rocket's center of gravity S. Its mass is referred to as m and the acceleration due to gravity is referred to as g. The rocket has a swiveling liquid-fuel propulsion whose thrust we can regulate. As a result, we have propulsive forces F_x and F_z in the directions x and z. We will now calculate the accelerations

$$\ddot{x} = \frac{F_x}{m},$$
$$\ddot{z} = \frac{F_z}{m} - g,$$

taking into account the earth's gravitational force.

The balance-of-torques equation can be formulated as

$$J\ddot{\Theta} = F_z d \sin(\Theta) - F_x d \cos(\Theta),$$

where d is the lever arm acted on by the forces F_x and F_z, meaning d is the distance between the forces' point of application and the center of gravity S, and Θ is the flight angle.

With the state vector

$$\begin{bmatrix} x_1 & x_2 & x_3 & x_4 & x_5 & x_6 \end{bmatrix} = \begin{bmatrix} \frac{x}{g} & \frac{\dot{x}}{g} & \frac{z}{g} & \frac{\dot{z}}{g} & \Theta & \dot{\Theta} \end{bmatrix},$$

the input variables

$$u_1 = \frac{F_x}{mg}, \qquad u_2 = \frac{F_z}{mg} - 1,$$

and the parameter

$$\varepsilon = \frac{J}{mgd},$$

we obtain the rocket model

$$\dot{\boldsymbol{x}} = \underbrace{\begin{bmatrix} x_2 \\ 0 \\ x_4 \\ 0 \\ x_6 \\ \dfrac{\sin(x_5)}{\varepsilon} \end{bmatrix}}_{\boldsymbol{a}} + \underbrace{\begin{bmatrix} 0 \\ 1 \\ 0 \\ 0 \\ 0 \\ -\dfrac{\cos(x_5)}{\varepsilon} \end{bmatrix}}_{\boldsymbol{b}_1} u_1 + \underbrace{\begin{bmatrix} 0 \\ 0 \\ 0 \\ 1 \\ 0 \\ \dfrac{\sin(x_5)}{\varepsilon} \end{bmatrix}}_{\boldsymbol{b}_2} u_2. \tag{5.117}$$

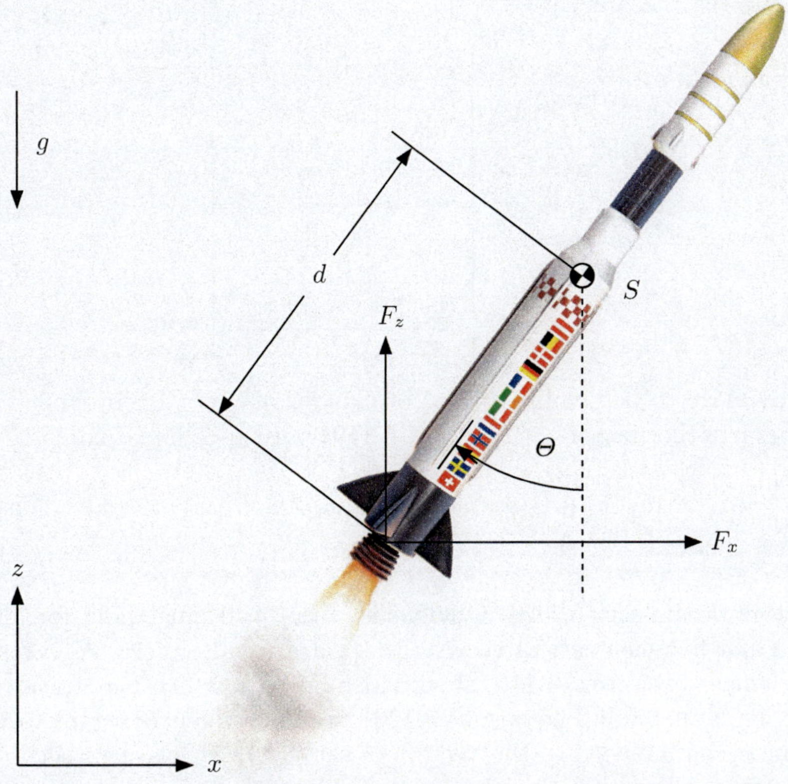

Fig. 5.20: Rocket with propulsion forces F_x and F_z and inclination angle Θ

To analyze the full-state linearizability of the system, we will apply Theorem 66 and show that its third condition is not fulfilled. To this end, we first determine the set of input vectors $\boldsymbol{b}_1$, $\boldsymbol{b}_2$ and the two Lie brackets

$$[\boldsymbol{a}, \boldsymbol{b}_1] = \mathrm{ad}_{\boldsymbol{a}}\, \boldsymbol{b}_1 \quad \text{and} \quad [\boldsymbol{a}, \boldsymbol{b}_2] = \mathrm{ad}_{\boldsymbol{a}}\, \boldsymbol{b}_2,$$

i. e.

$$\left\{\boldsymbol{b}_1, \boldsymbol{b}_2, \mathrm{ad}_{\boldsymbol{a}}\, \boldsymbol{b}_1, \mathrm{ad}_{\boldsymbol{a}}\, \boldsymbol{b}_2\right\} = \left\{ \begin{bmatrix} 0 \\ 1 \\ 0 \\ 0 \\ 0 \\ -\dfrac{\cos(x_5)}{\varepsilon} \end{bmatrix}, \begin{bmatrix} 0 \\ 0 \\ 0 \\ 1 \\ 0 \\ \dfrac{\sin(x_5)}{\varepsilon} \end{bmatrix}, \begin{bmatrix} -1 \\ 0 \\ 0 \\ 0 \\ \dfrac{\cos(x_5)}{\varepsilon} \\ \dfrac{x_6 \sin(x_5)}{\varepsilon} \end{bmatrix}, \begin{bmatrix} 0 \\ 0 \\ -1 \\ 0 \\ -\dfrac{\sin(x_5)}{\varepsilon} \\ \dfrac{x_6 \cos(x_5)}{\varepsilon} \end{bmatrix} \right\}. \tag{5.118}$$

Now we will determine whether the set (5.118) is involutive. For this, we will first calculate the Lie bracket

$$\left[\mathrm{ad}_{\boldsymbol{a}}\,\boldsymbol{b}_1, \boldsymbol{b}_2\right] = \begin{bmatrix} 0 \\ 0 \\ 0 \\ 0 \\ 0 \\ \dfrac{\cos^2(x_5) - \sin^2(x_5)}{\varepsilon^2} \end{bmatrix}. \tag{5.119}$$

For the set (5.118) to be involutive, i. e. for Condition (3) of Theorem 66 to be fulfilled, among other factors, the vector (5.119) must be a linear combination

$$\left[\mathrm{ad}_{\boldsymbol{a}}\boldsymbol{b}_1, \boldsymbol{b}_2\right] = \alpha_1(\boldsymbol{x})\cdot\boldsymbol{b}_1(\boldsymbol{x})+\alpha_2(\boldsymbol{x})\cdot\boldsymbol{b}_2(\boldsymbol{x})+\alpha_3(\boldsymbol{x})\cdot\mathrm{ad}_{\boldsymbol{a}}\,\boldsymbol{b}_1(\boldsymbol{x})+\alpha_4(\boldsymbol{x})\cdot\mathrm{ad}_{\boldsymbol{a}}\,\boldsymbol{b}_2(\boldsymbol{x}) \tag{5.120}$$

of the vectors of the set (5.118). Obviously, $\alpha_i(\boldsymbol{x}) = 0$ must hold for all α_i so that the first five elements of vector (5.119) are equal to zero. As a result, the sixth element of vector (5.119) should also be zero, which is not the case. Therefore, we cannot fulfill equation (5.120) and consequently vector (5.119) is not a linear combination of the vectors of set (5.118); thus, this set is not involutive. This means that Condition (3) of Theorem 66 is not fulfilled and the rocket model cannot be full-state linearized.

But it is flat, as already mentioned. We can prove this using the output

$$\boldsymbol{y} = \begin{bmatrix} y_1 \\ y_2 \end{bmatrix} = \begin{bmatrix} x_1 + \varepsilon\sin(x_5) \\ x_3 + \varepsilon\cos(x_5) \end{bmatrix}. \tag{5.121}$$

We can now calculate

$$\ddot{y}_1 = \sin(x_5)\cdot\left[u_1\sin(x_5) + (1+u_2)\cos(x_5) - \varepsilon\dot{x}_5^2\right], \tag{5.122}$$

$$\ddot{y}_2 = \cos(x_5)\cdot\left[u_1\sin(x_5) + (1+u_2)\cos(x_5) - \varepsilon\dot{x}_5^2\right] - 1. \tag{5.123}$$

Equation (5.122) and (5.123) yield

$$\frac{\ddot{y}_1}{\ddot{y}_2 + 1} = \tan(x_5)$$

or

$$x_5 = \arctan\left(\frac{\ddot{y}_1}{\ddot{y}_2+1}\right) = \Psi_{1,5}\left(\boldsymbol{y}, \dot{\boldsymbol{y}}, \ddot{\boldsymbol{y}}\right). \tag{5.124}$$

Inserting equation (5.124) into equation (5.121) results in

$$x_1 = y_1 - \frac{\varepsilon \ddot{y}_1}{\sqrt{\ddot{y}_1^2 + (\ddot{y}_2 + 1)^2}} = \Psi_{1,1}(\boldsymbol{y}, \dot{\boldsymbol{y}}, \ddot{\boldsymbol{y}}), \tag{5.125}$$

$$x_3 = y_2 - \frac{\varepsilon(\ddot{y}_2 + 1)}{\sqrt{\ddot{y}_1^2 + (\ddot{y}_2 + 1)^2}} = \Psi_{1,3}(\boldsymbol{y}, \dot{\boldsymbol{y}}, \ddot{\boldsymbol{y}}). \tag{5.126}$$

We have thus determined three elements of the function $\boldsymbol{\Psi}_1$. The missing coordinate representations of the flat system representation

$$\begin{aligned} x_2 &= \dot{x}_1 = \Psi_{1,2}(\boldsymbol{y}, \dot{\boldsymbol{y}}, \ddot{\boldsymbol{y}}, \dddot{\boldsymbol{y}}), \\ x_4 &= \dot{x}_3 = \Psi_{1,4}(\boldsymbol{y}, \dot{\boldsymbol{y}}, \ddot{\boldsymbol{y}}, \dddot{\boldsymbol{y}}), \\ x_6 &= \dot{x}_5 = \Psi_{1,6}(\boldsymbol{y}, \dot{\boldsymbol{y}}, \ddot{\boldsymbol{y}}, \dddot{\boldsymbol{y}}) \end{aligned}$$

are obtained by differentiating equations (5.124), (5.125), and (5.126).

To calculate the function $\boldsymbol{u} = \boldsymbol{\Psi}_2(\boldsymbol{y}, \dot{\boldsymbol{y}}, \ldots, \boldsymbol{y}^{(4)})$ of the flat system representation, we will use equation (5.122) and the relation

$$\ddot{x}_5 = \frac{1}{\varepsilon}\sin(x_5) - \frac{1}{\varepsilon}\cos(x_5)u_1 + \frac{1}{\varepsilon}\sin(x_5)u_2,$$

which results from equation (5.117), yielding the linear system of equations

$$\begin{bmatrix} \dfrac{\ddot{y}_1}{\sin(x_5)} + \varepsilon\dot{x}_5^2 - \cos(x_5) \\ \varepsilon\ddot{x}_5 - \sin(x_5) \end{bmatrix} = \begin{bmatrix} \sin(x_5) & \cos(x_5) \\ -\cos(x_5) & \sin(x_5) \end{bmatrix} \begin{bmatrix} u_1 \\ u_2 \end{bmatrix}.$$

From this, we can determine the control vector as follows:

$$\begin{bmatrix} u_1 \\ u_2 \end{bmatrix} = \begin{bmatrix} \sin(x_5) & -\cos(x_5) \\ \cos(x_5) & \sin(x_5) \end{bmatrix} \begin{bmatrix} \dfrac{\ddot{y}_1}{\sin(x_5)} + \varepsilon\dot{x}_5^2 - \cos(x_5) \\ \varepsilon\ddot{x}_5 - \sin(x_5) \end{bmatrix}.$$

Inserting equation (5.124) yields the function $\boldsymbol{u} = \boldsymbol{\Psi}_2(\boldsymbol{y}, \dot{\boldsymbol{y}}, \ldots, \boldsymbol{y}^{(4)})$ which we are looking for. Thus, we have proven that the system is flat.

5.4 Feedforward and Feedback Control of Flat Systems

5.4.1 Fundamentals

Based on the statements about flatness in Section 3.2 on p. 239 et seq., we will aim to design feedforward and feedback controllers for flat systems. The design of a feedforward controller for a flat system

$$\begin{aligned} \dot{\boldsymbol{x}} &= \boldsymbol{f}(\boldsymbol{x}, \boldsymbol{u}), \\ \boldsymbol{y} &= \boldsymbol{g}(\boldsymbol{x}) \end{aligned}$$

is obvious and immediately feasible if the real output

$$\boldsymbol{y} = \boldsymbol{g}(\boldsymbol{x}) = \boldsymbol{h}(\boldsymbol{x})$$

is flat; from now on, we will write $\boldsymbol{y} = \boldsymbol{h}(\boldsymbol{x})$ for the output function if $\boldsymbol{y}$ is a flat output and otherwise we will formulate it as $\boldsymbol{y} = \boldsymbol{g}(\boldsymbol{x})$. If $\boldsymbol{y}$ is flat, the flat system representation

$$\begin{aligned} \boldsymbol{x} &= \boldsymbol{\Psi}_1(\boldsymbol{y}, \dot{\boldsymbol{y}}, \ldots, \boldsymbol{y}^{(\beta)}), \\ \boldsymbol{u} &= \boldsymbol{\Psi}_2(\boldsymbol{y}, \dot{\boldsymbol{y}}, \ldots, \boldsymbol{y}^{(\beta+1)}) \end{aligned} \tag{5.127}$$

provides us with the feedforward control law (5.127) in dependence on the output variable vector $\boldsymbol{y}$ and its derivatives. Figure 5.21 shows the corresponding block diagram, where the reference variable is $\boldsymbol{y}_{\mathrm{ref}}$.

If a reference output signal $\boldsymbol{y}_{\mathrm{ref}}(t)$ is predefined, i. e. a signal $\boldsymbol{y}_{\mathrm{ref}}(t)$ and its derivatives

$$\dot{\boldsymbol{y}}_{\mathrm{ref}}(t), \ \ldots \ , \ \boldsymbol{y}_{\mathrm{ref}}^{(\beta+1)}(t)$$

in equation (5.127) are predetermined, and an exact model which is unaffected by a disturbance exists, the identity $\boldsymbol{y}(t) = \boldsymbol{y}_{\mathrm{ref}}(t)$ holds for the system's output signal. Note that not all reference variable signals $\boldsymbol{y}_{\mathrm{ref}}(t)$ are suitable. In fact, they must be differentiable at least $\beta+1$ times. Thus, reference variable steps are not allowed.

In practice, it will rarely or never be the case that an exact model exists and that there is an absence of disturbances. This means the output value $\boldsymbol{y}(t)$ will deviate from the reference value $\boldsymbol{y}_{\mathrm{ref}}(t)$. To solve this problem, a feedback control is usually superimposed on the feedforward control, as shown in Figure 5.22.

The feedback control eliminates disturbances affecting the plant and also compensates for the effects of discrepancies between the real process and the model on which the flat system representation is based. Feedback control is particularly important in the case of unstable plants. In the simplest case, PID controllers can be used. When complex nonlinearities occur, the use of nonlinear controllers such as gain-scheduling controllers may be necessary.

So far, we have only examined the case of a real flat output $\boldsymbol{y}$, in which the design of a flat feedforward controller is obvious. However, if the real output $\boldsymbol{y} = \boldsymbol{g}(\boldsymbol{x})$ is not flat, and we must use a fictitious flat output $\boldsymbol{y}_{\mathrm{f}} = \boldsymbol{h}(\boldsymbol{x})$ for the control, the design of a controller of this type is usually more complex. This is discussed in the following section.

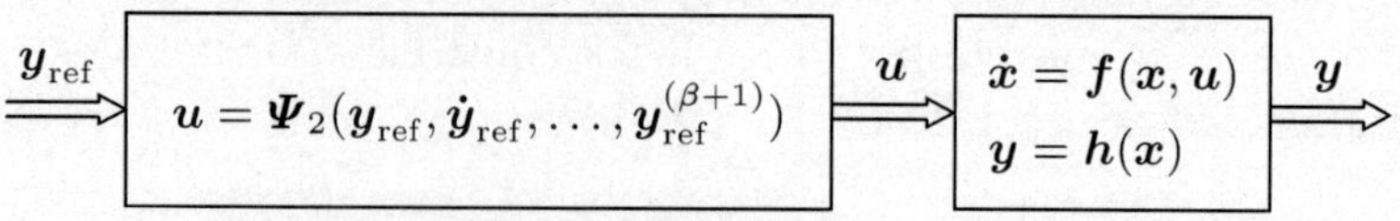

Fig. 5.21: Feedforward control of a flat system

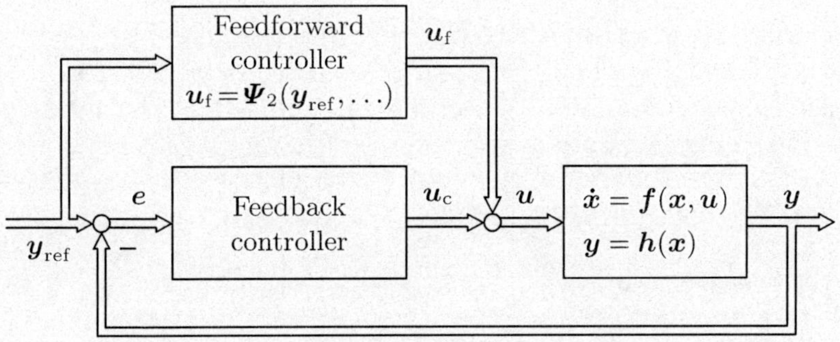

Fig. 5.22: Flatness-based feedforward and feedback control of a flat system

5.4.2 Feedforward Controls Using Fictitious Flat Outputs

Usually, for a flatness-based control, we use the reference signal $\boldsymbol{y}_{\mathrm{ref}}(t)$ of the real output variable $\boldsymbol{y}(t)$ as an input variable. If the real output $\boldsymbol{y}$ is not flat, we must convert it into a fictitious flat output $\boldsymbol{y}_{\mathrm{f}}$ or vice versa. In the following, we assume this situation and that we know a fictitious flat output $\boldsymbol{y}_{\mathrm{f}}$.

By inserting the flat system equation

$$\boldsymbol{x} = \boldsymbol{\Psi}_1(\boldsymbol{y}_{\mathrm{f}}, \dot{\boldsymbol{y}}_{\mathrm{f}}, \ldots, \boldsymbol{y}_{\mathrm{f}}^{(\beta)}) \tag{5.128}$$

into the equation $\boldsymbol{y} = \boldsymbol{g}(\boldsymbol{x})$ for the system's output, we obtain the desired conversion rule between $\boldsymbol{y}_{\mathrm{f}}$ and $\boldsymbol{y}$:

$$\begin{aligned} \boldsymbol{y} &= \boldsymbol{g}\left(\boldsymbol{\Psi}_1\left(\boldsymbol{y}_{\mathrm{f}}, \dot{\boldsymbol{y}}_{\mathrm{f}}, \ldots, \boldsymbol{y}_{\mathrm{f}}^{(\beta)}\right)\right) \\ &= \boldsymbol{\Gamma}\left(\boldsymbol{y}_{\mathrm{f}}, \dot{\boldsymbol{y}}_{\mathrm{f}}, \ldots, \boldsymbol{y}_{\mathrm{f}}^{(\gamma)}\right), \end{aligned} \tag{5.129}$$

with $\gamma \leq \beta$. The identity $\gamma = n - \delta$ holds in the SISO case, where δ denotes the relative degree of the system [169]. We can use the above equation (5.129), which is an implicit differential equation, in different ways. The obvious possibility is to specify the real output signal $\boldsymbol{y}(t)$ and to solve the differential equation (5.129) to calculate the progression of the fictitious flat output $\boldsymbol{y}_{\mathrm{f}}^{(\gamma)}$. This is usually a doable task if we can convert the differential equation (5.129) such that $\boldsymbol{y}_{\mathrm{f}}^{(\gamma)}$ is explicitly given. If this is not possible, the differential equation is implicit in $\boldsymbol{y}_{\mathrm{f}}^{(\gamma)}$, and calculating the fictitious flat output $\boldsymbol{y}_{\mathrm{f}}$ by solving the differential equation is often tricky.

In general, we can choose any set of initial values

$$\boldsymbol{y}_{\mathrm{f}}(0), \dot{\boldsymbol{y}}_{\mathrm{f}}(0), \ldots, \boldsymbol{y}_{\mathrm{f}}^{(\gamma-1)}(0).$$

For example, we can set all these values at zero and then use the progression $\boldsymbol{y}_{\mathrm{f}}(t)$ of the flat output to calculate the control variable

$$\boldsymbol{u} = \boldsymbol{\Psi}_2\left(\boldsymbol{y}_{\mathrm{f}}, \dot{\boldsymbol{y}}_{\mathrm{f}}, \ldots, \boldsymbol{y}_{\mathrm{f}}^{(\beta+1)}\right).$$

If we wish to specify an initial vector $\boldsymbol{x}(0)$ in addition to the signal $\boldsymbol{y}(t)$, we must select the initial values

$$\boldsymbol{y}_{\mathrm{f}}(0),\ \dot{\boldsymbol{y}}_{\mathrm{f}}(0),\ \ldots\ ,\ \boldsymbol{y}_{\mathrm{f}}^{(\beta)}(0)$$

so that the relation

$$\boldsymbol{x}(0) = \boldsymbol{\Psi}_1\left(\boldsymbol{y}_{\mathrm{f}}(0), \dot{\boldsymbol{y}}_{\mathrm{f}}(0), \ldots, \boldsymbol{y}_{\mathrm{f}}^{(\beta)}(0)\right) \tag{5.130}$$

resulting from equation (5.128) is fulfilled. In the SISO case, $\beta = n-1$ applies. Therefore, we have n values

$$y_{\mathrm{f}}(0), \ldots, y_{\mathrm{f}}^{(n-1)}(0),$$

i. e. the same number as states $x_i(0)$. Furthermore, a flat SISO-system is full-state linearizable using equation (5.128), which is the diffeomorphism required for this purpose. This means that the inverse function $\boldsymbol{\Psi}_1^{-1}$ always exists in the SISO case. The desired initial values

$$y_{\mathrm{f}}(0), \dot{y}_{\mathrm{f}}(0), \ldots, y_{\mathrm{f}}^{(n-1)}(0)$$

then result from the equation

$$\begin{bmatrix} y_{\mathrm{f}}(0) \\ \dot{y}_{\mathrm{f}}(0) \\ \vdots \\ y_{\mathrm{f}}^{(n-1)}(0) \end{bmatrix} = \boldsymbol{\Psi}_1^{-1}(\boldsymbol{x}(0)).$$

In the MIMO case, there may be more initial values $y_{\mathrm{f},i}^{(j)}(0)$, $i = 1, \ldots, r$ and $j = 0, \ldots, \beta$ than initial states $x_i(0)$. In this case, the system of equations (5.130) may be underdetermined.

Let us now examine how we can determine the fictitious flat output signal $\boldsymbol{y}_{\mathrm{f}}(t)$ and its derivatives

$$\dot{\boldsymbol{y}}_{\mathrm{f}}(t), \ldots, \boldsymbol{y}_{\mathrm{f}}^{(\gamma)}(t)$$

from a given real output signal $\boldsymbol{y}(t)$ using equation (5.129). As already stated, equation (5.129) is a differential equation in implicit form, i. e. the highest order derivative $\boldsymbol{y}_{\mathrm{f}}^{(\gamma)}$ is not explicitly given. However, if equation (5.129) can be transformed into the explicit form

$$\boldsymbol{y}_{\mathrm{f}}^{(\gamma)} = \boldsymbol{\Gamma}^{-1}\left(\boldsymbol{y}_{\mathrm{f}}, \dot{\boldsymbol{y}}_{\mathrm{f}}, \ldots, \boldsymbol{y}_{\mathrm{f}}^{(\gamma-1)}, \boldsymbol{y}\right), \tag{5.131}$$

we can solve this differential equation of order γ using standard numerical methods, such as the Runge-Kutta method, after the equation has been transformed into a first-order system of differential equations

$$\dot{\boldsymbol{z}} = \begin{bmatrix} \dot{\boldsymbol{z}}_1 \\ \dot{\boldsymbol{z}}_2 \\ \vdots \\ \dot{\boldsymbol{z}}_{\gamma-1} \\ \dot{\boldsymbol{z}}_\gamma \end{bmatrix} = \begin{bmatrix} \boldsymbol{z}_2 \\ \boldsymbol{z}_3 \\ \vdots \\ \boldsymbol{z}_\gamma \\ \boldsymbol{\Gamma}^{-1}(\boldsymbol{z}_1,\ldots,\boldsymbol{z}_\gamma,\boldsymbol{y}) \end{bmatrix}, \qquad \boldsymbol{z} = \begin{bmatrix} \boldsymbol{y}_\mathrm{f} \\ \dot{\boldsymbol{y}}_\mathrm{f} \\ \ddot{\boldsymbol{y}}_\mathrm{f} \\ \vdots \\ \boldsymbol{y}_\mathrm{f}^{(\gamma-1)} \end{bmatrix}.$$

By specifying the real output signal $\boldsymbol{y}(t)$, we can then determine the progression of the fictitious flat output $\boldsymbol{y}_\mathrm{f}(t)$. Subsequently inserting $\boldsymbol{y}_\mathrm{f}$ into the equation

$$\boldsymbol{u} = \boldsymbol{\Psi}_2\left(\boldsymbol{y}_\mathrm{f},\dot{\boldsymbol{y}}_\mathrm{f},\ldots,\boldsymbol{y}_\mathrm{f}^{(\beta+1)}\right) \tag{5.132}$$

yields the desired feedforward controller. If the differential equation (5.129) is unstable, the calculation of the control variable $\boldsymbol{u}$ using equation (5.132) can lead to problems. Very high values $\boldsymbol{y}_\mathrm{f}(t)$ and associated numerical difficulties can occur for the time interval $[0,T]$ in question. Furthermore, very high and possibly unrealizable control variable values $\boldsymbol{u}$ can result as well.

If equation (5.129) is not explicitly solvable for $\boldsymbol{y}_\mathrm{f}^{(\gamma)}$, the calculation of $\boldsymbol{y}_\mathrm{f}(t)$ for a given signal $\boldsymbol{y}(t)$ is more complex [131, 423]. However, the solution of the differential equation (5.129) and thus the calculation of the progression of the flat output $\boldsymbol{y}_\mathrm{f}(t)$ from the specification of the real output $\boldsymbol{y}$ need only be done once offline.

The flatness-based feedforward control has the structure shown in Figure 5.23. Here the feedforward controller is used to pilot the output signal $\boldsymbol{y}$ so that it follows a reference output signal $\boldsymbol{y}_\mathrm{f,ref}$ or $\boldsymbol{y}_\mathrm{ref}$. The feedback control is

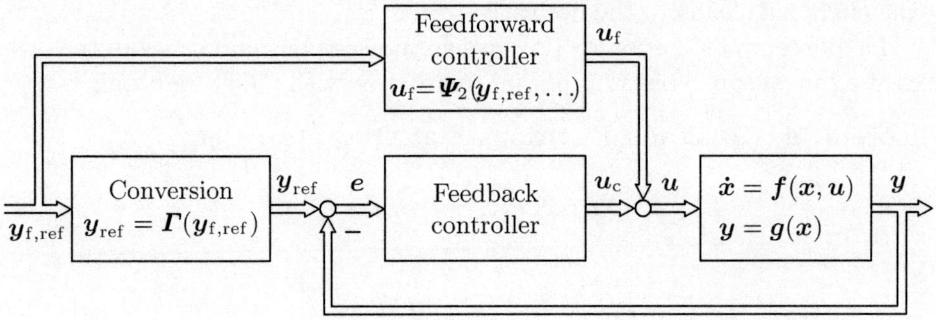

Fig. 5.23: Flatness-based feedforward and conventional feedback controller of a system with a fictitious flat output $\boldsymbol{y}_\mathrm{f}$

used to eliminate residual minor deviations from the reference output signal resulting from causes such as model inaccuracies or external disturbances.

If we do not succeed in solving the implicit differential equation (5.129), or if it is unstable and presents us with problems, we can proceed as follows. Instead of a real signal $\boldsymbol{y}(t)$, we can plan for a progression of the fictitious flat output $\boldsymbol{y}_{\mathrm{f}}(t)$, namely $\boldsymbol{y}_{\mathrm{f,ref}}(t)$. This is inserted into the implicit differential equation (5.129), and we then calculate the resulting reference course of the real output $\boldsymbol{y}_{\mathrm{ref}}(t)$. If it meets our requirements and possible constraints

$$y_{i,\min} \le y_i \le y_{i,\max}, \qquad i = 1, \ldots, r = \dim(\boldsymbol{y}), \tag{5.133}$$

we have found a feasible solution to the control problem. If it does not meet them, we must vary the reference progression of the fictitious output $\boldsymbol{y}_{\mathrm{f,ref}}(t)$ until $\boldsymbol{y}_{\mathrm{ref}}(t)$ is in the desired form and fulfills the constraints (5.133).

When selecting a fictitious output signal $\boldsymbol{y}_{\mathrm{f,ref}}(t)$, the initial value $\boldsymbol{y}_{\mathrm{f,ref}}(0)$ and the final value $\boldsymbol{y}_{\mathrm{f,ref}}(T)$ must correspond to the specified initial value $\boldsymbol{y}_{\mathrm{ref}}(0)$ and the specified final value $\boldsymbol{y}_{\mathrm{ref}}(T)$, i. e.

$$\boldsymbol{y}_{\mathrm{ref}}(0) = \boldsymbol{\Gamma}(\boldsymbol{y}_{\mathrm{f,ref}}(0), \mathbf{0}, \ldots, \mathbf{0}), \tag{5.134}$$

$$\boldsymbol{y}_{\mathrm{ref}}(T) = \boldsymbol{\Gamma}(\boldsymbol{y}_{\mathrm{f,ref}}(T), \mathbf{0}, \ldots, \mathbf{0}) \tag{5.135}$$

must apply. The elements of the derivatives $\boldsymbol{y}_{\mathrm{f,ref}}^{(j)}(0)$ and $\boldsymbol{y}_{\mathrm{f,ref}}^{(j)}(T)$ are set to zero for $j = 1, \ldots, \gamma$. For the calculation of $\boldsymbol{y}_{\mathrm{f,ref}}(0)$ and $\boldsymbol{y}_{\mathrm{f,ref}}(T)$, the implicit equations (5.134) and (5.135) may have to be solved numerically if necessary.

Note that the reference signal $\boldsymbol{y}_{\mathrm{ref}}(t)$ which is to be specified in equation (5.129) must be differentiable at least $\beta - \gamma + 1$ times. This condition results from the feedforward control law (5.132), in which the derivative $\boldsymbol{y}_{\mathrm{f}}^{(\beta+1)}$ is required. Utilizing equation (5.131), we can represent it as

$$\boldsymbol{y}_{\mathrm{f}}^{(\beta+1)} = \frac{\partial^{\beta-\gamma+1} \boldsymbol{y}_{\mathrm{f}}^{(\gamma)}}{\partial t^{\beta-\gamma+1}} = \frac{\partial^{\beta-\gamma+1} \boldsymbol{\Gamma}^{-1}(\boldsymbol{y}_{\mathrm{f}}, \dot{\boldsymbol{y}}_{\mathrm{f}}, \ldots, \boldsymbol{y}_{\mathrm{f}}^{(\gamma-1)}, \boldsymbol{y})}{\partial t^{\beta-\gamma+1}}.$$

This shows that the $(\beta - \gamma + 1)$-fold differentiation resulting from application of the chain rule leads to the derivative $\boldsymbol{y}^{(\beta-\gamma+1)}$.

The differential equation (5.129) has an interesting significance in the context of input-output linearization: in accordance with [169], we obtain

Theorem 71 (Real and Fictitious Flat Outputs). *Let*

$$\dot{\boldsymbol{x}} = \boldsymbol{f}(\boldsymbol{x}, u),$$
$$y = g(\boldsymbol{x})$$

with $\boldsymbol{x} \in D_{\boldsymbol{x},\mathrm{def}} \subseteq I\!R^n$ *and* $u \in D_{u,\mathrm{def}} \subseteq I\!R$ *be a system with the relative degree* δ*, and let*

$$y_{\mathrm{f}} = h(\boldsymbol{x})$$

be a fictitious flat output. In this case the differential equation

$$y = g\left(\boldsymbol{\Psi}_1\left(y_{\mathrm{f}}, \dot{y}_{\mathrm{f}}, \ldots, y_{\mathrm{f}}^{(n-1)}\right)\right) = \Gamma\left(y_{\mathrm{f}}, \dot{y}_{\mathrm{f}}, \ldots, y_{\mathrm{f}}^{(n-\delta)}\right)$$

represents the internal dynamics of the system.

Note that in the theorem above, the internal dynamics are not represented by a system of differential equations of order $n - \delta$; they are represented by a scalar differential equation. Here, the system's output variable y formally acts as an input variable. The order of the differential equation is $n - \delta$, not $\beta = n-1$. If we set $y = 0$, the differential equation describes the zero dynamics of the system.

The time-consuming computations for cases in which the real output is not flat and must be converted into a fictitious flat output make it clear how advantageous real flat outputs are.

5.4.3 Flatness-Based Feedforward Control of Linear Systems

Linear systems which are controllable are flat and vice versa. This has already been explained in Section 3.2.4, Theorem 49 on p. 251. In the following, we will assume that a controllable linear system is present in controller canonical form

$$\dot{\boldsymbol{x}} = \begin{bmatrix} 0 & 1 & 0 & \cdots & 0 \\ 0 & 0 & 1 & \cdots & 0 \\ \vdots & \vdots & \vdots & \ddots & \vdots \\ 0 & 0 & 0 & \cdots & 1 \\ -a_0 & -a_1 & -a_2 & \cdots & -a_{n-1} \end{bmatrix} \boldsymbol{x} + \begin{bmatrix} 0 \\ 0 \\ \vdots \\ 0 \\ 1 \end{bmatrix} u, \tag{5.136}$$

$$y = \begin{bmatrix} b_0 \; b_1 \; \cdots \; b_m \; 0 \cdots 0 \end{bmatrix} \boldsymbol{x}$$

or has been transformed into it. As we know from Section 3.2.5, p. 251 et seq., the real output y is only flat if

$$y = b_0 x_1$$

holds. All flat outputs y_{f} of the system can be stated as

$$y_{\mathrm{f}} = \alpha x_1$$

with $\alpha \in \mathbb{R}\backslash\{0\}$. Usually, $y_{\mathrm{f}} = x_1$ is used as a flat output.

It is possible to use the controller canonical form (5.136) to directly derive the flat system representation. The relation

$$\begin{aligned}
y_\mathrm{f} &= x_1, \\
\dot{y}_\mathrm{f} &= \dot{x}_1 = x_2, \\
\ddot{y}_\mathrm{f} &= \dot{x}_2 = x_3, \\
&\vdots \\
y_\mathrm{f}^{(n-1)} &= \dot{x}_{n-1} = x_n
\end{aligned}$$

applies; thus, for the first equation of the flat system representation, we obtain

$$\boldsymbol{x} = \boldsymbol{\Psi}_1\left(y_\mathrm{f}, \dot{y}_\mathrm{f}, \ldots, y_\mathrm{f}^{(n-1)}\right) = \begin{bmatrix} y_\mathrm{f} \\ \dot{y}_\mathrm{f} \\ \vdots \\ y_\mathrm{f}^{(n-1)} \end{bmatrix}.$$

From the controller canonical form (5.136), the second flat system equation also directly follows as

$$u = a_0 x_1 + a_1 x_2 + \ldots + a_{n-1} x_n + \dot{x}_n,$$

i. e.

$$u = a_0 y_\mathrm{f} + a_1 \dot{y}_\mathrm{f} + \ldots + a_{n-1} y_\mathrm{f}^{(n-1)} + y_\mathrm{f}^{(n)}. \tag{5.137}$$

The differential equation describing the relation between the real and fictitious output can also be simply determined as

$$\begin{aligned}
y &= b_0 x_1 + b_1 x_2 + \ldots + b_m x_{m+1} \\
&= b_0 y_\mathrm{f} + b_1 \dot{y}_\mathrm{f} + \ldots + b_m y_\mathrm{f}^{(m)}.
\end{aligned} \tag{5.138}$$

Therefore, $\gamma = m$ holds. Equation (5.138) is an example of the statement from Theorem 71 in the previous section, namely that equation (5.138) describes the internal dynamics of the system. In the case of linear systems, the internal dynamics are determined by the coefficients b_i and thus by the zeros of the numerator polynomial of the corresponding transfer function. We already noted this connection in Section 5.2.8 on p. 389 et seq.

For the design of a flat feedforward control for the system (5.136), as in the case of nonlinear systems, different scenarios exist. The simplest scenario is the one in which

$$y = b_0 y_\mathrm{f} = b_0 x_1$$

applies. The appropriate feedforward control law is given by equation (5.137).

Where

$$y \neq b_0 y_{\mathrm{f}}$$

holds, it is important for the transfer function of system (5.136) to have only zeros with a negative real part, because in this case the differential equation (5.138) has a stable solution. The fictitious flat output y_{f} required by the feedforward control law is determined from the differential equation (5.138) by specifying the real output signal $y(t)$ as the reference signal $y_{\mathrm{ref}}(t)$ and by solving the differential equation (5.138).

For the critical case in which $y \neq y_{\mathrm{f}}$ holds and the system (5.136) possesses zeros with a nonnegative real part, the differential equation (5.138) has unstable solutions. In this case, its solution y_{f} can only be used for a given reference signal $y_{\mathrm{ref}}(t)$ in the feedforward control law (5.137) if the signal $y_{\mathrm{f}}(t)$ does not assume values which are excessively high.

If the latter occurs, as in the nonlinear case, it is a valid procedure to plan a fictitious reference output signal $y_{\mathrm{f,ref}}(t)$. In this case, the initial value $y_{\mathrm{f,ref}}(0)$ and the final value $y_{\mathrm{f,ref}}(T)$ at time T are selected so that these two values at least correspond to the reference values $y_{\mathrm{ref}}(0)$ and $y_{\mathrm{ref}}(T)$. Suppose that, both at time $t = 0$ and $t = T$, all derivatives of $y_{\mathrm{f,ref}}$ and y_{ref} are identical to zero. This is typical of a change from one operating point to another. From equation (5.138) it then follows that

$$y_{\mathrm{f,ref}}(0) = \frac{1}{b_0} y_{\mathrm{ref}}(0)$$

and

$$y_{\mathrm{f,ref}}(T) = \frac{1}{b_0} y_{\mathrm{ref}}(T)$$

as well as from equation (5.137)

$$u(0) = a_0 y_{\mathrm{f,ref}}(0) = \frac{a_0}{b_0} y_{\mathrm{ref}}(0),$$
$$u(T) = a_0 y_{\mathrm{f,ref}}(T) = \frac{a_0}{b_0} y_{\mathrm{ref}}(T)$$

for the required control variables at the operating points.

5.4.4 Example: Propulsion-Based Aircraft Control

In rare but dramatic cases, the control hydraulics of an aircraft may fail. As a consequence, the elevator, rudder, and ailerons can no longer be operated by the pilots. In this case the rudders are freely mobile, free of torque, and therefore in an approximately neutral position. A similar situation occurs when the elevator, rudder, ailerons, and their trim tabs become stuck or damaged.

Examples of such emergencies are the flight of a Lockheed L-1011 TriStar passenger aircraft [317] and an Airbus A300 cargo aircraft [194] whose hydraulic systems and flight control failed due to the hit of a surface-to-air

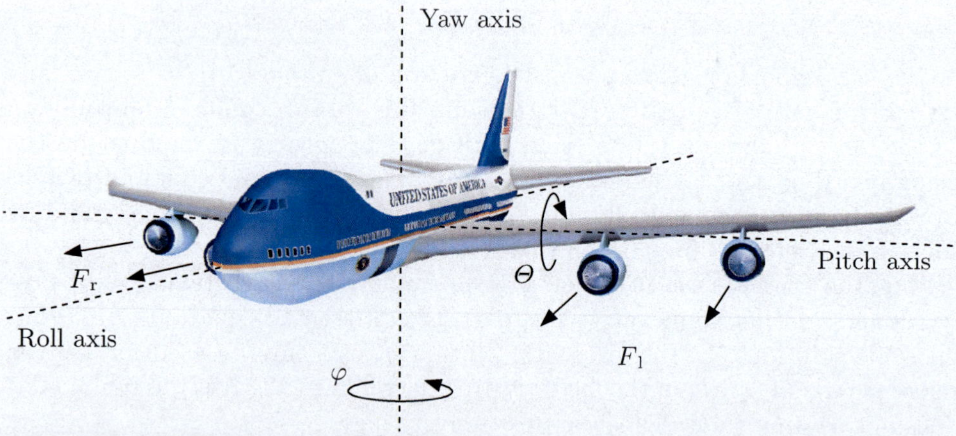

Fig. 5.24: An equal variation of the engine thrusts F_r and F_l causes a torque around the pitch axis and, hence, a change of the pitch angle Θ. The thrust of the inner engines of a Boeing 747-200 will act on the aircraft with a lever arm of 4.45 m and the outer engines act with just 1.63 m below the pitch axis. Similarly, different variations of the engine thrusts F_r and F_l cause a change in the yaw angle φ. The center of gravity S of the aircraft lies at the intersection of its three axes.

missile. Both flights became emergencies and it was only possible to land the aircraft due to the pilots' considerable flight experience. To prevent the impending crash, the pilots used the propulsion engines as the only remaining control variables. A change in engine thrust can influence the direction of flight, especially the angle of ascent ζ, i. e. the altitude of the aircraft. Figure 5.24 and Figure 5.25 illustrate how the angle of ascent can be influenced, based on the fact that the engines are usually below the center of gravity of the aircraft S and a torque around S can thus be generated. Furthermore, the yaw angle φ can be influenced by different engine thrusts F_r and F_l on the starboard and port sides, making a curved flight possible.

There have been a number of similar occurrences [62], not all of which have ended as well as the ones mentioned above. A McDonnell Douglas DC-10-10 was destroyed in an emergency landing and 111 people died [331]; a Boeing 747-100SR crashed into a mountain as a result of hydraulic failure and 520 people were killed [8]. As a response to these accidents and the fact that an aircraft with several engines can be controlled by means of the engines' thrusts, various controllers have been developed which support pilots in controlling the aircraft in emergencies such as this [14, 64, 176, 210, 211, 340].

As a concrete example, we will examine the failure of the control hydraulics in a Boeing 747-200 and design a flatness-based feedforward control of the angle of ascent using the four engines. We will limit ourselves to the longitudinal movement of the airplane shown in Figure 5.25, where we are interested in the velocity deviation x_1 from the intended flight velocity v_0, the descent velocity $x_2 = w$, the pitch rate $x_3 = \dot{\Theta}$, and the pitch angle $x_4 = \Theta$. We can influence

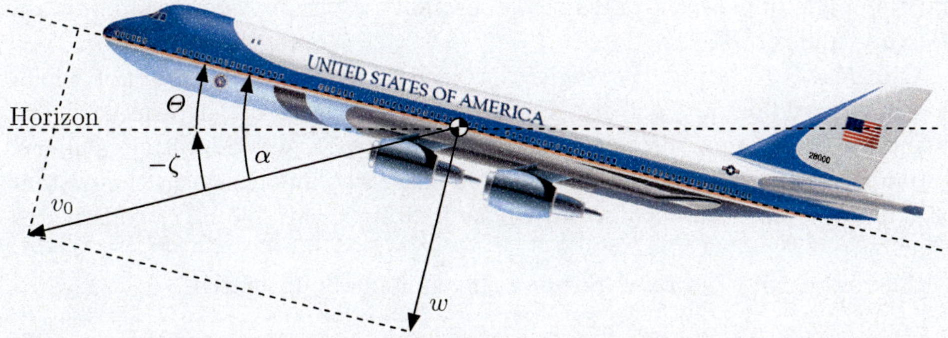

Fig. 5.25: Aircraft at its velocity v_0, its descent velocity w in the direction of the aircraft's yaw axis, its pitch angle $\boldsymbol{\Theta}$, and its angle of attack $\boldsymbol{\alpha}$. The angle of ascent is $\zeta = \Theta - \alpha$ which, in this case, is negative as the plane sinks.

these variables by changing the thrust x_5 of all four engines combined, which is regulated by a secondary control. Its reference value $x_{5,\mathrm{ref}}$ is used as control variable u. As a linearized model of longitudinal motion [340] we will use

$$\dot{\boldsymbol{x}} = \begin{bmatrix} -1.08\cdot 10^{-2} & 0.106 & -8.42 & -9.77 & 3.91\cdot 10^{-6} \\ -0.155 & -0.634 & 85.3 & -1.01 & -1.77\cdot 10^{-7} \\ 7.16\cdot 10^{-4} & -5.83\cdot 10^{-3} & -0.504 & 7.95\cdot 10^{-4} & 7\cdot 10^{-8} \\ 0 & 0 & 1 & 0 & 0 \\ 0 & 0 & 0 & 0 & -0.667 \end{bmatrix} \boldsymbol{x} + \begin{bmatrix} 0 \\ 0 \\ 0 \\ 0 \\ 0.667 \end{bmatrix} u, \tag{5.139}$$

which is valid for an altitude range of 0 m to 3000 m and a speed range of $v = 70\ldots 97\,\mathrm{m\,s^{-1}}$. Here we are assuming an operating point of $v_0 = 84.7\,\mathrm{m\,s^{-1}}$. The output value y is the angle of ascent ζ, i.e. the difference between the pitch angle Θ and the angle of attack α:

$$y = \Theta - \alpha \approx \Theta - \sin(\alpha) = \Theta - \frac{w}{v_0} = \begin{bmatrix} 0 & -v_0^{-1} & 0 & 1 & 0 \end{bmatrix} \boldsymbol{x}. \tag{5.140}$$

The eigenvalues of the system are

$$\begin{aligned} \lambda_1 &= -0.6670, \\ \lambda_{2/3} &= -0.5718 \pm j0.7084, \\ \lambda_{4/5} &= -0.0026 \pm j0.1265. \end{aligned}$$

The eigenvalue λ_1 reflects the dynamics of the underlying thrust control. The complex conjugate eigenvalues $\lambda_{2/3}$ cause damped fast oscillations of the angle of attack α. On the other hand, $\lambda_{4/5}$ cause slow, very weakly damped oscillations, so-called *phugoids*, where the angle of attack remains nearly constant and the angle of ascent ζ and speed v oscillate. Both types of oscillation are of course undesirable.

Our goal is to adjust the angle of ascent by means of an autopilot which uses a flat feedforward control. For this purpose we must determine a flat output of the system. From Theorem 49 on p. 251, we know that a linear system is flat if it is controllable. In this case, we can transform the system into the controller canonical form, whose first state variable $x_{\text{ccf},1}$ constitutes a flat output.

Therefore, we will first determine the controllability matrix

$$\boldsymbol{M}_{\text{contr}} = \begin{bmatrix} 0 & 2.6080 \cdot 10^{-6} & -2.1733 \cdot 10^{-6} & 1.5620 \cdot 10^{-6} & -1.1786 \cdot 10^{-6} \\ 0 & -1.1806 \cdot 10^{-7} & 3.7320 \cdot 10^{-6} & -6.5746 \cdot 10^{-6} & 6.0408 \cdot 10^{-6} \\ 0 & 4.6690 \cdot 10^{-8} & -5.2118 \cdot 10^{-8} & 2.3763 \cdot 10^{-8} & 1.3576 \cdot 10^{-8} \\ 0 & 0 & 4.6690 \cdot 10^{-8} & -5.2118 \cdot 10^{-8} & 2.3763 \cdot 10^{-8} \\ 0.667 & -0.44489 & 0.29674 & -0.19793 & 0.13202 \end{bmatrix}$$

whose determinant

$$\det(\boldsymbol{M}_{\text{contr}}) = -1.8623 \cdot 10^{-27}$$

is very low. This means that the aircraft is very difficult to steer using the engines alone, which is hardly surprising.

With

$$\boldsymbol{x}_{\text{ccf}} = \boldsymbol{T}^{-1}\boldsymbol{x}, \qquad \boldsymbol{T}^{-1} = \begin{bmatrix} \begin{bmatrix} 0 \cdots 0 \; 1 \end{bmatrix} \boldsymbol{M}_{\text{contr}}^{-1} \\ \begin{bmatrix} 0 \cdots 0 \; 1 \end{bmatrix} \boldsymbol{M}_{\text{contr}}^{-1}\boldsymbol{A} \\ \vdots \qquad \vdots \\ \begin{bmatrix} 0 \cdots 0 \; 1 \end{bmatrix} \boldsymbol{M}_{\text{contr}}^{-1}\boldsymbol{A}^{n-1} \end{bmatrix},$$

we can transform system (5.139), (5.140) into the controller canonical form

$$\begin{aligned} \dot{\boldsymbol{x}}_{\text{ccf}} &= \begin{bmatrix} 0 & 1 & 0 & 0 & 0 \\ 0 & 0 & 1 & 0 & 0 \\ 0 & 0 & 0 & 1 & 0 \\ 0 & 0 & 0 & 0 & 1 \\ -0.008843 & -0.02837 & -0.5901 & -1.617 & -1.816 \end{bmatrix} \boldsymbol{x}_{\text{ccf}} + \begin{bmatrix} 0 \\ 0 \\ 0 \\ 0 \\ 1 \end{bmatrix} u, \\ y &= 10^{-9} \begin{bmatrix} 3.825 & 32.53 & 5.159 & 1.394 & 0 \end{bmatrix} \boldsymbol{x}_{\text{ccf}}. \end{aligned} \tag{5.141}$$

We know that

$$y_{\mathrm{f}} = x_{\mathrm{ccf},1}$$

is a flat output, while y is not.

Next we will use the relation between y and y_{f} in the case of linear systems in controller canonical form, which was generally established in equation (5.138), to obtain the relation

$$\begin{aligned} y &= b_0 y_{\mathrm{f}} + b_1 \dot{y}_{\mathrm{f}} + \ldots + b_m \overset{(m)}{y_{\mathrm{f}}} \\ &= 10^{-9}(3.825 y_{\mathrm{f}} + 32.53 \dot{y}_{\mathrm{f}} + 5.159 \ddot{y}_{\mathrm{f}} + 1.394 \dddot{y}_{\mathrm{f}}), \end{aligned} \tag{5.142}$$

where $\gamma = m = 3$. This also follows directly from the identities

$$y_{\mathrm{f}} = x_{\mathrm{ccf},1}, \quad \dot{y}_{\mathrm{f}} = x_{\mathrm{ccf},2}, \quad \cdots, \quad \overset{(m)}{y_{\mathrm{f}}} = x_{\mathrm{ccf},m+1},$$

resulting from the first $m+1$ differential equations of the controller canonical form (5.141) after insertion into

$$y = \begin{bmatrix} b_0 \; b_1 \; \cdots \; b_m \; 0 \cdots 0 \end{bmatrix} \boldsymbol{x}_{\mathrm{ccf}}.$$

Since the differential equation (5.142) has the eigenvalues

$$\lambda_1 = -0.1198 \quad \text{and} \quad \lambda_{2/3} = -1.7905 \pm j4.4386,$$

i. e. it is stable, y_{f} can be utilized for the flat control without the described problems. With equation (5.137) and by calculating $y_{\mathrm{f}}(t)$ using equation (5.142) for a given signal $y(t)$, we obtain the control variable

$$\begin{aligned} u &= a_0 y_{\mathrm{f}} + a_1 \dot{y}_{\mathrm{f}} + a_2 \ddot{y}_{\mathrm{f}} + a_3 \dddot{y}_{\mathrm{f}} + a_4 \overset{(4)}{y_{\mathrm{f}}} + \overset{(5)}{y_{\mathrm{f}}} \\ &= 0.008843 y_{\mathrm{f}} + 0.02837 \dot{y}_{\mathrm{f}} + 0.5901 \ddot{y}_{\mathrm{f}} + 1.617 \, \dddot{y}_{\mathrm{f}} + 1.816 y_{\mathrm{f}}^{(4)} + y_{\mathrm{f}}^{(5)}. \end{aligned} \tag{5.143}$$

In this way, we can generate a control variable signal $u(t)$ for the feedforward controller by inserting

$$y_{\mathrm{f}}(t) = y_{\mathrm{f,ref}}(t)$$

into equation (5.143).

In the following, the angle of ascent should increase continuously from the value $\zeta = y = 0°$ to $-1°$, so that the aircraft sinks constantly in the end with $\zeta = -1° = -0.01745\,\mathrm{rad}$. For this purpose, we will choose

$$y_{\mathrm{ref}}(t) = \begin{cases} -0.008725\left(1 - \cos\left(\dfrac{\pi}{300} t\right)\right), & 0 \le t \le 300\,\mathrm{s}, \\ -0.01745, & t > 300\,\mathrm{s}, \end{cases}$$

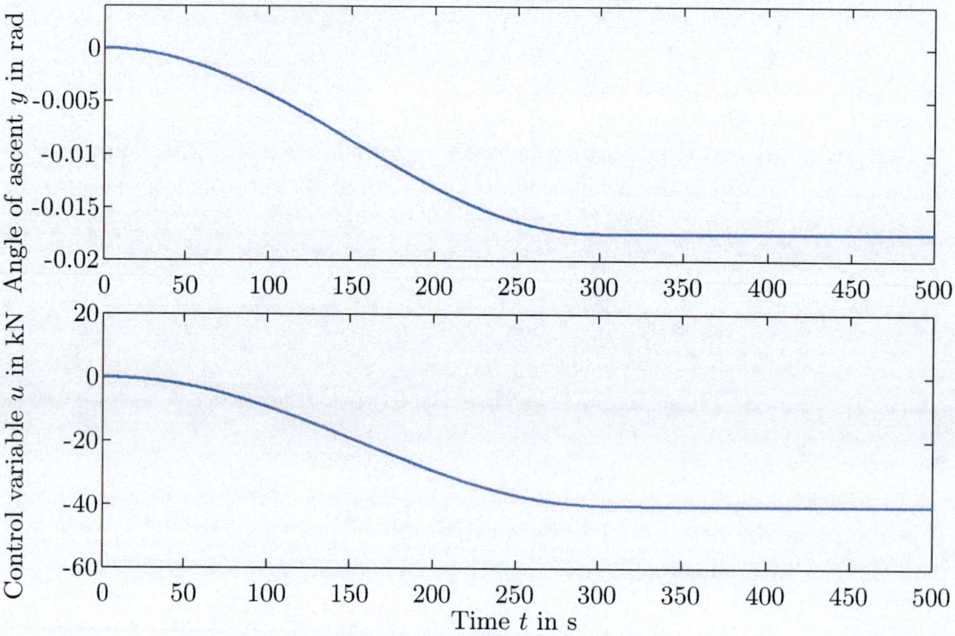

Fig. 5.26: Time courses of the angle of ascent y and the engine thrust u of the Boeing 747-200 for a transition of the angle of ascent from $y = 0$ to $y = -1° = -0.01745\,\mathrm{rad}$

as the reference signal of the angle of ascent $y = \zeta$. From equation (5.142) we can also calculate the flat output y_f using the initial values

$$y_\mathrm{f}(0) = 0,\ \dot{y}_\mathrm{f}(0) = 0,\ \ddot{y}_\mathrm{f}(0) = 0 \quad \text{and} \quad \dddot{y}_\mathrm{f}(0) = 0.$$

Figure 5.26 shows only the real output, i. e. the angle of ascent, since the flat output in this case has almost the same progression. Note that the reference signal $y_\mathrm{ref}(t)$, as required in Section 5.4.2, can be differentiated twice, meaning $\beta - \gamma + 1$ times with $\beta = 4$ and $\gamma = 3$. Its second derivative is not continuous, since this is not necessary here.

5.4.5 Flatness-Based Feedback Control of Nonlinear Systems

Theorem 69 on p. 430 states that a full-state linearizable control-affine SISO system is flat and vice versa, i. e. a flat SISO system is also full-state linearizable. We can extend this statement to general nonlinear SISO systems

$$\dot{\boldsymbol{x}} = \boldsymbol{f}(\boldsymbol{x}, u) \tag{5.144}$$

with $\boldsymbol{x} \in D_{\boldsymbol{x},\mathrm{def}} \subseteq I\!R^n$. First of all, we will assume that the system (5.144) can be full-state linearized and that

$$y = h(\boldsymbol{x}) \tag{5.145}$$

is a linearizing output. This means that the system (5.144), (5.145) has the relative degree $\delta = n$. For the linearization, we will again use the Lie derivative

$$L_{\boldsymbol{f}} h(\boldsymbol{x}) = \frac{\partial h(\boldsymbol{x})}{\partial \boldsymbol{x}} \boldsymbol{f}(\boldsymbol{x}, u)$$

and compute

$$\begin{aligned} y &= h(\boldsymbol{x}), \\ \dot{y} &= L_{\boldsymbol{f}} h(\boldsymbol{x}), \\ \ddot{y} &= L_{\boldsymbol{f}}^2 h(\boldsymbol{x}), \\ &\vdots \\ y^{(n-1)} &= L_{\boldsymbol{f}}^{n-1} h(\boldsymbol{x}), \\ y^{(n)} &= L_{\boldsymbol{f}}^{n} h(\boldsymbol{x}) = \varphi(\boldsymbol{x}, u), \end{aligned} \tag{5.146}$$

where all derivatives

$$y^{(i)} = L_{\boldsymbol{f}}^{i} h(\boldsymbol{x}), \qquad i = 1, \ldots, n-1, \tag{5.147}$$

of the linearizable system (5.144), (5.145) are independent of the control variable u. Only

$$y^{(n)} = L_{\boldsymbol{f}}^{n} h(\boldsymbol{x}) = \varphi(\boldsymbol{x}, u) \tag{5.148}$$

depends on u. In this way, using the new state variables

$$\boldsymbol{z} = \begin{bmatrix} z_1 \\ z_2 \\ \vdots \\ z_n \end{bmatrix} = \begin{bmatrix} y \\ \dot{y} \\ \vdots \\ y^{(n-1)} \end{bmatrix} = \begin{bmatrix} h(\boldsymbol{x}) \\ L_{\boldsymbol{f}} h(\boldsymbol{x}) \\ \vdots \\ L_{\boldsymbol{f}}^{n-1} h(\boldsymbol{x}) \end{bmatrix} = \boldsymbol{t}(\boldsymbol{x}),$$

we obtain the so-called *generalized controller canonical form* [123]

$$\dot{\boldsymbol{z}} = \begin{bmatrix} z_2 \\ \vdots \\ z_n \\ \varphi(\boldsymbol{x}, u) = \varphi(\boldsymbol{t}^{-1}(\boldsymbol{z}), u) \end{bmatrix}, \tag{5.149}$$

which is a generalization of the nonlinear controller canonical form. Note that if we introduce a new control variable

$$v = \varphi(\boldsymbol{x}, u),$$

the linearized system representation in Brunovsky canonical form is obtained from equation (5.149).

Since we assumed that the system is full-state linearizable, the function φ can be inverted, and

$$u = \varphi^{-1}(\boldsymbol{x}, y^{(n)}) = \varphi^{-1}\left(\boldsymbol{t}^{-1}(\boldsymbol{z}), y^{(n)}\right) = \Psi_2(y, \dot{y}, \ldots, y^{(n)}) \tag{5.150}$$

applies with $\boldsymbol{x} = \boldsymbol{t}^{-1}(\boldsymbol{z})$ and $v = y^{(n)}$. Furthermore, we know the expression

$$\boldsymbol{x} = \boldsymbol{t}^{-1}(\boldsymbol{z}) = \boldsymbol{\Psi}_1\left(y, \dot{y}, \ldots, y^{(n-1)}\right) \tag{5.151}$$

from equation (5.115) on p. 431. Equation (5.150) and equation (5.151) represent the flat system description of the system (5.144), (5.145). As a result, any full-state linearizable system (5.144) is flat.

Now we will take the opposite approach and show that any flat system (5.144), (5.145) can be state linearized. To this end, let us consider the first equation

$$\boldsymbol{x} = \boldsymbol{\Psi}_1\left(y, \dot{y}, \ldots, y^{(n-1)}\right).$$

This does not depend on the input signal u or one of its derivatives $u^{(i)}$, which means that all derivatives $y^{(i)}, i = 1, \ldots, n-1$, fulfill equation (5.147). Obviously, the identity

$$\boldsymbol{z} = \boldsymbol{\Psi}_1^{-1}(\boldsymbol{x}) = \boldsymbol{t}(\boldsymbol{x})$$

holds, where

$$\boldsymbol{z} = \begin{bmatrix} y \\ \dot{y} \\ \vdots \\ y^{(n-1)} \end{bmatrix}. \tag{5.152}$$

From the second equation of a flat system representation,

$$u = \Psi_2\left(y, \dot{y}, \ldots, y^{(n)}\right), \tag{5.153}$$

we can further conclude that $y^{(n)}$ is dependent on the input signal u, since the derivatives $y^{(i)}, i, \ldots, n-1$, in equation (5.146) are not. Thus, equation (5.153) can be transformed into an equation conforming to equation (5.148). As result, we obtain

$$\begin{aligned} y^{(n)} &= L_{\boldsymbol{f}}^n h(\boldsymbol{x}) = \varphi(\boldsymbol{x}, u) = \varphi(\boldsymbol{t}^{-1}(\boldsymbol{z}), u) \\ &= \hat{\varphi}(\boldsymbol{z}, u) = \tilde{\varphi}(y, \dot{y}, \ldots, y^{(n-1)}, u). \end{aligned} \tag{5.154}$$

With equation (5.152) and equation (5.154), we then obtain the generalized controller canonical form

$$\dot{\boldsymbol{z}} = \begin{bmatrix} z_2 \\ \vdots \\ z_n \\ \hat{\varphi}(\boldsymbol{z}, u) \end{bmatrix}, \qquad (5.155)$$
$$y = z_1$$

as a system representation equivalent to the flat system representation.

Because we assumed the system to be flat, equation (5.154) can be solved for u, yielding

$$u = \tilde{\varphi}^{-1}(y, \dot{y}, \ldots, y^{(n)}) = \Psi_2(y, \dot{y}, \ldots, y^{(n)}).$$

This also ensures that φ and $\tilde{\varphi}$ are invertible as well, from which the full-state linearizability of system (5.155) follows. Consequently, system (5.144) is full-state linearizable. So, as a generalization of Theorem 69, we can derive [420]

Theorem 72 (Flatness and Linearizability of SISO Systems). *A nonlinear system*

$$\dot{\boldsymbol{x}} = \boldsymbol{f}(\boldsymbol{x}, u), \quad \boldsymbol{x} \in D_{\boldsymbol{x},\text{def}} \subseteq I\!R^n, \quad u \in I\!R,$$

is flat if and only if it is full-state linearizable. If the system has the relative degree n with respect to an output y, then y is a flat output.

In addition, the transformation which follows from the above is important to the flatness-based controller design we aim to create. It can be stated as

Theorem 73 (Transformation into Brunovsky Canonical Form). *For a flat system*

$$\dot{\boldsymbol{x}} = \boldsymbol{f}(\boldsymbol{x}, u), \quad \boldsymbol{x} \in D_{\boldsymbol{x},\text{def}} \subseteq I\!R^n, \quad u \in I\!R,$$

the flat output y, and the flat system representation

$$\boldsymbol{x} = \boldsymbol{\Psi}_1(y, \dot{y}, \ldots, y^{(n-1)}),$$
$$u = \Psi_2(y, \dot{y}, \ldots, y^{(n)}),$$

the diffeomorphism which transforms the system into the generalized controller canonical form

$$\begin{bmatrix} \dot{z}_1 \\ \vdots \\ \dot{z}_{n-1} \\ \dot{z}_n \end{bmatrix} = \begin{bmatrix} z_2 \\ \vdots \\ z_n \\ \varphi(\boldsymbol{z}, u) \end{bmatrix}$$

is

$$z = \boldsymbol{\Psi}_1^{-1}(\boldsymbol{x})$$

with $\boldsymbol{z} = [y,\, \dot{y},\, \ldots,\, y^{(n-1)}]^T$. *Furthermore, the state-dependent mapping*

$$v = \varphi(\boldsymbol{z}, u) = \Psi_2^{-1}(y, \dot{y}, \ldots, y^{(n-1)}, u)$$

of the input variable u *transforms the system into Brunovsky canonical form.*

With the theorem above, we have also performed input-output linearization. This is because

$$\dot{z}_n = y^{(n)} = v \tag{5.156}$$

holds, so that the output variable y depends on the artificial input variable v via n integrators. From

$$u = \Psi_2(y, \dot{y}, \ldots, y^{(n)}) = \Psi_2(y, \dot{y}, \ldots, y^{(n-1)}, v),$$

we can obtain the real input or control variable u.

Theorem 73 now enables us to design a flatness-based feedback control system with respect to an output variable signal $y_{\text{ref}}(t)$ which we can select from a wide range. We can now insert the control law

$$v = y_{\text{ref}}^{(n)} - \sum_{i=0}^{n-1} a_i \left(y^{(i)} - y_{\text{ref}}^{(i)} \right) \tag{5.157}$$

into the state equation (5.156) of the Brunovsky canonical form. Thus, after inserting equation (5.157) into equation (5.156), the linear system dynamics

$$y^{(n)} - y_{\text{ref}}^{(n)} + a_{n-1} \left(y^{(n-1)} - y_{\text{ref}}^{(n-1)} \right) + \ldots + a_1 \left(\dot{y} - \dot{y}_{\text{ref}} \right) + a_0 \left(y - y_{\text{ref}} \right) = 0 \tag{5.158}$$

are obtained.

Provided that we choose all roots of the characteristic polynomial

$$P(s) = s^n + a_{n-1}s + \ldots + a_1 s + a_0$$

exclusively with negative real parts by selecting suitable coefficients a_i, the differential equation (5.158) is stable. Thus, we obtain

$$\lim_{t \to \infty} \left(y^{(i)}(t) - y_{\text{ref}}^{(i)}(t) \right) = 0, \quad i = 0, \ldots, n.$$

This means that the approximation

$$y^{(i)}(t) \approx y_{\text{ref}}^{(i)}(t), \quad i = 0, \ldots, n$$

applies such that the output variable $y(t)$ follows the reference variable $y_{\text{ref}}(t)$.

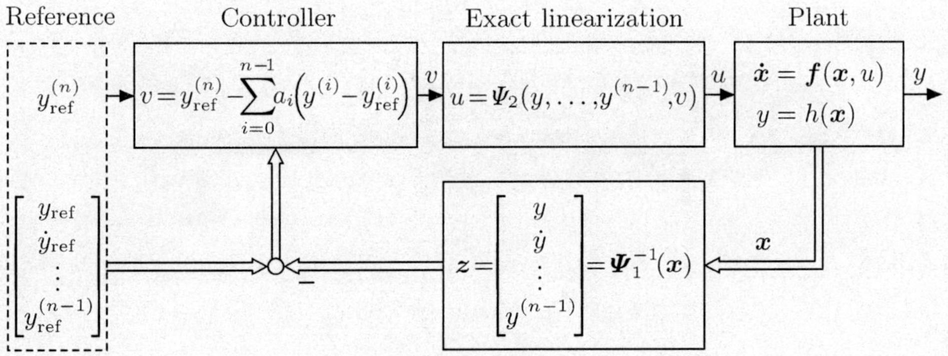

Fig. 5.27: Flatness-based feedback control system

The complete control law consists of the two equations

$$u = \Psi_2(y, \dot{y}, \ldots, y^{(n-1)}, v), \tag{5.159}$$

$$v = y_{\mathrm{ref}}^{(n)} - \sum_{i=0}^{n-1} a_i \left(y^{(i)} - y_{\mathrm{ref}}^{(i)} \right). \tag{5.160}$$

Figure 5.27 shows the resulting control-loop. The diffeomorphism $\boldsymbol{x} = \boldsymbol{\Psi}_1(\boldsymbol{z})$ and the transformation (5.159) linearizes the plant according to equation (5.156), and the control law (5.160) generates the linear dynamics (5.158). The coefficients a_i are freely selectable with the restriction that they must result in an asymptotically stable control loop. When specifying the time courses

$$y_{\mathrm{ref}}(t), \dot{y}_{\mathrm{ref}}(t), \ldots, y_{\mathrm{ref}}^{(n)}(t),$$

$y_{\mathrm{ref}}(t)$ must be differentiable at least n times.

As an alternative to the flatness-based feedforward control with superimposed feedback control, the flatness-based feedback control above is a second way to use the flatness of a system for the feedback control with respect to a given desired signal $y_{\mathrm{ref}}(t)$.

The flatness-based feedback control can also be implemented for MIMO plants. However, the design may be much more complicated in these cases. This is especially true if the controlled system is flat but not full-state linearizable. Information on this topic is provided in [1, 420, 485].

5.4.6 Example: Pneumatic Motor

We will examine a pneumatic motor used for the feed of a drill or for moving workpieces to grinding machines. Figure 5.28 shows the basic structure of such a motor. Air is blown through a hose or a pipe into a cylinder via a compressor. Due to the pressure P in the cylinder, a piston then moves distance s, and with it a load mass is shifted. This compresses a spring which brings the piston back to its initial position when the pressure in the cylinder drops. The spring

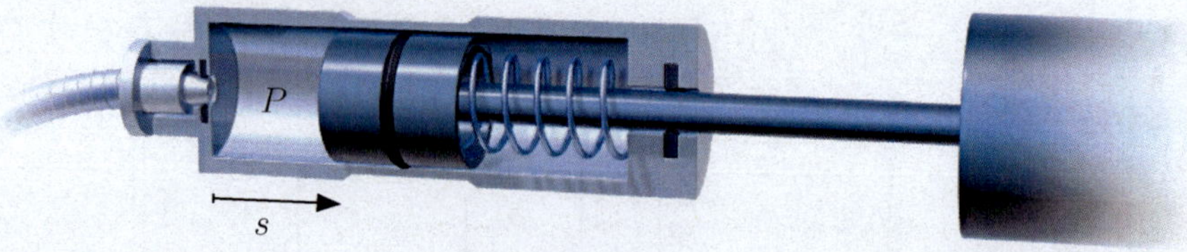

Fig. 5.28: Pneumatic motor

possesses the spring constant c, the piston the surface area A, and the moving parts, which are load, piston, and linkage, possess the mass m.

The equation of motion results from the spring force cs as well as from the force PA acting on the piston and can be stated as

$$m\ddot{s} = PA - cs. \tag{5.161}$$

The pressure P in turn can be determined by means of the ideal gas equation

$$\frac{1}{n}PV = RT \tag{5.162}$$

if we assume isothermal behavior. Here, $V = As$ is the cylinder volume, n is the number of gas molecules in moles, R is the ideal gas constant, and T is the temperature of the gas, which is assumed to be constant.

The next step is to differentiate the ideal gas equation (5.162) with respect to time, which yields

$$\frac{\dot{P}V}{n} + \frac{P\dot{V}}{n} - \frac{PV\dot{n}}{n^2} = 0$$

or

$$\dot{P} = \frac{P}{V}\left(\frac{V\dot{n}}{n} - \dot{V}\right). \tag{5.163}$$

With the air volume flow

$$q = \frac{V\dot{n}}{n},$$

which acts as the input variable u, and $V = As$, we obtain

$$\dot{P} = \frac{P}{s}\left(\frac{q}{A} - \dot{s}\right) \tag{5.164}$$

from equation (5.163). If we set

$$x_1 = s, \quad x_2 = \dot{s}, \quad x_3 = P, \quad \text{and} \quad u = q$$

as the state variables and the input variable, equations (5.161) and (5.164) provide the model of the pneumatic motor

$$\underbrace{\begin{bmatrix} \dot{x}_1 \\ \dot{x}_2 \\ \dot{x}_3 \end{bmatrix}}_{\dot{\boldsymbol{x}}} = \underbrace{\begin{bmatrix} x_2 \\ -\dfrac{c}{m}x_1 + \dfrac{A}{m}x_3 \\ -\dfrac{x_2 x_3}{x_1} \end{bmatrix}}_{\boldsymbol{a}(\boldsymbol{x})} + \underbrace{\begin{bmatrix} 0 \\ 0 \\ \dfrac{x_3}{Ax_1} \end{bmatrix}}_{\boldsymbol{b}(\boldsymbol{x})} u. \tag{5.165}$$

As the output variable of the system, we will select the load position

$$y = c(\boldsymbol{x}) = x_1 = s.$$

The system is full-state linearizable and $y = x_1$ is a linearizing output. This is easily verifiable, because the derivatives of y are

$$\begin{aligned} y &= x_1, \\ \dot{y} &= \dot{x}_1 = x_2, \\ \ddot{y} &= \dot{x}_2 = -\frac{c}{m}x_1 + \frac{A}{m}x_3, \\ \dddot{y} &= -\frac{c}{m}\dot{x}_1 + \frac{A}{m}\dot{x}_3 = -\frac{c}{m}x_2 - \frac{Ax_2x_3}{mx_1} + \frac{x_3}{mx_1}u. \end{aligned} \tag{5.166}$$

Thus, the relative degree $\delta = 3$ equals the system order and the system is full-state linearizable. As we recall from Theorem 69 and Theorem 72, this means that $y = x_1$ is a flat output.

Since y is a flat output, the flat system representation can also be quickly derived from equation (5.166). This is because

$$\begin{bmatrix} x_1 \\ x_2 \\ x_3 \end{bmatrix} = \begin{bmatrix} y \\ \dot{y} \\ \dfrac{m}{A}\left(\ddot{y} + \dfrac{c}{m}y\right) \end{bmatrix} = \boldsymbol{\Psi}_1(y, \dot{y}, \ddot{y})$$

and

$$u = A\frac{my\dddot{y} + m\dot{y}\ddot{y} + 2cy\dot{y}}{cy + m\ddot{y}} = \Psi_2(y, \dot{y}, \ddot{y}, \dddot{y}) \tag{5.167}$$

hold.

We can use equation (5.167) as a flat feedforward control law. An advantageous alternative to the feedforward control is a flatness-based feedback control, as described in the previous section and illustrated in Figure 5.27. For this purpose, we need the inverse function

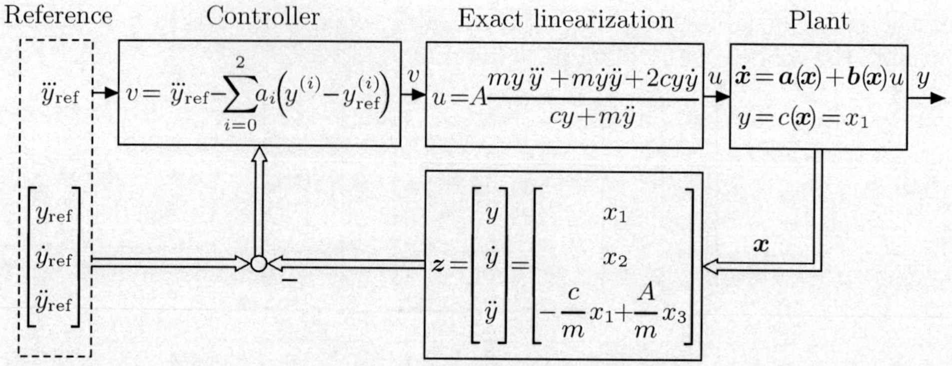

Fig. 5.29: Flatness-based feedback control of the pneumatic motor

$$\boldsymbol{z} = \begin{bmatrix} y \\ \dot{y} \\ \ddot{y} \end{bmatrix} = \boldsymbol{\Psi}_1^{-1}(\boldsymbol{x}) = \begin{bmatrix} x_1 \\ x_2 \\ -\dfrac{c}{m}x_1 + \dfrac{A}{m}x_3 \end{bmatrix}.$$

Recall that, according to Theorem 73, the mapping $\boldsymbol{\Psi}_1^{-1}$ is the diffeomorphism which transforms the system into the nonlinear controller canonical form. Furthermore, we need the controller

$$v = \dddot{y}_{\mathrm{ref}} - \sum_{i=0}^{2} a_i \left(y^{(i)} - {y_{\mathrm{ref}}}^{(i)}\right).$$

Figure 5.29 shows the structure of the flatness-based feedback controller for the pneumatic motor.

As a concrete numerical example, we can take

$$c = 20\,\mathrm{N\,cm}^{-1}, \quad A = 50\,\mathrm{cm}^2, \quad m = 1\,\mathrm{kg}.$$

The coefficients of the controller are selected as

$$a_0 = 1000, \quad a_1 = 300, \quad a_2 = 30,$$

so that the closed linear control loop has the threefold eigenvalue

$$\lambda = -10.$$

We will now move the pneumatic motor's piston from

$$y = x_1 = s = 1\,\mathrm{cm}$$

to

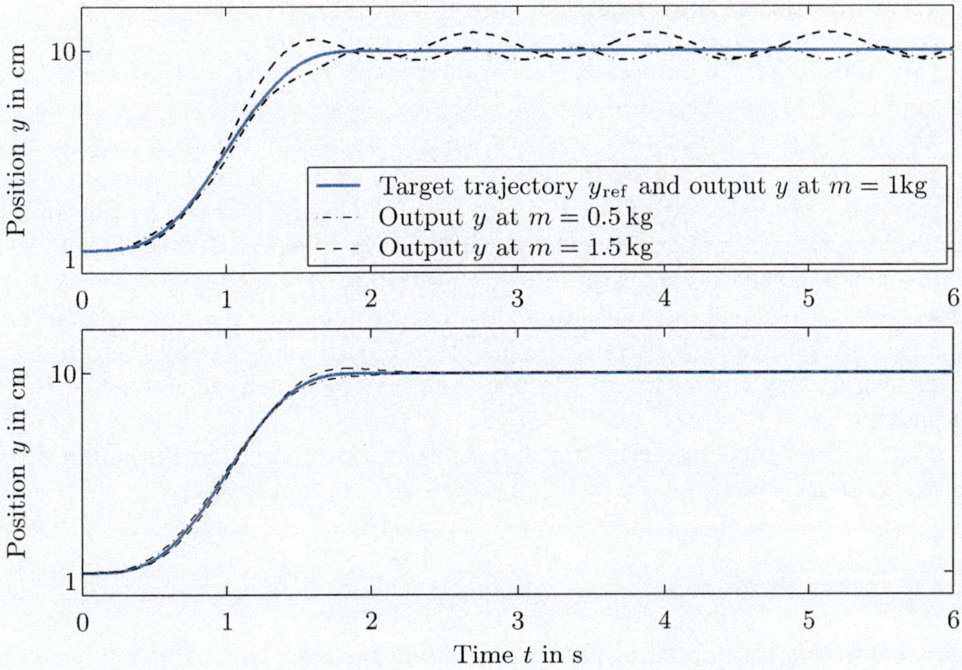

Fig. 5.30: Time courses of the position $y = x_1$ for the flat feedforward control in the upper diagram, and for the flatness-based feedback control in the lower diagram

$$y = 10\,\mathrm{cm}.$$

To do so, we specify

$$y_{\mathrm{ref}}(t) = \begin{cases} -1.406t^7 + 9.844t^6 - 23.630t^5 + 19.694t^4 + 1, & 0 \le t \le 2, \\ 10, & 2 < t, \end{cases}$$

with the reference values

$$\begin{aligned} y_{\mathrm{ref}}(0) &= 1\,\mathrm{cm}, & y_{\mathrm{ref}}(2) &= 10\,\mathrm{cm}, \\ \dot{y}_{\mathrm{ref}}(0) &= 0\,\mathrm{cm\,s^{-1}}, & \dot{y}_{\mathrm{ref}}(2) &= 0\,\mathrm{cm\,s^{-1}}, \\ \ddot{y}_{\mathrm{ref}}(0) &= 0\,\mathrm{cm\,s^{-2}}, & \ddot{y}_{\mathrm{ref}}(2) &= 0\,\mathrm{cm\,s^{-2}}, \\ \dddot{y}_{\mathrm{ref}}(0) &= 0\,\mathrm{cm\,s^{-3}}, & \dddot{y}_{\mathrm{ref}}(2) &= 0\,\mathrm{cm\,s^{-3}}. \end{aligned}$$

Figure 5.30 shows the time course of the load position $y = s$ of the piston for the nominal case $m = 1\,\mathrm{kg}$ as well as for the cases $m = 0.5\,\mathrm{kg}$ and $m = 1.5\,\mathrm{kg}$. In the upper diagram, the time course of the position $y(t)$ is shown for the flat feedforward control, and the lower diagram displays the course $y(t)$ for the flatness-based feedback control. As expected, unlike the feedforword control, the feedback control compensates for the changes in the load mass m.

5.4.7 Flat Inputs and Their Design

A prerequisite for the flatness-based design of a feedforward controller is a flat output. The ideal case is one in which the real output of the system is flat. If the real output is not flat, or if the system itself is not flat, it may make sense to design the input of the system in such a way that the measurable real output becomes flat. Thus the output is made flat by the creation of a suitable input. An input such as this is termed *flat input* [137, 138, 337, 466].

In general, the design engineer of the system to be controlled will have to select and implement the actuators required to integrate the flat input into the system. Since she or he must be informed of such an input by the control engineer, cooperation between the two is necessary at an early stage of the design.

We will proceed by designing flat inputs for the autonomous nonlinear SISO systems

$$\begin{aligned} \dot{\boldsymbol{x}} &= \boldsymbol{a}(\boldsymbol{x}), \\ y &= c(\boldsymbol{x}), \end{aligned} \tag{5.168}$$

i. e. vectors $\boldsymbol{b}(\boldsymbol{x})$, to obtain a flat control-affine system

$$\begin{aligned} \dot{\boldsymbol{x}} &= \boldsymbol{a}(\boldsymbol{x}) + \boldsymbol{b}(\boldsymbol{x}) \cdot u, \\ y &= c(\boldsymbol{x}) \end{aligned}$$

with y as a flat output. For this purpose, we will transform the system using

$$\boldsymbol{z} = \boldsymbol{q}(\boldsymbol{x}, u, \ldots, u^{(n-2)}) = \begin{bmatrix} y \\ \dot{y} \\ \ddot{y} \\ \vdots \\ y^{(n-1)} \end{bmatrix}$$

$$= \begin{bmatrix} c(\boldsymbol{x}) \\ L_{\boldsymbol{a}}c(\boldsymbol{x}) + L_{\boldsymbol{b}}c(\boldsymbol{x})u \\ L_{\boldsymbol{a}}^2 c(\boldsymbol{x}) + L_{\boldsymbol{b}}L_{\boldsymbol{a}}c(\boldsymbol{x})u + L_{\boldsymbol{a}}L_{\boldsymbol{b}}c(\boldsymbol{x})u + L_{\boldsymbol{b}}^2 c(\boldsymbol{x})u^2 + L_{\boldsymbol{b}}c(\boldsymbol{x})\dot{u} \\ \vdots \\ L_{\boldsymbol{a}}^{n-1}c(\boldsymbol{x}) + L_{\boldsymbol{b}}L_{\boldsymbol{a}}^{n-2}c(\boldsymbol{x})u + \ldots + L_{\boldsymbol{b}}^{n-1}c(\boldsymbol{x})u^{n-1} + L_{\boldsymbol{b}}c(\boldsymbol{x})u^{(n-2)} \end{bmatrix}. \tag{5.169}$$

For the time being, we will assume that the terms fulfill

$$L_{\boldsymbol{b}}L_{\boldsymbol{a}}^i c(\boldsymbol{x}) \neq 0, \quad i = 0, \ldots, n-2.$$

By differentiating the transformation equation above, we obtain

$$\begin{bmatrix} \dot{z}_1 \\ \dot{z}_2 \\ \vdots \\ \dot{z}_{n-1} \\ \dot{z}_n \end{bmatrix} = \begin{bmatrix} \dot{y} \\ \ddot{y} \\ \vdots \\ y^{(n-1)} \\ y^{(n)} \end{bmatrix} = \begin{bmatrix} L_{\boldsymbol{a}} c(\boldsymbol{x}) \\ L_{\boldsymbol{a}}^2 c(\boldsymbol{x}) \\ \vdots \\ L_{\boldsymbol{a}}^{n-1} c(\boldsymbol{x}) \\ L_{\boldsymbol{a}}^{n} c(\boldsymbol{x}) \end{bmatrix} + \begin{bmatrix} L_{\boldsymbol{b}} c(\boldsymbol{x}) \\ L_{\boldsymbol{b}} L_{\boldsymbol{a}} c(\boldsymbol{x}) \\ \vdots \\ L_{\boldsymbol{b}} L_{\boldsymbol{a}}^{n-2} c(\boldsymbol{x}) \\ L_{\boldsymbol{b}} L_{\boldsymbol{a}}^{n-1} c(\boldsymbol{x}) \end{bmatrix} u + \begin{bmatrix} 0 \\ L_{\boldsymbol{a}} L_{\boldsymbol{b}} c(\boldsymbol{x}) \\ \vdots \\ L_{\boldsymbol{a}}^{n-2} L_{\boldsymbol{b}} c(\boldsymbol{x}) \\ L_{\boldsymbol{a}}^{n-1} L_{\boldsymbol{b}} c(\boldsymbol{x}) \end{bmatrix} u + \ldots$$

$$+ \begin{bmatrix} 0 \\ L_{\boldsymbol{b}} c(\boldsymbol{x}) \dot{u} \\ \vdots \\ L_{\boldsymbol{b}} c(\boldsymbol{x}) u^{(n-2)} \\ L_{\boldsymbol{b}} c(\boldsymbol{x}) u^{(n-1)} \end{bmatrix}, \tag{5.170}$$

$$y = z_1.$$

We will design the input vector $\boldsymbol{b}(\boldsymbol{x})$ in such a way that

$$\begin{aligned} L_{\boldsymbol{b}} c(\boldsymbol{x}) &= \frac{\partial c(\boldsymbol{x})}{\partial \boldsymbol{x}} \boldsymbol{b}(\boldsymbol{x}) = 0, \\ L_{\boldsymbol{b}} L_{\boldsymbol{a}} c(\boldsymbol{x}) &= \frac{\partial L_{\boldsymbol{a}} c(\boldsymbol{x})}{\partial \boldsymbol{x}} \boldsymbol{b}(\boldsymbol{x}) = 0, \\ &\vdots \\ L_{\boldsymbol{b}} L_{\boldsymbol{a}}^{n-2} c(\boldsymbol{x}) &= \frac{\partial L_{\boldsymbol{a}}^{n-2} c(\boldsymbol{x})}{\partial \boldsymbol{x}} \boldsymbol{b}(\boldsymbol{x}) = 0 \end{aligned} \tag{5.171}$$

and

$$L_{\boldsymbol{b}} L_{\boldsymbol{a}}^{n-1} c(\boldsymbol{x}) = \frac{\partial L_{\boldsymbol{a}}^{n-1} c(\boldsymbol{x})}{\partial \boldsymbol{x}} \boldsymbol{b}(\boldsymbol{x}) = e(\boldsymbol{x}) \tag{5.172}$$

hold, where $e(\boldsymbol{x}) \neq 0$ is a freely selectable function. Combining equations (5.171) and (5.172), we obtain

$$\boldsymbol{Q}(\boldsymbol{x}) \boldsymbol{b}(\boldsymbol{x}) = \begin{bmatrix} 0 \\ \vdots \\ 0 \\ e(\boldsymbol{x}) \end{bmatrix} \quad \text{with} \quad \boldsymbol{Q}(\boldsymbol{x}) = \begin{bmatrix} \dfrac{\partial c(\boldsymbol{x})}{\partial \boldsymbol{x}} \\ \dfrac{\partial L_{\boldsymbol{a}} c(\boldsymbol{x})}{\partial \boldsymbol{x}} \\ \vdots \\ \dfrac{\partial L_{\boldsymbol{a}}^{n-1} c(\boldsymbol{x})}{\partial \boldsymbol{x}} \end{bmatrix}. \tag{5.173}$$

This choice of $\boldsymbol{b}(\boldsymbol{x})$ considerably simplifies system (5.170), because all terms containing

$$L_{\boldsymbol{b}} L_{\boldsymbol{a}}^{i} c(\boldsymbol{x}) = 0, \quad i = 0, \dots, n-2,$$

are now omitted, and in this case only, the system description

$$\dot{\boldsymbol{z}} = \begin{bmatrix} z_2 \\ z_3 \\ \vdots \\ z_n \\ L_{\boldsymbol{a}}^{n} c(\boldsymbol{x}) \end{bmatrix} + \begin{bmatrix} 0 \\ 0 \\ \vdots \\ 0 \\ L_{\boldsymbol{b}} L_{\boldsymbol{a}}^{n-1} c(\boldsymbol{x}) \end{bmatrix} u,$$

$$y = z_1$$

applies. The system is now in nonlinear controller canonical form and has the relative degree $\delta = n$.

Furthermore, the transformation equation (5.169) takes the simpler form

$$\boldsymbol{z} = \boldsymbol{q}(\boldsymbol{x}, u, \dot{u}, \dots, u^{(n-2)}) = \boldsymbol{t}(\boldsymbol{x}) = \begin{bmatrix} c(\boldsymbol{x}) \\ L_{\boldsymbol{a}} c(\boldsymbol{x}) \\ \vdots \\ L_{\boldsymbol{a}}^{n-1} c(\boldsymbol{x}) \end{bmatrix},$$

which we encountered in connection with input-output linearized systems with the maximum relative degree $\delta = n$ discussed in Section 5.2.1, p. 358 et seq., which do not depend on the derivatives

$$\dot{u}, \dots, u^{(n-2)}$$

or powers

$$u^2, \dots, u^{n-1}.$$

As we learned from Theorem 51 on p. 257, SISO systems are flat if and only if they are in controller canonical form or can be transformed into it. For this canonical form,

$$y = z_1$$

is a flat output. Furthermore, a system (5.168) is bijectively transformable into nonlinear controller canonical form if and only if the Jacobian matrix

$$\boldsymbol{Q}(\boldsymbol{x}) = \frac{\partial \boldsymbol{t}(\boldsymbol{x})}{\partial \boldsymbol{x}}$$

of the transformation $\boldsymbol{z} = \boldsymbol{t}(\boldsymbol{x})$ is regular. We know this from Theorem 54 on p. 262.

Based on the analysis above, a system is flat if the matrix $\boldsymbol{Q}(\boldsymbol{x})$ is regular and equation (5.173) applies. It then follows from equation (5.173) that the input vector

$$\boldsymbol{b}(\boldsymbol{x}) = \boldsymbol{Q}^{-1}(\boldsymbol{x}) \begin{bmatrix} 0 \\ \vdots \\ 0 \\ e(\boldsymbol{x}) \end{bmatrix} \tag{5.174}$$

creates a flat input u and thus a flat output $y = c(\boldsymbol{x})$. With the arbitrarily selectable function $e(\boldsymbol{x}) \neq 0$, equation (5.174) also generates all possible flat inputs, since firstly $\boldsymbol{Q}(\boldsymbol{x})$ is regular and secondly $\boldsymbol{b}(\boldsymbol{x})$ fulfills equations (5.171) and (5.172), which lead to the nonlinear controller canonical form and thus ensure the flatness.

This is summarized in [466]

Theorem 74 (Flat Inputs of Control-Affine Systems). *If an autonomous system*

$$\begin{aligned} \dot{\boldsymbol{x}} &= \boldsymbol{a}(\boldsymbol{x}), \quad \boldsymbol{x} \in D_{\boldsymbol{x},\mathrm{def}} \subseteq I\!R^n, \\ y &= c(\boldsymbol{x}) \end{aligned}$$

is extended to a system

$$\begin{aligned} \dot{\boldsymbol{x}} &= \boldsymbol{a}(\boldsymbol{x}) + \boldsymbol{b}(\boldsymbol{x}) \cdot u, \\ y &= c(\boldsymbol{x}) \end{aligned}$$

with an input u, this input is flat and y is a flat output if and only if the matrix

$$\boldsymbol{Q}(\boldsymbol{x}) = \begin{bmatrix} \dfrac{\partial c(\boldsymbol{x})}{\partial \boldsymbol{x}} \\ \dfrac{\partial L_{\boldsymbol{a}} c(\boldsymbol{x})}{\partial \boldsymbol{x}} \\ \vdots \\ \dfrac{\partial L_{\boldsymbol{a}}^{n-1} c(\boldsymbol{x})}{\partial \boldsymbol{x}} \end{bmatrix}$$

is regular and the input vector fulfills the equation

$$\boldsymbol{b}(\boldsymbol{x}) = \boldsymbol{Q}^{-1}(\boldsymbol{x}) \begin{bmatrix} 0 \\ \vdots \\ 0 \\ e(\boldsymbol{x}) \end{bmatrix}$$

for an arbitrary function $e(\boldsymbol{x}) \neq 0$.

Note that the matrix $\boldsymbol{Q}(\boldsymbol{x})$ is the observability matrix of a control-affine system with the relative degree n which will be discussed again in Chapter 7. For designing flat inputs of MIMO systems, see [337, 467].

5.4.8 Flat Inputs of Linear Systems

Now we will apply the results of the previous section to the special case of linear systems

$$\begin{aligned} \dot{\boldsymbol{x}} &= \boldsymbol{A}\boldsymbol{x}, \\ y &= \boldsymbol{c}^T\boldsymbol{x}. \end{aligned} \tag{5.175}$$

We are interested in determining when the system (5.175) has flat inputs and what form they take. Theorem 74 provides us with the answers to these questions. Firstly, the matrix

$$\boldsymbol{Q}(\boldsymbol{x}) = \begin{bmatrix} \dfrac{\partial c(\boldsymbol{x})}{\partial \boldsymbol{x}} \\ \dfrac{\partial L_{\boldsymbol{a}} c(\boldsymbol{x})}{\partial \boldsymbol{x}} \\ \vdots \\ \dfrac{\partial L_{\boldsymbol{a}}^{n-1} c(\boldsymbol{x})}{\partial \boldsymbol{x}} \end{bmatrix} = \begin{bmatrix} \boldsymbol{c}^T \\ \boldsymbol{c}^T\boldsymbol{A} \\ \vdots \\ \boldsymbol{c}^T\boldsymbol{A}^{n-1} \end{bmatrix}, \quad c(\boldsymbol{x}) = \boldsymbol{c}^T\boldsymbol{x},$$

which is the observability matrix of a linear system (5.175), must be regular. This means the system (5.175) has flat inputs if and only if it is observable. Secondly, the input vector $\boldsymbol{b}$ of the extended system

$$\begin{aligned} \dot{\boldsymbol{x}} &= \boldsymbol{A}\boldsymbol{x} + \boldsymbol{b}\cdot u, \\ y &= \boldsymbol{c}^T\boldsymbol{x} \end{aligned}$$

must take the form

$$\boldsymbol{b} = \boldsymbol{Q}^{-1} \begin{bmatrix} 0 \\ \vdots \\ 0 \\ e \end{bmatrix}, \quad e \in I\!R\backslash\{0\},$$

so that the output y of the linear system (5.175) is flat. We remember that the constant e can be selected arbitrarily.

As in the nonlinear case, we can design inputs which enable particularly efficient feedforward and feedback control of the system. It is also possible to analyze existing inputs in terms of their suitability for flat control.

5.4.9 Example: Economic Market Model

An example of an economic market is a simple model [34, 160, 268] of the dependence on the price P of a product on the supply quantity Q and the quantity Q_d demanded by customers. We can assume that the change in price is proportional to the difference between demand and supply quantity, i. e. that

$$\dot{P} = \lambda \left(Q_\mathrm{d}(P) - Q\right), \quad \lambda > 0, \tag{5.176}$$

holds, where λ is a constant of proportionality. If the demand Q_d is greater than the supply Q, the price P increases.

Furthermore, the manufacturer responds to the possible market price P, or, to put it more precisely, to the difference between the market price P and the manufacturer's supply price P_s. This can be modeled by

$$\dot{Q} = \mu \left(P - P_\mathrm{s}(Q)\right), \quad \mu > 0. \tag{5.177}$$

This means that if the market price P is higher than the supply price P_s and increases, more will be produced. As in equation (5.176), the relationship between $\dot{Q}$ and P is linear where μ is a constant of proportionality.

We can now assume that the demand Q_d

$$Q_\mathrm{d}(P) = \gamma - \kappa P, \quad \gamma,\, \kappa > 0 \tag{5.178}$$

depends linearly on the current market price P via a constant κ. A low price therefore leads to high demand.

The supply price P_s depends on the supply quantity Q via

$$P_\mathrm{s}(Q) = \delta Q^2 + \beta, \quad \delta,\, \beta > 0, \tag{5.179}$$

for example.

If we insert equation (5.178) and equation (5.179) into the differential equations (5.176) and (5.177), respectively, and select

$$x_1 = P \quad \text{and} \quad x_2 = Q$$

as state variables, we obtain

$$\begin{bmatrix} \dot{x}_1 \\ \dot{x}_2 \end{bmatrix} = \begin{bmatrix} a_0 - a_1 x_1 - a_2 x_2 \\ a_3 x_1 - a_4 - a_5 x_2^2 \end{bmatrix} = \boldsymbol{a}(\boldsymbol{x}),$$

where

$$a_0 = \lambda\gamma,\ a_1 = \lambda\kappa,\ a_2 = \lambda,\ a_3 = \mu,\ a_4 = \mu\beta,\ a_5 = \mu\delta.$$

Our goal now is to influence the market price $x_1 = P$ to increase it. To do this, we must identify an input with which we can exert influence on x_1, i. e. a flat input. Here, the price

$$y = x_1 = c(\boldsymbol{x})$$

is the system's output of interest.

Applying Theorem 74, we determine

$$\frac{\partial c(\boldsymbol{x})}{\partial \boldsymbol{x}} = [1\ 0]\,,$$
$$\frac{\partial L_{\boldsymbol{a}} c(\boldsymbol{x})}{\partial \boldsymbol{x}} = \frac{\partial (a_0 - a_1 x_1 - a_2 x_2)}{\partial \boldsymbol{x}} = [-a_1\ -a_2]$$

and hence

$$\boldsymbol{Q}(\boldsymbol{x}) = \begin{bmatrix} \dfrac{\partial c(\boldsymbol{x})}{\partial \boldsymbol{x}} \\ \dfrac{\partial L_{\boldsymbol{a}} c(\boldsymbol{x})}{\partial \boldsymbol{x}} \end{bmatrix} = \begin{bmatrix} 1 & 0 \\ -a_1 & -a_2 \end{bmatrix}.$$

With

$$\boldsymbol{b}(\boldsymbol{x}) = \boldsymbol{Q}^{-1}(\boldsymbol{x}) \begin{bmatrix} 0 \\ e(\boldsymbol{x}) \end{bmatrix} = \begin{bmatrix} 1 & 0 \\ -\dfrac{a_1}{a_2} & -\dfrac{1}{a_2} \end{bmatrix} \begin{bmatrix} 0 \\ e(\boldsymbol{x}) \end{bmatrix} = \begin{bmatrix} 0 \\ -\dfrac{e(\boldsymbol{x})}{a_2} \end{bmatrix}$$

and by choosing

$$e(\boldsymbol{x}) = -a_2,$$

we obtain the input vector

$$\boldsymbol{b}(\boldsymbol{x}) = \begin{bmatrix} 0 \\ 1 \end{bmatrix}$$

in the system description

$$\dot{\boldsymbol{x}} = \boldsymbol{a}(\boldsymbol{x}) + \boldsymbol{b}(\boldsymbol{x}) \cdot u.$$

In this way, the control variable u from the market model

$$\begin{bmatrix} \dot{x}_1 \\ \dot{x}_2 \end{bmatrix} = \begin{bmatrix} a_0 - a_1 x_1 - a_2 x_2 \\ a_3 x_1 - a_4 - a_5 x_2^2 \end{bmatrix} + \begin{bmatrix} 0 \\ 1 \end{bmatrix} u, \qquad (5.180)$$
$$y = x_1$$

becomes a flat input and $y = x_1$ is a flat output.

If we now wish to influence the current market price $y = x_1 = P$, we can use the flat system representation and a reference trajectory $y_{\text{ref}}(t)$ to specify the control variable

$$\begin{aligned} u &= \Psi_2\left(y_{\text{ref}}, \dot{y}_{\text{ref}}, \ddot{y}_{\text{ref}}\right) \\ &= a_4 - a_3 y_{\text{ref}} + a_5 \left(\frac{a_0 - a_1 y_{\text{ref}} - \dot{y}_{\text{ref}}}{a_2}\right)^2 - \frac{a_1 \dot{y}_{\text{ref}} + \ddot{y}_{\text{ref}}}{a_2}. \end{aligned}$$

From equation (5.180) it is clear that u affects the change in the supply quantity $\dot{x}_2 = \dot{Q}$. The above result is therefore simple to interpret, because the control variable u can directly influence the supply quantity $x_2 = Q$. The control variable u thus means either an increase in production or a product shortage.

5.5 Control Lyapunov Functions

5.5.1 Fundamentals

The direct Lyapunov method can be used for stability analysis as well as for controller synthesis. A controller synthesis method of this kind can be based on so-called *control Lyapunov functions* $V(\boldsymbol{x})$, often abbreviated as *CLF*. The basic concept of control Lyapunov functions is to use the control variable vector $\boldsymbol{u}$ of a plant

$$\dot{\boldsymbol{x}} = \boldsymbol{f}(\boldsymbol{x}, \boldsymbol{u})$$

which has the equilibrium point

$$\boldsymbol{x}_{\mathrm{eq}} = \boldsymbol{0}$$

such that the derivative

$$\dot{V}(\boldsymbol{x}) = \boldsymbol{f}^T(\boldsymbol{x}, \boldsymbol{u}) \cdot \operatorname{grad}(V(\boldsymbol{x}))$$

of a radially unbounded Lyapunov function $V(\boldsymbol{x})$ is minimized and the inequality

$$\dot{V}(\boldsymbol{x}) < 0$$

applies to all states $\boldsymbol{x} \neq \boldsymbol{0}$. For example, the control variable $\boldsymbol{u}$ is determined so that the infimum, i. e. the lower limit of $\dot{V}(\boldsymbol{x})$, satisfies the inequality

$$\inf_{\boldsymbol{u}} \left\{ \dot{V}(\boldsymbol{x}) = \boldsymbol{f}^T(\boldsymbol{x}, \boldsymbol{u}) \cdot \operatorname{grad}(V(\boldsymbol{x})) \right\} < 0 \tag{5.181}$$

for all $\boldsymbol{x} \neq \boldsymbol{0}$. The control variable $\boldsymbol{u}(\boldsymbol{x})$ thus determined constitutes a control law based on the Lyapunov stability theorem on p. 115, and yields a control loop

$$\dot{\boldsymbol{x}} = \boldsymbol{f}(\boldsymbol{x}, \boldsymbol{u}(\boldsymbol{x}))$$

with the globally asymptotically stable equilibrium point $\boldsymbol{x}_{\mathrm{eq}} = \boldsymbol{0}$. Note that it is also possible to aim for local stability only.

To ensure the stability of the control loop, it is not necessary to select the function $\boldsymbol{u}(\boldsymbol{x})$ as the control law which generates the infimum of $\dot{V}(\boldsymbol{x})$. In fact, a control law $\boldsymbol{u}(\boldsymbol{x})$ which yields $\dot{V}(\boldsymbol{x}) < 0$ for all $\boldsymbol{x}$ is sufficient. The choice of the controller $\boldsymbol{u}(\boldsymbol{x})$ which leads to the infimal $\dot{V}$ has the advantage that $V(\boldsymbol{x})$ decreases rapidly along the trajectories $\boldsymbol{x}(t)$ and the compensation is therefore rapid.

The term *control Lyapunov function* is derived from the fact that, using a Lyapunov function, we define a controller $\boldsymbol{u}(\boldsymbol{x})$ which ensures stability. Although the procedure above ensures the stability of the control loop being designed, it does not allow a direct, calculable influence on the control performance, i. e. the settling time. However, it is difficult or impossible to predict whether our choice of a Lyapunov function will positively or negatively affect the control performance. Namely, the faster the Lyapunov function decreases, the smaller the settling time becomes. Therefore, the choice of the Lyapunov function is critical for the control performance.

For time-invariant nonlinear systems, we will define the term *control Lyapunov function* as follows.

Definition 35 (Control Lyapunov Function). *Let*

$$\dot{\boldsymbol{x}} = \boldsymbol{f}(\boldsymbol{x}, \boldsymbol{u}), \quad \boldsymbol{x} \in I\!R^n, \quad \boldsymbol{u} \in I\!R^m,$$

be a system with the equilibrium point $\boldsymbol{x}_{\mathrm{eq}} = \boldsymbol{0}$ *for* $\boldsymbol{u} = \boldsymbol{0}$. *A continuously differentiable function* $V(\boldsymbol{x})$ *is called a control Lyapunov function for this system if it fulfills the following conditions:*

(1) $V(\boldsymbol{0}) = 0$,
(2) $V(\boldsymbol{x}) > 0$ *for all* $\boldsymbol{x} \neq \boldsymbol{0}$,
(3) $V(\boldsymbol{x}) \to \infty$ *for* $|\boldsymbol{x}| \to \infty$.
(4) A control law $\boldsymbol{u}(\boldsymbol{x})$ *exists such that* $\dot{V}(\boldsymbol{x}) < 0$ *applies to all* $\boldsymbol{x} \neq \boldsymbol{0}$.

If we apply the Barbashin and Krasovskii theorem, i. e. Theorem 14 on p. 116, we can define a control Lyapunov function somewhat less restrictively. We thus arrive at

Definition 36 (Extended Control Lyapunov Function). *Let*

$$\dot{\boldsymbol{x}} = \boldsymbol{f}(\boldsymbol{x}, \boldsymbol{u}), \quad \boldsymbol{x} \in I\!R^n, \quad \boldsymbol{u} \in I\!R^m,$$

denote a system with the equilibrium point $\boldsymbol{x}_{\mathrm{eq}} = \boldsymbol{0}$ *for* $\boldsymbol{u} = \boldsymbol{0}$. *A continuously differentiable function* $V(\boldsymbol{x})$ *is called an extended control Lyapunov function for this system if it satisfies the following conditions:*

(1) $V(\boldsymbol{0}) = 0$,
(2) $V(\boldsymbol{x}) > 0$ *for all* $\boldsymbol{x} \neq \boldsymbol{0}$,
(3) $V(\boldsymbol{x}) \to \infty$ *for* $|\boldsymbol{x}| \to \infty$.
(4) A control law $\boldsymbol{u}(\boldsymbol{x})$ *exists such that* $\dot{V}(\boldsymbol{x}) \leq 0$ *applies to all* $\boldsymbol{x} \in I\!R$.
(5) The set of states $\boldsymbol{x}$ *of the system controlled by* $\boldsymbol{u}(\boldsymbol{x})$, *to which* $\dot{V}(\boldsymbol{x}) = 0$ *applies, contains no trajectory* $\boldsymbol{x}(t)$ *except* $\boldsymbol{x}(t) = \boldsymbol{0}$.

Definition 36 makes the requirement $\dot{V}(\boldsymbol{x}) < 0$ from Definition 35 less restrictive. Now we can also address the case in which

$$\dot{V}(\boldsymbol{x}) \leq 0$$

holds, as long as, according to Condition (5), $\dot{V}(\boldsymbol{x}(t)) = 0$ does not apply along any trajectory $\boldsymbol{x}(t)$ of the controlled system.

Based on the Lyapunov and the Barbashin and Krasovskii theorems, Theorem 13 on p. 115 and Theorem 14 on p. 116, respectively, and the definitions above, we can directly derive

Theorem 75 (Existence of a Control Law). *Let there be a system*

$$\dot{\boldsymbol{x}} = \boldsymbol{f}(\boldsymbol{x}, \boldsymbol{u}), \quad \boldsymbol{x} \in I\!R^n, \quad \boldsymbol{u} \in I\!R^m,$$

with the equilibrium point $\boldsymbol{x}_{\text{eq}} = \boldsymbol{0}$ for $\boldsymbol{u} = \boldsymbol{0}$. If in this case a control Lyapunov function or an extended control Lyapunov function $V(\boldsymbol{x})$ exists for the system, a controller $\boldsymbol{u}(\boldsymbol{x})$ can always be found such that the control loop's equilibrium point $\boldsymbol{x}_{\text{eq}} = \boldsymbol{0}$ is globally asymptotically stable.

If Condition (3) of Definitions 35 and 36 is not fulfilled, only local asymptotic stability can be ensured.

As with the direct method in the case of stability analysis, the main difficulty with the above theorem on controller synthesis is finding a suitable Lyapunov function $V(\boldsymbol{x})$. Unfortunately, there is no general procedure for the design of a Lyapunov function or a control Lyapunov function. In many cases, we must rely on our intuition to find a control Lyapunov function or we must test different possible functions $V(\boldsymbol{x})$ to find out whether one of them fulfills the conditions of Definition 35 or 36.

5.5.2 Control Lyapunov Functions for Linear Systems

As a simple example to illustrate the method of control Lyapunov functions, we will examine the linear system

$$\dot{\boldsymbol{x}} = \boldsymbol{A}\boldsymbol{x} + \boldsymbol{b} \cdot u, \quad |u| \leq u_{\max},$$

which we assume to be Lyapunov stable, and

$$V(\boldsymbol{x}) = \boldsymbol{x}^T \boldsymbol{R}\, \boldsymbol{x}$$

as a possible control Lyapunov function which is positive definite and radially unbounded and therefore fulfills Conditions (1), (2), and (3) from Definition 35. Its derivative is

$$\dot{V}(\boldsymbol{x}) = \boldsymbol{x}^T \left(\boldsymbol{A}^T \boldsymbol{R} + \boldsymbol{R}\boldsymbol{A}\right) \boldsymbol{x} + 2\boldsymbol{b}^T \boldsymbol{R}\, \boldsymbol{x} \cdot u.$$

Obviously, $V(\boldsymbol{x}) = \boldsymbol{x}^T \boldsymbol{R} \boldsymbol{x}$ is a control Lyapunov function if the matrix

$$\boldsymbol{A}^T \boldsymbol{R} + \boldsymbol{R}\boldsymbol{A}$$

is negative definite or negative semidefinite, since the control law

$$u = -u_{\max} \operatorname{sgn}\left(\boldsymbol{b}^T \boldsymbol{R}\, \boldsymbol{x}\right) \tag{5.182}$$

yields

$$2\boldsymbol{b}^T \boldsymbol{R}\boldsymbol{x} \cdot u = -2u_{\max} \operatorname{sgn}\left(\boldsymbol{b}^T \boldsymbol{R}\, \boldsymbol{x}\right) \boldsymbol{b}^T \boldsymbol{R}\, \boldsymbol{x} = -2u_{\max}|\boldsymbol{b}^T \boldsymbol{R}\, \boldsymbol{x}|,$$

meaning that Condition (4) from Definition 35

$$\begin{aligned}\dot{V}(\boldsymbol{x}) &= \boldsymbol{f}^T(\boldsymbol{x},u)\,\mathrm{grad}(V(\boldsymbol{x}))\\ &= \boldsymbol{x}^T\left(\boldsymbol{A}^T\boldsymbol{R} + \boldsymbol{R}\boldsymbol{A}\right)\boldsymbol{x} - 2u_{\max}|\boldsymbol{b}^T \boldsymbol{R}\, \boldsymbol{x}| < 0\end{aligned}$$

is fulfilled.

As we have seen, a control Lyapunov function is simple to determine for a Lyapunov stable linear system. However, the control law (5.182) has the disadvantage of switching between $-u_{\max}$ and $+u_{\max}$. Only a few actuators can provide and endure such switching behavior. Furthermore, undesirable sliding modes can occur.

5.5.3 Control Lyapunov Functions for Control-Affine Systems

After this simple example, we will turn to a more general case. The plants in question are now the control-affine systems

$$\dot{\boldsymbol{x}} = \boldsymbol{a}(\boldsymbol{x}) + \boldsymbol{B}(\boldsymbol{x}) \cdot \boldsymbol{u}. \tag{5.183}$$

In this case Condition (4) from Definition 35 takes the form

$$\boldsymbol{a}^T(\boldsymbol{x}) \cdot V_{\boldsymbol{x}}(\boldsymbol{x}) + \boldsymbol{u}^T \boldsymbol{B}^T(\boldsymbol{x}) \cdot V_{\boldsymbol{x}}(\boldsymbol{x}) < 0, \tag{5.184}$$

where we have abbreviated the gradient of the control Lyapunov function $V(\boldsymbol{x})$ as

$$V_{\boldsymbol{x}}(\boldsymbol{x}) = \mathrm{grad}(V(\boldsymbol{x})) = \left(\frac{\partial V}{\partial \boldsymbol{x}}\right)^T.$$

From inequality (5.184) we can now derive

Theorem 76 (Control Lyapunov Function for Control-Affine Systems). *For a system*

$$\dot{\boldsymbol{x}} = \boldsymbol{a}(\boldsymbol{x}) + \boldsymbol{B}(\boldsymbol{x}) \cdot \boldsymbol{u}$$

a radially unbounded, positive definite function $V(\boldsymbol{x})$ is a control Lyapunov function if one of the conditions

(1) $\boldsymbol{B}^T(\boldsymbol{x})V_{\boldsymbol{x}}(\boldsymbol{x}) \neq \boldsymbol{0}$ for all $\boldsymbol{x} \neq 0$ or
(2) $\boldsymbol{a}^T(\boldsymbol{x})V_{\boldsymbol{x}}(\boldsymbol{x}) < 0$ for all $\boldsymbol{x} \neq 0$ with $\boldsymbol{B}^T(\boldsymbol{x})V_{\boldsymbol{x}}(\boldsymbol{x}) = \boldsymbol{0}$,
$\boldsymbol{a}^T(\boldsymbol{x})V_{\boldsymbol{x}}(\boldsymbol{x}) = 0$ for $\boldsymbol{x} = \boldsymbol{0}$

is fulfilled.

Condition (2) from the theorem above is particularly satisfied if

$$\begin{aligned} \boldsymbol{a}^T(\boldsymbol{x})V_{\boldsymbol{x}}(\boldsymbol{x}) &< 0 \quad \text{for} \quad \boldsymbol{x} \in I\!R^n \setminus \{\boldsymbol{0}\}, \\ \boldsymbol{a}^T(\boldsymbol{x})V_{\boldsymbol{x}}(\boldsymbol{x}) &= 0 \quad \text{for} \quad \boldsymbol{x} = \boldsymbol{0} \end{aligned} \tag{5.185}$$

hold. In this significant case, the function V is already a Lyapunov function for the free system $\dot{\boldsymbol{x}} = \boldsymbol{a}(\boldsymbol{x})$.

In the event that there is a known control Lyapunov function $V(\boldsymbol{x})$ for the plant (5.183), we use E. D. Sontag's control law, which globally asymptotically stabilizes the equilibrium point $\boldsymbol{x}_{\text{eq}} = \boldsymbol{0}$ of the system (5.183) [426, 431]. It is formulated in

Theorem 77 (Sontag's Control Law). *Let $V(\boldsymbol{x})$ be a control Lyapunov function for the system*

$$\dot{\boldsymbol{x}} = \boldsymbol{a}(\boldsymbol{x}) + \boldsymbol{B}(\boldsymbol{x}) \cdot \boldsymbol{u}.$$

In this case the control loop consisting of this system and the controller

$$\boldsymbol{u}(\boldsymbol{x}) = \begin{cases} -k_{\text{s}}(\boldsymbol{x})\boldsymbol{B}^T(\boldsymbol{x})V_{\boldsymbol{x}}(\boldsymbol{x}) & \text{for } \boldsymbol{B}^T(\boldsymbol{x})V_{\boldsymbol{x}}(\boldsymbol{x}) \neq \boldsymbol{0}, \\ \boldsymbol{0} & \text{for } \boldsymbol{B}^T(\boldsymbol{x})V_{\boldsymbol{x}}(\boldsymbol{x}) = \boldsymbol{0} \end{cases}$$

with

$$\begin{aligned} k_{\text{s}}(\boldsymbol{x}) &= \frac{\boldsymbol{a}^T(\boldsymbol{x})V_{\boldsymbol{x}}(\boldsymbol{x}) + \sqrt{\left(\boldsymbol{a}^T(\boldsymbol{x})V_{\boldsymbol{x}}(\boldsymbol{x})\right)^2 + h^2(\boldsymbol{x})}}{h(\boldsymbol{x})}, \\ h(\boldsymbol{x}) &= {V_{\boldsymbol{x}}}^T(\boldsymbol{x})\boldsymbol{B}(\boldsymbol{x})\boldsymbol{B}^T(\boldsymbol{x})V_{\boldsymbol{x}}(\boldsymbol{x}), \end{aligned}$$

has the globally asymptotically stable equilibrium point $\boldsymbol{x}_{\text{eq}} = \boldsymbol{0}$.

By inserting Sontag's control law into inequality (5.184), it can be proven that the inequality is satisfied, and thus the control loop is globally asymptotically stable. For systems (5.183) with bounded control inputs, a modified formula can be found in [283].

Further we will discuss two control laws for cases in which the control Lyapunov function fulfills condition (5.185) and thus Condition (2) from Theorem 76. In this case, the autonomous part $\dot{\boldsymbol{x}} = \boldsymbol{a}(\boldsymbol{x})$ of system (5.183) has a globally asymptotically stable equilibrium point $\boldsymbol{x}_{\text{eq}} = \boldsymbol{0}$. We can rewrite the system (5.183) as

$$\dot{\boldsymbol{x}} = \boldsymbol{a}(\boldsymbol{x}) + \sum_{i=1}^{m} \boldsymbol{b}_i(\boldsymbol{x}) \cdot u_i,$$

where the vectors $\boldsymbol{b}_i(\boldsymbol{x})$ are the column vectors of the matrix $\boldsymbol{B}(\boldsymbol{x})$. Let us assume that we have succeeded in identifying a control Lyapunov function for the system above. We then obtain

$$\dot{V}(\boldsymbol{x}) = \underbrace{\boldsymbol{a}^T(\boldsymbol{x})V_{\boldsymbol{x}}(\boldsymbol{x})}_{<0} + \sum_{i=1}^{m} V_{\boldsymbol{x}}{}^T(\boldsymbol{x})\boldsymbol{b}_i(\boldsymbol{x}) \cdot u_i. \tag{5.186}$$

It is reasonable to select all u_i in equation (5.186) in such a way that $\dot{V}(\boldsymbol{x})$ takes on the lowest possible negative values so that the compensation for a trajectory $\boldsymbol{x}(t)$ into the equilibrium point $\boldsymbol{x}_{\mathrm{eq}} = \boldsymbol{0}$ happens rapidly. In the case of control variable limitations $|u_i| \leq u_{i,\max}$, we can achieve this goal if we use the control law

$$u_i = -u_{i,\max}\,\mathrm{sgn}(V_{\boldsymbol{x}}{}^T(\boldsymbol{x})\boldsymbol{b}_i(\boldsymbol{x})), \quad i = 1,\ldots,m,$$

because then

$$\dot{V}(\boldsymbol{x}) = \boldsymbol{a}^T(\boldsymbol{x})V_{\boldsymbol{x}}(\boldsymbol{x}) - \sum_{i=1}^{m} u_{i,\max}\left|V_{\boldsymbol{x}}{}^T(\boldsymbol{x})\boldsymbol{b}_i(\boldsymbol{x})\right| < 0$$

and inequality (5.181) hold.

However, this switching control law has a discontinuous control variable signal and possibly undesirable sliding modes. Both are drawbacks which have already occurred with the control law (5.182).

The disadvantages above can be avoided if the control variable is selected as

$$u_i = -\,\mathrm{sat}(V_{\boldsymbol{x}}{}^T(\boldsymbol{x})\boldsymbol{b}_i(\boldsymbol{x})), \qquad \mathrm{sat}(v) = \begin{cases} v_{\max}, & v > v_{\max}, \\ v, & |v| \leq v_{\max}, \\ -v_{\max}, & v < -v_{\max}, \end{cases}$$

by means of a saturation function. As a result,

$$\dot{V}(\boldsymbol{x}) = \boldsymbol{a}^T(\boldsymbol{x})V_{\boldsymbol{x}}(\boldsymbol{x}) - \sum_{i=1}^{m} V_{\boldsymbol{x}}{}^T(\boldsymbol{x})\boldsymbol{b}_i(\boldsymbol{x})\,\mathrm{sat}(V_{\boldsymbol{x}}{}^T(\boldsymbol{x})\boldsymbol{b}_i(\boldsymbol{x})) < 0$$

holds and in this case V also decreases rapidly along any trajectory.

5.5.4 Illustrative Example

As an example, we will analyze the system

$$\dot{\boldsymbol{x}} = \underbrace{\begin{bmatrix} -x_1\left(1 - e^{-x_1^2 - x_2^2}\right) + x_2 e^{-x_1^2 - x_2^2} \\ -x_1 e^{-x_1^2 - x_2^2} - x_2\left(1 - e^{-x_1^2 - x_2^2}\right) \end{bmatrix}}_{\boldsymbol{a}(\boldsymbol{x})} + \underbrace{\begin{bmatrix} 1 & 0 \\ 0 & 1 \end{bmatrix}}_{\boldsymbol{B}(\boldsymbol{x})} \boldsymbol{u}$$

with the single equilibrium point $\boldsymbol{x}_{\mathrm{eq}} = \boldsymbol{0}$. Figure 5.31 shows the trajectories of the system for $\boldsymbol{u} = \boldsymbol{0}$. A possible control Lyapunov function is

$$V(\boldsymbol{x}) = \frac{1}{2}\left(x_1^2 + x_2^2\right).$$

We obtain

$$V_{\boldsymbol{x}}{}^T(\boldsymbol{x}) = [x_1 \;\; x_2]$$

for the gradient and hence

$$\boldsymbol{B}^T V_{\boldsymbol{x}}(\boldsymbol{x}) = \begin{bmatrix} x_1 \\ x_2 \end{bmatrix}.$$

Since $\boldsymbol{B}^T V_{\boldsymbol{x}}(\boldsymbol{x}) \neq \boldsymbol{0}$ is valid for all $\boldsymbol{x} \neq \boldsymbol{0}$, according to Theorem 76 the function V is a control Lyapunov function. This is also directly apparent from

$$\dot{V}(\boldsymbol{x}) = -\left(x_1^2 + x_2^2\right)\left(1 - e^{-x_1^2 - x_2^2}\right) + x_1 u_1 + x_2 u_2.$$

From Theorem 77, we obtain

$$k_{\mathrm{s}}(\boldsymbol{x}) = -\left(1 - e^{-x_1^2 - x_2^2}\right) + \sqrt{\left(1 - e^{-x_1^2 - x_2^2}\right)^2 + 1},$$

and Sontag's control law takes the form

$$\boldsymbol{u}(\boldsymbol{x}) = \left(\left(1 - e^{-x_1^2 - x_2^2}\right) - \sqrt{\left(1 - e^{-x_1^2 - x_2^2}\right)^2 + 1}\,\right)\boldsymbol{x}.$$

Figure 5.32 shows the trajectories of the controlled system and Figure 5.33 the time courses representing x_1 of the controlled system and the uncontrolled one, which both start at an initial value of $\boldsymbol{x}(0) = [0.2\;0.2]^T$. Compared to the markedly oscillating plant, the control loop no longer has a tendency to oscillate.

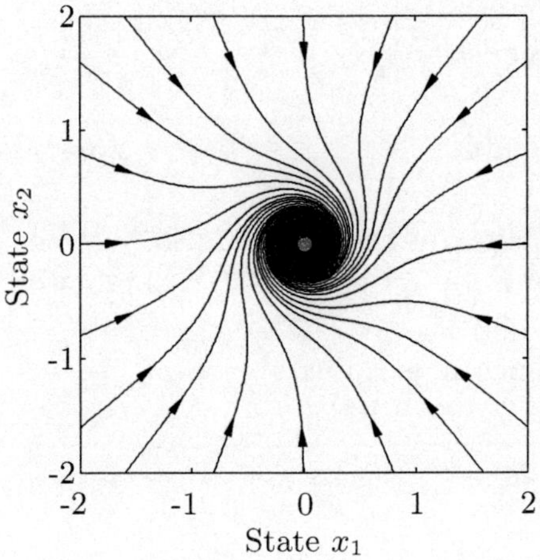

Fig. 5.31: Trajectories of the uncontrolled system

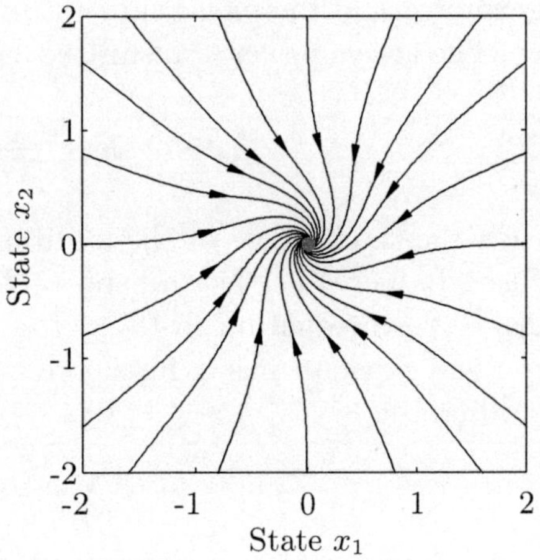

Fig. 5.32: Trajectories of the control loop

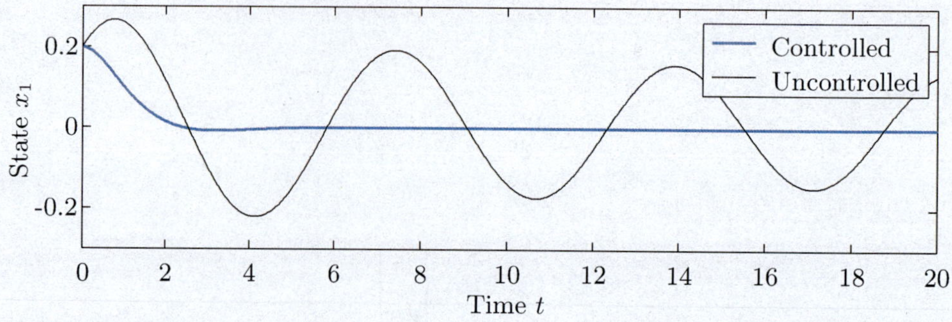

Fig. 5.33: Time courses of state variable x_1 in the controlled case and the uncontrolled one

5.5.5 Example: Power Plant with Grid Feed-In

Below we will describe a power plant generating electricity and connected to the power supply grid via a long high-voltage transmission line; we will examine its behavior in the case of a short circuit [11, 15, 254]. The power supply grid is assumed to be rigid, i. e. its frequency is constant and the stability of the grid is not affected by the power plant in question. For example, hydropower plants in remote regions are connected to the power grid by means of such long high-voltage transmission lines, as shown in Figure 5.34.

The transmission line is connected in series with an adjustable capacitive reactance X_C [150, 284, 437]. The sum of the transient inductive reactance of the generator, the inductive reactance of the transformer, and the reactance of the transmission line is denoted as X_L. Further, $\bar{E} = Ee^{j\delta}$ is the generator voltage, more precisely the pole-wheel voltage, and $\bar{V} = Ve^{j0}$ is the voltage of the rigid network. Figure 5.35 shows the equivalent circuit diagram of the system. The power angle δ consists of the load angle of the synchronous generator and the phase shift resulting from the transmission line.

The active power transmitted via the line is

$$P = \frac{E \cdot V}{X_L - X_C} \sin(\delta). \tag{5.187}$$

Now the significance of the additional capacitor in the system also becomes clear. It increases the transportable power P by reducing the denominator $X_L - X_C$ in equation (5.187).

The capacity has a further function. Since it is adjustably designed according to

$$X_C = X_{C_0} + \Delta X_C, \tag{5.188}$$

it can be used to correct system faults as quickly as possible after a transmission line malfunction such as a short circuit. With equation (5.188), we obtain

Fig. 5.34: A hydropower plant connected via a long transmission line to the rigid grid

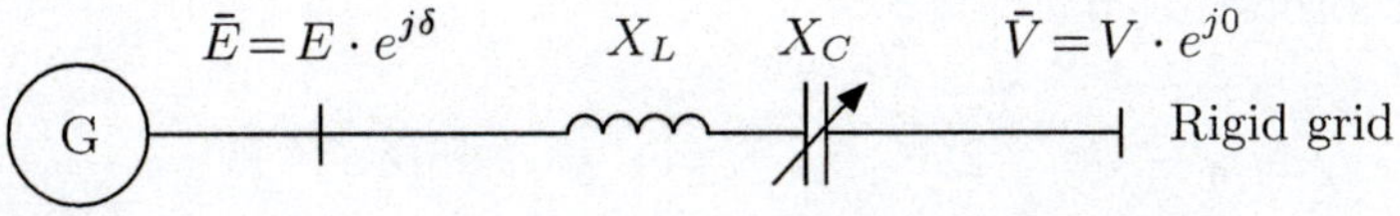

Fig. 5.35: Generator with long transmission line to the power grid

$$P = \frac{E \cdot V}{X_L - X_{C_0}}(1 + u)\sin(\delta) \tag{5.189}$$

with

$$u = \frac{\Delta X_C}{X_L - X_{C_0} - \Delta X_C}$$

for equation (5.187).

In stationary operation, the relation

$$\omega_{\mathrm{e}} = p\omega_{\mathrm{m}} \tag{5.190}$$

applies to the mechanical angular velocity ω_{m} of the pole wheel of the synchronous machine with the pole pair number p and to the frequency ω_{e} of the pole-wheel voltage $E\sin(\omega_{\mathrm{e}}t + \delta)$.

If temporal changes in the angle δ occur because of a short circuit or for other reasons, the frequency of the pole-wheel voltage $E\sin(\omega_{\mathrm{e}}t + \delta)$ and thus the mechanical frequency ω_{m} of the pole wheel changes according to

$$\omega_{\mathrm{e}} + \dot{\delta} = p(\omega_{\mathrm{m}} + \Delta\omega_{\mathrm{m}}).$$

Inserting equation (5.190), we obtain as the change in the mechanical frequency of the pole wheel

$$\Delta\omega_{\mathrm{m}} = \frac{\dot{\delta}}{p}.$$

We will now examine the system's power balance. The turbine power P_{T}, from which we substract the damping power loss

$$D\Delta\omega_{\mathrm{m}} = \frac{D}{p}\dot{\delta}$$

and the power accelerating or decelerating the rotor

$$J(\omega_{\mathrm{m}} + \Delta\omega_{\mathrm{m}})(\Delta\omega_{\mathrm{m}})^{\cdot} \approx J\omega_{\mathrm{m}}(\Delta\omega_{\mathrm{m}})^{\cdot} = \frac{J\omega_{\mathrm{e}}}{p^2}\ddot{\delta},$$

is equal to the electrical power P from equation (5.189) generated by the synchronous generator and transported via the transmission line. Here D is a damping constant and J is the moment of inertia of all the turbine's and generator's rotating parts. Thus,

$$P_{\mathrm{T}} - D\Delta\omega_{\mathrm{m}} - J\omega_{\mathrm{m}}(\Delta\omega_{\mathrm{m}})^{\cdot} = P_{\mathrm{E}}(1+u)\sin(\delta), \quad P_{\mathrm{E}} = \frac{E\cdot V}{X_L - X_{C_0}},$$

applies to the power balance, from which

$$P_{\mathrm{T}} - \frac{D}{p}\dot{\delta} - \frac{J\omega_{\mathrm{e}}}{p^2}\ddot{\delta} = P_{\mathrm{E}}(1+u)\sin(\delta)$$

and

$$\ddot{\delta} = \frac{p^2}{J\omega_{\mathrm{e}}}\left(P_{\mathrm{T}} - \frac{D}{p}\dot{\delta} - P_{\mathrm{E}}(1+u)\sin(\delta)\right) \tag{5.191}$$

follow.

The stationary operating point

$$\delta_{\text{eq}} = \arcsin\left(\frac{P_{\text{T}}}{P_{\text{E}}}\right)$$

of the system is obtained from equation (5.191) if we set $\ddot{\delta} = 0$ and $\dot{\delta} = 0$, as well as $u = 0$. The angle δ typically varies within narrow bounds around the operating point δ_{eq}. Only in the case of larger disturbances can higher values $\delta \in [-\pi, \pi]$ occur. Changes in the angle δ are undesirable because, according to equation (5.189), the transmitted electrical power P fluctuates. Furthermore, large changes in the angle δ can cause the synchronous generator to be out of step, which means the pole wheel no longer runs synchronously with the stator's rotating magnetic field.

We will define

$$\begin{bmatrix} x_1 \\ x_2 \end{bmatrix} = \begin{bmatrix} \delta - \delta_{\text{eq}} \\ \dot{\delta} \end{bmatrix}$$

as the state vector, and furthermore the abbreviations

$$a_1 = \frac{p^2 P_{\text{T}}}{J\omega_{\text{e}}}, \qquad a_2 = \frac{pD}{J\omega_{\text{e}}}, \qquad \text{and} \qquad a_3 = \frac{p^2 P_{\text{E}}}{J\omega_{\text{e}}}.$$

So from equation (5.191), we obtain the model of the power plant connected to a rigid grid as

$$\begin{bmatrix} \dot{x}_1 \\ \dot{x}_2 \end{bmatrix} = \begin{bmatrix} x_2 \\ a_1 - a_2 x_2 - a_3(1+u)\sin(x_1 + \delta_{\text{eq}}) \end{bmatrix}. \tag{5.192}$$

Here, u is the system's control variable. System (5.192) has the equilibrium points

$$\boldsymbol{x}_{\text{eq},2i} = \begin{bmatrix} \pm 2i\pi \\ 0 \end{bmatrix}, \quad \boldsymbol{x}_{\text{eq},(2i+1)} = \begin{bmatrix} \pm(2i+1)\pi - 2\delta_{\text{eq}} \\ 0 \end{bmatrix} \quad \text{for } i = 0, 1, 2, \ldots$$

for $u = 0$. Since the equilibrium points and the associated trajectories of indices $i > 0$ repeat periodically due to the sine function in equation (5.192), taking into account the equilibrium points

$$\boldsymbol{x}_{\text{eq}0} = \begin{bmatrix} 0 \\ 0 \end{bmatrix} \quad \text{and} \quad \boldsymbol{x}_{\text{eq}1} = \begin{bmatrix} \pi - 2\delta_{\text{eq}} \\ 0 \end{bmatrix}$$

is sufficient. The stability of the equilibrium point $\boldsymbol{x}_{\text{eq}0}$ and the instability of the equilibrium point $\boldsymbol{x}_{\text{eq}1}$ can be simply proven using Lyapunov's indirect method (Theorem 17 on p. 126), using the linearized model of system (5.192).

The aim is to quickly compensate for disturbances that occur because of line faults or other reasons, which corresponds to $\boldsymbol{x} \neq \mathbf{0}$. That is, we wish to quickly eliminate deviations from the equilibrium point $\boldsymbol{x}_{\mathrm{eq0}}$ by means of a controller.

To derive a control law for system (5.192), here we can employ the control Lyapunov function

$$V(\boldsymbol{x}) = \frac{1}{2}x_2^2 - a_1 x_1 + a_3(\cos(\delta_{\mathrm{eq}}) - \cos(\delta_{\mathrm{eq}} + x_1)). \tag{5.193}$$

The condition $V(\mathbf{0}) = 0$ is satisfied, and $V(\boldsymbol{x}) > 0$ is valid within a certain neighborhood of the equilibrium point $\boldsymbol{x}_{\mathrm{eq0}} = \mathbf{0}$. However, the control Lyapunov function does not fulfill the condition $V(\boldsymbol{x}) > 0$ for higher values of $x_1 > 0$. From a practical point of view, however, this is not necessary either, because no higher values of $x_1 = \delta - \delta_{\mathrm{eq}}$ occur. Figure 5.36 illustrates the graph of the function $V(\boldsymbol{x})$ depending on the state variable x_1. Here we will restrict ourselves to the case in which $x_2 = 0$, because only for the component dependent on x_1, the course of $V(\boldsymbol{x})$ is not directly evident. By contrast, the term $0.5x_2^2$ takes the form of a parabola.

With equation (5.192) and equation (5.193), we obtain

$$\dot{V}(\boldsymbol{x}) = -a_2 x_2^2 - u \cdot a_3 x_2 \sin(\delta_{\mathrm{eq}} + x_1) \tag{5.194}$$

for the derivative of the control Lyapunov function. We choose u in equation (5.194) as

$$u = k x_2 \sin(\delta_{\mathrm{eq}} + x_1), \tag{5.195}$$

so that

$$\dot{V}(\boldsymbol{x}) = -a_2 x_2^2 - a_3 k x_2^2 \sin^2(\delta_{\mathrm{eq}} + x_1) \le 0$$

is fulfilled. Only the inequality $\dot{V}(\boldsymbol{x}) \le 0$ and not the stricter condition $\dot{V}(\boldsymbol{x}) < 0$ holds here, because $\dot{V}(\boldsymbol{x}) = 0$ holds for the set $\{\boldsymbol{x} \mid x_1 \in I\!R,\, x_2 = 0\}$. However, since only the state vector $\boldsymbol{x} = \mathbf{0}$ in this set satisfies the differential

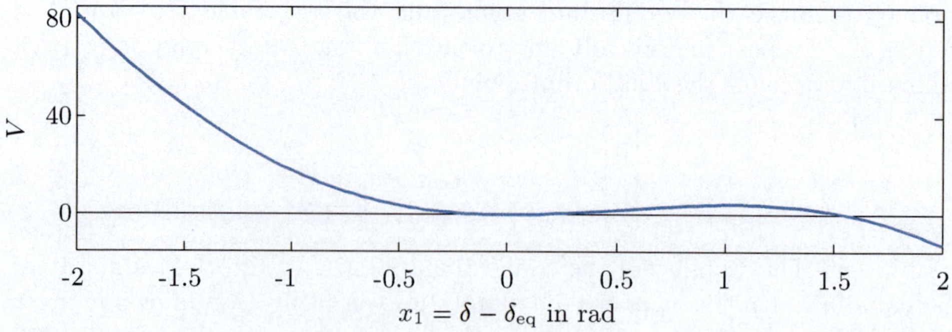

Fig. 5.36: Graph of the control Lyapunov function $V(\boldsymbol{x})$ for $x_2 = 0$ with $\delta_{\mathrm{eq}} = 1.05\,\mathrm{rad} = 60.41°$, $a_1 = 43.196$, and $a_3 = 49.676$

equation (5.192), $\dot{V}(\boldsymbol{x})$ is not identical to zero along any trajectory $\boldsymbol{x}(t)$ except for the trivial case in which $\boldsymbol{x}(t) = \mathbf{0}$. According to Definition 36, the function $V(\boldsymbol{x})$ is an extended control Lyapunov function, meaning that the stability of the equilibrium point $\boldsymbol{x}_{\text{eq}} = \mathbf{0}$ is ensured. The value $k > 0$ is a freely selectable parameter.

With its original coordinates δ and $\dot{\delta}$, using $x_1 = \delta - \delta_{\text{eq}}$ and $x_2 = \dot{\delta}$, the control law (5.195) takes the form

$$u = k\dot{\delta}\sin(\delta). \tag{5.196}$$

We will now present an example with $P_{\text{T}} = 540$ MW, $P_{\text{E}} = 621$ MVA, $D = 4.6889\,\text{MW\,s}$, $\omega_{\text{e}} = 2\pi \cdot 50\,\text{s}^{-1}$, $p = 1$, and $J = 39792\,\text{kg\,m}^2$, which results in

$$a_1 = \frac{P_{\text{T}}}{J\omega_{\text{e}}} = 43.196\,\text{s}^{-2}, \quad a_2 = \frac{D}{J\omega_{\text{e}}} = 0.375\,\text{s}^{-1}, \quad a_3 = \frac{P_{\text{E}}}{J\omega_{\text{e}}} = 49.676\,\text{s}^{-2}.$$

A three-phase short circuit from $t = 4\,\text{s}$ to $t = 4.06\,\text{s}$ can be viewed as a test case. In this time interval, the transmitted power is

$$P = P_{\text{E}}\sin(\delta) = 0.$$

For the controller constant, we will select the value $k = 0.075$.

Figure 5.37 shows the time courses of the angle $\delta = x_1 + \delta_{\text{eq}}$ and the trajectories $[\delta(t)\ \ \dot{\delta}(t)]^T$ for system (5.192), which is controlled by applying control law (5.195), or for system (5.191), which is controlled by applying control law (5.196), and for the free system. It is evident that both the free and the controlled system return to the equilibrium point, i.e. the operating point $\delta_{\text{eq}} = 1.05\,\text{rad} = 60.41°$, after the short circuit ends. The control system ensures that, in contrast to the free system, in this system the angle δ and thus the transmitted power $P = P_{\text{E}}\sin(\delta)$ barely oscillate.

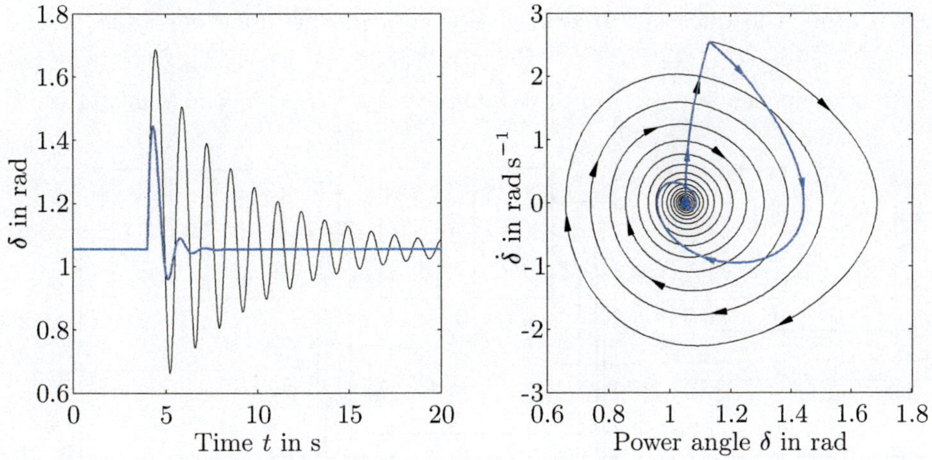

Fig. 5.37: Time courses of the power angle $\delta = x_1 + \delta_{\text{eq}}$ and the associated trajectories $[\delta(t)\ \ \dot{\delta}(t)]^T$ for the uncontrolled system (black) and the control loop (blue)

5.6 The Backstepping Method

5.6.1 Fundamentals

The backstepping procedure enables us to determine controllers and Lyapunov functions for nonlinear plants possessing the form

$$\begin{aligned}
\dot{\boldsymbol{x}}_1 &= \boldsymbol{f}_1(\boldsymbol{x}_1) + \boldsymbol{h}_1(\boldsymbol{x}_1) \cdot x_2, \\
\dot{x}_2 &= f_2(\boldsymbol{x}_1, x_2) + h_2(\boldsymbol{x}_1, x_2) \cdot x_3, \\
\dot{x}_3 &= f_3(\boldsymbol{x}_1, x_2, x_3) + h_3(\boldsymbol{x}_1, x_2, x_3) \cdot x_4, \\
&\ \vdots \\
\dot{x}_k &= f_k(\boldsymbol{x}_1, x_2, \ldots, x_k) + h_k(\boldsymbol{x}_1, x_2, \ldots, x_k) \cdot u
\end{aligned}$$

where $\boldsymbol{x}_1 \in \mathbb{R}^n$, $x_2, \ldots, x_k, u \in \mathbb{R}$ holds. The form of the above systems is called the *strict feedback form*. Such systems constitute a subclass of control-affine systems. Figure 5.38 illustrates the system structure of the plant.

If $\boldsymbol{x}_1$ is scalar, the output $y = x_1$ is flat, and consequently such systems in strict feedback form are flat. To prove this, we replace x_1 with y in all equations above and then insert the first equation into the second by eliminating x_2. In the second equation x_3 now only appears in dependence on $y, \dot{y}$, and $\ddot{y}$. When this is solved for x_3, we can insert the result into the third equation and so on, until we obtain both the state vector

$$\boldsymbol{x}^T = \begin{bmatrix} \boldsymbol{x}_1^T & x_2 & \cdots & x_k \end{bmatrix}$$

and the input variable u as functions of $y, \dot{y}, \ldots, y^{(k)}$. According to Theorem 50 on p. 251, the strict feedback form's controllability follows from its flatness in the SISO case.

As it is relatively simple to control control-affine systems using feedback linearization, the question arises: why would another method, such as the

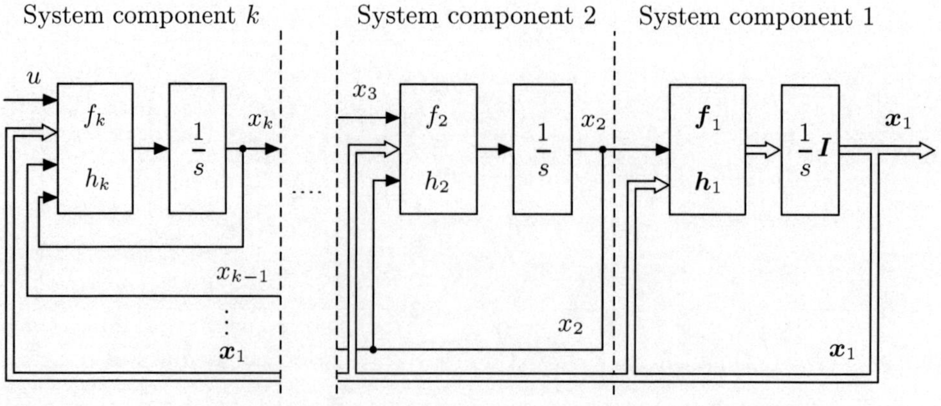

Fig. 5.38: Structure of a system in strict feedback form

backstepping method or the approaches based on control Lyapunov functions discussed in the previous section, be useful? One answer is that feedback linearization depends on an exact model of the plant. If there is a discrepancy between the plant and the model, we cannot achieve exact linearization. In this case, the control loop would be nonlinear and the feedback linearization method would fail in its objective. A second disadvantage of feedback linearization is that this method sometimes linearizes useful nonlinearities in the plant, involving an unnecessarily high actuating energy. The backstepping method and control Lyapunov functions both enable us to make suitable nonlinearities in the system usable for the control and also to render the control robust against inaccuracies in the plant model [251].

For the determination of the controller $u(\boldsymbol{x})$ and the Lyapunov function $V(\boldsymbol{x})$, we will begin our discussion with the special case

$$\dot{\boldsymbol{x}}_1 = \boldsymbol{f}(\boldsymbol{x}_1) + \boldsymbol{h}(\boldsymbol{x}_1) \cdot x_2, \tag{5.197}$$

$$\dot{x}_2 = u. \tag{5.198}$$

Figure 5.39 shows the structure of the system. The state variable x_2 is now taken as the input variable of system (5.197). A continuously differentiable control law

$$x_2 = \alpha(\boldsymbol{x}_1) \quad \text{with} \quad \alpha(\boldsymbol{0}) = 0 \tag{5.199}$$

is assumed to be known, so that $\boldsymbol{x}_1 = \boldsymbol{0}$ is an asymptotically stable equilibrium point in the α-controlled system. Here, x_2 is obviously not the real control variable. Rather, the state x_2 is used temporarily as a virtual control variable to derive the actual control law $u(\boldsymbol{x}_1, x_2)$.

Furthermore, a Lyapunov function $V(\boldsymbol{x}_1)$ is assumed to be known for system (5.197), which is controlled by the virtual control law (5.199); this means the inequality

$$\dot{V}(\boldsymbol{x}_1) = \frac{\partial V}{\partial \boldsymbol{x}_1}(\boldsymbol{f}(\boldsymbol{x}_1) + \boldsymbol{h}(\boldsymbol{x}_1) \cdot \alpha(\boldsymbol{x}_1)) < 0$$

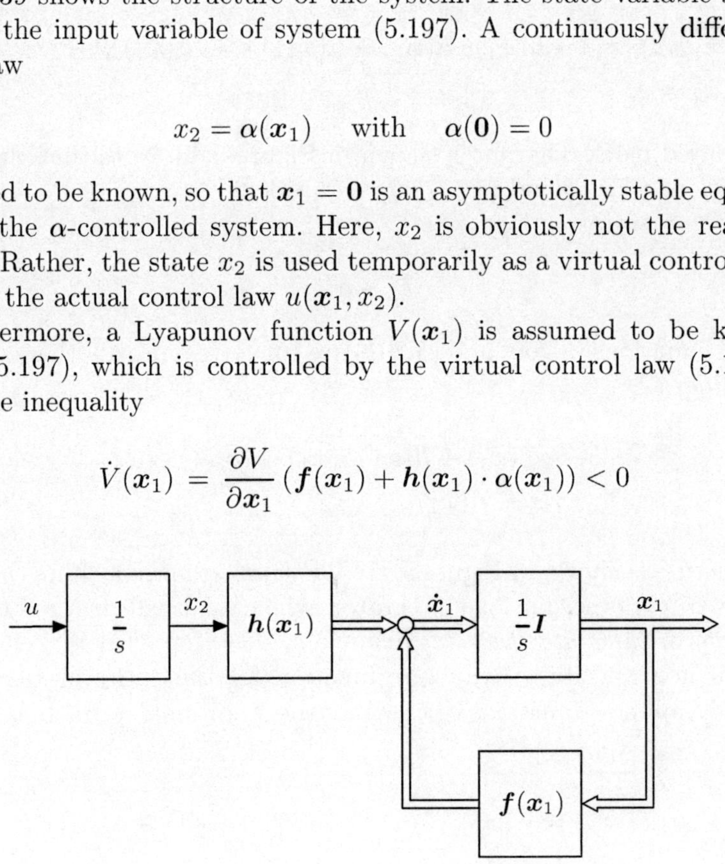

Fig. 5.39: Structure of the system $\dot{\boldsymbol{x}}_1 = \boldsymbol{f}(\boldsymbol{x}_1) + \boldsymbol{h}(\boldsymbol{x}_1) \cdot x_2$ with $\dot{x}_2 = u$

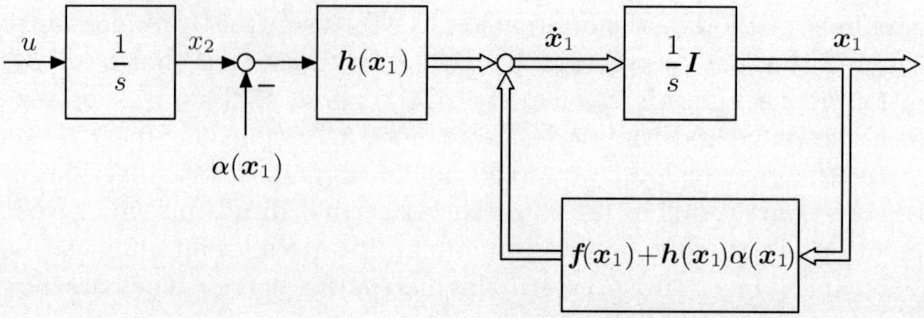

Fig. 5.40: System with a virtual control law $\alpha(\boldsymbol{x}_1)$, which is equivalent to the system from Figure 5.39

is fulfilled. To identify a Lyapunov function $V(\boldsymbol{x}_1)$ of this kind and a controller $\alpha(\boldsymbol{x}_1)$, we can use the method of control Lyapunov functions from the previous section, among other methods.

The system consisting of equations (5.197) and (5.198) can be represented in the form

$$\dot{\boldsymbol{x}}_1 = \boldsymbol{f}(\boldsymbol{x}_1) + \boldsymbol{h}(\boldsymbol{x}_1) \cdot \alpha(\boldsymbol{x}_1) + \boldsymbol{h}(\boldsymbol{x}_1)\,(x_2 - \alpha(\boldsymbol{x}_1)), \tag{5.200}$$

$$\dot{x}_2 = u. \tag{5.201}$$

The associated block diagram is shown in Figure 5.40. Note that the system representations (5.200) and (5.197) are equivalent.

Using the transformation

$$z = x_2 - \alpha(\boldsymbol{x}_1),$$

the system equations (5.200) and (5.201) are converted into another equivalent form yielding

$$\begin{aligned} \dot{\boldsymbol{x}}_1 &= \boldsymbol{f}(\boldsymbol{x}_1) + \boldsymbol{h}(\boldsymbol{x}_1)\alpha(\boldsymbol{x}_1) + \boldsymbol{h}(\boldsymbol{x}_1) \cdot z, \\ \dot{z} &= u - \dot{\alpha}(\boldsymbol{x}_1), \end{aligned} \tag{5.202}$$

whose structure is shown in Figure 5.41. The transformation shifts the virtual control law $\alpha(\boldsymbol{x}_1)$ preceding the integrator, which lends this method the name *backstepping* or *integrator backstepping*.

We will now specify a Lyapunov function for the entire system (5.202) using the Lyapunov function $V(\boldsymbol{x}_1)$ and the additional term $0.5z^2$, which yields

$$V_{\text{tot}}(\boldsymbol{x}_1, x_2) = V(\boldsymbol{x}_1) + \frac{1}{2}z^2 = V(\boldsymbol{x}_1) + \frac{1}{2}\,(x_2 - \alpha(\boldsymbol{x}_1))^2.$$

For the system (5.202), the derivative of this potential Lyapunov function is given by

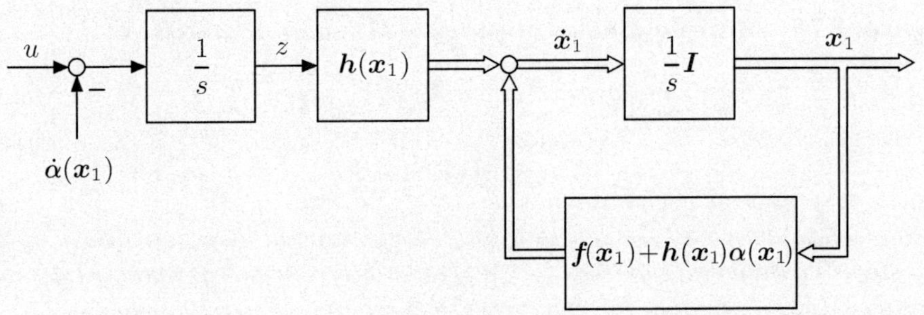

Fig. 5.41: System structure of Figure 5.40 where the virtual control law is shifted backward to the input of the integrator

$$\begin{aligned}\dot{V}_{\text{tot}}(\boldsymbol{x}_1, x_2) &= \frac{\partial V}{\partial \boldsymbol{x}_1}\dot{\boldsymbol{x}}_1 + z\dot{z} \\ &= \underbrace{\frac{\partial V}{\partial \boldsymbol{x}_1}(\boldsymbol{f}(\boldsymbol{x}_1)+\boldsymbol{h}(\boldsymbol{x}_1)\alpha(\boldsymbol{x}_1))}_{<0} + \frac{\partial V}{\partial \boldsymbol{x}_1}\boldsymbol{h}(\boldsymbol{x}_1)z + z\,(u - \dot{\alpha}(\boldsymbol{x}_1)).\end{aligned}$$

We note that the Lyapunov function's derivative $\dot{V}_{\text{tot}}$ depends on u.

We will now select

$$u = \dot{\alpha}(\boldsymbol{x}_1) - \frac{\partial V}{\partial \boldsymbol{x}_1}\boldsymbol{h}(\boldsymbol{x}_1) - k \cdot z, \quad k > 0, \tag{5.203}$$

as the control law such that

$$\dot{V}_{\text{tot}}(\boldsymbol{x}_1, x_2) = \frac{\partial V}{\partial \boldsymbol{x}_1}(\boldsymbol{f}(\boldsymbol{x}_1) + \boldsymbol{h}(\boldsymbol{x}_1)\alpha(\boldsymbol{x}_1)) - kz^2 < 0$$

holds for all $\boldsymbol{x}_1$. Further, with

$$\dot{\boldsymbol{x}}_1 = \boldsymbol{f}(\boldsymbol{x}_1) + \boldsymbol{h}(\boldsymbol{x}_1)x_2,$$

we next obtain

$$\begin{aligned}u &= \dot{\alpha}(\boldsymbol{x}_1) - \frac{\partial V}{\partial \boldsymbol{x}_1}\boldsymbol{h}(\boldsymbol{x}_1) - k \cdot z \\ &= \frac{\partial \alpha}{\partial \boldsymbol{x}_1}\dot{\boldsymbol{x}}_1 - \frac{\partial V}{\partial \boldsymbol{x}_1}\boldsymbol{h}(\boldsymbol{x}_1) - k\,(x_2 - \alpha(\boldsymbol{x}_1)) \\ &= \frac{\partial \alpha}{\partial \boldsymbol{x}_1}(\boldsymbol{f}(\boldsymbol{x}_1) + \boldsymbol{h}(\boldsymbol{x}_1)x_2) - \frac{\partial V}{\partial \boldsymbol{x}_1}\boldsymbol{h}(\boldsymbol{x}_1) - k\,(x_2 - \alpha(\boldsymbol{x}_1))\end{aligned}$$

from equation (5.203). Here, k is a freely selectable positive parameter which influences the dynamics of the control loop. High values of k lead to a fast decrease in V_{tot} and thus usually to a faster control.

With V_{tot}, we have identified a Lyapunov function, which happens to be a control Lyapunov function, and with the function u defined above a control law for the plant (5.197), (5.198) has been identified as well. We can summarize the above results in

Theorem 78 (Simple Backstepping). *Let there be a system*

$$\dot{\boldsymbol{x}}_1 = \boldsymbol{f}(\boldsymbol{x}_1) + \boldsymbol{h}(\boldsymbol{x}_1) \cdot x_2, \tag{5.204}$$

$$\dot{x}_2 = u. \tag{5.205}$$

We will assume that a virtual control law $x_2 = \alpha(\boldsymbol{x}_1)$ with $\alpha(\boldsymbol{0}) = 0$ is known for the subsystem (5.204) such that the virtual control loop has an asymptotically stable equilibrium point $\boldsymbol{x}_{1\text{eq}} = \boldsymbol{0}$. If the latter is the case and a Lyapunov function $V(\boldsymbol{x}_1)$ is known for the virtual control loop, then the control law

$$u = \frac{\partial \alpha(\boldsymbol{x}_1)}{\partial \boldsymbol{x}_1} \left(\boldsymbol{f}(\boldsymbol{x}_1) + \boldsymbol{h}(\boldsymbol{x}_1)\, x_2 \right) - \frac{\partial V}{\partial \boldsymbol{x}_1} \boldsymbol{h}(\boldsymbol{x}_1) - k \left(x_2 - \alpha(\boldsymbol{x}_1) \right)$$

with an arbitrary $k > 0$ stabilizes the system's equilibrium point

$$\left[\boldsymbol{x}_{1\text{eq}}^T \; x_{2\text{eq}} \right]^T = \boldsymbol{0}$$

asymptotically, and

$$V_{\text{tot}}(\boldsymbol{x}_1, x_2) = V(\boldsymbol{x}_1) + \frac{1}{2} \left(x_2 - \alpha(\boldsymbol{x}_1) \right)^2$$

is a Lyapunov function for the entire control loop.

If $V(\boldsymbol{x}_1) \to \infty$ also applies to $|\boldsymbol{x}_1| \to \infty$ in the theorem above, the equilibrium point of the control system is globally asymptotically stable.

The result above is only moderately useful in the form shown above, because for the theorem to be applied it is necessary to find the control law $\alpha(\boldsymbol{x}_1)$ and the Lyapunov function $V(\boldsymbol{x}_1)$ for the subsystem (5.204), which constitutes the essential part of the system (5.204), (5.205). Thus, we would certainly find a controller for this system even without the theorem. The result, however, assuming V and α are known, provides the starting point for a valuable result to be derived.

For this purpose, we will now deal with the system

$$\begin{aligned} \dot{\boldsymbol{x}}_1 &= \boldsymbol{f}_1(\boldsymbol{x}_1) + \boldsymbol{h}_1(\boldsymbol{x}_1) \cdot x_2, \\ \dot{x}_2 &= f_2(\boldsymbol{x}_1, x_2) + h_2(\boldsymbol{x}_1, x_2) \cdot u. \end{aligned} \tag{5.206}$$

By applying

$$u = \frac{u_2 - f_2(\boldsymbol{x}_1, x_2)}{h_2(\boldsymbol{x}_1, x_2)}, \qquad h_2(\boldsymbol{x}_1, x_2) \neq 0,$$

we transform it to

$$\begin{aligned} \dot{\boldsymbol{x}}_1 &= \boldsymbol{f}_1(\boldsymbol{x}_1) + \boldsymbol{h}_1(\boldsymbol{x}_1) \cdot x_2, \\ \dot{x}_2 &= u_2. \end{aligned}$$

For this transformed system, the control law $u_2(\boldsymbol{x}_1, x_2)$ and the Lyapunov function $V(\boldsymbol{x}_1)$ are known from Theorem 78. Therefore the control law u for the system (5.206) can also be directly specified by using

Theorem 79 (Backstepping). *Let there be a system*

$$\begin{aligned}\dot{\boldsymbol{x}}_1 &= \boldsymbol{f}_1(\boldsymbol{x}_1) + \boldsymbol{h}_1(\boldsymbol{x}_1) \cdot x_2,\\ \dot{x}_2 &= f_2(\boldsymbol{x}_1, x_2) + h_2(\boldsymbol{x}_1, x_2) \cdot u.\end{aligned} \tag{5.207}$$

We will assume that a virtual control law $x_2 = \alpha(\boldsymbol{x}_1)$ *with* $\alpha(\boldsymbol{0}) = 0$ *is known for the subsystem (5.207) such that the virtual control loop has an asymptotically stable equilibrium point* $\boldsymbol{x}_{1\mathrm{eq}} = \boldsymbol{0}$. *If the latter is the case and a Lyapunov function* $V(\boldsymbol{x}_1)$ *is known for the virtual control loop, then the control law*

$$\begin{aligned}u = \frac{1}{h_2(\boldsymbol{x}_1, x_2)} \cdot \Bigg(&\frac{\partial \alpha(\boldsymbol{x}_1)}{\partial \boldsymbol{x}_1} (\boldsymbol{f}_1(\boldsymbol{x}_1) + \boldsymbol{h}_1(\boldsymbol{x}_1) x_2)\\ &- \frac{\partial V}{\partial \boldsymbol{x}_1} \boldsymbol{h}_1(\boldsymbol{x}_1) - k\,(x_2 - \alpha(\boldsymbol{x}_1)) - f_2(\boldsymbol{x}_1, x_2)\Bigg)\end{aligned}$$

with an arbitrary $k > 0$ *stabilizes the system's equilibrium point*

$$[\boldsymbol{x}_{1\mathrm{eq}}^T \; x_{2\mathrm{eq}}]^T = \boldsymbol{0}$$

asymptotically, and

$$V_{\mathrm{tot}}(\boldsymbol{x}_1, x_2) = V(\boldsymbol{x}_1) + \frac{1}{2}\,(x_2 - \alpha(\boldsymbol{x}_1))^2$$

is a Lyapunov function for the entire control loop.

5.6.2 Recursive Scheme for the Controller Design

Based on the theorems above, for the systems described previously which take the form

$$\begin{aligned}\dot{\boldsymbol{x}}_1 &= \boldsymbol{f}_1(\boldsymbol{x}_1) + \boldsymbol{h}_1(\boldsymbol{x}_1) \cdot x_2,\\ \dot{x}_2 &= f_2(\boldsymbol{x}_1, x_2) + h_2(\boldsymbol{x}_1, x_2) \cdot x_3,\\ &\;\;\vdots\\ \dot{x}_k &= f_k(\boldsymbol{x}_1, \ldots, x_k) + h_k(\boldsymbol{x}_1, \ldots, x_k) \cdot u,\end{aligned}$$

we can now determine control laws $u(\boldsymbol{x})$ and Lyapunov functions $V(\boldsymbol{x})$. For this purpose, we will proceed with the following recursive design process in steps:

Step 1: Consider the subsystem

$$\begin{aligned}\dot{\boldsymbol{x}}_1 &= \boldsymbol{f}_1(\boldsymbol{x}_1) + \boldsymbol{h}_1(\boldsymbol{x}_1) \cdot x_2,\\ \dot{x}_2 &= f_2(\boldsymbol{x}_1, x_2) + h_2(\boldsymbol{x}_1, x_2) \cdot x_3,\end{aligned}$$

which we will call T_1. Design a virtual controller $x_3(\boldsymbol{x}_1, x_2)$ and a Lyapunov function $V_1(\boldsymbol{x}_1, x_2)$ using Theorem 79 or, if possible, Theorem 78.

Step 2: Summarize the subsystem T_1 to formulate a differential equation

$$\dot{\boldsymbol{x}}_1 = \begin{bmatrix} \dot{\boldsymbol{x}}_1 \\ \dot{x}_2 \end{bmatrix} = \begin{bmatrix} \boldsymbol{f}_1(\boldsymbol{x}_1) + \boldsymbol{h}_1(\boldsymbol{x}_1) \cdot x_2 \\ f_2(\boldsymbol{x}_1, x_2) + h_2(\boldsymbol{x}_1, x_2) \cdot x_3 \end{bmatrix},$$

and then formulate subsystem T_2:

$$\dot{\boldsymbol{x}}_1 = \begin{bmatrix} \boldsymbol{f}_1(\boldsymbol{x}_1) + \boldsymbol{h}_1(\boldsymbol{x}_1) \cdot x_2 \\ f_2(\boldsymbol{x}_1, x_2) \end{bmatrix} + \begin{bmatrix} 0 \\ h_2(\boldsymbol{x}_1, x_2) \end{bmatrix} x_3,$$

$$\dot{x}_3 = f_3(\boldsymbol{x}_1, x_2, x_3) + h_3(\boldsymbol{x}_1, x_2, x_3) \cdot x_4.$$

This subsystem T_2 consists of the first three differential equations of the total system. Its form corresponds to what we saw in Theorem 79. Since a virtual controller $x_3(\boldsymbol{x}_1, x_2)$ and a Lyapunov function $V_1(\boldsymbol{x}_1, x_2)$ are known from the first step, another virtual control law $x_4(\boldsymbol{x}_1, x_2, x_3)$ and a Lyapunov function $V_2(\boldsymbol{x}_1, x_2, x_3)$ for T_2 can be derived using Theorem 79.

Step 3: Summarize subsystem T_2 as a differential equation

$$\dot{\boldsymbol{x}}_1 = \begin{bmatrix} \dot{\boldsymbol{x}}_1 \\ x_2 \\ x_3 \end{bmatrix} = \begin{bmatrix} \boldsymbol{f}_1(\boldsymbol{x}_1) + \boldsymbol{h}_1(\boldsymbol{x}_1) \cdot x_2 \\ f_2(\boldsymbol{x}_1, x_2) + h_2(\boldsymbol{x}_1, x_2) \cdot x_3 \\ f_3(\boldsymbol{x}_1, x_2, x_3) + h_3(\boldsymbol{x}_1, x_2, x_3) \cdot x_4 \end{bmatrix}$$

and formulate subsystem T_3:

$$\dot{\boldsymbol{x}}_1 = \begin{bmatrix} \boldsymbol{f}_1(\boldsymbol{x}_1) + \boldsymbol{h}_1(\boldsymbol{x}_1)x_2 \\ f_2(\boldsymbol{x}_1, x_2) + h_2(\boldsymbol{x}_1, x_2)x_3 \\ f_3(\boldsymbol{x}_1, x_2, x_3) \end{bmatrix} + \begin{bmatrix} 0 \\ 0 \\ h_3(\boldsymbol{x}_1, x_2, x_3) \end{bmatrix} x_4,$$

$$\dot{x}_4 = f_4(\boldsymbol{x}_1, x_2, x_3, x_4) + h_4(\boldsymbol{x}_1, x_2, x_3, x_4) \cdot x_5.$$

Subsystem T_3 corresponds to the system in Theorem 79. Also, a virtual controller $x_4(\boldsymbol{x}_1, x_2, x_3)$ and a Lyapunov function $V_2(\boldsymbol{x}_1, x_2, x_3)$are known from Step 2, so that a control law $x_5(\boldsymbol{x}_1, x_2, x_3, x_4)$ and a Lyapunov function $V_3(\boldsymbol{x}_1, x_2, x_3, x_4)$ can be derived for T_3 using Theorem 79.

Step 4: ...

⋮

Step $k-1$: ...

Steps 2 and 3 are almost identical; Step 4 and all subsequent steps not listed here also correspond to their predecessors, so that it is possible to advance recursively from the subsystem T_1 to the subsystem T_2, from there to T_3, then to T_4, and so on. The sequence below illustrates this:

$$\left.\begin{array}{l}\left.\begin{array}{l}\left.\begin{array}{l}\left.\begin{array}{l}\dot{\boldsymbol{x}}_1 = \boldsymbol{f}_1(\boldsymbol{x}_1) + \boldsymbol{h}_1(\boldsymbol{x}_1)\cdot x_2\\ \dot{x}_2 = f_2(\boldsymbol{x}_1, x_2) + h_2(\boldsymbol{x}_1, x_2)\cdot x_3\end{array}\right\} T_1\\ \dot{x}_3 = f_3(\boldsymbol{x}_1, \dots, x_3) + h_3(\boldsymbol{x}_1, \dots, x_3)\cdot x_4\end{array}\right\} T_2\\ \dot{x}_4 = f_4(\boldsymbol{x}_1, \dots, x_4) + h_4(\boldsymbol{x}_1, \dots, x_4)\cdot x_5\end{array}\right\} T_3 \quad \dots\\ \quad\vdots\\ \dot{x}_k = f_k(\boldsymbol{x}_1, \dots, x_k) + h_k(\boldsymbol{x}_1, \dots, x_k)\cdot u\,.\end{array}\right\} T_{k-1}$$

Ultimately, after $k-1$ steps, we have derived the control law u we were seeking, and also a Lyapunov function of the entire control loop.

The advantage of the backstepping method is its systematic design technique, which leads to a stable and robust control loop. Its disadvantage is that the control performance can only be predicted or influenced to a limited extent. Extensions of and additions to the backstepping procedure can be found in [251, 252, 411, 500].

5.6.3 Illustrative Examples

As a preliminary example, we will examine the plant

$$\dot{x}_1 = x_1^2 - x_1 + x_2, \tag{5.208}$$

$$\dot{x}_2 = u. \tag{5.209}$$

For the design of a controller $u(x_1, x_2)$ and a Lyapunov function $V_{\text{tot}}(x_1, x_2)$, we will apply Theorem 78. First, a controller

$$x_2 = \alpha(x_1) = -x_1^2 - x_1$$

is designed for the subsystem (5.208). Clearly, this controller stabilizes the subsystem (5.208), because inserting it into equation (5.208) yields

$$\dot{x}_1 = -2x_1.$$

Furthermore,

$$V(x_1) = \frac{1}{2}x_1^2$$

is a Lyapunov function of the virtual control loop which proves that the loop is globally asymptotically stable, because

$$\dot{V}(x_1) = x_1\dot{x}_1 = -2x_1^2 < 0$$

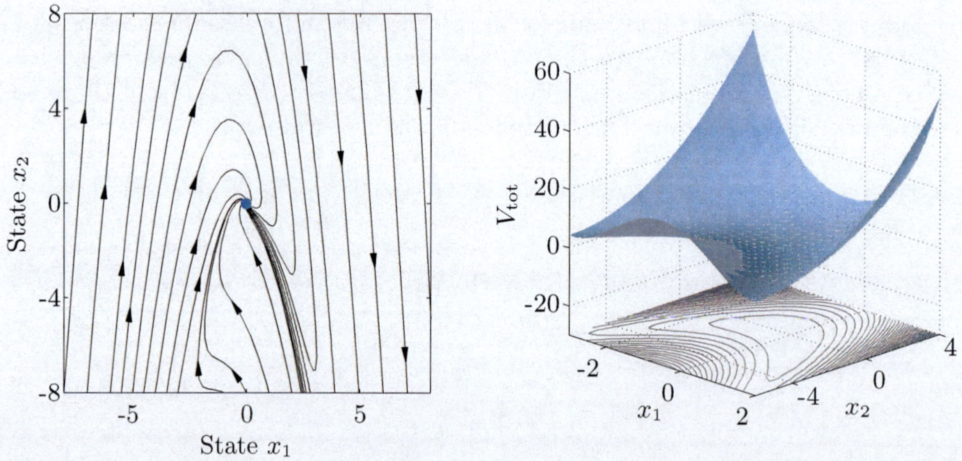

Fig. 5.42: Trajectories of the plant and Lyapunov function V_{tot} from the first example

holds for all points $x_1 \neq 0$. From Theorem 78, we now derive the control law

$$u = \frac{\partial \alpha}{\partial x_1}\left(x_1^2 - x_1 + x_2\right) - \frac{\partial V}{\partial x_1} - k(x_2 - \alpha(x_1))$$

for the entire system consisting of equations (5.208) and (5.209), where $k > 0$ holds. With

$$\frac{\partial \alpha}{\partial x_1} = \frac{\partial(-x_1^2 - x_1)}{\partial x_1} = -(2x_1 + 1)$$

and

$$\frac{\partial V}{\partial x_1} = x_1,$$

the control law

$$\begin{aligned} u &= -(2x_1 + 1)(x_1^2 - x_1 + x_2) - 2x_1 - x_1^2 - x_2 \\ &= -x_1 - 2x_1^3 - 2x_2 - 2x_1 x_2 \end{aligned} \tag{5.210}$$

is obtained for $k = 1$. For the entire control loop (5.208), (5.209), (5.210), the Lyapunov function is

$$V_{\text{tot}}(x_1, x_2) = V(x_1) + \frac{1}{2}\left(x_2 - \alpha(x_1)\right)^2 = \frac{1}{2}x_1^2 + \frac{1}{2}\left(x_1 + x_1^2 + x_2\right)^2.$$

Figure 5.42 shows the trajectories of the plant and the form of the Lyapunov function V_{tot}.

In the second example, we will begin with the controlled system from the first example and add an integrator to it, so that

$$\dot{x}_1 = x_1^2 - x_1 + x_2, \tag{5.211}$$
$$\dot{x}_2 = x_3, \tag{5.212}$$
$$\dot{x}_3 = u. \tag{5.213}$$

For the design of a control law $u(x_1, x_2, x_3)$, we can again apply Theorem 78. To design the controller, we first split the system denoted by (5.211), (5.212), (5.213) into two subsystems, i. e. into the subsystem T_1 consisting of equations (5.211) and (5.212), and the total system T_2 consisting of equations (5.211), (5.212), (5.213), such that we obtain

$$\left.\begin{array}{l} \left.\begin{bmatrix} \dot{x}_1 \\ \dot{x}_2 \end{bmatrix} = \underbrace{\begin{bmatrix} x_1^2 - x_1 + x_2 \\ 0 \end{bmatrix}}_{\boldsymbol{f}(x_1,x_2)} + \underbrace{\begin{bmatrix} 0 \\ 1 \end{bmatrix}}_{\boldsymbol{h}(x_1,x_2)} \cdot x_3, \quad \right\} T_1 \\ \dot{x}_3 = u. \end{array}\right\} T_2$$

For subsystem T_1, from equation (5.210) in the first example, we know that a controller exists, namely

$$x_3 = \alpha(x_1, x_2) = -x_1 - 2x_1^3 - 2x_2 - 2x_1 x_2,$$

as well as a Lyapunov function

$$V(x_1, x_2) = \frac{1}{2}x_1^2 + \frac{1}{2}(x_1 + x_1^2 + x_2)^2.$$

Thus Theorem 78 immediately yields the control law

$$\begin{aligned} u &= \frac{\partial \alpha(x_1, x_2)}{\partial [x_1 \; x_2]^T} \left(\boldsymbol{f}(x_1, x_2) + \boldsymbol{h}(x_1, x_2) \cdot x_3 \right) \\ &\quad - \frac{\partial V(x_1, x_2)}{\partial [x_1 \; x_2]^T} \cdot \boldsymbol{h}(x_1, x_2) - k\,(x_3 - \alpha(x_1, x_2)) \\ &= \frac{\partial \alpha}{\partial x_1}\left(x_1^2 - x_1 + x_2\right) + \frac{\partial \alpha}{\partial x_2} x_3 - \frac{\partial V}{\partial x_2} - k(x_3 - \alpha(x_1, x_2)) \end{aligned}$$

for the total system T_2 with its system equations (5.211), (5.212), (5.213). For $k = 1$ and

$$\begin{aligned} \frac{\partial \alpha(x_1, x_2)}{\partial x_1} &= -6x_1^2 - 2x_2 - 1, \\ \frac{\partial \alpha(x_1, x_2)}{\partial x_2} &= -2x_1 - 2, \\ \frac{\partial V(x_1, x_2)}{\partial x_2} &= x_2 + x_1 + x_1^2, \end{aligned}$$

the control law, of which we see a simulation in Figure 5.43, can now be formulated as

$$u = -x_1 - 2x_1^2 + 4x_1^3 - 6x_1^4 - 4x_2 - 8x_1^2x_2 - 2x_2^2 - 3x_3 - 2x_1x_3.$$

A simulation of the control loop is shown in Figure 5.43, where the vector of the initial values is

$$\boldsymbol{x}_0 = [1 \;\; 1.5 \;\; 1]^T.$$

For the Lyapunov function of the entire control loop, we obtain

$$\begin{aligned} V_{\text{tot}}(x_1, x_2, x_3) &= V(x_1, x_2) + \frac{1}{2}\left(x_3 - \alpha(x_1, x_2)\right)^2 \\ &= \frac{1}{2}x_1^2 + \frac{1}{2}\left(x_1 + x_1^2 + x_2\right)^2 \\ &+ \frac{1}{2}\left(x_3 + x_1 + 2x_1^3 + 2x_2 + 2x_1x_2\right)^2. \end{aligned}$$

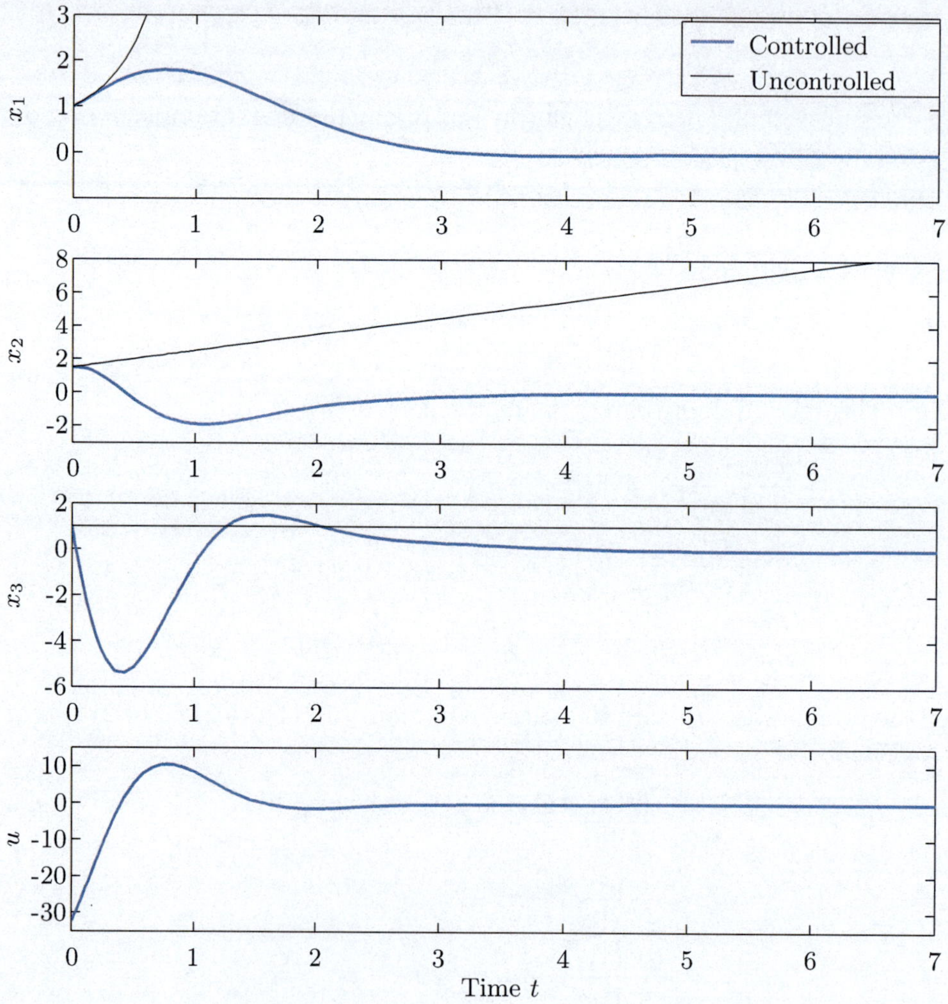

Fig. 5.43: Time courses of the state variables x_1, x_2, and x_3 and the control variable u from the second example

On the one hand, we recognize the integrating character of the plant: x_3 is constant, x_2 increases linearly, and x_1 increases at a greater than quadratic rate. Thus, the system is unstable. As can be seen from the simulation, and as expected from the design, the control stabilizes the system. In addition, we have achieved high control performance.

5.6.4 Example: Fluid System with Chaotic Behavior

We will design a backstepping controller for an experimental fluid system [87, 416, 470] consisting of a water-filled toroidal tube. The lower half of this torus is electrically heated using a filament. The filament is wrapped tightly around the torus and surrounded by insulating material so that the lower half of the torus is uniformly heated. The upper half of the torus is surrounded by a ring-shaped cooling jacket through which cooling water flows. The temperature of the torus wall of the entire upper half is thus maintained at a constant level. Figure 5.44 shows the configuration. The torus has a radius of $r_1 = 38\,\text{cm}$ and the radius of the torus tube is $r_2 = 1.5\,\text{cm}$.

Sensors are mounted at positions A, B, C, and D to measure the temperature of the water contained in the torus at these points. At a low heating power, meaning less than 190 W, there is a continuous flow of water inside the torus. The water flows clockwise or counterclockwise, depending on the

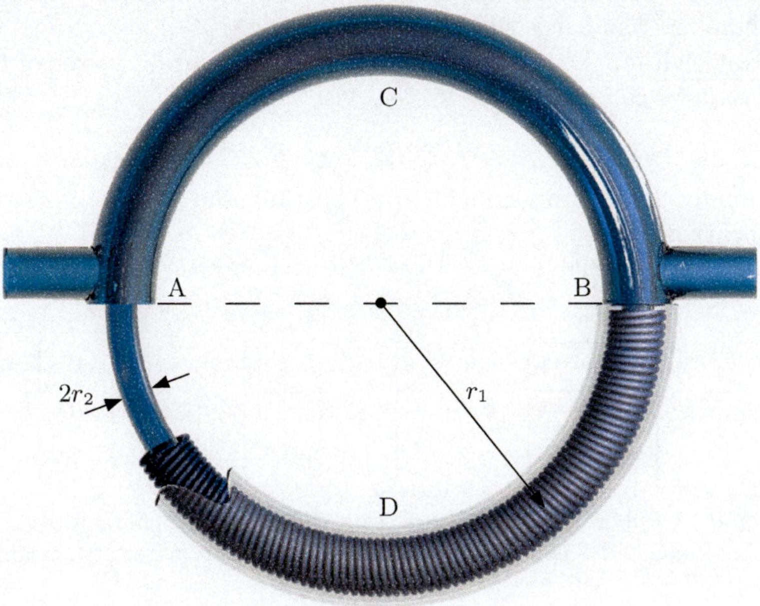

Fig. 5.44: Fluid system with chaotic behavior. The inner toroidal tube contains water. The upper half of the torus is cooled, the lower half heated.

initial values. If the heating power is now increased to 600 W, for example, there is no more continuous water flow within the torus. Under this condition, the water flow continually changes its direction. Sometimes the water flows clockwise, and sometimes it flows counterclockwise. The changes in direction of the water flow are chaotic, not predictable.

The emergence of this chaotic behavior can be explained as follows. If the speed of the water flow is reduced by a small irregularity, the water remains longer in the heating or cooling area of the torus. This increases the temperature difference between the cold water in the upper half of the torus and the hot water in the lower half. As a result, the flow velocity increases due to the increased temperature difference. This leads to a more rapid mixing of the hot water from the lower half with the cold water from the upper half. Thus, the temperature difference between the cold and hot water decreases again, causing a decrease in the flow velocity as well and, in some cases, changes in its direction.

The above behavior leads to oscillations and finally to the chaotic behavior mentioned above. It can be described by the Lorenz equations

$$\begin{aligned} \dot{z}_1 &= p(z_2 - z_1), \\ \dot{z}_2 &= -z_1 z_3 - z_2, \\ \dot{z}_3 &= z_1 z_2 - z_3 - R. \end{aligned} \tag{5.214}$$

Here, the variable z_1 is a measure of the mean flow velocity, z_2 is a measure of the temperature difference $T_{\mathrm{B}} - T_{\mathrm{A}}$ between points A and B, and z_3 is a measure of the temperature difference $T_{\mathrm{C}} - T_{\mathrm{D}}$. Furthermore, $p > 0$ is the Prandtl number[7] and R is the Rayleigh number[8].

The Rayleigh number R is proportional to the heating power of the filament, so that we can control the system via

$$R = R_0 + u.$$

The parameter R_0 is proportional to a constant and u to a superimposed, variable heating power.

For the control variable value $u = 0$, the system has the three equilibrium points

$$\boldsymbol{z}_{\mathrm{eq1}} = \begin{bmatrix} \sqrt{R_0 - 1} \\ \sqrt{R_0 - 1} \\ -1 \end{bmatrix}, \quad \boldsymbol{z}_{\mathrm{eq2}} = \begin{bmatrix} -\sqrt{R_0 - 1} \\ -\sqrt{R_0 - 1} \\ -1 \end{bmatrix}, \quad \boldsymbol{z}_{\mathrm{eq3}} = \begin{bmatrix} 0 \\ 0 \\ -R_0 \end{bmatrix}.$$

[7] The Prandtl number $p = \nu/a$ is a constant dimensionless characteristic number describing fluids and is equal to the quotient of kinematic viscosity ν and thermal diffusivity a.

[8] The temperature-dependent dimensionless Rayleigh number is an indicator of the type of heat transfer in a fluid. Below a critical Rayleigh number, the heat transfer in the fluid occurs primarily by heat conduction; above this value it occurs by convection. The Rayleigh number indicates the stability of thermal fluid layers.

We will now control the system (5.214) such that $\boldsymbol{z}_{\text{eq1}}$ is an asymptotically stable equilibrium point. For this purpose, the first step is to transform the coordinates of the system so that the equilibrium point $\boldsymbol{z}_{\text{eq1}}$ is shifted to the origin. This is done by means of the coordinate transformation $\boldsymbol{x} = \boldsymbol{z} - \boldsymbol{z}_{\text{eq1}}$, i. e.

$$\begin{bmatrix} x_1 \\ x_2 \\ x_3 \end{bmatrix} = \begin{bmatrix} z_1 - \sqrt{R_0 - 1} \\ z_2 - \sqrt{R_0 - 1} \\ z_3 + 1 \end{bmatrix}.$$

The transformed system has the equilibrium point $\boldsymbol{x}_{\text{eq}} = \boldsymbol{0}$, and with $\beta = \sqrt{R_0 - 1}$ it takes the form

$$\dot{x}_1 = -px_1 + px_2, \tag{5.215}$$

$$\dot{x}_2 = x_1 - x_2 - (x_1 + \beta)x_3, \tag{5.216}$$

$$\dot{x}_3 = x_1x_2 + \beta(x_1 + x_2) - x_3 - u. \tag{5.217}$$

The system is in strict feedback form, so we can directly apply the backstepping design theorems.

We will begin with the selection of the virtual control law

$$x_2 = \alpha_1(x_1) = 0,$$

which stabilizes subsystem (5.215). As a Lyapunov function, we choose

$$V_1(x_1) = \frac{1}{2}x_1^2.$$

With $x_2 = \alpha_1(x_1) = 0$ and equation (5.215), the inequality

$$\dot{V}_1(x_1) = -px_1^2 < 0$$

follows. We will now apply Theorem 79 to the subsystem (5.215), (5.216), i. e.

$$\begin{aligned} \dot{x}_1 &= \underbrace{-px_1}_{f_1(x_1)} + \underbrace{p}_{h_1(x_1)} \cdot x_2, \\ \dot{x}_2 &= \underbrace{x_1 - x_2}_{f_2(x_1,x_2)} \underbrace{-(x_1 + \beta)}_{h_2(x_1,x_2)} x_3, \end{aligned}$$

yielding

$$x_3 = \alpha_2(x_1, x_2) = -\frac{1}{x_1 + \beta}(-px_1 - k_1x_2 - x_1 + x_2)$$

as a virtual control law. We will now set $k_1 = 1$, resulting in

$$x_3 = \alpha_2(x_1, x_2) = \frac{(1 + p)x_1}{x_1 + \beta},$$

and

$$V_2(x_1, x_2) = \frac{1}{2}(x_1^2 + x_2^2)$$

for the Lyapunov function.

In the next step, Theorem 79 is applied again, but now the focus is on the complete system consisting of equations (5.215), (5.216), (5.217).

With the terms used in Theorem 79, we can formulate the system description

$$\begin{bmatrix} \dot{x}_1 \\ \dot{x}_2 \end{bmatrix} = \underbrace{\begin{bmatrix} -px_1 + px_2 \\ x_1 - x_2 \end{bmatrix}}_{\boldsymbol{f}_1(x_1,x_2)} + \underbrace{\begin{bmatrix} 0 \\ -(x_1+\beta) \end{bmatrix}}_{\boldsymbol{h}_1(x_1,x_2)} x_3,$$

$$\dot{x}_3 = \underbrace{x_1x_2 + \beta(x_1+x_2) - x_3}_{f_2(x_1,x_2,x_3)} \quad \underbrace{-1}_{h_2(x_1,x_2,x_3)} \cdot u.$$

After doing this, and including

$$\frac{\partial \alpha_2}{\partial [x_1\ x_2]^T} = \begin{bmatrix} \frac{(1+p)\beta}{(x_1+\beta)^2} & 0 \end{bmatrix}$$

as well as

$$\frac{\partial V_2}{\partial [x_1\ x_2]^T} = [x_1\ x_2],$$

we obtain the expression

$$u = -x_3 + k_2x_3 - k_2\frac{(1+p)x_1}{x_1+\beta} + \beta x_1 - \frac{p(1+p)\beta(x_2-x_1)}{(x_1+\beta)^2}$$

for the control law. With $k_2 = 1$, the control law can finally be written as

$$u = \beta x_1 - \frac{(1+p)x_1}{x_1+\beta} - \frac{\beta p(1+p)(x_2-x_1)}{(x_1+\beta)^2}.$$

Now we will simulate both the uncontrolled and the controlled process for the parameters

$$p = 10 \quad \text{and} \quad \beta = 6.$$

The initial values are

$$x_1 = 5, \quad x_2 = 5, \quad \text{and} \quad x_3 = 5.$$

The resulting time courses for the flow velocity x_1 are shown in Figure 5.45, in which we can clearly see that the control suppresses the chaotic behavior and compensates the process to the stationary value.

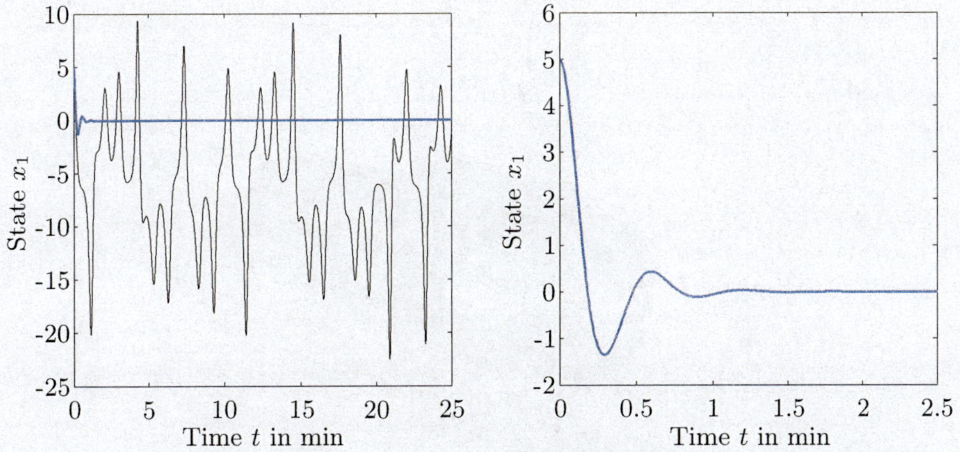

Fig. 5.45: Time courses of the deviation x_1 of the flow velocity from the equilibrium point of the chaotic process (black) and the control loop (blue). The diagram on the right shows an enlarged view of the control process.

5.7 Exercises

Exercise 5.1 Let us examine the longitudinal dynamics of a motor vehicle with the intention of designing a speed control using gain scheduling [332]. We obtain the motor vehicle's equation of motion based on the forces acting on it, which are shown in Figure 5.46. They are the air resistance

$$F_{\mathrm{d}} = \frac{1}{2}\rho c_{\mathrm{d}} A v^2,$$

where ρ is the air density, c_{d} is the drag coefficient of the vehicle, A is its front surface, and v its velocity. The rolling resistance is

$$F_{\mathrm{r}} = c_{\mathrm{r}} mg \cos(\varphi),$$

where c_{r} is the rolling resistance coefficient, m is the mass of the vehicle, g the gravitational acceleration, and φ the angle of the road's slope. Furthermore, we have to take into account the proportional gravitational force F_{s} acting on the vehicle due to the slope. In addition, the driving force or the braking force F_{a} act on the vehicle via the gas pedal or the brake.

(a) Formulate the state-space model for the vehicle, with $x = v$ being simultaneously the state variable and the output variable. The angle of the slope φ is a disturbance variable. The acceleration generated by the engine or brake is represented by the control variable $u = F_{\mathrm{a}}/m$.

(b) Calculate the parameterized linearization family with the velocity x as the scheduling parameter β. In doing this and in the steps that follow, set $\varphi = 0$.

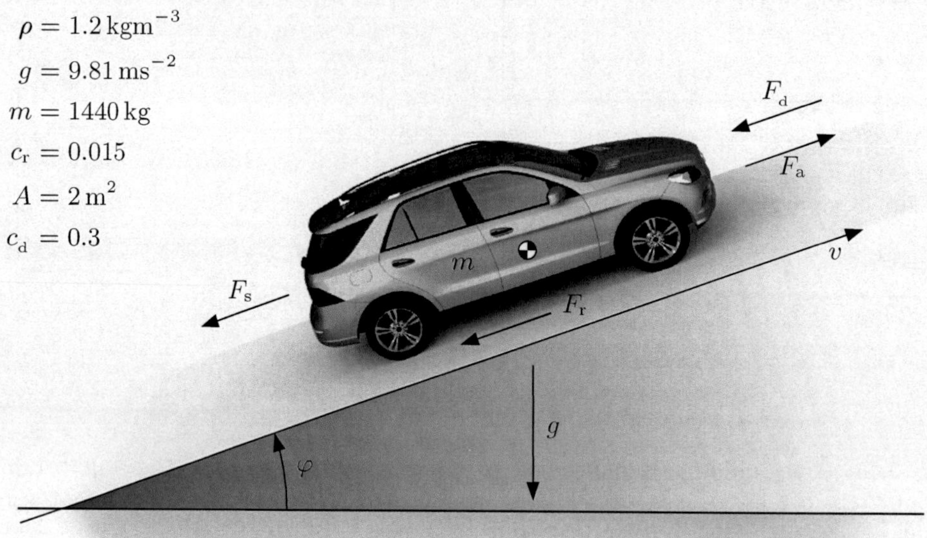

Fig. 5.46: Car on ascending road and average car parameters

(c) Now use the parameters from Figure 5.46 and determine three linearized models for the operating points $x_{\mathrm{op1}} = 10\,\mathrm{ms}^{-1}$, $x_{\mathrm{op2}} = 25\,\mathrm{ms}^{-1}$, and $x_{\mathrm{op3}} = 40\,\mathrm{ms}^{-1}$. For each of the three linearized models, calculate a PI controller

$$G_{\mathrm{PI}}(s) = P + \frac{I}{s}$$

such that all eigenvalues of each of the linear control loops are exactly ten times larger than the time constants of the linearized models in question.

(d) Identify the control law $u(x)$ based on the weighted average, using Gauss functions. Select the parameters σ_i of the Gauss functions such that σ_i in each case equals half the distance from one operating point $x_{\mathrm{op},i}$ to the next.

Exercise 5.2 Let us examine a spring-mass system with variable mass $m(t)$, as shown in Figure 5.47. The state-space representation is given by the following equations:

$$\begin{aligned}
\dot{x}_1 &= x_2,\\
\dot{x}_2 &= -\frac{c}{m(t)}x_1 + \frac{1}{m(t)}u,\\
y &= x_1.
\end{aligned}$$

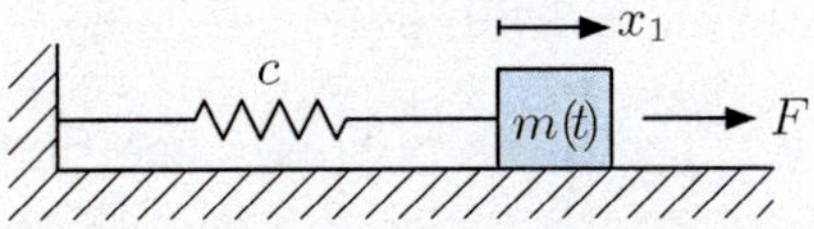

Fig. 5.47: Spring-mass system

Here, $c = 30\,\mathrm{Nm}^{-1}$ is the spring constant and u is the force being applied externally to the mass $m(t)$. Assume that we know the mass $m(t)$, which varies within the interval $[0.1\,\mathrm{kg}, 1.1\,\mathrm{kg}]$. The system is linear, but time-varying. Systems of this kind are called *linear parameter-varying systems (LPV systems).*

(a) For $m = 0.1\,\mathrm{kg}$ and $m = 1\,\mathrm{kg}$, calculate the eigenvalues of the system.

(b) Design a gain-scheduling controller based on the weighted average. As subcontrollers, select PID controllers with the transfer functions

$$H(s) = K_{\mathrm{P},i}\frac{T_{\mathrm{I},i}T_{\mathrm{D},i}s^2 + T_{\mathrm{I},i}s + 1}{T_{\mathrm{I},i}s}.$$

Use the mass $m(t)$ as a scheduling parameter β; as operating points, select $m_1 = 0.2\,\mathrm{kg}$, $m_2 = 0.4\,\mathrm{kg}$, $m_3 = 0.6\,\mathrm{kg}$, $m_4 = 0.8\,\mathrm{kg}$, and $m_5 = 1\,\mathrm{kg}$. The transfer function of the closed subcontrol loops should have the denominator $(s + 10)^3$ in all five cases. As weighting functions, select triangles and ramps whose sum is one.

(c) Design a single PID controller whose control parameters depend continuously on the mass $m(t)$ such that the control loop's transfer function has the denominator $(s + 10)^3$.

Exercise 5.3 Let us examine a model to calculate the human heartrate [449]:

$$\begin{aligned}
\dot{x}_1 &= -\frac{1}{\varepsilon}(x_1^3 + x_1x_2 + x_3), \quad 0 < \varepsilon \ll 1,\\
\dot{x}_2 &= -2x_1 - 2x_2,\\
\dot{x}_3 &= -x_2 - 1 + u,\\
y &= x_3.
\end{aligned}$$

Here, ε is a parameter, e. g. $\varepsilon = 0.2$, x_1 represents the current length of the fibers of the heart muscle, x_2 the tension of the fibers, and x_3 the electric potential. The latter is measurable and thus x_3 is also the output variable. The system also includes an input variable u with which the potential can be controlled using a pacemaker. As it pumps, the heart muscle contracts during the systolic phase, i.e. $x_1(t)$ decreases, and blood is pumped into the arteries. Subsequently, during the diastolic phase, the muscle relaxes, $x_1(t)$ increases, and the chambers of the heart take in fresh blood.

(a) Determine the relative degree of the system.

(b) Conduct an input-output linearization. To do so, identify the required diffeomorphism. If internal dynamics occur, select the diffeomorphism such that it is independent of u. Determine the transformed system and the zero dynamics.

(c) Formulate the controller and the prefilter such that the control loop has arbitrary linear dynamics.

Exercise 5.4 Assume that $\boldsymbol{f}(\boldsymbol{x})$ and $\boldsymbol{g}(\boldsymbol{x})$ are differentiable vector functions, $\mu(\boldsymbol{x})$ and $\lambda(\boldsymbol{x})$ are differentiable scalar functions, and c is a real number. Prove the following:

(a) $L_{\boldsymbol{f}} c = 0$,

(b) $L_{\mu \boldsymbol{f}} \lambda(\boldsymbol{x}) = \mu(\boldsymbol{x}) L_{\boldsymbol{f}} \lambda(\boldsymbol{x})$,

(c) $L_{\boldsymbol{f}+\boldsymbol{g}} \mu(\boldsymbol{x}) = L_{\boldsymbol{f}} \mu(\boldsymbol{x}) + L_{\boldsymbol{g}} \mu(\boldsymbol{x})$,

(d) $L_{\boldsymbol{f}}(\mu(\boldsymbol{x}) + \lambda(\boldsymbol{x})) = L_{\boldsymbol{f}} \mu(\boldsymbol{x}) + L_{\boldsymbol{f}} \lambda(\boldsymbol{x})$,

(e) $L_{\boldsymbol{f}}(\mu(\boldsymbol{x})\lambda(\boldsymbol{x})) = \mu(\boldsymbol{x}) L_{\boldsymbol{f}} \lambda(\boldsymbol{x}) + \lambda(\boldsymbol{x}) L_{\boldsymbol{f}} \mu(\boldsymbol{x})$,

(f) $L_{[\boldsymbol{f},\boldsymbol{g}]} \mu(\boldsymbol{x}) = L_{\boldsymbol{f}} L_{\boldsymbol{g}} \mu(\boldsymbol{x}) - L_{\boldsymbol{g}} L_{\boldsymbol{f}} \mu(\boldsymbol{x})$.

Exercise 5.5 Let the bilinear system[9]

$$\dot{\boldsymbol{x}} = \boldsymbol{A}\boldsymbol{x} + u\boldsymbol{B}\boldsymbol{x},$$
$$y = \boldsymbol{c}^T \boldsymbol{x}$$

with the relative degree $\delta = n$ be given.

(a) Conduct an input-output linearization. Determine the controller $r(\boldsymbol{x})$ and the prefilter $v(\boldsymbol{x})$.

(b) What disadvantage does the control law $u(\boldsymbol{x})$ have in this case?

(c) Identify the diffeomorphism $\boldsymbol{z} = \boldsymbol{t}(\boldsymbol{x})$ which transforms the system into the nonlinear controller canonical form.

(d) State the transformed differential equation $\dot{\boldsymbol{z}} = \boldsymbol{f}(\boldsymbol{z}, u)$, $y = g(\boldsymbol{z})$ of the plant.

Exercise 5.6 In automated arc welding, an electric arc is used to melt both the welding electrode and the part being worked on. As the electrode continues to move, the molten material cools and bonds the two parts together. During this process, the welding electrode is progressively consumed. To maintain the electric arc and thus continue the welding process, the electrode requires continual replenishment so that the length of the electric arc remains constant. A shielding gas protects the welding process from being influenced by the environment. Figure 5.48 shows the process. The welding process can be described by the equations [450]

$$\dot{\boldsymbol{x}} = \boldsymbol{a}(\boldsymbol{x}) + \boldsymbol{b}(\boldsymbol{x}) \cdot u$$
$$y = c(\boldsymbol{x})$$

with

[9] A system is called *bilinear* if it is linear both in the state $\boldsymbol{x}$ and the control variable $\boldsymbol{u}$, but is nonlinear overall.

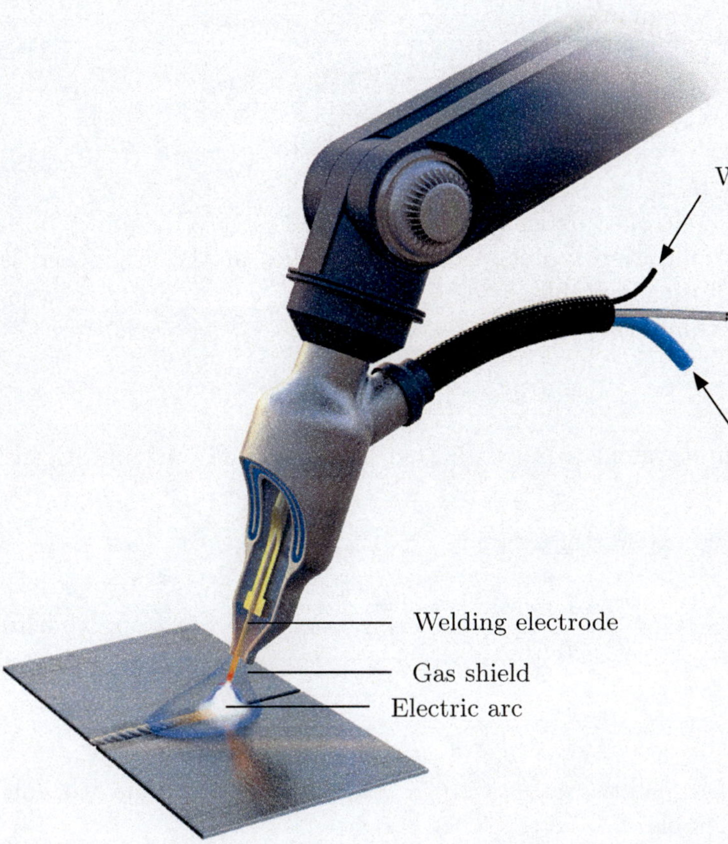

Fig. 5.48: Arc welding process

$$\boldsymbol{a}(\boldsymbol{x}) = \begin{bmatrix} -k_0 x_1 \\ k_1 x_1 + k_2 x_1^2 (l - x_2) - v_\mathrm{e} \end{bmatrix}, \quad \boldsymbol{b}(\boldsymbol{x}) = \begin{bmatrix} k_0 \\ 0 \end{bmatrix}, \quad c(\boldsymbol{x}) = x_2.$$

The electrical direct current in the electric arc will be represented by x_1 and the length of the electric arc by x_2. Here, due to the application of the solder, the length of the electric arc x_2 is somewhat smaller than the distance l between the tip of the electrode and the part being worked on. The control signal u is the reference current, i. e. the input signal, of an underlying current control. The process constants are the positive parameters k_0, k_1, k_2, and the wire feed speed v_e of the electrode. Below, our aim will be to design a control by means of input-output linearization.

(a) Determine the relative degree δ of the system.
(b) Determine the controller and the prefilter which linearize the system.
(c) Can the control be built as designed?

Exercise 5.7 Let us consider an input-output linearized system with the controlled external dynamics

$$\dot{x}_1 = -x_1 \tag{5.218}$$

and the internal dynamics

$$\dot{x}_2 = -x_2 + x_2^2 x_1. \tag{5.219}$$

(a) Show that the controlled external dynamics and the zero dynamics both have a single equilibrium point, each of which lies at the origin and is globally asymptotically stable.
(b) Show that the equilibrium point

$$[x_1\ x_2]^T = \mathbf{0}$$

of the composite system (5.218), (5.219) is not globally asymptotically stable.

Exercise 5.8 Let us extend the system

$$\begin{aligned} \dot{\boldsymbol{x}} &= \boldsymbol{f}(\boldsymbol{x}, u), \quad \boldsymbol{x} \in I\!R^{n-1}, \\ y &= g(\boldsymbol{x}) \end{aligned} \tag{5.220}$$

by an integrator

$$\dot{u} = v.$$

This process is called *dynamic extension*[10]. Use v as the new input variable and u as a state variable.
(a) Which system description does this yield?
(b) What is the lowest possible relative degree δ that the extended system can have?
(c) Conduct an input-output linearization of the extended system. Assume that the relative degree δ is maximum. State the transformed system description, i. e. the control loop's differential equation, and the control law $v(\boldsymbol{x}, u)$.
(d) State the control law $u(\boldsymbol{x})$ for the system (5.220).

Exercise 5.9 Prove the Jacobi identity

$$[\boldsymbol{f}, [\boldsymbol{g}, \boldsymbol{h}]] + [\boldsymbol{h}, [\boldsymbol{f}, \boldsymbol{g}]] + [\boldsymbol{g}, [\boldsymbol{h}, \boldsymbol{f}]] = \mathbf{0}.$$

To do this, use equation (5.105) on p. 416.

[10] The case described in this exercise is the simplest possible dynamic extension. In general, a dynamic system

$$\begin{aligned} \dot{\boldsymbol{\eta}} &= \boldsymbol{\rho}(\boldsymbol{x}, \boldsymbol{\eta}) + \boldsymbol{q}(\boldsymbol{x}, \boldsymbol{\eta})v, \\ u &= r(\boldsymbol{x}, \boldsymbol{\eta}) + s(\boldsymbol{x}, \boldsymbol{\eta})v \end{aligned}$$

is used to generate a dynamic extension [203].

Exercise 5.10 Let a system in the strict feedback form

$$\begin{aligned}
\dot{x}_1 &= f_1(x_1) + h_1(x_1) \cdot x_2, \\
\dot{x}_2 &= f_2(x_1, x_2) + h_2(x_1, x_2) \cdot x_3, \\
\dot{x}_3 &= f_3(x_1, x_2, x_3) + h_3(x_1, x_2, x_3) \cdot x_4, \\
&\vdots \\
\dot{x}_n &= f_n(x_1, x_2, \ldots, x_n) + h_n(x_1, x_2, \ldots, x_n) \cdot u
\end{aligned}$$

with $h_i(x_1, \ldots, x_i) \neq 0$ for $i = 1, \ldots, n$ be given.
(a) Determine the relative degree for each case $y = x_i$, $i = 1, \ldots, n$.
(b) For $y = x_2$, determine the internal dynamics of the system.

Exercise 5.11 Let us examine a macroeconomic model with the interest rate x_1, the investment demand x_2, and the price exponent x_3, which is a measure of the inflation rate [487]. The model takes the form

$$\begin{aligned}
\dot{x}_1 &= x_1(x_2 - a) + x_3, \\
\dot{x}_2 &= -x_1^2 - bx_2 + 1 + u, \\
\dot{x}_3 &= -x_1 - cx_3.
\end{aligned}$$

Here, the increase $\dot{x}_2$ in the investment demand can be influenced by the input variable u. The constants $a > 0$, $b > 0$, and $c > 0$ are the system parameters.
(a) Demonstrate that the model for $x_1 \neq 0$ is full-state linearizable.
(b) Identify a linearizing output y.
(c) Transform the system description into the Brunovsky canonical form with the new input variable v, and determine the corresponding transformations.
(d) Formulate the output variable y as a function dependent only on the input variable v of the transformed system.
(e) State a feedforward control law $v(t)$ such that

$$y(t) = \begin{cases} 0, & t < 0, \\ \sin^2(t), & 0 \leq t \leq \dfrac{\pi}{2}, \\ 1, & \dfrac{\pi}{2} < t. \end{cases}$$

Exercise 5.12 Let us examine the airship shown in Figure 5.49, which is also called a blimp and is often used for advertising purposes. We will limit ourselves to its dynamics in the horizontal plane, as shown in Figure 5.50. A blimp is typically powered by two main engines which allow it to move forward and backward. An additional engine is mounted at the back to enable the blimp to turn around its main axis and to fly in a curved trajectory. Larger blimps have additional tail units at the fin, while these are usually not present on smaller unmanned airships. Here, we will model the dynamics of a blimp without tail units.

Fig. 5.49: Airship

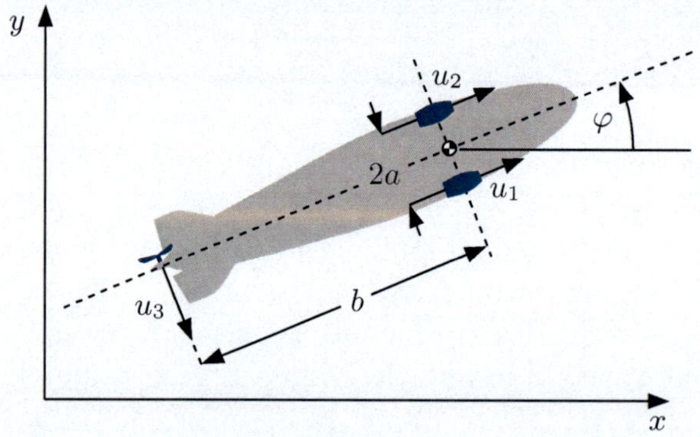

Fig. 5.50: Variables and parameters of the blimp's dynamics in the horizontal plane

Based on the horizontal positions $x_1 = x$ and $x_3 = y$, the course angle $x_5 = \varphi$, the propulsive thrusts u_1 and u_2 of the right and left main engines, and the propulsive thrust u_3 of the back engine, we obtain the dynamic model

$$
\begin{aligned}
\dot{x}_1 &= x_2, \\
\dot{x}_2 &= \frac{(u_1 + u_2)\cos(x_5) + u_3 \sin(x_5)}{m}, \\
\dot{x}_3 &= x_4, \\
\dot{x}_4 &= \frac{(u_1 + u_2)\sin(x_5) - u_3 \cos(x_5)}{m}, \\
\dot{x}_5 &= x_6, \\
\dot{x}_6 &= \frac{a(u_1 - u_2) + bu_3}{J},
\end{aligned}
\qquad
\boldsymbol{y} = \begin{bmatrix} x_1 \\ x_3 \\ x_5 \end{bmatrix}.
$$

Here, m is the mass of the blimp, J is its moment of inertia around the yaw axis, a is the lever arm of the right and the left main engine in relation to the yaw axis, and b is the lever arm of the back engine in relation to the yaw axis.
(a) Determine the vectorial relative degree $\boldsymbol{\delta}$ of the blimp.
(b) Is the vectorial relative degree $\boldsymbol{\delta}$ well-defined?
(c) Determine the feedback control law $\boldsymbol{u}(\boldsymbol{x})$ which completely decouples the input and output variables and stabilizes and linearizes the control loop.

Exercise 5.13 Show that the system

$$\dot{\boldsymbol{x}} = \begin{bmatrix} 1 \\ x_4 \\ x_5^2 \\ x_5 \\ 0 \end{bmatrix} u_1 + \begin{bmatrix} 0 \\ 0 \\ 0 \\ 0 \\ 1 \end{bmatrix} u_2$$

is globally controllable, but not full-state linearizable.

Exercise 5.14 Show that a driftless system $\dot{\boldsymbol{x}} = \boldsymbol{B}(\boldsymbol{x}) \cdot \boldsymbol{u}$ is not full-state linearizable, except in the special case in which it is omnidirectionally controllable.

Exercise 5.15 Is the model (5.44), p. 381, of the lunar module Eagle full-state linearizable?

Exercise 5.16 Let the plant

$$\begin{aligned} \dot{x}_1 &= \sqrt[3]{x_2}, \\ \dot{x}_2 &= -\sqrt[3]{x_2^2} \cdot x_3, \\ \dot{x}_3 &= x_1 x_3 + u \end{aligned}$$

with the flat output $y = x_1$ be given.
(a) State the flat system representation.
(b) Draw a picture of the structure of the flatness-based feedback control loop. This closed control loop should have three eigenvalues at $s = -5$. Formulate all the necessary equations in detail.

Exercise 5.17 The increase in a species is dependent on its population and, where resources are unlimited, it follows the simple law of growth

$$\dot{x} = \lambda x,$$

where λ is the growth coefficient. Thanks to progress in agriculture, medicine, science, engineering, etc., it has been possible for human beings to increase this growth coefficient λ and to make it dependent on x_2, the level of human knowledge. In a good approximation, the following applies to the growth $\dot{x}_1$ of the human population x_1 on Earth [238]:

$$\dot{x}_1 = \alpha(\beta x_2 - x_1)x_1.$$

Here, βx_2 is the maximum world population which receives a sufficient food supply based on a certain level of knowledge x_2 and the corresponding technologies. The level of human knowledge x_2 increases with time, along with the increasing human population x_1. Here,

$$\dot{x}_2 = 0.04093 x_1 x_2$$

applies. For a good approximation starting at $x_1(0) = 10^8$ humans in 500 B.C., we will set $x_2(0) = 0.01$. In addition, $\alpha = 1$ and $\beta = 1$ apply.

(a) If the population x_1 is the output of this growth model, identify a flat input of the system such that $y = x_1$ is a flat output and the output vector $\boldsymbol{b}(\boldsymbol{x})$ is not dependent on $\boldsymbol{x}$. In doing so, take into account that $x_1 > 0$ applies.
(b) State the flat system representation.
(c) What quantity can we influence using the flat input u?
(d) Calculate the control $u(y, \dot{y}, \ddot{y})$ such that $y = x_1$ remains constant.
(e) What statement can you make about the practical applicability of the flat input if you take the result from (c) into account?

Let us now assume that the growth rate of the human population on Earth can be directly influenced by means of a control variable u such that

$$\dot{x}_1 = u\alpha(\beta x_2 - x_1)x_1$$

applies. In practice, this can be achieved by means of improved education, improved social security, protection from unemployment, old age pension schemes, birth control, etc. Here, $\alpha = 1$ and $\beta = 1$ still apply.

(f) Calculate a flat output y_f.
(g) Determine the flat system representation.
(h) How should we select the flat control u in this case if we wish the human population on Earth

$$y = x_1 = x_{1,\infty}$$

to remain constant? In this case, what is the curve representing human knowledge x_2 over time?

Exercise 5.18 Is the model (5.44) of the lunar module Eagle, see Section 5.2.6 on p. 381, flat?

Exercise 5.19 Let us examine a general nonlinear system

$$\begin{aligned}\dot{\boldsymbol{x}} &= \boldsymbol{f}(\boldsymbol{x}, u),\\ y &= g(\boldsymbol{x}).\end{aligned} \tag{5.221}$$

Show that the dynamically extended system

$$\begin{aligned}\dot{\boldsymbol{x}} &= \boldsymbol{f}(\boldsymbol{x}, u),\\ u^{(k)} &= v, \qquad k \in I\!N,\\ y &= g(\boldsymbol{x})\end{aligned}$$

is flat and that y is a flat output if the system (5.221) is flat and has the flat output y.

Exercise 5.20 Like healthy tissue, cancerous tumors require a supply of blood in order to grow. The blood transports the nutrients and oxygen necessary for cell growth. If the supply of blood to the tumor is stopped by preventing the growth of blood vessels to the tumor, it can no longer grow. A medication with this effect prevents the new growth of blood vessels, or vascular angiogenesis, around the tumor and causes atrophy of already existing blood vessels. This starves the tumor, causing it to regress. This process can be described by the following model [101, 171]:

$$\begin{aligned}\dot{x}_1 &= -\lambda x_1 \ln\left(\frac{x_1}{x_2}\right),\\ \dot{x}_2 &= bx_1 - d \cdot x_2 \sqrt[3]{x_1^2} - \mu x_2 x_3,\\ \dot{x}_3 &= -a x_3 + u,\\ y &= x_1.\end{aligned}$$

Here, x_1 is the tumor volume, x_2 is the volume of the endothelium, i.e. the innermost cell layer of the blood vessels of the tumor, and x_3 is the concentration of the medication in the body. The input variable u represents the medication dose per day and per kilogram of the patient's body weight. The parameters λ, b, d, μ, and a are positive constants.

(a) State the transformation $\boldsymbol{z} = \boldsymbol{t}(\boldsymbol{x})$ which transforms the model into the nonlinear controller canonical form.

(b) Determine the inverse transformation $\boldsymbol{x} = \boldsymbol{t}^{-1}(\boldsymbol{z})$. To do this, use the flat system representation with $y = x_1$ as a flat output.

(c) State the flat feedforward control $u(y, \dot{y}, \ddot{y}, \dddot{y}\,)$.

(d) Our aim is to dose the medication to decrease the tumor volume y according to

$$y = e^{-\alpha t}, \qquad \alpha > 1.5a.$$

In this case, what is the flat feedforward control law $u(t)$?

(e) Sketch the curve of $u(t)$.

Exercise 5.21 Let us consider a system which has a real flat output. What is the difference between a flatness-based feedback control and a control of a flat system with input-output linearization, and what is identical between the two?

Exercise 5.22 Show that the function $V(\boldsymbol{x}) = \boldsymbol{x}^T \boldsymbol{R} \boldsymbol{x}$ with a positive definite matrix $\boldsymbol{R}$ is a control Lyapunov function for an omnidirectionally controllable system $\dot{\boldsymbol{x}} = \boldsymbol{a}(\boldsymbol{x}) + \boldsymbol{B}(\boldsymbol{x}) \cdot \boldsymbol{u}$.

Exercise 5.23 Let us examine the bilinear system

$$\dot{\boldsymbol{x}} = \boldsymbol{A}\boldsymbol{x} + u\boldsymbol{N}\boldsymbol{x}.$$

(a) What condition must be fulfilled such that the equilibrium point $\boldsymbol{x}_{\text{eq}} = \boldsymbol{0}$ is globally asymptotically stable for a constant input variable $u = u_{\text{c}}$?

Now we will suppose that the control variable u is limited by $|u| \leq u_{\text{max}}$.

(b) Assume that $\boldsymbol{A}$ has only eigenvalues with a negative real part. In this case, what condition is necessary for $V(\boldsymbol{x}) = \boldsymbol{x}^T \boldsymbol{R} \boldsymbol{x}$ to be a control Lyapunov function for all $\boldsymbol{x} \in I\!R^n$?

(c) What is the control law $u(\boldsymbol{x})$ that minimizes $\dot{V}(\boldsymbol{x})$?

(d) What practical disadvantage does the control law have, and how can this be remedied?

Exercise 5.24 Let us examine a system in the nonlinear controller canonical form

$$\dot{\boldsymbol{x}} = \begin{bmatrix} x_2 \\ \vdots \\ x_n \\ f(\boldsymbol{x}) \end{bmatrix} + \begin{bmatrix} 0 \\ \vdots \\ 0 \\ h(\boldsymbol{x}) \end{bmatrix} u,$$

which is a special form of the strict feedback form.

(a) Using the backstepping method, design a controller $u(\boldsymbol{x})$. In doing so, use a P controller $\alpha(x_1)$ as the virtual controller and

$$V_1(x_1) = \frac{1}{2} x_1^2$$

as the Lyapunov function in the first design step.

(b) State the differential equation of the control loop.

(c) How do the system descriptions of the above control loop and the controller differ from the descriptions we would obtain in the case of a feedback linearization?

Check for updates

6

Nonlinear Control of Linear and Nonlinear Systems

6.1 Model-Based Predictive Control

6.1.1 Basics and Functionality

Among all advanced control methods, model-based predictive controls (MPCs), or model predictive controls for short, are the most commonly used in industry [68, 120, 157, 193, 296, 378, 381]. In the processing industry, especially in refineries, chemical plants, cement mills, and paper and steel production, MPCs are well-established standard procedures. MPCs are nonlinear control methods which are generally suitable to linear plants with limited control and state variables, and to nonlinear plants as well. The reasons for their widespread and successful use are the good explainability of their basic principle and their applicability to complex high-order dynamical systems.

The operating principle of MPC is essentially based on the properties of mathematical process models. Process models generally have two tasks in control engineering. Firstly, a model should provide a deeper understanding of the process and its mode of operation. This knowledge is used for the design of controllers such as state controllers, time-optimal controllers, and variable structure controllers, among others. Secondly, a model allows for the prediction of future behavior. It is just this possibility of predicting the future which model predictive controls make use of. They calculate the model's output variable signals for different control input sequences online and then select the best one. The optimized control variable is used to control the actual process. Figure 6.1 illustrates the procedure.

Both continuous and discrete process models can be used for model predictive control. In most cases, however, discrete or discretized models

$$\begin{aligned}\boldsymbol{x}(k+1) &= \boldsymbol{f}(\boldsymbol{x}(k),\boldsymbol{u}(k)),\\ \boldsymbol{y}(k) &= \boldsymbol{g}(\boldsymbol{x}(k),\boldsymbol{u}(k))\end{aligned}$$

are employed, because in the case of continuous models, the optimization of the control variable signal is much more complex.

J. Adamy, *Nonlinear Systems and Controls*,
https://doi.org/10.1007/978-3-662-68690-4_6

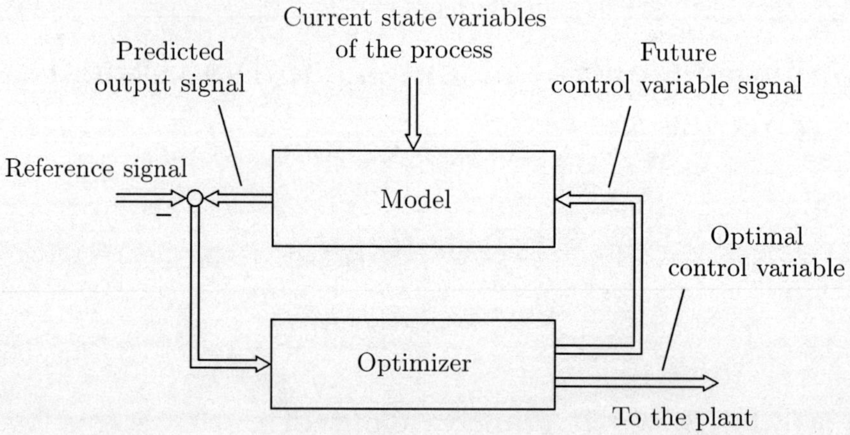

Fig. 6.1: Structure of a model predictive controller

We will begin with a known control variable $\boldsymbol{u}(k-1)$ at time instant $k-1$ and start with the optimization from this point on. The optimization varies the future control variable signal $\boldsymbol{u}(k+i)$ from the time k onward for a finite number $i=0,\ldots,n_\mathrm{c}-1$ of control variable steps in such a way that a given performance index J becomes minimal. The value n_c is called the *control horizon*. Commonly, the quadratic performance indices

$$J=\sum_{i=1}^{n_\mathrm{p}}|\boldsymbol{Q}_0(\boldsymbol{y}(k+i)-\boldsymbol{y}_\mathrm{ref}(k+i))|^2+\sum_{i=1}^{n_\mathrm{p}}|\boldsymbol{R}_0\boldsymbol{u}(k+i-1)|^2 \tag{6.1}$$

and

$$J=\sum_{i=1}^{n_\mathrm{p}}|\boldsymbol{Q}_0(\boldsymbol{y}(k+i)-\boldsymbol{y}_\mathrm{ref}(k+i))|^2+\sum_{i=1}^{n_\mathrm{c}}|\boldsymbol{R}_1(\boldsymbol{u}(k+i-1)-\boldsymbol{u}(k+i-2))|^2 \tag{6.2}$$

calculated over n_p time steps are used. The value n_p is called the *prediction horizon* since the future behavior is predicted for n_p time steps. The prediction horizon is always greater than the control horizon n_c or both are equal, i. e.

$$n_\mathrm{p}\geq n_\mathrm{c}.$$

The matrices $\boldsymbol{Q}_0$, $\boldsymbol{R}_0$, and $\boldsymbol{R}_1$ must be positive definite. They are often chosen as diagonal matrices. In the above performance indices the difference between the output variable sequence $\boldsymbol{y}(k+i)$ and a nominal or reference sequence $\boldsymbol{y}_\mathrm{ref}(k+i)$ is evaluated in the first term of the summation. In the second term, the squared control variables, or control variable differences, are added and weighted by the positive definite matrix $\boldsymbol{R}_0$ or $\boldsymbol{R}_1$. When minimizing J, this second term ensures that the control variables $\boldsymbol{u}(k+i)$ in equation (6.1) and

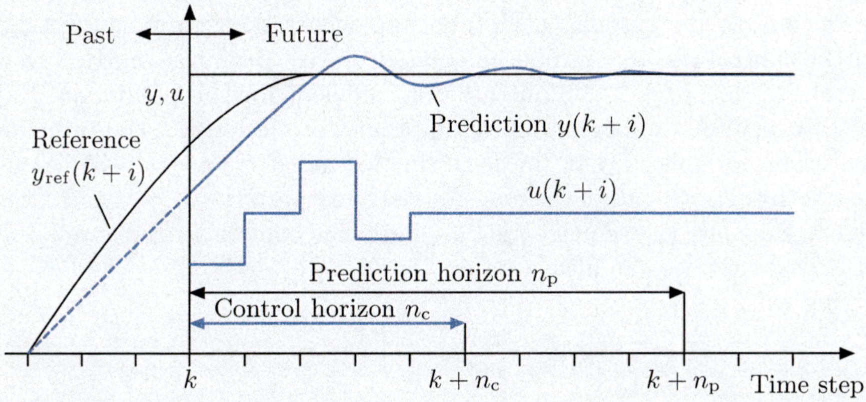

Fig. 6.2: Basic procedure of a model predictive control (MPC)

control variable differences in equation (6.2) do not assume values which are too high, either to conserve energy or for other reasons.

The prediction of $\boldsymbol{y}(k+i)$ is made for n_p time steps in the performance index J. The prediction horizon n_p is greater than or equal to the control horizon n_c, as mentioned above. To ensure for the case $n_p > n_c$ that control variables $\boldsymbol{u}(k+i)$ are available for the time range

$$i \geq n_c,$$

and to allow us to predict $\boldsymbol{y}(k+i)$, all control variables beyond the control horizon are constantly set at $\boldsymbol{u}(k+n_c-1)$. Figure 6.2 illustrates the procedure.

After the optimization has determined the control sequence $\boldsymbol{u}_{\text{opt}}(k+i)$ with $i = 0, \ldots, n_c - 1$, only the first value of this sequence, i. e.

$$\boldsymbol{u}_{\text{opt}}(k) = \boldsymbol{u}(k),$$

is applied to the real controlled system. This means the entire sequence $\boldsymbol{u}_{\text{opt}}(k+i)$ is not used for the control. Immediately after the activation of $\boldsymbol{u}(k)$, the prediction horizon and optimization process described above are shifted into the future by one step and a new optimal control variable sequence is calculated. This procedure is repeated after each optimization. With each repetition, the horizon is moved one time step further. Therefore, we refer to this as a *moving horizon*. The moving horizon and the repetitive optimization of the control sequence $\boldsymbol{u}(k+i)$ allow the MPC to react to disturbances and compensate for them.

We can compare the operating mode of the MPC to the behavior of a chess player – a frequently used simile. The player thinks through different sequences of moves in his mind, considering three, four, or more moves in advance. In reality, he then plays the first step of the combination that seems optimal to him. After his opponent's move, which may be regarded as a disturbance, he selects his next, newly optimized move, and so on.

The new prediction after each time step occurs based on the process model and the control variable signals calculated by the optimization. However, the control variables $\boldsymbol{u}(k+i)$ alone are not sufficient for the prediction. Rather, the state $\boldsymbol{x}(k)$ at the beginning of the prediction period must also be known. The reason for this is that the performance index J can only be calculated using both the sequence of the model's state vectors $\boldsymbol{x}(k+i)$, $i = 1, 2, \ldots n_{\mathrm{p}}$, in order to calculate the sequence $\boldsymbol{y}(k+i)$, and the control sequence $\boldsymbol{u}(k+i-1)$. In the best case, we can measure the plant's state vector $\boldsymbol{x}_{\mathrm{pl}}(k)$ and calculate, starting with

$$\boldsymbol{x}(k) = \boldsymbol{x}_{\mathrm{pl}}(k),$$

the model's state vectors

$$\boldsymbol{x}(k+i) = \boldsymbol{f}(\boldsymbol{x}(k+i-1), \boldsymbol{u}(k+i-1)), \quad i = 1, 2, \ldots, n_{\mathrm{p}},$$

and

$$\boldsymbol{y}(k+i) = \boldsymbol{g}(\boldsymbol{x}(k+i), \boldsymbol{u}(k+i)), \quad i = 1, 2, \ldots, n_{\mathrm{p}},$$

by iteration. If the measurement is not possible, $\boldsymbol{x}_{\mathrm{pl}}(k)$ must be estimated as in Figure 6.3 by $\tilde{\boldsymbol{x}}(k)$ using an observer based on the historical sequences

$$\begin{gathered}\boldsymbol{u}(k-1), \boldsymbol{u}(k-2), \ldots, \\ \boldsymbol{y}(k-1), \boldsymbol{y}(k-2), \ldots,\end{gathered}$$

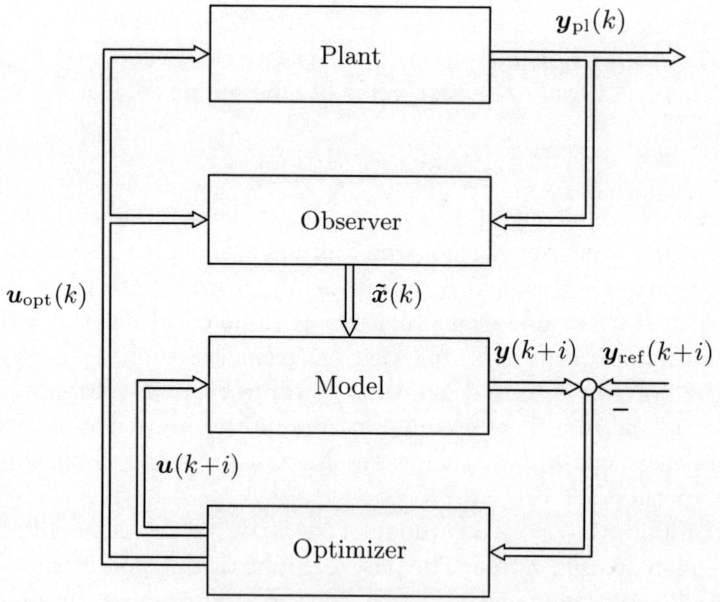

Fig. 6.3: Structure of an MPC with observer, model, and optimizer, where $\boldsymbol{y}_{\mathrm{pl}}(k)$ is the plant's output variable vector and $\tilde{\boldsymbol{x}}(k)$ is its estimated state vector

and

$$\boldsymbol{y}_{\mathrm{pl}}(k-1), \boldsymbol{y}_{\mathrm{pl}}(k-2), \ldots,$$

where $\boldsymbol{y}_{\mathrm{pl}}$ is the plant's output variable vector. Combining all elements of an MPC yields the structure that is shown in Figure 6.3.

6.1.2 Linear Model Predictive Control without Constraints

The MPCs most frequently used in industrial practice are based on linear process models, which usually possess constraints on the control variables and possibly on the state and output variables. MPCs of this kind are called *linear model predictive controls*, abbreviated LMPC. In this section, we will address the simple case of linear plants without constraints, which is not relevant to practical applications. However, the study of MPCs for these systems allows us firstly to gain a deeper understanding of the operation of MPC, and secondly it provides us with some basics for the MPC of linear systems with constraints. The latter is of greater interest in practice.

When using the quadratic performance index (6.1) or (6.2), the result is a linear controller, as we will see in the following. It is well known that linear controllers can also be designed using other, simpler methods. The true benefit of an LMPC is therefore only obtained if constraints must be included in the optimization problem. In this case, which we will discuss in the next section, the result is a nonlinear controller. However, the MPC is still referred to as an LMPC due to the linear process model.

Our starting point is the linear discrete-time model

$$\begin{aligned} \boldsymbol{x}(k+1) &= \boldsymbol{A}\boldsymbol{x}(k) + \boldsymbol{B}\boldsymbol{u}(k), \\ \boldsymbol{y}(k) &= \boldsymbol{C}\boldsymbol{x}(k) \end{aligned} \tag{6.3}$$

with $\boldsymbol{x} \in I\!R^n$, $\boldsymbol{u} \in I\!R^m$, and $\boldsymbol{y} \in I\!R^r$ without constraints on $\boldsymbol{u}$, $\boldsymbol{x}$, or $\boldsymbol{y}$, i. e. the linear case. The control variable vector $\boldsymbol{u}(k)$ is composed of the previous control variable vector $\boldsymbol{u}(k-1)$ and the stepwise increment $\Delta\boldsymbol{u}(k)$ at step k according to

$$\boldsymbol{u}(k) = \boldsymbol{u}(k-1) + \Delta\boldsymbol{u}(k) \tag{6.4}$$

For the system description (6.3) above, we thus obtain

$$\begin{aligned} \boldsymbol{x}(k+1) &= \boldsymbol{A}\boldsymbol{x}(k) + \boldsymbol{B}\boldsymbol{u}(k-1) + \boldsymbol{B}\Delta\boldsymbol{u}(k), \\ \boldsymbol{y}(k) &= \boldsymbol{C}\boldsymbol{x}(k). \end{aligned} \tag{6.5}$$

We will now perform the calculation for all output variable vectors $\boldsymbol{y}(k+i)$ from step $k+1$ up to step $k+n_{\mathrm{p}}$, i. e. up to the prediction horizon n_{p}. Equation (6.4) and equation (6.5) yield

$$
\begin{aligned}
\boldsymbol{y}(k+1) &= \boldsymbol{CAx}(k) + \boldsymbol{CBu}(k-1) + \boldsymbol{CB}\Delta\boldsymbol{u}(k),\\
\boldsymbol{y}(k+2) &= \boldsymbol{CA}^2\boldsymbol{x}(k) + \boldsymbol{C}(\boldsymbol{A}+\boldsymbol{I})\boldsymbol{Bu}(k-1) + \boldsymbol{C}(\boldsymbol{A}+\boldsymbol{I})\boldsymbol{B}\Delta\boldsymbol{u}(k)\\
&\quad + \boldsymbol{CB}\Delta\boldsymbol{u}(k+1),\\
&\ \ \vdots \\
\boldsymbol{y}(k+i) &= \boldsymbol{CA}^i\boldsymbol{x}(k) + \boldsymbol{C}\left(\boldsymbol{A}^{i-1}+\ldots+\boldsymbol{A}+\boldsymbol{I}\right)\boldsymbol{Bu}(k-1)\\
&\quad + \sum_{j=1}^{i} \boldsymbol{C}\left(\boldsymbol{A}^{i-j}+\ldots+\boldsymbol{A}+\boldsymbol{I}\right)\boldsymbol{B}\Delta\boldsymbol{u}(k+j-1),\\
&\ \ \vdots \\
\boldsymbol{y}(k+n_{\mathrm{c}}) &= \boldsymbol{CA}^{n_{\mathrm{c}}}\boldsymbol{x}(k) + \boldsymbol{C}\left(\boldsymbol{A}^{n_{\mathrm{c}}-1}+\ldots+\boldsymbol{A}+\boldsymbol{I}\right)\boldsymbol{Bu}(k-1)\\
&\quad + \sum_{j=1}^{n_{\mathrm{c}}} \boldsymbol{C}\left(\boldsymbol{A}^{n_{\mathrm{c}}-j}+\ldots+\boldsymbol{A}+\boldsymbol{I}\right)\boldsymbol{B}\Delta\boldsymbol{u}(k+j-1)
\end{aligned}
\tag{6.6}
$$

for the steps up to the control horizon n_{c}. Beyond the control horizon, i. e. $i > n_{\mathrm{c}}$, the control variable no longer changes, i. e.

$$
\Delta\boldsymbol{u}(k+i-1) = 0
$$

holds. Then we obtain

$$
\begin{aligned}
\boldsymbol{y}(k+n_{\mathrm{c}}+1) &= \boldsymbol{CA}^{n_{\mathrm{c}}+1}\boldsymbol{x}(k) + \boldsymbol{C}\left(\boldsymbol{A}^{n_{\mathrm{c}}}+\ldots+\boldsymbol{A}+\boldsymbol{I}\right)\boldsymbol{Bu}(k-1)\\
&\quad + \sum_{j=1}^{n_{\mathrm{c}}} \boldsymbol{C}\left(\boldsymbol{A}^{n_{\mathrm{c}}+1-j}+\ldots+\boldsymbol{A}+\boldsymbol{I}\right)\boldsymbol{B}\Delta\boldsymbol{u}(k+j-1),\\
&\ \ \vdots \\
\boldsymbol{y}(k+n_{\mathrm{p}}) &= \boldsymbol{CA}^{n_{\mathrm{p}}}\boldsymbol{x}(k) + \boldsymbol{C}\left(\boldsymbol{A}^{n_{\mathrm{p}}-1}+\ldots+\boldsymbol{A}+\boldsymbol{I}\right)\boldsymbol{Bu}(k-1)\\
&\quad + \sum_{j=1}^{n_{\mathrm{c}}} \boldsymbol{C}\left(\boldsymbol{A}^{n_{\mathrm{p}}-j}+\ldots+\boldsymbol{A}+\boldsymbol{I}\right)\boldsymbol{B}\Delta\boldsymbol{u}(k+j-1).
\end{aligned}
\tag{6.7}
$$

The vectors $\boldsymbol{y}(k+1)$ to $\boldsymbol{y}(k+n_{\mathrm{p}})$ are collected in the $rn_{\mathrm{p}} \times 1$ vector

$$
\bar{\boldsymbol{y}}(k+1) = \begin{bmatrix} \boldsymbol{y}(k+1) \\ \boldsymbol{y}(k+2) \\ \vdots \\ \boldsymbol{y}(k+n_{\mathrm{p}}) \end{bmatrix}
$$

and the control variable changes $\Delta\boldsymbol{u}(k)$ to $\Delta\boldsymbol{u}(k+n_{\mathrm{c}}-1)$ are lined up in the $mn_{\mathrm{c}} \times 1$ vector

$$\Delta\bar{\boldsymbol{u}}(k) = \begin{bmatrix} \Delta\boldsymbol{u}(k) \\ \Delta\boldsymbol{u}(k+1) \\ \vdots \\ \Delta\boldsymbol{u}(k+n_c-1) \end{bmatrix}.$$

Note that the vector $\Delta\bar{\boldsymbol{u}}(k)$ generally has a lower dimension than the vector $\bar{\boldsymbol{y}}(k+1)$ because $n_c \leq n_p$ holds.

Combining equation (6.6) and equation (6.7) results in

$$\bar{\boldsymbol{y}}(k+1) = \boldsymbol{F}\boldsymbol{x}(k) + \boldsymbol{G}\boldsymbol{u}(k-1) + \boldsymbol{H}\Delta\bar{\boldsymbol{u}}(k), \tag{6.8}$$

where

$$\boldsymbol{F} = \begin{bmatrix} \boldsymbol{CA} \\ \boldsymbol{CA}^2 \\ \boldsymbol{CA}^3 \\ \vdots \\ \boldsymbol{CA}^{n_p} \end{bmatrix}, \qquad \boldsymbol{G} = \begin{bmatrix} \boldsymbol{CB} \\ \boldsymbol{C}(\boldsymbol{A}+\boldsymbol{I})\boldsymbol{B} \\ \boldsymbol{C}(\boldsymbol{A}^2+\boldsymbol{A}+\boldsymbol{I})\boldsymbol{B} \\ \vdots \\ \boldsymbol{C}(\boldsymbol{A}^{n_p-1}+\ldots+\boldsymbol{I})\boldsymbol{B} \end{bmatrix},$$

$$\boldsymbol{H} = \begin{bmatrix} \boldsymbol{CB} & \boldsymbol{0} & \cdots & \boldsymbol{0} \\ \boldsymbol{C}(\boldsymbol{A}+\boldsymbol{I})\boldsymbol{B} & \boldsymbol{CB} & \cdots & \boldsymbol{0} \\ \boldsymbol{C}(\boldsymbol{A}^2+\boldsymbol{A}+\boldsymbol{I})\boldsymbol{B} & \boldsymbol{C}(\boldsymbol{A}+\boldsymbol{I})\boldsymbol{B} & \cdots & \boldsymbol{0} \\ \vdots & \vdots & \ddots & \vdots \\ \boldsymbol{C}(\boldsymbol{A}^{n_c-1}+\ldots+\boldsymbol{I})\boldsymbol{B} & \boldsymbol{C}(\boldsymbol{A}^{n_c-2}+\ldots+\boldsymbol{I})\boldsymbol{B} & \cdots & \boldsymbol{CB} \\ \boldsymbol{C}(\boldsymbol{A}^{n_c}+\ldots+\boldsymbol{I})\boldsymbol{B} & \boldsymbol{C}(\boldsymbol{A}^{n_c-1}+\ldots+\boldsymbol{I})\boldsymbol{B} & \cdots & \boldsymbol{C}(\boldsymbol{A}+\boldsymbol{I})\boldsymbol{B} \\ \vdots & \vdots & \ddots & \vdots \\ \boldsymbol{C}(\boldsymbol{A}^{n_p-1}+\ldots+\boldsymbol{I})\boldsymbol{B} & \boldsymbol{C}(\boldsymbol{A}^{n_p-2}+\ldots+\boldsymbol{I})\boldsymbol{B} & \cdots & \boldsymbol{C}(\boldsymbol{A}^{n_p-n_c}+\ldots+\boldsymbol{I})\boldsymbol{B} \end{bmatrix}$$

with $\boldsymbol{F} \in \mathbb{R}^{rn_p \times n}$, $\boldsymbol{G} \in \mathbb{R}^{rn_p \times m}$, and $\boldsymbol{H} \in \mathbb{R}^{rn_p \times mn_c}$ hold. The component

$$\boldsymbol{g}(k) = \boldsymbol{F}\boldsymbol{x}(k) + \boldsymbol{G}\boldsymbol{u}(k-1)$$

in equation (6.8) is determined by steps $0, \ldots, k-1$ of the control, which have already been performed. In contrast to the term $\boldsymbol{g}(k)$, which is constant, the term $\boldsymbol{H}\Delta\bar{\boldsymbol{u}}(k)$ in equation (6.8) contains the control variable sequence to be optimized, i. e.

$$\Delta\boldsymbol{u}(k), \ldots, \Delta\boldsymbol{u}(k+n_c-1).$$

Taking this into account, we write equation (6.8) as

$$\bar{\boldsymbol{y}}(k+1) = \boldsymbol{g}(k) + \boldsymbol{H}\Delta\bar{\boldsymbol{u}}(k). \tag{6.9}$$

The performance index

$$J(\Delta\bar{\boldsymbol{u}}(k)) = (\bar{\boldsymbol{y}}(k+1) - \bar{\boldsymbol{y}}_{\mathrm{ref}}(k+1))^T \boldsymbol{Q}\,(\bar{\boldsymbol{y}}(k+1) - \bar{\boldsymbol{y}}_{\mathrm{ref}}(k+1)) + \Delta\bar{\boldsymbol{u}}^T(k)\boldsymbol{R}\Delta\bar{\boldsymbol{u}}(k) \tag{6.10}$$

now needs to be minimized, where $\boldsymbol{Q}$ is a positive definite $rn_{\mathrm{p}} \times rn_{\mathrm{p}}$ matrix and $\boldsymbol{R}$ is a positive definite $mn_{\mathrm{c}} \times mn_{\mathrm{c}}$ matrix. The rn_{p} - dimensional vector

$$\bar{\boldsymbol{y}}_{\mathrm{ref}}(k+1) = \begin{bmatrix} \boldsymbol{y}_{\mathrm{ref}}(k+1) \\ \boldsymbol{y}_{\mathrm{ref}}(k+2) \\ \vdots \\ \boldsymbol{y}_{\mathrm{ref}}(k+n_{\mathrm{p}}) \end{bmatrix}$$

contains the progression of the reference variable $\boldsymbol{y}_{\mathrm{ref}}(k+i)$. Inserting equation (6.9) into the performance index (6.10) and introducing the abbreviation

$$\boldsymbol{e}(k) = \boldsymbol{g}(k) - \bar{\boldsymbol{y}}_{\mathrm{ref}}(k+1)$$

yields the quadratic form

$$\begin{aligned} J(\Delta\bar{\boldsymbol{u}}(k)) &= (\boldsymbol{e}(k) + \boldsymbol{H}\Delta\bar{\boldsymbol{u}}(k))^T \boldsymbol{Q}\,(\boldsymbol{e}(k) + \boldsymbol{H}\Delta\bar{\boldsymbol{u}}(k)) + \Delta\bar{\boldsymbol{u}}^T(k)\boldsymbol{R}\Delta\bar{\boldsymbol{u}}(k) \\ &= \Delta\bar{\boldsymbol{u}}^T(k)\left(\boldsymbol{H}^T\boldsymbol{Q}\boldsymbol{H} + \boldsymbol{R}\right)\Delta\bar{\boldsymbol{u}}(k) \\ &\quad + 2\Delta\bar{\boldsymbol{u}}^T(k)\boldsymbol{H}^T\boldsymbol{Q}\boldsymbol{e}(k) + \boldsymbol{e}^T(k)\boldsymbol{Q}\boldsymbol{e}(k), \end{aligned} \tag{6.11}$$

which is positive definite. The necessary condition for a minimum of J, which is

$$\frac{\partial J(\Delta\bar{\boldsymbol{u}}(k))}{\partial\Delta\bar{\boldsymbol{u}}(k)} = \boldsymbol{0}, \tag{6.12}$$

is also sufficient, because the performance index (6.11) is a positive definite form and thus is a convex function. For equation (6.12), we obtain

$$\left(\boldsymbol{H}^T\boldsymbol{Q}\boldsymbol{H} + \boldsymbol{R}\right)\Delta\bar{\boldsymbol{u}}(k) + \boldsymbol{H}^T\boldsymbol{Q}\boldsymbol{e}(k) = \boldsymbol{0}$$

in the present case. This linear equation results in the control variables

$$\Delta\bar{\boldsymbol{u}}(k) = -\left(\boldsymbol{H}^T\boldsymbol{Q}\boldsymbol{H} + \boldsymbol{R}\right)^{-1}\boldsymbol{H}^T\boldsymbol{Q}\boldsymbol{e}(k).$$

However, as mentioned earlier, only the first control variable $\Delta\boldsymbol{u}(k)$ in the vector

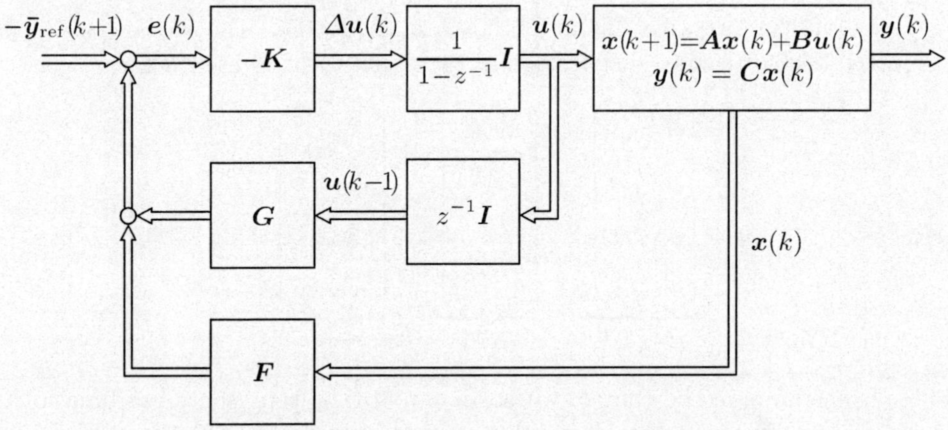

Fig. 6.4: MPC for a linear plant without limitations

$$\Delta\bar{\boldsymbol{u}}(k) = \begin{bmatrix} \Delta\boldsymbol{u}(k) \\ \Delta\boldsymbol{u}(k+1) \\ \Delta\boldsymbol{u}(k+2) \\ \vdots \\ \Delta\boldsymbol{u}(k+n_{\mathrm{c}}-1) \end{bmatrix}$$

is used for control. We will therefore address only the first element

$$\Delta\boldsymbol{u}(k) = -\boldsymbol{K}\boldsymbol{e}(k)$$

of this vector, where

$$\boldsymbol{K} = \begin{bmatrix} \boldsymbol{I} \; \boldsymbol{0} \cdots \boldsymbol{0} \end{bmatrix} \left[\left(\boldsymbol{H}^T\boldsymbol{Q}\boldsymbol{H} + \boldsymbol{R} \right)^{-1} \boldsymbol{H}^T\boldsymbol{Q} \right],$$

$\boldsymbol{I} \in \mathbb{R}^{m\times m}$ is the identity matrix. The control law is linear, has a dynamic, and can be summarized as

$$\begin{aligned} \boldsymbol{u}(k) &= \boldsymbol{u}(k-1) + \Delta\boldsymbol{u}(k) = \boldsymbol{u}(k-1) - \boldsymbol{K}\boldsymbol{e}(k), \\ \boldsymbol{e}(k) &= \boldsymbol{F}\boldsymbol{x}(k) + \boldsymbol{G}\boldsymbol{u}(k-1) - \bar{\boldsymbol{y}}_{\mathrm{ref}}(k+1). \end{aligned}$$

Figure 6.4 shows the corresponding structure of the linear MPC without constraints. An observer was not used. However, it can easily be inserted into the control loop to determine the state vector $\boldsymbol{x}(k)$ required for control.

6.1.3 LMPC with Constraints

As mentioned previously, linear MPCs without constraints have no great practical relevance and have been described here mainly to provide an easy-to-understand introduction to the topic, and to explain how they work in principle.

In industrial practice, linear MPCs are used when the variables to be optimized are subject to restrictions in the control variable $u_i(k)$, i. e.

$$\underbrace{\begin{bmatrix} u_{\min,1} \\ u_{\min,2} \\ \vdots \\ u_{\min,m} \end{bmatrix}}_{\boldsymbol{u}_{\min}} \le \underbrace{\begin{bmatrix} u_1(k) \\ u_2(k) \\ \vdots \\ u_m(k) \end{bmatrix}}_{\boldsymbol{u}(k)} \le \underbrace{\begin{bmatrix} u_{\max,1} \\ u_{\max,2} \\ \vdots \\ u_{\max,m} \end{bmatrix}}_{\boldsymbol{u}_{\max}}. \tag{6.13}$$

These constraints of the control variables can also be represented as functions of the changes in the control variables, so that equation (6.13) takes the form

$$\begin{aligned}
\boldsymbol{u}_{\min} \le \boldsymbol{u}(k) &= \boldsymbol{u}(k-1) + \Delta\boldsymbol{u}(k) \le \boldsymbol{u}_{\max},\\
\boldsymbol{u}_{\min} \le \boldsymbol{u}(k+1) &= \boldsymbol{u}(k-1) + \Delta\boldsymbol{u}(k) + \Delta\boldsymbol{u}(k+1) \le \boldsymbol{u}_{\max},\\
\boldsymbol{u}_{\min} \le \boldsymbol{u}(k+2) &= \boldsymbol{u}(k-1) + \Delta\boldsymbol{u}(k) + \Delta\boldsymbol{u}(k+1) + \Delta\boldsymbol{u}(k+2) \le \boldsymbol{u}_{\max},\\
&\;\;\vdots\\
\boldsymbol{u}_{\min} \le \boldsymbol{u}(k+n_{\mathrm{c}}-1) &= \boldsymbol{u}(k-1) + \Delta\boldsymbol{u}(k) + \ldots + \Delta\boldsymbol{u}(k+n_{\mathrm{c}}-1) \le \boldsymbol{u}_{\max}.
\end{aligned} \tag{6.14}$$

We define

$$\bar{\boldsymbol{u}}_{\min} = \begin{bmatrix} \boldsymbol{u}_{\min} \\ \vdots \\ \boldsymbol{u}_{\min} \end{bmatrix} \quad \text{and} \quad \bar{\boldsymbol{u}}_{\max} = \begin{bmatrix} \boldsymbol{u}_{\max} \\ \vdots \\ \boldsymbol{u}_{\max} \end{bmatrix}$$

as vectors of length mn_{c}. The inequalities (6.14) can then be represented in the matrix form

$$\bar{\boldsymbol{u}}_{\min} \le \boldsymbol{E} \cdot \boldsymbol{u}(k-1) + \boldsymbol{D} \cdot \Delta\bar{\boldsymbol{u}}(k) \le \bar{\boldsymbol{u}}_{\max} \tag{6.15}$$

with the $mn_{\mathrm{c}} \times m$ matrix

$$\boldsymbol{E} = \begin{bmatrix} \boldsymbol{I} \\ \vdots \\ \boldsymbol{I} \end{bmatrix}$$

and the lower triangular matrix of dimensions $mn_{\mathrm{c}} \times mn_{\mathrm{c}}$, given by

$$\boldsymbol{D} = \begin{bmatrix} \boldsymbol{I} & \boldsymbol{0} & \boldsymbol{0} & \cdots & \boldsymbol{0} \\ \boldsymbol{I} & \boldsymbol{I} & \boldsymbol{0} & \cdots & \boldsymbol{0} \\ \vdots & \vdots & \vdots & \ddots & \vdots \\ \boldsymbol{I} & \boldsymbol{I} & \boldsymbol{I} & \cdots & \boldsymbol{I} \end{bmatrix}.$$

The matrices $\boldsymbol{I}$ are identity matrices of dimension $m \times m$.

Constraints on the control variables' rate of change, i. e. constraints of the type

$$\Delta\bar{\boldsymbol{u}}_{\min} \leq \Delta\bar{\boldsymbol{u}}(k) \leq \Delta\bar{\boldsymbol{u}}_{\max}, \tag{6.16}$$

are also often relevant. The vectors $\Delta\bar{\boldsymbol{u}}_{\min}$ and $\Delta\bar{\boldsymbol{u}}_{\max}$ are similar in structure to the vectors $\bar{\boldsymbol{u}}_{\min}$ and $\bar{\boldsymbol{u}}_{\max}$, and are of the same dimensions.

In addition to control variables, the state variables x_i and output variables y_i can also be subject to constraints such as

$$\underbrace{\begin{bmatrix} y_{\min,1} \\ y_{\min,2} \\ \vdots \\ y_{\min,r} \end{bmatrix}}_{\boldsymbol{y}_{\min}} \leq \underbrace{\begin{bmatrix} y_1(k) \\ y_2(k) \\ \vdots \\ y_r(k) \end{bmatrix}}_{\boldsymbol{y}(k)} \leq \underbrace{\begin{bmatrix} y_{\max,1} \\ y_{\max,2} \\ \vdots \\ y_{\max,r} \end{bmatrix}}_{\boldsymbol{y}_{\max}}.$$

The constraints on the output variables can be converted into constraints on the control variable changes. With

$$\bar{\boldsymbol{y}}_{\min} \leq \bar{\boldsymbol{y}}(k+1) \leq \bar{\boldsymbol{y}}_{\max}$$

and with equation (6.9), we obtain

$$\bar{\boldsymbol{y}}_{\min} \leq \boldsymbol{g}(k) + \boldsymbol{H}\Delta\bar{\boldsymbol{u}}(k) \leq \bar{\boldsymbol{y}}_{\max}. \tag{6.17}$$

Here, the vector $\bar{\boldsymbol{y}}_{\min}$ consists of n_{p} vectors $\boldsymbol{y}_{\min}$ according to

$$\bar{\boldsymbol{y}}_{\min} = \begin{bmatrix} \boldsymbol{y}_{\min} \\ \vdots \\ \boldsymbol{y}_{\min} \end{bmatrix} \in I\!R^{rn_{\mathrm{p}} \times 1},$$

and for $\bar{\boldsymbol{y}}_{\max}$ the same applies. Constraints on x_i are of a similar form.

In summary, the inequalities (6.15), (6.16), and (6.17) can be represented as

$$\boldsymbol{W}\Delta\bar{\boldsymbol{u}}(k) \leq \boldsymbol{w}(k, k-1) \tag{6.18}$$

with

$$\boldsymbol{W} = \begin{bmatrix} -\boldsymbol{D} \\ \boldsymbol{D} \\ -\boldsymbol{I} \\ \boldsymbol{I} \\ -\boldsymbol{H} \\ \boldsymbol{H} \end{bmatrix} \quad \text{and} \quad \boldsymbol{w}(k, k-1) = \begin{bmatrix} -\bar{\boldsymbol{u}}_{\min} + \boldsymbol{E}\boldsymbol{u}(k-1) \\ \bar{\boldsymbol{u}}_{\max} - \boldsymbol{E}\boldsymbol{u}(k-1) \\ -\Delta\bar{\boldsymbol{u}}_{\min} \\ \Delta\bar{\boldsymbol{u}}_{\max} \\ -\bar{\boldsymbol{y}}_{\min} + \boldsymbol{g}(k) \\ \bar{\boldsymbol{y}}_{\max} - \boldsymbol{g}(k) \end{bmatrix}.$$

This results in the optimization problem

$$\min_{\Delta\bar{\boldsymbol{u}}(k)} J(\Delta\bar{\boldsymbol{u}}(k))$$

with the linear constraints (6.18).

We can now use the positive definite quadratic performance index (6.1) or (6.2) once again. In this case, we have a convex quadratic optimization problem with linear constraints, which has a unique minimum. Generally, this minimum cannot be determined analytically, only numerically. However, there are convergent standard procedures in quadratic programming which can be used to calculate the minimum [41, 68, 296]. At this point, the disadvantages of conventional model predictive control become clear: the control law cannot be explicitly specified and the control variable's calculation is time-consuming. Explicit MPC offers a way to mitigate this disadvantage [12, 156, 230, 256, 298]. Such an MPC solves the optimization problem offline. For this purpose, control variable vectors $\boldsymbol{u}_{\mathrm{opt},i}$ are calculated for a number of state vectors $\boldsymbol{x}_i$ and setpoints $\boldsymbol{y}_{\mathrm{ref},i}$ and summarized in a table. By interpolation between these values, a static controller

$$\boldsymbol{u}_{\mathrm{opt}} = \boldsymbol{h}(\boldsymbol{x}, \boldsymbol{y}_{\mathrm{ref}})$$

is obtained, which enables real-time control. However, the effort required by this method increases greatly as the system order increases, so that it is only applicable to systems with a low system order. Another way to mitigate the real-time problem is to use software and hardware specifically tailored to the problem [3]. Nonetheless, in many cases the advantages of MPCs, which are their high control performance, good comprehensibility, and simple adjustability, outweigh their disadvantages.

6.1.4 Example: Drainage System

For the drainage of areas with high groundwater levels, such as those found in the Netherlands or the American Everglades, sewer systems are used. Particularly during or after heavy rainfall, water must be pumped from the drainage sewers into neighboring rivers or the sea. Otherwise, excessive water levels would endanger dikes, towns, and agricultural land. However, low water levels in the canals are also problematic, as they can cause an embankment slide. Moreover, large fluctuations in the water level can cause loosening of the soil in the sewer walls due to pump effects, leading to damage of the sewers.

Therefore, water level in the sewers should be neither too low nor too high but should remain within a specified tolerance range. For this purpose, control systems are used to adjust the amount of water pumped out of the sewers. Figure 6.5 shows the structure of such a drainage system. In some cases where the facilities are near the sea, water is pumped into additional storage canals or reservoirs during high tide. At low tide, the water then flows into the sea without additional pumping.

The water level h in the sewer system is dependent on the rainwater flowing into the sewers and the amount of water that has been pumped out according to the relation

$$h(k+1) = h(k) + \frac{T}{A}(q_{\text{rain}}(k) - q_{\text{p}}(k - k_{\text{d}})). \tag{6.19}$$

Here, q_{rain} indicates the amount of water per second flowing into the canals due to rainfall or additional groundwater, and q_{p} indicates the amount of water pumped out per second. The area A comprises the total area of the sewers, T is the time span in seconds between the times k and $k+1$, and k_{d} is the number of time steps after which the control variable q_{p} acts on the water level in the sewers.

As an example we will describe the drainage system of the Delfland in the Netherlands [348]. Its parameter values are

Fig. 6.5: Drainage system

$$A = 7300000\,\mathrm{m}^2 = 7.3\,\mathrm{km}^2, \qquad T = 900\,\mathrm{s}, \qquad k_\mathrm{d} = 1.$$

The water level h and pump volume flow q_p are limited by

$$-0.55\,\mathrm{m} \le h \le -0.30\,\mathrm{m}$$

and

$$0\,\mathrm{m}^3\mathrm{s}^{-1} \le q_\mathrm{p} \le 75\,\mathrm{m}^3\mathrm{s}^{-1}.$$

The water level h in the sewers is indicated with respect to the sea level, i. e. for $h = 0$, the water level in the sewers is equal to sea level. In the case of the Delfland, it normally lies about half a meter below sea level. In our example, we will set its reference value at $h_\mathrm{ref} = -0.40\,\mathrm{m}$.

The system's state variables given by equation (6.19) are the water level

$$x_1(k) = h(k),$$

and the delayed control signal

$$x_2(k) = q_\mathrm{p}(k-1),$$

which is delayed here by one time step, i. e. $k_\mathrm{d} = 1$. With the parameter values above and the current control variable

$$u(k) = q_\mathrm{p}(k)$$

from equation (6.19), we obtain the state-space model

$$\begin{aligned} x_1(k+1) &= x_1(k) - 1.23 \cdot 10^{-4} x_2(k) + 1.23 \cdot 10^{-4} q_\mathrm{rain}(k), \\ x_2(k+1) &= u(k), \\ y(k) &= x_1(k). \end{aligned}$$

Here, q_rain acts as a disturbance.

The aim of a control is now to compensate for disturbances q_rain due to heavy precipitation in such a way that the water level x_1 is brought to the desired value $x_1 = -0.4$. During the regulation process, the state variable x_1 must not exceed the constraints

$$-0.55 \le x_1 \le -0.30$$

and the control variable u must not exceed the constraints

$$0 \le u \le 75.$$

It is evident that the above control task is well-suited to model predictive control. Firstly, the time $T = 900\,\mathrm{s}$ between two controller actions is sufficiently long for the computation, and secondly, the limitations of $y = x_1$ and u can very well be included in a model predictive control.

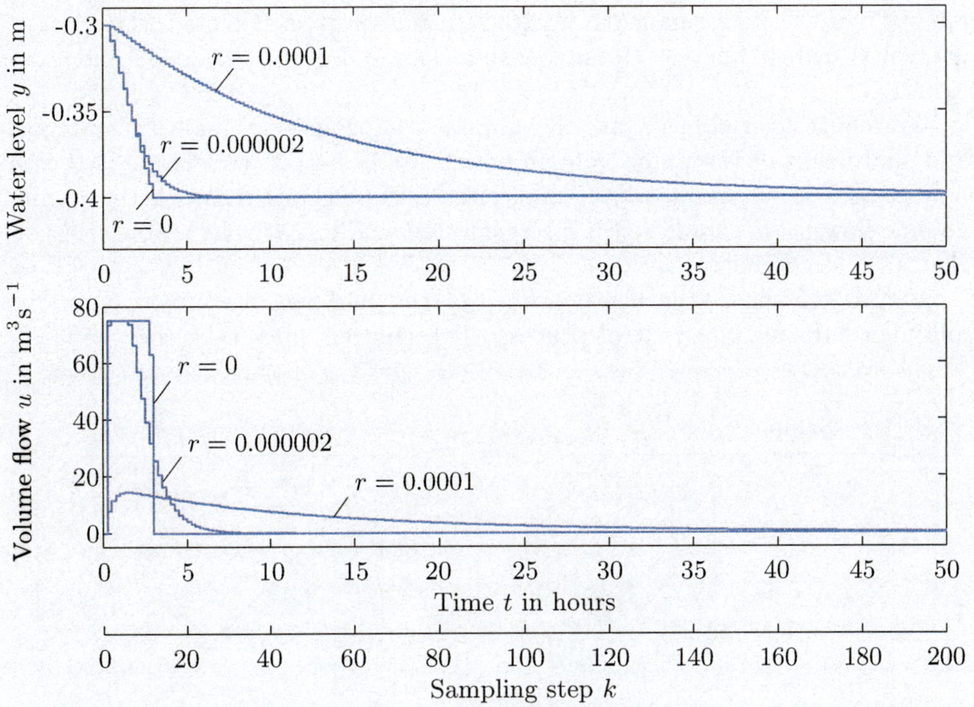

Fig. 6.6: Progression of the output variable y and progression of the control variable u depending on $t = kT$

For the design of the MPC, we now select the prediction horizon and the control horizon as

$$n_{\mathrm{p}} = 97 \quad \text{and} \quad n_{\mathrm{c}} = 2,$$

respectively. The long prediction horizon of 24 h 15 min allows the inclusion of predicted rainfall based on a weather forecast, among other factors. This amount of rain can be taken into account in the controller via q_{rain}.

For the performance index, we will choose

$$J = (\bar{\boldsymbol{y}}(k+1) - \bar{\boldsymbol{y}}_{\mathrm{ref}}(k+1))^T (\bar{\boldsymbol{y}}(k+1) - \bar{\boldsymbol{y}}_{\mathrm{ref}}(k+1)) + r\boldsymbol{\Delta}\bar{\boldsymbol{u}}^T(k)\boldsymbol{\Delta}\bar{\boldsymbol{u}}(k), \tag{6.20}$$

where the reference value of the water level is set constant at $y_{\mathrm{ref}} = h_{\mathrm{ref}} = -0.4\,\mathrm{m}$, i. e.

$$\bar{\boldsymbol{y}}_{\mathrm{ref}}(i) = [-0.4\ \mathrm{m} \quad -0.4\ \mathrm{m} \quad \cdots \quad -0.4\ \mathrm{m}\ ]^T$$

holds for all $i = 0, 1, \ldots$

As a practical example, let us take an elevated water level of $x_1(0) = -0.3\,\mathrm{m}$ following bad weather conditions. The second initial value is $x_2(0) = 0$. Figure 6.6 shows the courses of the control variable u and the water level y. Three different controllers have been used, resulting from different weighting factors r in the performance index (6.20). As expected, the model predictive control is fastest for $r = 0$ and becomes slower for $r = 0.000002$ and $r = 0.0001$.

The control variable u's rate of change also depends on r. If abrupt startups and shutdowns of the pump volume flow u are of no consequence, $r = 0$ can be selected. If we would rather make less abrupt changes Δu in the pump volume flow u, we should select a larger value r. Alternatively, a limitation of Δu can be introduced.

Figure 6.7 shows the contour lines of the performance index, which is minimized during the control process. The contour lines take the shape of

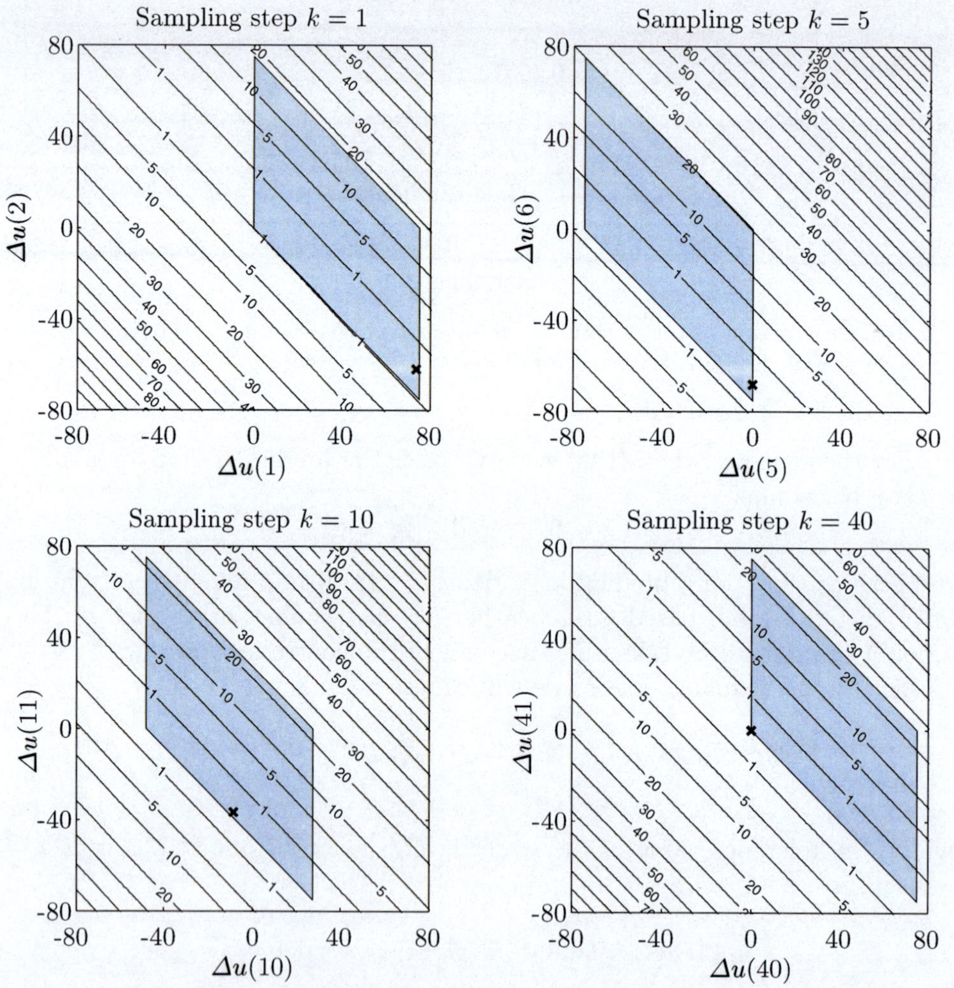

Fig. 6.7: Illustration of the optimization problems to be solved in the time steps $k = 1$, $k = 5$, $k = 10$, and $k = 40$ for the case $r = 0.000002$. The contour lines of J and the admissible optimization region (blue) are displayed. A cross marks the minimum in each case.

very elongated ellipses. Furthermore, the graph shows the admissible region for the optimization variables $\Delta u(k)$ and $\Delta u(k+1)$ as resulting from the constraints. The shapes of the contour lines of the performance index and the admissible region depend on the time step k, i. e. they change during the control process.

6.1.5 Nonlinear Model Predictive Control

For nonlinear plants

$$\begin{aligned}\dot{\boldsymbol{x}} &= \boldsymbol{f}(\boldsymbol{x}, \boldsymbol{u}),\\ \boldsymbol{y} &= \boldsymbol{g}(\boldsymbol{x})\end{aligned} \tag{6.21}$$

or

$$\begin{aligned}\boldsymbol{x}(k+1) &= \boldsymbol{f}(\boldsymbol{x}(k), \boldsymbol{u}(k)),\\ \boldsymbol{y}(k) &= \boldsymbol{g}(\boldsymbol{x}(k)),\end{aligned} \tag{6.22}$$

MPC can be used in a similar way as for linear plants. However, the implementation and calculation are usually much more complicated and complex than in the linear case. This is because, due to the nonlinear system dynamics (6.22), the optimization problem no longer takes a simple form. In particular, it is generally no longer convex. The nonlinearity causes a complicated dependence of the performance index J on the control values $\boldsymbol{u}(k+i)$ to be optimized. Such a nonconvex optimization problem often has several local minima, and there are no solution procedures which reliably converge to the global minimum. Despite the more complex numerics, the basic procedure is roughly the same as for the LMPC.

Numerical optimization methods are used to solve the constrained optimization problem [13, 94, 241]. One of the main problems is often solving the optimization problem in real time. It is particularly time-consuming to determine the solution of system (6.21) for each control variable and performance index calculation. Usually this is done by means of an integration procedure, such as the Runge-Kutta procedure in the continuous-time case (6.21). In the case of discrete systems, the system solution is calculated directly using the recursion equation (6.22).

Because of the real-time problem, MPC is mostly used in control systems with slow dynamics, such as in the processing industry. Another problem of nonlinear MPC, abbreviated to NMPC, is ensuring the stability of the control loop. Due to the nonlinearity of the plant and the fact that the control law cannot be explicitly specified, stability generally cannot be guaranteed.

Below we will take a closer look at the stability problem. The starting point of our analysis is system (6.22), where the input variable vector $\boldsymbol{u}$ is limited according to $\boldsymbol{u} \in W \subset \mathbb{R}^m$, and the state vector $\boldsymbol{x}$ is limited according to $\boldsymbol{x} \in X \subset \mathbb{R}^n$. Let us assume that $n_\mathrm{p} = n_\mathrm{c}$ holds. We will use the function

$$J(k) = \sum_{i=k}^{k+n_c-1} Q(\boldsymbol{x}(i), \boldsymbol{u}(i))$$

as the performance index to be minimized.

The function Q can have quadratic terms similar to those in equation (6.1). Of course, other forms are also possible. In any case,

$$Q(\boldsymbol{x}, \boldsymbol{u}) > 0$$

must hold for all $\boldsymbol{x} \neq \mathbf{0}$ and $\boldsymbol{u} \neq \mathbf{0}$. Furthermore, the controlled system must have an equilibrium point at

$$\boldsymbol{x}(k) = \mathbf{0}$$

for $\boldsymbol{u}(k) = \mathbf{0}$, and

$$Q(\mathbf{0}, \mathbf{0}) = 0$$

must apply. Due to this requirement,

$$J(k) = 0$$

also applies to $\boldsymbol{x}(k) = \mathbf{0}$ and to $\boldsymbol{u}(i) = \mathbf{0},\ i = k, k+1, \ldots$

Note that the performance index J indirectly depends on the state vector $\boldsymbol{x}$ only, since the control variable vector $\boldsymbol{u}$ depends on $\boldsymbol{x}$ via the NMPC. Thus, the performance index J seems to be a candidate for a Lyapunov function of the predictively controlled system, because at the origin $\boldsymbol{x} = \mathbf{0}$ it is identical to zero and everywhere else it is positive. As a third condition for a Lyapunov function (see Section 2.3.5, p. 125), J must fulfill the requirement

$$J(k) - J(k-1) < 0, \tag{6.23}$$

i. e. $J(k)$ must decrease in the course of each control process.

Condition (6.23) is also fulfilled if $J(k)$ is only slightly smaller than $J(k-1)$. In this case, however, the control occurs very slowly. Scokaert, Mayne, and Rawlings [408] have thus formulated condition (6.23) more strictly as

$$J(k) - J(k-1) < -\mu Q(\boldsymbol{x}(k-1), \boldsymbol{u}(k-1)), \tag{6.24}$$

where $\mu \in (0, 1)$ holds.

For values of the parameter μ close to zero, the optimization procedure will find a solution

$$\boldsymbol{u}(k), \ldots, \boldsymbol{u}(k + n_c - 1)$$

more easily than for greater values of μ, because there are more solutions in this case. The price to be paid for this is often a small decrease in the performance index J. Consequently, the compensation is slow again.

Choosing μ close to one causes a greater decrease in J from one step $k-1$ to the next. However, now the solution space is more limited, so that the calculation of a solution becomes more complicated and can take longer.

During the execution of the MPC, it must be constantly verified whether inequality (6.24) is fulfilled. Once a corresponding control sequence $\boldsymbol{u}(k), \ldots,$ $\boldsymbol{u}(k+n_{\mathrm{c}}-1)$ has been identified using the optimization procedure, two cases are possible. In the first, optimization time is still available and can be used to improve the result. In the second, the available optimization time is used up and the control variable $\boldsymbol{u}(k)$ is applied to the plant. The optimization then starts over again with $\tilde{k} = k+1$ in an attempt to identify a new combination of values $\boldsymbol{u}(\tilde{k}), \ldots, \boldsymbol{u}(\tilde{k}+n_{\mathrm{c}}-1)$. Conveniently, the old set of control variables $\boldsymbol{u}(k), \ldots, \boldsymbol{u}(k+n_{\mathrm{c}}-1)$ can be used as the starting point of the optimization.

Based on the above arguments, the performance index J seems to be suitable as a potential Lyapunov function, but we can only determine its decrease in the general case along a trajectory, i. e. the fulfillment of inequality (6.23), numerically during a control process. However, Lyapunov's stability theorem requires the verification of condition (6.23) in a neighborhood $U(\mathbf{0})$ of the equilibrium point $\boldsymbol{x}_{\mathrm{eq}} = \mathbf{0}$, i. e. for all $\boldsymbol{x} \in U(\mathbf{0}) \setminus \{\mathbf{0}\}$ and not only for individual trajectories, to prove stability. As long as we can only verify the decrease in J for individual trajectories and not for the entire neighborhood of $\boldsymbol{x} = \mathbf{0}$, the above procedure is not sufficient for the stability analysis.

The aforementioned problem can be illustrated by the following explanations. Let us analyze a nonlinear model predictive control loop of order two which has an equilibrium point at $\boldsymbol{x} = \mathbf{0}$, and a circular stable limit cycle with its center point at $\boldsymbol{x} = \mathbf{0}$. We will start the control procedure by means of the predictive control at a point $\boldsymbol{x}(0)$ outside the limit cycle. The trajectory $\boldsymbol{x}(k)$ of the control loop tends asymptotically from the point $\boldsymbol{x}(0)$ to the limit cycle. This means the performance index J decreases constantly, so that inequality (6.23) or (6.24) is fulfilled. The trajectory $\boldsymbol{x}(k)$ does not tend to the equilibrium point $\boldsymbol{x}_{\mathrm{eq}} = \mathbf{0}$ despite the decrease in J. Figure 6.8 illustrates this. In conclusion, it is important to note that, in the general case, we cannot deter-

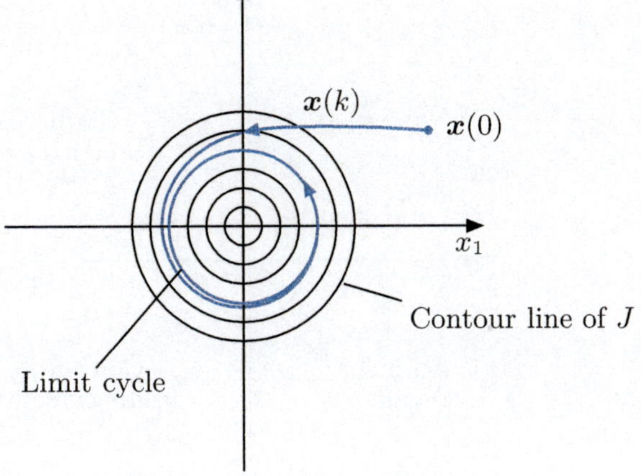

Fig. 6.8: Example of the continuous decrease in the performance index J along a trajectory where the trajectory $\boldsymbol{x}(k)$ does not tend to zero

mine whether the performance index J is a Lyapunov function. The decrease in J during a compensation operation is merely a practice-oriented condition and not sufficient to prove the stability of the model predictive control system.

Nonetheless, stability can be guaranteed if we succeed in proving that inequality (6.23) holds for all $\boldsymbol{x}$ in a neighborhood of $\boldsymbol{x}_{\text{eq}} = \mathbf{0}$. In specific cases, based on assumptions regarding the plant and the minimization problem, this is possible and has yielded some basic results for the stability of MPCs [51, 161, 205, 282, 315, 382]. A prerequisite for these is an infinite control horizon n_{c} or one of the following requirements.

To ensure that the equilibrium point $\boldsymbol{x}_{\text{eq}} = \mathbf{0}$ is actually reached, we require that

$$\boldsymbol{x}(k + n_{\text{c}}) = \mathbf{0} \tag{6.25}$$

holds for every time step k in addition to the decrease in J, i. e. that stability is given. Equation (6.25) is an additional constraint on the task of optimization. In order to fulfill it, the control horizon n_{c} must be sufficiently large. This can require very large values of n_{c}, meaning the amount n_{c} of control values

$$\boldsymbol{u}(k), \ldots, \boldsymbol{u}(k + n_{\text{c}} - 1)$$

to be optimized also becomes very large.

The latter problem is reduced if, instead of the constraint (6.25), we require

$$\boldsymbol{x}(k + n_{\text{c}}) \in U(\mathbf{0}),$$

where $U(\mathbf{0})$ is a neighborhood of the equilibrium point $\boldsymbol{x}_{\text{eq}} = \mathbf{0}$. The set $U(\mathbf{0})$ can be a tolerance range whose state values $\boldsymbol{x} \in U(\mathbf{0})$ are accepted as target values of the control because their deviation from $\boldsymbol{x} = \mathbf{0}$ is sufficiently small. The neighborhood $U(\mathbf{0})$ can also be a region in which a controller $\boldsymbol{u}_2(\boldsymbol{x})$ is used instead of the predictive controller $\boldsymbol{u}_1(\boldsymbol{x})$, as illustrated in Figure 6.9.

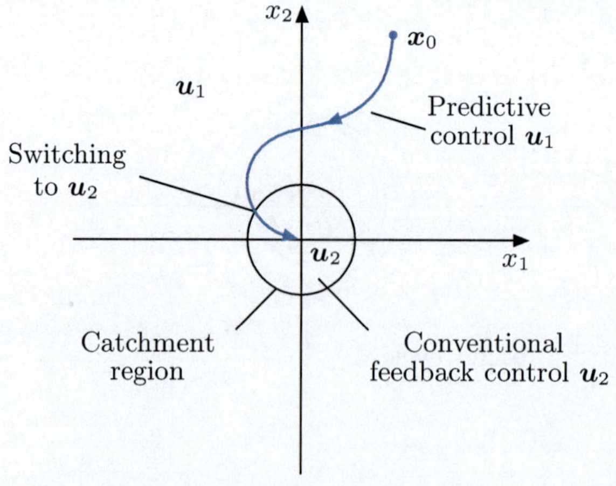

Fig. 6.9: Predictive dual-mode control system

In this case, the controller $\boldsymbol{u}_2(\boldsymbol{x})$ is designed in such a way that it leads to an asymptotically stable regulation to the equilibrium point $\boldsymbol{x}_{\mathrm{eq}} = \boldsymbol{0}$ for all $\boldsymbol{x} \in U(\boldsymbol{0})$. The neighborhood $U(\boldsymbol{0})$ is designed as a catchment region, i. e. no trajectory $\boldsymbol{x}(k)$ leaves $U(\boldsymbol{0})$ after entering this region. A compensation for an initial value $\boldsymbol{x}_0 \notin U(\boldsymbol{0})$ starts with the predictive control $\boldsymbol{u}_1(\boldsymbol{x})$ and switches to the second controller $\boldsymbol{u}_2(\boldsymbol{x})$ as soon as $\boldsymbol{x}(k) \in U(\boldsymbol{0})$ applies.

For example, the conventional control law $\boldsymbol{u}_2(\boldsymbol{x})$ can be designed linearly, based on a linearization of the plant around $\boldsymbol{x} = \boldsymbol{0}$. Controllers of this kind with two control laws $\boldsymbol{u}_1$ and $\boldsymbol{u}_2$ are called *dual-mode controllers*. We can summarize the above in the following dual-mode MPC algorithm proposed by Scokaert, Mayne, and Rawlings [408]:

Step 1: Choose $\mu \in (0,1)$.

Step 2: Let $k = 0$. If $\boldsymbol{x}(0) \in U(\boldsymbol{0})$, then set $\boldsymbol{u}(0) = \boldsymbol{u}_2(\boldsymbol{x}(0))$. Otherwise, find a control variable sequence

$$\boldsymbol{u}(0), \boldsymbol{u}(1), \ldots, \boldsymbol{u}(n_{\mathrm{c}} - 1)$$

by optimizing J and an associated state variable sequence

$$\boldsymbol{x}(0), \boldsymbol{x}(1), \ldots, \boldsymbol{x}(n_{\mathrm{c}}),$$

so that

$$\begin{aligned} &\boldsymbol{u}(i) \in W \quad \text{for all } i = 0, \ldots, n_{\mathrm{c}} - 1, \\ &\boldsymbol{x}(i) \in X \quad \text{for all } i = 0, \ldots, n_{\mathrm{c}}, \qquad \boldsymbol{x}(n_{\mathrm{c}}) \in U(\boldsymbol{0}) \end{aligned}$$

hold. As the active control variable value, use the vector $\boldsymbol{u}(0)$.

Step 3: Let $k \neq 0$. If $\boldsymbol{x}(k) \in U(\boldsymbol{0})$, then set $\boldsymbol{u}(k) = \boldsymbol{u}_2(\boldsymbol{x}(k))$. Otherwise, find a control variable sequence

$$\boldsymbol{u}(k), \ldots, \boldsymbol{u}(k + n_{\mathrm{c}} - 1)$$

by optimizing J and an associated state variable sequence

$$\boldsymbol{x}(k), \ldots, \boldsymbol{x}(k + n_{\mathrm{c}}),$$

so that

$$\begin{aligned} &\boldsymbol{u}(i) \in W \quad \text{for all } i = k, \ldots, k + n_{\mathrm{c}} - 1, \\ &\boldsymbol{x}(i) \in X \quad \text{for all } i = k, \ldots, k + n_{\mathrm{c}}, \qquad \boldsymbol{x}(k + n_{\mathrm{c}}) \in U(\boldsymbol{0}), \end{aligned}$$

and

$$J(k) - J(k-1) < -\mu Q(\boldsymbol{x}(k-1), \boldsymbol{u}(k-1))$$

holds. When optimizing J, the control variable sequence of the previous step is selected as the starting point. Use $\boldsymbol{u}(k)$ as the active control value. Repeat Step 3.

At this point, it must be mentioned that, in practice, stability analysis is often neglected for nonlinear MPC in industry. In order to be able to make a statement at all for an NMPC, stability is sometimes evaluated by simulating the control loop for selected cases or operating points. However, this does not provide a proof of stability. Thus, in addition to the extensive computational effort [121], the main disadvantage of real-world MPCs is that stability either cannot be rigorously proven, or this can be done only with great effort or only for special cases [51, 161, 205, 282, 315, 382].

6.1.6 Example: Evaporation Plant

As an example, let us view an evaporation plant as it is used for purposes such as the production of syrup in sugar factories [335, 458]. As shown in Figure 6.10, the system consists of an evaporator, a separator, and a condenser. The input material, raw juice from sugar beets, is fed into the evaporator which is designed as a heat exchanger. The evaporator is heated with steam of pressure P_{s}. The raw material heated in the evaporator at pressure P_{e} leaves the

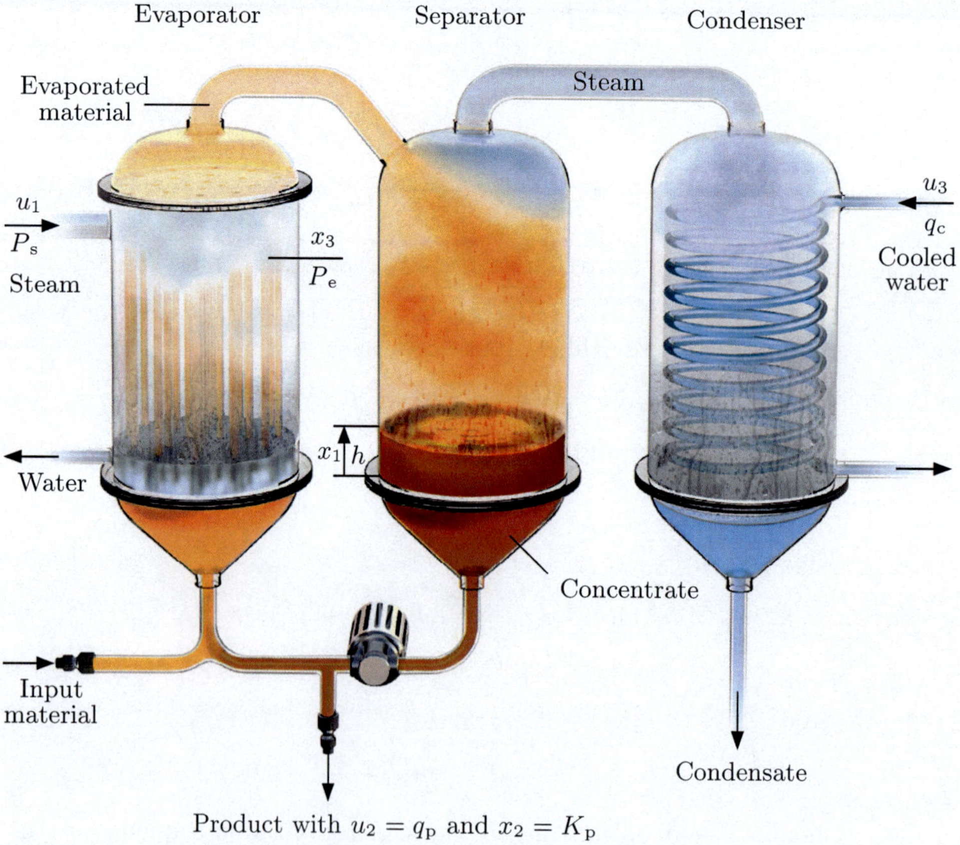

Fig. 6.10: Evaporation plant with evaporator, separator, and condenser

evaporator as a mixture of steam and liquid and then enters the separator. In the separator, the steam is separated and fed into a condenser. Cooled by water that flows into the condenser at rate q_c, the steam is condensed and extracted from the facility. The concentrate collected in the separator has a level of h. A part of this concentrate with concentration K_p is now removed as a product from the process at volume rate q_p. The much larger part, however, is mixed with the input material and returned to the evaporator.

The system can be described by a third-order nonlinear model, for which we will assume that the flow rate and the concentration of the input material are constant. Here,

$$x_1 = h, \quad x_2 = K_\mathrm{p}, \quad x_3 = P_\mathrm{e}$$

are the state variables, and

$$u_1 = P_\mathrm{s}, \quad u_2 = q_\mathrm{p}, \quad u_3 = q_\mathrm{c}$$

are the control variables. We will measure x_1 in m, x_2 in %, x_3 and u_1 in kPa, and u_2 and u_3 in kg min^{-1}. The model has the form

$$\begin{aligned}
\dot{x}_1 &= a_1 x_3 + a_2 x_2 - b_1 u_1 - b_2 u_2 - k_1, \\
\dot{x}_2 &= -a_3 x_2 u_2 + k_2, \\
\dot{x}_3 &= -a_4 x_3 - a_5 x_2 + b_3 u_1 - \frac{a_6 x_3 + b_4}{b_5 u_3 + k_3} u_3 + k_4.
\end{aligned} \tag{6.26}$$

The parameters of the system are

$$\begin{aligned}
a_1 &= 0.00751, & b_1 &= 0.00192, & k_1 &= 0.01061, \\
a_2 &= 0.00418, & b_2 &= 0.05, & k_2 &= 2.5, \\
a_3 &= 0.05, & b_3 &= 0.00959, & k_3 &= 6.84, \\
a_4 &= 0.03755, & b_4 &= 0.1866, & k_4 &= 2.5531, \\
a_5 &= 0.02091, & b_5 &= 0.14, \\
a_6 &= 0.00315.
\end{aligned}$$

The state variables are the output variables

$$y_1 = x_1, \quad y_2 = x_2, \quad y_3 = x_3$$

of the process. Both the state variables and the control variables are subject to constraints of the form

$$\begin{aligned}
0\,\mathrm{m} &\le x_1 \le 2\,\mathrm{m}, \\
0\,\% &\le x_2 \le 50\,\%, \\
0\,\mathrm{kPa} &\le x_3 \le 100\,\mathrm{kPa}, \\
0\,\mathrm{kPa} &\le u_1 \le 400\,\mathrm{kPa}, \\
0\,\mathrm{kg\,min}^{-1} &\le u_2 \le 4\,\mathrm{kg\,min}^{-1}, \\
0\,\mathrm{kg\,min}^{-1} &\le u_3 \le 400\,\mathrm{kg\,min}^{-1}.
\end{aligned} \tag{6.27}$$

The equilibrium point $\boldsymbol{x}_{\text{eq}}$ of system (6.26) can be specified via the control variables

$$u_{1\text{eq}} = \frac{1}{b_1}\left(a_1 x_{3\text{eq}} + a_2 x_{2\text{eq}} - k_1 - \frac{b_2 k_2}{a_3 x_{2\text{eq}}}\right),$$

$$u_{2\text{eq}} = \frac{k_2}{a_3 x_{2\text{eq}}},$$

$$u_{3\text{eq}} = \frac{k_3(-a_4 x_{3\text{eq}} - a_5 x_{2\text{eq}} + k_4 + b_3 u_{1\text{eq}})}{(a_6 + a_4 b_5) x_{3\text{eq}} + a_5 b_5 x_{2\text{eq}} - b_5 k_4 + b_4 - b_3 b_5 u_{1\text{eq}}}.$$

In the following, we will select

$$\boldsymbol{x}_{\text{eq}} = \begin{bmatrix} 1 & 15 & 70 \end{bmatrix}^T, \tag{6.28}$$

which yields

$$\boldsymbol{u}_{\text{eq}} = \begin{bmatrix} 214.13 & 3.33 & 65.40 \end{bmatrix}^T.$$

To make the design of a model predictive control as simple as possible, we discretize the model and use a discrete-time MPC. The duration of a control-value step is $T = 1\,\text{min}$. For the prediction horizon and the control horizon, we will choose $n_{\text{p}} = n_{\text{c}} = 5$, and for the performance index

$$J(k) = \sum_{i=k}^{k+4} (\boldsymbol{x}(i) - \boldsymbol{x}_{\text{eq}})^T \boldsymbol{Q} (\boldsymbol{x}(i) - \boldsymbol{x}_{\text{eq}}) + r\,(\boldsymbol{u}(i) - \boldsymbol{u}_{\text{eq}})^T (\boldsymbol{u}(i) - \boldsymbol{u}_{\text{eq}}).$$

Here, the weighting matrix is

$$\boldsymbol{Q} = \begin{bmatrix} 100 & 0 & 0 \\ 0 & 1 & 0 \\ 0 & 0 & 10 \end{bmatrix}$$

and $r = 0.1$. The performance index with the associated constraints (6.27) is converted using a barrier function so that an unconstrained performance index

$$\tilde{J}(k) = \begin{cases} J(k) & \text{if all } \boldsymbol{x}(i), \boldsymbol{u}(i), i = 1, \ldots, 4, \text{ fulfill inequality (6.27)}, \\ \infty & \text{if one } \boldsymbol{x}(i), \boldsymbol{u}(i), i = 1, \ldots, 4, \text{ does not fulfill inequality (6.27)}, \end{cases}$$

is to be minimized. Due to the Cartesian constraints, the performance index $\tilde{J}$ can be efficiently optimized with the optimization method developed by Hooke and Jeeves [189, 406]. The numerical solution of the system equation (6.26) is carried out using the Euler method with a step size of 2.4 s.

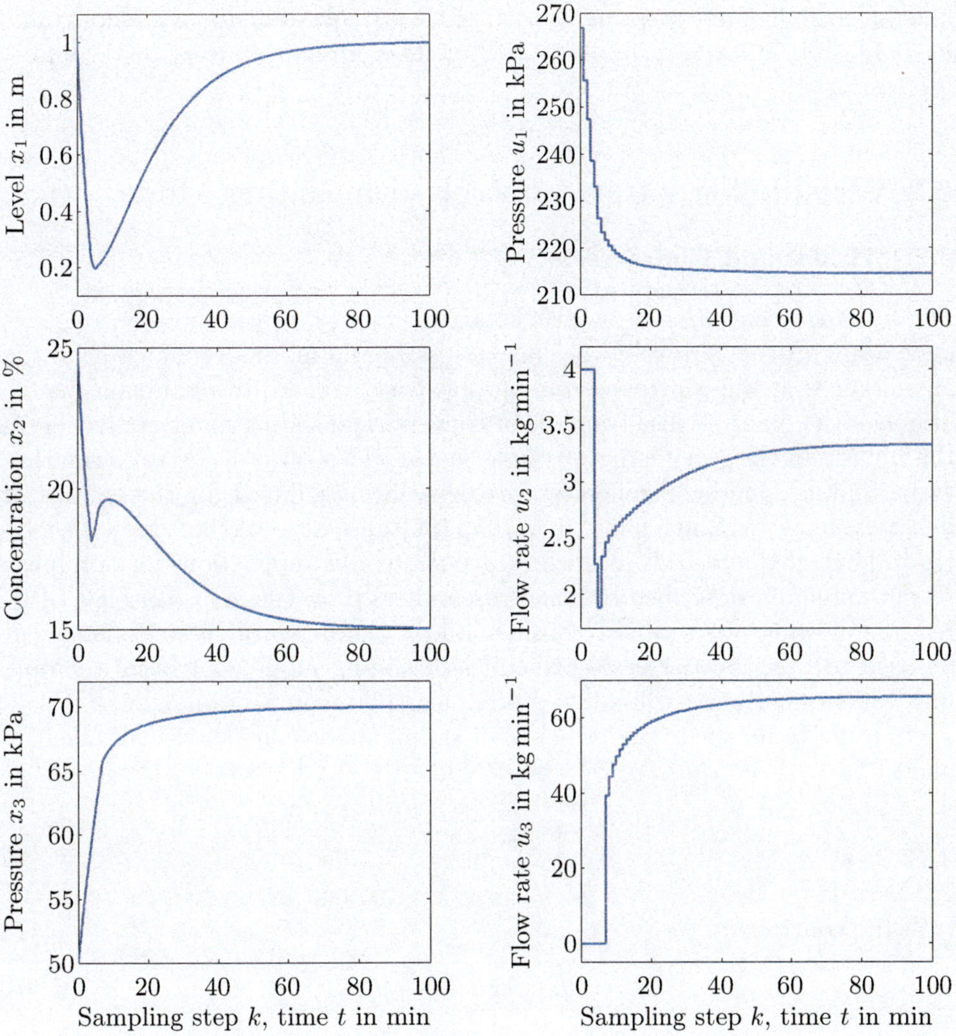

Fig. 6.11: Evaporation plant employing model predictive control with sampling time $T = 1$ min

A stability analysis for this NMPC would be complicated and very laborious, so we will proceed without it. Theoretically this is unsatisfactory, but in practice, as mentioned, this is often done.

Let us take the initial state

$$\boldsymbol{x}(0) = \begin{bmatrix} 1 & 25 & 50 \end{bmatrix}^T$$

to be regulated to the operating point (6.28). Figure 6.11 shows the variation in time of the control variables u_1, u_2, and u_3 and the controlled variables y_1, y_2, and y_3. In addition to the good compensation result, it is also evident that all constraints (6.27) for x_1, x_2, x_3, u_1, u_2, and u_3 are fulfilled. This especially holds for the flow rates u_2 and u_3, of which u_2 is held at its maximum value $u_2 = 4\,\mathrm{kg\,min}^{-1}$ for 4 min and u_3 is held at its minimum value $u_3 = 0\,\mathrm{kg\,min}^{-1}$ for 7 min.

6.2 Variable Structure Control with Sliding Mode

6.2.1 Basics and Characteristics

Sliding mode controllers[1] switch back and forth between two control laws depending on the state vector. This results in sliding modes which have the advantage that the control system is robust with regard to parameter fluctuations of the plant. A disadvantage of conventional sliding mode controllers is the high-frequency switching of the actuator. This often leads to premature wear. Sliding mode controllers were originally introduced for the control of linear systems by Emeljanov [104, 105, 106] and were further developed by Utkin [460, 461] and Itkis [204]. Improvements and applications for nonlinear plants were later described by other researchers [134, 196, 415, 421, 482, 501]. In the following, we will first consider linear plants, since these systems can be used to describe the basic principles of sliding mode control in a simple way. Subsequently, we will address nonlinear plants in Section 6.2.6.

In order to illustrate the principle of sliding mode control, we will take the plant

$$\begin{aligned} \dot{x}_1 &= x_2, \\ \dot{x}_2 &= a \cdot u, \quad a > 0, \end{aligned} \tag{6.29}$$

with the controller

$$u = -u_{\max} \operatorname{sgn}(x_2 + m x_1)\,. \tag{6.30}$$

Here, $u_{\max}$ and m are constants of the controller. The control law (6.30) separates the state space into two regions via the switching line

$$x_2 = -m x_1. \tag{6.31}$$

Below the switching line, we have

$$u = u_{\max},$$

while above it,

$$u = -u_{\max}$$

holds. Figure 6.12 shows the switching line (6.31) and the trajectories of the plant for the control variable values $u = \pm u_{\max}$, which are parabolas.

[1] In the literature, sliding mode controls are often referred to as *variable structure controls*. However, as described in Section 4.3, variable structure controls comprise a much larger class of controls.

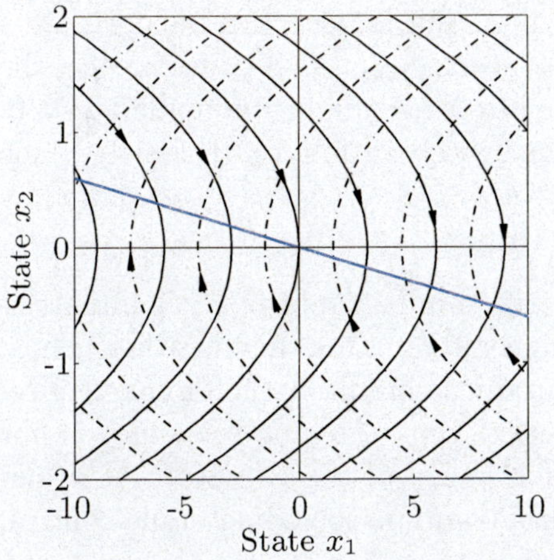

Fig. 6.12: System trajectories for actuator signals $u = -u_{\max}$ (solid line) and $u = u_{\max}$ (dashed line) as well as the switching line, shown in blue

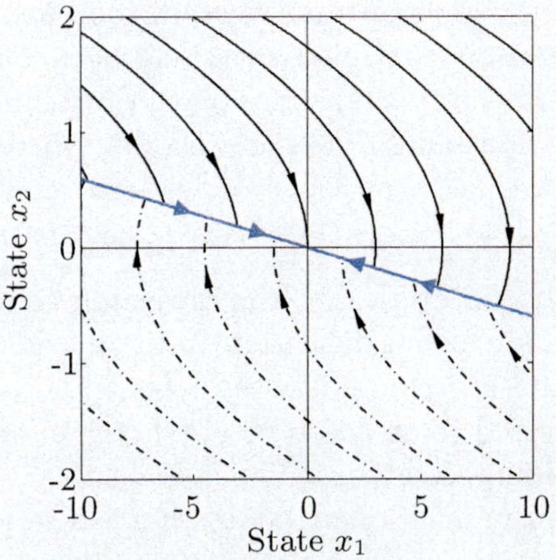

Fig. 6.13: Trajectories of the control loop with sliding mode shown in blue. Above the switching line, $u = -u_{\max}$ holds, while below it, $u = u_{\max}$ holds.

The trajectories $\boldsymbol{x}(t) = [x_1(t)\ x_2(t)]^T$ of the control loop consist of parts of the parabolas separated by the switching line. To determine them, we must calculate the solutions

$$x_2 = x_2(0) + \int_0^t a u_{\max}\, d\tau = a u_{\max} t + x_2(0), \tag{6.32}$$

$$x_1 = x_1(0) + \int_0^t x_2(\tau)\, d\tau = \frac{1}{2} a u_{\max} t^2 + x_2(0) t + x_1(0) \tag{6.33}$$

of the differential equations (6.29) for $u = u_{\max}$. From equation (6.32), we obtain

$$t = \frac{x_2 - x_2(0)}{a u_{\max}}.$$

Inserting this into equation (6.33) results in

$$x_1 = \frac{x_2^2}{2 a u_{\max}} - \frac{x_2^2(0)}{2 a u_{\max}} + x_1(0).$$

This equation describes the parabolas along which the trajectories run. Their branches are symmetrical with regard to the x_1-axis, i. e. their apexes lie on the x_1-axis. In the case of $u = -u_{\max}$, we obtain the parabolas

$$x_1 = -\frac{x_2^2}{2 a u_{\max}} + \frac{x_2^2(0)}{2 a u_{\max}} + x_1(0).$$

The control law (6.30) causes the trajectories to run toward the switching line $x_2 = -m x_1$ from both sides. Figure 6.13 illustrates this for the value

$m = 0.06$. After reaching the switching line, the trajectories on both sides cause a continual switching between $-u_{\max}$ and $u_{\max}$. As a consequence, the trajectory $\boldsymbol{x}(t)$ slides on the switching line into the equilibrium point $\boldsymbol{x}_{\text{eq}} = \mathbf{0}$. The sliding dynamics are described by the switching line (6.31) itself, which specifies them as

$$x_2 = \dot{x}_1 = -mx_1.$$

Theoretically, the trajectory changes at infinite rapidity with infinitesimally small deflection from one side of the switching line to the other. Thus, this switching between $-u_{\max}$ and $u_{\max}$ occurs at infinitely high frequency referred to as *chattering*. Of course, in practice, the switching frequency is not infinitely high; it depends on the maximum speed of the actuator. As previously mentioned, chattering is a serious disadvantage to sliding mode control, because mechanical actuators wear out quickly.

However, sliding mode control has a significant advantage: the control loop is robust with regard to plant variations in the sliding mode. This means that the control-loop dynamics are always the same, even if the plant changes. In the previous example, the dynamics are

$$\begin{aligned}\dot{x}_1 &= -mx_1,\\ \dot{x}_2 &= \ddot{x}_1 = -m\dot{x}_1 = m^2 x_1 = -mx_2\end{aligned}$$

in the sliding mode. These dynamics are therefore independent of a, i. e. the parameter of the plant (6.29).

6.2.2 Design for Linear Plants

The control process of a sliding mode control can be divided into three phases:

(1) the arrival phase, in which the trajectories approach the switching line or switching hyperplane and reach it in finite time;
(2) the sliding mode phase, in which the trajectory slides on the switching hyperplane into the equilibrium point; and
(3) the equilibrium point $\boldsymbol{x}_{\text{eq}} = \mathbf{0}$, at which the system remains.

For a globally stable equilibrium point $\boldsymbol{x}_{\text{eq}} = \mathbf{0}$, it must be ensured that all trajectories tend toward the switching hyperplane in finite time and then move to the equilibrium point $\boldsymbol{x}_{\text{eq}} = \mathbf{0}$ in the sliding phase.

Let us examine the linear plants

$$\dot{\boldsymbol{x}} = \boldsymbol{A}\boldsymbol{x} + \boldsymbol{b}u$$

with a switching plane

$$s(\boldsymbol{x}) = 0$$

and the associated control law

$$u(\boldsymbol{x}) = \begin{cases} u_+(\boldsymbol{x}) & \text{for } s(\boldsymbol{x}) > 0, \\ u_-(\boldsymbol{x}) & \text{for } s(\boldsymbol{x}) < 0. \end{cases}$$

In most cases, the switching function

$$s(\boldsymbol{x}) = \boldsymbol{r}^T \boldsymbol{x}$$

is used, i. e. a switching hyperplane $\boldsymbol{r}^T \boldsymbol{x} = 0$.

First, the accessibility of the switching hyperplane should be ensured for all trajectories of the state space. A necessary condition for this is that the control-loop trajectories tend toward the switching hyperplane from both sides, as shown in Figure 6.13. This means that

$$\dot{s} < 0 \quad \text{for} \quad s(\boldsymbol{x}) > 0$$

and

$$\dot{s} > 0 \quad \text{for} \quad s(\boldsymbol{x}) < 0$$

must hold. If both conditions are combined, we obtain

$$s\dot{s} < 0$$

as a condition for a sliding mode. Here

$$\dot{s} = \operatorname{grad}^T(s(\boldsymbol{x})) \cdot \dot{\boldsymbol{x}}, \qquad \operatorname{grad}^T(s(\boldsymbol{x})) = \frac{\partial s(\boldsymbol{x})}{\partial \boldsymbol{x}},$$

holds, and if $s(\boldsymbol{x}) = \boldsymbol{r}^T \boldsymbol{x}$, it follows that $\dot{s} = \boldsymbol{r}^T \dot{\boldsymbol{x}}$. Unfortunately, the condition $s\dot{s} < 0$ does not ensure that the trajectories reach the switching hyperplane $s(\boldsymbol{x}) = 0$ in finite time for every conceivable case. The condition is therefore necessary, but not sufficient for a functional sliding mode control system.

There are different approaches to ensure the accessibility of the switching hyperplane for all trajectories in finite time. A very common approach is that developed by Gao and Hung [145]. Here, the decrease $\dot{s}(\boldsymbol{x})$ in the switching function along the trajectories $\boldsymbol{x}(t)$ is preset to

$$\dot{s}(\boldsymbol{x}) = -q \operatorname{sgn}(s(\boldsymbol{x})) - ks(\boldsymbol{x}), \tag{6.34}$$

where q and k are positive constants. Obviously, in the case of equation (6.34), the switching hyperplane s fulfills the necessary condition

$$s\dot{s} = -q|s| - k \cdot s^2 < 0.$$

Because s has a decrease rate of $\dot{s} < -q$ (or an increase rate of $\dot{s} > q$) even for very small values of $|s|$ due to the choice of $\dot{s}$ in equation (6.34), the trajectories $\boldsymbol{x}(t)$ reach the switching hyperplane in finite time.

Using

$$\dot{s}(\boldsymbol{x}) = \operatorname{grad}^T(s(\boldsymbol{x})) \cdot \dot{\boldsymbol{x}} = \operatorname{grad}^T(s(\boldsymbol{x})) \cdot (\boldsymbol{A}\boldsymbol{x} + \boldsymbol{b}u)$$

the control law is obtained from equation (6.34) as

$$u(\boldsymbol{x}) = -\frac{\operatorname{grad}^T(s(\boldsymbol{x})) \cdot \boldsymbol{A}\boldsymbol{x} + q\operatorname{sgn}(s(\boldsymbol{x})) + ks(\boldsymbol{x})}{\operatorname{grad}^T(s(\boldsymbol{x})) \cdot \boldsymbol{b}}.$$

For the frequent case of a linear switching hyperplane

$$s(\boldsymbol{x}) = \boldsymbol{r}^T\boldsymbol{x} = 0,$$

we obtain the control law

$$u(\boldsymbol{x}) = -\frac{\boldsymbol{r}^T\boldsymbol{A}\boldsymbol{x} + q\operatorname{sgn}(\boldsymbol{r}^T\boldsymbol{x}) + k\boldsymbol{r}^T\boldsymbol{x}}{\boldsymbol{r}^T\boldsymbol{b}}.$$

The arbitrary positive parameters q and k allow for the creation of favorable control dynamics.

Note that sliding mode controllers can also be designed for systems with multiple input variables, i. e. for systems

$$\dot{\boldsymbol{x}} = \boldsymbol{A}\boldsymbol{x} + \boldsymbol{B}\boldsymbol{u};$$

see [196, 397, 460].

6.2.3 Dynamics in the Sliding Mode

When trajectories $\boldsymbol{x}(t)$ reach the switching plane and the sliding mode begins, the question arises: what dynamics does the control loop have in the sliding mode? When considering this question, the discontinuity of the differential equation

$$\dot{\boldsymbol{x}} = \boldsymbol{A}\boldsymbol{x} + \boldsymbol{b}u, \quad u(\boldsymbol{x}) = \begin{cases} u_+(\boldsymbol{x}) & \text{for } s(\boldsymbol{x}) > 0, \\ u_-(\boldsymbol{x}) & \text{for } s(\boldsymbol{x}) < 0 \end{cases} \tag{6.35}$$

of the closed control loop on the switching hyperplane

$$s(\boldsymbol{x}) = \boldsymbol{r}^T\boldsymbol{x} = 0$$

poses a problem.

It is evident that the differential equation is not defined on the switching hyperplane, i. e. the existence and uniqueness of its solution is not ensured on the hyperplane. This problem results from the switching between $u_+(\boldsymbol{x})$ and $u_-(\boldsymbol{x})$, which causes a discontinuous control signal $u(t)$. However, the right side of the differential equation (6.35) must be continuous, since the left side is continuous due to the differentiation of $\boldsymbol{x}$. There are various methods [196] to solve this problem, such as the Filippov method [119]. But this issue does not matter in practical applications, since the trajectory cannot run accurately on the switching hyperplane due to noise. Therefore, we do not need to take this circumstance into account below.

We can determine the system dynamics in the sliding mode by employing the following frequently-used method. The plant is in or has been transformed into the controller canonical form

$$\begin{aligned} \dot{x}_1 &= x_2, \\ &\vdots \\ \dot{x}_{n-1} &= x_n, \\ \dot{x}_n &= -a_0 x_1 - a_1 x_2 - \ldots - a_{n-1} x_n + u. \end{aligned} \tag{6.36}$$

Furthermore, the equation of the switching hyperplane is

$$r_1 x_1 + \ldots + r_{n-1} x_{n-1} + r_n x_n = 0. \tag{6.37}$$

Provided that $r_n \neq 0$ holds, we can choose $r_n = 1$ without loss of generality. Thus, we obtain

$$x_n = -r_1 x_1 - \ldots - r_{n-1} x_{n-1}. \tag{6.38}$$

Substituting the state variable x_n in the controller canonical form (6.36) of the plant by using equation (6.38) yields the differential equation system

$$\begin{aligned} \dot{x}_1 &= x_2, \\ &\vdots \\ \dot{x}_{n-2} &= x_{n-1}, \\ \dot{x}_{n-1} &= -r_1 x_1 - \ldots - r_{n-1} x_{n-1} \end{aligned} \tag{6.39}$$

which describes the dynamics of the control loop in the sliding mode. Note that in this case the state variable x_n is given by the algebraic equation (6.38) and the equation for $\dot{x}_n$ is therefore omitted. The differential equations (6.39) are no longer dependent on the parameters a_i of the plant. This means that the control loop (6.39) is robust with regard to parameter fluctuations of the plant. It is also noteworthy that the system order has been reduced by one degree to $n-1$ and the coefficients r_i of the switching hyperplane (6.37) are also the characteristic polynomial coefficients of the linear dynamics (6.39) in the sliding mode.

6.2.4 Verification of Robustness

The main advantage of sliding mode controllers is their robustness with regard to variations $\Delta\boldsymbol{A}$ in the plant parameters or external disturbances $\boldsymbol{d}(t)$ appearing in the system description

$$\dot{\boldsymbol{x}} = (\boldsymbol{A} + \Delta\boldsymbol{A})\,\boldsymbol{x} + \boldsymbol{b}u + \boldsymbol{d}.$$

This means that the closed-loop dynamics depend only on the parameters r_i of the switching hyperplane $s(\boldsymbol{x}) = \boldsymbol{r}^T\boldsymbol{x}$ in the sliding mode, as we saw in equation (6.39). The dynamics are independent of $\Delta\boldsymbol{A}$ and $\boldsymbol{d}$.

If both the following conditions are met [100], the robustness described above will indeed result:

(1) There is a vector $\boldsymbol{p}$, so $\Delta\boldsymbol{A} = \boldsymbol{b}\cdot\boldsymbol{p}^T$ applies.
(2) There is an $\alpha(t)$, so $\boldsymbol{d}(t) = \boldsymbol{b}\cdot\alpha(t)$ applies.

If, for example, the system is given in the controller canonical form

$$\dot{\boldsymbol{x}} = \begin{bmatrix} 0 & 1 & 0 & \cdots & 0 \\ 0 & 0 & 1 & \cdots & 0 \\ \vdots & \vdots & \vdots & \ddots & \vdots \\ 0 & 0 & 0 & \cdots & 1 \\ -a_0 & -a_1 & -a_2 & \cdots & -a_{n-1} \end{bmatrix} \boldsymbol{x} + \begin{bmatrix} 0 \\ 0 \\ \vdots \\ 0 \\ 1 \end{bmatrix} u,$$

the condition $\Delta\boldsymbol{A} = \boldsymbol{b}\cdot\boldsymbol{p}^T$ can obviously be fulfilled if only the coefficients a_i vary.

6.2.5 Example: DC-to-DC Converter

A DC-to-DC converter is used to convert a DC voltage into a lower or higher voltage. At the same time, it should operate with minimum losses. As an example we will describe a buck converter [441], i. e. a device that converts a voltage u_{in} into a lower voltage u_{out}. Figure 6.14 shows a converter of this type including the corresponding controller.

The circuit functions by setting the output voltage u_{out} based on a constant reference voltage u_{ref}, so that

$$\beta u_{\mathrm{out}} = \frac{R_2}{R_1 + R_2} u_{\mathrm{out}} = u_{\mathrm{ref}} \tag{6.40}$$

applies. For this purpose a power MOSFET, marked T in Figure 6.14, is used as a switch, and is turned on or off by the controller. A switching variable z indicates the state of the transistor T. If $z = 1$, the transistor conducts. If $z = 0$, the transistor is disabled. The transistor is switched using a pulse width modulation (PWM). This causes a continuous current flow i_{R}, alternately fed by i_{L} and i_{C}, so that the voltage βu_{out} over the resistor R_2 is equal to the reference voltage u_{ref}, i. e. equation (6.40) is fulfilled.

The derivative $\dot{u}_{\mathrm{out}}$ of the voltage across the load resistor R_{L} and the smoothing capacitor C is given by

$$\dot{u}_{\mathrm{out}} = \frac{1}{C} i_{\mathrm{C}} = \frac{1}{C}(i_{\mathrm{L}} - i_{\mathrm{R}}) = \frac{1}{C}\left(\int \frac{u_{\mathrm{in}} z - u_{\mathrm{out}}}{L}\, dt - \frac{u_{\mathrm{out}}}{R_{\mathrm{G}}}\right), \tag{6.41}$$

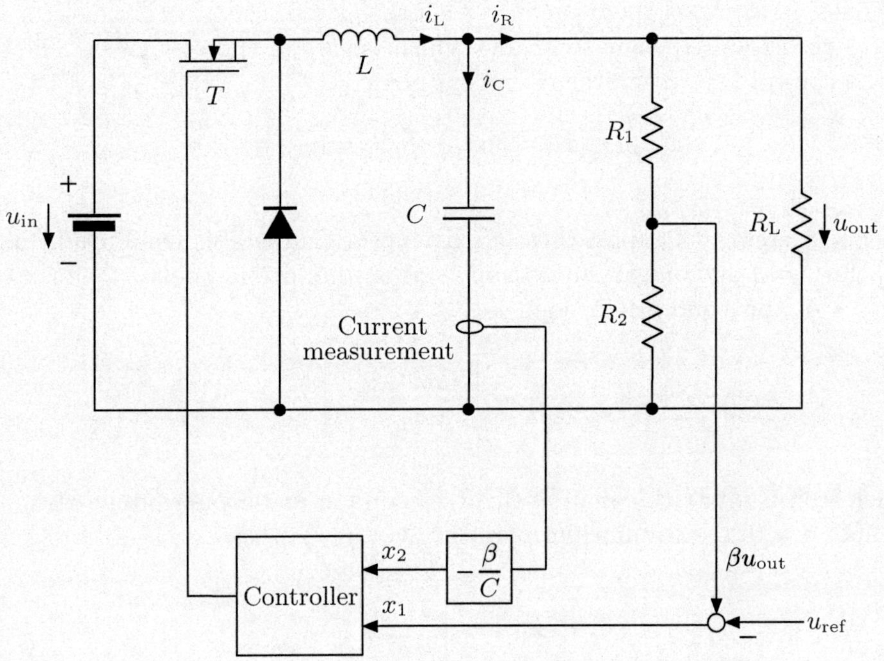

Fig. 6.14: DC-to-DC converter

where

$$R_G = \frac{(R_1 + R_2)R_L}{R_1 + R_2 + R_L}.$$

From equation (6.41), we obtain

$$\ddot{u}_{out} = -\frac{1}{LC}u_{out} - \frac{1}{R_G C}\dot{u}_{out} + \frac{u_{in}z}{LC}. \tag{6.42}$$

We will now define

$$\begin{aligned} x_1 &= \beta u_{out} - u_{ref}, \\ x_2 &= \dot{x}_1 = \beta \dot{u}_{out} \end{aligned} \tag{6.43}$$

as state variables. Equations (6.42) and (6.43) result in the state-space model

$$\begin{bmatrix} \dot{x}_1 \\ \dot{x}_2 \end{bmatrix} = \begin{bmatrix} 0 & 1 \\ -\dfrac{1}{LC} & -\dfrac{1}{R_G C} \end{bmatrix} \begin{bmatrix} x_1 \\ x_2 \end{bmatrix} + \begin{bmatrix} 0 \\ \dfrac{1}{LC} \end{bmatrix} u, \tag{6.44}$$

with the control variable u being

$$u = \beta u_{in} z - u_{ref}. \tag{6.45}$$

The system (6.44), (6.45) is ideal for the application of a sliding mode control. This is because, on the one hand, the digital control variable z already specifies a switching operation. On the other hand, the power transistor tolerates the high-frequency switching which occurs in sliding modes. The aim of the control is to reach the operating point

$$x_1 = \beta u_{\text{out}} - u_{\text{ref}} = 0, \tag{6.46}$$

i. e. $u_{\text{out}} = u_{\text{ref}}/\beta$.

The system (6.44) possesses one asymptotically stable equilibrium point $\boldsymbol{x}_{\text{eq1}}$ for $z = 1$ and one asymptotically stable equilibrium point $\boldsymbol{x}_{\text{eq2}}$ for $z = 0$. For $z = 1$, this equilibrium point is

$$\boldsymbol{x}_{\text{eq1}} = \begin{bmatrix} \beta u_{\text{in}} - u_{\text{ref}} \\ 0 \end{bmatrix},$$

which, consistent with equation (6.43), results in an output voltage of $u_{\text{out}} = u_{\text{in}}$. For $z = 0$, the equilibrium point is

$$\boldsymbol{x}_{\text{eq2}} = \begin{bmatrix} -u_{\text{ref}} \\ 0 \end{bmatrix},$$

which, according to equation (6.43), results in an output voltage $u_{\text{out}} = 0$. Note that the desired operating point (6.46) of the control loop is not identical to either of the two equilibrium points of the plant.

Let us now consider the converter's trajectories $\boldsymbol{x}(t)$. For $z = 1$, they can be calculated using equation (6.44) and are shown in Figure 6.15. For $z = 0$, the trajectories can also be obtained as solutions of equation (6.44), as long as $i_{\text{L}} > 0$ holds. If the energy stored in the coil is consumed, $i_{\text{L}} = 0$ holds. Then the diode blocks and the capacitor C discharges according to

$$\dot{u}_{\text{out}} + \frac{1}{R_{\text{G}}C} u_{\text{out}} = 0,$$

from which

$$u_{\text{out}} = u_{\text{out}}(0) e^{-t/(R_{\text{G}}C)}$$

follows. For the trajectories $\boldsymbol{x}(t)$, with equation (6.43) we obtain

$$x_1(t) = \beta u_{\text{out}}(0) e^{-t/(R_{\text{G}}C)} - u_{\text{ref}}, \tag{6.47}$$

$$x_2(t) = \frac{-\beta u_{\text{out}}(0)}{R_{\text{G}}C} e^{-t/(R_{\text{G}}C)}. \tag{6.48}$$

Inserting equation (6.47) into equation (6.48) results in

$$x_2 = -\frac{1}{R_{\text{G}}C} x_1 - \frac{u_{\text{ref}}}{R_{\text{G}}C}. \tag{6.49}$$

For $z = 0$, the trajectories $\boldsymbol{x}(t)$ are composed of two parts. As long as $i_{\text{L}} > 0$

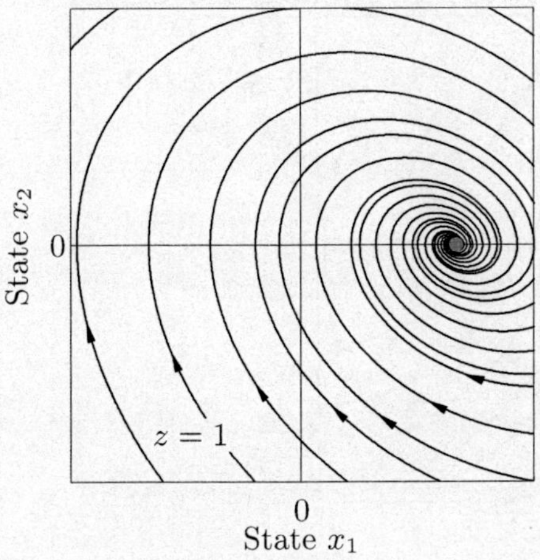

Fig. 6.15: Equilibrium point $\boldsymbol{x}_{\mathrm{eq}1}$ and trajectories for $z = 1$

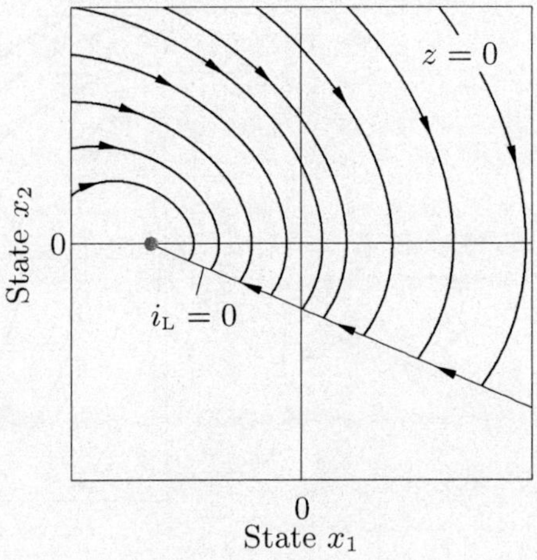

Fig. 6.16: Equilibrium point $\boldsymbol{x}_{\mathrm{eq}2}$ and trajectories for $z = 0$

holds, the trajectories result from the solutions of equation (6.44). If $i_{\mathrm{L}} = 0$ holds, the trajectories move on the straight line (6.49), and on this line they tend to the equilibrium point $\boldsymbol{x}_{\mathrm{eq}2}$, according to equations (6.47) and (6.48). Figure 6.16 shows the trajectories in question.

A switching line

$$s(\boldsymbol{x}) = r_1 x_1 + x_2 = 0, \quad r_1 > 0, \tag{6.50}$$

is now selected to separate the state space into two regions. In one region $z = 0$ applies, and in the other $z = 1$, according to

$$z = \begin{cases} 0 & \text{for } \; s(\boldsymbol{x}) > 0, \\ 1 & \text{for } \; s(\boldsymbol{x}) < 0. \end{cases}$$

Figure 6.17 shows both regions and the trajectories in question. These trajectories are composed of the trajectories shown in Figures 6.15 and 6.16. A sliding mode occurs on the switching line.

Next, the necessary condition for a sliding mode, $s\dot{s} < 0$, will be used to determine the parameter r_1 of the switching line (6.50). From

$$s\dot{s} = s(\boldsymbol{x}) \cdot \operatorname{grad}^T(s(\boldsymbol{x})) \cdot \dot{\boldsymbol{x}} < 0$$

and the system equations (6.44) and (6.45), we now obtain

$$s\dot{s} = s(\boldsymbol{x}) \left[\left(r_1 - \frac{1}{R_{\mathrm{G}} C} \right) x_2 - \frac{1}{LC} x_1 + \frac{\beta u_{\mathrm{in}} z - u_{\mathrm{ref}}}{LC} \right] < 0. \tag{6.51}$$

If we set

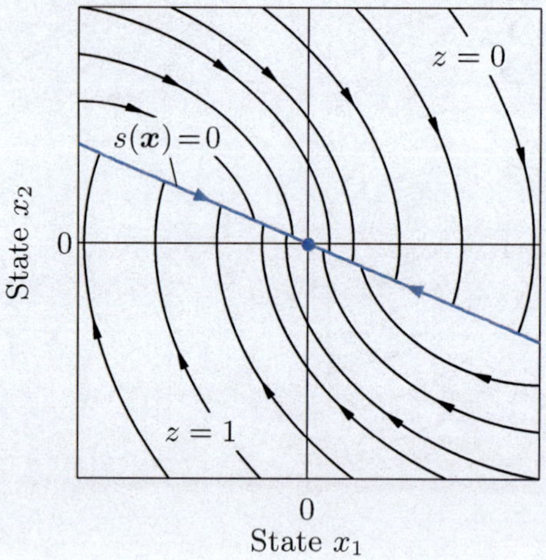

Fig. 6.17: Trajectories, switching line (blue), and sliding mode of the DC-to-DC converter

$$r_1 = \frac{1}{R_\mathrm{G}C},$$

the condition of existence (6.51) is independent of $x_2 = \beta \dot{u}_\mathrm{out}$. With equation (6.43), this results in

$$s\dot{s} = s(\boldsymbol{x})\frac{\beta}{LC}(u_\mathrm{in}z - u_\mathrm{out}) < 0. \tag{6.52}$$

From this condition, and in the case of $s(\boldsymbol{x}) < 0$ and $z = 1$, we obtain

$$u_\mathrm{in} > u_\mathrm{out}. \tag{6.53}$$

For $s(\boldsymbol{x}) > 0$ and $z = 0$, and with condition (6.52),

$$u_\mathrm{out} > 0 \tag{6.54}$$

follows. Conversely, if equations (6.53) and (6.54) are fulfilled, so is condition (6.52), which is necessary for the existence of the sliding mode. Furthermore, the trajectories of the control system also tend to the sliding mode in finite time, since

$$\dot{s} = \frac{\beta}{LC}(u_\mathrm{in}z - u_\mathrm{out})$$

generates a constant decrease rate of

$$\dot{s} = -u_\mathrm{out}\frac{\beta}{LC} < 0 \quad \text{for} \quad s(\boldsymbol{x}) > 0 \quad \text{and} \quad z = 0$$

or a constant increase rate of

$$\dot{s} = (u_{\text{in}} - u_{\text{out}})\frac{\beta}{LC} > 0 \quad \text{for} \quad s(\boldsymbol{x}) < 0 \quad \text{and} \quad z = 1.$$

The dynamics of the controlled converter in sliding mode are calculated using

$$s(\boldsymbol{x}) = r_1 x_1 + x_2 = 0 \quad \text{and} \quad x_2 = \dot{x}_1,$$

yielding

$$x_1(t) = x_1(0)e^{-r_1 t}.$$

As we can see, the dynamics in the sliding mode are independent of changes in the load resistance R_{L} and fluctuations in the input voltage, i. e. the battery voltage u_{in}.

Let us examine a concrete implementation with the following parameters:

$$\begin{aligned} R_{\text{L}} &= 6\,\Omega, & u_{\text{in}} &= 24\,\text{V}, \\ L &= 110\,\mu\text{H}, & u_{\text{out}} &= 12\,\text{V}, \\ C &= 100\,\mu\text{F}, & u_{\text{ref}} &= 3.3\,\text{V}. \end{aligned}$$

Hence,

$$\beta = \frac{u_{\text{ref}}}{u_{\text{out}}} = 0.275.$$

From

$$R_2 = \frac{\beta R_1}{1 - \beta}$$

and $R_1 = 870\,\Omega$ it follows that $R_2 = 330\,\Omega$. The time constant $1/r_1$ in sliding mode is 0.6 ms.

Figure 6.18 shows the output voltage in the event of a sudden decrease of 3 V in the input voltage u_{in}. As expected, the output voltage u_{out} does not change. In real devices a voltage drop of several 10 mV can be observed. Figure 6.19 shows the turning-on process. The output value of $u_{\text{out}} = 12\,\text{V}$ is reached after a time of $5/r_1 = 3\,\text{ms}$.

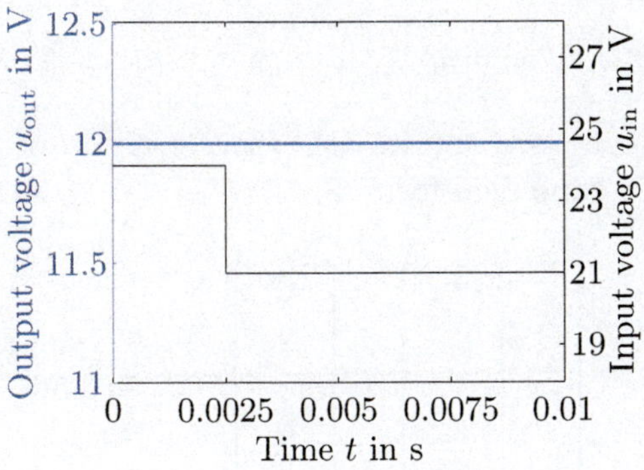

Fig. 6.18: Control of the output voltage u_{out} after a drop in the input voltage u_{in}

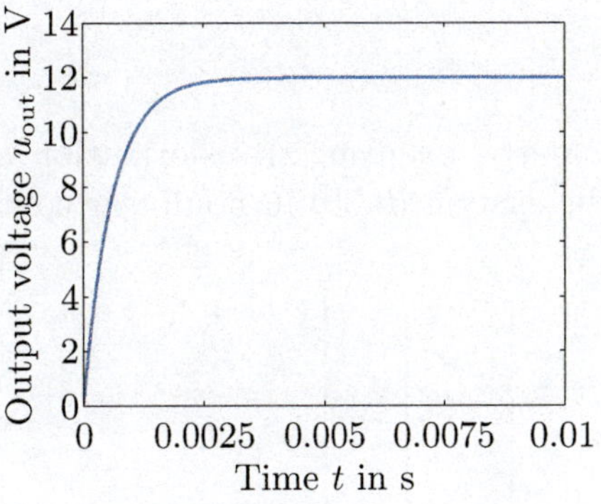

Fig. 6.19: Time course of the output voltage u_{out} after the converter is turned on

6.2.6 Design for Nonlinear Plants

In certain cases, the design of sliding mode controllers for nonlinear plants is no more difficult than for linear plants. Among other types of plants, this applies to the control-affine plants

$$\begin{aligned} \dot{\boldsymbol{x}} &= \boldsymbol{a}(\boldsymbol{x}) + \boldsymbol{b}(\boldsymbol{x}) \cdot u, \\ y &= c(\boldsymbol{x}) \end{aligned} \tag{6.55}$$

with a relative degree of $\delta = n$. The design is similar for $\delta < n$.

We can convert the system representation above into the nonlinear controller canonical form as described in Section 5.2.1, p. 358 et seq., for controller design by means of feedback linearization. For this purpose, we will use the Lie derivative

$$L_{\boldsymbol{f}} h(\boldsymbol{x}) = \frac{\partial h(\boldsymbol{x})}{\partial \boldsymbol{x}} \boldsymbol{f}(\boldsymbol{x}) = \operatorname{grad}^T(h(\boldsymbol{x})) \cdot \boldsymbol{f}(\boldsymbol{x}).$$

First, we will compute the derivatives

$$\begin{aligned} y &= c(\boldsymbol{x}), \\ \dot{y} &= \frac{\partial c(\boldsymbol{x})}{\partial \boldsymbol{x}} \dot{\boldsymbol{x}} = L_{\boldsymbol{a}} c(\boldsymbol{x}), \\ \ddot{y} &= \frac{\partial L_{\boldsymbol{a}} c(\boldsymbol{x})}{\partial \boldsymbol{x}} \dot{\boldsymbol{x}} = L_{\boldsymbol{a}}^2 c(\boldsymbol{x}), \\ &\vdots \\ y^{(n-1)} &= \frac{\partial L_{\boldsymbol{a}}^{n-2} c(\boldsymbol{x})}{\partial \boldsymbol{x}} \dot{\boldsymbol{x}} = L_{\boldsymbol{a}}^{n-1} c(\boldsymbol{x}), \\ y^{(n)} &= \frac{\partial L_{\boldsymbol{a}}^{n-1} c(\boldsymbol{x})}{\partial \boldsymbol{x}} \dot{\boldsymbol{x}} = L_{\boldsymbol{a}}^{n} c(\boldsymbol{x}) + L_{\boldsymbol{b}} L_{\boldsymbol{a}}^{n-1} c(\boldsymbol{x}) \cdot u. \end{aligned}$$

By using the new state coordinates

$$z_1 = y, \quad z_2 = \dot{y}, \quad \ldots, \quad z_n = y^{(n-1)},$$

and the resulting transformation $\boldsymbol{z} = \boldsymbol{t}(\boldsymbol{x})$, we obtain the representation of the system (6.55) in nonlinear controller canonical form

$$\begin{bmatrix} \dot{z}_1 \\ \vdots \\ \dot{z}_{n-1} \\ \dot{z}_n \end{bmatrix} = \begin{bmatrix} z_2 \\ \vdots \\ z_n \\ L_{\boldsymbol{a}}^{n} c(\boldsymbol{x}) + L_{\boldsymbol{b}} L_{\boldsymbol{a}}^{n-1} c(\boldsymbol{x}) \cdot u \end{bmatrix} = \underbrace{\begin{bmatrix} z_2 \\ \vdots \\ z_n \\ \alpha(\boldsymbol{z}) \end{bmatrix}}_{\tilde{\boldsymbol{a}}(\boldsymbol{z})} + \underbrace{\begin{bmatrix} 0 \\ \vdots \\ 0 \\ \beta(\boldsymbol{z}) \end{bmatrix}}_{\tilde{\boldsymbol{b}}(\boldsymbol{z})} u \tag{6.56}$$

with

$$\alpha(\boldsymbol{z}) = L_{\boldsymbol{a}}^{n} c(\boldsymbol{t}^{-1}(\boldsymbol{z})) \quad \text{and} \quad \beta(\boldsymbol{z}) = L_{\boldsymbol{b}} L_{\boldsymbol{a}}^{n-1} c(\boldsymbol{t}^{-1}(\boldsymbol{z})).$$

The design of a sliding mode controller with switching hypersurface

$$s(\boldsymbol{z}) = 0$$

is similar to the linear case. We will again use the approach

$$\dot{s}(\boldsymbol{z}) = -q\,\mathrm{sgn}(s(\boldsymbol{z})) - ks(\boldsymbol{z}) \tag{6.57}$$

developed by Gao and Hung. Recall that this ensures the trajectories $\boldsymbol{z}(t)$ will reach the switching plane in finite time. Again,

$$q > 0 \quad \text{and} \quad k > 0$$

apply.

Using

$$\dot{s}(\boldsymbol{z}) = \mathrm{grad}^T(s(\boldsymbol{z})) \cdot \dot{\boldsymbol{z}}$$

and equation (6.56) as well as equation (6.57) yields the control law

$$u(\boldsymbol{z}) = -\frac{\mathrm{grad}^T(s(\boldsymbol{z})) \cdot \tilde{\boldsymbol{a}}(\boldsymbol{z}) + q\,\mathrm{sgn}(s(\boldsymbol{z})) + ks(\boldsymbol{z})}{\mathrm{grad}^T(s(\boldsymbol{z})) \cdot \tilde{\boldsymbol{b}}(\boldsymbol{z})}.$$

In the case of a switching hyperplane $s(\boldsymbol{z}) = \boldsymbol{r}^T\boldsymbol{z} = 0$, this simplifies to

$$u(\boldsymbol{z}) = -\frac{\boldsymbol{r}^T\tilde{\boldsymbol{a}}(\boldsymbol{z}) + q\,\mathrm{sgn}(\boldsymbol{r}^T\boldsymbol{z}) + k\,\boldsymbol{r}^T\boldsymbol{z}}{\boldsymbol{r}^T\tilde{\boldsymbol{b}}(\boldsymbol{z})}. \tag{6.58}$$

If we set $\boldsymbol{r}^T = [r_1 \;\; r_2 \;\; \cdots \;\; r_{n-1} \;\; 1]$, for equation (6.58) we obtain

$$u(\boldsymbol{z}) = -\frac{r_1 z_2 + \ldots + r_{n-1} z_n + k\,\boldsymbol{r}^T\boldsymbol{z} + \alpha(\boldsymbol{z}) + q\,\mathrm{sgn}(\boldsymbol{r}^T\boldsymbol{z})}{\beta(\boldsymbol{z})}. \tag{6.59}$$

Inserted into the plant (6.56), this results in the control-loop dynamics

$$\begin{bmatrix} \dot{z}_1 \\ \vdots \\ \dot{z}_{n-1} \\ \dot{z}_n \end{bmatrix} = \begin{bmatrix} z_2 \\ \vdots \\ z_n \\ -kr_1 z_1 - (r_1 + kr_2) z_2 - \ldots - (r_{n-1} + k) z_n \end{bmatrix} - \begin{bmatrix} 0 \\ \vdots \\ 0 \\ q\,\mathrm{sgn}(\boldsymbol{r}^T\boldsymbol{z}) \end{bmatrix}.$$

This control loop is free of the nonlinearities of the plant. Further information on topics such as sliding mode control for nonlinear MIMO systems can be found in [89, 145, 360, 418].

6.2.7 Example: Optical Switch

In the following, we will design a sliding mode controller for an optical switch. This is implemented as a microelectromechanical system, abbreviated MEMS, and serves to switch the light path between different optical fibers [54, 349]. In this way it is possible to transmit an optical signal from one optical waveguide to another and back again, without conversion to an electrical signal.

Figure 6.20 shows the structure of this kind of optical switch. The comb drive converts an electrical voltage v, which serves as the switch's input signal, into a translational change of the position x_1 of a rod. A mirror is attached to the rod, enabling it to be pushed into and out of the optical path. The model

$$\begin{aligned} \dot{x}_1 &= x_2, \\ \dot{x}_2 &= -\frac{c}{m}x_1 - \frac{1}{m}(d_1 x_1 + d_0)\cdot x_2 + \frac{k_\mathrm{e}}{m}u + d \end{aligned} \tag{6.60}$$

of the optical switch is in nonlinear controller canonical form. Here, $u = v^2$ holds, while $y = x_1$ represents the system's output variable, and d is a mechanical, time-dependent disturbance. The available voltage v is limited by $0 \le v \le 35\,\mathrm{V}$.

The parameters are the moving mass $m = 2.35\cdot 10^{-9}\,\mathrm{kg}$, the spring constant of the suspension $c = 0.6\,\mathrm{N\,m^{-1}}$, the damping coefficients $d_1 = 0.0363\,\mathrm{kg\,s^{-1}m^{-1}}$ and $d_0 = 5.45\cdot 10^{-7}\,\mathrm{kg\,s^{-1}}$, and the converter constant $k_\mathrm{e} = 17.8\cdot 10^{-9}\,\mathrm{N\,V^{-2}}$ of the comb drive. The values of the damping coefficients d_0 and d_1 must be regarded as uncertain estimates. In the following,

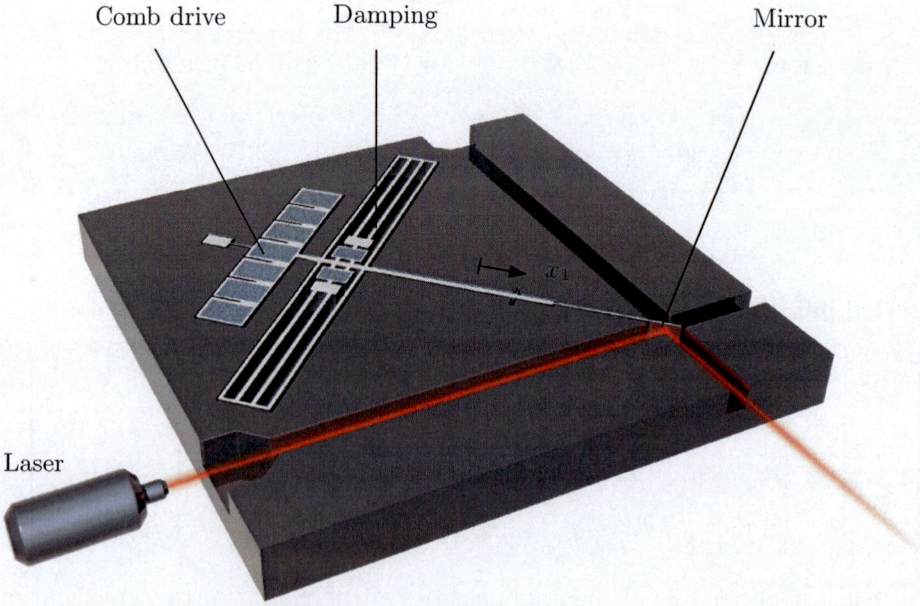

Fig. 6.20: Optical switch in microelectromechanical design

we will use the abbreviations

$$a_0 = \frac{c}{m} = 2.55 \cdot 10^8\,\mathrm{N\,m^{-1}kg^{-1}},$$
$$a_1 = \frac{d_1}{m} = 1.54 \cdot 10^7\,\mathrm{s^{-1}m^{-1}},$$
$$a_2 = \frac{d_0}{m} = 232\,\mathrm{s^{-1}},$$
$$b = \frac{k_e}{m} = 7.57\,\mathrm{N\,kg^{-1}V^{-2}}.$$

Now, our aim is to be able to adjust different positions $x_{1\text{eq}}$ of the mirror. These target positions $x_{1\text{eq}}$ are equal to the first element of the equilibrium points

$$\boldsymbol{x}_{\text{eq}} = \begin{bmatrix} \dfrac{b}{a_0} u_{\text{eq}} \\ 0 \end{bmatrix}, \tag{6.61}$$

where u_{eq} is the control variable. By using equation (6.61) and the transformation rule

$$x_1 = x_{1\text{eq}} + \Delta x_1 = \frac{b}{a_0} u_{\text{eq}} + \Delta x_1$$

with the new variable Δx_1, we obtain

$$\begin{aligned} \Delta\dot{x}_1 &= x_2, \\ \dot{x}_2 &= -(a_0 + a_1\,x_2)(x_{1\text{eq}} + \Delta x_1) - a_2\,x_2 + b\,u + d \end{aligned} \tag{6.62}$$

for the model (6.60).

Now we will take the linear switching line

$$s(\boldsymbol{x}) = r \cdot \Delta x_1 + x_2 = 0$$

with the parameter r yet to be determined. With the approach developed by Gao and Hung, we obtain

$$\dot{s}(\boldsymbol{x}) = -q\,\mathrm{sgn}(s(\boldsymbol{x})) - k\,s(\boldsymbol{x})$$

from equation (6.57). If we equate this with

$$\dot{s}(\boldsymbol{x}) = \mathrm{grad}^T(s(\boldsymbol{x})) \cdot (\boldsymbol{a}(\boldsymbol{x}) + \boldsymbol{b}(\boldsymbol{x})u),$$

the control law, which we had already derived in equation (6.59) for the general case, takes the form

$$u(\boldsymbol{x}) = -\frac{rx_2 - (a_0 + a_1x_2)(x_{1\text{eq}} + \Delta x_1) - a_2x_2 + q\,\mathrm{sgn}(s(\boldsymbol{x})) + ks(\boldsymbol{x})}{b} \tag{6.63}$$

for the present example. The parameters q and k are constants which need to be suitably chosen. Inserting the control law (6.63) into the system dynamics (6.62), the resulting expression for the control loop is

$$\begin{aligned}\Delta\dot{x}_1 &= x_2,\\ \dot{x}_2 &= -rx_2 - q\,\mathrm{sgn}(s(\boldsymbol{x})) - ks(\boldsymbol{x}) + d.\end{aligned}$$

Using $s(\boldsymbol{x}) = r\Delta x_1 + x_2$, it becomes

$$\begin{aligned}\Delta\dot{x}_1 &= x_2,\\ \dot{x}_2 &= -kr\Delta x_1 - (r+k)x_2 - q\,\mathrm{sgn}(r\Delta x_1 + x_2) + d.\end{aligned} \tag{6.64}$$

If the nonlinear part is ignored, i. e. $q = 0$, this control loop has linear dynamics with the characteristic polynomial

$$P(s) = s^2 + (r+k)s + rk.$$

The polynomial has the two zeros $s_1 = -r$ and $s_2 = -k$, which are identical to the two eigenvalues of the linear part of the control loop's system dynamics (6.64).

In the case of the sliding mode,

$$s(\boldsymbol{x}) = r\Delta x_1 + x_2 = 0$$

holds. Then we have the reduced system dynamics

$$\Delta\dot{x}_1 = -r\Delta x_1$$

and the algebraic equation

$$x_2 = -r\Delta x_1$$

for the second state variable.

We can simulate the control loop (6.64) for the parameters $k = 20000$, $r = 20000$, and $q = 1000$ for different damping values

$$\begin{aligned}d_1 &= d_{\mathrm{nom}} = 0.0363,\\ d_1 &= 0,\\ d_1 &= 10d_{\mathrm{nom}},\\ d_1 &= 100d_{\mathrm{nom}}.\end{aligned}$$

Here, we will take an initial displacement of $x_1(0) = 25\,\mu\mathrm{m}$ and an initial velocity of $x_2(0) = 0\,\mu\mathrm{m\,s^{-1}}$. The first element $x_{1\mathrm{eq}}$ of the equilibrium point $\boldsymbol{x}_{\mathrm{eq}}$ is at $10\,\mu\mathrm{m}$. Thus, $\Delta x_1(0) = 15\,\mu\mathrm{m}$. For the disturbance, we have assumed $d = 0$. Figure 6.21 shows the progressions of the deviation Δx_1 for the four different damping values. In all four cases, good control behavior becomes apparent. The control loop is in sliding mode for curves in blue, and is still

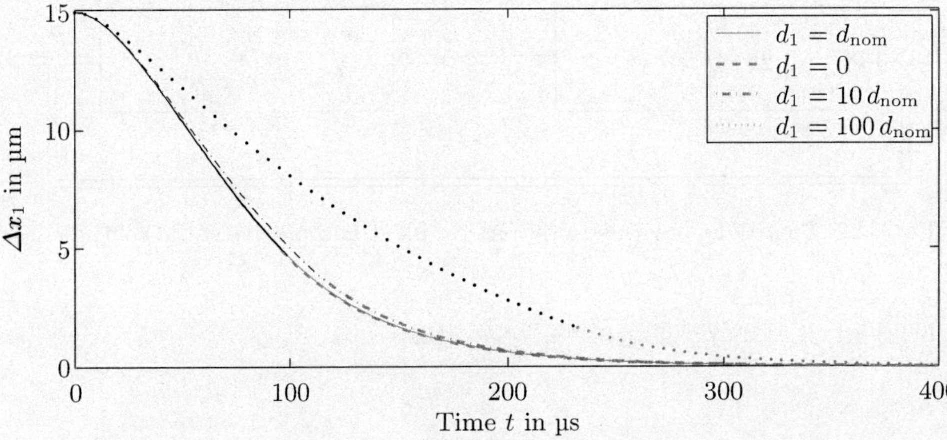

Fig. 6.21: Compensation of the displacement $\Delta x_1 = 15\,\mu\text{m}$ to the equilibrium point $\Delta x_1 = 0\,\mu\text{m}$ for different damping values d_1. For blue curve components the system is in sliding mode, for black curve components it is not.

moving toward the switching line for curves in black. The progressions for the values $d_1 = d_{\text{nom}}$, $d_1 = 10d_{\text{nom}}$, and $d_1 = 0$ are almost identical.

Note that a sliding mode control is only robust if it is in sliding mode. If the plant's parameters vary so that the sliding mode no longer exists or can no longer be achieved, the control loop is no longer robust in the way we have described. This is illustrated by the case of $d_1 \gg d_{\text{nom}}$. In this case, the sliding mode which generates robustness is reached very late.

6.3 Passivity-Based Control

6.3.1 Control of Passive Systems Using Static Controllers

In Section 2.4, we described passive systems and stated that their equilibrium point $\boldsymbol{x}_{\text{eq}} = \boldsymbol{0}$, provided that the systems have a positive definite storage function, is Lyapunov stable. In particular, strictly passive systems have an asymptotically stable equilibrium point at $\boldsymbol{x} = \boldsymbol{0}$. We are therefore interested in designing a control loop so that it is strictly passive.

First, we will look at a very simple passivity-based control loop. It is based on a strictly passive plant

$$\begin{aligned}\dot{\boldsymbol{x}} &= \boldsymbol{f}(\boldsymbol{x}, \boldsymbol{e}_2),\\ \boldsymbol{y}_2 &= \boldsymbol{g}(\boldsymbol{x})\end{aligned}$$

and a passive static controller

$$\boldsymbol{y}_1 = \boldsymbol{h}(\boldsymbol{e}_1, t),$$

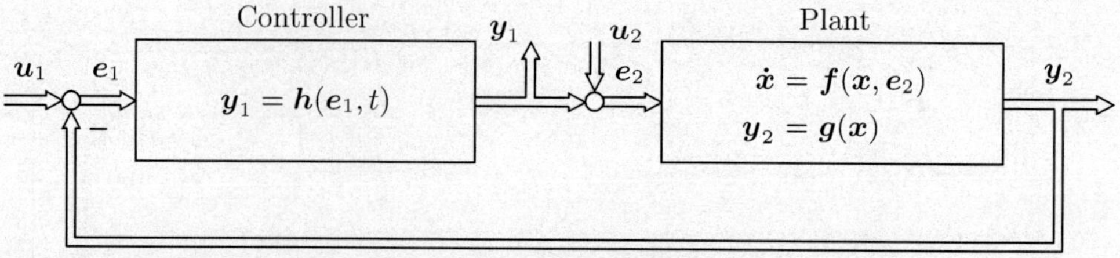

Fig. 6.22: Control loop with strictly passive plant and passive static controller

which may be time-variant. Along with

$$e_1 = u_1 - y_2,$$
$$e_2 = u_2 + y_1,$$

they result in the control loop shown in Figure 6.22. Remember Section 2.4.1, p. 143 et seq., where we found that the static controller

$$y_1 = h(e_1, t),$$

which may be time-variant, is passive if

$$0 \leq e_1^T y_1 \tag{6.65}$$

applies. In the SISO case, $0 \leq e_1 y_1$, this requirement means that the characteristic curve must lie in the first and the third quadrants. Such characteristics include two-position and three-position elements, among others.

In the following, it is assumed that $S(x)$ is the plant's storage function, and $R(x)$ is a positive definite function, so that the inequality

$$\dot{S}(x) + R(x) \leq e_2^T y_2 \tag{6.66}$$

for the strict passivity of the plant is fulfilled. We will also use $S(x)$ as a storage function of the control loop and the function $R(x)$ to prove the strict passivity of the control loop. With the composite input variable vector

$$u = \begin{bmatrix} u_1 \\ u_2 \end{bmatrix} = \begin{bmatrix} e_1 + y_2 \\ e_2 - y_1 \end{bmatrix}$$

and the output variable vector

$$y = \begin{bmatrix} y_1 \\ y_2 \end{bmatrix}$$

of the control loop, we obtain the inequality

$$\dot{S}(\boldsymbol{x}) + R(\boldsymbol{x}) \le \boldsymbol{u}^T\boldsymbol{y} = (\boldsymbol{e}_1 + \boldsymbol{y}_2)^T\boldsymbol{y}_1 + (\boldsymbol{e}_2 - \boldsymbol{y}_1)^T\boldsymbol{y}_2 = \boldsymbol{e}_1^T\boldsymbol{y}_1 + \boldsymbol{e}_2^T\boldsymbol{y}_2 \quad (6.67)$$

of strict passivity. Assuming that equation (6.65) applies, if the passivity inequality (6.66) of the plant is fulfilled, the passivity inequality (6.67) of the control loop is also. The control loop is therefore strictly passive and thus has an asymptotically stable equilibrium point. This is stated in

Theorem 80 (Control of Strictly Passive Plants Using Passive Static Controllers). *Let a passive static controller and a strictly passive plant, which are given as*

$$\boldsymbol{y}_1 = \boldsymbol{h}(\boldsymbol{e}_1, t) \quad \text{and} \quad \begin{aligned} \dot{\boldsymbol{x}} &= \boldsymbol{f}(\boldsymbol{x}, \boldsymbol{e}_2), \\ \boldsymbol{y}_2 &= \boldsymbol{g}(\boldsymbol{x}) \end{aligned}$$

with $\boldsymbol{x} \in I\!R^n$, $\boldsymbol{e}_1$, $\boldsymbol{e}_2$, $\boldsymbol{y}_1$, $\boldsymbol{y}_2 \in I\!R^m$, *and*

$$\boldsymbol{e}_1 = \boldsymbol{u}_1 - \boldsymbol{y}_2 \quad \text{and} \quad \boldsymbol{e}_2 = \boldsymbol{u}_2 + \boldsymbol{y}_1$$

with $\boldsymbol{u}_1$, $\boldsymbol{u}_2 \in I\!R^m$, *form the control loop*

$$\begin{aligned} \dot{\boldsymbol{x}} &= \boldsymbol{f}(\boldsymbol{x}, \boldsymbol{u}_2 + \boldsymbol{y}_1), \\ \boldsymbol{y}_1 &= \boldsymbol{h}(\boldsymbol{u}_1 - \boldsymbol{y}_2, t), \\ \boldsymbol{y}_2 &= \boldsymbol{g}(\boldsymbol{x}). \end{aligned}$$

This control loop possesses an asymptotically stable equilibrium point at $\boldsymbol{x} = \boldsymbol{0}$ *for* $\boldsymbol{u} = [\boldsymbol{u}_1^T \;\; \boldsymbol{u}_2^T]^T = \boldsymbol{0}$. *If the plant's storage function* $S(\boldsymbol{x})$ *is radially unbounded, the equilibrium point is globally asymptotically stable.*

Let us now consider the case in which the plant is only passive. Even then, an asymptotically stable control loop can be created using a passive static controller. However, additional conditions must be imposed in this case. For this purpose, we need the property of zero-state detectability, which we define in

Definition 37 (Zero-State Detectability). *A system*

$$\begin{aligned} \dot{\boldsymbol{x}} &= \boldsymbol{f}(\boldsymbol{x}, \boldsymbol{u}), \quad && \boldsymbol{x} \in I\!R^n, \quad \boldsymbol{u} \in I\!R^m, \\ \boldsymbol{y} &= \boldsymbol{g}(\boldsymbol{x}, \boldsymbol{u}), \quad && \boldsymbol{y} \in I\!R^r, \end{aligned}$$

is called zero-state detectable if

$$\lim_{t \to \infty} \boldsymbol{x}(t) = \boldsymbol{0}$$

results for the case that $\boldsymbol{u}(t) = \boldsymbol{0}$ *and* $\boldsymbol{y}(t) = \boldsymbol{0}$ *hold for all* $t \ge 0$.

Now we will start again from equation (6.67), which must be valid for the control loop in this case as well. However, the controlled system is only passive here, i. e.

$$\dot{S}(\boldsymbol{x}) \leq \boldsymbol{e}_2^T \boldsymbol{y}_2$$

applies instead of equation (6.66). Therefore,

$$\boldsymbol{e}_1^T \boldsymbol{y}_1 = \boldsymbol{e}_1^T \boldsymbol{h}(\boldsymbol{e}_1, t) > 0 \quad \text{for all } \boldsymbol{e}_1 \neq \boldsymbol{0}$$

must be valid for the fulfillment of equation (6.67). Furthermore, the limit

$$\lim_{t\to\infty} \boldsymbol{x}(t) = \boldsymbol{0} \quad \text{for all } t \geq 0$$

must follow from $\boldsymbol{e}_2(t) = \boldsymbol{0}$ and $\boldsymbol{y}_2(t) = \boldsymbol{0}$, i.e., for this case, the plant's trajectories tend to zero and remain there. The latter is equivalent to the controlled system being zero-state detectable. If this were not the case, limit cycles $\boldsymbol{x}(t)$ with $\dot{S}(\boldsymbol{x}) = 0$ could exist which cannot be observed based on the input variable vector $\boldsymbol{u}(t)$ and the output variable vector $\boldsymbol{y}(t)$.

Similar to Theorem 80[2], and with the statements above, we can now formulate the subsequent theorem [65].

Theorem 81 (Control of Passive Plants Using Passive Static Controllers). *Let the control loop from Theorem 80 be given, with the difference that the plant is passive and the passive static controller satisfies* $\boldsymbol{e}_1 \boldsymbol{h}(\boldsymbol{e}_1, t) > 0$ *for* $\boldsymbol{e}_1 \neq \boldsymbol{0}$. *If the plant has a positive definite storage function and is zero-state detectable, then the control loop has an asymptotically stable equilibrium point at* $\boldsymbol{x} = \boldsymbol{0}$ *for* $\boldsymbol{u} = \boldsymbol{0}$. *If the storage function is radially unbounded, the equilibrium point is globally asymptotically stable.*

We will now describe an example of the application of Theorem 81.

6.3.2 Example: Damping of Seismic Building Vibrations

Earthquakes trigger vibrations in buildings with steel girders, which can damage them and even cause them to collapse. These building vibrations can be attenuated using various damping methods. A possible approach is utilizing viscous fluid dampers that are mounted on one or more floors to compensate for seismic vibrations [434, 438, 439]. Figure 6.23 illustrates the case of a building vibrating due to an earthquake.

The dynamics of the building are those of a multi-mass oscillator. The walls act both as springs and dampers. We can therefore model two superimposed floors using their masses m_i and m_{i+1}, connected by a spring with the spring constant c_i and a damper with the damping constant d_i [350].

The building is set to vibration by the seismically induced acceleration $\ddot{x}_\mathrm{g}$ of the building's foundation. We can model this acceleration as acting immediately and with equal strength on all floors – except the foundation –

[2] In [65], Theorem 81 is only proven for time-invariant characteristics $\boldsymbol{h}(\boldsymbol{e}_1)$. However, the argument therein remains valid even with time-variant characteristics $\boldsymbol{h}(\boldsymbol{e}_1, t)$.

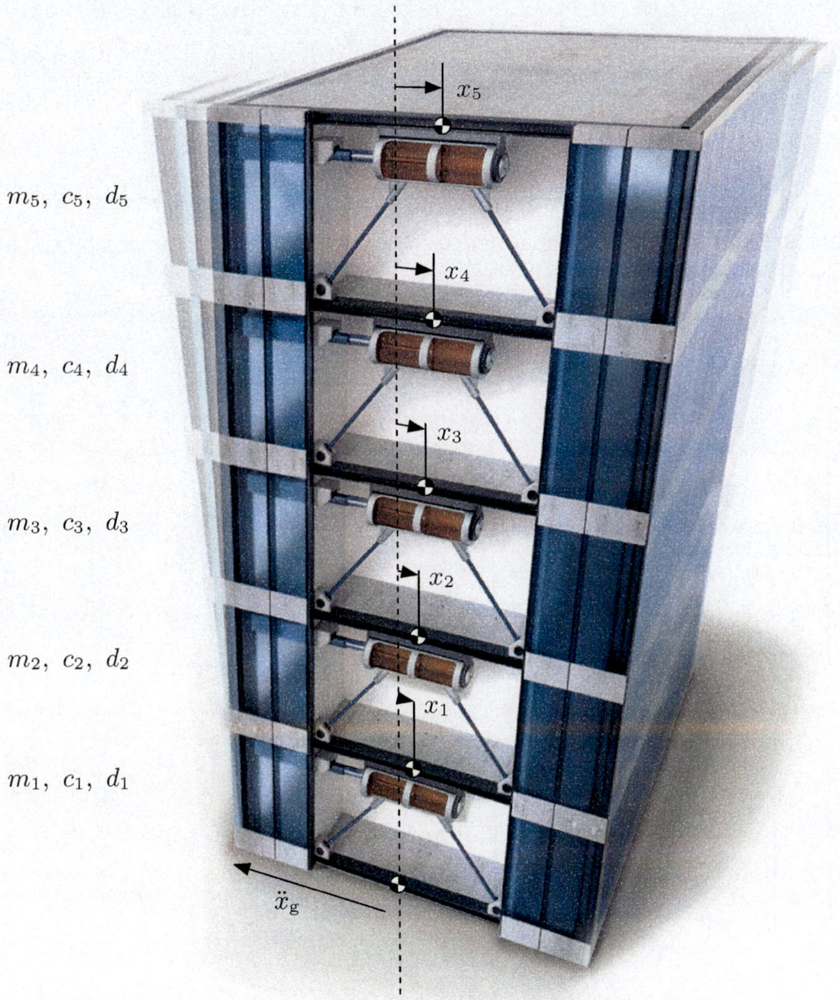

Fig. 6.23: A building with viscous fluid dampers on each floor. The parameters m_i, c_i, d_i, $i = 1, \ldots, 5$, indicate the masses, spring constants, and damping constants of the individual stories.

with the sign reversed. The damping forces $F_1, \ldots, F_5$ of the viscous fluid dampers counteract it. This results in the dynamics for the movement of the floor on the first story,

$$m_1\ddot{x}_1 = -d_1\dot{x}_1 - d_2(\dot{x}_1 - \dot{x}_2) - c_1x_1 - c_2(x_1 - x_2) + F_1 + m_1\ddot{x}_\text{g},$$

and for those on the second, third, and fourth stories,

$$m_2\ddot{x}_2 = -d_2(\dot{x}_2-\dot{x}_1) - d_3(\dot{x}_2-\dot{x}_3) - c_2(x_2-x_1) - c_3(x_2-x_3) + F_2 + m_2\ddot{x}_{\mathrm{g}},$$
$$m_3\ddot{x}_3 = -d_3(\dot{x}_3-\dot{x}_2) - d_4(\dot{x}_3-\dot{x}_4) - c_3(x_3-x_2) - c_4(x_3-x_4) + F_3 + m_3\ddot{x}_{\mathrm{g}},$$
$$m_4\ddot{x}_4 = -d_4(\dot{x}_4-\dot{x}_3) - d_5(\dot{x}_4-\dot{x}_5) - c_4(x_4-x_3) - c_5(x_4-x_5) + F_4 + m_4\ddot{x}_{\mathrm{g}},$$

as well as for the dynamics of the fifth and highest story,

$$m_5\ddot{x}_5 = -d_5(\dot{x}_5 - \dot{x}_4) - c_5(x_5 - x_4) + F_5 + m_5\ddot{x}_{\mathrm{g}}.$$

In summary, for the aforementioned five equations of motion in matrix notation, we obtain

$$\boldsymbol{M}\ddot{\boldsymbol{x}} = -\boldsymbol{D}\dot{\boldsymbol{x}} - \boldsymbol{C}\boldsymbol{x} + \boldsymbol{F} + \boldsymbol{M}\boldsymbol{1}\ddot{x}_{\mathrm{g}} \tag{6.68}$$

with

$$\boldsymbol{M} = \begin{bmatrix} m_1 & 0 & 0 & 0 & 0 \\ 0 & m_2 & 0 & 0 & 0 \\ 0 & 0 & m_3 & 0 & 0 \\ 0 & 0 & 0 & m_4 & 0 \\ 0 & 0 & 0 & 0 & m_5 \end{bmatrix}, \boldsymbol{D} = \begin{bmatrix} d_1+d_2 & -d_2 & 0 & 0 & 0 \\ -d_2 & d_2+d_3 & -d_3 & 0 & 0 \\ 0 & -d_3 & d_3+d_4 & -d_4 & 0 \\ 0 & 0 & -d_4 & d_4+d_5 & -d_5 \\ 0 & 0 & 0 & -d_5 & d_5 \end{bmatrix},$$

$$\boldsymbol{F} = \begin{bmatrix} F_1 \\ F_2 \\ F_3 \\ F_4 \\ F_5 \end{bmatrix}, \quad \boldsymbol{1} = \begin{bmatrix} 1 \\ 1 \\ 1 \\ 1 \\ 1 \end{bmatrix}, \quad \boldsymbol{C} = \begin{bmatrix} c_1+c_2 & -c_2 & 0 & 0 & 0 \\ -c_2 & c_2+c_3 & -c_3 & 0 & 0 \\ 0 & -c_3 & c_3+c_4 & -c_4 & 0 \\ 0 & 0 & -c_4 & c_4+c_5 & -c_5 \\ 0 & 0 & 0 & -c_5 & c_5 \end{bmatrix}.$$

We can now extend the position vector $\boldsymbol{x} = [x_1 \;\; x_2 \;\; x_3 \;\; x_4 \;\; x_5]^T$ by the derivatives $\dot{x}_1, \ldots, \dot{x}_5$ and obtain the ten-dimensional state vector

$$\begin{bmatrix} \boldsymbol{x} \\ \dot{\boldsymbol{x}} \end{bmatrix}.$$

In this way, from the five second-order differential equations (6.68), we obtain the ten first-order differential equations

$$\begin{bmatrix} \dot{\boldsymbol{x}} \\ \ddot{\boldsymbol{x}} \end{bmatrix} = \begin{bmatrix} \boldsymbol{0} & \boldsymbol{I} \\ -\boldsymbol{M}^{-1}\boldsymbol{C} & -\boldsymbol{M}^{-1}\boldsymbol{D} \end{bmatrix} \begin{bmatrix} \boldsymbol{x} \\ \dot{\boldsymbol{x}} \end{bmatrix} + \begin{bmatrix} \boldsymbol{0} \\ \boldsymbol{M}^{-1} \end{bmatrix} \boldsymbol{F} + \begin{bmatrix} \boldsymbol{0} \\ \boldsymbol{1} \end{bmatrix} \ddot{x}_{\mathrm{g}}. \tag{6.69}$$

As a five-dimensional output variable vector, we will use

$$\boldsymbol{y} = \dot{\boldsymbol{x}}. \tag{6.70}$$

System (6.69), (6.70) is passive, which we can prove with the storage function

$$S(\boldsymbol{x}, \dot{\boldsymbol{x}}) = \frac{1}{2}\dot{\boldsymbol{x}}^T \boldsymbol{M}\dot{\boldsymbol{x}} + \frac{1}{2}\boldsymbol{x}^T \boldsymbol{C}\boldsymbol{x} \tag{6.71}$$

by inserting it into the passivity inequality

$$\dot{S}(\boldsymbol{x}, \dot{\boldsymbol{x}}) \leq \underbrace{(\boldsymbol{F} + \boldsymbol{M}\boldsymbol{1}\ddot{x}_{\mathrm{g}})^T}_{\boldsymbol{u}^T} \boldsymbol{y},$$

where the force vector $\boldsymbol{F}$ and the disturbance $\boldsymbol{M}\boldsymbol{1}\ddot{x}_{\mathrm{g}}$ act as input variable vectors. Thus, inequality

$$\begin{aligned}\dot{S}(\boldsymbol{x}, \dot{\boldsymbol{x}}) &= \frac{1}{2}(\ddot{\boldsymbol{x}}^T \boldsymbol{M}\dot{\boldsymbol{x}} + \dot{\boldsymbol{x}}^T \boldsymbol{M}\ddot{\boldsymbol{x}} + \dot{\boldsymbol{x}}^T \boldsymbol{C}\boldsymbol{x} + \boldsymbol{x}^T \boldsymbol{C}\dot{\boldsymbol{x}})\\ &= -\dot{\boldsymbol{x}}^T \boldsymbol{D}\dot{\boldsymbol{x}} + (\boldsymbol{F} + \boldsymbol{M}\boldsymbol{1}\ddot{x}_{\mathrm{g}})^T \boldsymbol{y} \leq (\boldsymbol{F} + \boldsymbol{M}\boldsymbol{1}\ddot{x}_{\mathrm{g}})^T \boldsymbol{y}\end{aligned}$$

follows, which is fulfilled due to the positive definiteness of the matrix $\boldsymbol{D}$. Thus, system (6.69), (6.70) is passive. Furthermore, it is also zero-state detectable, because it is asymptotically stable, i. e. it holds that

$$\lim_{t\to\infty} \begin{bmatrix} \boldsymbol{x}(t) \\ \dot{\boldsymbol{x}}(t) \end{bmatrix} = \boldsymbol{0},$$

which we obtain if $\boldsymbol{F}(t) = \boldsymbol{0}$, $\ddot{x}_{\mathrm{g}} = 0$, and $\boldsymbol{y}(t) = \boldsymbol{0}$ hold for all $t \geq 0$.

In this example, viscous fluid dampers are mounted on each story of the building. They consist of a hydraulic cylinder and a moving piston. If the piston moves, it generates a force

$$\tilde{F} = \tilde{d} \cdot |v|^{\alpha} \operatorname{sgn}(-v)$$

due to viscous friction, which counteracts the movement of the piston and depends on its speed v, as shown in Figure 6.24. Here, $\tilde{d}$ is the damping coefficient and α is a design constant with $0 < \alpha \leq 1$ [190].

The viscous fluid dampers exert the forces

$$\begin{aligned}\tilde{F}_1 &= \tilde{d}_1|\dot{x}_1|^{\alpha} \operatorname{sgn}(-\dot{x}_1),\\ \tilde{F}_2 &= \tilde{d}_2|\dot{x}_2 - \dot{x}_1|^{\alpha} \operatorname{sgn}(\dot{x}_1 - \dot{x}_2),\\ \tilde{F}_3 &= \tilde{d}_3|\dot{x}_3 - \dot{x}_2|^{\alpha} \operatorname{sgn}(\dot{x}_2 - \dot{x}_3),\\ \tilde{F}_4 &= \tilde{d}_4|\dot{x}_4 - \dot{x}_3|^{\alpha} \operatorname{sgn}(\dot{x}_3 - \dot{x}_4),\\ \tilde{F}_5 &= \tilde{d}_5|\dot{x}_5 - \dot{x}_4|^{\alpha} \operatorname{sgn}(\dot{x}_4 - \dot{x}_5)\end{aligned}$$

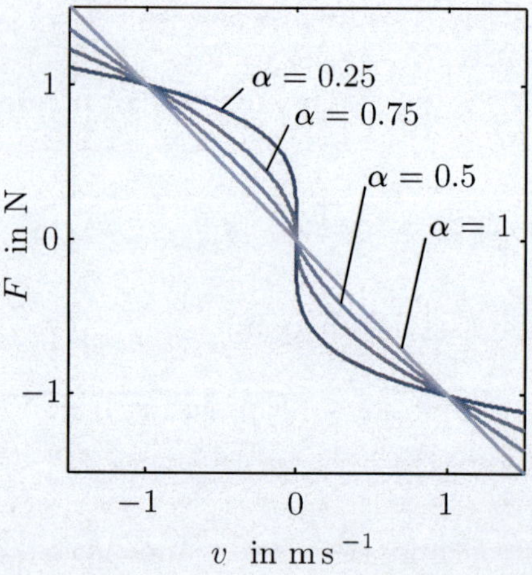

Fig. 6.24: Curve of the damping force $\tilde{F} = \tilde{d} \cdot |v|^{\alpha}\, \mathrm{sgn}\,(-v)$ for different values α and $\tilde{d} = 1$

on the masses m_i. Figures 6.25 and 6.26 illustrate these forces. For the total forces acting on the masses m_i, we obtain

$$\boldsymbol{F}(\dot{\boldsymbol{x}}) = \begin{bmatrix} F_1 \\ F_2 \\ F_3 \\ F_4 \\ F_5 \end{bmatrix} = \begin{bmatrix} \tilde{F}_1 - \tilde{F}_2 \\ \tilde{F}_2 - \tilde{F}_3 \\ \tilde{F}_3 - \tilde{F}_4 \\ \tilde{F}_4 - \tilde{F}_5 \\ \tilde{F}_5 \end{bmatrix} = \begin{bmatrix} 1 & -1 & 0 & 0 & 0 \\ 0 & 1 & -1 & 0 & 0 \\ 0 & 0 & 1 & -1 & 0 \\ 0 & 0 & 0 & 1 & -1 \\ 0 & 0 & 0 & 0 & 1 \end{bmatrix} \begin{bmatrix} \tilde{d}_1 |\dot{x}_1|^{\alpha}\, \mathrm{sgn}\,(-\dot{x}_1) \\ \tilde{d}_2 |\dot{x}_2 - \dot{x}_1|^{\alpha}\, \mathrm{sgn}\,(\dot{x}_1 - \dot{x}_2) \\ \tilde{d}_3 |\dot{x}_3 - \dot{x}_2|^{\alpha}\, \mathrm{sgn}\,(\dot{x}_2 - \dot{x}_3) \\ \tilde{d}_4 |\dot{x}_4 - \dot{x}_3|^{\alpha}\, \mathrm{sgn}\,(\dot{x}_3 - \dot{x}_4) \\ \tilde{d}_5 |\dot{x}_5 - \dot{x}_4|^{\alpha}\, \mathrm{sgn}\,(\dot{x}_4 - \dot{x}_5) \end{bmatrix}. \tag{6.72}$$

This nonlinearity is passive because it satisfies the inequality (6.65), i. e.

$$0 < \boldsymbol{e}_1^T \boldsymbol{y}_1 = \boldsymbol{y}_1^T \boldsymbol{e}_1,$$

for all $\boldsymbol{e}_1 \neq \boldsymbol{0}$. To prove this, we use

$$\boldsymbol{e}_1 = -\dot{\boldsymbol{x}} \quad \text{and} \quad \boldsymbol{y}_1 = \boldsymbol{F}(\dot{\boldsymbol{x}}) = \boldsymbol{F}(-\boldsymbol{e}_1)$$

and obtain

$$\boldsymbol{y}_1^T \boldsymbol{e}_1 = \boldsymbol{F}^T(-\boldsymbol{e}_1)\boldsymbol{e}_1 = -\boldsymbol{F}^T(\dot{\boldsymbol{x}})\dot{\boldsymbol{x}}.$$

From the latter, the passivity inequality

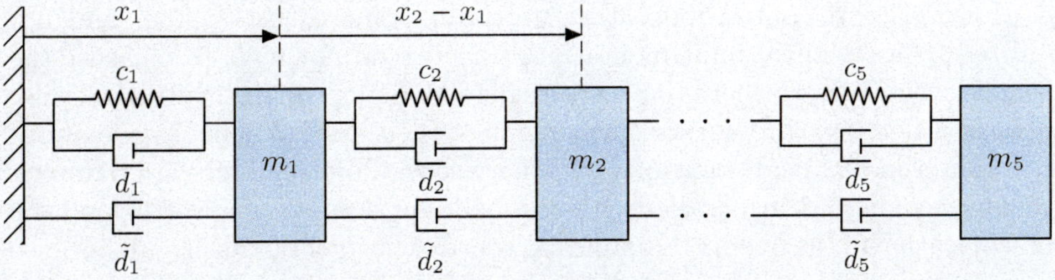

Fig. 6.25: Model of a building based on springs, masses, and dampers

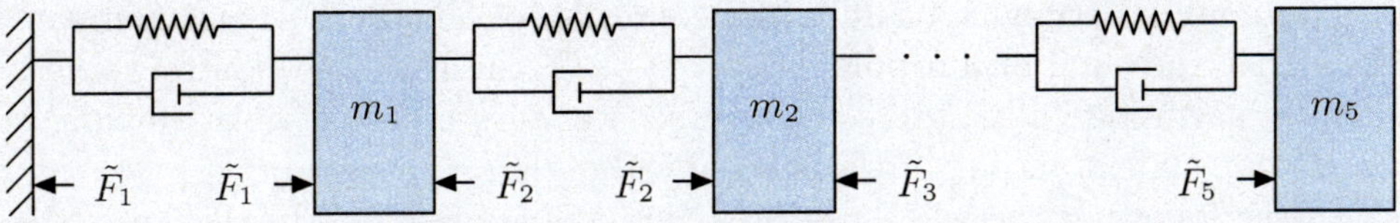

Fig. 6.26: Diagram of the damping forces $\tilde{F}_i$

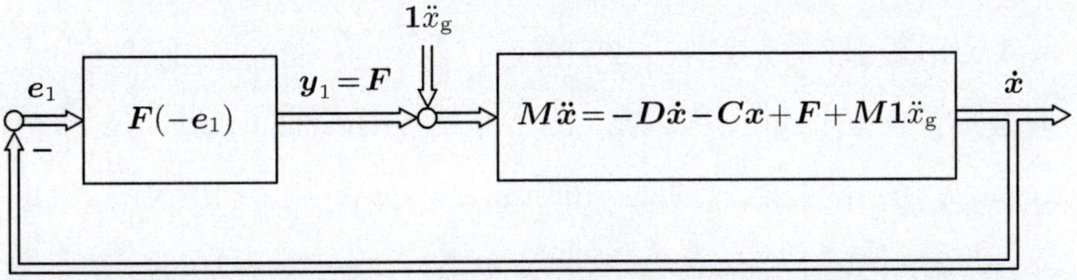

Fig. 6.27: Representation of the building model with viscous fluid dampers, shown as a control-loop structure

$$\boldsymbol{y}_1^T\boldsymbol{e}_1 = -\boldsymbol{F}^T(\dot{\boldsymbol{x}})\dot{\boldsymbol{x}} = \begin{bmatrix} -\tilde{d}_1|\dot{x}_1|^\alpha \operatorname{sgn}(-\dot{x}_1) \\ -\tilde{d}_2|\dot{x}_2 - \dot{x}_1|^\alpha \operatorname{sgn}(\dot{x}_1 - \dot{x}_2) \\ -\tilde{d}_3|\dot{x}_3 - \dot{x}_2|^\alpha \operatorname{sgn}(\dot{x}_2 - \dot{x}_3) \\ -\tilde{d}_4|\dot{x}_4 - \dot{x}_3|^\alpha \operatorname{sgn}(\dot{x}_3 - \dot{x}_4) \\ -\tilde{d}_5|\dot{x}_5 - \dot{x}_4|^\alpha \operatorname{sgn}(\dot{x}_4 - \dot{x}_5) \end{bmatrix}^T \begin{bmatrix} 1 & 0 & 0 & 0 & 0 \\ -1 & 1 & 0 & 0 & 0 \\ 0 & -1 & 1 & 0 & 0 \\ 0 & 0 & -1 & 1 & 0 \\ 0 & 0 & 0 & -1 & 1 \end{bmatrix} \begin{bmatrix} \dot{x}_1 \\ \dot{x}_2 \\ \dot{x}_3 \\ \dot{x}_4 \\ \dot{x}_5 \end{bmatrix}$$

$$= \tilde{d}_1|\dot{x}_1|^{\alpha+1} + \sum_{i=2}^{5} \tilde{d}_i|\dot{x}_i - \dot{x}_{i-1}|^{\alpha+1} > 0$$

results, which holds for all $\boldsymbol{e}_1 = -\dot{\boldsymbol{x}} \neq \mathbf{0}$.

We can represent the model (6.68) of the building and the model (6.72) of the viscous fluid dampers as a control-loop structure, as shown in Figure 6.27. This is a special case of the control-loop structure shown in Figure 6.22.

However, this control-loop structure is inherent to the system and not realized by a technical implementation, since we do not have to measure the velocity vector $\dot{\boldsymbol{x}}$. Rather, the viscous fluid dampers intrinsically make use of these velocities through the movement of their pistons with a velocity of $\dot{x}_i$. From a mechanical point of view, the viscous fluid dampers are damping elements connected in parallel with the building dampers. Nevertheless, the interpretation of the overall dynamics as a feedback structure is useful, because in this way we can prove the stability using Theorem 81.

Since the plant is a zero-state detectable passive system with the radially unbounded, positive definite storage function (6.71) and the controller is a passive nonlinearity which fulfills the condition of Theorem 81, we can conclude that the equilibrium point $\boldsymbol{x}_{\mathrm{eq}} = \boldsymbol{0}$ of the undisturbed control loop is globally asymptotically stable. Of course, we were aware of this previously due to our understanding of the mechanical design.

In the following, we will describe a concrete application. In this case, the building has five similar stories with the masses m_i, spring constants c_i, and damping coefficients d_i [350], which are in detail

$$
\begin{aligned}
m_1 &= 215.2 \cdot 10^3 \text{ kg}, & c_1 &= 147 \cdot 10^6 \,\mathrm{N\,m^{-1}}, & d_1 &= 113.8 \cdot 10^3 \,\mathrm{N\,s\,m^{-1}},\\
m_2 &= 209.2 \cdot 10^3 \text{ kg}, & c_2 &= 133 \cdot 10^6 \,\mathrm{N\,m^{-1}}, & d_2 &= 92.4 \cdot 10^3 \,\mathrm{N\,s\,m^{-1}},\\
m_3 &= 207.0 \cdot 10^3 \text{ kg}, & c_3 &= 99 \cdot 10^6 \,\mathrm{N\,m^{-1}}, & d_3 &= 81.0 \cdot 10^3 \,\mathrm{N\,s\,m^{-1}},\\
m_4 &= 204.8 \cdot 10^3 \text{ kg}, & c_4 &= 89 \cdot 10^6 \,\mathrm{N\,m^{-1}}, & d_4 &= 72.8 \cdot 10^3 \,\mathrm{N\,s\,m^{-1}},\\
m_5 &= 266.1 \cdot 10^3 \text{ kg}, & c_5 &= 84 \cdot 10^6 \,\mathrm{N\,m^{-1}}, & d_5 &= 68.7 \cdot 10^3 \,\mathrm{N\,s\,m^{-1}}.
\end{aligned}
$$

The damping coefficients of the viscous fluid dampers are

$$
\begin{aligned}
\tilde{d}_1 &= 6506 \cdot 10^3 \,\mathrm{N\,s\,m^{-1}},\\
\tilde{d}_2 &= 4343 \cdot 10^3 \,\mathrm{N\,s\,m^{-1}},\\
\tilde{d}_3 &= 3455 \cdot 10^3 \,\mathrm{N\,s\,m^{-1}},\\
\tilde{d}_4 &= 2914 \cdot 10^3 \,\mathrm{N\,s\,m^{-1}},\\
\tilde{d}_5 &= 2648 \cdot 10^3 \,\mathrm{N\,s\,m^{-1}},
\end{aligned}
$$

and their parameter α is

$$\alpha = 0.5125.$$

An example is the 1995 earthquake in Kōbe, Japan [140]. The quake lasted about 20 s and reached a strength of 6.9 on the moment-magnitude scale, destroying 170000 buildings and killing more than 6500 people. The elevated highway collapsed along a length of 5 km. The horizontal acceleration $\ddot{x}_{\mathrm{g}}$ caused by the Kōbe quake and the shift x_{g} resulting from the double integration of $\ddot{x}_{\mathrm{g}}$ are both shown in red in Figure 6.28. The graph also shows

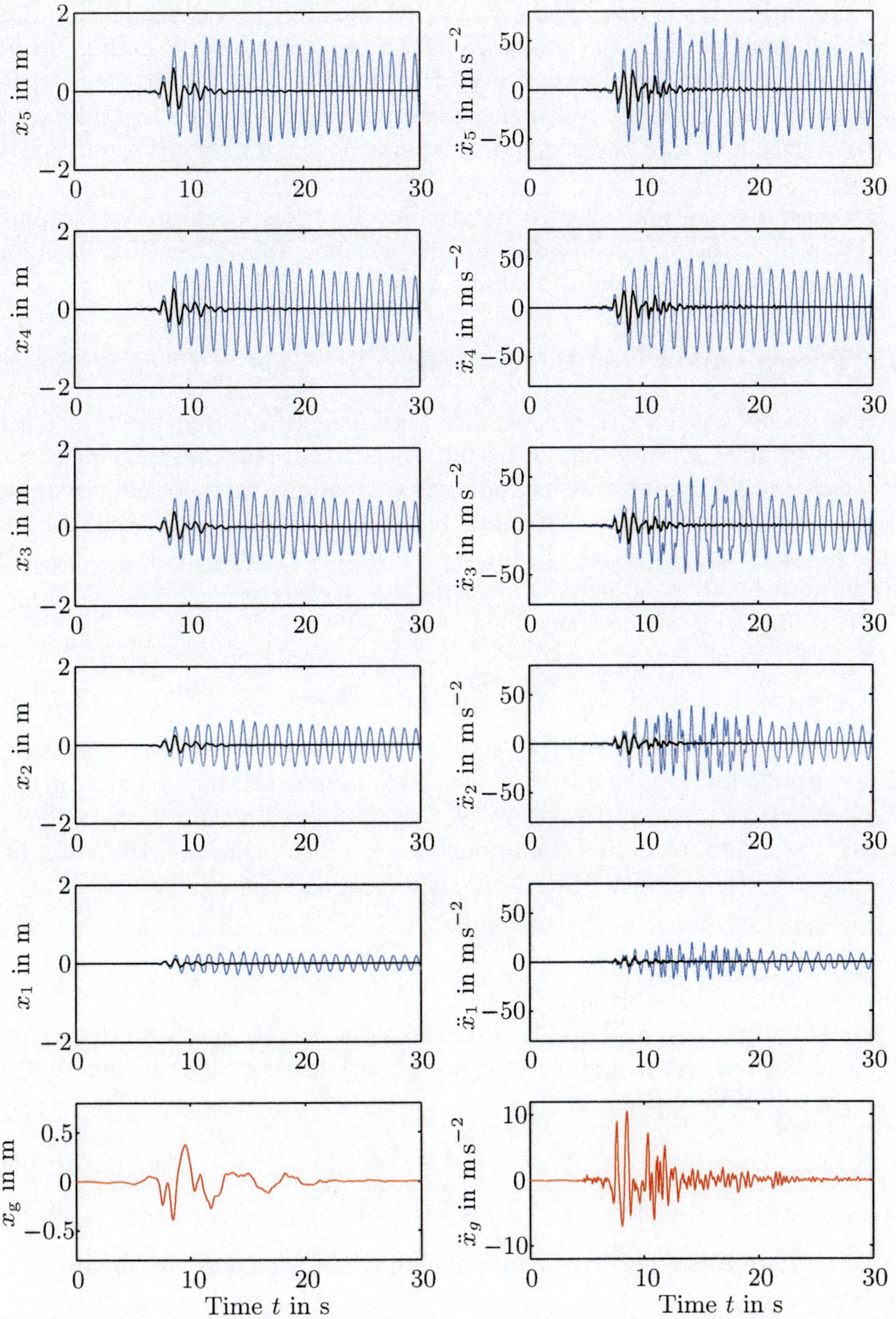

Fig. 6.28: The left column shows the shift x_g (red) in the foundation caused by the earthquake in Kōbe and the shift x_i of the floors. The right column shows the acceleration $\ddot{x}_g$ (red) of the foundation and the acceleration $\ddot{x}_i$ of the floors. The blue curves show the building vibrations without viscous fluid dampers, while the black curves show the building vibrations with these dampers.

the accelerations $\ddot{x}_i$ acting on the respective stories i of our model building and which, multiplied by the respective masses m_i, result in the acting forces. Also shown are the displacements x_i of the stories i, both with and without viscous fluid dampers. With these dampers, the load on the building is greatly reduced, so that the effect of the quake on the building is significantly decreased.

Because this method effectively reduces vibrations, additional damping devices, such as the viscous fluid dampers described above, are often used for large buildings, bridges, and tall towers located in earthquake-prone areas.

6.3.3 Passivation of Non-Passive Linear Systems

For non-passive systems, the controller design methods from Section 6.3.1 cannot be applied at this stage. A useful procedure in this case is to make the non-passive system passive by extending it with additional external circuitry, such as a system connected in parallel, or feedback via an additional system. After this passivation of the system, we can apply the theorems for passive systems and the appropriate controller design procedures.

Let us first address the controllable linear SISO systems

$$\begin{aligned} \dot{\boldsymbol{x}} &= \boldsymbol{A}\boldsymbol{x} + \boldsymbol{b} \cdot u, \\ y &= \boldsymbol{c}^T \boldsymbol{x} \end{aligned} \tag{6.73}$$

before turning to the general case of control-affine MIMO systems in the following section.

Now our aim is to derive conditions under which a linear system can be made passive using a static state feedback, i. e. a controller

$$u = -\boldsymbol{r}^T \boldsymbol{x} + V \cdot y_{\text{ref}}$$

with the reference variable y_{ref}.

For this purpose we will use Theorem 28 from p. 153, the theorem on passivity and the Kalman-Yakubovich-Popov equations. If a SISO system (6.73) is passive, the KYP equations

$$\boldsymbol{A}^T \boldsymbol{R} + \boldsymbol{R}\boldsymbol{A} = -\boldsymbol{Q}, \tag{6.74}$$

$$\boldsymbol{c}^T = \boldsymbol{b}^T \boldsymbol{R} \tag{6.75}$$

are valid, where $\boldsymbol{R}$ is a positive definite matrix and $\boldsymbol{Q}$ is a positive semidefinite matrix.

We now multiply equation (6.75) by vector $\boldsymbol{b}$ and obtain

$$\boldsymbol{c}^T \boldsymbol{b} = \boldsymbol{b}^T \boldsymbol{R} \boldsymbol{b} > 0 \tag{6.76}$$

as a necessary condition for the passivity of the system (6.73). We will also perform the same steps as for an exact input-output linearization, deriving the output variable y according to the equations

$$
\begin{aligned}
y &= \boldsymbol{c}^T \boldsymbol{x}, \\
\dot{y} &= \boldsymbol{c}^T \dot{\boldsymbol{x}} = \underbrace{\boldsymbol{c}^T \boldsymbol{A}\boldsymbol{x}}_{L_{\boldsymbol{a}}c(\boldsymbol{x})} + \underbrace{\boldsymbol{c}^T \boldsymbol{b}}_{L_{\boldsymbol{b}}c(\boldsymbol{x})} \cdot u, \\
\ddot{y} &= \ldots \\
&\;\vdots
\end{aligned}
$$

The first Lie derivative

$$L_{\boldsymbol{b}}c(\boldsymbol{x}) = \boldsymbol{c}^T \boldsymbol{b}$$

is already unequal to zero according to equation (6.76) if the system is passive. Consequently, a passive system (6.73) must have a relative degree of $\delta = 1$ due to $L_{\boldsymbol{b}}c(\boldsymbol{x}) > 0$.

We will derive a further necessary condition for the passivity of the system (6.73). For this purpose, we will transform system (6.73) using the diffeomorphism

$$\boldsymbol{z} = \boldsymbol{t}(\boldsymbol{x}) = \begin{bmatrix} \boldsymbol{c}^T \boldsymbol{x} \\ \boldsymbol{t}_2^T \boldsymbol{x} \\ \vdots \\ \boldsymbol{t}_n^T \boldsymbol{x} \end{bmatrix} = \boldsymbol{T}\boldsymbol{x} \tag{6.77}$$

with suitable linearly independent vectors $\boldsymbol{t}_i$, $i = 2, \ldots, n$. This will give us

$$\dot{\boldsymbol{z}} = \begin{bmatrix} \boldsymbol{c}^T \\ \boldsymbol{t}_2^T \\ \vdots \\ \boldsymbol{t}_n^T \end{bmatrix} \dot{\boldsymbol{x}} = \begin{bmatrix} \boldsymbol{c}^T \boldsymbol{A}\boldsymbol{x} + \boldsymbol{c}^T \boldsymbol{b} \cdot u \\ \boldsymbol{t}_2^T \boldsymbol{A}\boldsymbol{x} + \boldsymbol{t}_2^T \boldsymbol{b} \cdot u \\ \vdots \\ \boldsymbol{t}_n^T \boldsymbol{A}\boldsymbol{x} + \boldsymbol{t}_n^T \boldsymbol{b} \cdot u \end{bmatrix}, \tag{6.78}$$

$$y = z_1.$$

To eliminate the dependence of the terms

$$\boldsymbol{t}_i^T \boldsymbol{A}\boldsymbol{x} + \boldsymbol{t}_i^T \boldsymbol{b}\, u, \quad i = 2, \ldots, n,$$

from equation (6.78) on the control value u, we now select the vectors $\boldsymbol{t}_i$, $i = 2, \ldots, n$, so that

$$\begin{bmatrix} \boldsymbol{t}_2^T \\ \vdots \\ \boldsymbol{t}_n^T \end{bmatrix} \boldsymbol{b} = \boldsymbol{0}$$

applies. This means that all $\boldsymbol{t}_i$ are perpendicular to $\boldsymbol{b}$. Since according to equation (6.76), the inequality $\boldsymbol{c}^T \boldsymbol{b} \neq 0$ applies. If now the vector $\boldsymbol{b}$ is not a linear combination of the vectors $\boldsymbol{t}_i$, $i = 2, \ldots, n$, the vector $\boldsymbol{c}$ is also not a linear combination of the vectors $\boldsymbol{t}_i$, $i = 2, \ldots, n$, alone. Consequently, the vectors $\boldsymbol{c}$ and $\boldsymbol{t}_i$, $i = 2, \ldots, n$, form a base of $\mathbb{R}^n$ and the matrix $\boldsymbol{T}$ is regular. This results in

$$\dot{\boldsymbol{z}} = \begin{bmatrix} \boldsymbol{c}^T \boldsymbol{A} \boldsymbol{T}^{-1} \boldsymbol{z} + \boldsymbol{c}^T \boldsymbol{b} \cdot u \\ \boldsymbol{t}_2 \boldsymbol{A} \boldsymbol{T}^{-1} \boldsymbol{z} \\ \vdots \\ \boldsymbol{t}_n \boldsymbol{A} \boldsymbol{T}^{-1} \boldsymbol{z} \end{bmatrix}, \quad \begin{matrix} \} \text{ external dynamics} \\ \left.\begin{matrix} \\ \\ \\ \end{matrix}\right\} \text{ internal dynamics} \end{matrix} \qquad (6.79)$$

$$y = z_1.$$

To shorten this, we can define the vector

$$\boldsymbol{z}_{\text{int}} = \begin{bmatrix} z_2 & \cdots & z_n \end{bmatrix}^T$$

of internal states and obtain

$$\begin{aligned} \dot{z}_1 &= \Gamma_{11} z_1 + \boldsymbol{\Gamma}_{12}^T \boldsymbol{z}_{\text{int}} + \boldsymbol{c}^T \boldsymbol{b} \cdot u, && \} \text{ external dynamics} \\ \dot{\boldsymbol{z}}_{\text{int}} &= \boldsymbol{\Gamma}_{21} z_1 + \boldsymbol{\Gamma}_{22} \boldsymbol{z}_{\text{int}}, && \} \text{ internal dynamics} \\ y &= z_1, \end{aligned} \qquad (6.80)$$

where

$$\boldsymbol{z} = \begin{bmatrix} z_1 \\ \boldsymbol{z}_{\text{int}} \end{bmatrix}, \quad \begin{bmatrix} \Gamma_{11} & \boldsymbol{\Gamma}_{12}^T \end{bmatrix} = \boldsymbol{c}^T \boldsymbol{A} \boldsymbol{T}^{-1}, \quad \text{and} \quad \begin{bmatrix} \boldsymbol{\Gamma}_{21} & \boldsymbol{\Gamma}_{22} \end{bmatrix} = \begin{bmatrix} \boldsymbol{t}_2 \boldsymbol{A} \boldsymbol{T}^{-1} \\ \vdots \\ \boldsymbol{t}_n \boldsymbol{A} \boldsymbol{T}^{-1} \end{bmatrix}$$

hold.

Besides condition (6.75), equation (6.74) is the second requirement to be fulfilled. It is equivalent to saying that the equilibrium point $\boldsymbol{x}_{\text{eq}} = \boldsymbol{0}$ of the system (6.73) is Lyapunov stable. Because the linear transformation (6.77) is regular, equation (6.74) tells us that the transformed system (6.80) must also be Lyapunov stable. In particular, its internal dynamics must be Lyapunov stable, i. e. the matrix $\boldsymbol{\Gamma}_{22}$ may only have eigenvalues with a nonpositive real part. This is the second necessary condition we were attempting to identify.

Thus, we obtain

Theorem 82 (Passivity of Linear Systems). *A passive system*

$$\begin{aligned} \dot{\boldsymbol{x}} &= \boldsymbol{A}\boldsymbol{x} + \boldsymbol{b} \cdot u, \\ y &= \boldsymbol{c}^T \boldsymbol{x} \end{aligned}$$

with a positive definite storage function has

(1) a relative degree of $\delta = 1$ *and*
(2) internal dynamics which are Lyapunov stable.

For the internal dynamics, or equivalently the zero dynamics, to be Lyapunov stable, according to Theorem 16 from p. 121, it is sufficient and necessary for a Lyapunov function

$$V(\boldsymbol{z}_{\text{int}}) = \boldsymbol{z}_{\text{int}}{}^T \tilde{\boldsymbol{R}} \boldsymbol{z}_{\text{int}}$$

to exist. This means that a positive definite matrix $\tilde{\boldsymbol{R}}$ exists, so that the matrix $\tilde{\boldsymbol{Q}}$ from the Lyapunov equation

$$\boldsymbol{\Gamma}_{22}^T \tilde{\boldsymbol{R}} + \tilde{\boldsymbol{R}} \boldsymbol{\Gamma}_{22} = -\tilde{\boldsymbol{Q}} \tag{6.81}$$

is negative semidefinite. This condition will be required later.

At this point recall Section 5.2.8, p. 389 et seq., in which we defined zero dynamics as a special case of internal dynamics. Zero dynamics are derived from internal dynamics when we set the input variables of the internal dynamics, which are the external dynamics' state variables, to zero. For the stability analysis using Condition (2) of the previous theorem, we can therefore also use the system's zero dynamics and require that they be Lyapunov stable.

The two necessary conditions of the previous theorem are equivalent to Conditions (2) and (3) of Theorem 33, p. 161. The requirement is that a relative degree $\delta = 1$ be equal to a difference between the denominator degree n and the numerator degree m of the transfer function $G(s)$, i. e. $\delta = n - m = 1$. The requirement that the zero dynamics be Lyapunov stable is identical to the requirement that the real parts of all their eigenvalues be nonpositive with at most one at zero.

If we now use static feedback, i. e. a controller

$$u = -\boldsymbol{r}^T \boldsymbol{x} + V \cdot y_{\text{ref}} \tag{6.82}$$

with the reference variable or new input variable y_{ref}, to make a non-passive system (6.73) passive, we have to consider the following: based on equation (6.80), it is immediately apparent that the relative degree and stability of the internal dynamics cannot be changed by feedback (6.82). A relative degree of $\delta = 1$ and Lyapunov stable internal dynamics are thus necessary conditions for the passivability of the system (6.73).

Furthermore, these conditions are also sufficient, as shown below. We first notice that we can always stabilize the external dynamics in equation (6.80) using the control law (6.82). In addition, we can always passivate the system (6.80) by means of a suitable choice of the control law (6.82) as well. A suitable choice is

$$u = -\frac{1}{\boldsymbol{c}^T \boldsymbol{b}} \left(\Gamma_{11} z_1 + \boldsymbol{\Gamma}_{12}^T \boldsymbol{z}_{\text{int}} + 2 \boldsymbol{\Gamma}_{21}^T \tilde{\boldsymbol{R}} \boldsymbol{z}_{\text{int}} - y_{\text{ref}} \right). \tag{6.83}$$

We will prove this below. Inserting the control law (6.83) into the system representation (6.80) leads to the equations

$$\begin{aligned}
\dot{z}_1 &= -2\boldsymbol{\Gamma}_{21}^T \tilde{\boldsymbol{R}} \boldsymbol{z}_{\text{int}} + y_{\text{ref}}, \quad \} \text{ external dynamics}\\
\dot{\boldsymbol{z}}_{\text{int}} &= \boldsymbol{\Gamma}_{22} \boldsymbol{z}_{\text{int}} + \boldsymbol{\Gamma}_{21} z_1, \qquad \} \text{ internal dynamics}\\
y &= z_1
\end{aligned}$$

of the closed loop.

Its passivity can be verified by inserting the storage function

$$S(z_1, \eta) = \frac{1}{2} z_1^2 + \boldsymbol{z}_{\text{int}}{}^T \tilde{\boldsymbol{R}} \boldsymbol{z}_{\text{int}}$$

into the passivity inequality $\dot{S}(z_1, \boldsymbol{z}_{\text{int}}) \leq y_{\text{ref}}\, y$, i. e.

$$\begin{aligned}
\dot{S}(z_1, \boldsymbol{z}_{\text{int}}) &= z_1 \dot{z}_1 + \dot{\boldsymbol{z}}_{\text{int}}{}^T \tilde{\boldsymbol{R}} \boldsymbol{z}_{\text{int}} + \boldsymbol{z}_{\text{int}}{}^T \tilde{\boldsymbol{R}} \dot{\boldsymbol{z}}_{\text{int}}\\
&= \boldsymbol{z}_{\text{int}}{}^T (\boldsymbol{\Gamma}_{22}^T \tilde{\boldsymbol{R}} + \tilde{\boldsymbol{R}} \boldsymbol{\Gamma}_{22}) \boldsymbol{z}_{\text{int}} + y_{\text{ref}} z_1 \leq y_{\text{ref}}\, y.
\end{aligned}$$

Obviously, this inequality is fulfilled because $y = z_1$ and

$$\boldsymbol{z}_{\text{int}}{}^T (\boldsymbol{\Gamma}_{22}^T \tilde{\boldsymbol{R}} + \tilde{\boldsymbol{R}} \boldsymbol{\Gamma}_{22}) \boldsymbol{z}_{\text{int}} \leq 0$$

apply. The latter holds because the zero dynamics $\dot{\boldsymbol{z}}_{\text{int}} = \boldsymbol{\Gamma}_{22} \boldsymbol{z}_{\text{int}}$ are assumed to be Lyapunov stable. We stated this previously in equation (6.81). With equation (6.83), we can always specify a static control law

$$u = -\tilde{\boldsymbol{r}}^T \boldsymbol{z} + V \cdot y_{\text{ref}} = -\boldsymbol{r}^T \boldsymbol{x} + V \cdot y_{\text{ref}}, \quad \boldsymbol{r}^T = \tilde{\boldsymbol{r}}^T \boldsymbol{T},$$

which passivates the system (6.80) and its original representation (6.73).

Thus we arrive at

Theorem 83 (Passivability of Linear Systems). *A non-passive system*

$$\begin{aligned}
\dot{\boldsymbol{x}} &= \boldsymbol{A}\boldsymbol{x} + \boldsymbol{b} \cdot u,\\
y &= \boldsymbol{c}^T \boldsymbol{x}
\end{aligned}$$

can be converted into a passive control loop with a quadratic positive definite storage function by a static controller

$$u = -\boldsymbol{r}^T \boldsymbol{x} + V \cdot y_{\text{ref}}$$

if and only if the system

(1) has a relative degree of $\delta = 1$ and
(2) its internal dynamics are Lyapunov stable.

We have already established that we cannot change the relative degree or stability of the internal dynamics using static feedback. However, the stability of the external dynamics can certainly be influenced by such feedback. For example, we can stabilize an unstable system. The opposite situation occurs when we use a parallel arrangement with an additional system instead of the feedback. In this case we cannot influence the external dynamics, i. e. the locations of the poles of the system to be passivated. However, we can manipulate its relative degree and internal dynamics, i. e. its numerator degree and the location of the zeros, respectively. This becomes plausible if we take a system

$$G_1(s) = \frac{1}{D(s)}$$

which is non-passive but stable as an example, and combine it with a system

$$G_2(s) = \frac{N(s)}{D(s)}$$

to obtain

$$G_1(s) + G_2(s) = \frac{1}{D(s)} + \frac{N(s)}{D(s)} = \frac{1 + N(s)}{D(s)}$$

by connecting it in parallel. Obviously, using the numerator $N(s)$, we can change the numerator degree and thus the relative degree and the zeros. Consequently, the internal dynamics are changed as well.

In the following section, we will extend the results from this section to MIMO control-affine systems.

6.3.4 Passivation of Non-Passive Control-Affine Systems

Let us now examine control-affine systems

$$\begin{aligned} \dot{\boldsymbol{x}} &= \boldsymbol{a}(\boldsymbol{x}) + \boldsymbol{B}(\boldsymbol{x}) \cdot \boldsymbol{u}, \\ \boldsymbol{y} &= \boldsymbol{c}(\boldsymbol{x}). \end{aligned} \tag{6.84}$$

As in the preceding section, we will first clarify the conditions that must be fulfilled for these systems to be passive before we determine how such a system can be passivated. Similar to Theorem 82, the following theorem applies [58, 65].

Theorem 84 (Passivity of Control-Affine Systems). *A passive system*

$$\begin{aligned} \dot{\boldsymbol{x}} &= \boldsymbol{a}(\boldsymbol{x}) + \boldsymbol{B}(\boldsymbol{x}) \cdot \boldsymbol{u}, \\ \boldsymbol{y} &= \boldsymbol{c}(\boldsymbol{x}) \end{aligned}$$

with a two times continuously differentiable, positive definite storage function has

(1) the vector relative degree $\boldsymbol{\delta} = [1 \ \cdots \ 1]$ *and*
(2) internal dynamics with an equilibrium point which is Lyapunov stable.

In the following, we will address non-passive MIMO systems (6.84). Because we cannot change the relative degree of such a system using a static controller and a passive system (6.84) must have a relative degree

$$\boldsymbol{\delta} = [1 \ \cdots \ 1],$$

from the outset, we can limit our analysis to systems with this relative degree. We intend to make these systems passive using the static controller

$$\boldsymbol{u}(\boldsymbol{x}) = -\boldsymbol{r}(\boldsymbol{x}) + \boldsymbol{V}(\boldsymbol{x}) \cdot \boldsymbol{y}_{\text{ref}} = -\boldsymbol{D}^{-1}(\boldsymbol{x}) \left[\tilde{\boldsymbol{r}}(\boldsymbol{x}) - \tilde{\boldsymbol{V}} \cdot \boldsymbol{y}_{\text{ref}} \right], \tag{6.85}$$

$$\boldsymbol{D}(\boldsymbol{x}) = \begin{bmatrix} L_{\boldsymbol{b}_1} c_1(\boldsymbol{x}) & L_{\boldsymbol{b}_2} c_1(\boldsymbol{x}) & \cdots & L_{\boldsymbol{b}_m} c_1(\boldsymbol{x}) \\ L_{\boldsymbol{b}_1} c_2(\boldsymbol{x}) & L_{\boldsymbol{b}_2} c_2(\boldsymbol{x}) & \cdots & L_{\boldsymbol{b}_m} c_2(\boldsymbol{x}) \\ \vdots & \vdots & \ddots & \vdots \\ L_{\boldsymbol{b}_1} c_m(\boldsymbol{x}) & L_{\boldsymbol{b}_2} c_m(\boldsymbol{x}) & \cdots & L_{\boldsymbol{b}_m} c_m(\boldsymbol{x}) \end{bmatrix},$$

$$\tilde{\boldsymbol{r}}(\boldsymbol{x}) = \begin{bmatrix} L_{\boldsymbol{a}} c_1(\boldsymbol{x}) + a_{1,0}\, c_1(\boldsymbol{x}) \\ L_{\boldsymbol{a}} c_2(\boldsymbol{x}) + a_{2,0}\, c_2(\boldsymbol{x}) \\ \vdots \\ L_{\boldsymbol{a}} c_m(\boldsymbol{x}) + a_{m,0}\, c_m(\boldsymbol{x}) \end{bmatrix}, \quad \tilde{\boldsymbol{V}} = \begin{bmatrix} \tilde{V}_1 & 0 & \cdots & 0 \\ 0 & \tilde{V}_2 & \cdots & 0 \\ \vdots & \vdots & \vdots & \vdots \\ 0 & 0 & \cdots & \tilde{V}_m \end{bmatrix},$$

as previously derived in Section 5.2.11, p. 401 et seq.

In the SISO case, the $m \times m$ decoupling matrix $\boldsymbol{D}(\boldsymbol{x})$ with its elements

$$d_{ij} = L_{\boldsymbol{b}_j} c_i(\boldsymbol{x})$$

is identical to $L_{\boldsymbol{b}} c(\boldsymbol{x})$, i. e. the denominator of the control law (6.85). Also, recall

$$L_{\boldsymbol{b}} c(\boldsymbol{x}) \neq 0$$

must always apply to this denominator term. Accordingly,

$$\det(\boldsymbol{D}(\boldsymbol{x})) \neq 0$$

must hold in the MIMO case.

We obtain from [65]

Theorem 85 (Passivability of Control-Affine Systems). *A non-passive system*

$$\begin{aligned} \dot{\boldsymbol{x}} &= \boldsymbol{a}(\boldsymbol{x}) + \boldsymbol{B}(\boldsymbol{x}) \cdot \boldsymbol{u}, \\ \boldsymbol{y} &= \boldsymbol{c}(\boldsymbol{x}) \end{aligned}$$

with a regular decoupling matrix $\boldsymbol{D}(\boldsymbol{x})$ *can be converted into a passive control loop with a two times continuously differentiable, positive definite storage function using a static controller*

$$\boldsymbol{u}(\boldsymbol{x}) = -\boldsymbol{D}^{-1}(\boldsymbol{x})[\tilde{\boldsymbol{r}}(\boldsymbol{x}) - \tilde{\boldsymbol{V}} \cdot \boldsymbol{y}_{\text{ref}}]$$

if and only if the system

(1) has a vector relative degree $\boldsymbol{\delta} = [1 \ \cdots \ 1]$ *and*
(2) the internal dynamics have an equilibrium point which is Lyapunov stable.

Theorems 84 and 85 are generalizations of Theorems 82 and 83 from the preceding section. Since the systems we have addressed in Theorems 82 and 83 are linear, their Conditions (1) and (2) have global validity. This means that if they are fulfilled in any subset of $I\!R^n$, they are fulfilled in the whole $I\!R^n$. In contrast, it is possible that the Conditions (1) and (2) of Theorems 84 and 85 are only fulfilled in a proper subset of $I\!R^n$. Therefore, it may be that the passivity and passivability of system (6.84) are only local properties.

6.3.5 Passivity-Based Control with IDA

Passivity-based control with an interconnection and damping assignment (IDA) has the goal of providing a controller $\boldsymbol{u}(\boldsymbol{x})$ for a possibly, but not necessarily, passive system

$$\begin{aligned} \dot{\boldsymbol{x}} &= \boldsymbol{a}(\boldsymbol{x}) + \boldsymbol{B}(\boldsymbol{x}) \cdot \boldsymbol{u}, \\ \boldsymbol{y} &= \boldsymbol{c}(\boldsymbol{x}) \end{aligned} \tag{6.86}$$

so that the control loop takes the shape of a PCHD system

$$\begin{aligned} \dot{\boldsymbol{x}} &= (\boldsymbol{J}(\boldsymbol{x}) - \boldsymbol{D}(\boldsymbol{x})) \left(\frac{\partial V(\boldsymbol{x})}{\partial \boldsymbol{x}} \right)^T + \boldsymbol{B}(\boldsymbol{x}) \cdot \boldsymbol{y}_{\text{ref}}, \\ \boldsymbol{y}_{\text{PCHD}} &= \boldsymbol{B}^T(\boldsymbol{x}) \left(\frac{\partial V(\boldsymbol{x})}{\partial \boldsymbol{x}} \right)^T \end{aligned} \tag{6.87}$$

with the reference vector $\boldsymbol{y}_{\text{ref}}$[3] [344, 345, 346]. As we know from Section 2.4.10 on p. 179 et seq., a PCHD system is always passive. Furthermore, it is strictly passive if $\boldsymbol{D}(\boldsymbol{x})$ is positive definite. Our task is to design a controller $\boldsymbol{u}(\boldsymbol{x})$ that will generate a passive control loop (6.87) in interconnection with the plant (6.86) with good damping and an asymptotically stable equilibrium point. However, the output variables $\boldsymbol{y}$ and $\boldsymbol{y}_{\text{PCHD}}$ of the two systems are generally no longer identical.

We assume that the equilibrium point which is of interest to us has been transformed into the origin $\boldsymbol{x} = \boldsymbol{0}$ or is already there. Now, we obtain for the derivative of the positive definite function $V(\boldsymbol{x})$

[3] Note that the matrix $\boldsymbol{D}(\boldsymbol{x})$ in equation (6.87) is the damping matrix of a PCHD system and not the decoupling matrix $\boldsymbol{D}(\boldsymbol{x})$. Both matrices are designated $\boldsymbol{D}$ in the literature. We will retain these designations, as there is no risk of confusion in the following.

$$\dot{V}(\boldsymbol{x})=\frac{\partial V(\boldsymbol{x})}{\partial \boldsymbol{x}}(\boldsymbol{J}(\boldsymbol{x})-\boldsymbol{D}(\boldsymbol{x}))\left(\frac{\partial V(\boldsymbol{x})}{\partial \boldsymbol{x}}\right)^T=-\frac{\partial V(\boldsymbol{x})}{\partial \boldsymbol{x}}\boldsymbol{D}(\boldsymbol{x})\left(\frac{\partial V(\boldsymbol{x})}{\partial \boldsymbol{x}}\right)^T$$

for $\boldsymbol{y}_{\mathrm{ref}} = \boldsymbol{0}$, where $\dot{V}(\boldsymbol{x}) < 0$ holds if $\boldsymbol{D}(\boldsymbol{x})$ is positive definite. In the latter case, the function $V(\boldsymbol{x})$ is a strict Lyapunov function. Thus, the origin is an asymptotically stable equilibrium point. If $\boldsymbol{D}(\boldsymbol{x})$ is positive semidefinite, only $\dot{V}(\boldsymbol{x}) \le 0$ holds. Then Barbashin and Krasovskii's theorem on p. 116 or Theorem 23 on p. 147 can be used to ensure asymptotic stability. Although the following can be performed for the latter, we restrict ourselves, for simplicity, to the case of positive definite matrices $\boldsymbol{D}(\boldsymbol{x})$ since it is the most common.

Now, we will attempt to identify a control law $\boldsymbol{u}(\boldsymbol{x})$ such that the control loop with plant (6.86) as a controlled system is in PCHD form (6.87). For this purpose, as a first step, we split the control variable

$$\boldsymbol{u} = \boldsymbol{u}_1 + \boldsymbol{u}_2.$$

In the second step, we equate equation (6.86) and equation (6.87), obtaining

$$\boldsymbol{B}(\boldsymbol{x}) \cdot \boldsymbol{u}_1 + \boldsymbol{B}(\boldsymbol{x}) \cdot \boldsymbol{u}_2 = (\boldsymbol{J}(\boldsymbol{x}) - \boldsymbol{D}(\boldsymbol{x}))\left(\frac{\partial V}{\partial \boldsymbol{x}}\right)^T - \boldsymbol{a}(\boldsymbol{x}) + \boldsymbol{B}(\boldsymbol{x}) \cdot \boldsymbol{y}_{\mathrm{ref}}.$$

We then compensate for the influence of $\boldsymbol{B}(\boldsymbol{x}) \cdot \boldsymbol{y}_{\mathrm{ref}}$ by setting

$$\boldsymbol{u}_1 = \boldsymbol{y}_{\mathrm{ref}}.$$

This results in the equation

$$\boldsymbol{B}(\boldsymbol{x}) \cdot \boldsymbol{u}_2 = \underbrace{(\boldsymbol{J}(\boldsymbol{x}) - \boldsymbol{D}(\boldsymbol{x}))\left(\frac{\partial V}{\partial \boldsymbol{x}}\right)^T - \boldsymbol{a}(\boldsymbol{x})}_{\boldsymbol{v}}, \tag{6.88}$$

which we will examine in more detail below.

To this end, let us first identify the input variable vector $\boldsymbol{u}_2$ which satisfies this equation. Here, we will note that equation (6.88) is usually an overdetermined linear system of equations

$$\boldsymbol{B}(\boldsymbol{x}) \cdot \boldsymbol{u}_2 = \boldsymbol{v} \tag{6.89}$$

for a constant $\boldsymbol{x}$ with respect to $\boldsymbol{u}_2$. In this system of equations, $\boldsymbol{u}_2$ is an m - dimensional vector, $\boldsymbol{v}$ is an n - dimensional vector, and $\boldsymbol{B}$ is an $n \times m$ matrix. The system of equations (6.89) can only be solved if the vector $\boldsymbol{v}$ is a linear combination of the $\boldsymbol{B}$ columns. We can verify whether the latter is fulfilled or not by constructing $n - m$ linearly independent row vectors $\boldsymbol{b}_i^{\perp}$ which are perpendicular to all m columns $\boldsymbol{b}_j$ of the matrix $\boldsymbol{B}$, i. e.

$$\boldsymbol{b}_i^{\perp}\boldsymbol{b}_j = 0 \qquad \text{for all} \quad i = 1, \ldots, n - m \quad \text{and} \quad j = 1, \ldots, m$$

must hold. Together, the vectors $(\boldsymbol{b}_i^\perp)^T, i = 1,\ldots,n-m$, and the vectors $\boldsymbol{b}_j, j = 1,\ldots,m$, form a basis of $\mathbb{R}^n$. In matrix notation, we obtain

$$\boldsymbol{B}^\perp \boldsymbol{B} = \boldsymbol{0},$$

where

$$\boldsymbol{B}^\perp = \begin{bmatrix} \boldsymbol{b}_1^\perp \\ \vdots \\ \boldsymbol{b}_{n-m}^\perp \end{bmatrix}.$$

The $(n-m) \times n$ matrix $\boldsymbol{B}^\perp$ is called the *annihilator matrix* and its rank is $n-m$.

If

$$\boldsymbol{B}^\perp \boldsymbol{v} = \boldsymbol{0}$$

holds, i. e. if $\boldsymbol{v}$ is perpendicular to all vectors $\boldsymbol{b}_i^\perp$, then $\boldsymbol{v}$ is a linear combination of the columns of the matrix $\boldsymbol{B}$. This means that equation (6.89) is solvable.

The control-loop equation (6.88) can thus be fulfilled if and only if

$$\boldsymbol{B}^\perp(\boldsymbol{x}) \left[(\boldsymbol{J}(\boldsymbol{x}) - \boldsymbol{D}(\boldsymbol{x})) \left(\frac{\partial V}{\partial \boldsymbol{x}} \right)^T - \boldsymbol{a}(\boldsymbol{x}) \right] = \boldsymbol{0} \tag{6.90}$$

holds. In solving the previous equation, we have the choice between two different variables and approaches to aim at a control loop with an asymptotically stable equilibrium point:

(1) We can select a storage function $V(\boldsymbol{x})$ with its minimum at zero. Then the matrices $\boldsymbol{J}(\boldsymbol{x})$ and $\boldsymbol{D}(\boldsymbol{x})$ are to be determined, with the skew symmetry of $\boldsymbol{J}(\boldsymbol{x})$, i. e. $\boldsymbol{J}(\boldsymbol{x}) = -\boldsymbol{J}^T(\boldsymbol{x})$, and the positive definiteness of $\boldsymbol{D}(\boldsymbol{x})$ as properties to be satisfied. In this case, equation (6.90) is an algebraic equation with the variables $\boldsymbol{J}(\boldsymbol{x})$ and $\boldsymbol{D}(\boldsymbol{x})$.
(2) We can choose a skew-symmetric matrix $\boldsymbol{J}(\boldsymbol{x})$ and a positive definite matrix $\boldsymbol{D}(\boldsymbol{x})$. Then we have to determine the storage function $V(\boldsymbol{x})$ so that equation (6.90) is fulfilled. The constraints $V(\boldsymbol{0}) = 0$ and $V(\boldsymbol{x}) > 0$ must be complied with. In this case, equation (6.90) is a linear partial differential equation with respect to $V(\boldsymbol{x})$.

If we succeed in fulfilling the key equation (6.90), the system of equations (6.88) has a solution. Now we can determine the m-dimensional vector $\boldsymbol{u}_2$ by selecting m equations with linearly independent row vectors $\boldsymbol{b}_i^T$ of the $n \times m$ matrix $\boldsymbol{B}(\boldsymbol{x})$ from

$$\boldsymbol{B}(\boldsymbol{x}) \cdot \boldsymbol{u}_2 = \boldsymbol{v} \tag{6.91}$$

in the following manner. We will denote the indices of these linearly independent m row vectors $\boldsymbol{b}_i^T$ as $i_1, i_2, \ldots, i_m$, where $i_1, i_2, \ldots, i_m \in \{1, 2, \ldots, n\}$ applies. To select the corresponding equations

$$\boldsymbol{b}_{i_1}^T \boldsymbol{u}_2 = v_{i_1},$$
$$\vdots$$
$$\boldsymbol{b}_{i_m}^T \boldsymbol{u}_2 = v_{i_m}$$

from the system of equations (6.91), we now construct the $m \times n$ matrix

$$\boldsymbol{K} = \begin{bmatrix} \boldsymbol{e}_{i_1}^T \\ \vdots \\ \boldsymbol{e}_{i_m}^T \end{bmatrix},$$

where the n - dimensional vector $\boldsymbol{e}_{i_k}$ is the unit vector of the i_k-th coordinate direction of $I\!R^n$. This vector has zeros everywhere except in its i_k-th element, which is equal to one.

Now we multiply equation (6.91) by the matrix $\boldsymbol{K}$ from the left, selecting the m linearly independent equations

$$\boldsymbol{K}\boldsymbol{B}(\boldsymbol{x}) \cdot \boldsymbol{u}_2 = \boldsymbol{K}\boldsymbol{v}.$$

This yields

$$\boldsymbol{u}_2 = (\boldsymbol{K}\boldsymbol{B}(\boldsymbol{x}))^{-1}\boldsymbol{K}\boldsymbol{v} \tag{6.92}$$

and the control law we have been attempting to identify:

$$\boldsymbol{u} = \boldsymbol{u}_2 + \boldsymbol{u}_1 = (\boldsymbol{K}\boldsymbol{B}(\boldsymbol{x}))^{-1}\boldsymbol{K}\left[(\boldsymbol{J}(\boldsymbol{x}) - \boldsymbol{D}(\boldsymbol{x}))\left(\frac{\partial V}{\partial \boldsymbol{x}}\right)^T - \boldsymbol{a}(\boldsymbol{x})\right] + \boldsymbol{y}_{\text{ref}}. \tag{6.93}$$

Alternatively, we can also derive the control law (6.93) from the pseudoinverse of the matrix $\boldsymbol{B}(\boldsymbol{x})$. The pseudoinverse [40]

$$\boldsymbol{M}^+ = \left(\boldsymbol{M}^T\boldsymbol{M}\right)^{-1}\boldsymbol{M}^T$$

of an $n \times m$ matrix $\boldsymbol{M}$ is used to approximately solve an overdetermined system of the equations

$$\boldsymbol{M}\boldsymbol{z} = \boldsymbol{h}, \quad \boldsymbol{M} \in I\!R^{n\times m} \quad \text{and} \quad \boldsymbol{z} \in I\!R^m, \quad \boldsymbol{h} \in I\!R^n. \tag{6.94}$$

Its approximate solution is given by

$$\boldsymbol{z}_{\text{opt}} = \boldsymbol{M}^+\boldsymbol{h}$$

and yields

$$\boldsymbol{M}\boldsymbol{z}_{\text{opt}} \approx \boldsymbol{h}.$$

This approximation attains the smallest value of the square error

$$|\boldsymbol{M}\boldsymbol{z}-\boldsymbol{h}|^2 = (\boldsymbol{M}\boldsymbol{z}-\boldsymbol{h})^T\,(\boldsymbol{M}\boldsymbol{z}-\boldsymbol{h})\,. \tag{6.95}$$

If, however, the condition

$$\boldsymbol{M}^{\perp}\boldsymbol{h} = \boldsymbol{0}$$

is also fulfilled, the vector $\boldsymbol{h}$ is a linear combination of the $\boldsymbol{M}$ columns, and the system of equations (6.94) can be solved exactly. Hence, the error (6.95) is identical to zero for $\boldsymbol{z} = \boldsymbol{z}_{\text{opt}}$ and the solution is no longer approximate; it is exact, yielding

$$\boldsymbol{M}\boldsymbol{z}_{\text{opt}} = \boldsymbol{h} \quad \text{with} \quad \boldsymbol{z}_{\text{opt}} = \boldsymbol{M}^{+}\boldsymbol{h}.$$

With the approach using the pseudoinverse of $\boldsymbol{B}(\boldsymbol{x})$, we can calculate

$$\boldsymbol{u}_2 = \left(\boldsymbol{B}^T(\boldsymbol{x})\boldsymbol{B}(\boldsymbol{x})\right)^{-1}\boldsymbol{B}^T(\boldsymbol{x})\boldsymbol{v}$$

as an alternative to equation (6.92). This means we can represent the control law (6.93) as

$$\boldsymbol{u} = \left(\boldsymbol{B}^T(\boldsymbol{x})\boldsymbol{B}(\boldsymbol{x})\right)^{-1}\boldsymbol{B}^T(\boldsymbol{x})\left[(\boldsymbol{J}(\boldsymbol{x})-\boldsymbol{D}(\boldsymbol{x}))\left(\frac{\partial V}{\partial \boldsymbol{x}}\right)^T - \boldsymbol{a}(\boldsymbol{x})\right] + \boldsymbol{y}_{\text{ref}} \tag{6.96}$$

if condition (6.90) is fulfilled.

Now we will address the case in which

$$\boldsymbol{y} = \boldsymbol{y}_{\text{PCHD}} = \boldsymbol{B}^T(\boldsymbol{x})\left(\frac{\partial V}{\partial \boldsymbol{x}}\right)^T$$

is required. As mentioned earlier, this identity is generally not achievable. This is because the control law (6.96) is static in nature, and from Theorem 85 in the previous section, we know that we can make systems (6.86) passive using a static control law if and only if the system has the vector relative degree

$$\boldsymbol{\delta} = [1 \;\; \cdots \;\; 1]$$

and the internal dynamics have a Lyapunov stable equilibrium point. This will not always be the case.

However, since we are using a state control (6.96) to control the entire state vector and are mainly interested in the stability of the control loop, we can also classify the passivity with respect to an arbitrary output variable vector. Then it is irrelevant whether $\boldsymbol{y}$ is regarded as the output variable vector or $\boldsymbol{y}_{\text{PCHD}}$. Thus the above methodology provides a useful tool for controlling both control-affine and linear systems.

6.3.6 Example: Paper Machine

As an example, we will describe a paper machine whose basic structure is shown in Figure 6.29. From the stock preparation system (not shown here), a water-pulp suspension containing about 1% pulp is fed into a mixing tank via a valve which can be used to control the inflow q_f. The water stream q_b, generated during the manufacturing process and still containing pulp particles, is also fed into this mixing tank via a pump. The suspension mixed in this way is then fed into the headbox consisting of a long narrow nozzle gap through which the water-pulp suspension is sprayed onto a rotating mesh wire which functions as a sieve. This part of the machine is called the forming section. At this point, the water precipitates due to gravity. If necessary, it can also be suctioned off using a vacuum suction box. After dewatering, the precipitated water is collected in a container below the mesh; from there it is pumped back into the mixing tank. The pulp web formed on the mesh is then fed into a press section. At this point, the pulp content is approximately 20%. Next the pulp web is squeezed between two circulating press felts and thus further dewatered. As no further water can be removed from the pulp web by additional squeezing at the end of the press section, the pulp web is now dewatered by drying. For this purpose, the pulp web is guided meanderingly into the dryer section over steam-heated cylinders. After the dryer section, the paper web is finished and wound on a reel spool.

In a paper machine, there is a multitude of controllers. In the following, we will examine both the control of the water level

$$h_\mathrm{h} = h_{\mathrm{h,op}} + \Delta h_\mathrm{h}$$

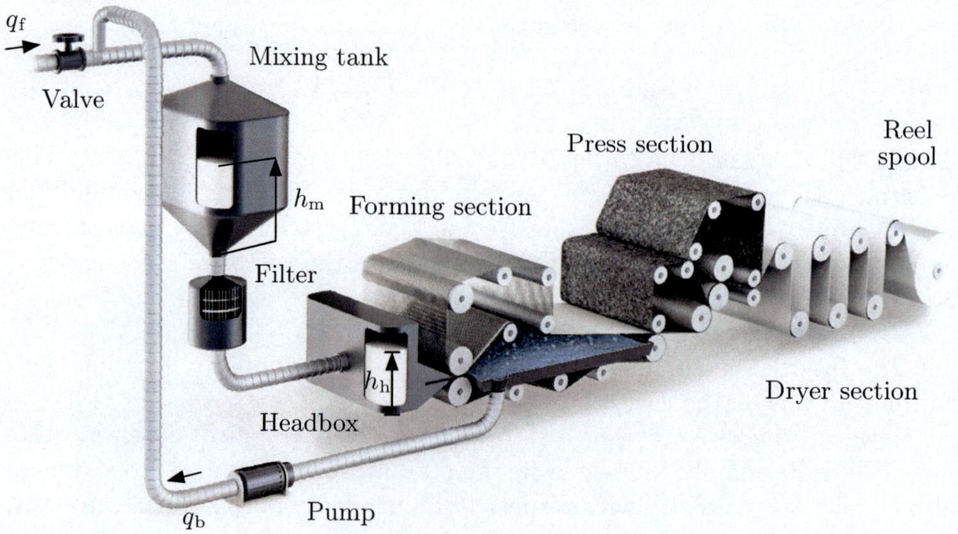

Fig. 6.29: Paper machine

and the pulp concentration

$$c_{\mathrm{h}} = c_{\mathrm{h,op}} + \Delta c_{\mathrm{h}}$$

of the suspension in the headbox. Here, Δh_{h} and Δc_{h} are the deviations from the operating point, whose elements are marked by the index op. These two variables, combined in the output variable vector

$$\boldsymbol{y}^T = [\Delta h_{\mathrm{h}} \;\; \Delta c_{\mathrm{h}}],$$

also make up two of the state variables in the vector

$$\boldsymbol{x}^T = [\Delta h_{\mathrm{m}} \;\; \Delta h_{\mathrm{h}} \;\; \Delta c_{\mathrm{m}} \;\; \Delta c_{\mathrm{h}}],$$

where

$$h_{\mathrm{m}} = h_{\mathrm{m,op}} + \Delta h_m$$

is the filling level and

$$c_{\mathrm{m}} = c_{\mathrm{m,op}} + \Delta c_{\mathrm{m}}$$

is the pulp concentration in the mixing tank. The variables Δh_{m} and Δc_{m} are once again the deviations from the operating point. The deviations Δq_{f} and Δq_{b} from the operating points of the two adjustable inflows

$$q_{\mathrm{f}} = q_{\mathrm{f,op}} + \Delta q_{\mathrm{f}} \quad \text{and} \quad q_{\mathrm{b}} = q_{\mathrm{b,op}} + \Delta q_{\mathrm{b}}$$

of the stock preparation system and the backflow serve as control variables. This means that the control variable vector is given by

$$\boldsymbol{u}^T = [\Delta q_{\mathrm{f}} \;\; \Delta q_{\mathrm{b}}]\,.$$

The concentrations of the pulp in the backflow and in the suspension coming from the stock preparation system, $c_{\mathrm{b}} + \Delta c_{\mathrm{b}}$ and $c_{\mathrm{f}} + \Delta c_{\mathrm{f}}$, fluctuate and act as disturbance variables. Here, the fluctuations are Δc_{b} and Δc_{f}.

The model [484] of a paper machine producing capacitor paper is given by

$$\dot{\boldsymbol{x}} = \underbrace{\underbrace{\begin{bmatrix} -1.93 & 0 & 0 & 0 \\ 0.394 & -0.426 & 0 & 0 \\ 0 & 0 & -0.63 & 0 \\ 0.82 & -0.784 & 0.413 & -0.426 \end{bmatrix}}_{\boldsymbol{A}} \boldsymbol{x}}_{\boldsymbol{a}(\boldsymbol{x})}$$

$$+ \underbrace{\begin{bmatrix} 1.274 & 1.274 \\ 0 & 0 \\ 1.34 - 0.327x_3 & -0.65 - 0.327x_3 \\ 0 & 0 \end{bmatrix}}_{\boldsymbol{B}(\boldsymbol{x})} \boldsymbol{u} + \underbrace{\begin{bmatrix} 0 \\ 0 \\ 0.203\Delta c_{\mathrm{f}} + 0.406\Delta c_{\mathrm{b}} \\ 0 \end{bmatrix}}_{\boldsymbol{d}(\Delta c_{\mathrm{f}}, \Delta c_{\mathrm{b}})},$$

an equation which is bilinear in the states x_i and the input variables u_i. The vector $\boldsymbol{d}(\Delta c_{\mathrm{f}}, \Delta c_{\mathrm{b}})$ represents the disturbances. Our goal will be to design a passivity-based control with IDA for the aforementioned system to compensate for the disturbances Δc_{f} and Δc_{b}, so that the control loop takes the form

$$\dot{\boldsymbol{x}} = (\boldsymbol{J}(\boldsymbol{x}) - \boldsymbol{D}(\boldsymbol{x})) \left(\frac{\partial V}{\partial \boldsymbol{x}}\right)^T + \boldsymbol{d}(\Delta c_{\mathrm{f}}, \Delta c_{\mathrm{b}})$$

with the reference vector $\boldsymbol{y}_{\mathrm{ref}} = \boldsymbol{0}$. As we know from the previous section, this is possible if and only if equation (6.90) in the form

$$\boldsymbol{B}^{\perp}(\boldsymbol{x}) \left[(\boldsymbol{J}(\boldsymbol{x}) - \boldsymbol{D}(\boldsymbol{x})) \left(\frac{\partial V}{\partial \boldsymbol{x}}\right)^T \right] = \boldsymbol{B}^{\perp}(\boldsymbol{x}) \boldsymbol{a}(\boldsymbol{x}) \tag{6.97}$$

is fulfilled. For the system in question, we will select

$$V(\boldsymbol{x}) = \frac{1}{2} \boldsymbol{x}^T \boldsymbol{x}.$$

As we can immediately see, the row vectors of the matrix

$$\boldsymbol{B}^{\perp} = \begin{bmatrix} 0 & 1 & 0 & 0 \\ 0 & 0 & 0 & 1 \end{bmatrix}$$

are perpendicular to all column vectors of the matrix $\boldsymbol{B}(\boldsymbol{x})$. We use these results in equation (6.97) and obtain

$$\begin{bmatrix} 0 & 1 & 0 & 0 \\ 0 & 0 & 0 & 1 \end{bmatrix} (\boldsymbol{J}(\boldsymbol{x}) - \boldsymbol{D}(\boldsymbol{x})) \boldsymbol{x} = \begin{bmatrix} 0.3940 & -0.4260 & 0 & 0 \\ 0.8200 & -0.7840 & 0.4130 & -0.4260 \end{bmatrix} \boldsymbol{x}. \tag{6.98}$$

The matrix

$$\boldsymbol{J}(\boldsymbol{x}) - \boldsymbol{D}(\boldsymbol{x}) = \begin{bmatrix} * & * & * & * \\ 0.3940 & -0.4260 & 0 & 0 \\ * & * & * & * \\ 0.8200 & -0.7840 & 0.4130 & -0.4260 \end{bmatrix}$$

has a number of arbitrarily selectable entries to fulfill the key equation (6.98), which are marked here with asterisks. We will now choose

$$\boldsymbol{J} - \boldsymbol{D} = \begin{bmatrix} -1.8322 & 0.0633 & -1.7247 & -0.8113 \\ 0.3940 & -0.4260 & 0 & 0 \\ -0.0039 & 0.7013 & -0.8526 & -0.1169 \\ 0.8200 & -0.7840 & 0.4130 & -0.4260 \end{bmatrix} = \boldsymbol{M},$$

and since each matrix $\boldsymbol{M}$ is composed of a skew-symmetric and a symmetric component according to

$$\boldsymbol{M} = \underbrace{\frac{1}{2}(\boldsymbol{M} - \boldsymbol{M}^T)}_{\text{skew-symmetric}} + \underbrace{\frac{1}{2}(\boldsymbol{M} + \boldsymbol{M}^T)}_{\text{symmetric}},$$

we obtain the skew-symmetric matrix

$$\boldsymbol{J} = \frac{1}{2}(\boldsymbol{M} - \boldsymbol{M}^T) = \begin{bmatrix} 0 & -0.1654 & -0.8604 & -0.8156 \\ 0.1654 & 0 & -0.3507 & 0.3920 \\ 0.8604 & 0.3507 & 0 & -0.2650 \\ 0.8156 & -0.3920 & 0.2650 & 0 \end{bmatrix}.$$

The symmetric component $\boldsymbol{D}$ takes the form

$$\boldsymbol{D} = -\frac{1}{2}(\boldsymbol{M} + \boldsymbol{M}^T) = \begin{bmatrix} 1.8322 & -0.2287 & 0.8643 & -0.0043 \\ -0.2287 & 0.4260 & -0.3507 & 0.3920 \\ 0.8643 & -0.3507 & 0.8526 & -0.1480 \\ -0.0043 & 0.3920 & -0.1480 & 0.4260 \end{bmatrix},$$

which is positive definite. We can now use equation (6.96) to define the control law

$$\boldsymbol{u} = (\boldsymbol{B}^T(\boldsymbol{x})\boldsymbol{B}(\boldsymbol{x}))^{-1}\boldsymbol{B}^T(\boldsymbol{x})(\boldsymbol{J} - \boldsymbol{D} - \boldsymbol{A})\boldsymbol{x} = \begin{bmatrix} u_1 \\ u_2 \end{bmatrix}$$

with

$$\begin{aligned} u_1 &= 0.0231x_1 + 0.3686x_2 - 0.5540x_3 - 0.2667x_4 + 0.0126x_1x_3 \\ &\quad + 0.0082x_2x_3 - 0.1046x_3x_4 - 0.2225x_3^2, \\ u_2 &= 0.0537x_1 - 0.3190x_2 - 0.7997x_3 - 0.3701x_4 - 0.0126x_1x_3 \\ &\quad - 0.0082x_2x_3 + 0.1046x_3x_4 + 0.2225x_3^2. \end{aligned}$$

Using this passivity-based control, we have obtained a linear control loop with the eigenvalues

$$\lambda_1 = -0.7918, \quad \lambda_2 = -0.8355, \quad \lambda_{3/4} = -0.9548 \pm j0.1803,$$

whose settling time is approximately half as long as that of the plant with its eigenvalues

$$\lambda_{1/2} = -0.426, \quad \lambda_3 = -0.63, \quad \lambda_4 = -1.93.$$

6.4 Fuzzy Control

6.4.1 Introduction

Simple nonlinear controllers, such as two- and three-position controllers, are often designed based on human experience. A good example of this is a simple temperature control as in a heater, as shown in Figure 6.30. Two simple rules describe the controller behavior of this control loop:

(1) *If the control error e is negative, then turn off the heater.*
(2) *If the control error e is positive, then turn on the heater.*

If the control error e is described by a Boolean variable ϑ with

$$\vartheta = 0 \quad \text{for} \quad e \leq 0$$

and

$$\vartheta = 1 \quad \text{for} \quad e > 0$$

and the status of the heater is described by $z = 0$ (heater off) and $z = 1$ (heater on), then the above rules can also be described by the simple Boolean logic expression

$$z = \vartheta.$$

The above control has some advantageous features: it is simple and inexpensive to implement. No process model is required for its design. The design is based on rules that are easy to formulate and result from human experience.

The advantages of this kind of design are so great that the question arises whether we can create a general design methodology for controllers with the same advantages. A design methodology such as this would obviously be very useful, because it could be used to design controllers that would be simple to implement based on human empirical knowledge, i. e. rules which can be verbally formulated. Ultimately, this means that we can mathematically imitate human behavior or causal knowledge using a computer to simulate the behavior of a process operator or for other purposes.

Fuzzy logic, developed by L. A. Zadeh in 1965 [490] (see also [70, 75, 102, 207, 249, 321, 490]), is a generalization of Boolean logic and makes it possible to apply the above procedure. The human behavior to be simulated by fuzzy logic must be describable by "if-then rules", such as the following:

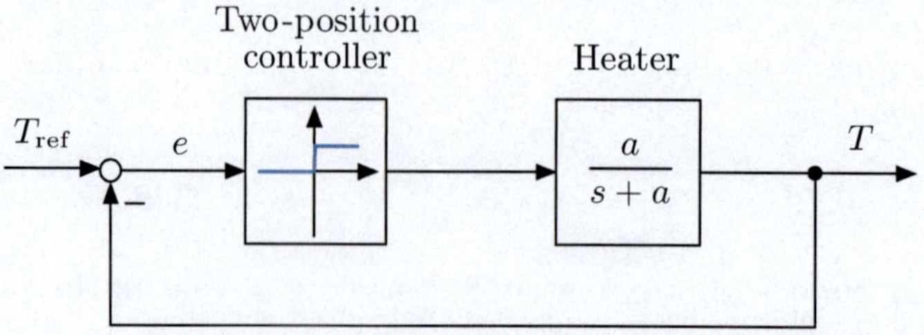

Fig. 6.30: Control with a two-position controller

If the boiler temperature T is very high or the boiler pressure P is very high, then set the valve opening V of the gas supply close to zero.

or

If the boiler temperature T is high and the boiler pressure P is low, then set the valve opening V of the gas supply to approximately half-open.

This raises the question of how such verbally formulated rules can be mathematized using fuzzy logic. We can answer this question by proceeding with the following five steps:

Step 1: We mathematize terms such as *very high* or *low* or *close to zero*.

Step 2: Logic operations such as AND or OR are defined by mathematical operators.

Step 3: The "if-then conclusion" is mathematized.

Step 4: The results of all conclusions, meaning of all rules, are combined.

Step 5: The result of the logical operation must be converted into a technically applicable numerical value.

Step 1 is referred to as *fuzzification*, Steps 2, 3, and 4 as *inference*, and Step 5 is *defuzzification*. They are discussed in detail in the next three sections.

6.4.2 Fuzzification

Let us review the following rule: if the boiler temperature T is *high* and the boiler pressure P is *low*, then set the valve opening of the gas supply to *approximately half-open*. In a rule of this type, variable quantities such as the *boiler temperature*, the *boiler pressure*, and the *valve opening* are called *linguistic variables*. Terms and expressions such as *high*, *low*, and *approximately half-open* are referred to as *linguistic values*. Unlike numerical values, linguistic values do not describe a single value, but a whole range of values.

In classical logic and classical set theory, a value range of this type can be stated as

$$M_{high} = \{T \in I\!R \mid 300\,°\mathrm{C} \leq T \leq 500\,°\mathrm{C}\},$$

for example. A temperature value T is either part of this quantity or not. Figure 6.31 shows the set M_{high}, where a membership function μ is used as a measure of the set membership. In this example, the membership function μ takes the form

$$\mu(T) = \begin{cases} 0 & \text{for } T \notin M_{high}, \\ 1 & \text{for } T \in M_{high}. \end{cases}$$

However, the above sharp separation into value ranges to which a temperature either belongs or does not belong has little in common with the human

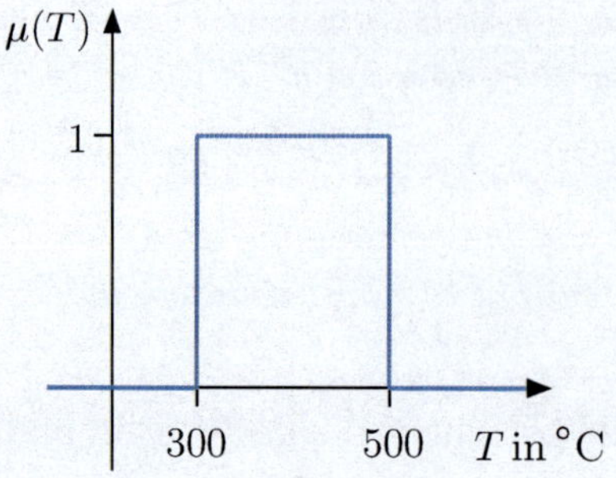

Fig. 6.31: Example of a discontinuous membership function of the set M_{high}

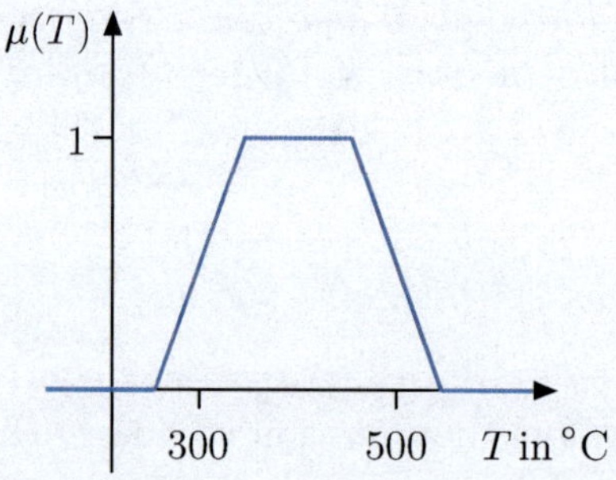

Fig. 6.32: Example of a continuous membership function of the set M_{high}

concept of the term *high*. For example, to a human being it would seem nonsensical that a temperature $T = 299.99\,°\text{C}$ would be assigned the membership value $\mu(299.99\,°\text{C}) = 0$, whereas $T = 300.01\,°\text{C}$ is assigned the membership value $\mu(300.01\,°\text{C}) = 1$.

In many cases, humans see membership in a set as a gradual shift and not as an abrupt transition. Figure 6.32 shows an example of a continuous membership function. A continuous membership function μ makes it possible to have membership values $\mu(T)$ with

$$0 \leq \mu(T) \leq 1,$$

which correspond to a linguistic value such as *high*. For example, $T = 300\,°\text{C}$ is classified as *high* with a value of only $\mu(300\,°\text{C}) = 0.45$.

The assignment to a set is no longer false or true, meaning $\mu = 0$ or $\mu = 1$, respectively; rather, the membership function μ has an infinite number of values between zero and one. An assignment such as this is called *fuzzy*. In this case, the term *fuzzy set* is used, which is defined as follows.

Definition 38 (Fuzzy Set). *A set*

$$F = \{(x, \mu(x)) \mid x \in G \quad \textit{and} \quad \mu(x) \in [0,1]\}$$

of pairs $(x, \mu(x))$ is called a fuzzy set in the universe G. The mapping

$$\mu \;:\; G \to [0,1]$$

is called the membership function of F. It assigns the degree of membership $\mu(x)$ to each element x of the universe G.

Linking a variable x with a degree of membership $\mu(x)$ to a linguistic value is called the *fuzzification* of x. The universe G is often identical to the set of real numbers $\mathbb{R}$.

The range of values for a linguistic variable which are of interest, such as the temperature T in the example above, is generally divided into several linguistic values, as shown in Table 6.1. A fuzzy set and a membership function

Table 6.1: Examples of linguistic values

Linguistic variable	Linguistic values			
temperature T	*zero*	*low*	*high*	*very high*

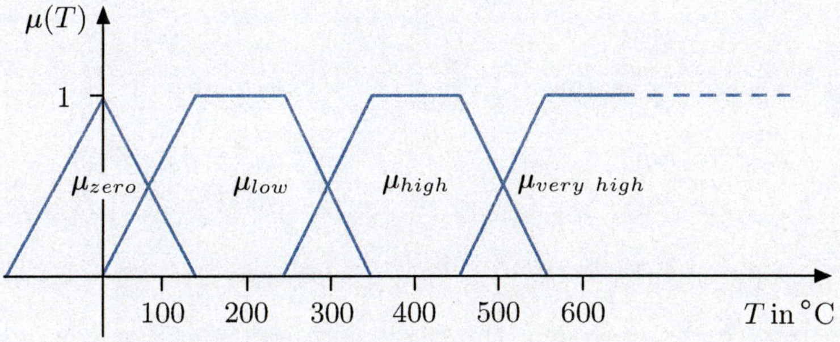

Fig. 6.33: Examples of membership functions

belong to each of the individual linguistic values. Figure 6.33 illustrates a possible selection made by a human and based on his or her intuitive assessment of the circumstances. Triangular, trapezoidal, and ramp-shaped functions are often used as membership functions. Furthermore, other functions such as the Gaussian function and the cosine-square function are also used.

If-then rules are often represented in the standardized form:

$$\textit{If } x_1 = LV_{1,j} \textit{ and } \dots \textit{ and } x_n = LV_{n,l}, \textit{ then } y = LV_p.$$

Here, $x_1, x_2, \dots, x_n$ are the n input variables of the rule, $LV_{i,1}, \dots, LV_{i,q}$ are the q linguistic values of the input variables x_i, the linguistic variable y is the output variable of the rule, and $LV_1, \dots, LV_r$ are the r linguistic values of the output variable y. The linguistic values $LV_{i,j}$ are numerically defined by the associated membership functions $\mu_{i,j}$. The expression $x_i = LV_{i,k}$ is therefore synonymous with the specification of a membership value of x_i to $LV_{i,k}$, i. e. $\mu_{i,k}(x_i)$.

6.4.3 Inference

In the next step, the AND or OR operations in the rules are to be implemented using mathematical operators, i. e. fuzzy operators. First, we will again consider the classical logic approach, which only includes the membership values zero and one. The AND and OR operations are known from Boolean logic and are shown in Tables 6.2 and 6.3. Obviously, we can implement the Boolean AND operation using the minimum operator

$$\mu_y = \min\{\mu_a, \mu_b\}$$

and the Boolean OR operation using the maximum operator

Table 6.2: Boolean AND

μ_a	μ_b	μ_y
0	0	0
0	1	0
1	0	0
1	1	1

$\mu_y = \mu_a \wedge \mu_b$

Table 6.3: Boolean OR

μ_a	μ_b	μ_y
0	0	0
0	1	1
1	0	1
1	1	1

$\mu_y = \mu_a \vee \mu_b$

$$\mu_y = \max\{\mu_a, \mu_b\}.$$

In the fuzzy logic approach, these logical operations are now adopted in such a manner that all intermediate values are allowed for membership, in addition to the values 0 and 1. Accordingly, the fuzzy AND and the fuzzy OR are defined as follows:

Definition 39 (Fuzzy AND and Fuzzy OR). *The fuzzy AND operation of two membership values $\mu_a, \mu_b \in [0,1]$ is defined by*

$$\mu_{a \wedge b} = \min\{\mu_a, \mu_b\}$$

while the fuzzy OR operation is defined by

$$\mu_{a \vee b} = \max\{\mu_a, \mu_b\}.$$

With the current results, the "if" part of a rule, called the *condition*, *antecedent*, or the *premise*, can be converted into a mathematical expression. For example, consider the "if" part of the rule

> *If the boiler temperature T is high and the boiler pressure P is low, then set the valve opening V of the gas supply to approximately half-open.*

From this "if" part, we obtain

$$\mu_{high \wedge low}(T, P) = \min\{\mu_{high}(T), \mu_{low}(P)\}.$$

This is illustrated in the following example. Inserting the temperature $T = 400\,°\mathrm{C}$ and the pressure $P = 1\,\mathrm{bar}$ into the rule above, the values

$$\mu_{high}(T = 400\,°\mathrm{C}) = 1$$

and

$$\mu_{low}(P = 1\,\mathrm{bar}) = 0.6$$

are obtained from the membership functions in Figure 6.34. For this single

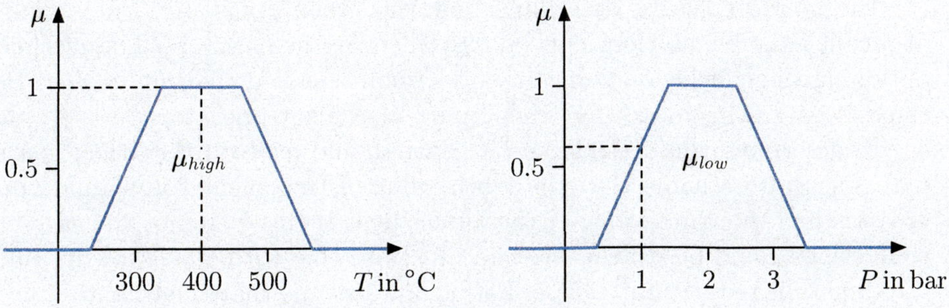

Fig. 6.34: Membership functions of temperature *high* and pressure *low*

point (400 °C, 1 bar), we obtain

$$\mu_{high \wedge low}(T = 400\,^\circ\text{C}, P = 1\,\text{bar}) = \min\{1, 0.6\} = 0.6$$

as the result of the fuzzy AND operation. In the general case, the mathematical evaluation of the "if" part of a rule, called the *antecedent aggregation* or the *aggregation* for short, is given by

$$\mu_{\text{agg}}(x_1, \ldots, x_n) = \min\{\mu_{LV1,k}(x_1) \ldots, \mu_{LVn,l}(x_n)\}$$

if it has only fuzzy AND operations.

Note that there are a number of operators, besides the minimum and the maximum operators, which can be used to implement the fuzzy AND operation and the fuzzy OR operation, respectively [75, 102, 207]. In industrial practice, the minimum and maximum operators are most frequently used. Other common fuzzy operators are the bounded product

$$\max\{0, \mu_a(x) + \mu_b(x) - 1\}$$

and the algebraic product

$$\mu_a(x) \cdot \mu_b(x)$$

for the fuzzy AND, and the bounded sum

$$\min\{1, \mu_a(x) + \mu_b(x)\}$$

and the algebraic sum

$$\mu_a(x) + \mu_b(x) - \mu_a(x)\mu_b(x)$$

for the fuzzy OR [304]. In contrast to the minimum and maximum operators, the algebraic product and the algebraic sum have the advantage of being differentiable. The differentiability of the operators is important if the parameters of the membership function are to be optimized using a gradient method.

The next step is to determine how the "then" part of a rule, which is also called the *conclusion*, can be described mathematically. The method of obtaining the conclusion from the antecedent is called the *implication*. Here, the "if" part is linked to the "then" part by a fuzzy operator.

To determine what effect the "if" part should have on the "then" part, it makes sense to assume that the truth value of the conclusion should not be greater than the truth value of the antecedent. If this were not the case, false conclusions could be drawn. We can illustrate this with the following rule: if the tomato is red, then it is ripe. Let us now assume that the tomato is green, i. e. the antecedent has the truth value $\mu_{\mathrm{agg}} = 0$. It would clearly be wrong to deduce that the tomato is nevertheless ripe, i. e. that the conclusion that the tomato is ripe has a truth value of $\mu = 1$.

A simple way to implement the above requirement is as follows: we can limit the "then" part, i. e. its membership function

$$\mu_{LV_p}(y),$$

to the result of the "if" part

$$\mu_{\mathrm{agg}}(x_1, \ldots, x_n).$$

This is easily possible using the minimum operator. The fuzzy implication is thus implemented using the minimum operator. For a rule k, this results in

$$\mu_k(x_1, \ldots, x_n, y) = \min\{\mu_{\mathrm{agg},k}(x_1, \ldots, x_n), \mu_{LV_p}(y)\}.$$

Note that, instead of the minimum operator, other operators can also be used for the fuzzy implication. For example, the algebraic product

$$\mu_{\mathrm{agg},k}(x_1, \ldots, x_n) \cdot \mu_{LV_p}(y)$$

is commonly used.

For an example, let us return to the following rule:

If the boiler temperature T is high and the boiler pressure P is low, then set the valve opening V of the gas supply to approximately half-open.

Converted into fuzzy logic using the above results, we obtain

$$\mu_k(T, P, V) = \min\{\mu_{\mathrm{agg}}(T, P), \mu_{half}(V)\}$$

with

$$\mu_{\mathrm{agg}}(T, P) = \min\{\mu_{high}(T), \mu_{low}(P)\}.$$

This is shown in graphical form in Figure 6.35.

For all possible value combinations (T, P, V), there would be a characteristic four-dimensional diagram which cannot be graphically depicted. Assuming two constant values T, P, e. g. $T = 400\,^\circ\mathrm{C}$ and $P = 1\,\mathrm{bar}$, the graph can be

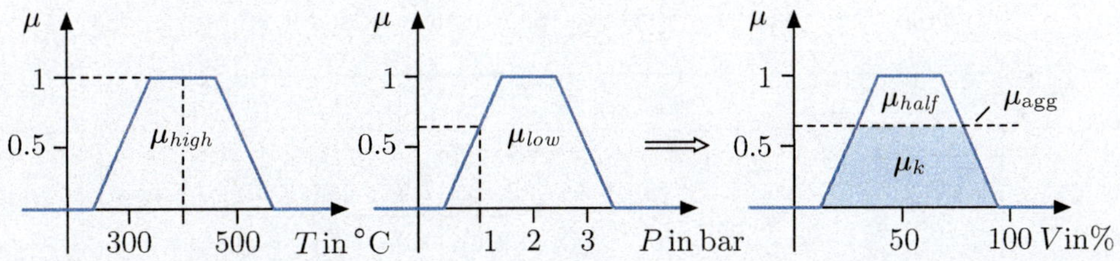

Fig. 6.35: Evaluation of a rule with the result μ_k (membership function highlighted in blue)

reduced to two dimensions, as shown in Figure 6.35. The example illustrates the limitation of μ_k to the result of the antecedent evaluation. The result μ_k is the membership function μ_{half} truncated to μ_{agg}.

In general, the description of a person's behavior will contain several rules. The next step is therefore to determine how the results of several rules are to be linked. This combination is called *accumulation*[4].

The entirety of all rules, the rule base, can be arranged either in a table or in a matrix. Table 6.4 and Table 6.5 show examples. The tabular form is particularly suitable for rule bases with many input variables. The arrangement as a matrix is similar to the Karnaugh map from Boolean logic. It is easier to keep track with the matrix than with the tabular form. However, the matrix form is only suitable for rule bases with few input variables x_i. An advantage of the matrix form is the fact that the rule base is automatically free of contradictions. This is because each field within the matrix represents exactly one antecedent. No antecedent can occur several times and therefore, no contradictory rules can occur due to different conclusions. Another advantage is that if no conclusion exists for an antecedent, this is immediately apparent.

Table 6.4: Rule base in tabular form

Rule 1: If $x_1 = LV_{1,i}$ and ... and $x_n = LV_{n,j}$, then $y = LV_s$.
Rule 2: If $x_1 = LV_{1,k}$ and ... and $x_n = LV_{n,l}$, then $y = LV_u$.
⋮
Rule m: If $x_1 = LV_{1,q}$ and ... and $x_n = LV_{n,r}$, then $y = LV_w$.

[4] In the literature, the term *rule aggregation* or briefly *aggregation* is used sometimes instead of the term *accumulation*. Here we adhere to the International Standard IEC 61131-7, which standardizes the terms of fuzzy systems, and use the term *accumulation*. This means there is no confusion between antecedent aggregation and rule aggregation.

Table 6.5: Rule base in matrix form. No rules exist for empty fields.

x_1	x_3	x_2: $LV_{2,1}$	x_2: $LV_{2,2}$	x_2: $LV_{2,3}$	x_2: $LV_{2,4}$
$x_1 = LV_{1,1}$	$LV_{3,1}$	LV_1	LV_3	LV_1	LV_1
	$LV_{3,2}$	LV_1	LV_2		
	$LV_{3,3}$	LV_2	LV_3		
$x_1 = LV_{1,2}$	$LV_{3,1}$	LV_1	LV_3	LV_1	LV_1
	$LV_{3,2}$	LV_1	LV_3	LV_1	LV_1
	$LV_{3,3}$	LV_3	LV_1	LV_2	LV_2
$x_1 = LV_{1,3}$	$LV_{3,1}$	LV_1	LV_3	LV_1	LV_3
	$LV_{3,2}$		LV_2	LV_1	LV_2
	$LV_{3,3}$	LV_1	LV_3	LV_2	LV_1

The accumulation can take place in different ways. The most common option is the union of all rule results. This is synonymous with the OR operation, i. e. all rules from the rule base are combined using the fuzzy OR operation.

As explained above, the fuzzy OR operation can be realized using the maximum operator. The accumulation result is then

$$\mu_{\text{res}}(x_1, \ldots, x_n, y) = \max\{\mu_1(x_1, \ldots, x_n, y), \ldots\ldots, \mu_m(x_1, \ldots, x_n, y)\}. \tag{6.99}$$

Again, this result can be illustrated graphically. For this purpose, consider the following two rules:

If the boiler temperature T is high and the boiler pressure P is low, then set the valve opening V of the gas supply to approximately half-open,

and

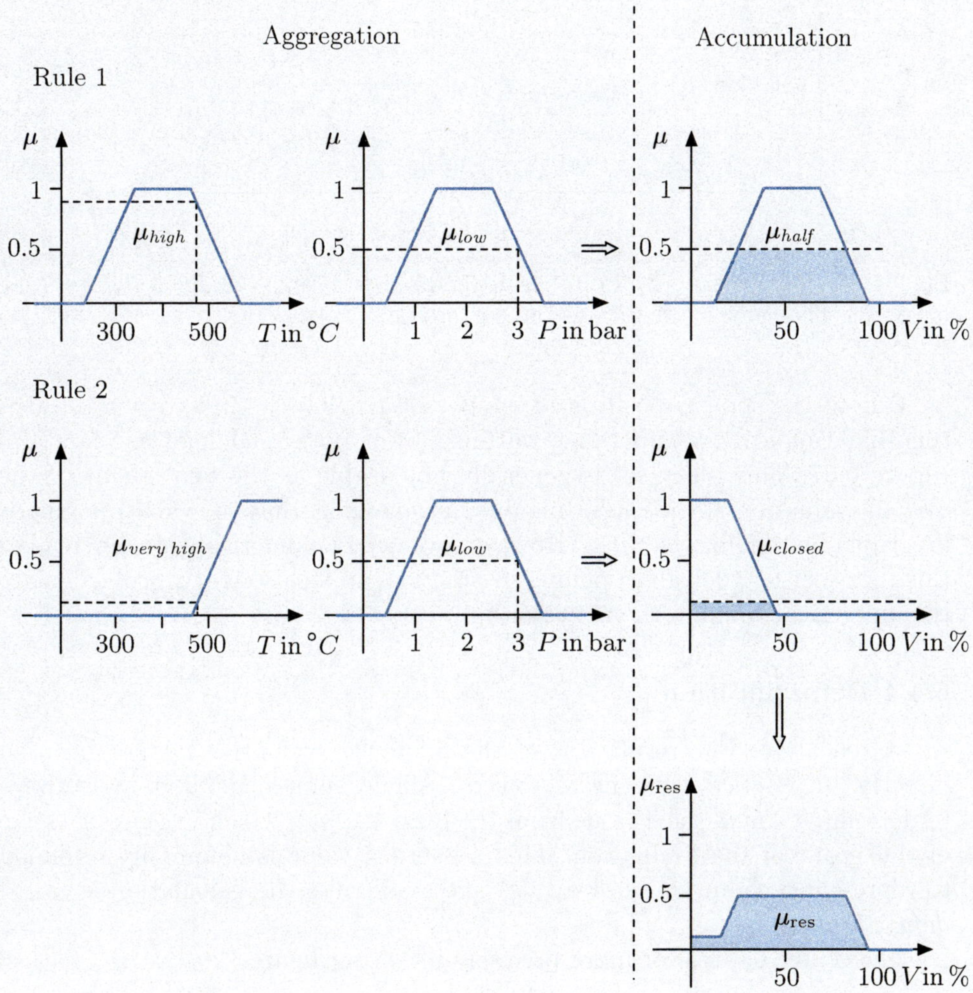

Fig. 6.36: Evaluation of a rule base using aggregation and accumulation

If the boiler temperature T is very high and the boiler pressure P is low, then set the valve opening V of the gas supply to almost closed.

These rules can then be evaluated for a situation (T, P) with the corresponding membership functions. Here, $(T, P) = (470\,°\text{C}, 3\,\text{bar})$ is selected. Figure 6.36 shows the resulting evaluation. It illustrates how the union μ_{res} of the accumulation is performed using the OR operator from the fuzzy sets or the membership values of the two rule outcomes, which are shown as blue areas.

The result of the accumulation is a superimposition of the results obtained from the individual rules. For a given situation $(x_1, \ldots, x_n)$, the accumulation result (6.99) is a function of y. The evaluation of all rules by aggregation, implication, and accumulation is called *inference*.

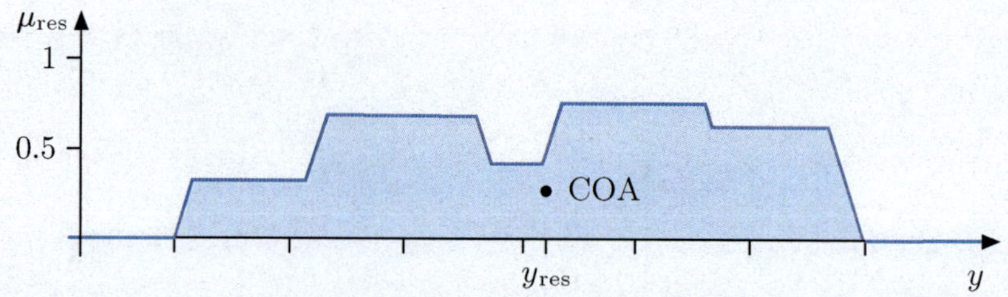

Fig. 6.37: Accumulation of all rule results and center of area (COA) of the resulting fuzzy set. The value y_{res} indicates the y coordinate of the COA.

Due to the superimposition of many rule results, the fuzzy set μ_{res} takes the shape of a mountain range outlined by a polygonal line, as shown in Figure 6.37. This fuzzy set is generally not usable as the final result of the overall evaluation of the rules. Imagine, for example, that the valve opening is described by the fuzzy set μ_{res}. However, we need a numerical value instead of this fuzzy set for the valve opening in practical applications. We will calculate this numerical value in the next section.

6.4.4 Defuzzification

As stated above, the result μ_{res} of the inference evaluation cannot be used directly. In practice, we generally need a single numerical value, so we have to determine a numerical value from the fuzzy set μ_{res} which is representative of the result of the evaluation. This numerical value is commonly intended to represent a compromise or an average of all rules. Its calculation is called *defuzzification.*

The center of area, or more precisely its y coordinate

$$y_{\mathrm{res}} = \frac{\displaystyle\int_{-\infty}^{\infty} y \cdot \mu_{\mathrm{res}}(x_1, \ldots, x_n, y)\, dy}{\displaystyle\int_{-\infty}^{\infty} \mu_{\mathrm{res}}(x_1, \ldots, x_n, y)\, dy}$$

is such a value. It is the final result of a fuzzy logic, i. e. the output value of the overall rule base evaluation. Figure 6.37 shows an example. The described method of defuzzification is the best-known and is known as the *center-of-area method* or *COA method.*

However, determining the center of area is complicated, since the membership function μ_{res} takes the shape of a polygonal line. The calculation of the center of area y_{res} can be simplified considerably by choosing singletons instead of trapezoids or triangles for the membership functions of the linguistic values LV_i of y. Singletons $\mu_{LV_i}(y)$ only take the function value one at one point y_{s}; everywhere else they take the function value zero, i. e.

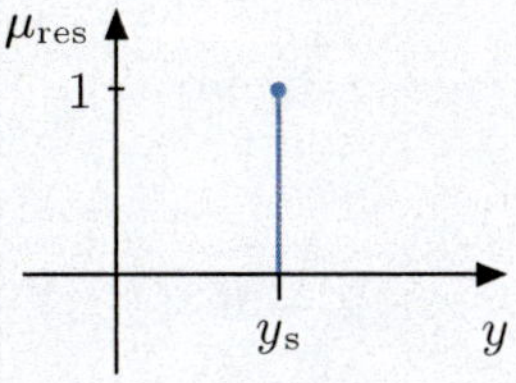

Fig. 6.38: Singleton

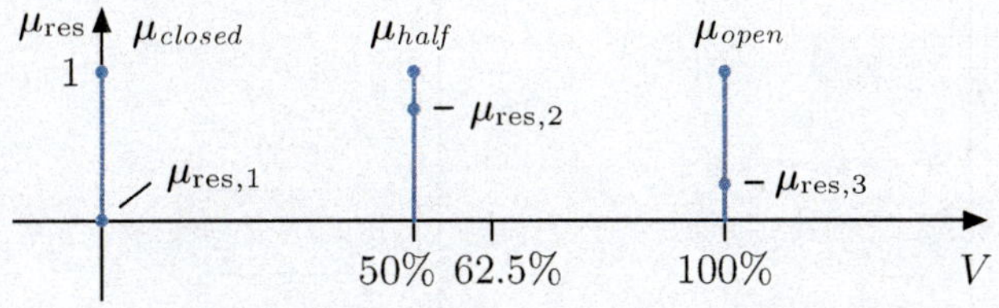

Fig. 6.39: Defuzzification with singletons

$$\mu_{LV_i}(y) = \begin{cases} 1 & \text{for } y = y_{\text{s}}, \\ 0 & \text{otherwise.} \end{cases}$$

Figure 6.38 shows an example of a singleton.

The result of the accumulation then takes on the form illustrated in Figure 6.39. In this case, the individual results of the accumulation no longer overlap to form a polygonal line. Now the membership function μ_{res} can be decomposed into

$$\mu_{\text{res}} = \mu_{\text{res},1} + \mu_{\text{res},2} + \mu_{\text{res},3} + \dots$$

The membership functions $\mu_{\text{res},i}$ are the accumulated results of all rules that affect the singleton i.

If all m output membership functions are singletons, we can replace the formula for the center of area, which is now no longer applicable, with a weighted average. This yields

$$y_{\text{res}} = \frac{\sum_{i=1}^{m} y_{\text{s},i} \cdot \mu_{\text{res},i}(x_1, \dots, x_n)}{\sum_{i=1}^{m} \mu_{\text{res},i}(x_1, \dots, x_n)},$$

which is much easier to calculate than the center of area. This variant of the COA method is called the *center-of-singletons method*, abbreviated as the *COS method*. The example of a valve position shown in Figure 6.39 with the three membership functions *closed*, *half*, and *open* illustrates this defuzzification method. The result of the defuzzification is

$$y_{\text{res}} = \frac{50 \cdot 0.75 + 100 \cdot 0.25}{0.75 + 0.25}\% = 62.5\%$$

with $\mu_{\text{res},1} = 0$, $\mu_{\text{res},2} = 0.75$, and $\mu_{\text{res},3} = 0.25$.

6.4.5 Fuzzy Systems and Fuzzy Controllers

By means of fuzzification, aggregation, implication, accumulation, and defuzzification, we can transform linguistically formulated rules into a mathematical function when using the results of the previous sections. Figure 6.40 illustrates this.

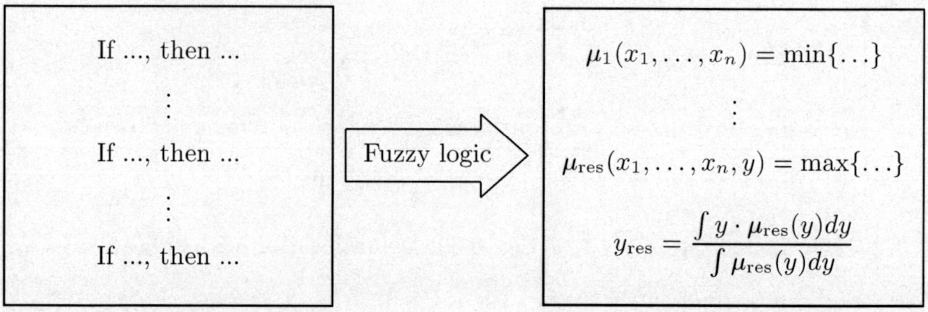

Fig. 6.40: Conversion of rules into equations using fuzzy logic

The function

$$y_{\text{res}} = f(x_1, \dots, x_n)$$

is composed several times, with the composition resulting from the following subfunctions and their associated calculation steps:

Fuzzification: determination of all membership values $\mu_{LV_{i,j}}(x_i)$,

Aggregation: $\mu_{\text{agg},k}(x_1,\dots,x_n) = \min\{\mu_{LV_{1,i}}(x_1),\dots,\mu_{LV_{n,l}}(x_n)\}$, $k = 1,\dots,m$,

Implication: $\mu_k(x_1,\dots,x_n,y) = \min\{\mu_{\text{agg},k}(x_1,\dots,x_n), \mu_{LV_p}(y)\}$, $k = 1,\dots,m$,

Accumulation: $\mu_{\text{res}}(x_1,\dots,x_n,y) = \max\{\mu_1(x_1,\dots,x_n,y),\dots, \mu_m(x_1,\dots,x_n,y)\}$,

Defuzzification:
$$y_{\text{res}} = \frac{\int\limits_{-\infty}^{\infty} y\cdot\mu_{\text{res}}(x_1,\dots,x_n,y)\,dy}{\int\limits_{-\infty}^{\infty} \mu_{\text{res}}(x_1,\dots,x_n,y)\,dy}.$$

When calculating y_{res} for a situation $(x_1,\dots,x_n)$, we go through the steps above. The variables $x_1,\dots,x_n$ can be interpreted as input variables and y as the output variable of a system, as shown in Figure 6.41.

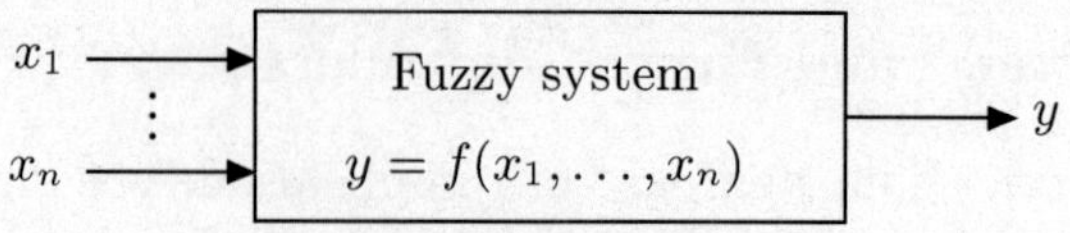

Fig. 6.41: Fuzzy system

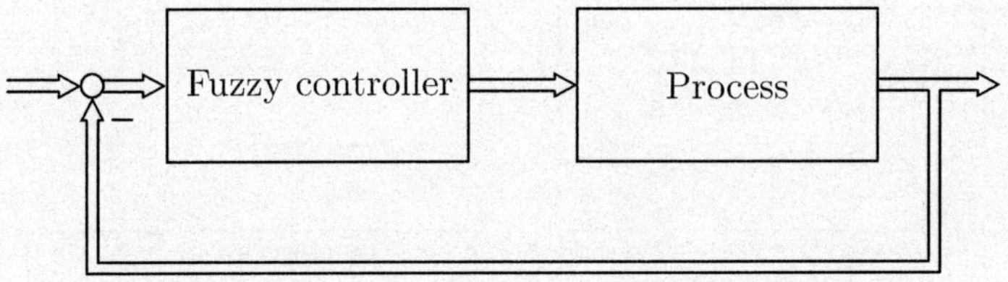

Fig. 6.42: Control loop with fuzzy controller

The possible applications of fuzzy systems are numerous. This is due to the fact that many human behaviors can be described by a mathematical function using fuzzy logic. Important areas of application are open- and closed-loop controls, as mentioned above. This branch of fuzzy logic is called *fuzzy control* [110, 177, 358, 497]. Figure 6.42 shows a control loop equipped with a fuzzy controller.

In many applications, human behavior can be reproduced in this way. Examples are the control of sewage plants [90, 326], ABS braking systems [117, 313, 314], the control of railway and subway trains [212, 447, 469], the control of paper machines [4, 5, 413], alarm systems for anesthesia [463], the autofocus control for cameras [95, 188, 262], washing machine controllers [7, 27, 379, 440], and others [70, 215, 226, 447, 463].

When designing fuzzy controllers, we proceed as follows:

Step 1: Determine the linguistic variables x_i, y_i, which make up the input and output variables of the fuzzy controller.

Step 2: Assign linguistic values $LV_{i,j}$ to each linguistic variable x_i, y_i.

Step 3: Select the membership functions $\mu_{i,j}$ for $LV_{i,j}$.

Step 4: Determine the rules.

In most cases, the fuzzy controller is used to imitate a person who performs or could perform the control. This could be the process operator of a chemical or biotechnological process, or the driver of a subway train, for example.

It is rare for us to be able to determine a feedforward or feedback controller with the desired behavior in a first design using the method above. This is because the first design will not be able to reproduce human behavior accurately enough. Consequently, the design is repeated with modified parameters and rules until satisfactory behavior is achieved. This procedure is shown in Figure 6.43. The design is based on trial and error and will require multiple optimizations.

Note that the fuzzy controller is a static controller. It is nonlinear, but without dynamics. Its nonlinearity entails the problem that the stability of the control loop is either not verifiable at all or only with great effort. This is one of the reasons why fuzzy controllers are mainly used in cases where

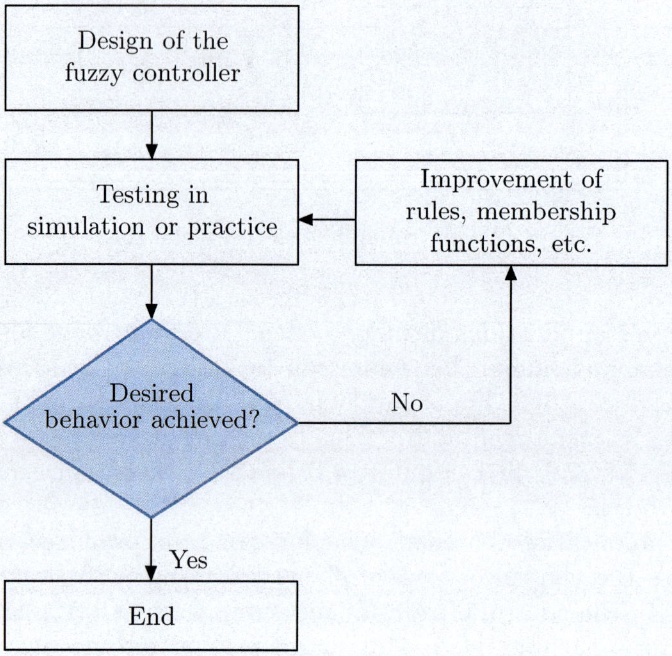

Fig. 6.43: Procedure for designing a fuzzy controller

classical controllers and the associated stability assurance of the control loop do not achieve satisfactory results.

6.4.6 Example: Distance Control for Automobiles

As an example application of a fuzzy controller, we will look at a distance control for automobiles, which is described in similar form in [149, 187]. The aim is to determine the speed v or acceleration a of an automobile so that the distance x_{rel} to a vehicle in front of the car takes a safe value, e. g. $x_{\mathrm{ref}} = 0.5v$ with x_{ref} in meters and v in kilometers per hour. Figure 6.44 illustrates the situation.

The following variables are relevant for distance control:

v_{lead}	: velocity of the vehicle in front,
v	: speed of the vehicle being controlled,
$v_{\mathrm{rel}} = v_{\mathrm{lead}} - v$	: relative velocity,
x_{rel}	: distance,
$\Delta x_{\mathrm{rel}} = x_{\mathrm{rel}} - x_{\mathrm{ref}}$	: distance deviation,
a_{ref}	: desired acceleration of the vehicle being controlled,
a	: acceleration of the vehicle being controlled.

Besides the speed v, the variables v_{rel} and x_{rel} can also be measured and thus used as control inputs. The latter are measured by radar or laser. The car's movement can be described using a double integrator with the acceleration

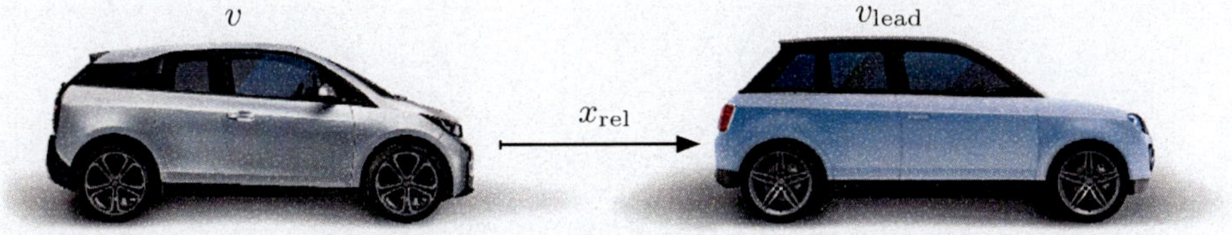

Fig. 6.44: Distance control

as the input variable and the distance as the output variable. In addition, there is a subordinate acceleration control loop, which can be represented by two first-order transfer elements. Therefore, the transfer function of the open control loop is

$$G(s) = \frac{1}{s^2(1 + T_1 s)(1 + T_2 s)}$$

with $T_1 = 0.074\,\text{s}$ and $T_2 = 0.14\,\text{s}$. This results in the control loop shown in Figure 6.45.

For the distance controller, it makes sense to imitate human behavior using a fuzzy controller. Humans also use Δx_{rel}, v_{rel}, and v as indicators that they need to reduce or increase the acceleration a_{ref} of their vehicle. In principle, a distance control of this kind could also be performed using a linear controller. However, a linear controller does not guarantee sufficiently safe driving behavior if the vehicles in question change lanes or encounter slow vehicles after rounding a bend. In contrast, the fuzzy distance controller shown in Figure 6.45 achieves this.

In the first design step, linguistic values and membership functions of Δx_{rel}, v, v_{rel}, and a_{ref} are defined. The membership functions of Figure 6.46

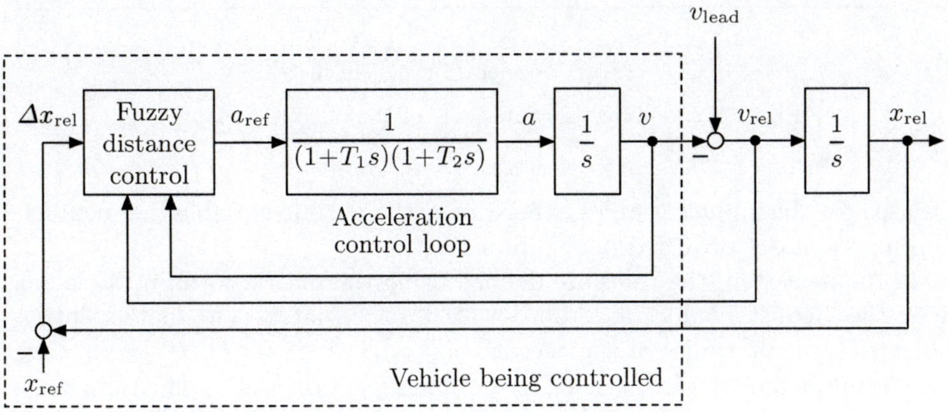

Fig. 6.45: Distance-control loop

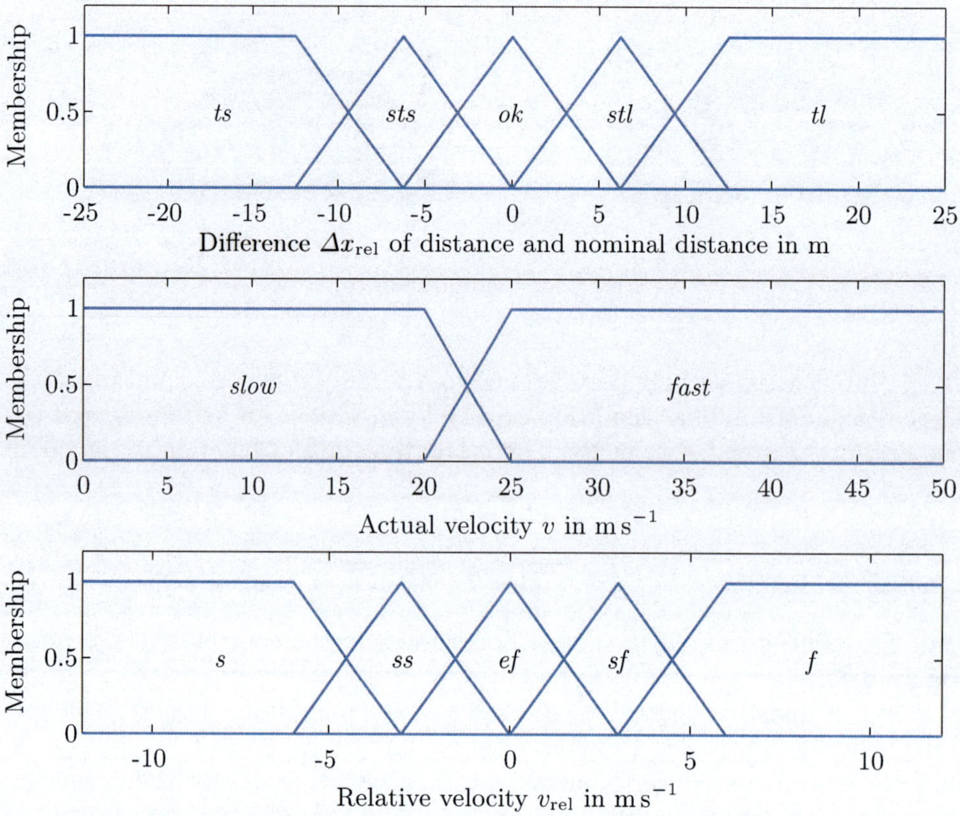

Fig. 6.46: Membership functions of the input variables

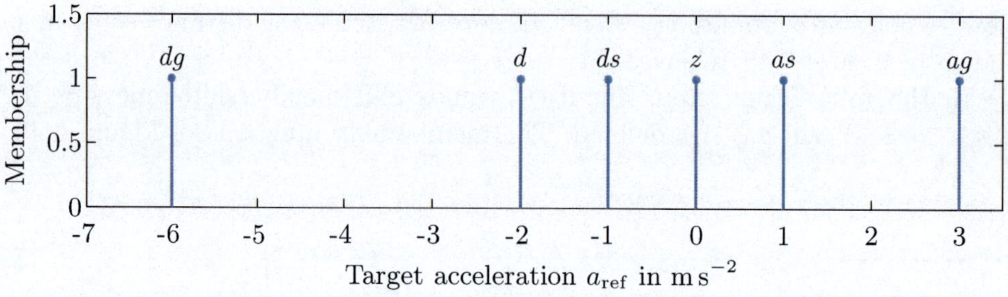

Fig. 6.47: Membership functions of the output variable

are used for the input variables, whereas for the output variable the membership functions are provided in Figure 6.47.

In the next step, the rules are defined using the matrix form in Table 6.6. Here, the linguistic values are appropriately abbreviated. The matrix entries indicate linguistic values of the acceleration a_{ref}.

The minimum operator is used for both aggregation and implication, while the maximum operator is used for accumulation. For defuzzification, we will apply the COS method. The above specifications have the advantage that they

Table 6.6: Fuzzy rules for distance control with the input variables v, Δx_{rel}, and v_{rel}, and the output variable a_{ref}

$v = slow$		Δx_{rel}: *ts*	*sts*	*ok*	*stl*	*tl*
v_{rel}	*s*	*dg*	*d*	*d*	*d*	*d*
	ss	*d*	*d*	*d*	*ds*	*as*
	ef	*ds*	*ds*	*z*	*as*	*ag*
	sf	*ds*	*ds*	*as*	*ag*	*ag*
	f	*z*	*z*	*as*	*ag*	*ag*

$v = fast$		Δx_{rel}: *ts*	*sts*	*ok*	*stl*	*tl*
v_{rel}	*s*	*d*	*d*	*d*	*d*	*z*
	ss	*d*	*d*	*d*	*ds*	*as*
	ef	*ds*	*ds*	*z*	*as*	*ag*
	sf	*ds*	*ds*	*as*	*ag*	*ag*
	f	*z*	*z*	*as*	*ag*	*ag*

Abbr.	Meaning
s	slower
ss	slightly slower
ef	equally fast
sf	slightly faster
f	faster
ts	too small
sts	slightly too small
ok	ok
stl	slightly too large
tl	too large
dg	decelerate greatly
d	decelerate
ds	decelerate slightly
z	zero
as	accelerate slightly
ag	accelerate greatly

are cost-effective and very simple to implement using a fixed-point processor, for example. Particularly the use of membership functions other than singletons or a defuzzification method other than the COS method would make the evaluation much more complex and costly. In the present case, there would also be no advantages, since the control behavior is very similar for both variants.

In total, 50 rules are obtained. The fuzzy controller is a static controller whose family of characteristics is four-dimensional. For the vehicle speeds $v = 10\,\text{m}\,\text{s}^{-1}$, i. e. for *slow*, and $v = 30\,\text{m}\,\text{s}^{-1}$, i. e. for *fast*, the corresponding family of characteristics are shown graphically in Figure 6.48. Five essential regions can be identified in this nonlinear controller. In region 1, the controller behavior is approximately the same as that of a linear controller. This region addresses the stationary state with $v_{\text{rel}} \approx 0\,\text{m}\,\text{s}^{-1}$ and the correct distance $x_{\text{rel}} \approx x_{\text{ref}}$, i. e. $\Delta x_{\text{rel}} \approx 0$. In region 2, the vehicle can catch up with the vehicle in front by accelerating at the maximum rate, as the distance is large and the vehicle in front is moving away. In region 3, maximum deceleration is required because the distance is much too small. In region 4, the approach

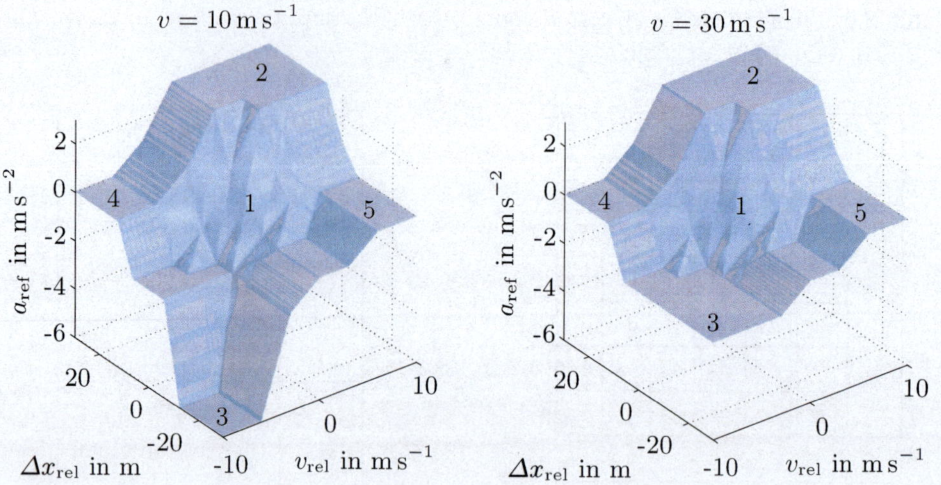

Fig. 6.48: Family of characteristics of the fuzzy distance controller at the velocities $v = 10\,\mathrm{m\,s^{-1}}$ and $v = 30\,\mathrm{m\,s^{-1}}$ of the vehicle being controlled

is at a negative relative speed v_{rel} from a large distance. The velocity is not changed in this case. Thus, $a_{\mathrm{ref}} = 0\,\mathrm{m\,s^{-2}}$ applies until the vehicle has caught up with the vehicle in front. In region 5, the vehicle in front is driving away at a distance which is too small. Therefore, it is not accelerated or decelerated until the correct distance is reached.

From the family of characteristics, it is also evident that it is only in the relatively small regions 4 and 5 that neither acceleration nor deceleration takes place. As a consequence, a distance control system is constantly accelerating or braking in dense traffic. In practice, a distance control is combined with cruise control, so that at low traffic density the speed control is performed by the cruise control. The deceleration and acceleration processes are then omitted.

We use for an example the same traffic situation as in [148]. Figure 6.49 shows the acceleration a_{ref}, the velocities v and v_{lead}, the distance x_{rel}, and the target distance x_{ref} between two vehicles during a drive on the highway of approximately 8 min in very dense traffic. The effective control behavior is clearly visible in all ranges. The vehicle at the back which is being controlled almost exactly maintains the distance prescribed by law. This also applies to the incident at $t = 200\,\mathrm{s}$, when a vehicle with a distance of about 18.5 m and a speed of $38\,\mathrm{m\,s^{-1}} \approx 137\,\mathrm{km\,h^{-1}} \approx 85$ mph enters the lane in front of the vehicle being controlled. The control restores the correct distance within 10 s.

Also noteworthy is the period between seconds 320 and 350, during which the traffic comes to a complete halt. Here, too, the control system succeeds in maintaining a safe distance which amounts to only centimeters when the vehicles are stationary. In practice, the nominal distance is limited to a minimum value of $x_{\mathrm{min}} = 2\,\mathrm{m}$, for example. The equation $x_{\mathrm{ref}} = 0.5v$ is then replaced by $x_{\mathrm{ref}} = 0.5v + x_{\mathrm{min}}$.

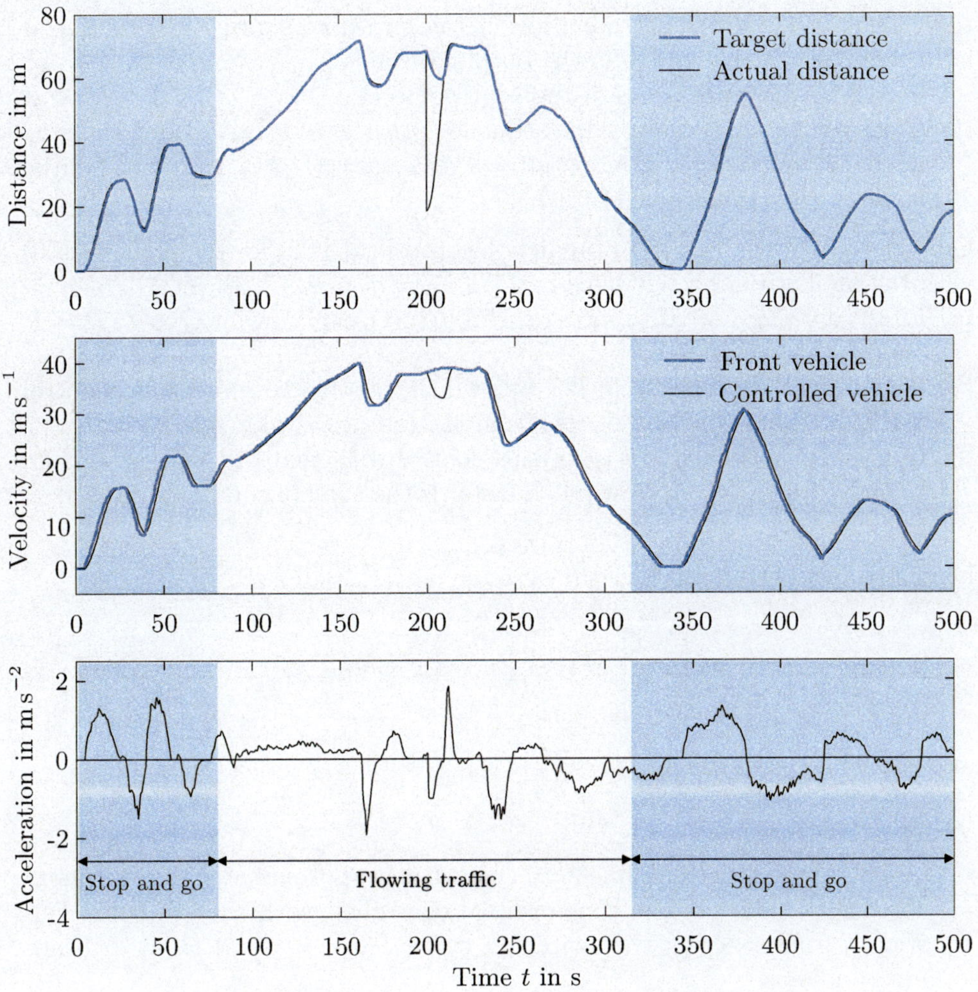

Fig. 6.49: Measurements during a simulated highway drive

6.5 Exercises

Exercise 6.1 Our aim is to design a model predictive control (MPC1) for the discrete-time system

$$\begin{aligned} x(k+1) &= x(k) + u(k), \\ y(k) &= x(k). \end{aligned} \tag{6.100}$$

(a) Calculate the matrices $\boldsymbol{F}$, $\boldsymbol{G}$, and $\boldsymbol{H}$ for the case in which the prediction and control horizons $n_\mathrm{p} = n_\mathrm{c} = 1$ are equally long.
(b) Formulate the quadratic performance index J with the values $Q = 1$ and $R = 1$ from equation (6.11) on p. 512 in dependence on the control variable $\Delta\boldsymbol{u}(k)$ and the control error $e(k)$.
(c) Calculate the control law $u(k)$ dependent solely on the previous output variables $y(k-i)$ and the reference variable $y_\mathrm{ref}(k+1)$.

(d) Calculate the difference equation of the closed control loop with y_{ref} as the input variable and y as the output variable.
(e) Calculate the eigenvalues of the control loop.
(f) Calculate the output variable sequence $y(k)$, $k = 0, \ldots, 18$, for the step in the reference value from $y_{\text{ref}}(0) = 0$ to $y_{\text{ref}}(k) = 1$, $k \geq 1$, and the initial values $y(0) = 0$ and $y(-1) = 0$.

Exercise 6.2 Design a model predictive control (MPC2) for the plant (6.100) with the predictive and reference horizon $n_{\text{p}} = n_{\text{c}} = 2$.
(a) First, solve Exercise 6.1, parts (a) through (f), but now for $\boldsymbol{Q} = \boldsymbol{I}$, $\boldsymbol{R} = \boldsymbol{I}$, and $n_{\text{p}} = n_{\text{c}} = 2$.
(b) Determine the time step $k_{3\%}$ at which the model predictive control (MPC2) deviates by 3% or less from the reference value $y_{\text{ref}}(k) = 1$.
(c) Compare the time step $k_{3\%}$ of this control with that of the control from Exercise 6.1 (MPC1). How much faster is the control compared to that in Exercise 6.1?

Exercise 6.3 Let us consider the system

$$\dot{\boldsymbol{x}} = \begin{bmatrix} 0 & 1 \\ 0 & 0 \end{bmatrix} \boldsymbol{x} + \begin{bmatrix} 0 \\ 1 \end{bmatrix} u, \qquad |u| \leq u_{\text{max}} = 1.$$

(a) Design a sliding mode controller $u(\boldsymbol{x})$ which has the switching line

$$s(\boldsymbol{x}) = r x_1 + x_2 = 0.$$

Calculate the switching line $s(\boldsymbol{x})$ such that it approximates the switching curve $S(x_2) = x_1$ of the time-optimal controller, with the switching line $s(\boldsymbol{x}) = 0$ intersecting the switching curve $S(x_2) = x_1$ at $x_1 = 2$. State $u(\boldsymbol{x})$ and $s(\boldsymbol{x})$.
(b) Draw the switching line and the trajectories of the control loop.
(c) On the switching line, where does a sliding mode occur and where does it not occur?
(d) Demonstrate that the sliding mode can be achieved in finite time.
(e) What are the dynamics of the control loop in the sliding mode?
(f) Identify and sketch the progression of $x_1(t)$ and $x_2(t)$ for the initial vector

$$\boldsymbol{x}^T(0) = \begin{bmatrix} -5 & 5 \end{bmatrix}. \tag{6.101}$$

(g) For the initial vector (6.101), calculate the switching times of the time-optimal control and draw the time-optimal progressions of $x_1(t)$ and $x_2(t)$.
(h) Given the initial vector

$$\boldsymbol{x}^T(0) = \begin{bmatrix} x_{10} & x_{20} = -x_{10} \end{bmatrix},$$

what is the maximum number k of switching intervals required to attain the sliding mode?

Exercise 6.4 For a plant

$$\dot{\boldsymbol{x}} = \boldsymbol{a}(\boldsymbol{x}) + \boldsymbol{b}(\boldsymbol{x})u$$

calculate a sliding mode controller $u(\boldsymbol{x})$ using Gao and Hung's method.

Exercise 6.5 A motor vehicle's brakes have an anti-lock braking system which prevents the wheels from locking as a result of overly abrupt braking, thus also preventing the driver from losing control over the vehicle. The measure for locking is the slip

$$\lambda = \frac{v - r\omega}{v}$$

of the wheel in question. It is the difference, in reference to the velocity v of the vehicle, between v and the circumferential velocity $r\omega$ of the wheel. Here, r is the radius of the wheel and ω is its angular velocity. The wheel has the moment of inertia J in relation to its rotational axis. The friction F_{f} between the wheel and the road is obtained from the weight mg (where m amounts to a quarter of the vehicle's mass and g is the gravitational acceleration) and the friction coefficient $\mu(\lambda)$, yielding

$$F_{\mathrm{f}} = mg\mu(\lambda).$$

The wheel is braked by the braking torque M_{b}. Figure 6.50 shows the process. The friction coefficient can vary depending on the slip λ between $\mu = 0$ for

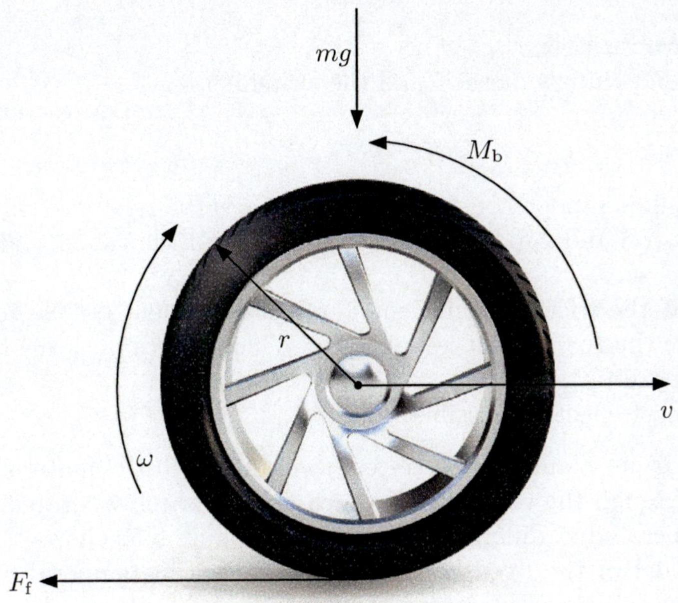

Fig. 6.50: Wheel and the forces acting on it

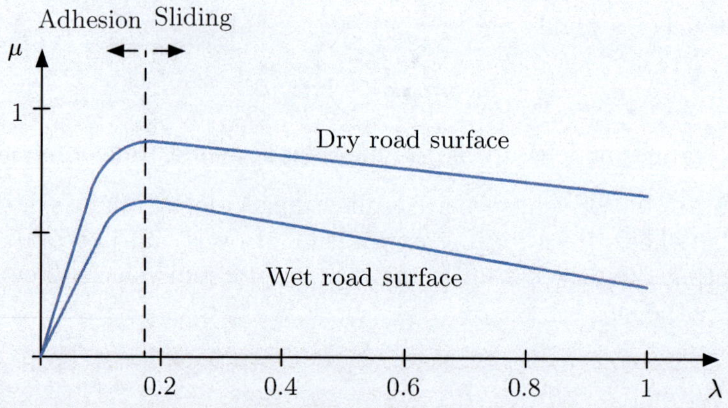

Fig. 6.51: Curve of the friction coefficient $\mu(\lambda)$ in varying road conditions

$\lambda = 0$ and $\mu \approx 1$ for $\lambda > 0$ [202]. Figure 6.51 shows a typical curve. If the road surface is dry and we have normal driving conditions, the friction coefficient is $\mu \approx 0.015$.

(a) Formulate the wheel's equations of motion and, in doing so, choose the state variables

$$x_1 = v \quad \text{and} \quad x_2 = \lambda,$$

and

$$u = M_\text{b}$$

as the control variable.

(b) Using Gao and Hung's method and the switching line

$$s(\boldsymbol{x}) = x_2 - x_{2,\text{ref}} = 0,$$

calculate a sliding mode controller. Here, $x_{2,\text{ref}}$ is the reference slip value.

(c) Determine $x_1(t)$ and $x_2(t)$ in the sliding mode and the time t_eq at which $x_1(t)$ reaches zero.

(d) Assume that the friction coefficient $\mu(x_2)$ is dependent on the material used to pave the road, i.e. it is unknown. Which values does the braking distance depend on?

(e) Select the most effective possible value $x_{2,\text{ref}}$.

Exercise 6.6 Let us examine a three-zone oven in which the material being heated passes through the three zones, which are separately operated. These zones may be operated at different temperatures. Figure 6.52 shows the basic design. The model of the oven describes the changes in temperature ΔT_i, $i = 1, 2, 3$, surrounding the operating point T_i, $i = 1, 2, 3$. Here, the changes in temperature ΔT_i are the output variables y_i. They depend on the changes u_i in the heating levels at the operating point via the model in the Laplace domain

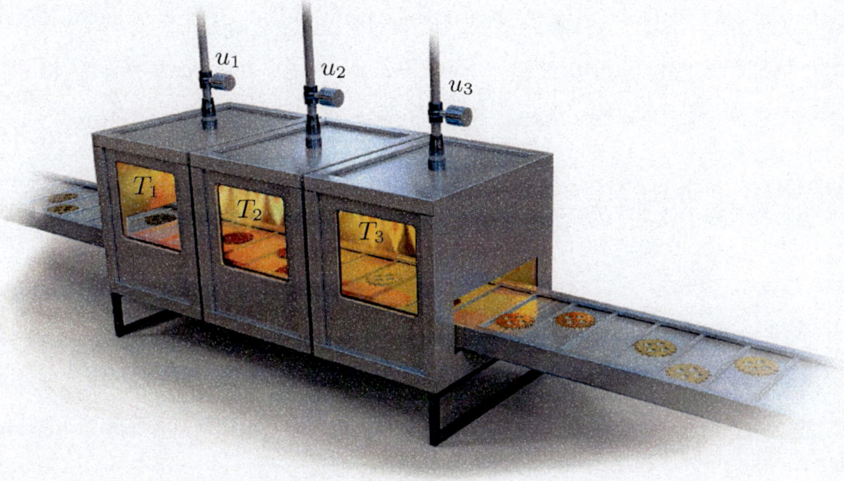

Fig. 6.52: Three-zone oven

$$\boldsymbol{Y}(s) = \boldsymbol{G}(s)\boldsymbol{U}(s)$$

with the transfer matrix

$$\boldsymbol{G}(s) = c \begin{bmatrix} \dfrac{1}{1+5s} & \dfrac{0.5}{1+6s} & \dfrac{0.2}{1+6s} \\[2ex] \dfrac{0.5}{1+6s} & \dfrac{1}{1+5s} & \dfrac{0.5}{1+6s} \\[2ex] \dfrac{0.2}{1+6s} & \dfrac{0.5}{1+6s} & \dfrac{1}{1+5s} \end{bmatrix}.$$

The parameter c depends on the mass and its specific thermal capacity.

(a) Show that the model of the three-zone oven is strictly positive real.

(b) For each of the three zones, a P controller is used which does not take into account the coupling of the zones, i. e.

$$\boldsymbol{u} = \begin{bmatrix} P_1 & 0 & 0 \\ 0 & P_2 & 0 \\ 0 & 0 & P_3 \end{bmatrix} \boldsymbol{e}, \qquad \boldsymbol{e} = -\boldsymbol{y}.$$

In addition, the control variables u_i, $i = 1, 2, 3$, are each subject to an actuator saturation

$$u_i = \begin{cases} u_{\max}, & u_i > u_{\max}, \\ \tilde{u}_i, & u_{\min} < u_i < u_{\max}, \\ u_{\min}, & u_i < u_{\min}, \end{cases}$$

where $u_{\max} > 0$ and $u_{\min} < 0$. Is the point at which $\Delta T_i = 0$, $i = 1, 2, 3$, applies, an asymptotically stable equilibrium point of the control loop?

Exercise 6.7 We will examine the so-called *Netushil backlash model* [334]

$$\dot{u} = D(e - \frac{u}{m}), \tag{6.102}$$

where $D(x)$ is the dead zone nonlinearity

$$D(x) = \begin{cases} \mu(x+a), & x \le -a, \\ 0, & |x| < a, \\ \mu(x-a), & x \ge a. \end{cases}$$

Here $\mu \gg 1$ is a very large value which tends to infinity in the ideal but unrealistic case.

(a) Demonstrate that the backlash model (2.6) on p. 75 has the same dynamic behavior as the Netushil backlash model (6.102).
(b) Show that the backlash nonlinearity is passive.

Exercise 6.8 Let us consider the nonlinear controller

$$\begin{aligned} u &= h(x, e), \\ \dot{x} &= e \end{aligned} \tag{6.103}$$

with the input variable e and the output variable u where the nonlinear function h has the properties

$$h(0,0) = 0$$

and

$$\begin{aligned} 0 &\le x[h(x,e) - h(0,e)] \qquad \text{for all } x, e \in I\!R, \\ 0 &\le e[h(x,e) - h(x,0)] \qquad \text{for all } x, e \in I\!R. \end{aligned}$$

(a) Which geometric properties does the characteristic diagram $u = h(x, e)$ have? For this purpose, examine the function $h(x, e)$, or more precisely its signs, in the four quadrants of the xe-plane.
(b) Show that the two properties

$$S(0) = 0 \quad \text{and} \quad S(x) \ge 0 \quad \text{for } x \ne 0$$

are fulfilled for the storage function

$$S(x) = \int_0^x h(\tilde{x}, 0) d\tilde{x}.$$

(c) Prove that the characteristic diagram controller (6.103) is passive.

Exercise 6.9 Given the system

$$\begin{aligned}\dot{x}_1 &= x_1^2 + u,\\ \dot{x}_2 &= -x_1 + x_2,\\ y &= x_1,\end{aligned}$$

why can't the system be converted into a passive control loop by means of a static controller $u(\boldsymbol{x})$?

Exercise 6.10 Use Theorem 80 on p. 549 to prove that the building with viscous fluid dampers from Section 6.3.2, p. 550 et seq., has a globally asymptotically stable equilibrium point.

Exercise 6.11 We would like to design a control with IDA for the hydropower plant with grid feed-in from Section 5.5.5, p. 472 et seq.

(a) State the model

$$\dot{\boldsymbol{x}} = \begin{bmatrix} x_2 \\ a_1 - a_2 x_2 - a_3(1+u)\sin(x_1 + \delta_{\mathrm{eq}}) \end{bmatrix}$$

from equation (5.192) in both the nonlinear controller canonical form and in the PCHD form. To do the latter, use the storage function

$$V(\boldsymbol{x}) = \frac{1}{2}x_2^2 - a_1 x_1 + a_3(\cos(\delta_{\mathrm{eq}}) - \cos(x_1 + \delta_{\mathrm{eq}})).$$

(b) Now, using an IDA passivity-based control (IDA PBC), we will attempt to design the control loop such that it has the dynamics

$$\dot{\boldsymbol{x}} = (\boldsymbol{J}_{\mathrm{CL}} - \boldsymbol{D}_{\mathrm{CL}})\frac{\partial V_{\mathrm{CL}}(\boldsymbol{x})}{\partial \boldsymbol{x}}$$

with

$$V_{\mathrm{CL}}(\boldsymbol{x}) = V(\boldsymbol{x}) + \frac{1}{2}k x_1^2.$$

What form must the matrix $\boldsymbol{J}_{\mathrm{CL}} - \boldsymbol{D}_{\mathrm{CL}}$ take for a controller design of this kind to be possible?

(c) Calculate the IDA control law based on the matrix

$$\boldsymbol{J}_{\mathrm{CL}} - \boldsymbol{D}_{\mathrm{CL}} = \begin{bmatrix} 0 & 1 \\ -1 & -a_2 - d \end{bmatrix}, \qquad d > 0.$$

Exercise 6.12 Let us examine Chen's chaotic system [76]

$$\begin{aligned}\dot{x}_1 &= -35x_1 + 35x_2,\\ \dot{x}_2 &= -7x_1 + 28x_2 - x_1x_3 + u,\\ \dot{x}_3 &= x_1x_2 - 3x_3,\\ y &= x_2.\end{aligned}$$

(a) What are the equilibrium points of the free system?
(b) State the system in PCHD form with the storage function

$$V_1(\boldsymbol{x}) = \frac{1}{10}x_1^2 + \frac{1}{2}x_2^2 + \frac{1}{2}x_3^2.$$

(c) Is the equilibrium point $\boldsymbol{x}_{\text{eq}} = \boldsymbol{0}$ stable? Give reasons for your answer.
(d) Convert the system into the Byrnes-Isidori canonical form and state the diffeomorphism $\boldsymbol{z} = \boldsymbol{t}(\boldsymbol{x})$ which is required for this transformation.
(e) Can we use a static controller $u(\boldsymbol{z})$ to design a passive control loop for the system in question, and if so, why?
(f) State the control law $u(\boldsymbol{z})$ such that the control loop in PCHD form with

$$V_2(\boldsymbol{z}) = \frac{1}{2}(z_1^2 + z_2^2 + z_3^2)$$

is strictly passive.
(g) State the equations of the control loop in the original coordinates $\boldsymbol{x}$.
(h) Is the equilibrium point $\boldsymbol{x}_{\text{eq}} = \boldsymbol{0}$ of the control loop globally asymptotically stable and if so, why?

Exercise 6.13 Show that De Morgan's laws

$$\overline{\mu_a \wedge \mu_b} = \bar{\mu}_a \vee \bar{\mu}_b$$
$$\overline{\mu_a \vee \mu_b} = \bar{\mu}_a \wedge \bar{\mu}_b$$

are valid even in fuzzy logic for the fuzzy operators min and max. In doing this, use the generally applicable equations

$$\max\{x_1, x_2\} = \frac{x_1 + x_2 + |x_1 - x_2|}{2},$$
$$\min\{x_1, x_2\} = \frac{x_1 + x_2 - |x_1 - x_2|}{2}.$$

Recall that the negation of a fuzzy set is defined by $\bar{\mu} = 1 - \mu$.

Exercise 6.14 In Boolean logic, $a \vee \bar{a} = 1$ applies. This is the law of the excluded middle, also known as *tertium non datur*. Determine whether this rule applies in fuzzy logic as well if we use
(a) the maximum operator,
(b) the algebraic sum, or
(c) the bounded sum
as fuzzy OR operators.

Exercise 6.15 Let us examine the distance control for automobiles from Section 6.4.6 on p. 588 et seq. We will assume that the dynamics of the acceleration control loop are so fast in comparison to the vehicle dynamics that they can be neglected. In this exercise, we would like to examine the distance control only for velocities of $v \geq 8.\bar{3}\,\text{ms}^{-1} = 10\,\text{kmh}^{-1} \approx 6.2\,\text{mph}$. In this case, it is sufficient to use only the laws for $v =$ *fast* in Table 6.6 on p. 591 for the fuzzy distance controller.

(a) Based on the statements made in Exercise 6.8, determine if the subsidiary system

$$\begin{aligned}\dot{x}_{\text{rel}} &= v_{\text{rel}},\\ a_{\text{ref}} &= f(x_{\text{rel}}, v_{\text{rel}}),\end{aligned}$$

which includes the fuzzy distance controller, is passive.

(b) Neglecting the acceleration control loop, i. e. $a = a_{\text{ref}}$, show that the distance control loop described in this problem has an asymptotically stable equilibrium point at $[x_{\text{rel}}\ v_{\text{rel}}] = \mathbf{0}^T$.

Exercise 6.16 To mimic human behavior in selecting the gear of a motor vehicle, a control system for an automatic transmission is to be developed using fuzzy logic [71, 417]. The output variable of the fuzzy logic is the upshifting rotational speed n_{up}, which can take values between 1500 rpm and 5500 rpm. The downshifting rotational speed is given by $n_{\text{down}} = 0.4 n_{\text{up}}$. The gear selection is dependent on the gas-pedal position P, the torque caused by the driving resistance M, and the driver's driving style F. The gas-pedal position P varies between 0% and 100% and can be assigned the three linguistic values small, large, or kickdown. The vehicle's driving resistance torque M is the difference between the motor torque M_{mot} and the torque M_{a} which actually accelerates the vehicle. The latter depends on the slope of the road, the air resistance, etc. For us to compare it to the motor torque M_{mot}, it is calculated back to the gearbox's input. This yields

$$M = M_{\text{mot}} - M_{\text{a}} = M_{\text{mot}} - \frac{\dot{v}}{R} rm$$

with the vehicle's acceleration $\dot{v}$, the wheel radius r, the vehicle mass m, and the transmission ratio R of the gear. The driving resistance torque M can vary between -200 Nm and 200 Nm and will be assigned the linguistic values negative, small, or positive.

The designation of the driver's driving style F as either defensive or aggressive can be calculated over the time interval $[t, t+T]$ using the number and degree of acceleration and braking processes by means of [306]

$$F = \frac{1}{T} \int\limits_{t}^{t+T} \frac{|\dot{v}|^4}{R} dt.$$

Here, F varies between $F = 0$ for a very defensive driving style and $F = 10$ for very aggressive drivers. The driving style F is allocated one of the linguistic values, defensive or aggressive.

Design a fuzzy control with the upshifting rotational speed n_{up} as the output variable and the gas-pedal position P, the vehicle's driving resistance torque M, and the driving style F as the input variables. For the input variables, use triangles and ramps as membership functions. For the output variable, which is the upshifting rotational speed n_{up}, choose nine singletons between the values 1500 rpm and 5500 rpm.

7

Observers for Nonlinear Systems

7.1 Observability of Nonlinear Systems

7.1.1 Definition of Observability

The following question arises in the case of nonlinear control loops with state controllers: how can we determine the plant's state variables x_i if it is either technically impossible to measure them or not desired due to cost or other reasons? This situation is comparable to the one we know from the case of linear systems, in which the state variables x_i are often not measurable or the measurement would be too expensive.

In the linear case, observers are used to solve this problem by estimating the state variables x_i. Figure 7.1 shows a state control loop of this kind with an observer. In particular, the separation principle applies in this case, i. e. the dynamics of the observer and those of the control loop can be specified independently of each other via the observer matrix $\boldsymbol{L}$ and the controller matrix $\boldsymbol{K}$, respectively.

Observers can also be used for nonlinear systems. However, this is not as straightforward as in the linear case. This is because the design of nonlinear observers can be complicated and the stability of nonlinear control systems with observers is often impossible or very difficult to verify. One reason for this is that the separation principle does not normally apply to nonlinear control loops with observers. However, the structure of a nonlinear control loop with an observer, as shown in Figure 7.2, equals that of the linear one.

The type of nonlinear observer to be selected depends on the characteristics of the plant. The following sections describe some of the common types. Supplementary literature can be found in [42, 47, 107, 180, 229]. However, before an observer can be designed, it must be ensured that the nonlinear system is observable [47, 146, 180]. In practice, this step is sometimes neglected, since it is often laborious to examine a nonlinear system's observability. Furthermore, most of the real-world systems, particularly technical systems, are observable.

J. Adamy, *Nonlinear Systems and Controls*,
https://doi.org/10.1007/978-3-662-68690-4_7

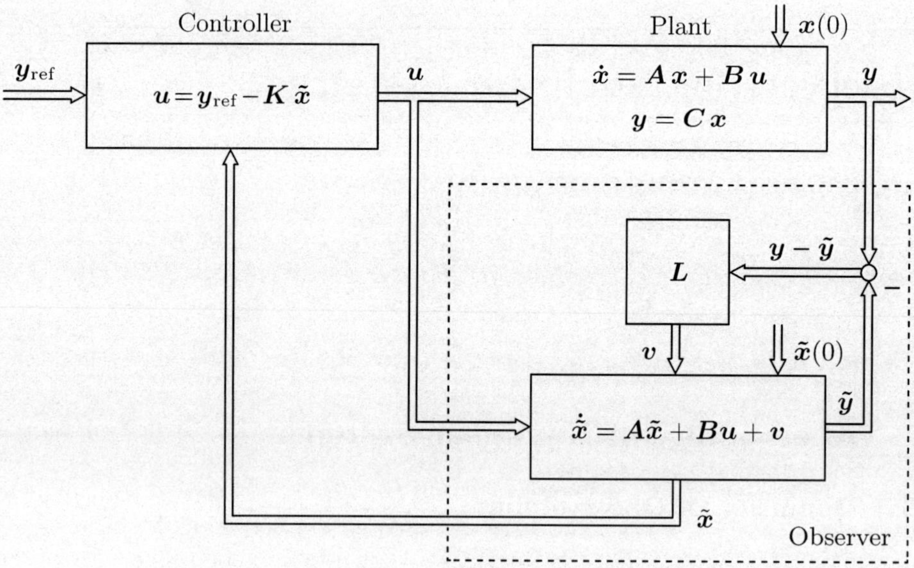

Fig. 7.1: Linear state control loop with observer

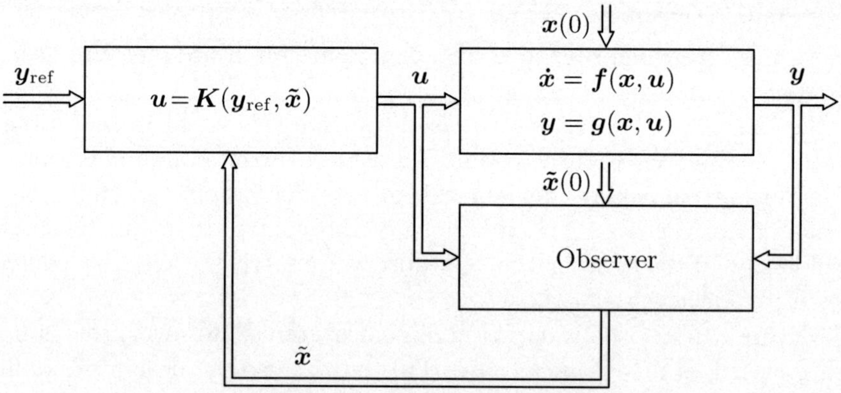

Fig. 7.2: Nonlinear state control loop with observer

We want to be thorough, so we will first define the concept of *observability*. In this context, we will make a distinction between observability and weak observability[1].

Definition 40 (Observability). *Let a system*

$$\begin{aligned} \dot{x} &= f(x, u) \qquad \textit{with} \;\; x(t_0) = x_0, \\ y &= g(x, u) \end{aligned}$$

[1] There are different concepts of observability. In addition, identical concepts are sometimes referred to by different names [47, 182]. Here we will only describe the two concepts which are particularly meaningful from a practical point of view in control engineering.

be defined for $\boldsymbol{x} \in D_{\boldsymbol{x},\text{def}} \subseteq I\!R^n$ *and* $\boldsymbol{u} \in D_{\boldsymbol{u},\text{def}} \subseteq I\!R^m$*, and let* $\boldsymbol{y} \in I\!R^r$*. If all initial vectors* $\boldsymbol{x}_0 \in D_{\boldsymbol{x}} \subseteq D_{\boldsymbol{x},\text{def}}$ *are uniquely determinable from the knowledge of* $\boldsymbol{u}(t)$ *and* $\boldsymbol{y}(t)$ *in a finite time interval* $[t_0, t_1]$ *for all* $\boldsymbol{u} \in D_{\boldsymbol{u}} \subseteq D_{\boldsymbol{u},\text{def}}$*, then the system is called observable.*

Similarly, although with fewer system requirements, we can define the concept of weak observability[2].

Definition 41 (Weak Observability). *Let a system*

$$\begin{aligned} \dot{\boldsymbol{x}} &= \boldsymbol{f}(\boldsymbol{x}, \boldsymbol{u}) \quad \text{with } \boldsymbol{x}(t_0) = \boldsymbol{x}_0, \\ \boldsymbol{y} &= \boldsymbol{g}(\boldsymbol{x}, \boldsymbol{u}) \end{aligned}$$

be defined for $\boldsymbol{x} \in D_{\boldsymbol{x},\text{def}} \subseteq I\!R^n$ *and* $\boldsymbol{u} \in D_{\boldsymbol{u},\text{def}} \subseteq I\!R^m$*, and let* $\boldsymbol{y} \in I\!R^r$*. If all initial vectors* $\boldsymbol{x}_0 \in D_{\boldsymbol{x}} \subseteq D_{\boldsymbol{x},\text{def}}$ *within a neighborhood*

$$U = \{\boldsymbol{x}_0 \in I\!R^n \mid |\boldsymbol{x}_0 - \boldsymbol{x}_\text{p}| < \rho\}$$

of a point $\boldsymbol{x}_\text{p} \in D_{\boldsymbol{x}}$ *known to us are uniquely determinable from the knowledge of* $\boldsymbol{u}(t)$ *and* $\boldsymbol{y}(t)$ *in a finite time interval* $[t_0, t_1]$ *for all* $\boldsymbol{u} \in D_{\boldsymbol{u}} \subseteq D_{\boldsymbol{u},\text{def}}$*, and if this is possible for all* $\boldsymbol{x}_\text{p} \in D_{\boldsymbol{x}}$*, then the system is called weakly observable.*

As with the system properties *stability* and *controllability*, a distinction has to be made between global and local observability. In this context, globally observable means that the system property applies to the system's domain of definition $D_{\boldsymbol{x},\text{def}} \times D_{\boldsymbol{u},\text{def}}$; locally observable[3] means that it only applies to a proper subset $D_{\boldsymbol{x}} \times D_{\boldsymbol{u}}$ of the domain. Since the observability of a nonlinear system may depend on both the state $\boldsymbol{x}$ and the input variable vector $\boldsymbol{u}$, we will highlight this dependence by the notion *observable on* $D_{\boldsymbol{x}}$ *for* $D_{\boldsymbol{u}}$.

In many cases, weakly observable systems are also observable. This is always the case for linear systems. A weakly observable but not observable example is the nonlinear autonomous system

$$\begin{aligned} \dot{x} &= -\frac{1}{x}, \\ y &= x^2, \end{aligned} \tag{7.1}$$

where the domain of definition is $D_{x,\text{def}} = I\!R\backslash\{0\}$. Obviously, it is not possible to determine the initial value x_0 from the knowledge of $y(t)$ uniquely, because the equation $y = x^2$ has the two solutions

$$x_1 = -\sqrt{y} \quad \text{and} \quad x_2 = \sqrt{y}.$$

[2] The concept of *weak observability* results from the analysis by R. Hermann and A. J. Krener [182].

[3] Local observability is often defined differently in the literature [182]. In some of these cases, however, the observability is not local in the common sense of the word but refers to a stronger property than observability.

Therefore, system (7.1) is not observable. However, it is weakly observable and even globally weakly observable, because for all values x_0 from a suitably selected neighborhood

$$U = \{x_0 \in D_{x,\mathrm{def}} \mid |x_0 - x_\mathrm{p}| < \rho\}$$

of each point $x_\mathrm{p} \in D_{x,\mathrm{def}}$ the value x_0 can be determined uniquely from $y = x^2$. On the one hand, we obtain

$$x_0 = -\sqrt{y(t_0)} \qquad \text{for} \quad x_\mathrm{p} < 0,$$

and on the other

$$x_0 = \sqrt{y(t_0)} \qquad \text{for} \quad x_\mathrm{p} > 0.$$

However, determining x_0 requires knowledge of x_p.

As another example, let us examine the system

$$\begin{aligned} \dot{x}_1 &= \alpha(x_2), \qquad \alpha(x_2) = \begin{cases} 0, & x_2 \le 0, \\ x_2^3, & x_2 > 0, \end{cases} \\ \dot{x}_2 &= u, \\ y &= x_1, \end{aligned}$$

whose domain of definition is $D_{\boldsymbol{x},\mathrm{def}} = \mathbb{R}^2$. Using

$$\dot{y} = \dot{x}_1 = \alpha(x_2)$$

for $x_2 > 0$, we obtain

$$\begin{aligned} x_1 &= y, \\ x_2 &= \sqrt[3]{\dot{y}} \end{aligned}$$

for the vectors $\boldsymbol{x}$ of the set

$$D_{\boldsymbol{x},\mathrm{part1}} = \{\boldsymbol{x} \in \mathbb{R}^2 \mid x_1 \in \mathbb{R},\ x_2 > 0\}.$$

So we can conclude that the system is observable for all $\boldsymbol{x} \in D_{\boldsymbol{x},\mathrm{part1}}$. In contrast to this result, we obtain

$$\begin{aligned} x_1 &= y, \\ x_2 &= x_2(0) + \int_0^t u(\tau)d\tau \end{aligned}$$

for $D_{\boldsymbol{x},\mathrm{part2}} = D_{\boldsymbol{x},\mathrm{def}} \setminus D_{\boldsymbol{x},\mathrm{part1}}$. We can immediately conclude that the system is not observable for all $x \in D_{\boldsymbol{x},\mathrm{part2}}$, since we cannot calculate $x_2(t)$ using

$u(t)$ and $y(t)$ due to the unknown initial value $x_2(0)$. Therefore, the system is only locally observable.

As an additional example, we will now describe a locally weakly observable system. It is given by

$$\dot{x}_1 = \beta(x_2), \qquad \beta(x_2) = \begin{cases} 0, & x_2 < 0, \\ \sin^2(x_2), & x_2 \geq 0, \end{cases}$$
$$\dot{x}_2 = u,$$
$$y = x_1.$$

This system is weakly observable, not observable, for the subset

$$D_{\boldsymbol{x},\text{part1}} = \{\boldsymbol{x} \in \mathbb{R}^2 \mid x_1 \in \mathbb{R},\ \frac{\pi}{2}i \leq x_2 < \frac{\pi}{2}(i+1),\ i = 0, 1, 2, \ldots\},$$

since the equation

$$\dot{y} = \dot{x}_1 = \sin^2(x_2)$$

has an infinite number of solutions x_2 for a value $\dot{y}$ but can be solved uniquely in each interval $[\frac{\pi}{2}i, \frac{\pi}{2}(i+1)),\ i = 0, 1, 2, \ldots$ For the set

$$D_{\boldsymbol{x},\text{part2}} = \{\boldsymbol{x} \in \mathbb{R}^2 \mid x_1 \in \mathbb{R},\ x_2 < 0\},$$

we have

$$x_1 = y,$$
$$x_2 = x_2(0) + \int_0^t u(\tau)d\tau$$

with the unknown initial value $x_2(0)$. From this we can immediately see that the system is only locally weakly observable for $\boldsymbol{x} \in D_{\boldsymbol{x},\text{part2}}$.

For nonlinear systems, a distinction has to be made between observability and weak observability, according to the definitions above. In contrast, linear systems are always observable if they are weakly observable and vice versa, which is equivalent to saying that they are always globally observable. For nonlinear systems it is possible that their observability depends on the input variable vector $\boldsymbol{u}$. The observability of a linear system, however, is always completely independent of the input variable vector $\boldsymbol{u}$.

7.1.2 Observability of Autonomous Systems

First of all, let us view the autonomous systems

$$\dot{\boldsymbol{x}} = \boldsymbol{f}(\boldsymbol{x}),$$
$$y = g(\boldsymbol{x}),$$

i. e. systems which are time-invariant and are not dependent on an input variable vector $\boldsymbol{u}$. In order to develop an observability criterion for them, we use the Lie derivative

$$L_{\boldsymbol{f}} g(\boldsymbol{x}) = \frac{\partial g(\boldsymbol{x})}{\partial \boldsymbol{x}} \boldsymbol{f}(\boldsymbol{x})$$

and the multiple Lie derivative

$$L_{\boldsymbol{f}}^k g(\boldsymbol{x}) = L_{\boldsymbol{f}} L_{\boldsymbol{f}}^{k-1} g(\boldsymbol{x}) = \frac{\partial L_{\boldsymbol{f}}^{k-1} g(\boldsymbol{x})}{\partial \boldsymbol{x}} \boldsymbol{f}(\boldsymbol{x})$$

for the determination of

$$\begin{aligned} y &= L_{\boldsymbol{f}}^0 g(\boldsymbol{x}) = g(\boldsymbol{x}), \\ \dot{y} &= L_{\boldsymbol{f}}^1 g(\boldsymbol{x}) = \frac{\partial g(\boldsymbol{x})}{\partial \boldsymbol{x}} \boldsymbol{f}(\boldsymbol{x}), \\ &\vdots \\ y^{(n-1)} &= L_{\boldsymbol{f}}^{n-1} g(\boldsymbol{x}) = L_{\boldsymbol{f}} L_{\boldsymbol{f}}^{n-2} g(\boldsymbol{x}). \end{aligned}$$

The above Lie derivatives are now combined in the vector

$$\boldsymbol{q}(\boldsymbol{x}) = \begin{bmatrix} L_{\boldsymbol{f}}^0 g(\boldsymbol{x}) \\ \vdots \\ L_{\boldsymbol{f}}^{n-1} g(\boldsymbol{x}) \end{bmatrix}.$$

Using the vector

$$\boldsymbol{z} = \begin{bmatrix} y \\ \dot{y} \\ \vdots \\ y^{(n-1)} \end{bmatrix}$$

with the new state variable $z_1, \ldots, z_n$, we obtain

$$\boldsymbol{z} = \boldsymbol{q}(\boldsymbol{x}).$$

If the inverse function $\boldsymbol{q}^{-1}(\boldsymbol{z})$ exists, $\boldsymbol{x}$ can be determined using the knowledge of $z_1 = y, z_2 = \dot{y}, \ldots, z_n = y^{(n-1)}$. The knowledge of $y(t)$ within an interval $[t_0, t_1]$ therefore leads to the knowledge of the state vector $\boldsymbol{x}(t_0)$. We can summarize this in

Theorem 86 (Observability of Autonomous Systems). *A system*

$$\begin{aligned} \dot{\boldsymbol{x}} &= \boldsymbol{f}(\boldsymbol{x}), \\ y &= g(\boldsymbol{x}) \end{aligned}$$

defined on $D_{\boldsymbol{x},\mathrm{def}} \subseteq \mathbb{R}^n$ *is observable if the inverse function* $\boldsymbol{q}^{-1}(\boldsymbol{z})$ *to determine all* $\boldsymbol{x} \in D_{\boldsymbol{x}} \subseteq D_{\boldsymbol{x},\mathrm{def}}$ *exists.*

The condition of Theorem 86 is only sufficient, since it is possible that a nonlinear system is observable even though the inverse function $\boldsymbol{q}^{-1}(\boldsymbol{z})$ does not exist. An example of such a system can be found in Exercise 7.6.

For many nonlinear systems, the inverse function $\boldsymbol{q}^{-1}$ is either impossible or very difficult to determine. We will therefore derive a criterion for weak observability which is easier to apply. For this purpose, we will expand $\boldsymbol{q}$ in a Taylor series around a point $\boldsymbol{x}_{\mathrm{p}}$:

$$\boldsymbol{z} = \boldsymbol{q}(\boldsymbol{x}_{\mathrm{p}}) + \left.\frac{\partial \boldsymbol{q}(\boldsymbol{x})}{\partial \boldsymbol{x}}\right|_{\boldsymbol{x}=\boldsymbol{x}_{\mathrm{p}}} \cdot (\boldsymbol{x} - \boldsymbol{x}_{\mathrm{p}}) + \text{remainder}.$$

Neglecting the remainder yields the equation

$$\boldsymbol{z} - \boldsymbol{q}(\boldsymbol{x}_{\mathrm{p}}) \approx \left.\frac{\partial \boldsymbol{q}(\boldsymbol{x})}{\partial \boldsymbol{x}}\right|_{\boldsymbol{x}=\boldsymbol{x}_{\mathrm{p}}} \cdot (\boldsymbol{x} - \boldsymbol{x}_{\mathrm{p}}).$$

From this equation, it is obvious that the state $\boldsymbol{x}$ can be reconstructed based on the knowledge of $\boldsymbol{z} - \boldsymbol{q}(\boldsymbol{x}_{\mathrm{p}})$ if the Jacobian matrix

$$\boldsymbol{Q}(\boldsymbol{x}_{\mathrm{p}}) = \left.\frac{\partial \boldsymbol{q}(\boldsymbol{x})}{\partial \boldsymbol{x}}\right|_{\boldsymbol{x}=\boldsymbol{x}_{\mathrm{p}}} = \left.\begin{bmatrix} \dfrac{\partial L_{\boldsymbol{f}}^{0} g(\boldsymbol{x})}{\partial \boldsymbol{x}} \\ \vdots \\ \dfrac{\partial L_{\boldsymbol{f}}^{n-1} g(\boldsymbol{x})}{\partial \boldsymbol{x}} \end{bmatrix}\right|_{\boldsymbol{x}=\boldsymbol{x}_{\mathrm{p}}}$$

is of rank n. This is equivalent to the fact that the system of linear equations

$$\boldsymbol{z} - \boldsymbol{q}(\boldsymbol{x}_{\mathrm{p}}) = \boldsymbol{Q}(\boldsymbol{x}_{\mathrm{p}})(\boldsymbol{x} - \boldsymbol{x}_{\mathrm{p}})$$

can be solved uniquely for $\boldsymbol{x}$. In this case, within a neighborhood

$$U = \{\boldsymbol{x} \in I\!R^{n} \mid |\boldsymbol{x} - \boldsymbol{x}_{\mathrm{p}}| < \rho,\ \rho > 0\}$$

of $\boldsymbol{x}_{\mathrm{p}}$ we can determine whether the system in question is observable. This yields the following theorem [42] on weak observability.

Theorem 87 (Weak Observability of Autonomous Systems). *A system*

$$\begin{aligned} \dot{\boldsymbol{x}} &= \boldsymbol{f}(\boldsymbol{x}), \\ y &= g(\boldsymbol{x}) \end{aligned}$$

which is defined on $D_{\boldsymbol{x},\mathrm{def}} \subseteq I\!R^{n}$ *is weakly observable if*

$$\operatorname{rank}\left(\frac{\partial \boldsymbol{q}(\boldsymbol{x})}{\partial \boldsymbol{x}}\right) = \operatorname{rank}\left(\begin{bmatrix} \dfrac{\partial L_{\boldsymbol{f}}^{0} g(\boldsymbol{x})}{\partial \boldsymbol{x}} \\ \vdots \\ \dfrac{\partial L_{\boldsymbol{f}}^{n-1} g(\boldsymbol{x})}{\partial \boldsymbol{x}} \end{bmatrix}\right) = n$$

holds for all $\boldsymbol{x} \in D_{\boldsymbol{x}} \subseteq D_{\boldsymbol{x},\mathrm{def}}$.

The matrix

$$\boldsymbol{Q}(\boldsymbol{x}) = \frac{\partial \boldsymbol{q}(\boldsymbol{x})}{\partial \boldsymbol{x}}$$

is also called the *observability matrix*. The application of the criterion above can be somewhat complicated, since the matrix $\boldsymbol{Q}$ depends on $\boldsymbol{x}$. However, determining the rank of $\boldsymbol{Q}$ is usually easier than determining the inverse function $\boldsymbol{q}^{-1}$ in Theorem 86.

Theorem 87 can also be derived directly from the transformation equation $\boldsymbol{z} = \boldsymbol{q}(\boldsymbol{x})$. According to the implicit function theorem, the inverse function $\boldsymbol{q}^{-1}$ exists in a neighborhood of a point $\boldsymbol{x}$ if the Jacobian matrix $\boldsymbol{Q}(\boldsymbol{x})$ has rank n at that point.

To illustrate Theorem 87, let us take the linear systems

$$\begin{aligned} \dot{\boldsymbol{x}} &= \boldsymbol{A}\boldsymbol{x}, \\ y &= \boldsymbol{c}^T \boldsymbol{x}. \end{aligned}$$

With $g(\boldsymbol{x}) = \boldsymbol{c}^T \boldsymbol{x}$, we obtain

$$\boldsymbol{q}(\boldsymbol{x}) = \begin{bmatrix} L_{\boldsymbol{f}}^0 g(\boldsymbol{x}) \\ L_{\boldsymbol{f}}^1 g(\boldsymbol{x}) \\ L_{\boldsymbol{f}}^2 g(\boldsymbol{x}) \\ \vdots \\ L_{\boldsymbol{f}}^{n-1} g(\boldsymbol{x}) \end{bmatrix} = \begin{bmatrix} \boldsymbol{c}^T \boldsymbol{x} \\ \boldsymbol{c}^T \boldsymbol{A}\boldsymbol{x} \\ \boldsymbol{c}^T \boldsymbol{A}^2 \boldsymbol{x} \\ \vdots \\ \boldsymbol{c}^T \boldsymbol{A}^{n-1} \boldsymbol{x} \end{bmatrix}$$

and hence the Jacobian matrix

$$\boldsymbol{Q}(\boldsymbol{x}) = \frac{\partial \boldsymbol{q}(\boldsymbol{x})}{\partial \boldsymbol{x}} = \begin{bmatrix} \boldsymbol{c}^T \\ \boldsymbol{c}^T \boldsymbol{A} \\ \boldsymbol{c}^T \boldsymbol{A}^2 \\ \vdots \\ \boldsymbol{c}^T \boldsymbol{A}^{n-1} \end{bmatrix}. \tag{7.2}$$

If the matrix in equation (7.2), i. e. the observability matrix, has rank n, the linear system is observable. For linear systems, this requirement is not only sufficient; it can be shown [218] that it is also necessary. This is the well-known observability criterion for linear systems, which is obviously a special case of the above criterion for the observability of nonlinear systems.

7.1.3 Example: Synchronous Generator

Let us examine a synchronous generator [325] as shown in Figure 7.3. The load angle x_1, which is a measure of the stability of the generator and whose sine is proportional to the generated power, is one of the state variables and is also the output variable. The other state variables are the frequency deviation

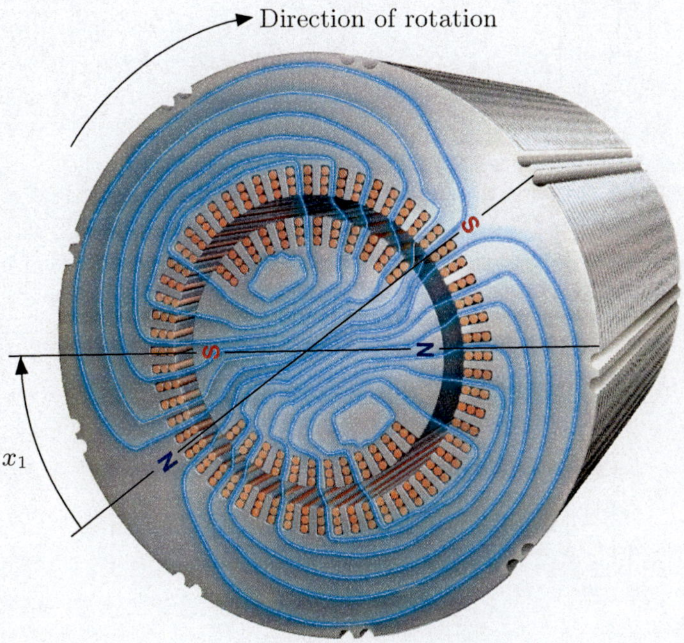

Fig. 7.3: Synchronous generator with magnetic field shown in blue and load angle x_1

x_2 of the rotor with respect to the power grid and the field flux linkage x_3 of the magnetic field. The flux linkage x_3 is either impossible or very difficult to measure. Therefore, we will estimate it using an observer, for example to perform a load angle control. The prerequisite for this is the observability of the system, which we will analyze subsequently.

The state-space model of the generator is given by

$$\begin{aligned}\dot{x}_1 &= x_2,\\ \dot{x}_2 &= b_1 - a_1 x_2 - a_2 x_3 \sin(x_1) - \frac{b_2}{2}\sin(2x_1),\\ \dot{x}_3 &= -c_1 x_3 + c_2\cos(x_1) + c_3,\\ y &= x_1.\end{aligned}$$

Here, a_1, a_2, b_1, b_2, c_1, c_2, and c_3 are all constant machine parameters. We can now determine

$$\begin{aligned}\dot{y} &= \dot{x}_1 = x_2,\\ \ddot{y} &= \dot{x}_2 = b_1 - a_1 x_2 - a_2 x_3 \sin(x_1) - \frac{b_2}{2}\sin(2x_1)\end{aligned}$$

and thus arrive at

$$\begin{bmatrix} z_1 \\ z_2 \\ z_3 \end{bmatrix} = \begin{bmatrix} y \\ \dot{y} \\ \ddot{y} \end{bmatrix} = \begin{bmatrix} x_1 \\ x_2 \\ b_1 - a_1 x_2 - a_2 x_3 \sin(x_1) - \dfrac{b_2}{2}\sin(2x_1) \end{bmatrix} = \boldsymbol{q}(\boldsymbol{x}).$$

For the relevant region of state vectors

$$D_{\boldsymbol{x}} = \{\boldsymbol{x} \in I\!R^3 \mid x_1 \in (0,\pi), x_2 \in I\!R, x_3 \in I\!R\},$$

according to Theorem 86, the synchronous machine is observable. This is because the mapping

$$\boldsymbol{z} = \boldsymbol{q}(\boldsymbol{x})$$

is invertible, and the inverse mapping is

$$\begin{bmatrix} x_1 \\ x_2 \\ x_3 \end{bmatrix} = \begin{bmatrix} z_1 \\ z_2 \\ \dfrac{b_1 - a_1 z_2 - 0.5 b_2 \sin(2z_1) - z_3}{a_2 \sin(z_1)} \end{bmatrix}$$

for all $\boldsymbol{x} \in D_{\boldsymbol{x}}$.

7.1.4 Observability of General Nonlinear Systems

Next we will examine the nonlinear systems

$$\begin{aligned} \dot{\boldsymbol{x}} &= \boldsymbol{f}(\boldsymbol{x}, \boldsymbol{u}), \\ y &= g(\boldsymbol{x}, \boldsymbol{u}) \end{aligned}$$

with an input variable vector $\boldsymbol{u}$. As in the case of autonomous systems, we will again calculate the $n-1$ total time derivatives

$$\begin{aligned} \dot{y} &= \frac{\partial g}{\partial \boldsymbol{x}}\boldsymbol{f}(\boldsymbol{x},\boldsymbol{u}) + \frac{\partial g}{\partial \boldsymbol{u}}\dot{\boldsymbol{u}} = h_1(\boldsymbol{x},\boldsymbol{u},\dot{\boldsymbol{u}}), \\ \ddot{y} &= \frac{\partial h_1}{\partial \boldsymbol{x}}\boldsymbol{f}(\boldsymbol{x},\boldsymbol{u}) + \frac{\partial h_1}{\partial \boldsymbol{u}}\dot{\boldsymbol{u}} + \frac{\partial h_1}{\partial \dot{\boldsymbol{u}}}\ddot{\boldsymbol{u}} = h_2(\boldsymbol{x},\boldsymbol{u},\dot{\boldsymbol{u}},\ddot{\boldsymbol{u}}), \\ \dddot{y} &= \frac{\partial h_2}{\partial \boldsymbol{x}}\boldsymbol{f}(\boldsymbol{x},\boldsymbol{u}) + \frac{\partial h_2}{\partial \boldsymbol{u}}\dot{\boldsymbol{u}} + \frac{\partial h_2}{\partial \dot{\boldsymbol{u}}}\ddot{\boldsymbol{u}} + \frac{\partial h_2}{\partial \ddot{\boldsymbol{u}}}\dddot{\boldsymbol{u}} = h_3(\boldsymbol{x},\boldsymbol{u},\dot{\boldsymbol{u}},\ddot{\boldsymbol{u}},\dddot{\boldsymbol{u}}), \\ &\;\;\vdots \\ y^{(n-1)} &= \frac{\partial h_{n-2}}{\partial \boldsymbol{x}}\boldsymbol{f}(\boldsymbol{x},\boldsymbol{u}) + \sum_{i=1}^{n-1} \frac{\partial h_{n-2}}{\partial \boldsymbol{u}^{(i-1)}}\boldsymbol{u}^{(i)} = h_{n-1}(\boldsymbol{x},\boldsymbol{u},\dot{\boldsymbol{u}},\ldots,\boldsymbol{u}^{(n-1)}). \end{aligned}$$

Similar to the observability of autonomous systems, we now define

$$\boldsymbol{z} = \begin{bmatrix} y \\ \dot{y} \\ \ddot{y} \\ \vdots \\ y^{(n-1)} \end{bmatrix} = \begin{bmatrix} g(\boldsymbol{x}, \boldsymbol{u}) \\ h_1(\boldsymbol{x}, \boldsymbol{u}, \dot{\boldsymbol{u}}) \\ h_2(\boldsymbol{x}, \boldsymbol{u}, \dot{\boldsymbol{u}}, \ddot{\boldsymbol{u}}) \\ \vdots \\ h_{n-1}(\boldsymbol{x}, \boldsymbol{u}, \dot{\boldsymbol{u}}, \ldots, \boldsymbol{u}^{(n-1)}) \end{bmatrix} = \boldsymbol{q}(\boldsymbol{x}, \boldsymbol{u}, \dot{\boldsymbol{u}}, \ldots, \boldsymbol{u}^{(n-1)}).$$

Again, the existence of the inverse function of

$$\boldsymbol{z} = \boldsymbol{q}(\boldsymbol{x}, \boldsymbol{u}, \dot{\boldsymbol{u}}, \ldots, \boldsymbol{u}^{(n-1)})$$

is essential for the observability of the system. From the calculations above, we obtain

Theorem 88 (Observability of Nonlinear Systems). *A system*

$$\dot{\boldsymbol{x}} = \boldsymbol{f}(\boldsymbol{x}, \boldsymbol{u}),$$
$$y = g(\boldsymbol{x}, \boldsymbol{u})$$

which is defined on $D_{\boldsymbol{x},\mathrm{def}} \subseteq I\!R^n$ and for all $\boldsymbol{u}(t) \in C_m^{n-1}$ is observable if the inverse function $\boldsymbol{q}^{-1}(\boldsymbol{z}, \boldsymbol{u}, \dot{\boldsymbol{u}}, \ldots, \boldsymbol{u}^{(n-1)})$ to determine all $\boldsymbol{x} \in D_{\boldsymbol{x}} \subseteq D_{\boldsymbol{x},\mathrm{def}}$ exists for all $\boldsymbol{u}(t) \in C_{\boldsymbol{u}} \subseteq C_m^{n-1}$.

Here, C_m^{n-1} is the space of $(n-1)$ - times continuously differentiable m - dimensional vector functions. If $\boldsymbol{q}$ is a global diffeomorphism, the system is globally observable.

In Section 7.1.1, we noted that the observability of a nonlinear system, in contrast to that of a linear system, can depend on the input variable vector $\boldsymbol{u}$. Theorem 88 substantiates this statement, because the mapping $\boldsymbol{z} = \boldsymbol{q}(\boldsymbol{x})$ and its solvability for $\boldsymbol{x}$, i.e. the calculation of its inverse, can also depend on $\boldsymbol{u}$. The simple example

$$\dot{x}_1 = -x_2 \cdot u,$$
$$\dot{x}_2 = -x_1 - x_2,$$
$$y = x_1$$

illustrates this fact. For $u = 0$ the system is not observable, while for $u \neq 0$ it is.

The above theorem is difficult to apply in practice. In simple cases only, it is possible to determine the inverse function

$$\boldsymbol{x} = \boldsymbol{q}^{-1}(\boldsymbol{z}, \boldsymbol{u}, \dot{\boldsymbol{u}}, \ldots, \boldsymbol{u}^{(n-1)}).$$

Again, it may be easier to verify weak observability than to verify observability. Analogous to the case of autonomous systems, we can derive

Theorem 89 (Weak Observability of Nonlinear Systems). *A system*

$$\dot{\boldsymbol{x}} = \boldsymbol{f}(\boldsymbol{x}, \boldsymbol{u}),$$
$$y = g(\boldsymbol{x}, \boldsymbol{u})$$

which is defined on $D_{\boldsymbol{x},\text{def}} \subseteq I\!R^n$ and for all $\boldsymbol{u}(t) \in C_m^{n-1}$ is weakly observable if the condition

$$\operatorname{rank}\left(\frac{\partial \boldsymbol{q}(\boldsymbol{x}, \boldsymbol{u}, \dot{\boldsymbol{u}}, \ldots, \boldsymbol{u}^{(n-1)})}{\partial \boldsymbol{x}}\right) = \operatorname{rank}\left(\begin{bmatrix} \dfrac{\partial g(\boldsymbol{x}, \boldsymbol{u})}{\partial \boldsymbol{x}} \\ \dfrac{\partial h_1(\boldsymbol{x}, \boldsymbol{u}, \dot{\boldsymbol{u}})}{\partial \boldsymbol{x}} \\ \dfrac{\partial h_2(\boldsymbol{x}, \boldsymbol{u}, \dot{\boldsymbol{u}}, \ddot{\boldsymbol{u}})}{\partial \boldsymbol{x}} \\ \vdots \\ \dfrac{\partial h_{n-1}(\boldsymbol{x}, \boldsymbol{u}, \dot{\boldsymbol{u}}, \ldots, \boldsymbol{u}^{(n-1)})}{\partial \boldsymbol{x}} \end{bmatrix}\right) = n$$

is fulfilled for all $\boldsymbol{x} \in D_{\boldsymbol{x}} \subseteq D_{\boldsymbol{x},\text{def}}$ and $\boldsymbol{u}(t) \in C_{\boldsymbol{u}} \subseteq C_m^{n-1}$.

In contrast to Theorem 88, Theorem 89 can be applied at least for single points, if the analytical verification of the rank condition fails, by calculating the rank or the determinant of the observability matrix

$$\boldsymbol{Q}(\boldsymbol{x}, \boldsymbol{u}, \dot{\boldsymbol{u}}, \ldots, \boldsymbol{u}^{(n-1)}) = \frac{\partial \boldsymbol{q}(\boldsymbol{x}, \boldsymbol{u}, \dot{\boldsymbol{u}}, \ldots, \boldsymbol{u}^{(n-1)})}{\partial \boldsymbol{x}}$$

at given grid points $(\boldsymbol{x}, \boldsymbol{u}, \dot{\boldsymbol{u}}, \ldots, \boldsymbol{u}^{(n-1)})$.

7.1.5 Nonlinear Observability Canonical Form

Let us consider an observable system

$$\begin{aligned} \dot{\boldsymbol{x}} &= \boldsymbol{f}(\boldsymbol{x}, \boldsymbol{u}), \\ y &= g(\boldsymbol{x}, \boldsymbol{u}). \end{aligned} \tag{7.3}$$

We assume that we can determine the mapping

$$\boldsymbol{z} = \boldsymbol{q}(\boldsymbol{x}, \boldsymbol{u}, \dot{\boldsymbol{u}}, \ldots, \boldsymbol{u}^{(n-1)}) \tag{7.4}$$

and the inverse mapping

$$\boldsymbol{x} = \boldsymbol{q}^{-1}(\boldsymbol{z}, \boldsymbol{u}, \dot{\boldsymbol{u}}, \ldots, \boldsymbol{u}^{(n-1)}). \tag{7.5}$$

In this case, it is possible to convert the system representation (7.3) into the representation

$$\begin{bmatrix} \dot{z}_1 \\ \dot{z}_2 \\ \vdots \\ \dot{z}_{n-1} \\ \dot{z}_n \end{bmatrix} = \begin{bmatrix} \dot{y} \\ \ddot{y} \\ \vdots \\ y^{(n-1)} \\ y^{(n)} \end{bmatrix} = \begin{bmatrix} z_2 \\ z_3 \\ \vdots \\ z_n \\ h_n\left(\boldsymbol{q}^{-1}(\boldsymbol{z}, \boldsymbol{u}, \dot{\boldsymbol{u}}, \dots, \boldsymbol{u}^{(n-1)}), \boldsymbol{u}, \dot{\boldsymbol{u}}, \dots, \boldsymbol{u}^{(n)}\right) \end{bmatrix},$$

$$y = z_1$$

by means of the mapping (7.5) and

$$y^{(n)} = h_n(\boldsymbol{x}, \boldsymbol{u}, \dot{\boldsymbol{u}}, \dots, \boldsymbol{u}^{(n)}) = \frac{\partial h_{n-1}}{\partial \boldsymbol{x}} \boldsymbol{f}(\boldsymbol{x}, \boldsymbol{u}) + \sum_{i=1}^{n} \frac{\partial h_{n-1}}{\partial \boldsymbol{u}^{(i-1)}} \boldsymbol{u}^{(i)}.$$

Finally, by abbreviating

$$\varphi\left(\boldsymbol{z}, \boldsymbol{u}, \dot{\boldsymbol{u}}, \dots, \boldsymbol{u}^{(n)}\right) = h_n\left(\boldsymbol{q}^{-1}(\boldsymbol{z}, \boldsymbol{u}, \dot{\boldsymbol{u}}, \dots, \boldsymbol{u}^{(n-1)}), \boldsymbol{u}, \dot{\boldsymbol{u}}, \dots, \boldsymbol{u}^{(n)}\right),$$

we obtain

$$\dot{\boldsymbol{z}} = \begin{bmatrix} z_2 \\ \vdots \\ z_n \\ \varphi\left(\boldsymbol{z}, \boldsymbol{u}, \dot{\boldsymbol{u}}, \dots, \boldsymbol{u}^{(n)}\right) \end{bmatrix}, \tag{7.6}$$

$$y = z_1.$$

The system representation (7.6) is referred to as the *nonlinear observability canonical form* [493]. Figure 7.4 shows the corresponding block diagram. If a system is represented in this form, the observability can be inferred directly from the system equations. This is because all states z_i affect the output variable $y = z_1$ via the integrator chain in equation (7.6). Furthermore, all state values z_i can be determined from the output variable y and its derivatives $y^{(i)}$. Hence, a system that is represented in nonlinear observability canonical form is always observable. Furthermore, any globally observable system can be transformed into the observability canonical form.

As an example, let us take the system

$$\begin{aligned} \dot{x}_1 &= -x_2 + u, \\ \dot{x}_2 &= -x_3, \\ \dot{x}_3 &= -x_1^3, \\ y &= x_1. \end{aligned} \tag{7.7}$$

In this case, we obtain

$$\boldsymbol{q}(\boldsymbol{x}, u, \dot{u}) = \begin{bmatrix} z_1 \\ z_2 \\ z_3 \end{bmatrix} = \begin{bmatrix} y \\ \dot{y} \\ \ddot{y} \end{bmatrix} = \begin{bmatrix} x_1 \\ -x_2 + u \\ x_3 + \dot{u} \end{bmatrix}$$

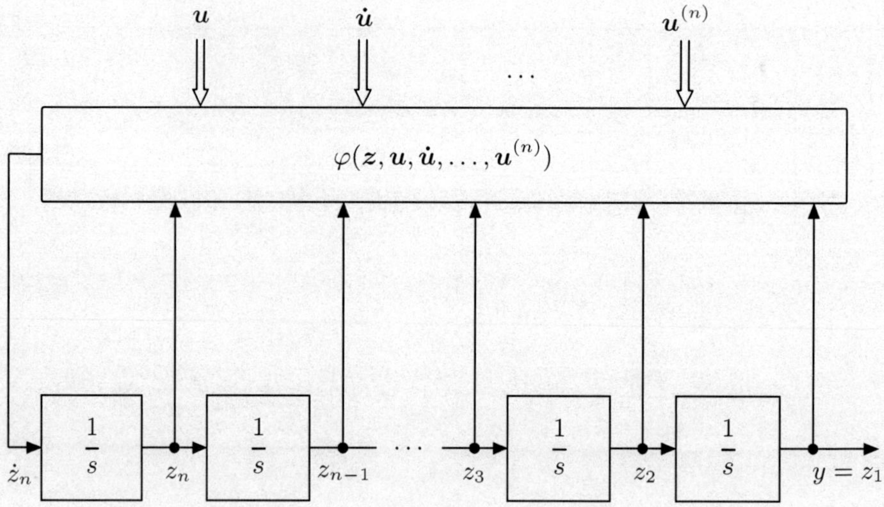

Fig. 7.4: System in nonlinear observability canonical form

for the mapping (7.4), while its inverse mapping is given by

$$\boldsymbol{q}^{-1}(\boldsymbol{z}, u, \dot{u}) = \begin{bmatrix} x_1 \\ x_2 \\ x_3 \end{bmatrix} = \begin{bmatrix} z_1 \\ -z_2 + u \\ z_3 - \dot{u} \end{bmatrix}. \tag{7.8}$$

Since the inverse mapping $\boldsymbol{q}^{-1}$ exists for all $\boldsymbol{x} \in I\!R^3$, system (7.7) is observable. With the derivatives

$$\begin{aligned} \dot{z}_1 &= z_2, \\ \dot{z}_2 &= z_3, \\ \dot{z}_3 &= \dddot{y} = \dot{x}_3 + \ddot{u} = -x_1^3 + \ddot{u}, \end{aligned}$$

we obtain the nonlinear observability canonical form

$$\dot{\boldsymbol{z}} = \begin{bmatrix} z_2 \\ z_3 \\ -z_1^3 + \ddot{u} \end{bmatrix}. \tag{7.9}$$

This result can also be obtained more formally by using the differential equation (3.95) of the transformed system that we derived in Section 3.3.1 on p. 263. To refresh our memory, this differential equation is derived once again below. Inserting the diffeomorphism

$$\boldsymbol{x} = \boldsymbol{q}^{-1}(\boldsymbol{z}, \boldsymbol{u}, \ldots, \boldsymbol{u}^{(n-1)})$$

into the system equation (7.3) yields

$$\begin{aligned}\frac{d\boldsymbol{q}^{-1}(\boldsymbol{z},\boldsymbol{u},\ldots,\boldsymbol{u}^{(n-1)})}{dt} &= \frac{\partial \boldsymbol{q}^{-1}(\boldsymbol{z},\boldsymbol{u},\ldots,\boldsymbol{u}^{(n-1)})}{\partial \boldsymbol{z}}\cdot\dot{\boldsymbol{z}}\\ &\quad+\sum_{j=0}^{n-1}\frac{\partial \boldsymbol{q}^{-1}(\boldsymbol{z},\boldsymbol{u},\ldots,\boldsymbol{u}^{(n-1)})}{\partial \boldsymbol{u}^{(j)}}\cdot\boldsymbol{u}^{(j+1)}\\ &= \boldsymbol{f}(\boldsymbol{q}^{-1}(\boldsymbol{z},\boldsymbol{u},\ldots,\boldsymbol{u}^{(n-1)}),\boldsymbol{u}),\end{aligned}$$

from which

$$\begin{aligned}\dot{\boldsymbol{z}} &= \left(\frac{\partial \boldsymbol{q}^{-1}(\boldsymbol{z},\boldsymbol{u},\ldots,\boldsymbol{u}^{(n-1)})}{\partial \boldsymbol{z}}\right)^{-1}\\ &\quad\cdot\left(\boldsymbol{f}(\boldsymbol{q}^{-1}(\boldsymbol{z},\boldsymbol{u},\ldots,\boldsymbol{u}^{(n-1)}),\boldsymbol{u})-\sum_{j=0}^{n-1}\frac{\partial \boldsymbol{q}^{-1}(\boldsymbol{z},\boldsymbol{u},\ldots,\boldsymbol{u}^{(n-1)})}{\partial \boldsymbol{u}^{(j)}}\cdot\boldsymbol{u}^{(j+1)}\right)\end{aligned}$$

follows. After inserting diffeomorphism (7.8) from our exemplary system, we obtain the transformed differential equation

$$\dot{\boldsymbol{z}} = \begin{bmatrix}1 & 0 & 0\\ 0 & -1 & 0\\ 0 & 0 & 1\end{bmatrix}\left(\begin{bmatrix}z_2\\ -z_3+\dot{u}\\ -z_1^3\end{bmatrix}-\begin{bmatrix}0\\1\\0\end{bmatrix}\dot{u}-\begin{bmatrix}0\\0\\-1\end{bmatrix}\ddot{u}\right)=\begin{bmatrix}z_2\\ z_3\\ -z_1^3+\ddot{u}\end{bmatrix},$$

and thus equation (7.9) follows once again.

7.1.6 Observability of Control-Affine Systems

We will now address the control-affine systems

$$\begin{aligned}\dot{\boldsymbol{x}} &= \boldsymbol{a}(\boldsymbol{x})+\boldsymbol{b}(\boldsymbol{x})\cdot u,\\ y &= c(\boldsymbol{x}).\end{aligned}\tag{7.10}$$

If a system of this kind has a relative degree of n, it can be transformed into nonlinear controller canonical form

$$\begin{bmatrix}\dot{z}_1\\ \vdots\\ \dot{z}_{n-1}\\ \dot{z}_n\end{bmatrix}=\begin{bmatrix}z_2\\ \vdots\\ z_n\\ L_{\boldsymbol{a}}^n c(\boldsymbol{t}^{-1}(\boldsymbol{z}))+L_{\boldsymbol{b}}L_{\boldsymbol{a}}^{n-1}c(\boldsymbol{t}^{-1}(\boldsymbol{z}))u\end{bmatrix}\tag{7.11}$$
$$y = z_1$$

using the diffeomorphism

$$\boldsymbol{z}=\boldsymbol{q}(\boldsymbol{x})=\boldsymbol{t}(\boldsymbol{x})=\begin{bmatrix}c(\boldsymbol{x})\\ L_{\boldsymbol{a}}c(\boldsymbol{x})\\ \vdots\\ L_{\boldsymbol{a}}^{n-1}c(\boldsymbol{x})\end{bmatrix},\tag{7.12}$$

as we know from Section 5.2.1, p. 359 et seq. Since the nonlinear controller canonical form is identical to the nonlinear observability canonical form in the case of control-affine systems (7.11), we obtain

Theorem 90 (Observability of Control-Affine Systems). *A system*

$$\begin{aligned}\dot{\boldsymbol{x}} &= \boldsymbol{a}(\boldsymbol{x}) + \boldsymbol{b}(\boldsymbol{x}) \cdot u, \qquad \boldsymbol{x} \in D_{\boldsymbol{x},\text{def}} \subseteq I\!R^n,\\ y &= c(\boldsymbol{x})\end{aligned}$$

is weakly observable if its relative degree is n.

Note that the theorem above only guarantees weak observability and not observability. This is because the mapping (7.12) may be a diffeomorphism only in a subset of the region of interest $D_{\boldsymbol{x}}$ and thus does not guarantee bijectivity everywhere in $D_{\boldsymbol{x}}$. However, if the diffeomorphism holds everywhere in the region $D_{\boldsymbol{x}}$ or is even global, the system is observable.

Theorem 90 is only sufficient, but not necessary. The reverse conclusion, that a control-affine system with a relative degree $\delta < n$ is not observable, generally does not apply. As an example, let us take the system

$$\begin{aligned}\dot{\boldsymbol{x}} &= \boldsymbol{A}\boldsymbol{x} + \boldsymbol{b}u,\\ y &= \boldsymbol{c}^T\boldsymbol{x}\end{aligned}$$

with

$$\boldsymbol{A} = \begin{bmatrix} 0 & -1 \\ 1 & 0 \end{bmatrix}, \quad \boldsymbol{b} = \begin{bmatrix} 1 \\ 0 \end{bmatrix}, \quad \boldsymbol{c}^T = \begin{bmatrix} 1 & 0 \end{bmatrix}.$$

It is of relative degree $\delta = 1$ and the regular observability matrix is given by

$$M_{\text{obs}} = \begin{bmatrix} \boldsymbol{c}^T \\ \boldsymbol{c}^T\boldsymbol{A} \end{bmatrix} = \begin{bmatrix} 1 & 0 \\ 0 & -1 \end{bmatrix}.$$

Thus, this is an example of an observable system with a relative degree of $\delta < n$.

Now we will examine the case with a relative degree less than n. As in the more general case $\dot{\boldsymbol{x}} = \boldsymbol{f}(\boldsymbol{x}, \boldsymbol{u})$, $y = g(\boldsymbol{x}, \boldsymbol{u})$, which we discussed in Section 7.1.4, we can calculate

$$\begin{aligned}
y &= c(\boldsymbol{x}),\\
\dot{y} &= L_{\boldsymbol{a}}c(\boldsymbol{x}) + L_{\boldsymbol{b}}c(\boldsymbol{x})u,\\
\ddot{y} &= L_{\boldsymbol{a}}^2c(\boldsymbol{x}) + L_{\boldsymbol{b}}L_{\boldsymbol{a}}c(\boldsymbol{x})u + L_{\boldsymbol{a}}L_{\boldsymbol{b}}c(\boldsymbol{x})u + L_{\boldsymbol{b}}^2c(\boldsymbol{x})u^2 + L_{\boldsymbol{b}}c(\boldsymbol{x})\dot{u},\\
&\ \vdots\\
y^{(n-1)} &= L_{\boldsymbol{a}}^{n-1}c(\boldsymbol{x}) + L_{\boldsymbol{b}}L_{\boldsymbol{a}}^{n-2}c(\boldsymbol{x})u + \ldots + L_{\boldsymbol{b}}^{n-1}c(\boldsymbol{x})u^{(n-1)} + L_{\boldsymbol{b}}c(\boldsymbol{x})u^{(n-2)}.
\end{aligned}$$

In this way, the transformation equation

$$\boldsymbol{z} = \begin{bmatrix} y \\ \dot{y} \\ \vdots \\ y^{(n-1)} \end{bmatrix} = \begin{bmatrix} c(\boldsymbol{x}) \\ L_{\boldsymbol{a}}c(\boldsymbol{x}) + L_{\boldsymbol{b}}c(\boldsymbol{x})u \\ \vdots \\ L_{\boldsymbol{a}}^{(n-1)}c(\boldsymbol{x}) + \ldots + L_{\boldsymbol{b}}c(\boldsymbol{x})u^{(n-2)} \end{bmatrix} = \boldsymbol{q}(\boldsymbol{x}, u, \dot{u}, \ldots, u^{(n-2)})$$

is obtained. As discussed in Section 5.2.4, p. 372 et seq., the terms $L_{\boldsymbol{b}}L_{\boldsymbol{a}}^{k-1}c(\boldsymbol{x})$ are identical to zero for all $k < \delta$.

We can examine the observability and the weak observability of system (7.10) using Theorems 88 and 89, respectively. In using the latter, we have to verify whether the Jacobian matrix

$$\frac{\partial \boldsymbol{q}}{\partial \boldsymbol{x}} = \begin{bmatrix} \dfrac{\partial c(\boldsymbol{x})}{\partial \boldsymbol{x}} \\ \dfrac{\partial L_{\boldsymbol{a}}c(\boldsymbol{x})}{\partial \boldsymbol{x}} + \dfrac{\partial L_{\boldsymbol{b}}c(\boldsymbol{x})}{\partial \boldsymbol{x}}u \\ \dfrac{\partial L_{\boldsymbol{a}}^2c(\boldsymbol{x})}{\partial \boldsymbol{x}} + \dfrac{\partial L_{\boldsymbol{b}}L_{\boldsymbol{a}}c(\boldsymbol{x})}{\partial \boldsymbol{x}}u + \dfrac{\partial L_{\boldsymbol{a}}L_{\boldsymbol{b}}c(\boldsymbol{x})}{\partial \boldsymbol{x}}u + \dfrac{\partial L_{\boldsymbol{b}}^2c(\boldsymbol{x})}{\partial \boldsymbol{x}}u^2 + \dfrac{\partial L_{\boldsymbol{b}}c(\boldsymbol{x})}{\partial \boldsymbol{x}}\dot{u} \\ \vdots \\ \dfrac{\partial L_{\boldsymbol{a}}^{n-1}c(\boldsymbol{x})}{\partial \boldsymbol{x}} + \dfrac{\partial L_{\boldsymbol{b}}L_{\boldsymbol{a}}^{n-2}c(\boldsymbol{x})}{\partial \boldsymbol{x}}u + \ldots + \dfrac{\partial L_{\boldsymbol{b}}c(\boldsymbol{x})}{\partial \boldsymbol{x}}u^{(n-2)} \end{bmatrix}$$

has rank n. Note that the rank condition of Theorem 89 again depends on u. Evidently, there may be input signals u, so that the rank condition is not fulfilled and the system is not weakly observable. However, no statement about weak observability can be made using Theorem 89 if the rank condition is not fulfilled. This is because Theorem 89 provides a sufficient condition only.

Let us view the simple example

$$\begin{aligned} \dot{x}_1 &= -x_2 + x_2 \cdot u, \\ \dot{x}_2 &= -x_1, \\ y &= x_1. \end{aligned} \tag{7.13}$$

Here,

$$\frac{\partial \boldsymbol{q}}{\partial \boldsymbol{x}} = \begin{bmatrix} \dfrac{\partial c(\boldsymbol{x})}{\partial \boldsymbol{x}} \\ \dfrac{\partial L_{\boldsymbol{a}}c(\boldsymbol{x})}{\partial \boldsymbol{x}} + \dfrac{\partial L_{\boldsymbol{b}}c(\boldsymbol{x})}{\partial \boldsymbol{x}}u \end{bmatrix} = \begin{bmatrix} 1 & 0 \\ 0 & u-1 \end{bmatrix}.$$

For $u \neq 1$, we obtain

$$\operatorname{rank}\left(\frac{\partial \boldsymbol{q}}{\partial \boldsymbol{x}}\right) = 2.$$

Therefore the system is weakly observable. If $u = 1$, it follows that $\partial \boldsymbol{q}/\partial \boldsymbol{x}$ only has rank one. In fact, system (7.13) is not observable for $u = 1$, since in this case

$$\begin{aligned} \dot{x}_1 &= 0, \\ \dot{x}_2 &= -x_1, \\ y &= x_1 \end{aligned} \tag{7.14}$$

holds and, for this system, we cannot determine the value of x_2 from the knowledge of y and u. This can also be shown by applying the necessary and sufficient observability criterion for linear systems, because the observability matrix from equation (7.2) for system (7.14),

$$\begin{bmatrix} \boldsymbol{c}^T \\ \boldsymbol{c}^T \boldsymbol{A} \end{bmatrix} = \begin{bmatrix} 1 & 0 \\ 0 & 0 \end{bmatrix},$$

only has rank one.

7.2 Canonical Forms and the Canonical Form Observer

We have already noted the usefulness of system representations in special coordinates, meaning canonical forms, for various tasks or concepts such as controllability, controller design using feedback linearization, and observability. For example, these coordinates and the canonical forms enable us to easily assess the controllability and observability of a system.

The best known canonical forms are the *controller canonical form*

$$\dot{\boldsymbol{x}} = \begin{bmatrix} 0 & 1 & 0 & \cdots & 0 \\ 0 & 0 & 1 & \cdots & 0 \\ \vdots & \vdots & \vdots & \ddots & \vdots \\ 0 & 0 & 0 & \cdots & 1 \\ -a_0 & -a_1 & -a_2 & \cdots & -a_{n-1} \end{bmatrix} \boldsymbol{x} + \begin{bmatrix} 0 \\ 0 \\ \vdots \\ 0 \\ 1 \end{bmatrix} u,$$

$$y = \begin{bmatrix} b_0 \; b_1 \; \cdots \; b_m \; 0 \cdots 0 \end{bmatrix} \boldsymbol{x},$$

and the *observer canonical form*

$$\dot{\boldsymbol{x}} = \begin{bmatrix} 0 & 0 & 0 & \cdots & 0 & -a_0 \\ 1 & 0 & 0 & \cdots & 0 & -a_1 \\ 0 & 1 & 0 & \cdots & 0 & -a_2 \\ \vdots & \vdots & \vdots & \ddots & \vdots & \vdots \\ 0 & 0 & 0 & \cdots & 1 & -a_{n-1} \end{bmatrix} \boldsymbol{x} + \begin{bmatrix} b_0 \\ \vdots \\ b_m \\ 0 \\ \vdots \\ 0 \end{bmatrix} u,$$

$$y = x_n$$

of linear systems[4]. The term *controller canonical form* is derived from the fact that the design of a state controller

$$u = -\boldsymbol{k}^T \boldsymbol{x}$$

is particularly simple for systems in this representation. In the case of the observer canonical form, the name has a similar origin, since this form makes the design of a Luenberger observer easily possible.

For nonlinear systems as well, the names of the canonical forms were coined because of these possibilities. For example, when designing a controller using feedback linearization, we used the nonlinear controller canonical form in Section 5.2.1, p. 359 et seq. Table 7.1 gives us an overview of some nonlinear canonical forms used in control theory.

In the case of the nonlinear observer canonical form [248, 494, 495]

$$\dot{\boldsymbol{x}} = \begin{bmatrix} -\,a_1(x_n, u, \dot{u}, \ldots, u^{(n)}) \\ x_1 - a_2(x_n, u, \dot{u}, \ldots, u^{(n-1)}) \\ \vdots \\ x_{n-2} - a_{n-1}(x_n, u, \dot{u}, \ddot{u}) \\ x_{n-1} - a_n(x_n, u, \dot{u}) \end{bmatrix} = \boldsymbol{A}\boldsymbol{x} - \boldsymbol{a}(x_n, \boldsymbol{u}^{0\ldots n}), \tag{7.15}$$
$$y = g(x_n), \quad \text{and} \quad x_n = g^{-1}(y)$$

with

$$\boldsymbol{A} = \begin{bmatrix} 0 & 0 & \cdots & 0 & 0 \\ 1 & 0 & \cdots & 0 & 0 \\ 0 & 1 & \cdots & 0 & 0 \\ \vdots & \vdots & \ddots & \vdots & \vdots \\ 0 & 0 & \cdots & 1 & 0 \end{bmatrix}, \quad \boldsymbol{u}^{0\ldots n} = \left[u \; \dot{u} \; \cdots \; u^{(n)}\right]^T,$$
$$\boldsymbol{a}(x_n, \boldsymbol{u}^{0\ldots n}) = \left[a_1(x_n, \boldsymbol{u}^{0\ldots n}) \; \cdots \; a_n(x_n, \boldsymbol{u}^{0\ldots n})\right]^T,$$

we are able to easily design an observer which is termed a *canonical form observer* or, synonymously, a *normal form observer*. The dynamics of this type of an observer are given by

$$\dot{\tilde{\boldsymbol{x}}} = \boldsymbol{A}\tilde{\boldsymbol{x}} - \boldsymbol{a}(g^{-1}(y), \boldsymbol{u}^{0\ldots n}) + \boldsymbol{l}(g^{-1}(y) - \boldsymbol{c}^T\tilde{\boldsymbol{x}}),$$
$$\boldsymbol{c}^T = \left[0 \cdots 0 \; 1\right],$$

consist of a model of system (7.15) and the observer gain $\boldsymbol{l}$. These are similar

[4] The nomenclature is not consistent in the literature. Although the one used here is the most common, the terms *controllable canonical form*, *control canonical form*, and *controllability canonical form* are used synonymously. Similarly, the term *observable canonical form* is used instead of *observer canonical form*. Furthermore, the term *normal form* is used synonymously with the term *canonical form*.

Table 7.1: Canonical forms

1 Nonlinear controller canonical form

$$\dot{\boldsymbol{x}} = \begin{bmatrix} x_2 \\ x_3 \\ \vdots \\ x_n \\ \alpha(\boldsymbol{x}) \end{bmatrix} + \begin{bmatrix} 0 \\ 0 \\ \vdots \\ 0 \\ \beta(\boldsymbol{x}) \end{bmatrix} u, \qquad y = g(\boldsymbol{x})$$

Characteristics	:	controllable if $\beta(\boldsymbol{x}) \neq 0$, control-affine
Usage	:	designing controllers, verifying controllability
Equivalence	:	to the nonlinear observability canonical form in the case of control-affine systems if $y = x_1$
Special cases	:	linear controller canonical form, Brunovsky canonical form

2 Generalized controller canonical form

$$\dot{\boldsymbol{x}} = \begin{bmatrix} x_2 \\ x_3 \\ \vdots \\ x_n \\ f(\boldsymbol{x}, u, \dot{u}, \ldots, u^{(n)}) \end{bmatrix}, \qquad y = g(\boldsymbol{x})$$

Usage	:	designing controllers
Special cases	:	(nonlinear) controller canonical form

3 Nonlinear observability canonical form

$$\dot{\boldsymbol{x}} = \begin{bmatrix} x_2 \\ x_3 \\ \vdots \\ x_n \\ f(\boldsymbol{x}, u, \dot{u}, \ldots, u^{(n)}) \end{bmatrix}, \qquad y = x_1$$

Characteristics	:	observable
Usage	:	verifying observability, designing high-gain observers
Equivalence	:	to the nonlinear controller canonical form in the case of control-affine systems and to the generalized controller canonical form if $g(x) = x_1$
Special cases	:	Brunovsky canonical form

Table 7.1: Canonical forms - continued

4 Brunovsky canonical form

$$\dot{\boldsymbol{x}} = \begin{bmatrix} x_2 \\ x_3 \\ \vdots \\ x_n \\ 0 \end{bmatrix} + \begin{bmatrix} 0 \\ 0 \\ \vdots \\ 0 \\ 1 \end{bmatrix} u, \qquad \text{Characteristics : controllable, observable if } y = x_1$$

5 Byrnes-Isidori and generalized Byrnes-Isidori canonical form

$$\dot{\boldsymbol{x}} = \begin{bmatrix} x_2 \\ x_3 \\ \vdots \\ x_\delta \\ \alpha(\boldsymbol{x}) \\ q_{\delta+1}(\boldsymbol{x}) \\ \vdots \\ q_n(\boldsymbol{x}) \end{bmatrix} + \begin{bmatrix} 0 \\ 0 \\ \vdots \\ 0 \\ \beta(\boldsymbol{x}) \\ 0 \\ \vdots \\ 0 \end{bmatrix} u, \quad y = x_1; \qquad \dot{\boldsymbol{x}} = \begin{bmatrix} x_2 \\ x_3 \\ \vdots \\ x_\delta \\ \varphi(\boldsymbol{x}, u) \\ q_{\delta+1}(\boldsymbol{x}) \\ \vdots \\ q_n(\boldsymbol{x}) \end{bmatrix}, \quad y = x_1$$

Usage : designing controllers

6 Nonlinear controllability canonical form

$$\dot{\boldsymbol{x}} = \begin{bmatrix} -a_1(x_n) \\ x_1 - a_2(x_n) \\ \vdots \\ x_{n-1} - a_n(x_n) \end{bmatrix} + \begin{bmatrix} 1 \\ 0 \\ \vdots \\ 0 \end{bmatrix} u, \qquad y = g(\boldsymbol{x})$$

Characteristics : controllable Usage : verifying controllability

7 Nonlinear observer canonical form

$$\dot{\boldsymbol{x}} = \begin{bmatrix} -a_1(x_n, u, \dot{u}, \ldots, u^{(n)}) \\ x_1 - a_2(x_n, u, \dot{u}, \ldots, u^{(n-1)}) \\ \vdots \\ x_{n-2} - a_{n-1}(x_n, u, \dot{u}, \ddot{u}) \\ x_{n-1} - a_n(x_n, u, \dot{u}) \end{bmatrix}, \quad y = g(x_n) \text{ and } x_n = g^{-1}(y)$$

Characteristics : observable

Usage : designing canonical form observers

Special cases : linear observer canonical form

to a linear observer. The constant vector

$$\boldsymbol{l}^T = \left[l_1 \cdots l_n\right]$$

is freely selectable. Note that we have used the system's output value y instead of the observer's output value $\tilde{y}$ for calculating the observer's dynamics. This is possible because we can measure the output variable y.

In order to calculate the estimation error

$$\boldsymbol{e} = \boldsymbol{x} - \tilde{\boldsymbol{x}},$$

we will first determine its time derivative

$$\begin{aligned}\dot{\boldsymbol{e}} &= \dot{\boldsymbol{x}} - \dot{\tilde{\boldsymbol{x}}}\\ &= \boldsymbol{A}\boldsymbol{x} - \boldsymbol{a}(g^{-1}(y), \boldsymbol{u}^{0\cdots n}) - \boldsymbol{A}\tilde{\boldsymbol{x}} + \boldsymbol{a}(g^{-1}(y), \boldsymbol{u}^{0\cdots n}) - \boldsymbol{l}(g^{-1}(y) - \boldsymbol{c}^T\tilde{\boldsymbol{x}})\\ &= \boldsymbol{A}\boldsymbol{x} - \boldsymbol{A}\tilde{\boldsymbol{x}} - \boldsymbol{l}(\boldsymbol{c}^T\boldsymbol{x} - \boldsymbol{c}^T\tilde{\boldsymbol{x}}).\end{aligned}$$

The terms $\boldsymbol{a}(g^{-1}(y), \boldsymbol{u}^{0\cdots n})$ are identical in the system dynamics and the observer dynamics above, so they cancel each other out. Thus we obtain the linear error dynamics

$$\dot{\boldsymbol{e}} = (\boldsymbol{A} - \boldsymbol{l}\boldsymbol{c}^T)\boldsymbol{e}.$$

This approach to designing a nonlinear observer is known as *exact error linearization*. The linear dynamics and the fact that we are free to choose the eigenvalues of $\boldsymbol{A} - \boldsymbol{l}\boldsymbol{c}^T$ makes the canonical form observer so effective. However, normally we have to transform a system representation into the observer canonical form to design an observer of this type. This is generally a laborious procedure, since we have to solve some partial differential equations to obtain the appropriate diffeomorphism [43, 385, 386, 494].

7.3 Luenberger Observers for Nonlinear Control Loops

Luenberger's observer theory was developed for linear systems. It can be extended to nonlinear systems for some special cases. For example, a Luenberger observer can be used [289] in the case of a control loop with a linear plant

$$\begin{aligned}\dot{\boldsymbol{x}} &= \boldsymbol{A}\boldsymbol{x} + \boldsymbol{B}\boldsymbol{u},\\ \boldsymbol{y} &= \boldsymbol{C}\boldsymbol{x}\end{aligned}$$

and a nonlinear controller

$$\boldsymbol{u} = \boldsymbol{h}(\boldsymbol{x}, \boldsymbol{y}_{\text{ref}}),$$

i. e. a nonlinear control loop

$$\dot{\boldsymbol{x}} = \boldsymbol{A}\boldsymbol{x} + \boldsymbol{B}\boldsymbol{h}(\boldsymbol{x}, \boldsymbol{y}_{\text{ref}}).$$

As in the linear case, the observer

$$\underbrace{\dot{\tilde{\boldsymbol{x}}} = \boldsymbol{A}\tilde{\boldsymbol{x}} + \boldsymbol{B}\boldsymbol{u}}_{\text{Model}} \quad + \underbrace{\boldsymbol{L}(\boldsymbol{y} - \boldsymbol{C}\tilde{\boldsymbol{x}})}_{\text{Feedback system}}$$

consists of a system model and a feedback system which acts like a controller.

The task of the feedback system

$$\boldsymbol{L}\,(\boldsymbol{y} - \boldsymbol{C}\tilde{\boldsymbol{x}})$$

is to asymptotically reduce the estimation error

$$\boldsymbol{e} = \boldsymbol{x} - \tilde{\boldsymbol{x}},$$

which is the difference between the system state $\boldsymbol{x}$ and the observer state $\tilde{\boldsymbol{x}}$, to zero. Figure 7.5 shows the structure of the control system with an observer and the nonlinear control law $\boldsymbol{u} = \boldsymbol{h}(\tilde{\boldsymbol{x}}, \boldsymbol{y}_{\text{ref}})$.

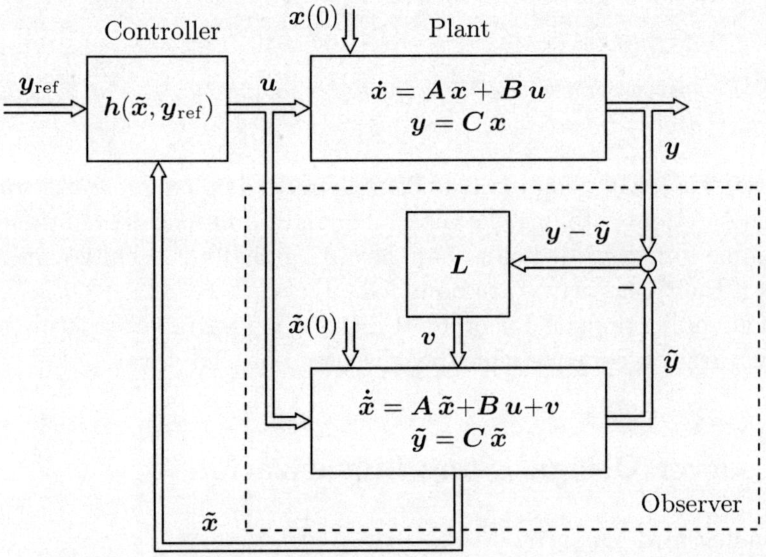

Fig. 7.5: Structure of the system with linear plant, nonlinear controller, and linear observer

In the case of a constant reference vector $\boldsymbol{y}_{\text{ref}}$, it is possible to transform the control system in such a way that $\boldsymbol{y}_{\text{ref}} = \boldsymbol{0}$ applies with no loss of generality. Thus, a controller of the form

$$\boldsymbol{u} = \boldsymbol{h}(\tilde{\boldsymbol{x}})$$

is assumed below.

In order to ensure the stability of the control system in question, we can use the following theorem, which was proven by D. G. Luenberger [289].

Theorem 91 (Luenberger Observer for Nonlinear Control Loops). *Let there be a control loop*

$$\begin{aligned}\dot{\boldsymbol{x}} &= \boldsymbol{A}\boldsymbol{x} + \boldsymbol{B}\boldsymbol{u}, \quad \boldsymbol{y} = \boldsymbol{C}\boldsymbol{x},\\ \boldsymbol{u} &= \boldsymbol{h}(\boldsymbol{x})\end{aligned}$$

with a globally (locally) asymptotically stable equilibrium point $\boldsymbol{x}_{\mathrm{eq}} = \boldsymbol{0}$. *Further, let the vector function* $\boldsymbol{h}$ *fulfill the Lipschitz condition*

$$|\boldsymbol{h}(\boldsymbol{x}_1) - \boldsymbol{h}(\boldsymbol{x}_2)| \le k \cdot |\boldsymbol{x}_1 - \boldsymbol{x}_2|$$

for all $\boldsymbol{x}_1, \boldsymbol{x}_2 \in \mathbb{R}^n$ *(for all* $\boldsymbol{x}_1, \boldsymbol{x}_2$ *within a neighborhood* $U \subset \mathbb{R}^n$ *of* $\boldsymbol{x}_{\mathrm{eq}} = \boldsymbol{0}$*) and for a real number* $k > 0$. *By inserting an asymptotically stable, linear observer into the control loop, we obtain the overall system*

$$\begin{aligned}\dot{\boldsymbol{x}} &= \boldsymbol{A}\boldsymbol{x} + \boldsymbol{B}\boldsymbol{u}, \quad \boldsymbol{y} = \boldsymbol{C}\boldsymbol{x},\\ \boldsymbol{u} &= \boldsymbol{h}(\tilde{\boldsymbol{x}}),\\ \dot{\tilde{\boldsymbol{x}}} &= \boldsymbol{A}\tilde{\boldsymbol{x}} + \boldsymbol{B}\boldsymbol{u} + \boldsymbol{L}(\boldsymbol{y} - \boldsymbol{C}\tilde{\boldsymbol{x}}).\end{aligned}$$

This overall system has a globally (locally) asymptotically stable equilibrium point at $\left[\boldsymbol{x}^T \; \tilde{\boldsymbol{x}}^T\right] = \boldsymbol{0}^T$.

In this context, recall Chapter 1, p. 39, where we defined *Lipschitz continuity* and called functions fulfilling the Lipschitz condition Lipschitz continuous. For differentiable functions, it holds that they are globally Lipschitz continuous if and only if their derivatives are bounded.

Typical applications of Theorem 91 include soft variable structure controls and the saturation controls which we discussed in Chapter 4.

7.4 Observer Design Using Linearization

7.4.1 Basics and Design

In this section, we will continue to address the Luenberger observer as we aim to leverage its general concept to develop an observer for nonlinear systems.

If we consider a linear system

$$\begin{aligned}\dot{\boldsymbol{x}} &= \boldsymbol{A}\boldsymbol{x} + \boldsymbol{B}\boldsymbol{u},\\ \boldsymbol{y} &= \boldsymbol{C}\boldsymbol{x}\end{aligned}$$

with the Luenberger observer

$$\dot{\tilde{\boldsymbol{x}}} = \boldsymbol{A}\tilde{\boldsymbol{x}} + \boldsymbol{B}\boldsymbol{u} + \boldsymbol{L}(\boldsymbol{y} - \boldsymbol{C}\tilde{\boldsymbol{x}}),$$

the idea of applying this scheme to the nonlinear systems

$$\dot{\boldsymbol{x}} = \boldsymbol{f}(\boldsymbol{x}, \boldsymbol{u}),$$
$$\boldsymbol{y} = \boldsymbol{g}(\boldsymbol{x})$$

seems reasonable. The observer is then given by

$$\dot{\tilde{\boldsymbol{x}}} = \boldsymbol{f}(\tilde{\boldsymbol{x}}, \boldsymbol{u}) + \boldsymbol{L}(\boldsymbol{y} - \boldsymbol{g}(\tilde{\boldsymbol{x}})). \tag{7.16}$$

For the observer error

$$\boldsymbol{e} = \boldsymbol{x} - \tilde{\boldsymbol{x}},$$

we obtain the differential equation

$$\dot{\boldsymbol{e}} = \boldsymbol{f}(\tilde{\boldsymbol{x}} + \boldsymbol{e}, \boldsymbol{u}) - \boldsymbol{f}(\tilde{\boldsymbol{x}}, \boldsymbol{u}) - \boldsymbol{L}(\boldsymbol{y} - \boldsymbol{g}(\tilde{\boldsymbol{x}})).$$

In contrast to the linear case, a nonlinear differential equation for the estimation error is obtained. Replacing

$$\boldsymbol{y} = \boldsymbol{g}(\boldsymbol{x}) = \boldsymbol{g}(\tilde{\boldsymbol{x}} + \boldsymbol{e})$$

in this equation results in

$$\dot{\boldsymbol{e}} = \boldsymbol{f}(\tilde{\boldsymbol{x}} + \boldsymbol{e}, \boldsymbol{u}) - \boldsymbol{f}(\tilde{\boldsymbol{x}}, \boldsymbol{u}) - \boldsymbol{L}(\boldsymbol{g}(\tilde{\boldsymbol{x}} + \boldsymbol{e}) - \boldsymbol{g}(\tilde{\boldsymbol{x}})). \tag{7.17}$$

For the estimation error $\boldsymbol{e}$ to decay asymptotically, the design task is to manipulate the nonlinear system equation (7.17) using $\boldsymbol{L}$ in such a way that an asymptotically stable equilibrium point at $\boldsymbol{e} = \boldsymbol{0}$ results.

The simplest way to design the observer matrix $\boldsymbol{L}$ is to linearize equation (7.17). A linearization around a fixed point $(\tilde{\boldsymbol{x}}_{\mathrm{p}}, \boldsymbol{u}_{\mathrm{p}})$ is performed by applying the Taylor series expansion, resulting in

$$\boldsymbol{f}(\tilde{\boldsymbol{x}}_{\mathrm{p}} + \Delta\tilde{\boldsymbol{x}} + \boldsymbol{e}, \boldsymbol{u}) = \boldsymbol{f}(\tilde{\boldsymbol{x}}_{\mathrm{p}}, \boldsymbol{u}) + \underbrace{\left.\frac{\partial \boldsymbol{f}}{\partial \tilde{\boldsymbol{x}}}\right|_{\substack{\tilde{\boldsymbol{x}} = \tilde{\boldsymbol{x}}_{\mathrm{p}} \\ \boldsymbol{u} = \boldsymbol{u}_{\mathrm{p}}}}}_{\boldsymbol{A}} \cdot (\Delta\tilde{\boldsymbol{x}} + \boldsymbol{e}) + \ldots$$

Simultaneously, the relation

$$\boldsymbol{f}(\tilde{\boldsymbol{x}}_{\mathrm{p}} + \Delta\tilde{\boldsymbol{x}}, \boldsymbol{u}) = \boldsymbol{f}(\tilde{\boldsymbol{x}}_{\mathrm{p}}, \boldsymbol{u}) + \boldsymbol{A} \cdot \Delta\tilde{\boldsymbol{x}} + \ldots$$

holds. For the output vector function $\boldsymbol{g}$, we obtain the similar results

$$\boldsymbol{g}(\tilde{\boldsymbol{x}}_{\mathrm{p}} + \Delta\tilde{\boldsymbol{x}} + \boldsymbol{e}) = \boldsymbol{g}(\tilde{\boldsymbol{x}}_{\mathrm{p}}) + \underbrace{\left.\frac{\partial \boldsymbol{g}}{\partial \tilde{\boldsymbol{x}}}\right|_{\substack{\tilde{\boldsymbol{x}} = \tilde{\boldsymbol{x}}_{\mathrm{p}} \\ \boldsymbol{u} = \boldsymbol{u}_{\mathrm{p}}}}}_{\boldsymbol{C}} \cdot (\Delta\tilde{\boldsymbol{x}} + \boldsymbol{e}) + \ldots$$

and

$$\boldsymbol{g}(\tilde{\boldsymbol{x}}_{\text{p}} + \Delta\tilde{\boldsymbol{x}}) = \boldsymbol{g}(\tilde{\boldsymbol{x}}_{\text{p}}) + \boldsymbol{C} \cdot \Delta\tilde{\boldsymbol{x}} + \ldots$$

We will neglect the terms of the Taylor series after the first derivative. Inserting this reduced Taylor series into the observer equation (7.17) yields

$$\dot{\boldsymbol{e}} \approx (\boldsymbol{A} - \boldsymbol{LC}) \cdot \boldsymbol{e}. \tag{7.18}$$

The latter is obviously the estimation equation of the linear observer. Note that equation (7.18) is only an approximation, due to the omission of higher derivatives in the Taylor series.

If we design the matrix $\boldsymbol{A} - \boldsymbol{LC}$ of the observer so that it is asymptotically stable, the estimation error $\boldsymbol{e}$ will also converge to zero for $t \to \infty$. The situation is similar to that of linear systems with linear observers. However, due to the above approximation, this statement of stability only applies within a neighborhood of unknown size surrounding the linearization point $\tilde{\boldsymbol{x}}_{\text{p}}$.

To achieve better, more general results, it is possible to represent the linearization described above as a function of $\tilde{\boldsymbol{x}}$ and $\boldsymbol{u}$. Then the system matrix

$$\boldsymbol{A}(\tilde{\boldsymbol{x}}, \boldsymbol{u}) = \frac{\partial \boldsymbol{f}(\tilde{\boldsymbol{x}}, \boldsymbol{u})}{\partial \tilde{\boldsymbol{x}}}$$

and the output matrix

$$\boldsymbol{C}(\tilde{\boldsymbol{x}}) = \frac{\partial \boldsymbol{g}(\tilde{\boldsymbol{x}})}{\partial \tilde{\boldsymbol{x}}}$$

are no longer constant; they are dependent on the operating point.

Furthermore, the observer matrix $\boldsymbol{L}$ is now selected as a function of $\tilde{\boldsymbol{x}}$ and $\boldsymbol{u}$. Thus, instead of the linear differential equation

$$\dot{\boldsymbol{e}} \approx (\boldsymbol{A} - \boldsymbol{LC}) \cdot \boldsymbol{e} \tag{7.19}$$

for the estimation error e, a nonlinear, more precise equation

$$\dot{\boldsymbol{e}} \approx (\boldsymbol{A}(\tilde{\boldsymbol{x}}, \boldsymbol{u}) - \boldsymbol{L}(\tilde{\boldsymbol{x}}, \boldsymbol{u})\boldsymbol{C}(\tilde{\boldsymbol{x}})) \cdot \boldsymbol{e}, \tag{7.20}$$

dependent on $\tilde{\boldsymbol{x}}$ and $\boldsymbol{u}$, results and incorporates a continuum of operating points. In contrast to a linear observer with its error dynamics (7.19), a nonlinear observer with its error equation (7.20) can better reproduce the plant's dynamics, and thus achieve a lower error $\boldsymbol{e}$ than the linear observer.

The aim now is to design the system matrix

$$\boldsymbol{F}(\tilde{\boldsymbol{x}}, \boldsymbol{u}) = \boldsymbol{A}(\tilde{\boldsymbol{x}}, \boldsymbol{u}) - \boldsymbol{L}(\tilde{\boldsymbol{x}}, \boldsymbol{u})\boldsymbol{C}(\tilde{\boldsymbol{x}})$$

of the estimation error equation to be independent of $\tilde{\boldsymbol{x}}$ and $\boldsymbol{u}$, which requires a suitable choice of $\boldsymbol{L}(\tilde{\boldsymbol{x}}, \boldsymbol{u})$. Moreover, this should be done in such a way that all eigenvalues of $\boldsymbol{F}$ have negative real parts. In this case,

$$\dot{\boldsymbol{e}} = \boldsymbol{F}\boldsymbol{e}$$

applies, and the estimation error $\boldsymbol{e}$ decays asymptotically.

To obtain a constant matrix $\boldsymbol{F}$ of this kind, we can attempt to calculate $\boldsymbol{L}(\tilde{\boldsymbol{x}}, \boldsymbol{u})$ for a given constant matrix $\boldsymbol{F}$ from

$$\boldsymbol{L}(\tilde{\boldsymbol{x}}, \boldsymbol{u}) = (\boldsymbol{A}(\tilde{\boldsymbol{x}}, \boldsymbol{u}) - \boldsymbol{F})\,\boldsymbol{C}^{-1}(\tilde{\boldsymbol{x}}).$$

However, this is only possible for an invertible matrix $\boldsymbol{C}(\tilde{\boldsymbol{x}})$. A prerequisite for the invertibility of $\boldsymbol{C}$ is that

$$\boldsymbol{C}(\tilde{\boldsymbol{x}}) = \frac{\partial \boldsymbol{g}(\tilde{\boldsymbol{x}})}{\partial \tilde{\boldsymbol{x}}}$$

is a quadratic matrix. This in turn requires that $\boldsymbol{g}(\tilde{\boldsymbol{x}})$ is an n-dimensional function, i. e. there must be as many output variables as there are states. In this case, we are able to determine the state vector by means of

$$\boldsymbol{x} = \boldsymbol{g}^{-1}(\boldsymbol{y})$$

and consequently do not need an observer. However, the existence of the same number of states and output variables is rather rare in practice; therefore it is usually not relevant.

For this reason, we will use an alternative method, albeit not one which generally leads to a constant matrix $\boldsymbol{F}$: the characteristic polynomial

$$\begin{aligned} P(s) &= \det(s\boldsymbol{I} - \boldsymbol{F}(\tilde{\boldsymbol{x}}, \boldsymbol{u})) \\ &= \det(s\boldsymbol{I} - \boldsymbol{A}(\tilde{\boldsymbol{x}}, \boldsymbol{u}) + \boldsymbol{L}(\tilde{\boldsymbol{x}}, \boldsymbol{u})\boldsymbol{C}(\tilde{\boldsymbol{x}})) \\ &= \prod_{i=1}^{n}(s - \lambda_i) \end{aligned}$$

is calculated and our requirement is that the real parts of all eigenvalues λ_i are negative. Here it must be ensured that the eigenvalues λ_i and thus the coefficients of $P(s)$ are independent of both $\tilde{\boldsymbol{x}}$ and $\boldsymbol{u}$.

To ensure that all $\mathrm{Re}\{\lambda_i\} < 0$, the observer matrix $\boldsymbol{L}(\tilde{\boldsymbol{x}}, \boldsymbol{u})$ must be designed accordingly. We succeed in this if we choose the matrix $\boldsymbol{L}(\tilde{\boldsymbol{x}}, \boldsymbol{u})$ in such a way that the characteristic polynomial

$$P(s) = s^n + a_{n-1}(\boldsymbol{L}, \tilde{\boldsymbol{x}}, \boldsymbol{u})s^{n-1} + \ldots + a_0(\boldsymbol{L}, \tilde{\boldsymbol{x}}, \boldsymbol{u}) \qquad (7.21)$$

is identical to the freely selectable polynomial

$$\hat{P}(s) = s^n + \hat{a}_{n-1}s^{n-1} + \ldots + \hat{a}_0 \qquad (7.22)$$

which has only zeros with negative real parts and constant coefficients a_i. By equating the coefficients of the polynomials (7.21) and (7.22), we obtain

$$\begin{aligned} a_{n-1}(\boldsymbol{L}, \tilde{\boldsymbol{x}}, \boldsymbol{u}) &= \hat{a}_{n-1}, \\ a_{n-2}(\boldsymbol{L}, \tilde{\boldsymbol{x}}, \boldsymbol{u}) &= \hat{a}_{n-2}, \\ &\vdots \\ a_0(\boldsymbol{L}, \tilde{\boldsymbol{x}}, \boldsymbol{u}) &= \hat{a}_0. \end{aligned}$$

This is a system with n nonlinear equations and $n \cdot r$ unknowns. The latter are the elements of the $n \times r$ matrix $\boldsymbol{L}$. Remember that n is the system order and r is the dimension of the output variable vector $\boldsymbol{y}$.

The solution of the above system of equations ensures that the eigenvalues of $\boldsymbol{F}$ are constant and correspond to the zeros of the freely selectable polynomial $\hat{P}(s)$. However, this does not guarantee fulfillment of the original requirement that the system matrix

$$\boldsymbol{F}(\tilde{\boldsymbol{x}}, \boldsymbol{u}) = \boldsymbol{A}(\tilde{\boldsymbol{x}}, \boldsymbol{u}) - \boldsymbol{L}(\tilde{\boldsymbol{x}}, \boldsymbol{u})\boldsymbol{C}(\tilde{\boldsymbol{x}}) \tag{7.23}$$

of the linearized observer equation is constant, i. e. independent of $\tilde{\boldsymbol{x}}$ and $\boldsymbol{u}$.

Rather, the dependence of $\boldsymbol{F}$ on $\tilde{\boldsymbol{x}}$ and $\boldsymbol{u}$ must be verified by inserting the matrix $\boldsymbol{L}(\tilde{\boldsymbol{x}}, \boldsymbol{u})$ into equation (7.23). If it turns out that $\boldsymbol{F}$ is not independent of $\tilde{\boldsymbol{x}}$ and $\boldsymbol{u}$, the observer

$$\dot{\tilde{\boldsymbol{x}}} = \boldsymbol{f}(\tilde{\boldsymbol{x}}, \boldsymbol{u}) + \boldsymbol{L}(\tilde{\boldsymbol{x}}, \boldsymbol{u}) \cdot (\boldsymbol{y} - \boldsymbol{g}(\tilde{\boldsymbol{x}}))$$

can still be used if the elements of $\boldsymbol{F}$ do not vary widely. In the latter case, it is essential to assess the stability and performance by means of simulations or other methods.

7.4.2 Control Loop with Observer

Inserting the observer developed above into a nonlinear control loop results in a similar structure as in the case of the Luenberger observer for the linear control loop. This is shown in Figure 7.6. However, $\boldsymbol{L}$ is not constant in the nonlinear case; rather, it is a function of $\tilde{\boldsymbol{x}}$ and $\boldsymbol{u}$.

The corresponding equations for the control loop with the observer are

$$\begin{aligned} \left.\begin{aligned} \dot{\boldsymbol{x}} &= \boldsymbol{f}(\boldsymbol{x}, \boldsymbol{u}), \\ \boldsymbol{y} &= \boldsymbol{g}(\boldsymbol{x}), \end{aligned}\right\} &\quad \text{plant} \\ \left.\boldsymbol{u} = \boldsymbol{h}(\tilde{\boldsymbol{x}}, \boldsymbol{y}_{\text{ref}}),\right\} &\quad \text{controller} \\ \left.\dot{\tilde{\boldsymbol{x}}} = \boldsymbol{f}(\tilde{\boldsymbol{x}}, \boldsymbol{u}) + \boldsymbol{L}(\tilde{\boldsymbol{x}}, \boldsymbol{u}) \cdot (\boldsymbol{y} - \boldsymbol{g}(\tilde{\boldsymbol{x}})).\right\} &\quad \text{observer} \end{aligned} \tag{7.24}$$

As an alternative to this notation, the equations for the overall system can also be formulated in terms of the estimation error

$$\boldsymbol{e} = \boldsymbol{x} - \tilde{\boldsymbol{x}}.$$

In this case, we obtain the differential equations of the control loop with the approximate observer dynamics

$$\begin{aligned} \dot{\boldsymbol{x}} &= \boldsymbol{f}(\boldsymbol{x}, \boldsymbol{u}), \\ \boldsymbol{y} &= \boldsymbol{g}(\boldsymbol{x}), \\ \boldsymbol{u} &= \boldsymbol{h}(\boldsymbol{x} - \boldsymbol{e}, \boldsymbol{y}_{\text{ref}}), \\ \dot{\boldsymbol{e}} &\approx \boldsymbol{F}(\boldsymbol{x} - \boldsymbol{e}, \boldsymbol{u})\boldsymbol{e}. \end{aligned}$$

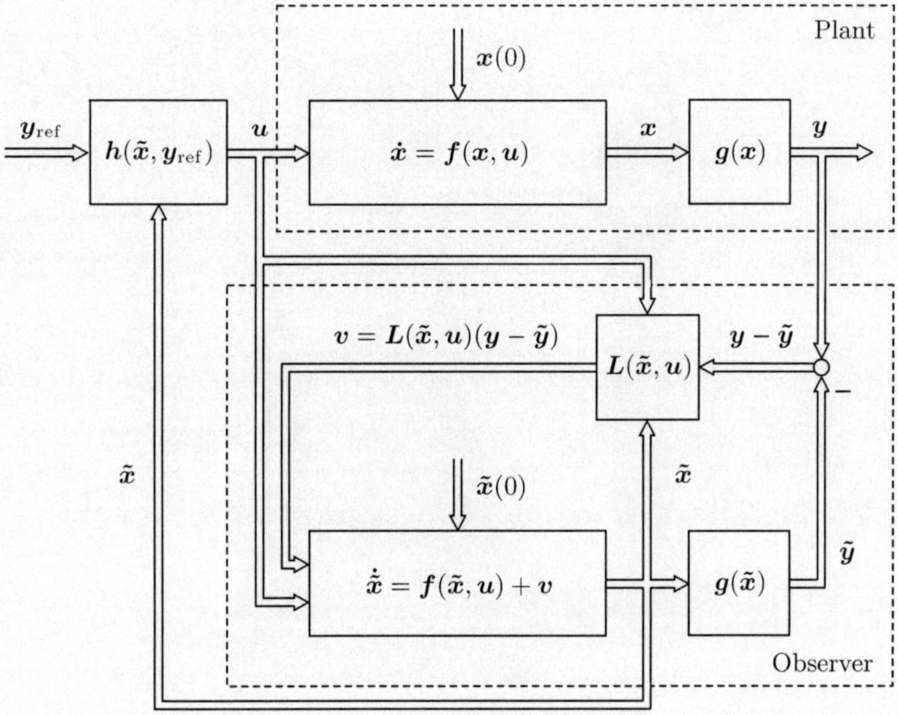

Fig. 7.6: Structure of the plant with observer and controller $\boldsymbol{h}(\tilde{\boldsymbol{x}}, \boldsymbol{y}_{\text{ref}})$

In contrast to the control loop's description (7.24), this notation is only suitable for analysis and not for implementation, since the estimation error $\boldsymbol{e}$ cannot be calculated with $\boldsymbol{x}$ being unknown.

But the equations of the latter system allow for a plausibility analysis with regard to stability. Since $\boldsymbol{F}$ is approximately constant and has eigenvalues with a negative real part,

$$\boldsymbol{e} \to \boldsymbol{0} \qquad \text{for} \qquad t \to \infty$$

applies. This means the control law for large values t becomes

$$\boldsymbol{u} = \boldsymbol{h}(\boldsymbol{x} - \boldsymbol{e}, \boldsymbol{y}_{\text{ref}}) \approx \boldsymbol{h}(\boldsymbol{x}, \boldsymbol{y}_{\text{ref}}).$$

If the control loop is stable without an observer, it is plausible to assume, although it is not proven, that the control loop with an observer is also stable. However, to ensure that the control loop is stable, it is necessary to provide a rigorous proof of stability using the Lyapunov method, for example.

7.4.3 Example: Bioreactor

Let us view a bioreactor [175] as an example. Bioreactors are used for the production of vitamins and medicines. In the first step, a cell culture is cultivated and used for the production of the target substance. We will model the growth phase of the cell culture as follows.

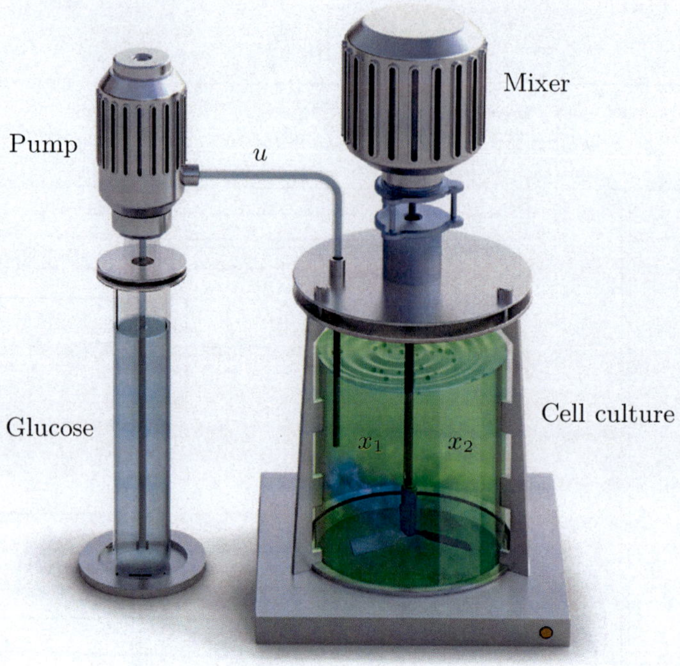

Fig. 7.7: Bioreactor

The reactor has a constant volume. Glucose is pumped into the bioreactor as a substrate. The cell culture reproduces by consuming the substrate and thereby increases its biomass. A mixer ensures even blending. The mixture is finally removed from the reactor for further processing steps. Figure 7.7 illustrates the process.

The biomass concentration x_1 of the cell culture, measured in $\mathrm{g\,l^{-1}}$, increases proportionally to its population according to

$$\dot{x}_1 = \mu(x_2) \cdot x_1 .$$

In other words, it obeys a classic law of growth. The growth constant μ depends on the concentration x_2 of the growth substrate, i.e. the glucose, according to the growth kinetics

$$\mu(x_2) = \frac{\mu_0 \cdot x_2}{k_1 + x_2 + k_2 x_2^2}$$

with the associated growth rate $\mu_0 = 1\ \mathrm{h}^{-1}$ and the two affinity constants $k_1 = 0.03\ \mathrm{g\,l^{-1}}$ and $k_2 = 0.5\ \mathrm{l\,g^{-1}}$. The substrate concentration x_2 in the cell culture is measured in $\mathrm{g\,l^{-1}}$.

The inflow relative to the reactor's volume is the input variable u and measured in h^{-1}. It dilutes the biomass by the value $-x_1 \cdot u$. This results in a change in the biomass concentration over time according to

$$\dot{x}_1 = \mu(x_2) \cdot x_1 - x_1 \cdot u.$$

The substrate mass in the reactor is consumed by the biomass and therefore decreases proportionally to the cell culture amount x_1. The substrate inflow increases the substrate concentration x_2 in the reactor proportionally to the inflow u and the difference between the concentration K of glucose in the inflow and the concentration x_2 in the reactor. Hence,

$$\dot{x}_2 = -\frac{1}{\alpha}\mu(x_2) \cdot x_1 + (K - x_2) \cdot u$$

holds, where $\alpha = 0.5$ is the yield coefficient of the bioreactor and $K = 10\,\mathrm{g\,l^{-1}}$ is the glucose feed concentration. In summary, the state-space model

$$\begin{aligned} \dot{\boldsymbol{x}} &= \boldsymbol{a}(\boldsymbol{x}) + \boldsymbol{b}(\boldsymbol{x}) \cdot u = \begin{bmatrix} \mu(x_2) \cdot x_1 \\ -\dfrac{1}{\alpha}\,\mu(x_2) \cdot x_1 \end{bmatrix} + \begin{bmatrix} -x_1 \\ K - x_2 \end{bmatrix} u, \\ y &= g(\boldsymbol{x}) = \begin{bmatrix} 1 \; 0 \end{bmatrix} \boldsymbol{x} \end{aligned}$$

results.

In the following, we will design an observer using linearization and the state-dependent observer matrix $\boldsymbol{L}(\tilde{\boldsymbol{x}}, u)$. Our goal is to estimate the concentrations x_1 and x_2. For the observer design, the Jacobian matrices

$$\boldsymbol{A}(\tilde{\boldsymbol{x}}, u) = \frac{\partial \boldsymbol{f}(\tilde{\boldsymbol{x}}, u)}{\partial \tilde{\boldsymbol{x}}} \quad \text{and} \quad \boldsymbol{C}(\tilde{\boldsymbol{x}}) = \frac{\partial \boldsymbol{g}(\tilde{\boldsymbol{x}})}{\partial \tilde{\boldsymbol{x}}}$$

must first be computed for the linearization of the system. This results in

$$\boldsymbol{A}(\tilde{\boldsymbol{x}}, u) = \begin{bmatrix} \mu(\tilde{x}_2) - u & \mu'(\tilde{x}_2)\tilde{x}_1 \\ -\dfrac{1}{\alpha}\mu(\tilde{x}_2) & -\dfrac{1}{\alpha}\mu'(\tilde{x}_2)\tilde{x}_1 - u \end{bmatrix}, \quad \mu'(\tilde{x}_2) = \frac{\partial \mu(\tilde{x}_2)}{\partial \tilde{x}_2},$$

and

$$\boldsymbol{C}(\tilde{\boldsymbol{x}}) = \begin{bmatrix} 1 \; 0 \end{bmatrix}.$$

In the next design step, the characteristic polynomial of the observer matrix is calculated as

$$P(s) = \det(s\boldsymbol{I} - \underbrace{[\boldsymbol{A}(\tilde{\boldsymbol{x}}, u) - \boldsymbol{L}(\tilde{\boldsymbol{x}}, u)\boldsymbol{C}(\tilde{\boldsymbol{x}})]}_{\text{observer matrix } \boldsymbol{F}(\tilde{\boldsymbol{x}}, u)}),$$

where

$$\boldsymbol{L}(\tilde{\boldsymbol{x}}, u) = \begin{bmatrix} l_1(\tilde{\boldsymbol{x}}, u) \\ l_2(\tilde{\boldsymbol{x}}, u) \end{bmatrix}.$$

For the sake of clarity, we will use the abbreviations

$$l_1 = l_1(\tilde{\boldsymbol{x}}, u), \qquad l_2 = l_2(\tilde{\boldsymbol{x}}, u),$$
$$\mu = \mu(\tilde{x}_2), \qquad \mu' = \mu'(\tilde{x}_2) = \frac{\partial \mu(\tilde{x}_2)}{\partial \tilde{x}_2}.$$

This yields

$$P(s) = s^2 + \left(l_1 - \mu + \frac{\mu'}{\alpha}\tilde{x}_1 + 2u\right)s + \mu'\tilde{x}_1\left(l_2 + \frac{l_1 + u}{\alpha}\right) + l_1 u - \mu u + u^2$$

for the characteristic polynomial of $\boldsymbol{F}$.

Comparing the coefficients with those of the polynomial we intend to obtain,

$$\hat{P}(s) = s^2 + \hat{a}_1 s + \hat{a}_0,$$

yields

$$\hat{a}_1 = l_1 - \mu + \frac{\mu'}{\alpha}\tilde{x}_1 + 2u\,,$$
$$\hat{a}_0 = \mu'\tilde{x}_1\left(l_2 + \frac{l_1 + u}{\alpha}\right) + l_1 u - \mu u + u^2.$$

Solving this system of equations, we obtain

$$l_1 = \hat{a}_1 + \mu - \frac{\mu'}{\alpha}\tilde{x}_1 - 2u\,,$$
$$l_2 = \frac{\hat{a}_0 - \hat{a}_1 u + u^2}{\mu'\tilde{x}_1} - \frac{\hat{a}_1 + \mu - \alpha^{-1}\mu'\tilde{x}_1 - 2u}{\alpha}.$$

We can now calculate the matrix $\boldsymbol{F}$ of the observer's dynamics

$$\dot{\boldsymbol{e}} \approx \boldsymbol{F}(\tilde{\boldsymbol{x}}, u)\boldsymbol{e}$$

by inserting the matrix $\boldsymbol{L}(\tilde{\boldsymbol{x}}, u)$ into equation (7.23). This leads to

$$\boldsymbol{F}(\tilde{\boldsymbol{x}}, u) = \boldsymbol{A}(\tilde{\boldsymbol{x}}, u) - \boldsymbol{L}(\tilde{\boldsymbol{x}}, u)\,\boldsymbol{C}(\tilde{\boldsymbol{x}})$$
$$= \begin{bmatrix} -\hat{a}_1 + \dfrac{\mu'}{\alpha}\tilde{x}_1 + u & \mu'\tilde{x}_1 \\ -\dfrac{\hat{a}_0 - \hat{a}_1 u + u^2}{\mu'\tilde{x}_1} + \dfrac{\hat{a}_1 - \alpha^{-1}\mu'\tilde{x}_1 - 2u}{\alpha} & -\dfrac{\mu'}{\alpha}\tilde{x}_1 - u \end{bmatrix}.$$

Note that although $\boldsymbol{F}$ depends on $\tilde{\boldsymbol{x}}$ and u, the characteristic polynomial

$$P(s) = \hat{P}(s) = s^2 + \hat{a}_1 s + \hat{a}_0$$

of the matrix $\boldsymbol{F}$ does not depend on $\tilde{\boldsymbol{x}}$ and u. We will design the observer with $\hat{a}_0 = 100$ and $\hat{a}_1 = 20$, so that the eigenvalues of $\boldsymbol{F}(\tilde{\boldsymbol{x}}, u)$, i. e. the zeros of the characteristic polynomial P, are $s_{1/2} = -10$.

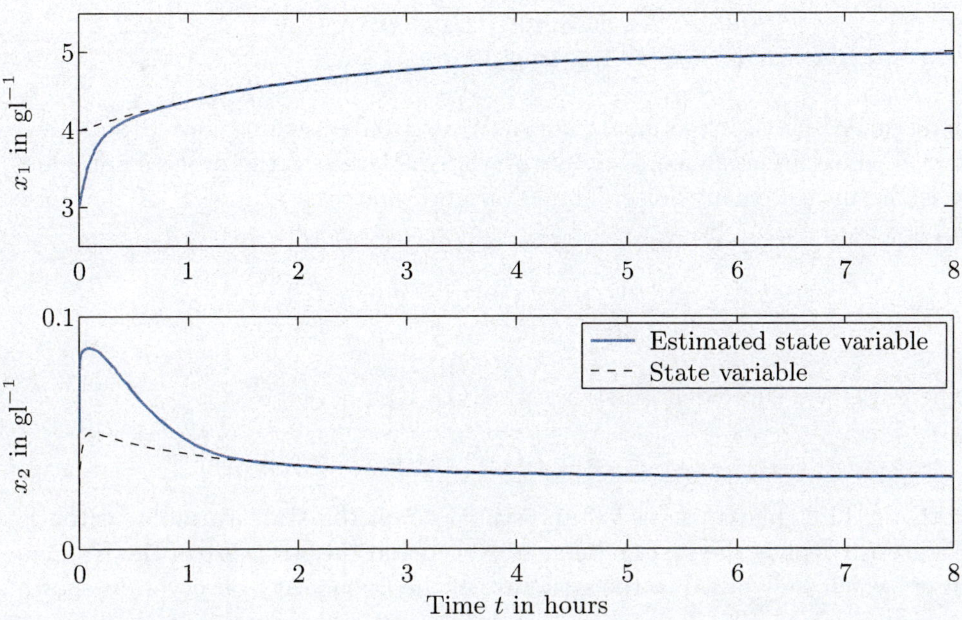

Fig. 7.8: Time courses of actual and estimated biomass concentration x_1 and substrate concentration x_2

The diagrams in Figure 7.8 show the progression of the biomass and substrate concentration in the bioreactor for the initial state

$$\boldsymbol{x}(0) = [4.0 \; 0.02]^T \; \mathrm{g\,l}^{-1}.$$

As a comparison, the system states of the observer are given, starting from

$$\tilde{\boldsymbol{x}}(0) = [3.0 \; 0.05]^T \; \mathrm{g\,l}^{-1}.$$

For an input variable value of $u = 0.5$, the system states approach the final values $x_1 = 4.94 \; \mathrm{g\,l}^{-1}$ and $x_2 = 0.03 \; \mathrm{g\,l}^{-1}$. The difference between the actual system states of the reactor and those observed are rapidly and steadily eliminated.

7.5 The Extended Kalman Filter

7.5.1 Kalman Filter for Linear Systems

In practice, the extended Kalman filter, abbreviated EKF, is the most commonly used type of observer for nonlinear systems. It is based on a linearized representation of the nonlinear system dynamics. To explain how it works, we will first briefly describe the Kalman filter [133, 219] for linear systems.

The system to be observed,

$$\dot{\boldsymbol{x}} = \boldsymbol{A}\boldsymbol{x} + \boldsymbol{B}\boldsymbol{u} + \boldsymbol{\mu},$$
$$y = \boldsymbol{C}\boldsymbol{x} + \boldsymbol{\rho},$$

is disturbed by two zero-mean, normally distributed, white-noise processes $\boldsymbol{\mu}$ and $\boldsymbol{\rho}$ which are assumed to be uncorrelated. Here $\boldsymbol{\mu}$ is the process noise and $\boldsymbol{\rho}$ is the measurement noise. The covariance matrices $\boldsymbol{Q}$ and $\boldsymbol{S}$ of the noise processes are given by

$$\operatorname{cov}\{\boldsymbol{\mu}(t_1), \boldsymbol{\mu}(t_2)\} = \boldsymbol{Q} \cdot \delta(t_1 - t_2),$$
$$\operatorname{cov}\{\boldsymbol{\rho}(t_1), \boldsymbol{\rho}(t_2)\} = \boldsymbol{S} \cdot \delta(t_1 - t_2),$$

where δ is the Dirac delta distribution. The Kalman filter

$$\dot{\tilde{\boldsymbol{x}}} = (\boldsymbol{A} - \boldsymbol{L}\boldsymbol{C})\,\tilde{\boldsymbol{x}} + \boldsymbol{B}\boldsymbol{u} + \boldsymbol{L}\boldsymbol{y}$$

with the filter matrix $\boldsymbol{L}$ yields an estimate $\tilde{\boldsymbol{x}}$ of the state variable vector $\boldsymbol{x}$. Figure 7.9 depicts the system being observed and the structure of the Kalman filter, which is identical to the structure of the Luenberger observer, as shown in Figure 7.1, p. 604.

In fact, according to linear systems theory [133, 181], Kalman filters and Luenberger observers are identical in their equations. They differ only in the calculation of the matrix $\boldsymbol{L}$. To design a Luenberger observer, appropriate eigenvalues of the matrix $\boldsymbol{A} - \boldsymbol{L}\boldsymbol{C}$ must be chosen via $\boldsymbol{L}$. In contrast to this approach, the Kalman matrix $\boldsymbol{L}$ is designed so that the influence of the process noise $\boldsymbol{\mu}$ and the measurement noise $\boldsymbol{\rho}$ on the estimation error $\boldsymbol{e}$ is minimal. For this purpose, we will use the performance index

$$J = \sum_{i=1}^{n} \mathrm{E}\left\{e_i^2\right\},$$

which is based on the estimation error

$$\boldsymbol{e} = [e_1 \ \cdots \ e_n]^T = \boldsymbol{x} - \tilde{\boldsymbol{x}}$$

and the expected values

$$\mathrm{E}\left\{e_i^2\right\} = \lim_{T\to\infty} \frac{1}{2T} \int_{-T}^{T} e_i^2(t)dt.$$

Minimizing J in dependence on $\boldsymbol{L}$ leads to the matrix

$$\boldsymbol{L} = \boldsymbol{P}\boldsymbol{C}^T\boldsymbol{S}^{-1} \tag{7.25}$$

which we were attempting to determine. The matrix $\boldsymbol{P}$ is yielded by the algebraic Riccati equation

$$\boldsymbol{A}\boldsymbol{P} + \boldsymbol{P}\boldsymbol{A}^T - \boldsymbol{P}\boldsymbol{C}^T\boldsymbol{S}^{-1}\boldsymbol{C}\boldsymbol{P} = -\boldsymbol{Q}. \tag{7.26}$$

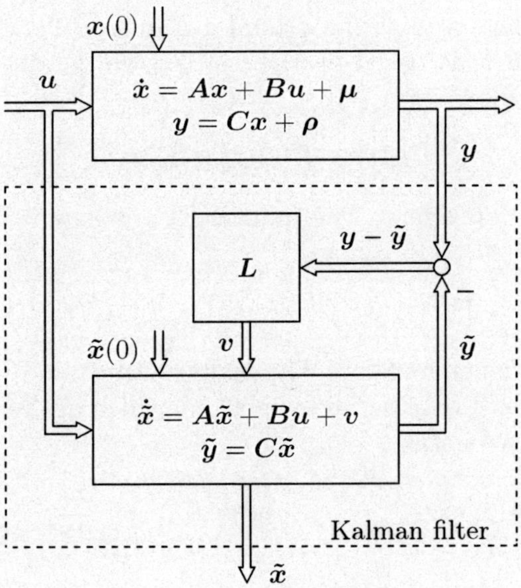

Fig. 7.9: Structure of the linear Kalman filter

The matrices $\boldsymbol{S}$ and $\boldsymbol{Q}$ are generally unknown and are often assumed to be identity matrices. In many cases, however, only subsequent, iterative testing or optimization by trial and error using other matrices $\boldsymbol{S}$ and $\boldsymbol{Q}$ leads to a satisfactory design result.

7.5.2 The EKF for Nonlinear Systems

In the case of a nonlinear system

$$\begin{aligned} \dot{\boldsymbol{x}} &= \boldsymbol{f}(\boldsymbol{x}, \boldsymbol{u}) + \boldsymbol{\mu}, \\ \boldsymbol{y} &= \boldsymbol{g}(\boldsymbol{x}) + \boldsymbol{\rho}, \end{aligned} \tag{7.27}$$

we will utilize the estimation equation of the linear Kalman filter

$$\dot{\tilde{\boldsymbol{x}}} = \boldsymbol{f}(\tilde{\boldsymbol{x}}, \boldsymbol{u}) + \boldsymbol{L}(\boldsymbol{y} - \boldsymbol{g}(\tilde{\boldsymbol{x}}))$$

once again. We used this equation, which equals the equation (7.16) of the Luenberger observer in Section 7.4.1, to design an observer using linearization. However, the choice of a constant matrix $\boldsymbol{L}$ was disadvantageous, because the observer will only perform well in the neighborhood of a given operating point. The same applies here. Therefore, here we select a time-dependent observer matrix $\boldsymbol{L}$ so that we can adapt it to the nonlinearity of the system which depends on the trajectories $\boldsymbol{x}(t)$ or $\tilde{\boldsymbol{x}}(t)$. Thus, the observer equation takes the form

$$\dot{\tilde{\boldsymbol{x}}} = \boldsymbol{f}(\tilde{\boldsymbol{x}}, \boldsymbol{u}) + \boldsymbol{L}(t)(\boldsymbol{y} - \boldsymbol{g}(\tilde{\boldsymbol{x}})). \tag{7.28}$$

This observer is referred to as the *extended Kalman filter.*

The design of $\boldsymbol{L}(t)$ is based on the well-known design equation (7.25) of the linear Kalman filter

$$\boldsymbol{L}(t) = \boldsymbol{P}(t)\boldsymbol{C}^T(t)\boldsymbol{S}^{-1},$$

which is now time-dependent. The matrix $\boldsymbol{P}(t)$ is again calculated from the Riccati equation (7.26)

$$\boldsymbol{A}(t)\boldsymbol{P}(t) + \boldsymbol{P}(t)\boldsymbol{A}^T(t) - \boldsymbol{P}(t)\boldsymbol{C}^T(t)\boldsymbol{S}^{-1}\boldsymbol{C}(t)\boldsymbol{P}(t) = -\boldsymbol{Q},$$

which is now time-dependent too. The matrices $\boldsymbol{A}(t)$ and $\boldsymbol{C}(t)$ are the result of linearizations, i. e. Taylor series expansions which are terminated after the first derivative. So we obtain

$$\boldsymbol{A}(t) = \left.\frac{\partial \boldsymbol{f}}{\partial \boldsymbol{x}}\right|_{\tilde{\boldsymbol{x}}(t)} \quad \text{and} \quad \boldsymbol{C}(t) = \left.\frac{\partial \boldsymbol{g}}{\partial \boldsymbol{x}}\right|_{\tilde{\boldsymbol{x}}(t)}$$

at the current estimated point $\tilde{\boldsymbol{x}}(t)$.

In practice, the algebraic Riccati equation above is solved using the Riccati differential equation

$$\dot{\boldsymbol{P}}(t) = \boldsymbol{A}(t)\boldsymbol{P}(t) + \boldsymbol{P}(t)\boldsymbol{A}^T(t) + \boldsymbol{Q} - \boldsymbol{P}(t)\boldsymbol{C}^T(t)\boldsymbol{S}^{-1}\boldsymbol{C}(t)\boldsymbol{P}(t) \tag{7.29}$$

as in the linear case. The covariance matrix

$$\boldsymbol{P}(0) = \operatorname{cov}\{\boldsymbol{x}_0 - \tilde{\boldsymbol{x}}_0, \boldsymbol{x}_0 - \tilde{\boldsymbol{x}}_0\} = \mathrm{E}\{(\boldsymbol{x}_0 - \tilde{\boldsymbol{x}}_0)(\boldsymbol{x}_0 - \tilde{\boldsymbol{x}}_0)^T\}$$

of the initial estimation error

$$\boldsymbol{e}_0 = \boldsymbol{x}_0 - \tilde{\boldsymbol{x}}_0, \quad \text{with} \quad \boldsymbol{x}_0 = \boldsymbol{x}(0), \; \tilde{\boldsymbol{x}}_0 = \tilde{\boldsymbol{x}}(0)$$

is used as the initial value for the matrix $\boldsymbol{P}(t)$ we are attempting to determine.

Note that the Riccati differential equation (7.29) is not solved offline to determine the stationary value of $\boldsymbol{P}$. This would not be possible because $\boldsymbol{A}(t)$ and $\boldsymbol{C}(t)$ vary continuously. Thus, the estimation equation (7.28) and the Riccati differential equation (7.29) are solved simultaneously; the Jacobian matrices $\boldsymbol{A}(t)$ and $\boldsymbol{C}(t)$ must also be calculated continuously. This dynamic adjustment becomes apparent in the block diagram of the extended Kalman filter in Figure 7.10.

The stability and accuracy of the estimation generally cannot be guaranteed for the extended Kalman filter [32, 39, 53]; they have to be evaluated using simulations. The downside of this is that no general statement can be made and the stability and the accuracy of the estimation only hold for the specific simulations which have been carried out.

Designing the Kalman filter, i. e. choosing suitable values of $\boldsymbol{S}$ and $\boldsymbol{Q}$, involves trial and error and requires experience. In summary, for the system (7.27) being observed, the equations of the extended Kalman filter are given by

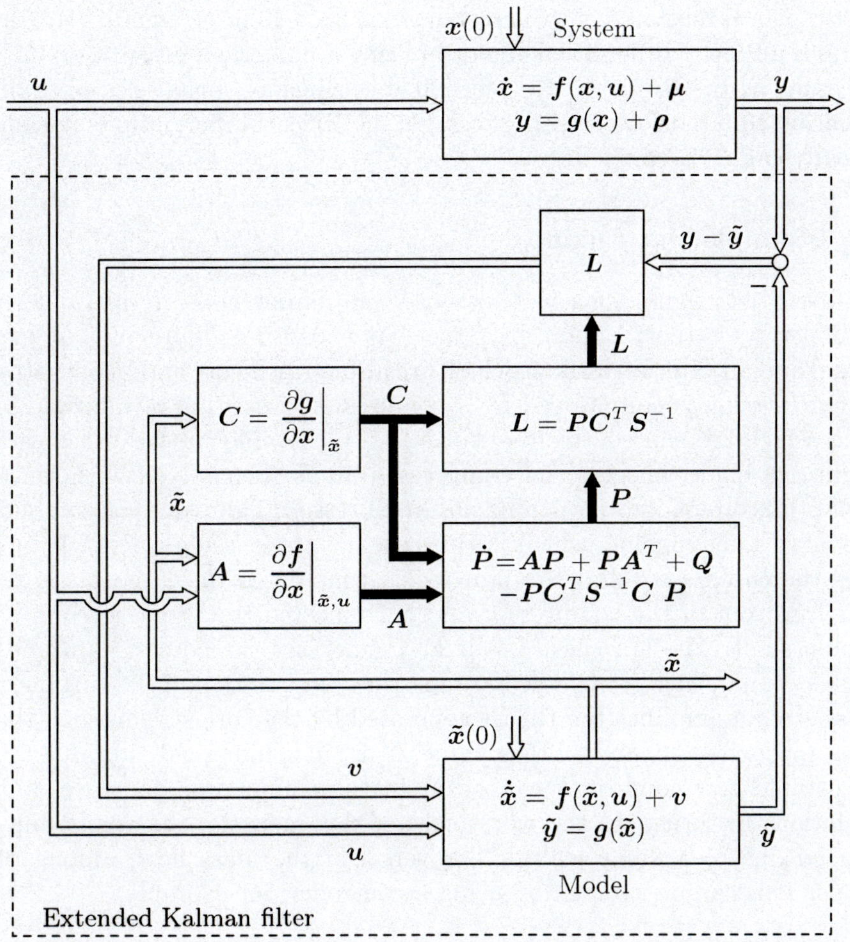

Fig. 7.10: System being observed and extended Kalman filter

$$
\begin{aligned}
\dot{\tilde{\boldsymbol{x}}} &= \boldsymbol{f}(\tilde{\boldsymbol{x}},\boldsymbol{u}) + \boldsymbol{L}(\boldsymbol{y} - \boldsymbol{g}(\tilde{\boldsymbol{x}})),\\
\boldsymbol{A} &= \left.\frac{\partial \boldsymbol{f}}{\partial \boldsymbol{x}}\right|_{\tilde{\boldsymbol{x}}(t),\boldsymbol{u}(t)},\\
\boldsymbol{C} &= \left.\frac{\partial \boldsymbol{g}}{\partial \boldsymbol{x}}\right|_{\tilde{\boldsymbol{x}}(t)},\\
\dot{\boldsymbol{P}} &= \boldsymbol{AP} + \boldsymbol{PA}^T + \boldsymbol{Q} - \boldsymbol{PC}^T\boldsymbol{S}^{-1}\boldsymbol{CP},\\
\boldsymbol{L} &= \boldsymbol{PC}^T\boldsymbol{S}^{-1}.
\end{aligned}
\tag{7.30}
$$

The complete structure of the extended Kalman filter and the mutual dependencies of the individual equations are shown in Figure 7.10. Here the bold arrows represent the forwarding of matrices.

With the exception of very simple cases, practical implementation requires solving the Riccati differential equation using a numerical integration method [323], such as the Runge-Kutta method. In principle, this also applies to the estimation equation. Attempting to solve the Riccati differential equation numerically can cause difficulties.

7.5.3 Example: Jet Engine

Consider a jet engine such as those used in commercial aircraft. The most common type is the turbofan or fanjet. As shown in Figure 7.11, it has an internal and an external airflow, which are also termed core and bypass stream. The core stream is compressed by a compressor consisting of multiple blade wheels which are consecutively arranged. The example shown in Figure 7.11 depicts five blade wheels. The compressed air is then mixed with kerosene and this mixture is ignited behind the compressor. The resulting combustion gases drive a gas turbine, which in this case has three blade wheels. It in turn drives the compressor and the fan, which consists of the large blade wheel located in front of the compressor and is connected to the gas turbine via a common shaft. The fan sucks in air and generates the core stream and the bypass stream, which is bypassed around the compressor and the turbine. The bypass stream amplifies the thrust generated by the core stream such that it causes approximately 80% of the engine thrust in commercial aircraft engines.

Unstable flow conditions can occur in the compressor due to a stall or fluctuations in thrust. In the worst case of this scenario, the engine may be damaged due to a flame leak or a reversal of the mass flow, among other reasons. This can be avoided by using a controller, for example.

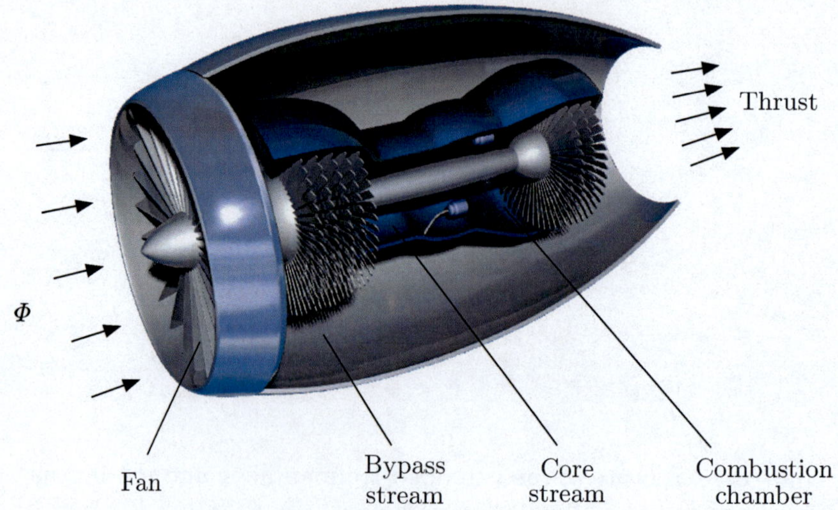

Fig. 7.11: Jet engine

The conditions in the compressor are described by the Moore-Greitzer equations [250, 299]

$$\begin{aligned}
\dot{\Phi} &= -\Psi + \Psi_{\text{co}} + 1 + \frac{3}{2}\Phi - \frac{1}{2}\Phi^3 - 3\Phi R,\\
\dot{\Psi} &= \frac{1}{\beta^2}(\Phi - \gamma\sqrt{\Psi} + 1),\\
\dot{R} &= \sigma R(1 - \Phi^2 - R).
\end{aligned} \tag{7.31}$$

Here, Φ represents the mass flow through the compressor, Ψ is the pressure rise, $R \geq 0$ is a measure of the stall, and β, σ, and Ψ_{co} are constants. The variable γ acts as the control variable u. It can represent bleed air, which is taken from the core or bypass stream and is fed into the compressor in such a way that neither a flow stall nor a reversal of the direction of the mass flow occurs.

The aim is to stabilize the engine, meaning its compressor, at the equilibrium point

$$R_{\text{eq}} = 0, \ \Phi_{\text{eq}} = 1 \ \text{ and } \ \Psi_{\text{eq}} = \Psi_{\text{co}} + 2 \tag{7.32}$$

at

$$\gamma = 2/\sqrt{\Psi_{\text{co}} + 2}.$$

This can be done with a nonlinear state controller. However, all three state variables Φ, Ψ, and R are required for a controller of this type. The pressure increase Ψ is measurable, but the quantities Φ and R are not. Therefore, they must be estimated using an observer.

Before we design an extended Kalman filter as an observer, we will transform system (7.31) so that the equilibrium point (7.32) is at the origin. For this purpose, we will introduce the new state variables

$$\begin{aligned}
x_1 &= \Phi - 1,\\
x_2 &= \Psi - \Psi_{\text{co}} - 2,\\
x_3 &= R
\end{aligned}$$

and thus obtain the transformed system equations

$$\begin{bmatrix} \dot{x}_1 \\ \dot{x}_2 \\ \dot{x}_3 \end{bmatrix} = \boldsymbol{f}(\boldsymbol{x}, u) = \begin{bmatrix} -x_2 - \frac{3}{2}x_1^2 - \frac{1}{2}x_1^3 - 3x_1x_3 - 3x_3 \\ \frac{1}{\beta^2}\left(x_1 - u\sqrt{x_2 + \Psi_{\text{co}} + 2} + 2\right) \\ -\sigma x_3^2 - \sigma x_3(2x_1 + x_1^2) \end{bmatrix} \tag{7.33}$$

using the control variable $u = \gamma$. The output variable y is a measure of the pressure rise Ψ and is given by

$$y = x_2 = \Psi - \Psi_{\text{co}} - 2.$$

At the equilibrium point $\boldsymbol{x}_{\text{eq}} = \boldsymbol{0}$, the control variable is

$$u_{\text{eq}} = \frac{2}{\sqrt{\Psi_{\text{co}} + 2}}.$$

We will now linearize system (7.33) around the estimated state $\tilde{\boldsymbol{x}}$, resulting in

$$\boldsymbol{A}(t) = \left.\frac{\partial \boldsymbol{f}}{\partial \boldsymbol{x}}\right|_{\tilde{\boldsymbol{x}},\boldsymbol{u}} = \begin{bmatrix} -3\tilde{x}_1 - \frac{3}{2}\tilde{x}_1^2 - 3\tilde{x}_3 & -1 & -3\tilde{x}_1 - 3 \\ \beta^{-2} & \dfrac{-\beta^{-2}u}{2\sqrt{(\tilde{x}_2 + \Psi_{\text{co}} + 2)}} & 0 \\ -2\sigma\tilde{x}_3(1 + \tilde{x}_1) & 0 & -2\sigma\tilde{x}_3 - \sigma(2\tilde{x}_1 + \tilde{x}_1^2) \end{bmatrix}$$

and

$$\boldsymbol{C}(t) = \left.\frac{\partial g}{\partial \boldsymbol{x}}\right|_{\tilde{\boldsymbol{x}}} = \left.\frac{\partial x_2}{\partial \boldsymbol{x}}\right|_{\tilde{\boldsymbol{x}}} = [0 \ 1 \ 0].$$

The linearized system matrix $\boldsymbol{A}(t)$, the linearized output vector $\boldsymbol{C}(t)$, and the nonlinear system dynamics $\boldsymbol{f}(\boldsymbol{x}, u)$ are now inserted into equations (7.30) of the extended Kalman filter. For the observer, we thus obtain

$$\begin{aligned} \dot{\tilde{\boldsymbol{x}}} &= \boldsymbol{f}(\tilde{\boldsymbol{x}}, u) + \boldsymbol{L}(y - \tilde{x}_2), \\ \dot{\boldsymbol{P}} &= \boldsymbol{AP} + \boldsymbol{PA}^T + \boldsymbol{Q} - \boldsymbol{PC}^T\boldsymbol{S}^{-1}\boldsymbol{CP}, \\ \boldsymbol{L} &= \boldsymbol{PC}^T\boldsymbol{S}^{-1}. \end{aligned}$$

Note that $\boldsymbol{C}$ is a 1×3 matrix, i. e. a row vector, and $\boldsymbol{L}$ is a 3×1 matrix, i. e. a column vector.

For the covariance matrices $\boldsymbol{Q}$ and $\boldsymbol{S}$, with $\boldsymbol{S}$ being scalar in this case, we select

$$\boldsymbol{Q} = \begin{bmatrix} 0.45 & 1.15 & 0.88 \\ 1.15 & 8.65 & 1.77 \\ 0.88 & 1.77 & 3.04 \end{bmatrix}, \qquad \boldsymbol{S} = 0.12.$$

The system parameters are

$$\Psi_{\text{co}} = 0.72, \ \sigma = 4, \ \text{ and } \ \beta = 0.71.$$

Figure 7.12 shows the progression of the original state variables Φ, Ψ, and R for the initial state vector

$$[\Phi(0) \ \Psi(0) \ R(0)] = [1.5 \ \ 1.2 \ \ 0.5]$$

of the plant and for the initial state vector

$$[\tilde{\Phi}(0) \ \tilde{\Psi}(0) \ \tilde{R}(0)] = [0.5 \ \ 0.2 \ \ 0]$$

of the Kalman filter with the constant control variable

$$u = u_{\text{eq}} = 2/\sqrt{2.72}.$$

At $t = 0$, the matrix $\boldsymbol{P}(t)$ is the identity matrix.

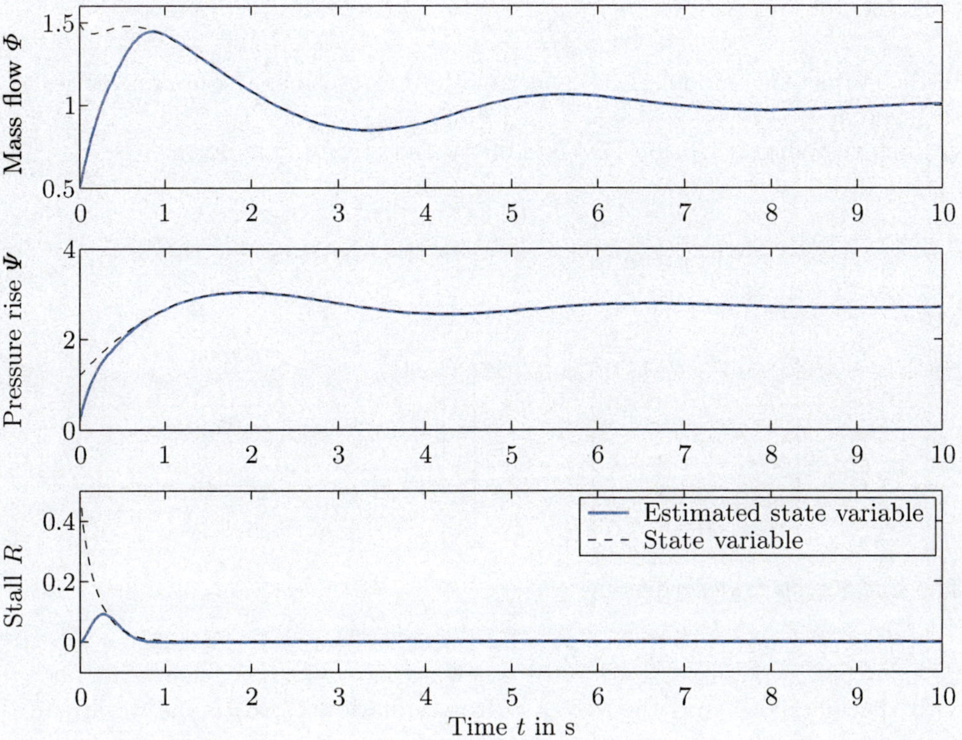

Fig. 7.12: Time courses of the real and estimated mass flow Φ, pressure increase Ψ, and stall R of the flow

7.6 High-Gain Observer

7.6.1 Concept and Design

In order to explain the operating principle of a high-gain observer, we will first assume that the nonlinear SISO system

$$\begin{aligned}\dot{\boldsymbol{x}} &= \boldsymbol{f}(\boldsymbol{x}, u),\\ y &= g(\boldsymbol{x})\end{aligned}$$

is formulated in the nonlinear observability canonical form

$$\begin{aligned}\dot{\boldsymbol{z}} &= \begin{bmatrix} z_2 \\ z_3 \\ \vdots \\ z_n \\ \varphi(\boldsymbol{z}, u, \dot{u}, \ldots, u^{(n-1)}) \end{bmatrix},\\ y &= z_1,\end{aligned} \tag{7.34}$$

or has been transformed into this form by means of

$$\boldsymbol{z} = \boldsymbol{q}(\boldsymbol{x}, u, \dot{u}, \ldots, u^{(n-1)}).$$

In the state-space model (7.34), the function φ describes the nonlinearities of the system.

The system description (7.34) is often also given in the form

$$\begin{aligned} \dot{\boldsymbol{z}} &= \boldsymbol{A}\boldsymbol{z} + \boldsymbol{b}\varphi(\boldsymbol{z}, u, \dot{u}, \ldots, u^{(n-1)}), \\ y &= \boldsymbol{c}^T\boldsymbol{z}, \end{aligned} \tag{7.35}$$

where

$$\boldsymbol{A} = \begin{bmatrix} 0 & 1 & 0 & \cdots & 0 \\ 0 & 0 & 1 & \cdots & 0 \\ \vdots & \vdots & \vdots & \ddots & \vdots \\ 0 & 0 & 0 & \cdots & 1 \\ 0 & 0 & 0 & \cdots & 0 \end{bmatrix}, \quad \boldsymbol{b} = \begin{bmatrix} 0 \\ 0 \\ \vdots \\ 0 \\ 1 \end{bmatrix}, \quad \boldsymbol{c}^T = \begin{bmatrix} 1 \; 0 \cdots 0 \end{bmatrix}. \tag{7.36}$$

The linear system description

$$\dot{\boldsymbol{z}} = \boldsymbol{A}\boldsymbol{z} + \boldsymbol{b} \cdot \hat{u}$$

with the matrix $\boldsymbol{A}$ and the vector $\boldsymbol{b}$ from equation (7.36) is, as we already know, referred to as the *Brunovsky canonical form*. It consists of n integrators arranged in series. Since the output variable is

$$y = z_1 = \boldsymbol{c}^T\boldsymbol{z},$$

the nonlinear observability canonical form is always observable, as is directly apparent from the integration chain.

Much like the Luenberger observer for linear systems, the high-gain observer for the above system is set to

$$\dot{\tilde{\boldsymbol{z}}} = \boldsymbol{A}\tilde{\boldsymbol{z}} + \boldsymbol{b}\varphi(\tilde{\boldsymbol{z}}, u, \dot{u}, \ldots, u^{(n-1)}) + \boldsymbol{\ell}(\varepsilon)\big(y - \boldsymbol{c}^T\tilde{\boldsymbol{z}}\big) \tag{7.37}$$

with the estimation vector $\tilde{\boldsymbol{z}}$ or, using $\tilde{y} = \boldsymbol{c}^T\tilde{\boldsymbol{z}}$, to

$$\dot{\tilde{\boldsymbol{z}}} = \boldsymbol{A}\tilde{\boldsymbol{z}} + \boldsymbol{b}\varphi(\tilde{\boldsymbol{z}}, u, \dot{u}, \ldots, u^{(n-1)}) + \boldsymbol{\ell}(\varepsilon)(y - \tilde{y}). \tag{7.38}$$

Here,

$$\boldsymbol{\ell}(\varepsilon) = \begin{bmatrix} \varepsilon^{-1} & 0 & \cdots & 0 \\ 0 & \varepsilon^{-2} & \cdots & 0 \\ \vdots & \vdots & \ddots & \vdots \\ 0 & 0 & \cdots & \varepsilon^{-n} \end{bmatrix} \begin{bmatrix} l_1 \\ l_2 \\ \vdots \\ l_n \end{bmatrix} = \boldsymbol{D}^{-1}(\varepsilon) \cdot \boldsymbol{l}$$

is the observer gain, which can be adjusted by its parameter $\epsilon > 0$. The matrix $\boldsymbol{D}^{-1}(\varepsilon)$ is a diagonal matrix which has the values $\varepsilon^{-i} > 0$ as elements d_{ii}. In contrast to $\boldsymbol{\ell}(\varepsilon)$, the vector

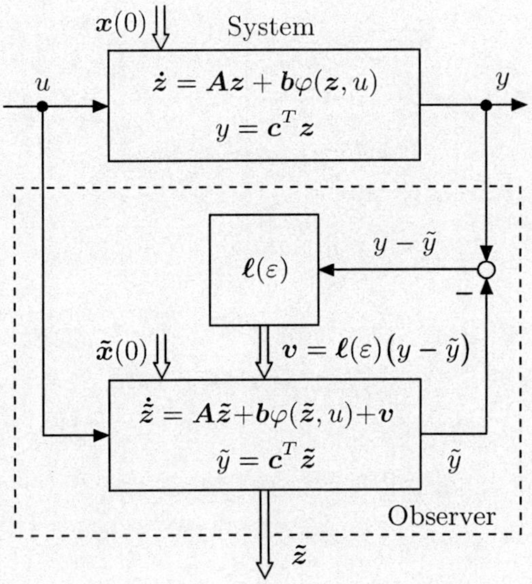

Fig. 7.13: Structure of the high-gain observer using the abbreviated notation $\varphi(\tilde{\boldsymbol{z}}, u) = \varphi(\tilde{\boldsymbol{z}}, u, \dot{u}, \ldots, u^{(n-1)})$

$$\boldsymbol{l} = [l_1 \;\; l_2 \;\; \cdots \;\; l_n]^T$$

is constant but can be freely selected to influence the observer's dynamic behavior. The structure of the high-gain observer for system (7.35) is shown in Figure 7.13.

Based on the estimation error

$$\boldsymbol{e} = \boldsymbol{z} - \tilde{\boldsymbol{z}},$$

the description of system (7.35), and observer (7.38), the error dynamics of the observer are given by

$$\begin{aligned}\dot{\boldsymbol{e}} &= \left(\boldsymbol{A} - \boldsymbol{\ell}(\varepsilon)\boldsymbol{c}^{\mathrm{T}}\right)\boldsymbol{e} + \boldsymbol{b}\Big(\varphi(\boldsymbol{z}, u, \dot{u}, \ldots, u^{(n-1)}) - \varphi(\tilde{\boldsymbol{z}}, u, \dot{u}, \ldots, u^{(n-1)})\Big)\\ &= (\boldsymbol{A} - \boldsymbol{D}^{-1}(\varepsilon)\boldsymbol{l}\boldsymbol{c}^{\mathrm{T}})\boldsymbol{e} + \boldsymbol{b}\Big(\varphi(\boldsymbol{z}, u, \dot{u}, \ldots, u^{(n-1)}) - \varphi(\tilde{\boldsymbol{z}}, u, \dot{u}, \ldots, u^{(n-1)})\Big).\end{aligned} \quad (7.39)$$

If we first assume that

$$\varphi(\tilde{\boldsymbol{z}}, u, \dot{u}, \ldots, u^{(n-1)}) = \varphi(\boldsymbol{z}, u, \dot{u}, \ldots, u^{(n-1)})$$

holds, the error dynamics are linear, i. e.

$$\dot{\boldsymbol{e}} = \left(\boldsymbol{A} - \boldsymbol{D}^{-1}(\varepsilon)\boldsymbol{l}\boldsymbol{c}^{T}\right)\boldsymbol{e}.$$

To calculate the eigenvalues of the observer's system matrix

$$\tilde{\boldsymbol{F}}(\varepsilon) = \boldsymbol{A} - \boldsymbol{D}^{-1}(\varepsilon)\boldsymbol{l}\boldsymbol{c}^T = \begin{bmatrix} -l_1\varepsilon^{-1} & 1 \cdots 0 \\ -l_2\varepsilon^{-2} & 0 \cdots 0 \\ \vdots & \vdots \ddots \vdots \\ -l_{n-1}\varepsilon^{-(n-1)} & 0 \cdots 1 \\ -l_n\varepsilon^{-n} & 0 \cdots 0 \end{bmatrix},$$

we determine its characteristic polynomial

$$\begin{aligned} P(s) &= s^n + \frac{l_1}{\varepsilon}\, s^{n-1} + \frac{l_2}{\varepsilon^2}\, s^{n-2} + \ldots + \frac{l_{n-1}}{\varepsilon^{n-1}}\, s + \frac{l_n}{\varepsilon^n} \\ &= \left(s - \frac{\lambda_1}{\varepsilon}\right)\left(s - \frac{\lambda_2}{\varepsilon}\right)\ldots\left(s - \frac{\lambda_n}{\varepsilon}\right). \end{aligned}$$

Depending on ε, the eigenvalues

$$\tilde{\lambda}_i = \frac{\lambda_i}{\varepsilon}$$

of $\tilde{\boldsymbol{F}}(\varepsilon)$ thus move along rays which emanate from the origin of the complex plane. The real parts of the eigenvalues $\tilde{\lambda}_i$ become smaller with decreasing ε, which means that they shift more and more to the left side of the complex plane. Eventually, the limit

$$\lim_{\varepsilon \to 0} \operatorname{Re}\{\tilde{\lambda}_i\} = \lim_{\varepsilon \to 0} \frac{\operatorname{Re}\{\lambda_i\}}{\varepsilon} = -\infty$$

applies, provided that all eigenvalues λ_i are chosen such that $\operatorname{Re}\{\lambda_i\} < 0$ holds. Figure 7.14 illustrates this dependence of the eigenvalues $\tilde{\lambda}_i$ on ε. For the case in question,

$$\varphi(\tilde{\boldsymbol{z}}, u, \dot{u}, \ldots, u^{(n-1)}) = \varphi(\boldsymbol{z}, u, \dot{u}, \ldots, u^{(n-1)}),$$

the estimation error $\boldsymbol{e}$ converges to zero due to the linear error dynamics. The convergence becomes more rapid the smaller we choose the parameter ε.

In the following, we will analyze the case

$$\varphi(\tilde{\boldsymbol{z}}, u, \dot{u}, \ldots, u^{(n-1)}) \neq \varphi(\boldsymbol{z}, u, \dot{u}, \ldots, u^{(n-1)}),$$

starting with a transformation of the estimation error $\boldsymbol{e}$ and selecting

$$\boldsymbol{e} = \begin{bmatrix} e_1 \\ e_2 \\ \vdots \\ e_n \end{bmatrix} = \begin{bmatrix} \hat{e}_1 \\ \varepsilon^{-1}\,\hat{e}_2 \\ \vdots \\ \varepsilon^{-(n-1)}\,\hat{e}_n \end{bmatrix} = \begin{bmatrix} 1 & 0 & \cdots & 0 \\ 0 & \varepsilon^{-1} & \cdots & 0 \\ \vdots & \vdots & \ddots & \vdots \\ 0 & 0 & \cdots & \varepsilon^{-(n-1)} \end{bmatrix} \hat{\boldsymbol{e}} = \varepsilon\boldsymbol{D}^{-1}(\varepsilon)\hat{\boldsymbol{e}}$$

as the transformation rule. Applying this, we obtain

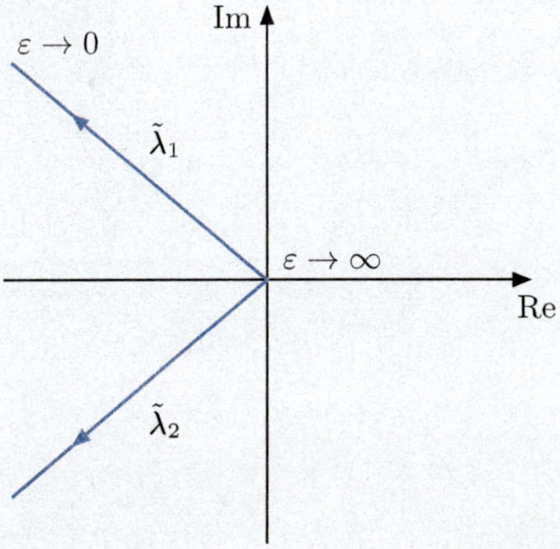

Fig. 7.14: Course of the eigenvalues depending on ϵ

$$\begin{aligned}\varepsilon \boldsymbol{D}^{-1}(\varepsilon)\dot{\hat{\boldsymbol{e}}} = \left(\boldsymbol{A} - \boldsymbol{D}^{-1}(\varepsilon)\boldsymbol{l}\boldsymbol{c}^T\right) \cdot \varepsilon \boldsymbol{D}^{-1}(\varepsilon)\hat{\boldsymbol{e}} \\ + \boldsymbol{b}\left(\varphi(\boldsymbol{z}, u, \dot{u}, \ldots, u^{(n-1)}) - \varphi(\tilde{\boldsymbol{z}}, u, \dot{u}, \ldots, u^{(n-1)})\right)\end{aligned}$$

for the error equation (7.39) of the observer. After multiplication by $\boldsymbol{D}(\varepsilon)$, this results in

$$\begin{aligned}\varepsilon\dot{\hat{\boldsymbol{e}}} = \left(\varepsilon \boldsymbol{D}(\varepsilon)\boldsymbol{A}\boldsymbol{D}^{-1}(\varepsilon) - \boldsymbol{l}\boldsymbol{c}^T \cdot \varepsilon \boldsymbol{D}^{-1}(\varepsilon)\right)\hat{\boldsymbol{e}} \\ + \boldsymbol{D}(\varepsilon)\boldsymbol{b}\left(\varphi(\boldsymbol{z}, u, \dot{u}, \ldots, u^{(n-1)}) - \varphi(\tilde{\boldsymbol{z}}, u, \dot{u}, \ldots, u^{(n-1)})\right).\end{aligned} \tag{7.40}$$

Because

$$\varepsilon \boldsymbol{D}(\varepsilon)\boldsymbol{A}\boldsymbol{D}^{-1}(\varepsilon) = \varepsilon \begin{bmatrix} \varepsilon & 0 & \cdots & 0 \\ 0 & \varepsilon^2 & \cdots & 0 \\ \vdots & \vdots & \ddots & \vdots \\ 0 & 0 & \cdots & \varepsilon^n \end{bmatrix} \begin{bmatrix} 0 & 1 & 0 & \cdots & 0 \\ 0 & 0 & 1 & \cdots & 0 \\ \vdots & \vdots & \vdots & \ddots & \vdots \\ 0 & 0 & 0 & \cdots & 0 \end{bmatrix} \begin{bmatrix} \varepsilon^{-1} & 0 & \cdots & 0 \\ 0 & \varepsilon^{-2} & \cdots & 0 \\ \vdots & \vdots & \ddots & \vdots \\ 0 & 0 & \cdots & \varepsilon^{-n} \end{bmatrix} = \boldsymbol{A},$$

$$\boldsymbol{l}\boldsymbol{c}^T \cdot \varepsilon \boldsymbol{D}^{-1}(\varepsilon) = \varepsilon \begin{bmatrix} l_1 & 0 & \cdots & 0 \\ l_2 & 0 & \cdots & 0 \\ \vdots & \vdots & \ddots & \vdots \\ l_n & 0 & \cdots & 0 \end{bmatrix} \begin{bmatrix} \varepsilon^{-1} & 0 & \cdots & 0 \\ 0 & \varepsilon^{-2} & \cdots & 0 \\ \vdots & \vdots & \ddots & \vdots \\ 0 & 0 & \cdots & \varepsilon^{-n} \end{bmatrix} = \boldsymbol{l}\boldsymbol{c}^T,$$

and

$$\boldsymbol{D}(\varepsilon)\boldsymbol{b} = \varepsilon^n \boldsymbol{b}$$

hold, we obtain the transformed error dynamics

$$\varepsilon\dot{\hat{\boldsymbol{e}}} = (\boldsymbol{A} - \boldsymbol{l}\boldsymbol{c}^T)\hat{\boldsymbol{e}} + \varepsilon^n \boldsymbol{b}\left(\varphi(\boldsymbol{z}, u, \dot{u}, \ldots, u^{(n-1)}) - \varphi(\tilde{\boldsymbol{z}}, u, \dot{u}, \ldots, u^{(n-1)})\right) \tag{7.41}$$

for equation (7.40).

Now, the time t is rescaled according to

$$\tau = \varepsilon^{-1} \cdot t,$$

resulting in a new time variable denoted by τ. This yields

$$\dot{\hat{\boldsymbol{e}}} = \frac{d\hat{\boldsymbol{e}}}{dt} = \frac{d\hat{\boldsymbol{e}}}{\varepsilon d\tau}$$

and

$$\frac{d\hat{\boldsymbol{e}}}{d\tau} = (\boldsymbol{A} - \boldsymbol{l}\boldsymbol{c}^T)\hat{\boldsymbol{e}} + \varepsilon^n \boldsymbol{b}\Big(\varphi(\boldsymbol{z}, u, \dot{u}, \ldots, u^{(n-1)}) - \varphi(\tilde{\boldsymbol{z}}, u, \dot{u}, \ldots, u^{(n-1)})\Big)$$

for equation (7.41). From this equation, it is directly apparent that the smaller ε is, the less important the nonlinear term

$$\varepsilon^n \boldsymbol{b}\left(\varphi(\boldsymbol{z}, u, \dot{u}, \ldots, u^{(n-1)}) - \varphi(\tilde{\boldsymbol{z}}, u, \dot{u}, \ldots, u^{(n-1)})\right)$$

becomes. In contrast to this, the linear part of the error dynamics of the high-gain observer

$$\frac{d\hat{\boldsymbol{e}}}{d\tau} = (\boldsymbol{A} - \boldsymbol{l}\boldsymbol{c}^T)\hat{\boldsymbol{e}}$$

has a constant system matrix

$$\boldsymbol{F} = \boldsymbol{A} - \boldsymbol{l}\boldsymbol{c}^T = \begin{bmatrix} -l_1 & 1 & 0 & \cdots & 0 \\ -l_2 & 0 & 1 & \cdots & 0 \\ \vdots & \vdots & \vdots & \ddots & \vdots \\ -l_{n-1} & 0 & 0 & \cdots & 1 \\ -l_n & 0 & 0 & \cdots & 0 \end{bmatrix},$$

whose eigenvalues λ_i are arbitrarily selectable via $l_1, \ldots, l_n$, and is dominant. If a sufficiently small value of ε is selected, the observer will be asymptotically stable, and the estimation error $\hat{\boldsymbol{e}}$ and thus

$$\boldsymbol{e} = \boldsymbol{z} - \tilde{\boldsymbol{z}}$$

both converge to zero. Consequently, the estimated value $\tilde{\boldsymbol{z}}$ will converge to the system state $\boldsymbol{z}$. The term *high-gain observer* results from the fact that for small values of ε, the elements of the observer gain

$$\boldsymbol{\ell}(\varepsilon) = \boldsymbol{D}^{-1}(\varepsilon)\boldsymbol{l}$$

assume large values. In this context, we note that the eigenvalues $\tilde{\lambda}_i = \lambda_i/\varepsilon$ of the high-gain observer with the system matrix $\tilde{\boldsymbol{F}}(\varepsilon)$ are larger by a factor of $1/\varepsilon$ than the eigenvalues of the transformed observer with the system matrix $\boldsymbol{F}$.

7.6.2 High-Gain Observers in General Form

In the preceding section, we explained the design of high-gain observers for systems which were already formulated in or were transformed into nonlinear observability canonical form. In practice, however, the system description will rarely be in this canonical form. For us to design a high-gain observer, we must transform the system

$$\begin{aligned} \dot{\boldsymbol{x}} &= \boldsymbol{f}(\boldsymbol{x}, u), \\ y &= g(\boldsymbol{x}) \end{aligned} \tag{7.42}$$

into the observability canonical form (7.34) using mapping (7.4) from Section 7.1.5, p. 614, i. e. by means of the diffeomorphism

$$\boldsymbol{z} = \boldsymbol{q}(\boldsymbol{x}, u, \dot{u}, \dots, u^{(n-1)}). \tag{7.43}$$

This cannot always be easily done. Note that state transformations of the above type were already discussed in Section 3.3.1, p. 259 et seq.

Let us assume that we have succeeded in designing a high-gain observer (7.38) for the system in canonical form, as described in the previous section. One way to obtain the original state vector $\tilde{\boldsymbol{x}}$ being observed is to formulate the observer in such a way that its states are consistent with the original coordinates $\boldsymbol{x}$ of system (7.42). To this end, the diffeomorphism (7.43) must be inserted into the observer equation (7.38), i. e. into

$$\dot{\tilde{\boldsymbol{z}}} = \boldsymbol{A}\tilde{\boldsymbol{z}} + \boldsymbol{b}\varphi(\tilde{\boldsymbol{z}}, u, \dot{u}, \dots, u^{(n-1)}) + \boldsymbol{\ell}(\varepsilon)(y - \tilde{y}). \tag{7.44}$$

Using the diffeomorphism

$$\tilde{\boldsymbol{z}} = \boldsymbol{q}(\tilde{\boldsymbol{x}}, u, \dot{u}, \dots, u^{(n-1)}), \tag{7.45}$$

we obtain

$$\dot{\tilde{\boldsymbol{z}}} = \frac{d\boldsymbol{q}(\tilde{\boldsymbol{x}}, u, \dot{u}, \dots, u^{(n-1)})}{dt} = \frac{\partial \boldsymbol{q}}{\partial \tilde{\boldsymbol{x}}}\dot{\tilde{\boldsymbol{x}}} + \sum_{i=0}^{n-1} \frac{\partial \boldsymbol{q}}{\partial u^{(i)}} u^{(i+1)}, \tag{7.46}$$

where the vector $\tilde{\boldsymbol{x}}$ denotes the estimate of the vector $\boldsymbol{x}$. Inserting equation (7.46) into equation (7.44) yields

$$\frac{\partial \boldsymbol{q}}{\partial \tilde{\boldsymbol{x}}}\dot{\tilde{\boldsymbol{x}}} + \sum_{i=0}^{n-1} \frac{\partial \boldsymbol{q}}{\partial u^{(i)}} u^{(i+1)} = \boldsymbol{A}\tilde{\boldsymbol{z}} + \boldsymbol{b}\varphi(\tilde{\boldsymbol{z}}, u, \dot{u}, \dots, u^{(n-1)}) + \boldsymbol{\ell}(\varepsilon)(y - \tilde{y}),$$

from which

$$\begin{aligned} \dot{\tilde{\boldsymbol{x}}} = {} & \left(\frac{\partial \boldsymbol{q}}{\partial \tilde{\boldsymbol{x}}}\right)^{-1} \left(\boldsymbol{A}\tilde{\boldsymbol{z}} + \boldsymbol{b}\varphi(\tilde{\boldsymbol{z}}, u, \dot{u}, \dots, u^{(n-1)}) - \sum_{i=0}^{n-1} \frac{\partial \boldsymbol{q}}{\partial u^{(i)}} u^{(i+1)} \right) \\ & + \left(\frac{\partial \boldsymbol{q}}{\partial \tilde{\boldsymbol{x}}}\right)^{-1} \boldsymbol{\ell}(\varepsilon)(y - \tilde{y}) \end{aligned} \tag{7.47}$$

follows. Transforming the plant model

$$\dot{\tilde{\boldsymbol{x}}} = \boldsymbol{f}(\tilde{\boldsymbol{x}}, u)$$

of the observer using the diffeomorphism (7.45) leads to

$$\left(\frac{\partial \boldsymbol{q}}{\partial \tilde{\boldsymbol{x}}}\right)^{-1}\left(\boldsymbol{A}\tilde{\boldsymbol{z}} + \boldsymbol{b}\varphi(\tilde{\boldsymbol{z}}, u, \dot{u}, \ldots, u^{(n-1)}) - \sum_{i=0}^{n-1} \frac{\partial \boldsymbol{q}}{\partial u^{(i)}} u^{(i+1)}\right) = \boldsymbol{f}(\tilde{\boldsymbol{x}}, u).$$

Inserting this into equation (7.47), we obtain the general equation

$$\dot{\tilde{\boldsymbol{x}}} = \boldsymbol{f}(\tilde{\boldsymbol{x}}, u) + \left(\frac{\partial \boldsymbol{q}(\tilde{\boldsymbol{x}}, u, \dot{u}, \ldots, u^{(n-1)})}{\partial \tilde{\boldsymbol{x}}}\right)^{-1} \boldsymbol{\ell}(\varepsilon)(y - \tilde{y}) \tag{7.48}$$

for the high-gain observer. Note that

$$\tilde{y} = \boldsymbol{c}^T \tilde{\boldsymbol{z}} = g(\tilde{\boldsymbol{x}})$$

holds.

The observer not only requires the measured variable y and the control variable u; it also requires the derivatives $\dot{u}, \ldots, u^{(n-1)}$ which are also arguments of the vector function $\boldsymbol{q}$. Figure 7.15 shows the structure of the observer.

The design of an observer according to equation (7.44) or equation (7.48) normally requires calculating the transformation

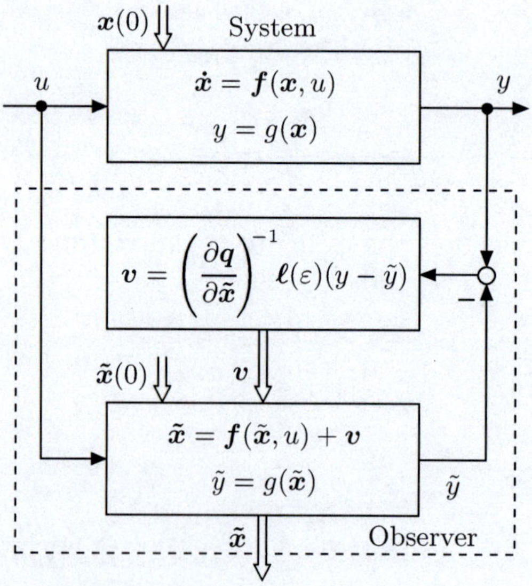

Fig. 7.15: General high-gain observer using the abbreviated notation $\boldsymbol{q}$ for $\boldsymbol{q}(\tilde{\boldsymbol{x}}, u, \dot{u}, \ldots, u^{(n-1)})$

$$z = q(x, u, \dot{u}, \dots, u^{(n-1)})$$

or the corresponding inverse transformation q^{-1}. In practice, this can be a very challenging task.

In principle, there are two ways to implement a high-gain observer. On the one hand, the observer equation (7.44), the input variable u, and the output variable y can be used to determine the estimated state $\tilde{z}$ of the transformed system (7.35). The actual estimation vector $\tilde{x}$ is calculated by applying the inverse transformation formula

$$\tilde{x} = q^{-1}(\tilde{z}, u, \dot{u}, \dots, u^{(n-1)}).$$

On the other hand, the observer can also be realized in the original coordinates $\tilde{x}$ using the general equation (7.48) of the high-gain observer. In any case, as mentioned earlier, the crux of the method lies in the handling of the function q. This is because either the inverse function q^{-1} or the inverse of the Jacobian matrix $\partial q / \partial \tilde{x}$ must be determined. Both are often difficult.

7.6.3 Example: Chemical Reactor

An example is a stirred tank reactor [107, 229]. This reactor is continuously supplied with substance A at a volume rate of q_{in}. Substance A has the input concentration c_0 and the input temperature T_0. In the reactor, the substance is catalytically decomposed, releasing heat. For this reason, the reactor is equipped with a cooling jacket through which cooling water flows at the temperature T_{c}. In the reactor, an agitator ensures that mixing is homogeneous at all times. The decomposition products finally leave the reactor at a volume rate q_{out}. In the reactor, the mixture has the temperature T, and the concentration of substance A in the mixture is denoted by c. Figure 7.16 illustrates the process.

The reactor can be modeled by a second-order nonlinear system. The state variables are defined as the normalized quantities

$$x_1 = \frac{c_0 - c}{c_0} \qquad \text{and} \qquad x_2 = \frac{T - T_0}{T_0}.$$

In contrast to the normalized temperature x_2, the state variable x_1 cannot be measured. The input or control variable u is the normalized difference between the coolant temperature T_{c} and the input temperature T_0, i. e.

$$u = \frac{T_{\text{c}} - T_0}{T_0}.$$

The model equations are

$$\begin{aligned} \dot{x} &= f(x, u), \\ y &= x_2 \end{aligned} \qquad (7.49)$$

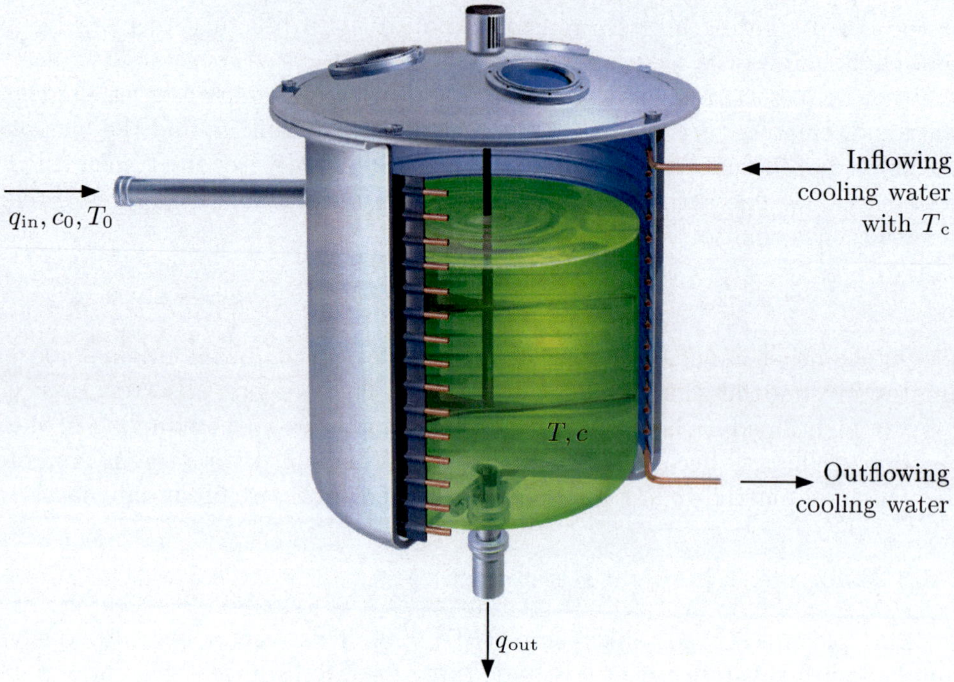

Fig. 7.16: Chemical reactor

with

$$\boldsymbol{f}(\boldsymbol{x}, u) = \begin{bmatrix} -ax_1 + k(1-x_1)e^{-\alpha/(1+x_2)} \\ -\beta x_2 + kd(1-x_1)e^{-\alpha/(1+x_2)} + bu \end{bmatrix}$$

and the parameters

$$\begin{aligned} c_0 &= 0.848 \text{ mol l}^{-1}, & T_0 &= 308.5 \text{ K}, \\ k &= 1.05 \cdot 10^{14} \text{ min}^{-1}, & \alpha &= 34.2583, \\ a &= 0.2674 \text{ min}^{-1}, & \beta &= 1.815 \text{ min}^{-1}, \\ d &= 0.4682, & b &= 1.5476 \text{ min}^{-1}. \end{aligned}$$

The time unit is assumed to be one minute.

The first step is determining the diffeomorphism

$$\boldsymbol{z} = \boldsymbol{q}\left(\boldsymbol{x}, u, \dot{u}, \ldots, u^{(n-1)}\right),$$

which transforms system (7.49) into the nonlinear observability canonical form. This yields

$$\boldsymbol{z} = \begin{bmatrix} z_1 \\ z_2 \end{bmatrix} = \begin{bmatrix} y \\ \dot{y} \end{bmatrix} = \begin{bmatrix} x_2 \\ -\beta x_2 + kd(1-x_1)e^{-\alpha/(1+x_2)} + bu \end{bmatrix} = \boldsymbol{q}(\boldsymbol{x}, u). \quad (7.50)$$

From this, the relation

$$\boldsymbol{x} = \begin{bmatrix} x_1 \\ x_2 \end{bmatrix} = \begin{bmatrix} 1 - \dfrac{1}{kd}(z_2 + \beta z_1 - bu)e^{\alpha/(1+z_1)} \\ z_1 \end{bmatrix} = \boldsymbol{q}^{-1}(\boldsymbol{z}, u) \quad (7.51)$$

follows for $\boldsymbol{q}^{-1}$. Note that $\boldsymbol{q}^{-1}$ in equation (7.51) is defined for all

$$z_1 = x_2 \neq -1.$$

Because $T > 0$ and thus $x_2 > -1$, the system can be considered observable for all relevant values of $\boldsymbol{x}$ and $\boldsymbol{z}$. To calculate the nonlinear observability canonical form, equation (7.51) is inserted into equation (7.49), resulting in

$$\begin{bmatrix} \dot{z}_1 \\ \dot{z}_2 \end{bmatrix} = \begin{bmatrix} z_2 \\ \varphi(\boldsymbol{z}, u, \dot{u}) \end{bmatrix},$$

$$y = \boldsymbol{c}^T \boldsymbol{z}, \quad \boldsymbol{c}^T = [1 \;\; 0],$$

with

$$\varphi(\boldsymbol{z}, u, \dot{u}) = \frac{\alpha \cdot z_2(z_2 + \beta \cdot z_1 - bu)}{(1+z_1)^2} - \left(a + \beta + ke^{-\alpha/(1+z_1)}\right) z_2$$
$$- k(\beta \cdot z_1 - bu - ad)e^{-\alpha/(1+z_1)} - a(\beta \cdot z_1 - bu) + b\dot{u}.$$

In this way, according to equations (7.44) and (7.36), it is possible to directly determine the high-gain observer for the chemical reactor as

$$\dot{\tilde{\boldsymbol{z}}} = \begin{bmatrix} \tilde{z}_2 \\ \varphi(\tilde{\boldsymbol{z}}, u, \dot{u}) \end{bmatrix} + \boldsymbol{\ell}(\varepsilon)(y - \tilde{y}). \quad (7.52)$$

By retransforming the observer dynamics (7.52) via equation (7.50) or by calculating them using the general equation (7.48), we obtain the observer dynamics in the original coordinates

$$\begin{aligned} \dot{\tilde{\boldsymbol{x}}} &= \boldsymbol{f}(\tilde{\boldsymbol{x}}, u) + \left(\frac{\partial \boldsymbol{q}}{\partial \tilde{\boldsymbol{x}}}\right)^{-1} \boldsymbol{\ell}(\varepsilon)(y - \boldsymbol{c}^T \boldsymbol{q}(\tilde{\boldsymbol{x}}, u)) \\ &= \boldsymbol{f}(\tilde{\boldsymbol{x}}, u) + \begin{bmatrix} r_{11} & r_{12} \\ 1 & 0 \end{bmatrix} \boldsymbol{\ell}(\varepsilon)(y - \tilde{y}) \end{aligned}$$

using the acronyms

$$r_{11} = -\frac{\beta(1+\tilde{x}_2)^2 e^{\alpha/(1+\tilde{x}_2)} + kd\alpha(\tilde{x}_1 - 1)}{kd(1+\tilde{x}_2)^2},$$

$$r_{12} = -\frac{e^{\alpha/(1+\tilde{x}_2)}}{kd}.$$

The observer is dimensioned using

$$\varepsilon = 0.1, \quad \lambda_1 = -1.69, \quad \text{and} \quad \lambda_2 = -2.6.$$

We assume the initial state vector

$$\boldsymbol{x}^T(0) = [0.504 \;\; 0.02]$$

for the simulation shown in Figure 7.17. Since we can measure the temperature x_2, we will set $\tilde{x}_2(0) = x_2(0) = 0.02$. In contrast to x_2, we cannot measure x_1. Having no accurate knowledge of the reactor's initial values, we will select $\tilde{x}_1(0) = 0$ which yields

$$\tilde{\boldsymbol{x}}^T(0) = [0 \;\; 0.02].$$

The input value is

$$u(t) = -0.01.$$

Figure 7.17 shows the concentration x_1, the temperature x_2, and the estimated values $\tilde{x}_1$ and $\tilde{x}_2$. The estimated state variable $\tilde{x}_1$ reaches the actual value of the state variable x_1 after about 25 seconds. As is to be expected, there is almost no difference in the progression between the measured value x_2 and the estimated value $\tilde{x}_2$ after this time period.

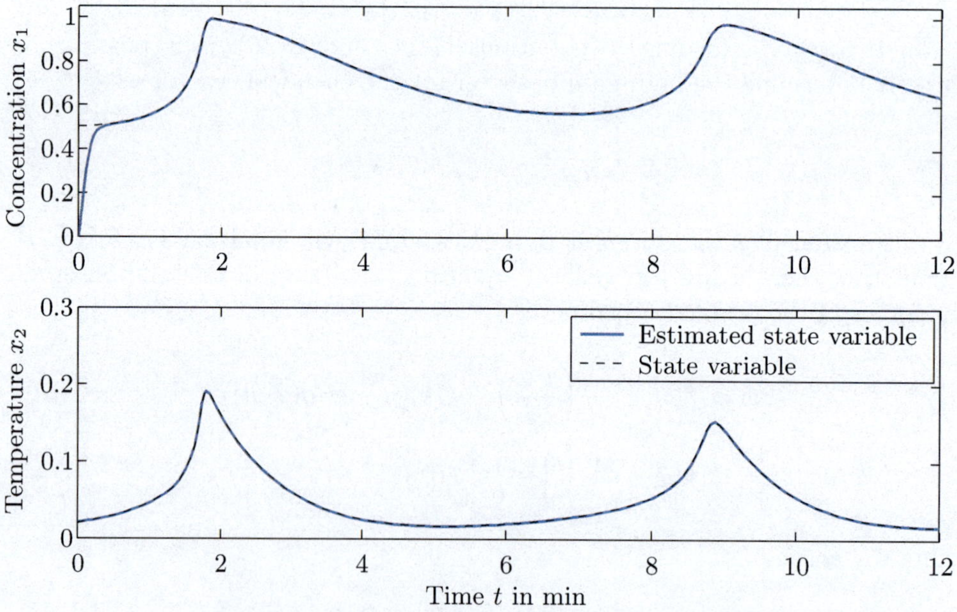

Fig. 7.17: Progressions of the observed and true states of the chemical reactor, i. e. the normalized concentration x_1 and the normalized temperature x_2

7.6.4 The Case of Control-Affine Systems

For the design of high-gain observers, we had previously assumed that the plant is in the nonlinear observability canonical form

$$\dot{\boldsymbol{x}} = \begin{bmatrix} x_2 \\ x_3 \\ \vdots \\ x_n \\ \varphi(\boldsymbol{x}, u, \dot{u}, \ldots, u^{(n-1)}) \end{bmatrix}, \qquad (7.53)$$
$$y = x_1$$

or can be transformed into this canonical form. As has been mentioned several times, this will only rarely be the case. Fortunately, many control-affine systems

$$\dot{\boldsymbol{x}} = \boldsymbol{a}(\boldsymbol{x}) + \boldsymbol{b}(\boldsymbol{x}) \cdot u,$$
$$y = c(\boldsymbol{x})$$

can be bijectively transformed into the form of equation (7.53) by means of a transformation

$$\boldsymbol{z} = \boldsymbol{q}(\boldsymbol{x}) = \boldsymbol{t}(\boldsymbol{x}). \qquad (7.54)$$

This transformation was already discussed in Section 5.2, p. 359 et seq. and in Section 7.1.6, p. 617 et seq. For the sake of clarity and convenience, we will allow ourselves to repeat the derivation of the transformation equation (7.54) in the following. Note that the transformation $\boldsymbol{t}$ in Section 5.2 is a special case of the transformation $\boldsymbol{q}$ in Section 7.1.4, p. 612.

The mapping $\boldsymbol{t}$ is determined by means of the Lie derivative

$$L_{\boldsymbol{a}} c(\boldsymbol{x}) = \frac{\partial c(\boldsymbol{x})}{\partial \boldsymbol{x}} \boldsymbol{a}(\boldsymbol{x})$$

and the multiple Lie derivatives

$$L_{\boldsymbol{a}}^{k} c(\boldsymbol{x}) = L_{\boldsymbol{a}} L_{\boldsymbol{a}}^{k-1} c(\boldsymbol{x})$$
$$= \frac{\partial L_{\boldsymbol{a}}^{k-1} c(\boldsymbol{x})}{\partial \boldsymbol{x}} \boldsymbol{a}(\boldsymbol{x}).$$

Now, the output function

$$y = c(\boldsymbol{x})$$

will be differentiated multiple times with respect to time t, yielding the equations

$$
\begin{aligned}
y &= c(\boldsymbol{x}),\\
\dot{y} &= \frac{\partial c\,(\boldsymbol{x})}{\partial \boldsymbol{x}}\dot{\boldsymbol{x}} = \frac{\partial c\,(\boldsymbol{x})}{\partial \boldsymbol{x}}\boldsymbol{a}(\boldsymbol{x}) + \frac{\partial c\,(\boldsymbol{x})}{\partial \boldsymbol{x}}\boldsymbol{b}(\boldsymbol{x})u = L_{\boldsymbol{a}}c(\boldsymbol{x}) + \underbrace{L_{\boldsymbol{b}}c(\boldsymbol{x})}_{=0}\,u,\\
\ddot{y} &= \frac{\partial L_{\boldsymbol{a}}c(\boldsymbol{x})}{\partial \boldsymbol{x}}\dot{\boldsymbol{x}} = \frac{\partial L_{\boldsymbol{a}}c(\boldsymbol{x})}{\partial \boldsymbol{x}}\boldsymbol{a}(\boldsymbol{x}) + \frac{\partial L_{\boldsymbol{a}}c(\boldsymbol{x})}{\partial \boldsymbol{x}}\boldsymbol{b}(\boldsymbol{x})u = L_{\boldsymbol{a}}^2c(\boldsymbol{x}) + \underbrace{L_{\boldsymbol{b}}L_{\boldsymbol{a}}c(\boldsymbol{x})}_{=0}\,u,\\
&\;\;\vdots\\
y^{(\delta-1)} &= \frac{\partial L_{\boldsymbol{a}}^{\delta-2}c(\boldsymbol{x})}{\partial \boldsymbol{x}}\dot{\boldsymbol{x}} = L_{\boldsymbol{a}}^{\delta-1}c(\boldsymbol{x}) + \underbrace{L_{\boldsymbol{b}}L_{\boldsymbol{a}}^{\delta-2}c(\boldsymbol{x})}_{=0}\,u,\\
y^{(\delta)} &= \frac{\partial L_{\boldsymbol{a}}^{\delta-1}c(\boldsymbol{x})}{\partial \boldsymbol{x}}\dot{\boldsymbol{x}} = L_{\boldsymbol{a}}^{\delta}c(\boldsymbol{x}) + \underbrace{L_{\boldsymbol{b}}L_{\boldsymbol{a}}^{\delta-1}c(\boldsymbol{x})}_{\neq 0}\,u.
\end{aligned}
$$

In the process above, the output value y is differentiated until the term which is multiplied by the control variable u becomes

$$L_{\boldsymbol{b}}L_{\boldsymbol{a}}^{\delta-1}c(\boldsymbol{x}) \neq 0$$

for the first time. As mentioned previously, the above sequence of derivatives corresponds to that calculated in controller design using input-output linearization, as described in Section 5.2. The order δ of the derivative, at which $L_{\boldsymbol{b}}L_{\boldsymbol{a}}^{\delta-1}c(\boldsymbol{x}) \neq 0$ holds for the first time, is called the *relative degree* δ.

In the following, we will only address systems in which the relative degree δ is equal to the system order n, i. e. $\delta = n$. The sequence of derivatives discussed previously takes the form

$$
\begin{aligned}
y &= c(\boldsymbol{x}),\\
\dot{y} &= L_{\boldsymbol{a}}c(\boldsymbol{x}),\\
\ddot{y} &= L_{\boldsymbol{a}}^2c(\boldsymbol{x}),\\
&\;\;\vdots\\
y^{(n-1)} &= L_{\boldsymbol{a}}^{n-1}c(\boldsymbol{x}),\\
y^{(n)} &= L_{\boldsymbol{a}}^{n}c(\boldsymbol{x}) + L_{\boldsymbol{b}}L_{\boldsymbol{a}}^{n-1}c(\boldsymbol{x})\cdot u\,.
\end{aligned}
\tag{7.55}
$$

New state variables

$$
\boldsymbol{z} = \begin{bmatrix} z_1\\ z_2\\ z_3\\ \vdots\\ z_n \end{bmatrix} = \begin{bmatrix} y\\ \dot{y}\\ \ddot{y}\\ \vdots\\ y^{(n-1)} \end{bmatrix} = \begin{bmatrix} c(\boldsymbol{x})\\ L_{\boldsymbol{a}}c(\boldsymbol{x})\\ L_{\boldsymbol{a}}^2c(\boldsymbol{x})\\ \vdots\\ L_{\boldsymbol{a}}^{n-1}c(\boldsymbol{x}) \end{bmatrix} = \boldsymbol{t}(\boldsymbol{x}) \tag{7.56}
$$

are now defined, and the mapping $\boldsymbol{t}(\boldsymbol{x})$ we wish to calculate is directly obtained. Along with equations (7.55) and (7.56), the system description we are attempting to determine follows as

$$\begin{bmatrix} \dot{z}_1 \\ \vdots \\ \dot{z}_{n-1} \\ \dot{z}_n \end{bmatrix} = \begin{bmatrix} z_2 \\ \vdots \\ z_n \\ L_{\boldsymbol{a}}^n c(\boldsymbol{x}) + L_{\boldsymbol{b}} L_{\boldsymbol{a}}^{n-1} c(\boldsymbol{x}) \cdot u \end{bmatrix},$$
$$y = z_1 \,.$$

Now the original state $\boldsymbol{x}$ in the last row must be replaced by $\boldsymbol{z}$ using

$$\boldsymbol{x} = \boldsymbol{t}^{-1}(\boldsymbol{z}).$$

So we finally obtain

$$\begin{bmatrix} \dot{z}_1 \\ \vdots \\ \dot{z}_{n-1} \\ \dot{z}_n \end{bmatrix} = \begin{bmatrix} z_2 \\ \vdots \\ z_n \\ \varphi(\boldsymbol{z}, u) \end{bmatrix}, \qquad (7.57)$$
$$y = z_1$$

with

$$\varphi(\boldsymbol{z}, u) = L_{\boldsymbol{a}}^n c(\boldsymbol{t}^{-1}(\boldsymbol{z})) + L_{\boldsymbol{b}} L_{\boldsymbol{a}}^{n-1} c(\boldsymbol{t}^{-1}(\boldsymbol{z})) \cdot u.$$

We can formulate equation (7.57) as

$$\dot{\boldsymbol{z}} = \boldsymbol{A}\boldsymbol{z} + \boldsymbol{b} \cdot \varphi(\boldsymbol{z}, u),$$
$$y = \boldsymbol{c}^T \boldsymbol{z},$$

where

$$\boldsymbol{A} = \begin{bmatrix} 0 & 1 & 0 \cdots 0 \\ 0 & 0 & 1 \cdots 0 \\ \vdots & \vdots & \vdots \ddots \vdots \\ 0 & 0 & 0 \cdots 1 \\ 0 & 0 & 0 \cdots 0 \end{bmatrix}, \quad \boldsymbol{b} = \begin{bmatrix} 0 \\ 0 \\ \vdots \\ 0 \\ 1 \end{bmatrix}, \quad \boldsymbol{c}^T = \begin{bmatrix} 1 \; 0 \cdots 0 \end{bmatrix}.$$

The associated high-gain observer

$$\dot{\tilde{\boldsymbol{z}}} = \boldsymbol{A}\tilde{\boldsymbol{z}} + \boldsymbol{b}\varphi(\tilde{\boldsymbol{z}}, u) + \boldsymbol{\ell}(\varepsilon)\big(y - \boldsymbol{c}^T \tilde{\boldsymbol{z}}\big) \qquad (7.58)$$

can be transformed using

$$\tilde{\boldsymbol{z}} = \boldsymbol{t}(\tilde{\boldsymbol{x}}) \qquad (7.59)$$

so that it is formulated in the coordinates of the original system

$$\begin{aligned} \dot{\boldsymbol{x}} &= \boldsymbol{a}(\boldsymbol{x}) + \boldsymbol{b}(\boldsymbol{x}) \cdot u, \\ y &= c(\boldsymbol{x}). \end{aligned} \tag{7.60}$$

Inserting equation (7.59) into equation (7.58), we obtain

$$\frac{d\boldsymbol{t}(\tilde{\boldsymbol{x}})}{dt} = \boldsymbol{A}\,\boldsymbol{t}(\tilde{\boldsymbol{x}}) + \boldsymbol{b}\,\varphi(\boldsymbol{t}(\tilde{\boldsymbol{x}}), u) + \boldsymbol{\ell}(\varepsilon)\big(y - \mathbf{c}^T \boldsymbol{t}(\tilde{\boldsymbol{x}})\big)$$

and, consequently,

$$\frac{\partial \boldsymbol{t}(\tilde{\boldsymbol{x}})}{\partial \tilde{\boldsymbol{x}}}\dot{\tilde{\boldsymbol{x}}} = \boldsymbol{A}\,\boldsymbol{t}(\tilde{\boldsymbol{x}}) + \boldsymbol{b}\,\varphi(\boldsymbol{t}(\tilde{\boldsymbol{x}}), u) + \boldsymbol{\ell}(\varepsilon)\,(y - c(\tilde{\boldsymbol{x}}))$$

for the estimation vector $\tilde{\boldsymbol{x}}$ in the original coordinates. In the last equation, we multiply by the inverse of the Jacobian matrix $\partial \boldsymbol{t}(\boldsymbol{x})/\partial \boldsymbol{x}$ from the left. Thus, we obtain the equation of the high-gain observer

$$\dot{\tilde{\boldsymbol{x}}} = \boldsymbol{a}(\tilde{\boldsymbol{x}}) + \boldsymbol{b}(\tilde{\boldsymbol{x}}) \cdot u + \left(\frac{\partial \boldsymbol{t}(\tilde{\boldsymbol{x}})}{\partial \tilde{\boldsymbol{x}}}\right)^{-1} \boldsymbol{\ell}(\varepsilon)\,(y - \tilde{y}) \tag{7.61}$$

in the original coordinates, where $\tilde{y} = c(\tilde{\boldsymbol{x}})$ applies. Figure 7.18 shows the structure of the high-gain observer for the original system (7.60). Equation (7.61) also immediately follows from the general equation (7.48) of the high-gain observer.

Finally, note that high-gain observers can also be designed if the relative degree δ is smaller than the system order n, i. e. if $\delta < n$ holds. For more on this topic, see [208, 387].

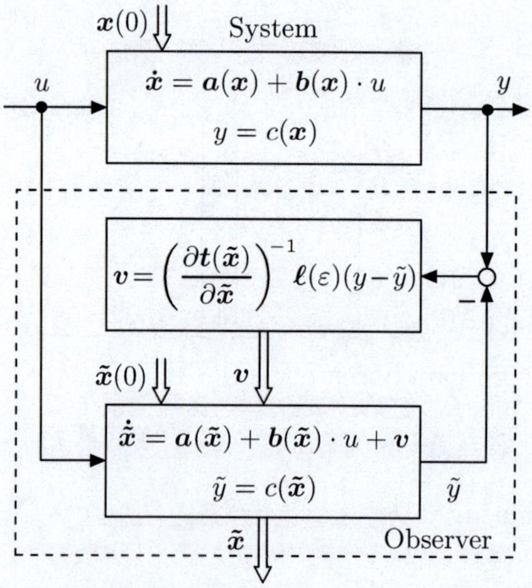

Fig. 7.18: Structure of the high-gain observer for control-affine systems

7.7 Exercises

Exercise 7.1 Show that a linear system

$$\dot{\boldsymbol{x}} = \boldsymbol{A}\boldsymbol{x} + \boldsymbol{b}u,$$
$$y = \boldsymbol{c}^T\boldsymbol{x}$$

with the relative degree $\delta = n$ is always observable.

Exercise 7.2 Let us examine the satellite model (1.8) from p. 8. Let

$$J_x < J_y \quad \text{and} \quad J_x < J_z$$

apply. For the torques, we will assume

$$M_x = M_y = M_z = 0.$$

Due to a malfunction in the satellite's sensor system, we can only measure ω_x. In which cases is it possible to prove that the satellite is weakly observable using Theorem 87 on p. 609?

Exercise 7.3 The dynamics of the glucose and insulin concentrations in diabetes patients can be described in a simplified way by the model [38, 242]

$$\begin{aligned}
\dot{x}_1 &= 0.028x_1 - x_2(x_1 + 110),\\
\dot{x}_2 &= -0.025x_2 + 0.00013x_3,\\
\dot{x}_3 &= 0.093(x_3 - 1.5) + 0.0083u,\\
y &= x_1.
\end{aligned}$$

Here, x_1 is the glucose concentration in the blood plasma in milligrams per deciliter (mg/dL), x_2 is the insulin concentration in the tissue fluids in milliunits per deciliter (mU/dL), x_3 is the insulin concentration in the blood plasma in mU/dL, and u is the intravenously injected amount of insulin in mU/min.

(a) Determine whether the diabetes model is weakly observable or even observable.

(b) Is the type of observability found in (a) globally valid for $\boldsymbol{x} \in D_{\boldsymbol{x},\text{def}} = I\!R^3$ or only locally valid?

Exercise 7.4 Show that the strict feedback form

$$\begin{aligned}
\dot{x}_1 &= f_1(x_1) + h_1(x_1)x_2,\\
\dot{x}_2 &= f_2(x_1, x_2) + h_2(x_1, x_2)x_3,\\
\dot{x}_3 &= f_3(x_1, x_2, x_3) + h_3(x_1, x_2, x_3)x_4,\\
&\vdots\\
\dot{x}_n &= f_n(x_1, \dots, x_n) + h_n(x_1, \dots, x_n)u,\\
y &= x_1
\end{aligned}$$

is observable and state the sufficient conditions under which this applies.

Exercise 7.5 Demonstrate that a system

$$\begin{aligned}
\dot{x}_1 &= x_2 + b_1(x_1)u \\
\dot{x}_2 &= x_3 + b_2(x_1, x_2)u \\
\dot{x}_3 &= x_4 + b_3(x_1, x_2, x_3)u \\
&\vdots \\
\dot{x}_{n-1} &= x_n + b_{n-1}(x_1, \ldots, x_{n-1})u \\
\dot{x}_n &= \alpha(\boldsymbol{x}) + \beta(\boldsymbol{x})u \\
y &= x_1
\end{aligned}$$

is observable for all $\boldsymbol{x} \in I\!R^n$ and $\boldsymbol{u} \in \mathbb{C}^{n-1}$.

Exercise 7.6 Let us examine the system [48]

$$\begin{aligned}
\dot{x}_1 &= 1, \\
\dot{x}_2 &= x_1^2, \\
y &= x_2
\end{aligned}$$

with $\boldsymbol{x} \in D_{\boldsymbol{x}} = I\!R^2$.

(a) Show that with the diffeomorphism

$$\boldsymbol{z} = \begin{bmatrix} y \\ \dot{y} \end{bmatrix} = \boldsymbol{q}(\boldsymbol{x})$$

the observability of the system via Theorem 86 on p. 608 is not provable for all $\boldsymbol{x} \in D_{\boldsymbol{x},\text{def}} = I\!R^2$.

(b) Demonstrate that with the addition of $\ddot{y}$, the state vector $\boldsymbol{x}$ for all $\boldsymbol{x} \in I\!R^2$ can be calculated from the knowledge of y and $\ddot{y}$, and that the system is therefore globally observable.

(c) Which property of Theorem 86 on p. 608 becomes clearly evident with the above results?

Exercise 7.7 Let us examine the system

$$\begin{aligned}
\dot{x}_1 &= x_2 - x_1 + u, \\
\dot{x}_2 &= -x_1^3, \\
y &= x_1.
\end{aligned}$$

(a) Determine the diffeomorphism

$$\boldsymbol{x} = \boldsymbol{q}(\boldsymbol{z})$$

which transforms the system into the nonlinear observer canonical form.

(b) Formulate the system in the nonlinear observer canonical form.

(c) Design a canonical form observer in z-coordinates whose error dynamics have the eigenvalues $s_{1,2} = -10$.
(d) Transform the canonical form observer into the original coordinates $\boldsymbol{x}$.

Exercise 7.8 Show that a system in nonlinear controllability canonical form and $y = x_n$ can be transformed into the nonlinear controller canonical form with $\beta(\boldsymbol{x}) = 1$.

Exercise 7.9 Cable-operated elevators in very tall buildings involve a number of problems, such as cable elongation and vertical oscillations. These problems can be avoided by using cableless elevators which are operated by an electrical linear motor. The elevator cage runs along tracks without touching them; these tracks are part of the linear motor and also serve to maintain the contactless distancing of the elevator cage from the tracks. Here, the distance x_1 between the elevator cage and the track is maintained using electromagnets mounted in the elevator cage. Figure 7.19 shows the general design.

The system's state variables are the distance x_1, the derivative $x_2 = \dot{x}_1$ of the distance, and the magnetic flux density x_3 of the magnets. The simplified

Fig. 7.19: Cableless elevator

system dynamics are given by the differential equations

$$\begin{aligned}
\dot{x}_1 &= x_2, \\
\dot{x}_2 &= -a\left(\frac{x_3}{x_1}\right)^2, \\
\dot{x}_3 &= -cx_3 + d\frac{x_2 x_3}{x_1^2} + bu, \\
y &= x_1.
\end{aligned}$$

Here, the inductor voltage u is the input variable. The constants $a, b, c,$ and d are parameters dependent on the construction.

(a) To calculate an observer

$$\dot{\tilde{\boldsymbol{x}}} = \boldsymbol{f}(\tilde{\boldsymbol{x}}, u) + \boldsymbol{L}(\tilde{\boldsymbol{x}}, u) \cdot (y - \tilde{y})$$

using linearization, first determine the characteristic polynomial $P(s, \tilde{\boldsymbol{x}}, u)$ of the matrix

$$\boldsymbol{F}(\tilde{\boldsymbol{x}}, u) = \boldsymbol{A}(\tilde{\boldsymbol{x}}, u) - \boldsymbol{L}(\tilde{\boldsymbol{x}}, u)\boldsymbol{C}(\tilde{\boldsymbol{x}})$$

of the error dynamics.

(b) Which conditions must the matrix $\boldsymbol{L}(\tilde{\boldsymbol{x}}, u)$ fulfill for the characteristic polynomial $P(s, \tilde{\boldsymbol{x}}, u)$ of the matrix $\boldsymbol{F}(\tilde{\boldsymbol{x}}, u)$ to be dependent on neither $\tilde{\boldsymbol{x}}$ or u?

Exercise 7.10 We will now attempt to design a high-gain observer for the pneumatic engine

$$\dot{\boldsymbol{x}} = \begin{bmatrix} x_2 \\ -\dfrac{c}{m}x_1 + \dfrac{A}{m}x_3 \\ -\dfrac{x_2 x_3}{x_1} \end{bmatrix} + \begin{bmatrix} 0 \\ 0 \\ \dfrac{x_3}{Ax_1} \end{bmatrix} u,$$

$$y = x_1$$

which we encountered in equation (5.165) in Section 5.4.6, p. 455.

(a) Show that the system is globally observable for all $\boldsymbol{x} \in I\!R^3$.

(b) Transform the system into the nonlinear observability canonical form.

(c) Calculate a high-gain observer in the original coordinates $\tilde{\boldsymbol{x}}$.

8

Solutions to the Exercises

In this chapter the solutions to the exercises are presented. Due to the extent of the calculations, the solutions to the exercises do not contain any calculation steps, but only the final results. For the same reason, where a proof is to be given or an assumption is to be verified, the proof or the solution method is not included here.

1.1 $\ddot{y}y - \dot{y}^2 + 2\dot{y}^2y^3 = 2\dot{y}yu$

1.2 Equilibrium points: $\boldsymbol{x}_{\text{eq1}} = \boldsymbol{0}, \quad \boldsymbol{x}_{\text{eq2}} = \begin{bmatrix}1 & 0\end{bmatrix}^T, \quad \boldsymbol{x}_{\text{eq3}} = \begin{bmatrix}0 & 0.5\end{bmatrix}^T$

1.3 (a) The system has no equilibrium points.

(b) The system has the set of equilibrium points $E = \{\boldsymbol{x} \in I\!R^2 \mid x_1 = 0, x_2 \in I\!R\}$ and the infinite number of equilibrium points $\boldsymbol{x}_{\text{eq},k}^T = k\pi\begin{bmatrix}1 & \pm 1\end{bmatrix}$, $k \in \mathbb{Z}$.

(c) $\boldsymbol{x}_{\text{eq1}} = \begin{bmatrix}0 & 0\end{bmatrix}^T, \quad \boldsymbol{x}_{\text{eq2}} = \begin{bmatrix}\sqrt{6} & 0\end{bmatrix}^T, \quad \boldsymbol{x}_{\text{eq3}} = \begin{bmatrix}-\sqrt{6} & 0\end{bmatrix}^T$

(d) $\boldsymbol{x}_{\text{eq1}} = \begin{bmatrix}0 & 0\end{bmatrix}^T, \quad \boldsymbol{x}_{\text{eq2}} = \begin{bmatrix}-5-\sqrt{6} & -5-\sqrt{6}\end{bmatrix}^T, \quad \boldsymbol{x}_{\text{eq3}} = \begin{bmatrix}-5+\sqrt{6} & -5+\sqrt{6}\end{bmatrix}^T$

(e) $\boldsymbol{x}_{\text{eq}} = \begin{bmatrix}0 & 0\end{bmatrix}^T$ (f) $\boldsymbol{x}_{\text{eq}} = \begin{bmatrix}0 & 0\end{bmatrix}^T$

(g) The system has the equilibrium point $\boldsymbol{x}_{\text{eq1}} = \boldsymbol{0}$ and the set $E = \{\boldsymbol{x} \in I\!R^2 \mid x_1^2 + x_2^2 = 1\}$ of equilibrium points.

(h) $\boldsymbol{x}_{\text{eq1,2}} = \begin{bmatrix}\frac{1}{2}(\sqrt{5}-1) & \pm\sqrt{\frac{1}{2}(\sqrt{5}-1)} & 0\end{bmatrix}^T, \quad \boldsymbol{x}_{\text{eq3,4}} = \begin{bmatrix}0 & 0 & \pm 1\end{bmatrix}^T$

1.4 (a)

$$\boldsymbol{x}_{\text{eq1}} = \begin{bmatrix} N\dfrac{\alpha}{\mu} \\ 0 \\ 0 \end{bmatrix}, \quad \boldsymbol{x}_{\text{eq2}} = \begin{bmatrix} N\dfrac{\gamma+\mu}{r\beta} \\ N\dfrac{r\beta\alpha - \mu(\gamma+\mu)}{r\beta(\gamma+\mu)} \\ N\gamma\dfrac{r\beta\alpha - \mu(\gamma+\mu)}{\mu r\beta(\gamma+\mu)} \end{bmatrix}$$

J. Adamy, *Nonlinear Systems and Controls*,
https://doi.org/10.1007/978-3-662-68690-4_8

(b) The equilibrium point $\boldsymbol{x}_{\text{eq1}}$ is reached if no pathogens are present in the population. The equilibrium point $\boldsymbol{x}_{\text{eq2}}$ represents the case of illness.

(c) $\dot{x}_2 = \dfrac{r\beta}{N}(N - x_2)x_2$ (d) $\boldsymbol{x}_{\text{eq1}} = \begin{bmatrix}0 & N\end{bmatrix}^T$, $\boldsymbol{x}_{\text{eq2}} = \begin{bmatrix}N & 0\end{bmatrix}^T$

(e) $x_2(t) = \dfrac{x_2(0)N}{(N - x_2(0))e^{-r\beta t} + x_2(0)}$ (f) $\lim\limits_{t\to\infty} x_2(t) = N$

(g) At the end of an incurable infectious illness such as AIDS, the entire population will be infected in the case of the simplified model. This is not realistic, because in reality non-infected newborns are born and infected patients die. These factors must be taken into account in the model for the situation in reality to be described more accurately.

1.5 (d) $G_{\text{ROA}} = \{\boldsymbol{x} \in I\!R^2 \mid x_{10}^3 x_{20}^2 \neq 5\}$ (e) $G_{\text{RAS}} = \{\boldsymbol{x} \in I\!R^2 \mid x_{10}^3 x_{20}^2 < 5\}$

1.6 (a)

$$i(u_{\text{C1}}) = \begin{cases} m_2 u_{\text{C1}} - \dfrac{u_{\text{d}}}{R_3}, & u_{\text{C1}} \geq u_0, \\ m_1 u_{\text{C1}}, & |u_{\text{C1}}| < u_0, \\ m_2 u_{\text{C1}} + \dfrac{u_{\text{d}}}{R_3}, & u_{\text{C1}} \leq -u_0, \end{cases} \quad \text{with} \quad \begin{aligned} u_0 &= \frac{R_2}{R_2 + R_3} u_{\text{d}}, \\ m_1 &= -\frac{1}{R_1}, \\ m_2 &= -\frac{1}{R_1} + \frac{1}{R_2} + \frac{1}{R_3}, \end{aligned}$$

$$\dot{u}_{\text{C1}} = -\frac{u_{\text{C1}}}{RC_1} + \frac{u_{\text{C2}}}{RC_1} - \frac{1}{C_1} i(u_{\text{C1}}), \quad \dot{u}_{\text{C2}} = \frac{u_{\text{C1}}}{RC_2} - \frac{u_{\text{C2}}}{RC_2} + \frac{i_{\text{L}}}{C_2}, \quad \dot{i}_{\text{L}} = -\frac{u_{\text{C2}}}{L}$$

(b) Point-symmetric polygon line with bends at two points

(c)

$$\begin{aligned} \dot{x}_1 &= \alpha(-x_1 + x_2 - g(x_1)) \\ \dot{x}_2 &= x_1 - x_2 + x_3 \\ \dot{x}_3 &= -\beta x_2 \end{aligned} \quad \text{with} \quad g(x_1) = \begin{cases} bx_1 + a - b, & x_1 \geq 1, \\ ax_1, & |x_1| < 1, \\ bx_1 + b - a, & x_1 \leq -1. \end{cases}$$

(d) The state variables x_1, x_2, and x_3 become dimensionless as a result of the normalization. Thus the equations become generally applicable and no longer apply only to the electric circuit; in this form they can be applied to other projects as well, such as mechanical ones.

1.7 (a)

$$\boldsymbol{x}_{\text{eq1}} = \boldsymbol{0}, \quad \boldsymbol{x}_{\text{eq2}} = \begin{bmatrix} \dfrac{b-a}{b+1} & 0 & \dfrac{a-b}{b+1} \end{bmatrix}^T, \quad \boldsymbol{x}_{\text{eq3}} = \begin{bmatrix} \dfrac{a-b}{b+1} & 0 & \dfrac{b-a}{b+1} \end{bmatrix}^T$$

(b) The system, which is the normalized version of Chua's circuit, consists of three subsystems, each with linear dynamics. Each of these subsystems has one of the three equilibrium points. The characteristic polynomials for the subsystems are

$$P_a(s) = s^3 - \frac{9}{35}s^2 + \frac{173}{35}s - \frac{132}{7} \quad \text{for } \boldsymbol{x}_{\text{eq1}} \quad \text{and}$$

$$P_b(s) = s^3 + \frac{123}{35}s^2 + \frac{61}{7}s + \frac{264}{7} \quad \text{for } \boldsymbol{x}_{\text{eq2}} \text{ and } \boldsymbol{x}_{\text{eq3}}.$$

Using the Routh criterion, we determine that both polynomials and thus all three equilibrium points are unstable.

(c) The system is chaotic.

1.8 The output signal y_1 is a square wave; y_2 is a triangular wave.

1.9 (a) The output signal is shown in Figure 8.1 below.

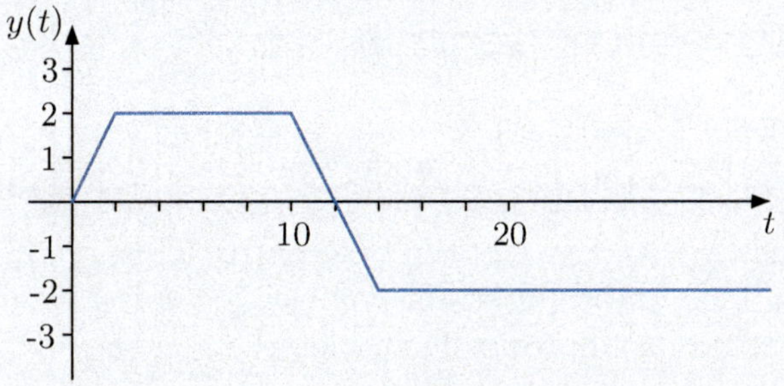

Fig. 8.1: Output signal $y(t)$ from solution 1.9

(b) The remaining control error for $y_{\mathrm{ref}} = 3$ is $e_\infty = 1$; for $y_{\mathrm{ref}} = -3$ it is $e_\infty = -1$.

1.10 (a) $\lambda_{1/2} = \pm 1$

(b) The paths of the trajectories shown in Figure 8.2 are described by

$$x_2 = \pm\sqrt{(x_1+1)^2 + c_1}, \quad c_1 = x_2^2(0) - (x_1(0)+1)^2, \quad \text{for} \quad x_1 < 0,$$

$$x_2 = \pm\sqrt{(x_1-1)^2 + c_2}, \quad c_2 = x_2^2(0) - (x_1(0)-1)^2, \quad \text{for} \quad x_1 > 0.$$

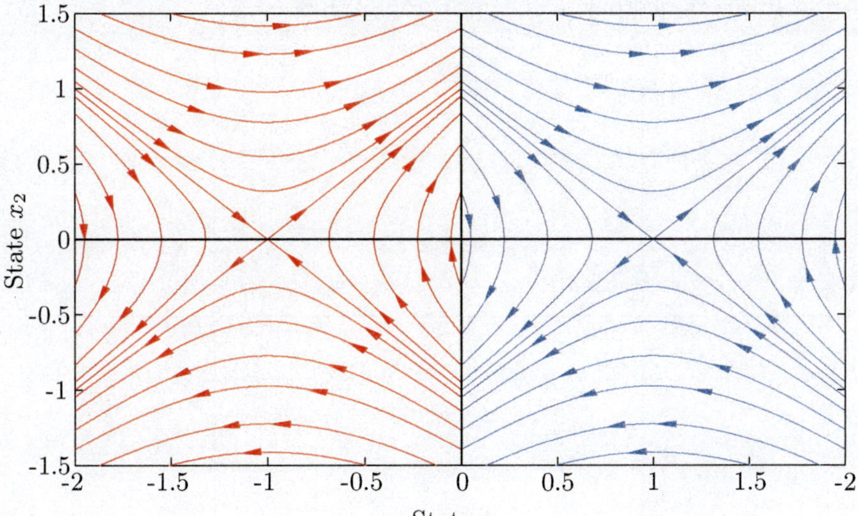

Fig. 8.2: Trajectories of the control loop; red trajectories for $u = 1$ and blue ones for $u = -1$

Because $\dot{x}_1 = x_2$ applies, all trajectories in the upper half-plane tend to the right and those in the lower half-plane to the left.

(c) There are three equilibrium points in the system. The equilibrium point $\boldsymbol{x}_{\text{eq1}} = \mathbf{0}$ is Lyapunov stable, but not asymptotically stable; $\boldsymbol{x}_{\text{eq2}} = \begin{bmatrix}-1 & 0\end{bmatrix}^T$ and $\boldsymbol{x}_{\text{eq3}} = \begin{bmatrix}1 & 0\end{bmatrix}^T$ are strictly unstable.

(d) Stable trajectories exist only for the initial vectors

$$\boldsymbol{x}(0) \in \{\boldsymbol{x} \in I\!R \mid |x_1| + |x_2| \leq 1\}.$$

1.11 (a) $k=4\colon \begin{bmatrix}\dot{x}_1\\ \dot{x}_2\end{bmatrix} = \begin{bmatrix}0 & 1\\ -4 & 0\end{bmatrix}\begin{bmatrix}x_1\\ x_2\end{bmatrix}$, and $k=0.25\colon \begin{bmatrix}\dot{x}_1\\ \dot{x}_2\end{bmatrix} = \begin{bmatrix}0 & 1\\ -0.25 & 0\end{bmatrix}\begin{bmatrix}x_1\\ x_2\end{bmatrix}$

(b) The eigenvalues $\lambda_{1,2} = \pm j\sqrt{k}$ of the two control loops are $\lambda_{1,2} = \pm j2$ and $\lambda_{1,2} = \pm j0.5$. Thus the equilibrium point $\boldsymbol{x}_{\text{eq}} = \mathbf{0}$ is Lyapunov stable in both cases, but not asymptotically stable.

(c)
$$\boldsymbol{x}(t) = \begin{bmatrix}\dfrac{x_2(0)}{\sqrt{k}}\sin(\sqrt{k}t)\\ x_2(0)\cos(\sqrt{k}t)\end{bmatrix} \quad \text{for} \quad k = 4 \quad \text{and} \quad k = 0.25$$

The trajectories are ellipses

$$\frac{x_1^2}{x_2^2(0)k^{-1}} + \frac{x_2}{x_2^2(0)} = 1$$

with semiaxes $a = x_{20}/\sqrt{k}$ and $b = x_{20}$, as shown in Figure 8.3.

(d) There is a single equilibrium point at $\boldsymbol{x} = \mathbf{0}$, which is asymptotically stable. Figure 8.4 shows an example trajectory.

(e) There is a single equilibrium point at $\boldsymbol{x} = \mathbf{0}$, which is strictly unstable. An example trajectory is shown in Figure 8.5.

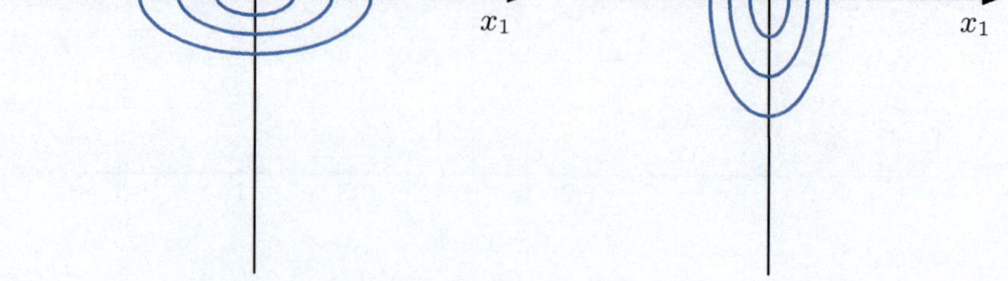

Fig. 8.3: Trajectories of the two subsystems

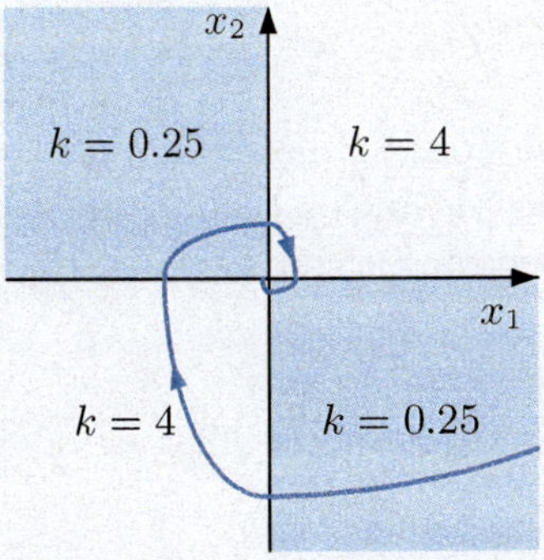

Fig. 8.4: An example trajectory of the stable control loop

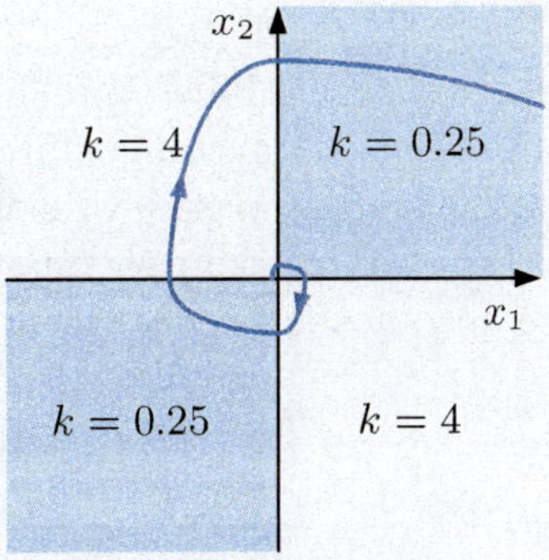

Fig. 8.5: An example trajectory of the unstable control loop

1.13 (a)
$$\boldsymbol{x}_{\mathrm{eq1}}^T = \begin{bmatrix} \dfrac{a_3 + b_2 u_2}{a_4} & \dfrac{a_1 - b_1 u_1}{a_2} \end{bmatrix}, \quad \boldsymbol{x}_{\mathrm{eq2}} = \boldsymbol{0},$$
and for the case in which $a_1 = b_1 u_1$: $\boldsymbol{x}_{\mathrm{eq3}}^T = [x_1 \in \mathbb{R} \;\; 0]$

(b) $(\boldsymbol{\Delta x})^{\boldsymbol{\cdot}} = \boldsymbol{A}\boldsymbol{\Delta x} + \boldsymbol{B}\boldsymbol{\Delta u}$,
$$\boldsymbol{A} = \begin{bmatrix} 0 & -\dfrac{a_2}{a_4}\left(a_3 + \dfrac{b_2}{2}\right) \\ \dfrac{a_4}{a_2}\left(a_1 - \dfrac{b_1}{2}\right) & 0 \end{bmatrix}, \quad \boldsymbol{B} = \begin{bmatrix} -\dfrac{b_1}{a_4}\left(a_3 + \dfrac{b_2}{2}\right) & 0 \\ 0 & -\dfrac{b_2}{a_2}\left(a_1 - \dfrac{b_1}{2}\right) \end{bmatrix}$$

(c) $\lambda_{1,2} = \pm j\sqrt{(a_1 - 0.5b_1)(a_3 + 0.5b_2)}$ (d) Yes

1.14 (a)
$$\dot{V}_1 = \left(z - \frac{5}{9}\right)\dot{V}_{\mathrm{in}}, \quad \dot{V}_2 = \left(\frac{4}{9} - z\right)\dot{V}_{\mathrm{in}}, \quad z = \begin{cases} 1 & \text{if } V_1 = 0.1, \text{ or } \dot{V}_1 > 0 \text{ and } V_2 > 0.1, \\ 0 & \text{if } V_2 = 0.1, \text{ or } \dot{V}_2 > 0 \text{ and } V_1 > 0.1. \end{cases}$$

(b)
$$\Delta t_i = 4\left(\frac{4}{5}\right)^{i-1}, \quad i = 1, 2, \ldots$$
(c)
$$t_i = 20\left[1 - \left(\frac{4}{5}\right)^i\right]$$
(d)
$$t_{\mathrm{tot}} = \sum_{i=0}^{\infty} \Delta t_i = 20$$

(e) The more the time approaches the time $t_{\mathrm{tot}} = 20$ at which both silos reach a volume of $V_1 = V_2 = 0.1V$, the shorter the duration Δt_i of the switching intervals become. This means that the switching frequency increases and eventually becomes infinitely high for $t \to t_{\mathrm{tot}}$. The control obviously cannot cope with this, because each element of the control system, including the swiveling crane, has a limited switching speed. For practical reasons as well, high-frequency switching back and forth is undesirable in this situation. The effect of the increasingly shorter switching intervals whose duration Δt_i becomes infinitely small for $t \to t_{\mathrm{tot}}$ is referred to as Zeno behavior, named after the Greek philosopher Zeno and his paradox of the race between Achilles and the tortoise.

(f) If we simulate a system exhibiting Zeno behavior, an integration procedure with a fixed step size will of necessity become inexact, and one with a step-size adaptation will never attain a time $t \geq t_{\text{tot}}$, because the step size becomes infinitely small for $t \to t_{\text{tot}}$.

1.15 (a) $\dot{z} = (1-\alpha)(p(t)z + q(t))$ (b) Linear differential equation

1.16 $x(t) = \dfrac{bx(0)}{x(0) + (b - x(0))e^{-at}}$ (logistic function)

1.17 $x(t) = h^{-1}\Big(h(x(0)) + \int\limits_0^t g(\tau)d\tau\Big), \quad h(x) = \int \frac{1}{f(x)}dx$

1.18 (a) $x(t) = t^n$ (b) $x(t) = \sqrt[3]{(x^3(0)+1)e^{-2t} - 1}$

(c) $x(t) = \dfrac{x(0)}{x(0) + (1 - x(0))e^t}$ (d) $x(t) = \sin(\frac{1}{3}t^3 + \arcsin(x(0)))$

(e) $x(t) = x(0) + \cosh(t + \operatorname{arsinh}(\dot{x}(0))) - \cosh(\operatorname{arsinh}(\dot{x}(0)))$

1.19 (c) $t_{\text{eq}} = 2\sqrt{|x_0|}$

1.20 (a) We obtain $m_i = q_i$, because $\lg(\varepsilon^{\text{rel}}) = q\lg(h) + \lg(\tilde{\alpha})$ follows from $\varepsilon^{\text{rel}} = \tilde{\alpha}h^q$. See also Table 8.1.

(b) $h_i = \sqrt[m_i]{\varepsilon^{\text{rel}}/\tilde{\alpha}_i}$, see also Table 8.1. (c) $N_i = n_i \cdot 10/h_i$, see also Table 8.1.

(d) The selection of the integration method does not depend only on its order of accuracy. It also depends on the calculation effort, i. e. on the time required for the simulation. In terms of both accuracy and the time required, useful approaches are the 4th-order Runge-Kutta method and the 5th-order Adams-Moulton procedure.

Table 8.1: Parameters for the Euler method (1), modified Euler method (2), Simpson method (3), Runge-Kutta method (4), Adams-Bashforth method (5), Adams-Moulton method with one iteration (6), and the Adams-Moulton method with 10 iterations (7).

Number of the procedure i	1	2	3	4	5	6	7
Order of accuracy q_i	1	2	3	4	4	5	5
Slope of the error curve m_i	1	2	3	4	4	5	5
Constant $\tilde{\alpha}_i$	480	190	46	10	460	120	20
Step size h_i for $\varepsilon = 10^{-6}$	$2 \cdot 10^{-9}$	$7 \cdot 10^{-5}$	$3 \cdot 10^{-3}$	$2 \cdot 10^{-2}$	$7 \cdot 10^{-3}$	$2 \cdot 10^{-2}$	$3 \cdot 10^{-2}$
Function calculations n_i per step	1	2	3	4	1	2	11
Total function calculations N_i	$4.8 \cdot 10^9$	$2.8 \cdot 10^5$	$1.1 \cdot 10^4$	$2.2 \cdot 10^3$	$1.5 \cdot 10^3$	$8.3 \cdot 10^2$	$3.2 \cdot 10^3$

1.21 See Table 1.3 on p. 54. In the case of the Gear method, the stability range of the step size h is $h\lambda < \infty$.

2.1 (a),(b) See Table 2.1, p. 78 et seq.
(c)
$$N(A) = m_2 + \frac{2}{\pi}(m_1 - m_2)\left[\arcsin\left(\frac{a}{A}\right) + \frac{a}{A}\sqrt{1 - \frac{a^2}{A^2}}\right], \quad A > a$$

(d)
$$N(A) = \frac{4b}{\pi A}\sum_{i=1}^{k}\sqrt{1 - \left(\frac{2i-1}{2A}a\right)^2}, \quad \frac{2k-1}{2}a < A < \frac{2k+1}{2}a, \quad k \in \mathbb{N}$$

2.2 (a)
$$N(A) = k_1 + \frac{3k_2}{4}A^2 + \frac{5k_3}{8}A^4$$
(b)
$$N(A) = \sum_{i=1}^{n} k_i \left(\frac{A}{2}\right)^{2(i-1)} \binom{2i-1}{i-1}$$

2.3 (a) $G(s) = \dfrac{s^2 + s + 15}{s^3 + 9.8s^2 + 15s + 132}$

(b) Chua's circuit is shown in Figure 8.6.

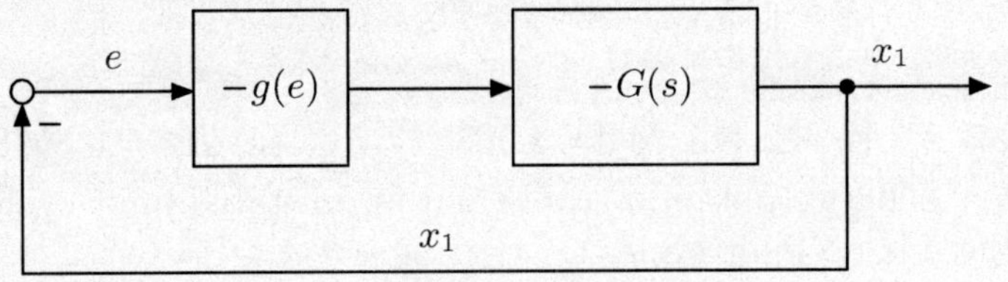

Fig. 8.6: Chua's circuit represented as a nonlinear standard control loop

(c)
$$N(A) = \frac{5}{7} + \frac{6}{7\pi}\left[\arcsin\left(\frac{1}{A}\right) + \frac{1}{A}\sqrt{1 - \frac{1}{A^2}}\right], \quad A > a = 1$$

(d) The nonlinear locus $-1/N(A)$ for $A = a = 1$ starts at $-7/8$ on the negative real axis and follows the negative real axis for $A \to \infty$ to $-7/5$. The Nyquist plot of $-G(j\omega)$ intersects the real axis in several places; the furthest to the left is at
$$\mathrm{Re}\,\{G(j3.6196)\} = -0.5268.$$
This means no point of intersection exists between the plots of $-G(j\omega)$ and $-1/N(A)$, and thus there is probably no limit cycle either.

2.4 $b < 2\pi a$

2.5 (a)
$$N(A) = \begin{cases} 0, & A \le a, \\ m\left[1 - \frac{2}{\pi}\arcsin\left(\frac{a}{A}\right) - \frac{2}{\pi}\cdot\frac{a}{A}\sqrt{1 - \frac{a^2}{A^2}}\right], & a < A \le b, \\ \frac{2m}{\pi}\left[\arcsin\left(\frac{b}{A}\right) - \arcsin\left(\frac{a}{A}\right) + \frac{b}{A}\sqrt{1 - \frac{b^2}{A^2}} - \frac{a}{A}\sqrt{1 - \frac{a^2}{A^2}}\right], & b < A, \end{cases}$$

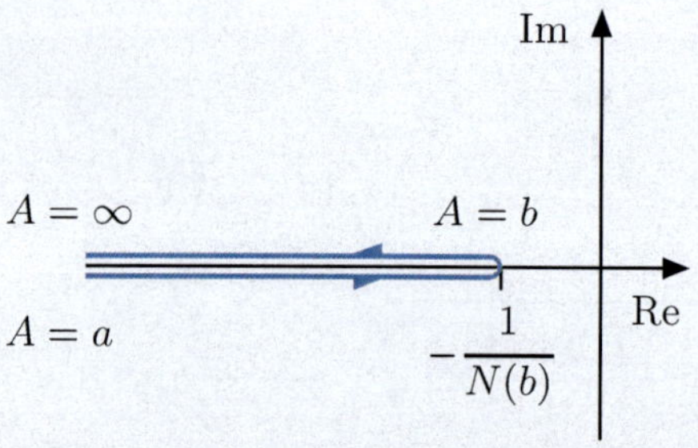

Fig. 8.7: Nonlinear locus

with $m = c/(b-a)$. Figure 8.7 shows the nonlinear locus $-1/N(A)$. Its turning point is $-1/N(b)$, where

$$N(b) = m\left[1 - \frac{2}{\pi}\arcsin\left(\frac{a}{b}\right) - \frac{2}{\pi}\frac{a}{b}\sqrt{1-\frac{a^2}{b^2}}\right]$$

is reached at $A = b$.

(b) $\omega_1 = \omega_2 = 2,\ A_1 = 2.475,\ A_2 = 4.505$ (c) $K_\mathrm{s} = 0.8568$

(d) For $K > 0.8568$ we obtain a stable and an unstable limit cycle; for $K < 0.8568$ there is no limit cycle.

2.6 (a)

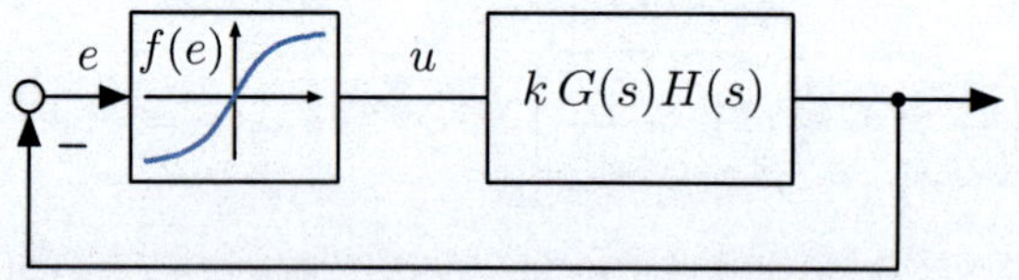

Fig. 8.8: Blood-pressure control loop

$$f(e) = b\arctan(me)$$

$$kG(s)H(s) = k\frac{e^{-(\tau_\mathrm{e}+\tau_\mathrm{a})s}}{1+\tau_\nu s} = \frac{1.65e^{-2.99s}}{1+5s}$$

(b) $$N(A) = \frac{2b}{\pi A}\int_0^\pi \arctan(mA\sin(\alpha))\sin(\alpha)d\alpha = \frac{2b}{\pi A}\int_0^\pi \frac{mA\cos^2(\alpha)}{1+m^2A^2\sin^2(\alpha)}d\alpha$$

$$= \frac{2bm}{\pi}\int_0^\pi \frac{1}{1+m^2A^2\sin^2(\alpha)}d\alpha - \frac{2b}{\pi m A^2}\int_0^\pi\left(1-\frac{1}{1+m^2A^2\sin^2(\alpha)}\right)d\alpha$$

$$= \frac{2b}{mA^2}\left(\sqrt{m^2A^2+1}-1\right)$$

(c) One single limit cycle exists with $\omega = 0.2\pi$, i. e. $f = 0.1\,\mathrm{Hz}$.

(d) The limit cycle is stable.

2.7 (a) The nonlinear curve can lie in the sector $[0,\ 45.5)$.

(b) A linear curve (proportional control) can lie in the sector $(-81,\ 140.5)$.

(c) $|e| < 6.74$

2.8 (a) $T_{\mathrm{I}}=T_{\mathrm{e}}=0.2\,\mathrm{s}$, $K_{\mathrm{P}}=\dfrac{c}{125}$ (b) $c=\dfrac{\pi}{12}$, $K_{\mathrm{P}}=\dfrac{\pi}{1500}$

2.9 (a) [0, 27.94] (b) [0, 60)
(c) Aizerman's conjecture is correct in this case and leads to the sector (−6, 60).

2.10 (a) The Nyquist plot of $G(j\omega)$ is shown in Figure 8.9.
(b) Circle with its midpoint on the real axis between −3 and −5
(c) Midpoint $c = -4.39$, radius $r = 0.23$, sector [0.22, 0.24]
(d) Hurwitz sector: $(0.2, 0.\overline{3})$

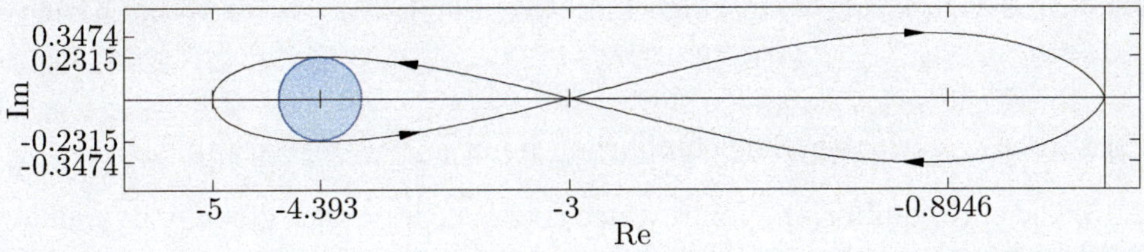

Fig. 8.9: Nyquist plot

2.11 (a) $\Omega = \pi$
(b)

$$[K_1, K_2] = \left[\frac{\sqrt{8}}{\sqrt{7}+\sqrt{8}} + \epsilon,\ \frac{\sqrt{8}}{\sqrt{8}-\sqrt{7}} - \epsilon\right] \quad \text{with arbitrarily small} \quad \epsilon > 0$$

(c) Hurwitz sector: $(0, \infty)$

2.12 Solution for $G_1(s)$: (a) The plant has two poles at $s = 0$, i. e. $\mu = 2$. Thus we use the general circle criterion, Theorem 10. The continuous angle variation is $\Omega = \pi$. The sector of absolute stability $[K_1, K_2]$ is dependent on the circle selected, as we can see from Figure 8.10. (b) Hurwitz sector: $(0, \infty)$

Solution for $G_2(s)$: (a) Using Theorem 9, we obtain $(1, \infty)$ as a sector of absolute stability, because $\nu = 2$ (2 unstable poles) and the circular Nyquist plot of $G_2(j\omega)$ encircles each circle $D\left(1/(1-\epsilon), 1/\varepsilon\right)$, $0.5 \geq \varepsilon > 0$, twice counterclockwise. See Figure 8.11. (b) Hurwitz sector: $K \in (1, \infty)$

Solution for $G_3(s)$: (a) The general circle criterion, Theorem 10, must be applied, because there are three poles $s_{1,2,3} = 0$. No sector of absolute stability can be identified using the criterion, because the continuous angle variation for all circles with $K_1 > 0$, $K_2 > 0$ takes the value $\Omega = \pi/2$. See Figure 8.12. However, to prove absolute stability, $\Omega = \nu\pi + \mu\pi/2 = 3\pi/2$ would have to apply. See Figure 8.12. (b) There is no Hurwitz sector, i. e. the plant cannot be stabilized by a P controller. That is why there is no sector of absolute stability at all.

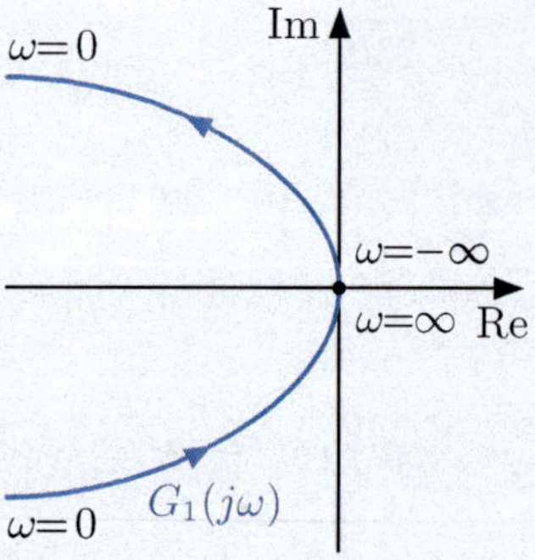

Fig. 8.10: Nyquist plot of $G_1(j\omega)$

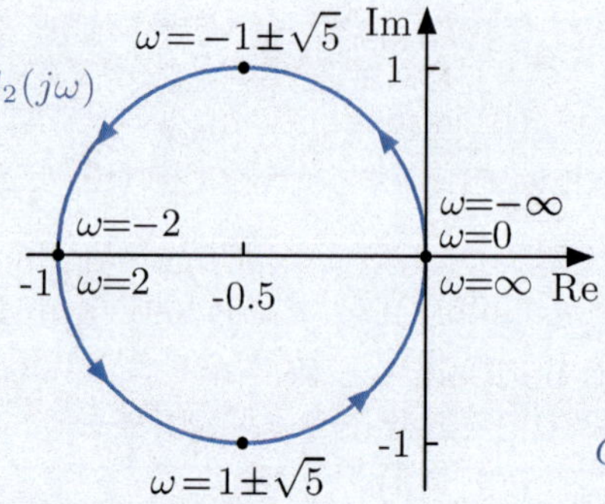

Fig. 8.11: Nyquist plot of $G_2(j\omega)$

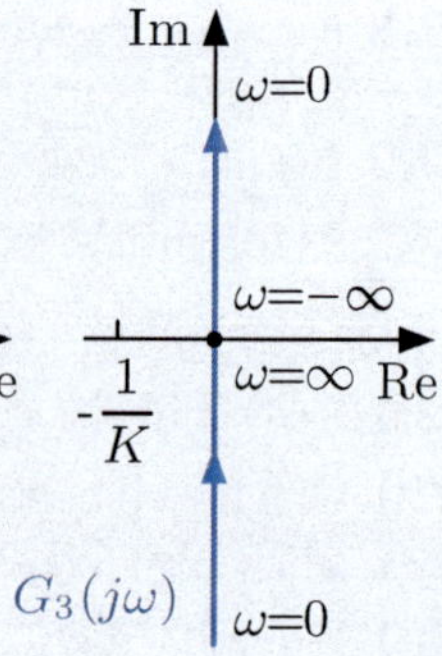

Fig. 8.12: Nyquist plot of $G_3(j\omega)$

2.14 (a) $\boldsymbol{x}_{\text{eq}} = \mathbf{0}$ (b) The equilibrium point $\boldsymbol{x}_{\text{eq}} = \mathbf{0}$ is unstable.

2.15 The only equilibrium point, which is $[x_{\text{eq}}\ \dot{x}_{\text{eq}}]^T = \mathbf{0}$, is globally Lyapunov stable. Thus the system itself is globally Lyapunov stable.

2.16 (c) $G = \{\boldsymbol{x} \in \mathbb{R}^2 \mid x_1^2 + x_2^2 < 4\}$.

2.17 (b) The contour lines of $V(\boldsymbol{x})$ are circles.
(c) Figure 8.13 shows the plot of the Lyapunov function $V(\boldsymbol{x})$ for $x_2 = 0$.

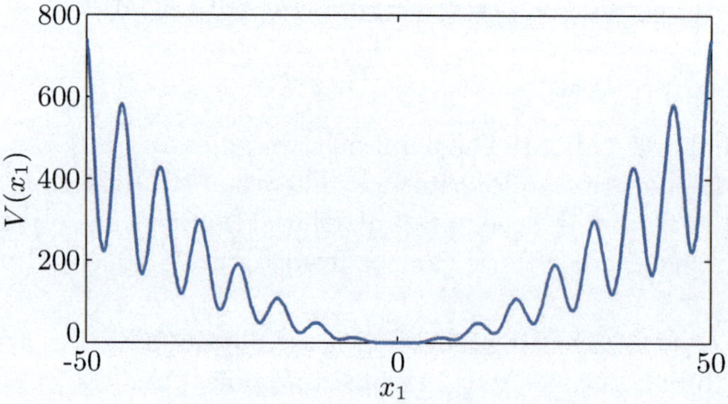

Fig. 8.13: Plot of the rotationally symmetric Lyapunov function $V(\boldsymbol{x})$

(d) The trajectories are circles, because the system is a harmonic oscillator.

2.18 (a)
$$\dot{\boldsymbol{x}} = \begin{bmatrix} -4 & -15 \\ -4 & -23 \end{bmatrix} \boldsymbol{x} - \begin{bmatrix} 1 \\ 1 \end{bmatrix} (3x_1 + 21x_2)^3$$
(b) There is a single equilibrium point at $\boldsymbol{x}_{\text{eq}} = [0,\ 0]^T$.

2.19 (a) Equilibrium points: $\boldsymbol{x}_{\text{eq1}} = \mathbf{0}$ and all $\boldsymbol{x}_{\text{eq}} \in \{\boldsymbol{x} \in \mathbb{R}^3 \mid x_1^2 + x_2^2 + x_3^2 = 1\}$
(b) 1. With $V(\boldsymbol{x}) = \boldsymbol{x}^T\boldsymbol{x}$, $\dot{V}(\boldsymbol{x}) > 0$ applies for all $\boldsymbol{x} \in \{\boldsymbol{x} \in \mathbb{R}^3 \mid x_1^2 + x_2^2 + x_3^2 < 1\}$. This means that $\boldsymbol{x}_{\text{eq1}} = \mathbf{0}$ is unstable (instability theorem, p. 142).

2. With

$$V_{\text{LaSalle}}(\boldsymbol{x})=(x_1^2+x_2^2+x_3^2-1)^2,$$

$\dot{V}_{\text{LaSalle}}(\boldsymbol{x}) = 0$ applies for

$$\boldsymbol{x}\in M=N=\left\{\boldsymbol{x}\in I\!R^3 \mid x_1^2+x_2^2+x_3^2=1\right\}.$$

In addition, $\dot{V}_{\text{LaSalle}}(\boldsymbol{x}) < 0$ applies for all $\boldsymbol{x} \in I\!R^3\backslash\{M \cup \{\boldsymbol{0}\}\}$. This means that all equilibrium points $\boldsymbol{x}_{\text{eq},i} \in M$ are Lyapunov stable and all trajectories with the exception of $\boldsymbol{x}(t) = \boldsymbol{0}$ tends to M.

2.21 (a) $S(\boldsymbol{\omega}) = 0.5\left(J_x\omega_x^2 + J_y\omega_y^2 + J_z\omega_z^2\right)$

(b) The controller matrix $\boldsymbol{K}$ must be positive definite.

(c) As new coordinates, select the angular momentum values $L_x = J_x\omega_x$, $L_y = J_y\omega_y$, and $L_z = J_z\omega_z$. Use the storage function

$$H(\boldsymbol{L}) = \frac{1}{2}\left(\frac{L_x^2}{J_x} + \frac{L_y^2}{J_y} + \frac{L_z^2}{J_z}\right) = S(\boldsymbol{\omega}).$$

The PCH model is

$$\dot{\boldsymbol{L}} = \boldsymbol{J}(\boldsymbol{L})\frac{\partial H(\boldsymbol{L})}{\partial \boldsymbol{L}} + \boldsymbol{M} \qquad \text{with} \qquad \boldsymbol{J}(\boldsymbol{L}) = \begin{bmatrix} 0 & -L_z & L_y \\ L_z & 0 & -L_x \\ -L_y & L_x & 0 \end{bmatrix}.$$

(d) The control law is $M(\boldsymbol{L}) = -\boldsymbol{D}\dfrac{\partial H(\boldsymbol{L})}{\partial \boldsymbol{L}}$, with $\boldsymbol{D}$ being positive definite.

2.22 A PID controller is passive, but not strictly passive.

2.24 (a) strictly positive real (b) strictly passive

(c) All characteristic curves in the sector $[0, \infty]$ result in a control loop with an asymptotically stable equilibrium point.

(d) Based on Aizerman's conjecture, the largest sector of absolute stability corresponds to the Hurwitz sector, which is $(-1, \infty]$ in the case of the system considered.

2.25 $P \cdot \text{Re}\{G(j\omega)\} \geq -\dfrac{I}{\omega}\text{Im}\{G(j\omega)\}$

2.26 (a) $L = E_{\text{kin}} - E_{\text{pot}} = \dfrac{1}{2}ml^2\dot{\varphi}^2 - mgl(1 - \cos(\varphi))$

(b) $\dfrac{d}{dt}\dfrac{\partial L}{\partial \dot{\varphi}} - \dfrac{\partial L}{\partial \varphi} = ml\ddot{\varphi} + mlg\sin(\varphi) = 0 \;\Rightarrow\; l\ddot{\varphi} + g\sin(\varphi) = 0$

(c) $H(q,p) = \dfrac{p^2}{2ml^2} + mgl(1 - \cos(q))$ with $q = \varphi$ and $p = ml^2\dot{\varphi}$

(d) $\dot{q} = \dfrac{p}{ml^2}, \qquad \dot{p} = -mgl\sin(q)$

3.1 The solution to the differential equation $\dot{x} = xu$ is

$$x(t) = x(0)e^{\int_0^t u(\tau)d\tau}.$$

Depending on the sign of the initial value $x(0)$, $x(t)$ is either positive for all $t \in \mathbb{R}$ or it is negative. Thus the system is not globally controllable. However, it is locally controllable for $x(0) > 0$ on $\mathbb{R}^+$ and for $x(0) < 0$ on $\mathbb{R}^-$. It is not controllable if $x(0) = 0$.

3.2 The system is not controllable, because the state variable x_3 cannot be influenced by the control variable u.

3.3 (a) $\begin{bmatrix} \dot{x}_1 \\ \dot{x}_2 \\ \dot{x}_3 \end{bmatrix} = \begin{bmatrix} \cos(x_3) & -\sin(x_3) & 0 \\ \sin(x_3) & \cos(x_3) & 0 \\ 0 & 0 & 1 \end{bmatrix} \begin{bmatrix} v_1 \\ v_2 \\ v_3 \end{bmatrix} = \boldsymbol{A}(\boldsymbol{x}) \cdot \boldsymbol{v}$

(c) The overall model

$$\begin{bmatrix} \dot{\boldsymbol{x}} \\ \dot{\boldsymbol{v}} \end{bmatrix} = \begin{bmatrix} \boldsymbol{A}(\boldsymbol{x})\boldsymbol{v} \\ -\boldsymbol{M}^{-1}\boldsymbol{D}\boldsymbol{v} \end{bmatrix} + \begin{bmatrix} \boldsymbol{0} \\ \boldsymbol{M}^{-1} \end{bmatrix} \boldsymbol{u}$$

is not omnidirectionally controllable. It is controllable, however, because it is flat.

(d) A flat output is $\boldsymbol{y} = \boldsymbol{x}$. The flat system description is

$$\begin{bmatrix} \boldsymbol{x} \\ \boldsymbol{v} \end{bmatrix} = \begin{bmatrix} \boldsymbol{y} \\ \boldsymbol{A}^{-1}(\boldsymbol{y})\dot{\boldsymbol{y}} \end{bmatrix}, \qquad \boldsymbol{A}^{-1}(\boldsymbol{y}) = \begin{bmatrix} \cos(y_3) & \sin(y_3) & 0 \\ -\sin(y_3) & \cos(y_3) & 0 \\ 0 & 0 & 1 \end{bmatrix},$$

$$\boldsymbol{u} = \boldsymbol{M}\boldsymbol{A}^{-1}(\boldsymbol{y}) \begin{bmatrix} \ddot{y}_1 + \dot{y}_2\dot{y}_3 \\ \ddot{y}_2 - \dot{y}_1\dot{y}_3 \\ \ddot{y}_3 \end{bmatrix} + \boldsymbol{D}\boldsymbol{A}^{-1}(\boldsymbol{y})\dot{\boldsymbol{y}}.$$

3.5 (a) The fuel cell model is locally omnidirectionally controllable for

$$\boldsymbol{x} \in \left\{ \boldsymbol{x} \in \mathbb{R}^3 \mid x_1 \neq p \text{ and } x_2 \neq p \text{ and } x_3 \neq \frac{2}{3}p \right\}.$$

(b) The fuel cell model is not globally omnidirectionally controllable.

(c) The fuel cell model is locally small-time locally controllable for

$$\boldsymbol{x} \in \left\{ \boldsymbol{x} \in \mathbb{R}^3 \mid x_1 \neq p \text{ and } x_2 \neq p \right\}.$$

(d) The fuel cell model is not globally small-time locally controllable.

3.6 (a)

$$\boldsymbol{z} = \begin{bmatrix} y \\ \dot{y} \\ \ddot{y} \end{bmatrix} = \begin{bmatrix} x_1 \\ f_1(x_1) + h_1(x_1)x_2 \\ \mu(x_1, x_2) + h_1(x_1)h_2(x_1, x_2)x_3 \end{bmatrix}$$

$$\mu(x_1, x_2) = [f_1(x_1) + h_1(x_1)x_2] \left[\frac{\partial f_1(x_1)}{\partial x_1} + x_2 \frac{\partial h_1(x_1)}{\partial x_1} \right] + h_1(x_1)f_2(x_1, x_2)$$

(b) $\boldsymbol{b}(\boldsymbol{z}) = [0 \;\; 0 \;\; h_1(x_1)h_2(x_1, x_2)h_3(x_1, x_2, x_3)]^T$

with $x_1 = z_1, \quad x_2 = \dfrac{z_2 - f_1(z_1)}{h_1(z_1)}, \quad x_3 = \dfrac{z_3 - \mu(z_1, x_2)}{h_1(z_1)h_2(z_1, x_2)}$

(c) The system is controllable if the inequalities $h_1(x_1) \neq 0$, $h_2(x_1, x_2) \neq 0$, and $h_3(x_1, x_2, x_3) \neq 0$ hold.

3.7 (a) Flat output: $y = x_1$; $\boldsymbol{x} = \begin{bmatrix} y \ \dot{y} \end{bmatrix}^T$, $\quad u = \ddot{y} + y \qquad$ (b) $u = 1 - 2te^{-t}$

3.8 (a)

$$\boldsymbol{y} = \begin{bmatrix} \omega_{\mathrm{x}} \\ \omega_{\mathrm{y}} \\ \omega_{\mathrm{z}} \end{bmatrix}, \quad \boldsymbol{u} = \begin{bmatrix} M_{\mathrm{x}} \\ M_{\mathrm{y}} \\ M_{\mathrm{z}} \end{bmatrix} = \begin{bmatrix} J_{\mathrm{x}}\dot{y}_1 + (J_{\mathrm{z}} - J_{\mathrm{y}})y_2 y_3 \\ J_{\mathrm{y}}\dot{y}_2 + (J_{\mathrm{x}} - J_{\mathrm{z}})y_1 y_3 \\ J_{\mathrm{z}}\dot{y}_3 + (J_{\mathrm{y}} - J_{\mathrm{x}})y_1 y_2 \end{bmatrix}$$

(b)

$$\boldsymbol{y} = \begin{bmatrix} x_1 \\ x_2 \end{bmatrix}, \quad \boldsymbol{x} = \begin{bmatrix} y_1 \\ y_2 \\ \arctan\left(\dfrac{\dot{y}_2}{\dot{y}_1}\right) \end{bmatrix}, \quad \boldsymbol{u} = \begin{bmatrix} \dfrac{\dot{y}_1\ddot{y}_2 - \dot{y}_2\ddot{y}_1}{\dot{y}_1^2 + \dot{y}_2^2} \\ \sqrt{\dot{y}_1^2 + \dot{y}_2^2} \end{bmatrix}$$

(c)

$$\boldsymbol{y} = \begin{bmatrix} x_1 \\ x_2 \\ x_3 \end{bmatrix}, \quad \boldsymbol{u} = \begin{bmatrix} 0 & 0 & 1 \\ \cos(y_3) & \sin(y_3) & 0 \\ -\sin(y_3) & \cos(y_3) & 0 \end{bmatrix} \dot{\boldsymbol{y}}$$

3.9 (a) Flat output: $\boldsymbol{y} = \begin{bmatrix} y_1 \ y_2 \end{bmatrix}^T = \begin{bmatrix} x_1 \ x_3 \end{bmatrix}^T$

(b) Flat system representation:

$$\boldsymbol{x} = \begin{bmatrix} y_1 \\ \dfrac{J\dot{y}_2 + dy_2 + M_{\mathrm{L}}}{ky_1} \\ y_2 \end{bmatrix}, \quad \boldsymbol{u} = \begin{bmatrix} u_{\mathrm{r}} \\ u_{\mathrm{s}} \end{bmatrix},$$

$$u_{\mathrm{r}} = L_{\mathrm{r}}\dot{y}_1 + R_{\mathrm{r}}y_1 + y_2\frac{J\dot{y}_2 + dy_2 + M_{\mathrm{L}}}{y_1},$$

$$u_{\mathrm{s}} = \frac{J(y_1\ddot{y}_2 - \dot{y}_1\dot{y}_2) + d(y_1\dot{y}_2 - \dot{y}_1 y_2) - M_{\mathrm{L}}\dot{y}_1}{ky_1^2} + \frac{R_{\mathrm{s}}}{L_{\mathrm{s}}}\frac{J\dot{y}_2 + dy_2 + M_{\mathrm{L}}}{ky_1}$$

3.10 (a)

$$x_1 = \frac{1}{m_{\mathrm{c}}}y + \frac{m_{\mathrm{s}}}{m_{\mathrm{c}}c_{\mathrm{t}}}\ddot{y} - \frac{m_{\mathrm{s}}}{m_{\mathrm{c}}}x_{\mathrm{d}}, \qquad x_2 = \frac{1}{m_{\mathrm{c}}}\dot{y} + \frac{m_{\mathrm{s}}}{m_{\mathrm{c}}c_{\mathrm{t}}}\dddot{y} - \frac{m_{\mathrm{s}}}{m_{\mathrm{c}}}\dot{x}_{\mathrm{d}},$$

$$x_3 = -\frac{1}{c_{\mathrm{t}}}\ddot{y} + x_{\mathrm{d}}, \qquad x_4 = -\frac{1}{c_{\mathrm{t}}}\dddot{y} + \dot{x}_{\mathrm{d}},$$

$$u = \frac{m}{c_{\mathrm{t}}}y^{(4)} + \ddot{y} + F_{\mathrm{s}}(x_1, x_3) + F_{\mathrm{d}}(x_2, x_4) - m_{\mathrm{s}}\ddot{x}_{\mathrm{d}}$$

(b) The system is controllable, because flat systems are controllable.

(c) The transformation $\boldsymbol{z} = \boldsymbol{t}(\boldsymbol{x})$ and the transformed system $\dot{\boldsymbol{z}} = \dot{\boldsymbol{t}}(\boldsymbol{x})$ with $\boldsymbol{x} = \boldsymbol{t}^{-1}(\boldsymbol{z})$, which is in the nonlinear controller canonical form, are

$$\boldsymbol{z} = \boldsymbol{t}(\boldsymbol{x}) = \begin{bmatrix} m_{\mathrm{c}}x_1 + m_{\mathrm{s}}x_3 \\ m_{\mathrm{c}}x_2 + m_{\mathrm{s}}x_4 \\ -c_{\mathrm{t}}x_3 + c_{\mathrm{t}}x_{\mathrm{d}} \\ -c_{\mathrm{t}}x_4 + c_{\mathrm{t}}\dot{x}_{\mathrm{d}} \end{bmatrix}, \quad \dot{\boldsymbol{z}} = \begin{bmatrix} z_2 \\ z_3 \\ z_4 \\ \alpha(\boldsymbol{z}) \end{bmatrix} + \begin{bmatrix} 0 \\ 0 \\ 0 \\ \beta \end{bmatrix} u$$

with

$$\alpha(\boldsymbol{z}) = -\frac{c_\mathrm{t}}{m_\mathrm{s}}\left[z_3 + \hat{F}_\mathrm{s}(\boldsymbol{z}) + \hat{F}_\mathrm{d}(\boldsymbol{z}) - m_\mathrm{s}\ddot{x}_\mathrm{d}\right], \quad \beta = \frac{c_\mathrm{t}}{m_\mathrm{s}},$$

$$\hat{F}_\mathrm{s}(\boldsymbol{z}) = c_\mathrm{s} g(\boldsymbol{z}) + c_\mathrm{nls} g^3(\boldsymbol{z}), \quad g(\boldsymbol{z}) = \frac{z_1}{m_\mathrm{c}} + \frac{m_\mathrm{s}+m_\mathrm{c}}{m_\mathrm{c}}\left(\frac{z_3}{c_\mathrm{t}} - x_\mathrm{d}\right),$$

$$\hat{F}_\mathrm{d}(\boldsymbol{z}) = d\cdot h(\boldsymbol{z}) + d_\mathrm{nl} h^2(\boldsymbol{z})\,\mathrm{sgn}(h(\boldsymbol{z})), \quad h(\boldsymbol{z}) = \frac{z_2}{m_\mathrm{c}} + \frac{m_\mathrm{s}+m_\mathrm{c}}{m_\mathrm{c}}\left(\frac{z_4}{c_\mathrm{t}} - \dot{x}_\mathrm{d}\right).$$

3.11

$$\boldsymbol{x} = \boldsymbol{q}^{-1}(\boldsymbol{z}) = \begin{bmatrix} \ln\left(\sqrt{z_1^2 + z_2^2}\right) \\ \arctan\left(\frac{z_2}{z_1}\right) \end{bmatrix}$$

3.12 (a) $x = p(z) = z^{-1}$ (b) $x(t) = \dfrac{x_0}{x_0 + (1 - x_0)e^{-t}}$

3.13 (a)

$$\boldsymbol{x} = \begin{bmatrix} z_1 \\ \arcsin(z_2) \end{bmatrix} = \boldsymbol{q}^{-1}(\boldsymbol{z})$$

(b)

$$\dot{\boldsymbol{z}} = \begin{bmatrix} z_2 \\ -z_1\sqrt{1 - z_2^2} \end{bmatrix} + \begin{bmatrix} 0 \\ \sqrt{1 - z_2^2} \end{bmatrix} u$$

(c) The transformed system description $\dot{\boldsymbol{z}} = \boldsymbol{f}(\boldsymbol{z}, u)$ takes the nonlinear controller canonical form.

3.14 (a) Local diffeomorphism: $z = \sqrt[3]{x}$, $M_z = I\!R$, $M_x = I\!R$; although the diffeomorphism bijectively maps $I\!R^n$ into $I\!R^n$ it is not a global diffeomorphism, since $z = \sqrt[3]{x}$ is not continuously differentiable at $x = 0$.
(b) Global diffeomorphism if $\boldsymbol{T} \in I\!R^{n\times n}$ is regular;
$\boldsymbol{z} = \boldsymbol{T}^{-1}\boldsymbol{x} - \boldsymbol{T}^{-1}\boldsymbol{a}$, $M_{\boldsymbol{z}} = I\!R^n$, $M_{\boldsymbol{x}} = I\!R^n$.
(c) Local diffeomorphism:

$$\boldsymbol{z} = \begin{bmatrix} \sqrt{x_1^2 + x_2^2} \\ \kappa(x_2)\arccos\left(\dfrac{x_1}{\sqrt{x_1^2 + x_2^2}}\right) \end{bmatrix}, \quad \kappa(x_2) = \begin{cases} 1, & x_2 \geq 0, \\ -1, & x_2 < 0, \end{cases}$$

$$M_{\boldsymbol{z}} = \{\boldsymbol{z} \in I\!R \mid z_1 \neq 0,\ z_2 \in (-\pi, \pi]\}, \quad M_{\boldsymbol{x}} = I\!R^2 \setminus \{\boldsymbol{0}\}.$$

(d) Global diffeomorphism; the inverse is not analytically determinable.
(e) Not a diffeomorphism
(f) Local diffeomorphism if $z_1 \neq 0$ and $z_1 < R$;

$$\boldsymbol{z} = \begin{bmatrix} \sqrt{x_3^2 + \left(\sqrt{x_1^2 + x_2^2} - R\right)^2} \\ \kappa(x_3)\arccos\left(\dfrac{\sqrt{x_1^2 + x_2^2} - R}{\sqrt{x_3^2 + \left(\sqrt{x_1^2 + x_2^2} - R\right)^2}}\right) \\ \kappa(x_2)\arccos\left(\dfrac{x_1}{\sqrt{x_1^2 + x_2^2}}\right) \end{bmatrix}, \quad \kappa(x_i) = \begin{cases} 1, & x_i \geq 0, \\ -1, & x_i < 0, \end{cases}$$

$$M_{\boldsymbol{z}} = \left\{\boldsymbol{z} \in \mathbb{R}^3 \mid 0 < z_1 < R,\ -\pi < z_2 \le \pi,\ -\pi < z_3 \le \pi\right\},$$

$$M_{\boldsymbol{x}} = \left\{\boldsymbol{x} \in \mathbb{R}^3 \mid 0 < x_3^2 + \left(\sqrt{x_1^2 + x_2^2} - R\right)^2 < R^2\right\}.$$

3.15 (a) $\dot{z}_1 = z_1(1 - z_1), \qquad \dot{z}_2 = \sin^2(0.5z_2)$

(b) The system has two equilibrium points at $\boldsymbol{z}_{\text{eq1}} = \boldsymbol{0}$ and $\boldsymbol{z}_{\text{eq2}} = [1 \quad 0]^T$. Both are unstable, because all values z_1 with $1 > z_1 > 0$ in a neighborhood of $\boldsymbol{z}_{\text{eq1}}$ lead to $\dot{z}_1 > 0$ and, at $\boldsymbol{z}_{\text{eq2}}$, arbitrarily small values $z_2 > 0$ lead to $\dot{z}_2 > 0$. In addition, the radius z_1 is constant for $z_1 = 1$, because $\dot{z}_1 = 0$. This yields a circular trajectory if $\dot{z}_2 > 0$. This trajectory tends to the equilibrium point $\boldsymbol{z}_{\text{eq2}}$. From outside, i.e. $z_1 > 1$, all trajectories tend to the unit circle, because $\dot{z}_1 < 0$ applies here. From within, for $z_1 < 1$ all trajectories also tend to the unit circle, because $\dot{z}_1 > 0$ applies.

3.16 (a) $\dot{z} = 2\,\text{sgn}(z) \qquad$ (b) $x(t) = \text{sgn}(x(0))\sqrt{2t + x^2(0)}$

(c) Both diffeomorphisms transform the nonlinear differential equation into a linear differential equation – but into two different ones. So it may be that there are several diffeomorphisms linearizing a differential equation.

(d) It is both a Bernoulli differential equation and a separable one.

4.1 (a)

$$Q(s) = \frac{N_{\text{PID}}(s)}{s + mK_{\text{I}}}, \qquad R(s) = \frac{N_{\text{PID}}(s)N(s) - mK_{\text{I}}D(s)}{(s + mK_{\text{I}})N(s)},$$

$$G(s)R(s) = \frac{N_{\text{PID}}(s)N(s) - mK_{\text{I}}D(s)}{(s + mK_{\text{I}})D(s)}$$

(b) $G(s)G_{\text{PI}}(s) = \dfrac{2s + 1}{s^2}, \qquad G(s)R(s) = \dfrac{-998s + 1}{s(s + 1000)}$

(c) Both criteria are applicable, provided that excessively high values of the saturation block's input variable cannot occur. This is always the case in real world applications. In this context, see the remarks concerning the modified saturation characteristic in Section 2.2.4 on p. 99 et seq.

(d) $G(j\omega)G_{\text{PI}}(j\omega) = -\dfrac{1}{\omega^2} - j\dfrac{2}{\omega}, \quad G(j\omega)R(j\omega) = -\dfrac{998 \cdot 10^3 + 1}{\omega^2 + 10^6} + j\dfrac{998\omega^2 - 10^3}{\omega(\omega^2 + 10^6)}$

(e) For the control loop without anti-windup, i. e. the plant's transfer function is $G(s)G_{\text{PI}}(s)$, we can identify no sector of absolute stability which proves stability, because the Popov criterion returns no result and the general circle criterion yields no circles whose corresponding sectors of absolute stability contain the saturation curve $u = \text{sat}(v)$. This can be seen in Figure 8.14 for the circle criterion and in the Figure 8.16 for the Popov criterion.

In the case of the control loop with anti-windup, i. e. the plant's transfer function is $G(s)R(s)$, there is a circle with an infinite radius which intersects the point $-1/k_2 = -1/1.0020$. This yields the sector $[\epsilon,\ 1.0020]$ of absolute stability. Figure 8.15 illustrates this. The Popov criterion yields the same result, as can be seen in Figure 8.17.

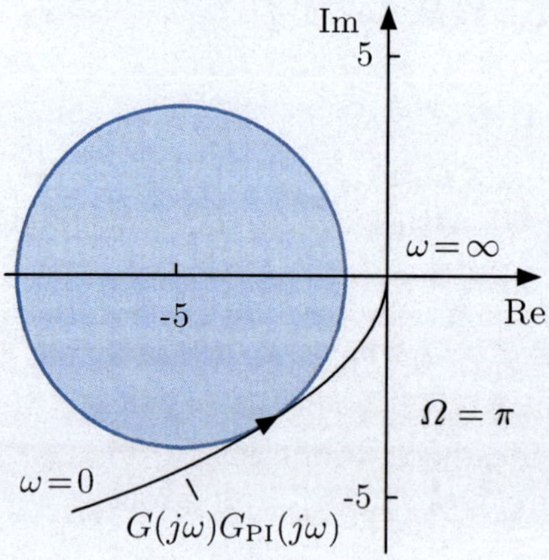

Fig. 8.14: Nyquist plot of $G(s)G_{\mathrm{PI}}(s)$

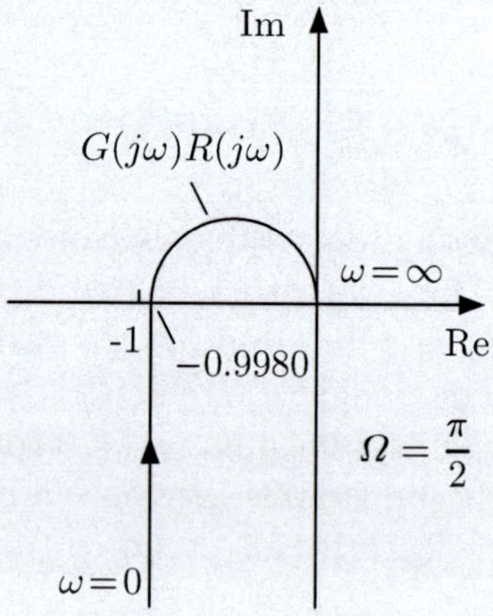

Fig. 8.15: Nyquist plot of $G(s)R(s)$

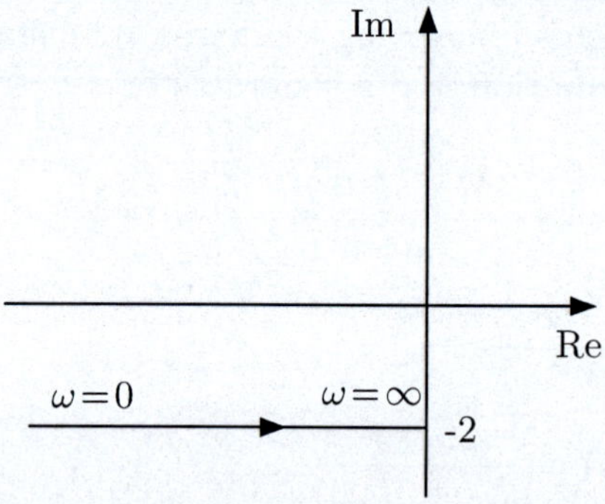

Fig. 8.16: Popov plot of $G(s)G_{\mathrm{PI}}(s)$

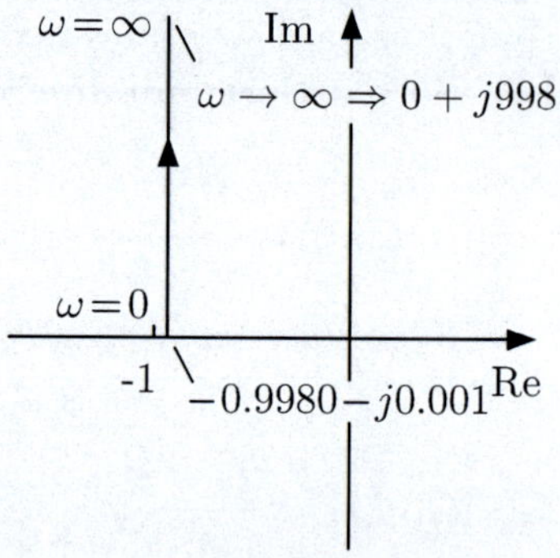

Fig. 8.17: Popov plot of $G(s)R(s)$

(f) The Hurwitz sectors for the control loop with and without anti-windup are $[\epsilon, 1000/998 \approx 1.0020]$ and $[\epsilon, \infty)$, respectively, where $0 < \epsilon \ll 1$. Aizerman's conjecture is true in both cases.

4.3 (a) $x_1(t) = \dfrac{\alpha}{2u_{\max}}\left(x_2^2(t) - x_2^2(0)\right), \quad \alpha = \pm 1, \quad x_1(0) = 0$

(b) $x_1(t) = \dfrac{x_2^2(t)}{2u_{\max}}$ for $x_2 < 0$ and $x_1(t) = -\dfrac{x_2^2(t)}{2u_{\max}}$ for $x_2 > 0$

(c)

$$u(\boldsymbol{x}) = \begin{cases} u_{\max}, & x_1 < S(x_2), \\ -u_{\max}, & x_1 > S(x_2), \\ -\operatorname{sgn}(x_2)u_{\max}, & x_1 = S(x_2), \end{cases} \qquad S(x_2) = -\frac{x_2|x_2|}{2u_{\max}}$$

4.4 $X_{\text{null}} = \left\{ x \in I\!R \mid |x| < \dfrac{u_{\max}|b|}{\lambda} \right\}$

4.5 (a) $t_1 = \dfrac{|x(0)|}{u_{\max}}, \quad u(x) = -u_{\max}\,\mathrm{sgn}(x)$

(b) $t_1 = -\dfrac{1}{\lambda}\ln\left(1 - \dfrac{\lambda|x(0)|}{u_{\max}b}\right), \quad u(x) = -u_{\max}\,\mathrm{sgn}(x);$

for $\lambda < 0$ all initial values $x(0) \in \mathbb{R}$ can be compensated to zero, for $\lambda > 0$ only initial values $x(0)$ restricted by $|x(0)| < u_{\max}b/\lambda$ can be compensated to zero.

4.6 $u = -u_{\max}\,\mathrm{sgn}(S(x_2) - x_1), \quad S(x_2) = -\dfrac{x_2}{\lambda} + \dfrac{x_2 + u_{\max}\,\mathrm{sgn}(x_2)}{\lambda}\ln\left(1 + \dfrac{|x_2|}{u_{\max}}\right)$

4.7 $u = -u_{\max}\,\mathrm{sgn}\left(\hat{a}|\hat{a}| + \dfrac{\hat{b}^3}{u_{\max}}\right),$

$$\hat{a} = x_1 + \frac{x_3^3}{3u_{\max}{}^2} + \frac{x_2x_3}{u_{\max}}\,\mathrm{sgn}\left(x_2 + \frac{x_3|x_3|}{2u_{\max}}\right), \quad \hat{b} = x_2 + \frac{x_3^2}{2u_{\max}}\,\mathrm{sgn}\left(x_2 + \frac{x_3|x_3|}{2u_{\max}}\right)$$

4.8 (a) $$\begin{bmatrix} \dot{i}_a \\ \dot{i}_v \end{bmatrix} = \begin{bmatrix} -1.2988 & 3.4199 \\ 46.7567 & 178.7697 \end{bmatrix} \begin{bmatrix} i_a \\ i_v \end{bmatrix} + \begin{bmatrix} 37.1085 \\ -1335.9062 \end{bmatrix} u_a,$$

$$z = \begin{bmatrix} -575 \cdot 10^{-6} & -16 \cdot 10^{-6} \end{bmatrix} \begin{bmatrix} i_a \\ i_v \end{bmatrix}$$

Eigenvalues: $\lambda_1 = 179.6534, \quad \lambda_2 = -2.1825$

(b) $$\boldsymbol{i} = \begin{bmatrix} -0.1389 & -28.4326 \\ -7.3468 & 7.3468 \end{bmatrix} \boldsymbol{x}$$

(c) $u = u_{\max}\,\mathrm{sgn}(S(x_2) - x_1), \quad S(x_2) = u_{\max}\,\mathrm{sgn}(x_2)\left[\left(1 + \dfrac{|x_2|}{u_{\max}}\right)^{\lambda_1/\lambda_2} - 1\right]$

(d) $G = \{\boldsymbol{x} \in \mathbb{R}^2 \mid |x_1| < u_{\max},\, x_2 \in \mathbb{R}\}$

4.9 (a)
$$u_{\mathrm{to}}(\boldsymbol{x}) = \begin{cases} u_{\max}, & x_1 < S(x_2), \\ -u_{\max}, & x_1 > S(x_2), \\ \mathrm{sgn}\,(x_1)u_{\max}, & x_1 = S(x_2), \end{cases} \quad S(x_2) = -\frac{x_2|x_2|}{2u_{\max}}$$

$$= u_{\max}\Big[\mathrm{sgn}(S(x_2) - x_1) + \mathrm{sgn}(x_1)\Big(1 - |\mathrm{sgn}(S(x_2) - x_1)|\Big)\Big]$$

(b) $u = -\begin{bmatrix} 1 & 2 \end{bmatrix}\boldsymbol{x}$

(c) Using Barbashin and Krasovskii's theorem on p. 116, we obtain

$$V(\mathbf{0}) = 0, \quad V(\boldsymbol{x}) = x_1^2 + x_2^2 > 0 \quad \text{for} \quad x \neq 0, \quad \text{and} \quad \dot{V}(\boldsymbol{x}) = -4x_2^2 \leq 0.$$

The set of states for which $\dot{V}(\boldsymbol{x}) = 0$ applies (except for $\boldsymbol{x}(t) = \mathbf{0}$) is $M = \{\boldsymbol{x} \in \mathbb{R}^2 \mid x_2 = 0,\, x_1 \in \mathbb{R}\backslash\{0\}\}$. No trajectory $\boldsymbol{x}(t)$ of the linear control loop lies within M, because a trajectory of this kind would have to take the form

$$\boldsymbol{x}_M(t) = \begin{bmatrix} x_1(t) \\ 0 \end{bmatrix}$$

but would not fulfill the differential equation

$$\dot{\boldsymbol{x}} = \begin{bmatrix} 0 & 1 \\ -1 & -2 \end{bmatrix} \boldsymbol{x}$$

of the linear control loop.

(d) The disc with radius 0.1, shown in Figure 8.18, indicates the region in which the linear controller $u = -\boldsymbol{k}^T\boldsymbol{x}$ is activated. Outside the disc, the time-optimal controller regulates the plant.

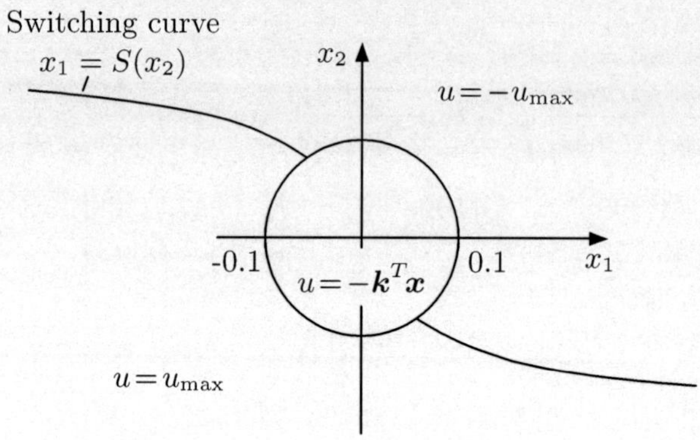

Fig. 8.18: The dual-mode controller illustrated in the phase plane

(e) All trajectories of the time-optimal control loop which start outside the disc tend to it. After entering the disc, the linear controller is active such that all these trajectories tend to the origin. The reason for this is that the disc is a catchment region. Thus all trajectories are asymptotically stable.

(f) In practice, all control loops have output and state noise. Thus, when the equilibrium point is reached and afterwards, the time-optimal control switches continually back and forth between $u_{\max}$ and $-u_{\max}$. The linear controller in the area around the equilibrium point makes it possible to avoid this switching.

4.10 (a) $x(k) = a^k x(0) + \sum\limits_{i=0}^{k-1} a^{k-i-1} bu(i)$

(c) $N = \operatorname{ceil}\left(\dfrac{\ln(bu_{\max}) - \ln(|x(0)|(1-a) + bu_{\max})}{\ln(a)}\right)$

(d)

$$u(i) = \begin{cases} -u_{\max}\operatorname{sgn}(x(0)), & 0 \le i \le N-2 \\ u_{\max}\operatorname{sgn}(x(0))\dfrac{a - a^N}{1-a} - \dfrac{a^N}{b}x(0), & i = N-1 \\ 0, & i \ge N \end{cases}$$

4.11 (b) The switching system is globally asymptotically stable if and only if the matrix $e^{\boldsymbol{A}_2 \Delta t} e^{\boldsymbol{A}_1 \Delta t}$ has only eigenvalues λ_i with $|\lambda_i| < 1$.

(c) For $\Delta t = \ln(1.4)$ the system is stable; for $\Delta t = \ln(3)$ it is unstable.

4.12 (a) $u_1 = 2, \quad u_2 = -1$

(b) No values u_1, u_2 exist for the case in which $t_1 = \pi$, $t_2 = 2\pi$.

(c) In order for there to be a control sequence $u_1, \ldots, u_n$ which transforms one arbitrary state to another, it is necessary and sufficient for the matrix

$$\boldsymbol{M} = \left[\int_0^{t_1} e^{-\boldsymbol{A}\tau} \boldsymbol{b} d\tau \quad \int_{t_1}^{t_2} e^{-\boldsymbol{A}\tau} \boldsymbol{b} d\tau \quad \cdots \quad \int_{t_{N-1}}^{t_N} e^{-\boldsymbol{A}\tau} \boldsymbol{b} d\tau \right]$$

to have rank n.

4.13 (f) Design the procedure of a piecewise linear control:

Step 1: Choose eigenvalues λ_{1i} and thus the coefficients $\hat{a}_i$ of the control loop's characteristic polynomial.

Step 2: Determine the matrix $\boldsymbol{R}_1$ from

$$\hat{\boldsymbol{A}}_1^T \boldsymbol{R}_1 + \boldsymbol{R}_1 \hat{\boldsymbol{A}}_1 = -\boldsymbol{Q}$$

for a positive definite matrix $\boldsymbol{Q}$.

Step 3: Check whether all catchment regions

$$G_i = \left\{ \boldsymbol{x} \in I\!R^n \mid e(q) \boldsymbol{x}^T \boldsymbol{D}^{-1}(q) \boldsymbol{R}_1 \boldsymbol{D}^{-1}(q) \boldsymbol{x} \le 1 \right\}$$

are nested within one another. This is the case if

$$\frac{\partial e(q)}{\partial q} \le 0$$

and

$$\boldsymbol{N} \boldsymbol{R}_1 + \boldsymbol{R}_1 \boldsymbol{N} = -\boldsymbol{S}$$

is negative definite.

Step 4: Select l controllers $k_i = k(q_i), \ i = 1, 2, \ldots, l$.

(g) Control law:

$$u = -\boldsymbol{k}^T(q) \boldsymbol{x} = -\left(\boldsymbol{D}^{-1}(q) \hat{\boldsymbol{a}} - \boldsymbol{a} \right) \boldsymbol{x}$$

Implicit equation for calculating q:

$$e(q) \boldsymbol{x}^T \boldsymbol{D}^{-1}(q) \boldsymbol{R}_1 \boldsymbol{D}^{-1}(q) \boldsymbol{x} = 1$$

Calculate $\boldsymbol{R}_1$ by satisfying

$$\hat{\boldsymbol{A}}_1^T \boldsymbol{R}_1 + \boldsymbol{R}_1 \hat{\boldsymbol{A}}_1 = -\boldsymbol{Q}, \quad \boldsymbol{N} \boldsymbol{R}_1 + \boldsymbol{R}_1 \boldsymbol{N} = -\boldsymbol{S}, \quad \text{and} \quad \frac{\partial e(q)}{\partial q} \le 0$$

simultaneously.

(h) For the parameter q, the following applies:

1. For $\boldsymbol{x} \to \boldsymbol{0}$, the volume of G tends to 0; thus q does as well. Therefore $q(\boldsymbol{x} = \boldsymbol{0}) = 0$ applies.
2. For all $\boldsymbol{x} \in I\!R^n \setminus \{0\}$, $q > 0$.

3. The selected parameter q decreases as the size of the catchment regions G_i becomes smaller. Because all G_i are nested within one another and all trajectories of larger catchment regions tend to smaller ones, q decreases along each trajectory. Thus $\dot{q} < 0$ applies.

From 1, 2, and 3 above, it follows that q is a Lyapunov function.

4.14 The catchment region G lies between the hyperplanes $|\boldsymbol{h}^T\boldsymbol{x}| = u_{\max}$. Consequently, the asymptotically stable regulation with $u = -\operatorname{sat}(\boldsymbol{h}^T\boldsymbol{x})$ is linear for $\boldsymbol{x} \in G$.

4.15 (a) $\lambda_{1,2} = \pm j$ (b) $\boldsymbol{h}^T = [0\ 2]$ (c) $\boldsymbol{R} = \begin{bmatrix} 3 & 1 \\ 1 & 1 \end{bmatrix}$

(d) $G = \{\boldsymbol{x} \in I\!R^2 \mid \boldsymbol{x}^T\boldsymbol{R}\boldsymbol{x} < 6\}$ (e) $\boldsymbol{k}^T = [3\ 4]$

(f) In accordance with Hu and Lin's theorem, the equilibrium point $\boldsymbol{x}_{\text{eq}} = \boldsymbol{0}$ is asymptotically stable for $\boldsymbol{x} \in G$, because the matrices $\boldsymbol{Q}_{\text{h}}$ and $\boldsymbol{Q}_{\text{k}}$ are both positive definite.

5.1 (a) $\dot{x} = -\dfrac{1}{2}\dfrac{\rho c_{\text{d}} A}{m}x^2 - g(c_{\text{r}}\cos(\varphi) + \sin(\varphi)) + u, \quad y = x$

(b) $(\Delta x)^{\cdot} = -\dfrac{\rho c_{\text{d}} A}{m}\beta\Delta x + \Delta u = -0.0005\beta\Delta x + \Delta u, \quad \beta = x, \quad \Delta y = \Delta x$

(c) $x_{\text{op1}} = 10\,\text{ms}^{-1}: \quad u_{\text{op1}} = 0.17215,\ (\Delta x)^{\cdot} = -0.005\Delta x + \Delta u$

$x_{\text{op2}} = 25\,\text{ms}^{-1}: \quad u_{\text{op2}} = 0.3034,\ (\Delta x)^{\cdot} = -0.0125\Delta x + \Delta u$

$x_{\text{op3}} = 40\,\text{ms}^{-1}: \quad u_{\text{op3}} = 0.54715,\ (\Delta x)^{\cdot} = -0.02\Delta x + \Delta u$

$P_1 = 0.095,\ I_1 = 0.0025, \quad P_2 = 0.2375,\ I_2 = 0.015625, \quad P_3 = 0.38,\ I_3 = 0.04$

(d)

$$u = \frac{\sum\limits_{i=1}^{3}(u_{\text{op},i} + \Delta u_i)e^{-(\Delta x_i^2)/(2\sigma_i^2)}}{\sum\limits_{i=1}^{3} e^{-(\Delta x_i^2)/(2\sigma_i^2)}}, \qquad \Delta x_i = x - x_{\text{op},i}, \quad \sigma_1 = \sigma_2 = \sigma_3 = 7.5$$

$$\Delta u_i = P_i\Delta e_i + I_i\int_0^t \Delta e_i d\tau = P_i e + I_i\int_0^t e d\tau$$

$$\Delta e_i = \Delta x_{\text{ref}} - \Delta x_i = (x_{\text{ref}} - x_{\text{op},i}) - (x - x_{\text{op},i}) = x_{\text{ref}} - x = e$$

5.2 (a) $m = 0.1\,\text{kg}: \ s_{1,2} = \pm j\sqrt{300}$, and $m = 1\,\text{kg}: \ s_{1,2} = \pm j\sqrt{30}$

(b) $u = \sum\limits_{i=1}^{5} d_i(m(t))u_i, \ u_i = K_{\text{P}i}\left(y - y_{\text{ref}} + \dfrac{1}{T_{\text{I}i}}\displaystyle\int_0^t (y - y_{\text{ref}})d\tau + T_{\text{D}i}\dfrac{d(y - y_{\text{ref}})}{dt}\right)$

Table 8.2 shows the parameters of the subcontrollers and Figure 8.19 the weighting functions.

(c) $u = K_{\text{P}}(m)\left(y - y_{\text{ref}} + \dfrac{1}{T_{\text{I}}(m)}\displaystyle\int_0^t (y - y_{\text{ref}})d\tau + T_{\text{D}}(m)\dfrac{d(y - y_{\text{ref}})}{dt}\right)$

$$K_{\mathrm{P}}(m) = 300m - c, \quad T_{\mathrm{I}}(m) = \frac{300m - c}{1000m}, \quad T_{\mathrm{D}}(m) = \frac{30m}{300m - c}, \quad c = 30\,\mathrm{Nm}^{-1}$$

Table 8.2: Parameters of the subcontrollers

Controller	1	2	3	4	5
m_i	0.2 kg	0.4 kg	0.6 kg	0.8 kg	1 kg
$K_{\mathrm{P}i}$	30	90	150	210	270
$T_{\mathrm{I}i}$	0.15	0.225	0.25	0.2625	0.27
$T_{\mathrm{D}i}$	0.2	2/15	0.12	4/35	1/9

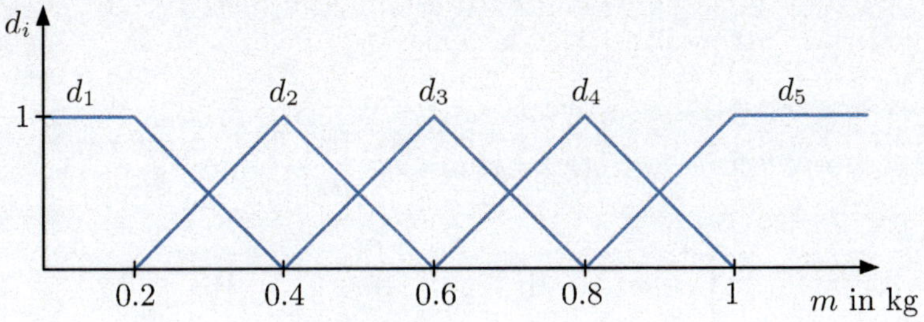

Fig. 8.19: Weighting functions

5.3 (a) $\delta = 1$

(b)
$$\boldsymbol{z} = \begin{bmatrix} x_3 \\ x_1 \\ x_2 \end{bmatrix}, \quad \dot{\boldsymbol{z}} = \begin{bmatrix} -z_3 - 1 \\ -\dfrac{1}{\varepsilon}(z_2^3 + z_2 z_3 + z_1) \\ -2z_2 - 2z_3 \end{bmatrix} + \begin{bmatrix} 1 \\ 0 \\ 0 \end{bmatrix} u, \quad \dot{\boldsymbol{z}}_{\text{zero}} = \begin{bmatrix} -\dfrac{1}{\varepsilon}(z_2^3 + z_2 z_3) \\ -2z_2 - 2z_3 \end{bmatrix}$$

(c) $u = x_2 + 1 - a_0 x_3 + V y_{\text{ref}}, \quad \dot{y} + a_0 y = V y_{\text{ref}}$

5.5 (a) $u(\boldsymbol{x}) = -\dfrac{\boldsymbol{c}^T\boldsymbol{A}^n\boldsymbol{x} + a_{n-1}\boldsymbol{c}^T\boldsymbol{A}^{n-1}\boldsymbol{x} + \ldots + a_1\boldsymbol{c}^T\boldsymbol{A}\boldsymbol{x} + a_0\boldsymbol{c}^T\boldsymbol{x}}{\boldsymbol{c}^T\boldsymbol{A}^{n-1}\boldsymbol{B}\boldsymbol{x}} + \dfrac{V y_{\text{ref}}}{\boldsymbol{c}^T\boldsymbol{A}^{n-1}\boldsymbol{B}\boldsymbol{x}}$

(b) If, during regulation, the trajectory $\boldsymbol{x}(t)$ approaches the equilibrium point $\boldsymbol{x}_{\text{eq}} = \boldsymbol{0}$, the term $\boldsymbol{c}^T\boldsymbol{A}^{n-1}\boldsymbol{B}\boldsymbol{x}$ becomes increasingly small and tends towards zero. As a consequence, the control signal u becomes greater and greater and tends to infinity. This is, of course, not feasible.

(c)
$$\boldsymbol{z} = \begin{bmatrix} y \\ \dot{y} \\ \vdots \\ y^{(n-2)} \\ y^{(n-1)} \end{bmatrix} = \begin{bmatrix} \boldsymbol{c}^T \\ \boldsymbol{c}^T\boldsymbol{A} \\ \vdots \\ \boldsymbol{c}^T\boldsymbol{A}^{n-2} \\ \boldsymbol{c}^T\boldsymbol{A}^{n-1} \end{bmatrix} \boldsymbol{x} = \boldsymbol{T}\boldsymbol{x}$$

(d)

$$\dot{\boldsymbol{z}} = \begin{bmatrix} z_2 \\ z_3 \\ \vdots \\ z_n \\ \boldsymbol{c}^T\boldsymbol{A}^n\boldsymbol{T}^{-1}\boldsymbol{z} \end{bmatrix} + \begin{bmatrix} 0 \\ 0 \\ \vdots \\ 0 \\ \boldsymbol{c}^T\boldsymbol{A}^{n-1}\boldsymbol{B}\boldsymbol{T}^{-1}\boldsymbol{z} \end{bmatrix} u, \quad y = \boldsymbol{c}^T\boldsymbol{T}^{-1}\boldsymbol{z}$$

5.6 (a) $\delta = 2$

(b) $u(\boldsymbol{x}) = -\dfrac{L_{\boldsymbol{a}}^2 c(\boldsymbol{x}) + a_1 L_{\boldsymbol{a}} c(\boldsymbol{x}) + a_0 c(\boldsymbol{x})}{L_{\boldsymbol{b}} L_{\boldsymbol{a}} c(\boldsymbol{x})} + \dfrac{V y_{\mathrm{ref}}}{L_{\boldsymbol{b}} L_{\boldsymbol{a}} c(\boldsymbol{x})}, \quad y = c(\boldsymbol{x}) = x_2$

$L_{\boldsymbol{a}} c(\boldsymbol{x}) = k_1 x_1 + k_2 x_1^2 (l - x_2) - v_{\mathrm{e}}$

$L_{\boldsymbol{a}}^2 c(\boldsymbol{x}) = -k_0 \left[k_1 + 2k_2 x_1 (l - x_2)\right] x_1 - k_2 \left[k_1 x_1 + k_2 x_1^2 (l - x_2) - v_{\mathrm{e}}\right] x_1^2$

$L_{\boldsymbol{b}} L_{\boldsymbol{a}} c(\boldsymbol{x}) = k_0 \left[k_1 + 2k_2 x_1 (l - x_2)\right]$

(c) The control can be implemented if $L_{\boldsymbol{b}} L_{\boldsymbol{a}} c(\boldsymbol{x}) = k_0[k_1 + 2k_2(l - x_2)x_1] \neq 0$ is fulfilled. The latter is the case, because $k_0 > 0$, $k_1 > 0$, $k_2 > 0$, $l > x_2 > 0$, and $x_1 > 0$.

5.8 (a) $\dot{\tilde{\boldsymbol{x}}} = \boldsymbol{a}(\tilde{\boldsymbol{x}}) + \boldsymbol{b}(\tilde{\boldsymbol{x}})v, \; y = c(\tilde{\boldsymbol{x}}) = g(\boldsymbol{x}), \; \tilde{\boldsymbol{x}} = \begin{bmatrix} \boldsymbol{x} \\ u \end{bmatrix}, \; \boldsymbol{a}(\tilde{\boldsymbol{x}}) = \begin{bmatrix} \boldsymbol{f}(\boldsymbol{x}, u) \\ 0 \end{bmatrix}, \; b(\tilde{\boldsymbol{x}}) = \begin{bmatrix} 0 \\ 1 \end{bmatrix}$

(b) $\delta \geq 2$

(c)

$$\dot{\boldsymbol{z}} = \begin{bmatrix} \dot{y} \\ \ddot{y} \\ \vdots \\ y^{(n-1)} \\ y^{(n)} \end{bmatrix} = \begin{bmatrix} 0 & 1 & 0 & \dots & 0 \\ 0 & 0 & 0 & \dots & 0 \\ \vdots & \vdots & \vdots & \ddots & \vdots \\ 0 & 0 & 0 & \dots & 1 \\ -a_0 & -a_1 & -a_2 & \dots & -a_{n-1} \end{bmatrix} z + \begin{bmatrix} 0 \\ 0 \\ \vdots \\ 0 \\ 1 \end{bmatrix} y_{\mathrm{ref}}, \quad y = c(\tilde{\boldsymbol{x}}),$$

$$v = -\frac{L_{\boldsymbol{a}}^n c(\tilde{\boldsymbol{x}}) + a_{n-1} L_{\boldsymbol{a}}^{n-1} c(\tilde{\boldsymbol{x}}) + \ldots + a_1 L_{\boldsymbol{a}} c(\tilde{\boldsymbol{x}}) + a_0 c(\tilde{\boldsymbol{x}})}{L_{\boldsymbol{b}} L_{\boldsymbol{a}}^{n-1} c(\tilde{\boldsymbol{x}})} + \frac{V y_{\mathrm{ref}}}{L_{\boldsymbol{b}} L_{\boldsymbol{a}}^{n-1} c(\tilde{\boldsymbol{x}})}$$

(d) $u = \displaystyle\int_0^t v \, d\tau$

5.10 (a) $\delta = n + 1 - i \quad$ for $y = x_i\,, \quad i = 1, \ldots, n$

(b) Internal dynamics: $\dot{z}_n = f_1(z_n) + h(z_n) \cdot z_1$ with $z_1 = x_2$ and $z_n = x_1$

5.11 (b) A linearizing output is $y = x_3$.

(c)

$$z = \begin{bmatrix} y \\ \dot{y} \\ \ddot{y} \end{bmatrix} = \begin{bmatrix} x_3 \\ -x_1 - c x_3 \\ x_1(a + c - x_2) + (c^2 - 1)x_3 \end{bmatrix}$$

$$u = \frac{\left[x_1(x_2 - a) + x_3\right](a + c - x_2) + x_1(x_1^2 + b x_2 - c^2) - (c^3 - c)x_3}{x_1} + \frac{v}{x_1}$$

(d) $\dddot{y} = v \quad \Rightarrow \quad y = \displaystyle\int_0^t \int_0^\tau \int_0^{\tilde{\tau}} v \, d\hat{\tau} \, d\tilde{\tau} \, d\tau$

(e)

$$v = \dddot{y} = \begin{cases} 0, & t < 0, \\ -8\sin(t)\cos(t), & 0 \le t \le \dfrac{\pi}{2}, \\ 0, & \dfrac{\pi}{2} < t \end{cases}$$

5.12 (a) $\boldsymbol{\delta} = \begin{bmatrix} 2 & 2 & 2 \end{bmatrix}$

(b) Since $\det(\boldsymbol{D}(\boldsymbol{x})) = -2a/(Jm^2) \neq 0$ holds, the vectorial relative degree $\boldsymbol{\delta}$ is well-defined.

(c)

$$\boldsymbol{u}(\boldsymbol{x}) = -\boldsymbol{D}^{-1}(\boldsymbol{x}) \begin{bmatrix} a_{1,1}x_2 + a_{1,0}x_1 - V_1 y_{\mathrm{ref},1} \\ a_{2,1}x_4 + a_{2,0}x_3 - V_2 y_{\mathrm{ref},2} \\ a_{3,1}x_6 + a_{3,0}x_5 - V_3 y_{\mathrm{ref},3} \end{bmatrix}$$

$$\boldsymbol{D}^{-1}(\boldsymbol{x}) = \frac{m}{2a} \begin{bmatrix} a\cos(x_5) - b\sin(x_5) & b\cos(x_5) + a\sin(x_5) & J/m \\ b\sin(x_5) + a\cos(x_5) & a\sin(x_5) - b\cos(x_5) & -J/m \\ 2a\sin(x_5) & -2a\cos(x_5) & 0 \end{bmatrix}$$

5.15 No, because $\operatorname{rank}(\begin{bmatrix} \boldsymbol{b} & \mathrm{ad}_{\boldsymbol{a}}\boldsymbol{b} & \mathrm{ad}_{\boldsymbol{a}}^2\,\boldsymbol{b} \end{bmatrix}) \neq 3$. See Theorem 64 on p. 418.

5.16 (a) $x_1 = y, \quad x_2 = \dot{y}^3, \quad x_3 = -3\ddot{y}, \quad u = 3(y\ddot{y} - \dddot{y})$

(b) $P(s) = (s+5)^3 = s^3 + 15s^2 + 75s + 125 \quad \Rightarrow \quad a_2 = 15,\ a_1 = 75,\ a_0 = 125$

$$\boldsymbol{\Psi}_1^{-1}(\boldsymbol{x}) = \begin{bmatrix} y \\ \dot{y} \\ \ddot{y} \end{bmatrix} = \begin{bmatrix} x_1 \\ \sqrt[3]{x_2} \\ -\dfrac{1}{3}x_3 \end{bmatrix}, \qquad \boldsymbol{\Psi}_2(y,\dot{y},\dddot{y}) = 3(y\ddot{y} - \dddot{y}) = 3(y\ddot{y} - v)$$

Figure 8.20 shows the block diagram of the control.

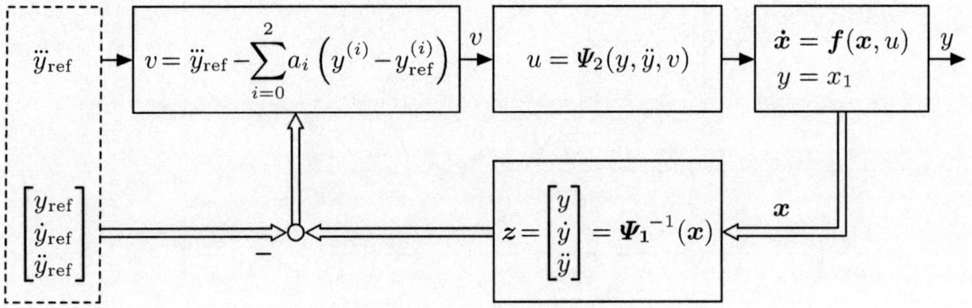

Fig. 8.20: Flatness-based feedback control

5.17 (a)

$$\dot{\boldsymbol{x}} = \begin{bmatrix} (x_2 - x_1)x_1 \\ 0.04093x_1x_2 \end{bmatrix} + \begin{bmatrix} 0 \\ 1 \end{bmatrix} u, \quad y = x_1$$

(b) $x_1 = y, \quad x_2 = \dfrac{\dot{y} + y^2}{y}, \quad u = \dfrac{\ddot{y}y - \dot{y}^2}{y^2} + \dot{y} - 0.04093(\dot{y} + y^2)$

(c) The flat input can be used to influence the growth of human knowledge. Good education and research funding, for example, can do this.
(d) $u = -0.04093y^2$
(e) Using the flat input, the population x_1 can only be kept constant if human knowledge and human technological possibilities x_2 are held constant, i. e. if they stagnate. This can be seen directly from $\dot{x}_1 = (x_2 - x_1)x_1 = 0$.
(f) The flat output is $y_f = x_2$.
(g) $x_1 = \dfrac{1}{0.04093} \cdot \dfrac{\dot{y}_f}{y_f}, \quad x_2 = y_f, \quad u = 0.04903\dfrac{\ddot{y}_f y_f - \dot{y}_f^2}{(0.04903y_f^2 - \dot{y}_f)\dot{y}_f}$
(h) If the population size $x_1(t) = x_{1,\infty}$ is kept constant by $u = 0$, then

$$y_f(t) = x_2(t) = x_2(0)e^{0.04093x_{1,\infty}t}$$

represents the increase in human knowledge. The population can thus be kept constant while the amount of knowledge continues to grow exponentially.

5.18 No, because it is not full-state linearizable (Theorem 69, on p. 430).

5.20 (a)

$$\begin{bmatrix} z_1 \\ z_2 \\ z_3 \end{bmatrix} = \begin{bmatrix} y \\ \dot{y} \\ \ddot{y} \end{bmatrix} = \begin{bmatrix} x_1 \\ -\lambda x_1 \ln\left(\dfrac{x_1}{x_2}\right) \\ \lambda x_1 \left[\lambda\left(\ln\left(\dfrac{x_1}{x_2}\right)+1\right)\ln\left(\dfrac{x_1}{x_2}\right)+b\dfrac{x_1}{x_2} - dx_1^{2/3} - \mu x_3\right] \end{bmatrix} = \boldsymbol{t}(\boldsymbol{x})$$

(b)

$$\boldsymbol{x} = \boldsymbol{\Psi}_1(y, \dot{y}, \ddot{y}) = \begin{bmatrix} y \\ ye^{\dot{y}/(\lambda y)} \\ \dfrac{1}{\mu}\left(be^{-\dot{y}/(\lambda y)} - dy^{2/3} - \dfrac{\dot{y}}{y} - \dfrac{\ddot{y}}{\lambda y} + \dfrac{\dot{y}^2}{\lambda y^2}\right) \end{bmatrix} = \boldsymbol{t}^{-1}(\boldsymbol{z})$$

(c)
$$u = \frac{b}{\lambda\mu}e^{-\dot{y}/(\lambda y)}\left(\lambda a + \frac{\dot{y}^2}{y^2} - \frac{\ddot{y}}{y}\right) - \frac{d}{\mu}y^{2/3}\left(a + \frac{2}{3}\frac{\dot{y}}{y}\right)$$
$$+ \frac{1}{\mu}\frac{\dot{y}}{y}\left(\frac{\dot{y}}{y} - \frac{2}{\lambda}\frac{\dot{y}^2}{y^2} + \frac{3}{\lambda}\frac{\ddot{y}}{y} + \frac{a}{\lambda}\frac{\dot{y}}{y} - a\right) - \frac{1}{\mu y}\left(\frac{a\ddot{y}}{\lambda} + \ddot{y} + \frac{\dddot{y}}{\lambda}\right)$$

(d) $u(t) = \left(\dfrac{2d\alpha}{3\mu} - \dfrac{da}{\mu}\right)e^{-2\alpha t/3} + \dfrac{a}{\mu}\left(be^{\alpha/\lambda} + \alpha\right)$

(e) Figure 8.21 shows the medication $u(t)$.

5.21 Both controls are based on a full-state linearization which converts the nonlinear system into the Brunovsky canonical form. In this aspect they are identical. Furthermore, both use the output variable y and its derivatives for control. The difference between the two is the use of the reference variable

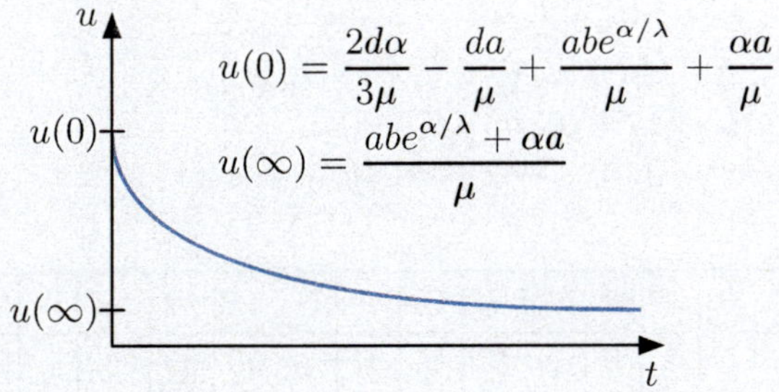

Fig. 8.21: Progression of the medication u

y_{ref}. In the case of a full-state linearization, the control law depends on a reference signal $y_{\text{ref}}(t)$, which may be constant. In the case of the flatness-based feedback control, the control law depends not only on a reference signal $y_{\text{ref}}(t)$ but also on its derivatives $\dot{y}_{\text{ref}}(t), \ldots, y_{\text{ref}}^{(n)}(t)$. In this context, see also Fig. 5.9 on p. 365 and Fig. 5.27 on p. 453.

5.23 (a) The matrix $\boldsymbol{A} + u_c\boldsymbol{N}$ must possess only eigenvalues with a negative real part.

(b) The matrix $\boldsymbol{A}^T\boldsymbol{R} + \boldsymbol{R}\boldsymbol{A}$ must be negative definite.

(c) $u = -u_{\max}\,\text{sgn}(\boldsymbol{x}^T(\boldsymbol{N}^T\boldsymbol{R} + \boldsymbol{R}\boldsymbol{N})\boldsymbol{x})$

(d) The controller from (c) switches perpetually with a high frequency between $-u_{\max}$ and $u_{\max}$. This also happens near the equilibrium point $\boldsymbol{x}_{\text{eq}} = \boldsymbol{0}$. Of course, this is undesirable. We can avoid the high-frequency switching using the modified control law

$$u = -\,\text{sat}(\mu\boldsymbol{x}^T(\boldsymbol{N}\boldsymbol{R} + \boldsymbol{R}\boldsymbol{N})\boldsymbol{x}).$$

The control parameter μ requires adjustment according to the application in question. It determines for which values of $\boldsymbol{x}$ the control variable u is smaller than $u_{\max}$.

5.24 (a)

$$u = -\frac{f(\boldsymbol{x}) + \sum\limits_{i=1}^{n} a_{i-1}x_i}{h(\boldsymbol{x})}, \qquad a_i,\ i = 1, ..., n-1, \quad \text{freely selectable}$$

(b)

$$\dot{\boldsymbol{x}} = \begin{bmatrix} 0 & 1 & 0 & \ldots & 0 & 0 \\ 0 & 0 & 1 & \ldots & 0 & 0 \\ \vdots & \vdots & \vdots & \ddots & \vdots & \vdots \\ 0 & 0 & 0 & \ldots & 0 & 1 \\ -a_0 & -a_1 & -a_2 & \ldots & -a_{n-2} & -a_{n-1} \end{bmatrix} \boldsymbol{x}$$

(c) The two control loops are identical, with the difference that no reference variable y_{ref} is present in the backstepping control. However, its control law can be easily completed by adding $Vy_{\text{ref}}/h(\boldsymbol{x})$.

6.1 (a) $F = 1,\ G = 1,\ H = 1$
(b) $J(\Delta u(k)) = 2\Delta u^2(k) + 2\Delta u(k)e(k) + e^2(k)$
(c) $u(k) = -0.5y(k-1) + 0.5y_{\text{ref}}(k+1)$
(d) $y(k+1) = y(k) - 0.5y(k-1) + 0.5y_{\text{ref}}(k+1)$
(e) $z_{1/2} = 0.5 \pm j0.5$
(f)

k	0	1	2	3	4	5	6	7	8	9	10	11	12	13	14	15	16	17	18
$y(k)$	0	$\frac{1}{2}$	1	$\frac{5}{4}$	$\frac{5}{4}$	$\frac{9}{8}$	1	$\frac{15}{16}$	$\frac{15}{16}$	$\frac{31}{32}$	1	$\frac{65}{64}$	$\frac{65}{64}$	$\frac{129}{128}$	1	$\frac{255}{256}$	$\frac{255}{256}$	$\frac{511}{512}$	1

6.2 (a) $\boldsymbol{F} = \begin{bmatrix}1\\1\end{bmatrix},\quad \boldsymbol{G} = \begin{bmatrix}1\\2\end{bmatrix},\quad \boldsymbol{H} = \begin{bmatrix}1&0\\2&1\end{bmatrix},$

$$J(\Delta\bar{\boldsymbol{u}}(k)) = \Delta\bar{\boldsymbol{u}}^T(k)\begin{bmatrix}6&2\\2&2\end{bmatrix}\Delta\bar{\boldsymbol{u}}(k) + 2\Delta\bar{\boldsymbol{u}}^T(k)\begin{bmatrix}1&2\\0&1\end{bmatrix}\boldsymbol{e}(k) + \boldsymbol{e}^T(k)\boldsymbol{e}(k),$$

$$u(k) = -\frac{1}{4}\big[y(k) + y(k-1) - y_{\text{ref}}(k+1) - y_{\text{ref}}(k+2)\big],$$

$$y(k+1) = \frac{1}{4}\big[3y(k) - y(k-1) + y_{\text{ref}}(k+1) + y_{\text{ref}}(k+2)\big],$$

$$z_{1/2} = \frac{3}{8} \pm j\frac{1}{8}\sqrt{7}$$

k	0	1	2	3	4	5	6	7	8
$y(k)$	0	0.5	0.875	1.0313	1.0547	1.0332	1.0112	1.0001	0.9973

k	9	10	11	12	13	14	15	16 ... 18
$y(k)$	0.9979	0.9991	0.9999	1.0001	1.0001	1.0001	1.0000	1.0000

(b) MPC2: $k_{3\%} = 6$
(c) MPC1: $k_{3\%} = 10$; the MPC2 is faster than the MPC1 by a factor of $1.\overline{6}$.

6.3 (a) $s(\boldsymbol{x}) = x_1 + x_2,\quad u(\boldsymbol{x}) = -\operatorname{sgn}(s(\boldsymbol{x}))$
(b) Figure 8.22 shows the trajectories of the control loop.
(c) The sliding mode occurs between the points $\boldsymbol{x}^T = \begin{bmatrix}-1 & 1\end{bmatrix}$ and $\boldsymbol{x}^T = \begin{bmatrix}1 & -1\end{bmatrix}$ on the switching line, and nowhere else.
(e) $x_1(t) = -x_2(t) = x_1(0)e^{-t}$
(f)

$$x_1(t) = \begin{cases} -5 + 5t - 0.5t^2, & 0 \le t < 8, \\ 59 - 11t + 0.5t^2, & 8 \le t < 12, \\ -e^{-(t-12)}, & 12 \le t, \end{cases} \quad x_2(t) = \begin{cases} 5 - t, & 0 \le t < 8, \\ -11 + t, & 8 \le t < 12, \\ e^{-(t-12)}, & 12 \le t; \end{cases}$$

see Figure 8.23.

(g) $t_1 = 5 + \sqrt{\dfrac{15}{2}} \approx 7.7386,\quad t_2 = 5 + 2\sqrt{\dfrac{15}{2}} \approx 10.4772$; see Figure 8.24.

(h) $k = \operatorname{ceil}\left(\dfrac{|x_1(0)| - 1}{2}\right)$

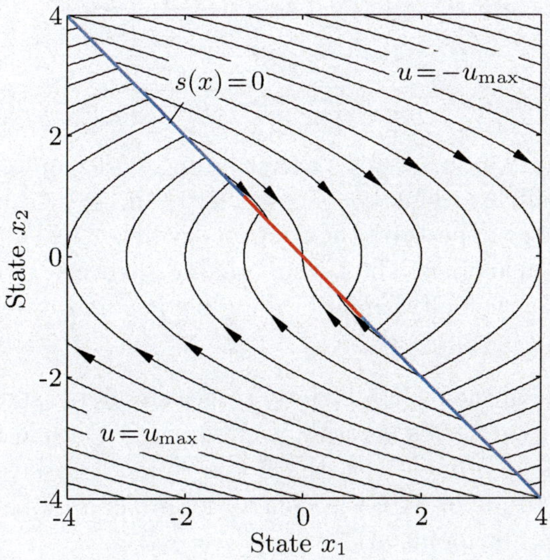

Fig. 8.22: Approximation of the time-optimal controller's switching curve using a switching line. On this switching line a sliding mode occurs, which is shown in red.

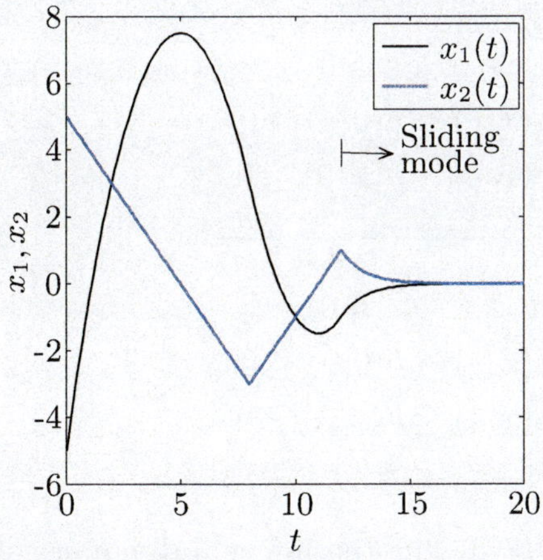

Fig. 8.23: Time courses of the sliding mode control loop's state variables

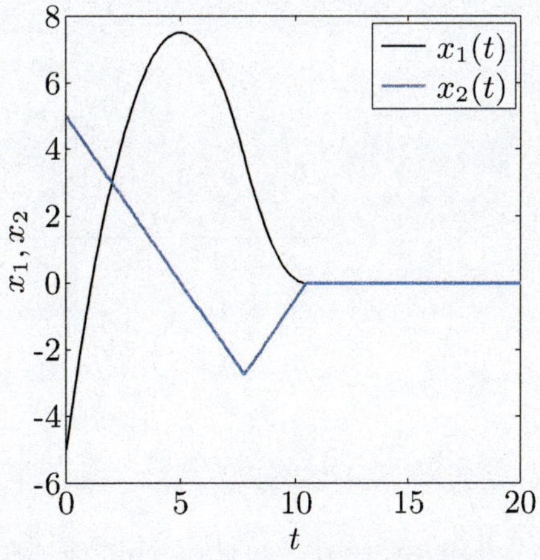

Fig. 8.24: Time courses of the time-optimal control loop's state variables

6.4 $u = \dfrac{-q\,\mathrm{sgn}(s(\boldsymbol{x})) - ks(\boldsymbol{x}) - \mathrm{grad}^T(s(\boldsymbol{x}))\cdot \boldsymbol{a}(\boldsymbol{x})}{\mathrm{grad}^T(s(\boldsymbol{x}))\cdot \boldsymbol{b}(\boldsymbol{x})}$

6.5 (a) $\dot{x}_1 = -g\mu(x_2), \quad \dot{x}_2 = -\dfrac{g\mu(x_2)}{x_1}\left(1 - x_2 + \dfrac{mr^2}{J}\right) + \dfrac{r}{Jx_1}u$

(b) $u(\boldsymbol{x}) = -\dfrac{Jx_1}{r}\Big(q\,\mathrm{sgn}(x_2 - x_{2,\mathrm{ref}}) + k(x_2 - x_{2,\mathrm{ref}})\Big) + \dfrac{Jg\mu(x_2)}{r}\left(1 - x_2 + \dfrac{mr^2}{J}\right)$

(c)

$$x_1(t) = \begin{cases} x_1(0) - g\mu(x_{2,\mathrm{ref}})t, & t < t_{\mathrm{eq}} = \dfrac{x_1(0)}{g\mu(x_{2,\mathrm{ref}})}, \\ 0, & t \geq t_{\mathrm{eq}} = \dfrac{x_1(0)}{g\mu(x_{2,\mathrm{ref}})}, \end{cases} \qquad x_2(t) = x_{2,\mathrm{ref}}$$

(d) The slip $x_2 = \lambda$ is constant in the sliding mode, more precisely $x_2(t) = x_{2,\mathrm{ref}}$. The vehicle's velocity $x_1(t) = x_1(0) - g\mu(x_{2,\mathrm{ref}})t$ in the sliding mode during braking depends on the friction coefficient. The latter is a function of the road surfacing. Thus, the braking distance $0.5x_1(0)/(g\mu(x_{2,\mathrm{ref}}))$ depends on the velocity when braking begins, and on the road surface.

(e) $x_{2,\mathrm{ref}} = \lambda_{\mathrm{ref}} \approx 0.16$

6.6 (b) The model of the oven is strictly positive real, i.e. strictly passive. Since it is linear, it possesses a radially unbounded storage function. The saturation curve combined with the P controller is passive. Thus, according to Theorem 80 on p. 549, the control loop has a globally asymptotically stable equilibrium point at $\Delta T_i = 0$, $i = 1, 2, 3$.

6.8 (a) The characteristic diagram passes through the origin. Furthermore, the following inequalities apply:

$x \leq 0, \quad e \geq 0$ $h(x,e) \leq h(0,e) \geq 0$ $h(x,e) \geq h(x,0) \leq 0$	$x \geq 0, \quad e \geq 0$ $h(x,e) \geq h(0,e) \geq 0$ $h(x,e) \geq h(x,0) \geq 0$
$x \leq 0, \quad e \leq 0$ $h(x,e) \leq h(0,e) \leq 0$ $h(x,e) \leq h(x,0) \leq 0$	$x \geq 0, \quad e \leq 0$ $h(x,e) \geq h(0,e) \leq 0$ $h(x,e) \leq h(x,0) \geq 0$

6.9 The system has the relative degree $\delta = 1$, but the internal dynamics are unstable. Thus it is impossible to generate a passive control loop using a static controller $u(\boldsymbol{x})$. See Theorem 85 on p. 564.

6.10 This exercise is out of the ordinary, because the task is not solvable. (It is the only unsolvable exercise in this book.) It is meant to illustrate that you may be faced with mathematical problems and tasks in engineering which are not solvable. You may recognize this immediately with some problems, but not with others. Examples are the derivation of a general criterion for the controllability of nonlinear systems or the construction of a perpetual motion machine. In the example considered here, the plant (6.69), (6.70) is not strictly passive because there is no positive definite matrix $\boldsymbol{R}$ which satisfies the Kalman-Yakubovich-Popov equations

$$A^T R + RA = -\mu P,$$
$$RB = C^T$$

of Theorem 29 on p. 154 such that $\boldsymbol{P}$ is positive definite.

6.11 (a) Nonlinear controller canonical form $\dot{\boldsymbol{x}} = \boldsymbol{a}(\boldsymbol{x}) + \boldsymbol{b}(\boldsymbol{x})u$:

$$\dot{\boldsymbol{x}} = \begin{bmatrix} x_2 \\ a_1 - a_2 x_2 - a_3 \sin(x_1 + \delta_{\text{eq}}) \end{bmatrix} + \begin{bmatrix} 0 \\ -a_3 \sin(x_1 + \delta_{\text{eq}}) \end{bmatrix} u$$

PCHD form:

$$\dot{\boldsymbol{x}} = (\boldsymbol{J} - \boldsymbol{D}) \frac{\partial V}{\partial \boldsymbol{x}} + \boldsymbol{b}(\boldsymbol{x})u, \qquad \boldsymbol{J} = \begin{bmatrix} 0 & 1 \\ -1 & 0 \end{bmatrix},$$

$$\boldsymbol{b}(\boldsymbol{x}) = \begin{bmatrix} 0 \\ -a_3 \sin(x_1 + \delta_{\text{eq}}) \end{bmatrix} \qquad \boldsymbol{D} = \begin{bmatrix} 0 & 0 \\ 0 & a_2 \end{bmatrix}.$$

(b) $\boldsymbol{J}_{\text{CL}} - \boldsymbol{D}_{\text{CL}} = \begin{bmatrix} 0 & 1 \\ * & * \end{bmatrix}$, $*$ = arbitrary, but such that $\boldsymbol{J}_{\text{CL}} - \boldsymbol{D}_{\text{CL}}$ is negative definite.

(c) $u = \dfrac{kx_1 + dx_2}{a_3 \sin(x_1 + \delta_{\text{eq}})}$

6.12 (a) $\boldsymbol{x}_{\text{eq1}} = \boldsymbol{0}$, $\boldsymbol{x}_{\text{eq2}}^T = \begin{bmatrix} \sqrt{63} & \sqrt{63} & 21 \end{bmatrix}$, $\boldsymbol{x}_{\text{eq3}}^T = \begin{bmatrix} -\sqrt{63} & -\sqrt{63} & 21 \end{bmatrix}$

(b)
$$\dot{\boldsymbol{x}} = (\boldsymbol{J} - \boldsymbol{D}) \frac{\partial V_1}{\partial \boldsymbol{x}} + \boldsymbol{b}u, \; \boldsymbol{J} = \begin{bmatrix} 0 & 35 & 0 \\ -35 & 0 & -x_1 \\ 0 & x_1 & 0 \end{bmatrix}, \; \boldsymbol{D} = \begin{bmatrix} 175 & 0 & 0 \\ 0 & -28 & 0 \\ 0 & 0 & 3 \end{bmatrix}, \; \boldsymbol{b} = \begin{bmatrix} 0 \\ 1 \\ 0 \end{bmatrix}$$

(c) The equilibrium point $\boldsymbol{x}_{\text{eq}} = \boldsymbol{0}$ is unstable, because the linearized system has an unstable eigenvalue (indirect Lyapunov method, Theorem 17, p. 126).

(d) The Byrnes-Isidori canonical form and diffeomorphism $\boldsymbol{t}(\boldsymbol{x})$ are

$$\dot{\boldsymbol{z}} = \begin{bmatrix} 28z_1 - 7z_2 - z_2 z_3 \\ 35z_1 - 35z_2 \\ z_1 z_2 - 3z_3 \end{bmatrix} + \begin{bmatrix} 1 \\ 0 \\ 0 \end{bmatrix} u, \qquad \boldsymbol{z} = \boldsymbol{t}(\boldsymbol{x}) = \begin{bmatrix} x_2 \\ x_1 \\ x_3 \end{bmatrix}.$$

(e) According to Theorem 85 on p. 564 we can construct a passive control loop using a static controller, because the relative degree is $\delta = 1$ and the internal dynamics

$$\dot{\boldsymbol{z}}_{\text{int}} = \begin{bmatrix} \dot{z}_2 \\ \dot{z}_3 \end{bmatrix} = \begin{bmatrix} -35 & 0 \\ 0 & -3 \end{bmatrix} \begin{bmatrix} z_2 \\ z_3 \end{bmatrix} + \begin{bmatrix} 35 \\ z_2 \end{bmatrix} z_1$$

are stable.

(f) The control law is $u(\boldsymbol{z}) = -28z_1 - kz_1 - 28z_2 + y_{\text{ref}}$ and the control loop in PCHD form $\dot{\boldsymbol{z}} = (\boldsymbol{J}_{\text{cl}} - \boldsymbol{D}_{\text{cl}}) \dfrac{\partial V_2(\boldsymbol{z})}{\partial \boldsymbol{x}} + \boldsymbol{b}_{\text{cl}} u$, where

$$\boldsymbol{J}_{\text{cl}} = \begin{bmatrix} 0 & -35 & -z_2 \\ 35 & 0 & 0 \\ z_2 & 0 & 0 \end{bmatrix}, \quad \boldsymbol{D}_{\text{cl}} = \begin{bmatrix} k & 0 & 0 \\ 0 & 35 & 0 \\ 0 & 0 & 3 \end{bmatrix}, \quad \boldsymbol{b}_{\text{cl}} = \begin{bmatrix} 1 \\ 0 \\ 0 \end{bmatrix}.$$

(g) $\dot{\boldsymbol{x}} = (\boldsymbol{J}_{\text{org}} - \boldsymbol{D}_{\text{org}}) \dfrac{\partial V_2(\boldsymbol{x})}{\partial \boldsymbol{x}} + \boldsymbol{b} y_{\text{ref}},$

$$\boldsymbol{J}_{\text{org}} = \begin{bmatrix} 0 & 35 & 0 \\ -35 & 0 & -x_1 \\ 0 & x_1 & 0 \end{bmatrix}, \quad \boldsymbol{D}_{\text{org}} = \begin{bmatrix} 35 & 0 & 0 \\ 0 & k & 0 \\ 0 & 0 & 3 \end{bmatrix}, \quad \boldsymbol{b} = \begin{bmatrix} 0 \\ 1 \\ 0 \end{bmatrix}$$

(h) The equilibrium point $\boldsymbol{x}_{\text{eq1}} = \boldsymbol{0}$ is globally asymptotically stable, because the constant matrix $\boldsymbol{D}_{\text{org}}$ is positive definite and V_2 is a Lyapunov function with $V(\boldsymbol{x}) \to \infty$ for $|\boldsymbol{x}| \to \infty$.

6.14 The bounded sum satisfies $a \vee \bar{a} = 1$, but the maximum operator and the algebraic sum do not.

6.15 (a) The subsidiary system is passive.

6.16 The membership functions and the rule base are shown in Figure 8.25 and Table 8.3, respectively. Probably your solution is slightly different due to different possible solutions.

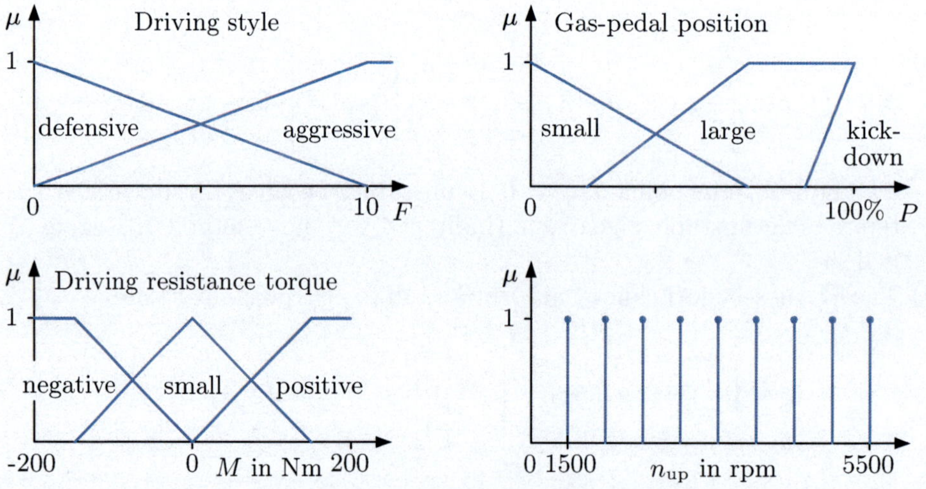

Fig. 8.25: Membership functions of the input variables and the output variable n_{up}

7.2 The weak observability can be proven for all $\omega \in \mathbb{R}^3$, except for the cases (1) $J_z = J_y$, (2) $\omega_x = 0$, and (3) $\omega_y = \omega_z = 0$.

7.3 (a) The system is control-affine and has the relative degree $\delta = 3$. It is thus weakly observable (see Theorem 90 on p. 618). Furthermore, the system is even observable since the inverse $\boldsymbol{q}^{-1}(\boldsymbol{z})$ of the function $\boldsymbol{z} = \boldsymbol{q}(\boldsymbol{x})$ exists with the exception of $x_1 = z_1 = -110$.

Table 8.3: Rule base of the gear selection

Driving style = defensive

Driving resistance M	Gas-pedal position P: small	large	kickdown
positive	2500	3000	5500
small	2000	2500	5000
negative	1500	2000	4500

Driving style = aggressive

Driving resistance M	Gas-pedal position P: small	large	kickdown
positive	4500	5500	5500
small	4000	5000	5000
negative	3500	4500	4500

(b) For $x_1 = -110$, $\dot{x}_1 = 0.028x_1$ applies. In this case, x_2 and x_3 do not act on $y = x_1$. Thus the system is not observable at this point, i.e. it is only locally observable and only locally weakly observable.

7.4 $h_1(x_1) \neq 0, \quad h_2(x_1, x_2) \neq 0, \quad \ldots, \quad h_{n-1}(x_1, \ldots, x_{n-1}) \neq 0$

7.6 (a) With the inverse function $\boldsymbol{q}^{-1}$ yielding

$$\begin{bmatrix} x_1 \\ x_2 \end{bmatrix} = \begin{bmatrix} \pm\sqrt{\dot{y}} \\ y \end{bmatrix} \quad \text{of} \quad z = \begin{bmatrix} y \\ \dot{y} \end{bmatrix} = \begin{bmatrix} x_2 \\ x_1^2 \end{bmatrix} = \boldsymbol{q}(\boldsymbol{x}),$$

not all $\boldsymbol{x} \in I\!R^2$ can be uniquely determined.

(b) We use $\ddot{y}$ in addition. Now, with the function

$$\tilde{\boldsymbol{q}}^{-1}(y, \ddot{y}) = \begin{bmatrix} 0.5\ddot{y} \\ y \end{bmatrix} = \begin{bmatrix} x_1 \\ x_2 \end{bmatrix} = \boldsymbol{x}$$

and if y and $\ddot{y}$ are known, all $\boldsymbol{x} \in I\!R^2$ can uniquely determined.

(c) The example shows that Theorem 86 on p. 608 is only sufficient. A nonlinear system can thus be observable although no formula

$$\boldsymbol{x} = \boldsymbol{q}^{-1}(y, \dot{y}, \ldots, y^{(n-1)})$$

exists with which all $\boldsymbol{x} \in D_{\boldsymbol{x}}$ are uniquely determinable.

7.7 (a) $x_1 = z_2, \quad x_2 = z_1$ (b) $\dot{z}_1 = -z_2^3, \quad \dot{z}_2 = z_1 - z_2 + u, \quad y = z_2$

(c)
$$\dot{\tilde{\boldsymbol{z}}} = \begin{bmatrix} 0 & 0 \\ 1 & 0 \end{bmatrix} \tilde{\boldsymbol{z}} - \begin{bmatrix} y^3 \\ y - u \end{bmatrix} + \begin{bmatrix} l_1 \\ l_2 \end{bmatrix} (y - \tilde{z}_2), \quad l_1 = 100, \quad l_2 = 20$$

(d)
$$\dot{\tilde{\boldsymbol{x}}} = \begin{bmatrix} 0 & 1 \\ 0 & 0 \end{bmatrix} \tilde{\boldsymbol{x}} - \begin{bmatrix} y - u \\ y^3 \end{bmatrix} + \begin{bmatrix} l_2 \\ l_1 \end{bmatrix} (y - \tilde{x}_1), \quad y = x_1$$

7.9 (a)
$$P(s) = s^3 + \left(l_1 + c - d\frac{x_2}{x_1^2}\right)s^2 + \left[l_2 + l_1\left(c - d\frac{x_2}{x_1^2}\right) + \frac{2ax_3^2}{x_1^3}\left(\frac{d}{x_1} - 1\right)\right]s$$
$$+ \frac{2ax_3}{x_1^2}\left[-l_3 - 2d\frac{x_2x_3}{x_1^3} + l_1 d\frac{x_3}{x_1^2} + \left(c - d\frac{x_2}{x_1^2}\right)\left(\frac{l_2x_1^2}{2ax_3} - \frac{x_3}{x_1}\right)\right]$$

(b) $P(s)$ is independent of $\boldsymbol{x}$ and u, i.e. $P(s) = s^3 + \hat{a}_2s^2 + \hat{a}_1s + \hat{a}_0$, if the following apply:

$$
\begin{aligned}
l_1 &= \hat{a}_2 - c + d\frac{x_2}{x_1^2}, \\
l_2 &= \hat{a}_1 + \frac{2ax_3^2}{x_1^3}\left(1 - \frac{d}{x_1}\right) - l_1\left(c - d\frac{x_2}{x_1^2}\right), \\
l_3 &= -\frac{\hat{a}_0 x_1^2}{2ax_3} + d\frac{x_3}{x_1^2}\left(l_1 - \frac{2x_2}{x_1}\right) + \left(\frac{l_2 x_1^2}{2ax_3} - \frac{x_3}{x_1}\right)\left(c - d\frac{x_2}{x_1^2}\right).
\end{aligned}
$$

7.10 (b)

$$
\dot{\boldsymbol{z}} = \begin{bmatrix} z_2 \\ z_3 \\ -\dfrac{2cz_1z_2 + mz_2z_3}{mz_1} \end{bmatrix} + \begin{bmatrix} 0 \\ 0 \\ \dfrac{cz_1 + mz_3}{mAz_1} \end{bmatrix} u, \quad y = z_1
$$

(c)

$$
\dot{\tilde{\boldsymbol{x}}} = \begin{bmatrix} \tilde{x}_2 \\ -\dfrac{c}{m}\tilde{x}_1 + \dfrac{A}{m}\tilde{x}_3 \\ -\dfrac{\tilde{x}_2\tilde{x}_3}{\tilde{x}_1} \end{bmatrix} + \begin{bmatrix} 0 \\ 0 \\ \dfrac{\tilde{x}_3}{A\tilde{x}_1} \end{bmatrix} u + \begin{bmatrix} l_1\varepsilon^{-1} \\ l_2\varepsilon^{-2} \\ \dfrac{cl_1\varepsilon^{-1} + ml_3\varepsilon^{-3}}{A} \end{bmatrix} (y - \tilde{y})
$$

A

Appendix

A.1 Proof of the General Circle Criterion

Below we will prove the general circle criterion, Theorem 10, p. 109.

According to Theorem 80, p. 549, the equilibrium point $\boldsymbol{x}_{\mathrm{eq}} = 0$ of the nonlinear standard control loop

$$\begin{aligned} Y(s) &= G(s)U(s), \\ e &= -y, \\ u &= f(e,t) \end{aligned} \tag{A.1}$$

with the open-loop transfer function $G(s)$ and the nonlinear characteristic $f(e,t)$ is globally asymptotically stable if $f(e,t)$ lies in the sector $[0,\infty]$, i.e. the nonlinearity is passive, and $G(s)$ is strictly positive real. In the following, we will restrict ourselves to the sector $[K_1, K_2]$ with $0 < K_1 < K_2$, since all other cases are identical to those of the circle criterion without poles on the imaginary axis. For the nonlinearity,

$$K_1 \leq \frac{f(e,t)}{e} \leq K_2$$

holds. We will transform the sector $[K_1, K_2]$ into sector $[0,\infty]$. For this purpose, we will convert the control loop (A.1) in the following three steps.

Step 1: We add a negative feedback with gain K_1 to the plant $G(s)$ for

$$Y(s) = G(s)(U(s) - K_1 Y(s)) \;\Leftrightarrow\; G_1(s) = \frac{Y(s)}{U(s)} = \frac{G(s)}{1 + K_1 G(s)}$$

to apply. So that the control loop is not changed in its dynamics, we compensate this feedback by a gain K_1 connected in parallel to the nonlinearity

$$u = f(e,t) + K_1 y = f(e,t) - K_1 e, \quad e = -y. \tag{A.2}$$

J. Adamy, *Nonlinear Systems and Controls*,
https://doi.org/10.1007/978-3-662-68690-4

Step 2: We extend the transfer function of Step 1 by the gain $K_2 - K_1$ and obtain

$$G_2(s) = \frac{Y(s)}{U_2(s)} = \frac{(K_2 - K_1)G(s)}{1 + K_1 G(s)}.$$

Again, the control loop's dynamics must not change. Therefore, we multiply the nonlinearity (A.2) by $1/(K_2 - K_1)$ which results in

$$u_2 = \frac{f(e,t) - K_1 e}{K_2 - K_1} \quad \text{and} \quad U_2(s) = \frac{U(s)}{K_2 - K_1}.$$

Step 3: Now we add a parallel branch with gain 1 between the input of the transfer function $G_2(s)$ and its output. So we obtain

$$G_3(s) = G_2(s) + 1 = \frac{1 + K_2 G(s)}{1 + K_1 G(s)}. \tag{A.3}$$

This parallel branch is compensated by a positive feedback from the output of the nonlinearity of Step 2 (which is the input of $G_3(s)$) to its input (which was the negated output of $G_3(s)$ in Step 2). This results in the modified nonlinearity

$$\begin{aligned}
u_2 &= \frac{f(z,t) - K_1 z}{K_2 - K_1}, \quad z = u_2 + e, \; e = -y \\
\Leftrightarrow (K_2 - K_1)u_2 &= \frac{f(z,t) - K_1 z}{z}(u_2 + e) \\
\Leftrightarrow K_2 u_2 - \frac{f(z,t)}{z} u_2 &= \frac{f(z,t)}{z} e - K_1 e \\
\Leftrightarrow u_2 &= \frac{\dfrac{f(z,t)}{z} - K_1}{K_2 - \dfrac{f(z,t)}{z}} e = \tilde{f}(e,t), \quad z = u_2 + e.
\end{aligned} \tag{A.4}$$

The modified nonlinearity (A.4) transforms the sector $[K_1, K_2]$ into the sector $[0, \infty]$. We can easily see this if we insert $f(z,t) = K_1 z$ and $f(z,t) = K_2 z$ into equation (A.4).

Figure A.1 illustrates these mutually canceling additions to the control loop (A.1). Consequently, the resulting control loop

$$\begin{aligned}
Y(s) &= G(s)U(s), \\
e &= -y, \\
u &= \tilde{f}(e,t)
\end{aligned}$$

has the same dynamic behavior as the control loop (A.1).

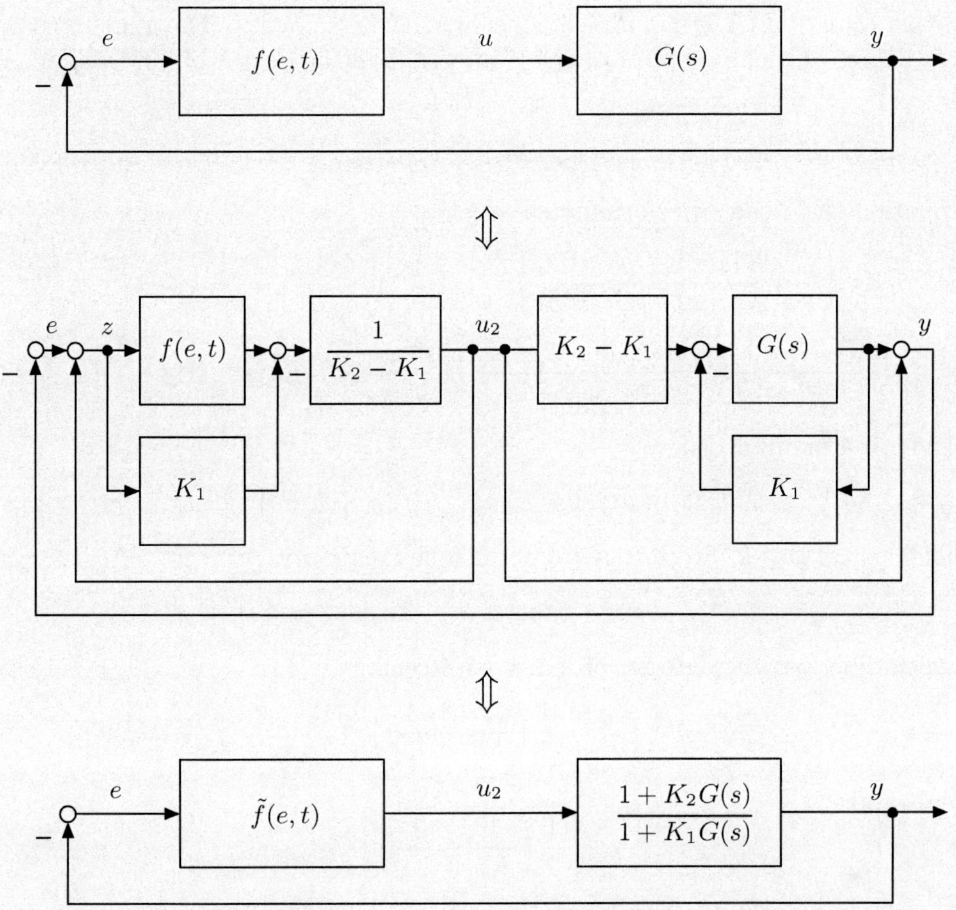

Fig. A.1: Transformation of sector $[K_1, K_2]$ into sector $[0, \infty]$

Thus, we can conclude that the nonlinear standard control loop is absolutely stable in the sector $[K_1, K_2]$, $0 < K_1 < K_2$, if the transfer function (A.3) is strictly positive real. According to Theorem 32 on p. 160, the latter is fulfilled if

$$\operatorname{Re}\left\{\frac{1 + K_2 G(j\omega)}{1 + K_1 G(j\omega)}\right\} > 0 \tag{A.5}$$

holds for all $\omega \geq 0$, and if $G_3(s)$ is asymptotically stable, which is the case if

$$\frac{G(s)}{1 + K_1 G(s)} \tag{A.6}$$

is asymptotically stable, and if

$$G_3(\infty) > 0 \tag{A.7}$$

holds. Equation (A.7) is fulfilled, since we have assumed in Theorem 10 that the degree of the numerator of $G(s)$ is lower than the degree of its denominator.

Using

$$G(j\omega) = a(\omega) + jb(\omega),$$

equation (A.5) can be reformulated as

$$\begin{aligned} &\operatorname{Re}\left\{\frac{1+K_2a(\omega)+jK_2b(\omega)}{1+K_1a(\omega)+jK_1b(\omega)}\right\} \\ &= \frac{(1+K_1a(\omega))(1+K_2a(\omega))+K_1K_2b^2(\omega)}{(1+K_1a(\omega))^2+K_1^2b^2(\omega)} > 0 \quad \text{for all} \quad \omega \geq 0. \end{aligned} \tag{A.8}$$

The latter is fulfilled if

$$(1+K_1a(\omega))(1+K_2a(\omega))+K_1K_2b^2(\omega) > 0$$

holds for all $\omega \geq 0$. From this, we obtain the inequality

$$(a(\omega)+c)^2 + b^2(\omega) > r^2 \quad \text{for all} \quad \omega \geq 0,$$

which describes the exterior of a disc with center

$$c = \frac{1}{2}\left(\frac{1}{K_1} + \frac{1}{K_2}\right)$$

and radius

$$r = \frac{1}{2}\left(\frac{1}{K_1} - \frac{1}{K_2}\right).$$

The circle lies symmetrically on the real axis, passing through the points $-1/K_1$ and $-1/K_2$. Inequality (A.8) and thus inequality (A.5) are fulfilled if and only if the Nyquist plot of $G(j\omega)$ neither intersects nor touches the circle.

So far, we have followed the proofs of the circle criterion published in the literature [168, 232, 308, 402, 422, 465]. Now our task is to ensure that equation (A.6) is satisfied. In the literature, equation (A.6) is verified using the well-known Nyquist criterion [98, 133, 465]. In contrast, we will use a different form of the Nyquist criterion [130, 483], based on the continuous angle variation Ω, to handle systems which may also have poles on the imaginary axis. For the sake of convenience, we will repeat this Nyquist criterion within our proof:

Theorem 92 (Nyquist criterion). *The feedback control loop*

$$\begin{aligned} Y(s) &= G(s)U(s) \\ U(s) &= -KY(s), \quad K \in \mathbb{R}, \end{aligned}$$

with the coprime open-loop transfer function $KG(s)$ whose numerator degree is less than its denominator degree is asymptotically stable if and only if the continuous angle variation with respect to the point $-1/K + j0$ is

$$\Omega = \nu\pi + \mu\frac{\pi}{2}.$$

Condition (1) of the general circle criterion results from this Nyquist criterion and the conditions above. Conditions (2), (3), and (4) are identical to the circle criterion for systems without poles on the imaginary axis. □

A.2 Parameters of the Container Crane Control

Subcontrollers $u = -\boldsymbol{k}_i^T \boldsymbol{x}$ of the crane control described in Section 4.3.3:

$$
\begin{aligned}
\boldsymbol{k}_1^T &= \begin{bmatrix} 1.9245 & 16.8960 & -67.7493 & -120.7999 & 0.3651 \end{bmatrix},\\
\boldsymbol{k}_2^T &= \begin{bmatrix} 5.3581 & 39.3339 & 287.3235 & -181.1081 & 0.8983 \end{bmatrix},\\
\boldsymbol{k}_3^T &= \begin{bmatrix} 7.4166 & 51.0577 & 325.7613 & -306.2776 & 1.1650 \end{bmatrix},\\
\boldsymbol{k}_4^T &= \begin{bmatrix} 11.7940 & 72.2878 & 138.0207 & -535.5579 & 1.6982 \end{bmatrix},\\
\boldsymbol{k}_5^T &= \begin{bmatrix} 17.3572 & 94.5922 & -118.3572 & -678.5956 & 2.2315 \end{bmatrix},\\
\boldsymbol{k}_6^T &= \begin{bmatrix} 23.9695 & 117.0157 & -417.4605 & -708.2384 & 2.7648 \end{bmatrix},\\
\boldsymbol{k}_7^T &= \begin{bmatrix} 31.4013 & 138.7914 & -730.7397 & -602.6913 & 3.2980 \end{bmatrix},\\
\boldsymbol{k}_8^T &= \begin{bmatrix} 49.8302 & 188.8935 & -1005.7853 & -405.2965 & 4.2313 \end{bmatrix},\\
\boldsymbol{k}_9^T &= \begin{bmatrix} 72.0268 & 241.9377 & -1118.9194 & 54.1054 & 5.1645 \end{bmatrix},\\
\boldsymbol{k}_{10}^T &= \begin{bmatrix} 97.0202 & 297.4944 & -954.7146 & 787.7226 & 6.0977 \end{bmatrix}.
\end{aligned}
$$

Matrices $\boldsymbol{R}_i$ of the catchment regions $G_i = \{\boldsymbol{x} \in I\!R^5 |\ \boldsymbol{x}^T \boldsymbol{R}_i \boldsymbol{x} \leq 1\}$:

$$
\boldsymbol{R}_1 = \begin{bmatrix}
4.1107\cdot10^{-4} & 3.3656\cdot10^{-3} & -4.0166\cdot10^{-3} & -2.4248\cdot10^{-2} & 6.6442\cdot10^{-5}\\
3.3656\cdot10^{-3} & 3.1180\cdot10^{-2} & -2.8011\cdot10^{-2} & -2.3607\cdot10^{-1} & 6.2148\cdot10^{-4}\\
-4.0166\cdot10^{-3} & -2.8011\cdot10^{-2} & 4.1307 & 1.7240\cdot10^{-1} & -2.7836\cdot10^{-3}\\
-2.4248\cdot10^{-2} & -2.3607\cdot10^{-1} & 1.7240\cdot10^{-1} & 5.4049 & -2.9691\cdot10^{-3}\\
6.6442\cdot10^{-5} & 6.2148\cdot10^{-4} & -2.7836\cdot10^{-3} & -2.9691\cdot10^{-3} & 1.4478\cdot10^{-5}
\end{bmatrix},
$$

$$
\boldsymbol{R}_2 = \begin{bmatrix}
3.7979\cdot10^{-3} & 2.6079\cdot10^{-2} & 1.8755\cdot10^{-1} & -2.2657\cdot10^{-1} & 5.6711\cdot10^{-4}\\
2.6079\cdot10^{-2} & 1.8712\cdot10^{-1} & 1.2913 & -1.6330 & 4.0165\cdot10^{-3}\\
1.8755\cdot10^{-1} & 1.2913 & 2.8190\cdot10 & -1.1024\cdot10 & 1.9722\cdot10^{-2}\\
-2.2657\cdot10^{-1} & -1.6330 & -1.1024\cdot10 & 3.0710\cdot10 & -2.7863\cdot10^{-2}\\
5.6711\cdot10^{-4} & 4.0165\cdot10^{-3} & 1.9722\cdot10^{-2} & -2.7863\cdot10^{-2} & 9.5314\cdot10^{-5}
\end{bmatrix},
$$

$$
\boldsymbol{R}_3 = \begin{bmatrix}
7.0254\cdot10^{-3} & 4.5741\cdot10^{-2} & 3.1248\cdot10^{-1} & -4.1397\cdot10^{-1} & 1.0098\cdot10^{-3}\\
4.5741\cdot10^{-2} & 3.0763\cdot10^{-1} & 2.0342 & -2.7809 & 6.7269\cdot10^{-3}\\
3.1248\cdot10^{-1} & 2.0342 & 3.3884\cdot10 & -1.8433\cdot10 & 3.6004\cdot10^{-2}\\
-4.1397\cdot10^{-1} & -2.7809 & -1.8433\cdot10 & 4.2920\cdot10 & -5.3480\cdot10^{-2}\\
1.0098\cdot10^{-3} & 6.7269\cdot10^{-3} & 3.6004\cdot10^{-2} & -5.3480\cdot10^{-2} & 1.5701\cdot10^{-4}
\end{bmatrix},
$$

$$\boldsymbol{R}_4=\begin{bmatrix} 1.8355\cdot10^{-2} & 1.0811\cdot10^{-1} & 4.7125\cdot10^{-1} & -9.4012\cdot10^{-1} & 2.4504\cdot10^{-3} \\ 1.0811\cdot10^{-1} & 6.5416\cdot10^{-1} & 2.8488 & -5.5883 & 1.4831\cdot10^{-2} \\ 4.7125\cdot10^{-1} & 2.8488 & 3.8658\cdot10 & -2.5478\cdot10 & 5.1777\cdot10^{-2} \\ -9.4012\cdot10^{-1} & -5.5883 & -2.5478\cdot10 & 7.2404\cdot10 & -1.1723\cdot10^{-1} \\ 2.4504\cdot10^{-3} & 1.4831\cdot10^{-2} & 5.1777\cdot10^{-2} & -1.1723\cdot10^{-1} & 3.5057\cdot10^{-4} \end{bmatrix},$$

$$\boldsymbol{R}_5=\begin{bmatrix} 5.2217\cdot10^{-2} & 2.7366\cdot10^{-1} & 4.0373\cdot10^{-1} & -2.3095 & 6.2049\cdot10^{-3} \\ 2.7366\cdot10^{-1} & 1.4687 & 2.4757 & -1.2133\cdot10 & 3.3474\cdot10^{-2} \\ 4.0373\cdot10^{-1} & 2.4757 & 4.2570\cdot10 & -2.1409\cdot10 & 3.7964\cdot10^{-2} \\ -2.3095 & -1.2133\cdot10 & -2.1409\cdot10 & 1.3541\cdot10^{2} & -2.6235\cdot10^{-1} \\ 6.2049\cdot10^{-3} & 3.3474\cdot10^{-2} & 3.7964\cdot10^{-2} & -2.6235\cdot10^{-1} & 7.8876\cdot10^{-4} \end{bmatrix},$$

$$\boldsymbol{R}_6=\begin{bmatrix} 1.1949\cdot10^{-1} & 5.6923\cdot10^{-1} & -5.1097\cdot10^{-1} & -4.2345 & 1.3063\cdot10^{-2} \\ 5.6923\cdot10^{-1} & 2.7826 & -1.6116 & -2.0252\cdot10 & 6.4387\cdot10^{-2} \\ -5.1097\cdot10^{-1} & -1.6116 & 6.7149\cdot10 & 9.3161 & -6.8998\cdot10^{-2} \\ -4.2345 & -2.0252\cdot10 & 9.3161 & 2.0442\cdot10^{2} & -4.4193\cdot10^{-1} \\ 1.3063\cdot10^{-2} & 6.4387\cdot10^{-2} & -6.8998\cdot10^{-2} & -4.4193\cdot10^{-1} & 1.5430\cdot10^{-3} \end{bmatrix},$$

$$\boldsymbol{R}_7=\begin{bmatrix} 2.3786\cdot10^{-1} & 1.0378 & -2.8921 & -5.9433 & 2.4280\cdot10^{-2} \\ 1.0378 & 4.6715 & -1.1067\cdot10 & -2.6154\cdot10 & 1.1058\cdot10^{-1} \\ -2.8921 & -1.1067\cdot10 & 1.5204\cdot10^{2} & 5.1281\cdot10 & -3.2462\cdot10^{-1} \\ -5.9433 & -2.6154\cdot10 & 5.1281\cdot10 & 2.5478\cdot10^{2} & -5.6212\cdot10^{-1} \\ 2.4280\cdot10^{-2} & 1.1058\cdot10^{-1} & -3.2462\cdot10^{-1} & -5.6212\cdot10^{-1} & 2.7258\cdot10^{-3} \end{bmatrix},$$

$$\boldsymbol{R}_8=\begin{bmatrix} 5.8841\cdot10^{-1} & 2.2140 & -8.4857 & -7.0109 & 4.9746\cdot10^{-2} \\ 2.2140 & 8.6981 & -2.9546\cdot10 & -2.7779\cdot10 & 1.9994\cdot10^{-1} \\ -8.4857 & -2.9546\cdot10 & 3.2425\cdot10^{2} & 6.6341\cdot10 & -7.8731\cdot10^{-1} \\ -7.0109 & -2.7779\cdot10 & 6.6341\cdot10 & 3.0685\cdot10^{2} & -5.3622\cdot10^{-1} \\ 4.9746\cdot10^{-2} & 1.9994\cdot10^{-1} & -7.8731\cdot10^{-1} & -5.3622\cdot10^{-1} & 4.8279\cdot10^{-3} \end{bmatrix},$$

$$\boldsymbol{R}_9=\begin{bmatrix} 1.0563 & 3.5328 & -1.3691\cdot10 & -2.9545 & 7.6710\cdot10^{-2} \\ 3.5328 & 1.2558\cdot10 & -4.2423\cdot10 & -1.2953\cdot10 & 2.8163\cdot10^{-1} \\ -1.3691\cdot10 & -4.2423\cdot10 & 5.5357\cdot10^{2} & -1.1448\cdot10 & -1.1622 \\ -2.9545 & -1.2953\cdot10 & -1.1448\cdot10 & 4.6012\cdot10^{2} & -9.4034\cdot10^{-2} \\ 7.6710\cdot10^{-2} & 2.8163\cdot10^{-1} & -1.1622 & -9.4034\cdot10^{-2} & 6.7514\cdot10^{-3} \end{bmatrix},$$

$$\boldsymbol{R}_{10}=\begin{bmatrix} 1.6062 & 4.8038 & -1.3282\cdot10 & 6.2810 & 9.7536\cdot10^{-2} \\ 4.8038 & 1.5756\cdot10 & -3.3740\cdot10 & 1.2281\cdot10 & 3.3568\cdot10^{-1} \\ -1.3282\cdot10 & -3.3740\cdot10 & 9.2668\cdot10^{2} & -1.0862\cdot10^{2} & -1.1734 \\ 6.2810 & 1.2281\cdot10 & -1.0862\cdot10^{2} & 1.0134\cdot10^{3} & 6.5813\cdot10^{-1} \\ 9.7536\cdot10^{-2} & 3.3568\cdot10^{-1} & -1.1734 & 6.5813\cdot10^{-1} & 7.9645\cdot10^{-3} \end{bmatrix}.$$

B

List of Symbols

$\mathbb{R}$	: Set of real numbers
$\mathbb{R}^+$	: Set of positive real numbers
$\mathbb{R}^-$	: Set of negative real numbers
$\mathbb{R}^n$	: Real coordinate space of dimension n
$D_{\boldsymbol{x},\mathrm{def}}$	: A function's or a free system's domain of definition
$D_{\boldsymbol{x}}$	: Subset of $D_{\boldsymbol{x},\mathrm{def}}$, i. e. $D_{\boldsymbol{x}} \subseteq D_{\boldsymbol{x},\mathrm{def}}$
$D_{\boldsymbol{u},\mathrm{def}}$	: Subspace of a non-free system's domain of definition $D_{\boldsymbol{x},\mathrm{def}} \times D_{\boldsymbol{u},\mathrm{def}}$
$D_{\boldsymbol{u}}$	: Subset of $D_{\boldsymbol{u},\mathrm{def}}$, i. e. $D_{\boldsymbol{u}} \subseteq D_{\boldsymbol{u},\mathrm{def}}$
$D_{t,\mathrm{def}}$	: Subspace of a system's domain of definition $D_{\boldsymbol{x},\mathrm{def}} \times D_{t,\mathrm{def}}$
$\mathbb{C}$	: Set of complex numbers
$\mathbb{C}^n$	: Complex coordinate space of dimension n
$\mathbb{N}$	: Set of natural numbers
$\mathbb{N}_0$	: Set of natural numbers and zero
$\mathbb{Z}$	: Set of integers
C^k	: Set of k-times continuously differentiable functions
C^k_m	: Set of k-times continuously differentiable, m-dimensional vector functions
$T_1 \Leftrightarrow T_2$	: T_1 is equivalent to T_2.
$a \approx b$	: a is approximately equal to b.
$a \ll b$	: a is much smaller than b.
$a \gg b$	: a is much greater than b.
$x \to a$	: x tends to a.
j	: Imaginary unit $j = \sqrt{-1}$
$\underline{z}$	: Complex numbers are marked by an underscore in certain cases

J. Adamy, *Nonlinear Systems and Controls*,
https://doi.org/10.1007/978-3-662-68690-4

$\overline{z}$	: Complex conjugate number of $z = a + jb$, i. e. $\overline{z} = a - jb$
$\mathrm{Re}\{a + jb\} = a$	: Real part of a complex number
$\mathrm{Im}\{a + jb\} = b$	: Imaginary part of a complex number
$[a, b]$	: Closed interval
$[a, b)$ and $(a, b]$	: Half-open interval
(a, b)	: Open interval
$\lvert x\rvert$	: Absolute value of a real or complex number x
$\boldsymbol{a}, \boldsymbol{b}, \ldots$	: Vectors are represented by bold lower case letters
$\boldsymbol{a}^T$	: Transposed vector of $\boldsymbol{a}$
$\boldsymbol{a}^*$	: Conjugate transpose vector $\boldsymbol{a}$
$\mathbf{1}$	: Vector of ones, $\mathbf{1} = [1\ 1\ \cdots\ 1]^T$
$\mathbf{0}$	: Vector of zeros, $\mathbf{0} = [0\ 0\ \cdots\ 0]^T$
$\dim(\boldsymbol{x})$	: Dimension of a vector $\boldsymbol{x}$, i. e. the number of the vector's elements x_i
$\lVert\boldsymbol{x}\rVert$	: Euclidean norm of a vector $\boldsymbol{x}$, $\lVert\boldsymbol{x}\rVert = (x_1^2 + \ldots + x_n^2)^{1/2}$
$\lvert\boldsymbol{x}\rvert$	: Absolute value of a vector $\boldsymbol{x}$, identical to $\lVert\boldsymbol{x}\rVert$
$\boldsymbol{a} \times \boldsymbol{b}$	: Cross product of vectors $\boldsymbol{a}$ and $\boldsymbol{b}$
$\boldsymbol{A}, \boldsymbol{B}, \ldots$	: Matrices and some vectors of physical quantities, e. g. the torque vector $\boldsymbol{M}$, are represented by bold capital letters.
$\boldsymbol{I}$	: Identity matrix
$\boldsymbol{A}^{-1}$	: Inverse matrix of $\boldsymbol{A}$
$\boldsymbol{A}^T$	: Transposed matrix of $\boldsymbol{A}$
$\overline{\boldsymbol{A}}$	: Conjugate matrix $\overline{\boldsymbol{A}}$ of a matrix $\boldsymbol{A}$, i. e. each element a_{ij} is replaced by its complex conjugate $\overline{a}_{ij}$
$\boldsymbol{A}^*$	: Conjugate transpose $\overline{\boldsymbol{A}}$ of a complex matrix $\boldsymbol{A}$, also called an adjoint matrix
$\mathrm{rank}(\boldsymbol{A})$	: Rank of a matrix $\boldsymbol{A}$; the rank equals the number of linearly independent columns of $\boldsymbol{A}$.
$\det(\boldsymbol{A})$	: Determinant of a matrix $\boldsymbol{A}$
$\mathrm{diag}(a_{11}, a_{22}, \ldots, a_{nn})$	: $n \times n$ diagonal matrix with the elements $a_{11}, a_{22}, \ldots, a_{nn}$
$\{a, b, c, \ldots\}$	: Set with the elements $a, b, c, \ldots$
$a \in M$	: a is an element of the set M.
$a \notin M$	: a is not an element of the set M.

$M_1 \times M_2$	: Cartesian product $M_1 \times M_2 \in I\!R^{n+m}$ of the sets $M_1 \in I\!R^n$ and $M_2 \in I\!R^m$
$M_1 = M_2$	: Set M_1 equals set M_2.
$M_1 \cup M_2$	: Union of set M_1 and set M_2
$M_1 \cap M_2$	: Intersection of set M_1 and set M_2
$M_1 \subset M_2$	: M_1 is a proper (also called strict) subset of the set M_2, i. e. $M_1 \neq M_2$ holds.
$M_1 \subseteq M_2$	: M_1 is a subset of the set M_2, i. e. $M_1 = M_2$ is possible.
$M_1 \setminus M_2$	: Set M_1 without set M_2, also called the set difference
$U(\boldsymbol{x})$	: Neighborhood of a point $\boldsymbol{x}$; if it is unambiguous which point $\boldsymbol{x}$ is meant, only the designation U is used.
$U_\delta(\boldsymbol{x})$	: Neighborhood with $U_\delta(\boldsymbol{x}) = \{\boldsymbol{x} \in I\!R^n \mid \lvert\boldsymbol{x}\rvert < \delta\}$
$\inf\{M\}$	: Infimum of a set M
$\sup\{M\}$	: Supremum of a set M
$\min\{M\}$	: Minimum of a set M
$\max\{M\}$	: Maximum of a set M
$\boldsymbol{f}^{-1}(\boldsymbol{z})$	: Inverse function of $\boldsymbol{z} = \boldsymbol{f}(\boldsymbol{x})$, i. e. $\boldsymbol{x} = \boldsymbol{f}^{-1}(\boldsymbol{z})$
$\min f(x)$	: Minimum of a function f
$\min\limits_i x_i$	: Minimum of the values x_i
$\arg\min\limits_i x_i$	: The index i that identifies the minimum x_i
$\mathrm{sat}(u) = \begin{cases} u_{\max}, & u > u_{\max} \\ u, & u_{\min} \leq u \leq u_{\max} \\ u_{\min}, & u < u_{\min} \end{cases}$	: Saturation function
$\mathrm{sgn}(x) = \begin{cases} 1, & x > 0 \\ 0, & x = 0 \\ -1, & x < 0 \end{cases}$	: Signum function
$H(x) = \begin{cases} 0, & x < 0 \\ 1, & x \geq 0 \end{cases}$	: Heaviside function[1], step function
$\mathrm{ceil}(x)$	: Next higher whole number of a real number x, e. g. $\mathrm{ceil}(3.34) = 4$
$\ln(x)$	: Natural logarithm, logarithm to the base e

[1] There are different ways of defining the Heaviside function at $x = 0$. Here we use $H(0) = 1$. Furthermore, $H(0) = 0$ and $H(0) = 0.5$ are in use.

$\lg(x)$	: Common logarithm, logarithm to the base 10
$\arcsin(x)$	: Inverse of the sine function
$\arccos(x)$	: Inverse of the cosine function
$\arctan(x)$	: Inverse of the tangent function
$\operatorname{arsinh}(x)$	: Area hyperbolic sine, inverse hyperbolic sine
$\operatorname{arcosh}(x)$	: Area hyperbolic cosine, inverse hyperbolic cosine
$X(s) = \mathcal{L}\{x(t)\}$	: Laplace transform
$x(t) = \mathcal{L}^{-1}\{X(s)\}$	: Inverse Laplace transform
$G(s)$	: Laplace transfer function
$G(z)$	: z-transfer function
$f'(x) = \dfrac{df(x)}{dx}$	: Derivative of the function f concerning the variable x
$\left.\dfrac{df(x)}{dx}\right\rvert_{x=a}$ or $\left.\dfrac{df(x)}{dx}\right\rvert_{a}$	: Derivative of f at the point $x = a$
$\dot{x} = \dfrac{dx}{dt}$	: Time derivative of x
$\ddot{x},\ \dddot{x}$	: Two- and threefold derivative of x with respect to time t
$x^{(i)}$	: i-th derivative of x with respect to time t
$\dfrac{df(\boldsymbol{x})}{dt} = \sum_{i=1}^{n} \dfrac{\partial f(\boldsymbol{x})}{\partial x_i} \cdot \dfrac{dx_i}{dt} = \dfrac{\partial f(\boldsymbol{x})}{\partial \boldsymbol{x}} \cdot \dot{\boldsymbol{x}}$	: Total derivative, i. e. derivation of all variables with respect to t
$\dfrac{\partial f(\boldsymbol{x})}{\partial x_i}$	: Partial derivative of a function f with several variables, i. e. derivative with respect to the variable x_i
$\dfrac{\partial f(\boldsymbol{x})}{\partial \boldsymbol{x}} = \left[\dfrac{\partial f(\boldsymbol{x})}{\partial x_1}\ \dfrac{\partial f(\boldsymbol{x})}{\partial x_2} \cdots \dfrac{\partial f(\boldsymbol{x})}{\partial x_n}\right]$	: Derivative of a function f with respect to the vector $\boldsymbol{x}$
$\operatorname{grad}(f(\boldsymbol{x})) = \begin{bmatrix} \dfrac{\partial f(\boldsymbol{x})}{\partial x_1} \\ \vdots \\ \dfrac{\partial f(\boldsymbol{x})}{\partial x_n} \end{bmatrix} = \left(\dfrac{\partial f(\boldsymbol{x})}{\partial \boldsymbol{x}}\right)^T$	: Gradient, i. e. directional derivative, of a function f (the gradient specifies a direction in space and is therefore a column vector)

$$\frac{\partial \boldsymbol{f}(\boldsymbol{x})}{\partial \boldsymbol{x}} = \begin{bmatrix} \frac{\partial f_1(\boldsymbol{x})}{\partial x_1} & \frac{\partial f_1(\boldsymbol{x})}{\partial x_2} & \cdots & \frac{\partial f_1(\boldsymbol{x})}{\partial x_n} \\ \frac{\partial f_2(\boldsymbol{x})}{\partial x_1} & \frac{\partial f_2(\boldsymbol{x})}{\partial x_2} & \cdots & \frac{\partial f_2(\boldsymbol{x})}{\partial x_n} \\ \vdots & \vdots & \ddots & \vdots \\ \frac{\partial f_n(\boldsymbol{x})}{\partial x_1} & \frac{\partial f_n(\boldsymbol{x})}{\partial x_2} & \cdots & \frac{\partial f_n(\boldsymbol{x})}{\partial x_n} \end{bmatrix}$$

: Derivative of a vector function with respect to a vector $\boldsymbol{x}$; the result is the Jacobian matrix.

$$L_{\boldsymbol{f}} h(\boldsymbol{x}) = \frac{\partial h(\boldsymbol{x})}{\partial \boldsymbol{x}} \boldsymbol{f}(\boldsymbol{x})$$

: Lie derivative of h with respect to $\boldsymbol{f}$

$$L^i_{\boldsymbol{f}} h(\boldsymbol{x}) = L_{\boldsymbol{f}} L^{i-1}_{\boldsymbol{f}} h(\boldsymbol{x})$$

: i-th Lie derivative of h with respect to $\boldsymbol{f}$

$$\mathring{\boldsymbol{c}}(\boldsymbol{x}) = \begin{bmatrix} L^{\delta_1}_{\boldsymbol{a}} c_1(\boldsymbol{x}) \\ L^{\delta_2}_{\boldsymbol{a}} c_2(\boldsymbol{x}) \\ \vdots \\ L^{\delta_m}_{\boldsymbol{a}} c_m(\boldsymbol{x}) \end{bmatrix}$$

: m-dimensional vector consisting of Lie derivatives

$$[\boldsymbol{f}, \boldsymbol{g}] = \frac{\partial \boldsymbol{g}(\boldsymbol{x})}{\partial \boldsymbol{x}} \boldsymbol{f}(\boldsymbol{x}) - \frac{\partial \boldsymbol{f}}{\partial \boldsymbol{x}} \boldsymbol{g}(\boldsymbol{x})$$

: Lie bracket of the vector functions $\boldsymbol{f}$ and $\boldsymbol{g}$

$$\mathrm{ad}^i_{\boldsymbol{f}} \boldsymbol{g} = [\boldsymbol{f}, \mathrm{ad}^{i-1}_{\boldsymbol{f}} \boldsymbol{g}]$$

: i-times nested Lie bracket

$$\mathrm{E}\{x(t)\} = \lim_{T \to \infty} \frac{1}{2\mathrm{T}} \int_{-\mathrm{T}}^{\mathrm{T}} x(t)\, dt$$

: Expected value, i. e. time average of a signal $x(t)$

$$\mathrm{cov}\{\boldsymbol{x}_1(t), \boldsymbol{x}_2(t)\}$$

: Covariance matrix of two signals $\boldsymbol{x}_1(t)$ and $\boldsymbol{x}_2(t)$

References

1. Abdallah, M. B., Ayadi, M., and Benrejeb, M. (2009): Flatness-Based Control of MIMO Linear Systems. Proceedings of the 6th IEEE Multi-Conference on Systems, Signals and Devices, Djerba, 1–6.
2. Abdulgalil, F., and Siguerdidjane, H. (2004): Input-state linearisation technique and nonlinear control in oil well drillstrings. 6th IFAC Symposium on Nonlinear Control Systems (NOLCOS), Stuttgart. In IFAC Proceedings Volumes 37(13):1425–1430.
3. Abughalieh, K. M., and Alawneh, S. G. (2019): A Survey of Parallel Implementations for Model Predictive Control. IEEE Access 7:34348–34360.
4. Adamy, J. (1997): Adaptation of Cross-Direction Basis-Weight Control in Paper Machines Using Fuzzy Decision Logic. International Journal of Approximate Reasoning 16(1):25–42.
5. Adamy, J. (1997): Method for actuator identification during the transverse profile control of a continuous material web. U. S. Patent 5,691,908.
6. Adamy, J., and Flemming, A. (2004): Soft variable-structure controls: a survey. Automatica 40(11):1821–1844.
7. Ahmed, T., and Toki, A. (2016): A Review on Washing Machine Using Fuzzy Logic Controller. International Journal of Emerging Trends in Engineering Research 4(7):64–67.
8. Aircraft Accident Investigation Commission, Japanese Ministry of Transport (1987): Japan Air Lines Co., Ltd, Boeing 747 SR-100, JA8119, Gunma Prefecture, Japan, August 12, 1985. Aircraft Accident Investigation Report.
9. Айзерман, М. А. (1949): Об одной проблеме, касающейся устойчивости „в большом“ динамических систем. Успехи Математических Наук 4(4):187–188. (Aizerman, M. A. (1949): On a problem concerning the stability "in the large" of dynamical systems. Uspehi Matematicheskih Nauk 4(4):187–188, in Russian).
10. Aizerman, M. A., and Gantmacher, F. R. (1964): Absolute Stability of Regulator Systems, Holden-Day.
11. Alberto, L. F. C., and Bretas, N. G. (1998): Damping Estimation for Multi-Swing Transient Stability Analysis: the OMIB case. Proceedings of the IEEE International Conference on Power System Technology (POWERCON'98) (2):1383–1387, Beijing.

J. Adamy, *Nonlinear Systems and Controls*,
https://doi.org/10.1007/978-3-662-68690-4

12. Alessio, A., and Bemporad, A. (2009): A Survey on Explicit Model Predictive Control. In Magni, L., Raimondo, D. M., and Allgöwer, F. (Eds.): Nonlinear Model Predicitve Control, 345–369, Springer.
13. Allgöwer, F., and Zheng, A. (Eds.) (2000): Nonlinear Model Predictive Control, Birkhäuser.
14. Alwi, H., and Edwards, C. (2009): Propulsion Control of a Large Civil Aircraft using On-line Control Allocation. IEEE Proceedings of the American Control Conference, St. Louis, 4581–4586.
15. Ametani, A., Nagaoka, N., Baba, Y., Ohno, T., and Koichi, Y. (2016): Power System Transients: Theory and Applications, Taylor & Francis.
16. Anderson, B. D. O. (1968): A Simplified Viewpoint of Hyperstability. IEEE Transactions on Automatic Control 13(3):292–294.
17. Anderson, B. D. O., and Vongpanitlerd, S. (2006): Network Analysis and Synthesis: A Modern Systems Theory Approach, Dover Publications.
18. Anderson, R. L., and Ibragimov, N. H. (1979) Lie-Bäcklund Transformations in Applications, SIAM, Philadelphia.
19. Andrievskii, B. R., and Fradkov, A. L. (2003): Control of Chaos: Methods and Applications I. Methods. Automation and Remote Control 64(5):673–713.
20. Andrievskii, B. R., and Fradkov, A. L. (2003): Control of Chaos: Methods and Applications II. Applications. Automation and Remote Control 65(4):505–533.
21. Apkarian, P., and Adams, R. J. (1998): Advanced Gain-Scheduling Techniques for Uncertain Systems. IEEE Transactions on Control Systems Technology 6(1):21–32.
22. Apollo Program Office - MAO (1969): Apollo 11 [AS-506] Mission. NASA, Mission Operation Report No. M-932-69-11.
23. Atkinson, K., Han, W., and Stewart, D. (2009): Numerical Solutions of Ordinary Differential Equations, Wiley & Sons.
24. Athans, M., and Falb, P. L. (1966): Optimal Control, McGraw-Hill.
25. Auernig, J. W., and Troger, H. (1987): Time optimal control of overhead cranes with hoisting of the load. Automatica 23(4):437–447.
26. Bacciotti, A., and Rosier, L. (2010): Liapunov Functions and Stability in Control Theory, Springer.
27. Badami, V. V., and Dausch, M. E. (1994): Fuzzy logic control method for reducing water consumption in a machine for washing articles. U. S. Patent 5,284,523.
28. Barabanov, N. E. (1988): On the Kalman problem. Siberian Mathematical Journal 29(3):333–341.
29. Барбашин, Е. А. и Красовский, Н. Н. (1952): Об устойчивости движения в целом. Доклад Академии Наук СССР 86(3):453–456 (Barbashin, E. A., and Krasovskii, N. N. (1952): On the stability of motion in the large. Doklady of Akademii Nauk SSSR 86(3):453–456, in Russian).
30. Barnett, S., and Storey, C. (1970): Matrix Methods in Stability Theory, Barnes & Noble.
31. Barraquand, J., and Latombe, J.-C. (1993): Nonholonomic Multibody Mobile Robots: Controllability and Motion Planning in the Presence of Obstacles. Algorithmica 10(2-3-4):121–155.
32. Barrau, A., and Bonnabel, S. (2017): The Invariant Extended Kalman Filter as a Stable Observer. IEEE Transactions on Automatic Control 62(4):1797–1812.
33. Bashein, G. (1971): A Simplex Algorithm for On-Line Computation of Time Optimal Controls. IEEE Transactions on Automatic Control 16(5):479–482.

34. Beckmann, M. J., and Ryder, H. E. (1969): Simultaneous Price and Quantity Adjustment in a Single Market. Econometrica 37(3):470–484.
35. Ben-Asher, J., Burns, J. A., and Cliff, E. M. (1992): Time-optimal slewing of flexible spacecraft. Journal of Guidance, Control and Dynamics 15(2):360–367.
36. Benz, K., Rech, C., and Mercorelli, P. (2019): Sustainable Management of Marine Fish Stocks by Means of Sliding Mode Control. IEEE Proceedings of the Federated Conference on Computer Science and Information Systems, Leipzig, 907–910.
37. Bergen, A. R., and Williams, I. J. (1962): Verification of Aizerman's Conjecture for a Class of Third-Order Systems. IRE Transactions on Automatic Control 7(3):42–46.
38. Bergman, R. N., Phillips, L. S., and Cobelli, C. (1981): Physiologic Evaluation of Factors Controlling Glucose Tolerance in Man: Measurement of Insulin Sensitivity and Beta-cell Glucose Sensitivity from the Response to Intravenous Glucose. The Journal of Clinical Investigation 68(6):1456–1467.
39. Bernard, P., Andrieu, V., and Astolfi, D. (2022): Observer design for continuous-time dynamical systems. Annual Reviews in Control 53:224–248.
40. Bernstein, D. S. (2018): Scalar, Vector, and Matrix Mathematics: Theory, Facts, and Formulas, Princeton University Press.
41. Bertsekas, D. P. (1999): Nonlinear Programming, Athena Scientific.
42. Besançon, G. (Ed.) (2007): Nonlinear Observers and Applications, Springer.
43. Bestle, D., and Zeitz, M. (1983): Canonical form observer design for non-linear time-variable systems. International Journal of Control 38(2):419–431.
44. Bhatia, N. P., and Szegö, G. P. (2002): Stability Theory of Dynamical Systems, Springer.
45. Bhatta, P. (2006): Nonlinear Stability and Control of Gliding Vehicles, PhD thesis, Princeton University.
46. Bhatta, P., and Leonard, N. E. (2008): Nonlinear gliding stability and control for vehicles with hydrodynamic forcing. Automatica 44(5):1240–1250.
47. Birk, J. (1992): Rechnergestützte Analyse and Lösung nichtlinearer Beobachtungsaufgaben. VDI Fortschritt-Berichte, Reihe 8, Nr. 294, VDI-Verlag, (Computer aided analysis and solution of nonlinear observation tasks; in German).
48. Birk, J., and Zeitz, M. (1992): Computer-aided analysis of nonlinear observation problems. 2nd IFAC Symposium on Nonlinear Control Systems (NOLCOS), Bordeaux. In IFAC Proceedings Volumes 25(3):257–262.
49. Blanchini, F. (1992): Minimum-Time Control for Uncertain Discrete-Time Linear Systems. Proceedings of the IEEE Conference on Decision and Control, Tucson (3):2629–2634.
50. Blümel, R., and Reinhardt, W. P. (1997): Chaos in Atomic Physics, Cambridge University Press.
51. Boccia, A., Grüne, L., and Wortmann, K. (2014): Stability and feasibility of state constrained MPC without stabilizing terminal constraints. System & Control Letters 72:14–21
52. Bode, H. W. (1945): Network Analysis and Feedback Amplifier Design, D. Van Nostrand.
53. Bonnabel, S., and Slotine, J.-J. E. (2014): A Contraction Theory-Based Analysis of the Stability of the Deterministic Extended Kalman Filter. IEEE Transactions on Automatic Control 60(2):565–569.

54. Borovic, B., Hong, C., Liu, A. Q., Xie, L., and Lewis, F. L. (2004): Control of a MEMS Optical Switch. Proceedings of the 43rd IEEE Conference on Decision and Control, Atlantis (3):3039–3044.
55. Boyd, S., El Ghaoui, L., Feron, E., and Balakrishnan, V. (1994): Linear Matrix Inequalities in System and Control Theory, SIAM.
56. Bradley, T.-D., Bryson, T., and Terilla, J. (2020): Topology: A Categorical Approach. MIT Press.
57. Bragin, V. O., Vagaitsev, V. I., Kuznetsov, N. V., and Leonov, G. A. (2011): Algorithms for Finding Hidden Oscillations in Nonlinear Systems. The Aizerman and Kalman Conjectures and Chua's Circuits. Journal of Computer and Systems Sciences International 50(4):511–543.
58. Brogliato, B., Lozano. R., Maschke, B., and Egeland, O. (2007): Dissipative Systems Analysis and Control, Springer
59. Brown, C. J., and Ma, J. T. (1968): Time-Optimal Control of a Moving-Coil Linear Actuator. IBM Journal of Research and Development 12(5):372–379.
60. Brunovský, P. (1970): A Classification of Linear Controllable Systems. Kybernetika 6(3):173–188.
61. Buhl, M., and Lohmann, B. (2009): Control with Exponentially Decaying Lyapunov Functions and Its Use for Systems with Input Saturation. IEEE Proceedings of the 10th European Control Conference, Budapest, 3148–3153.
62. Burcham, F. W., Trindel, A., Fullerton, C. G., and Webb, L. D. (1996): Development and Flight Evaluation of an Emergency Digital Flight Control System Using Only Engine Thrust on an F-15 Airplane. NASA Technical Paper 3627.
63. Burden, R. L., Faires, J. D., and Burden, A. M. (2015): Numerical Analysis, Cengage Learning.
64. Burken, J. J., Burcham, F. W., Maine, T. A., Feather, J., Goldthorpe, S., and Kahler, J. A. (1996): Flight Test of a Propulsion-Based Emergency Control System on the MD-11 Airplane With Emphasis on the Lateral Axis. NASA Technical Memorandum 4746.
65. Byrnes, C. I., Isidori, A., and Willems, J. C. (1991): Passivity, Feedback Equivalence, and the Global Stabilization of Minimum Phase Nonlinear Systems. IEEE Transactions on Automatic Control 36(11):1228–1240.
66. Byrnes, C. I., and Isidori, A. (1991): Asymptotic Stabilization of Minimum Phase Nonlinear Systems. IEEE Transactions on Automatic Control 36(10):1122–1137.
67. Camacho, E. F., Berenguel, M., and Rubio, F. R. (1994): Application of a gain scheduling generalized predictive controller to a solar power plant. Control Engineering Practice 2(2):227–238.
68. Camacho, E. F., and Bordons, C. (2007): Model Predictive Control, Springer.
69. Canudas-de-Wit, C., Rubio, F. R., and Corchero, M. A. (2008): D-OSKIL: A New Mechanism for Controlling Stick-Slip Oscillations in Oil Well Drillstrings. IEEE Transactions on Control Systems Technology 16(6):1177–1191.
70. Carter, J., Chiclana, F., Khuman, A. S., and Chen, T. (Eds.) (2021): Fuzzy Logic: Recent Applications and Developments, Springer.
71. Casavola, A., Prodi, G., and Rocca, G. (2010): Efficient Gear Shifting Strategies for Green Driving Policies. IEEE Proceedings of the American Control Conference, Baltimore, 4331–4336.
72. Charlet, B., Lévine, J., and Marino, R. (1991): Sufficient conditions for dynamic state feedback linearization. SIAM Journal on Control and Optimization 29(1):38–57.

73. Chen, C.-T. (2014): Linear System Theory and Design, Oxford University Press.
74. Chen, D., Bako, L., and Lecoeuche, S. (2012): The minimum-time problem for discrete-time linear systems: a non-smooth optimization approach. Proceedings of the IEEE International Conference on Control Applications, Dubrovnik, 196–201.
75. Chen, G., and Pham, T. T. (2000): Introduction to Fuzzy Sets, Fuzzy Logic, and Fuzzy Control Systems, Taylor & Francis.
76. Chen, G., and Ueta, T. (1999): Yet another chaotic attractor. International Journal of Bifurcation and Chaos 9(7):1465–1466.
77. Chen, G. (2023): Nonlinear Systems: Stability, Dynamics and Control, World Scientific.
78. Chen, M. Z. Q., and Smith, M. C. (2009): A Note on Tests for Positive-Real Functions. IEEE Transactions on Automatic Control 54(2):390–393.
79. Cheng, P., Vasić, M., García, O, Oliver, J. Á., Alou, P., and Cobos, J. A. (2014): Minimum Time Control for Multiphase Buck Converter: Analysis and Application. IEEE Transactions on Power Electronics 29(2):958–967.
80. Chern, S. S. und Shokurov, V. V. (Eds.) (2002): The Collected Papers of Wei-Liang Chow, World Scientific.
81. Chow, W.-L. (1940): Über Systeme von linearen partiellen Differentialgleichungen erster Ordnung. Mathematische Annalen 117(1):98–105, (On systems of linear partial differential equations of first order; in German).
82. Consolini, L., and Piazzi, A. (2009): Generalized bang-bang control for feedforward constrained regulation. Automatica 45(10):2234–2243.
83. Corless, M., and Shorten, R. (2009): A correct characterization of strict positive realness for MIMO systems. IEEE Proceedings of the American Control Conference, St. Louis, 469–475.
84. Corless, M., and Shorten, R. (2010): On the Characterization of Strict Positive Realness for General Matrix Transfer Functions. IEEE Transactions on Automatic Control 55(8):1899–1904.
85. Corless, M., and Shorten, R. (2011): On a class of generalized eigenvalue problems and equivalent eigenvalue problems that arise in systems and control theory. Automatica 47(3):431–442.
86. Cragon, S. D. (1991): Time-Optimal Control Application for an Exoatmospheric Interceptor. Proceedings of the IEEE Telesystems Conference (1):31–36.
87. Creveling, H. F., De Paz, J. F., Baladi, J. Y., and Schoenhals, R. J. (1975): Stability characteristics of a single-phase free convection loop. Journal of Fluid Mechanics 67(1):65–84.
88. Davison, E., and Constantinescu, D. (1971): A Describing Function Technique for Multiple Nonlinearities in a Single-Loop Feedback System. IEEE Transactions on Automatic Control 16(1):56–60.
89. DeCarlo, R. A., Zak, S. H., and Matthews, G. P. (1988): Variable Structure Control of Nonlinear Multivariable Systems: A Tutorial. Proceedings of the IEEE 76(3):212–232.
90. Demirci, Y, Pekel, L. C., Altinten, A., and Alpbaz, M. (2015): Application of fuzzy control on the electrocoagulation process to treat textile wastewater. Environmental Technology 36(24):3243–3252.

91. de Paor, A., and Ringwood, J. (2006): A Simple Soft Limiter Describing Function for Biomedical Applications. IEEE Transactions on Biomedical Engineering 53(7):1233–1240.
92. Devaney, R. L. (2003): An Introduction to Chaotic Dynamical Systems, Westview Press Inc.
93. Dewey, A. G., and Jury, E. I. (1965): A Note on Aizerman's Conjecture. IEEE Transactions on Automatic Control 10(4):482–483.
94. Diehl, M., Findeisen, R., Schwarzkopf, S., Uslu, I., Allgöwer, F., Bock, H. G., Gilles, E. D., and Schlöder, J. P. (2002): An Efficient Algorithm for Nonlinear Model Predictive Control of Large-Scale Systems - Part I: Description of the Method. Automatisierungstechnik 50(12):557–567.
95. Di Martino, F., and Sessa, S. (2020): Fuzzy Transforms for Image Processing and Data Analysis, Springer.
96. Dodds, S. J. (2015): Feedback Control: Linear, Nonlinear and Robust Techniques and Design with Industrial Applications, Springer.
97. Döring, D. (2003): Ein Beitrag zur Synthese von Gain-Schedulingreglern unter Nutzung normierter Gaußscher Radial-Basisfunktionen, PhD thesis, Universität Magdeburg, (A Contribution to the Synthesis of Gain-Scheduling Controllers Using Normalized Gaussian Radial Basis Functions; in German).
98. Dorf, R. C., and Bishop, R. H. (2016): Modern Control Systems, Prentice Hall.
99. Dormand, J. R., and Prince, P. J. (1980): A family of embedded Runge-Kutta formulae. Journal of Computational and Applied Mathematics 6(1):19–26.
100. Draženóvic, B. (1969): The invariance conditions in variable structure systems. Automatica 5(3):287–295.
101. Drexler, D., Sápi, J., Szeles, A., Harmati, I., Kovács, A., and Kovács, L. (2012): Flat control of tumor growth with angiogenic inhibition. Proceedings of the 7th IEEE International Symposium on Applied Computational Intelligence and Informatics, Timişoara, 179–183.
102. Driankov, D., Hellendoorn, H., and Reinfrank, M. (1996): An Introduction to Fuzzy Control, Springer.
103. Edwards, C., and Postlethwaite, I. (1998): Anti-windup and bumpless-transfer schemes. Automatica 34(2):199–210.
104. Емельянов, С. В. (1967): Системы автоматического управления с переменной структурой, Наука (Russian original publication of [105]).
105. Emel'yanov, S. V. (1967): Variable-Structure Control Systems, Nauka.
106. Emel'yanov, S. V. (2007): Theory of variable-structure control systems: Inception and initial development. Computational Mathematics and Modeling 18(4):321–331.
107. Engell, S. (Ed.) (1995): Entwurf nichtlinearer Regelungen, R. Oldenbourg, (Nonlinear control design; in German).
108. Engeln-Müllges, G., and Uhlig, F. (1996): Numerical Algorithms with C, Springer.
109. Fadden, E. J., and Gilbert, E. G. (1964): Computational Aspects of the Time-Optimal Control Problem. In Balakrishnan, A.V., and Neustadt, L.W. (Eds.): Computing Methods in Optimization Problems: 167–192, Academic Press.
110. Farinwata, S. S., Filev, D. P., and Langari, R. (2000): Fuzzy Control: Synthesis and Analysis, Wiley & Sons.
111. Фельдбаум, А. А. (1955): О синтезе оптимальных систем с помощью фазового пространства. Автоматика и телемеханика 16(2):129–149

(Fel'dbaum, A. A. (1955): On the synthesis of optimal systems with the aid of the phase space. Automatica y Telemehanika 16(2):129–149, in Russian).

112. Фельдбаум, А. А. (1959): Вычислительные устройства в автоматических системах, Физматгиз, Москва (Russian original publication of [113]).
113. Feldbaum, A. A. (1962): Rechengeräte in automatischen Systemen, R. Oldenbourg, (Computers in Automatic Systems; in German).
114. Fel'dbaum, A. A. (1965): Optimal Control Systems, Academic Press.
115. Ferfera, A., and Hammami, M. A. (1995): Growth Conditions for Global Stabilization of Cascade Nonlinear Systems. 3rd IFAC Conference on System Structure and Control, Nantes. In IFAC Proceedings Volumes 28(8):471–474.
116. Fernández-Anaya, G., Martínez-García, J. G., and Kučera, V. (2006): Characterizing families of positive real matrices by matrix substitution on scalar rational functions. System & Control Letters 55(11):871–878.
117. Fernández, J. P., Vargas, M. A., García, J. M. V., Carrillo, J. A. C, and Aguilar, J. J. C. (2021): Coevolutionary Optimization of a Fuzzy Logic Controller for Antilock Braking Systems Under Changing Road Conditions. IEEE Transactions on Vehicular Technology 70(2):1255–1268.
118. Ferrante, A., Lanzon, A., and Brogliato, B. (2020): A Direct Proof of the Equivalence of Side Conditions of Strictly Positive Real Transfer Functions. IEEE Transactions on Automatic Control 65(1):450–452.
119. Filippov, A. F. (1988): Differential Equations with Discontinuous Righthand Sides, Kluwer Academic Publishers.
120. Findeisen, R., Imsland, L., Allgöwer, F., and Foss, B. A. (2003): State and Output Feedback Nonlinear Model Predictive Control: An Overview. European Journal of Control 9(2-3):190–206.
121. Findeisen, R., and Allgöwer, F. (2004): Computational delay in nonlinear model predictive control. Proceedings of the 7th IFAC International Symposium on Advanced Control of Chemical Processes (ADCHEM), Hong Kong. In IFAC Proceedings Volumes 37(1):427–432.
122. Fitts, R. (1966): Two Counterexamples to Aizerman's Conjecture. IEEE Transactions on Automatic Control 11(3):553–556.
123. Fliess, M. (1990): Generalized Controller Canonical Form for Linear and Nonlinear Dynamics. IEEE Transactions on Automatic Control, 35(9):994–1001.
124. Fliess, M., Lévine, J., Martin, P., and Rouchon, P. (1992): Sur les systèmes non linéaires différentiellement plats. Comptes Rendus de l'Académie des Science, Serie I-315:619–624, (On differentially flat nonlinear systems; in French).
125. Fliess, M., Lévine, J., Martin, P., and Rouchon, P. (1992): On differentially flat nonlinear systems. 2nd IFAC Symposium on Nonlinear Control Systems (NOLCOS), Bordeaux. In IFAC Proceedings Volumes 25(13):159–163.
126. Fliess, M., Lévine, J., Martin, P., and Rouchon, P. (1994): Nonlinear control and Lie-Bäcklund transformations: Towards a new differential geometric standpoint. Proceedings of the 33rd IEEE Conference on Decision and Control, Lake Buena Vista, 1:339–344.
127. Fliess, M., Lévine, J., Martin, P., and Rouchon, P. (1995): Flatness and defect of non-linear systems: introductory theory and examples. International Journal of Control 61(6):1327–1361.
128. Fliess, M., Lévine, J., Martin, P., Ollivier, F., and Rouchon, P. (1997): Controlling Nonlinear Systems by Flatness. In Byrnes, C. I., Datta, B. N., Gilliam, D. S., and Martin, C. F. (Eds.): Systems and Control in the Twenty-First Century, 137–154, Birkhäuser.

129. Fliess, M., Lévine, J., Martin, P., and Rouchon, P. (1995): A Lie-Bäcklund Approach to Equivalence and Flatness of Nonlinear Systems. IEEE Transactions on Automatic Control 44(5):922–937.
130. Föllinger, O. (2022): Regelungstechnik, VDE-Verlag, (Control Engineering; in German).
131. Fox, L., and Mayers, D. F. (1987): Numerical Solution of Ordinary Differential Equations for Scientists and Engineers, Chapman and Hall.
132. Franke, D. (1982): Ausschöpfung von Stellgrößenbeschränkungen mittels weicher strukturvariabler Regelung. Regelungstechnik 30(10):348–355, (Exhausting bounds on control by means of soft variable structure control; in German).
133. Franklin, G. F., Powell, J. D., Emami-Naeini, A. (2018): Feedback Control of Dynamic Systems, Prentice-Hall.
134. Fridman, L., Moreno, J., and Iriarte, R. (2011): Sliding Modes after the First Decade of the 21st Century, Springer.
135. Friedland, B. (2005): Control System Design: An Introduction to State-Space Methods, Dover Publications.
136. Frik, M. (1966): Zeitoptimale Steuerung des Einstellvorganges bei Kreiselkompassen. Archive of Applied Mechanics 35(4):262–268, (Time-optimal control of the adjustment process of gyrocompasses; in German).
137. Fritzsche, K., and Röbenack, K.(2018): On the Computation of Differentially Flat Inputs. IEEE Proceedings of the 22nd International Conference on System Theory, Control, and Computing (ICSTCC), Sinaia, 12–19.
138. Fritzsche, K., and Röbenack, K.(2020): On a generalized flat input definition and physical realizability. 21st IFAC World Congress, Berlin. In IFAC Papers-Online 53(2):5994–5999.
139. Fujii, K, and Shoji, K. (1972): Verification of the Aizerman and/or Kalman Conjecture. IEEE Transactions on Automatic Control 17(3):406–408.
140. Fukushima, Y., Irikura, K., Uetake, T., and Matsumoto, H. (2000): Characteristics of Observed Peak Amplitude for Strong Ground Motion from the 1995 Hyogoken Nanbu (Kobe) Earthquake. Bulletin of the Seismological Society of America 90(3):545–565.
141. Fuller, A (1974): Simplification of Some Time-Optimal Switching Functions. IEEE Transactions on Automatic Control 19(1):65–65.
142. Galeani, S., Tarbouriech, S., Turner, M., and Zaccarian, L. (2009): A Tutorial on Modern Anti-Windup Design. IEEE Proceedings of the 10th European Control Conference, Budapest, 306–323.
143. Galor, O. (2007): Discrete Dynamical Systems, Springer.
144. Gantmacher, F. R. (1984): The Theory of Matrices, American Mathematical Society.
145. Gao, W., and Hung, J. C. (1993): Variable Structure Control of Nonlinear Systems: A New Approach. IEEE Transactions on Industrial Electronics 40(1):45–55.
146. Gauthier, J. P., and Bornard, G. (1981): Observability for any $u(t)$ of a Class of Nonlinear Systems. IEEE Transactions on Automatic Control 26(4):922–926.
147. Gelb, A., and Vander Velde, W. E. (1968): Multiple-Input Describing Functions and Nonlinear System Design, McGraw-Hill.
148. Germann, S., and Isermann, R. (1995): Nonlinear distance and cruise control for passenger cars. IEEE Proceedings of the American Control Conference, Washington, 3081–3085.

149. Germann, S. (1997): Modellbildung und modellgestützte Regelung der Fahrzeuglängsdynamik. VDI Fortschritt-Berichte, Reihe 12, Nr. 309, VDI-Verlag, (Modeling and Model-Based Control of Vehicle Longitudinal Dynamics; in German).
150. Ghandhari, M., Andersson, G., Pavella, M., and Ernst, D. (2001): A control strategy for controllable series capacitor in electric power systems. Automatica 37(10):1575–1583.
151. Giesl, P., and Hafstein, S. (2015): Review on computational methods for Lyapunov functions. Discrete and Continuous Dynamical Systems 20(8):2291–2331.
152. Glad, T., and Ljung, L. (2000): Control Theory: Multivariable and Nonlinear Methods, Taylor & Francis.
153. Glattfelder, A. H., and Schaufelberger, W. (2003): Control Systems with Input and Output Constraints, Springer.
154. Gordon, W. B. (1972): On the Diffeomorphisms of Euclidean Spaces. The American Mathematical Monthly 79(7):755–759.
155. Gowers, T., Barrow-Green, J., and Leader, I. (Eds.) (2008): The Princeton Companion to Mathematics, Princeton University Press.
156. Grancharova, A., and Johansen, T. A. (2012): Explicit Nonlinear Model Predictive Control, Springer.
157. Grimble, M J., and Majecki, P. (2020): Nonlinear Industrial Control Systems, Springer.
158. Grognard, F., Sepulchre, R., and de Dona, J. (2001): Dynamical Systems that Compute Time-Optimal Switchings. 5th IFAC Symposium on Nonlinear Control Systems (NOLCOS), St. Petersburg. In IFAC Proceedings Volumes 34(6):1189–1194.
159. Grognard, F., and Sepulchre, R.. (2001): Global analysis of a continuous-time flow which computes time-optimal switchings. Proceedings of the 40th IEEE Conference on Decision and Control, Orlando, 3826–3831.
160. Grossman, H. I. (1974): The Nature of Quantities in Market Disequilibrium. The American Economic Review 64(3):509–514.
161. Grüne, L., and Pannek, J. (2011): Nonlinear Model Predictive Control, Springer.
162. Gruyitch, L., Richard, J.-P., Borne, P., and Gentina, J.-C. (2004): Stability Domains, Chapman & Hall.
163. Gupta, S., and Joshi, S. M. (1995): State Space Characterization and Robust Stabilization of Dissipative LTI Systems. IEEE Proceedings of the American Control Conference, Seattle, 3616–3619.
164. Gurel, O., and Lapidus, L. (1969): A Guide to the Generation of Liapunov Functions. Industrial and Engineering Chemistry 61(3):30–41.
165. Gußner, T., Jost, M., and Adamy, J. (2012): Controller design for a class of nonlinear systems with input saturation using convex optimization. Systems & Control Letters 61(1):258–265.
166. Gutman, P.-O., and Hagander, P. (1985): A New Design of Constrained Controllers for Linear Systems. IEEE Transactions on Automatic Control 30(1):22–33.
167. Guzman, A. (2003): Derivatives and Integrals of Multivariable Functions, Birkhäuser.
168. Haddad, W. M., and Chellaboina, V. (2008): Nonlinear Dynamical Systems and Control: A Lyapunov-Based Approach, Princeton University Press.

169. Hagenmeyer, V., and Zeitz, M. (2004): Internal dynamics of flat nonlinear SISO systems with respect to a non-flat output. Systems & Control Letters 52(3-4):323–327.
170. Hahn, W. (1967): Stability of Motion, Springer.
171. Hahnfeldt, P., Panigrahy, D., Folkman, J., and Hlatky, L. (1999): Tumor Development under Angiogenic Signaling: A Dynamical Theory of Tumor Growth, Treatment Response, and Postvascular Dormancy. Cancer Research 59(19):4770–4775.
172. Hairer, E., Nørsett, S. P., and Wanner, G. (1993): Solving Ordinary Differential Equations I: Nonstiff Problems, Springer.
173. Hairer, E., and Wanner, G. (2004): Solving Ordinary Differential Equations II: Stiff and Differential-Algebraic Problems, Springer.
174. Hamaya, K., Waita, H., Gomi, H., and Shin, J. (2014): Inverted Pendulum type vehicle. U. S. Patent 8,763,733 B2.
175. Hangos, K. M., Bokor, J., and Szederkényi, G. (2004): Analysis and Control of Nonlinear Process Systems, Springer.
176. Härefors, M., and Bates, D. G. (2003): Integrated propulsion based flight control system design for a civil transport aircraft. International Journal of Turbo and Jet Engines 20(2):95–114.
177. Harris, J. (2005): Fuzzy Logic Applications in Engineering Science, Springer.
178. Haxel, J. H., Matsumoto, H., Meining, C., Kalbach, G., Lau, T.-K., Dziak, R. P., and Stalin, S. (2019): Ocean sound levels in the northeast Pacific recorded from an autonomous underwater glider. PLoS One 14(11):1–20.
179. Haynes, G. W., and Hermes, H. (1970): Nonlinear Controllability via Lie Theory. SIAM Journal on Control 8(4):450–460.
180. Hedrick, J. K., and Misawa, E. A. (1989): Nonlinear Observers – A State-of-the-Art Survey. Transactions of the ASME, Journal of Dynamic Systems, Measurement, and Control 111(3):344–352.
181. Hendricks, E., Jannerup, O., and Sørensen, P. H. (2008): Linear Systems Control: Deterministic and Stochastic Methods, Springer.
182. Hermann, R., and Krener, A. J. (1977): Nonlinear Controllability and Observability. IEEE Transactions on Automatic Control 22(5):728–740.
183. He, R., Woods, W., Zambon, J. B., and Xue, Z. (2016): Monitoring the Gulf Stream and shelf environment in the South Atlantic Bight through integrated autonomous underwater glider observations and data assimilative ocean model predictions. IEEE Proceedings of Oceans 2016, Shanghai, 1-4.
184. Hibbeler, R. C. (2016): Engineering Mechanics: Dynamics, Pearson.
185. Hippe, P. (2006): Windup in Control, Springer.
186. Hocking, L. M. (1991): Optimal Control, Oxford University Press.
187. Holzmann, H., Germann, S., Halfmann, C., and Isermann, R. (1998): Intelligent fuzzy distance and cruise control for passenger cars. Journal of Intelligent and Fuzzy Systems 6(3):315–327.
188. Honma, H. (1996): Automatic focusing system with response control using fuzzy interference. U. S. Patent 5,566,380.
189. Hooke, R., and Jeeves, T. A. (1961): „Direct Search“ Solution of Numerical and Statistical Problems. Journal of the Association of Computing Machinery (JACM) 8(2):212–229.
190. Hou, C. (2008): Fluid Dynamics and Behavior of Nonlinear Viscous Fluid Dampers. Journal of Structural Engineering 134(1):56–63.

191. Hu, T., and Lin, Z. (2001): Control Systems with Actuator Saturation: Analysis and Design, Birkhäuser.
192. Hu, T., and Lin, Z. (2002): On Improving the Performance with Bounded Continuous Feedback Laws. IEEE Transactions on Automatic Control 47(9):1570–1575.
193. Huang, S., Tan, K. K., and Lee, T. H. (2001): Applied Predictive Control, Springer.
194. Hughes, D., and Dornheim, M. A. (2003): No Flight Controls. Aviation Week & Space Technology 159(23):42–43.
195. Hughes, P. C. (2004): Spacecraft Attitude Dynamics, Dover Publications.
196. Hung, J. Y., Gao, W., and Hung, J. C. (1993): Variable Structure Control: A Survey. IEEE Transactions on Industrial Electronics 40(1):2–22.
197. Hunt, L. R., Su, R., Meyer, G., and Lin, Z. (1983): Design for multi-input nonlinear systems. In Brockett, R. W., Millman, R. S., and Sussman, H. J. (Eds.): Differential Geometric Control Theory. Proceedings of the Conference Held at Michigan University, Volume 27 of Progress in Mathematics, Birkhäuser, 268–297.
198. Hunt, L. R., Su, R., and Meyer, G. (1983): Global Transformations of Nonlinear Sytems. IEEE Transactions of Automatic Control 28(1):24–31.
199. Ilchmann, A., and Wirth, F. (2013): On Minimum Phase. Automatisierungstechnik 61(12):805–817.
200. Isermann, R. (1989): Digital Control Systems: Fundamentals, Deterministic Control, Springer.
201. Isermann, R. (1991): Digital Control Systems: Stochastic Control, Multivariable Control, Adaptive Control, Applications, Springer.
202. Isermann, R. (2022): Automotive Control: Modeling and Control of Vehicles, Springer.
203. Isidori, A. (1995): Nonlinear Control Systems, Springer.
204. Itkis, U. (1976): Control Systems of Variable Structure, Wiley & Sons.
205. Jadbabaie, A., and Hauser, (2005): On the Stability of Receding Horizon Control with a General Terminal Cost. IEEE Transactions on Automatic Control 50(5):674–678.
206. Jakubczyk, B., and Respondek, W. (1980): On linearization of control systems. Bulletin de L'Académie Polonaise des Sciences, Série des Sciences Mathématiques 28(9-10):517–522.
207. Jin, Y. (2003): Advanced Fuzzy Systems Design and Applications, Physica-Verlag.
208. Jo, N. H., and Seo, J. H. (2000): Input Output Linearization Approach to State Observer Design for Nonlinear System. IEEE Transactions on Automatic Control 45(12):2388–2393.
209. Johansen, T. A., Hunt, K. J., and Petersen, I. (2000): Gain-scheduled control of a solar power plant. Control Engineering Practice 8(9):1011–1022.
210. Jonckheere, E. A., and Yu, G.-R. (1998): H_∞ longitudinal control of crippled trijet aircraft with throttles only. Control Engineering Practice 6(5):601–613.
211. Jonckheere, E. A., and Yu, G.-R. (1999): Propulsion Control of Crippled Aircraft by H_∞ Model Matching. IEEE Transaction on Control Systems Technology 7(2):142–159.
212. Jordan, L. B. (1995): Locomotive traction control system using fuzzy logic. U. S. Patent 5,424,948.

213. Joshi, S. M. (1989): Out of Control. IEEE Control Systems Magazine 9(5):50.
214. Joshi, S. M., and Gupta, S. (1996): On a Class of Marginally Stable Positive-Real Systems. IEEE Transactions on Automatic Control 41(1):152–155.
215. Kahraman, C. (Ed.) (2006): Fuzzy Applications in Industrial Engineering, Springer.
216. Kailath, T. (1980): Linear Systems, Prentice Hall.
217. Kalman, R. E. (1957): Physical and Mathematical Mechanisms of Instability in Nonlinear Automatic Control Systems. Transactions of the American Society of Mechanical Engineers (ASME) 79(3):553–563.
218. Kalman, R. E. (1960): On the General Theory of Control Systems. 1st International IFAC Congress on Automatic and Remote Control, Moscow. In IFAC Proceedings Volumes 1(1):491–502.
219. Kalman, R. E., and Bucy, R. S. (1961): New Results in Linear Filtering and Prediction Theory. Transactions of the ASME, Journal of Basic Engineering 83(1):95–108.
220. Kamen, D. L., Arling, R. W., Field, J. D., Morell, J. B., Pompa, J. B., and Heinzmann, R. K. (2002): Personal Transporter. U. S. Patent 6,796,396 B2.
221. Kane, T. R., Likins, P. W., and Levinson, D. A. (1983): Spacecraft Dynamics, McGraw Hill.
222. Kapila, V., and Grigoriadis, K. M. (2002): Actuator Saturation Control, Marcel Dekker.
223. Kapoor, N., and Daoutidis, P. (2000): Stabilization of nonlinear processes with input constraints. Computers and Chemical Engineering 24:9–21.
224. Kaps, P., and Rentrop, P. (1979): Generalized Runge-Kutta methods of order four with stepsize control for stiff ordinary differential equations. Numerische Mathematik 33(1):55–68.
225. Karnopp, D. (2001): Motorized power steering system, German Patent DE 40 31 316 C2.
226. Kaur, R., and Singh, A. (2019): Fuzzy logic: an overview of different application areas. Advances and Applications in Mathematical Sciences 18(8):677–689.
227. Keerthi, S., and Gilbert, E. (1987): Computation of Minimum-Time Feedback Control Laws for Discrete-Time Systems with State-Control Constraints. IEEE Transactions on Automatic Control 32(5):432–435.
228. Kelkar, A. G., and Joshi, S. M. (1996): Control of Nonlinear Multibody Flexible Space Structures, Springer.
229. Keller, H. (1986): Entwurf nichtlinearer Beobachter mittels Normalformen. VDI Fortschritt-Berichte, Reihe 8, Nr. 124, VDI-Verlag, (Design of Nonlinear Observers Using Normal Forms; in German).
230. Kellett, C. M., and Braun, P. (2023): Introduction to Nonlinear Control: Stability, Control Design, and Estimation, Princeton University Press.
231. Kelley, W. G., and Peterson, A. C. (2001): Difference Equations: An Introduction with Applications, Academic Press.
232. Khalil, H. K. (2002): Nonlinear Systems, Prentice-Hall.
233. Kiel, L. D., and Elliott, W. E. (Eds.) (1997): Chaos Theory in the Social Sciences: Foundations and Applications, University of Michigan Press.
234. Kiendl, H., and Schneider, G. (1972): Synthese nichtlinearer Regler für die Regelstrecke $const/s^2$ aufgrund ineinandergeschachtelter abgeschlossener Gebiete beschränkter Stellgröße. Regelungstechnik und Prozeß-Datenverarbeit-

ung 20(7):289–296, (Synthesis of nonlinear controllers for the controlled system $const/s^2$ due to nested closed sets of limited actuator variable; in German).
235. Kiendl, H. (1972): Suboptimale Regler mit abschnittweise linearer Struktur, Springer, (Suboptimal Controllers with Sectional Linear Structure; in German).
236. Kim, M.-H., and Engell, S. (1994): Speed-up of Linear Programming for Time-Optimal Control. IEEE Proceedings of the American Control Conference, Baltimore, 3:2667–2670.
237. Kirk, D. E. (2004): Optimal Control Theory: An Introduction, Dover Publications.
238. Korotayev, A. (2005): A Compact Macromodel of World System Evolution. Journal of World-Systems Research 11(1):79–93.
239. Kothare, M. V., Campo, P. J., Morari, M., and Nett, C. N. (1994): A unified framework for the study of anti-windup designs. Automatica 30(12):1869–1883.
240. Kottenstette, N., and Antsaklis, P. J. (2010): Relationships Between Positive Real, Passive Dissipative, & Positive Systems. IEEE Proceedings of the American Control Conference, Baltimore, 409–416.
241. Kouvaritakis, B., and Cannon, M. (Eds.) (2001): Nonlinear Predictive Control: theory and practice, The Institution of Engineering and Technology.
242. Kovács, L., György, A., Szalay, P., Benyó, B., Benyó, Z., Hann, C. E., and Chase, J. G. (2010): Investigating the Applicability of qALPV Modeling to ICU Models for Glycaemic Control. IEEE Proceedings of the UKACC International Conference on Control, Coventry, 1–6.
243. Krämer, W. (2006): Grenzzyklen in einem Servolenksystem. Arbeitsberichte der FH Ingolstadt, Nr. 12, (Limit cycles in a servo steering systems; in German).
244. Krämer, W. (2007): Vermeidung von Grenzzyklen in einem Servolenksystem. In Mechatronik 2007 - Innovative Produktentwicklung: Maschinenbau, Elektrotechnik, Informationstechnik, VDI-Berichte (2007), Nr. 1971, S. 811–818, VDI-Verlag, (Avoiding limit cycles in a servo steering system; in German).
245. Krantz, S. G., and Parks, H. R. (2013): The Implicit Function Theorem. Birkhäuser.
246. Красовский, H. H. (1952): Теоремы об устойчивости движений, опрделяемых системой двух уравнений. Прикладная математика и маханика 16(5):547–554 (Krasovskii, N. N. (1952): Theorems on stability of motions based on a system of two equations. Prikladnaia matematika i mekhanika 16(5):547–554, in Russian).
247. Krause, P. C., Wasynczuk, O., and Sudhoff, S. D. (2013): Analysis of Electric Machinery and Drive Systems, Wiley & Sons.
248. Krener, A. J., and Isidori, A. (1983): Linearization by output injection and nonlinear observers. Systems & Control Letters 3(1):47–52.
249. Kroll, A. (2013): Computational Intelligence, R. Oldenbourg.
250. Krstić, M., Fontaine, D., Kokotović, P. V., and Paduano, J. D. (1998): Useful Nonlinearities and Global Stabilization of Bifurcations in a Model of Jet Engine Surge and Stall. IEEE Transactions on Automatic Control 43(12):1739–1745.
251. Krstić, M., Kanellakopoulos, I., and Kokotović, P. V. (1995): Nonlinear and Adaptive Control Design, Wiley & Sons.

252. Krstić, M., and Smyshlyaev, A. (2008): Boundary Control of PDEs: A Course on Backstepping Designs, SIAM.
253. Kugi, A. (2001): Non-linear Control Based on Physical Models, Springer.
254. Kundur, P. (1994): Power System Stability and Control, McGraw-Hill.
255. Kuznetsov, N. V., Kuznetsova, O. A., Koznov, D., Mokaev, R. N., and Andrievsky, B. (2018): Counterexamples to the Kalman Conjectures. 5th IFAC Conference on Analysis and Control of Chaotic Systems (CHAOS), Eindhoven. In IFAC PapersOnLine 51(33):138–143.
256. Kvasnica, M., Jones, C. N., Pejcic, I., Holaza, J., Korda, M., and Bakaráč, P. (2019): Real-Time Implementation of Explicit Model Predictive Control. In Raković, S. V., and Levine, W. S. (Eds.): Handbook of Model Predictive Control, 387–412, Birkhäuser.
257. Lafferriere, G., and Sussmann, H. (1991): Motion planning for controllable systems without drift. Proceedings of the IEEE International Conference on Robotics and Automation, Sacramento, 1148–1153.
258. Lancaster, P., and Tismenetsky, M. (1985): The Theory of Matrices, Academic Press.
259. Landau, I. D., and Zito, G. (2006): Digital Control Systems, Springer.
260. Larsen, M., and Kokotović, P. V. (2001): A brief look at the Tsypkin criterion: from analysis to design. International Journal of Adaptive Control and Signal Processing 15(2):121–128.
261. LaSalle, J. P. (1960): Some Extensions of Liapunov's Second Method. IRE Transactions on Circuit Theory 7(4):520–527.
262. Lee, Y., Jang, S.-I., Chung, K., Lee, D., Kim, W., and Lee, C.-W. (1994): A Fuzzy-Control Processor for Automatic Focusing. IEEE Transactions on Consumer Electronics 40(2):138–144.
263. Lee, J.-H., Hyeon, J.-W., Jung, S.-K., Lee, Y.-K., and Ko, S.-H.(2020): Observations of Temperature and Salinity Mesoscale Variability off the East Coast of Korea using an Underwater Glider: Comparison with Ship CTD Survey Data. Journal of Coastal Research 95(Special Issue):1167–1171.
264. Lein, M. (2022): A Mathematical Journey Through Differential Equations Of Physics, World Scientific.
265. Leith, D. J., and Leithead, W. E. (1998): Gain-scheduled controller design: an analytic framework directly incorporating non-equilibrium plant dynamics. International Journal of Control 70(2):249–269.
266. Leith, D. J., and Leithead, W. E. (2000): Survey of gain-scheduling analysis and design. International Journal of Control 73(11):1001–1025.
267. Lens, H., Adamy, J., and Domont-Yankulova, D. (2011): A fast nonlinear control method for linear systems with input saturation. Automatica 47(4):857–860.
268. Lentz, W. (1993): Neuere Entwicklungen in der Theorie dynamischer Systeme and ihre Bedeutung für die Agrarökonomie. Duncker & Humblot, (Recent developments in the theory of dynamic systems and their relevance for agricultural economics; in German).
269. Leonhard, W. (1976): Zeitoptimale Scherenregelung. Archiv für Elektrotechnik 58(1):61–67, (Time-optimal shear control; in German).
270. Leonov, G. A., Burkin I. M., and Shepeljavyi, A. I. (1996): Frequency Methods in Oscillation Theory, Kluwer Academic Publishers.

271. Leonov, G. A., Bragin, V. O., and Kuznetsov, N. V. (2010): Algorithm for constructing counterexamples to the Kalman problem. Doklady Mathematics 82(1):540–542.
272. Lévine, J. (1997): Static and dynamic feedback linearization. In Fossard, A. J., and Normand-Cyrot, D. (Eds.): Nonlinear Systems - Vol. 3, Control: 93–126, Chapman & Hall.
273. Lévine, J. (2009): Analysis and Control of Nonlinear Systems - A Flatness-based Approach, Springer.
274. Levine, W. S. (2011): The Control Handbook, Taylor & Francis.
275. Li, A., and Ando, K. (2013): Measuring the Acceptability of Self-Balancing Two-Wheeled Personal Mobility Vehicles. Journal of the Eastern Asia Society for Transportation Studies 10:444–453.
276. Li, T.-Y., and Yorke, J. A. (1975): Period Three Implies Chaos. The American Mathematical Monthly 82(10):985–992.
277. Li, X.-J. (1963): On the absolute stability of systems with time lags. Chinese Mathematics 3(4):609–626.
278. Liao, X., and Yu, P. (2008): Absolute Stability of Nonlinear Control Systems, Springer.
279. Liapounoff, A. M. (1907): Problème générale de la stabilité du mouvement. Translation of the original publication [292] by M. É. Davaux. Annales de la Faculté des Sciences de l'Université de Toulouse, 2e série, 9:203–474.
280. Liberzon, D. (2003): Switching in Systems and Control, Birkhäuser.
281. Liberzon, M. R. (2006): Essays on the Absolute Stability Theory. Automation and Remote Control 67(10):1610–1644.
282. Limon, D., Alamo, T., Salas, F., and Camacho, E. F. (2006): On the Stability of Constrained MPC without Terminal Constraint. IEEE Transactions on Automatic Control 51(5):832–836.
283. Lin, Y., and Sontag, E. D. (1991): A universal formula for stabilization with bounded controls. System & Control Letters 16(6):393–397.
284. Llamas, A., De La Ree Lopez, J., Mili, L., Phadke, A. G., and Thorp, J. S. (1995): Clarifications of the BCU Method for Transient Stability Analysis. IEEE Transactions on Power Systems 10(1):210–219.
285. Lobry, C. (1970): Contrôlabilité des systèmes non linéaires. SIAM Journal on Control 8(4):573–605, (Controllability of nonlinear systems; in French).
286. Locatelli, A. (2017): Optimal Control of a Double Integrator: A Primer on Maximum Principle, Springer.
287. Loughman, M., de Paor, A., Ringwood, J. V. (2016): The Effect of Blood Pressure Pulsatility on the Baroreflex Gain. IEEE Proceedings of the 27th Irish Signals and Systems Conference, Londonderry, 1–6.
288. Lozano-Leal, R., and Joshi, S. M. (1990): Strictly Positive Real Transfer Functions Revisited. IEEE Transactions on Automatic Control 35(11):1243–1245.
289. Luenberger, D. G. (1966): Observers for Multivariable Systems. IEEE Transactions on Automatic Control 11(2):190–197.
290. Лурье, А. И. (1951): Некоторые нелинейные задачи теории автоматического регулирования, Государственое издательство технико-теоретической литературы, Москва (Russian original publication of [291]).
291. Lur'e, A. I. (1957): Some Non-linear Problems in the Theory of Automatic Control, Her Majesty's Stationery Office, London.

292. Ляпунов, А. М. (1892): Общая задача объ устойчивости движенія, Разсуждение, Харьковъ (Ljapunov, A. M (1892): The general problem of the stability of motion. PhD thesis, University of Kharkiv; in Russian).
293. Lyapunov, A. M. (1992): The general problem of the stability of motion. Translation of the French version [279] by A. T. Fuller. International Journal of Control 55(3):531–534.
294. Ma, W., Wang, Y., Yang, S., Wang, S., and Xue, Z. (2018): Observation of Internal Solitary Waves Using an Underwater Glider in the Northern South China Sea. Journal of Coastal Research 34(5):1188–1195.
295. Ma, Z., and Zou, S. (2021): Optimal Control Theory, Springer.
296. Maciejowski, J. M. (2002): Predictive Control with Constraints, Pearson.
297. Madeira, D., and Adamy, J. (2016): On the Equivalence between Strict Positive Realness and Strict Passivity of Linear Systems. IEEE Transactions on Automatic Control 61(10):3091–3095.
298. Maeder, U., Cagienard, R., and Morari, M. (2007): Explicit Model Predicitve Control. In Tarbouriech, S., Garcia, G., and Glattfelder, A. H. (Eds.): Advanced Strategies in Control Systems with Input and Output Constraints, 237–271, Springer.
299. Maggiore, M., and Passino, K. M. (2003): A Separation Principle for Non-UCO Systems: The Jet Engine Stall and Surge Example. IEEE Transactions on Automatic Control 48(7):1264–1269.
300. Малкин, И. Г. (1952): Теория устойчивости движения, Издательство Наука. (Russian original publication of [301]).
301. Malkin, I. G. (1958): Theory of stability of motion, University of Michigan Library.
302. Mancilla Aguilar, J. L., García, R. A., and D'Attellis, C. E. (1996): Exact linearization of nonlinear systems: trajectory tracking with bounded controls and state constraints. International Journal of Control 65(3):455–467.
303. Margono, B. S. (2004): Optimierung von Bewegungsabläufen mit schwingungsfreien Endpositionen zur Verkürzung der Arbeitszyklen von Container-Schnellumschlaganlagen, PhD thesis, Universität Duisburg-Essen, (Optimization of motion sequences with oscillation-free end positions for shortening the operating cycles of container handling systems; in German).
304. Mizumoto, M., and Tanaka, K. (1981): Fuzzy Sets and Their Operations. Information and Control 48(1):30–48.
305. Marino, R. (1986): On the largest feedback linearizable subsystem. Systems & Control Letters 6(1):345–351.
306. Maron, C. (1992): Automatische Gangwahl bei PKW – Automatikgetrieben mit Hilfe von Fuzzy-Logik. VDE-Fachtagung Technische Anwendungen von Fuzzy-Systemen, Dortmund, 278–287, (Automatic gear selection in automatic transmission of passenger cars using fuzzy logic. VDE Symposium Technical Applications of Fuzzy Systems; in German).
307. Marquez, H. J., and Damaren, G. J. (1996): Analysis and Synthesis of Strictly Positive Real Transfer Functions, Journal of the Franklin Institute 333 B(2):245–256.
308. Marquez, H. J. (2003): Nonlinear Control Systems, Wiley & Sons.
309. Martin, P., Devasia, S., and Paden, B. (1996): A Different Look at Output Tracking: Control of a VTOL Aircraft. Automatica 32(1):101–107.
310. Martin, P., Murray, R. M., and Rouchon, P. (2009): Flatness based design. Control Systems, Robotics and Automation XIII, 65–89.

311. Massera, J. L. (1956): Contributions to Stability Theory. Annals of Mathematics 64(1):182–206.
312. Massera, J. L. (1958): Erratum: Contributions to Stability Theory. Annals of Mathematics 68(1):202.
313. Matsumoto, N., Kuraoka, H., Ohoka, N., and Ohba, M. (1991): Servo Control System. U. S. Patent 5,001,640.
314. Maurer, G. F. (1995): A Fuzzy Logic Controller for an ABS Braking System. IEEE Transactions on Fuzzy Systems 3(4):381–388.
315. Mayne, D. Q., Rawlings, J. B., Rao, C. V., and Scokaert, P. O. M. (2000): Constrained model predictive control: Stability and optimality. Automatica 36(6):789–814.
316. Mazenc, F., de Queiroz, M. S., Malisoff, M., and Gao, F. (2006): Further Results on Active Magnetic Bearing Control with Input Saturation. IEEE Transactions on Control Systems Technology 14(5):914–919.
317. McMahan, J. (1978): Flight 1080. Air Line Pilot 47(7):6–10.
318. Meditch, J. S. (1964): On the Problem of Optimal Thrust Programming For a Lunar Soft Landing. IEEE Transactions on Automatic Control 9(4):477–484.
319. Mees, A. I. (1984): Describing Functions: Ten Years On. IMA Journal of Applied Mathematics 32(1-3):221–233.
320. Michel, A. N., Hou, L., and Liu, D. (2015): Stability of Dynamical Systems, Birkhäuser.
321. Michels, K., Klawonn, F., Kruse, R., and Nürnberger, A. (2006): Fuzzy Control: Fundamentals, Stability and Design of Fuzzy Controllers, Springer.
322. Miller, S., Kennedy, J. F., Molino, J., Emo, A., Rousseau, G., Tan, G., and Do, A. (2010): Operating Characteristics of the Segway Human Transporter. Technical Report of the U. S. Department of Transportation, No. FHWA-HRT-10-025.
323. Mohammadi, F., and Hosseini, M. M. (2011): A comparative study of numerical methods for solving quadratic Riccati differential equations. Journal of the Franklin Institute 348(2):156–164.
324. Mokhtari, F., Sicard, P., and Lechevin, N. (2007): Stabilizing Winding Systems by Injection Damping Control on Control Based on Controlled Hamiltonian Systems. IEEE Proceedings of the International Electric Machines and Drives Conference (IEMDC), Antalya, 95–100.
325. Mukhopadhyay, B. K., and Malik, O. P. (1972): Optimal control of synchronous-machine excitation by quasilinearisation techniques. IEE Proceedings - Electric Power Applications 119(1):91–98.
326. Müller, A., Marsili-Libelli, S., Aivasidis, A., Lloyd, T., Kroner, S., and Wandrey, C. (1997): Fuzzy control of disturbances in a wastewater treatment process. Water Research 31(12):3157–3167.
327. Murray, F. J., and Miller, K. S. (2007): Existence Theorems for Ordinary Differential Equations, Dover Publications.
328. Na, W. K., Gou, B, and Diong, B. (2007): Nonlinear Control of PEM Fuel Cells by Exact Linearization. IEEE Transactions on Industry Applications 43(6):1426–1433.
329. Narendra, K. S., and Taylor, J. H. (1973): Frequency Domain Criteria for Absolute Stability, Academic Press.
330. Narendra, K. S., and Goldwyn, R. M. (1964): A Geometrical Criterion for the Stability of Certain Nonlinear Nonautonomous Systems. IEEE Transactions on Circuit Theory 11(3):406–408.

331. National Transportation Safety Board (1989): Aircraft Accident Report - United Airlines Flight 232, McDonnell Douglas DC-10-10, Sioux Gateway Airport, Sioux City, Iowa, July 19, 1989. Report No. NTSB/AAR-90/06.
332. Naus, G. J. L. (2010): Model-based control for automotive applications, PhD Thesis, University of Technology Eindhoven.
333. Nayfeh, A. H., and Mook, D. T. (1995): Nonlinear Oscillations, Wiley & Sons.
334. Netushil, A. (Ed.) (1978): Theory of Automatic Control, MIR.
335. Newell, R. B., and Lee, P. L. (1989): Applied Process Control: A Case Study, Prentice-Hall.
336. Newman, W. S. (1990): Robust Near Time-Optimal Control. IEEE Transactions on Automatic Control 35(7):841–844.
337. Nicolau, F., Respondek, W., and Barbot, J.-P. (2020): Flat Inputs: Theory and Applications. SIAM Journal on Control and Optimization 58(6):3293–3321.
338. Nijmeijer, H., and van der Schaft, A. (2016): Nonlinear Dynamical Control Systems, Springer.
339. Nikravesh, S. K. Y. (2013): Nonlinear Systems Stability Analysis: Lyapunov-Based Approach, Taylor & Francis.
340. Ochi, Y. (2005): Flight control system design for propulsion-controlled aircraft. Proceedings of the Institution of Mechanical Engineers - Part G: Journal of Aerospace Engineering 219(4):329–340.
341. Ogata, K. (1987): Discrete-Time Control Systems, Prentice-Hall.
342. Oppenheim, A. V., and Schafer, R. W. (2010): Discrete-Time Signal Processing, Pearson.
343. Orloff, R. W., and Harland, D. M. (2006): Apollo, Springer.
344. Ortega, R., and Garcia-Canseco, E. (2004): Interconnection and Damping Assignment Passivity-Based Control: A Survey. European Journal of Control 10(5):432–450.
345. Ortega, R., van der Schaft, A., Maschke, B., and Escobar, G. (2002): Interconnection and damping assignment passivity-based control of port-controlled Hamiltonian systems. Automatica 38(4):585–596.
346. Ortega, R., van der Schaft, A., Mareels, J., and Maschke, B. (2001): Putting Energy Back in Control. IEEE Control Systems Magazine 21(2):18–33.
347. Outrata J, V. (1975): On the Minimum Time Problem in Linear Discrete Systems with the Discrete Set of Admissible Controls. Kybernetika 11(5):368–374.
348. van Overloop, P.-J., Weijs, S., and Dijkstra, S. (2008): Multiple model predictive control on a drainage canal system. Control Engineering Practice 16(5):531–540.
349. Owusu, K. O., Lewis, F. L., Borovic, B., and Liu, A. Q. (2006): Nonlinear Control of a MEMS Optical Switch. Proceedings of the IEEE Conference on Decision and Control, San Diego, 597–602.
350. Palacios-Quiñonero, F., Rubió-Massegú, J., Rossell, J. M., and Karimi, H. R. (2012): Optimal passive-damping design using a decentralized velocity-feedback H_∞ approach. Modelling, Identification and Control 33(3):87–97.
351. Panteley, E., Loría, A., and Sokolov, A. (1999): Global Uniform Asymptotic Stability of Cascaded Non-autonomous Non-linear Systems: Application to Stabilisation of a Diesel Engine. European Journal of Control 5(1):107–115.
352. Pappas, G. J., Lygeros, J., and Godbole, D. N. (1995): Stabilization and Tracking of Feedback Linearizable Systems under Input Constraints. Proceedings of the 34th IEEE Conference on Decision and Control, New Orleans, 596–601.

353. Park, M.-H., and Won, C.-Y. (1991): Time Optimal Control for Induction Motor Servo System. IEEE Transactions on Power Electronics 6(3):514–524.
354. Parker, T. S., and Chua, L. O. (1989): Practical Numerical Algorithms for Chaotic Systems, Springer.
355. Parks, P. C. (1992): A. M. Lyapunov's stability theory – 100 years on. IMA Journal of Mathematical Control & Information 9(4):275–303.
356. Parks, P. C., and Hahn, V. (1993): Stability Theory, Prentice Hall.
357. Pavliotis, G. A., and Stuart, A. M. (2000): Multiscale Methods, Springer.
358. Peckol, J. K. (2021): Introduction to Fuzzy Logic, Wiley.
359. Peng, Y., Vrancic, D., and Hanus, R. (1996): Anti-Windup, Bumpless, and Conditioned Transfer Techniques for PID Controllers. IEEE Control Systems Magazine 16(4):48–57.
360. Perruquetti, W., and Barbot, J. P. (Eds.) (2002): Sliding Mode Control in Engineering, Marcel Dekker Inc.
361. Pfiffner R., and Guzella, L. (1999): Feedback Linearization Idle-Speed Control: Design and Experiments. Kybernetika 35(4):441–458.
362. Плисс, В. А. (1958): Некоторые проблемы теории устоичивости в целом, Издательство Ленинградского университета (Pliss, V. A. (1958): Certain problems in the theory of stability of motion in the whole. Leningrad University Press; in Russian).
363. Pliss, V. A. (1965): Certain problems in the theory of stability of motion in the whole. English translation of the original publication [362], NASA Technical Translation F-280.
364. Polyanin, A. D., and Zaitsev, V. F. (2018): Handbook of Ordinary Differential Equations: Exact Solutions, Methods, and Problems, Chapman & Hall.
365. Popov, V. M. (1959): Criterii de stabilitate pentru sistemele neliniare de reglare automată, bazate pe utilizarea transformatei Laplace. Studii şi Cercetări de Energetică 9(1):119–135, (Stability criteria for nonlinear systems with automatic control based on the use of the laplace transformation; in Romanian).
366. Popov, V. M. (1959): Criterii suficiente de stabilitate asimptotică in mare pentru sistemele automate neliniare cu mai multe organe de execuţie. Studii şi Cercetări de Energetică 9(4):647–680, (Sufficient criteria for asymptotic stability of general nonlinear systems with multiple actuating variables; in Romanian).
367. Popov, V. M. (1960): Criterion of Quality for Non-linear Controlled Systems. 1st International IFAC Congress on Automatic and Remote Control, Moscow. In IFAC Proceedings Volumes 1(1):183–187.
368. Popov, V.M. (1960): Noi criterii de stabilitate pentru sistemele automate neliniare. Studii şi Cercetări de Energetică 10(1):159–174, (New stability criteria for nonlinear control systems; in Romanian).
369. Popov, V. M. (1960): Nouveaux critériums de stabilité pour les systèmes automatiques nonlinéaires. French translation of the original publication [368]. Revue d'Électrotechnique et Énergétique 5(1):73–88.
370. Popov, V. M. (1960): Noi criterii grafice pentru stabilitatea stării staţionare a sistemelor automate neliniare. Studii şi Cercetări de Energetică 10(3):601–612 (Romanian original publication of [371]).
371. Popov, V. M. (1961): New graphical criteria for the stability of the steady state of nonlinear control systems. Revue d'Électrotechnique et Énergétique 6(1):25–34.

372. Popov, V. M. (1961): Absolute Stability of Nonlinear Systems of Automatic Control. Automation and Remote Control 22(8):857–875.
373. Popov, V. M., and Halanay, A. (1962): On the Stability of Nonlinear Automatic Control Systems with Lagging Argument. Automation and Remote Control 23(7):783–786.
374. Popov, V. M. (1963): The Solution of a New Stability Problem for Controlled Systems. Automation and Remote Control 24(1):1–23.
375. Popov, V. M. (1973): Hyperstability of Control Systems, Springer.
376. Powell, B. K., and Cook, J. A. (1987): Nonlinear Low Frequency Phenomenological Engine Modeling and Analysis. IEEE Proceedings of the American Control Conference, Minneapolis, 332–340.
377. Pyatnitskii, E. S. (1973): Existence of Absolutely Stable Systems for Which the Popov Criterion Fails. Automation and Remote Control 34(1):22–29.
378. Quin, S. J., and Badgwell, T. A. (2003): A survey of industrial model predictive control technology. Control Engineering Practice 11(7):733–764.
379. Raja, K., and Ramathilagam, S. (2021): Washing machine using fuzzy logic controller to provide wash quality. Soft Computing 25(15):9957–9965.
380. Răsvan, V., and Popescu, D. (2012): On the Problem of Aizerman, IEEE Proceedings of the 16th International Conference on System Theory, Control and Computing (ICSTCC), Sinaia, 1–6.
381. Rawlings, J. B. (2000): Tutorial Overview of Model Predictive Control. IEEE Control Systems Magazine 20(3):38–52.
382. Rawlings, J. B., and Mayne, D. Q. (2009): Model Predictive Control: Theory and Design, Nob Hill Publishing.
383. Ray, K. S., and Majumder, D. D. (1985): Fuzzy Logic Control of a Nonlinear Multivariable Steam Generating Unit Using Decoupling Theory. IEEE Transactions on Systems, Man, and Cybernetics 15(4):539–558.
384. Ringwood, J. V., and Malpas, S. C. (2001): Slow oscillations in blood pressure via a nonlinear feedback model. American Journal of Physicol Regulatory Integrative and Comperative Physiology 280(4):R1105–R1115.
385. Röbenack, K. (2004): An Approximation of Normal Form Observer Design: Convergence and Computation. 6th IFAC Symposium on Nonlinear Control Systems (NOLCOS), Stuttgart. In IFAC Proceedings Volumes 37(13):1353–1358.
386. Röbenack, K. (2007): Direct approximation of observer error linearization for nonlinear forced systems. IMA Journal of Mathematical Control and Information 24(4):551–566.
387. Röbenack, K. (2007): Observer Design for a Class of Nonlinear Systems with Non-Full Relative Degree. Nonlinear Dynamics and Systems Theory 7(4):399–408.
388. Röbenack, K. (2017): On the Direct Computation of the Byrnes-Isidori Normal Form. Proceedings in Applied Mathematics and Mechanics 17(1):811–812.
389. Rouche, N., Habets, P., and Laloy, M. (1977): Stability Theory by Liapunov's Direct Method, Springer.
390. Rouchon, P. (1994): Necessary Condition and Genericity of Dynamic Feedback Linearization. Journal of Mathematical Systems, Estimation, and Control 4(2):1–14.
391. Rozenvasser, E. N. (1963): The Absolute Stability of Nonlinear Systems. Automation and Remote Control 24(3):283–291.

392. Rubio, F. R., Camacho, E. F., and Berenguel, M. (2006): Control de Campos de Colectores Solares. Revista iberoamericana de automática e informática industrial 3(4):26–45, (Control of solar collector fields; in Spanish).
393. Rugh, W. J., and Shamma, J. S. (2000): Research on gain scheduling. Automatica 36(10):1401–1425.
394. Ruzhansky, M., and Sugimoto, M. (2015): On global inversion of homogeneous maps. Bulletin of Mathematical Sciences 5(1):13–18.
395. Ryan, E. P. (1974): Time-optimal feedback control laws for certain third-order relay control systems. International Journal of Control 20(6):881–911.
396. Ryan, E. P. (1977): Time-optimal feedback control of certain fourth-order systems. International Journal of Control 26(5):675–688.
397. Sabanovic, A., Fridman, L. M., and Spurgeon, S. K. (Eds.) (2004): Variable Structure Systems: From Principles to Implementation, The Institution of Engineering and Technology (IET).
398. Sadegh, N, Finney, J. D., and Heck, B. S. (1997): An explicit method for computing the positive real lemma matrices. International Journal of Robust and Nonlinear Control 7(12):1057–1069.
399. Sage, A. P., and White, C. C. (1977): Optimum Systems Control, Prentice Hall.
400. Sager, S., Bock, H. G., Diehl M., Reinelt, G., and Schloder, J. P. (2006): Numerical Methods for Optimal Control with Binary Control Functions Applied to a Lotka-Volterra Type Fishing Problem. In Seeger A. (Ed.): Recent Advances in Optimization, Springer.
401. Sandberg, I. W. (1964): A Frequency-Domain Condition for the Stability of Feedback Systems Containing a Single Time-Varying Nonlinear Element. The Bell System Technical Journal 43(4):1601–1608.
402. Sastry, S. (1999): Nonlinear Systems: Analysis, Stability and Control, Springer.
403. Savkin, A. V., and Evans, R. J. (2002): Hybrid Dynamical Systems: Controller and Sensor Switching Problems, Birkhäuser.
404. Schmidt, G., and Preusche, G. (1967): Popovs Stabilitätssatz als Mittel zur teilweisen Bestätigung von Aizermans Vermutung. Regelungstechnik 15(1):20–25, (Popov's stability theorem as a means of partially confirming Aizerman's conjecture; in German).
405. Schwartz, C. A., and Yan, A. (1995): Systematic Construction of Lyapunov Functions for Nonlinear Systems in Critical Cases. Proceedings of the 34th IEEE Conference on Decision and Control, New Orleans, 3779–3784.
406. Schwefel, H.-P. (1995): Evolution and Optimum Seeking, Wiley.
407. Scibile, L. (1997): Non-linear Control of the Plasma Vertical Position in a Tokamak, PhD Thesis, University of Oxford.
408. Scokaert, P. O. M., Mayne, D. Q., and Rawlings, J. B. (1999): Suboptimal Model Predictive Control (Feasibility Implies Stability). IEEE Transactions on Automatic Control 44(3):648–654.
409. Scott, M. (1986): Time/fuel optimal control of constrained linear discrete systems. Automatica 22(6):711–715.
410. Seibert, P., and Suarez, R. (1990): Global stabilization of nonlinear cascade systems. Systems & Control Letters 14(4):347–352.
411. Sepulchre, R., Janković, M., and Kokotović, P. V. (1997): Constructive Nonlinear Control, Springer.

412. Serrarens, A. F. A., van de Molengraft, M. J. G., Kok, J. J., and van den Steen, L. (1998): H_∞ Control for Suppressing Stick-Slip in Oil Well Drillstrings. IEEE Control Systems Magazine 18(2):19–30.
413. Sharma, S. C., Sharma, C., Kumar, C., Rajawat, P. S., and Mohindra, M. (2008): Fuzzy approach to control moisture content on a paper machine. Paper Technology 49(4):33–38.
414. Shen, Z., Huang, P., and Andersson, S. B. (2013): Calculating switching times for the time-optimal control of single-input, single-output second-order systems. Automatica 49(5):1340–1347.
415. Shtessel, Y., Edwards, C., Fridman, L., and Levant, A. (2014): Sliding Mode Control and Observation, Birkhäuser.
416. Singer, J., Wang, Y.-Z., and Bau, H. H. (1991): Controlling a Chaotic System. Physical Review Letters 66(9):1123–1125.
417. Singh, D. S., Kumari, S., Abhishek, V., and Sathyaraj, R. (2017): Optimal Gear Indication for Green Environment. International Journal of Engineering Research 6(5):271–274.
418. Sira-Ramírez, H. (1989): Nonlinear Variable Structure Systems in Sliding Mode: The General Case. IEEE Transactions on Automatic Control 34(11):1186–1188.
419. Sira-Ramírez, H. (1998): A general canonical form for feedback passivity of non-linear systems. International Journal of Control 71(5):891–905.
420. Sira-Ramírez, H., and Agrawal, S. K. (2004): Differentially Flat Systems, Marcel Dekker.
421. Sira-Ramírez, H. (2015): Sliding Mode Control, Birkhäuser.
422. Slotine, J.-J. E., and Li, W. (1991): Applied Nonlinear Control, Prentice-Hall.
423. Soetaert, K., Cash, J., and Mazzia, F. (2012): Solving Differential Equations in R, Springer.
424. Sofrony, J. (2009): Anti-windup Compensation of Input Constrained Systems: Synthesis using Riccati Equations, VDM Verlag.
425. Sommer, R. (1980): Control design for multivariable nonlinear time-varying systems. International Journal of Control 31(5):883–891.
426. Sontag, E. D. (1989): A 'universal' construction of Artstein's theorem on nonlinear stabilization. Systems & Control Letters 13(2):117–123.
427. Sontag, E. D. (1989): Smooth Stabilization Implies Coprime Factorization. IEEE Transactions on Automatic Control 34(4):435–443.
428. Sontag, E. D. (1989): Remarks on Stabilization and Input-to-State Stability. IEEE Proceedings on the 28th Conference on Decision and Control, Tampa, 1376–1378.
429. Sontag, E. D. (1990): Further Facts about Input to State Stabilization. IEEE Transactions on Automatic Control 35(4):473–476.
430. Sontag, E. D., and Wang, Y. (1995): On characterizations of the input-to-state stability property. Systems & Control Letters 24(5):351–359.
431. Sontag, E. D. (1998): Mathematical Control Theory, Springer.
432. Sontag, E. D., and Wang, Y. (1999): Notions of input to output stability. Systems & Control Letters 38(4-5):235–248.
433. Sontag, E. D. (2008): Input to State Stability: Basic Concepts and Results. In Nistri, P., and Stefani, G. (Eds.): Nonlinear and Optimal Control Theory, Springer.
434. Spencer, B. F., and Nagarajaiah, S. (2003): State of the Art of Structural Control. Journal of Structural Engineering 129(7):845–856.

435. Stefan, P. (1973): Two Proofs of Chow's Theorem. In Mayne, D. Q., and Brockett, R. W. (Eds.): Geometric Methods in System Theory, NATO Advanced Study Institutes Series 3:159–164.
436. Strogatz, S. H. (2001): Nonlinear Dynamics and Chaos, Westview Press Inc.
437. Sun, Y., and Peng, J. (2002): A New Lyapunov Function for Transient Stability Analysis of Power Systems with Emergency Control. Proceedings of the IEEE International Conference on Power System Technology (PowerCon 2002) 3:1540–1544, Kunming.
438. Symans, M. D., and Constantinou, M. C. (1997): Experimental Testing and Analytical Modeling of Semi-Active Fluid Dampers for Seismic Protection. Journal of Intelligent Material Systems and Structures 8(8):644–657.
439. Symans, M. D., and Constantinou, M. C. (1999): Semi-active control systems for seismic protection of structures: a state-of-the-art review. Engineering Structures 21(6):469–487.
440. Takagi, H. (1992): Application of neural networks and fuzzy logic to consumer products. Proceedings of the IEEE International Conference on Industrial Electronics, Control, Instrumentation, and Automation 3:1629–1633, San Diego.
441. Tan, S.-C., Lai, Y. M., Cheung, M. K. H., and Tse, C. K. (2005): On the Practical Design of a Sliding Mode Voltage Controlled Buck Converter. IEEE Transactions on Power Electronics 20(2):425–437.
442. Tao, G., and Ioannou, P. A. (1988): Strictly Positive Real Matrices and the Lefschetz-Kalman-Yakubovich Lemma. IEEE Transactions on Automatic Control 33(12):1183–1185.
443. Tarbouriech, S., and Turner, M. (2009): Anti-windup design: an overview of some recent advances and open problems. IET Control Theory and Applications 3(1):1–19.
444. Tarbouriech, S., Garcia, G., Gomes da Silva, J. M., and Queinnec, J. (2011): Stability and Stabilization of Linear Systems with Saturating Actuators, Springer.
445. Taylor, J. H. (1974): Strictly Positive-Real Functions and the Lefschetz-Kalman-Yakubovich (LKY) Lemma. IEEE Transactions on Circuits and Systems 21(2):310–311.
446. Taylor, J. H. (1999): Describing Functions. In Webster, J. G. (Ed.): Wiley Encyclopedia of Electrical and Electronics Engineering, Wiley-Interscience.
447. Terano, T., Asai, K., and Sugeno, M. (1994): Applied Fuzzy Systems, Academic Press.
448. Terrell, W. J. (2009): Stability and Stabilization: An Introduction, Princeton University Press.
449. Thanom, W., and Loh, R. N. K. (2012): Observer-Based Nonlinear Feedback Controls for Heartbeat ECG Tracking Systems. Intelligent Control and Automation 3(3):251–261.
450. Thomsen, J. S. (2005): Feedback Linearization Based Arc Length Control for Gas Metal Arc Welding. IEEE Proceedings of the American Control Conference, Portland, 3568–3573.
451. Thompson, J. M. T., and Bishop, S. R. (Eds.) (1994): Nonlinearity and Chaos in Engineering Dynamics, Wiley & Sons.
452. Thompson, J. M. T., and Stewart, H. B. (Eds.) (2002): Nonlinear Dynamics and Chaos, Wiley & Sons.

453. Tolle, H. (1975): Optimization Methods, Springer.
454. Tsiotras, P., and Wilson, B. C. (2003): Zero- and Low- Bias Control Designs for Active Magnetic Bearings. IEEE Transactions on Control Systems Technology 11(6):889–904.
455. Tsypkin, Ya. S. (1964): A criterion for absolute stability of automatic pulse systems with monotonic characteristics of the nonlinear element. Soviet Physics-Doklady 9(4):262–266.
456. Tsypkin, Ya. S. (1964): Frequency Criteria for the Absolute Stability of Nonlinear Sampled-Data Systems. Automation and Remote Control 25(3):261–267.
457. Turner, M. C., and Bates, D. G. (Eds.) (2007): Mathematical Methods for Robust and Nonlinear Control, Springer.
458. Tyagunov, A. A. (2004): High-Performance Model Predictive Control for Process Industry, PhD thesis, Technische Universität Eindhoven.
459. Unbehauen R. (1998): Systemtheorie 2: Mehrdimensionale, adaptive und nichtlineare Systeme. R. Oldenbourg, (System Theory 2: Multidimensional, Adaptive, and Nonlinear Systems; in German).
460. Utkin, V.I. (1992): Sliding Modes in Control and Optimization, Springer.
461. Utkin, V.I. (1977): Variable Structure Systems with Sliding Modes. IEEE Transactions on Automatic Control 22(2):212–222.
462. Valmorbida, G., and Anderson, J. (2017): Region of attraction estimation using invariant sets and rational Lyapunov functions. Automatica 75(1):37–45.
463. Verbruggen, H. B., and Babuška, R. (Eds.) (1999): Fuzzy Logic Control: Advances in Applications, World Scientific.
464. Vidyasagar, M. (1980): Decomposition Techniques for Large-Scale Systems with Nonadditive Interactions: Stability and Stabilization. IEEE Transactions on Automatic Control 25(4):773–779.
465. Vidyasagar, M. (2002): Nonlinear Systems Analysis, Society for Industrial & Applied Mathematics.
466. Waldherr, S., and Zeitz, M. (2008): Conditions for the existence of a flat input. International Journal of Control 81(3):439–443.
467. Waldherr, S., and Zeitz, M. (2010): Flat inputs in the MIMO case. 8th IFAC Symposium on Nonlinear Control Systems (NOLCOS), Bologna. In IFAC Proceedings Volumes 43(14):695–700.
468. Wang, J., and Rugh, W. (1987): Feedback Linearization Families for Nonlinear Systems. IEEE Transactions on Automatic Control 32(10):935–940.
469. Wang, X., Li, S., Su, S., and Tang, T. (2019): Robust Fuzzy Predictive Control for Automatic Train Regulation in High-Frequency Metro Lines. IEEE Transactions on Fuzzy Systems 27(6):1295–1308.
470. Wang, Y., Singer, J., and Bau, H. H. (1992): Controlling chaos in a thermal convection loop. Journal of Fluid Mechanics 237:479–498.
471. Weerasooriya, S., Low, T. S., and Huang, Y. H. (1994): Adaptive Time Optimal Control of a Disk Drive Actuator. IEEE Transactions on Magnetics 30(6):4224–4226.
472. Wei, G., Wang, Z., Li, W., and Ma, L. (2014): A Survey on Gain-Scheduled Control and Filtering for Parameter-Varying Systems. Discrete Dynamics in Nature and Society, Volume 2014 (April):1–10.
473. Weinberg, L., and Slepian, P. (1960): Positive real matrices. Indiana University Mathematics Journal 9(1):71–83.

474. Wen, J., and Desrochers, A. A. (1985): A minimum time control algorithm for linear and nonlinear systems. Proceedings of the 24th IEEE Conference on Decision and Control, Fort Lauderdale, 1441–1446.
475. Wen, J. T. (1988): Time Domain and Frequency Domain Conditions for Strict Positive Realness. IEEE Transactions on Automatic Control 33(10):988–922.
476. Wichner, W., Wilharm, H., and Mahr, J. (1997): Crane with a travel drive for the horizontal movement of a load hanging on a rope. German Patent DE 44 05 525 C2.
477. Willems, J. C. (1972): Dissipative dynamical systems: Part II: Linear Systems with quadratic supply rates. Archive for Rational Mechanics and Analysis 45(5):352–393.
478. Willems, J. L. (1968): Further comments on a paper by G. Schmidt and G. Preusche: „Popovs Stabilitätssatz als Mittel zur teilweisen Bestätigung von Aizermans Vermutung". Regelungstechnik 16(2):74–75.
479. Willems, J. L. (1970): Stability Theory of Dynamical Systems, Thomas Nelson and Sons.
480. Wredenhagen, G. F., and Bélanger, P. R. (1994): Piecewise-linear LQ control for systems with input constraints. Automatica 30(3):403–416.
481. Wu, F. F., and Desoer, C. A (1972): Global Inverse Function Theorem. IEEE Transactions on Circuit Theory 19(2):199–201.
482. Yan, X.-G., Spurgeon, S. K., and Edwards, C. (2017): Variable Structure Control of Complex Systems, Springer.
483. Yeung, K. S. (1985): A Reformulation of Nyquist's Criterion. IEEE Transactions on Education 28(1):58–60.
484. Ying, Y., Rao, M., and Sun, Y. (1992): Bilinear control strategy for paper-making process. Chemical Engineering Communications 111(1):13–28.
485. Yong, S, Z., Paden, B., and Frazzoli, E. (2015): Computational Methods for MIMO Flat Linear Systems: Flat Output Characterization, Test and Tracking Control. IEEE Proceedings of the American Control Conference, Chicago, 3898–3904.
486. Yoshizawa, T. (1966): Stability Theory by Liapunov's Second Method, The Mathematical Society of Japan.
487. Yu, H., Cai, G., and Li, Y. (2012): Dynamic analysis and control of a new hyperchaotic finance system. Nonlinear Dynamics 67(3):2171–2182.
488. Yulin, X. (2009): Modeling and LPV Control of Web Winding System with Sinusoidal Tension Disturbance. IEEE Proceedings of the Chinese Control Conference (CCDC), Guilin, 3815–3820.
489. Zaccarian, L., and Teel, A. R. (2011): Modern Anti-Windup Synthesis: Control Augmentation for Actuator Saturation, Princeton University Press.
490. Zadeh, L. A. (1965): Fuzzy Sets. Information and Control 8(3):338–353.
491. Zames, G. (1966): On the Input-Output Stability of Time-Varying Nonlinear Feedback Systems-Part I: Conditions Derived Using Concepts of Loop Gain, Conicity, and Positivity. IEEE Transactions on Automatic Control 11(2):228–238.
492. Zames, G. (1966): On the Input-Output Stability of Time-Varying Nonlinear Feedback Systems-Part II: Conditions Involving Circles in the Frequency Plane and Sector Nonlinearities. IEEE Transactions on Automatic Control 11(3):465–476.
493. Zeitz, M. (1983): Controllability canonical (phase-variable) form for non-linear time-variable systems. International Journal of Control 37(6):1449–1457.

494. Zeitz, M. (1987): The extended Luenberger observer for nonlinear systems. Systems & Control Letters 9(2):149–156.
495. Zeitz, M. (1989): Canonical Forms for Nonlinear Systems. IFAC Symposium on Nonlinear Control Systems Design (NOLCOS), Capri. In IFAC Proceedings Volumes 22(3):33–38.
496. Zeitz, M. (2014): Minimalphasigkeit – keine relevante Eigenschaft für die Regelungstechnik! Automatisierungstechnik 62(1):3–10, (The property to be minimum phase is not a relevant property for control engineering; in German).
497. Zhang, H., and Liu, D. (2006): Fuzzy Modeling and Fuzzy Control, Birkhäuser.
498. Zhang, J., Tugal, H., Carrasco, J., and Heath, W. P. (2018): Absolute Stability of Systems With Integrator and/or Time Delay via Off-Axis Circle Criterion. IEEE Control Systems Letters 2(3):411–416.
499. Zhao, Z.-Y., Tomizuka, M., and Isaka, S. (1993): Fuzzy Gain Scheduling of PID Controllers. IEEE Transactions on System, Man & Cybernetics 23(5):1392–1398.
500. Zhou, J., and Wen, C. (2008): Adaptive Backstepping Control of Uncertain Systems: Nonsmooth Nonlinearities, Interactions of Time-Variations, Springer.
501. Zinober, A. S. I. (Ed.) (1994): Variable Structure and Lyapunov Control, Springer.
502. Zypkin, Ja. S. (1963): Die absolute Stabilität nichtlinearer Impulsregelsysteme. Regelungstechnik 11(4):145–148, (On the Global Stability of Nonlinear Automatic Sampled-Data Systems; in German).
503. Цыпкин, Я. З. (1977): Основы теории автоматических систем, Наука (Russian original publication of [504])
504. Zypkin, Ja. S. (1981): Grundlagen der Theorie automatischer Systeme, R. Oldenbourg, (Fundamentals of the Theory of Automatic Control Systems; in German).

Index

A

B

J. Adamy, *Nonlinear Systems and Controls*,
https://doi.org/10.1007/978-3-662-68690-4

D

F

I

J

M

N

Q

R

S

T